AS and A LEVEL

CHEMISTRY

Erik Lewise and Martyn Berry

Pearson Education
Edinburgh Gate
Harlow
Essex

ISBN 0582 33733 X
Third impression 2004

Design concept by Moondisks Ltd, Cambridge.
Typeset by Techset Ltd, Gateshead.
Printed in China
SWTC/03

Acknowledgements
The authors would like to thank many people for their help, support and encouragement in writing this book. In particular: Dr Sandra Evans, for so much helpful criticism and comment; Mrs Olga Trebell, Senior Technician at Cranbrook School; colleagues and students at Cranbrook School, especially Dr N P Vinall, for their encouragement; and Mr Allan Masson and Ms Penny Johnson for their advice and assistance. A very special thanks goes to Margaret and Jill for their understanding and forbearance throughout the making of this book.

We are grateful to the following for permission to reproduce photographs:

FRONT COVER: Ballard Power Systems *top right*; Science Photo Library *top left*; Science Photo Library/Philippe Plailly *bottom*; Science Photo Library/ Andrew Syred *background*.

Professor J F Allen, School of Physics and Astronomy, University of St Andrews, page 239; Art Directors & TRIP, pages 79 (Eric Smith), 168 (M Barlow), 670 (H Rogers); A-Z Botanical Collection, page 458 *left* (Adrian Thomas); Ballard Power Systems, page 629 *bottom*; Centre for Alternative Technology, page 98; De Beers, page 80 *top and bottom*; Mary Evans Picture Library, pages 7, 32 *top*, 32 *centre*, 32 *bottom*; GSF Picture Library, page 431; Ken Harris, New Lisbon Schools, WI, USA, page 565; ICI, page 329; Image Bank, page 172 (Marvin E Newman); Life Science Images, page 164; Andrew Lambert, pages 93, 195, 219, 775; Paul Mulcahy, pages 82, 94, 178, 458 *right*; Salt Union Limited/Weston Point Studios, page 160; Science Photo Library, pages 1 *left* (Philippe Plailly), 1 *right*, 84 (Andrew Syred), 96 *right* (Mehau Kulyk), 100 *top* (James King-Holmes/OCMS), 100 *bottom* (James King-Holmes/ OCMS), 127 *bottom* (Novosti), 169 (Sheila Terry), 179 *top and bottom* (Ken Edward), 311 (J C Revy), 441 (Lawrence Livermore, National Library), 812 (John Greim), 815 (Alfred Pasieka); Stock Shot, page 96 left (Gary Pearl); Stone, pages 15 (Dennis Coleman), 73 (Daniel J Cox), 381; Thames Water, page 213; Weston Point Studios, page 170.

All other photography by Gareth Boden.

Contents

Contents

Contents

v

We start from these ideas:

- **Elements** are the basic building blocks of the Universe. By analysing the light from stars we can tell what they are made of. Wherever we look, matter is made of the same ninety or so elements that we meet here on Earth. (More than a hundred elements are known, but most of the additional elements can only be obtained in tiny quantities artificially.)

- Other than the hydrogen and some helium made in the Big Bang, elements are made by nuclear reactions inside stars, and are scattered into space when giant stars explode as supernovae. (All the atoms in our bodies, other than hydrogen, were made in this way. We are, quite literally, stardust.)

- Elements consist of **atoms**.

- The atoms of a particular element all behave in the same way chemically, but may differ slightly in mass. They are different from the atoms of any other element. So another way of thinking about elements is this: *an element is a substance which consists of one kind of atom only.*

Scanning tunnelling electron micrograph of gold atoms on a graphite, substrate

An atom is mostly empty space. This means that any form of matter which we know here on Earth is mostly empty space, including us.

Whirlpool galaxy

In this chapter

We look at the internal structure of the atoms which make our Universe. Unfortunately, there is no room to do more than summarise briefly the history of the growth of our knowledge of the atom.

1.1 **The particles which make an atom**
1.2 **Atomic number and mass number; isotopes**
1.3 **Relative atomic mass (A_r); the mass spectrometer**
1.4 **Electron configurations and the Periodic Table**
1.5 **Ionisation energies and electron affinities**
1.6 **Atomic radius and ionic radius**

1
atomic structure

1.1 The particles which make an atom

Chemists work with energies which are very small compared to those which hold atomic nuclei together. (Compare even the biggest chemical explosion with those of atom bombs or hydrogen bombs.) High-energy physicists know of many sub-atomic particles. These are created at huge temperatures, like those thought to have existed very soon after the Big Bang. Such high-energy particles can be studied on Earth only by using very large and expensive equipment.

For the chemist, the temperatures available range from −273 °C to about 4000 °C; and, in that range, the three stable particles from which all atoms are built exist. The important properties of these three particles are briefly summarised in *Table 1.1.1*. In this table, relative mass means in relation to each other. Notice that, for ordinary uses, the masses of protons and neutrons can be taken as equal and the masses of electrons can be neglected.

The slight difference in mass of protons and neutrons, which is shown in the fourth column, is very important when explaining what happens in radioactive decay and nuclear reactions. Notice also that the charges on protons and electrons cancel each other out exactly. We shall see the importance of this fact very soon.

Name	Symbol	Approximate relative mass	More accurate mass (relative to the mass of the proton)	Relative charge
Proton	p	1	1.0000	+1
Neutron	n	1	1.0016	0
Electron	e	almost zero	0.000547	−1

Table 1.1.1 The important sub-atomic particles

The absolute temperature scale

Temperature can be measured on a number of different scales. Our common scale is the Celsius scale, where water melts at 0 °C and boils at 100 °C.

The British scientist Lord Kelvin, in the nineteenth century, suggested an alternative temperature scale, the **absolute temperature scale**, in which the point at which the average kinetic energy of all particles should be zero is the zero point on the scale – **absolute zero**.

Most scientific relationships involving temperature use this absolute temperature scale, in which the unit is the kelvin (named after Lord Kelvin), so it is essential that you can convert degrees Celsius into kelvin.

Kelvin temperature = degrees Celsius + 273
(K = °C + 273)

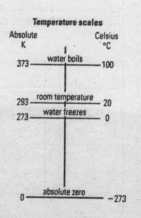

Temperature scales

Absolute K		Celsius °C
373	water boils	100
293	room temperature	20
273	water freezes	0
0	absolute zero	−273

Dalton, Thomson, Rutherford, Bohr, and Chadwick: What do we mean by an 'atom'?
Greek philosophers of the fifth century BC, notably Leucippus and Democritus, were the first people to think seriously about the nature of matter and to introduce the idea of atoms. (Indeed, the word 'atom' comes from the Greek 'atomos', meaning something that cannot be cut up into smaller pieces.)

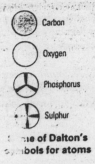

Carbon

Oxygen

Phosphorus

Sulphur

ne of Dalton's
ɔɯbols for atoms

John Dalton (1766–1844), a Quaker schoolteacher from Manchester, published his atomic theory of matter in 1803 and 1808. He was the first to provide a rational basis for measurements in chemistry. The great advances in understanding which occurred during the next hundred years can be traced back to his work. Briefly, Dalton emphasised:

- that atoms cannot be broken up by any chemical process;
- that atoms of the same element are identical, but different from atoms of any other element;
- that atoms of different elements have different masses;
- that compounds are formed by atoms of different elements joining together in simple ratios, e.g. 1:1, 1:2, 2:3, etc; and
- that the masses of elements which combine together can be used to find the relative masses of atoms.

This final point means that measurement of the combining masses of the elements can enable us to find out properties of individual atoms.

Joseph John (always known as J J) Thomson (1856–1940) was for many years head of the famous Cavendish Laboratory at Cambridge. It was here, around the end of the nineteenth century, that he and his research team established the reality of electrons. (The word 'electron' had first been used by Johnstone Stoney in 1874.) The 'discovery' of the electron is usually said to date from Thomson's publications of 1897. Thomson won the Nobel Prize for Physics in 1906.

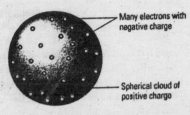

Many electrons with negative charge

Spherical cloud of positive charge

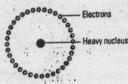

Electrons

Heavy nucleus

Ernest Rutherford (1871–1937) was a New Zealander who worked in Manchester and Canada before becoming Cavendish Professor at Cambridge. He announced in 1911 that the atom consisted of a heavy, small nucleus surrounded by electrons. The initial experimental observations had been made in 1909 by his co-workers Ernest Marsden and Hans Geiger. Rutherford won the Nobel Prize for Chemistry in 1908 for his work on radioactivity and atomic structure.

Neils Bohr (1885–1962) was born and lived most of his life in Copenhagen. He was the first to bring together a great deal of experimental observation on atomic structure, and he used this to state, in 1913, that electrons orbit the nucleus in fixed orbits, and absorb or emit definite amounts of energy when they move from one orbit to another. He won the Nobel Prize for Physics in 1922. His son, Aage, won the same prize in 1975 for work on the structure of the nucleus.

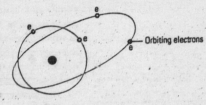

Orbiting electrons

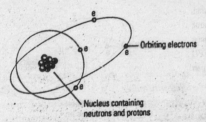

Orbiting electrons

Nucleus containing neutrons and protons

James Chadwick (1891–1974) was the first to discover the neutron, so solving the problem of the structure of the nucleus. This work was also carried out at the Cavendish Laboratory and was announced in 1932. Chadwick and Maurice Goldhaber went on to measure the mass of the neutron (1934); they found that it was greater than the mass of (proton + electron), so establishing that neutrons are separate and distinct particles. Chadwick won the Nobel Prize for Physics in 1935.

1.2 Atomic number and mass number; isotopes

Atomic number

The work of Rutherford and many others showed clearly that an atom consists of a nucleus surrounded by electrons. The nucleus contains protons and neutrons (except in ordinary hydrogen, in which the nucleus is composed of a single proton). So, from *Table 1.1.1*, we see that the nucleus is where the mass of the atom is concentrated.

Because an atom is electrically neutral, the number of protons in an atom *must* equal the number of electrons, because the positive charge on the proton is exactly cancelled by the negative charge on the electron. As we shall see in Ch 1.4, the number of electrons in an atom fixes the way in which the electrons are arranged in that atom. It is the arrangement of the outermost electrons in the atom which is responsible for the chemical properties of an element.

So, if you know (or can read off from the Periodic Table, which can be found inside the back cover of this book) the atomic number of an element and hence its electron structure, you can predict quite accurately how it will react. This point will be reinforced frequently in this book!

One of the implications of the nuclear model of the atom is that an atom is mostly empty space.

For example, the diameter of a fluorine atom is about 1.4×10^{-10} m. The diameter of a fluorine nucleus is about 10^{-14} m. The mass of a fluorine atom is about 3×10^{-23} g, so the density of a fluorine nucleus can easily be calculated to be about 60 million tonnes per cm^3.

If electrons are stripped off atoms to leave bare nuclei, as we believe happens at the high temperatures reached inside stars, matter with such startling densities – and even higher! – can be obtained.

At the start of this chapter we stated that one way of thinking about the term 'element' is that 'an element is a substance which consists of one kind of atom only'. You can now see what this means. All the atoms of a given element contain the same number of protons, and hence the same number of electrons; therefore, they have the same chemical properties. If the number of protons is different, it must be a different element.

If the way an element behaves is fixed by its atomic number, it would obviously help you to get control of a massive amount of information about the elements and their reactions if the elements could be put in some sort of order. This is achieved in the Periodic Table (Ch 1.4 and Ch 5.1), in which the elements are arranged in order of increasing atomic number. When you have learned the important parts of the Periodic Table and the trends within it (see Ch 5), you will have a great deal of information stored in your mind and ready for use in making predictions and in helping to remember details.

The atomic number of an element is written as a subscript in front of the symbol. So, for carbon: $_6$C. And for uranium and iodine: $_{92}$U, $_{53}$I, respectively.

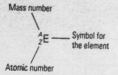

Mass number

$$^A_ZE$$ —— Symbol for the element

Atomic number

**Figure 1.2.1
Atomic notation**

Mass number and isotopes

Since protons and neutrons are the particles found in the nucleus, they are often known as *nucleons*; so, the mass number may also be called the *nucleon number*.

The mass number of an atom is written as a superscript in front of the symbol, thus: ^{12}C, ^{238}U, ^{127}I. We can now describe a particular atom (or *nuclide*) in terms of both its mass number and atomic number, thus: $^{12}_6$C, $^{238}_{92}$U, $^{127}_{53}$I (*Figure 1.2.1*).

Notice that *the number of neutrons in an atom is easily found by subtracting the atomic number (= number of protons) from the mass number (= number of protons + neutrons)*.

All the atoms of a particular element *must* contain the same number of protons and the same number of electrons, but need not necessarily contain the same number of neutrons.

Isotopes are atoms of the same element, which have different numbers of neutrons (*Figure 1.2.2*).

So, arsenic, $^{75}_{33}$As, with only a single naturally-occurring isotope, has 33 protons in the nucleus, 33 electrons outside the nucleus and $(75 - 33) = 42$ neutrons in the nucleus. In the same way, the major isotope of copper, $^{63}_{29}$Cu, has 29 protons, 29 electrons, and $(63 - 29) = 34$ neutrons.

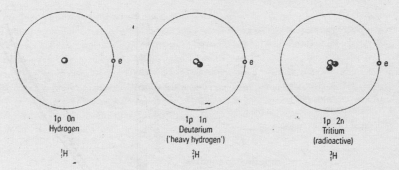

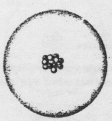

$^{12}_{6}C$ – 6 protons, 6 neutrons

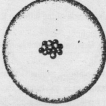

$^{13}_{6}C$ – 6 protons, 7 neutrons

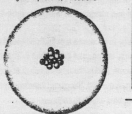

$^{14}_{6}C$ – 6 protons, 8 neutrons

Figure 1.2.3 Isotopes of carbon

Figure 1.2.2   Isotopes of hydrogen

Carbon has more than one naturally occurring isotope. Practically all of the carbon atoms found in nature (98.9%) can be represented as $^{12}_{6}C$, i.e. they contain $(12 - 6) = 6$ neutrons (*Figure 1.2.3*). The remaining 1.1% consists almost entirely of $^{13}_{6}C$, with $(13 - 6) = 7$ neutrons. A very small proportion of carbon atoms contain eight neutrons. These $^{14}_{6}C$ atoms are unstable and subject to radioactive decay. This property is used in radiocarbon dating of archaeological finds.

Most uranium (99.3%) occurs as $^{238}_{92}U$; the 0.7% which occurs as $^{235}_{92}U$ is radioactive and is used in the nuclear power industry.

The number of protons, neutrons and electrons in an atom – worked example

Calculating the number of protons, neutrons and electrons in an atom of any isotope is straightforward – as long as the mass number and atomic number are known. The atomic number can always be found from the Periodic Table inside the back cover of the book – it is the smaller of the two numbers quoted for each element!

Examples:

Element, $^{A}_{Z}E$	Number of protons $= Z$	Number of electrons $= Z$	Number of neutrons $= A - Z$
Helium, $^{4}_{2}He$	2	2	2
Aluminium, $^{27}_{13}Al$	13	13	14
Nickel, $^{58}_{28}Ni$	28	28	30

Radioactivity

This is a very large topic which is beyond the scope of this book. However, the following points might be useful.

- Very large nuclei are always unstable. The largest completely stable nucleus is that of lead, $_{82}Pb$. Protons are positively charged and so will repel each other; neutrons in the nucleus help to keep nuclei together. As the atomic number of the element increases, so the number of neutrons needed to keep the nucleus together begins to increase more and more rapidly, e.g.: $^{16}_{8}O$ – 8p, 8n; $^{40}_{20}Ca$ – 20p, 20n; $^{56}_{26}Fe$ – 26p, 30n; $^{107}_{47}Ag$ – 47p, 60n; $^{208}_{82}Pb$ – 82p, 126n (*Figure 1.2.4*). Eventually, even with this increasing ratio of neutrons to protons, stability is impossible.
- Some smaller nuclei which occur in Nature are also unstable. We have already met $^{14}_{6}C$, used in radiocarbon dating. The much more slowly-decaying potassium isotope, $^{40}_{19}K$, is very useful for dating rocks.

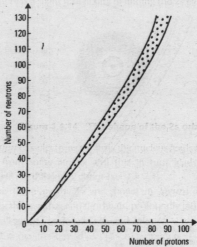

Figure 1.2.4   The region of stable neutron/proton ratios

- Radioactive decay involves nuclei breaking up, so that other elements – often with smaller nuclei – *always* result. The break-up causes α particles (helium nuclei, $^4_2$He) or β particles (fast nuclear electrons) to be emitted, and γ radiation (very short wavelength, high energy, electromagnetic radiation) to be given off.

An example of α decay: $^{226}_{88}$Ra $\longrightarrow$ $^{222}_{86}$Rn + $^4_2\alpha$ together with some γ radiation

An example of β decay: $^{14}_6$C $\longrightarrow$ $^{14}_7$N + $^0_{-1}\beta$ together with some γ radiation

It is worth noting that the mass numbers and the sum of the atomic numbers on each side of the arrow in these equations must each be equal.

All three – α and β particles, and γ radiation – are highly dangerous for living tissue. Neutrons may also be emitted during decay.

- When heavy nuclei break up, the total mass of the products (new smaller atoms, α or β particles, neutrons) is always slightly less than the original mass. This mass loss always appears as energy, according to the famous Einstein equation, $E = mc^2$. In this, E = energy in Joules; m = mass in kg, and c = velocity (speed) of light in m s^{-1}. Since $c = 3 \times 10^8$ m s^{-1}, the loss of a tiny amount of mass releases a vast amount of energy. This is the basis of the use of uranium-235 ($^{235}_{92}$U) in the nuclear power industry, and of the abuse of uranium and plutonium in 'atomic' bombs. This break up of heavy nuclei is known as '**nuclear fission**'.

- At very high temperatures, very light nuclei (usually hydrogen) have enough energy to overcome the repulsion caused by their positive charges to crash into each other and fuse together. In this case also, the mass of the product is less than that of the original particles! So, once again, huge amounts of energy are released. This is the process which happens in so-called 'hydrogen bombs', and which is going on inside every 'live' star, such as our Sun. Our planet intercepts a tiny fraction of this energy, which keeps us warm and also – via photosynthesis – enables all life to continue.

This '**thermonuclear fusion**' process in stars now long dead is what made all the atoms, other than hydrogen, in the whole of the Earth and all that lives on it.

- A radioactive isotope decays in such a way that the time taken for half of it to decay is equal to the time taken for the remaining half to decay to one-quarter of the original material, and so on. In other words, whatever amount of the isotope is present, its 'half-life' is constant. *The half-life of a radioactive isotope, $t_{1/2}$, is the time required for the radioactivity of that particular isotope to reduce to half the original amount.* The situation is often complicated by the fact that many decay products are themselves radioactive. For example, natural $^{238}_{92}$U decays through thirteen intermediate radioactive species to end up as $^{206}_{82}$Pb.

$^{14}_6$C has a $t_{1/2}$ of 5600 years and is found in all living things; because of this long $t_{1/2}$ it is used in radiocarbon dating. $^{40}_{19}$K has a $t_{1/2}$ of 1.3 *billion* years; it is therefore useful for dating very old rocks.

Some half-lives are extremely short, e.g. for $^{215}_{85}$At, $t_{1/2} = 10^{-4}$ s.

Writing nuclear equations – worked example

Some examinations require you to be able to complete nuclear equations to show how a particular radioactive isotope decays into another element, with the emission of an α or β particle.

An α particle is a helium nucleus, written either as $^4_2\alpha$ or $^4_2\text{He}^{2\oplus}$

A β particle is an electron ejected from the nucleus of the atom, written $^0_{-1}\beta$ or $^0_{-1}\text{e}^{\ominus}$. (Overall, you might like to consider that a neutron has become a proton plus an electron.)

What you have to remember is that the sum of the mass numbers and the sum of the atomic numbers on each side of the equation must be the same.

If we use the examples we have already quoted in the box on radioactivity above, the principles become clear.

$$^{226}_{88}\text{Ra} \longrightarrow \ldots\ldots + ^4_2\alpha$$

$$^{14}_{6}\text{C} \longrightarrow \ldots\ldots + ^0_{-1}\beta$$

In the first equation, radium decays with the loss of an α particle, so radium has lost 4 mass units and 2 protons. The mass number decreases from 226 to 222 and the atomic number from 88 to 86. Using the Periodic Table at the back of the book, we can see that the element with atomic number 86 is radon, Rn.

$$^{226}_{88}\text{Ra} \longrightarrow ^{222}_{86}\text{Rn} + ^4_2\alpha$$

In the second equation the radioactive isotope of carbon, ^{14}C (which is used for radiocarbon dating of antiquities), decays by losing a β particle. The mass number does not change, since an electron has very little mass compared to the other particles present in the atom. As the atomic number is effectively -1 for an electron, the atomic number of the carbon atom must increase by 1, forming the next element in the Periodic Table, nitrogen, N.

$$^{14}_{6}\text{C} \longrightarrow ^{14}_{7}\text{N} + ^0_{-1}\beta$$

Marie Curie

Becquerel and the Curies

In 1896, Antoine-Henri Becquerel discovered that untreated uranium compounds could fog a photographic plate wrapped in black paper, so must be emitting some kind of penetrating radiation. This was the first observation of natural radioactivity. In 1898, Polish-born Marie Curie and her husband Pierre noticed that an ore of uranium, pitchblende, was more radioactive than would be expected from the amount of uranium in it. Using elementary methods of chemical separation, and with immense labour, the Curies separated two new elements from pitchblende – polonium (named in honour of Poland) and radium (because of its exceptional radioactivity).

Radium was soon being used inside metal needles which were inserted into cancer tumours to destroy them; a drastic treatment, but better than anything else then available and leading to some cures. (One of the major cancer charities in Britain is *Marie Curie Cancer Care*, which, of course, uses totally modern methods.)

Becquerel and the Curies were jointly awarded the 1903 Nobel Prize for Physics for their discoveries about radioactivity. After the death of Pierre in a traffic accident in 1906, Marie Curie took over her husband's professorship at the Sorbonne in Paris, one of the first women ever to be a professor. She carried on working, and in 1911 became the first person ever – and still the only woman – to win a second Nobel Prize, this time for Chemistry.

Irène Joliot-Curie, daughter of Pierre and Marie, with her husband Frédéric won the Nobel Prize for Chemistry in 1935 for the synthesis of new radioactive elements. So far, Marie and Irène Curie are the only mother and daughter to have both won Nobel Prizes.

The dangers of working with radioactive materials were at first not understood. Marie Curie died in 1934 from radiation-induced disease. Her laboratory notebook is still locked away – it is too radioactive to handle. Irène died in 1956 from leukaemia, most probably caused by her work.

SELF-ASSESSMENT
QUESTIONS
1.2

1 Work out the number of protons, electrons and neutrons present in the following atoms:

a $^{7}_{3}Li$ **b** $^{40}_{18}Ar$

c $^{127}_{53}I$ **d** $^{56}_{26}Fe$

2 Write the usual notation for the following isotopes.

a Magnesium, $A = 24$

b Hydrogen, $A = 1$

c Hydrogen, $A = 3$

d Silicon, $A = 28$

3 Define the terms:

a atomic number

b isotope

c mass number

4 The most abundant element on the surface of the Earth is oxygen. The major isotope is $^{16}_{8}O$, with much smaller quantities of ^{17}O and ^{18}O. One of the artificial isotopes of oxygen is $^{20}_{8}O$.

a What is the mass number of the major isotope?

b What is the atomic number of the major isotope?

c How many protons and electrons are present in each of these isotopes?

d How many neutrons are present in the nuclei of each of these isotopes?

e The artificial isotope of oxygen, $^{20}_{8}O$, is radioactive. Explain what radioactive means.

f This isotope decays with the emission of a β particle to form an isotope of fluorine, $^{20}_{9}F$. Write an equation to illustrate this radioactive decay.

1.3 Relative atomic mass (A_r); the mass spectrometer

Some elements consist entirely of only one isotope, e.g. all fluorine atoms are $^{19}_{9}F$, and all aluminium atoms $^{27}_{13}Al$. Tin holds the record for stable isotopes with ten, ranging from $^{112}_{50}Sn$ to $^{124}_{50}Sn$. Chlorine has two isotopes, $^{35}_{17}Cl$ and $^{37}_{17}Cl$.

The internationally agreed scale for relative atomic masses is based on a value of 12.0000 for ^{12}C. (We shall use 12.00, i.e. to four significant figures.) On this scale, the relative atomic masses of fluorine and aluminium should be very nearly 19.00 and 27.00 respectively. The symbol for relative atomic mass is A_r.

But what about tin, with mass numbers ranging from 112 to 124? And what about chlorine? Its isotopes have mass numbers of 35 and 37. The *average* atomic mass is therefore 36, so there is a temptation to say that its A_r is 36. But this fails to allow for the fact that in Nature there is far more of one isotope of chlorine than the other: 75.53% $^{35}_{17}Cl$ and only 24.47% $^{37}_{17}Cl$ (*Figure 1.3.1*), i.e. approximately $\frac{3}{4}$ ^{35}Cl and $\frac{1}{4}$ ^{37}Cl.

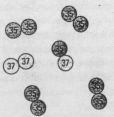

$$\text{Weighted mean} = \frac{(9 \times 35) + (3 \times 37)}{12} = 35.5$$

Figure 1.3.1

'Averaging' isotopic masses

To get the A_r of chlorine we must find what the mathematicians call the 'weighted mean' of the isotopic masses, which allows for the different percentages of isotopes (*Figure 1.3.2*).

'Weighted mean' of mass numbers for chlorine:

$$= \frac{(75.53 \times 35) + (24.47 \times 37)}{100} = 35.49$$

∴ A_r of chlorine should be 35.49. (But see the bottom of this page!)

Just remember

Number of protons + neutrons

Number of electons

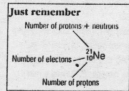

Number of protons

Relative atomic mass

From isotope abundances of an element which consists of isotopes A, B and C:

$$A_r = \frac{(\%A \times \text{mass number of A}) + (\%B \times \text{mass number of B}) + (\%C \times \text{mass number of C})}{100}$$

Figure 1.3.2

The nucleus – worked example

The naturally occurring isotopes of neon are $^{20}_{10}Ne$, $^{21}_{10}Ne$ and $^{22}_{10}Ne$. The abundances of these isotopes in a sample of neon were found to be ^{20}Ne (90.92%), ^{21}Ne (0.26%), and ^{22}Ne (8.82%).

a *Show, in the form of a table, the number of protons, neutrons and electrons present in each isotope.*

b *Calculate the relative atomic mass of this sample of neon.*

a

	Protons	Neutrons	Electrons
$^{20}_{10}Ne$	10	10	10
$^{21}_{10}Ne$	10	11	10
$^{22}_{10}Ne$	10	12	10

b The A_r is found from the sum of (% of the isotope × mass number)/100

So, for this sample of neon,

$$A_r = \frac{(90.92 \times 20) + (0.26 \times 21) + (8.82 \times 22)}{100}$$

$$= 18.18 + 0.06 + 1.94$$

$$= 20.18$$

Accurate relative atomic masses

However, accurate measurement shows that the A_r of chlorine is in fact 35.46. This difference may seem very small, but is nevertheless important. So how does it happen?

The answer is that the mass of an atomic nucleus – other than hydrogen, of course – is always slightly less than the sum of the accurate masses of the protons and neutrons in its nucleus (see *Table 1.1.1*). This difference in mass (or *mass defect*) represents the *binding energy* of the nucleus – i.e. the energy released when the protons and neutrons are brought together to form the nucleus. We have already seen (Ch 1.2) that energy and mass are connected through the Einstein equation, $E = mc^2$.

So, for accurate work, we should take the actual isotope masses as: ^{35}Cl, 34.969; ^{37}Cl, 36.966. If we substitute these figures into the weighted mean mass expression above, we get the correct value of 35.46 for the A_r of chlorine.

The usual way of writing a list of relative atomic masses – e.g. for a calculation question – is like this: $A_r(C) = 12.00$; $A_r(O) = 16.00$; $A_r(H) = 1.01$, etc.

(You shouldn't write C = 12.00, O = 16.00 etc!)

We can now give a precise definition for the relative atomic mass, A_r, of an element.

For most elements which have stable isotopes, the 'mix' of isotopes is exactly the same for any sample of the elements from any source. So, any sample of chlorine will consist of 75.5% $^{35}_{17}Cl$ and 24.5% $^{37}_{17}Cl$. The situation is different for elements which are radioactive or result from radioactive decay. For example, lead is the final, stable element at the end of some of the natural radioactive decay series. Uranium-238, $^{238}_{92}U$, decays eventually to $^{206}_{82}Pb$, whereas $^{235}_{92}U$ ends up as $^{207}_{82}Pb$. So, samples of lead made from ores mined in different areas will differ slightly in their isotope ratios, and this can be used to identify where the lead was mined.

The names of elements and compounds, and the basic standards and definitions used in chemistry, are decided by the International Union of Pure and Applied Chemistry (IUPAC) and its various committees and working parties. Every country in which chemists are active is represented in IUPAC. The British National Committee for Chemistry, which consists of representatives of all the major organisations connected with chemistry in Britain, is the British 'branch' of IUPAC.

Early tables of relative atomic masses were based on the hydrogen atom having a mass of 1.0, later ones on oxygen having a mass of 16.0. For many years chemists and physicists used slightly different values of relative atomic masses. This obviously unsatisfactory situation was sorted out in 1961, when IUPAC and the other international scientific unions agreed that the relative atomic mass scale should be based on the isotope $^{12}_{6}C$ having a mass of 12.0000.

The mass spectrometer

When a charged particle is fired into a magnetic field, it is deflected by the field. The path which the particle follows depends on its mass-to-charge ratio, m/e. The modern mass spectrometer was developed from Francis Aston's mass spectrograph, which he first used in 1919 and which led to him being awarded the Nobel Prize for Chemistry in 1922. The way it works is shown in *Figure 1.3.3*.

The inside of the apparatus is pumped down to as near a perfect vacuum as possible. The sample is vaporised, then ionised by a stream of electrons which knocks one or more electrons out of some of the sample atoms or molecules, making them into positively charged ions, e.g. $A(g) + e^{\ominus} \longrightarrow A^{\oplus}(g) + 2e^{\ominus}$ for an atom of element A. Any positive ions go through a slit and are accelerated by plates which are negatively charged to several thousand volts. The ions, now moving fast, are sent through another slit into an intense magnetic field, and so are deflected.

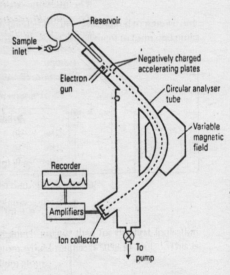

Figure 1.3.3   The mass spectrometer

Assuming that the ions are singly charged (higher charges will cause much more deflection), the amount of deflection will depend on the relative masses of the ions, with heavy ions being deflected least. In old instruments, the ions were detected by means of a long photographic plate, on which the beams of separated ions left a trace.

Newer instruments use various methods, including varying the field strength to bring the beams separately to an electronic detector. The relative intensities are proportional to the amount of each isotope present. The abundance of each isotope can therefore be measured, the weighted mean calculated and an accurate value found for the A_r. Typical mass spectra for two elements are shown in *Figures 1.3.4* and *1.3.5*. The major peak is always scaled to 100.

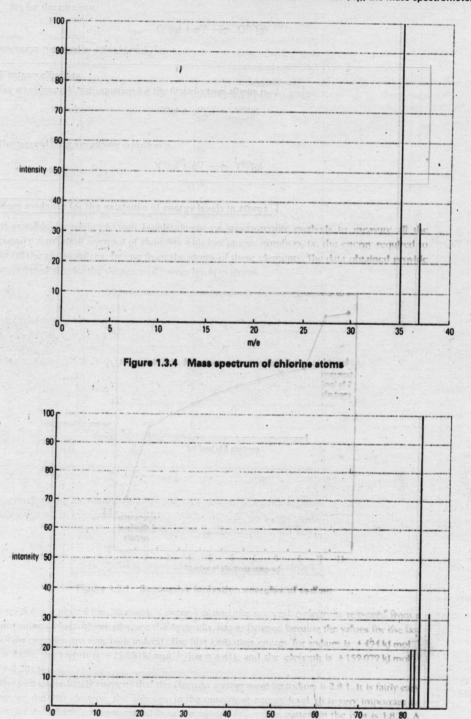

Figure 1.3.4   Mass spectrum of chlorine atoms

Figure 1.3.5   Mass spectrum of krypton atoms

The mass spectrum of Cl_2

Since the chlorine molecule contains two atoms, Cl_2, and there are two isotopes of chlorine: ^{35}Cl and ^{37}Cl, the mass spectrum of Cl_2 gas could be expected to show five m/e peaks, for the following reason.

The individual atoms at 35 and 37 would show the 3:1 ratio. The molecule could be $^{35}Cl^{35}Cl$, $^{35}Cl^{37}Cl$ or $^{37}Cl^{37}Cl$, so peaks at 70, 72 and 74 would appear, in a ratio 9:6:1.

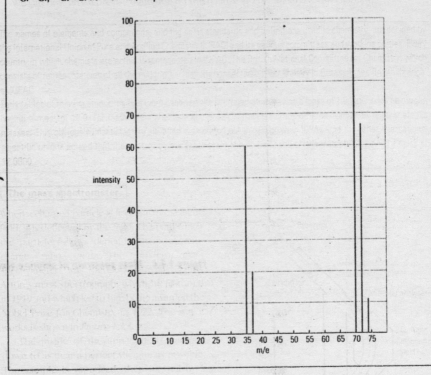

Mass spectrometry (now usually known by the initials MS) is now rarely used for measuring isotopic ratios. The major uses are in the accurate determination of the relative masses of new compounds and of the structures of organic compounds (see Ch 30.5).

Points to remember about the mass spectrometer.

- High vacuum;
- vaporised sample;
- sample ionised using high-energy electrons;
- $E(g) \rightarrow E^{\oplus}(g) + e^{\ominus}$;
- ions are accelerated by an intense electric field;
- these ions are deflected in a magnetic field;
- the amount of deflection depends on the mass/charge ratio (m/e), and hence, for all singly-charged ions, on the relative mass;
- the heavier ions are deflected the least;
- deflected ions are detected electronically;
- relative intensities and relative masses are measured;
- VIADD – Vaporise, Ionise, Accelerate, Deflect, Detect.

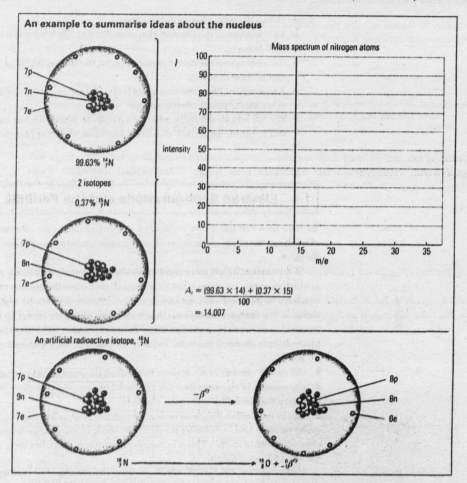

An example to summarise ideas about the nucleus

7p
7n
7e

99.63% $^{14}_{7}N$

2 isotopes

0.37% $^{15}_{7}N$

7p
8n
7e

Mass spectrum of nitrogen atoms

intensity

m/e

$$A_r = \frac{(99.63 \times 14) + (0.37 \times 15)}{100}$$
$$= 14.007$$

An artificial radioactive isotope, $^{16}_{7}N$

7p
9n
7e

$-\beta^{0}$

8p
8n
8e

$^{16}_{7}N \longrightarrow {}^{16}_{8}O + {}^{0}_{-1}\beta^{0}$

Q

EXAMINATION
QUESTIONS ON THE
NUCLEUS

1 (Remember there is a Periodic Table inside the back cover.)

   a Outline the principles of the simple mass spectrometer.

   b Why must the atoms be ionised in the mass spectrometer?

   c Show how the mass spectrometer can be used to determine the relative abundances of the isotopes in an element.

   d In the mass spectrum of gallium there are two peaks on the mass spectrum at m/e values of 69 (60.2%) and 71 (39.8%). Calculate the relative atomic mass of gallium to three significant figures.

2 **a** Draw up a table to summarise the relative masses and charges of protons, electrons and neutrons.

   b Work out the numbers of protons, neutrons and electrons present in $^{59}_{27}Co$ and in $^{39}_{19}K$.

   c Write, using conventional notation, the symbol for the isotope of caesium containing 78 neutrons.

3 Complete the following nuclear equations.

   a $^{24}_{11}Na \longrightarrow {}^{24}_{12}Mg + \ldots$ **b** $^{220}_{86}Rn \longrightarrow \ldots + {}^{4}_{2}\alpha$

   c $^{239}_{92}U \longrightarrow \ldots + {}^{0}_{-1}\beta$ **d** $^{238}_{92}U \longrightarrow {}^{234}_{90}Th + \ldots$

4 **a** Define the term *relative atomic mass*.

  b Two isotopes of the element boron are $^{10}_{5}B$ and $^{11}_{5}B$. What name is given to the number 5 in each isotope?

  c Naturally-occurring boron contains 19.7% of ^{10}B and 80.3% of ^{11}B. Calculate the relative atomic mass of boron.

  d A man-made radioactive isotope of boron has a mass number of 12. How many protons and neutrons are present in this isotope?

  e The $^{12}_{5}B$ isotope decays by emitting a β particle. Write a nuclear equation to show this decay, and so decide which new element is formed in the decay process.

1.4 Electron configurations and the Periodic Table

We have seen that the atomic number of an element means the number of protons in the nucleus of *all* the atoms of the element, and that the atomic number must equal the number of electrons in the atom.

In this section we will show that the *number* of electrons is all that is needed to give you the *way* in which electrons are arranged in an atom, and particularly the number of outer electrons – the ones which are furthest from the nucleus. It is these outermost electrons which get involved in chemical changes. The energies needed to disturb electrons which are nearer to the nucleus, and therefore attracted more firmly by it, are *far larger* than the energies met with in ordinary chemical reactions. (And 'ordinary chemical reactions' includes, for example, the explosions of Semtex or TNT!)

So:

- the atomic number of an element determines the number of electrons in its atom;
- the number of electrons fixes the arrangement – or *configuration* – of these electrons;
- hence the number of outermost electrons;
- that in turn makes the element react chemically in specific ways.

Someone once said: 'Chemistry is merely the study of the arrangement and re-arrangement of the outer electrons of atoms'. This is certainly not the whole truth, but there's enough truth in it to give us a basis for our thinking.

Understanding the way in which electrons are arranged in atoms will provide you with the power to make chemical predictions.

Provided you know where elements are situated in the Periodic Table, you will be able to predict quite well how they will react, what compounds they will form, and what the formulae and properties of these compounds will be.

But first we must look at the evidence for the existence of electron energy levels within atoms.

The evidence for electron energy levels

Electromagnetic radiation

Visible light forms only a tiny part of the whole spectrum of electromagnetic radiation (*Figure 1.4.1*).

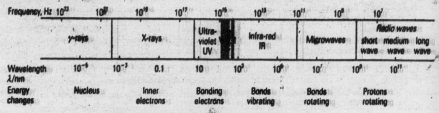

Figure 1.4.1 The electromagnetic spectrum

The shorter the wavelength of the radiation, the higher its frequency and the greater its energy. Ultra-violet (UV) radiation can damage the retina of the eye, and X-rays and γ-radiation can damage living tissue (severely so in the case of γ-radiation).

Two simple equations are needed.

$$c = \nu\lambda \tag{1}$$

where c, the speed of light, is a universal constant with a value of $3.00 \times 10^8 \, m \, s^{-1}$,
   ν (in Hertz, Hz) is the frequency, i.e. the number of oscillations per second, and
   λ (in metres, m) is the wavelength of the radiation.

> ν is the Greek letter 'nu';
> λ is the Greek letter 'lambda'.

This equation applies to all kinds of electromagnetic radiation. Note carefully that the equation clearly states that high frequency means short wavelength and low frequency means long wavelength.

$$E = h\nu \tag{2}$$

where E is energy (in joules, J),
   ν is frequency (units Hz), and
   h is Planck's constant = $6.63 \times 10^{-34} \, J \, s$.

Note that this equation clearly shows that high frequency (and hence short wavelength) means high energy.

Energy is not continuous and can only come in tiny packets. Such a packet of energy, $h\nu$, is known as a **quantum** (plural '**quanta**'). The higher the frequency, the bigger the quantum of energy. A quantum of light energy is called a *photon*.

> Max Planck was, in 1900, the first person to suggest that energy was not continuous and to develop the idea that energy can only be emitted or absorbed in small packets, called **quanta**. This led to developments in almost every branch of science and increasingly in technology as well. Planck received the Nobel Prize for Physics in 1918. One of Albert Einstein's first contributions to science was to realise, in 1905, that energy can be propagated through space, at the velocity of light, as quanta or photons with energy equal to $h\nu$. Einstein was given the Nobel Prize for Physics in 1921.

Emission spectra

When a solid is heated it eventually starts to glow red – it becomes red-hot. Even before then, you can sense that it is hot because of the infra-red (IR) radiation it is emitting. At an even higher temperature it becomes 'white hot'. If it appears white, an object must be emitting radiation in all parts of the visible spectrum.

Radiation of this kind – involving all wavelengths over a wide range – is known as *continuous radiation*. It is typical of *incandescent* objects – such as a tungsten light bulb filament, or the Sun. If such radiation is passed through a prism, or undergoes total internal reflection in water droplets so that a rainbow is formed, the familiar violet-to-red *continuous spectrum* is seen.

Reminder: Continuous radiation consists of a very large number of very small quanta. The spectrum appears to be continuous but is made up of quantised packets of energy.

The spectacular colours of fireworks are due to excited electrons and their emission spectra

When a gas is heated the radiation emitted is *not* continuous. You will have seen such *emission spectra* in sodium vapour (yellow) and mercury vapour (blue-green) street lights, neon advertising signs (red), and also in the flame tests that you may have done at school in order to identify certain metals, especially sodium and potassium (see Ch 6.3). Why do gases behave in this way?

The spectrum of atomic hydrogen

Radiation is given off from matter which has been *excited* by a high temperature (as in a flame test) or an electric discharge (as in a sodium vapour street lamp). As we have seen, a white-hot solid emits continuous radiation, and passing this through a prism in a spectrometer gives the usual continuous violet-to-red visible spectrum. However, if the excited material is a gas or vapour (e.g. in a hydrogen discharge tube), a *line spectrum* is the result (*Figure 1.4.2*).

The main point to remember is this: *each line in a line spectrum corresponds to a definite wavelength, hence to a definite frequency (equation (1) above), and therefore also to a definite quantum of energy (equation (2) above).*

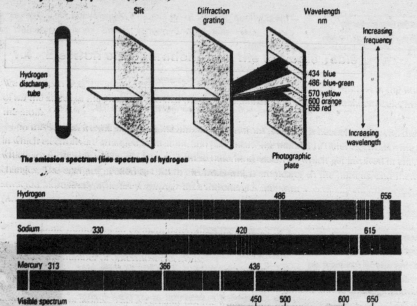

The emission spectrum (line spectrum) of hydrogen

Figure 1.4.2   Examples of emission spectra

The simplest emission spectrum is, not surprisingly, that of the simplest atom – hydrogen. The spectrum is obtained by passing an electric discharge through hydrogen at very low pressure. The H_2 molecules are torn apart to give excited H atoms. The electron in the hydrogen atom can be excited to higher energies and *light energy is given out as it drops back to lower energy levels (Figure 1.4.3).*

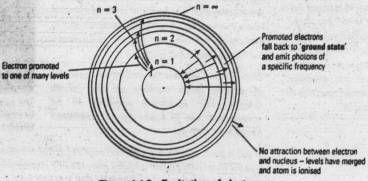

Figure 1.4.3   Excitation of electrons

Visible spectra can be observed not only by using a prism but also by using a 'diffraction grating'. This consists of many fine lines scratched parallel and very close together on a flat surface. Real diffraction gratings are very expensive, but you can get the same effect with a CD or an old-fashioned LP record. The CD surface gives the spectral colours in daylight. With the LP, hold it so that light from a clear filament bulb reflects off it at a low angle into your eye. Tilt it to and fro slightly to get the best effect. You should be able to see two complete spectra and part of a third. If you have sodium or mercury vapour street lamps or neon advertising signs nearby, try the same with them.

Figure 1.4.4 is a simplified diagram of the spectrum of atomic hydrogen. Note particularly that there are several series of lines, and that in each series the lines *converge* – they become closer together – at the high frequency/short wavelength/high energy end of the series.

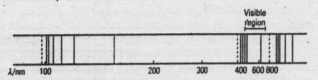

Figure 1.4.4   **Part of the spectrum of atomic hydrogen**

Study of the spectrum of atomic hydrogen led to the conclusion that electrons in an atom can exist only in definite energy levels. Energy can only be absorbed in definite packets – quanta – to excite the electrons in the hydrogen atom to a higher energy level. As the excited electron falls back to a lower energy level, radiation is emitted. The frequency of this radiation is given by a modification of equation (2) above, $\Delta E = h\nu$, where ΔE is the difference in energy between the energy level the electron started from and the level it lands in. Obviously, the greater the energy difference the higher the frequency – and hence the shorter the wavelength – of the radiation given out.

As the electron gets further away from the nucleus, it is influenced less and less by it. The difference between successive energy levels becomes smaller (*Figure 1.4.5*).

This means that if an electron is dropping back from higher energy levels into a particular lower level, the differences in energy between successive lines towards the higher-energy end of the series will get less. This then means that the lines in a series of lines become closer together – they *converge* – towards the high-energy end.

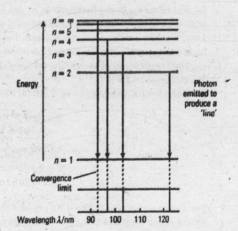

Figure 1.4.5   **'Falling electrons' and line spectra**

When the lines eventually run together and the series therefore comes to an end at what is called the '*convergence limit*', the electron must be so far away from the nucleus that the nucleus no longer has any attraction for it. The hydrogen atom has lost its electron; the atom has become a hydrogen ion.

The series of lines in the hydrogen spectrum

Now look at *Figure 1.4.6*.

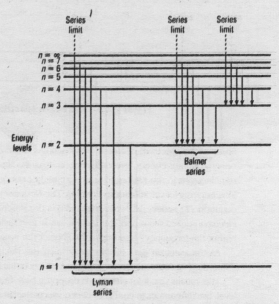

Figure 1.4.6 Some of the electron energy transitions in the hydrogen atom

Excited electrons in the hydrogen atom can fall back into lower energy levels. One series of lines (called the Lyman series, after its discoverer) is obtained when electrons fall into the lowest energy level, i.e. the one closest to the nucleus. The energy differences are large, so the frequencies are high. Consequently, the Lyman series is in the ultra-violet (UV) part of the spectrum.

The Lyman series of lines in the UV spectrum of atomic hydrogen reaches its convergence limit at a frequency of $3.3 \times 10^{15}\, s^{-1}$.

Using $E = h\nu$, where $h = 6.63 \times 10^{-34}\, J\, s$ (see equation (2) above), we see that:

$$E = 6.63 \times 10^{-34} \times 3.31 \times 10^{15}\, J$$
$$= 2.195 \times 10^{-18}\, J \text{ per atom}$$

∴ for one **mole** of atoms

$$E = 2.195 \times 10^{-18} \times 6 \times 10^{23}\, J\, mol^{-1}$$
$$= 1320\, kJ\, mol^{-1}$$

> A **mole** is the standard unit of the amount of substance (see Ch 9.3). One mole of a substance contains 6×10^{23} of the specified particles.

The Lyman series (see *Figure 1.4.6*) results from electrons falling back into the ground state – the lowest energy level – of hydrogen. So +1320 kJ is the energy needed to remove the electron from one mole of hydrogen atoms and to convert them into $H^{\oplus}$ ions – i.e. *the ionisation energy of hydrogen is +1320 kJ mol⁻¹*.

$$H(g) \longrightarrow H^{\oplus}(g) + e^{\ominus}$$

The Balmer series results from electrons falling into the second energy level; the energy differences are less and the lines in this series are in the visible region of the spectrum. The series which result from electrons falling to the third, fourth, fifth, etc., levels are linked with progressively smaller energy changes, and therefore are found in the infra-red (IR) part of the spectrum.

Energy levels and electron configurations

When better spectrometers became available, it was found that most lines in the hydrogen spectrum could, in their turn, be analysed into a series of fine lines. It became apparent that each energy level, except the lowest, consisted of two or more sub-levels. The lines corresponding to the sub-levels could themselves often be split in the presence of a magnetic field. Also, when they were examined at very high resolution in the absence of a magnetic field, many lines were found to consist of two lines very close together.

It was also found that a beam of electrons could be diffracted, just like a beam of light. Such experiments showed that electrons – indeed, *all* particles – have a 'wave-particle' duality: which means that in some experiments electrons show the properties of particles, and in others they behave like waves. The Austrian physicist Erwin Schrödinger and others developed a mathematical treatment known as 'wave mechanics' which was applied to the spectroscopy results to build the foundations of our ideas about electron configurations in atoms.

The ideas which summarise our understanding of all these observations are as follows on the next page and are summarised in *Figure 1.4.7*, which shows the electron arrangement in an atom of sulphur.

One of the nicest possible illustrations of the dual nature of the electron is that, as we have seen, J J Thomson won the Nobel Prize for Physics in 1906 for showing the reality of the electron as a *particle*, J J's son, George Paget (G P) Thomson, jointly with Clinton Davisson of America, was awarded the Nobel Prize for Physics in 1937 for their demonstration that electrons could behave as *waves!* They showed that a beam of electrons could be diffracted by crystals, in the same way that a beam of X-rays can.

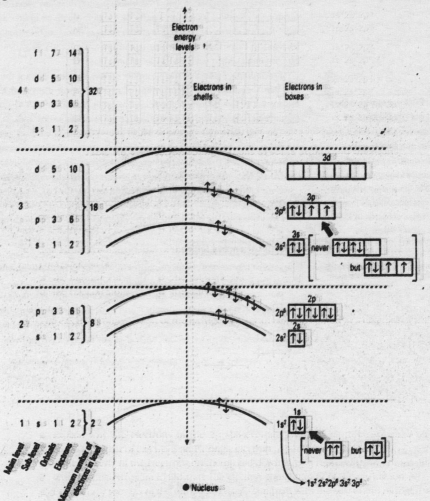

Figure 1.4.7 The electron arrangement for sulphur

- The *main energy levels* are numbered 1 (the lowest energy level), 2, 3, etc.
- Each main energy level consists of *sub-levels*. The number of sub-levels is equal to the number used for the main energy level. So, the first main energy level has only one sub-level, the second has two, and so on.

  The sub-levels (thanks to the rather odd and unhelpful names given to the groups of lines by the old-time spectroscopists) are known by the letters *s, p, d,* and *f* (for sharp, principle, diffuse and fundamental ... well, we did describe them as odd and unhelpful!)
- Each sub-level consists of one or more *orbitals*. Each sub-level contains an odd number of orbitals. The s-sub-level consists of only one orbital, the p-sub-level has three orbitals, the d- five and the f- seven.
- Each orbital can contain a maximum of *two* electrons.

  If there are two electrons in one orbital, they must have opposite *spins*. (No-one knows exactly what is meant by the spin of an electron, but the idea works!)

shown as ⏐↑↓⏐

Spinning electrons?

The energy sub-levels

Two main consequences follow from these points. First, any electron in an atom can be described by using four so-called 'quantum numbers' – which define its main energy level, sub-level, orbital and spin. No two electrons in any one atom can have the same four quantum numbers. This is the **Pauli Exclusion Principle**. Second, we can now work out the *maximum* number of electrons which can be contained within each energy level and sub-level.

From points 3 and 4 in the summary above, it is obvious that the s-sub-level has only one orbital and can therefore hold a maximum of two electrons; the p-sub-level has three orbitals, hence a maximum of six electrons; for d, five orbitals, maximum 10 electrons; and for f, seven orbitals, maximum 14 electrons. (Note the simple arithmetical series 2, 6, 10, 14, ...).

The first main energy level has only the s-sub-level, and so can only hold a maximum of two electrons. The second energy level has both s- and p-sub-levels, so can hold up to eight electrons. The third has s, p, and d, so can hold up to 18 electrons. The fourth has s, p, d, and f, so can hold up to 32 electrons. The further the main energy levels are from the nucleus, the more electrons they can hold.

If we know the atomic number of an atom (see Ch 1.2), we know the number of electrons in an atom; we can now easily work out its ground state electron (or electronic) configuration. Simply put the electrons into the atom and fill the energy levels from the bottom up.

> At GCSE you may have been taught that electrons are placed in 'shells' with the third shell only holding eight electrons. As the atoms become larger, the sequence of filling the energy levels and sub-levels is slightly more complex, but perfectly logical, as we will shortly see.

Filling the energy levels

The easiest way to remember the order in which electrons fill the energy levels in atoms is shown in *Figure 1.4.8*, which is a very useful aid for your memory!

The parallel diagonal lines give the order of filling, beginning at the lowest energy level, 1s. For example, boron has atomic number 5 and therefore five electrons. The electron arrangement can be written as 2.3, showing two electrons in the first energy level and three in the second. In more detail, we could write this as $1s^2 2s^2 2p^1$, showing how many electrons there are in each sub-level.

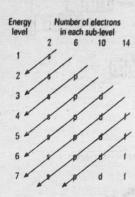

Figure 1.4.8 Filling electron energy levels

Remember: the notation for electron configuration is:

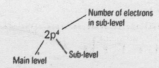

$$2p^4$$

Number of electrons in sub-level

Main level   Sub-level

A more complicated example is arsenic, atomic number 33, which can be written as 2.8.18.5, or in more detail: $1s^2 2s^2 2p^6 3s^2 3p^6 4s^2 3d^{10} 4p^3$.

(Usually, however, once the 'filling-up' sequence has been worked out, the sequence is tidied up and the sub-levels in each main level are put together: so for arsenic $1s^2 2s^2 2p^6 3s^2 3p^6 3d^{10} 4s^2 4p^3$.) Now try some for yourself.

Electron configurations

$_{11}Na$ *has 11 electrons.*

So: 2.8.1 in the main energy levels (the figures must add up to 11) and $1s^2 2s^2 2p^6 3s^1$ (going through the grid in *Figure 1.4.8* until the superscripts add up to 11).

$_{20}Ca$ *has 20 electrons.*

So: 2.8.8.2 in the main energy levels (the figures must add up to 20) and $1s^2 2s^2 2p^6 3s^2 3p^6 4s^2$ (going through the grid in *Figure 1.4.8* until the superscripts add up to 20).

An alternative way to show the electrons filling the available energy levels uses the energy levels and sub-levels. This is illustrated for chlorine – 2.8.7:

Each electron, represented by an arrow, is placed in the lowest available level, following the rules outlined in the last section.

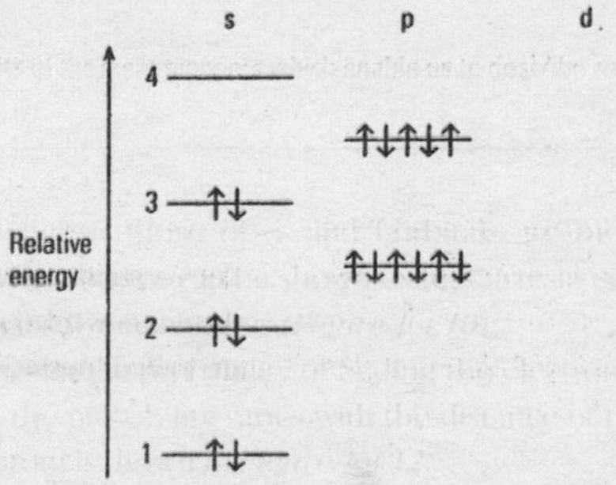

The electron arrangement for chlorine

Electrons in boxes

This electron configuration for arsenic can be used to bring out another point. The detailed structure ends ... $4s^2 4p^3$. We can represent orbitals by little boxes – one box for the 4s orbital, three for the 4p orbitals. We can represent the electron spins by arrows. Remember, for two electrons to occupy the same orbital/box, they must have opposite spins. So $4s^2 4p^3$ looks like this:

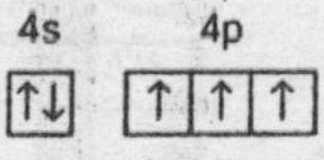

Why does each 4p orbital only have one electron in it? (That this is actually so can be shown by careful examination of the emission spectrum of arsenic vapour.) The answer is that electrons don't pair up unless they have to, which is not so very surprising. Under normal circumstances, the orbitals in a sub-level all have the same energy. One electron will go into each orbital, and only when each of the orbitals is half-filled with one electron will additional electrons start pairing up to fill the orbitals one by one, showing the electrons with opposite spins as ↑↓.

An *exactly* half-filled p- or d-sub-level appears to possess extra stability, as does a filled sub-level. (The reasons for this are complicated.) Some consequences of this extra stability are found in Ch 25.3.

It is for this reason that the ground state configurations of chromium and copper are not exactly as predicted by *Figure 1.4.8*. That of chromium ends ... $3d^5 4s^1$ instead of ...$3d^4 4s^2$, and copper ends ...$3d^{10} 4s^1$ instead of ...$3d^9 4s^2$. These are the only common exceptions to the 'filling up' rules.

Electron energy levels and the Periodic Table

The result of some hard work is that now we can start to link the structure of the Periodic Table to the structure of the atom. This linkage was one of the great triumphs of early twentieth-century chemistry and physics, and opened the way to many of the developments of materials which have influenced our lives.

The first 'row' of the Periodic Table corresponds to the 1s level and is complete at helium, atomic number 2. Thereafter, each row corresponds to the s- and p-sub-levels of the successive main energy levels (see *Figure 1.4.9* and *Table 1.4.1*), with the d-block and f-block breaking up the later rows in accordance with the order of filling of sub-levels shown in *Figure 1.4.8*.

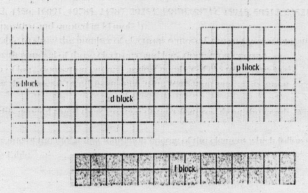

Figure 1.4.9   The block structure of the Periodic Table

The vertical columns in the Periodic Table are called 'groups' and the group number of many elements is related to the electron configuration of the outermost energy level. The horzontal 'rows' are called 'periods'. The Periodic Table will be discussed in detail in Ch 5.

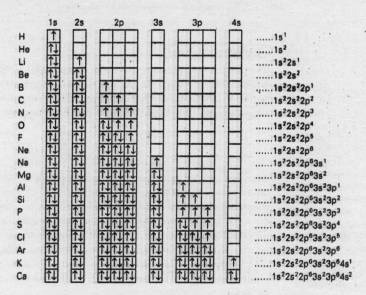

	1s	2s	2p	3s	3p	4s	
H	↑						......$1s^1$
He	↑↓						......$1s^2$
Li	↑↓	↑					......$1s^2 2s^1$
Be	↑↓	↑↓					......$1s^2 2s^2$
B	↑↓	↑↓	↑				......$1s^2 2s^2 2p^1$
C	↑↓	↑↓	↑ ↑				......$1s^2 2s^2 2p^2$
N	↑↓	↑↓	↑ ↑ ↑				......$1s^2 2s^2 2p^3$
O	↑↓	↑↓	↑↓ ↑ ↑				......$1s^2 2s^2 2p^4$
F	↑↓	↑↓	↑↓ ↑↓ ↑				......$1s^2 2s^2 2p^5$
Ne	↑↓	↑↓	↑↓ ↑↓ ↑↓				......$1s^2 2s^2 2p^6$
Na	↑↓	↑↓	↑↓ ↑↓ ↑↓	↑			......$1s^2 2s^2 2p^6 3s^1$
Mg	↑↓	↑↓	↑↓ ↑↓ ↑↓	↑↓			......$1s^2 2s^2 2p^6 3s^2$
Al	↑↓	↑↓	↑↓ ↑↓ ↑↓	↑↓	↑		......$1s^2 2s^2 2p^6 3s^2 3p^1$
Si	↑↓	↑↓	↑↓ ↑↓ ↑↓	↑↓	↑ ↑		......$1s^2 2s^2 2p^6 3s^2 3p^2$
P	↑↓	↑↓	↑↓ ↑↓ ↑↓	↑↓	↑ ↑ ↑		......$1s^2 2s^2 2p^6 3s^2 3p^3$
S	↑↓	↑↓	↑↓ ↑↓ ↑↓	↑↓	↑↓ ↑ ↑		......$1s^2 2s^2 2p^6 3s^2 3p^4$
Cl	↑↓	↑↓	↑↓ ↑↓ ↑↓	↑↓	↑↓ ↑↓ ↑		......$1s^2 2s^2 2p^6 3s^2 3p^5$
Ar	↑↓	↑↓	↑↓ ↑↓ ↑↓	↑↓	↑↓ ↑↓ ↑↓		......$1s^2 2s^2 2p^6 3s^2 3p^6$
K	↑↓	↑↓	↑↓ ↑↓ ↑↓	↑↓	↑↓ ↑↓ ↑↓	↑	......$1s^2 2s^2 2p^6 3s^2 3p^6 4s^1$
Ca	↑↓	↑↓	↑↓ ↑↓ ↑↓	↑↓	↑↓ ↑↓ ↑↓	↑↓	......$1s^2 2s^2 2p^6 3s^2 3p^6 4s^2$

Table 1.4.1   'Electrons in boxes' for the first 20 elements

The shapes of orbitals

This may seem a rather over-the-top subject, but an understanding of these shapes will help you greatly when we come to consider chemical bonding and why molecules have the shapes they do.

In 1926 Erwin Schrödinger developed an equation to describe the behaviour of electrons in atoms. This equation can only be solved exactly for the hydrogen atom, which has only two particles – a proton and an electron. Whenever there are three or more bodies involved – as, for example, with a planet under the gravitational pulls of a double star – exact solutions of equations are not possible. However, computers can now be used to obtain increasingly better approximate solutions.

An atomic orbital is best thought of as a solution of the Schrödinger equation. It is quite in order to talk about the shapes of *vacant* orbitals – and, indeed the exact solutions for hydrogen are used to show the orbital shapes. Orbitals indicate the *probability* of finding an electron at any point. When an electron, or at most two electrons, have suitable energies and locations for an orbital, they are said to *occupy* that orbital.

Because different orbitals represent how the probability of finding the electrons is distributed in space around the nucleus, different orbitals correspond to different energies. An electron in a particular orbital in a particular atom therefore has a definite energy.

If all this sounds a bit vague, so in a sense it is. Complex and abstract mathematics is used to try to describe the behaviour of *real* systems. One of the basic ideas in physics is summed up by *Heisenberg's Uncertainty Principle* (1927) – *It is not possible to know both the velocity and position of any particle at the same time.* The smaller the particle the more impossible it is ... and electrons are very small indeed.

For our purposes in this book, all that matters is to use those results of the great pioneers which enable us to describe *real* materials.

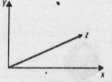

Figure 1.4.10 Cartesian co-ordinates

The shapes of s- and p-orbitals are quite simple, whereas those of d- and f-orbitals are more complicated. We shall discuss only s- and p-orbitals. Their shapes can be shown using Cartesian co-ordinates – that is, using x, y, and z axes all at right angles to one another (*Figure 1.4.10*).

All s-orbitals are spherical in shape (*Figure 1.4.11*). But the probability of finding the electron is not equal in all parts of the sphere. The way in which the probability varies with the distance of the electron from the nucleus in an unexcited hydrogen atom is shown in *Figure 1.4.12*.

Figure 1.4.11 The shape of the 1s orbital

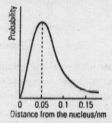

Figure 1.4.12 Electron probability profile – 1s orbital in the hydrogen atom

Notice that the maximum probability is at a very precisely known distance from the nucleus, 0.0529 nm, but the electron could be further in or further out. The orbital (indeed, the main energy level) involved in the unexcited hydrogen atom is 1s. The 2s orbital is again spherical, but the variation of probability with distance from the nucleus – this time for a slightly excited hydrogen atom, with the electron possessing enough energy to occupy the 2s orbital – is shown in *Figure 1.4.13*.

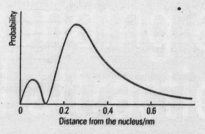

Figure 1.4.13 Electron probability profile of the 2s orbital in the hydrogen atom

Notice that this time there is a slight probability of finding the 2s electrons quite near to the nucleus (*Figure 1.4.14*).

Figure 1.4.14   The shape of the 2s orbital

This so-called *penetration* of s-electrons towards the nucleus leads to a slight increase in difficulty in removing s-electrons from some atoms, and this in turn leads to some interesting chemical properties which you will need to know about (see Ch 7.3).

The shapes of the three p-orbitals are shown in *Figure 1.4.15*. These shapes, and the consequences of mixing them with s- and d-orbitals, profoundly influence the shapes and behaviour of molecules (as we shall see in Chs 2, 3, 14, 25, and 26!)

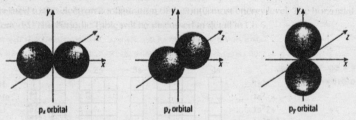

Figure 1.4.15   The shapes of the p-orbitals

Note that under *normal circumstances*, all the orbitals in a sub-level are at the same energy. However, in certain compounds, the five orbitals of the d-sub-level can be split, and this also has some important consequences in terms of both chemical behaviour and colour (see Ch 25.3).

Electron structures – worked example

The element phosphorus has 15 electrons in the neutral atom. On the diagram (Figure 1.4.16) show the electron structure in the various energy levels and sub-levels, using arrows to represent the electrons. Justify the inclusion of phosphorus in the p-block of the Periodic Table. Use this diagram to show how the emission (or line) spectrum of phosphorus is produced when the electrons are excited.

Start with the 1s level and fill the levels from the lowest level until the 15 electrons have been accounted for. Remember to put the electrons as pairs with two electrons in an s-sub-level and six electrons (three pairs) in a p-sub-level. Put the electrons into the p-sub-levels as three single electrons before they start to pair up. So the 3p sub-level has three single electrons.

Figure 1.4.16

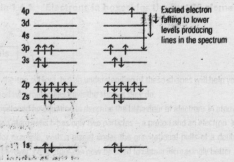

Werner Heisenberg won the Nobel Prize for Physics in 1932 for his work in quantum mechanics. One consequence of his Uncertainty Principle was used by Neils Bohr as a basis for Bohr's idea of 'complementarity'; if you observe a system, you can interact with it and disturb it. This only becomes important for very small particles, such as the electron.

In the few years from 1928, Bristol-born Paul Dirac succeeded in linking Schrödinger's and Heisenberg's work with Einstein's theory of relativity and even predicted the existence of antimatter. Dirac and Schrödinger were jointly awarded the Nobel Prize for Physics in 1933.

Since the last electron is placed into a p-sub-level, this element must be in the p-block in the Periodic Table.

When the electrons are excited (by putting energy into the atom) the electrons move up to higher energy levels. For example, one of the 3p electrons could move up to a 4p sub-level. When this electron falls back to a lower level, the extra energy is emitted as a series of energy values, producing a series of lines in the energy spectrum.

While not required in this answer, it might help you to appreciate that this electron configuration of phosphorus can also be written as:

$1s^2 2s^2 2p^6 3s^2 3p^3$, or in electron boxes as:

1s	2s	2p	3s	3p
↑↓	↑↓	↑↓ ↑↓ ↑↓	↑↓	↑ ↑ ↑

An overall picture of an atom? – a number of orbitals each containing up to two electrons

SELF-ASSESSMENT QUESTIONS 1.4

1　Write out the ground state electron configuration, using the s and p notation, of the following atoms.
　a　Carbon　　　　　b　Argon
　c　Phosphorus　　　d　Oxygen

2　Using electrons in boxes, show how the electrons are arranged in the following atoms.
　a　Lithium　　　　　b　Magnesium
　c　Silicon　　　　　 d　Boron

3　In which block and group of the Periodic Table would you place each of the elements with the following electron configurations?
　a　$1s^2 2s^1$　　　　　　　b　$1s^2 2s^2 2p^5$
　c　$1s^2 2s^2 2p^6 3s^2$　　　d　$1s^2 2s^2 2p^6 3s^2 3p^6 3d^{10} 4s^2 4p^4$

Definition: An ion is an atom or group of atoms which carries an electric charge.

It is incorrect to say that the charge on the nucleus has increased or decreased. The atom may carry a charge, but the nucleus is untouched.

1.5　Ionisation energies and electron affinities

It takes energy to remove an electron from an atom of any element. Even the outermost electron of a very big atom is still attracted quite strongly to the nucleus, and work will have to be done against that attractive force.

Once an electron *is* removed, the balance of protons and electrons in the atom is upset. Protons outnumber electrons, and a positively charged **ion** is left.

Once the atom becomes an ion with one positive charge, it becomes much harder to remove a second electron. This is because the electron must now be removed from a positive ion. The energy needed to remove the last electrons from even a fairly small atom like sodium is so great that the process is probably easier to observe (using a spectroscope) in stars than it is to make it happen on Earth.

It is easy to remember that there is *attraction* between the positive nucleus in an atom and the electrons around it. It is equally easy to forget that there is *repulsion* between the negatively charged electrons. The size of an atom is determined by the balance between these attractions and repulsions. If an electron is removed from an atom, the electron–electron repulsion is reduced and the ion becomes smaller in size. This means that the new outermost electrons are *closer* to the nucleus than was the first electron which was removed; so the second electron is *always* far harder to remove, and the second ionisation energy always much higher than the first – even if the two electrons were in the same sub-shell or even orbital in the neutral atom.

Therefore, a positive ion is always smaller than its parent atom.

And what is more, *a negative ion is always bigger than its parent atom*, because any extra electron(s) will increase the repulsions inside the atom.

Ionisation energies

Obviously, the amount of energy needed to remove an electron depends on the atomic number (= number of protons in the nucleus) and the electron configuration of atoms of the element involved. And the mole is, as always, taken as the standard quantity of substance.

So, for example, for the process:

$$Na(g) \longrightarrow Na^{\oplus}(g) + e^{\ominus}$$

the energy needed is $+494\ kJ\ mol^{-1}$. This is the **first ionisation energy** of sodium. Note that this process of removing electrons against attraction by the nucleus *always* requires energy and is *endothermic*.

The **second ionisation energy** of an element is defined as the energy required to remove one electron from each ion in one mole of gaseous monopositive ions of the element to form one mole of gaseous di-positive ions.

> **Definition:** The molar first ionisation energy of an element is the energy required to remove one electron from each atom in one mole of gaseous atoms of the element to form one mole of gaseous monopositive ions.

Ionisation energies

For an element, E, the equation for the first ionisation energy is:

$$E(g) \longrightarrow E^{\oplus}(g) + e^{\ominus}$$

The second ionisation energy equation is:

$$E^{\oplus}(g) \longrightarrow E^{2\oplus}(g) + e^{\ominus}$$

For the process:

$$Na^{\oplus}(g) \longrightarrow Na^{2\oplus}(g) + e^{\ominus}$$

the energy required is $+4564\ kJ\ mol^{-1}$ – which is very much greater than for the first ionisation energy. So to remove two electrons from a sodium atom would require $+5058\ kJ\ mol^{-1}$. This is more than double the first ionisation energy of the helium atom.

And so on for the third and successive ionisation energies.

Electron affinities

Electron affinities are the opposite of ionisation energies!

For example, for the process:

$$O(g) + e^{\ominus} \longrightarrow O^{\ominus}(g)$$

the energy released is $-142\ kJ\ mol^{-1}$. Note that this process usually releases energy, i.e. it is usually *exothermic*. However, second and successive electron affinities *always* require energy, e.g. they are *endothermic*. This is because once the first electron is added to the atom, a negative ion is formed; this ion then *repels* incoming electrons.

> **Definition:** The molar first electron affinity of an element is the energy change when one electron is added to each atom in one mole of gaseous atoms of the element to form one mole of gaseous mononegative ions.

So, for the process:

$$O^{\ominus}(g) + e^{\ominus} \longrightarrow O^{2\ominus}(g)$$

the energy required is $+844$ kJ mol^{-1}.

Electron affinities

For an element Y, the equation for the first electron affinity is:

$$Y(g) + e^{\ominus} \longrightarrow Y^{\ominus}(g)$$

The second electron affinity equation is:

$$Y^{\ominus}(g) + e^{\ominus} \longrightarrow Y^{2\ominus}(g)$$

A large majority of the first electron affinities of the elements are exothermic. The twenty or so elements for which the first electron affinity is endothermic include the elements in Group 2 of the Periodic Table plus the noble gases.

More evidence for the existence of energy levels in atoms

It is possible by using electron bombardment or spectroscopic methods to measure all the *successive ionisation energies* of elements with low atomic numbers, i.e. the energy required to strip off the electrons one by one from the atoms of these elements. The data obtained provide powerful evidence for the existence of energy levels in atoms.

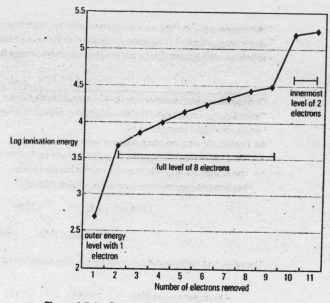

Figure 1.5.1 Successive ionisation energies of sodium

Figure 1.5.1 is a plot of $\log_{10}$(ionisation energy) against the successive electrons removed from a sodium atom. (A logarithmic plot, see the Appendix, has to be used because the values for the last ionisation energies are very high indeed. The first ionisation energy for sodium is $+494$ kJ mol^{-1} ($\log = 2.69$); the eighth is $+25\,500$ kJ mol^{-1} ($\log = 4.41$); and the eleventh is $+159\,079$ kJ mol^{-1} ($\log = 5.20$).)

The plot immediately suggests that the electron arrangement in sodium is 2.8.1. It is fairly easy to remove the first electron, the only one in the outermost energy level. (It is very important to remember that electrons are removed *from the outside in* – so the pattern in the plot is 1.8.2). A group of eight comes next, with roughly linear variation. Finally come the two innermost electrons, where very high energies are needed to remove them from the attraction of the nucleus.

Figure 1.5.1 shows clearly the three main energy levels in sodium. We know from Ch 1.4 that the detailed structure of sodium is $1s^2 2s^2 2p^6 3s^1$; using the boxes notation:

1s	2s	2p	3s
↑↓	↑↓	↑↓ ↑↓ ↑↓	↑

Once the 3s electron is removed, there is a big 'jump' in the graph. This reflects the increased difficulty of breaking into the next energy level, which is closer to the nucleus. There is another big jump once the 2p and 2s electrons have been removed.

Examination of the successive ionisation energies for any element reveals straight away the number of electrons in its outer energy level. For example, the first eight ionisation energies for magnesium and silicon are given in *Table 1.5.1*.

	1st	2nd	3rd	4th	5th	6th	7th	8th
Magnesium	+738	+1451	+7733	+10541	+13629	+17995	+21704	+25657
Silicon	+789	+1577	+3232	+4356	+16091	+19785	+23787	+29253

Table 1.5.1   Successive ionisation energies/kJ mol^{-1}

For magnesium, the big increase comes after two electrons have been removed; consequently the outer energy level of magnesium must contain two electrons. With silicon, it comes after four electrons: so silicon has four electrons in the outer energy level.

Recognising the new energy level – worked example

With a string of numbers for the successive ionisation energies, one way of recognising the break into a new main energy level is:

- Look at the first number only in each pair of numbers.
- What factor do you need to change the first number into the second?
- The largest 'jump' will then involve the largest factor.

For example: successive ionisation energies (kJ mol^{-1})

Lithium: +520   +7298   +11815

about ×14   about ×1½

The new level starts at the second figure – therefore there must be one outer electron.

Aluminium: +578   +1817   +2745   +11578   +14831

about ×3   about ×1½   about ×5   about ×1

The new level starts at the fourth figure – therefore there must be three outer electrons.

The plot of successive ionisation energies for one element does not give much information about sub-levels, although with careful plotting it is possible to spot a 3.3.2 structure within the group of eight electrons in the plot for sodium (*Figure 1.5.2*). This corresponds to the removal of the paired electrons in each 3p orbital, then the unpaired electrons, and finally the electrons in the 3s orbital.

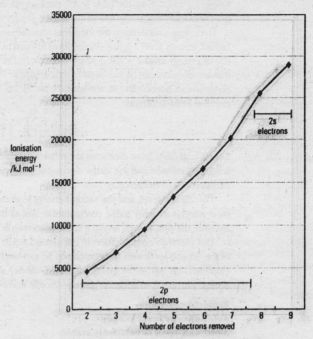

Figure 1.5.2   Sodium 2nd level ionisation energies

The first ionisation energies of the elements

A much clearer indication of the sub-levels is given by a plot of the molar first ionisation energies of the elements against their atomic numbers (*Figure 1.5.3*).

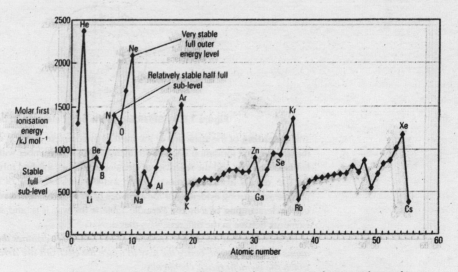

Figure 1.5.3   Plot of molar first ionisation energy against atomic number

Reference to the Periodic Table inside the back cover reveals that each peak in the plot corresponds to a chemically inactive noble gas, and each trough to a very reactive alkali metal.

Hydrogen and helium, the elements with the simplest atoms, form an initial grouping of two. At helium, energy level 1 is full. Energy level 2 begins at lithium (Li). At beryllium (Be) the first sub-level (2s) of energy level 2 is full. A single electron can be removed from a half-full 2p orbital more easily than an electron can be removed from a complete 2s sub-level; this is why there is a slight dip from beryllium to boron (B). At nitrogen (N; $1s^2 2s^2 2p^3$) the 2p sub-level is exactly half full, with one electron in each orbital:

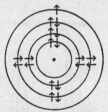

Exactly half-filled p- or d-orbitals are relatively stable (see Ch 1.4). With one more electron, stability is slightly reduced and the 'extra' electron is more easily removed; so the plot shows a slight dip from nitrogen (N) to oxygen (O).

The 2p sub-level, and the second energy level overall, is full at neon (Ne). Eight electrons in the outer shell is a highly stable configuration, and all the noble gases have it (except for helium which has only the first energy level); it is the main reason for their exceptionally low chemical reactivity.

The electron configuration of calcium is 2.8.8.2. The next ten elements correspond to the filling of the 3d sub-level with the completed 3d sub-level (and hence completed third energy level of 18 electrons – $3s^2 3p^6 3d^{10}$) giving increased stability and hence a slightly higher value for zinc. The normal pattern is resumed at gallium (atomic number 31).

Remember:
A full energy level is stable.
A full sub-level is relatively stable.
A half-full sub-level is relatively stable.

Shielding electrons

It is worth looking once more at the peaks (the noble gases, He, Ne, Ar, Kr, Xe) and the troughs (the alkali metals, Li, Na, K, Rb, Cs) in *Figure 1.5.3*. Notice that as for each of these families of elements, as indeed for *any* such family, the first ionisation energy decreases with increasing atomic number. The more complete energy levels of electrons there are between the outermost electrons (negative) and the nucleus (positive), the easier it is to remove an electron from the outermost level (*Figure 1.5.4*).

Figure 1.5.4   Shielding of the outer electron in potassium

This **shielding** of the attraction of the nucleus by complete electron energy levels is the reason for the drop in first ionisation energy with increasing atomic number within a family of elements. The link between this and the reactivity of elements will be established further in Ch 5.4.

The results from both spectroscopy and the measurement of ionisation energies together provide very strong evidence for *the existence of electron energy levels and sub-levels*, and so help our understanding of *why* the Periodic Table is the way it is, and so our understanding of why elements behave in the different ways they do.

Remember that *the atomic number of an element determines the number of electrons in the atoms of that element, hence the way in which the electrons are arranged, and hence the way in which the element reacts chemically*.

SELF-ASSESSMENT
QUESTIONS
1.5

1　Write an equation to show:
　a　the first ionisation energy of beryllium;
　b　the third ionisation energy of neon;
　c　the first electron affinity of fluorine.

2　Justify the following statements, using the data provided.
　a　Caesium has one outer electron in its atom, since the first three ionisation energies of caesium are $+376$, $+2420$ and $+3300$ kJ mol^{-1}.
　b　The element with the electron configuration $1s^2 2s^2 2p^6 3s^2 3p^4$ is a p-block element.
　c　The element with successive ionisation energies $+789$, $+1577$, $+3232$, $+4356$, $+16091$, $+19785$, $+23787$ and $+29253$ kJ mol^{-1} has four outer electrons in its atom.

3　a　Define the term *molar first ionisation energy*.
　b　Write an equation for the first ionisation energy of helium.
　c　Explain why helium has the highest first ionisation energy of all the elements.
　d　How does the first ionisation energy of the noble gases change with increasing atomic number?
　e　Why does it change in this way?

4　a　The first ionisation energies of the elements of atomic number 4 to 9 are (in kJ mol^{-1}):
　　$+900$, $+801$, $+1086$, $+1402$, $+1314$, $+1681$.
　　Plot a graph of these figures against the atomic number of the element.
　b　On your graph show where you would expect the values of element number 3 and element number 10 to appear.
　c　How would you expect the first ionisation energy of element number 11 to compare with that of element number 3?
　d　Although there is a general increase in ionisation energy from elements 4 to 9, explain why there is a 'blip' between element number 7 and element number 8.

5　a　Write out the electron configuration, as electrons in boxes, for sulphur.
　b　Sketch the graph you might expect to see for the successive ionisation energies of sulphur.
　c　Explain the general trend and comment on any important points in your sketched graph.
　d　How many outer electrons are there in an atom of sulphur?
　e　Write an equation for the first electron affinity of sulphur.
　f　Explain why the molar first electron affinity of sulphur is negative (-200 kJ mol^{-1}), whereas the second electron affinity for sulphur is positive ($+640$ kJ mol^{-1}).

1.6　Atomic radius and ionic radius

The sizes of individual atoms and ions can be found by several methods, the most well known of which is X-ray diffraction as applied to crystalline solids (see Ch 3.9). This method does not involve shining a beam of X-rays of a single wavelength through the material on to a photographic plate, as you might have experienced if you have ever broken a bone or accidentally swallowed something unfortunate. Instead, because X-rays have a wavelength comparable with the distance between atoms or ions in a crystal, the crystal can *diffract* the X-rays as they reflect at suitable angles from layers of particles near the surface of the crystal. (Compare the formation of spectra by reflection from CDs or LP records – see Ch 1.4.)

Dorothy Hodgkin

Sir William Bragg

Lawrence Bragg

The technique of X-ray crystallography has been used to solve the structures of increasingly complex substances. The method was developed by Max Von Laue in Germany (who won the Nobel Prize for Physics in 1914) and by the father and son team, Sir William and Lawrence Bragg in Britain. (The Braggs won the Nobel Prize for Physics in 1915. They are the only father and son to win a Nobel Prize at the same time; J J and G P Thomson won theirs with an interval of more than 30 years between them. In 1915, Lawrence Bragg was a junior officer in the trenches in France. He was just 25. Some of us are behind schedule in our bid to win a Nobel …)

Laue and the Braggs worked mostly with simple salts like sodium chloride. In 1949, Dorothy Hodgkin of Oxford University was able to work out the structure of the antibiotic penicillin; by 1956, she had progressed to the much more complicated vitamin B_{12}, for which work she was awarded the Nobel Prize for Chemistry in 1964.

The 1950s and 1960s were a good time for X-ray crystallographers. In 1962, Max Perutz and John Kendrew of Cambridge University won the Nobel Prize for Chemistry for their determination of the complex structures of haemoglobin and related compounds. In the same year Francis Crick, also of Cambridge, and his colleague James Watson of America were given the Nobel Prize for Biology for their brilliant deduction of the molecular structure of the most exciting molecule of all, DNA; their success depended on the X-ray crystallographic work of Rosalind Franklin, who had unfortunately died of cancer in 1958, and of Maurice Wilkins, who shared the prize with them.

X-ray diffraction can measure the distance between the centres of atoms, but cannot measure from the middle of an atom to its edge because an atom doesn't suddenly stop. The data from X-ray measurements are represented in the form of an electron density map (see *Figure 1.6.1*).

Figure 1.6.1   Electron density map for H_2

In metals, the particles are arranged in a regular lattice, so we can take the radius of a metallic atom as half the distance between the centres of atoms next to each other. In simple molecules of non-metals such as H_2, Cl_2, N_2, etc., the radius of the atom is again half the distance between the atomic centres (*Figure 1.6.2*).

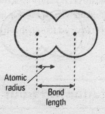

Atomic radius
Bond length

Figure 1.6.2   Atomic radius

With ions, the situation is more complicated, and many measurements on different compounds had to be made before a self-consistent table of ionic radii could be produced. Tables of radii from different sources often give slightly different values.

The most important trends in atomic radii were soon apparent. Within any family, or group, of elements in the Periodic Table, atomic radius increases with atomic number; for example, in the alkali metals (*Table 1.6.1*).

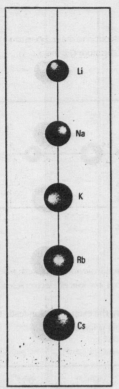

Element	Symbol	Atomic number	Electron structure	Atomic radius / nm
Lithium	Li	3	2.1	0.157
Sodium	Na	11	2.8.1	0.191
Potassium	K	19	2.8.8.1	0.235
Rubidium	Rb	37	2.8.18.8.1	0.250
Caesium	Cs	55	2.8.18.18.8.1	0.272

Table 1.6.1   The atomic radii of the alkali metals

The atoms get bigger as there are more complete electron energy levels within the atom, but not in proportion to the number of electron shells, otherwise caesium would have an atomic radius of about 0.800 nm – *Figure 1.6.3*. This is because the more complete energy levels there are in an atom, the more protons there are in the nucleus attracting those electrons towards the nucleus.

This effect of increased nuclear attraction can also be seen in the decrease in atomic radius across a typical row, or period, in the Periodic Table (*Table 1.6.2*).

Element	Symbol	Atomic number	Electron structure	Atomic radius / nm
Sodium	Na	11	2.8.1	0.191
Magnesium	Mg	12	2.8.2	0.160
Aluminium	Al	13	2.8.3	0.130
Silicon	Si	14	2.8.4	0.118
Phosphorus	P	15	2.8.5	0.110
Sulphur	S	16	2.8.6	0.102
Chlorine	Cl	17	2.8.7	0.099
Argon	Ar	18	2.8.8	0.095

Figure 1.6.3   Relative atomic radii – Li to Cs

Table 1.6.2   The atomic radii of elements in the row sodium to argon

Although more electrons are being added to the atoms across the period, so that electron–electron repulsion is increased, the atomic radius of argon is only half that of sodium. This is because all the electrons are going into the same energy level. The increasing charge on the nucleus therefore dominates – *Figure 1.6.4*.

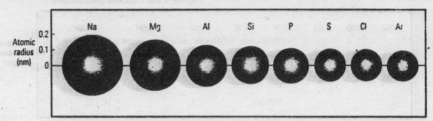

Figure 1.6.4   Atomic radii – Na to Ar

With the next element after argon, which is potassium, the new electron goes into the fourth energy level – which is shielded from the nucleus by three complete levels, so that this electron can be thought of as much further from the nucleus. Compare the atomic radius of potassium, 0.235 nm, with that of sodium, 0.191 nm, and especially that of argon, 0.095 nm. Opening up the new energy level has more than doubled the atomic radius.

The relative sizes of atoms are very important, as we shall see when we consider, for example, bonding (Ch 2), the strengths of bonds (Ch 11.5), the relative reactivities of certain compounds (Ch 17.4) and the formation of alloys (Ch 25.2).

Ionic radius

From what we have already discussed, it is obvious that removal of an electron to make a positive ion will reduce the repulsion between electrons and so reduce the size of the atom (*Figure 1.6.5*).

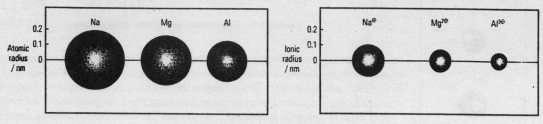

Figure 1.6.5   Atoms into cations

Removal of more than one will cause drastic shrinkage. The reduction is most marked when electrons are completely removed from the outermost energy level, but any loss of electrons leads to reduced electron–electron repulsions and a decrease in size.

Conversely, when atoms gain an electron to become a negative ion, the extra repulsion leads to an increase in size. Addition of more than one can cause drastic swelling (*Figure 1.6.6*).

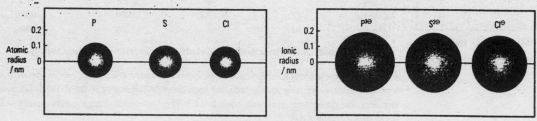

Figure 1.6.6   Atoms into anions

Some atomic and ionic radii are compared in *Table 1.6.3*.

Element	Symbol	Atomic radius / nm	Ion	Ionic radius / nm
Lithium	Li	0.134	$Li^{\oplus}$	0.074
Sodium	Na	0.154	$Na^{\oplus}$	0.102
Magnesium	Mg	0.145	$Mg^{2\ominus}$	0.072
Calcium	Ca	0.197	$Ca^{2\oplus}$	0.100
Aluminium	Al	0.130	$Al^{3\oplus}$	0.053
Fluorine	F	0.071	$F^{\ominus}$	0.133
Chlorine	Cl	0.099	$Cl^{\ominus}$	0.180
Oxygen	O	0.073	$O^{2\ominus}$	0.140
Sulphur	S	0.102	$S^{2\ominus}$	0.185
Nitrogen	N	0.075	$N^{3\ominus}$	0.171

Table 1.6.3   The radii of some atoms and ions

The sizes of ions have very important consequences. We shall meet the idea when we consider, for example, how ions can fit together in crystals of ionic compounds (Ch 2.3), how ions interact with water and the solubility of ionic compounds (Ch 3.4), and why some ionic compounds are more stable to heat than others or give rise to acid or alkaline solutions (Ch 23.1).

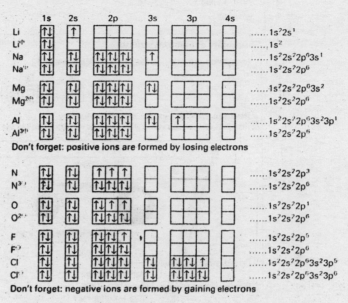

Don't forget: positive ions are formed by losing electrons

Don't forget: negative ions are formed by gaining electrons

Table 1.6.4   Electron configurations of some atoms and their ions

1 In each of the following pairs of atoms/ions which is the larger?
   a Na and Na$^{\oplus}$.   **b** Si and P.
   c Br and Br$^{\ominus}$.   **d** F and I.

2 The electron configuration of the lithium ion is $1s^2$. Using a similar notation, write the electron configuration of:
   a the fluoride ion   **b** the sulphide ion
   c the calcium atom   **d** the calcium ion

3 The electron configuration of the nitrogen atom can be represented in a diagram of the form:

2p   ↑ ↑ ↑
2s   ↑↓

1s   ↑↓

   a Explain why the electrons in the 2p sub-level are arranged as three separate electrons instead of a pair of electrons and a single electron.
   b What property of the electrons do the up and down arrows represent?
   c Write a similar diagram to illustrate the electron configuration of the nitride ion, N$^{3\ominus}$.
   d Explain how the size of the nitride ion would compare with the size of the nitrogen atom.
   e How would the size of the oxygen atom compare with the size of the nitrogen atom?

4 **a** Sketch a graph to show how the atomic radii of atoms vary across the row lithium to fluorine.

b Sketch a graph to show how the ionic radii vary from $Na^{\oplus}$ to $Mg^{2\oplus}$ to $Al^{3\oplus}$.

c Sketch a graph to show how the atomic radii vary from fluorine to chlorine to bromine to iodine.

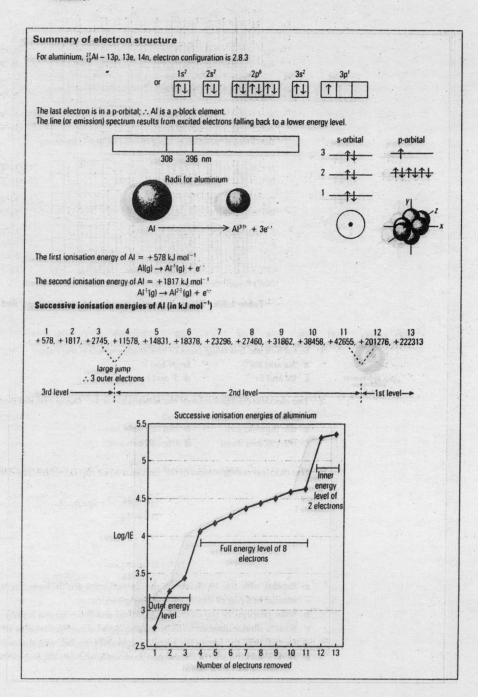

Summary of electron structure

For aluminium, $^{27}_{13}Al$ – 13p, 13e, 14n, electron configuration is 2.8.3

The last electron is in a p-orbital; ∴ Al is a p-block element.
The line (or emission) spectrum results from excited electrons falling back to a lower energy level.

308   396 nm

Radii for aluminium

$$Al \longrightarrow Al^{3+} + 3e^-$$

The first ionisation energy of Al = +578 kJ mol⁻¹
$$Al(g) \rightarrow Al^+(g) + e^-$$
The second ionisation energy of Al = +1817 kJ mol⁻¹
$$Al^+(g) \rightarrow Al^{2+}(g) + e^-$$

Successive ionisation energies of Al (in kJ mol⁻¹)

1	2	3	4	5	6	7	8	9	10	11	12	13
+578	+1817	+2745	+11578	+14831	+18378	+23296	+27460	+31862	+38458	+42655	+201276	+222313

large jump
∴ 3 outer electrons

3rd level ——— | 2nd level ——— | 1st level

Successive ionisation energies of aluminium

Inner energy level of 2 electrons

Full energy level of 8 electrons

Outer energy level

Log/IE

Number of electrons removed

S
SUMMARY
1

In this chapter the following new ideas have been covered. *Have you understood and learned them?* Could you explain them clearly to a friend? The page references will help.

- An atom consists of a heavy nucleus surrounded by electrons. p3
- The number of protons in the nucleus (the atomic number) equals the number of electrons in the atom. p4
- All atoms of the same element contain the same number of protons, but the number of neutrons may vary. p4
- Radioactivity involves mostly very large nuclei, and some smaller ones such as $^{14}_{6}C$. p5
- The relative masses of atoms can be measured and expressed on a scale for which $^{12}_{6}C = 12.0000$ is the standard. p8
- The mass spectrometer can be used to measure isotopic ratios for an element. p10
- The atomic number of an element determines the arrangement of electrons in its atoms. p14
- How an element reacts depends on its outermost electrons. p14
- Emission spectra give information about electron configurations in atoms. p15
- Electron configurations can be discussed in terms of main energy levels, sub-levels and orbitals. p19
- The structure of the Periodic Table of the elements is directly related to electron energy levels and sub-levels. p21
- Ionisation energies provide further evidence for electron energy levels and sub-levels. p25
- Variations in atomic radii within the Periodic Table can be explained in terms of attractions and repulsions within the atom. p31
- Positive ions are always smaller than their parent atoms and negative ions are always larger. p34

Q
EXAMINATION
QUESTIONS
1

1 **a** Define the terms *mass number* and *relative atomic mass*.
   b In the mass spectrum of the element silver (the atomic number, Z, is 47) there are two peaks corresponding to m/e values of 107 (51.35%) and 109 (48.65%). What does m/e signify?
   c Write the number of protons in ^{107}Ag.
   d Write the number of neutrons in ^{109}Ag.
   e Write the number of electrons in the neutral ^{107}Ag atom.
   f Calculate the relative atomic mass of silver, to four significant figures, showing all your working.
   g What difference would you expect in the chemical reactions of ^{107}Ag and ^{109}Ag?

2 The molar first ionisation energies for the first twelve elements are listed below:
   1312, 2372, 520, 900, 801, 1086, 1402, 1314, 1681, 2081, 496, 738
   (all values are positive and quoted in kJ mol^{-1}).
   a Plot a graph of molar first ionisation energy against atomic number of the element.
   b Write an equation for the first ionisation energy of carbon.
   c Why is the first ionisation energy of sodium less than that of neon?
   d Why is the first ionisation energy of boron less than that of beryllium?
   e How would you expect the second ionisation energy of lithium to compare with its first ionisation energy?
   f Write an equation to illustrate the second ionisation energy of lithium.
   g Write an equation for the molar first electron affinity of oxygen.
   h Explain why the second electron affinity of oxygen has a positive value ($+798$ kJ mol^{-1}) while the first electron affinity is negative (-141 kJ mol^{-1}).

3 The successive ionisation energies for silicon are listed below:
789, 1577, 3232, 4356, 16091, 19785, 23787, 29253, 33878, 38734, 45935, 50512, 235211, 257928
(all values are positive and quoted in kJ mol^{-1}).

a Draw up a table to show the number of electrons removed, the successive ionisation energy associated with removal of those electrons, and $\log_{10}$(ionisation energy).

b Plot a graph of $\log_{10}$(ionisation energy) against number of electrons removed.

c Explain why the general trend is an increasing ionisation energy as more electrons are removed.

d In spite of this general trend there are two distinct features about the graph. Why do they occur?

e Predict the value of the molar first ionisation energy of the element which follows silicon in the Periodic Table.

- The meaning of **element** and **atom**. (Ch 1)
- The atom consists of a **nucleus** and surrounding **electrons**. (Ch 1.2)
- The particles in the nucleus are **protons** and **neutrons**. (Ch 1.2)
- How electrons are arranged in atoms in **energy levels**, **sub-levels** and **orbitals**. (Ch 1.4)
- The link between electron energy levels and sub-levels and the layout of the **Periodic Table**. (Ch 1.4)
- Atoms can **gain or lose electrons** to form **ions**. (Ch 1.5)
- **Positive ions** are smaller and **negative ions** larger than their parent atoms. (Ch 1.6)

In this chapter

We shall discuss how particles can cling together to form the huge variety of materials which we see around us, and of which we ourselves are built.

2.1 **Introduction to bonding**
2.2 **Electron pairs and electronegativity**
2.3 **Ionic bonding**
2.4 **Covalent bonding and dative covalent bonding**
2.5 **Polar bonds, polarising power and polarisability**
2.6 **The shapes of molecules and ions; polar compounds**
2.7 **Metallic bonding**
2.8 **Intermolecular attractions**
2.9 **Hydrogen bonding**

2
chemical bonding and chemical attractions

2.1 Introduction to bonding

> **Definition:** A compound consists of two or more elements chemically combined together in fixed proportions by mass.

Compounds are made when atoms of two or more elements combine together. An essential difference between a *mixture* and a *compound* is that a mixture can be separated by purely physical methods (e.g. filtration, distillation, chromatography, etc) whereas a compound *never* can be. It takes a chemical reaction to make any compound; another chemical reaction is needed in order to break any compound up.

In the introduction to Chapter 1 we saw that Dalton stated in his atomic theory that compounds are formed by atoms of different elements joining together in simple ratios: 1:1; 1:2, 2:1, etc (e.g. $NaCl$, CO_2, H_2O). Since atoms of different elements have different relative atomic masses (see Ch 1.3), it will be seen that all samples of a particular compound must contain the same elements in the same proportion by mass. In every sample of water which has ever been analysed, the mass ratio of hydrogen to oxygen is found by experiment to be 1:8; in every sample of CO_2, the mass ratio of carbon to oxygen is 3:8 (*Figure 2.1.1*).

Figure 2.1.1   Mass ratios in simple compounds

As well as forming compounds, the atoms of elements can often join with each other – e.g. in the familiar H_2, O_2, Cl_2, N_2, but also in P_4 and S_8, as well as the billions and billions of carbon atoms all linked together in any sample of diamond or graphite.

Once the nature of the electron arrangements in atoms was understood, it became possible to interpret the formation of compounds in terms of electrons. It came to be realised that in even the most violent chemical reactions, involving the making and breaking of compounds and the chemical bonds which hold them together, it is only a few of the outermost electrons of each atom which are involved.

Gilbert Newton Lewis and Linus Pauling

The ideas which are developed in this chapter are based largely on the work of these two American chemists. G N Lewis, in 1916, explained how molecules are constructed in terms of shared electrons; in 1926 he coined the word *photon* for the quantum of light energy (see Ch 1.4); in 1933 he was the first to obtain 'heavy water', D_2O, containing the heavy isotope of hydrogen (Ch 1.2). Obviously, a wide-ranging scientist! Pauling's classic book *The Nature of the Chemical Bond* was first published in 1939. Pauling used Lewis' ideas and the results of X-ray crystallography to develop a comprehensive theory of bonding in molecules and ionic materials. In 1951, Pauling and Elias Corey determined the α-helix structure of proteins, and Pauling narrowly missed working out the structure of DNA before Crick and Watson did so in 1953. His consolation was the Nobel Prize for Chemistry in 1954, for his work on chemical bonding. He later was given a Nobel Peace Prize, only the third person (after Marie Curie and John Bardeen) to gain two Nobel Prizes. (The fourth was Fred Sanger: see Ch 28.5.)

The richness and diversity of all the material world around us and within us is a result of the very few different ways in which a fairly small number of elements can join together. An understanding of how matter hangs together and of how bonds are broken and made is so fundamental that a good knowledge of chemistry is demanded of, for example, all doctors, vets, environmental scientists, agricultural scientists, food scientists and many kinds of engineers.

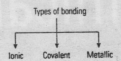

The three main types of bonding between atoms are **ionic, covalent** and **metallic**. Each of these types of bonding is – as we shall see – linked to a different set of physical and chemical properties. The much weaker – but still very important – forces of attraction which exist *between*, rather than *inside*, clumps of atoms are also discussed in this chapter. These interactions, too, give rise to typical physical properties in materials.

When you have read this chapter, you should be able to predict what sort of bond will be formed between any two common elements, and also how many bonds can be formed by any atom of those elements.

2.2 Electron pairs and electronegativity

Electronegativity

In all but a relatively few compounds, the formation of a bond between two atoms can be described as occurring between a *pair* of electrons, and on the influence of each of those atoms on the position of the electron pair. The idea of **electronegativity** has proved very useful.

An atom of a very electronegative element will attract electrons away from the influence of an atom of lower electronegativity. *Table 2.2.1* is based on the Periodic Table and shows the electronegativity values given to the elements by Linus Pauling. They are *relative* values, not absolute ones; they show how electronegative an element is compared with other elements.

Group														3	4	5	6	7	
																H 2.1			He –
1	2													3	4	5	6	7	
Li 1.0	Be 1.5													B 2.0	C 2.5	N 3.0	O 3.5	F 4.0	Ne –
Na 0.9	Mg 1.2													Al 1.5	Si 1.8	P 2.1	S 2.5	Cl 3.0	Ar –
K 0.8	Ca 1.0	Sc 1.3	Ti 1.5	V 1.6	Cr 1.6	Mn 1.5	Fe 1.8	Co 1.8	Ni 1.8	Cu 1.9	Zn 1.6			Ga 1.6	Ge 1.8	As 2.0	Se 2.4	Br 2.8	Kr –
Rb 0.8	Sr 1.0									Ag 1.9	Cd 1.7			In 1.7	Sn 1.8	Sb 1.9	Te 2.1	I 2.5	Xe –
Cs 0.7	Ba 0.9									Au 2.4	Hg 1.9			Tl 1.8	Pb 1.8	Bi 1.9	Po 2.0	At 2.2	Rn –

Table 2.2.1 The relative electronegativity values of some elements (Pauling)

Notice that – in general, and certainly for the most reactive families of elements (groups 1, 2, 6 and 7) – electronegativity *decreases* within a family of elements as atomic number increases. Also, electronegativity *increases* along the table from left to right. We therefore find the most electronegative elements at the top right of the table, and the least electronegative (i.e. the most *electropositive*) elements at the bottom left.

The most electronegative elements are non-metals with small atoms; it is no accident that fluorine, F, is the most reactive non-metal. Conversely, the least electronegative (and therefore the most electropositive) elements are metals with large atoms; and caesium, Cs, is the most reactive non-radioactive metal.

There is no need to remember electronegativity values, but it would certainly help to remember the trends. These will be discussed further in Ch 5.2.

Electrons are negatively charged and normally repel one another; but, as we saw in Ch 1.4, they are able to pair up. Imagine a pair of electrons between the nuclei of two atoms. The nuclei will repel each other, but the negative electrons between them attract the positive nuclei of each atom and help keep them together. **And that, basically, is what a chemical bond is.**

Covalent bonds

Let's imagine further that a bond exists between two atoms of elements A and B, and that (as is usually the case) each atom has provided one electron to the pair which make up the bond. If the electronegativities of A and B are similar, the attraction of both A and B for the pair of electrons will be similar, and the electrons will stay roughly midway between the nuclei of the atoms of A and B. This is a normal, non-polar *covalent bond*, as we shall meet in Ch 2.4.

$$A \overset{\bullet}{\bullet} B$$

Polarised bonds

If B is more electronegative than A, the electrons will be drawn towards B. This will mean that B has slightly more than its fair share of electron density and will be slightly negatively charged; on the other hand, A has lost some of its electron density and will be slightly positive. These *partial charges* can be represented by the symbols $\delta\ominus$ and $\delta\oplus$. The bond between A and B is said to be *polarised*: we shall discuss polar bonds of this type in Ch 2.5.

$$^{\delta+}A \overset{\circ}{\bullet} B^{\delta-}$$

Ionic bonds

If B is *much* more electronegative than A, the pair of electrons will be associated almost completely with the atom of B. B has not only got back its own electron but gained one from A. The atom of B is therefore fully negatively charged – it is a *negative ion*; and the atom of A, having lost an electron, is a *positive ion*. This type of bond is called an *ionic bond* (see Ch 2.3).

$$A^+ \overset{\circ}{\bullet} B^{-}$$

Bonds which are completely or almost 100% ionic or 100% covalent are very rare. Most compounds contain bonds which are somewhere between the two extremes, but which are sufficiently biased to one side or the other for us to be able to classify them as either covalent or ionic (*Figure 2.2.1*). The next two sections (Ch 2.3 and Ch 2.4) will concentrate on these two types of bond.

Spectrum of bonding

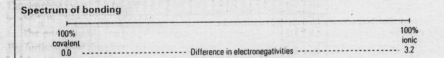

| 100% covalent 0.0 | - - - - - - - - - - Difference in electronegativities - - - - - - - - - - | 100% ionic 3.2 |

As a rough guide:
- a difference greater than 2.0 will lead to essentially ionic bonding, and
- a difference that is less than 1.0 will lead to bonding that is mainly covalent.

Figure 2.2.1

Predicting types of bonds

An oversimplified but very helpful way of predicting what kind of bond is formed between any two elements is as follows.

There are only *two kinds of elements*:
  METALS and NON-METALS (although a few elements are near the border between them).
So there should be only *three kinds of bonding*:
  METAL to NON-METAL,
  NON-METAL to NON-METAL and
  METAL to METAL.

Metals are generally electropositive (i.e. they have low electronegativity values) and non-metals are usually electronegative. The differences in electronegativity between most metals and most non-metals are such that **bonding between a metal and a non-metal is usually ionic**. When a non-metal bonds with another non-metal, differences in electronegativity are usually low; so **bonding between two non-metals is covalent**.

The third and final possibility is metal-to-metal bonding. There are very few compounds which you are likely to meet at this level which contain any metal–metal bonds; and compounds containing *only* two or more metals combined together in fixed proportion by mass are very rare indeed.

It is, however, possible to get 'mixtures' of metals – sometimes with one or more non-metals also present – called **alloys**, and some of these are extremely important in everyday life. Examples are cupronickel (for coins), brass, bronze and the many kinds of steel. The type of bonding which holds metals together is discussed in Ch 2.7.

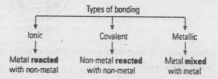

If metal-to-metal compounds are rare, then in almost all compounds involving metals they are bonded to non-metals. From what we have discussed, it follows logically that **if any compound contains a metal, it is very likely to be ionic.**

SELF-ASSESSMENT QUESTIONS 2.2

1   What type of bonding would you expect in the following compounds?
   a Potassium bromide.   **b** Sulphur dioxide.
   c Copper chloride.   **d** Magnesium aluminium alloy.

2   Draw a diagram to represent the position of the pair of electrons in the bond between:
   a Sodium and iodine.   **b** Rubidium and chlorine.
   c Hydrogen and chlorine.   **d** Chlorine atoms in the Cl_2 molecule.

3   Use the difference in electronegativity values (in *Table 2.2.1*) to decide if the following bonds will be ionic or covalent:
   a Caesium and fluorine.   **b** Carbon and hydrogen.
   c Oxygen and oxygen.   **d** Chlorine and bromine.

2.3 Ionic bonding

What's in a name?

- If a name of a compound ends in *-ide*, it is a simple binary compound – i.e. it consists of *only two elements*. This is true even if the compound contains no metal – e.g. H_2S, hydrogen sulphide.
- If the name ends in *-ate*, oxygen is also present. So calcium sulphide is CaS, and calcium sulphate is $CaSO_4$.
- A name ending in *-ite* shows that oxygen is present, but not to the same extent as in the corresponding *-ate* compounds.
   Thus, sodium sulph*ide* is Na_2S, sodium sulph*ate* is Na_2SO_4, and sodium sulph*ite* is Na_2SO_3.

In simple discussions of bonding it is easier to draw three-dimensional atoms as flat, two-dimensional objects, and to show electron energy levels as circles with electrons orbiting the nucleus. We *know* this isn't a true picture (see Ch 1.4) but, as often in science, we use a simplified model to help our understanding. The important thing is to accept the model for what it is and what it does, and no more than that!

The atomic number of sodium, Na, is eleven. We saw in Ch 1.4 that we can write its electron configuration as 2.8.1. The atomic number of chlorine, Cl, is 17; its configuration is therefore 2.8.7. We can draw sodium and chlorine atoms as shown below. (Please note that the dots and crosses used to represent electrons are only there to help us follow what happens!)

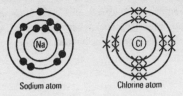

Sodium atom   Chlorine atom

Since chemical reactions involve the electrons in the outer energy levels, we usually only show the outer electrons:

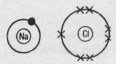

Now chlorine is *much* more electronegative than sodium (see *Table 2.2.1*). So any pair of electrons forming a bond between an atom of sodium and an atom of chlorine would be pulled on to the chlorine atom. In other words, *sodium will have lost an electron to the chlorine*. The result is a sodium ion and a chlor*ide* ion (note: not a chlor*ine* ion), and a compound, sodium chloride, is formed.

The attraction between the oppositely charged ions is what holds the ionic compound together (*Figure 2.3.1*). (A full discussion of the structure of ionic crystalline materials such as sodium chloride can be found in Ch 3.3. This includes an explanation of why, in an ionic crystal, the *attractions* between oppositely charged ions are always greater than the *repulsions* between ions with the same charge!)

An ion in a solid ionic compound is always surrounded by ions of the opposite charge; there are electrostatic attractions, but no actual bonds and therefore no molecules. However, the *ratios* between the number of positive and negative ions are always simple whole numbers, and this we shall now explain.

Note that electron density measurements from X-ray diffraction (see Ch 1.6) show that this way of describing the bonding in sodium chloride agrees with observed facts (*Figure 2.3.2*).

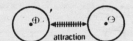

attraction

**Figure 2.3.1
The ionic bond**

Definition: Ionic bonding is the result of electrostatic attraction between oppositely charged ions. The positive ions are usually formed from metals and the negative ions usually formed from non-metals.

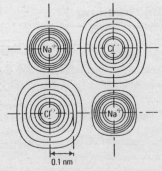

0.1 nm

Figure 2.3.2   Electron density map of sodium chloride

Dot-and-cross diagrams for ionic bonding

The reaction between atoms of sodium and chlorine can be represented like this. (N.B. It is a good idea always to draw 'before' and 'after' diagrams to show clearly what happens to electrons during the reaction. Alternatively, and taking up less space, the electron configurations can be shown as numbers.)

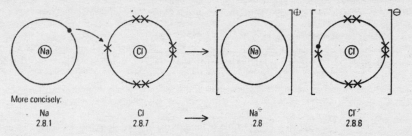

More concisely:

Na	Cl		Na⁺	Cl⁻
2.8.1	2.8.7	⟶	2.8	2.8.8

Look again at the diagrams representing the reaction of sodium and chlorine atoms to form ions. The electron arrangement in the Na⁺ ion is 2.8, and in the Cl⁻ ion is 2.8.8. These are the electron arrangements in the chemically 'dead' noble gases neon and argon (*Figure 2.3.3*).

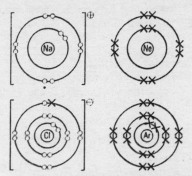

Figure 2.3.3   Electron configurations

Sodium and chlorine haven't turned into noble gases; *but their ions now have full outer electron shells and are chemically very unreactive*. They are now **iso-electronic** with neon and argon respectively (see *Figure 2.3.3*). As we shall see, it takes considerable energy to force an electron back on to a sodium ion or to drag one off a chloride ion.

Remember: When elements react their atoms gain, lose or share electrons to reach a more stable electron arrangement, which is usually that of a noble gas with a full outer energy level (shell) of electrons. Because the full outer 'shell' contains eight electrons, it is often called the 'octet'.

The electron configuration of a noble gas atom plays a crucial part in any discussion on bonding, and these noble gas configurations should be learned.

Lithium oxide

Consider now the formation of the compound lithium oxide.

The electron configuration of the lithium atom is 2.1 and of oxygen 2.6. Lithium must lose one electron to arrive at the configuration of the noble gas helium, and oxygen must gain two electrons to get to the 2.8 structure of neon. The resulting lithium ions must therefore carry one positive charge each, and the oxide ion two negative charges.

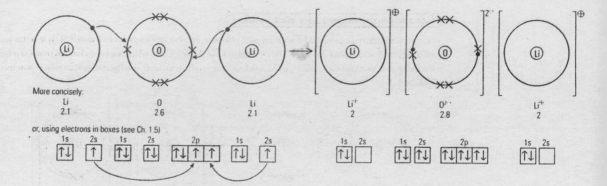

More concisely:

Li	O	Li	Li⁺	O²⁻	Li⁺
2.1	2.6	2.1	2	2.8	2

or, using electrons in boxes (see Ch. 1.5)

The formula of lithium oxide must therefore be Li_2O. Two atoms of lithium join with each atom of oxygen so that all three atoms have lost or gained electrons to end up with ions which have noble gas electron configurations. If the configuration of the magnesium atom is 2.8.2 and of the sulphur atom 2.8.6, can you see why the formula of magnesium sulphide must be MgS?

By now you should have noticed that **you can work out the formula of a simple ionic compound between two elements if you know the electron configurations of the metal and the non-metal.**

'Criss-cross' formulae

Since the ionic compound formed must be electrically neutral overall, the charges on the ions must balance out. The formula of lithium oxide must show this. You can work out the formula using a *criss-cross technique* – crossing over the number of charges on each of the ions – the 1 of Li⁺ becomes the number of oxygen atoms needed in the formula. The number 2 of the O²⁻ ion becomes the number of lithium ions required in the formula. As a diagram it looks like:

So we write Li_2O for the formula of lithium oxide – the figure 1 is not needed in a formula.

Don't forget:

- If you know the atomic number of an element (given by its position in the Periodic Table) you know the number of electrons in each atom of the element.
- If you know the number of electrons in an atom of the element, you know its electron configuration (see Ch 1.4).
- If you know that, you know how many electrons the atom has in its outer energy level.
- And that tells you how the element will behave when it reacts chemically.

So metals are metals because they only need to lose one, two or at most three electrons, and reactive non-metals are such because they only have to gain one or two electrons, to form in each case the electron configuration of the nearest noble gas.

1 Draw dot-and-cross diagrams to show how the ionic bond is formed in:

    a sodium fluoride, NaF

    b calcium oxide, CaO

    c potassium oxide, K_2O

    d magnesium bromide, $MgBr_2$.

2 Write the electron configuration, in terms of s-, p-, and d-energy sub-levels, of:

    a a chlorine atom

    b a bromide ion

    c a sulphur atom

    d an oxide ion.

3 Work out the formulae of the following ionic compounds:

    a sodium sulphide

    b calcium bromide

    c aluminium oxide

    d magnesium nitride.

2.4 Covalent bonding and dative covalent bonding

Molecules are groups of atoms joined together by covalent bonds (see Ch 2.2). It is rather unfair to draw most molecules in two dimensions, but dot-and-cross diagrams are useful in helping our understanding of what happens when two or more non-metal atoms join together. Remember – nearly all covalent bonds are between atoms of non-metals; therefore, almost all molecules which you will meet contain only non-metallic elements.

The simplest possible molecule is H_2; this is the particle found in hydrogen gas. The 1s orbitals (see Ch 1.4) of the two hydrogen atoms overlap; each hydrogen atom now has two electrons in its outer (indeed only) energy level, and therefore has the electron configuration of the noble gas helium. *Note that each hydrogen atom has provided one electron for the bond*, and in return shares the pair.

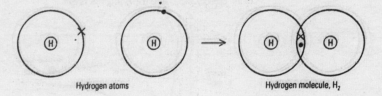

Hydrogen atoms   Hydrogen molecule, H_2

Another way of looking at it is that the atomic orbitals of the hydrogen atoms merge together to form a **molecular orbital**. Note that, as in an atomic orbital (see Ch 1.4), electrons, which normally repel one another, can only pair up if they have opposite spin.

Hydrogen atomic orbitals   Hydrogen molecule, H_2

And this is the picture we see with the electron density map for hydrogen (*Figure 2.4.1*).

Figure 2.4.1 Electron density map for hydrogen, H₂

The idea of molecular orbitals will become very useful later on, but for the moment dot-and-cross diagrams will do.

What holds a covalent bond together? Any such bond consists of a pair of electrons between two positive nuclei. The nuclei repel each other, but are attracted by the pair of negatively charged electrons (*Figure 2.4.2*).

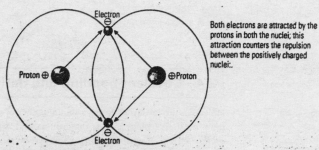

Both electrons are attracted by the protons in both the nuclei; this attraction counters the repulsion between the positively charged nuclei:.

Figure 2.4.2 The covalent bond

This is how an ordinary covalent bond is formed. As we shall see later on in this section, it is also possible for an atom or molecule to provide *both* electrons for a covalent bond. Such a bond is called a *dative* bond.

A covalent bond involves a balance between this attraction and repulsion, and the atoms in the bond are – at any normal temperature – vibrating continuously. The situation in the H₂ molecule is summed up in *Figure 2.4.3*. Note that if enough energy is given to the molecule it will be lifted out of its 'energy well', and the bond will break.

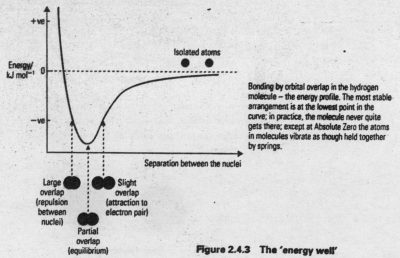

Bonding by orbital overlap in the hydrogen molecule – the energy profile. The most stable arrangement is at the lowest point in the curve; in practice, the molecule never quite gets there; except at Absolute Zero the atoms in molecules vibrate as though held together by springs.

Figure 2.4.3 The 'energy well'

The usual way of representing a covalent bond is to draw a line between the symbols of the elements, e.g. H—H for the hydrogen molecule, H_2 (see *Figure 2.4.4* for representations of the hydrogen molecule).

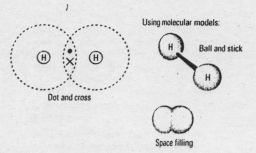

Figure 2.4.4   **Different models of the hydrogen molecule**

Dot-and-cross diagrams for covalent bonding

For hydrogen chloride, HCl, the before and after diagram is:

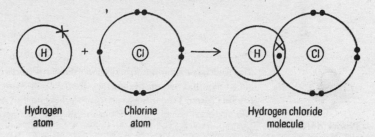

| Hydrogen atom | Chlorine atom | | Hydrogen chloride molecule |

By sharing a pair of electrons, the hydrogen atom has reached the helium structure, and the chlorine atom now has the argon structure. Once you are familiar with writing the dot-and-cross diagrams for covalent bonding you can even leave out the 'circles' that represent the electron energy levels – but you must be careful and tidy whenever you do so.

**Figure 2.4.5
Hydrogen chloride –
space filling model**

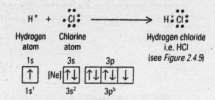

The number of outer electrons must still be obvious in your diagram, and with covalent bonding, it must be clear which pair(s) of electrons are being shared between two atoms. For chlorine, Cl_2, the covalent dot and cross diagram is:

**Figure 2.4.6
Chlorine molecule model**

Methane is the major component of natural gas, and is the simplest hydrocarbon (see Ch 15.2). The configuration of the carbon atom is 2.4. You can probably already see *why* the formula of methane is CH₄!

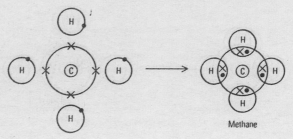

Methane

Showing just the outer electrons:

$$H \cdot \; \overset{\displaystyle H}{\underset{\displaystyle H}{\times C \times}} \; \cdot H \longrightarrow H \times \overset{\displaystyle H}{\underset{\displaystyle H}{\cdot C \cdot}} \times H$$

Methane contains four bonding pairs of electrons, and is often written as:

$$\begin{array}{c} H \\ | \\ H - C - H \\ | \\ H \end{array}$$

**Figure 2.4.7
Model of methane**

This tendency to compress three-dimensional structures on to a two-dimensional page is, as we have already pointed out, misleading (see *Figure 2.4.7*). As we shall soon see, although some molecules *are* flat ('planar') the majority are not. Working on the basis that pairs of electrons repel each other, you may even be able to work out the shape of the methane molecule before you get to the next section.

Multiple covalent bonds

It is possible to have molecules which contain double or even triple covalent bonds. These are often met with in organic chemistry (e.g. the alkenes, Ch 16; the carbonyls, Ch 19; and the nitriles, Ch 28), but also occur in simple inorganic molecules. For example, oxygen atoms have the electron configuration 2.6. Each atom therefore shares two pairs of electrons in order to complete its outer electron energy levels. Showing only the outer electrons:

$$\overset{\cdot\cdot}{\underset{\cdot\cdot}{O}} \; \overset{\times\times}{\underset{\times\times}{O}} \longrightarrow \; \overset{\cdot\cdot}{\underset{\cdot\cdot}{O}} \overset{\times}{\underset{\times}{\vdots}} \overset{\times\times}{\underset{\times\times}{O}}$$

Oxygen atoms   Oxygen molecule
ie, O₂, O = O

Note that the symbol for a double bond is two parallel lines, e.g. O=O. Nearly all of the ordinary oxygen we breathe exists as O_2 molecules. There is a very small amount of trioxygen, O_3 ('ozone'), particularly in the upper atmosphere (Ch 14.6) – and *you might already know why it is just as well there **is** some ozone there!*

The behaviour of oxygen in a strong magnetic field shows that it contains two unpaired electrons per molecule. This can be explained in terms of molecular orbital theory, but need not concern us now. We can account for all the reactions of oxygen which we need in this book by using the O=O picture. However, the precise reactions which occur in the transport of oxygen around your body and during respiration could not occur if O_2 was simply O=O! To put it bluntly, if that *were* the case, it is unlikely that complex animals such as ourselves could exist.

As you might guess, it is hard to break the strong triple bond between the atoms in nitrogen molecules in air. And that too, is a very good thing from the point of view of our breathing or, indeed, our control of burning – although it makes the manufacture of fertiliser more difficult (see the Haber process for making ammonia, Ch 13.3).

Incidentally, the carbon monoxide molecule, CO, contains 14 electrons, just like N_2. The formula CO is hard to justify with C having a 2.4 configuration and O having 2.6. Molecular orbital theory shows that the bonding in CO is very similar to that in N_2; so the structure should probably be written C≡O!

In a similar way, nitrogen atoms – with configuration 2.5 – share three pairs of electrons. Showing the outer electrons only:

$$ \overset{\cdot\cdot}{\underset{\cdot}{\text{N}}} \cdot \quad \overset{\cdot\cdot}{\underset{\times}{\text{N}}} \times \longrightarrow \overset{\cdot\cdot}{\underset{\cdot\cdot}{\text{:N:N:}}} $$

Nitrogen atoms

So the molecule N_2, which dominates the atmosphere of our planet, can be represented as N≡N.

Iso-electronic structures are possible with covalent molecules. Dot-and-cross diagrams for nitrogen and carbon monoxide show how. The atoms in the nitrogen molecule sharing three pairs of electrons, and a similar six electrons being shared in the carbon monoxide molecule – although now oxygen provides four of the six electrons.

$$ \text{:N:N:} \qquad \text{:C:O:} $$

Double bonds

σ bonds

The behaviour of many organic compounds can be explained in terms of two types of covalent bonds between carbon atoms. The shared electron pair may be caused by overlap of orbitals, so that there is maximum probability of the electron pair being found directly between the nuclei of the two atoms involved.

Region of maximum probability

Such bond are called **sigma bonds** (or **σ bonds**, using the Greek letter *sigma*). These bonds can be formed by overlap of any directionally suitable orbitals (*Figure 2.4.8*), e.g. p with p or p with s orbitals (see Ch 1.4).

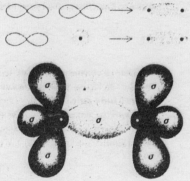

Figure 2.4.8   The σ bond – there are seven σ bonds in this diagram of ethane C_2H_6

π bonds

The other type of covalent bond is caused by sideways overlap of (usually) p orbitals (see *Figure 2.4.9* and Ch 1.4 for orbital shapes).

Region of maximum probability

2 overlapping probabilities
producing a π bond

Figure 2.4.9   The π bond

Such bonds are called **pi bonds** (or **π bonds**, using the Greek letter *pi*).

You will usually meet π bonds in carbon–carbon double bonds; the electron pair of the π bond is distributed both above and below the σ bond. Such a double bond is usually written as C=C, but the two bonds are not the same. A better picture would be:

C ———σ bond——— C

← p orbital

C ———σ bond——— C   π bond – both these 'areas' make up the π bond,
showing where the electron pair probably
is (95% probability)

π bonds are, not surprisingly, weaker than σ bonds. You will meet such bonds, and study the properties of the compounds which contain them, particularly in Chapters 16 (the alkenes) and 26 (arenes). Further discussion of these points will be left until then.

Dative covalent bonding

This is also known as **dative bonding** or **co-ordinate bonding**. The word 'dative' comes from the Latin word which means 'to give' or 'donate'. In an ordinary covalent bond one electron in the shared pair comes from each of the atoms which are joined by the bond; in a dative bond, *both* electrons are provided by one of the atoms.

Consider the ammonia molecule, NH_3. The nitrogen atom has the configuration 2.5. It has room for only three more electrons in its outer energy level and can therefore join with three hydrogen atoms to form three shared pairs of electrons and hence three covalent bonds.

The pairs of electrons which are used in bonds are, not surprisingly, known as **bonding pairs**. But one pair of electrons in the ammonia molecule is not used in bonding. This is called a **non-bonding pair** or a **lone pair**. Such a pair of electrons can be donated to another molecule, atom, or ion which has 'room' to accommodate it. For example, an important reaction of ammonia is that it can very readily accept a proton from an acid (see Ch 23.1) to form the ammonium ion, $NH_4^{\oplus}$ (as in ammonium chloride, $NH_4^{\oplus}Cl^{\ominus}$). We can represent what happens in this way:

Ammonia   Hydrogen
                     ion

Ammonium ion
$NH_4^{\oplus}$

Once this dative bond is formed, it cannot be distinguished from the other covalent bonds.

The ability of ammonia, water and similar molecules and ions to form dative bonds is vitally important in the chemistry of the transition metals (see Ch 25.6).

A dative bond is often written as a line with an arrow to show where the donated pair of electrons came from, e.g.

$$\left[\begin{array}{c} H \\ | \\ H-N\rightarrow H \\ | \\ H \end{array} \right]^{+}$$

If the radioactive isotope of hydrogen, tritium, ^3H, is used to make ammonium chloride, all the radioactivity of the tritium is incorporated into the product (H* indicates the tritium).

$$\begin{array}{c} H \\ | \\ H-N+H^*Cl \\ | \\ H \end{array} \rightarrow \left[\begin{array}{c} H \\ | \\ H-N-H^* \\ | \\ H \end{array}\right]^{\oplus} Cl^{\ominus}$$

If the ammonium chloride is warmed with an alkali such as sodium hydroxide, $Na^{\oplus}OH^{\ominus}$, the ammonia released contains about *three-quarters* of the original radioactive tritium.

$$NH_4Cl + OH^{\ominus} \rightarrow NH_3 + H_2O + Cl^{\ominus}$$

This is because the NH_3 molecule has a 3-in-4 chance of incorporating the tritium which was put into the ammonium ion, $NH_4^{\oplus}$. This shows that there is nothing special about the dative bond once it has been formed.

Ammonia also forms a compound with boron trifluoride, BF_3. Like most compounds of boron, BF_3 is unusual because in it the boron does not possess the electron configuration of a noble gas, although the fluorine atoms do. Boron has the configuration 2.3, so BF_3 can be written:

$$\begin{array}{ccc} F & & F \\ \cdot\!\!^\circ & & | \\ B & & B \\ ^\times\!\!^\times & & / \quad \backslash \\ F \quad F & & F \qquad F \end{array}$$

With only six electrons in its outer energy level, the boron atom in BF_3 has a vacant slot into which a pair of electrons can be donated from a suitable molecule such as ammonia. The reaction is immediate.

Using 'shorthand', the reaction looks like this:

$$\begin{array}{ccc} \begin{array}{c} H \\ | \\ H-N\!\!\times \\ | \\ H \end{array} & \begin{array}{c} H \\ | \\ B-F \\ | \\ H \end{array} & \longrightarrow \begin{array}{c} H \quad H \\ | \quad | \\ H-N\!\rightarrow\!B-F \\ | \quad | \\ H \quad H \end{array} \end{array}$$

Ammonia   Boron   Ammonia – boron
          trifluoride   trifluoride addition
                          compound

Covalent radius

When discussing the sizes of atoms and ions in Ch 1.6, we referred to the *atomic radius* of elements. Strictly, the term 'atomic radius' is not adequate, as it depends on the environment the atom is in.

The **covalent radius** of an element is defined as half the distance between the centres of two atoms of that element which are joined together by a single covalent bond. This is easy enough to determine for molecules such as H_2, Cl_2, etc, but harder for some other non-metallic elements.

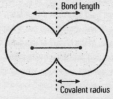

The **metallic radius** of an element is defined as half the distance between the centres of two adjacent atoms in the giant metallic structures (see Ch 3.5) of that element. The term metallic radius applies only to metals, but a covalent radius can also be found for metals. There is some difference between the two values: for example, the covalent radius of aluminium is 0.125 nm whereas the metallic radius is 0.143 nm.

Molecules which break the 'octet' rule

As we noted earlier, many atoms lose, gain or share electrons during chemical reactions in order to finish up with a full outer electron energy level containing eight electrons – an 'octet', similar to the electron configuration of an atom of a noble gas.

This is quite a sound rule, but it is only true *as a maximum* for elements in the row (Period 2 of the Periodic Table) which runs from lithium (Li) to neon (Ne). We have already seen that in the molecule BF_3, the boron atom only contains six electrons in its outer energy level; the beryllium atom in $BeCl_2$ is even worse off, with four electrons .

$$\overset{\times\times}{\underset{\times\times}{\times}}Cl\overset{\bullet}{\bullet}Be\overset{\bullet}{\bullet}Cl\overset{\times\times}{\underset{\times\times}{\times}}$$

But in *no* compound of these Period 2 elements is the octet exceeded.

With larger atoms there are more energy levels and sub-levels, and there is less of an energy gap between them, so that extra orbitals may be used. For example, although an oxygen atom (configuration 2.6) is limited to eight electrons in its outer energy level and can therefore form only two covalent bonds (as in H_2O), sulphur (configuration 2.8.6) can go up to 12 electrons, as in the compound SF_6.

Phosphorus can form the expected chloride, PCl_3, but with excess chlorine PCl_5 is produced – with ten electrons in the outer levels of the phosphorus atom. (It is perhaps worth commenting that although SF_6 is stable, PCl_5, when warmed, falls apart to PCl_3 and Cl_2: $PCl_5 \longrightarrow PCl_3 + Cl_2$). Even the noble gases with larger atoms can form compounds, for example the xenon fluorides XeF_2, XeF_4 and even XeF_6. Iodine forms a series of fluorides – IF, IF_3, IF_5 and IF_7. You will see that compounds such as XeF_6 and IF_7 don't so much *break* the octet rule as *smash* it, the limit being the number of atoms that can fit around the central atom. For example, there is a compound BrF_5 but not BrF_7, because the bromine atom is too small to fit seven fluorine atoms around it.

Even so, the octet rule is still very useful in predicting the likely existence of stable compounds and in working out the formulae of simple ionic and covalent compounds.

Theory vs. fact

Until 1962 everyone thought the noble gases – helium, neon, argon, krypton and xenon – really *were* totally chemically 'dead'. They were called the 'inert gases'. Neil Bartlett, a British chemist working at the University of British Columbia, had a supply of the powerful gaseous oxidising agent platinum(VI) fluoride, PtF_6. He found it would even drag an electron off an oxygen molecule to form the solid compound '$O_2^{\oplus}PtF_6^{\ominus}$'. While looking through a data book for other materials to try his new reagent on, he noticed that the energy needed to remove an electron from a xenon atom was less than that needed to take one from the O_2 molecule. So he mixed xenon with PtF_6 – and immediately obtained orange crystals of '$XePtF_6$', the first compound ever made of one of the 'inert gases'. (The compound was later shown to have a more complicated formula).

The news was published in *Proceedings of the Chemical Society* within a few weeks. All over the world, laboratories which had supplies of xenon and equipment for the safe handling of fluorine dropped everything else and leapt into this new research field. When the first International Conference on Compounds of the Noble Gases (notice the rapid renaming!) was held less than a year later, more than 400 research papers had already been published, xenon fluorides and oxides were well-known and some compounds of krypton had been made. One eminent chemist described it as the biggest gold rush in the history of inorganic chemistry.

But the biggest and best effect was probably to remind the chemical community of a basic scientific principle – **never** accept *theory* as *fact*. Everyone was taught, and everyone believed, that the 'inert gases' were 'inert' – and every competent chemistry student could explain *why* they were inert. But the existing theoretical structure was shattered by just one experiment, confirmed and extended by other experimenters, and a new and better theory had to be constructed. In this book we must often simplify and sometimes – without meaning to – we may state theory in such a way as to make it appear to be fact. We hope very much that you will develop the ability to think critically and spot any such error, in this book and elsewhere. And our dearest wish is that some readers may be inspired to do in some field of science what Neil Bartlett did to noble gas chemistry and bonding theory – demonstrate that existing ideas *don't* provide an adequate explanation, and make everyone think again.

---

1 Draw dot-and-cross diagrams to illustrate the bonding in:

   a Methane, CH_4.

   b Hydrogen, H_2.

   c Hydrogen sulphide, H_2S.

   d Hydrogen iodide, HI.

2 a Carbon forms a range of compounds with the halogens, the simplest of which have the formulae CCl_4, CBr_4 and CI_4. Draw a dot-and-cross diagram for the CBr_4 molecule.

   b Explain what is meant by the term covalent radius, using Cl_2 as an example.

   c Explain the variation in covalent radii for the halogen group of elements.

   d Why is the term covalent radius meaningless for the element neon?

3 a Draw dot-and-cross diagrams to illustrate the bonds present in H_2O and $H_3O^{\oplus}$.

   b Using H_2O to illustrate your answer, describe how a covalent bond is formed between the hydrogen and oxygen atoms.

   c Although, once they are formed, all the bonds present in $H_3O^{\oplus}$ are identical, the oxygen atom has provided four electrons to the three covalent bonds. Explain how oxygen has provided more than three electrons to form three bonds.

2.5 Polar bonds, polarising power and polarisability

Purely ionic and purely covalent bonds are ends of a spectrum, and the majority of bonds are probably somewhere in between – depending on the difference in electronegativity between the two atoms in the bond (*Figure 2.5.1*).

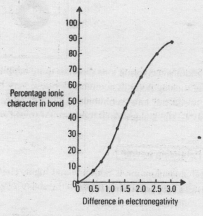

Figure 2.5.1 Ionic vs. covalent

In, for example, a molecule of hydrogen chloride, chlorine is more electronegative than hydrogen. The bonding pair of electrons is therefore displaced towards the chlorine, which means that the chlorine end of the bond has increased electron density and the hydrogen end has less. The chlorine has a *partial* negative charge, and the chlorine a partial positive charge. The symbol for 'partial' is the Greek letter 'delta', δ.

The polarity in a bond such as H—Cl can be represented like this (*Figure 2.5.2*):

Figure 2.5.2   Distortion of the electron cloud in hydrogen chloride

This distortion of the electron cloud is what is called a polar bond. The molecule is then said to have a **dipole** – a measure of the charge separation in the molecule.

$$H \rightarrow Cl \qquad \text{i.e. } H^{\delta \oplus}—Cl^{\delta \ominus}$$

Representing bonds

A—B   shows a simple covalent bond involving a pair of electrons

A=B   shows a double bond, with a σ and a π bond

A→B   shows a dative covalent bond, where A donates a pair of electrons to B

A⇢B   shows the polarity in a simple covalent bond, where the bonding pair is closer to B, i.e. it represents $A^{\delta \oplus}—B^{\delta \ominus}$

Table 2.2.1 gave the values of electronegativities, which were assigned to elements by Linus Pauling. You can see that in carbon–halogen bonds, polarity increases in the order C—I < C—Br < C—Cl < C—F.

Bond polarity is extremely important in determining how compounds react, and, it can also greatly affect the physical properties of compounds.

Potentially ionic bonds can also be greatly affected by the capacity of positive ions (**cations**) to attract electrons back to themselves and of negative ions to be affected by this. We saw in Ch 1.6 that when metal atoms lose electrons to become positive ions, they become smaller. The more electrons that are lost, the smaller the ion becomes. Conversely, when non-metal atoms gain electrons to become negative ions, they get larger.

Compare the relative sizes of sodium, $Na^{\oplus}$, aluminium, $Al^{3\oplus}$, and chloride, $Cl^{\ominus}$, ions.

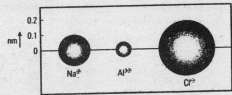

Sodium and chloride ions can exist quite happily in crystals of common salt, sodium chloride, NaCl. The melting point of sodium chloride is very high, 801 °C, showing that there is strong attraction between the ions. Aluminium chloride, $AlCl_3$, sublimes (i.e. vaporises without melting) at only 180 °C. This suggests that there are no strong attractions between particles in $AlCl_3$.

Polarising power

The aluminium ion is both tiny and highly charged. It therefore has a high **charge density**, which gives it high **polarising power** – the ability to attract electrons back from negative ions (*Figure 2.5.3*).

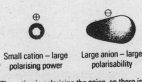

Small cation – large   Large anion – large
polarising power   polarisability

The cation is polarising the anion, so there is
some covalent character in the ionic bond

Figure 2.5.3   Polarisation of ions

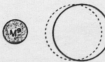

Figure 2.5.4 $M^\oplus$ and $M^{2\oplus}$ polarising power (M = metal)

So the bonding in aluminium chloride is far from *ionic*; it is nearly *covalent*. The larger the charge density of a positive ion (a *cation*) the greater its polarising power (*Figure 2.5.4*).

Conversely, the larger the negative ion (or *anion*), the easier it is for a suitable cation to pull electron density out of it – in other words, the greater its **polarisability**. The iodide ion, with a radius of 0.216 nm, is therefore more polarisable than the chloride ion. Larger, compound ions such as the carbonate ion, $CO_3^{2\ominus}$, are very easily polarised.

Any attempt to prepare aluminium carbonate, which would have the formula $Al_2(CO_3)_3$, is doomed to failure. The compound does not exist. If carbonate ions get close to $Al^{3\oplus}$ ions, the anions are broken up and carbon dioxide is given off.

Fully ionic bonds are favoured by:
- Large cations
- Small anions
- Small charge

SELF-ASSESSMENT QUESTIONS 2.5

1 Using the symbols $\delta\oplus$ and $\delta\ominus$, show the polarity on the bonds in each of the following molecules. (See *Table 2.2.1* for relevant data).
   a HCl
   b H_2S
   c SCl_2
   d NF_3

2 By referring to the electronegativity values for each of the elements (see *Table 2.2.1*), identify the most polar molecule in each of the following groups.
   a H_2O, H_2S
   b HBr, HCl, HF
   c NH_3, NF_3
   d CH_4, NH_3, H_2O

3 Identify the following compounds as ionic, covalent or polar covalent.
   a Potassium chloride, KCl.
   b Silane, SiH_4.
   c Hydrogen iodide, HI.
   d Calcium oxide, CaO.

2.6 The shapes of molecules and ions; polar compounds

As we shall see in the next chapter, the way in which ions fit together in regular patterns in ionic crystals is determined by the relative sizes of the ions and their charges. In this section, we shall concentrate on the shapes of simple molecules, and show how their shapes can be explained and predicted from the simple statement that **pairs of electrons repel one another.**

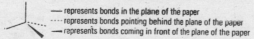

—— represents bonds in the plane of the paper
----- represents bonds pointing behind the plane of the paper
—— represents bonds coming in front of the plane of the paper

The electron pair in any π bond only has a slight effect on the shape of a molecule. It is only the repulsion between σ bonding pairs that needs to be considered. Consequently, the shapes of molecules are determined by the number of σ bonds and the number of lone pairs of electrons around a particular atom.

But not all pairs of electrons are the same. In Ch 2.4 we saw that the ammonia molecule contained three *bonding pairs* of electrons and one *non-bonding pair* (*Figure 2.6.1*).

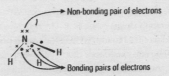

Figure 2.6.1   The ammonia molecule

A bonding pair is shared between two atoms. It may not be shared *equally*, and the resulting bond may be polar, but at least it is attracted by two nuclei. A non-bonding pair only has the attraction from its parent nucleus acting on it, and therefore it has a greater probability of being closer to that nucleus than do the bonding pairs. This means that **non-bonding pairs repel more strongly than do bonding pairs** (*Figure 2.6.2*). (But *don't* call them 'more repulsive'!)

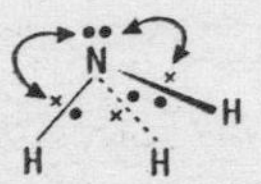

Figure 2.6.2   Electron pair repulsion

> **Definition:** The shape of a molecule is determined by the Electron-Pair Repulsion Theory. The outer pairs of electrons around a covalently bonded atom minimise the repulsions between them by moving as far apart as possible.

Bonding pair–bonding pair repulsion is less than bonding pair–non-bonding pair repulsion, which is less than non-bonding pair–non-bonding pair repulsion.

Bonding pair	–	Bonding pair
Bonding pair	–	Non-bonding pair
Non-bonding pair	–	Non-bonding pair

↓ Increasing repulsion

The shapes of non-metal hydride molecules

Having established the principles, we shall use them on the hydrides of the first elements in Groups 4–7 of the Periodic Table: methane, CH_4; ammonia, NH_3; water, H_2O; and hydrogen fluoride, HF.

Methane

First, the 'flat' picture of the bonding in methane:

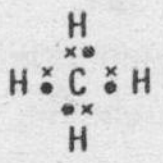

We can see that there are *four bonding pairs*, and no non-bonding pairs. All four pairs of electrons will repel each other equally. It can be shown mathematically that the furthest they can get away from each other is if the carbon is at the centre of gravity of a perfect tetrahedron and the hydrogen atoms are at the corners (*Figure 2.6.3*).

The angle made by any two H–C–H bonds is 109.5°. We say that *the tetrahedral bond angle is 109.5°*.

Any molecule which has four bonding pairs to identical atoms and no non-bonding pairs will have a perfect tetrahedral shape and a bond angle of 109.5° (*Figure 2.6.4*).

Figure 2.6.3   Model of methane

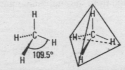

Figure 2.6.4   The tetrahedral shape

Slight distortions will be caused if different-sized atoms, or atoms of different electronegativity, are at the corners.

Ammonia

Next, ammonia. The dot-and-cross diagram is:

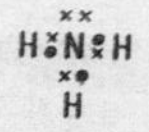

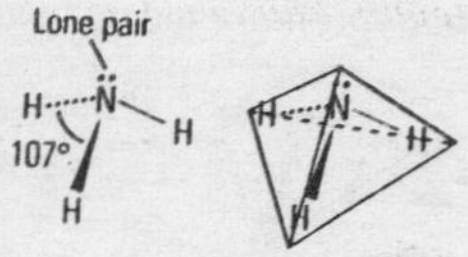

Figure 2.6.5
The shape of ammonia

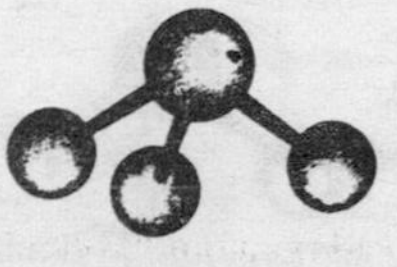

Ammonia contains *three bonding pairs* and *one non-bonding pair*. Repulsion will ensure that the electron pairs will still be distributed towards the corners of a tetrahedron. But, remember, the non-bonding pair repels more strongly, so that the others are pushed together slightly. The $^H\diagdown_N\diagup^H$ bond angle is therefore less than the tetrahedral bond angle of 109.5°; it is in fact 107° (*Figure 2.6.5*).

There is a temptation to describe the shape of the ammonia molecule as tetrahedral. However, the shape is determined by the position of the *atoms*, not the electron pairs. The shape is described as *trigonal pyramidal*, i.e. a pyramid based on an equilateral triangle (*Figure 2.6.6*).

Any molecule which contains three bonding pairs (to identical atoms) and one non-bonding pair will have a regular trigonal pyramidal shape, and the bond angle will be slightly less than the tetrahedral angle. The angle may not be the same as that in ammonia (107°) because of different bond lengths and bond polarities.

Figure 2.6.6
Model of ammonia

Water

Figure 2.6.7 The shape of water

The unique properties of water, which will be met frequently in this book, are linked to its shape. With only three atoms, the molecule must be 'flat' in the sense that all its atoms must be in one plane. The dot-and-cross picture is:

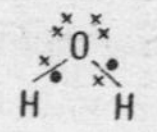

Figure 2.6.8 Model of water molecule

We see that the water molecule contains *two bonding pairs* and *two non-bonding pairs*. Again, the distribution in space of these four pairs will be roughly tetrahedral. But now there are *two* non-bonding pairs, which will not only repel the bonding pairs but also each other. The $^H\diagdown_O\diagup^H$ bond angle is therefore even less than the bond angle in ammonia, and is in fact 104.5° (*Figure 2.6.7*).

Any molecule which contains two bonding pairs and two non-bonding pairs will have a crooked shape, and the bond angle will be significantly less than the tetrahedral angle. (Look at the shape of the molecule in *Figure 2.6.8*).

Hydrogen halides

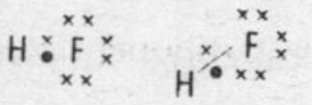

Figure 2.6.9 Hydrogen fluoride

Hydrogen fluoride (and any other hydrogen halide), only has two atoms and must therefore be a linear molecule.

But a glance at the dot-and-cross diagram shows that it contains *one bonding pair* and *three non-bonding pairs*, so the distribution of electron pairs around the fluorine atom is still roughly tetrahedral (*Figure 2.6.9*).

Remember: the principles for determining the shape around an atom are:
- electron pairs repel each other;
- π bonds do not need to be considered;
- the order of the effect of repulsion is:
  bonding pair–bonding pair < non-bonding pair–bonding pair < non-bonding pair–non-bonding pair.

All the molecules discussed so far have four electron pairs and all the actual bonds in them are **single** bonds – i.e. they involve only one pair of electrons. We shall now look at the various shapes and introduce some common examples.

Linear

The simplest case would be beryllium chloride, $BeCl_2$; there are only *two bonding pairs* and *no non-bonding pairs* on the central beryllium (*Figure 2.6.10*).

Carbon dioxide   Beryllium chloride

Figure 2.6.10   Carbon dioxide and beryllium chloride

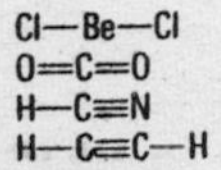

But carbon dioxide, CO_2, has four bonding pairs, two on each side of the carbon atom in double bonds which are repelling each other (*Figure 2.6.10*). The $O{=}C{=}O$ bond angle is therefore also 180°. Other linear molecules include hydrogen cyanide, HCN, and the gas ethyne (acetylene), C_2H_2 (*Figure 2.6.11*).

Figure 2.6.11   Linear molecules

Crooked

Species which have the same basic shape as the water molecule, $H{-}O{-}H$, include the foul-smelling gas hydrogen sulphide, H_2S, the corrosive acidic gas sulphur dioxide, SO_2 (*Figure 2.6.12*) (which is *not* like CO_2, because there is an extra non-bonding pair on sulphur), and the nitrite ion, $NO_2^{\ominus}$.

Figure 2.6.12   Crooked molecules

Trigonal planar

The simplest example is boron trifluoride, BF_3, which we met in Ch 2.4 as an example of a compound in which the 'octet' rule is broken (*Figure 2.6.13*).

The molecule BF_3 contains *three bonding pairs* and *no non-bonding pairs* on the boron atom. The furthest apart the electron pairs can get is towards the corners of an equilateral triangle, with an $F{-}B{-}F$ bond angle of 120°. Species with this or a very similar shape include sulphur trioxide, SO_3 (*Figure 2.6.14*), and the carbonate and nitrate ions, $CO_3^{2\ominus}$ and $NO_3^{\ominus}$.

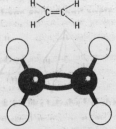

Figure 2.6.13   Boron trifluoride

Note that these contain, on average, more than a simple single bond; their existence (as shown by various techniques including X-ray diffraction) suggests that bonds of a type intermediate between single and double bonds are possible. We shall meet more of these later, especially in the chapter on benzene and related compounds (Ch 26).

The shape around each of the carbon atoms in ethene, C_2H_4, is also basically trigonal planar (*Figure 2.6.15*).

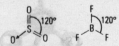

Figure 2.6.14   Trigonal planar molecules

Figure 2.6.15   Ethene

Trigonal pyramidal

Species which have the same basic shape as ammonia include, not surprisingly, the phosphorus analogue of ammonia, phosphine, PH_3. Rather more surprisingly, the presence of an extra non-bonding pair has the same effect when the other bonds are not simple single bonds, as in the sulphite ion, $SO_3^{2\ominus}$ (*Figure 2.6.16*).

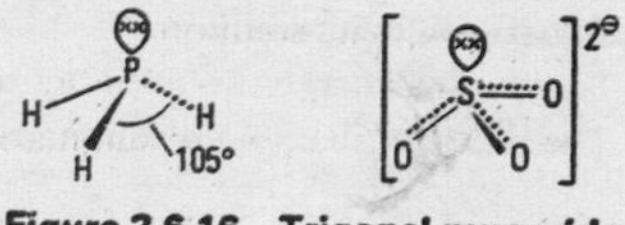

Figure 2.6.16   Trigonal pyramids

Tetrahedral

All that is required for the species to have the same shape as methane is that the central atoms be connected with four other identical atoms or groups of atoms by bonds of the same kind. Mutual repulsion will then make sure that the resulting shape is symmetrically tetrahedral (*Figure 2.6.17*).

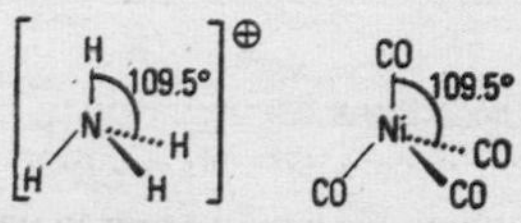

Figure 2.6.17   Tetrahedra

Examples include the ammonium ion, $NH_4^{\oplus}$ (which contains the same number of electrons as CH_4), the sulphate and phosphate ions, $SO_4^{2\ominus}$ and $PO_4^{3\ominus}$, and the unusual molecule tetracarbonylnickel(0), $Ni(CO)_4$, which is made by passing carbon monoxide over warm nickel and is used in purifying nickel.

Square planar

This shape is met in some complexes of d-block metals, but is otherwise not very common. For simple molecules or ions, *four bonding pairs* and *two non-bonding pairs* are required. The non-bonding pairs repel each other strongly, and so get as far apart as possible – i.e. at 180°. The four bonding pairs can get as far apart as possible by pointing towards the corners of a square (*Figure 2.6.18*).

Figure 2.6.18   Square planar

Examples include xenon(IV) fluoride, XeF_4, and the tetrachloroiodate(III) ion, $ICl_4^{\ominus}$.

Trigonal bipyramid

This shape consists of one trigonal pyramid on top of another, upside-down, pyramid. *Five bonding pairs* and *no non-bonding pairs* are needed. The only common example is PCl_5. Notice that there are two bond angles, 90° and 120° (see *Figure 2.6.19*).

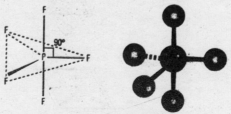

Figure 2.6.19   Trigonal bipyramid

61

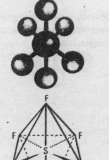

The molecule forming an octahedron

Figure 2.6.20   Octahedron

Octahedral

This requires *six bonding pairs* and *no non-bonding pairs*. Minimum repulsion between electron pairs is achieved with a bond angle of 90° throughout (*Figure 2.6.20*).

Examples are sulphur(VI) fluoride, SF_6, and a very large number of d-block complex ions, as we shall see in Ch 25. An example is the familiar $[Cu(H_2O)_6]^{2\oplus}$, which causes the blue colour of copper(II) sulphate solution.

Some other shapes do occur in various compounds but they are not important.

The different shapes are summarised in *Table 2.6.1*.

Shape		Bonding pairs	Non-bonding pairs	Bond angle	Examples
Linear	Cl—Hg—Cl	2	0	180	$BeCl_2$, CO_2, HCN, C_2H_2 (ethyne)
Trigonal planar		3	0	120	BF_3, SO_3, NO_3^{-}, CO_3^{2-}, C_2H_4 (ethene)
Tetrahedral		4	0	109.5	CH_4, $NH_4^{\oplus}$, SO_4^{2-}, PO_4^{3-}, $Ni(CO)_4$
Trigonal pyramidal		3	1	107	NH_3, PH_3, SO_3^{2-}
Crooked		2	2	105	H_2O, H_2S, SO_2, NO_2^{-}
Trigonal bipyramidal		5	0	120, 90	PCl_5
Octahedral		6	0	90	SF_6, $[Cu(H_2O)_6]^{2\oplus}$
Square planar		4	2	90	XeF_4, ICl_4^{-}

Table 2.6.1   A summary of the shapes of molecules and ions

Since δ means 'a little' is it necessary to balance δ⊕ and δ⊖ charges? You do not need to balance them, as long as you realise that in the molecule overall they must, of course, balance out.

> **Do polar bonds make a polar molecule?**

Oxygen is much more electronegative than hydrogen, so the O—H bond is polarised thus: $^{\delta\ominus}$O—H$^{\delta\oplus}$. If the water molecule was linear, i.e. straight, this wouldn't matter as the two polarities would cancel out; $^{\delta\oplus}$H—O$^{\delta\ominus}$—H$^{\delta\oplus}$. But water is a crooked molecule (and without it being so life would not be possible on this planet, as we shall shortly see), and so it is polar:

So the molecule of water has a partially-positive end, and a partially-negative end; it is a **polar molecule**. Water molecules will respond to an electric field. In carbon dioxide, on the other hand, although the individual *bonds* are polar, the molecule is linear; and it is because of this symmetry that the O=C=O *molecule* is not polar.

A
ACTIVITY

The polarity of water
The polarity of water can easily be demonstrated. All that is needed is a *dry* rod or ruler made from polythene or other plastic, a *dry* cloth (cat fur used to be used, but this is not now recommended!) and a *very thin* continuous stream of water from a tap. Rub the rod *very* vigorously with the cloth and hold it close alongside the thin stream of water. (It must *not* get wet.) You will see that the stream is deflected.

Rubbed (charged) polythene rod

Thin stream of falling water

If some electrons are removed as you rub the rod, so that it becomes positively charged, the water molecules will align with their negative ends towards the rod and be attracted to it.

Working on the same principle, you should be able to see why the symmetrically octahedral sulphur(VI) fluoride, SF$_6$ (*Figure 2.6.21*), is *not* a polar molecule although it contains polar S—F bonds.

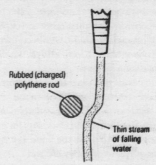

Figure 2.6.21   Sulphur(VI) fluoride

Tetrachloromethane (carbon tetrachloride), CCl$_4$, is not polar; trichloromethane, CHCl$_3$, *is* polar. Both are tetrahedral molecules; can you explain the difference?

1 Use diagrams to show the shapes of the following molecules, marking in the appropriate bond angles.

   a PH_3 b CH_4 c H_2S d SF_6

2 Draw dot-and-cross diagrams for each of the following molecules and explain their three-dimensional shape in terms of the electron pair repulsion theory.

   a NH_3 b CO_2 c PCl_3 d H_2Se

3 Decide if the following compounds will have ionic or covalent bonding present. Give your reasoning in each case.

   a C_2H_6 b I_2 c RbCl   d $CaBr_2$

2.7 Metallic bonding

In Ch 2.2 we pointed out that it is rare in common compounds to find genuine metal-to-metal chemical bonds, and even rarer to find compounds which contain just metals combined with one another and with no non-metallic element present. There are plenty of compounds which have a metal as part of the negative ion: potassium manganate(VII) (potassium permanganate), $K^{\oplus}MnO_4^{\ominus}$ and potassium dichromate(VI), $K_2Cr_2O_7$, are two examples of common laboratory reagents which fall into this category.

A relatively recently-made example of a metal-only compound is caesium auride, CsAu, which appears to behave as an ionic solid; but note the large difference in electronegativity of the two components – and their rarity and expense!

What metals can do, within certain limits and particularly if their atomic radii are similar, is mix with one another to form **alloys**. But these are *not* compounds as they do not have a rigorously fixed composition by mass.

In this section we shall give a simple picture of bonding in metals. A more detailed discussion of metallic structures will follow in the next chapter.

Metals may be wildly different in terms of melting point (e.g. mercury, −39 °C; tungsten, 3410 °C) and hardness. Compare, for example sodium, which can be cut with a table knife, and iron which is used to make the knife. But two of the properties of *all* metals which make them so useful in industry and everyday life are:

- they conduct electricity;
- as solids they can be bent, rolled into sheets, drawn into wires or otherwise shaped *without breaking*.

Aluminium mixed with between 1% and 10% magnesium produces a range of alloys with improved mechanical properties – particularly for use in the aircraft and satellite industries. The atom ratios vary from $Al_{20}Mg$ to Al_8Mg. Lithium-aluminium alloys have lower density than even aluminium and are resistant to high-temperature corrosion, so are often used as components in high-temperature engines.

The 'sea of electrons'

The explanation for these two properties is the same. The outer electrons of each metal atom (*Figure 2.7.1*) are free to move through the whole of the piece of metal.

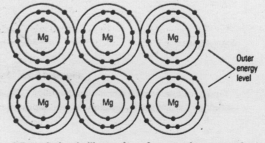

Figure 2.7.1 A simple illustration of magnesium atoms in the metal

A very helpful picture is of a lattice of metal ions (*ions* because they have lost their outermost electrons: if only temporarily) immersed in a sea of mobile electrons (*Figure 2.7.2*). Because the outermost electrons are not in fixed positions, they are said to be 'delocalised'.

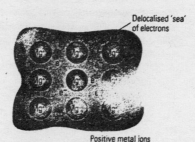

Figure 2.7.2   **A more useful picture of magnesium metal**

Because electrons repel, if electrons are fed into one end of the piece of metal (e.g. a wire), the resulting repulsion is transmitted across to the other end where an equal number of electrons leave. Hence electrical conduction.

Next consider what holds a metal together. The positive ions repel one another, so do the electrons; but there is a strong attraction between ions and electrons.

Now look at *Figure 2.7.3*, and mentally slide the layers of ions *above* the dotted line to the left or right on top of the layer below. The electrons may be in constant movement, *but there will always be some electrons between the moving layers, attracting them and keeping them attached to one another* (*Figure 2.7.3*). As a result, the piece of metal can be bent or otherwise distorted without breaking. Pure metals are *not* brittle!

Figure 2.7.3   **Movement of layers in metals**

Contrast this with sliding the layers in an ionic crystal – repulsion between the opposite charges leads to break up of the crystal lattice (see Chapter 3, *Figure 3.4.1*).

Of all the non-metals, only one form of carbon – graphite – can conduct electricity almost as well as a metal. We shall see in the next chapter how it can do this. In addition, we shall briefly explain how **semi-conductors** work – and hence the basis of the silicon chips which are all around us, in computers, calculators, watches and control mechanisms.

Q

SELF-ASSESSMENT
QUESTIONS
2.7

1   Aluminium is an excellent conductor of electricity but at least one structure of the next element in the Periodic Table (carbon) has no conduction properties at all. Explain why this should be so.

2   Draw diagrams to explain the bonding in:
   a   potassium chloride   b   ammonia
   c   hydrogen bromide   d   the Mg–Al alloy.

2.8   Intermolecular attractions

We saw in Ch 2.4 that molecules consist of two or more atoms held together by covalent bonds. These bonds are usually strong, so it is often hard to get molecules to break up. (Water only starts to break up into hydrogen and oxygen at about 1800 °C). It is tempting to think that when a liquid boils, the molecules are breaking up: they are in fact *separating* from one another, but are *not* breaking up (*Figure 2.8.1*).

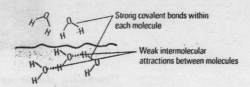

Strong covalent bonds within each molecule

Weak intermolecular attractions between molecules

Figure 2.8.1   Boiling water

You must be absolutely clear on this point.

The forces which attract molecules to each other (i.e. intermolecular attractions) are often very weak, so that the melting and boiling points of simple covalent compounds are generally low. If there were no attraction at all between particles, gases would never condense to liquids. But even helium atoms will eventually come together to form a liquid, albeit not until −269 °C (4K).

The evidence for such attraction is that whenever a gas condenses to a liquid, or a liquid freezes to a solid, heat energy is given out although the temperature does not change. Conversely, heat energy has to be taken in to melt a solid or vaporise a liquid.

Energy is given out whenever particles come together; energy is needed to separate any particles. So there must be attractive forces between the particles. Any change of state of a material is accompanied by an energy change. You have probably already met terms such as 'latent heat of fusion of ice' or 'latent heat of vaporisation of water'. Well, you now have a simple explanation in terms of inter-particle attractions.

Although *all* particles attract one another, these attractions are very weak in comparison with the electrostatic attraction between oppositely charged ions and with the attractions involved in a covalent bond. We shall discuss here only *attractions between molecules*. And please, do not think of attractive forces as *bonds* – they are far too weak for that.

There are three types of attractive force which concern us (*Figure 2.8.2*).

Remember: The dipole is a measure of the charge separation in the molecule.

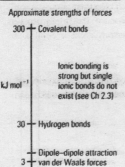

Approximate strengths of forces

300 ┼ Covalent bonds

Ionic bonding is strong but single ionic bonds do not exist (see Ch 2.3)

kJ mol⁻¹

30 ┼ Hydrogen bonds

┼ Dipole–dipole attraction
3 ┼ van der Waals forces

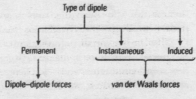

Type of dipole

Permanent   Instantaneous   Induced

Dipole–dipole forces   van der Waals forces

Figure 2.8.2

The so-called **van der Waals forces** are common to *all* particles These forces result from the interaction of *instantaneous dipoles* and *induced dipoles* possible in every atom and molecule. The third type is called **permanent dipole–dipole attraction**.

Compounds whose molecules contain polar bonds involving hydrogen atoms can also experience the more intense attractions known as **hydrogen bonding** (see Ch 2.9) – although we must emphasise again that anything approaching the strength of an ordinary covalent bond is very rare.

A useful guide to remembering the relative magnitudes of van der Waals forces, hydrogen bonds and covalent bonds is 1:10:100. Typical values for van der Waals forces are a few kJ mol^{-1}, for hydrogen bonds, 15–40 kJ mol^{-1}; and for single covalent bonds, 200–400 kJ mol^{-1}. (If you want the units explained see Ch 11.2.)

Permanent dipole–dipole forces

As we have seen in Ch 2.5, a molecule such as hydrogen chloride is *polar*, because the chlorine atom is much more electronegative than the hydrogen atom. Each end of the molecule is *partially charged*. As it has two ends, it is said to possess a **dipole**. The positive end of one molecule will attract the negative end of an adjacent molecule, and so on:

$$\delta\oplus \; \widehat{H-Cl} \; \delta\ominus \quad \delta\oplus \; \widehat{H-Cl} \; \delta\ominus \quad \delta\oplus \; \widehat{H-Cl} \; \delta\ominus$$

> Johannes van der Waals was a Dutch physicist who was given the Nobel Prize for Physics in 1910 for his equation which described the behaviour of *real* gases rather than ideal ones.

These permanent dipole–dipole forces usually contribute surprisingly little to the attractions between molecules. Even for hydrogen chloride, by far the most important attractive forces are those called **instantaneous dipole-induced dipole forces**, which are the ones more usually called **van der Waals forces**.

Instantaneous dipole-induced dipole forces

To explain these forces of attraction, we have to accept that the electrons in covalent bonds in molecules (and even the electrons in noble gas atoms) are not completely fixed in their positions but can move to some extent. Alternatively, we can consider there to be temporary distortions in the electron orbitals. A covalent bond is always vibrating. So the distribution of electron density in a molecule does not stay constant (*Figure 2.8.3*).

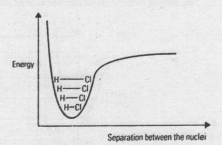

Figure 2.8.3   A vibrating molecule

Even a symmetrical molecule, or an atom like helium, can develop, for a tiny fraction of a second, ends that are slightly positive and negative – i.e. an *instantaneous dipole* (*Figure 2.8.4*).

Figure 2.8.4   Instantaneous dipole in a helium atom

Once again, the basic law of attraction and repulsion will take over. The electron movement in one molecule and consequent instantaneous dipole will induce an opposite movement of electrons and consequent opposite dipole in a nearby molecule; the polar molecules then attract one another (*Figure 2.8.5*).

67

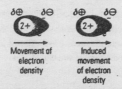

Figure 2.8.5 Induced dipoles

A split second later the dipoles in the molecules may all be round the other way.

So our picture of a *molecular solid* like candle wax, or even solid hydrogen (melting point −259 °C), is that the molecules are held together – weakly – in the solid by fluctuations in electron density *within* the molecules.

Obviously, the more electrons there are in the particles the greater the possibility of instantaneous dipole-induced dipole (**van der Waals**) attractions between them, and the higher will be their melting and boiling points. Noble gases exist as single atoms; the simplest such atom, that of helium, has only two electrons. Helium is the material which is the most difficult of all to liquefy – its boiling point is only 4K (−269 °C). The boiling points of the noble gases are given in *Table 2.8.1* and *Figure 2.8.6.*

Name	Symbol	Atomic number	Boiling point / °C
Helium	He	2	−269
Neon	Ne	10	−246
Argon	Ar	18	−186
Krypton	Kr	36	−152
Xenon	Xe	54	−107
Radon	Rn	86	−62

Table 2.8.1 Boiling points of noble gases

You can see that the more electrons there are in the particle, the higher the boiling point.

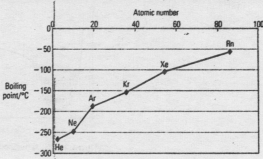

Figure 2.8.6 Boiling points of noble gases

The same effect can be seen (*Table 2.8.2*) in the diatomic molecules of the halogen elements, Group 7 of the Periodic Table.

Name	Molecular formula	Number of electrons in molecule	Boiling point / °C
Fluorine	F_2	18	−188
Chlorine	Cl_2	34	−35
Bromine	Br_2	70	59
Iodine	I_2	106	184
Astatine	At_2	170	337 (estimated)

Table 2.8.2 Boiling points of halogen elements

Do the shapes of molecules influence intermolecular bonding?

The *number* of van der Waals attractions between molecules also affects the physical properties such as melting and boiling points. If we consider the simple alkane hydrocarbons – methane (CH_4), ethane (C_2H_6), propane (C_3H_8), butane (C_4H_{10}) and pentane (C_5H_{12}), as the number of carbon atoms in the molecule increases, so does the boiling point of the compound. For example, methane, CH_4, boils at $-164\,°C$; but pentane, C_5H_{12}, boils at $36\,°C$. Two other molecules have the molecular formula C_5H_{12} – 2-methylbutane and 2,2-dimethylpropane. (Don't be put off by the names – all will be made clear in Ch 14.3). Their structures and boiling points are shown in *Figure 2.8.7*.

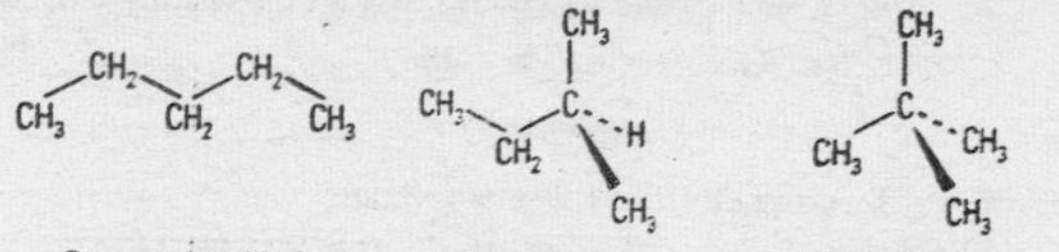

Pentane, b.p. 36 °C

2-Methylbutane b.p. 28 °C

2-2-Dimethylpropane b.p. 10 °C

Figure 2.8.7   Structures and boiling points of molecules with the formula C_5H_{12}

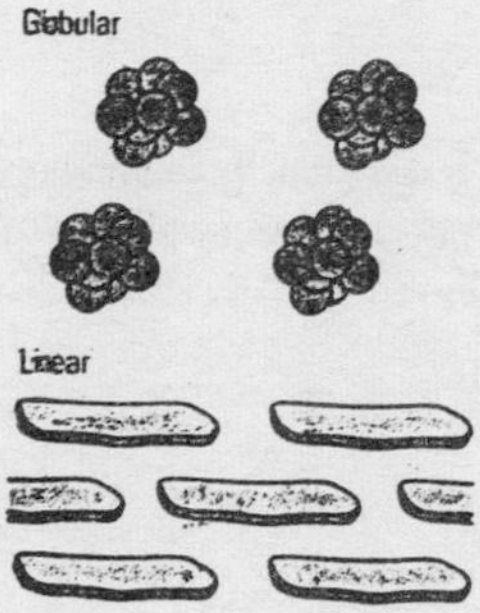

Figure 2.8.8   Globular and linear molecules

These three C_5H_{12} molecules contain the same number of atoms of each kind – and thus the same number of electrons – and the compounds would therefore be expected to boil at the same temperature. But notice that the more globular the molecule, the lower the boiling point. This is basically because there are fewer points of contact between 'round' molecules than there are between 'long' molecules in the liquid, and so there are fewer chances for van der Waals interactions between the 'round' molecules (*Figure 2.8.8*).

Very long molecules exist in many polymers, such as Polythene (see Ch 16.7 if you wish). As Polythene film for wrapping is produced it is stretched to partially orientate the long molecular chains so that they lie close and relatively parallel to one another. This increases the van der Waals interactions between the chains, and makes the film much stronger. Polymer fibres are strengthened by 'pulling' in the same way. The molecules may be only held together by weak van der Waals forces, but there are so many of these between long parallel molecules that overall the material is quite strong (*Figure 2.8.9*).

Figure 2.8.9   Polythene strands

Q

SELF-ASSESSMENT
QUESTIONS
2.8

1 The boiling points of four related compounds are shown in the table.

Compound	Relative molecular mass	Boiling point / °C
CH_4	16	−164
SiH_4	32	−112
GeH_4	77	−89
SnH_4	123	−52

a Plot a graph of relative molecular mass of the compound against the boiling point of the compound.

b Explain the trend in boiling point as the relative molecular mass increases.

2 Which molecule has the largest dipole in each of the following groups of molecules?

   a HF, HCl, HBr, HI.

   b CH_4, HF, NH_3.

3 Which of the following, in the liquid state, will have only van der Waals forces holding the atoms or molecules together?

   a Kr   **b** SO_2 **c** CH_4 **d** HBr

4 Which of the following molecules would you expect to possess a permanent dipole?

   a CCl_4 **b** BF_3 **c** PH_3 **d** H_2S

5 Consider the following data:

	Relative molecular mass	Boiling point / °C
Argon	40	−186
LiCl	42.5	1340
C_2H_6	30	−88

Although the relative masses are not very different, there is a wide range in the boiling points of the three liquids. By discussing the interparticle forces that exist in each case, explain why there is such a difference in the boiling points.

2.9 Hydrogen bonding

As we have just seen, van der Waals (instantaneous dipole-induced dipole) interactions increase with the number of electrons in molecules, so that for similar molecules melting points and boiling points should increase as the relative molecular masses increase. For a given family of elements – i.e. for a particular Group in the Periodic Table – we would expect an increase in the boiling points of the hydrides of the elements in a Group as the atomic number of the elements increases. What actually happens for the hydrides of Groups 4, 5, 6 and 7 is shown in *Figure 2.9.1*. The noble gases are included for comparison.

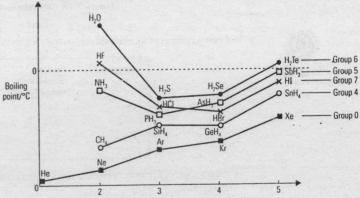

Figure 2.9.1   The boiling points of hydrides of Groups 4–7 in the Periodic Table

N 3.0	O 3.5	F 4.0	
	S 2.5	Cl 3.0	
		Br 2.8	
		I 2.5	

**Figure 2.9.2
Electronegative
elements**

For the hydrides of Group 4 (CH_4, SiH_4, GeH_4, SnH_4) the variation in boiling point is as expected, and similar to – though the values are higher than – the variation in the noble gases. But in Groups 5, 6 and 7, the hydride of the first member of each group (NH_4, H_2O, and HF) has a *much* higher boiling point than expected. *

The three most electronegative elements are fluorine, oxygen and nitrogen (see Ch 2.2), with chlorine being similar to nitrogen (*Figure 2.9.2*).

This means that the bonds which these elements form with hydrogen are *very* polar, and therefore the molecules of the hydrides are also very polar. Electron density is pulled away from the hydrogen, so leaving it with a larger partial positive charge than is usual in a polar bond. This can then attract non-bonding pairs of electrons (see Ch 2.4) on nearby atoms; and fluorine, oxygen and nitrogen all have such 'lone pairs'. The situation in water is shown in *Figure 2.9.3*.

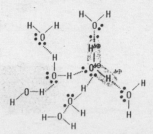

Figure 2.9.3   Hydrogen bonding in water

Hydrogen bonds are *directional*. They are usually shown as a dotted or interrupted line to indicate that a hydrogen bond is *not* the same as a covalent bond and that – as we have seen – it is usually much weaker than covalent bonding. The hydrogen atom – or, more accurately, the proton – is, in effect, shared between two adjacent electronegative atoms. The situation is *dynamic*, not static. The proton is not firmly attached to either atom, but can be thought of as 'flickering' between them.

Remember: Hydrogen bonds are formed by **electronegative elements with a non-bonding pair** of electrons available to bond with a **hydrogen** attached to an **electronegative element**.

Because of hydrogen bonding, the molecules in a polar liquid such as water or ethanol clump together. It therefore takes more energy than usual to separate the particles in the liquid and so to form the vapour; which means that the liquid has a higher boiling point than expected.

An example of an exceptionally strong hydrogen bond is found in the hydrogen fluoride ion, $HF_2^{\ominus}$, $(F\text{—}H\text{⸺}F)^{\ominus}$. In this ion, the distance between the two fluorine nuclei is shown by X-ray diffraction to be *less than two ionic radii of fluorine*. The only sensible explanation is that considerable overlap of the atomic orbitals of the fluoride ions is achieved because of the positively-charged proton embedded in them.

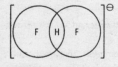

But this is very rare – the usual H-bond is many times weaker than the normal covalent bond.

Some consequences of hydrogen bonding in water

The hydrogen bond may be only a weak form of bonding, but its consequences are immense – for example, in shaping the surface of the Earth and in enabling all living things to exist and reproduce.

Melting point of water

Because water molecules cling together through hydrogen bonding, the melting (m.p.) and boiling points (b.p.) of water are very high for such a small molecule. (For methane, CH_4, a molecule of comparable mass, the m.p. is $-182\,°C$ and b.p. $-164\,°C$; water, H_2O, has a m.p. of $0\,°C$ and b.p. of $100\,°C$ – a huge difference!) This means that on most of the surface of this planet water can exist as a liquid.

Without water, life as we know it would be impossible. Living cells are largely water, and are bathed in a watery fluid; they depend on water for the transport of materials in and out. Water is an extraordinarily good solvent. Because it has such a polar molecule, many ionic salts such as sodium chloride are soluble in water; and because of their capacity for forming hydrogen bonds, so are carbohydrates and amino acids. All these materials are essential for life.

Surface tension of water

Water has an unusually high surface tension and viscosity. The high surface tension caused by hydrogen bonding leads to a surface layer on water which is capable of supporting insects such as the 'water boatman'. Water's comparatively high viscosity, caused by hydrogen bonding of molecules, means that it will not flow as easily as comparable non-polar liquids, and it is harder to stir.

In Ch 2.8, we noted that when a solid melts or a liquid vaporises, energy is taken in; when a gas condenses or a liquid freezes, energy is given out (*Figure 2.9.4*).

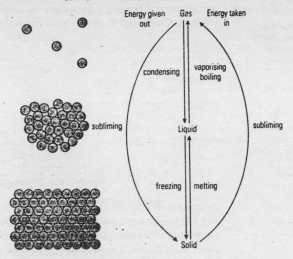

Figure 2.9.4

In each case (provided the process isn't done too rapidly), the temperature stays the same until the change of state is complete.

Breaking bonds or overcoming attractions between particles always requires energy, whereas forming bonds or attractions of any kind always gives out energy. You can now see why these so-called 'latent heats' (Ch 2.8) exist – and why they are greater for materials possessing hydrogen bonding (like water) rather than just van der Waals forces (such as methane).

The energy changes involved in evaporation and condensation of water have a major influence on the world's climate. In addition, water has a very high *heat capacity* – it takes about ten times as much energy to increase the temperature of a given mass of water by 1 °C than it takes to do the same to the same mass of rock. The reluctance of the ocean to warm up in summer and cool down in winter also has a massive effect on climate.

Density of water

One almost unique property of water is that the solid form, ice, is *less dense* than the liquid. No other natural material has this property. This is because the crystal of ice is based on an open, hydrogen-bonded structure (*Figure 2.9.5*).

The low density of ice means that it floats on water. (So the *Titanic* disaster and at least three very expensive films were caused by hydrogen bonding in water!)

Another consequence is that water expands when it freezes. When ice melts, the open lattice breaks up and the molecules can get closer together, so water is denser than ice. The ice structure persists for some time after melting, so that the density of water continues to increase above 0 °C. But of course, water also expands on heating, like any other material. These conflicting tendencies explain why water is at its densest at 4 °C.

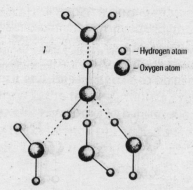

Figure 2.9.5   Structure of ice

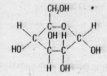

About 10³⁵ hydrogen bonds hold this iceberg together

The consequences of the freezing and melting of water in Nature are immense. For example, when the temperature drops, ice forms on the *surface* of seas, lakes, rivers and ponds – and the ice layer insulates the remaining water from the cold. The densest water, at 4 °C, will be at the bottom of the river or lake. If water behaved like a normal liquid, any solid formed would sink, and the lake, etc would rapidly freeze *from the bottom up*. But, as it is, the main body of the lake will stay liquid and the living creatures in it will survive.

The Earth has passed through several Ice Ages (with more to come?). The consequences for evolution of the low density of solid water must be large. A more practical consequence is that water in pipes (or in car engines when someone has forgotten to add antifreeze!) expands on freezing and can easily fracture metal; the leak only becomes obvious when the ice melts.

Much of the weathering of rock occurs for the same reason. Water gets into cracks in rocks, freezes and expands. The force of expansion is very great. Lumps of rock are loosened and eventually fall away. In ranges such as the Alps and the Himalayas, south-facing slopes and gullies are notorious for rockfall as the sun releases already-loosened rock. Much of the mountain scenery of Britain was rough-hewn by glaciers during the more recent Ice Ages. The flow of glaciers depends on the ability of ice to deform under pressure.

Figure 2.9.6   Glucose

Hydrogen bonding in Nature

Sugars, such as glucose (*Figure 2.9.6*), have plenty of polar —OH groups and their structures and properties depend on hydrogen bonding. Starch (as found in, for example, wheat, maize and potatoes) and cellulose are formed by large numbers of molecules of glucose joined by covalent bonds.

Wood is largely cellulose; hydrogen bonding gives wood its strength. Paper is largely cellulose fibres; it owes its strength to hydrogen bonds. Untreated paper loses its strength when wet because the fibres form hydrogen bonds with water rather than with each other. Cotton is another form of cellulose. Spun cotton thread and fabrics owe their strength to hydrogen bonding.

Many of the molecules necessary for life are proteins (see Ch 28.5). What a protein can do depends on its detailed structure, which in turn depends to a large extent on **intramolecular** hydrogen bonds (i.e. *hydrogen bonds between atoms in the same molecule*).

This is a major reason why nearly all living organisms are so sensitive to temperature.

Hydrogen bonds are weak; high temperatures rapidly disrupt them; so protein structure is destroyed by increased temperature. (This *denaturing* of proteins is essentially what happens when an egg is boiled). Mammals have evolved a marvellous mechanism to keep body temperature constant. This not only ensures that reactions in living cells occur at a reasonable rate (see Ch 12.2), but also keeps their proteins intact.

Intermolecular – forces *between* different molecules.

Intramolecular – forces *within* the one molecule.

Finally, hydrogen bonding is a crucial factor in the structure of the material essential for the continuation of life – DNA. The two strands in DNA are joined by precise hydrogen bonds between adenine and thiamine and between cytosine and guanine (see Ch 28.5). These are strong enough to hold the strands together under normal conditions, but weak enough to allow them to 'unzip' during cell division. It is hydrogen bonding which enables each strand to reform a new partner strand for the DNA in the new cells.

SELF-ASSESSMENT QUESTIONS 2.9

1 Which of the following molecules can be expected to exhibit hydrogen bonding?
   a SiH_4
   b HCl
   c NF_3
   d NH_3

2 What are the requirements for hydrogen bonds to be formed?

3 When ethanoic acid is dissolved in benzene, two molecules join together in a structure shown in the diagram.

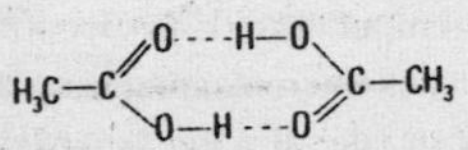

$$H_3C-C \overset{O---H-O}{\underset{O-H---O}{}} C-CH_3$$

   a Calculate the apparent relative molecular mass of this structure.
   b Name the type of bonding represented by the dotted lines in the diagram.
   c When dissolved in water the relative molecular mass of ethanoic acid is very close to half of that measured in benzene. Explain why this is so.

4 The boiling points of the compounds formed between the Group 6 elements and hydrogen are listed in the table.

Formula	Boiling point / °C
H_2O	100
H_2S	−61
H_2Se	−42
H_2Te	−2

   a Plot a graph of boiling point of the hydride against atomic number of the Group 6 elements.
   b What trend does your graph show for the compounds H_2S, H_2Se, and H_2Te?
   c Explain this trend in terms of intermolecular attractions.
   d Why should the value for H_2O not fall in with this trend?

5 Methane, CH_4, ammonia, NH_3 and neon, Ne, have a wide range of boiling points, although their relative molecular masses are very similar (see table).

Formula	Relative molecular mass	Boiling point / °C
CH_4	16	−164
NH_3	17	−33
Ne	20	−246

Explain, in terms of the type and strength of the intermolecular forces involved, why there should be such a large difference between the boiling points.

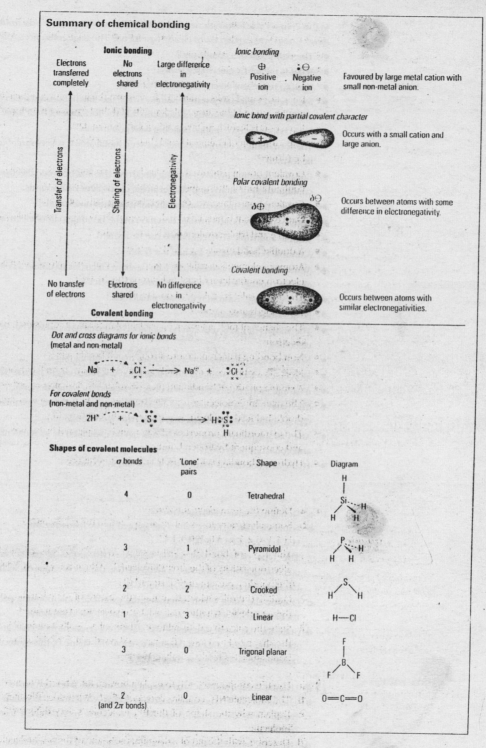

Summary of chemical bonding

Ionic bonding

| Electrons transferred completely | No electrons shared | Large difference in electronegativity |

Ionic bonding

Positive ion   Negative ion

Favoured by large metal cation with small non-metal anion.

Ionic bond with partial covalent character

Occurs with a small cation and large anion.

Polar covalent bonding

$\delta\oplus$ $\delta\ominus$

Occurs between atoms with some difference in electronegativity.

Covalent bonding

Occurs between atoms with similar electronegativities.

Transfer of electrons   Sharing of electrons   Electronegativity

| No transfer of electrons | Electrons shared | No difference in electronegativity |

Covalent bonding

Dot and cross diagrams for ionic bonds
(metal and non-metal)

$$Na \; + \; \overset{\times\times}{\underset{\times\times}{\times}Cl\times} \longrightarrow Na^{+} \; + \; \overset{\times\times}{\underset{\times\times}{\times}Cl\times}$$

For covalent bonds
(non-metal and non-metal)

$$2H \; + \; \overset{}{\underset{}{\bullet}S\bullet} \longrightarrow H\overset{\times}{\bullet}S\overset{\times}{\bullet}$$

Shapes of covalent molecules

σ bonds	'Lone' pairs	Shape	Diagram
4	0	Tetrahedral	
3	1	Pyramidal	
2	2	Crooked	
1	3	Linear	H—Cl
3	0	Trigonal planar	
2 (and 2π bonds)	0	Linear	O=C=O

SUMMARY 2

In this chapter the following new ideas have been covered. *Have you understood and learned them?* Could you explain them clearly to a friend? The page references will help.

- The definition of a compound. — p40
- What is meant by electronegativity. — p41
- How to represent electron pairs in bonds. — p42
- Ionic bonding involves transfer of electrons; the link between the electronic configuration of the atoms of elements and the formulae of their ionic compounds. — p43
- Ionic bonds involve metals bonding with non-metals. — p43
- The formation of chemical bonds between metal atoms is very rare; but alloys can be formed. — p43
- Covalent bonding involves sharing electrons; the link between electronic configuration of atoms and the formulae of covalent compounds. — p47
- Covalent bonds involve non-metals bonding with non-metals. — p43
- Dative covalent bonds involve donation of the bonding pair of electrons by one atom. — p52
- Double and triple covalent bonds can be formed. — p50
- A double bond involves a σ and a π bond. — p51
- Atoms of elements usually lose, gain or share electrons to obtain a complete outer electron energy level of eight electrons (the 'octet'). — p45
- The 'octet rule' is broken in some molecules. — p54
- Polar bonds are common. — p55
- The shapes of molecules can be explained in terms of repulsion between pairs of electrons. — p57
- Non-bonding pairs repel more strongly than bonding pairs. — p58
- Molecules which contain polar bonds need not have overall polarity themselves. — p63
- A simple picture of bonding in metals explains their main properties. — p64
- The main intermolecular forces are dipole–dipole attractions, instantaneous dipole-induced dipole attractions, and hydrogen bonding. — p66
- The extraordinary properties of water can be explained by its molecular structure and consequent hydrogen bonding. — p71
- Hydrogen bonding is immensely important in Nature. — p73

EXAMINATION QUESTIONS 2

1 a Define the term *electronegativity*.
   b Some electronegativity values are quoted in the following list:
      H 2.1; C 2.5; Cl 3.0; K 0.8; I 2.5.
      The type of bond that exists between two atoms is partly due to the difference in electronegativity of the atoms involved. Use the given values of electronegativities to suggest the bonding present in KCl, HI and CH_4.
   c Outline the rules which determine the extent of ionic bonding present in a solid, and use them to decide whether CsF will be a typical ionic compound.
   d State the rules that determine the shape of a covalent molecule, and use them to work out the shape of PF_3. Draw a diagram to show the shape of this molecule, and explain why the bond angle is slightly less than 109.5°.

2 a Use the compound $CaCl_2$ to explain how an ionic bond is formed.
   b The molecule NH_3 contains *covalent bonds*. What does this mean?
   c Explain why the shape of the BF_3 molecule is very different from the shape of the NF_3 molecule.
   d Describe, with the aid of a diagram, the shape of the SF_6 molecule.

3 a Describe, with diagrams, the bonding present in KBr, Br_2 and potassium.

 b Explain the trends in both molar first ionisation energies and electronegativities as atomic number increases in Group 1 of the Periodic Table.

 c Across the Periodic Table, from sodium to chlorine, the covalent radius of the atoms decreases. Why should this be so? Why is there no quoted value for the covalent radius for the argon atom?

4 a Explain what is meant by hydrogen bonding.

 b Using ethanol, C_2H_5—O—H, as an example, show how hydrogen bonds might be formed between at least three ethanol molecules.

 c The molecule with the formula CH_3—O—CH_3 has a formula similar to ethanol. Explain why this molecule is not able to hydrogen bond.

 d Explain why ethanol is soluble in water, whereas CH_3—O—CH_3 does not dissolve in water.

 e Which of the two compounds would you expect to have the higher boiling point: C_2H_5—O—H or CH_3—O—CH_3? Explain your answer.

Do you remember ...?

- The meaning of **atom** (Ch 1.1) and **ion**. (Ch 1.5)
- **Ionic bonding**. (Ch 2.3)
- **Covalent bonding** and the meaning of **molecule**. (Ch 2.4)
- The **shapes** of common molecules. (Ch 2.6)
- **Metallic bonding**. (Ch 2.7)
- The **forces** between molecules. (Ch 2.8)
- How **hydrogen bonds** are formed. (Ch 2.9)

The urge to understand *why* materials behave in the way they do is one of the driving forces behind chemistry. *Chemistry enables us to explain the behaviour of large, visible chunks of material in terms of the invisibly small particles from which they are built up*. If we understand enough about the behaviour of their particles, and how they are arranged, we can predict how the materials will behave. We may even (as has been done many times) put the particles together in such a way as to make a new material which behaves as we want it to behave.

In this chapter

We shall attempt to show how our knowledge of atoms, ions, molecules and chemical bonds can be used to account for the observed properties of materials.

3.1 **Types of structure**
3.2 **Giant atomic structures**
3.3 **Giant ionic structures**
3.4 **Properties of ionic materials**
3.5 **Giant metallic structures**
3.6 **Properties of metallic materials**
3.7 **Molecular structures and their properties**
3.8 **Amorphous materials**
3.9 **How structures are determined**

3

structure

3.1 Types of structure

A selection of crystalline solids from Hans Stern's jewellery factory

In crystalline solids, particles are arranged in a regular arrangement known as a **lattice**. There are four main types of solid structure. Three of these types are **giant structures**. This means that they contain very large numbers of units in a regular, repeating arrangement, usually with strong forces holding the structure together. The fourth type is a **molecular structure**, which consists of large numbers of molecules held together in the solid by comparatively weak forces.

The four types are summarised in *Figure 3.1.1*. The structures, and the properties linked to each of them, are summarised in *Table 3.1.1*. The information in this table will be invaluable later on. Much of the rest of this chapter will involve *explaining* the very strong connections between the structures and their properties.

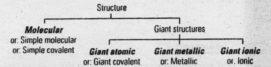

We shall use the highlighted terms in this book – just realise that they might be called something slightly different.

Figure 3.1.1 Types of structure

Type	Particles	Bonding between particles	Melting point	Electrical conductivity	Solubility in water
Giant					
Atomic	Atoms	Covalent	Very high	Poor, except graphite	Poor
Ionic	Ions	Ionic	High	Poor when solid, but good when molten	Variable, depending on forces holding the lattice together, but generally good
Metallic	Metal ions with free electrons	Metallic	Ranges from low (mercury) to very high (tungsten)	Good	Poor, but very reactive metals react to release H_2
Molecular	Molecules	Weak forces only (see Ch 2.8)	Usually low	Poor	Generally poor, unless the substance reacts with water

Table 3.1.1 Types of structures and their properties

3.2 Giant atomic structures

The bonding in these is covalent, so they are often called 'giant covalent structures'; but the particles of which they are built are *atoms*, so the term **giant atomic structures** is preferred.

The two most familiar giant atomic structures are diamond and graphite, which are both crystalline forms of carbon. If an element can exist in more than one form in the same state of matter it is said to be **polymorphic** or **allotropic**. (The word 'polymorphic' comes from the Greek for 'able to exist in many shapes'.) The recently discovered third crystalline form of carbon, the fullerenes, will be discussed at the end of this section.

In both polymorphs of carbon, there are covalent bonds between the atoms. The two different arrangements are shown in *Figure 3.2.1*.

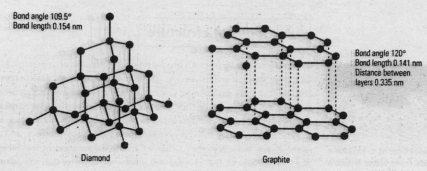

Bond angle 109.5°
Bond length 0.154 nm

Bond angle 120°
Bond length 0.141 nm
Distance between
layers 0.335 nm

Diamond

Graphite

Figure 3.2.1 Arrangements of carbon atoms

'The Great Star of Africa'

The Great Star of Africa
diamond – over 530
carats – contains
5×10^{24} atoms.

Diamond tipped drill bit

Diamond

In diamond, each carbon atom in the body of the solid is joined by four strong covalent bonds to four other carbon atoms. The four bonding pairs around each carbon atom point towards the corners of a tetrahedron – exactly as in the molecule of methane, CH_4 (see Ch 2.6). The result of linking these tetrahedral units is an immensely strong, rigid structure with a very high melting point (above 3550 °C). Diamond is the hardest known natural substance. It is 'used' in expensive jewellery. Of far greater benefit to us is its use in drilling through rock and in cutting glass.

Artificial diamond can be made by subjecting graphite to extremes of temperature and pressure (~1600 °C and 50 000 atmospheres) using metal catalysts. It is now possible to coat objects with diamond film; the potential benefits from this, if the process can be applied commercially, would include scratchproof glass, machinery bearings which do not wear out, etc.

Silicon is a non-metal with a giant atomic structure like that of diamond. When pure, it does not conduct electricity. Tiny precise amounts of specific impurities convert it into a **semi-conductor** (see Ch 3.6); and this is the basis of the 'silicon chip', billions of which are now used in communication, control, calculating and timing mechanisms throughout the world.

Graphite

In graphite, each carbon atom is joined to three others in the same plane. As there are three bonding pairs, the bond angle is 120° (see Ch 2.6), so the carbon atoms can form hexagonal (i.e. six-membered) rings. These hexagonal rings connect up to form planes – flat sheets – of carbon atoms. The bonding in the planes is very strong, and comparable with the bonding in diamond.

The melting point of graphite is very high, about 3700 °C. Each carbon atom has four electrons in its outer electron energy level; three of these are used in the three bonds to other carbon atoms in the plane. The fourth electron is free to move. So, one electron from each carbon atom is free to move throughout the plane; they are not fixed in place as they would be in a normal covalent bond. The word used to describe such electrons is **delocalised**; delocalised electrons enable metals to conduct electricity, as we shall see in Ch 3.5 and 3.6.

In diamond and the other giant atomic structures, all the outer electrons in the atoms are firmly fixed in covalent bonds, so they do not conduct electricity. They are 'localised' – fixed in one spot!

The unique properties of graphite

Carbon in the form of graphite is the only non-metallic element which is a reasonably good conductor of electricity. A perfect crystal of graphite conducts well *along* the planes of carbon atoms, but *not across* the planes, at right angles to them. This shows the delocalisation of electrons in the planes, but the lack of electron flow across the gaps between them.

The distance between carbon atoms *in* the planes is 0.142 nm, but the distance between the layers is much greater: 0.335 nm. So only relatively weak forces (see Ch 2.8) hold the planes together. This means that the layers of atoms can easily slip over one another. One result of this is that graphite can be used in 'lead' pencils. As the pencil moves across the paper, friction causes successive planes of carbon atoms to slide off the graphite core of the pencil on to the paper. The 'hardness' of the pencil lead (compare a 4H pencil with a 4B) is controlled by mixing graphite with clay in different amounts before baking. Graphite is also used as a lubricant for heavy machinery.

Because of its very high melting point, graphite is used in containers for molten metals and in the exhausts of rockets. All ordinary forms of carbon, including charcoal, consist of microcrystals of graphite. This means that the orientation of the carbon planes is random, and all normal samples of carbon (except diamond) conduct electricity in any direction. Carbon is used for the electrodes in industrial electrolysis, as in the manufacture of aluminium, and in electric furnaces. Because it is both a conductor and a lubricant, graphitic carbon is ideal material for the 'brushes' in electric motors, dynamos, alternators, etc.

In recent years it has been possible to make long fibres or 'whiskers' of graphite. These are very strong and of low density and, mixed with other materials, have been used to make objects as varied as turbine blades for jet engines, Formula One racing car brakes, and tennis and squash racquets.

> Interestingly, the planes of graphite do not slide easily in a vacuum. Graphite is useless as a lubricant in deep space.

> With diamond and graphite we have a very dramatic illustration of the influence of structure on the properties of matter. Identical atoms of the same element, with the same type of bond between them but arranged in different ways, give, on the one hand, the hardest material found in Nature and on the other a commercial lubricant.

Quartz

The only other giant structure we shall consider here is that of quartz (*Figure 3.2.2*).

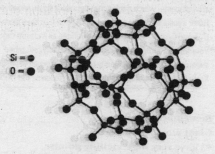

Si = ⊙
O = ●

Figure 3.2.2   Structure of quartz

This consists of silicon dioxide, with the empirical formula SiO_2. Unlike carbon dioxide, CO_2, which is a well-known gas, silicon dioxide forms a giant atomic network. The basic arrangement of silicon atoms is similar to that of carbon in diamond, but with an oxygen atom between the silicon atoms. As you can see in *Figure 3.2.2*, this results in a 1:2 ratio between silicon and oxygen atoms. However, in quartz crystals the tetrahedrons are arranged in spirals, so its structure is complex.

You will be familiar with the usual form of impure quartz – which is sand. The rigid giant structure of quartz ensures that it is very hard and has a high melting point.

Because of the large amount of energy needed to break huge numbers of strong covalent bonds, *all* materials which have giant atomic structures have high melting points and are insoluble in water.

Fullerenes

Diamond and graphite have been known for many centuries, but a third form of carbon has been discovered very recently. Although the fullerenes are not 'giant' structures, they are certainly large. And their discovery happened because a group of chemists was interested in what happened to carbon atoms in atmospheres of giant red stars.

Harry Kroto of Sussex University and Richard Smalley and Robert Curl of Rice University (Houston, Texas) thought that under such conditions in stars carbon atoms would link up in long chains. They blasted samples of graphite to atoms using high-powered lasers and found that long carbon chains were indeed formed. They also found that the mass spectrum (see Ch 1.3) also gave a large peak at m/e = 720, corresponding to a molecule containing exactly sixty carbon atoms. Furthermore, this molecule appeared to be very stable.

In trying to visualise possible structures, Kroto remembered the 'geodesic domes' designed by the American architect Buckminster Fuller, and also remembered building a spherical star map for his children by using pentagons and hexagons. Smalley built a model using 12 pentagons and 20 hexagons which had the exact shape of a good quality football. The new molecule was called 'buckminsterfullerene'.

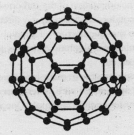 

The next development came in 1988 when it was found that soot containing up to 10% C_{60} could be made by striking an electric arc between two carbon electrodes. In 1990, Kroto and his team found that some of the soot dissolved in benzene to give a red solution containing C_{60}. The C_{60} could be purified by column chromatography (see Ch 29.2) and gave a carbon-13 n.m.r. spectrum (see Ch 30.5) with only a single peak, showing that all 60 carbon atoms must be in exactly the same environment – which means that they must be arranged in a sphere.

Other similar molecules, including C_{70} and C_{76}, have been isolated, and also a series of 'giant' fullerenes, beginning with C_{240} and C_{540}. (A 'fullerene' is now defined as a closed cage of carbon atoms consisting solely of pentagons and hexagons.) Very thin but strong 'buckytubes' – essentially 'stretched' fullerenes – can be grown. Potential uses of fullerenes involve superconduction, exceptionally rapid darkening on exposure to a flash, and even cancer treatment.

Kroto (now Sir Harold), Smalley and Curl received the Nobel Prize for Chemistry in 1996.

Remember:

Giant atomic structures:

- have very high melting points;
- are insoluble in any solvent;
- never conduct electricity (except graphite).

**SELF-ASSESSMENT
QUESTIONS
3.2**

1 Explain each of the following terms, using diamond and graphite to illustrate your answers.

  a Giant atomic structure.

  b Polymorph (allotrope).

  c Delocalised electrons.

2 Why would rubbing a pencil lead along a stiff zip fastener prove an acceptable alternative to using liquid machine oil?

3 Although carbon and silicon are close to each other in the Periodic Table, the oxides formed when these elements burn in oxygen have very different structures.

   a Write an equation for the complete combustion of each of these elements to form CO_2 and SiO_2 respectively.

   b Describe the physical state of each of the oxides at room temperature and pressure.

   c The temperature at which each of these oxides changes state is:

CO_2	−78 °C (sublimes)
SiO_2	above 1600 °C (melts)

   What does *sublime* mean?

   d Explain why these apparently similar compounds should behave so differently when heated.

3.3 Giant ionic structures

As we saw in Ch 2.3, there is no such thing – at least in the solid state – as a single ionic bond. A sodium chloride crystal consists of a huge number of sodium ions and an equal number of chloride ions, held together by the attraction between opposite charges. The attractions between ions of opposite charge are non-directional – each sodium ion attracts the negative ions all around it, and similarly for the chloride ions (*Figure 3.3.1*).

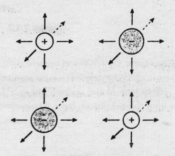

Figure 3.3.1 Non-directional attractions

Forces such as gravitation and electrostatic attraction or repulsion follow an 'inverse square' law. The force between two objects or ions is inversely proportional to the square of the distance, D, between them, i.e. force $\propto 1/D^2$.
For electric charges this is known as Coulomb's Law. If the distance doubles, the new force is only one-quarter of the magnitude of the original force.

And that causes the first difficulty. If there *are* equal numbers of positive and negative ions, why don't the repulsions between like charges cancel out the attractions between unlike charges, so that the crystal can't hang together at all?

We shall start with a very simple model – oversimplified, as we shall see, but it does explain what is happening in an ionic crystal, and it is based on that extremely important rule found in all sciences: *like charges or poles repel, unlike ones attract*, and the force of repulsion or attraction becomes less as the distance between them becomes greater.

The sodium chloride lattice

Consider a positively charged sodium ion. Which ion will be next to it? Obviously, a negatively-charged chloride ion. Next will come a sodium ion, then a chloride ion, and so on.

So, in one dimension the arrangement of ions in a sodium chloride crystal lattice (remember – 'lattice' simply means a giant, regularly repeating array) will be:

A similar argument leads to the arrangement in two dimensions:

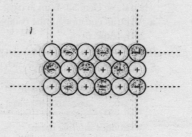

and finally to the arrangement in three dimensions (*Figure 3.3.2*).

○ – Sodium ion, Na⁺

⊛ – Chloride ion, Cl⁻

Models of the sodium chloride lattice

Figure 3.3.2 The structure of sodium chloride

<table>
<tr><td>**Definition:** The number of nearest neighbours around a given atom or ion is known as its *co-ordination number.*</td></tr>
</table>

From *Figure 3.3.2*, you can see that each sodium ion has, as its nearest neighbours, six chloride ions; in turn, each chloride ion has six sodium ions as *its* nearest neighbours.

Both ions in the sodium chloride lattice have a co-ordination number of 6; we say that the sodium chloride lattice has 6:6 *co-ordination* (*Figure 3.3.3*).

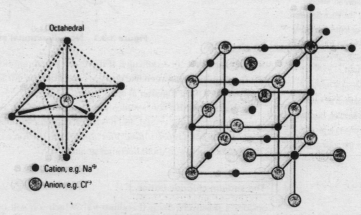

Octahedral

● Cation, e.g. Na⁺

⊛ Anion, e.g. Cl⁻

Figure 3.3.3 Co-ordination number

NaCl crystals

Looking again at *Figure 3.3.2*; it is probably no surprise by now that the shape of a sodium chloride crystal is **cubic**, which does not necessarily mean that all the crystals are perfect cubes, but that all the faces of the crystals are at 90° to each other.

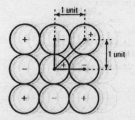

Figure 3.3.4 Attraction and repulsion in an idealised ionic lattice

If sodium chloride crystals have a high melting point (it is in fact 801 °C), the attractive forces between positive and negative ions in the lattice must be *far greater* than the repulsive forces between ions of the same charge. How does this happen?

Again, a simplified model can help us to understand the essential ideas. Assume that all the ions are the same size and have a diameter of 1 unit. Consider a small two-dimensional part of the lattice (*Figure 3.3.4*).

The distance between the centres of oppositely-charged ions is only one unit, but the distance between the centres of similarly charged ions is, using Pythagoras' Theorem, $\sqrt{2} = 1.414$ units. So repulsion occurs at a larger distance than attraction; and since both attraction and repulsion obey Coulomb's Law and decrease in proportion to the *square* of the distance (i.e. force $\propto 1/D^2$), the attractions greatly outweigh the repulsions.

So far, we have been using a simplified model, in which we consider the sodium and chloride ions to be the same size. But, as we discussed in Ch 1.6, the two ions are *not* the same size. The radius of the sodium ion is 0.102 nm and that of the chloride ion is 0.180 nm. The **radius ratio** is $0.102/0.180 = 0.567$.

The sodium chloride crystal lattice (*Figure 3.3.2*) can be thought of as an open **face-centred cubic** (f.c.c.) lattice of chloride ions (notice that the chloride ions do *not* touch each other) with the smaller sodium ions filling the holes between the chloride ions (*Figure 3.3.5*). 'Face-centred' simply means that there is a particle in the centre of each face. The sodium chloride crystal contains two interpenetrating f.c.c. lattices of chloride and sodium ions.

The unit cell

The simplest repeating unit in a crystal – the unit which, repeated many millions of times, makes up the crystal we can see – is called the **unit cell**. *Figure 3.3.5* shows a single face of a unit cell of sodium chloride.

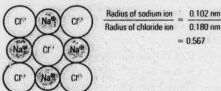

$$\frac{\text{Radius of sodium ion}}{\text{Radius of chloride ion}} = \frac{0.102 \text{ nm}}{0.180 \text{ nm}}$$
$$= 0.567$$

Figure 3.3.5 A single face of a unit cell of sodium chloride

The usual way of drawing the unit cell of sodium chloride is shown in *Figure 3.3.2*, but *Figure 3.3.6* is a *scale drawing* of the same unit cell.

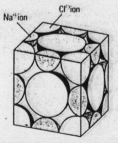

Figure 3.3.6 A scale drawing of a unit cell of sodium chloride

These figures of $\frac{1}{2}$, $\frac{1}{4}$ and $\frac{1}{8}$ can easily be verified by using suitably marked, identical, cubic children's building blocks. Try it.

It also helps us to see how many ions are actually in a unit cell. Only the $Na^\oplus$ ion at the centre is not shared with other unit cells. Each ion at the centre of a face is shared with another unit cell, so counts only as $\frac{1}{2}$; ions on an edge are shared with three other unit cells, so count as $\frac{1}{4}$; and the ions at corners are shared with seven(!) other unit cells and count only as $\frac{1}{8}$.

So, looking again at *Figure 3.3.2* or *Figure 3.3.6*, the sum total of ions in the unit cell is:

Chloride ions: $(6 \times \frac{1}{2}) + (8 \times \frac{1}{8}) = 4$

Sodium ions: $1 + (12 \times \frac{1}{4}) = 4$

The unit cell of sodium chloride therefore contains four sodium ions and four chloride ions.

Remember:
In the unit cell:
- an ion in the centre $= 1$
- an ion in the centre of a face $= \frac{1}{2}$
- an ion in the edge of a face $= \frac{1}{4}$
- an ion at the corner of a face $= \frac{1}{8}$

The caesium chloride lattice

But what happens if the metal ions are *too big* to fit into the holes in the lattice of chloride ions? This happens with ions of the alkali metal caesium, $Cs^{\oplus}$. In this case, the chloride ions are arranged in a more open simple cubic arrangement, with the $Cs^{\oplus}$ ions entering the holes between them. The result is two interpenetrating simple cubic lattices (*Figure 3.3.7*).

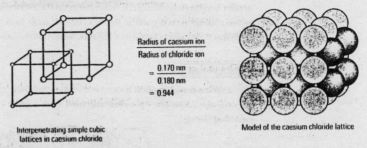

$$\frac{\text{Radius of caesium ion}}{\text{Radius of chloride ion}} = \frac{0.170 \text{ nm}}{0.180 \text{ nm}} = 0.944$$

Interpenetrating simple cubic lattices in caesium chloride

Model of the caesium chloride lattice

Figure 3.3.7   Caesium chloride structure

Each $Cs^{\oplus}$ ion has eight $Cl^{\ominus}$ ions as its nearest neighbours, and each $Cl^{\ominus}$ has eight $Cs^{\oplus}$ ions around it. Both the $Cs^{\oplus}$ and $Cl^{\ominus}$ ions therefore have a co-ordination number of 8; we say that caesium chloride has 8:8 co-ordination, unlike the 6:6 co-ordination in the face-centred cubic lattice of sodium chloride.

Note that the radius ratio in caesium chloride is 0.944. In general, if in a simple binary ionic compound, i.e. one with the general formula MX, the ratio

$$\frac{\text{radius of cation}}{\text{radius of anion}} > 0.73$$

the result is a structure like that of caesium chloride. If the radius ratio is between 0.41 and 0.73, the compound will have the sodium chloride structure. A scale drawing of the unit cell of caesium chloride is shown in *Figure 3.3.8*.

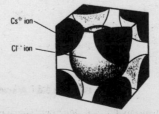

$Cs^{\oplus}$ ion

$Cl^{\ominus}$ ion

Figure 3.3.8   The caesium chloride unit cell

It is much simpler than for sodium chloride. There is a single $Cl^{\ominus}$ ion in the middle, and $(8 \times \frac{1}{8}) = 1$ $Cs^{\oplus}$ ion at the corners. So the unit cell contains only one ion of each kind.

SELF-ASSESSMENT QUESTIONS 3.3

1 When an electric current is passed through molten sodium bromide, metallic sodium and bromine gas are formed. Describe, with appropriate diagrams, how the movement of electrons between sodium and bromine produces the bonding found in sodium bromide.

2 The ionic radii, in nm, of some ions are given below:

K^+	0.138	O^{2-}	0.140
Cs^+	0.170	S^{2-}	0.102
Ca^{2+}	0.100	Br^-	0.195
Ba^{2+}	0.136		

Use these data to calculate the cation/anion ratio in each of the following compounds. Decide if the structure adopts a sodium chloride or caesium chloride type lattice. In each example state the co-ordination number of each ion.

a Barium oxide.

b Caesium bromide.

c Calcium sulphide.

d Potassium bromide.

3 Sodium oxide is primarily an ionic solid.

a Draw a dot-and-cross diagram to show how the bonding electrons in sodium oxide are distributed.

b Explain why sodium oxide is held together in a giant structure even though the charge on two sodium ions will be exactly cancelled out by the charges on one oxide ion.

4 Solid magnesium oxide has a structure very similar to that of sodium chloride. Complete the diagram to illustrate the structure of the ionic compound MgO, representing $Mg^{2\oplus}$ ions by ● and $O^{2\ominus}$ ions with o.

3.4 Properties of ionic materials

● Because of the attractions between oppositely charged ions, ionic crystals are *always* solids at room temperature, and usually have high melting and boiling points. (For sodium chloride, $Na^{\oplus}Cl^{\ominus}$, melting point (m.p.) is 801 °C, boiling point (b.p.) is 1412 °C.) In general, a double charge on both ions should result in greater attraction and so even higher m.p. and b.p. (e.g. magnesium oxide, $Mg^{2\oplus}O^{2\ominus}$, has a m.p. of 2852 °C and b.p. of 3600 °C).

But it is hard to make a general statement, as a lot depends on the sizes of the ions, and on the polarising power of the positive ion and the polarisability of the negative ion (see Ch 2.5). For example, all nitrates (containing the ion $NO_3^{\ominus}$) and nearly all carbonates (containing $CO_3^{2\ominus}$) decompose before they boil, and in many cases before they melt.

● Ionic materials are hard, but – unlike giant covalent structures such as diamond – are also very brittle. They shatter easily when dropped or struck. To break a diamond involves breaking a huge number of very strong carbon–carbon covalent bonds. But consider two planes of ions in an ionic crystal. An impact has only to shift one plane the tiny inter-ionic distance relative to the other plane (see *Figure 3.4.1*), and instead of opposite charges attracting there are now like charges repelling.

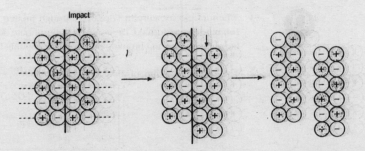

Figure 3.4.1 A possible effect of an impact on an ionic crystal

- Ionic materials do not conduct electricity when solid, because the ions are in fixed positions and cannot move. However, as soon as the ions become free to move – either by melting the crystal or dissolving it in water – conduction can occur. This type of conduction is called **electrolysis**. Whereas conduction in a metal is caused by *electrons* moving, electrolysis always involves movement of **ions**. Positive ions go to the negative electrode (the cathode) and gain electrons; they are called **cations** (note – *not* cathions). Negative ions go to the positive electrode (the anode) and give up electrons; they are called **anions**. For example, molten lead bromide will give lead at the cathode and bromine at the anode (*Figure 3.4.2*).

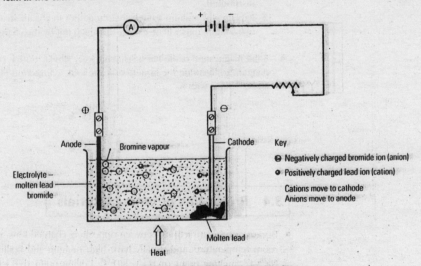

Figure 3.4.2 Electrolysis circuit

At the anode
$$2Br^{\ominus} \longrightarrow Br_2(g) + 2e^{\ominus}$$

At the cathode
$$Pb^{2\oplus} + 2e^{\ominus} \longrightarrow Pb(l)$$

Whereas there is no chemical change in a metal when it conducts electricity, electrolysis **always** results in the breaking-up of the conducting material. A material which conducts electricity when molten or in aqueous solution is called an **electrolyte**. Electrolysis is discussed further in Ch 24.8.

- Ionic materials are very often soluble in water and some other polar solvents. This is because the charged ions and the polar water molecules (see Ch 2.5) are attracted to each other. This helps to loosen ions at the surface of the crystal and ensures that any dissolved ion is surrounded by water molecules (*Figure 3.4.3*) – it is 'hydrated'.

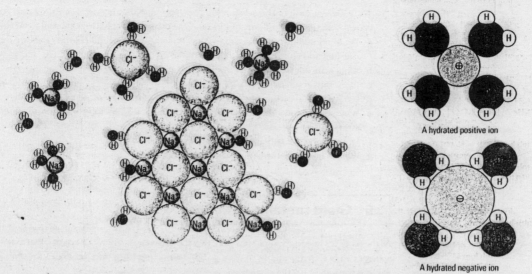

A hydrated positive ion

A hydrated negative ion

Figure 3.4.3   Dissolving ionic materials

The hydration of ions releases energy (see Ch 11.3). If the ions in an ionic solid are strongly attracted to each other – e.g. if the energy needed to break up the ionic lattice is high (see *lattice energy*, Ch 20.4) – the material will not be soluble. Ionic materials are insoluble in non-polar solvents such as hexane, benzene and ethoxyethane (ether).

Remember:
Giant ionic structures:
- **have high melting points;**
- **are hard and brittle;**
- **are usually soluble in polar solvents such as water;**
- **are electrolytes – they conduct electricity when molten or dissolved in water, and there are chemical reactions at the electrodes.**

SELF-ASSESSMENT QUESTIONS 3.4

1   Using the following data, identify which of the following compounds are likely to involve ionic bonding.
   a   RbBr, m.p. 693 °C, solubility in water is 1160 g dm^{-3}.
   b   PH$_3$, m.p. −133 °C, solubility in water is 0.3 g dm^{-3}.
   c   NO, m.p. −163 °C, solubility in water is 0.06 g dm^{-3}.
   d   SrO, m.p. 2430 °C, solubility in water is 8.6 g dm^{-3}.

2   **a**   Using appropriate diagrams, describe the way in which the movement of electrons between the atoms produces ionic bonding in potassium fluoride.
   Explain why potassium fluoride:
   b   has a high melting point;
   c   dissolves readily in water;
   d   does **not** conduct electricity at room temperature;
   e   **does** conduct electricity at 800 °C.

3 · The different radii of the sodium ion (0.95 nm) and the caesium ion (0.169 nm) result in the crystal structures of sodium chloride and caesium chloride having different ionic lattices.

    a Draw a dot-and-cross diagram to show how the caesium and chloride ions are formed from caesium and chlorine atoms.

    b What is meant by the term *ionic lattice*?

    c Draw the crystal lattice of sodium chloride.

    d What is the co-ordination number of the sodium ions in this structure?

    e Why should the differences in size of the ions lead to the formation of different lattice structures?

    f Draw the crystal lattice of caesium chloride.

    g Explain how an ionic lattice is held together.

3.5 Giant metallic structures

All the particles inside a sample of a pure metal are identical, and we can treat them as spheres. Such spheres can fit together tightly in what is called **close packing**; this minimises the amount of space which is wasted.

ACTIVITY

Bubble rafts

Blow a bubble raft (blow bubbles, using a large syringe, into a dish of water with a little washing-up liquid). Notice the bubbles take up a close packed arrangement – *never* an open packed arrangement.

A close packed arrangement   An open packed arrangement

If you have at least 20 identical marbles, table tennis balls or snooker balls you can investigate the two possible close packing arrangements for yourself. Resist looking at the figures until you have tried it!

ACTIVITY

Close packing of identical spheres

Arrange pieces of wood or books to form a flat triangular space of suitable size, and then put in ten balls or spheres to make a 4:3:2:1 triangle. The spheres should be packed as tightly as possible. Notice that the only sphere in this arrangement which is completely surrounded is in contact with six other spheres in the same plane. Mark this sphere in some way. Now put a second layer of spheres on top of the first. They will sit in the hollows. Six spheres can be placed in this way. Three of these are in contact with the marked sphere. Obviously, if there were another layer of spheres underneath, three of those would also be in contact with the marked sphere. This is a characteristic of close packing – each sphere is in contact with 12 others. The co-ordination number is 12.

    The interest really begins with the third layer. If a sphere is placed dead centre, it is directly above the marked sphere in the bottom layer. Three small 'tunnels' through to the bottom of the pile can be seen around the top sphere. (These can best be seen if the whole arrangement is built on an overhead projector protected with a piece of clear acetate.) You can see that the spheres in the *third* layer are directly above those which are in the *first*. The 'ababab…' arrangement gives rise to what is called **hexagonal close packing** (h.c.p.)

Now roll that single top sphere gently towards a corner of the triangle. It will fit into a hollow which will allow two other spheres to be fitted into the top layer. Your final sphere can now be placed on top to make a fourth layer. Notice that this sphere is directly above the marked sphere in the bottom layer, and that the 'tunnels' through the arrangement no longer exist: if it is on an OHP lens, light can no longer get through. The fourth layer of spheres is directly above the first. We now have an abcabcabc…' arrangement. This is called **cubic close packing** (c.c.p.).

All right, you can now look at the figures. You can see (*Figure 3.5.1*) how the ababab… arrangement gives rise to a hexagonal structure, and how (*Figure 3.5.2*) the abcabcabc… arrangement (tilted slightly towards you for a better view) gives the cubic structure.

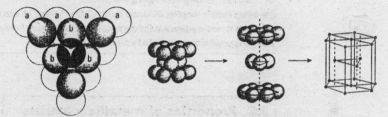

Figure 3.5.1 Hexagonal close packing – ababab

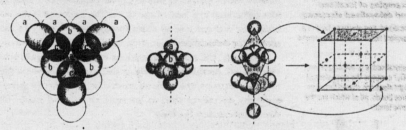

Figure 3.5.2 Cubic close packing – abcabc

Examples of metals with h.c.p. structures are zinc and magnesium. Those with c.c.p. structures include aluminium, copper and silver. A close-packed structure often, but not always, means a comparatively tough metal.

In some metals there is a less tightly-packed structure. This is the **body-centred cubic** (b.c.c.) structure (*Figure 3.5.3*).

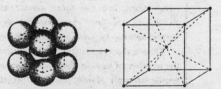

Figure 3.5.3 Body-centred cubic structure

A b.c.c. structure wastes more space. It is no surprise that the alkali metals (e.g. sodium and potassium), which are so soft that they can easily be cut with an ordinary knife and of so low a density that they float on water while reacting with it, have the b.c.c. structure; but it is rather a surprise that so do manganese and (at temperatures below 910 °C) iron.

As with several other metals, iron can change its crystalline form depending on the temperature. This is an example of **polymorphism** (which used to be called 'allotropy'). Diamond and graphite are **polymorphs** of carbon; both are stable at room temperature, although graphite is marginally the more stable. Iron freezes at 1535 °C into the b.c.c. form. At 1400 °C, the lattice changes to c.c.p. and at 910 °C it changes back to b.c.c.

There seems to be no clear link between the position of a metal in the Periodic Table and its crystal structure at room temperature.

The polymorphism of tin has generated several stories, which may or may not be true. When very cold, ordinary shiny tin certainly does crumble to a grey powder. The transition temperature between the two forms is 13 °C, but in ordinary cold weather the process is very slow. It occurs most rapidly at − 48 °C.

When Napoleon attacked Russia in 1812 and almost reached Moscow before his Grande Armée was forced to retreat though a bitter Russian winter, it is said that the buttons on the greatcoats and trousers of his soldiers were made of tin … and their crumbling made fighting very difficult.

The famous Russian composer Alexander Borodin, whose full-time job was as a Professor of Chemistry, is said to have been asked to resolve a dispute in St Petersburg. A merchant deposited expensive shiny ingots of tin in the basement of a bank and returned in the spring to find a pile of grey powder.

3.6 Properties of metallic materials

Examples of localised and delocalised electrons

Localised: two electrons held in a covalent bond between two atoms

A $\overset{\bullet}{\bullet}$ Y

Delocalised: e.g. the nitrate ion, $NO_3^{\ominus}$ contains the equivalent of eight electrons shared between three bonds, all of which are the same length

$$\left[O = \overset{\underset{\displaystyle O^{\ominus}}{|}}{N} \overset{\displaystyle O}{\underset{\displaystyle O}{\diagdown}} \right]^{\ominus}$$

Many of the properties of metals can be explained by thinking of them as a regular lattice of metal ions embedded in a 'sea' of electrons. The electrons in the outer energy level of the metal atoms break free and are able to roam throughout the piece of metal. These electrons are said to be **mobile**, or **delocalised** – i.e. they are not fixed in one position.

Obviously, the positive metal ions in the lattice repel each other. What holds a piece of metal together, i.e. **metallic bonding**, is the attraction between the positive ions in the lattice and the negative electrons which are all around and in between them (see Ch 2.7). So metallic bonding is even less directional than in an ionic crystal, and totally different from the almost rigorously directional covalent bonding.

Evidence for this model of metallic structure is given by the properties of sodium, magnesium and aluminium. These metals form a sequence in the Periodic Table, and have the electron structures 2.8.1, 2.8.2, 2.8.3 respectively. Sodium is very soft and – even if it didn't react with water – it would be impossible to build a bridge with it. Magnesium is harder, but magnesium ribbon can easily be broken between finger and thumb. Aluminium is tough enough to be used in vehicle engines. The atoms of the elements get smaller in the sequence sodium – magnesium – aluminium, but the more convincing explanation for this increase in hardness is that aluminium has *three* electrons per atom available for holding the lattice together, whereas magnesium has two and sodium only one.

Electrical conductivity increases from sodium to aluminium, again as one would expect from the number of mobile electrons. Because of its low density, good conductivity and fairly good tensile strength, aluminium (reinforced with a strand of steel) is the main component of the high-voltage cables which are slung between the pylons of the National Grid for distributing electricity.

It is the delocalised electrons which give metals their typical physical properties and make them so useful. An electric current in a metal is a flow of electrons. As electrons repel other electrons, a simple picture of electrical conduction is that electrons fed in at one end of a metal wire will cause repulsion along the wire, and an equal number of electrons will come out of the other end (see Ch 2.7). **Metals are good conductors of electricity.**

Heating a metal will cause mobile electrons to move around faster. This motion, and the energy associated with it, can be rapidly transmitted through the lattice causing it to vibrate more strongly. **Metals are therefore also good conductors of heat.**

When light of sufficient energy strikes the surface of a metal, electrons are given off from the metal. Light of lower energy, i.e. visible light, excites the mobile electrons near the surface of the metal into a vast number of energy levels (see Ch 1.4). This means that as the electrons return to lower energy levels, light of all visible wavelengths is re-emitted. In effect, the incident light is totally reflected, which is why metals are shiny and (with the exception of copper and gold) silvery in appearance. The reflecting surface of a mirror is silver or aluminium metal behind the glass.

Metal crystals

The beautiful structure of metal crystals is easily demonstrated by growing crystals of silver on a copper tree. The shape of the tree, cut out of thin copper sheet, is suspended in a dilute solution of silver nitrate. After a day or two, the silver metal, displaced from the solution appears as fine crystals attached to the copper tree.

Even though the outer electrons of the atoms in a metal are delocalised, they still occupy definite energy levels – but there are a great number of such levels, which form an **energy band**. The number of electrons in each of the energy levels in the band can vary. In the absence of light or other exciting agents, the higher energy levels will mostly be empty (*Figure 3.6.1*).

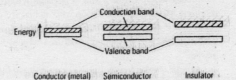

Figure 3.6.1

Semiconductor device

In many non-metal materials, a gap appears in the energy band. This means that there are now two bands – a lower *valence band* and an upper *conduction band*. In pure silicon and the diamond form of carbon, each atom has four outer electrons, and this results in a completely full valence band and an empty conduction band. The difference between the top of the valence band and the bottom of the conduction band is known as the *band-gap* energy. If electrons can be excited up into the conduction band, conduction is possible. Such a material is a **semiconductor**. A few substances, such as germanium and grey tin, have a sufficiently narrow band-gap to be semiconductors without further treatment.

Highly purified silicon, however, has tiny but very precise amounts of impurities added to it to give it useful semiconductor properties. There will now be additional narrow *impurity bands* between the valence and conduction bands, either unfilled levels just above a filled band or filled levels just below an empty band. 'Doping' silicon with phosphorus provides extra electrons and forms an 'n-type' (for negative, because of the extra electrons) semiconductor. Doping with, say, aluminium, leaves the lattice one electron short for each aluminium atom present, resulting in a 'p-type' (for positive) semiconductor. Many silicon-based devices depend on 'p–n' junctions between slices of suitably doped silicon.

Almost the whole of IT and communications, together with timekeeping, the control of machines, and devices for the direct conversion of solar energy to electricity, now depend on manipulating the band-gaps in slices of deliberately impure silicon. The band-gap energy in diamond is large, corresponding to radiation in the ultra-violet part of the spectrum; diamond is therefore colourless. Materials with large band-gap energies are good insulators. A small band-gap means that all light is absorbed and the material is black, as with, for example, graphite or the lead ore galena (lead(II) sulphide). The dramatic red colour of the pigment vermilion (mercury(II) sulphide) is caused by its fairly small band-gap; it absorbs all visible light except the lower energy red end of the spectrum.

Strength of metals

The reason why metals are so vitally important (indeed, whole Ages of the Western world have been named after one alloy and one metal!) is not only that some of them are very tough and chemically resistant, but also that *they can be shaped* – hammered, rolled, bent, stretched, drawn out into wire – *without breaking or losing strength*. They are said to be **malleable** (i.e. they can be hammered into shapes) and **ductile** (i.e. they can be drawn into wires). You can probably name whole industries which are largely based on forcing bits of metal into desired shapes. (Railway track, for a start.)

The explanation for this behaviour of metals is in terms of their structure, and depends once again on the mobile electrons. The ions in the metal lattice are arranged in layers, as we have already noted (Ch 2.7); there are no real bonds holding the layers together, only the attraction for the mobile electrons. The layers can slip over each other and *still* be attracted by the electrons in between them (*Figure 3.6.2*).

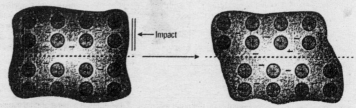

←— Impact

Figure 3.6.2 Deformation of metal (schematic)

In other words, the metal can be deformed without breaking.

Metals when stressed will bend slightly and return to their original shape when the stress is released. This is another reason why metals are used in structures where a certain amount of 'give' is necessary – e.g. bridges, tall chimneys, aircraft wings. However, if the stress is too great, slip occurs between layers in the metal structure – which therefore becomes permanently deformed, or even breaks.

A metal can be strengthened against slip in two ways. First, close examination of a piece of metal shows that it consists of *grains* – small irregular crystals tightly packed together (*Figure 3.6.3*). The metal lattice *within* the grain is regular, but slip does not easily cross grain boundaries. The result is that metal samples with small grain size are harder to deform than those with large grains.

Metallurgical techniques such as *annealing, tempering* and *quenching* can be used to control the grain size and hence the toughness of the metal.

The second way of strengthening a metal is by making an **alloy**. An alloy is a metal containing one or more deliberate impurities, often other metals (*Figure 3.6.4*).

> Some metals are more easily deformed than others. Gold is unique in that it can be beaten out into extremely thin gold sheets. Gold leaf is only a few microns thick.
> (1 micron, $\mu = 10^{-6}$ m)

**Figure 3.6.3
Microcrystalline
structure**

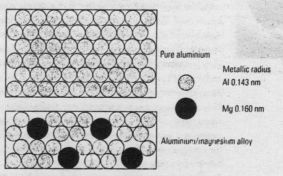

Pure aluminium

Metallic radius
Al 0.143 nm

Mg 0.160 nm

Aluminium/magnesium alloy

Figure 3.6.4 Alloy structure

Alloys are not compounds, since a particular alloy can vary to a small extent in composition and there is no genuine bonding between atoms.

Some approximate alloy compositions

Bronze	90% Cu	10% Sn		
Brass	90% Cu	10% Zn		
Stainless steel	73% Fe	18% Cr	8% Ni	1% C
Duralumin	94% Al	4% Cu	1% Mn	1% Mg

Ordinary 'mild steel' contains about 1% carbon, and is much tougher than pure iron. The extra strength is because the small carbon atoms are inserted between the iron ions and 'lock' the structure to some extent by preventing layers slipping. Brass is an alloy of zinc and copper and is much stronger than either. A particular approximate composition is necessary in which a suitable electron/atom ratio gives the extra strength.

Remember:
Many metals:
- **have excellent structural properties;**
- **have high melting points;**
- **are insoluble in any solvent (although several react with water at different rates).**

All metals:
- **conduct electricity with no obvious chemical changes.**

SELF-ASSESSMENT QUESTIONS 3.6

1 a Although sodium is very reactive and fairly soft for a metal (compared to more common metals such as iron) it still conducts electricity. Why?

   b A metallic lattice is often described as *metal ions in a sea of mobile electrons*. Explain how such a description is used to account for the physical properties of metals.

2 Caesium bromide, gold and graphite all conduct electricity, although not necessarily under the same conditions. By describing the bonding and structures of the three materials, explain why this is so.

3 Duralumin is an alloy containing aluminium and small amounts of copper, manganese and magnesium.

   a Draw a diagram, to show why aluminium can be quite a malleable metal, but the alloy is much less malleable.

   b Using a diagram, explain why the electrical conductivity of aluminium is not appreciably affected by the formation of the alloy.

   c Aluminium oxide also conducts electricity, but only at temperatures above 2100 °C. Explain why this should only happen at this high temperature.

3.7 Molecular structures and their properties

Most known compounds, together with many non-metal elements, exist as molecules rather than as giant structures. The bonding *within* the molecules is covalent (see Ch 2.4) and usually *strong*. The forces holding the separate molecules together in a solid or liquid are usually *weak* (see Ch 2.8 and 2.9).

The point about the bonds *inside* molecules being strong and the forces *between* molecules being weak is one that is very often missed by examination candidates, and accuracy of language is particularly important here. We have quite often, in examination answers, seen statements such as: 'when water boils, the molecules break up'. Can you analyse *why* this is wrong, and suggest what the candidate should have written? (*Figure 2.8.1* might help you.)

Because of these *weak* forces, it is comparatively easy to free the molecules in such a solid; simple molecular solids are usually soft and have low melting points.

Artificial snow

There is demand for winter sports and the 'right' type of snow on the pistes. The fine snowflake crystals are formed naturally around dust or pollution particles in the air. In making artificial snow the growth centre for the flake is a protein from a freeze-dried strain of a bacterium. This protein attracts water molecules and allows them to build up into the myriads of beautiful snowflake shapes (*Figure 3.7.1*).

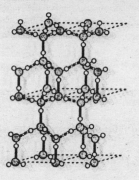

A snow gun in action

The beauty of a single snowflake

Figure 3.7.1

Figure 3.7.2 Iodine molecules (I$_2$) in a molecular crystal lattice

Similarly, the molecules in a liquid are relatively easy to separate, so simple molecular materials also tend to have low boiling points. (Large molecules, of course, have more points of attraction between them, and many polymers – see Ch 16.7 and Ch 28.1 – have quite high melting and boiling points).

Any material can be obtained as a solid if the temperature is low enough. Studies on simple molecular solids reveal that, although the packing together of individual molecules can often be complicated (see Ch 28.5), a pure molecular material can always form a regular crystalline lattice. A fairly simple lattice is found in solid iodine, I$_2$ (*Figure 3.7.2*).

Properties of molecular materials

As already stated, molecular materials usually have low melting and boiling points. Because they contain no free electrons, they do not conduct electricity when solid. Because they contain no ions, they cannot conduct even when molten or in solution. Normally, molecular materials are almost insoluble in water – unless, like hydrogen chloride or ammonia, they react with it, or like sugar, form extensive hydrogen bonds with it. Molecular materials tend to be soluble in non-polar solvents such as hexane or benzene.

Remember:
Materials with molecular structures:
- **often have low melting and boiling points (may be gas, liquid *or* solid at room temperature);**
- **are often not soluble in polar solvents such as water, but do dissolve in non-polar solvents such as hexane;**
- **do not normally conduct electricity.**

Q

SELF-ASSESSMENT
QUESTIONS
3.7

1 (See Ch 2 if you need help with this question.)
  a Draw a diagram to show the anticipated shape of the PCl_5 molecule, including approximate bond angles.
  b In the gas phase, PCl_5 consists of five chlorine atoms around a single phosphorus atom. However, in the solid state the structure contains $PCl_4^{\oplus}$ and $PCl_6^{\ominus}$ ions. Using electron pair repulsion theory predict the shapes of these two ions in solid phosphorus chloride.

2 i) Write the electron configurations of the following atoms:
     a hydrogen
     b nitrogen
     c silicon
     d sulphur
     e bromine.
  ii) Describe the movement of electrons required to form the following simple molecular structures from their atoms.
     a NH_3
     b H_2S
     c HBr
     d SiH_4.

3 Crystals of sodium chloride and diamond both have transparent structures. Using your knowledge of bonding and the structures of these two materials, explain why one is hard, insoluble and a poor conductor of electricity, whereas the other is soluble in water and conducts electricity under certain circumstances.

4 Classify the materials A–D, by the type of bonding and structure likely to be present when their properties are taken into account.

Substance	m.p.	Electrical conductivity	Solubility in water
A	Very high	Poor	Poor
B	High	Good	Poor
C	Low	Poor	Poor
D	High	Good when molten	Good

5 Sulphur can exist as two polymorphs (allotropes), rhombic sulphur (m.p. 113 °C) and monoclinic sulphur (m.p. 119 °C).
  a What is meant by the term polymorph or allotrope?
  b Why should there be a difference in the melting points of the two forms of sulphur?
  c Give a reason why sulphur in the rhombic form melts at 113 °C whereas the element above sulphur in the Periodic Table, oxygen, melts at −218 °C.

Figure 3.8.1
Amorphous solid

3.8 Amorphous materials

The best known example of a non-crystalline solid is glass. Such materials are said to be **amorphous** (from the Greek 'without shape'). The particles in glass (usually, in ordinary glass, sodium and calcium positive ions and silicate negative ions) are arranged more or less at random, and it is probably best considered as a *supercooled liquid* – a liquid which has been cooled so quickly that the particles in it have had no chance to form a regular lattice (*Figure 3.8.1*).

97

If subjected to stress or sudden changes in temperature, glass may crystallise. When it does, it becomes opaque, as light reflects off the crystal boundaries rather than going straight through. However, some glassy materials of this kind – known as *glass ceramics* – have very desirable properties. Some are almost unbreakable and have a low coefficient of thermal expansion. These may be used in ovenware, and even in vehicle engines.

The use of cheaper, 'glassy' semi-conductors rather than crystalline ones has been extensively investigated. Comparatively cheap amorphous silicon devices are now being used for direct conversion of sunlight into electricity, and banks of these devices are now frequently seen alongside, or on the roofs of, relay transmitters, lighthouses, weather stations, etc. The roof of the shop at the Centre for Alternative Technology, Machynlleth, Wales, consists almost entirely of such devices. It can produce 13 kW; any electricity not required in the Centre is sold to the National Grid.

Solar panels at Machynlleth

3.9 How structures are determined

In this chapter and in Ch 2.6 we have discussed many structures and molecular shapes, and you may have wondered whether there is evidence (other than the properties of the materials) for these structures and shapes. There are several methods available for gaining such information. The most common are summarised below. Details of all the methods will be found in standard degree-course textbooks.

Mass spectrometry has already been introduced (Ch 1.3). Here, we shall consider only X-ray crystallography. Infra-red and nuclear magnetic resonance spectroscopy will be discussed in Ch 30, as will the application of mass spectrometry to organic molecules.

Type	Method	Application
Diffraction methods	X-ray diffraction	Patterns of atoms and ions in crystals and distances between particles
	Electron diffraction	Bond angles and interatomic distances in gas molecules
	Neutron diffraction	Position of hydrogen atoms in crystals
Spectroscopic methods	Visible and ultra-violet (UV) spectroscopy	Structures of some organic molecules, and of complexes of transition metals
	Infra-red (IR) spectroscopy	Identification of particular bonds and groups of atoms
	Nuclear magnetic resonance (NMR) spectroscopy	Arrangement of certain atoms, especially hydrogen, in molecules
Other methods	Mass spectrometry	Molar masses; structures of organic molecules
	Magnetic effects	Compounds containing transition metals

Table 3.9.1   Methods of determining structure

X-ray crystallography

The idea of diffraction has already been met (Ch 1.4), as has the use of X-ray diffraction in finding the size of atoms and ions, together with a brief history of X-ray crystallography (Ch 1.6). The basic requirements for X-ray diffraction are a pure crystal of the substance, which contains all the atoms or ions in a regular lattice consisting of a huge number of repetitions of the unit cell (see Ch 3.3), and a monochromatic beam of X-rays. 'Monochromatic' means that they must all be of the same wavelength. X-rays are used because their wavelength is so short that it is comparable with the tiny distances between particles in a crystal.

The technique depends on the reinforcement or destruction of electromagnetic waves. You may have seen the effect demonstrated using water waves in a ripple tank, and also perhaps when sea waves enter a partially enclosed harbour or bay.

If two waves are 'in phase', i.e. the peaks and troughs coincide, they reinforce each other and the resulting waves have greater amplitude.

If they are exactly 'out of phase', so that the peaks of one set of waves coincide with the troughs of the other, they will 'interfere', i.e. destroy each other as the waves are cancelled out.

Electromagnetic radiation, including visible light and X-rays can be treated as a wave motion; hence the term 'wavelength'. *Figure 3.9.1* represents a beam of monochromatic X-rays interacting with a single crystal made up of regular layers of particles, e.g. sodium chloride.

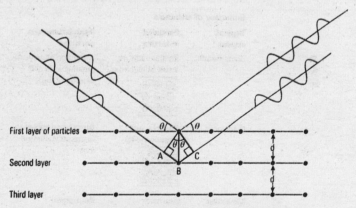

Figure 3.9.1   The principle of X-ray diffraction

Some of the radiation will go straight through, but some will be reflected. All the waves are in phase as they reach the crystal. Some will be reflected off the top layer, but others will penetrate and be reflected off the second and lower layers.

Note that the ray which is reflected off the second layer has travelled further than the ray which is reflected off the surface. If the waves are to be in phase after reflection, the 'path difference' must equal a whole number of wavelengths. With the help of a little trigonometry we can calculate this path difference.

If the wavelength of the X-rays is λ, then for reinforcement the path difference = nλ, where n is a whole number.

∴ In *Figure 3.9.1*, AB + BC = nλ

but where d = separation between the layers in the crystal,

AB = BC = d sinθ

∴ nλ = 2d sinθ.

This equation is known as the Bragg equation, after William and Lawrence Bragg (see Ch 1.6). It is the basis of all X-ray structural work. From this equation, if λ is known and θ can be measured, d – the distance between the layers – can be worked out.

In the classical method of X-ray crystallography, a perfect single crystal of the substance is held in a narrow beam of monochromatic X-rays and rotated slowly through 360°.

As the different planes of particles in the crystal reach the correct orientation for reinforcement, the beam is diffracted towards photographic film placed on a drum around the crystal. When the film is developed a complicated pattern of spots is seen. The angle of diffraction, the position of the spots and their relative brightness provide the information from which a structure can be worked out.

X-ray diffraction/ crystallography apparatus

It is important to realise that X-rays are reflected (strictly, 'scattered') most from regions of high electron density, so that large atoms such as iodine are easily 'seen' while hydrogen is very hard to detect. In the early decades of X-ray crystallography, the calculations had to be done by hand or by using primitive mechanical or electronic machines – a terribly laborious job. Modern computers have made the crystallographer's life much less arduous. The first complex structure to be solved with significant help from an electronic computer was that of vitamin B_{12}, completed by Dorothy Hodgkin and her group in 1956 (see box in Ch 1.6).

Summary of structure

Type of crystal	Particles in lattice	Forces between particles	Examples	Properties
Giant metallic	Positive ions in a sea of electrons	Attractions between positive ions and electron cloud	Na, Cu, Fe, Mg, Al	Wide range of hardness and melting points. Good electrical and thermal conductivity. Malleable and ductile
Giant ionic	Positive and negative ions	Attractions between oppositely charged ions	NaCl, CaF_2, KNO_3	Hard, high melting points. Non-conductors of electricity in solid state, but conduct when molten or in solution. Brittle
Molecular	Atoms or molecules	Permanent dipole–permanent dipole interactions, van der Waals forces, hydrogen bonds	HCl, Ar, H_2O, sugar	Soft, low melting points. Non-conductors of electricity in solid and liquid states
Giant atomic	Atoms	Covalent bonds between atoms	Diamond, graphite, SiC, SiO_2 (sand)	Very hard, high melting points. Non-conductors of electricity, except graphite

SUMMARY 3

In this chapter the following new ideas have been covered. *Have you understood and learned them?* Could you explain them clearly to a friend? The page references will help.

- There are four main types of structure: giant atomic, ionic, metallic and molecular.   p79
- The structure of a material greatly affects its physical properties, especially melting and boiling points, electrical conductivity, and solubility.   p79
- Diamond and graphite are polymorphs of carbon.   p80
- Sodium chloride and caesium chloride have different crystal structures.   p85
- There are two ways of close-packing identical spheres, as in metals.   p90
- The usefulness of metals lies largely in their ability to be shaped without breaking, and in their electrical conductivity.   p92
- Metals can have their properties modified, e.g. they can be strengthened by the formation of alloys or by control of grain size.   p94
- The properties of molecular structures arise from weak intermolecular forces.   p95
- Some solid materials have no regular structure; they are amorphous, e.g. glass.   p97
- Various methods can be used for determining solid structures and the shapes of molecules.   p98

EXAMINATION QUESTIONS 3

1   Materials in common use have widely different physical properties. Explain how the different physical properties of sodium chloride (used for flavouring food), chlorine (used for keeping swimming pools hygienic), graphite (used as a high temperature lubricant) and copper (used for electrical wiring and cooking utensils) are related to their structures. In particular you should consider the melting point and electrical conductivity of each of these materials.

2   Draw dot-and-cross diagrams to show the bonding in an ionic and in a covalent oxide. Explain how the electrical conductivity of each structure can be explained in terms of its bonding.

3   Decide on the likely structures of the compounds AX, BY, and CZ, using the data provided. Draw suitable diagrams to illustrate the structure in each case.

	m.p. / °C	Electrical conductivity	Solubility in water
AX	27	Very poor	Decomposes
BY	693	Good when molten	1000 g dm^{-3}
CZ	3027	Insulator	Insoluble

4   Consider a metal, L (with an outer energy level electron structure ns^1), and two non-metals, M and N (both with an outer energy level electronic structure ns^2np^5).

   a   Explain the meaning of the terms ionic, covalent and metallic bonding by drawing dot-and-cross diagrams for L, M and/or N and appropriate combinations of these elements.

   b   Consider and compare the electrical conductivity of each of the materials L, N, LN and MN, and explain any differences.

5   The three compounds detailed below were kept at $-50\,°C$ and then tested by placing each in a test tube and heating it if needed. Explain the observations in terms of the structures of the materials. Do not forget to include discussions about the bonding present in each of these compounds.

   a CO_2 was a white solid. As soon as the solid was placed in the test tube, it changed into a white mist, which soon cleared, apparently leaving nothing in the tube.

   b   CsCl crystals were white and when gently heated, did not appear to change. When they were heated strongly they changed into a clear, colourless liquid.

   c SiO_2 was a white powder, which did not change even when it was strongly heated over a period of some hours.

6 a Iodine, I_2, is an almost black solid which sublimes, at 114 °C, to produce a purple vapour. Describe the type of bonding present in the I_2 molecules.

  b Describe the forces holding the I_2 molecules together in the solid state.

  c Other common materials which sublime are carbon dioxide and graphite. What does the term *sublime* mean?

  d Graphite is thought to sublime at around 3370 °C. Why should graphite sublime at a much higher temperature than iodine?

  e Describe briefly the changes in structure which occur when iodine sublimes.

- What is meant by **element**, **compound**, and **molecule**. (Ch 1.1, 2.1, 2.4)
- **Ionic** compounds and **ionic bonding**. (Ch 2.3)
- **Covalent** compounds and **covalent bonding**. (Ch 2.4)
- How atoms lose and gain electrons to form full energy levels. (Ch 1.6)
- **Hydrogen bonding**. (Ch 2.9)

Figure 4.1.1   Short Periodic Table

H_2

N_2, O_2, F_2

Cl_2

Br_2

I_2

**Figure 4.1.2
The diatomic gases (at room temperature bromine is mostly liquid and iodine solid)**

It's time now to look at the language that we use in chemistry. We have already come across the symbols for the atoms of many of the elements. Quite a number of them are familiar. We write the symbol for an element using a capital letter, and if the symbol contains two letters, the second letter is always written as a small letter; the symbol for boron is B, the symbol for beryllium is Be. Take care with the small letters. Co is the symbol for the very useful metal cobalt, whereas CO is the formula for the extremely toxic gas carbon monoxide.

You should, by now, be familiar with the symbols for the first twenty elements in the Periodic Table (*Figure 4.1.1*).

It is worth remembering that a few of the gaseous elements always exist as **diatomic** molecules, so that the outer electron energy levels of each of their atoms are completed by forming covalent bonds between the atoms (*Figure 4.1.2*).

In this chapter

We are deliberately starting from a very simple level. You may feel that much of the material is familiar to you from your GCSE course. If that is so, and you can answer all the questions straight off without working through the text, fair enough. *But please be totally honest with yourself.* Our advice is to read through the material and make sure you understand it *thoroughly*!

4.1   Chemical formulae
4.2   Chemical equations
4.3   Ideas about redox chemistry

4

chemical equations and introduction to redox

4.1 Chemical formulae

Atoms form molecules or ions by combining in a variety of ratios and the information about the precise ratio in which they combine gives us the formula of a particular compound. This is where the language of chemistry really starts.

The combinations of chemical symbols provide a shorthand for writing formulae. It is, however, much more than a shorthand. Far more importantly, it provides us with detailed information about the atoms which are combined to form the compound (e.g. the formula of glucose is $C_6H_{12}O_6$). It also provides us with a precise notation about what is going on in a particular chemical reaction, and as we shall see later (Ch 9.2) it lets us calculate the masses of material involved.

Atoms combine in a variety of ratios to form compounds; one of the simplest – and one of the most familiar – must be water, H_2O. Two atoms of hydrogen combine with one atom of oxygen. The resulting compound is water; it has none of the properties of the elements from which it is made.

It is important to realise the significance of the numbers used with chemical symbols in chemistry. As John Dalton stated in 1808 (see Ch 1.1) the atoms of elements always combine in simple *whole number ratios*. The numbers of atoms of each element in the formula of a particular compound are signified by the small numbers, as a subscript, appearing after the symbol of the element: e.g. H_2O, NH_3, CO_2 and CH_3OH (*Figure 4.1.3*).

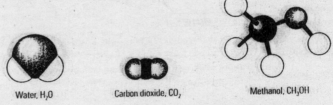

Water, H_2O Carbon dioxide, CO_2 Methanol, CH_3OH

Figure 4.1.3   Simple formulae

For *simple molecules* the idea of a formula representing the molecule is sufficient – although it is worth appreciating that the formula must represent the whole molecule. H_2O contains two hydrogen atoms attached to an oxygen atom, and CH_4 means that there are four hydrogen atoms attached to one carbon atom. But H_2O_2 contains two hydrogen atoms bonded to two oxygen atoms. The formula for hydrogen peroxide must be written as H_2O_2, and not simplified to HO, which is meaningless as the formula for a molecule since oxygen must form two bonds and hydrogen only one. It is also incorrect to write 2OH for hydrogen peroxide, as this would mean that there are two molecules of the formula OH (if it existed!).

Giant structures are built up of a very large number of particles – atoms or ions. In this case, the simplest formula is shown for the structure – a perfect diamond is a single molecule of carbon, but we write the formula as just C (and not $C_{60000000000000000000000}$ – which would be a very small diamond).

A sodium chloride crystal can contain huge but equal numbers of sodium and chloride ions depending on the size of the crystal. To make life simpler we just write the simplest formula, NaCl, which represents the real 1:1 ratio of sodium to chloride ions. Once you are familiar with these ideas – the use of the true formula in simple structures, and the simplest formula for any giant structure – they pose little difficulty. But do try to remember that giant structures are always represented by the simplest formula for the structure (*Figure 4.1.4*).

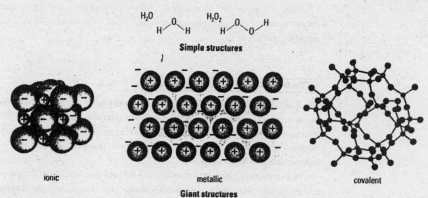

Figure 4.1.4   Simple and giant structures

Brackets in formulae

More complex formulae might have a special group of atoms, most commonly as an ion, e.g. $SO_4^{2\ominus}$, somewhere in the formula. We often enclose the formula of such a polyatomic ion in brackets if there is more than one in the formula. We put a number outside the bracket to indicate how many of the ions are included in the formula. So, calcium sulphate is $CaSO_4$, and sodium sulphate is Na_2SO_4. But aluminium sulphate contains three sulphate ions for every two aluminium ions, so we write the formula as $Al_2(SO_4)_3$, instead of writing $Al_2S_3O_{12}$. The first formula shows us immediately that the compound contains the sulphate ion, $SO_4^{2\ominus}$. This allows us to appreciate which ions are present in a particular compound.

Ions that are sometimes found enclosed in brackets (polyatomic ions) are listed in *Table 4.1.1*.

Formula	Ion
$NH_4^{\oplus}$	ammonium
$OH^{\ominus}$	hydroxide
$HCO_3^{\ominus}$	hydrogencarbonate
$NO_3^{\ominus}$	nitrate
$SO_4^{2\ominus}$	sulphate
$CO_3^{2\ominus}$	carbonate
$PO_4^{3\ominus}$	phosphate

Table 4.1.1   Formulae of some polyatomic ions

The only positive ion of this type which we are likely to come across is the ammonium ion, $NH_4^{\oplus}$; the others are anions. In Ch 25 we shall meet compounds called complexes in which brackets are used for both ions and neutral molecules. (N.B. brackets are also used to enclose organic groups, e.g. $(CH_3)_3Cl$ – see Ch 14.1.)

All the common compounds that form the basis of chemistry are **stoichiometric** compounds, i.e. compounds in which the ratios of atoms or ions are in simple whole numbers. For any of these compounds, a relatively simple formula can be written.

Some compounds contain groups of atoms which are linked to the rest by hydrogen bonds or dative covalent bonds. These 'loosely bonded' groups are shown in the formula with a 'dot'. The familiar blue crystalline copper sulphate is just such a compound. The formula, $CuSO_4.5H_2O$, shows that each $CuSO_4$ is linked with five molecules of *'water of crystallisation'*. This form of copper sulphate is the **hydrated** compound. If the compound is heated, the water molecules are removed, and white **anhydrous** copper sulphate is formed.

$$CuSO_4.5H_2O(s) \xrightarrow{\text{heat}} CuSO_4(s) + 5H_2O(l)$$

This 'dot' does not represent the whole truth of the structure in such compounds. In Ch 25.4 a more meaningful representation of such formulae will be considered. But for the moment the 'dot' notation is sufficient.

The formulae of ionic compounds

We have already seen (Ch 2.3) that ionic compounds are formed when electrons transfer from one atom or group of atoms to another to form charged particles – *ions*. The numbers of ions involved in a compound have to balance out the charges if the compound is to remain electrically neutral. Sodium chloride is therefore $Na^{\oplus}Cl^{\ominus}$ (one positive charge and one negative charge), and magnesium chloride is $Mg^{2\oplus}Cl^{\ominus}_2$.

But remember that the formula NaCl represents only a tiny fragment of a salt crystal. If we wish to emphasise that the compound is ionic, the charges of the ions are usually included in the formula – $Na^{\oplus}Cl^{\ominus}$, rather than simply NaCl.

The formulae of some of the more common ions are shown in *Table 4.1.2*.

Positive ions (cations)			Negative ions (anions)		
Charge	Cation	Symbol	Charge	Anion	Symbol
1⊕	Ammonium	$NH_4^{\oplus}$	1⊖	Bromide	$Br^{\ominus}$
	Copper(I)	$Cu^{\oplus}$		Chloride	$Cl^{\ominus}$
	Hydrogen	$H^{\oplus}$		Hydroxide	$OH^{\ominus}$
	Potassium	$K^{\oplus}$		Iodide	$I^{\ominus}$
	Silver	$Ag^{\oplus}$		Nitrate	$NO_3^{\ominus}$
	Sodium	$Na^{\oplus}$			
2⊕	Calcium	$Ca^{2\oplus}$	2⊖	Carbonate	$CO_3^{2\ominus}$
	Copper(II)	$Cu^{2\oplus}$		Oxide	$O^{2\ominus}$
	Iron(II)	$Fe^{2\oplus}$		Sulphate	$SO_4^{2\ominus}$
	Magnesium	$Mg^{2\oplus}$		Sulphide	$S^{2\ominus}$
	Zinc	$Zn^{2\oplus}$		Sulphite	$SO_3^{2\ominus}$
3⊕	Aluminium	$Al^{3\oplus}$	3⊖	Nitride	$N^{3\ominus}$
	Iron (III)	$Fe^{3\oplus}$		Phosphate	$PO_4^{3\ominus}$

Table 4.1.2   Some common ions

This list will help you work out the formula of any simple ionic compound.

To work out such a formula, given the formulae of the ions it contains, all you have to remember is that the electrical charges *must* balance. So, for the formula of sodium oxide, which contains the ions $Na^{\oplus}$ and $O^{2\ominus}$, two sodium ions are needed to cancel out each oxide ion. The formula is therefore Na_2O.

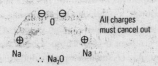

All charges must cancel out

∴ Na_2O

Similarly, the formula for ammonium sulphate can be worked out if we ensure that the number of positive charges on the ammonium ions equals the number of negative charges on the sulphate ion. The formula is therefore $(NH_4^{\oplus})_2SO_4^{2\ominus}$.

Criss cross

We have already introduced one method of working out the formula of an ionic compound using a criss-cross rule (Ch 2.3). You can see the *number* in front of the positive charge on the cation appears behind the anions. And similarly the *number* in front of the negative charge appears behind the cation part of the molecule. In the example you should be able to justify the formula $MgBr_2$. Similarly, the formula of aluminium oxide can be worked out to be Al_2O_3.

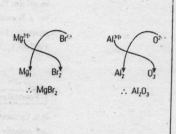

$$\therefore MgBr_2 \qquad \therefore Al_2O_3$$

$$\therefore Al_2(SO_4)_3$$

Figure 4.1.5 Aluminium sulphate

Don't forget to put the larger ions, where necessary, in brackets before including the number – so that the formula of aluminium sulphate is $Al_2(SO_4)_3$ (*Figure 4.1.5*) although the formula of magnesium sulphate is simply $MgSO_4$. It is worth remembering that ionic compounds are usually formed between a metal and one or more non-metals. For any simple compound that contains a metal and a non-metal in the name, you should be able to work out the formula.

You can work out the charges on ions formed by the first twenty elements by looking at the number of outer electrons. This will tell you the number of electrons needed to be lost or gained in order to gain a stable electron configuration. Therefore, one outer electron, $1^{\oplus}$ (Group 1); two outer electrons, $2^{\oplus}$ (Group 2); six outer electrons, $2^{\ominus}$ (Group 6); seven outer electrons, $1^{\ominus}$ (Group 7). So you now have access to the formulae of any ionic compound made from these elements. Things become a little more complicated in compounds of the d-block elements – see Ch 25, and remember (Ch 2.4) that atoms of non-metallic elements bond covalently.

It is worth learning the charges associated with all the common polyatomic ions in *Table 4.1.1*; and also worth remembering that:

- Group 1 elements all have a $1^{\oplus}$ charge on their ions;
- Group 2 elements all have a $2^{\oplus}$ charge on their ions;
- Group 6 elements often have a $2^{\ominus}$ charge on their ions;
- Group 7 elements often have a $1^{\ominus}$ charge on their ions.

The formulae of about 100 ionic compounds can easily be worked out as shown in *Figure 4.1.6*.

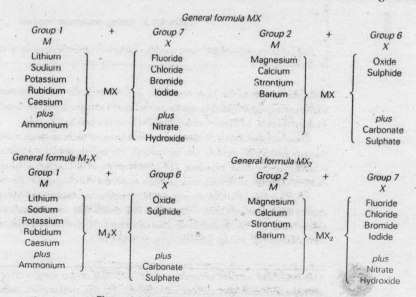

Figure 4.1.6 Formulae of simple ionic compounds

Some elements can form a number of ions with different charges. For example, iron readily forms both the $Fe^{2\oplus}$ and $Fe^{3\oplus}$ ions. In such circumstances the charge on the ion is indicated in the name by a Roman numeral, in brackets, after the name of the metal. Iron chloride, $Fe^{2\oplus}Cl^{\ominus}_2$ is called iron(II) chloride; and when iron reacts with moderately concentrated nitric acid, iron(III) nitrate, $Fe(NO_3)_3$, is formed. Naming compounds of this kind is discussed in Ch 4.3.

Formulae of covalent compounds

It is far more difficult to generalise about the formulae of covalent compounds. Covalent compounds are usually formed between two or more non-metallic elements. For simple **binary compounds**, i.e. compounds containing only two elements, the formula can be worked out by thinking of the number of 'spaces' in the outer electron energy level. Elements other than hydrogen share electrons to form an 'octet'. Hydrogen atoms need one electron to complete the only energy level of two electrons, and the other non-metal atoms share electrons to achieve the full outer level of eight electrons.

Picture the atoms as those used in making molecular models, with 'holes' in the atoms which are used to form the model covalent bonds (*Table 4.1.3*).

Element	Colour	'Bonds'	Element	Colour	'Bonds'
Hydrogen	White	1	Oxygen	Red	2
Fluorine	Green	1	Sulphur	Yellow	2
Chlorine	Green	1	Nitrogen	Blue	3
Bromine	Green	1	Carbon	Black	4
Iodine	Green	1			

Table 4.1.3   Using molecular models

Molecular models usually have specific coloured balls to represent atoms.

Hydrogen model atoms have one hole, and so form one bond. Fluorine, chlorine, bromine and iodine can be represented by similar model atoms and also form one bond; oxygen and sulphur can form two bonds, nitrogen forms three bonds and carbon four bonds.

The covalent formula of any binary combination of these atoms is now straightforward – think of them as molecular models. But be careful, this only applies for some of the very simple binary combinations of these elements. The combinations of just carbon and hydrogen provides us with a range of different molecules wide enough to need three chapters in this book!

In Ch 2.4 we found that the non-metals phosphorus, sulphur and chlorine (among others) could expand their outer electron energy level. This means that combinations such as SF_6 and PCl_5 are all known. Also, several apparently obvious molecules (SO, I_2O) do not exist, and others are quite stable with what seems 'wrong' numbers (e.g. SO_2, CO, NO).

ACTIVITY

Use molecular models to work out the formulae of the simple compounds formed between nitrogen and hydrogen, carbon and oxygen, hydrogen and bromine, sulphur and hydrogen.

Naming chemical compounds

The names of some of the elements have been used for a long time, and the elements discovered in the last few centuries were initially given Latin names. The symbols used for some of these elements are still based on the Latin name (*Table 4.1.4*).

Element	Old name	Symbol
All known to the ancients		
Iron	Ferrum	Fe
Copper	Cuprum	Cu
Silver	Argentum	Ag
Gold	Aurum	Au
Tin	Stannum	Sn
Lead	Plumbum	Pb
Discovered in the nineteenth century		
Sodium	Natrium	Na
Potassium	Kalium	K

Table 4.1.4   'Old' names of elements

The very new man-made elements (see Ch 5.1) are initially given a systematic name which is related to the atomic number of the new element – the recently discovered element, atomic number 113, is, for the moment, called ununtrium, Uut.

The name of a chemical compound should, with the exception of a few common names of simple compounds such as water and ammonia (NH_3) and the wide range of organic compounds (see Ch 14.3), tell you which elements are present in the compound.

With new compounds being discovered at the rate of twenty each hour it is important that naming is done systematically. The systematic way of naming **organic compounds** (those compounds which contain carbon and hydrogen and often a limited number of other elements) is discussed in Ch 14.3. For **inorganic compounds** there are some relatively simple rules.

- Binary compounds – containing only two elements – have a name which ends in – *ide* (potassium oxide, K_2O).
- Compounds containing three elements, one of which is oxygen in the negative (anion) part of the compound – i.e. compounds with *oxoanions* – end with – *ate* (sodium nit*rate*, $NaNO_3$); similar compounds with less oxygen may end in – *ite* (sodium nit*rite*, $NaNO_2$).
- In simple ionic compounds the cation comes first in the name (the metal, or ammonium). The anion appears at the end of the name or formula.
- In simple covalent compounds the names of the elements are arranged in reverse order of electronegativity, i.e. with the more *electropositive* element first (e.g. nitrogen oxide, hydrogen sulphide).

When there is more than one possible combination of the two elements, e.g. CO and CO_2, the number of atoms that can vary will usually have a prefix indicating the number of atoms in the molecule (*Table 4.1.5*).

mono-	one – only used when necessary to avoid confusion	carbon monoxide, CO
di-	two	carbon dioxide, CO_2
tri-	three	sulphur trioxide, SO_3
tetra-	four	silicon tetrachloride, $SiCl_4$
penta-	five	phosphorus pentachloride, PCl_5
hexa-	six	sulphur hexafluoride, SF_6

Table 4.1.5   Naming covalent compounds

CO is carbon *monoxide* (not mono-oxide), whereas CO_2 is carbon *dioxide*. Another common example is SO_2, sulphur *dioxide*, and SO_3, sulphur *trioxide*.

As always there are a number of exceptions, especially with compounds containing hydrogen: ammonia is NH_3, but water is H_2O. With Groups 6 and 7 the hydrogen appears first (hydrogen sulphide, H_2S; hydrogen bromide, HBr). Otherwise the hydrogen is at the end of the formula: e.g. NaH is sodium hydride and PH_3 is phosphorus hydride (known as phosphine).

Terminology in naming compounds

-ide – only two elements;

-ate – three elements, one of which is oxygen;

-ite – same as *-ate* but with less oxygen.

e.g.

sodium sulph*ide*, Na_2S;

sodium sulph*ate*, Na_2SO_4

sodium sulph*ite*, Na_2SO_3

Remember:

- **In a chemical formula, each subscript figure tells you the number of atoms of that element present in the formula.**
- **More complicated ions may be enclosed in brackets in the formula, and the number of such ions will then appear as a subscript after the brackets.**

Relative molecular mass

We have already come across the relative atomic mass of an element, A_r (Ch 1.4).

Some A_r values

H, 1.01

N, 14.01

Na, 23.00

Cl, 35.45

The Periodic Table at the back of the book gives you the A_r of each element to four significant figures. This takes into account the normal abundances of the naturally occurring isotopes. They are the figures you would normally use in calculation work.

The **relative molecular mass** (M_r) is the sum of all the atomic masses of the atoms shown in the formula of a compound. M_r is based on the carbon-12 scale (see Ch 1.3) and, being a *relative* mass, has no units. (We shall meet molar mass, M, in Ch 9.3; molar mass is numerically the same as M_r, but has the units g mol^{-1}.)

The term relative molecular mass is straightforward. The formula of ethanol is CH_3CH_2OH. The relative molecular mass is the sum of the relative atomic masses, i.e. 46.08, and is the relative mass of one molecule of ethanol (*Figure 4.1.7*).

However sodium chloride is more of a problem. There is no such thing as a molecule of sodium chloride – it is a giant ionic structure, and a single crystal of sodium chloride will consist of many millions of $Na^{\oplus}$ and $Cl^{\ominus}$ ions. Consequently, rather than write the formula NaCl with a large number, we use the simplest formula we can, NaCl. The sum of the relative atomic masses in the formula gives the relative formula mass. This principle applies to any giant structure – ionic, atomic or metallic.

Diamond is another giant structure, each crystal consisting of many billions of carbon atoms. There are no individual molecules at all. So the simplest formula is C, and the relative formula mass is the same as the relative atomic mass. It should always be made clear, by putting the formula after the M_r symbol, which substance is being referred to: thus $M_r(NaCl) = 58.45$; $M_r(CaH_2) = 42.10$; $M_r(Cl_2) = 70.90$.

CH_3CH_2OH

$2 \times C$	2×12.01	24.02
$1 \times O$	1×16.00	16.00
$6 \times H$	6×1.01	6.06
	$M_r(CH_3CH_2OH)$	**46.08**

$CaCl_2$

$1 \times Ca$	1×40.08	40.08
$2 \times Cl$	2×35.45	70.90
	$M_r(CaCl_2)$	**110.98**

$(NH_4)_2SO_4 = N_2H_8SO_4$

$2 \times N$	2×14.01	28.02
$8 \times H$	8×1.01	8.08
$1 \times S$	1×32.06	32.06
$4 \times O$	4×16.00	64.00
	$M_r((NH_4)_2SO_4)$	**132.16**

Figure 4.1.7 Relative molecular masses

The M_r is worked out by adding up all the relative atomic masses (*Figure 4.1.7*). It is worth taking care over the layout of these simple calculations and checking that the formula you are using is correct! It is so easy to make a simple mistake in the first step of any calculation. It is also worth showing how you have calculated the relative mass. In an exam, an obvious error in the calculation will only be penalised once if the mistake can be recognised. If the mistake is not recognised (because you have simply put a string of numbers into a calculator without any explanation), then you are likely to lose more marks.

Don't forget that with any formulae containing brackets everything inside the brackets has to be multiplied by the number outside the brackets. So, $Ca(NO_3)_2$ means one calcium atom, but two nitrogen atoms and six oxygen atoms.

If the formula contains 'water of crystallisation', then the relative mass of that has also to be included in the calculation of the value of M_r.

Worked examples

Calculate the relative formula mass of strontium chloride.

Decide in which groups of the Periodic Table the elements belong.

Strontium is in Group 2; chlorine is in Group 7.

Is the compound ionic or covalent (metal/non-metal or non-metal/non-metal)?

Ionic, since there is a metal and a non-metal.

Decide on the charges for each of the ions.

The charge on a Group 2 ion is $^{2\oplus}$ and on a Group 7 ion is $^{1\ominus}$.

So, using criss-cross:

or simply by balancing charges so that they cancel out:

the formula is $SrCl_2$.

Using the Periodic Table inside the back cover of the book, the relative atomic masses are: $A_r(Sr) = 87.62$; $A_r(Cl) = 35.45$.

For $SrCl_2$:

$$\begin{array}{lll} Sr & 1 \times 87.62 = & 87.62 \\ Cl & 2 \times 35.45 = & 70.90 \\ \hline M_r(SrCl_2) & & 158.52 \end{array}$$

What is the formula and relative molecular mass of the simplest compound formed from arsenic and bromine?

Decide in which groups of the Periodic Table the elements belong.

Arsenic is in Group 5; bromine is in Group 7.

Is the compound ionic or covalent (metal/non-metal or non-metal/non-metal)?

Covalent since there is a non-metal and a non-metal.

Decide on the bonds normally found with each atom.

There are normally three bonds for a Group 5 atom and one bond for a Group 7 atom.

So, to use up all the bonds:

$$ \begin{matrix} | \\ -As- \end{matrix} \quad Br- \quad \longrightarrow \quad \begin{matrix} Br \\ | \\ Br-As-Br \end{matrix} $$

the formula should be $AsBr_3$.

Using the Periodic Table inside the back cover of the book, the relative atomic masses are: $A_r(As) = 74.90$; $A_r(Br) = 79.91$.

For $AsBr_3$

$$\begin{array}{lll} As & 1 \times 74.90 = & 74.90 \\ Br & 3 \times 79.91 = & 239.73 \\ \hline M_r(AsBr_3) & & 314.63 \end{array}$$

Calculate the relative formula mass of barium nitrate.

It is an ionic compound.

Barium is in Group 2 and therefore $^{2\oplus}$; the nitrate group has one $^\ominus$ charge.

So, using criss-cross (don't forget to use brackets for a group of atoms forming one ion), or simply by balancing charges so that they cancel out:

$$Ba^{2\oplus} \quad NO_3{}^\ominus$$
$$Ba_1(NO_3)_2$$

the formula is $Ba(NO_3)_2$.

Using the Periodic Table inside the back cover of the book, the relative atomic masses are:
$A_r(Sr) = 137.3$; $A_r(N) = 14.01$; $A_r(O) = 16.00$

For $Ba(NO_3)_2$, which is equivalent to BaN_2O_6:

Ba	$1 \times 137.3 =$	137.30
N	$2 \times 14.01 =$	28.02
O	$6 \times 16.00 =$	96.00
$M_r(Ba(NO_3)_2)$		261.32

Once you are familiar with the ideas of working out formulae there should be no need to repeat every step in these examples – but do take care in calculating the relative masses. Do not just put a string of figures into a calculator and hope the answer is correct – show which figures you have used clearly every time.

SELF-ASSESSMENT QUESTIONS 4.1

1 How many of each type of atom are there in the formulae of each of the following compounds?
  a $CaBr_2$
  b B_2O_3
  c $Ca(NO_3)_2$
  d $(NH_4)_2SO_4$
  e C_2H_5OH

2 Work out the formulae of the following ionic compounds.
  a Magnesium sulphide
  b Sodium oxide
  c Potassium iodide
  d Calcium chloride
  e Ammonium phosphate

3 Calculate M_r for:
  a sodium bromide;
  b potassium sulphide;
  c silicon oxide;
  d ammonium sulphate;
  e magnesium sulphate;
  f sulphur dioxide.

4.2 Chemical equations

Chemistry is all about changing one material into another. Chemical equations are a shorthand way of describing what goes on when we change these materials. It is possible to write a **word equation** for the materials being reacted together. Don't be afraid to use word equations. They are very useful in starting off the understanding of any chemical reaction. At least, in any chemical situation, you would normally know the **reactants** you are putting together. Linking **products** to reactants is what chemistry is all about.

$$\text{reactants} \longrightarrow \text{products}$$

When magnesium burns with a very bright flare, it is just the reaction of magnesium and oxygen.

$$\text{magnesium + oxygen} \longrightarrow \text{magnesium oxide}$$

The arrow between the reactants and products does have a specific meaning; it is certainly *not* an 'equals' sign. In mathematics the 'equals' sign tells us that one side is exactly equivalent to the other side – and, as we shall see later on, this is not the case in chemistry.

Formulae in equations

Once we have the word equation we can start to put in the formulae that we know. If the equation involves ionic compounds we can certainly write in their formulae. Once the correct formulae have been put into the equation, *the formulae cannot be altered* (see *Figure 4.2.1*). They must not be altered just so that the equation can be balanced.

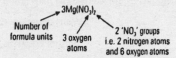

Figure 4.2.1   What does a formula tell us?

The formula of a compound has always been found by experiment, and shows precisely the number of atoms of each kind in a molecule of a compound or the ratio of atoms or ions in a giant structure. A correct formula is a fact, and nothing can ever change it. For example, if water is produced in a chemical reaction and we know that its formula is H_2O, to alter the formula to H_2O_2, or anything other than H_2O, in order to balance the equation, cannot be right.

Let's look a little more closely at the reaction of hydrogen burning in oxygen – the 'pop' test for hydrogen. Hydrogen burns quickly if a lighted splint is put anywhere near it.

$$\text{hydrogen} + \text{oxygen} \longrightarrow \text{water}$$

We have already seen that hydrogen and oxygen are both diatomic gases; that is, the gases have the formulae H_2 and O_2; and we know that the formula for water is H_2O.

So putting these formulae under the names in the word equation, we get:

$$\text{hydrogen and oxygen} \longrightarrow \text{water}$$
$$H_2 \quad + \quad O_2 \quad \longrightarrow \quad H_2O$$

Or using molecular models,

But there are different numbers of atoms on each side of the arrow – and this cannot happen in the chemistry laboratory! **Atoms can never be created or destroyed in any chemical reaction.** Only the way in which they are joined together can be changed.

Balancing the equation

We can see there are two hydrogen atoms on both sides of this equation, but only one oxygen appears to be formed from two oxygen atoms – and this is the kind of inequality that we cannot accept in chemical equations. *The numbers of atoms on each side of the arrow have to be exactly the same.*

The subscript numbers within the formulae cannot be altered – the formulae of the elements or compounds are already fixed. Certainly, in the equation above, the formation of hydrogen peroxide, H_2O_2, would not be correct. We *know* that water is formed!

The only way of balancing this equation is to alter the numbers of molecules involved in the reaction. If two molecules of water were formed – written as $2H_2O$ (with a number in front of the formula telling us that there are two molecules of H_2O) – the number of oxygen atoms would balance, but the number of hydrogen atoms is now incorrect.

$$H_2 + O_2 \longrightarrow 2H_2O$$

So we need to have two hydrogen molecules to balance the number of hydrogen atoms.

$$2H_2 + O_2 \longrightarrow 2H_2O$$

We can see that now the numbers of atoms on each side of the equation are both the same – we have a balanced equation. We can show this with molecular models:

Another example is the fact that carbon burns in a limited supply of oxygen to form carbon monoxide.

Word equation: carbon + oxygen $\longrightarrow$ carbon monoxide

Symbol equation: $C + O_2 \longrightarrow CO$

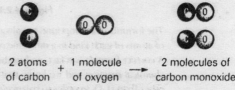

2 atoms   1 molecule   2 molecules of
of carbon + of oxygen $\longrightarrow$ carbon monoxide

Balanced equation: $2C + O_2 \longrightarrow 2CO$

Further worked examples in balancing equations

The reaction of hydrogen and bromine produces hydrogen bromide. What is the balanced equation for the reaction?

Write a word equation.

hydrogen + bromine $\longrightarrow$ hydrogen bromide

Write the formulae for the elements and compounds (remember that some gases are diatomic).

$$H_2 + Br_2 \longrightarrow HBr$$

Or, if you prefer to see it as a 'model picture':

Obviously, the equation is not balanced.

Start to count up the atoms.

$$H_2 \quad + \quad Br_2 \quad \longrightarrow \quad HBr$$

2 H atoms   2 Br atoms $\longrightarrow$ 1 H atom   1 Br atom

Alter the number of molecules (don't forget, you can only insert numbers in front of the formulae). In this case try 2HBr.

$$H_2 \quad + \quad Br_2 \longrightarrow 2HBr$$

2 H atoms   2 Br atoms   2 H atoms   2 Br atoms

The same number of each kind of atom is now on both sides of the equation – so the balanced equation is:

$$H_2 \qquad + \qquad Br_2 \qquad \longrightarrow \qquad 2HBr$$

1 molecule of hydrogen + 1 molecule of bromine $\longrightarrow$ 2 molecules of hydrogen bromide

Zinc metal reacts with dilute hydrochloric acid (HCl) to produce zinc chloride and liberate hydrogen gas. What is the balanced equation for the reaction?

Write a word equation.

$$zinc + hydrochloric\ acid \longrightarrow zinc\ chloride + hydrogen$$

Write the formulae for the elements and compounds (remember that some gases are diatomic).

$$Zn + HCl \longrightarrow ZnCl_2 + H_2$$

Or, if you prefer to see it as a 'model picture':

Obviously, the equation is not balanced.

Start to count up the atoms.

Zn	+	HCl	$\longrightarrow$	$ZnCl_2$	+	H_2
1 Zn atom		1 H atom 1 Cl atom	$\longrightarrow$	1 Zn atom 2 Cl atoms		2 H atoms

Alter the number of molecules (don't forget, you can only insert numbers in front of the formulae). In this case try 2HCl.

Zn	+	2HCl	$\longrightarrow$	$ZnCl_2$	+	H_2

1 Zn atom	2 H atoms 2 Cl atoms	$\longrightarrow$	1 Zn atom 2 Cl atoms	2 H atoms

The same number of each kind of atom is now on both sides of the equation – so the balanced equation is:

Zn	+	2HCl	$\longrightarrow$	$ZnCl_2$	+	H_2
1 atom of zinc	+	2 molecules of hydrochloric acid	$\longrightarrow$	1 "molecule" of zinc chloride	+	1 molecule of hydrogen

Zinc chloride is an ionic compound, but we can analyse equations in this way for both covalent and ionic materials.

Remember:

to balance a chemical equation:

- **write a word equation;**
- **write in the formulae of each species;**
- **don't forget to put in any diatomic gases correctly;**
- **check that all the formulae are correct;**
- **count the numbers of atoms of each type on each side of the equation;**
- **alter the numbers of 'molecules' (with numbers in front of the formulae) to achieve a balanced equation.**

Why is there such a need to be precise with these chemical equations? Chemists use balanced equations to tell us much more than just a shorthand representation of what is happening in the reaction. At a molecular or ionic level, the equation can show us what bonds are changing during the reaction, how some of the electrons are moving during the reaction, and what chemical changes are taking place.

It is worth pointing out that every equation has, at some time, been 'discovered' through experiment. It is only after all these equations had been worked out by actually reacting the chemicals together that we could start to make chemical generalisations.

On an industrial scale, equations can give us information about the quantities involved in the reaction. In, for example, the blast furnace used in one method of making zinc metal from zinc ore, it is essential to know how many tons of coke (carbon) are needed to form the desired amount of zinc. Too little and we have wasted the zinc ore; too much and the coke is wasted.

$$ZnO(s) + C(s) \longrightarrow Zn(s) + CO(g)$$

This emphasises two fields of human activity which are covered by the 'umbrella' of chemistry: the wish to understand and explain changes in the material world by means of theories about atoms, molecules and ions, and the need to convert materials in the world into new useful products to make life easier!

What else does the equation tell us? For the moment we will stick to the information it provides – the chemicals that react and what is formed. In Ch 9.5 we will see how we can use a balanced equation to calculate the masses of reactants and products in a reaction. But for now we will just use the equation to show the ratios of molecules, atoms, and ions which are involved in the reaction.

State symbols

Further information about a chemical reaction is often included in the equation. The physical state of the reactants is usually included as a symbol (called the **state symbol**) in brackets after the formula.

So the equation for the reaction of hydrogen and oxygen becomes:

$$2H_2(g) + O_2(g) \longrightarrow 2H_2O(l)$$

The water formed would be a gas at the temperature of the flame produced, but unless conditions are specified the states of the reactants and products are shown as they are at room temperature.

The reaction of zinc with hydrochloric acid which we considered earlier becomes, with state symbols:

$$Zn(s) + 2HCl(aq) \longrightarrow ZnCl_2(aq) + H_2(g)$$

This shows that the zinc is a solid, and that it is added to an aqueous solution of hydrochloric acid. The resulting solution contains the zinc chloride, and we might expect to see bubbles as the hydrogen gas is liberated.

Ionic equations

As we go through the chemistry course we will find that there are often groups of reactions which produce very similar results. For example, the reaction of sodium chloride and silver nitrate solutions produces a dense white precipitate of silver chloride. A similar dense white precipitate is produced if solutions of potassium chloride, magnesium chloride or zinc chloride are mixed with silver nitrate solution. Can we put all these reactions into a common group?

For the reaction of sodium chloride and silver nitrate solutions, the word equation would start:

silver nitrate + sodium chloride ⟶ silver chloride + ...

Simple inspection allows us to guess that the other product is sodium nitrate.

silver nitrate + sodium chloride ⟶ silver chloride + sodium nitrate

Now we can insert the formulae of each of the compounds in the equation. Since these compounds are essentially ionic it is worth putting in the charges associated with each of the ions.

$$Ag^{\oplus}NO_3^{\ominus} + Na^{\oplus}Cl^{\ominus} \longrightarrow Ag^{\oplus}Cl^{\ominus} + Na^{\oplus}NO_3^{\ominus}$$

When an ionic solid dissolves in water the ions are hydrated; they move around and have a separate existence in the solution. When these hydrated ions collide – as they frequently do – if the attraction between two oppositely charged ions is greater than the attractions of the ions for the solvent, a **precipitate** is formed. The ions combine and form a solid. In this reaction, the reactants were in solution and the silver chloride appeared as a precipitate – a solid. When the solid silver chloride is precipitated, the separate silver and chloride ions are no longer around (*Figure 4.2.2*).

So, putting in the state symbols as well:

$$Ag^{\oplus}NO_3^{\ominus}(aq) + Na^{\oplus}Cl^{\ominus}(aq) \longrightarrow AgCl(s) + Na^{\oplus}NO_3^{\ominus}(aq)$$

If we show the separate hydrated ions in the solution, the equation becomes:

$$Ag^{\oplus}(aq) + NO_3^{\ominus}(aq) + Na^{\oplus}(aq) + Cl^{\ominus}(aq) \longrightarrow AgCl(s) + Na^{\oplus}(aq) + NO_3^{\ominus}(aq)$$

State symbols

(g)	gas
(l)	liquid
(s)	solid
(aq)	aqueous – dissolved in water – from the Latin for water (aqua)

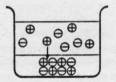

Figure 4.2.2   Precipitation

We can now see that the $Na^\oplus$ ions appear as separate ions on both sides of the equation. *They do not take any part in the reaction* – they are **spectator ions**. As far as this particular equation is concerned they can be left out. The same applies to the nitrate ions. They are also spectator ions, and do not contribute to the chemical reaction.

We are now left with silver ions reacting with chloride ions to form a solid precipitate of silver chloride.

$$Ag^\oplus(aq) + Cl^\ominus(aq) \longrightarrow AgCl(s)$$

This equation is called an **ionic equation**. The numbers of atoms on each side of the arrow still have to balance, and the *number of charges on each side of the arrow also have to balance* – they must cancel out exactly (as in this equation). The sum of the positive and negative ions on each side must be the same.

Why bother with ionic equations? They allow us to generalise about reactions. In our example, any soluble chloride (for example, potassium chloride or magnesium chloride) will give the same reaction – a dense white precipitate with silver nitrate solution.

$$\text{e.g.} \quad KCl(aq) + AgNO_3(aq) \xrightarrow{\text{room} \atop \text{temperature}} KNO_3(aq) + AgCl(s)$$

$$MgCl_2(aq) + 2AgNO_3(aq) \xrightarrow{\text{room} \atop \text{temperature}} Mg(NO_3)_2(aq) + 2AgCl(s)$$

The effective part of the reaction in all cases will be represented by the ionic equation.

$$Ag^\oplus(aq) + Cl^\ominus(aq) \longrightarrow AgCl(s)$$

Some typical ionic reactions are shown in *Figure 4.2.3*.

Reactions with acids:
$$H^\oplus(aq) + OH^\ominus(aq) \longrightarrow H_2O(l)$$
$$2H^\oplus(aq) + O^{2\ominus}(s) \longrightarrow H_2O(l)$$
$$2H^\oplus(aq) + M(s) \longrightarrow M^{2\oplus}(aq) + H_2(g)$$
$$2H^\oplus(aq) + CO_3^{2\ominus} \longrightarrow H_2O(l) + CO_2(g)$$
(Note: M refers to any metal able to form a $2^\oplus$ ion)

Precipitation reactions:
$$Ba^{2\oplus}(aq) + SO_4^{2\ominus}(aq) \longrightarrow BaSO_4(s)$$
$$Ag^\oplus(aq) + Cl^\ominus(aq) \longrightarrow AgCl(s)$$
$$Ag^\oplus(aq) + Br^\ominus(aq) \longrightarrow AgBr(s)$$
$$Ag^\oplus(aq) + I^\ominus(aq) \longrightarrow AgI(s)$$

Figure 4.2.3   Ionic equations

| Further information |

Finally, we usually also want to know the following about a reaction:
- what conditions are used;
- what other materials are needed for the reaction to work (for example, a catalyst or other material which may be needed for the reaction to work at all or to occur at a reasonable rate).

The conditions are usually included above the arrow of the reaction – even if the reaction is carried out at room temperature (R.T.). Reaction conditions that might be included in the equation are:
- the temperature of the reaction (including room temperature);
- the pressure of the reaction, if there are gases involved;
- the solvent used, particularly if the solvent is not water;
- the catalyst or other material used.

1 Write word equations for the reaction of:

a caesium and oxygen;

b aluminium and chlorine;

c magnesium sulphate and barium chloride.

2 Balance the following equations and insert the state symbols.

a $Sr + O_2 \longrightarrow SrO$

b $Al + Cr_2O_3 \longrightarrow Cr + Al_2O_3$

c $CH_4 + O_2 \longrightarrow CO_2 + H_2O$

d $Fe_2O_3 + C \longrightarrow CO + Fe$

3 Write ionic equations, including the spectator ions, for the following reactions in aqueous solution. Clearly mark any spectator ions that appear in the equation.

a Barium nitrate and magnesium sulphate.

b Silver nitrate and calcium chloride.

4.3 Ideas about redox chemistry

One of the principle aims of scientific work is to try to classify the material world around us. Biologists classify living organisms in different ways and try to collect a body of knowledge in order to develop an understanding of the living world. In chemistry, one method of classification is to look at chemical reactions and to see if there are any patterns we can observe.

The reactions of many materials with oxygen, for example the burning of coal or the rusting of iron, are examples of a process called oxidation. It is not surprising that one of the first reaction types we need to look at involves oxidation.

The reaction of materials directly with oxygen is really of limited use, except for the combustion of fossil fuels (see Ch 15.4). You might have seen magnesium react with oxygen – it burns with a very bright flare – in the laboratory or as a distress signal.

$$2Mg(s) + O_2(g) \xrightarrow{\text{heat}} 2MgO(s)$$

Redox as electron transfer

The development of the idea of redox

Oxidation is defined as:

- adding oxygen
- removing hydrogen
- adding an electronegative element
- removing electrons

Reduction is defined as:

- removing oxygen
- adding hydrogen
- removing an electronegative element
- adding electrons

Look more closely at this reaction of magnesium with oxygen. What has happened to the magnesium and oxygen atoms during the course of the reaction? The magnesium has lost the two electrons in the outermost energy level to form the $Mg^{2\oplus}$ ion, and each oxygen atom has gained these two electrons to complete the outer octet in the oxygen atom. (Notice that the covalent bonds in the oxygen molecule, O_2, have to break before this can happen.)

$$Mg\!:\;\; + \;\;\overset{\times\times}{\underset{\times\times}{O}}\!: \longrightarrow Mg^{2\oplus} \;\; :\!\overset{\times\times}{\underset{\times\times}{O}}\!:^{2\ominus}$$

The result is the formation of the ionic compound magnesium oxide. So during the course of the reaction the magnesium has lost electrons and oxygen has gained electrons. This is a common feature of all oxidation reactions – **an element loses electrons in the process of oxidation.**

But if an element has to lose electrons then there must be another atom present to pick up the electrons. Otherwise the reaction would not proceed. What about the oxygen atom in this reaction? The oxygen atom gains the two electrons from the magnesium atom. **The process where an element gains electrons is called reduction** – the opposite of oxidation.

A species is:
- **oxidised** if it loses electrons
- a **reducing agent** (or reductant) when it loses electrons

A species is:
- **reduced** if it gains electrons
- an **oxidising agent** (or oxidant) when it gains electrons

> **Definition:** A redox reaction is a chemical reaction in which there is an oxidation reaction and a reduction reaction occurring simultaneously. One element is reduced by gaining electrons, and another element is oxidised by losing electrons.

The point about such oxidation reactions is that if **an element is oxidised there must be another element that is reduced**. If we combine the two words, the whole process is called a **redox** reaction – the reaction consists of a **red**uction and an **ox**idation reaction occurring *simultaneously*.

One way to remember this is to remember **OILRIG**.

Oxidation Is Loss, Reduction Is Gain – of electrons

The idea is straightforward with ionic compounds, where the transfer of one or more electrons completely from one atom to another is the basis for the formation of a compound or for a displacement reaction (*Figure 4.3.1*).

> OILRIG is an example of a mnemonic. A **mnemonic** is a word or saying which helps you remember something.

OILRIG

Oxidation is loss } of electrons
Reduction is gain

copper sulphate + zinc $\longrightarrow$ zinc sulphate + copper

$Cu^{2+}(aq) + Zn(s) \longrightarrow Zn^{2+}(aq) + Cu(s)$

Zinc has lost electrons

$Zn(s) \longrightarrow Zn^{2+}(aq) + 2e^-$

so zinc has been oxidised.

Copper has gained electrons

$Cu^{2+}(aq) + 2e^- \longrightarrow Cu(s)$

so copper ions have been reduced.

Figure 4.3.1

However, the concept of electron transfer, even if the transfer is only partial, is very useful in understanding and explaining many other redox reactions. The idea of polar covalent bonds has already been met in Ch 2.5, and we use this idea when discussing redox reactions involving molecules. After all, electronegative elements are the ones that will 'have control' of the electrons in a covalent bond.

Oxidation state

In redox reactions, elements compete for electrons. This section is all about accounting for the transfer of electrons during such a reaction. We need a consistent way of assigning the electrons to each of the atoms in a compound. If an atom has completely or partly gained electrons, we say it has a *negative oxidation state*; and, conversely, atoms that have lost electrons have a *positive oxidation state*. **Oxidation state** is often called *oxidation number* – the two terms mean exactly the same. **Oxidation states are always written with the sign before the number** – unlike ionic charges.

Oxidation state in ionic compounds

Ionic compounds are formed when electrons are transferred from one element to another. And the movement of electrons is what redox is all about. For simple ions, i.e. ions formed from just one atom, the *charge on the ion is the oxidation state* of the element. So in the Na^+ ion, the oxidation state of sodium is $+1$. In the sulphide ion, S^{2-}, the oxidation state of sulphur is -2. The oxidation state of iron in the ion Fe^{3+} is $+3$.

Don't forget, **the oxidation state of any neutral atom of an element is zero; the total oxidation state of a compound must be zero; and the oxidation state of a metal in its compounds is almost always positive.**

Using a few simple ionic compounds as examples, and remembering that the oxidation state is the same as the charge on the ion:

$CaCl_2 - Ca^{2\oplus}Cl^{\ominus}{}_2$; oxidation states; calcium +2, chlorine −1;

$Al_2O_3 - Al^{3\oplus}{}_2O^{2\ominus}{}_3$; oxidation states: aluminium +3, oxygen −2.

In the reaction of iron with copper sulphate solution:

$$Fe(s) + Cu^{2\oplus}(aq) \longrightarrow Fe^{2\oplus}(aq) + Cu(s)$$

The oxidation states at the start and finish are: 0   +2   +2   0

In the reaction of iron(III) sulphate solution with potassium iodide solution, the oxidation states are shown straightforwardly by writing the ionic equation:

$$2Fe^{3\oplus}(aq) + 2I^{\ominus}(aq) \longrightarrow 2Fe^{2\oplus}(aq) + I_2(aq)$$

The oxidation states are: +3   −1   +2   0

Notice here that for each $Fe^{3\oplus}$ ion which gains an electron and so drops from the +3 to the +2 oxidation state, an iodide ion in the −1 state must lose an electron and end up as an iodine atom. We shall meet this balancing of electron gain with electron loss in Ch 10.3.

Oxidation state in covalent compounds

In covalent compounds, deciding on which atom has 'responsibility' for the electrons in the bond is more difficult – because the electron pair of any bond must be shared in the covalent bond. With a polar bond, it is possible to think in terms of the electron pair being 'controlled' by the more electronegative of the two atoms. With hydrogen chloride, for example, we assign an oxidation state of −1 to the electronegative chlorine atom, and so the hydrogen atom has an oxidation state of +1.

Even in molecules without such obvious polarity, the same principle can be used – the more electronegative of the two atoms in the bond is said to have a negative oxidation state (*Figure 4.3.2*).

| | | H | | |
| | | 2.1 | | |

C	N	O	F	Ne
2.5	3.0	3.5	4.0	–
Si	P	S	Cl	Ar
1.8	2.1	2.5	3.0	–
	As	Se	Br	Kr
	2.0	2.4	2.8	–

Figure 4.3.2   Electronegativities (Pauling Scale)

In methane, CH_4, carbon has an oxidation state of −4, and each of the four hydrogen atoms an oxidation state of +1.

Oxidation states in covalent molecules

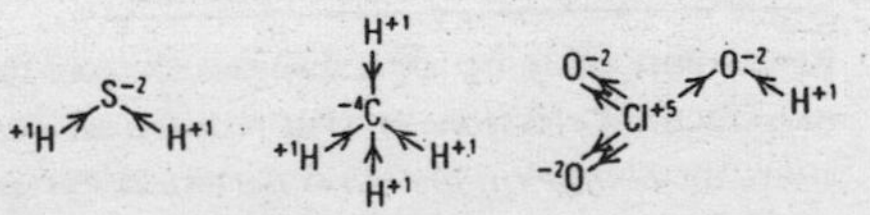

N.B. The arrow head on a bond points towards the electronegative atom. Each arrow head pointing to an atom contributes −1 to the oxidation state of the atom. Remember → represents a dative covalent bond. By using →⟶ we are trying to indicate that the electron pair is closer to the more electronegative atom.

If there are double bonds in the compound, then both the electrons in the double bond have been 'gained' by the electronegative element – one element has gone up by 2 and the other down by 2. So, in carbon dioxide, CO_2 ($O=C=O$), the oxidation state of carbon is $+4$ and of each oxygen is -2.

The only real difficulty arises if an element which can exist in different oxidation states is found *in* those different states in the same compound. Can you show that the average oxidation state of sulphur in the compound $Na_2S_4O_6$ is $+2\frac{1}{2}$? As oxidation states must be whole numbers (because it is not possible to have *half* an electron!), what do you think are the *real* oxidation states of sulphur in this compound?

Rules for assigning oxidation states

To help decide the oxidation states of elements in their compounds, a number of facts about oxidation states are worth remembering.

- Unreacted elements always have an oxidation state of zero.
- Group 1 elements always have an oxidation state of $+1$ in their compounds.
- Group 2 elements always have an oxidation state of $+2$ in their compounds.
- Aluminium always has an oxidation state of $+3$.
- Fluorine always has an oxidation state of -1.

The list so far is always true for substances we shall meet at this level. The following elements usually have a particular oxidation state, but there are specific exceptions.

- Hydrogen is usually $+1$, except in metal hydrides containing Group 1 and 2 elements, where it is -1.
- Oxygen is usually -2, except in peroxides and in compounds with fluorine.
- Chlorine is usually -1, except in compounds that also contain fluorine or oxygen.

Remember the sign is *always* quoted when discussing an oxidation state, and the sign comes *before* the number (*Table 4.3.1*).

Element	Oxidation state
Unreacted elements	0
Na, K	+1
Mg, Ca	+2
Al	+3
F	−1
H	usually +1
O	usually −2
Cl	usually −1

Table 4.3.1   Assigning oxidation states

(For the *charge* on ions, we say and write that, for example, the oxide ion is 2 minus, i.e. O^{2-}.)

The sum of the oxidation states in a neutral compound must be zero, and in any ion the sum of the oxidation states will be equal to the charge on the ion.

When there is more than one possible oxidation state for an element, the oxidation state is quoted as the Roman numeral, in brackets. So, for example, iron(III) chloride contains the iron ion Fe^{3+}. Copper(II) sulphate contains Cu^{2+}. A little more complex, but on the same principle, sodium chlorate(VII) ($NaClO_4$) has chlorine in the $+7$ oxidation state. (Notice that oxidation states can get numerically far larger than ionic charges! Compounds containing elements in $+6$ and $+7$ states are well known; stable ions with more than *three* charges are almost unknown.)

Working out oxidation states

By adopting a set routine it is possible to calculate oxidation states in simple compounds without any difficulty. Try it this way:

Under the formula:

- Enter zero for any uncombined element (e.g. Na or Cl_2).
- Put in the oxidation state(s) of the elements you are sure of – from the list given in *Table 4.3.1.*
- Multiply this figure by the number of atoms of that element that are present.
- Take this new number, change the sign, and assign it to the element with the unknown oxidation state.
- Divide by the number of atoms of that element in the compound.
- This final figure is the oxidation state of the element.

	What is the oxidation state of:				
	S in K_2S	P in PCl_5	N in Mg_3N_2	S in SO_2	S in SO_3
known oxidation state:	of K $+1$ 2 K atoms $\times 2$	of Cl -1 5 Cl atoms $\times 5$	of Mg $+2$ 3 Mg atoms $\times 3$	of O -2 2 O atoms $\times 2$	of O -2 3 O atoms $\times 3$
∴ total oxidation state: must balance with	$+2$ -2	-5 $+5$	$+6$ -6	-4 $+4$	-6 $+6$
oxidation state:	1 S atom $\div 1$ of S $\overline{-2}$	1 P atom $\div 1$ of P $\overline{+5}$	2 N atom $\div 2$ of N $\overline{-3}$	1 S atom $\div 1$ of S $\overline{+4}$	1 S atom $\div 1$ of S $\overline{+6}$

The oxidation states of all the elements except one must be known, if we need to work out the oxidation state of the one unknown element. The charge on the ion is part of the first total, and needs to be considered in the calculation.

	What is the oxidation state of:		
	Cl in $KClO_2$	Fe in $Fe_2(SO_4)_3$	N in $NH_4^{\oplus}$
known oxidation states:	of K $+1$ of O -2 1 K atoms $\times 1$ 2 O atoms $\times 2$ $+1$ -4	of S $+6$ of O -2 3 S atoms $\times 3$ 12 O atoms $\times 12$ $+18$ -24	of H $+1$ 4 H atoms $\times 4$ $+4$ with $1^{\oplus}$
∴ total oxidation state must balance with	$= -3$ $+3$	$= +6$ $+6$	$= +3$ -3
oxidation state:	1 Cl atom $\div 1$ of Cl $\overline{+3}$	2 Fe atoms $\div 2$ of Fe $\overline{+3}$	1 N atom $\div 1$ of N $\overline{-3}$

Half equations

The redox process, as we have been emphasising, involves gaining and losing electrons. One element gains electrons at the expense of another element. These processes occur simultaneously.

$$Mg(s) + Cl_2(g) \longrightarrow MgCl_2(s)$$

The separate processes of oxidation and reduction can be written as separate equations. For this reaction, the oxidation of magnesium is represented as:

$$Mg \longrightarrow Mg^{2\oplus} + 2e^{\ominus}$$

and the reduction of chlorine is represented as:

$$Cl_2 + 2e^{\ominus} \longrightarrow 2Cl^{\ominus}$$

These equations are called **half equations** – they represent only half of the overall reaction, and show very clearly the oxidation of magnesium (by losing electrons) and the reduction of chlorine (by accepting the electrons).

The two equations together will add up to the overall equation, **as long as we make sure that the number of electrons on each side of the arrow cancel out**.

Now, if we add these equations together:

$$Mg + Cl_2 + 2e^\ominus \longrightarrow Mg^{2\oplus} + 2Cl^\ominus + 2e^\ominus$$

the two electrons on each side of the arrow cancel out and we are left with

$$Mg + Cl_2 \longrightarrow Mg^{2\oplus} + 2Cl^\ominus \text{ (which is } MgCl_2)$$

We shall use half equations of this kind in Ch 10.3 and Ch 24.1.

SELF-ASSESSMENT
QUESTIONS
4.3

1 Identify the element oxidised and the element reduced in the following reactions.

   a $2Al + 3Cl_2 \longrightarrow 2AlCl_3$

   b $2Ba + O_2 \longrightarrow 2BaO$

   c $2HgO \longrightarrow 2Hg + O_2$

2 Work out the oxidation states of the indicated element in the following compounds.

   a P in PCl_5 b C in CH_4

   c S in SO_3 d S in Na_2SO_4

   e P in K_3PO_4

3 Which of the following half equations represent oxidation and which represent reduction?

   a $Ca \longrightarrow Ca^{2\oplus} + 2e^\ominus$

   b $O_2 + 4e^\ominus \longrightarrow 2O^{2\ominus}$

   c $Fe^{2\oplus} \longrightarrow Fe^{3\oplus} + e^\ominus$

4 Which species is oxidised and which reduced in the following redox reactions?

   a $2K + Cl_2 \longrightarrow 2KCl$

   b $Fe + CuCl_2 \longrightarrow FeCl_2 + Cu$

Summary of formulae and equations

Ionic formulae

   e.g. $Ca^{2\oplus}$ and $Cl^\ominus$ gives $CaCl_2$ (calcium chloride)

   $Ca^{2\oplus}$ $PO_4^{3\ominus}$ gives $Ca_3(PO_4)_2$ (calcium phosphate)

Covalent formulae

   e.g. $-N- + H$ gives NH_3 $H-N-H$ with N above

Balancing equations

   e.g. $N_2(g) + 3H_2(g) \longrightarrow 2NH_3(g)$

   2N   6H   2N

                       6H

Ionic equations

   e.g. $Ag^\oplus(aq) + Cl^\ominus(aq) \longrightarrow AgCl(s)$

Oxidation state

   e.g.

   gain $e^\ominus$, reduction

   $2Al + 3I_2 \longrightarrow 2AlI_3$

   0   0   +3 −1

   lose $e^\ominus$, oxidation

(overall $6e^\ominus$ lost by 2Al, $6e^\ominus$ gained by 6I)

SUMMARY 4

In this chapter the following new ideas have been covered. *Have you understood and learned them?* Could you explain them clearly to a friend? The page references will help.

- Writing the formulae of simple ionic compounds. p106
- Writing the formulae of covalent compounds. p108
- Naming simple binary compounds, and the meaning of the endings *-ide*, *-ate* and *-ite* in the names of compounds. p109
- Calculating the M_r of a compound. p110
- Balancing chemical equations. p11
- Writing ionic equations and understanding the role of spectator ions. p11
- Appreciating that redox reactions involve electron transfer. p11
- Working out the oxidation state of elements in ionic and covalent compounds. p11

EXAMINATION QUESTIONS 4

1 The manufacture of ammonium nitrate as a fertiliser starts with the fixation of atmospheric nitrogen and with the preparation of hydrogen from methane and steam. The reactions, from nitrogen and hydrogen to ammonium nitrate, are:

A $CH_4 + H_2O \longrightarrow H_2 + CO_2$ with a catalyst (unbalanced)
B $N_2 + H_2 \longrightarrow NH_3$ with iron catalyst at 400 °C (unbalanced)
C $4NH_3 + 5O_2 \longrightarrow 4NO + 6H_2O$ with Pt catalyst at 800 °C
D $2NO + O_2 \longrightarrow 2NO_2$
E $3NO_2 + H_2O \longrightarrow 2HNO_3 + NO$
F $HNO_3 \longrightarrow H^{\oplus} + NO_3^{\ominus}$
G $NH_3 + H^{\oplus} \longrightarrow NH_4^{\oplus}$
H $NH_4^{\oplus} + NO_3^{\ominus} \longrightarrow NH_4NO_3$

a Which of the reactions C to H are redox reactions?
b What are the oxidation states of nitrogen as reactant and product in each of reactions D and F?
c How does the oxidation state of nitrogen change as a product in the reactions B to C to D to E?
d What is the oxidation state of hydrogen on each side of the reaction B?
e Balance the equations for reactions A and B.

2 From the first twenty elements of the Periodic Table choose the element or elements which:
a has the largest value for the first ionisation energy;
b forms a $1^{\ominus}$ ion with the same electron configuration as neon;
c forms a $2^{\oplus}$ ion which is iso-electronic with argon (the word iso-electronic is explained in Ch 2.3);
d is a metal which forms an oxide with the formula X_2O;
e forms an ionic chloride with the formula XCl;
f forms a giant atomic oxide, XO_2;
g forms a $2^{\oplus}$ ion with the electron configuration $1s^2 2s^2 2p^6$;
h fits the equation:

$$XCl_5 \longrightarrow XCl_3 + Cl_2$$

3 M is a very reactive metal. The mass spectrum of M is shown in the diagram opposite.
a What are the relative quantities of the two isotopes in M?
b Define relative atomic mass, A_r.
c Calculate the value of A_r for this particular sample of M.

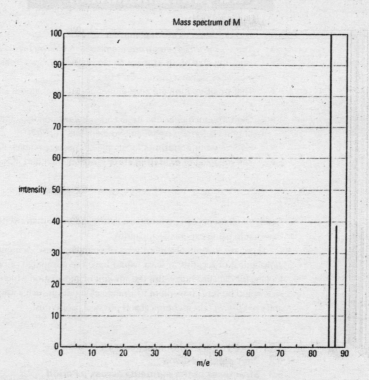

Mass spectrum of M

d The first four ionisation energies of M (in kJ mol^{-1}) are +403, +2632, +3900, +5080. In which Group of the Periodic Table does the element M belong? What is the evidence for your answer?

e M reacts with chlorine to form a chloride, MCl. Write a balanced equation for this reaction.

f What is the formula of the sulphate formed when the hydroxide of M reacts with sulphuric acid?

g M reacts violently with water to produce hydrogen gas and the hydroxide. Write a word equation for the reaction.

h What is the relative formula mass, M_r, of the nitrate of the element M?

4 Lanthanum is a metal which occurs naturally, mainly as the isotope $^{138}_{57}La$.

a What is meant by an isotope?

b How many protons, electrons and neutrons are there in this isotope of lanthanum?

c What is the oxidation state of lanthanum in the compound, LaF_3?

d Lanthanum reacts with water. The unbalanced equation is:

$$La + H_2O \longrightarrow La(OH)_3 + H_2$$

Balance this equation.

e When the hydroxide formed in the reaction with water is heated, lanthanum oxide is formed. Construct a balanced equation for the decomposition.

f Lanthanum is prepared by reacting the trichloride with calcium metal. Construct an equation for the reaction, and identify the changes in oxidation state that occur in the reaction.

- The definition of **atomic number**. (Ch 1.2)
- The way in which **electrons** are placed in **energy levels** and **sub-levels**. (Ch 1.4)
- How the **size of an atom** of an element contracts across a row in the Periodic Table and increases down a Group. (Ch 1.6)
- That **positive ions** are smaller than their parent atoms and **negative ions** are larger (Ch 1.6)
- The difference between **ionic** and **covalent** bonding, and the rules for deciding which type of bond is formed between two elements. (Ch 2.2)
- How to write **equations** for simple chemical reactions. (Ch 4.2)
- The meaning of **oxidation state** (oxidation number). (Ch 4.3)

Nearly 90 elements can be found on Earth, and more than twenty others – all very radioactive – have been made using expensive apparatus.

There are, potentially, many ways of arranging these elements. One way would be to put them in alphabetical order; this at least would have the advantage of being able to find the element when looking for data in a hurry. But the amount of information which you will need to know is quite large, so it would be very convenient to arrange elements in such a way that will help you learn, remember, and even *predict* their chemical and physical properties.

5.1 The Periodic Table and electron structure
5.2 Physical periodicity
5.3 Structures of the elements across a Period
5.4 Chemical periodicity – elements
5.5 Chemical periodicity – compounds

5
the periodic table and periodic trends

5.1 The Periodic Table and electron structure

The American satirist and singer Tom Lehrer – a maths professor at Harvard University! – discovered that the names of all the elements known in the 1950s fitted exactly into the Major-General's song from Gilbert and Sullivan's *Pirates of Penzance*. Tapes and CDs of his songs are still around; see if you can find one, and play *The Elements Song* track. It will help to familiarise you with the names.

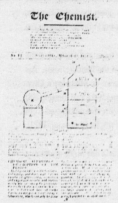

The Chemist – first issue March 1824

The development of the Periodic Table

During the first half of the nineteenth century many new elements were discovered.

If you had been trying to learn chemistry at that time, you would have had to learn a vast number of apparently unconnected facts. There was no framework of ideas on which to hang all the observed chemical facts in order to find a pattern and make sense of it all.

There had been earlier attempts at arranging elements into sequences that grouped elements with similar properties together (see below), but it was a Russian, Dmitri Mendeléev, who in 1869 developed the Periodic Table in almost its modern form.

Early attempts to organise the elements

Early in the nineteenth century, John Dalton in Britain and Jöns Jacob Berzelius in Sweden firmly established the idea of relative atomic mass (see Ch 1.3). One problem was that, because of uncertainty about formulae (for example, was water HO or H_2O?), some published values of atomic mass were wrong. Another problem for those trying to seek order among the elements was that many were still undiscovered.

In 1829, Wolfgang Döbereiner of Germany noticed that in some groups of three chemically similar elements – for example, Cl, Br, I; Ca, Sr, Ba; S, Se, Te – the relative atomic mass of the middle element is nearly midway between those of the first and last, and its properties are also roughly 'in the middle'. These 'triads' were probably the first attempt at systematisation.

In order to try to sort things out, Friedrich Kekulé organised the First International Chemical Congress at Karlsruhe in Germany in 1860. This was the first ever international gathering of scientists. The most important lecture was given by an Italian, Stanislao Cannizzaro, who persuaded his listeners of the importance of accurate values for atomic masses.

In the next few years several chemists began to list the known elements in the order of their relative atomic masses. A few of them noticed that, certainly to begin with, every eighth element had similar properties. First to publish his results in 1862 was a Frenchman, Alexandre-Emil Beguyer de Chancourtois, but the argument was not well presented. Next was a Londoner, John Alexander Newlands, who proposed his 'Law of Octaves' at a meeting of The Chemical Society in 1863. His choice of a musical metaphor was unfortunate and provoked ridicule from fellow chemists at the meeting, which discouraged him from further work on the idea.

In 1998, the centenary of his death, a plaque commemorating Newlands was unveiled on the wall of his family home near the Elephant and Castle in South London by Professor Edward Abel, who had just finished his term as President of The Royal Society of Chemistry – the modern successor to the old Chemical Society. On behalf of the RSC, Professor Abel apologised for the way in which Newlands had been treated. Fourteen of Newlands' descendants were there to hear the apology, including a Professor of Chemistry who had flown from Barbados especially for the ceremony.

In 1869 Dmitri Mendeléev and in 1870 Lothar Meyer published full versions of the Periodic Table. Mendeléev gained most of the credit because in 1871 he stated that the gaps in the Table would be filled by the discovery of new elements. Although Mendeléev lived until 1907, and the Nobel awards started in 1901, he was never given the Prize for Chemistry. It is possible that his Table was not thought worthy of a Prize until after the discovery of its link with atomic structure, from about 1910 onwards.

Dmitri Mendeléev

Mendeléev arranged the elements, in order of increasing atomic mass, in horizontal rows – called **Periods** – so that elements with similar properties were in the same vertical column. These families of similar elements were called **Groups**. So confident was Mendeléev that his idea was correct that he left gaps where his arrangement would not comfortably fit, and predicted the physical and chemical properties of each of these missing elements from those of the other elements in their Group. His predictions were found to be very accurate when the elements were eventually discovered.

N.B. A good test for the soundness of any scientific theory is that it can be used to make predictions which are then confirmed by experiment.

If Mendeléev had stuck *strictly* to the order of relative atomic masses, he would not have succeeded. Nickel (A_r 58.70) should come before cobalt (58.93), and iodine (126.9) before tellurium

(127.6). But their chemical properties quite clearly showed otherwise. Mendeléev trusted in his judgement, kept to chemical principles, and assumed that the anomalies were caused either by inaccurate measurement of relative atomic mass or to some effect not yet understood.

We now know that the **Periodic Table** is based on **atomic number** (see Ch 1.2), the number of protons in the nucleus of an atom of the element, rather than on atomic mass. The anomalies are caused by the isotopic mixture (Ch 1.2) of the atoms of the elements concerned. Iodine consists mainly of relatively 'light' isotopes and tellurium mainly of 'heavy' ones; so their relative atomic masses (Ch 1.3) are in the wrong order.

Mendeléev knew nothing of the noble gases, which were not discovered until the 1890s. Argon was found to have a relative atomic mass of 39.95, compared with potassium's 39.10. To put them the other way around in the Table would be stupid!

Ramsay and Moseley

The reclusive English aristocrat Henry Cavendish, studying air in 1785, found that when he had removed all the nitrogen and oxygen from a sample there was always a small bubble of gas left. He estimated this to be rather less than 1% of the original air – a remarkably accurate figure. We now know that he had obtained argon. But because argon does nothing, nobody else noticed it or knew anything about it, although 0.93% by volume of air consists of argon – nearly thirty times as much as the amount of carbon dioxide!

Argon was eventually isolated in 1892 by a Scotsman, Sir William Ramsay. He was following up an observation of Lord Rayleigh's that 'nitrogen' obtained from air was slightly denser than nitrogen obtained from a chemical reaction. In 1895 Ramsay isolated helium, and in 1898 completed the set with krypton, xenon and neon. He was given the Nobel Prize for Chemistry in 1904 for discovering the inert gases (as they were then called) and placing them in the Periodic Table. In the same year Rayleigh received the Nobel Prize for Physics.

In 1913 Henry Moseley generated X-rays by using more than 30 metals as targets in an X-ray tube, and found that the X-ray spectrum changed regularly from element to element, exactly in order of their position in the Periodic Table. The relation between an element's characteristic X-ray frequency and its atomic number is now known as **Moseley's Law**. He was the first to link the atomic nucleus to the Periodic Table. He established that six elements were still missing and predicted their properties. During the next twenty years or so these elements were hunted down and discovered. Lord Rutherford called Moseley 'a born experimenter'. Rutherford's colleague, Frederic Soddy (Nobel Prize for Chemistry, 1921), said that Moseley had 'called the roll of the elements'.

In 1915, 28-year-old Henry Moseley was shot through the head by a sniper during the battle of Suvla Bay in the disastrous Gallipoli campaign.

A new element had been observed in the spectrum of the sun in 1868 by Janssen and Lockyer and named helium from 'helios' – the Greek for 'sun'.

With the discovery of the noble gases and other elements to fill the gaps, the Periodic Table was completed. In recent years the Table has been extended by using high energy machines to create elements with very high atomic numbers, all of them radioactive and most with very short half lives (see Ch 1.2). The highest atomic number of an element found in any quantity in nature is 92 – the element with this atomic number is uranium.

A reasonable knowledge of the Periodic Table (*Figure 5.1.1*) – and the trends within it – is essential if you wish to have any mastery of inorganic chemistry. It will be immensely helpful if you can immediately place any common element in its correct block, Period and Group; or, given an atomic number, state immediately what the element is and what is the electron configuration of its atom. A copy of the full modern Table is printed inside the back cover of this book. Study it well and often. It contains a lot of information, some of it obvious and some that – by the end of this chapter – you will be able to deduce.

Building the Table: blocks, Periods, Groups

In Ch 1.4 we showed how electrons in atoms can exist in main energy levels, sub-levels, and orbitals. The sub-levels – known as s, p, d, and f and containing a maximum of 2, 6, 10 and 14 electrons respectively – show up as 'blocks' of elements in the Periodic Table (*Figure 5.1.2*).

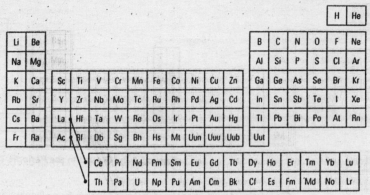

Figure 5.1.1   Periodic Table of the elements

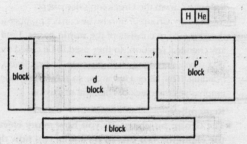

Figure 5.1.2   Outline of the Periodic Table

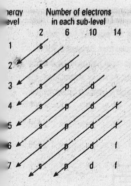

Figure 1.4.8   Filling electron energy levels

The first and smallest block contains the first two elements, hydrogen and helium. With helium the first main energy level is filled. Each row across the Table is called a **Period**. After the first Period ends with the noble gas helium, each successive row or Period corresponds to the filling of the s- and p-sub-levels of the successive main energy levels.

After the third Period, the d- and eventually the f- blocks break up the Periods (after the s-block). *Figure 1.4.8* (Ch 1.4) shows an easily-remembered way of sorting out the order in which the sub-levels are filled with electrons. Once you know the atomic number of an element it is easy to find its electron configuration, and especially the number of electrons in its outer energy level – which determines how the element will behave chemically.

So, the idea of 'blocks' is important. Many properties of the elements can be related to the type of sub-level containing their outermost electrons.

Groups in the Periodic Table

Within each block there are columns of elements which possess similar properties. These vertical columns are called **Groups** (*Figure 5.1.3*).

You will probably be familiar with a few of these Groups from your GCSE course. The Groups in the s- and p-blocks are numbered from 1 to 7 and finish with 0. Group 1 contains the very reactive metals lithium, sodium and potassium and, because these react with water to form an alkaline solution, the Group is known as the **alkali metals**.

With the exception of beryllium, Group 2 contains metals which are almost as reactive as Group 1. Magnesium and calcium occur in the Earth's crust as their carbonates, which can, on strong heating,

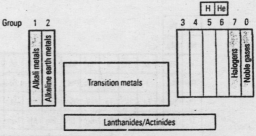

Figure 5.1.3 Groups in the Periodic Table

form the oxides. These are very basic and in turn form slightly soluble alkaline hydroxides. Group 2 is therefore known as the **alkaline earth metals**. The last but one column in the p-block consists of the very reactive non-metals fluorine, chlorine, bromine and iodine. Group 7 is known as the **halogens**, from the Greek for 'salt makers'.

But why Group 0? Surely the correct number to use in the sequence is Group 8? The answer is that this Group consists of the **noble gases**. Until 1962 no one had found any of them capable of any chemical reaction, so they used to be called the 'inert gases'. You can look at their structures in two ways: they have 'full' electron energy levels, or they have no outer electrons available for bonding. The latter view won, so they were called Group 0; and in any case, the title Group 0 emphasises their lack of chemical reactivity.

Remember:

The Group number tells you how many electrons there are in the outer energy level of the element. And this in turn tells you how the elements in the Group are likely to react and what kinds of bonds they can form (see Ch 2.3 and 2.4).

So already you can use the Periodic Table to predict the properties of an element!

See Ch 2.4 ('Theory vs. fact') for a reminder of how compounds of xenon were discovered – and how they made people think hard!

Some time ago the International Union of Pure and Applied Chemistry (IUPAC) decided to use an 18-Group Periodic Table. This was because older Periodic Tables had used A and B sub-Groups – so that manganese, for example, was placed in Group 7A and chlorine in 7B. The problem was that Americans used A and B the other way round from Europeans. IUPAC's solution was the 18-Group arrangement. They were apparently unaware that most sensible people had given up A and B sub-Groups many years earlier and referred to Groups 1–7 and 0 and to the d-block elements. At senior school level, the 8-Group labelling is still used, to the greater convenience of both teacher and student – *because it links the Group number directly to the number of outer electrons*, and so to chemical behaviour.

To summarise

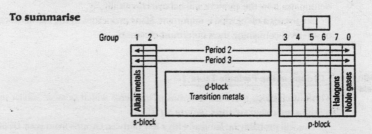

A number of different forms of the Periodic Table have been published – but they all contain the elements arranged using the principles developed in Mendeléev's Periodic Table. Circular, spiral, pyramidal and cubic forms of the Table are all possible, including one drawn up by one of our students some years ago (see page 131)! But they all show the same blocks, the same Periods and the same Groups.

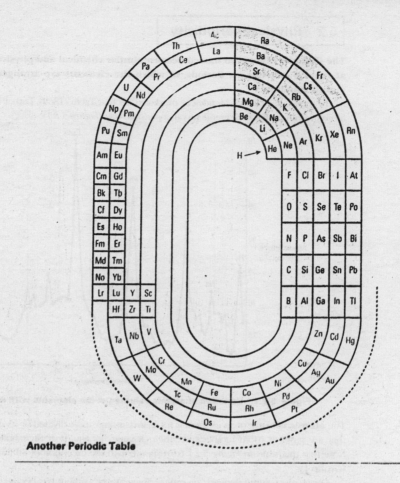

Another Periodic Table

SELF-ASSESSMENT
QUESTIONS
5.1

1 Using the information in this section and any other sources (such as a CD-ROM, the Internet, other books or an encyclopaedia), summarise the important stages in the development of the modern Periodic Table.

2 The formulae of the chlorides in the Period lithium to neon are:
$LiCl$, $BeCl_2$, BCl_3, CCl_4, NCl_3, Cl_2O and ClF.

   a What is the oxidation state of each of the elements in each of these compounds?

   b The equation for the formation of lithium chloride is:

$$2Li(s) + Cl_2(g) \xrightarrow{\text{heat}} 2LiCl(s)$$

   Explain, in terms of electrons, what happens to the oxidation states of lithium and chlorine during this reaction.

3 The electron configurations of four elements are listed below. Decide to which block and which Group in the Periodic Table they belong. Then use the Table to identify each element.

   a $1s^2 2s^2 2p^4$

   b $1s^2 2s^2 2p^6 3s^2 3p^6$

   c $1s^2 2s^1$

   d $1s^2 2s^2 2p^6 3s^2 3p^6 3d^4 4s^2$

5.2 Physical periodicity

The Periodic Table is so named because similar chemical and physical properties occur at regular intervals, or Periods, in which the elements are arranged in order of their atomic numbers.

In the same year that Mendeléev produced his first Table (1869), Lothar Meyer noted that the atomic volumes of the elements vary in a periodic way (*Figure 5.2.1*).

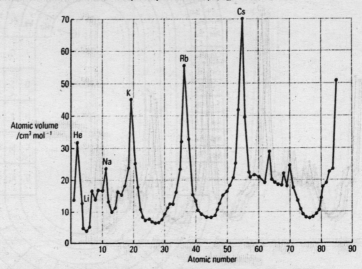

Figure 5.2.1 **Variation of atomic volumes of the elements with atomic number**

The **atomic volume** of an element is its relative atomic mass divided by its density. (For elements that are gases at room temperature, their densities as liquids at their boiling points are used.) Note that the peaks in *Figure 5.2.1* correspond, with the exception of lithium, to the alkali metals (Group 1).

Obviously, volume depends on radius. Since Meyer's time it has become possible, using X-ray diffraction, to measure atomic radii (see Ch 3.9). In Ch 1.6 we saw how **atomic radius** increased down a Group (*Table 1.6.1*) and decreased across the Period sodium to argon (*Table 1.6.2*). The increase within the Group can be explained by the increasing number of completed electron energy levels; the decrease across the Period is because extra protons are in the nucleus, but no new electron energy level is being filled.

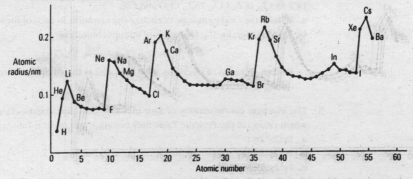

Figure 5.2.2 **Variation of atomic radius of the elements with atomic number**

Figure 5.2.2 shows a graph of atomic radius plotted against atomic number; not surprisingly, it is similar to Lothar Meyer's graph (*Figure 5.2.1*), and again the peaks correspond to the alkali metals.

The more the attraction from the nucleus is 'shielded' by complete electron energy levels, the easier it is to remove an electron from the atom to make it into a positive ion. As a result, the first ionisation energies of elements decrease down a Group (see Ch 1.5).

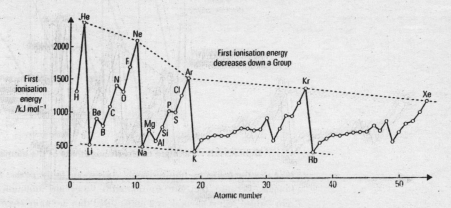

Figure 5.2.3   Variation of first ionisation energy of the elements with atomic number

Figure 5.2.3 shows the variation in **first ionisation energy** with atomic number. This shows clear periodicity, with the peaks corresponding to the noble gases and the troughs to the alkali metals. In Ch 1.5, we discussed how the detailed structure in the first ionisation energy/atomic number graph (*Figure 1.5.3*) gave powerful evidence about electron energy levels and sub-levels.

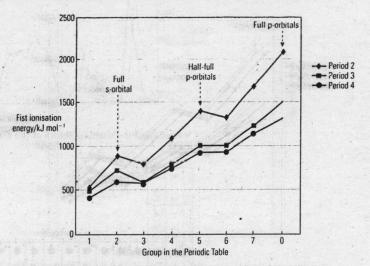

Figure 5.2.4   Periodicity of first ionisation energies of the elements

Figure 5.2.4 shows the variation in first ionisation energies across Periods 2, 3 and 4; the shapes provide evidence for the existence of s- and p-orbitals and for the slight extra stability of the half-filled p-sub-level.

Atomic radius, as we noted above, decreases across a Period and increases down a Group. In Ch 2.2 we discussed how this led to an **increase in electronegativity** across a Period but a **decrease** down a Group. These trends are clearly seen in *Figure 5.2.5*.

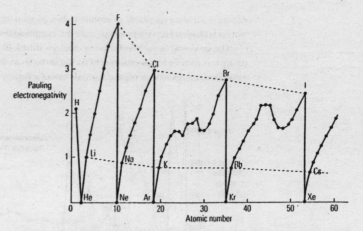

Figure 5.2.5   Variation of electronegativity of the elements with atomic number

Note that the halogens, unsurprisingly, correspond to the peaks. The noble gases are found in the troughs. Because they are so unreactive, they are arbitrarily given an electronegativity of zero – which seems illogical, as it implies that they are more *electropositive* than the alkali metals, and so should behave as extremely reactive metals!

The **melting and boiling points** of elements do not show such clear periodicity (*Figures 5.2.6 and 5.2.7*), largely because of the variety of structures (see Ch 5.3). Note, however, that the boiling point graph is roughly similar in shape to that for melting point.

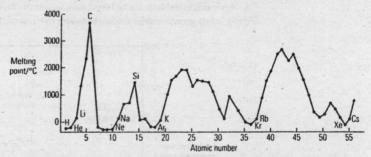

Variation of the melting points of the elements with atomic number

2 Li 181	Be 1278											**1** H −259	He −270	B 2300	C 3700s	N −210	O −218	F −220	Ne −248

Figure 5.2.6   Melting points of the elements / °C

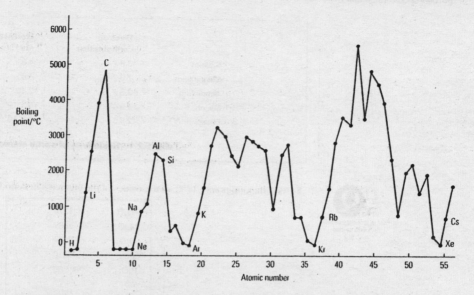

Figure 5.2.7   Variation of the boiling point of the elements with atomic number

For Groups in which all the members have the same structure – notably the noble gases and the alkali metals – there is some regularity down the Group. For Group 0, the noble gases, melting and boiling points *increase* with increasing atomic number. These gases exist as single atoms. The bigger the atom, the more electrons, and so the greater the interparticle forces (see Ch 2.8), resulting in increased melting and boiling points. So why do the values for Group 1, the alkali metals, actually *decrease* with increasing atomic number? The answer lies in the nature of metallic bonding (see Ch 2.7).

All the elements in Group 1 have one outer electron. It is these mobile outer electrons and the attraction between them and the lattice of metal ions which hold a piece of metal together. Potassium has a larger atom than sodium; consequently there are fewer atoms in a given volume of potassium than there are in the same volume of sodium, and so fewer electrons to hold the lump together. This means that potassium melts at a lower temperature than sodium (63 °C compared with 98 °C).

In metals, the metallic bonding has a much greater influence than the simple instantaneous dipole-induced dipole forces (van der Waals forces, Ch 2.8), though these are of course present in metals, as in any other materials.

As we shall see in Ch 5.4, elements become *less metallic* in their behaviour across a Period, but *more metallic* down a Group. It is difficult to find reliable data on the **conductivity** (or, rather, **electrical resistivity**) of pure elements – but, obviously, metals are all conductors whereas only one non-metal (carbon in the form of graphite) conducts well. For reasons similar to those given in the last paragraph, one would expect conductivity to *decrease*, and the resistivity to increase, down a Group of metals as the number of mobile electrons per unit volume decreases. Conversely, so long as the elements in a Period *are* metals, one would expect an *increase in conductivity*, and hence a *decrease in electrical resistivity*, from left to right in the Period. The atoms are getting smaller but the number of outer electrons is increasing, so that the number of mobile electrons per unit volume increases. This is borne out by the data for the first three elements in Period 3 (see *Table 5.2.1*).

	Electron configuration	Electrical resistivity /10^{-8} Ω cm
Sodium	2.8.1	4.8
Magnesium	2.8.2	3.9
Aluminium	2.8.3	2.45
Silicon	2.8.4	3×10^6
Phosphorus	2.8.5	1×10^{17}

Table 5.2.1 Resistivity of some elements

SELF ASSESSMENT QUESTIONS 5.2

1 The melting points, in °C, of the oxides of the elements in Period 3 are listed below.

Na_2O	1275
MgO	2852
Al_2O_3	2072
SiO_2	1610
P_4O_{10}	580
SO_3	17
Cl_2O_7	−92

a Rewrite these formulae to show the number of oxygen atoms that combine with one atom of the other element; e.g. for sodium, the formula would be $NaO_{0.5}$.

b Plot a graph of melting point against the number of oxygen atoms present in these simplified formulae.

c Explain any trends in melting points in terms of the structures of the oxides.

2 The boiling points, in °C, of a number of binary hydrides are listed below.

AlH_3	Not known
HBr	−67
HCl	−85
HF	20
HI	−35
H_2S	−61
MgH_2	m.p. 280 decomposes
NaH	m.p. 800 decomposes
PH_3	−88
SiH_4	−112

Using these data, describe the trends in boiling points of the hydrides across a Period and down a Group in the Periodic Table. Make sure to illustrate any trends with appropriate graphs. Explain in terms of the structures of the compounds, the general trends and anomalies that are observed.

3 The first two molar ionisation energies, in kJ mol^{-1}, for three *successive* elements in the Periodic Table are:

A	1170	2047
B	376	2420
C	503	965

a In which Group of the Periodic Table is element A?

b Write an equation for the third ionisation energy of element B.

c What would you expect the value of the third ionisation energy for element B to be?

d Explain why the first ionisation energies are considered to be periodic.

5.3 Structures of the elements across a Period

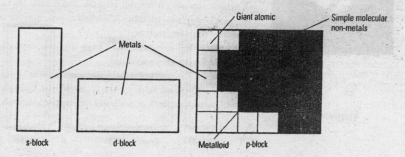

Figure 5.3.1 Structures in the Periodic Table

As shown in *Figure 5.3.1*, most of the elements are metals; their structures have been sufficiently discussed in Ch 2.7. The non-metals consist mostly of simple molecules, although some – notably carbon and boron – have giant atomic structures (see Ch 3.2). It is, of course, the structures and the bonding within them which are largely responsible for the variations in physical properties discussed in the previous section.

In a few Groups, the structures are the same for all the members. So, in Group 1, all the elements from lithium to caesium have body-centred cubic metallic structures. Presumably francium as well, although it can only be obtained in tiny quantities and is extremely radioactive, so is very difficult to investigate. In Group 0, all the elements from neon to radon are monatomic gases – they consist of single atoms. For all members of Group 7 from fluorine to astatine, the structure is diatomic molecules, X_2, whether the element exists as a gas (F_2, Cl_2), liquid (Br_2), or solid (I_2, At_2) at room temperature.

Across a Period, as shown in *Figure 5.3.1*, the general trend is from giant metal structures to simple molecular non-metals. But there is a wide variation within the Groups other than 1, 7 and 0. Group 2 begins with an element – beryllium – which is a good conductor of electricity but has few of the chemical properties of a metal; and Group 3 contains an obvious non-metal, boron, which has a very high melting point and a giant atomic structure. But the next member in each case is an obvious metal (magnesium and aluminium respectively).

Group 4

In Group 4, carbon is a non-metal with two polymorphs (see Ch 3.2), both of which possess giant atomic structures. Silicon is also a non-metal, though with some metallic properties; the same is true of germanium, where metallic properties are more noticeable. These two elements are best described as **metalloids**, sometimes known as 'semi-metals'. The next elements, tin and lead, are obvious metals.

Group 5

Group 5 is the most interesting group because of the variety in the elements. It starts with the third most electronegative element, nitrogen, which exists at room temperature as a diatomic gas, N_2. The next, phosphorus, is a polymorphic non-metal. The molecular unit in all forms of phosphorus is P_4, with four phosphorus atoms at the points of a tetrahedron, the tetrahedrons linking together.

Next is arsenic, a polymorphic metalloid with quite good electrical conductivity. Antimony is another polymorphic metalloid, but with a greater tendency to metallic behaviour and also with quite good conductivity. The final member, bismuth, is a reasonably straightforward metal with good conductivity.

Phosphorus, P_4

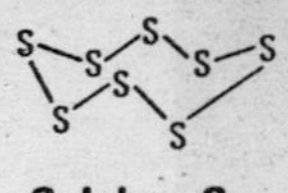

Sulphur, S_8

Group 6

Group 6 begins with the second most electronegative element, oxygen, which normally exists as O_2, but which can also form the unstable, but vitally important, triatomic molecule ozone, O_3 (see Ch 17.2). The second member is sulphur, a polymorphic non-metal whose molecule in the solid is a crown-like ring structure, S_8.

Selenium and tellurium are metalloids with increasing electrical conductivity. Polonium is a radioactive silvery-grey metal.

Figure 5.3.1 summarises the major variations of structure within the Periodic Table. *Table 5.3.1* shows the part of the Table in which simple molecular structures occur, and the normal molecular states of those elements. *You should make sure you remember these.*

Group	5	6	7	0
	N_2	O_2, (O_3)	F_2	Ne
	P_4	S_8	Cl_2	Ar
			Br_2	Kr
			I_2	Xe

Table 5.3.1   Simple molecular structures of elements in the Periodic Table

1 Are the elements with the following electron configurations metals or non-metals?
  a $1s^2 2s^2 2p^5$
  b $1s^2$
  c $1s^2 2s^2 2p^6 3s^2 3p^6 3d^7 4s^2$
  d $1s^2 2s^2 2p^6 3s^2 3p^6 4s^1$

2 What type of bonding would you expect to occur between the elements with the electron configurations shown below?
  a $1s^2 2s^1$ and $1s^2 2s^2 2p^4$
  b $1s^2 2s^2 2p^5$ and $1s^2 2s^2 2p^6 3s^1$
  c $1s^2 2s^2 2p^2$ and $1s^2 2s^2 2p^5$
  d $1s^2 2s^2 2p^6 3s^2 3p^5$ and $1s^2 2s^2$

3 The melting and boiling points of the Period 3 elements are shown in Table 5.3.2.

	Na	Mg	Al	Si	P_4	S_8	Cl_2	Ar
m.p./°C	98	649	660	1410	44	119	−101	−189
b.p./°C	883	1107	2467	2355	280	445	−35	−186
Electrical resistivity /$10^{-8} \Omega$ cm	4.8	3.9	2.45	3×10^6	1×10^{17}			

Table 5.3.2

  a Explain why the melting point of phosphorus is lower than that of sulphur.
  b Magnesium and aluminium both have metallic structures. Why should the resistivity of aluminium be lower (i.e. conductivity be higher) than that of magnesium?
  c Why should the first four elements in the Period have significantly higher boiling points than the last four elements in the Period?
  d The relative atomic mass of chlorine is 35.45 and that of argon is 39.95. Although these values are similar, the boiling points are very different. Why do you think this is so?

5.4   Chemical periodicity – elements

A look at the Periodic Table shows that the great majority of elements – nearly 70 of the first 90 – are metals. You can see that the metals are on the left of the Table, the most reactive of them at the bottom left. Non-metals are on the right, with the most reactive at the top right – not counting the noble gases.

It would be useful to know the reasons for this distribution of types of elements within the Table.

Typical metals are good conductors of heat and electricity, easily lose electrons to form positive ions, form basic oxides, and form ionic, solid chlorides. An example of the structure of metals is shown in *Figure 5.4.1* and discussed in Ch 2.7.

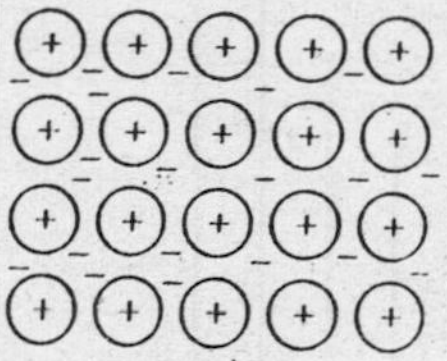

Typical non-metals are non-conductors, easily gain electrons to form negative ions, and form acidic or neutral oxides and volatile, often liquid, covalent chlorides. The different types of oxides and chlorides are discussed in Ch 5.5.

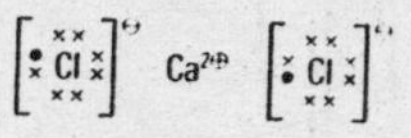

- Metals typically form compounds by ionic bonding with non-metals (*Figure 5.4.2*; see also Ch 2.3).

- They can also often form alloys with each other (see Ch 2.7).

- Non-metals either bond ionically with metals or covalently with each other (*Figure 5.4.3*; see also Ch 2.4).

- The properties of some elements are neither completely metallic nor non-metallic, and the bonding in many compounds is neither clearly ionic nor covalent (see Ch 2.5).

- Elements, other than those in Groups 1 and 2, are often capable of existing in more than one **oxidation state** (see Ch 4.3). Such metals can sometimes, judging from the properties of their compounds, show both metallic and non-metallic properties.

> **Oxidation state** (or oxidation number) **is the charge on an atom if the electrons in its bonds in a compound (ionic or covalent) are assigned to the more electronegative element** (see Ch 4.3).

**Figure 5.4.1
Metallic bonding**

CaCl₂

**Figure 5.4.2
Ionic bonding**

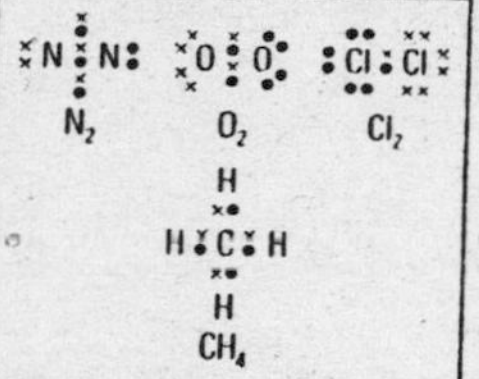

N₂   O₂   Cl₂

CH₄

**Figure 5.4.3
Covalent bonding**

Trends within a Period

Any Period in the Table (other than the first one, which finishes abruptly at helium!) follows the same general pattern, but the best to study is Period 3, Na → Ar. Late Periods are interrupted by the d-block, and eventually the f-block as well; and Period 2 contains the first members of each Group, which, as we have seen already, often behave oddly.

A Period starts with an extremely reactive metal in Group 1. The elements gradually become more non-metallic across the Period. Non-metallic character is greatest at Group 7, and the Period ends with a chemically-inert noble gas.

This trend corresponds to the electron configurations of the atoms of the elements. In Group 1, the outer electron energy level contains just one electron, which is shielded from the attraction of the nucleus by one or more completed energy levels. So this electron is fairly easy to remove; look again at *Figure 5.2.3*. In any Period, the element in Group 1 has the lowest first ionisation energy. There is a general increase in the first ionisation energy across any Period.

Metals in general have low first ionisation energies, and can readily have electrons removed so that their atoms become positive ions. As this tendency decreases from left to right across a Period, so **metallic character decreases across a Period.** Alternatively, **non-metallic character increases across a Period.**

The reason for the slight hiccups in the plot of first ionisation energy against atomic number is given in Ch 1.5.

Oxidation state

The **maximum oxidation state** (see Ch 4.3) which can be achieved by an element is equal to the Group number, except for the first elements in Groups 6 and 7. The maximum and also the common oxidation states are shown in *Table 5.4.1*, and the maxima and minima (together with some others) for the first 20 elements are shown graphically in *Figure 5.4.4*.

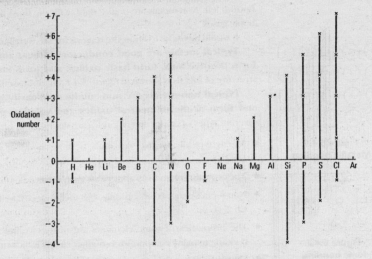

Figure 5.4.4   Maximum and minimum oxidation states

Group	1	2	3	4	5	6	7	0
Maximum oxidation state								
Period Li → Ne	+1	+2	+3	+4	+5	−2	−1	0
Period Na → Ar	+1	+2	+3	+4	+5	+6	+7	0
Common oxidation states								
	+1	+2	+3	+4	+5	+6	+7	0
				+2	+3	+4	+5	
					−3	−2	−1	

Table 5.4.1   Oxidation states of some elements

Trends within a Group

Down a Group there is an:
- increase in atomic number
- increase in atomic radius
- increase in ionic radius
- increase in electropositive character
- increase in 'metallic' character
- decrease in first ionisation energy.

Once more, look at the data on first ionisation energies in *Figure 5.2.3*. The Group 1 elements, the alkali metals, are found in the troughs. But, as atomic number increases down the Group, the energy needed to remove the outermost electron becomes less. We can therefore say that elements become more typically metallic – more reactive as metals, in other words more **electropositive** – down the Group. And this is true for all other Groups as well. Because both atomic radius and the shielding caused by completed energy levels increases down *any* Group of the Periodic Table, **down any Group there will be an increase in metallic (electropositive) character**.

Group 5 shows this best. There is a gradual change from the highly electronegative gaseous non-metal, nitrogen, to the fairly typical metal, bismuth. The increase in metallic character even applies to the halogens, Group 7; iodine has a few slightly metallic properties (see Ch 8.4).

One can stand the above argument on its head. Small atoms, with few completed electron energy levels and therefore less shielding of the attraction from the positive nucleus, are more likely to accept extra electron(s) and become negative ions; i.e. they should be more **electronegative**. And in fact, the four most electronegative elements are, in order, fluorine, oxygen and (third equal) nitrogen and chlorine. Fluorine, oxygen and nitrogen are at the top of their respective Groups.

The two major trends within the Periodic Table are therefore:

- an increase in non-metallic character across a Period from Group 1 to Group 7;
- an increase in metallic character with increasing atomic number down a Group.

The result of these trends is that **the most reactive metals are found at the bottom left of the Table**, and (ignoring the noble gases) **the most reactive non-metals are at the top right of the Table.**

The diagonal relationship

The two trends discussed above, which appear to be (as the Table is arranged) at right angles to each other, produce the so-called **diagonal relationship**. Lithium has some similarities to magnesium. For example, its fluoride and carbonate are insoluble, unlike those of the other Group 1 metals, but like those compounds of magnesium. Also, on heating, lithium carbonate and lithium nitrate decompose in the same way as the corresponding compounds of magnesium and unlike those of sodium.

Beryllium has some similarities to aluminium, and boron to silicon. But, in general, the diagonal relationship is not very important and only worth noticing for elements which possess fairly small atoms towards the left of Periods 2 and 3 (*Figure 5.4.5*).

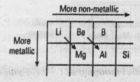

Figure 5.4.5   The diagonal relationship

Anomalous behaviour of the first elements in each Group

The difference in properties between the first and second member of a Group is often **greater than** that between the other members of the Group altogether. This is because of the small atomic radius and consequent electronegativity of the first element. The odd behaviour of lithium and beryllium (s-block), carbon (Group 4) and fluorine (Group 7) is discussed where appropriate during the next three chapters.

Should you ever be asked to discuss the trends in properties and behaviour across a Period, *always* (unless instructed otherwise) pick the elements in Period 3, sodium to argon.

One consequence of their small size is that, as indicated in *Table 5.4.1*, small non-metal atoms cannot achieve the Group's normal maximum oxidation state. For example, in Group 6 oxygen is so small that it has no d-orbitals available for bonding. Its electron configuration is $1s^2 2s^2 2p^4$

(see Ch 1.4 for how to write electron configurations). The second main energy level, remember, has no d-orbitals. Sulphur's electron configuration is $1s^2 2s^2 2p^6 3s^2 3p^4$. The 3d-orbitals are empty but can be used for bonding, so that sulphur can form compounds such as SF_6 and SO_3, and ions such as sulphate, $SO_4^{2\ominus}$.

The 'inert pair' effect

In a Group where the elements can exist in more than one oxidation state, as atomic number increases down the Group the lower oxidation state tends to become more stable. For example, in Group 4 the common oxidation states are +2 and +4. Tin(II) oxide, SnO, is very unstable in air, whereas tin(IV) oxide, SnO_2, is stable – it is in fact the main ore of tin, cassiterite. However, lead(II) oxide, PbO, is stable whereas lead(IV) oxide, PbO_2, is unstable and a powerful oxidising agent.

The effect is most noticeable in thallium, lead and bismuth, at the bottom of Groups 3, 4 and 5, respectively. These

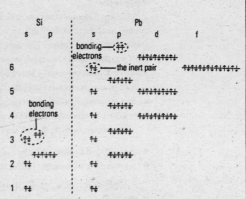

Figure 5.4.6 The 'inert pair' effect

elements follow immediately after the d-block, and after the f-block as well. It is thought that the shielding due to electrons in d-orbitals is comparatively weak. In any case, the two s-electrons from the outermost level penetrate towards the nucleus (see Ch 1.4), and are therefore exposed to greater attraction from it (*Figure 5.4.6*).

So, particularly in Period 6, the result is a decreased tendency for this pair of s-electrons to get involved in bonding: they are therefore known as the **inert pair**.

SELF-ASSESSMENT QUESTIONS 5.4

1 From the following elements, K, Mg, P, and S, select:
   a the most electronegative element;
   b the element which will lose electrons most readily;
   c the element(s) which will conduct electricity as a solid;
   d the element which is most electropositive;
   e the elements which will form a basic oxide.

2 In the following questions the elements are described by their electron configuration.
   a What will be the maximum oxidation state in the compounds formed by the element with configuration $1s^2 2s^2 2p^6 3s^2 3p^2$?
   b Will the element with configuration $1s^2 2s^2 2p^6 3s^2$ form an ionic or a covalent compound when reacted with chlorine?
   c Is the chloride formed by the element with configuration $1s^2 2s^2 2p^6 3s^2 3p^4$ a solid with a high melting point?
   d Which of the elements with configuration $1s^2 2s^2 2p^6 3s^2$ and $1s^2 2s^2$ is the more electropositive?
   e Which of the elements with configuration $1s^2 2s^2 2p^5$ and $1s^2 2s^2 2p^6 3s^2$ will form positive ions when reacted with chlorine?

5.5 Chemical periodicity – compounds

As we discussed above, the best Period for studying trends is Period 3, sodium → argon.

Period 3 elements   Na Mg Al Si P S Cl Ar

The most instructive compounds to study are the oxides and chlorides.

But first, how do the Period 3 elements react with three common reagents, water, oxygen and chlorine? The information is given in *Table 5.5.1*.

Reagent	Sodium	Magnesium	Aluminium	Silicon	Phosphorus	Sulphur	Chlorine	Argon
Water	Reacts violently with cold water. $2Na(s) + 2H_2O(l) \xrightarrow{RT} 2NaOH(aq) + H_2(g)$ Strongly alkaline solution formed.	Reacts with hot water or steam. $Mg(s) + H_2O(g) \xrightarrow{100°C} MgO(s) + H_2(g)$ MgO is only very slightly soluble in water, alkaline solution.	Thin, tough surface layer of oxide prevents reaction (hence use in cooking pans).	No reaction.	No reaction (white phosphorus is stored under water).	No reaction.	Soluble in water with some reaction. $Cl_2(g) + H_2O(l) \rightleftharpoons HClO(aq) + HCl(aq)$ Acidic solution.	No reaction.
Oxygen	Burns when ignited. $4Na(s) + O_2(g) \xrightarrow{ignite} 2Na_2O(s)$	Burns with intense light when ignited. $2Mg(s) + O_2(g) \xrightarrow{ignite} 2MgO(s)$	Needs > 1000°C to ignite in air, but burns readily when ignited in oxygen. $4Al(s) + 3O_2(g) \xrightarrow[in 0_2]{ignite} 2Al_2O_3(s)$	Amorphous form more reactive than crystalline form, burns readily when ignited in oxygen. $Si(s) + O_2(g) \xrightarrow[in 0_2]{ignite} SiO_2(s)$	White form more reactive than red form; can ignite in air at body temperature. Burns fiercely in oxygen when ignited. $4P(s) + 5O_2(g) \xrightarrow{ignite} P_4O_{10}(s)$ Forms P_4O_6 if oxygen supply is limited	Burns when ignited in oxygen to form a colourless choking gas, SO_2, and white fumes, SO_3. $S(s) + O_2(g) \xrightarrow{ignite} SO_2(g)$ $2S(s) + 3O_2(g) \xrightarrow{ignite} 2SO_3(l)$	No direct reaction. (Unstable chlorine oxides can be made by other methods.)	No reaction.
Chlorine	Burns when ignited. $2Na(s) + Cl_2(g) \xrightarrow{ignite} 2NaCl(s)$	Burns when ignited. $Mg(s) + Cl_2(g) \xrightarrow{ignite} MgCl_2(s)$	Heat in dry chlorine and sublime off the product. $2Al(s) + 3Cl_2(g) \xrightarrow{heat} 2AlCl_3(s)$	Heat in dry chlorine. $Si(s) + 2Cl_2(g) \xrightarrow{heat} SiCl_4(l)$	Warm in dry chlorine. $2P(s) + 3Cl_2(g) \xrightarrow{heat} 2PCl_3(l)$ (For PCl_5, react PCl_3 with chlorine in a cooled apparatus.) $PCl_3(s) + Cl_2(g) \rightarrow PCl_5(s)$	Pass dry chlorine over molten sulphur. $S(l) + Cl_2(g) \xrightarrow{heat} S_2Cl_2(l)$ (SCl_2 and SCl_4 can be made by action of chlorine on S_2Cl_2 in cooled apparatus.)		No reaction.

Table 5.5.1   The reactions of the elements of Period 3 with water, oxygen and chlorine

The oxides of the elements of Period 3

Firstly, some vocabulary.

A basic oxide or hydroxide is one which reacts with an acid to form a salt and water only.

Metal oxides are generally basic. If a metal oxide can react with water to form a solution of a hydroxide, the solution is called an **alkali**. For example, sodium oxide, Na_2O is basic; it can react with water to form an alkali, and directly with an acid to form a salt.

$$Na_2O(s) + H_2O(l) \xrightarrow{\text{room temperature}} 2NaOH(aq)$$

$$Na_2O(s) + 2HCl(aq) \xrightarrow{\text{room temperature}} 2NaCl(aq) + H_2O(l)$$

Magnesium oxide, MgO, is almost insoluble in water, but is still basic.

$$MgO(s) + H_2SO_4(aq) \xrightarrow{heat} MgSO_4(aq) + H_2O(l)$$

An acidic oxide is one which reacts with an alkali or insoluble basic metal oxide to form a salt.

The pH scale – a reminder

Oxides of non-metals are generally acidic (*Figure 5.5.1*).

If an acidic non-metal oxide can react with water, it forms an **acidic solution**; e.g. sulphur trioxide, SO_3, can react with water to form sulphuric acid.

$$SO_3(l) + H_2O(l) \xrightarrow[stir]{cool} H_2SO_4(aq)$$

Silicon oxide, SiO_2, is insoluble in water, but is still acidic.

$$SiO_2(s) + 2NaOH(aq) \xrightarrow[\substack{concentrated \\ solution}]{heat} Na_2SiO_3(aq) + H_2O(l)$$

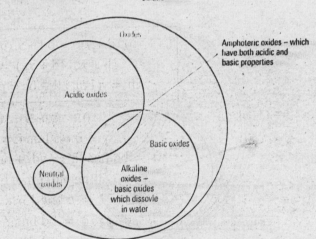

Figure 5.5.1   Classification of oxides

An amphoteric oxide or hydroxide is one which can react with both acids and alkalis.

The oxides and hydroxides of aluminium, zinc, lead and some d-block metals are able to do this; e.g. aluminium oxide, Al_2O_3, is usually amphoteric. (Interestingly, if heated to above 900 °C its relative density increases from 2.8 to 4.0 as it undergoes a change in structure, after which it is unaffected by acids.)

> The word *amphoteric* comes from the same Greek or Latin word-roots as *amphibian*, *amphitheatre* and even *ambiguous* and *ambidextrous*. Can you see the connection?

$$Al_2O_3(s) + 3H_2SO_4(aq) \xrightarrow{warm} Al_2(SO_4)_3(aq) + 3H_2O(l)$$

$$Al_2O_3(s) + 6NaOH(aq) + 3H_2O(l) \xrightarrow{warm} 2Na_3[Al(OH)_6](aq)$$

A few oxides of non-metals are **neutral oxides**, reacting with neither acid nor alkali. Examples are NO, N_2O and (except under extreme conditions) CO. Water may be thought to be a neutral oxide but, as we shall see in Ch 23.2, it can itself act as both an acid and a base.

The oxides change across Period 3 from strongly basic through amphoteric to strongly acidic (*Figure 5.5.2*). This correlates well with the change in the elements from electropositive on the left to electronegative on the right, and the change in the structures of their oxides from giant ionic to molecular. The details are shown in *Table 5.5.2*. The different types of oxides can be summarised in the Venn diagram shown in *Figure 5.5.1*.

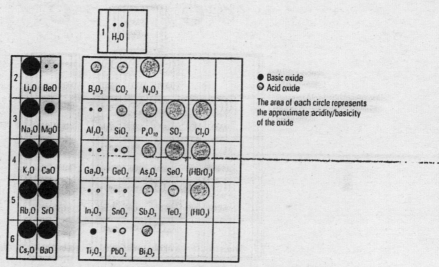

Figure 5.5.2   Acid–base properties of some principal oxides

	Sodium	Magnesium	Aluminium	Silicon	Phosphorus	Sulphur	Chlorine	Argon
Formula	Na_2O	MgO	Al_2O_3	SiO_2	P_4O_{10}	SO_3	Cl_2O_7	None formed
Other oxides	Na_2O_2				P_4O_8	SO_2	Cl_2O	
Physical state at room temperature	Solid	Solid	Solid	Solid	Solid	Liquid	Liquid	
m.p/°C	1275	2852	2027	1610	Sublimes at 300	17	−92	
b.p/°C	Sublimed at m.p.	3600	2980	2230		45	80	
Structure	Giant ionic	Giant ionic	Giant ionic	Giant atomic	Simple covalent molecules	Simple covalent molecules	Simple covalent molecules	
Electrical conductivity when molten	Good	Good	Good	Very poor	None	None	None	
Reaction with water	Forms NaOH(aq), strongly alkaline solution	Forms $Mg(OH)_2$, only slightly soluble	Does not react	Does not react	P_4O_{10} reacts to form H_3PO_4, a strong acid	SO_3 forms H_2SO_4, a strong acid; SO_2 forms H_2SO_3	Cl_2O_7 forms $HClO_4$, an acidic solution	
Nature of oxide	Alkaline	Basic	Amphoteric	Acidic	Acidic	Acidic	Acidic	
Equations	$Na_2O(s) + H_2O(l)$ $\xrightarrow{RT}$ $2NaOH(aq)$	$MgO(s) + H_2O(l)$ $\xrightarrow{RT}$ $Mg(OH)_2(aq)$ only very slight reaction.			$P_4O_{10}(s) + 6H_2O(l)$ $\xrightarrow{heat}$ $4H_3PO_4(aq)$	$SO_3(l) + H_2O(l)$ $\xrightarrow{RT}$ $H_2SO_4(aq)$ $SO_2(g) + H_2O(l)$ $\xrightarrow{RT}$ $H_2SO_3(aq)$	$Cl_2O_7(l) + H_2O(l)$ $\xrightarrow{RT}$ $2HClO_4(aq)$	

Table 5.5.2   Oxides of the elements in Period 3

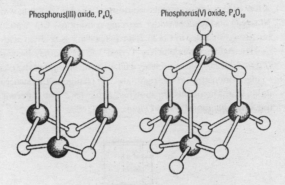

Phosphorus(III) oxide, P_4O_6 Phosphorus(V) oxide, P_4O_{10}

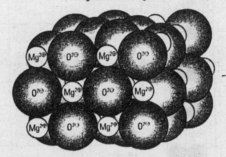

Magnesium oxide, MgO

The structures of the oxides of phosphorus and magnesium

As electropositivity increases down a Group, so does the basicity of the oxides. However, if a metal can exist in more than one oxidation state, the higher the oxidation state in its compounds the less metallic the metal appears to be (see Ch 5.4). So, **the higher the oxidation state of the metal, the more acidic the oxide.**

The chlorides of the elements of Period 3

As with the oxides, there is a change from ionic to covalent across the Period as the elements become more electronegative. The ionic chlorides dissolve in water. Sodium chloride forms a solution of pH7.

$$NaCl(s) \xrightarrow[\text{room temperature}]{H_2O} Na^{\oplus}(aq) + Cl^{\ominus}(aq)$$

There is an earlier switch from giant to molecular structures. Whereas aluminium oxide shows by its electrolysis when molten that it has some ionic character, aluminium chloride is volatile and reacts violently with water – typically covalent behaviour.

$$2AlCl_3(s) + 6H_2O(l) \xrightarrow[\text{temperature}]{\text{room}} 2Al(OH)_3(s) + 6HCl(g)$$

This is because of the high charge and small size, and consequent polarising power (see Ch 2.5), of the $Al^{3\oplus}$ ion.

All the non-metal chlorides in Period 3 react in the same way to give hydrogen chloride and acidic solutions. Details of the Period 3 chlorides are given in *Table 5.5.3.*

	Sodium	Magnesium	Aluminium	Silicon	Phosphorus	Sulphur	Chlorine	Argon
Formula	NaCl	$MgCl_2$	$AlCl_3$	$SiCl_4$	PCl_3	S_2Cl_2	Cl_2	None formed
Other chlorides					PCl_5	SCl_2, SCl_4		
Physical state at room temperature	Solid	Solid	Solid	Liquid	Liquid	Liquid	Gas	
m.p. (°C)	801	714	178	−70	112	−80	−101	
b.p. (°C)	1413	1412	sublimed to Al_2Cl_6	58	76	136	−35	
Structure	Giant ionic	Giant ionic	Largely covalent	Simple covalent molecules	Simple covalent molecules	Simple covalent molecules	Simple covalent molecules	
Electrical conductivity when molten	Good	Good	Poor	None	None	None	None	
Reaction with water	Dissolves easily	Dissolves easily	Fumes of HCl produced	Fumes of HCl produced	Fumes of HCl produced	Fumes of HCl produced	Some reaction with water	
Nature of aqueous solution	Neutral	Weakly acidic	Acidic	Acidic	Acidic	Acidic	Acidic	
Equations	$NaCl(s)$ $\xrightarrow[RT]{H_2O}$ $Na^+(aq)$ $+ Cl^-(aq)$	$MgCl_2 \xrightarrow[RT]{H_2O}$ $Mg^{2+}(aq)$ $+ 2Cl^-(aq)$	$2AlCl_3(s)$ $+ 6H_2O(l)$ $\xrightarrow{RT}$ $2Al(OH)_3(s)$ $+ 6HCl(aq)$	$SiCl_4 + 4H_2O(l)$ $\xrightarrow{RT}$ $Si(OH)_4(s)$ $+ 4HCl(aq)$	$PCl_3(l) + 3H_2O(l)$ $\xrightarrow{RT}$ $H_3PO_3(aq)$ $+ 3HCl(aq)$ $PCl_5(s) + 5H_2O(l)$ $\xrightarrow{heat}$ $H_3PO_4(aq)$ $+ 5HCl(aq)$	Variety of products e.g. H_2S, SO_2, H_2SO_3, H_2SO_4	$Cl_2(g) + H_2O(l)$ $\xrightleftharpoons{RT}$ $HClO(aq)$ $+ HCl(aq)$	

Table 5.5.3   Chlorides of the elements in Period 3

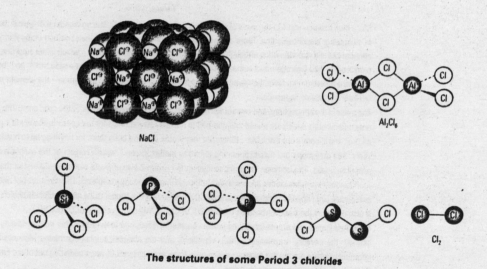

NaCl

Al_2Cl_6

Cl_2

The structures of some Period 3 chlorides

Hydrogen

Hydrogen is difficult to classify in the Periodic Table. By being able to ionise by loss of an electron, it resembles the alkali metals of Group 1; but the $H^\oplus$ ion is a bare proton, and as such could only exist uncombined for a tiny fraction of a second if anything else is around. In aqueous solution, for example, it is *always* hydrated (see Ch 23.1).

147

Hydrogen can also take up one electron to gain the structure of helium, either by forming the hydride ion, $H^{\ominus}$, by combining with a very reactive metal such as calcium, or by covalent bonding.

e.g.

$$Ca(s) + H_2(g) \xrightarrow[\text{temperature}]{\text{room}} CaH_2(s) \quad (\text{i.e. } Ca^{2+}H^{\ominus}{}_2)$$

$$H_2(g) + Cl_2(g) \xrightarrow[\text{light}]{\text{diffused}} 2HCl(g) \quad (\text{i.e. } H\!:\!\overset{\times\times}{\underset{\times\times}{Cl}}\!\times)$$

In these reactions, hydrogen resembles the halogens of Group 7. But the similarities with Group 1 and Group 7 are very limited; and hydrogen is best placed with helium at the head of the Table, and associated with no particular Group.

The Earth's crust and the elements

The abundances of the most common elements in the Earth's crust (including the atmosphere and the oceans) are shown in *Table 5.5.4*.

Element	Mass per cent	Element	Mass per cent
Oxygen	49.5	Magnesium	1.9
Silicon	25.7	Hydrogen	0.88
Aluminium	7.5	Titanium	0.58
Iron	4.7	Chlorine	0.19
Calcium	3.4	Phosphorus	0.12
Sodium	2.6	Carbon	0.09
Potassium	2.4	All others	0.42

Table 5.5.4

More than three-quarters of the mass of the crust is due to three elements. It is no wonder that typical rocks consist largely of silicates or aluminosilicates. There seems to be little correlation between their position in the Periodic Table and the abundances of these elements, although most are among the 'lighter' elements – which is not surprising, since almost all atoms other than hydrogen and some helium were built up by nuclear reactions inside stars, and 'heavier' atoms are therefore less common. Note, however, that the s-block is well represented but carbon – the element on which all life is based – is relatively uncommon.

The mass of the atmosphere, and even of the oceans, is small compared with that of the solid crust. The crust contains the ores and minerals which are mined and quarried to provide raw materials for the needs of the world's population. Nearly all ores and minerals are insoluble – otherwise they would not have been there for anything up to hundreds of millions of years. The exceptions are those containing the alkali metals, Group 1, which consist of the evaporites – the remains of evaporated seas. The common minerals containing the Group 2 elements are carbonates. Most of the other metals are found as their insoluble oxides and sulphides, although there are many exceptions. The major ores of zinc, for example, are zinc blende (ZnS) and the familiar calamine ($ZnCO_3$). The main ore of copper is the mixed sulphide chalcopyrite, $CuFeS_2$, but it also occurs as the basic carbonates malachite ($CuCO_3.Cu(OH)_2$) and azurite ($2CuCO_3.Cu(OH)_2$). Malachite (green) and azurite (blue), the iron ore haematite (Fe_2O_3, red-brown) and other iron-containing minerals (such as ochres), the mercury ore cinnabar (HgS, vermilion), and the complex sulphur-containing aluminosilicate ultramarine (glorious blue) are among the minerals which have for many thousands of years been dug out of the Earth's crust, ground finely, and used as pigments for painting.

**SELF-ASSESSMENT
QUESTIONS
5.5**

1 Describe what you would observe in the following cases. Write relevant equations to show what happens to the chemicals involved.
   a Sodium oxide is added to a dilute solution of pH indicator.
   b Sulphur dioxide is bubbled through a solution of pH indicator in water.

2 Sodium chloride is a non-volatile solid which dissolves readily in water, whereas phosphorus(V) chloride is a volatile solid which reacts violently with water. Explain these facts for the two compounds in terms of their structure and bonding. Write balanced equations for any reactions that might occur.

3 Consider the following list of compounds:

Na_2O, $NaCl$, MgO, $SiCl_4$, P_4O_6.

a Which compound(s) will dissolve in water to form a neutral solution?

b Which compound(s) will react with water to form hydrogen chloride?

c Which of the compound(s) is virtually insoluble in water?

d Which compound(s) will react with water to produce a strongly alkaline solution?

e Which compound(s) will produce a white precipitate when added to water?

f Which compound(s) will react with water to produce a strongly acidic solution?

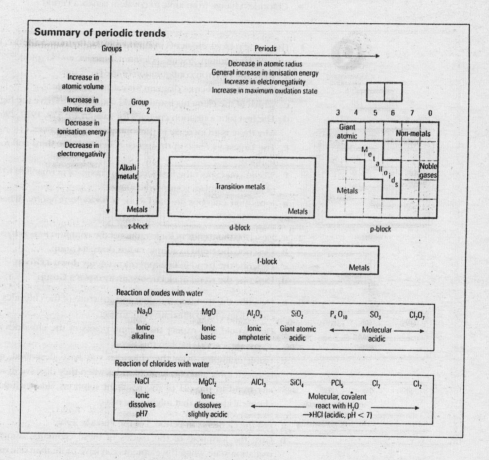

Summary of periodic trends

Groups / Periods

Decrease in atomic radius
General increase in ionisation energy
Increase in electronegativity
Increase in maximum oxidation state

Increase in atomic volume
Increase in atomic radius
Decrease in ionisation energy
Decrease in electronegativity

Group 1 2 / 3 4 5 6 7 0

Alkali metals / Transition metals / Giant atomic / Non-metals / Metalloids / Noble gases / Metals

Metals / Metals

s-block / d-block / p-block

f-block / Metals

Reaction of oxides with water

Na_2O	MgO	Al_2O_3	SiO_2	P_4O_{10}	SO_3	Cl_2O_7
Ionic alkaline	Ionic basic	Ionic amphoteric	Giant atomic acidic	← Molecular →	acidic	

Reaction of chlorides with water

$NaCl$	$MgCl_2$	$AlCl_3$	$SiCl_4$	PCl_5	Cl_2	Cl_2
Ionic dissolves pH7	Ionic dissolves slightly acidic	← Molecular, covalent react with H_2O →		$\rightarrow$HCl (acidic, pH < 7)		

SUMMARY 5

In this chapter the following new ideas have been covered. *Have you understood and learned them?* Could you explain them clearly to a friend? The page references will help.

- The Periodic Table contains elements arranged in order of their atomic numbers.   p128
- The horizontal rows in the Table are called Periods.   p129
- Elements with similar chemical properties occur in vertical columns called Groups.   p129

- The structure of the Table corresponds to the arrangement of electrons in electron energy levels and sub-levels, and consists of s-, p-, d- and f-blocks. p129
- The position of an element in the Table (i.e. its atomic number) determines its chemical behaviour. p130
- Metals are found to the left and middle of the Table and non-metals to the right; most of the elements are metals. p137
- Across a Period, elements change from very reactive metal to very reactive non-metal; the Period ends with a noble gas. p139
- Down any Group, elements become more metallic. p140
- There is periodicity of chemical properties within the Table. p139
- There is also periodicity of physical properties. p132
- Oxides can be acidic, basic, amphoteric or neutral. p144
- Across a Period, oxides change from basic to acidic. p145
- Chlorides change from ionic to covalent across a Period. p146

EXAMINATION QUESTIONS 5

1 The s- and p-block elements in Period 4 of the Periodic Table are, in order, potassium, calcium, gallium, germanium, arsenic, selenium, bromine and krypton.
   a Write the electron configuration of the $Ga^{3\oplus}$ ion.
   b Draw a dot-and-cross diagram for calcium bromide.
   c Which of the elements listed would you expect to have the highest first ionisation energy?
   d The first four ionisation energies for gallium are 579, 1979, 2963 and 6200 kJ mol^{-1}. Explain why there is an increase in the ionisation energy values.
   e The largest increase in this series is between the third and fourth values. Why should this happen?
   f Would you expect the first ionisation energy of germanium to be larger or smaller than that of gallium? Explain your prediction.
   g Ionisation energies are said to be a *periodic* property. What does periodic mean in this context?

2 a Describe the trends in ionic radius of the common ions across Period 3 from Na to Cl.
   b Describe the trend in atomic radius down a Group.
   c Describe the trend in first ionisation energy down a Group.
   d Describe the trend in electronegativity down a Group.

3 a Describe the trends in bonding and structure of the chlorides across a period. Use dot-and-cross diagrams to illustrate your answer.
   b How would you expect the boiling points of the chlorides of these elements to change across the Period? Explain this trend.
   c Using examples from the chlorides you have described, comment on the differences between ionic and covalent chlorides when they dissolve in water. Make sure you mention any trend in the pH of the resultant solutions, and include balanced equations for any chemical changes that might take place.

4 a List the elements in Period 3 of the Periodic Table.
   b Give the formulae of the oxides of these elements, using the element in the highest oxidation state when the elements can have more than one oxidation state.
   c How do the maximum oxidation states of the elements change across the Period?
   d Draw a dot-and-cross diagram for an ionic oxide and a covalent oxide from the formulae you have already given.
   e Describe how these two oxides react with water. What would be the approximate pH of each of the resulting solutions?
   f Write a balanced equation for the reaction of a Group 1 oxide with water.

6.1 **Introduction to Groups 1 and 2 – the s-block**
6.2 **Trends in physical properties of the s-block elements**
6.3 **Flame colours**
6.4 **The reactions of Group 1 elements and compounds**
6.5 **The reactions of Group 2 elements and compounds**
6.6 **The s-block in Nature and industry**

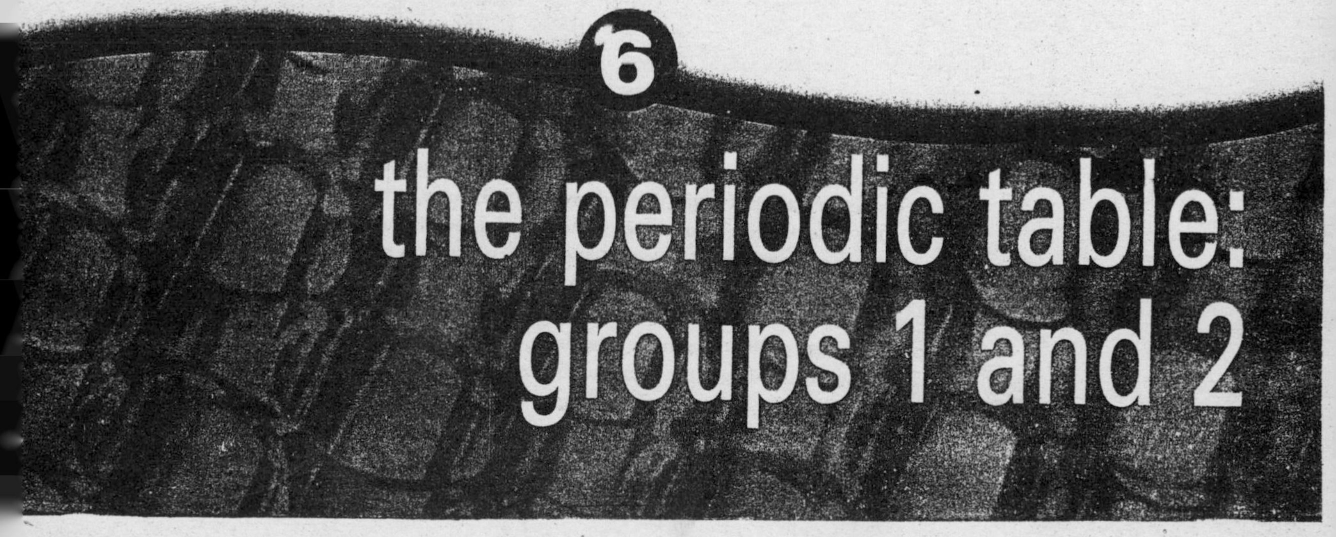

6

the periodic table: groups 1 and 2

6.1 Introduction to Groups 1 and 2 – the s-block

As we saw in Ch 5.1, the s-block of the Periodic Table consists of those elements whose atoms have their outermost electrons in the s-sub-level (see *Table 6.1.1*).

Group 1 Element	Atomic number	Group 2 Element	Atomic number
Li	3	Be	4
Na	11	Mg	12
K	19	Ca	20
Rb	37	Sr	38
Cs	55	Ba	56
Fr	87	Ra	88

Table 6.1.1 The s-block elements

This means that the atoms have one or two electrons outside the stable noble gas configuration. These electrons are therefore able to move through any piece of the element – they are *delocalised* and can move under the influence of an electric field (see Ch 3.5). The s-block elements, with the exception of beryllium (Ch 5.2), can therefore readily conduct electricity; they are *metals*.

The energy required to remove electrons from atoms of these elements is relatively low; they have low (first or first and second) ionisation energies (see Ch 1.5). This is because the one or two electrons in the outer energy level are shielded from the attraction of the positive nucleus by complete energy levels of electrons; the atomic volume of the atoms is high (Ch 5.2). So, because it is comparatively easy to remove these 'way-out' electrons, the metals of the s-block react very readily with anything which can accept electrons – e.g. atoms of non-metals such as chlorine, or hydrogen ions in water or acids.

The electron configurations (in terms of main energy levels) of the s-block elements are shown in *Figure 6.1.1*.

Group 1	
Li	2.1
Na	2.8.1
K	2.8.8.1
Rb	2.8.18.8.1
Cs	2.8.18.18.8.1
Group 2	
Be	2.2
Mg	2.8.2
Ca	2.8.8.2
Sr	2.8.18.8.2
Ba	2.8.18.18.8.2

Figure 6.1.1 Electron configurations

Obviously, with more completed levels, the atomic sizes increase down each Group, and the outer electrons are more completely shielded from the attraction of the nucleus. As would be expected, because reactivity of metals depends mainly on the ease with which they lose electrons, the reactivity of the s-block elements increases with increasing atomic number within the Group.

As is common in the Periodic Table, the first elements in a Group behave oddly. As we shall see, lithium and beryllium have such small atoms that their properties – particularly those of beryllium – are significantly different from those of their fellow group members. This will be discussed in Ch 6.4.

Because of their electronic structures, with the exception of beryllium and to a lesser extent lithium, the bonding in the compounds of s-block elements can be assumed to be largely ionic. The oxidation state (see Ch 4.3) of all Group 1 elements in their compounds is always +1, and of Group 2 elements always +2. The oxidation states correspond to the charges on the metal ions.

Remember, oxidation states are *always* written with the sign before the number, e.g. +2; charges on ions are always written with the number first: e.g. $Ca^{2\oplus}$.

All the s-block elements are far too reactive to be obtained at realistic temperatures by reduction of their oxides with carbon (see Ch 4.3). Because (except for Be) they exist as positive ions in all their compounds, the only way to obtain the metals from their compounds is by forcing electrons on to the ions during electrolysis (see Ch 24.8); and, because the metals react with water, this electrolysis has to use the molten salts rather than their solutions.

It is possible to argue that the metal elements were, for the most part, discovered in reverse order to their chemical reactivity. Unreactive metals – especially gold and silver – have been known and used for thousands of years. Copper and tin are a little more difficult to extract from their ores. Together, they make the alloy bronze. Iron is much more difficult to extract, requiring very strong heating of the oxide ore with carbon. Hence, the Bronze Age came before the Iron Age. (It has been suggested jokingly that history is chemistry in reverse!)

Before the s-block elements could be obtained, a source of continuous electric current was needed to make electrolysis possible. Alessandro Volta in 1800 announced his invention of the Voltaic pile, the first electric battery. In 1807, Sir Humphry Davy, working at the Royal Institution in London, used such batteries to isolate sodium and potassium in relatively pure form, and in 1808, he obtained impure magnesium and calcium. Sodium and potassium were at first called *natrium* and *kalium*, which is how the symbols Na for sodium and K for potassium originated.

6.2   Trends in physical properties of the s-block elements

As we have seen (Ch 1.6), within any Group atomic radius increases with increasing atomic number and hence increasing numbers of complete electron energy levels (*Figure 6.2.1*).

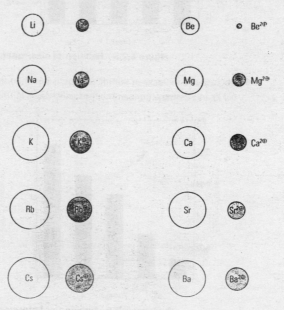

Figure 6.2.1   Relative atomic and ionic radii of the elements of Groups 1 and 2

This has the following results, as atomic number increases within a Group (see Ch 5.2).

- Outer electrons are more easily lost, because of increased 'shielding' of the outer electrons from the attraction of the nucleus; as a result, *first ionisation energies decrease* (*Figure 6.2.2*; see also Ch 2.2).

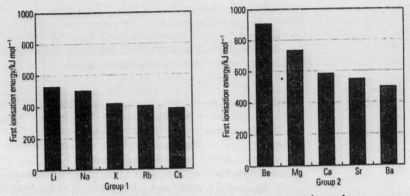

Figure 6.2.2   Relation of first ionisation energy to atomic number

- The tendency for an atom of the element to attract pairs of electrons to itself decreases; *electronegativity decreases* (*Figure 6.2.3*; see also Ch 2.2). Or, to put it another way, the elements within a group become more *electropositive* with increasing atomic number.

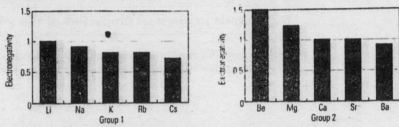

Figure 6.2.3   Relation of electronegativity to atomic number

- Because the increase in atomic volume is less than the increase in relative atomic mass (*Figure 6.2.4*), increasing atomic number usually results in *increasing density* (*Figure 6.2.5*).

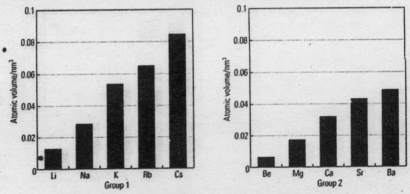

Figure 6.2.4   Relation of atomic volume to atomic number

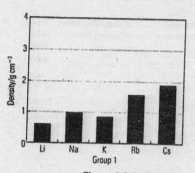 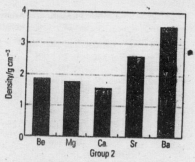

Figure 6.2.5   Relation of density to atomic number

- The increase in atomic volume means that there are fewer atoms of an element in a given volume, and therefore fewer mobile electrons in that volume to hold the metal lattice together (see Ch 3.5). Consequently, both *melting point and hardness usually decrease* with increasing atomic number, as does electrical conductivity (*Figure 6.2.6*).

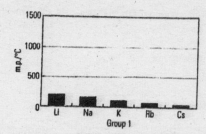 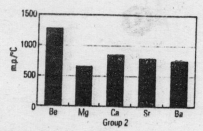

Figure 6.2.6   Relation of melting point to atomic number

The variation in hardness is best seen with the alkali metals. Lithium can be cut with a scalpel, albeit with some effort; sodium can easily be cut; and potassium can be sliced as easily as salami. (But you are strongly advised not to try it! – leave it to a teacher who is taking all precautions. See Ch 6.4 for the reason why.)

---

anomalous = odd, and unlike the other members of the Group.

The statements above hold well for Group 1, although the density of potassium is anomalous, but not so well for Group 2, where beryllium behaves very differently, and the melting point of magnesium and density of calcium are also anomalous.

In Group 1, the metals all have the same body-centred cubic structure, but the Group 2 metals do not all have the same structures. The reason for beryllium's failure to conform has already been explained in terms of the small size of the beryllium atom (Ch 5.2).

As expected for elements with an extra proton but no new main electron energy level, the elements of Group 2 have smaller atomic volumes than their immediate predecessors in Group 1. This and the possession of *two* electrons in the outer energy level largely determine the properties of the elements of Group 2.

Some physical properties of the s-block elements are given in *Table 6.2.1*.

Element	Atomic number Z	Electron structure	Electronegativity (Pauling scale)	Atomic radius /nm	m.p. /°C	Molar first ionisation energy /kJ mol⁻¹	Density at 25 °C /g cm⁻³
Group 1							
Lithium	3	2.1	1.0	0.157	181	+520	0.53
Sodium	11	2.8.1	0.9	0.191	98	+496	0.97
Potassium	19	2.8.8.1	0.8	0.235	63	+419	0.86
Rubidium	37	2.8.18.8.1	0.8	0.250	39	+403	1.53
Caesium	55	2.8.18.18.8.1	0.7	0.272	29	+376	1.88
Group 2							
Beryllium	4	2.2	1.5	0.112	1278	+900	1.85
Magnesium	12	2.8.2	1.2	0.160	649	+738	1.74
Calcium	20	2.8.8.2	1.0	0.197	839	+590	1.54
Strontium	38	2.8.18.8.2	1.0	0.215	769	+550	2.60
Barium	56	2.8.18.18.8.2	0.9	0.224	725	+503	3.51

Table 6.2.1 Physical properties of the s-block elements

SELF ASSESSMENT QUESTIONS 6.2

1 Write the electron configurations of the following s-block elements or ions.
a Na
b Ca
c K⊕
d Be

2 Explain why the potassium ion carries a 1⊕ charge.

3 State how first ionisation energies change with increasing atomic number:
a in the Group 1 elements;
b in the Group 2 elements;
c in neighbouring atoms in the same Period in the s-block.

4 The energy change for the process

$$Rb(g) \longrightarrow Rb^{\oplus}(g) + e^{\ominus}$$

is +403 kJ mol⁻¹.
a What is this energy change called?
b Why is the value lower than for the corresponding process for sodium?
c How does the volume of the Rb⊕ ion compare to the volume of the Rb atom?
d Explain any differences you have mentioned in part (c).

6.3 Flame colours

We have seen (in Ch 1.4) that excited atoms give out radiation as electrons fall back from higher energy levels to lower ones. This produces, for each element, a pattern of lines in the emission spectrum. This pattern is unique to each element, and can be used to identify it.

Because of their electron configurations – with the one or two electrons in the s-sub-level of the outermost energy level comparatively easy to excite – part of the emission spectrum of each of most of the s-block elements is in the visible region. It is this fact which is used in the familiar **flame test.**

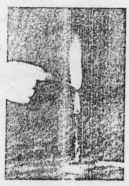
Lithium

Sodium

Potassium

Calcium

Barium

In the flame test, the tip of a clean nichrome or platinum wire is moistened with concentrated hydrochloric acid, and dipped into some powdered compound of the metal so that a tiny amount of the solid sticks to the wire. This is then held to the side of the cone in a roaring Bunsen flame. The volatilisation of a tiny amount of the material is helped by the presence of the hydrochloric acid, and it is excitation of the elements in the gas phase which gives the characteristic flame colour.

All the useful flame colours are given in Ch 30.3. Those of the s-block elements are shown in *Table 6.3.1.*

All the Group 1 elements give a characteristic flame. Indeed, Robert Bunsen and Gustav Kirchoff discovered rubidium (named for its red flame colour) and caesium (named for its blue colour) through flame tests on their compounds in 1861. Of the Group 2 elements, beryllium and magnesium give no flame colour. Sodium vapour excited *electrically* is used in many street lamps; some compounds of the other elements are used to colour fireworks, particularly barium and strontium.

The potassium colour is quite faint and often invisible in the intense yellow colour caused by sodium impurities. This colour should be observed through cobalt blue glass, which absorbs the yellow (sodium) light but not the lilac (potassium) colour. The sodium colour is so intense that a flame test can detect as little as one nanogram (1×10^{-9}g).

Element	Colour
Group 1	
Lithium	Rich scarlet
Sodium	Yellow/gold
Potassium	Lilac/violet
Rubidium	Red
Caesium	Blue
Group 2	
Calcium	Brick-red
Strontium	Crimson
Barium	Yellow-green

Table 6.3.1

157

1 When a very small piece of sodium is dropped on to the surface of some water, there is a vigorous reaction and, under the right conditions, the hydrogen gas evolved ignites and burns with a yellow flame, characteristic of the sodium flame test. The resulting solution is strongly alkaline.

   a Write a word equation for the reaction of sodium and water.
   b Insert the formulae of the reactants and products.
   c Balance the equation for the reaction of sodium and water and include the state symbols.
   d Write the electron configuration for sodium.
   e Explain, using an energy level diagram, why sodium should give a characteristic flame colour.

2 Calcium reacts readily with oxygen and burns with a brick-red colour.
   a Explain why the brick-red colour seen in this reaction gives a series of lines in the visible spectrum.
   b Write a word equation for this reaction.
   c Now write this equation as a balanced equation.
   d Explain which element has been oxidised and which element reduced in the reaction.

6.4 The reactions of Group 1 elements and compounds

Compounds of sodium and potassium are very common; those of lithium, rubidium and caesium are much rarer, and francium is too radioactive to exist for any time in nature. Francium was not discovered until 1939.

The elements of Group 1 are often called the **alkali metals**, because of their reaction with water to give an alkaline solution. Almost all of the reactions of Group 1 metals involve loss of the one electron in the outer energy level, which is shielded from the nucleus by completed energy levels. The first ionisation energies are very low, and these outer electrons are lost more easily the more shielded they are.

In other words, *reactivity increases with increasing atomic number within the Group*. All Group 1 metals are far too reactive to be found in nature other than in ionic compounds with non-metals. Lithium, sodium and potassium are kept in oil, in airtight containers, to avoid being in contact with air or moisture.

The $Li^{\oplus}$ ion is so tiny (see *Figure 6.4.1*) that it has a high charge density, and so considerable polarising power (see Ch 2.5). Consequently, some lithium compounds possess less ionic character than is usual for Group 1, and behave oddly.

Sodium in oil

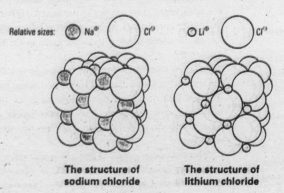

Relative sizes: $Na^{\oplus}$ $Cl^{\ominus}$ $Li^{\oplus}$ $Cl^{\ominus}$

The structure of sodium chloride   **The structure of lithium chloride**

Figure 6.4.1

The main similarities between the elements of Group 1

- In all their compounds the elements are in an oxidation state of +1 (see Ch 4.3). All atoms of Group 1 elements have one electron in the outer energy level which is easily lost to a non-metal atom, so that the Group 1 atom achieves the very stable noble gas configuration as the monopositive ion.

$$\text{e.g.} \qquad \begin{array}{ccccc} Na & \longrightarrow & Na^{\oplus} & + & e^{\ominus} \\ 2.8.1 & & 2.8 & & \text{donated to} \\ & & & & \text{e.g. chlorine} \end{array}$$

- All the Group 1 metals burn fiercely when heated in air. The ordinary oxides are very strongly basic, and they react violently with water to form hydroxides. All the hydroxides are freely soluble in water to give alkalis.

$$\text{e.g.} \qquad Na_2O(s) + H_2O(l) \xrightarrow{\text{room temperature}} 2NaOH(aq)$$

As well as the ordinary oxide, Na_2O, sodium can form a peroxide, Na_2O_2. This contains the peroxide ion, $O_2^{2\ominus}$. Sodium peroxide is a dangerously powerful oxidising agent. It can also react with carbon dioxide to produce oxygen, which makes it potentially useful in confined spaces, e.g. submarines. (It is mentioned in Jules Verne's *20000 Leagues Under the Sea*!)

$$2Na_2O_2(s) + 2CO_2(g) \longrightarrow 2Na_2CO_3(s) + O_2(g)$$

As well as the ordinary oxide, K_2O, and peroxide, K_2O_2, potassium can form a yellow superoxide, KO_2. The superoxide ion, $O_2^{\ominus}$ is essentially an oxygen molecule with an extra electron. Of lithium, sodium and potassium only potassium has an ion large enough to form a reasonably stable ionic lattice with $O_2^{\ominus}$ ions. If potassium is taken out of storage in oil for a piece to be cut off, it must always be treated with *great care* if it has a yellow coating. Explosions have been caused by a blunt knife forcing the superoxide into the body of the metal.

- All the Group 1 metals react with water to give hydrogen gas and the hydroxide of the metal.

$$\text{e.g.} \quad 2Na(s) + 2H_2O(l) \xrightarrow{\text{room temperature}} 2NaOH(aq) + H_2(g)$$

The violence of the reaction increases with increasing atomic number. A small piece of sodium placed on water melts to a silver coloured globule, dashes about the surface fizzing, and may explode with a yellow flash. Potassium appears to catch fire instantly; it is in fact the hydrogen given off by the reaction which burns. The potassium globules crackle, spit, and finally explode.

Rubidium and caesium explode at once, with no messing about; and since they are denser than water, a depth charge effect is produced!

- All form salt-like hydrides, e.g. $Na^{\oplus}H^{\ominus}$, which give hydrogen when water is added.

$$\text{e.g.} \quad NaH(s) + H_2O(l) \xrightarrow{\text{room temperature}} NaOH(aq) + H_2(g)$$

- All burn fiercely when heated in chlorine, to give ionic chlorides.

Potassium in oil

$$\text{e.g.} \qquad 2Na(s) + Cl_2(g) \xrightarrow{\text{heat}} 2Na^{\oplus}Cl^{\ominus}(s)$$

Group 1 reactions – summary

Oxidation state, +1

With oxygen

$$4M(s) + O_2(g) \xrightarrow{heat} 2M_2O(s)$$

With water

$$2M(s) + 2H_2O(l) \xrightarrow[temperature]{room} 2MOH(aq) + H_2(g)$$

With hydrogen

$$2M(s) + H_2(g) \xrightarrow{heat} 2MH(s)$$

Hydrides with water

$$MH(s) + H_2O(l) \xrightarrow[temperature]{room} MOH(aq) + H_2(g)$$

With chlorine

$$2M(s) + Cl_2(g) \xrightarrow{heat} 2MCl_2(s)$$

All common Group I salts are soluble in water, except for a few lithium salts.

(Note: in chemical formulae, M refers to any Group 1 metal).

- Except for a few salts of lithium (e.g. lithium carbonate, Li_2CO_3, and lithium fluoride, LiF), all the common salts of Group 1 metals are soluble in water (*Figure 6.4.2*).

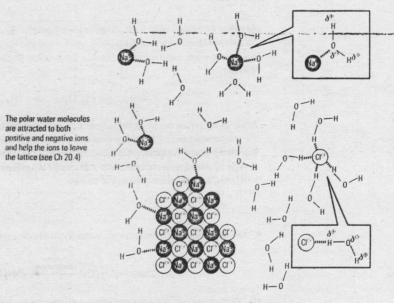

The polar water molecules are attracted to both positive and negative ions and help the ions to leave the lattice (see Ch 20.4)

Figure 6.4.2 Dissolving sodium chloride in water

A very useful slogan to remember (and it is *very* nearly true!) is: **all salts of sodium and potassium are soluble** (see Ch 3.4).

The high solubility of sodium and potassium salts means that the ores of the alkali metals are usually found as salt deposits left by the evaporation of an inland sea (e.g. the deposits at Stassfurt in Germany; and the rock salt beds mined by ICI in Cheshire, used for the treatment of icy roads). The sea is salty largely because of the alkali metal compounds washed out of soil by rain and carried down by rivers over many millions of years. Ordinary seawater contains about 3% by mass of sodium chloride. In the Dead Sea, however, which is the lowest point on the Earth's surface, the situation is different. The River Jordan carries down dissolved material via Tiberias and Galilee to the Dead Sea – and water can only leave the Dead Sea by evaporation. The concentration of salts by mass in the Dead Sea is about 25%; the density is so great that normal swimming is impossible.

A salt mine

The unusual behaviour of the nitrates, carbonates and hydroxides of Group 1

Nearly all metal nitrates break up (decompose) when heated to give the metal oxide plus the gases nitrogen dioxide and oxygen:

$$\text{e.g.} \quad 2Mg(NO_3)_2(s) \xrightarrow{\text{heat}} 2MgO(s) + 4NO_2(g) + O_2(g)$$

Sodium and potassium nitrates behave very differently; they decompose to give the nitrite and oxygen:

$$\text{e.g.} \quad 2NaNO_3(s) \xrightarrow{\text{heat}} 2NaNO_2(s) + O_2(g)$$

Lithium nitrate behaves like magnesium nitrate, and not like its colleagues in Group 1. This is one piece of evidence for the 'diagonal relationship' between lithium and magnesium (see Ch 5.4).

Again, nearly all metal carbonates decompose on heating to give the metal oxide and carbon dioxide, and are very insoluble in water:

$$\text{e.g.} \quad CaCO_3(s) \xrightarrow{\text{heat}} CaO(s) + CO_2(g)$$

The carbonates of sodium and potassium, Na_2CO_3 and K_2CO_3, only decompose very slowly at very high temperatures – they are said to be *thermally stable* – and they are freely soluble in water. Lithium carbonate, Li_2CO_3, however, is once again unusual. It is much less stable to heat than the sodium and potassium carbonates, and it is only slightly soluble. Once more, the lithium compound resembles the magnesium compound.

The hydroxides of the Group 1 metals merely melt when they are heated. The hydroxides of other metals decompose to the oxide plus water.

$$\text{e.g.} \quad Mg(OH)_2(s) \xrightarrow{\text{heat}} MgO(s) + H_2O(l)$$

Summary:

Effect of heat on nitrates	Effect of heat on carbonates	Effect of heat on hydroxides
Nitrate	Carbonate	Hydroxide
↓ heat	(only Na, K, Rb, Cs are stable)	(only Na, K, Rb, Cs are stable)
Oxide + O_2 + NO_2	↓ heat	↓ heat
(For Na, K, Rb, Cs only	Oxide + CO_2	Oxide + H_2O
$\xrightarrow{\text{heat}}$ nitrite + O_2)		

A summary of the anomalous behaviour of lithium

Remember, the basic cause of these differences is that the lithium atom – and, even more so, the lithium ion – is very small. As a result, the lithium ion has relatively high polarising power (see Ch 2.5) and causes large ions such as carbonates to become unstable (see *Figure 6.4.3*).

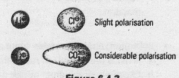

Slight polarisation

Considerable polarisation

Figure 6.4.3

At the same time, the lithium ion can fit in strong lattices with relatively small anions such as fluoride and nitride.

Some of the ways in which lithium differs from the rest of the elements in Group 1 are:

- The fluoride and carbonate are only slightly soluble.
- The nitrate and carbonate behave 'normally' on heating, just like the compounds of magnesium:

$$4LiNO_3(s) \xrightarrow{heat} 2Li_2O(s) + 4NO_2(g) + O_2(g)$$

$$Li_2CO_3(s) \xrightarrow{heat} Li_2O(s) + CO_2(g)$$

and unlike the nitrates of sodium and potassium, which decompose to the nitrite and oxygen, and their carbonates which are almost unaffected by heat.

- Lithium is the only Group 1 metal which can form a simple nitride with nitrogen; again, this behaviour is like that of magnesium:

$$6Li(s) + N_2(g) \xrightarrow{heat} 2Li_3N(s)$$

- lithium iodide, LiI, once again, demonstrates the high polarising power of the $Li^{\oplus}$ ion and the polarisability of the comparatively large $I^{\ominus}$ ion: it has so much covalent character that it is soluble in organic solvents such as methanol and propanone.

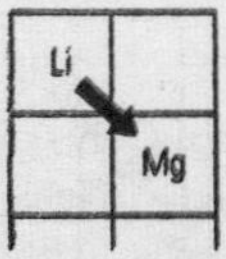

In the Periodic Table, lithium is to the upper left of magnesium (*Figure 6.4.4*); its diagonal relationship with magnesium has been briefly discussed (see Ch 5.4), but the idea is only of limited use.

Figure 6.4.4 Diagonal relationship

SELF-ASSESSMENT QUESTIONS 6.4

1 Write word equations for the following reactions.
   a Burning rubidium in chlorine.
   b Adding potassium hydride to water.
   c Gently heating lithium carbonate.
   d Gently heating sodium carbonate.

2 Write balanced chemical equations for:
   a the reaction of potassium with water;
   b the reaction when lithium nitrate is heated;
   c the burning of caesium in oxygen;
   d the effect of heating potassium nitrate.

3 a Define relative atomic mass.
   b Potassium readily burns in chlorine. What colour might you expect to see during the reaction?
   c Write a balanced equation for the reaction.
   d Draw a dot-and-cross diagram to show the bonding present in potassium chloride.
   e Calculate M_r for potassium chloride.
   f Explain, in terms of the movement of electrons, which element has been oxidised in this reaction.

4 Rubidium nitrate is a stable white solid at room temperature.
   a What is the formula of rubidium nitrate?
   b Explain how rubidium nitrate dissolves in water.
   c What would you *observe* if rubidium nitrate is heated strongly.
   d How would you test any gas that might be given off in this reaction?
   e Write a balanced equation for any reaction in c.

6.5 The reactions of Group 2 elements and compounds

Group 2 elements are often referred to as the 'alkaline earth metals'. (Be careful not to confuse them with Group 1, the 'alkali metals'.)

Beryllium is very much the odd one out. If it existed, the beryllium ion would be very tiny indeed (ionic radius 0.027 nm), and as it would be doubly charged the charge density would be extremely high (*Figure 6.5.1*).

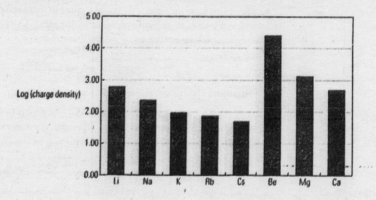

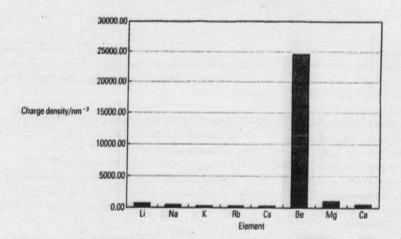

Figure 6.5.1 Charge density of M$^{\oplus}$ and M$^{2\oplus}$ ions

Consequently its ability to drag electrons off negative ions (its polarising power, see Ch 2.5) would be far, far higher than that of lithium. The bonding in lithium compounds is often largely ionic.

The extremely high charge density of the beryllium ion is very evident when charge density is plotted on a graph (*Figure 6.5.1*). By taking a log value of charge density the trend within each group can be seen – as the atom increases in size the charge density of the ion decreases. Beryllium compounds are almost entirely covalent. We shall not discuss beryllium any further here.

At the other end of the group, radium is far too radioactive for general use. The story of its isolation by the Curies and early use in cancer treatment (see Ch 1.2) is one of the great sagas of science, but we shall also not discuss radium here.

The remaining Group 2 elements – magnesium, calcium, strontium and barium – form a very closely related family. As with the Group 1 elements, all of these Group 2 elements are too reactive to be found uncombined in Nature; and, also like Group 1, these metals can only be obtained by electrolysis of molten salts.

The main similarities between magnesium, calcium, strontium and barium

- In all their compounds they are in an oxidation state of $+2$ (see Ch 4.3), which is to be expected as they have two electrons in their outer energy level.
- They burn in air or oxygen to give the 'normal' oxides, of the general formula MO, which react with cold water to give strongly basic hydroxides, $M(OH)_2$. The vigour of the reaction increases from magnesium to barium.

$$\text{e.g.} \qquad CaO(s) + H_2O(l) \xrightarrow{\text{room temperature}} Ca(OH)_2(s)$$

Strontium and barium also form peroxides, MO_2.

- Magnesium reacts with cold water to give hydrogen only very slowly, but reacts rapidly with boiling water. Calcium reacts very rapidly with cold water to give hydrogen and a white deposit of calcium hydroxide. The violence of the reaction increases with strontium and barium.

$$\text{e.g.} \qquad Ca(s) + 2H_2O(l) \xrightarrow{\text{room temperature}} Ca(OH)_2(s) + H_2(g)$$

- All form salt-like hydrides, MH_2, which contain the hydride ion H^-. These give hydrogen when water is added.

$$\text{e.g.} \quad CaH_2(s) + 2H_2O(l) \xrightarrow{\text{room temperature}} Ca(OH)_2(s) + 2H_2(g)$$

The old name for calcium hydride was 'hydrolith'. It is reported that it was carried by Victorian balloonists for replenishing the hydrogen in leaking balloons. The calcium hydroxide was then thrown overboard to lighten the balloon. What the Victorian accounts carefully fail to mention is that to save weight it sometimes wasn't water which was used on the hydrolith!

- The metals are all strongly electropositive, with reactivity increasing from magnesium to barium. All react fiercely when heated in chlorine to give the chloride, MCl_2. Their salts are all essentially ionic although – as expected – the smallest of the ions, Mg^{2+}, has sufficient charge density to cause a slightly covalent character in some of its salts.
- The solubility of the hydroxides *increases* from magnesium to barium, but the solubility of the sulphates *decreases* (*Table 6.5.1*). The reason for this will be discussed in Ch 20.4. Note that limewater, a saturated solution of calcium hydroxide, contains very little calcium hydroxide.

Magnesium sulphate crystals, $MgSO_4.7H_2O$, are known as 'Epsom salts' and are widely used in mild laxatives. Barium compounds are normally extremely poisonous, but barium sulphate is so insoluble that it is used in large quantities – either swallowed or as an enema – to show up detail and provide contrast when X-rays of the alimentary canal are needed.

X-ray of the alimentary canal after taking a barium meal

Compound	Solubility /mol/100 g water at 25 °C
$Mg(OH)_2$	2×10^{-5}
$Ca(OH)_2$	0.0015
$Sr(OH)_2$	0.0034
$Ba(OH)_2$	0.015
$MgSO_4$	0.183
$CaSO_4$	0.0047
$SrSO_4$	7.1×10^{-6}
$BaSO_4$	9.4×10^{-7}

Table 6.5.1 Solubilities of Group 2 hydroxides and sulphates

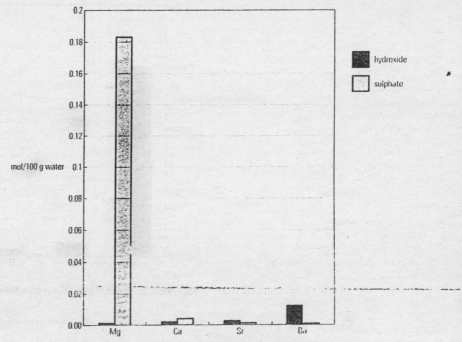

Figure 6.5.2   Solubility of Group 2 hydroxides and sulphates

---

Group 2 reactions – summary

Oxidation state, +2

With oxygen

$$2M(s) + O_2(g) \longrightarrow 2MO(s)$$

With water

$$M(s) + 2H_2O(l) \longrightarrow M(OH)_2 + H_2(g)$$

With chlorine

$$M(s) + Cl_2(g) \longrightarrow MCl_2(s)$$

(Note: in chemical formulae, M refers to any Group 2 metal).

When limestone or chalk (both calcium carbonate, $CaCO_3$) is heated it produces calcium oxide, CaO, and carbon dioxide. The old name for calcium oxide was *quicklime* ('quick' has an ancient meaning of 'lively', as in 'the quick and the dead'). When water is added to quicklime, it swells, steams, becomes very hot and crumbles. The resulting calcium hydroxide, $Ca(OH)_2$ was known as 'slaked lime' or just 'lime'. ('Slaked' because the quicklime had absorbed water, i.e. 'slaked its thirst…'!) Limestone was so called because it produced lime, for use in agriculture (see below).

- All of the Group 2 metal carbonates are insoluble (*Figure 6.5.3*). The usual occurrence of these Group 2 metals in Nature is as their carbonates. Calcium carbonate, $CaCO_3$, occurs as chalk, limestone and marble, and also in coral reefs; it is one of the commonest compounds on Earth. Chalk was formed, often in deposits hundred of metres thick (e.g. the 'white cliffs of Dover' and Beachy Head), from the exoskeletons of small marine organisms. The seas must have persisted in one place for many millions of years for such thick deposits to be found.

Limestone, and the harder metamorphic form marble, are the result of great pressure and high temperature being applied to chalk over long geological periods. Limestone often contains fossils of shells and other larger sea animals; such fossils have been found near the summit of Mount Everest. Dolomite, $MgCO_3.CaCO_3$, is a particularly hard form of limestone and is an excellent rock for climbing.

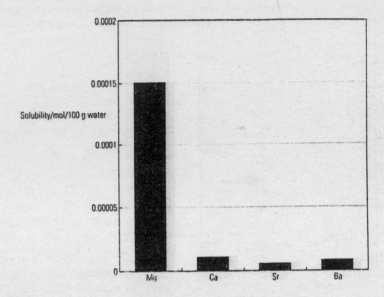

Figure 6.5.3 Solubility of Group 2 carbonates

● The thermal stability of the Group 2 carbonates increases from magnesium carbonate to barium carbonate (*Figure 6.5.4*).

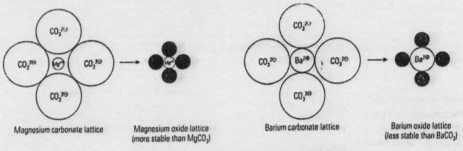

Figure 6.5.4 Thermal stability of carbonates

The small $Mg^{2\oplus}$ ion is capable of polarising and destabilising the carbonate ion; magnesium carbonate decomposes to the oxide and carbon dioxide fairly readily on heating. Calcium carbonate needs strong heating (industrially, a limekiln is used); strontium and barium carbonates require very strong heating.

Q

SELF-ASSESSMENT
QUESTIONS
6.5

1   Write balanced chemical equations for:
   a   strontium reacting with water;
   b   barium reacting with oxygen;
   c   magnesium reacting with chlorine;
   d   barium oxide reacting with water;
   e   the effect of heating calcium nitrate;
   f   the effect of heating barium carbonate;
   g   the effect of heating calcium hydroxide.

2 **a** How do the solubilities of the Group 2 sulphates change with increasing atomic number of the metal? Use this information to explain why, although barium salts are toxic, barium sulphate can be given to people when taking X-rays of the alimentary tract.

  b Does barium react with water more or less vigorously than calcium? Explain your answer and write a balanced equation for the reaction.

  c Write the formula of strontium carbonate and calculate its M_r. Is this compound more or less thermally stable than calcium carbonate?

6.6 The s-block in Nature and industry

No attempt is made in this section to provide a comprehensive coverage of all the naturally and industrially important compounds of the s-block elements. The selection is limited to the most important and to those which are specified in examination syllabuses. This principle will also be followed in the next two chapters.

The biological importance of the s-block elements

Sodium and potassium ions are found in the cells of all animals, and play an essential part in the conduction of impulses along nerve fibres. It is the uneven distribution of $Na^{\oplus}$ and $K^{\oplus}$ ions across nerve cell boundaries that starts off the impulses along nerve fibres. A diet which lacks sodium and potassium leads to collapse and death.

Potassium compounds are widely used in agriculture and horticulture, often mixed with nitrogen and phosphorus compounds in general (NPK) fertilisers. Some lithium compounds are used in medicine for treating certain forms of mental illness, and for gout.

Magnesium is found in chlorophyll, the green pigment in all plants and phytoplankton, which traps energy from the Sun and so enables photosynthesis to occur (*Figure 6.6.1*).

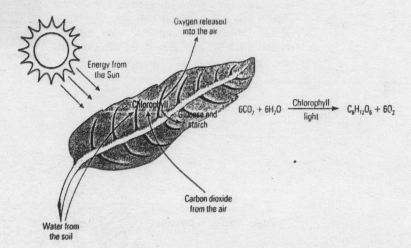

Figure 6.6.1 Photosynthesis

Magnesium is therefore deeply involved in the most important reaction for all living things on Earth, the one which underpins their continued existence. When we eat, we eat either plant material, or animals or fish which in the end depend on plants or phytoplankton for food. Either way, the energy which maintains our bodies and minds and enables all our physical activity comes from the Sun; and magnesium is essential for trapping and fixing that energy.

Calcium phosphate, $Ca_3(PO_4)_2$, is a major component of bones and teeth. Bone contains a helical protein, collagen, 'cemented' together with hydroxyapatite, approximately $Ca_5(PO_4)_3(OH)$. Teeth also contain hydroxyapatite. In the presence of fluoride ions, $F^{\ominus}$, some of the hydroxy groups are replaced forming fluoroapatite.

Sodium fluoride, in small amounts, is added to toothpaste, and to some drinking water supplies in order to encourage the formation of fluoroapatite in the teeth. Fluoroapatite is resistant to acid attack from decaying food. Calcium phosphate also occurs as a mineral, apatite. Calcium phosphate is, as should be obvious from these facts, insoluble in water. Calcium is also involved in muscle contraction, cell division, hormone regulation and the clotting of blood.

Hard water

'Hard water' is water which does not lather easily with soap. It is caused by calcium and magnesium ions dissolved in the water. When soap is mixed with hard water an insoluble scum is formed. This not only wastes expensive soap but would be an obvious nuisance in baths, laundries, etc.

As rain water trickles through soil and rock it dissolves some calcium or magnesium compounds, e.g. the sulphates. These compounds cause so-called 'permanent' hardness – i.e. hardness which cannot be cured by boiling the water. Much more interesting is 'temporary' hardness, which can be removed by boiling.

Rainwater falling through air dissolves some carbon dioxide. This very dilute and unstable solution of 'carbonic acid' dissolves chalk or limestone, but only extremely slowly. In limestone areas this can give rise to spectacular caves and gorges, and to what geographers call 'karst' scenery. When a film of this solution runs down rock and some water evaporates, calcium carbonate is formed once again. It is this reaction which makes stalactites and stalagmites and other beautiful formations in caves (*Figure 6.6.2*).

Typical 'karst' scenery in the Ikola river gorge, Dalmatia, Croatia

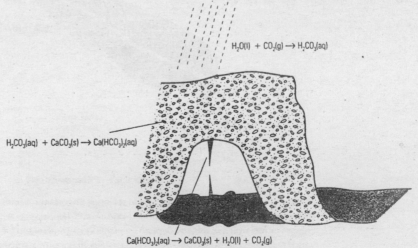

$$H_2O(l) + CO_2(g) \rightarrow H_2CO_3(aq)$$

$$H_2CO_3(aq) + CaCO_3(s) \rightarrow Ca(HCO_3)_2(aq)$$

$$Ca(HCO_3)_2(aq) \rightarrow CaCO_3(s) + H_2O(l) + CO_2(g)$$

Figure 6.6.2 Stalactite formation

A quicker way of re-forming the calcium carbonate is by boiling the solution. If you live in a chalk or limestone area, look inside your kettle (unless you have a 'water softener' attached to your water supply!). The formation of this *limescale* can cause havoc in boilers and heating systems; blocked pipes in boilers have caused fatal explosions.

$$H_2O(l) + CO_2(g) \rightleftharpoons H_2CO_3(aq)$$
very dilute carbonic acid (unstable)

$$H_2CO_3(aq) + CaCO_3(s) \underset{\text{evaporation or boil}}{\overset{\text{room temperature}}{\rightleftharpoons}} Ca(HCO_3)_2(aq)$$

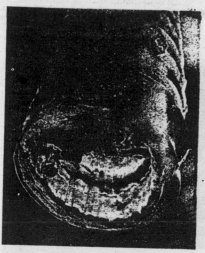

Scale inside a pipe

> Some chemical reactions can go in either direction – they are **reversible** (see Ch 13.1). In such reactions a double arrow $\rightleftharpoons$ is used to show that, by changing the conditions, the reaction can be 'made' to go in either direction.

Note that, like carbonic acid itself, calcium hydrogen carbonate can only exist in solution and breaks down when the solution is warmed or allowed to evaporate.

An obvious way of 'softening' water is to distil it. This is far too expensive for household use. A common method is to add *washing soda* (hydrated sodium carbonate, $Na_2CO_3.10H_2O$); this removes calcium and magnesium ions as their insoluble carbonates.

$$\text{e.g.} \qquad Ca^{2\oplus}(aq) + CO_3{}^{2\ominus}(aq) \longrightarrow CaCO_3(s)$$
calcium carbonate precipitates

The offending calcium or magnesium ions are exchanged for sodium ions:

$$\text{Overall:} \qquad Ca(HCO_3)_2(aq) + Na_2CO_3(aq) \longrightarrow 2NaHCO_3(aq) + CaCO_3(s)$$

Because virtually all sodium salts are soluble in water, they do not make water 'hard'. Soap itself is a sodium salt – sodium stearate, $C_{17}H_{35}COO^\ominus Na^\oplus$.

Household water supply pipes can now be fitted with rechargeable or replaceable canisters containing ion-exchange materials in order to soften the water on a continuous basis.

Uses of Group 1 elements

Because the alkali metals are so reactive, the use of the elements themselves is limited. Sodium is the only one used in large amounts.

- It is used in the familiar yellow sodium vapour street lamps; the high-pressure version gives a light which is much less yellow. The light is emitted from vaporised sodium excited by high voltage electricity.
- A sodium/lead alloy is used in making tetraethyl-lead(IV), $Pb(C_2H_5)_4$. For many years this was added to petrol to improve its octane rating (see Ch 15.4), but because its combustion products are toxic – especially for growing children – and 'poison' the catalysts in vehicle exhaust systems, it is now banned as a petrol additive in countries of the European Union and in many other places.

Uses of compounds of Group 1 elements

We shall mention only a few sodium compounds (*Figure 6.6.3*).

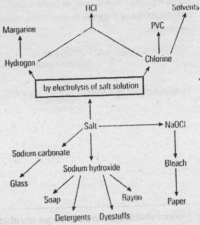

Figure 6.6.3   Uses of sodium compounds

Sodium chloride, NaCl, 'common salt'

This occurs in seawater, and in the deposits of rock salt which have formed by evaporation of inland seas long ago. Salt solution, known as *brine*, is the raw material for making the vitally important materials sodium hydroxide (see below) and chlorine (see Ch 8.8). The salt can be extracted from below ground by *solution mining* – holes are drilled down into the rock salt, water is pumped down, and brine is forced up to the surface.

Rock salt is also mined for spreading on roads to clear them from ice and snow.

Purified sodium chloride ('salt') is used in foodstuffs for flavour and as a preservative. Molten sodium chloride is electrolysed to give sodium metal and chlorine.

Solution mining

Sodium carbonate, Na_2CO_3

As mentioned above, hydrated sodium carbonate, $Na_2CO_3.10H_2O$ is used as a household water softener (*washing soda*). Ordinary glass is made by melting anhydrous (i.e. with no water in the solid) sodium carbonate, Na_2CO_3, with sand (impure silicon dioxide, SiO_2) and calcium carbonate $CaCO_3$.

Some communities bake 'soda bread' or 'soda scones', using only sodium hydrogencarbonate with no tartaric acid or citric acid. These are enjoyed by many people, but others find the taste resulting from the sodium carbonate which is formed far too strong.

Natural fats and oils consist largely of esters. Many oven cleaners contain 'caustic alkali', in the form of sodium hydroxide, to destroy fat deposits. If you rub your fingers with a little dilute alkali, it will feel soapy. This is because the alkali is hydrolysing the fat in your skin to form a soap. Given a chance it will start to hydrolyse the proteins as well, causing severe damage. That is why strong alkalis are called 'caustic' (from the Greek word for 'burnt'). As with acids, any strong alkali spilled on flesh or clothes should be washed off immediately with plenty of cold water. Eyes are especially vulnerable to caustic alkalis, and oven cleaners should not be used without eye protection.

Sodium hydrogencarbonate, NaHCO₃

This is known to cooks as 'bicarbonate of soda' or 'baking soda'. When heated, it decomposes to give off carbon dioxide and water. It is therefore used in baking powder to 'enable' cakes, scones, etc, to 'rise'.

$$2NaHCO_3(s) \xrightarrow{heat} Na_2CO_3(s) + H_2O(g) + CO_2(g)$$

However, the sodium carbonate formed would give an odd flavour. So, in baking powder a solid weak acid, such as tartaric acid or citric acid, is mixed with sodium hydrogencarbonate. The acid reacts with the sodium carbonate which is formed when the mixture is heated; this reaction produces more carbon dioxide – thus making the baking powder more efficient – and leaves an almost tasteless compound, sodium tartrate or sodium citrate.

Sodium hydroxide, NaOH

This very important industrial chemical is made, together with chlorine, by electrolysis of a saturated solution of sodium chloride, 'brine'. The brine is usually obtained by solution mining of rock salt (see above). A solution of sodium hydroxide, known as 'caustic soda', is a very powerful alkali which we shall often meet in this book. It reacts directly with halogens to give salts (see Ch 8.6). It is also neutralised by acids to give salts (see Ch 10.2 and Ch 23.1). It is widely used in organic chemistry for **hydrolysis.**

One example of hydrolysis is the breaking up of naturally-occurring compounds called esters to give *soaps* (see Ch 27.10).

A word which ends in *-lysis* means that something is being broken up. In *electrolysis*, the breaking up is done by electricity; in **hydrolysis**, water is the agent. In organic chemistry, hydrolysis involves the breaking of a bond in such a way that the products contains elements of an added molecule of water, H_2O: usually – OH goes with one product and – H with the other.

Sodium chlorate(I) ('sodium hypochlorite'), NaOCl

This compound is a powerful oxidising agent. A solution of it is used as a sterilising agent, bleach and disinfectant. It is made by allowing the sodium hydroxide and chlorine produced during electrolysis of brine to react together.

$$2NaOH(aq) + Cl_2(g) \longrightarrow NaOCl(aq) + NaCl(aq) + H_2O(l)$$

Sodium hydroxide and sodium chlorate(I)

The manufacture of these two sodium compounds, together with chlorine and hydrogen, by electrolysis of brine in a diaphragm cell is described in Ch 8.8.

Uses of Group 2 elements

Magnesium metal is used in low density alloys for making aircraft and specialist vehicles, in the extraction of titanium, and also for removing sulphur from molten iron during steel making. It is also used in fireworks and flares.

Uses of compounds of Group 2 metals

Magnesium oxide, MgO

This has an exceptionally high melting point (2850 °C) and is therefore used as a refractory lining in furnaces.

'Refractory' means stubborn, rebellious, hard to melt; or, as now, a substance very resistant to heat.

Calcium carbonate, CaCO₃

Various types of limestone and marble are widely used as building materials. They are, however, liable to attack by 'acid rain'. Limestone is the starting point for many other materials (*Figure 6.6.4*).

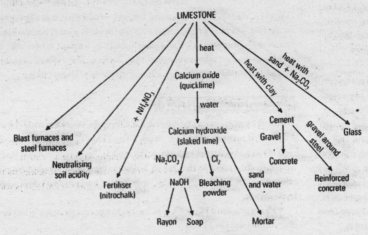

Figure 6.6.4   Uses of limestone

Michelangelo's David – carved from a single block of marble (calcium carbonate)

As discussed in Ch 6.5, strong heating of limestone gives calcium oxide, CaO ('quicklime') which reacts violently with water to give calcium hydroxide, $Ca(OH)_2$ ('slaked lime', or just 'lime').

$$CaCO_3(s) \xrightarrow{heat} CaO(s) + CO_2(g)$$

$$CaO(s) + H_2O(l) \longrightarrow Ca(OH)_2(s)$$

'Lime' is used extensively, especially in moorland areas, to correct soil pH. Many upland soils are acidic; the lime neutralises the acid. Some reactions of chalk and limestone are summarised in *Figure 6.6.5*.

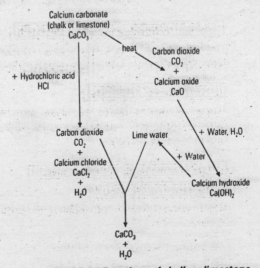

Figure 6.6.5   Reactions of chalk or limestone

Mortar, cement and concrete

These materials are fundamental to the construction industries and are based on calcium compounds. The detailed chemistry of the setting and hardening processes is very complex, and we shall not discuss it.

Mortar consists of a paste of slaked lime (calcium hydroxide), 1 part, with sand (silica), 4 parts, and water. It slowly sets and hardens.

Cement is made by heating a mixture of limestone and clay in a sloping rotating cylinder which is heated on the inside by burning coal-dust in a blast of air. Lumps of 'clinker' are formed, which are ground to a powder. This is cement. When it is thoroughly mixed with sand and water, the slurry so formed sets in a few hours.

A mixture of cement, sand and water with gravel or well-broken stone is called concrete. When mixed, it can be poured into moulds. It is usually vibrated or 'puddled' to make sure there are no cavities. It sets rock-hard in a day or two. It is not only used for buildings or roads; the moulds can be any shape. Boats or the roof of Sydney Opera House, for example, have been made of concrete. For 'reinforced concrete' ordinary concrete is poured around frameworks of steel rods.

SELF-ASSESSMENT QUESTIONS 6.6

1 **a** Read the section on 'the biological importance of the s-block elements' and, as you read it, write the important words or ideas on a sheet of paper.

   b Summarise the biological importance of the s-block using the words and ideas you have already written down.

2 Chalk, calcium carbonate, is found in many areas as a white rock which is insoluble in water. In some hard water areas it is the chalk that is said to be responsible for the temporary hardness in the water.

   a Write a balanced equation to show how chalk breaks up when heated.

   b Write a balanced equation to show how the solid product in the reaction in part (a) reacts with water. What is the pH of the resulting solution?

   c What is meant by hard water?

   d How does the chalk get into the water supply to cause temporary hardness.

Summary

Down the group		Group 1 (Li and Be are anomalous)	Group 2
	Li Be	$4Na(s) + O_2(g) \longrightarrow 2Na_2O(s)$	$2Ca(s) + O_2(g) \longrightarrow 2CaO(s)$
Lose electrons more easily		$2Na(s) + 2H_2O(l) \longrightarrow 2NaOH(aq) + H_2(g)$	$Ca(s) + 2H_2O(l) \longrightarrow Ca(OH)_2(s) + H_2(g)$
Increase size of atoms	Na Mg	$2Na(s) + Cl_2(g) \longrightarrow 2NaCl(s)$	$Ca(s) + Cl_2(g) \longrightarrow CaCl_2(s)$
		$2NaNO_3(s) \longrightarrow 2NaNO_2(s) + O_2(g)$	$Ca(NO_3)_2(s) \longrightarrow 2CaO(s) + 4NO_2(g) + O_2(g)$
First ionisation energy decreases	K Ca	$Na_2CO_3(s) \longrightarrow$ no reaction	$CaCO_3(s) \longrightarrow CaO(s) + CO_2(g)$
		$NaOH(s) \longrightarrow$ no reaction	$Ca(OH)_2(s) \longrightarrow CaO(s) + H_2O(l)$
Electronegativity decreases	Rb Sr		
	Cs Ba		

In this chapter the following new ideas have been covered. *Have you understood and learned them?* Could you explain them clearly to a friend? The page references will help.

- The s-block of the Periodic Table consists of the elements in Groups 1 and 2.   p152
- The first members of these Groups, lithium and beryllium, behave differently from the other members of their Groups.   p152
- This difference in behaviour can be explained by the small size of the atoms of lithium and beryllium.   p152
- There are trends in physical properties and chemical properties within each Group; these trends can be explained in terms of atomic size and electron structure.   p153
- Most of the s-block elements give characteristic colours to a Bunsen flame; these colours can be used in analysis.   p156
- Except for those of lithium, the nitrates and carbonates of Group 1 elements behave unusually when heated.   p161
- Several s-block elements, and especially magnesium, play vital roles in living systems.   p167
- Some compounds of the s-block are of major industrial importance.   p170

EXAMINATION QUESTIONS 6

1 a Write the electron configurations of sodium and magnesium.
  b Sodium forms an ion, $Na^{\oplus}$, and magnesium, $Mg^{2\oplus}$. These ions are said to be isoelectronic. What does *isoelectronic* mean?
  c Write equations to show the first molar ionisation energy for sodium and the second molar ionisation energy for magnesium.
  d The ionic radii of $Na^{\oplus}$ and $Mg^{2\oplus}$ are 0.102 nm and 0.072 nm respectively. Why should these ionic radii be so very different?

2

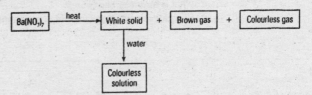

  a Complete the electron configuration of barium:
    $1s^2 2s^2 2p^6 3s^2 3p^6 3d^{10} 4s^2 4p^6 4d^{10} 5s^2 5p^6 \ldots\ldots$
  b Write the formulae of the white solid, the brown gas and the colourless gas formed in the reaction.
  c Why should barium nitrate decompose at a higher temperature than magnesium nitrate?
  d What would you expect the approximate pH of the colourless solution to be?

3 The tests on a number of s-block elements and their compounds are summarised in the table. Suggest, with reasons, possible identities for the metals A to D using this information.

	Reaction with H$_2$O	Formula of chloride	Solubility of carbonate in water	Effect of heat on nitrate
A	Violent	ACl	Soluble	Nitrite + O$_2$ formed
B	Rapid	BCl$_2$	Insoluble	all produce a brown gas and oxygen
C	Rapid	CCl	Insoluble	
D	Very rapid	DCl$_2$	Insoluble	

Do you remember ...?

- How **atomic radius** increases down a Group. (Ch 1.6)
- **Ionic bonding**. (Ch 2.3)
- **Covalent bonding**. (Ch 2.4)
- Electron **energy levels** and **sub-levels**. (Ch 1.4)
- The meaning of **oxidation state** and the rules for assigning oxidation states. (Ch 4.3)
- The general trends in **physical properties** down a Group. (Ch 5:2)
- The general trends in **chemical properties** down a Group. (Ch 5.4)
- The '**inert pair**' effect. (Ch 5.4)
- The general properties of **oxides** and **chlorides**. (Ch 5.5)
- Why molecules have particular **shapes**. (Ch 2.6)
- **Giant structures**. (Ch 3.2)
- **Molecular structures**. (Ch 3.7)

7.1 **Introduction to Group 4**
7.2 **Physical properties**
7.3 **Chemical properties; chlorides, oxides, hydrides**
7.4 **Some special aspects of carbon chemistry**

7
the periodic table: group 4

7.1 Introduction to Group 4

Group 4 is in the p-block of the Periodic Table. It consists of the elements carbon, silicon, germanium, tin and lead; all these have four electrons in the outermost electron energy level, with the general electron configuration ns^2np^2.

The Group clearly shows an increasing tendency towards metallic behaviour as atomic number increases. Carbon is an obvious non-metal (although graphite conducts electricity) with its most common oxide, CO_2, being acidic. Silicon and germanium show semiconductor behaviour (see Ch 3.6); they are **metalloids**. Tin and lead are good conductors, and fairly normal silvery-grey metals.

The reason for this increasing metallic character down a Group was discussed in Ch 5.4. Because there is an increase down the Group in both atomic radius *and* the shielding from the attraction of the nucleus caused by completed electron energy levels, electrons can be removed increasingly easily. In other words, there is an increase in *metallic* (i.e. electropositive) character.

From their outer electron configuration, ns^2np^2, the expected oxidation state of the Group 4 elements is +4. All of them, however, also form compounds in which they are in the +2 state. (Carbon and silicon, as non-metals, can also have negative oxidation states, as low as −4.) The +2 oxidation state becomes more stable as atomic number increases down the Group. Carbon monoxide, CO, in which carbon is in the +2 state, is a reducing agent and is easily oxidised to carbon dioxide, CO_2, where carbon is in the +4 state, e.g.

> Massive metal-extraction industries, including iron-making using blast furnaces, are based on the reducing power of carbon monoxide.

$$Fe_2O_3(s) + 3CO(g) \longrightarrow 2Fe(s) + 3CO_2(g)$$

Even tin(II) is a reducing agent.

On the other hand, lead(IV) oxide (PbO_2) is a very powerful oxidising agent. This can be explained in terms of the so-called 'inert pair' effect (Ch 5.4). Of the s^2p^2 outer electrons, the pair of s-electrons can penetrate closer to the nucleus (see Ch 1.4) and so is less likely to become involved in bonding.

7.2 Physical properties

Carbon

The giant structures of the two common forms of carbon (diamond and graphite) have already been discussed (Ch 3.2). Their melting points are immensely high, above 3500 °C (*Figure 7.2.1*).

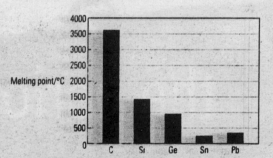

Figure 7.2.1   Melting points of Group 4 elements

Graphite is the only form of a non-metal element which conducts electricity well, and under normal conditions can act as a lubricant.

Diamond, on the other hand, is a non-conductor and the hardest known natural material. Because of its very high refractive index, diamond appears to sparkle when cut into highly faceted shapes. (Some people find this irresistible!) Impure diamonds can now be synthesised using very high temperatures and pressures (2000 °C and 5.5×10^5 atmospheres pressure).

Because it both conducts electricity and is a lubricant, graphite is used in almost all electrical machines which involve rotations – dynamos, motors, alternators, etc. Diamond's major industrial use is in drills for rock and other hard materials, and also in cutting glass.

Silicon

Silicon occurs in two forms: amorphous or 'glassy' ('without shape', i.e. non-crystalline – see *Figure 7.2.2*) and crystalline. The melting point of the crystalline form is 1410 °C, which points to a giant structure. This is essentially a giant atomic structure, but crystalline silicon is a semiconductor (see Ch 3.6) and is best considered as a metalloid. For making computer chips, silicon is rigorously purified, and then precisely controlled tiny amounts of impurities are added to give it the necessary properties. Huge numbers of chips are used in computers and control devices. Silicon can also be used in the direct conversion of sunlight – solar energy – to electricity (see Ch 3.8); cheaper glassy or amorphous silicon wafers are usually used.

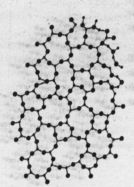

**Figure 7.2.2
Glassy silicon**

The purification of silicon uses a method known as *zone melting*. This is based on the fact that when a liquid containing a small amount of impurity starts to freeze, the impurity tends to stay in the liquid. (This is why people who walk to the North Pole have to carry fuel but not water. To make a drink, all they have to do is chip some of the sea ice and melt it. If the sea was reasonably calm when it froze, the ice will contain very little salt.)

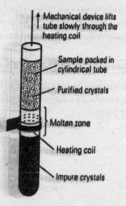

Mechanical device lifts tube slowly through the heating coil

Sample packed in cylindrical tube

Purified crystals

Molten zone

Heating coil

Impure crystals

A rod of impure silicon is passed very slowly through a very intense electrical furnace, known as an eddy current furnace. The part of the rod which is in the furnace at any moment is largely molten. As it passes through, the impurities tend to stay in the molten zone. When the zone gets to the end of the rod it is allowed to solidify and is sawn off. This process is repeated several times until the required degree of purity is reached. The rod is now much shorter – but the silicon is as pure a material as any industrial process can achieve.

Germanium

Look at the Periodic Table: how many other elements are named after countries, states, or cities and towns? And, of course, people?

Germanium is a grey metalloid, melting point 937 °C. It is also often used in semiconductor devices. When Mendeléev devised his Periodic Table, germanium was unknown. He named the missing element *eka-silicon*, left a gap for it in the Periodic Table, and predicted its physical and chemical properties (see Ch 5.1). It was first isolated by C A Winkler at Freiberg in Germany in 1886, who named the new element in honour of his country.

You can tell an old window by looking through it at an angle and moving your head from side to side. Because the glass is not flat, you will notice distortion of the objects on the other side of the window. Old-style plate glass was made by blowing large cylinders of glass, re-softening them, cutting the cylinder lengthways, and then rolling the glass out flat. Except that it never was quite flat!

Lead used for roofing

A *plumber* used to be someone who worked with lead. Why 'plumber'? Because the Romans know and used lead a great deal, and the Latin word for it is *plumbum*.

Tin

Tin has two polymorphic forms. Ordinary shiny white tin can collapse to a grey powder below about 13 °C, with the rate being fastest at −48 °C (see Ch 3.5 for two historical stories related to this fact). The melting point of tin is low, only 232 °C. A band of sheet steel can be passed through a bath of molten tin to become coated with a thin layer of tin on both sides. This is *tin plate*, which is used in very large amounts for canning food ('tin cans').

The modern method of making glass ('float glass') involves pouring molten glass on to a bath of molten tin. The glass cools and solidifies while the tin is still liquid. This means that both of the glass surfaces will be almost perfectly flat.

Tin is used in several important alloys, including solder (tin, lead and bismuth) and bronze (tin and copper).

Lead

Lead also has a low melting point, 328 °C, but a high density. Both tin and lead (and mercury) have unusually low melting points which are difficult to explain using simple metallic bonding theory. Lead is also rather soft and pliable. Being so malleable and also chemically fairly unreactive, lead sheeting was used for roofing (look at the roofs of old churches or cathedrals) and also for piping.

Lead water piping in soft water areas – i.e. those in which water contains only small amounts of dissolved calcium and magnesium compounds (see Ch 6.6) – very slowly dissolves. Lead poisoning is cumulative; once lead gets into the body it is hard to get rid of it. Consequently, people who lived in such areas tended to suffer from lead poisoning. Lead pipes should not be used for drinking water.

Lead is used in car batteries (see Ch 24.7), in several alloys (e.g. solder), and in making tetraethyl-lead(IV), which for many years was used as an additive (see Ch 6.6) to increase the octane rating of petrol. Because of its high density lead is used in weights, clock pendulums, plumb bobs, etc. It absorbs X-rays, and lead aprons and lead glass are used to shield hospital radiographers.

Summary of physical properties

The physical properties of the Group 4 elements are summed up in *Table 7.2.1*. As usual, there are one or two anomalies in the data, but in general, the expected trends are well shown (*Figures 7.2.3* and *7.2.4*).

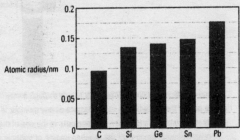

Figure 7.2.3   Atomic radius of Group 4 elements

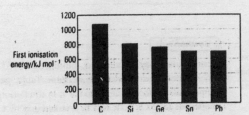

Figure 7.2.4   First ionisation energy of Group 4 elements

Element	Atomic number Z	Electron structure	Electronegativity (Pauling scale)	Atomic radius /nm	m.p. /°C	Molar first ionisation energy/kJ mol^{-1}	Density at 25 °C /g cm^{-1}	Electrical resistivity* /Ω m (20 °C)
Carbon (graphite) (diamond)	6	2.4	2.5	0.092	~ 3675 (sub) >3550	+1086	~2.25 3.51	1.4×10^{-5} 10.0×10^{11}
Silicon	14	2.8.4	1.8	0.132	1410	+789	2.33	0.001
Germanium	32	2.8.18.4	1.8	0.139	937	+762	5.35	0.46
Tin (white) (grey)	50	2.8.18.18.4	1.8	0.158	732	+709	7.28 5.75	1.1×10^{-7}
Lead	82	2.8.18.32.18.4	1.8	0.175	328	+716	11.34	2.1×10^{-7}

*Note. a lower resistivity value means a higher conductivity.

Table 7.2.1   Physical properties of Group 4 elements

SELF-ASSESSMENT QUESTIONS 7.2

1. a Sketch a graph to show how the atomic volume of the atom changes in Group 4 as the atomic number of the element increases. Explain the underlying trend in your graph.
   b How would you expect the size of the Pb^{2+} ion to compare with the size of the lead atom?
   c Explain why the first ionisation energies of the Group 4 elements generally decrease as the atomic number of the element increases. Write equations to represent the first and second ionisation energies of lead.
   d Considering the density of lead, would you expect the structure of lead to be an open or close-packed structure? Explain your answer.

7.3   Chemical properties: chlorides, oxides, hydrides

Reactions of the elements with water, air, acids and alkalis

Carbon

Graphite is more reactive than diamond, because it has a more accessible structure (*Figure 7.3.1*).

Chemical attack usually occurs initially between the planes of carbon atoms. Ordinary forms of carbon (charcoal, etc) consist of microcrystals of graphite.

Graphite will ignite if strongly heated in air. Diamond is harder to ignite. Both burn to carbon dioxide in a plentiful supply of air.

$$C(s) + O_2(g) \xrightarrow{\text{ignite}} CO_2(g)$$

Carbon does not react with liquid water. At 1000 °C and above, carbon reacts with steam to produce a 1:1 mixture of carbon monoxide and hydrogen known as *water gas*. This has been, in the past, a valuable industrial fuel, and also a starting material for processes such as the manufacture of ammonia by the Haber Process (see Ch 13.3). The production of water gas absorbs energy, so the carbon cools down. Alternative bursts of steam and air (to heat the carbon up again) are used.

$$C(s) + H_2O(g) \overset{>1000\,°C}{\rightleftharpoons} CO(g) + H_2(g)$$

Both diamond and graphite are extremely resistant to acids.

Silicon

Amorphous silicon burns in *oxygen* (not air) at red heat.

$$Si(s) + O_2(g) \xrightarrow{~900\,°C} SiO_2(g)$$

(a)

A view of the planes of atoms in the structure of graphite. Within the horizontal planes, the carbon atoms are arranged in interconnected hexagons

(b)

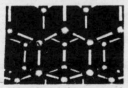

A view through the structure of diamond. Notice the tetrahedral arrangement of carbon atoms

Figure 7.3.1

It is attacked slowly by steam at red heat.

$$Si(s) + 2H_2O(g) \xrightarrow{\sim 900\,°C} SiO_2(s) + 2H_2(g)$$

It is very resistant to acids but reacts with hot, concentrated alkali to give hydrogen and the silicate.

$$Si(s) + 2OH^{\ominus}(aq) + H_2O(l) \xrightarrow[concentrated]{hot} SiO_3^{2\ominus}(aq) + 2H_2(g)$$

Crystalline silicon behaves in much the same way as amorphous silicon, but is less reactive.

Germanium

This resembles silicon in behaviour. It is unreactive with air and water except under extreme conditions. It is not normally affected by acids, but reacts with alkalis in the same way as silicon.

Tin

Tin does not react with water or air under normal conditions (hence its use in tin plate and 'tin cans') but burns to form tin(IV) oxide, SnO_2, at about 1500 °C. It reacts rapidly with concentrated hydrochloric acid to give tin(II) chloride and hydrogen.

$$Sn(s) + 2H^{\oplus}(aq) \xrightarrow{concentrated} Sn^{2\oplus}(aq) + H_2(g)$$

Hot concentrated sulphuric acid, however, gives a reaction in which the main products are tin(IV) sulphate and sulphur dioxide.

$$Sn(s) + 4H_2SO_4(l) \xrightarrow[concentrated]{hot} Sn(SO_4)_2(aq) + 4H_2O(l) + 2SO_2(g)$$

Note that in this reaction, sulphuric acid is partly reduced to sulphur dioxide (sulphur in the +6 state in $SO_4^{2\ominus}$ is reduced to the +4 state in SO_2 – see Ch 4.3), and the tin is oxidised to its highest oxidation state, +4.

Tin, like aluminium and zinc, reacts with alkali to give hydrogen plus an anion which contains the metal. It is, however, less reactive than aluminium and zinc, so hot, concentrated alkali is needed.

$$Sn(s) + 2OH^{\ominus}(aq) + H_2O(l) \xrightarrow[concentrated]{hot} SnO_3^{2\ominus}(aq) + 2H_2(g)$$

> Aluminium powder reacts rapidly with dilute aqueous alkali to give hydrogen; zinc also does so, but less rapidly. This behaviour is very uncommon for metals.

Lead

Lead is quickly coated in air by a thin layer of hydroxide and carbonate, which protects it from further corrosion (hence its use in roofing). If heated just above its melting point of 323 °C, lead forms yellow lead(II) oxide, PbO. If heated to 400–450 °C, PbO then forms 'red lead', Pb_3O_4, dilead(II)lead(IV) oxide.

$$2Pb(l) + O_2(g) \xrightarrow{heat} 2PbO(s)$$

then:

$$6PbO(s) + O_2(g) \xrightarrow{heat} 2Pb_3O_4(s)$$

Lead is not attacked by water which has no air in it. But ordinary water slowly reacts with lead to form the slightly soluble 'lead hydroxide'.

$$2Pb(s) + 2H_2O(l) + O_2(aq) \longrightarrow 2Pb(OH)_2(s)$$

This is the reason for the unsuitability of lead piping for drinking water in 'soft water' areas (see Ch 7.2).

Lead, because of the general insolubility of its salts, and so the formation of an insoluble layer on the surface of the metal, is generally not attacked by acids. There is a complex reaction with nitric acid.

Because lead(II) chloride is soluble at high temperatures, hot concentrated hydrochloric acid does react to give hydrogen. As with tin, hot concentrated sulphuric acid liberates sulphur dioxide in a redox reaction. As with tin, lead is slowly attacked by hot concentrated alkali to give hydrogen. Tin forms the ion $SnO_3^{2\ominus}$, but lead – because it is more stable as lead(II) (see Ch 7.1) – forms $PbO_2^{2\ominus}$.

	With oxygen	With water	With acid	With alkali
Carbon	$\xrightarrow{\text{ignite}}$ CO_2 $\left(\xrightarrow{\text{limited air}} CO\right)$	$\xrightarrow{1000°C} CO + H_2$		
Silicon	$\xrightarrow[\text{heat}]{\text{red}}$ SiO_2	$\xrightarrow[\text{heat}]{\text{red}}$ $SiO_2 + 2H_2$		$\xrightarrow[\text{conc}]{\text{hot}}$ $SiO_3^{2-} + 2H_2$
Germanium	$\xrightarrow[\text{temperature}]{\text{very high}}$ GeO_2	$\xrightarrow[\text{temperature}]{\text{very high}}$ $GeO_2 + 2H_2$		$\xrightarrow[\text{conc}]{\text{hot}}$ $GeO_3^{2-} + 2H_2$
Tin	$\xrightarrow{1500°C}$ SnO_2	$\xrightarrow[\text{temperature}]{\text{very high}}$ $SnO_2 + 2H_2$	$\xrightarrow[\text{RT}]{\text{conc HCl}}$ $SnCl_2 + H_2$ $\xrightarrow[\text{conc } H_2SO_4]{\text{heat}}$ $Sn(IV) + SO_2$	$\xrightarrow[\text{conc}]{\text{hot}}$ $SnO_3^{2-} + 2H_2$
Lead	$\xrightarrow{330°C}$ PbO $\xrightarrow{450°C}$ Pb_3O_4	$\xrightarrow[+ O_2]{H_2O}$ $Pb(OH)_2$ (slow) slow at room temperature	$\xrightarrow{\text{conc HCl}}^{\text{heat}}$ slow $PbCl_2 + H_2$ $\xrightarrow[\text{conc } H_2SO_4]{\text{hot}}$ $Pb(SO_4)_2 + SO_2$	$\xrightarrow[\text{conc}]{\text{hot}}$ $PbO_2^{2-} + 2H_2$

Table 7.3.1   Reactions of Group 4 elements

The chlorides of Group 4

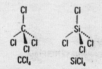

Figure 7.3.2   CCl₄ and SiCl₄

Tetrachloromethane (carbon tetrachloride), CCl_4, and silicon tetrachloride, $SiCl_4$, are liquids at room temperature. They have simple molecules which, because they contain four bonding pairs only, are tetrahedral in shape (*Figure 7.3.2*; see also Ch 2.6).

$GeCl_2$, $SnCl_2$ and $PbCl_2$ are solids at room temperature, but $GeCl_4$, $SnCl_4$ and $PbCl_4$ are liquids (*Figure 7.3.3*).

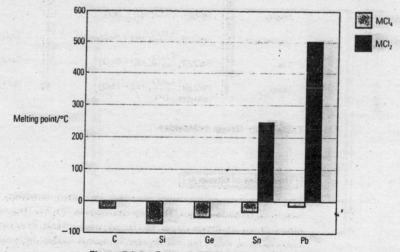

Figure 7.3.3   Group 4 chlorides – melting points

This difference in physical properties is because the chlorides of formula MCl_4 contain these elements in their +4 oxidation state and are *covalent*; those of formula MCl_2 are at least partly ionic, and $PbCl_2$ is largely ionic.

Note that, as always, if a metal has more than one oxidation state, the higher the oxidation state the more covalent its compounds are. The reason for the increasing stability of the +2 state down the Group is given in Ch 7.1.

All the covalent MCl_4 compounds, except CCl_4, are attacked rapidly by water to give hydrogen chloride and the oxide or hydrated oxide.

e.g.
$$SiCl_4(l) + 4H_2O(l) \longrightarrow SiO_2.2H_2O(s) + 4HCl(g)$$

Hydrated oxides

Many oxides (other than for the s-block elements) precipitate from aqueous solution as hydrated oxides. Their formulae can be represented in a number of ways.

$$Si(OH)_4 \longrightarrow SiO_2.2H_2O \longrightarrow SiO_2 + 2H_2O$$

Tetrachloromethane is unique among covalent chlorides in being unaffected by water. This is basically because, with four chlorine atoms around the small carbon atom, there is no way for a water molecule to get in to begin the attack.

A better way of looking at it is that, whereas silicon and the larger atoms Ge, Sn, and Pb all have d-orbitals available (3d orbitals in the case of silicon) into which a lone pair of electrons from the water can be donated to start the reaction going, there is no such thing as a 2d orbital in carbon. There are simply no low energy orbitals available for the water molecule to 'latch on to', and CCl_4 is therefore inert chemically.

So, in the Group 4 chlorides, we note that all of the formulae MCl_4 are *covalent* and are easily hydrolysed (except CCl_4); those of formula MCl_2 only start at germanium, and become more ionic in character for Ge → Sn → Pb.

> Tetrachloromethane is nonetheless a dangerous material and now banned for public use. It is soluble in body fat. For many years it was used in very large quantities for dry-cleaning.

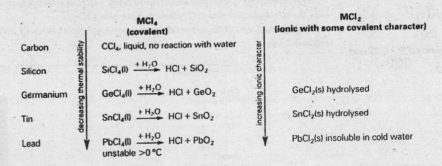

Table 7.3.2   Group 4 chlorides

The oxides of Group 4

The two most important oxides of carbon are carbon monoxide, CO, and carbon dioxide, CO_2, which are colourless gases. They are discussed in Ch 7.4. They are both acidic, carbon dioxide obviously so – it will turn damp blue litmus paper reddish, if not actually red. Carbon monoxide is not soluble in water, but will dissolve in molten sodium hydroxide under pressure to give the sodium salt of methanoic acid, $HCOO^{\ominus}Na^{\oplus}$; which shows that, under extreme conditions, carbon monoxide is *acidic*. However, under more normal conditions, carbon monoxide is considered to be a *neutral* oxide.

$$CO(g) + Na^{\oplus}OH^{\ominus}(l) \xrightarrow{heat} HCOO^{\ominus}Na^{\oplus}$$

The usual oxide of silicon, SiO_2, is better known as silica, and occurs in quartz, sand, etc. Whereas the carbon oxides have simple molecular structures, SiO_2 has a giant atomic structure (*Figure 7.3.4*; see also Ch 3.2), and is therefore a very hard solid.

Concentrated sodium silicate solution is known as water glass. Before the days of fridges and intensive poultry farming, eggs were preserved for winter use by putting them in a bucket of water glass, which sealed the shells and prevented entrance of bacteria. You might even have seen a 'crystal garden' grown in water glass.

Even so, it *is* acidic – it reacts with hot alkalis to give silicate salts; e.g. hot concentrated sodium hydroxide solution with sand eventually gives a sticky solution of sodium silicate.

$$SiO_2(s) + 2OH^\ominus(aq) \xrightarrow{heat} SiO_3^{2\ominus}(aq) + 2H_2O(l)$$

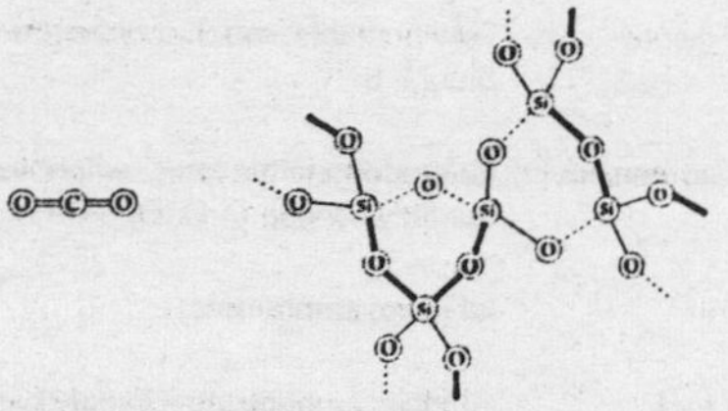

Figure 7.3.4 CO_2 and SiO_2

Silicones

One of the reasons for the stability of silicon dioxide, SiO_2, is the strength of the silicon–oxygen bond. The stability of O—Si—O systems has been developed to form a range of large molecules containing a backbone of silicon and oxygen atoms. This backbone also contains two carbon groups attached to each silicon atom.

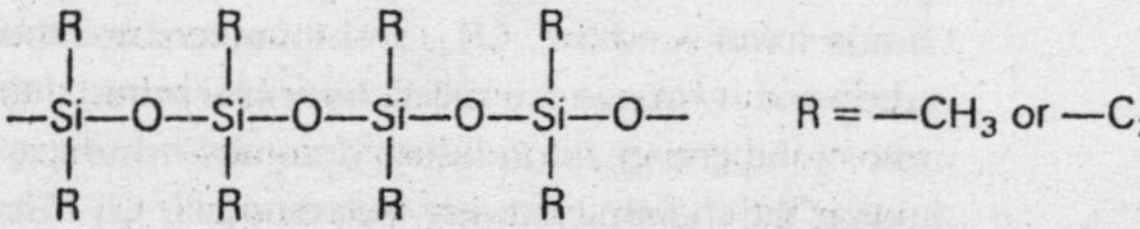

$$\begin{array}{ccccccc} & R & & R & & R & & R \\ & | & & | & & | & & | \\ -Si&-O&-Si&-O&-Si&-O&-Si&-O- \\ & | & & | & & | & & | \\ & R & & R & & R & & R \end{array} \qquad R = -CH_3 \text{ or } -C_2H_5$$

These silicones are non-toxic and very stable over a wide temperature range. By varying the lengths of the silicon–oxygen chain and the side chains, the properties of the silicone can be altered to provide liquids or low melting solids. These are used as lubricants, as polishes, as water repellents and for coating paper to provide a non-stick surface.

Other than GeO_2, which is an acidic solid, all the oxides of Ge, Sn and Pb are **amphoteric** (i.e. they will react with both acids and alkalis, see Ch 5.5). Tin(II) oxide, SnO, if freshly made actually *glows* in air as it is oxidised to the more stable tin(IV) oxide, SnO_2. The usual ore of tin, cassiterite, is the stable SnO_2. Lead(II) oxide, PbO, is also *amphoteric* but mostly basic. Lead(IV) oxide, PbO_2, is more acidic and, as noted above, a *very* powerful oxidising agent.

This switch of stable oxidation state from +4 for tin to +2 for lead is usually explained in terms of the 'inert pair' effect (see Ch 7.1 and Ch 5.4).

As noted above, gentle heating of PbO in air gives 'red lead', Pb_3O_4. This can be thought of as a 'mixed oxide', dilead(II)lead(IV) oxide, $2PbO.PbO_2$ (which adds up to Pb_3O_4). If red lead is treated with dilute nitric acid, the chocolate-coloured PbO_2 is left and can be filtered off. As the lead is in its higher oxidation state, PbO_2 is more acidic and covalent than the more basic PbO. The PbO reacts with the nitric acid, whereas the PbO_2 cannot.

$$\underset{\text{'red lead'}}{2PbO.PbO_2(s)} + 4HNO_3(aq) \longrightarrow 2Pb(NO_3)_2(aq) + PbO_2(s) + 2H_2O(l)$$

Tin(II) can be used as a reducing agent; lead(IV) oxide is used as a powerful oxidising agent (see Ch 7.1). For example, lead(IV) oxide, PbO_2, has been used to oxidise concentrated hydrochloric acid to chlorine.

$$PbO_2(s) + 4HCl(aq) \longrightarrow PbCl_2(s) + Cl_2(g) + 2H_2O(l)$$

An interesting comparison is seen in the ores of tin and lead. Tin occurs as cassiterite, SnO_2, because any tin(II) compounds in nature would be oxidised to tin(IV), whereas lead is typically found as lead(II) compounds, e.g. galena, PbS.

	MO		MO$_2$
Carbon	(g) covalent; reducing agent; neutral; (acidic under extreme conditions)	$\xrightarrow{+O_2}$	(g) covalent; acidic
Silicon	very unstable; rapidly converted to SiO_2 + Si		(s) giant covalent; 'acidic'
Germanium	(s) predominantly ionic; amphoteric; readily changes to GeO_2 +Ge	$\xrightarrow{+O_2}$	(s) acidic/amphoteric; covalent/ionic
Tin	(s) ionic; amphoteric	$\xrightarrow{+O_2}$	(s) covalent/ionic; amphoteric
Lead	(s) ionic; amphoteric – largely basic		(s) covalent/ionic; oxidising agent; amphoteric – more acidic

Table 7.3.3   Group 4 oxides

The hydrides of Group 4

Carbon forms methane, CH_4, and innumerable other hydrides. Compounds which contain *only* carbon and hydrogen are called **hydrocarbons**; they occur widely in fossil fuels, and so provide most of the energy for industry, domestic heating and transport, as well as starting materials for much of the chemical industry (see especially Ch 15 and Ch 16).

At room temperature, hydrocarbons range from gases through liquids to waxy solids, depending on the size of the molecules. They are insoluble in water, and neither acidic nor basic. Although they can be ignited in air or oxygen, most hydrocarbons are otherwise very stable to heat. In sufficient air or oxygen, hydrocarbons burn completely to form carbon dioxide and water.

e.g. $$CH_4(g) + 2O_2(g) \xrightarrow{\text{ignite}} CO_2(g) + H_2O(l)$$

In insufficient air, e.g. in a badly ventilated room or a badly adjusted petrol engine, the dangerous gas carbon monoxide is formed (see Ch 7.4).

$$2CH_4(g) + 3O_2(g) \xrightarrow{\text{ignite}} 2CO(g) + 4H_2O(l)$$

If there is a serious shortage of air or the hydrocarbon and air are not fully mixed, e.g. in a candle flame or a malfunctioning diesel engine, particles of carbon – i.e. *soot* – are also formed.

Silicon can form silane, SiH_4, and several other hydrides. They are dangerously unstable in air, and decompose on heating.

The same is true of germane, GeH_4. Stannane, SnH_4, is even less stable to heat, and plumbane, PbH_4, is highly unstable even at room temperature.

The simple hydrides, MH_4, of the Group 4 elements become increasingly unstable, when heated, with increasing size of the atom of the Group 4 elements.

When magnesium powder is heated with sand (impure silica, SiO_2), silicon is released from its oxide, but some magnesium silicide, Mg_2Si, is also formed. If the result of the experiment is thrown into dilute acid, any magnesium silicide reacts rapidly to form bubbles of silane. These explode as they come into contact with air, sometimes producing white vortex rings of very fine SiO_2 particles.

$$Mg_2Si(s) + 4HCl(aq) \longrightarrow 2MgCl_2(aq) + SiH_4(g)$$

$$SiH_4(g) + 2O_2(g) \longrightarrow SiO_2(s) + 2H_2O(l)$$

---

1 **a** Silicon tetrachloride reacts violently with water. Write an equation for this reaction. What would be the likely pH of the resultant solution?

b Silane is the binary compound containing silicon and hydrogen (b.p. −112 °C). What type of bond exists between silicon and hydrogen in the silane molecule?

c Draw a dot-and-cross diagram for the silane molecule, and a diagram to show the shape of the molecule.

d Impure silane ignites as soon as it comes into contact with oxygen in the air. Construct a balanced equation for the reaction of silane with oxygen.

2 Using the elements in Group 4 give the name and formula of:

a a gaseous acidic oxide;

b a neutral oxide;

c an amphoteric oxide;

d a mainly basic oxide;

e a covalent chloride unaffected by water;

f a chloride with a giant ionic lattice;

g an oxide with a simple molecular structure.

3 In the following test tube reactions, describe what you would expect to see and explain these observations. Where possible write balanced equations for the reactions.

a Germanium(IV) chloride and water.

b Lead(II) oxide and nitric acid.

c Lead heated with oxygen until it melts.

7.4 Some special aspects of carbon chemistry

Reducing action

Carbon and carbon monoxide are used in huge quantities to obtain some metals from their oxides.

If powdered charcoal – which is a quite pure form of carbon – is heated with the oxide of a fairly unreactive metal, the metal is obtained.

e.g. $$2CuO(s) + C(s) \xrightarrow{heat} 2Cu(s) + CO_2(g)$$

The metal must not be too reactive, i.e. it must not require too much energy to force electrons back on to the metal ions in the oxide. A classic industrial example is the blast furnace, in which hot air is blown through a charge of iron ore (usually haematite, iron(III) oxide, Fe_2O_3), coke and limestone.

The coke, which is almost completely carbon, burns to carbon dioxide and releases a great deal of heat energy. It is this reaction which keeps the furnace hot enough.

$$C(s) + O_2(g) \xrightarrow{ignite} CO_2(g)$$

At high temperatures, carbon dioxide reacts with more coke to form carbon monoxide.

$$CO_2(g) + C(s) \xrightarrow{high\ temperature} 2CO(g)$$

Although there is some direct reduction of iron oxide by carbon, a large part of the reduction is done by the carbon monoxide.

$$Fe_2O_3(s) + 3CO(g) \xrightarrow{heat} 2Fe(l) + 3CO_2(g)$$

The molten iron runs to the bottom of the furnace and is tapped off at intervals.

The limestone is added to remove silica impurities in the iron ore. When heated, limestone breaks up to carbon dioxide and the very basic calcium oxide. This reacts at high temperature with the acidic oxide of the non-metal, silicon.

$$CaCO_3(s) \xrightarrow{\text{high temperature}} CaO(s) + CO_2(g)$$

$$CaO(s) + SiO_2(s) \xrightarrow{\text{high temperature}} CaSiO_3(l)$$

The product is calcium silicate. This forms a large proportion of the waste product from a blast furnace, which is known as *slag*. This slag floats above the molten iron (because it is both less dense than iron and immiscible with it) and can also be tapped off.

Carbon monoxide

Carbon monoxide has a well-deserved reputation as a poisonous gas, and in spite of precautions and much publicity it causes deaths in Britain every year.

Although it is not very poisonous – much less so than ozone, hydrogen sulphide or chlorine – because it has no smell and is colourless, it cannot easily be detected, and is very dangerous indeed. It is formed when hydrocarbons or other fossil fuels burn in limited air, and is present in the exhaust fumes of vehicles.

In the body, oxygen is carried from the lungs, by attaching itself to the iron part of the haemoglobin in blood, to where it is needed in the muscles, brain and elsewhere. It can then easily leave the haemoglobin, which gets circulated round to the lungs again to pick up another load of oxygen.

Carbon monoxide attaches itself firmly to the iron of the haemoglobin, and only leaves it very slowly. So, if you breathe an atmosphere containing more than a small amount of carbon monoxide, your haemoglobin becomes more and more clogged up with carbon monoxide, your blood becomes less and less capable of transporting oxygen, and you eventually die.

The compound formed by carbon monoxide and haemoglobin is bright red, much redder than normal oxygenated arterial blood. A symptom of carbon monoxide poisoning is therefore cherry-red cheeks and other regions where blood vessels are near the surface.

The first aid treatment is to keep the victim breathing at all costs and apply oxygen, preferably under pressure, as soon as possible. The carbon monoxide will eventually be released from the haemoglobin. (One reason why people who smoke a lot tend to suffer from heart disease is that carbon monoxide from burning cigarettes reduces oxygen transport in their blood, making the heart work harder.)

A rather unexpected use of carbon monoxide is in the purification of nickel. If carbon monoxide is passed over impure nickel at about 60 °C, the volatile nickel carbonyl (tetracarbonylnickel(0), $Ni(CO)_4$), is formed. This highly poisonous gas decomposes back to carbon monoxide and pure nickel at about 200 °C. The process is known as the Mond Process, was discovered in 1889, and in Britain is carried out at Clydach near Swansea.

$$Ni(s) + 4CO(g) \underset{\sim 200\,°C}{\overset{\sim 60\,°C}{\rightleftharpoons}} Ni(CO)_4(g)$$

Carbon dioxide

Carbon dioxide can be made by the action of any acid on almost any carbonate:

$$2H^{\oplus}(aq) + CO_3^{2\ominus} \longrightarrow CO_2(g) + H_2O(g)$$

It is usually made by reacting dilute hydrochloric acid with marble chips. Marble is a particularly hard natural form of calcium carbonate, and in lump form reacts with the acid at a reasonable rate. The reaction with powdered chalk – a softer form of calcium carbonate – is uncontrollably fast.

$$CaCO_3(s) + 2HCl(aq) \longrightarrow CaCl_2(aq) + H_2O(l) + CO_2(g)$$

Electronic carbon monoxide detectors are now available and should be fitted in rooms with gas fires and stoves.

With a formula of CO, carbon monoxide still does obey the 'rules' of covalent bonding (see Ch 2.4). It contains the same number of electrons as the nitrogen molecule, N_2, and can be thought of as having very similar bonding, i.e. C≡O.

As noted in Ch 1.6, Max Perutz and John Kendrew won the Nobel Prize for Chemistry in 1962 for their solution, using X-ray crystallography, of the complicated structures of haemoglobin and the related compound myoglobin. Perutz afterwards went on to work out the detailed changes which occur in the haemoglobin molecule when it takes up and loses oxygen, and also in the mutant forms of haemoglobin which are characteristic of some inherited diseases such as thalassaemia and sickle-cell anaemia.

Carbon dioxide is produced in large quantities in industry as a by-product of fermentation, e.g. in the brewing of beer. It is moderately soluble in cold water, especially when under pressure; it gives the fizz and the tingly taste to fizzy drinks. The highly unstable *carbonic acid* is formed.

$$H_2O(l) + CO_2(g) \underset{}{\overset{\text{room temperature}}{\rightleftharpoons}} H_2CO_3(aq)$$

Note that, like sulphuric acid, H_2SO_4, carbonic acid has two replaceable hydrogens – it is **dibasic** (see Ch 13.4). It can form two families of salts, the carbonates (with the ion, $CO_3^{2\ominus}$) and the hydrogencarbonates (with the ion $HCO_3^{\ominus}$).

The solubility of atmospheric carbon dioxide in rain water has very important geographical and economic consequences – see 'hard water' in Ch 6.6.

Because it is much more dense than air and is usually unreactive (except with burning metals, such as magnesium alloys used in some aircraft and racing cars), carbon dioxide is very widely used in fire extinguishers. It 'blankets' the fire and prevents oxygen reaching it. But the gas has little cooling action, and hot objects (e.g. wood) may re-ignite as the carbon dioxide drifts away.

A test for carbon dioxide

A saturated solution of calcium hydroxide ('lime water') contains so little of the solute that it is totally useless for *absorbing* carbon dioxide; but, because calcium carbonate is so insoluble, lime water provides a sensitive method for *detecting* carbon dioxide (see Ch 30.3).

$$Ca(OH)_2(aq) + CO_2(g) \xrightarrow{\text{room temperature}} CaCO_3(s) + H_2O(l)$$

If excess carbon dioxide is present, the precipitate dissolves and a clear, colourless solution of calcium hydrogencarbonate results. (This is the same reaction which is responsible for hard water, limescale, stalactites, etc – see Ch 6.6).

$$CaCO_3(s) + CO_2(g) + H_2O(l) \underset{\text{heat}}{\overset{\text{room temperature}}{\rightleftharpoons}} Ca(HCO_3)_2(aq)$$

The properties of metal carbonates

As discussed in Ch 6.4, all carbonates except those of sodium and potassium – which are hardly affected by heat – decompose on heating to give the metal oxide and carbon dioxide if the gas is allowed to escape (see Ch 22.2).

$$\text{e.g.} \qquad CaCO_3(s) \xrightarrow{\text{heat}} CaO(s) + CO_2(g)$$

Also, apart from those of sodium and potassium, all metal carbonates are insoluble in water.

An American biochemist, Melvin Calvin, studied the early stages of photosynthesis in single-celled green algae by using radioactive CO_2 and then identifying the products formed. He discovered a cycle of reactions which now bears his name, and was given the Nobel Prize for Chemistry in 1961.

The carbon cycle

The carbon cycle and its two main processes, together with the 'greenhouse effect' (see Ch 17.2), can only be dealt with very briefly here. (These are major topics influencing all living things, and large books have been written about them.) Our discussion will ignore the huge amounts of carbon dioxide locked away in carbonate rocks and released into the atmosphere by volcanic activity, and further huge amounts dissolved in deep ocean water.

Only about 0.035% (or 350 parts per million, ppm) of the atmosphere consists of carbon dioxide. It is constantly taken out by the process of **photosynthesis**, common to all green plants and the basis of all life on Earth, and replaced by the processes of **respiration** and decay.

Photosynthesis, beginning billions of years ago with cyanobacteria and blue-green algae and accelerating several hundreds of millions of years ago with the first plants, is the main reason why our atmosphere – unique among the planets in the Solar System – is rich in oxygen. The overall equation for photosynthesis is:

$$6CO_2(g) + 6H_2O(l) \xrightarrow[\text{light}]{\text{chlorophyll}} \underset{\text{glucose}}{C_6H_{12}O_6(s)} + 6O_2(g)$$

This process is **endothermic** (see Ch 11.2) – it requires energy. The necessary energy is captured from sunlight by the chlorophyll in any green plant leaf or phytoplankton; the equation summarises a long series of reactions which occur in the leaf. The glucose can then be further converted in the plant to a wide range of other materials, including starch and cellulose.

The overall equation for **respiration** is the exact opposite of that for photosynthesis.

$$C_6H_{12}O_6(s) + 6O_2(g) \longrightarrow 6CO_2(g) + 6H_2O(l)$$

This equation summarises a long series of reactions for which enzymes (see Ch 28.5) are required. Respiration is an **exothermic** process – it gives out energy; indeed, *it provides the energy needed by all living animals.*

By looking at these two equations, it seems clear that if the rate at which animals respire worldwide is equal to the rate at which plants photosynthesise, the amount of carbon dioxide in the atmosphere should remain constant.

However, combustion of coal and hydrocarbons (see Ch 15.4) releases extra carbon dioxide into the atmosphere – carbon dioxide that was trapped through photosynthesis many millions of years ago and locked underground in fossil fuels. If, at the same time, large areas of forest – especially equatorial forest in South America and Africa – are being cut down, the proportion of carbon dioxide in the atmosphere is bound to rise. This, considerably oversimplified, explanation is the main reason for concern about the so-called 'greenhouse effect'.

How does the effect occur? Briefly, sunlight heats up the Earth's surface. (If there was *no* greenhouse effect caused by water vapour and carbon dioxide in the atmosphere, the surface would be about 20 °C cooler). Any warm surface gives off infra-red (IR) radiation.

Normally, there is a balance between the heating effect of the Sun on the surface and the loss of heat energy from the surface by radiation (as IR radiation) back into space. But carbon dioxide is a good absorber of IR radiation, so extra carbon dioxide in the atmosphere absorbs more of the energy radiated away from the Earth's surface. The CO_2 molecules transfer this energy to other molecules by collision; so the atmosphere heats up, with potentially very serious effects on climate, agriculture, sea levels, etc.

Methane, CH_4, and chlorofluorocarbons (CFCs) (see Ch 17.2) are also 'greenhouse gases', but the main effect is due to increased CO_2 levels.

Primo Levi was an Italian Jew, an industrial chemist by profession, who was captured while fighting against the fascists in Italy during the Second World War and sent to the Nazi extermination camp at Auschwitz – Birkenau. He was one of the few (out of more than a million) to survive, mainly because he was used as slave labour in the artificial rubber plant near the camp. During the hell of Auschwitz he conceived the idea of perhaps his most extraordinary book, among several, *The Periodic Table*. This consists of 21 chapters, each having as title the name of an element; most are factual, some are fiction. As the book review in *New Scientist* put it: 'Nominally it is prose; in actuality, it is a narrative poem of magical quality.'

Here is a short extract from the chapter entitled *Carbon*. 'Carbon dioxide … this gas which constitutes the raw material of life, the permanent store upon which all that grows draws, and the ultimate destiny of all flesh, is not one of the principal components of air but, rather, a ridiculous remnant, an "impurity", thirty times less abundant than argon, which nobody even notices … [It is] from this ever-renewed impurity of the air [that] we come, we animals and we plants, and we the human species, with our four billion discordant opinions, our millenniums of history, our wars and shames, nobility and pride …'

SELF-ASSESSMENT QUESTIONS 7.4

1 What would you expect to observe when the following reactions were carried out? Write balanced equations for any chemical changes that occur.
 a Impure silane is allowed to come into contact with air.
 b Carbon dioxide is bubbled into calcium hydroxide solution over a long period of time.
 c Silica is added to water.
 d Tin(IV) chloride is added dropwise to water.

2 Carbon and silicon are elements in the same Group of the Periodic Table. By discussing the differences in bonding and structure, explain why CO_2 and SiO_2 have such different melting points and CCl_4 and $SiCl_4$ have such different reactivities. How would you expect the simplest hydrides of these two elements to react with water?

3 Write equations for:
 a the oxidation of carbon monoxide;
 b the reduction of lead(IV) oxide when reacting with concentrated hydrochloric acid;
 c the reaction when carbon dioxide dissolves in water.

A summary of the chlorides, oxides and hydrides of Group 4

	Carbon, C	Silicon, Si	Germanium, Ge	Tin, Sn	Lead, Pb
Chlorides reaction with water MCl_2			$GeCl_2$; solid; hydrolysed.	$SnCl_2$; solid; partly ionic; slowly hydrolysed.	$PbCl_2$; solid; ionic; insoluble in cold water; soluble in hot.
MCl_4	CCl_4; liquid; unaffected by water.	$SiCl_4$; liquid; rapidly hydrolysed $SiCl_4(s) + 4H_2O(g)$ $\longrightarrow Si(OH)_4(s) + 4HCl(g)$	$GeCl_4$; liquid; hydrolysed. $GeCl_4(l) + 4H_2O(l)$ $\longrightarrow Ge(OH)_4(s) + 4HCl(g)$	$SnCl_4$; liquid; rapidly hydrolysed. $SnCl_4(l) + 2H_2O(l)$ $\longrightarrow SnO_2(s) + 4HCl(g)$	$PbCl_4$; liquid; hydrolysed. $PbCl_4(l) + 2H_2O(l)$ $\longrightarrow PbO_2(s) + 4HCl(g)$
	Other chlorocarbons exist, e.g. C_2Cl_6.			Soluble in organic solvents.	$PbCl_4(l) \xrightarrow{heat} PbCl_2(s)$ $+ Cl_2(g)$
Oxides MO	CO; gas; neutral; (acidic only under extreme conditions).	SiO; solid; unstable $2SiO(s)$ $\longrightarrow SiO_2(s) + Si(s)$	GeO; solid; amphoteric. $2GeO(s)$ $\longrightarrow GeO_2(s) + Ge(s)$	SnO; solid; amphoteric; oxidised by air.	PbO; solid; amphoteric but normally basic.
MO_2	CO_2; gas; acidic. $CO_2(s) + H_2O(l)$ $\longrightarrow H_2CO_3(aq)$ (plus C_3O_2 and several others).	SiO_2; solid; very stable; acidic.	GeO_2; solid; acidic.	SnO_2; solid; amphoteric; very stable.	PbO_2; solid; amphoteric; powerful oxidising agent. Pb_3O_4; solid; a 'mixed' oxide.
Hydrides stability of MH_4	CH_4 and a huge number of other hydrocarbons; generally very stable.	SiH_4 and several other silanes, e.g. Si_2H_6; moderately stable to heat; very reactive in air.	GeH_4 and several other germanes; decompose on heating.	SnH_4; stannane; decomposes on very gentle heating.	PbH_4; plumbane; decomposes rapidly at room temperature.

SUMMARY
7

In this chapter the following new ideas have been covered. *Have you understood and learned them?* Could you explain them clearly to a friend? The page references will help.

- The elements in Group 4 become more metallic down the Group (i.e. as atomic number increases).   p176
- The usual oxidation state in compounds is $+4$, but the $+2$ oxidation state becomes more stable down the Group.   p176
- With a slight irregularity at tin, the melting point and molar first ionisation energies of the elements increase down the Group.   p178
- Density generally increases down the Group, with irregularities for diamond and grey tin.   p179
- Electrical resistivity decreases (i.e. electrical conduction increases) down the Group, with an irregularity at germanium.   p179
- All the elements form oxides and chlorides, and general reactivity increases down the group.   p181
- Chlorides of the form MCl_2 increase in stability and ionic character down the Group; chlorides of the form MCl_4 are covalent liquids.   p181
- All the MCl_4 chlorides react with water to give hydrogen chloride and the oxide or hydrated oxide, other than CCl_4.   p182
- Oxides of the form MO_2 are acidic, although the acidity gets less down the Group.   p182
- Tin in the $+2$ oxidation state is a reducing agent while lead $+4$ is a powerful oxidising agent.   p183
- The hydrides, MH_4, become less stable down the Group. Carbon forms innumerable hydrides.   p184
- Carbon monoxide is poisonous and a reducing agent; carbon dioxide is deeply involved in the carbon cycle.   p186
- The combustion of fossil fuels contributes to the 'greenhouse effect'.   p188

EXAMINATION
QUESTIONS
7

1 One of the chlorides of lead, $PbCl_4$, is thermally unstable, and once it warms above $0\,°C$ it decomposes to $PbCl_2$ and chlorine gas.
   a Write a balanced equation for the reaction.
   b Explain the term *redox* in terms of electrons.
   c In the equation for the decomposition of $PbCl_4$, identify the element that has been oxidised and the element which has been reduced by showing the oxidation states of the elements involved.
   d Carbon reacts with very hot concentrated sulphuric acid according to the equation:

$$C(s) + H_2SO_4(l) \xrightarrow[\text{concentrated}]{\text{hot}} CO(g) + SO_2(g) + H_2O(l)$$

   What is the oxidation state of the carbon at the start of the reaction and at the end of the reaction?
   e What has been reduced to enable the carbon to be oxidised?

2 When a compound of lead, A, is heated gently it decomposes to produce a yellow powder, B, and a colourless gas, C, is given off. The gas, C, is slightly acidic when tested with damp pH paper. The yellow solid, B, on heating more strongly in air turns a reddish colour forming a new solid, D.
   If this new solid, D, is treated with dilute nitric acid, some of the solid appears to dissolve leaving a brown sediment, E.
   This brown sediment, if filtered off and dried, can be used to convert concentrated hydrochloric acid to chlorine gas.

a Suggest likely identities for the compounds A to E, explaining your reasoning for each decision.

b Copy and complete the following table.

	C	Si	Sn	Pb
Oxidation state	+4	+4	+4	+2
Formula of oxide				
Formula of chloride				

c Write an equation for the reaction of the chloride of silicon with water.

d Write an equation for the reaction of lead with water in the presence of oxygen.

e How does the oxide of silicon react with water?

f Identify the species oxidised and reduced in the reaction of tin with concentrated sulphuric acid.

$$Sn(s) + 4H_2SO_4(l) \longrightarrow Sn(SO_4)_2(aq) + 4H_2O(l) + 2SO_2(g)$$

Do you remember ...?

- **Electronegativity**. (Ch 2.2)
- **Ionic bonding.** (Ch 2.3)
- **Covalent bonding**. (Ch 2.4)
- **Intermolecular forces**. (Ch 2.8)
- **Oxidation states** and the rules for them. (Ch 4.3)
- **Redox reactions**. (Ch 4.3)
- The general trends in **physical properties** down a Group. (Ch 5.2)
- The general trends in **chemical properties** down a Group. (Ch 5.4)
- The **shapes** of molecules. (Ch 2.6)

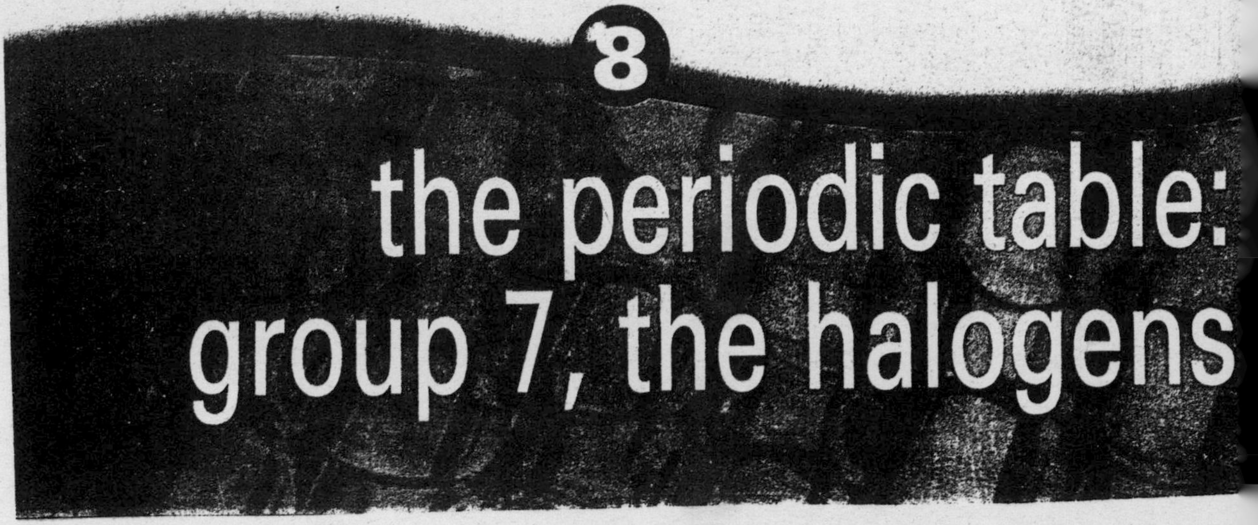

8
the periodic table: group 7, the halogens

8.1   Introduction to Group 7

The elements of Group 7 are fluorine, chlorine, bromine, iodine and the rare and radioactive astatine. They are called collectively, 'the halogens'. The word **halogen** comes from the Greek word for 'salt-former'. All the halogens have the outer electron configuration ns^2np^5, i.e. with seven electrons in the outer energy level.

All the halogens are electronegative non-metals and, since they easily gain the extra electron needed to complete the outer energy level, are good oxidising agents. As would be expected with their electron structure, the common oxidation state for halogens in their compounds is -1.

The ions, with complete outer levels and carrying a single negative charge, are known as **halide ions** – fluor*ide*, chlor*ide*, brom*ide*, and iod*ide*.

Group 7

Fluorine, F_2 – fluoride, $F^{\ominus}$

Chlorine, Cl_2 – chloride, $Cl^{\ominus}$

Bromine, Br_2 – bromide, $Br^{\ominus}$

Iodine, I_2 – iodide, $I^{\ominus}$

Fluorine is the most electronegative element of all; in all its compounds fluorine is found only in the -1 oxidation state. The other halogens can all have oxidation states of $+1$, $+3$, $+5$ and even $+7$, usually when in compounds involving oxygen.

In contrast to the metals of Groups 1 and 2, in Group 7 reactivity *decreases* down the Group. Halogen atoms *gain* electrons, by transfer or by sharing, when they react. In a metal atom more full energy levels leads to increased shielding of the outermost electrons from the attraction of the nucleus, and hence leads to increased reactivity as electrons are more easily lost. With the halogens and other non-metals, the less the shielding, the more ready the atom to react by electron gain.

Fluorine is the most reactive halogen and is the most violently reactive non-metal. It will ignite wood, steel wool and rubber when it comes into contact with them. Fluorine would be expected, therefore, to have the most negative first electron affinity of the Group 7 elements; but in fact the value is slightly less than that for chlorine (see *Figure 8.2.2*). This is because the small size of the fluorine atom leads to some repulsion of the incoming electron needed to form the $F^{\ominus}$ ion. The high reactivity is associated with the comparatively low energy needed to break the F—F bond in gaseous F_2 molecules, and the very high energies released when new bonds between fluorine and another element are formed (see Ch 11.5).

Fluorine is certainly the most electronegative element of all. Electronegativity *decreases* down the Group (*Figure 8.1.1*).

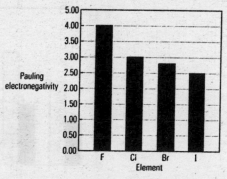

Figure 8.1.1   Electronegativities of the halogens

It can even be argued that iodine has a few slightly metallic properties; e.g. a salt, $Na^+I_9^-$, exists, in which an electron is delocalised among nine iodine atoms in the I_9^- ion. Astatine (from the Greek word for 'unstable') is an intensely radioactive artificial element, first made in 1940; the most stable of its isotopes has a half-life of only 8 hours.

SELF ASSESSMENT QUESTIONS 8.1

1 Using the symbols $\delta\oplus$ and $\delta\ominus$ indicate the polarity in the following bonds.
  a P—Br
  b Br—Cl
  c Cl—F
  d H—F

2 The halogens form a wide range of compounds with almost all the other elements in the Periodic Table. What type of bonding would you expect to find in the following halogen-containing compounds?
  a CsCl
  b PCl_3
  c XeF_4
  d $CuCl_2$
  e Cl_2O_7

3 An element, X, has the electron configuration $1s^2 2s^2 2p^6 3s^2 3p^5$.
  a What is the name of this element?
  b Write the formula of the binary compound formed between element X and sodium.
  c What type of bonding would you expect in this compound?
  d Draw a dot-and-cross diagram to show the bonding in the compound formed between X and sodium.

8.2 Physical properties

As usual, the *atomic radius increases* with increasing atomic number down a Group because there are more completed electron energy levels (*Figure 8.2.1*).

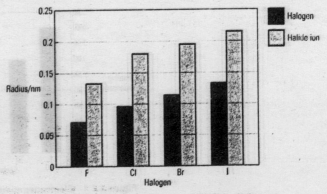

Figure 8.2.1 Atomic and ionic radii of the halogens

As stated above, more full levels means that outer electrons can more easily be removed; so *first ionisation energies decrease* down a Group (*Figure 8.2.2*).

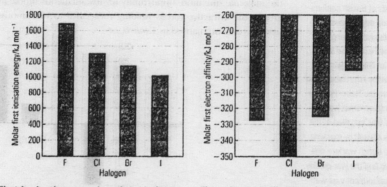

First ionisation energies of the halogens First electron affinities of the halogens

Figure 8.2.2

There is an excellent gradation of the more visible properties within the Group (*Figure 8.2.3*).

Figure 8.2.3 Halogens

All the halogens exist as diatomic molecules. Fluorine, F_2, is a pale yellow gas, b.p. $-188\,°C$; chlorine, Cl_2, is a dense yellow-green gas, b.p. $-35\,°C$ (its name comes from the Greek for green); bromine, Br_2, is a dense reddish-brown liquid with a deep brown vapour, b.p. $59\,°C$; iodine, I_2, is a black solid with violet-purple vapour (the name comes from the Greek for 'violet') which melts at $114\,°C$, and very easily sublimes. All the gases and vapours have intense and very unpleasant smells (bromine is named from the Greek word *bromos*, which means stench), corrode the tissue of lungs and eyes, and are highly poisonous.

Chlorine was first used as a war gas at Ypres in 1915. Thousands of British soldiers died. It and other gases were used extensively during the rest of the First World War. This use of chlorine was suggested by Fritz Haber, who discovered the Haber Process for making ammonia (see Ch 13.3). He was appalled at the slaughter caused by machine guns and trench warfare and wanted to end the war quickly. His mental torment was well displayed in Tony Harrison's verse play *Square Rounds*, first produced at the Royal National Theatre in 1992.

The differences in volatility, and hence melting points (*Figure 8.2.4*) and boiling points, among the halogens are easily explained in terms of intermolecular forces (see Ch 2.8). The more electrons in the molecule, the more opportunity for instantaneous dipole-induced dipole interactions, and the greater the attractive forces holding molecules to one another. The physical properties of the halogens are summarised in *Table 8.2.1*.

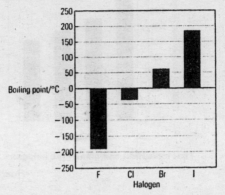

Figure 8.2.4   Boiling points of the halogens

Element	Atomic number Z	Electron structure	Electronegativity (Pauling scale)	Atomic radius /nm	b.p. /°C	Molar first ionisation energy /kJ mol⁻¹	Molar first electron affinity /kJ mol⁻¹	Density /g cm⁻³
Fluorine	9	2.7	4.0	0.071	−188	+1681	−328	1.51 (at −188 °C)
Chlorine	17	2.8.7	3.0	0.099	−35	+1251	−349	1.56 (at −35 °C)
Bromine	35	2.8.18.7	2.8	0.114	59	+1140	−325	3.12 (at 20 °C)
Iodine	53	2.8.18.18.7	2.5	0.133	184	+1008	−295	4.93 (at 25 °C)
Astatine	85	2.8.18.32.18.7	2.2			+930		

Table 8.2.1   Physical properties of the halogens

SELF-ASSESSMENT QUESTIONS 8.2

1 **a** The covalent radius of bromine is 0.114 nm. Explain what is meant by covalent radius.
   b Explain why the ionic radius of the bromide ion is more difficult to measure.
   c How does the ionic radius of the bromide ion compare with the covalent radius of bromine?
   d Why should there be this difference in the two radii?

2 Describe and explain the trends in the values of the first ionisation energies and the first electron affinities in the halogens.

3 Astatine is an element with several known unstable isotopes. One of the most stable isotopes is $^{210}_{85}At$.
   a Explain what is meant by an isotope.
   b Predict the physical state of astatine at room temperature. Suggest a melting point for astatine.
   c What colour would astatine be?
   d Write an ionic equation for the formation of the astatide ion.
   e How would the ionic radius of the astatide ion compare with the covalent radius of astatine?
   f Write an equation which represents the change occurring when the first ionisation energy is measured.
   g Explain why the first ionisation energy for astatine is the lowest in the halogen Group.

8.3 The hydrogen halides

All the hydrogen halides are colourless. With the important exception of hydrogen fluoride, HF, they are all gases with formulae of the type HX, and are highly soluble in water to give **strong acids,** i.e. acids which are completely dissociated into hydrated hydrogen ions and halide anions in dilute solution in water (see Ch 23.1)

$$HX(g) + H_2O(l) \rightleftharpoons H^{\oplus}(aq) + X^{\ominus}(aq)$$

Concentrated hydrochloric acid is a saturated solution of hydrogen chloride in water. At room temperature it contains about 35% by mass of hydrogen chloride. Most of the gas can be driven from the solution by warming.

The gaseous HX molecules are covalent but polar (see Ch 2.5). Their interaction with the polar water molecules helps them to ionise.

e.g. $\delta^{\ominus}Cl{-}H^{\delta\oplus}$ + $\delta^{\ominus}O \overset{H^{\delta\oplus}}{\underset{H^{\delta\oplus}}{\big<}} \rightleftharpoons Cl^{\ominus} \quad \overset{\oplus}{\underset{H \quad H}{O}}{\cdots}H$

> HCl is extremely soluble in water: 1 dm³ of water will dissolve over 200 dm³ of hydrogen chloride gas at room temperature and pressure!

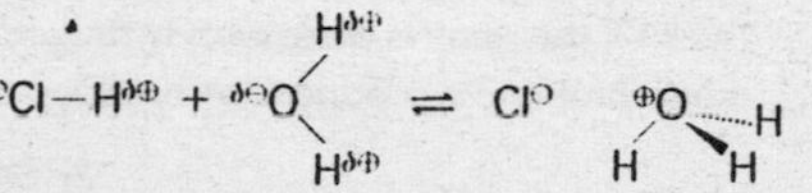

So far, we have discussed the molecules HX, where X is Cl, Br or I. Hydrogen fluoride is *very* different! The boiling point of HF is 20 °C; the boiling points of the other hydrogen halides are: HCl, −85 °C, HBr, −67 °C, and HI, −35 °C (*Figure 8.3.1*).

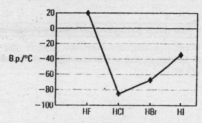

Figure 8.3.1 Boiling points of hydrogen halides

There is the expected change in boiling point from HCl through HBr to HI. Why is HF different?

The answer is that HF molecules are *so very polar* that they hydrogen bond strongly to each other (see *Figure 8.3.2*).

HF gas is a mixture of monomers and cyclic hexamers (HF)₆, although some *chain* dimers can exist at some temperatures and pressures. The solid is:

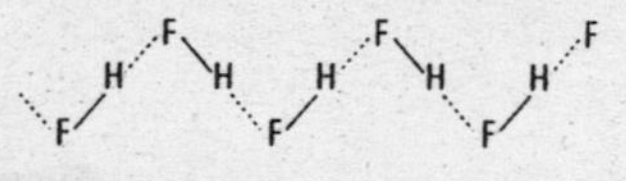

The zigzag chains show how HF₂⁻ ions can be formed.
The structure of the ion is linear (F—H—F)⁻ as in KHF₂, with hydrogen essentially midway between the two fluorine atoms.

Figure 8.3.2 Hydrogen bonding in hydrogen fluoride

In solution in water, hydrogen fluoride is only a **weak acid** – i.e. it is only slightly dissociated into hydrated hydrogen ions and fluoride anions in dilute solution. This is mostly because of the tendency of HF molecules to hydrogen bond strongly with each other rather than interact with water molecules.

Although only a weak acid, hydrogen fluoride is extremely reactive and corrosive (*and* highly poisonous). It attacks glass rapidly and is therefore used in etching. The problem of storage was solved by the discovery of Teflon (poly(tetrafluoroethene), see Ch 16.7).

The thermal stability of hydrogen halides

On the general principle that **long bonds are weak bonds** (*Table 8.3.1*; see also Ch 11.5), w should expect that hydrogen fluoride would be the most stable to heat of the hydrogen halides, and hydrogen iodide the least stable.

	nm
H—F	0.092
H—Cl	0.128
H—Br	0.141
H—I	0.160

Table 8.3.1 Covalent bond lengths

If a hot metal wire (e.g. Nichrome) is placed in a test tube of hydrogen chloride no change is seen. (It is no use trying this with hydrogen fluoride: it attacks not only the metal wire but the test tube as well.) If the same is done with hydrogen bromide, the contents of the tube become pale brown as some bromine is produced by break-up of the HBr molecules.

$$2HBr(g) \rightleftharpoons H_2(g) + Br_2(g)$$

With hydrogen iodide, the purple colour of iodine vapour is very clearly seen.

$$2HI(g) \rightleftharpoons H_2(g) + I_2(g)$$

These observations demonstrate that the stability of the hydrogen halides to heat is as expected.

The redox behaviour of hydrogen halides

When concentrated sulphuric acid is reacted with a fluoride or chloride, the hydrogen halide is produced; this is the standard way of making hydrogen fluoride and hydrogen chloride (*Figure 8.3.3*).

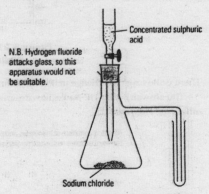

Concentrated sulphuric acid

N.B. Hydrogen fluoride attacks glass, so this apparatus would not be suitable.

Sodium chloride

Figure 8.3.3   Preparation of HCl

$$NaCl(s) + H_2SO_4(l) \xrightarrow{\text{room temperature}} NaHSO_4(s) + HCl(g)$$

Notice that the hydrogensulphate, $NaHSO_4$, is formed; a very high temperature is needed before the normal sulphate, Na_2SO_4, is a product.

When concentrated sulphuric acid, H_2SO_4, is added to a bromide, brown fumes of bromine are seen. Some of the hydrogen bromide which is formed in the reaction is oxidised to bromine by the sulphuric acid. As with all redox reactions (see Ch 4.3), this must mean a corresponding reduction, and some sulphuric acid is reduced to sulphur dioxide, SO_2.

The equations for the complex series of reactions involved when concentrated sulphuric acid reacts with sodium bromide can be written as:

$$NaBr(s) + H_2SO_4(l) \longrightarrow HBr(g) + NaHSO_4(s)$$

$$2HBr(g) + H_2SO_4(l) \longrightarrow Br_2(g) + SO_2(g) + 2H_2O(l)$$

Or overall:

$$2NaBr(s) + 3H_2SO_4(l) \longrightarrow Br_2(g) + SO_2(g) + 2NaHSO_4(s) + 2H_2O(l)$$

Do not try to remember these equations.

Can you recognise which elements have been oxidised and reduced in these reactions, and can you work out the oxidation states of these elements before and after the redox reaction has occurred?

When concentrated sulphuric acid is added to an iodide, however, the result is chaos. Most of the hydrogen iodide produced is oxidised at once to iodine; the acid is reduced not only to sulphur dioxide, but to sulphur and even to the evil-smelling and poisonous hydrogen sulphide, H_2S. What you get in the test tube is a foul mixture of black and yellow solids, purple vapour, choking fumes and a nauseating bad-egg smell. This experiment is NOT recommended!

The equations are not recommended either! In terms of half equations, the reactions involved are:

$$2NaI(s) \longrightarrow I_2(s) + 2Na^{\oplus} + 2e^{\ominus}$$

$$H_2SO_4(l) + 2H^{\oplus} + 2e^{\ominus} \longrightarrow SO_2(g) + 2H_2O(l)$$

Two iodide(−1) ions are *oxidised* to iodine(0) as one sulphur(+6) is *reduced* to sulphur(+4).

$$6NaI(s) \longrightarrow 3I_2(s) + 6Na^{\oplus} + 6e^{\ominus}$$

$$H_2SO_4(l) + 6H^{\oplus} + 6e^{\ominus} \longrightarrow S(s) + 4H_2O(l)$$

Six iodide(−1) go to iodine(0) as one sulphur(+6) goes to sulphur(0).

$$8NaI(s) \longrightarrow 4I_2(s) + 8Na^{\oplus} + 8e^{\ominus}$$

$$H_2SO_4(l) + 8H^{\oplus} + 8e^{\ominus} \longrightarrow H_2S(g) + 4H_2O(l)$$

Eight iodide(−1) go to iodine(0) as one sulphur(+6) goes to sulphur(−2).

A much more effective way of producing the hydrogen halides is to react the sodium or potassium salt with 100% phosphoric acid, H_3PO_4. This works because it is a non-volatile acid and has no oxidising properties.

$$\text{e.g.} \quad 2KBr(s) + H_3PO_4(s) \longrightarrow 2HBr(g) + K_2HPO_4(s)$$

1 The boiling points of hydrogen, hydrogen chloride and hydrogen fluoride are, respectively, −253 °C, −85 °C and 20 °C.

   a Write the formulae of each of the molecules mentioned, showing all the bonds between the atoms involved.

   b Explain, in terms of intermolecular attractions, why there is such a large difference between the boiling points of these substances.

   c If a hot glass rod is placed in a test tube of hydrogen iodide, purple fumes are seen. Explain what has happened and write a balanced equation for the reaction. What would you expect to see if the experiment was repeated with a test tube of hydrogen chloride?

2 Concentrated sulphuric acid, H_2SO_4, is an oxidising acid. In the course of an oxidising reaction at least three reduction products of sulphuric acid are possible.

a Write the formulae of three common reduction products of concentrated sulphuric acid.

b When NaCl, NaBr and NaI are reacted with concentrated sulphuric acid, distinctive reactions occur. Explain how you would identify the three sodium compounds using concentrated sulphuric acid. Be certain to mention what you might expect to see in each of the reactions.

c Write equations for any redox changes in each of the halide reactions with sulphuric acid.

3 a Describe the bonding and structure in sodium fluoride.

b What would you expect to see when sodium fluoride is mixed with water?

c Predict the reaction which would occur between concentrated sulphuric acid and sodium fluoride. Write a balanced equation for this reaction.

8.4 Chemical properties of the halogens

Preparation of the elements

A metal is obtained from its compounds by **reduction**, i.e. by forcing electrons back on to the positive metal ions to re-form metal atoms. The more reactive the metal, the harder this is to do. It should therefore be possible to produce non-metal elements, particularly from their ionic compounds, by **oxidation** – removal of electrons from the negative ions to re-form the atoms.

$$X^{\ominus} \longrightarrow X + e^{\ominus}$$
$$\text{negative ion} \qquad \text{non-metal atom}$$

This is the basic idea behind the preparation and production of the halogen elements.

You will remember, however, that fluorine is the *most electronegative* element. Once a fluorine atom has gained an electron to become a fluoride ion, *it is impossible to oxidise it using any chemical reagent*. The only available method involves electrolysis. One commercial process uses a mixture of potassium hydrogen fluoride, $K^{\oplus}HF_2^{\ominus}$ and anhydrous hydrogen fluoride, in the proportions of about 7:1 by mass, with a small amount of lithium fluoride to lower the melting temperature to below 100 °C.

Early attempts by Sir Humphry Davy and others to isolate fluorine were always unsuccessful and occasionally fatal. Success was achieved in 1886 by Ferdinand Frédéric Henri Moissan in France; he electrolysed a solution of potassium fluoride in anhydrous hydrofluoric acid in an apparatus made from fluorspar (calcium fluoride) and platinum. For some time fluorine was arguably the most expensive element on Earth, as Moissan's method meant that several grams of platinum were lost for every gram of fluorine obtained.

Moissan was the first to make boron. He devised a very high temperature carbon arc furnace (~3500 °C), which enabled him to make synthetic gemstones and other useful materials as well as to study the uncommon metals such as tantalum and niobium. He received the Nobel Prize for Chemistry in 1906 for his isolation of fluorine. He died the next year, aged 54; fluorine poisoning may have contributed to his early death.

Although chlorine is produced industrially by electrolysis of salt solution (see Ch 8.8), it can also be prepared by reacting concentrated hydrochloric acid with a suitably powerful chemical oxidising agent – for example, by dripping the acid onto crystals of potassium manganate(VII), $KMnO_4$.

$$16HCl(aq) + 2KMnO_4(s) \xrightarrow{\text{room temperature}} 5Cl_2(g) + 2MnCl_2(aq) + 2KCl(aq) + 8H_2O(l)$$

As chlorine is much denser than air it can be collected by downward delivery in the apparatus shown in *Figure 8.3.3*. A test for chlorine gas is to hold a piece of damp blue litmus paper in the suspected chlorine. If chlorine is present, the litmus goes red *and is then bleached white* almost immediately.

It was the density of chlorine which enabled it to be used as a war gas. Cylinders of chlorine gas were opened when a gentle breeze blew towards the enemy trenches. The dense gas stayed in reasonable concentration near ground level for long enough to do damage before diffusion, wind and convection currents dispersed it. Accounts written by survivors of the first attack exist, in which they tell how the green cloud slowly approached their positions, and of what happened when it got there. Read Wilfred Owen's poem *Dulce et decorum est*.

Bromide and iodide ions are easier to oxidise than chloride, and bromine and iodine may be prepared by heating sodium bromide or iodide with manganese(IV) oxide, MnO_2, which is a moderately powerful oxidising agent, and concentrated sulphuric acid. The half equations for the oxidation and reduction reactions are:

$$2Br^\ominus \longrightarrow Br_2(g) + 2e^\ominus$$
$$MnO_2(s) + 4H^\oplus + 2e^\ominus \longrightarrow Mn^{2\oplus} + 2H_2O(l)$$

Industrially, bromine is obtained from bromides in sea water by using chlorine – see below. Iodine is obtained in a similar way from concentrated solutions made from certain salt deposits, especially in Japan. It is also obtained by sulphur dioxide reduction of the sodium iodate(V), $NaIO_3$, which occurs as an impurity in sodium nitrate, $NaNO_3$, deposits of which are found in Chile.

Redox reactions of the halogens

Halogens as oxidising agents

As we have already seen, because of their readiness to gain one electron to complete the outer energy level in their atoms, the halogens are oxidising agents. The element fluorine is the most powerful oxidising agent among the halogens. Iodine is the least powerful.

When they act as *oxidising agents*, halogen atoms are themselves *reduced* to halide ions:

$$X + e^\ominus \longrightarrow X^\ominus$$

i.e. Reduction as the atom gains an electron.

These halide ions can in turn be oxidised back to halogen atoms. As we suggested above, iodide ions are the easiest to oxidise and fluoride by far the hardest:

$$X^\ominus \longrightarrow X + e^\ominus$$

i.e. Oxidation as the ion loses an electron.

But since the halogens are all diatomic:

$$2X^\ominus \longrightarrow X_2 + 2e^\ominus$$

If iodide ions are easy to oxidise, this must mean that they can easily give up electrons. *They are therefore quite powerful reducing agents.*

e.g. $$2Fe^{3\oplus}(aq) + 2I^\ominus(aq) \longrightarrow I_2(aq) + Fe^{2\oplus}(aq)$$

The situation is summed up in *Figure 8.4.1*.

F		F$^\ominus$	
Cl	increasing	Cl$^\ominus$	increasing
Br	oxidising	Br$^\ominus$	reducing
I	power	I$^\ominus$	power

Figure 8.4.1   The oxidising power of halogens and reducing power of halide ions

Displacement reactions

In general, a more reactive metal can *give up* electrons to the positive ions of a less reactive metal. Examples which you may have seen include putting zinc into blue copper(II) sulphate solution (the solution goes colourless, and a pink deposit of copper metal is obtained), and the exciting – but potentially very dangerous! – thermit reaction between iron(III) oxide and aluminium powder. (The energy released is so much that molten iron is formed.)

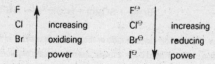

$$Cu^{2\oplus}(aq) + Zn(s) \xrightarrow{\text{room temperature}} Cu(s) + Zn^{2\oplus}(aq)$$

Each zinc atom loses two electrons; each copper ion gains two.

Putting zinc into copper(II) sulphate solution

Thermit reaction between iron(III) oxide and aluminium powder

$$Fe_2O_3(s) + 2Al(s) \xrightarrow{\text{ignite}} Al_2O_3(s) + 2Fe(s)$$

Each aluminium atom loses three electrons; each iron(III) ion gains three.

In the same way, a more reactive halogen can *remove* electrons from the negative ions of a less reactive one. For example, chlorine bubbled through a solution of potassium iodide causes a black precipitate of iodine to form:

$$Cl_2(g) + 2KI(aq) \longrightarrow 2KCl(aq) + I_2(aq)$$

There is a competition for electrons, and the stronger oxidising agent wins. So chlorine can displace both bromine and iodine, but bromine displaces only iodine.

> Fluorine is disqualified from this competition because it doesn't bother much with halide ions dispersed in the solution: it oxidises the water immediately, forming a mixture which includes oxygen difluoride, OF_2, and even some ozone, O_3!

$$Cl_2(g) + 2Br^-(aq) \longrightarrow Br_2(l) + 2Cl^-(aq)$$

$$Cl_2(g) + 2I^-(aq) \longrightarrow I_2(s) + 2Cl^-(aq)$$

$$Br_2(aq) + 2I^-(aq) \longrightarrow I_2(s) + 2Br^-(aq)$$

Bromine is produced commercially, often from sea water, by displacement with chlorine. Sea water only contains 65 ppm (parts per million) of bromine by mass (0.0065%); which doesn't seem much until we realise each cubic kilometre of sea water must therefore contain about 65 000 tonnes of bromine.

Filtered sea water is acidified to about pH 3.5; chlorine is then passed in:

$$2Br^-(aq) + Cl_2(g) \longrightarrow Br_2(aq) + 2Cl^-(aq)$$

Air is blown through; this removes bromine from the mixture as its vapour. Sulphur dioxide and water are mixed with the bromine vapour to re-form a solution:

$$Br_2(g) + SO_2(g) + H_2O(l) \longrightarrow 4H^+(aq) + SO_4^{2-}(aq) + 2Br^-(aq)$$

So far, the bromide concentration in the solution has been increased to 1500 times more than it is in sea water. This solution is then dripped down a tower against a stream of chlorine, steam being blown up the tower. The mixture, which emerges from the top of the tower, condenses to two layers of liquid – an aqueous layer on top of impure bromine. The bromine is purified by distillation.

The reactions of halogens with alkalis

Acids and alkalis

An acid is a proton donor; if soluble, it will give rise to a solution containing a high concentration of hydrated protons, $H^+(aq)$. It is this solution in water which is usually described as 'acid'.

The hydrated proton is often written as $H_3O^\oplus$. This is misleading, as it implies that the proton is stuck to one water molecule. It is, in fact, shared among a hydrogen-bonded clump of water molecules; the most frequent protonated species in water is actually $H_9O_4^\oplus$ – i.e. a proton shared among a group of four H_2O molecules. But we will still use $H^\oplus(aq)$ or $H_3O^\oplus(aq)$ when it is appropriate.

An example is hydrogen chloride – or, indeed, any hydrogen halide other than hydrogen fluoride (see Ch 8.3).

$$HCl(aq) + aq \rightleftharpoons H^\oplus(aq) + Cl^\ominus(aq)$$

A base is a proton acceptor. An alkali is a soluble base; the solution contains a high concentration of hydroxide ions, $OH^\ominus$. The best known example is sodium hydroxide ('caustic soda'), NaOH.

Acids and bases mutually neutralise each other. An acid and an alkali always react to form a *salt* and *water*.

e.g.
$$Na^\oplus OH^\ominus(aq) + H^\oplus Cl^\ominus(aq) \longrightarrow Na^\oplus Cl^\ominus(aq) + H_2O(l)$$

If we eliminate the *spectator ions* – the ions which do not actually take part in the reaction – the equation for *any* acid–alkali reaction becomes:

$$H^\oplus(aq) + OH^\ominus(aq) \longrightarrow H_2O(l)$$

We will meet acid–alkali reactions again in Ch 10 and Ch 13 and they are discussed fully in Ch 23.

The halogens and water

As mentioned above, fluorine attacks water violently. Chlorine is moderately soluble. At 0 °C, one dm^3 of water (one litre) dissolves about 4.5 dm^3 of chlorine gas. This may sound a lot, but is only about 13 g of chlorine. The solution is faintly green (from the dissolved chlorine) and contains the chloride and chlorate(I) ions, $Cl^\ominus$ and $ClO^\ominus$, as well as molecules of HCl and HClO.

Overall:

$$Cl_2(g) + H_2O(l) \rightleftharpoons HCl(aq) + HClO(aq) \rightleftharpoons 2H^\oplus(aq) + ClO^\ominus(aq) + Cl^\ominus(aq)$$

In dilute solution, HCl(aq) ionises fully but HClO is still mostly molecular. Bromine dissolves to about the same extent as chlorine, giving a similar mix of ions ($Br^\ominus$ and $BrO^\ominus$) and acid molecules (HBr and HBrO). The solution is brown. Iodine is only slightly soluble; the faint brown solution again contains $I^\ominus$, $IO^\ominus$, HI and HIO.

The halogens and alkalis – disproportionation

We shall only consider the reaction of chlorine with sodium hydroxide solution. (Virtually *all* the following reactions happen for *any* halogen – always excepting fluorine! – with *any* alkali).

The equation for the reaction which happens when chlorine is bubbled into *cold, dilute* alkali to form chloride and chlorate(I) ions is:

$$Cl_2(g) + 2OH^\ominus(aq) \xrightarrow{\text{room temperature}} Cl^\ominus(aq) + ClO^\ominus(aq) + H_2O(l)$$

Cl oxidation state   0   −1   +1

The oxidation states of chlorine in the molecules and ions are shown (see Ch 4.3).

All the halogens, other than fluorine, form a number of anions containing oxygen with the halogen atom in positive oxidation states. Remember the name of the compound has the oxidation state of the halogen atom in brackets after the name of the anion (ending in -ate).
e.g.
    sodium iodate(I) is NaIO;
    sodium chlorate(III) is $NaClO_2$;
    potassium bromate(V) is $KBrO_3$;
    potassium chlorate(VII) is $KClO_4$.

Notice that we started with chlorine as an element, and therefore with its oxidation state as zero. During the reaction, half the chlorine atoms have been *oxidised* (to a higher oxidation state) and half have been *reduced* (to a lower oxidation state). This is an example of **disproportionation**. Here, $Cl(0)$ ends up as $Cl(-1)$ and $Cl(+1)$.

> Another example of disproportionation we shall meet is copper(I) changing to copper metal and copper(II) ions:
>
>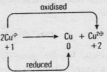
>
> The copper $+1$ acts as both the oxidising agent and the reducing agent, forming Cu^0 and Cu^{+2}.

> **Definition:** Disproportionation occurs when, during a reaction, an element in an intermediate oxidation state is simultaneously oxidised and reduced.

If the solution is now heated, further disproportionation occurs:

$$3ClO^\ominus(aq) \xrightarrow{heat} 2Cl^\ominus(aq) + ClO_3^\ominus(aq)$$

Cl oxidation state $\quad +1 \qquad\qquad -1 \quad\ +5$

(Note here that $3 \times (+1) = 2 \times (-1) + 5$. The sum of the oxidation states of chlorine stays constant. This must happen in *any* disproportionation reaction.)

Obviously, if chlorine is bubbled directly into hot, concentrated sodium hydroxide solution, the two reactions happen simultaneously; so, overall, the equation is:

$$3Cl_2(g) + 6OH^\ominus(aq) \longrightarrow 5Cl^\ominus(aq) + ClO_3^\ominus(aq) + 3H_2O(l)$$

Cl oxidation state $\quad\ 0 \qquad\qquad\qquad -1 \quad\ +5$

Here we have shown the oxidation states of chlorine below the symbols in the equation. Note that $6 \times 0 = 5 \times (-1) + 5$ – once again we see confirmation that this reaction is a true disproportionation: the sum of the oxidation states of chlorine on both sides of the equation is the same.

The $ClO_3^\ominus$ ion is known as chlorate(V). If sodium or potassium chlorate(V) is heated gently, yet more disproportionation occurs – this time, to chloride and chlorate(VII):

$$4ClO_3^\ominus \xrightarrow{heat} Cl^\ominus + 3ClO_4^\ominus$$

Cl oxidation state $\qquad +5 \qquad\quad -1 \quad\ +7$

checking $\qquad\qquad 4 \times (+5) = -1 + 3 \times (+7) = 20$

Interhalogen compounds

The halogens can react directly with each other to form a number of compounds, including ICl and ICl_3. An interesting series is the fluorides of chlorine, bromine and iodine – all form a monofluoride, but chlorine also forms ClF_3, bromine BrF_3 and BrF_5, and iodine IF_3, IF_5 and IF_7. The maximum number of fluorine atoms capable of joining with the other halogen atom seems to be determined by the size of that other atom and therefore the space for the fluorine atoms to fit round it. Normally, fluorine brings out the maximum oxidation state of elements it combines with.

SELF-ASSESSMENT QUESTIONS 8.4

1 Sodium chlorate(I) is used in solution as a commercial bleach. It is formed when chlorine gas is bubbled through a cold dilute solution of an alkali. In the reaction, chlorine disproportionates to chlorate(I) and chloride ions.
 a What is meant by disproportionation?
 b Write an equation for the reaction.
 c In hot concentrated alkali solution a different disproportionation reaction occurs. What are the formulae of the two chlorine-containing products of this reaction?
 d If the chlorine is bubbled through water instead of alkali, a pale green solution is formed. What do you think gives rise to the green solution?

2 Write balanced equations for the following reactions.
   a Bromine dissolving in water.
   b Iodine reacting with potassium hydroxide solution.
   c Chlorine reacting with potassium hydroxide at 20 °C.
   d Chlorine reacting with potassium hydroxide at 80 °C.

3 The oxidising power of the halogen decreases as the atomic number of the halogen increases.
   a How would you use this fact, and a solution of chlorine in water, to distinguish between potassium chloride, potassium bromide and potassium iodide solutions? Write balanced ionic equations for any chemical changes that occur.
   b In one of these reactions, identify the species being oxidised and the species being reduced. Make sure you show the oxidation states of the elements involved in the redox reaction.

4 Iodine monochloride is formed when iodine is treated with chlorine gas. In excess chlorine the reaction goes further and iodine trichloride is formed.
   a Write an equation for the formation of iodine monochloride.
   b Draw a dot-and-cross diagram to represent the bonding in iodine monochloride.
   c What are the oxidation states of iodine in iodine monochloride and in iodine trichloride?
   d Iodine monochloride readily decomposes to the elements. Write a balanced equation for this reaction.

8.5 Reactions of the halates

Except for fluorine, all the halogens can exist in positive oxidation states. These are usually found in compounds which also include oxygen. Such compounds include oxides, oxoacids and the salts of such acids; the salts are usually called halates. The positive oxidation states most frequently met with are $+1$, $+3$, $+5$ and even $+7$. These oxidation states occur for chlorine, bromine and iodine. *Table 8.5.1* shows all the possible oxidation states of chlorine, illustrated by known compounds.

The oxidation states of chlorine

$+7$	Cl_2O_7	$HClO_4$	$Na^{\oplus}ClO_4^{\ominus}$	
$+6$	Cl_2O_6			
$+5$		$HClO_3$	$Na^{\oplus}ClO_3^{\ominus}$	
$+4$	ClO_2			
$+3$		$HClO_2$	$Na^{\oplus}ClO_2^{\ominus}$	ClF_3
$+2$				
$+1$	Cl_2O	$HClO$	$Na^{\oplus}ClO^{\ominus}$	
0	Cl_2			
-1		HCl	$Na^{\oplus}Cl^{\ominus}$	PCl_3

Table 8.5.1   The oxidation states of chlorine

All the oxides of chlorine are unstable, some dangerously so. Chlorine(IV) oxide, ClO_2, is nonetheless an industrial oxidising agent which is used, among other things, for bleaching flour.

Chloric(I) acid, $HClO$, is a weak acid; the more oxygen, the stronger the acid, so that $HClO_4$ is relatively strong and ionises well in dilute solution. This is because the anion formed during ionisation is more stable if there are more electronegative oxygen atoms over which the negative charge can be spread; i.e. the extra electron can move over these atoms, and is to some extent **delocalised.** The shapes of the halate ions are shown in *Figure 8.5.1*.

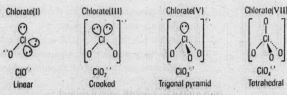

Figure 8.5.1   Shapes of halate ions

(See Ch 2.6 if you need reminding of *why* they have these particular shapes).

All the halate compounds are powerful **oxidising agents**. A halogen in a positive oxidation state is essentially unstable, and will react whenever possible to get to the more stable −1 oxidation state.

The chlorate(I) ion, $ClO^{\ominus}$, is only stable in solution. Such solutions are used in household and industrial bleaches and disinfectants. (The disinfectant action is because pathogenic bacteria are sensitive to oxidising agents – including atmospheric oxygen.)

Sodium or potassium chlorate(V) breaks down when heated:

	$2NaClO_3(s)$	$\xrightarrow{heat}$	$2NaCl(s)$	+	$3O_2(g)$
oxidation state	+5 −2		−1		0
checking	$[2 \times (+5)] + [6 \times (-2)]$	=	$2 \times (-1)$	+	$6 \times (0) = -2$

This reaction is catalysed by manganese(IV) oxide, MnO_2; potassium manganate(VII) is formed as an intermediate (see Ch 12.4). Notice that here the oxidation state of the chlorine changes from +5 to −1. To balance the two chlorines changing from +5 to −1, six oxygens must change from −2 to 0; i.e.

for chlorine: $2 \times (+5) = +10$ to $2 \times (-1) = -2$

for oxygen: $6 \times (-2) = \dfrac{-12}{-2}$ to $6 \times (0) = \dfrac{0}{-2}$

As already mentioned in Ch 8.4, gentle heating of chlorate(V) gives a mixture of chlorate(VII) and chloride by disproportionation.

$$4ClO_3^{\ominus} \xrightarrow{heat} Cl^{\ominus} + 3ClO_4^{\ominus}$$

Chlorate(VII) salts are extremely powerful oxidising agents and are used in rocket propellants.

Sodium chlorate(V), $NaClO_3$, was widely used as a weed killer. However, it is a powerful oxidising agent which also assists combustion. It is reported that one unfortunate man decided to kill all the grass on his lawn in order to pave it over. He watered the lawn with weed killer solution; and a fortnight of drought followed during which the grass died and dehydrated, and the dried cellulosic material was left mixed with crystals of sodium chlorate(V) … When he walked out to inspect his handiwork, the friction of his shoes caused ignition, the whole area flared up, and he eventually died of burns. Halates must never be treated carelessly!

SELF-ASSESSMENT
QUESTIONS
8.5

1   Work out the oxidation states of the halogen atom in the following compounds:
   a $CsClO_4$
   b $HBrO$
   c HIO_3
   d $Ca(ClO_2)_2$

2   Write a systematic name for each of the following compounds:
   a $NaClO_3$
   b HIO_4
   c I_2O_5
   d $KBrO_3$

Bleaching occurs where a dyestuff or other colouring material is oxidised so that its molecular structure no longer absorbs light in the visible region of the spectrum – i.e. it becomes colourless. For some dyestuff structures, see Ch 28.4.

8.6   Metal halides

Nearly all metals form compounds with the halogens, which are usually present as the halide ion, $X^\ominus$. The bonding in compounds such as potassium chloride, KCl, is very largely ionic.

However, the smaller the metal ion and the greater the charge that would be on it, the higher its **charge density** and consequent **polarising power** (see Ch 2.5), and the more covalent the resulting metal–chlorine bond (*Figures 8.6.1* and *8.6.2*).

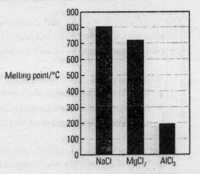

Figure 8.6.1   **Melting points of some metal halides**

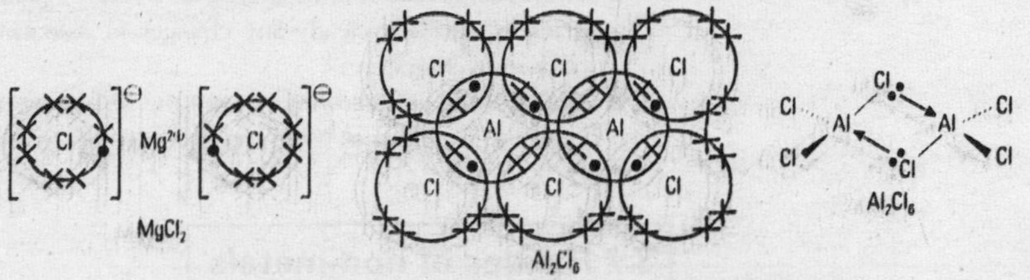

Figure 8.6.2   Structures of some metal chlorides

The result is that, whereas NaCl and $MgCl_2$ have the reasonably high melting points associated with giant ionic structures, $AlCl_3$ sublimes at 180 °C and must therefore be largely covalent (see Ch 5.5, *Table 5.5.3*). Its structure in the vapour phase is Al_2Cl_6, as shown (*Figure 8.6.2*).

Because of its largely covalent nature and because of the powerful effect that the tiny, highly-charged $Al^{3\oplus}$ ions have on water molecules, $AlCl_3$ reacts rapidly with water to give hydrogen chloride gas. It is **hydrolysed**, i.e. broken up by water, in a way very similar to that of the non-metal chlorides $SiCl_4$ and PCl_5.

$$AlCl_3(s) + 3H_2O(l) \xrightarrow{\text{room temperature}} Al(OH)_3(s) + 3HCl(aq)$$

For metals which possess more than one oxidation state, compounds of the higher states tend to be more covalent in character. For example, iron(II) chloride, $FeCl_2$, is easily oxidised in aqueous solution. If iron(III) chloride, $FeCl_3$, is placed in water, a brown sludge of the hydrated iron(III) oxide forms and the solution becomes highly acidic, containing hydrochloric acid. Such solutions will have a pH much lower than 7.

$$FeCl_3(s) + 3H_2O(l) \rightleftharpoons Fe(OH)_3(s) + 3HCl(aq)$$

Many of the metal chlorides are soluble in water; e.g. sodium chloride readily dissolves in water to form a solution of pH7.

$$NaCl(s) \xrightarrow[\text{room temperature}]{H_2O} Na^{\oplus}(aq) + Cl^{\ominus}(aq)$$

207

Exceptions which you may meet include silver chloride, AgCl, and lead(II) chloride, $PbCl_2$. The lead chloride is, however, soluble in hot water.

The silver halides show an interesting gradation. Silver fluoride is freely soluble in cold water; silver chloride is only very slightly soluble, and silver bromide and silver iodide are virtually insoluble. This is the basis of the detection of chloride, bromide and iodide ions in solution by using silver nitrate solution acidified with dilute *nitric* acid – see Ch 8.10. (It has to be nitric acid: the other common acids would form precipitates.)

$$X^{\ominus}(aq) + AgNO_3(aq) \xrightarrow[\text{room temperature}]{HNO_3(aq)} AgX(s) + NO_3^{\ominus}(aq)$$

The reactions of sodium and potassium halides with concentrated sulphuric acid have been discussed in Ch 8.3. Metal chlorides give hydrogen chloride gas; the bromides and iodides give more complex mixtures because of redox reactions involving HBr and HI as well as the sulphuric acid.

SELF-ASSESSMENT
QUESTIONS
8.6

1 What would you expect to see if the following aqueous solutions were mixed together?
  a Potassium chloride and silver nitrate.
  b Aqueous chlorine and methyl orange indicator.
  c Caesium bromide and silver nitrate acidified with dilute nitric acid.

2 On strong heating, potassium chlorate(V) decomposes to potassium chloride and oxygen.
  a What is the formula of potassium chlorate(V)?
  b Write an equation for the reaction described.
  c The decomposition is described as a redox reaction. Explain what is meant by a redox reaction, showing clearly any changes in oxidation states during the decomposition of potassium chlorate(v).
  d After heating and allowing to cool the resultant solid is added to water. What might you expect to *see* happen? Write an equation to show any changes that occur.

8.7 Halides of non-metals

The discussion here will mostly be limited to chlorides, as these are the most common of the halides of non-metals.

The chlorides of silicon, phosphorus and sulphur can be prepared by direct combination with chlorine in an air-free apparatus. Silicon requires heating; sulphur should be molten; white phosphorus need only be warmed.

$$Si(s) + 2Cl_2(g) \xrightarrow{heat} SiCl_4(l)$$
$$P_4(s) + 6Cl_2(g) \longrightarrow 4PCl_3(l)$$
$$S_8(l) + 4Cl_2(g) \longrightarrow 4S_2Cl_2(l)$$

The usual compounds obtained are $SiCl_4$, PCl_3 and – rather unexpectedly – S_2Cl_2. All these are liquid at room temperature. With excess chlorine, phosphorus can also form PCl_5, in which the outer electron energy level of the phosphorus has expanded to hold ten electrons. Phosphorus pentachloride dissociates into phosphorus trichloride and chlorine when heated.

$$PCl_5(s) \underset{cool}{\overset{heat}{\rightleftharpoons}} PCl_3(l) + Cl_2(g)$$

The properties of the chlorides of the elements of Period 3 (Na → Ar) have already been summarised in Ch 5.5, *Table 5.5.3*. **All chlorides of non-metals react rapidly with cold water** (i.e. they are **hydrolysed) to release hydrogen chloride gas**. The only exception to this rule is tetrachloromethane, CCl_4. The reason *why* the CCl_4 molecule is unaffected by water is fully discussed in Ch 7.3.

The reactions of the silicon, phosphorus and sulphur chlorides could be represented as follows:

$$SiCl_4(l) + 4H_2O(l) \xrightarrow[\text{temperature}]{\text{room}} SiO_2.2H_2O(s) + 4HCl(g)$$

$$PCl_3(l) + 3H_2O(l) \xrightarrow[\text{temperature}]{\text{room}} \underset{\text{phosphonic acid}}{H_3PO_3(s)} + 3HCl(g)$$

$$2S_2Cl_2(l) + 3H_2O(l) \xrightarrow[\text{temperature}]{\text{room}} H_2SO_3(aq) + 3S(s) + 4HCl(g)$$

With phosphorus pentachloride, the reaction can happen in two stages.

- With the correct mass of water:

$$PCl_5(s) + H_2O(l) \xrightarrow[\text{temperature}]{\text{room}} \underset{\text{phosphorus trichloride oxide}}{PCl_3O(l)} + 2HCl(g)$$

- With excess water and warming:

$$PCl_3O(l) + 3H_2O(l) \xrightarrow{\text{warm}} \underset{\text{phosphoric(V) acid}}{H_3PO_4(aq)} + 3HCl(aq)$$

Both the phosphorus chlorides react with the –OH group in alcohols and the –COOH in carboxylic acids to form chloroalkanes and acyl chlorides (see Ch 18.4 and Ch 27.7).

Phosphorus can form the bromides PBr_3 and PBr_5, and also PI_3 but not PI_5.

Sulphur reacts with fluorine to form sulphur hexafluoride, SF_6. In this, the sulphur atom has *twelve* electrons in its outer level. The molecule is perfectly octahedral and extremely unreactive (*Figure 8.7.1*). It is used as an insulator in high voltage electrical gear.

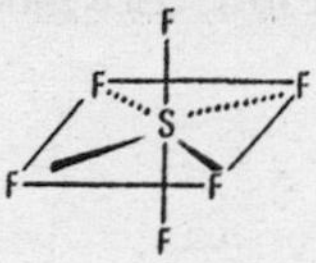

Figure 8.7.1

In Period 2 (Li → Ne) the chloride of carbon, CCl_4, is very stable, as already mentioned. The chloride of nitrogen, NCl_3, and the chlorides of oxygen (i.e. the chlorine oxides, Cl_2O, ClO_2, Cl_2O_6, Cl_2O_7) are notoriously unstable and explosive.

Nitrogen tri-iodide, NI_3, is a dark solid which, when dry, is so sensitive to detonation that one of the authors has used it as a fly killer. The damp crude material was spread thinly on a windowsill with fragments of food. When it dried, if a fly landed on it there was a sharp crack, a puff of purple vapour, and no fly!

It is now realised that both the preparation and storage of nitrogen tri-iodide, even in small amounts, are full of hazards and should never be attempted.

The related compound, nitrogen trichloride, NCl_3, is a yellow liquid which is equally dangerous and has caused fatal accidents.

SELF-ASSESSMENT QUESTIONS 8.7

1 Phosphorus can form a wide range of compounds with halogen atoms:
   PF_3, PF_5, PCl_3, PCl_5, PBr_3, PBr_5, PI_3.

   a Draw a dot-and-cross diagram, showing only the outer energy level electrons, for PBr_3.

   b Explain why this molecule does not have a trigonal planar structure. What is the shape of the PBr_3 molecule?

   c What are the oxidation states of phosphorus in PBr_5 and in PCl_3?

   d Although PF_5, PCl_5 and PBr_5 all exist, PI_5 is not known. Can you suggest a reason why not?

   e When a bottle containing PCl_5 is opened, white misty fumes are formed as the compound reacts with the moisture in the air. Write an equation for this reaction.

2 The boiling points of three fluorides are as follows:

$$SiF_4 \qquad -86\,°C$$
$$PF_5 \qquad -75\,°C$$
$$SF_6 \qquad -64\,°C$$

a Explain in terms of intermolecular attractions why these compounds are all gases at room temperature and pressure.
b Draw diagrams to show the structures and bond angles in each of these compounds.
c What is the oxidation state of sulphur in SF_6?
d Although oxygen is in the same Group of the Periodic Table as sulphur, the compound OF_6 does not exist. Why not?

8.8 The chlor-alkali industry

Brine – a saturated solution of sodium chloride, containing about 25% by mass of sodium chloride – can be obtained by solution mining of rock salt deposits, e.g. in Cheshire. Water is forced down a pipe which is bored down to the deposits, and brine is pumped up. The brine is then electrolysed to give a range of useful products.

The electrolysis can be carried out in a variety of cells, depending on which product is most required and in what degree of purity. Only the **diaphragm cell** will be discussed here.

The basic cell is shown in *Figure 8.8.1*. In practice, many such cells would be combined in one unit.

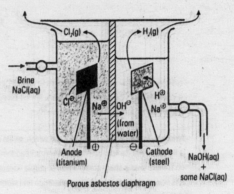

Figure 8.8.1 The diaphragm cell

Electrolysis will be treated in detail in Ch 24.8. You may have met some of the ideas during your GCSE course. We shall use here only those ideas which are required to understand how the cell works.

Water is always very slightly ionised:

$$H_2O(l) \rightleftharpoons H^{\oplus}(aq) + OH^{\ominus}(aq)$$

Sodium chloride, $Na^{\oplus}Cl^{\ominus}$, is an ionic compound and therefore fully ionises when in solution.

$$NaCl(s) \xrightarrow{H_2O} Na^{\oplus}(aq) + Cl^{\ominus}(aq)$$

The ions present in the brine are therefore:

In electrolysis, electrical conduction is caused by a flow of ions. A material which, when molten or in solution, allows conduction is called an **electrolyte**. When conduction occurs in an electrolyte, it is **always decomposed**. The breaking-up of a substance by conduction of electricity through it is termed **electrolysis**.

Any **positive** ions in the electrolyte travel towards the **negative** electrode – the **cathode** – where they *pick up* electrons to neutralise their positive charge. Such ions are called **cations** (not cathions, as we might expect).

e.g. With molten sodium chloride:

$$Na^{\oplus} + e^{\ominus} \longrightarrow Na(l)$$

Any **negative** ions travel towards the **positive** electrode – the **anode** – where they *lose* electrons in order to become neutral. These negative ions are called **anions**.

$$2Cl^{\ominus} \longrightarrow Cl_2(g) + 2e^{\ominus}$$

If more than one ion of each kind is present, the order in which they are **discharged** (i.e. lose their positive or negative charge) on an electrode is determined not only by how hard it is to force electrons on to a positive ion or pull the electrons off a negative ion, but also by the *relative concentrations* of the ions in the solution.

That should be enough for the moment.

Remember:
Conduction in metals involves a flow of *electrons*; and *no* chemical changes occur in the metal (see Ch 3.6).

In the cell, brine is fed into the anode compartment, which is separated from the cathode compartment by the porous asbestos diaphragm.

A diaphragm is essential is because the chlorine and sodium hydroxide produced would otherwise react immediately, and then we would be back to where we started from!

The level in the anode compartment is held slightly higher, so there is a small positive pressure towards the cathode.

The anode is made of titanium – expensive, but it resists attack from the chlorine.

- The two ions which could discharge at the anode are hydroxide, $OH^{\ominus}$, and chloride, $Cl^{\ominus}$. Hydroxide discharges to form oxygen, and chloride to form – of course! – chlorine.

$$4OH^{\ominus}(aq) \longrightarrow O_2(g) + 2H_2O(l) + 4e^{\ominus}$$
$$2Cl^{\ominus}(aq) \longrightarrow Cl_2(g) + 2e^{\ominus}$$

By using saturated brine solution, chlorine is produced rather than oxygen – although the chlorine always contains a little oxygen and may require purification.

- The ions which could discharge at the cathode are $H^{\oplus}$ and $Na^{\oplus}$. As expected, it is far harder to force an electron back on to the sodium ion, and hydrogen gas is produced. A steel cathode can be used.

$$2H^{\oplus}(aq) + 2e^{\ominus} \longrightarrow H_2(g)$$

As $H^{\oplus}$ ions are removed by discharge on the cathode, more water molecules ionise, so producing more $OH^{\ominus}$ ions. These, together with the $Na^{\oplus}$ ions already present, form a solution of sodium hydroxide, $Na^{\oplus}OH^{\ominus}$.

- The solution which is led off from the cathode compartment contains about 10% sodium hydroxide and 15% sodium chloride by mass. If this is evaporated to about one-fifth of its original volume, most of the sodium chloride crystallises (it is much less soluble than sodium hydroxide). The resulting solution contains 50% sodium hydroxide by mass and less than 1% sodium chloride.

● The products from the cell are hydrogen, chlorine, and aqueous sodium hydroxide.

Hydrogen and chlorine can be reacted together to form hydrogen chloride and hence hydrochloric acid.

$$H_2(g) + Cl_2(g) \xrightarrow{\text{catalyst}} 2HCl(g)$$

$$HCl(g) \xrightarrow[\text{room temperature}]{H_2O} HCl(aq)$$

● Chlorine and cold sodium hydroxide solution react to form a mixture of aqueous sodium chloride and sodium chlorate(I) (see Ch 8.4). Sodium chlorate(I) is a powerful oxidant, and is sold as a sterilising agent, some of it under the trade name 'Milton'.

$$Cl_2(g) + 2NaOH(aq) \xrightarrow[\text{temperature}]{\text{room}} NaCl(aq) + NaClO(aq) + H_2O(l)$$

Ionically: $Cl_2(g) + 2OH^\ominus(aq) \longrightarrow Cl^\ominus(aq) + ClO^\ominus(aq) + H_2O(l)$

The British production of chlorine is about 2 million tonnes per annum. World production is about 40 million tonnes..

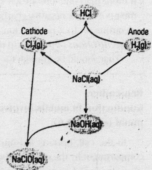

A summary of the electrolysis of brine

1 Brine is an important commercial feedstock for the chemical industry.
   a Explain the principles of the diaphragm cell for the electrolysis of brine.
   b What are the two main gaseous products in the electrolysis process?
   c Which gas appears at the cathode? Write an equation to represent the discharge of ions at the cathode.
   d Which gas is formed at the anode? Write an equation for the reaction at the anode.
   e The solution in the cell and the gas produced at the anode can be reacted together. What are the products in this reaction if the reaction is carried out at room temperature?
   f How does the reaction change if the reaction in e is carried out at 80 °C?

8.9 Uses of halogens and halogen compounds

The halogens in human biology

Although all the halogens and – to some extent – halide ions are toxic, fluoride, chloride and iodide are essential components in the human diet. The human body contains about 2 g of fluoride. Laboratory animals kept on fluoride-free diets became anaemic and infertile and failed to grow properly.

Chloride is usually taken as salt, sodium chloride; the average Briton eats almost 10 g of salt a day, which is roughly three times more than is needed.

Iodine is concentrated in the body in the thyroid gland; it is incorporated in the thyroid hormone, thyroxine, which partly controls the body's metabolic rate.

A deficiency of iodine in the diet – as used to be common in some Alpine valleys – leads to a great enlargement of the thyroid gland and consequent swellings around and hanging from the neck. This condition is known as goitre and is often accompanied by mental incapacity; the unfortunate sufferer was known as a 'cretin'. So, a word which is now a rather silly term of mild abuse should only be used for those who suffer from the effects of iodine deficiency.

One of the products of nuclear fission, e.g. in atomic bomb explosions, is the radioactive isotope ^{131}I. During the Cold War local authorities held stocks of tablets of potassium iodate(V), KIO_3. These were to be issued to children in the event of nuclear warfare. Their thyroid glands would then have plenty of 'healthy' iodine and be less likely to absorb the ^{131}I released, which would cause severe radiation damage and mutation. Thankfully, they have never had to be used.

In adults, thyroid cancers can often be successfully treated using controlled doses of ^{131}I, which concentrates in the tumour and destroys it.

Bromine appears to have no essential rôle in the human diet, although bromides were once used as sedatives.

Fluorine

Addition of fluoride to drinking water in very low concentrations (1–2 ppm) has been shown to reduce tooth decay, as has the use of 'fluoride toothpaste', which contains almost 1000 ppm of fluoride in the form of sodium monofluorophosphate.

Hydrogen fluoride (hydrofluoric acid) attacks glass, and is used for etching and 'frosting' glass. It has also been used in the production of chlorofluorocarbons (CFCs), e.g. CCl_2F_2, used as refrigerants and aerosol propellants. CFCs are excellent for these uses, but because of their chemical inertness they can diffuse unchanged to the upper atmosphere and cause potentially major problems on a global scale by weakening the Earth's protective ozone layer (see Ch 17.2)

Hydrogen fluoride can be used to make artificial cryolite, Na_3AlF_6, an important material which is used in the production of aluminium by electrolysis. It is also used to make the volatile uranium compound, UF_6, which is the key to enriching natural uranium with the isotope ^{235}U for use in nuclear power stations. Hydrogen fluoride is also used in making the almost frictionless, chemically inert, non-stick material poly(tetrafluoroethene) – better known as PTFE or 'Teflon'.

Chlorine

Water treatment works

A vital use is in sterilising drinking water.

This has undoubtedly saved many tens of millions of lives throughout the world, by eliminating the micro-organisms responsible for water-borne diseases such as dysentery, cholera and typhoid fever wherever it is used properly. (But, of course, as we saw in Ch 8.2, chlorine also killed many thousands of soldiers in the First World War. Chlorine illustrates how scientific knowledge can be used for good or evil, depending on the attitudes and prejudices of, and the pressures on, the people involved in making decisions.)

More than a quarter of all the chlorine produced in the UK is used in making the versatile plastic, poly(chloroethene), better known as PVC (polyvinyl chloride – see Ch 16.7). Much is also used in making chlorinated solvents for industrial use or for dry-cleaning. Chlorine is also used directly as a bleaching agent, and for making other bleaching agents such as 'bleaching powder' (calcium hydroxide – 'slaked lime' – treated with chlorine) and sodium chlorate(I).

Bromine

Silver bromide is used in large quantities in photographic emulsions. Bromine is still used to make 1,2-dibromoethane, $BrCH_2—CH_2Br$ (see Ch 16.5), a petrol additive used together with tetraethyllead(IV) ('tetraethyl lead', TEL) – which is now being removed from petrol to reduce lead pollution.

Bromine is also used in making dyes, drugs, pesticides and inhaled anaesthetics such as halothane, $CF_3CHBrCl$.

Additives are put into petrol to minimise the formation of unwanted free radicals. (Free radicals are highly reactive species with unpaired electrons – see Ch 14.6.)

Tetraethyllead(IV) used to be added since it breaks up readily to form ethyl radicals.

$$(C_2H_5)_4Pb \longrightarrow 4C_2H_5\cdot + Pb$$

(The dot is used to show the unpaired electron)

The ethyl free radicals react with any other free radicals formed during combustion and stop unwanted reactions proceeding. The 1,2-dibromethane was added to remove lead from the engine ...

| Iodine |

Iodine, dissolved in ethanol or in potassium iodide solution, used to be widely used as an antiseptic. It is still used in the manufacture of other antiseptics. Iodine is also used to detect the presence of starch (and starch is used to detect iodine); the two react together to form a deep blue colour. Silver iodide is used to some extent in photography.

SELF-ASSESSMENT QUESTIONS 8.9

1 Chlorine reacts slightly when it is bubbled through water producing two halogen-containing compounds.
  a Write an equation for the reaction.
  b This reaction forms the basis of the use of chlorine to purify drinking water supplies. Which of the two compounds produced is likely to have oxidising properties?
  c Which of the chemicals in your equation is likely to be responsible for the removal of bacteria in the water supplies?

2 a Silver bromide is used in photography because it is light sensitive – i.e. it decomposes in sunlight. Write an equation for the chemical change that occurs in sunlight.
  b Iodine reacts when added to potassium iodide solution as follows.

$$KI(aq) + I_2(s) \rightleftharpoons KI_3(aq)$$

  Explain why the solubility of iodine in water is very much lower than the solubility of KI_3.
  c Give two ways in which fluorine is not a typical halogen.

8.10 Chemical tests for halogens and halide compounds

These tests are based on properties and reactions which have already been met in this chapter. They are summarised in *Tables 8.10.1* and *8.10.2*.

	Appearance	pH paper	With water
Chlorine Cl_2	Pale green gas: smells of 'swimming pool'	Bleaches damp pH paper	Gives very pale green solution
Bromine Br_2	Dark brown liquid with brown vapour clearly seen at room temperature	Slowly bleaches damp pH paper	Dissolves slightly to give pale brown solution
Iodine I_2	Dark solid; gives purple vapour on heating	No pH change observed	Slightly soluble – pale brown solution.

Table 8.10.1   Tests for halogens

With silver nitrate acidified with dilute nitric acid, chloride gives a white precipitate, bromide a precipitate described as 'off-white' or 'cream', and iodide a pale yellow precipitate.

$$Cl^{\ominus}(aq) + Ag^{\oplus}(aq) \xrightarrow[\text{room temperature}]{HNO_3(aq)} AgCl(s) \quad \text{white}$$

$$Br^{\ominus}(aq) + Ag^{\oplus}(aq) \xrightarrow[\text{room temperature}]{HNO_3(aq)} AgBr(s) \quad \text{cream}$$

$$I^{\ominus}(aq) + Ag^{\oplus}(aq) \xrightarrow[\text{room temperature}]{HNO_3(aq)} AgI(s) \quad \text{yellow}$$

It is hard to distinguish between the chloride and bromide colours, and in any case they darken rapidly in strong daylight (see their use in photography, Ch 8.9).

$$\text{e.g.} \quad 2AgCl(s) \xrightarrow{\text{sunlight}} 2Ag(s) + Cl_2(g)$$

So the analysis concludes by testing the silver halide precipitate with ammonia solution. Silver chloride dissolves easily in *dilute* aqueous ammonia; the bromide is unaffected by normal quantities of dilute ammonia but dissolves if concentrated ammonia is used; the iodide is not affected by ammonia. The reason why the silver halide precipitates behave as they do with ammonia is given in Ch 25.4. The tests for anions are summarised in Ch 30.3.

	With silver nitrate solution	With aqueous chlorine (chlorine water)
Chloride $Cl^{\ominus}$	Dense white precipitate – darkens in sunlight – dissolves in dilute aqueous ammonia	No change to normally colourless solutions
Bromide $Br^{\ominus}$	Dense cream precipitate – dissolves slightly in dilute aqueous ammonia and readily in concentrated ammonia	Light brown colour due to formation of free bromine $2Br^-(aq) + Cl_2(g) \longrightarrow Br_2(aq) + 2Cl^-(aq)$
Iodide $I^{\ominus}$	Dense pale yellow precipitate – insoluble in ammonia solution	Dark 'blackish' colour due to formation of free iodine $2I^-(aq) + Cl_2(g) \longrightarrow I_2(s) + 2Cl^-(aq)$

Table 8.10.2   Tests for halide ions

SELF-ASSESSMENT QUESTIONS 8.10

1   What are the formulae of the compounds that should be included in the empty boxes in this reaction sequence?

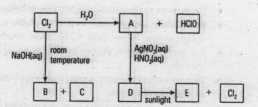

2   How would you distinguish between solutions of caesium chloride, caesium bromide and caesium iodide using solutions of silver nitrate, nitric acid and ammonia?

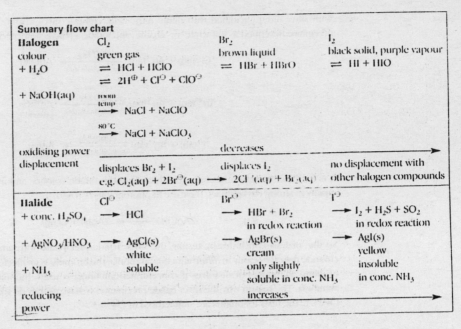

Summary flow chart

Halogen	Cl_2	Br_2	I_2
colour	green gas	brown liquid	black solid, purple vapour
+ H_2O	$\rightleftharpoons HCl + HClO$ $\rightleftharpoons 2H^{\oplus} + Cl^{\ominus} + ClO^{\ominus}$	$\rightleftharpoons HBr + HBrO$	$\rightleftharpoons HI + HIO$
+ NaOH(aq)	room temp $\longrightarrow$ NaCl + NaClO 80°C $\longrightarrow$ NaCl + NaClO$_3$		

oxidising power   decreases →

displacement	displaces $Br_2 + I_2$ e.g. $Cl_2(aq) + 2Br^{\ominus}(aq) \longrightarrow 2Cl^{\ominus}(aq) + Br_2(aq)$	displaces I_2	no displacement with other halogen compounds

Halide	$Cl^{\ominus}$	$Br^{\ominus}$	$I^{\ominus}$
+ conc. H_2SO_4	$\longrightarrow$ HCl	$\longrightarrow HBr + Br_2$ in redox reaction	$\longrightarrow I_2 + H_2S + SO_2$ in redox reaction
+ $AgNO_3$/HNO_3	$\longrightarrow AgCl(s)$ white soluble	$\longrightarrow AgBr(s)$ cream only slightly soluble in conc. NH_3	$\longrightarrow AgI(s)$ yellow insoluble in conc. NH_3
+ NH_3			

reducing power   increases →

SUMMARY 8

In this chapter the following new ideas have been covered. *Have you understood and learned them?* Could you explain them clearly to a friend? The page references will help.

- The halogens become less reactive as atomic number increases (i.e. 'down the Group'). p193
- There is definite gradation of physical properties down the Group. p193
- The hydrogen halides are colourless gases which (except for HF) dissolve in water to give strongly acid solutions. p197
- Hydrogen fluoride is a weak acid because HF molecules hydrogen bond preferentially to each other. p197
- The thermal stability of the hydrogen halides decreases from HF → HI, as does their stability towards oxidising agents. p198
- Fluorine is exceptionally difficult to prepare as an element. p200
- All the halogens are oxidising agents; oxidising power increases from iodine → fluorine. p201
- A more reactive halogen will displace a less reactive one from a solution of its ions. p202
- Except for fluorine, halogens react with alkalis to give halates with the halogens in positive oxidation states. p203
- The reactions with alkalis can involve disproportionation. p203
- The halates are themselves powerful oxidising agents. p206
- Metal halides are generally ionic and soluble in water; those in which the metal is in an oxidation state higher than +2 are more covalent and tend to be hydrolysed by water. p207
- Non-metal chlorides, except for CCl_4, react rapidly with cold water to give hydrogen chloride. p208
- The chlor-alkali industry is extremely important – electrolysis of brine gives chlorine and sodium hydroxide. p210
- Chlorine is required in very large amounts for the manufacture of PVC, for water sterilisation, etc. p213
- Halogens are important in human biology; halogens and their compounds have many other uses. p212

1 The elements in the halogen Group are all electronegative.

  a What does the term electronegative mean?

  b How does electronegativity change within the halogen group?

  c What will be the polarity in the bond between bromine and chlorine?

  d Potassium bromate(V) contains the bromate(V) ion. What is the formula of this anion?

  e What is the shape of the bromate(V) ion, and what is the Br—O—Br bond angle in such a molecule?

2 The boiling points (in °C) of the simplest compounds between the halogens and chlorine are:

$$
\begin{array}{ll}
\text{F—Cl} & -101 \\
\text{Cl—Cl} & -35 \\
\text{Br—Cl} & 5 \\
\text{I—Cl} & 97 \\
\end{array}
$$

  a Plot a graph of these values.

  b Explain the trend within the group in terms of intermolecular forces.

  c The graph for the hydrogen halides follows a similar trend except that the boiling point of HF is anomalous. In what way is the boiling point of HF anomalous?

  d The anomalous behaviour is explained in terms of hydrogen bonding between the HF molecules. What are the requirements for hydrogen bonding to occur?

  e Draw a diagram to show the hydrogen bonding in hydrogen fluoride.

3 The oxoacids of chlorine are $HClO$, $HClO_2$, $HClO_3$, and $HClO_4$.

  a HClO is formed when chlorine is bubbled into cold water. Write an equation for the reaction.

  b The sodium salt of $HClO_3$ is formed if chorine gas is bubbled into hot sodium hydroxide solution. Write an equation for this reaction.

  c Work out all the oxidation states of chlorine in the species involved in this last reaction.

  d What type of reaction has occurred in **c**?

- That the **mass spectrometer** is used to measure the **m/e ratio** of the atoms of an element. (Ch 1.3)
- The definition of the **relative mass** of an **isotope**. (Ch 1.2)
- The **relative atomic mass** of an element, A_r, is the **weighted average** of the relative masses of the isotopes of that element. (Ch 1.3)
- The molecular **formula** of a simple molecular compound represents the number of atoms of each kind in one molecule of the compound. (Ch 4.1)
- In **giant structures**, the formula represents the simplest ratio of atoms or ions in the giant structure. (Ch 4.1)
- **Chemical equations** summarise what happens during a chemical reaction. (Ch 4.2)
- The numbers of each type of atom on each side of a chemical equation must be equal. (Ch 4.2)

In this chapter

We will look at some of the basic types of calculations used in chemistry. The topic is dealt with from very first principles. If you find it easy, check with the questions at the end of each section. *If you are happy that you can do them without difficulty, then go on to the next section.*

If, however, you are not so confident in calculation work, do read the text and work through the examples. This will help you understand how to approach calculations in chemistry, and give you confidence in dealing with them. Once you have developed a technique, and keep using that technique, calculations at this level should not pose any difficulty. What we hope to do is to establish methods which you can use confidently and accurately. We do not expect you to know how to solve any particular problem at first sight!

9.1 **More mass spectrometry**
9.2 **Composition calculations**
9.3 **Avogadro and the mole**
9.4 **Empirical and molecular formulae**
9.5 **Calculations in chemical equations**
9.6 **The gaseous state**

9

introduction to
chemical calculations

> **Definition:** Relative atomic mass, A_r, of an element is defined as the ratio of the weighted mean of the isotopic masses of the element to 1/12 of the mass of an atom of ^{12}C.

Earlier in this book (see Ch 1.3) we have talked about the reason for using relative masses in chemistry – the particles are so small that it is far easier to compare the masses (of atoms, molecules, ions or formula units) with a standard mass. The standard is 1/12 of the mass of the ^{12}C isotope.

The relative atomic mass of an element would be written as, for example $A_r(Na) = 23.0$; it is a ratio and as such has no units. Other terms have been used to describe the relative atomic mass (RAM, atomic mass, 'atomic weight').

Since M_r is also a ratio, it has no units. Other terms used to describe the quantity are 'relative molecular mass' (RMM), 'relative formula mass' (RFM), and 'molecular weight'. **Note that M_r is not confined to molecules**.

> **Definition:** The relative molecular mass, M_r, is defined as the ratio of the mass of an entity (molecule, formula of ionic species, formula of an ion) to 1/12 of the mass of an atom of ^{12}C.

As we have just said, the relative atomic and relative molecular masses do not have units. One other definition that we need is the **molar mass, M.** This quantity does have a unit – g mol^{-1} – and it is the mass of one **mole** of any entity (atom, molecule, ion, formula) in grams. Numerically it is identical to the A_r or M_r of the entity.

e.g. $M(Cl) = 35.45$ g mol^{-1}; $M(NaCl) = 58.45$ g mol^{-1}; $M(CCl_4) = 153.8$ g mol^{-1}.

> **Definition:** The molar mass, M, is the mass, in grams, of one mole of the entity.

9.1 More mass spectrometry

In the first chapter you met the basic principles of mass spectrometry. A particle is given a positive charge by bombarding it with high energy electrons and knocking an electron off an atom. The positively charged particle is accelerated by an electrostatic field before being deflected in an electromagnetic field. The deflection depends on the mass and the charge of the charged particle – its m/e ratio (*Figure 9.1.1*).

A modern mass spectrometer

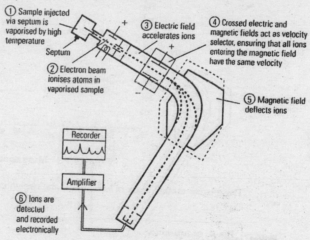

Figure 9.1.1 Outline of a mass spectrometer

As far as we are concerned, the charge is always $1\oplus$, and so the amount of deflection depends on the mass. Any $2\oplus$, $3\oplus$ or even $\ominus$ ions produced by electron bombardment of the sample will not reach the detector – they will have been deflected too little or too much to reach the detector at the end of the apparatus. The instrument provides a method for determining the mass of the charged particle.

If the mass spectrometer is calibrated using a suitable sample, the mass of the particle compared with the mass of the ^{12}C isotope can be measured. A mass spectrometer is a powerful way of finding A_r for elements and M_r for molecular compounds, and it is values of A_r and M_r which are essential for most chemistry calculations

The mass spectrum of sodium contains only a single peak at m/e = 23. This shows that there is only one type of sodium atom present (see *Figure 9.1.2*).

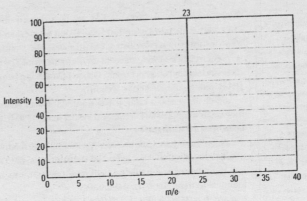

Figure 9.1.2   Mass spectrum of sodium

So, as sodium only has one isotope, the relative atomic mass of sodium is 23. On a more accurate machine the value can be measured as 22.9898. (Remember, the relative atomic mass of ^{12}C is by definition 12.00000 – see Ch 1.3).

However, the mass spectrum of bromine has two peaks, because bromine has two isotopes of relative masses 79 and 81 (*Figure 9.1.3*).

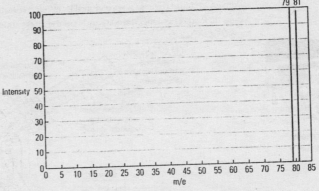

Figure 9.1.3   Mass spectrum of bromine

The relative atomic mass of bromine is the 'weighted mean' of these two isotopic masses (see Ch 1.3).

The A_r of bromine

The relative atomic mass of bromine is the 'weighted mean' of the two isotopes of bromine – ^{79}Br and ^{81}Br.

$$A_r(Br) = \frac{(\%\,^{79}Br \times 79) + (\%\,^{81}Br \times 81)}{100}$$

$$= \frac{(50.52 \times 79) + (49.48 \times 81)}{100}$$

$$= \frac{3991 + 4008}{100}$$

$$= 79.99$$

The mass spectrometer is not limited just to the measurement of the relative atomic masses of the elements. Molecules when bombarded with high energy electrons will also change, and one consequence might be the loss of an electron from the molecule.

$$M(g) \longrightarrow M^{\oplus}(g) + e^{\ominus}$$

The positively charged molecular ion will, after acceleration through an electric field, be deflected in the electro-magnetic field, and the extent of deflection will again depend on the mass to charge (m/e) ratio. The spectrum will now contain a peak at a reading which represents the M_r value of the molecule being studied. This peak – usually the highest value of m/e – is called the *molecular ion peak*, and provides an accurate (although expensive) way of determining M_r for a compound.

Also, it is worth remembering that the technique requires the sample to be vaporised, so the mass spectrum is only used for compounds which can be vaporised.

> Note: zeros at the start of a decimal number *do not* count as significant figures, but zeros at the end may count. So 0.0037 is given to 2 significant figures; 3.000 is given to 4 significant figures. If you are asked to give a volume to 3 significant figures, 25 cm³ is wrong but 25.0 cm³ is correct

Significant figures

'*Significant figures*' refers to the total number of figures you use – nothing to do with the number of *decimal places* in your result.

Always use the same number of significant figures as there are in the data given in the question, or which can be measured with the apparatus, e.g:

23.56 – 4 significant figures;

64830 – may still be 4 significant figures;

0.0037 – 2 significant figures.

Modern instruments are capable of routinely measuring masses extremely accurately, to seven or more significant figures. *Table 9.1.1* shows a few values of A_r with up to 7 significant figures.

Accurate values of A_r	
^{12}C	12.00000 by definition
^1H	1.00782
^{14}N	14 00307
^{16}O	15.99492

Table 9.1.1

It is possible to identify particular combinations of atoms, even though they add up to the same approximate relative molecular mass. Using *high resolution mass spectrometers*, it is possible to identify a particular molecule, even though there might be a number of different possibilities using a value of M_r with only two significant figures.

A peak in the mass spectrum at m/e = 44 could result from at least three simple combinations of atoms. Some possibilities are CO_2, C_2H_4O or CN_2H_4. All these formulae have an M_r of 44 (*Figure 9.1.4*), but only the $C_2H_4O^{\oplus}$ ion will produce a peak at exactly 44.02620 (*Figure 9.1.5*).

	CO_2			C_2H_4O			CH_4N_2	
$1 \times C$	1×12.0	12.0	$2 \times C$	2×12.0	24.0	$1 \times C$	1×12.0	12.0
$2 \times O$	2×16.0	32.0	$4 \times H$	4×1.0	4.0	$4 \times H$	4×1.0	4.0
	$M_r(CO_2) =$	44.0	$1 \times O$	1×16.0	16.0	$2 \times N$	2×14.0	28.0
				$M_r(C_2H_4O) =$	44.0		$M_r(CH_4N_2) =$	44.0

Figure 9.1.4   Adding A_r values to get M_r for three samples

	CO_2			C_2H_4O			CH_4N_2	
$1 \times C$	1×12.00000	12.00000	$2 \times C$	2×12.00000	24.00000	$1 \times C$	1×12.00000	12.00000
$2 \times O$	2×15.99492	31.98984	$4 \times H$	4×1.00782	4.03128	$4 \times H$	4×1.00782	4.03128
	$M_r(CO_2) =$	43.98984	$1 \times O$	1×15.99492	15.99492	$1 \times N$	2×14.00307	28.00614
				$M_r(C_2H_4O) =$	44.02620		$M_r(CH_4N_2) =$	44.03742

Figure 9.1.5   Accurate additions for three molecules

But where can we go from here? The conditions inside the mass spectrometer are very vigorous. After all, considerable energy is needed to knock off an electron from an atom or molecule which is stable at ordinary temperatures. As the electron is knocked off, it is not surprising that the molecule can break up in other ways. The molecule fragments result in a series of peaks corresponding to their m/e ratios. It is these **fragmentation patterns** that often help in identifying the structures of particular molecules.

Figure 9.1.6 shows the mass spectrum for ethanol, C_2H_5OH.

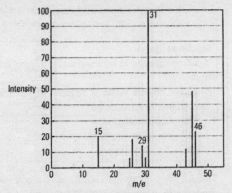

Figure 9.1.6   Mass spectrum of ethanol

The molecular ion peak at m/e = 46 is obvious, but it is by no means the only peak. It is not even the most intense peak in the spectrum. To make a comparison of the relative intensities of the peaks in the spectrum, the most intense peak is standardised to 100% (in the case of ethanol, the peak at m/e = 31). All the other peaks are then compared with the intensity of this peak.

In the spectrum of ethanol the molecular ion peak at 46 is produced by the $CH_3CH_2OH^{\oplus}$ ion. But where do the other peaks come from? If we draw the structure of the ethanol molecule, we can see how some of the fragments are formed by the molecule fragmenting in a number of ways.

The mass spectrum fragments in the ethanol spectrum

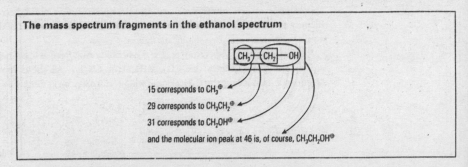

15 corresponds to $CH_3^{\oplus}$

29 corresponds to $CH_3CH_2^{\oplus}$

31 corresponds to $CH_2OH^{\oplus}$

and the molecular ion peak at 46 is, of course, $CH_3CH_2OH^{\oplus}$

The fragmentation patterns will be used in looking at organic molecules, but their use is by no means confined to organic chemistry. The mass spectrum of aluminium chloride is a good example. Firstly, it shows a molecular ion peak at 267. This shows that the molecular formula is Al_2Cl_6. Further, the loss of Cl atoms from the structure is shown by the peaks at intervals of 35 and 37.

Remember:
All the fragments and the molecular ion have a positive charge. Without the charge they would not be accelerated and deflected, and therefore would not be observed in the spectrum.

**SELF-ASSESSMENT
QUESTIONS
9.1**

1 Explain the following terms, which are used in describing how a mass spectrometer works.
  a Electro-magnetic field.
  b Bombarding electrons.
  c Molecular ion peak.

2 The mass spectra of a number of different compounds are listed below. Only the four most intense peaks are shown (the relative intensity is shown in brackets after each value). What is the value of M_r for each of these molecules?
  a 18(44), 29(95), 45(81), 46(100).
  b 29(22), 43(100), 57(56), 72(72).
  c 15(78), 28(100), 43(27), 44(51).
  d 27(55), 28(83), 29(64), 64(100).

3 In the mass spectrum of iodomethane, CH_3I, the following peaks are seen: 15, 127, 142, 141. Identify the fragment responsible for each of the peaks.

4 Calculate A_r for:
  a antimony, containing 57.25% ^{121}Sb and 42.74% ^{123}Sb;
  b copper, containing 69.1% ^{63}Cu and 30.9% ^{65}Cu;
  c lithium, containing 7.42% ^6Li and 92.58% ^7Li.

5 Identify the following fragments from peaks obtained using a high resolution mass spectrometer. You will need to use the figures from *Table 9.1.1*.
  a Is the fragment with m/e ratio 30.04692 due to $C_2H_6^{\oplus}$ or $H_2CO^{\oplus}$?
  b Is $C_4H_{10}^{\oplus}$ or $C_3H_6O^{\oplus}$ responsible for a peak at 58.04184?

9.2 Composition calculations

How much nitrogen by mass is there in potassium nitrate, KNO_3? The meaning of the 'formula of a compound' has already been covered (Ch 4.1). But what if we wish to know *how much of a particular element* is present in a compound?

The question is very relevant when preparing a fertiliser. The percentage of nitrogen present in a fertiliser is of vital importance to the farmer or gardener trying to improve the yield of the crop. Knowing the percentage by mass of an element in a compound helps in working out which compound is best value for money.

So, how do we work out the information from the formula of a compound?

First of all, let us briefly review the work we did on the formulae of compounds. In Ch 4.1, we learnt how it is possible to work out the formulae of ionic and some covalent compounds. By knowing the charges on the ions, or by knowing how many bonds each atom normally has, it is possible to derive a formula (see e.g. *Figure 9.2.1*).

Once its formula is known, the value of M_r for the compound is easily worked out. M_r is the sum of the A_r values of the elements in the formula. The A_r values of the correct number of atoms of each kind are added together. But *the formula must be correct*, including any 'extras' such as water of crystallisation. By using the relative atomic masses (found in the Periodic Table at the back of the book) in the correct proportions, M_r can be calculated (see examples in *Figure 9.2.2*).

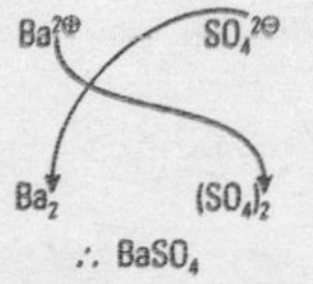

Figure 9.2.1 Formulae and criss-cross diagrams

```
M_r
K₂O
2 × K      2 × 39.10    78.20
1 × O      1 × 16.00    16.00
           M_r(K₂O) =   94.20

H₂S
2 × H      2 ×  1.01     2.02
1 × S      1 × 32.06    32.06
           M_r(H₂S) =   34.08

CuSO₄.5H₂O
1 × Cu     1 × 63.55    63.55
1 × S      1 × 32.06    32.06
4 × O      4 × 16.00    64.00
5 × H₂O    5 × 18.02    90.10
           M_r(CuSO₄.5H₂O) =  249.71
```

Figure 9.2.2

The percentage composition by mass for a compound is found as follows:

- First, work out M_r for the compound using the A_r values for the elements.
- Then find the relative mass of one element in a mole of the compound.
- Finally (relative mass of element/M_r × 100) gives the percentage by mass of that element in the compound. (Note that the molar mass, M, is used at this stage if the mass of the element is wanted and the units are to be consistent.)

This can be done for each element in a compound, or even for parts of the compound – i.e. for water molecules or ions such as sulphate, $SO_4^{2\ominus}$.

Worked examples

A *Calculate the percentage by mass of sodium present in sodium nitrate.*

Formula: $NaNO_3$
```
1 × Na     1 × 23.00    23.00
1 × N      1 × 14.01    14.01
3 × O      3 × 16.00    48.00
           M_r(NaNO₃) =  85.01
```

$$\therefore \%Na = \frac{\text{relative mass Na} \times 100}{M_r} = \frac{23.00 \times 100}{85.01} = 27.06\%$$

B *Calculate the percentage by mass of phosphorus present in ammonium phosphate.*

Formula: $(NH_4)_3PO_4$
```
 3 × N      3 × 14.01    42.03
12 × H     12 ×  1.01    12.12
 1 × P      1 × 30.97    30.97
 4 × O      4 × 16.00    64.00
           M_r((NH₄)₃PO₄) =  149.12
```

$$\therefore \%P = \frac{\text{relative mass P} \times 100}{M_r} = \frac{30.97 \times 100}{149.12} = 20.77\%$$

C *Calculate the percentage by mass of phosphorus present in phosphorus trichloride.*

Formula: PCl_3
```
1 × P      1 × 30.97    30.97
3 × Cl     3 × 35.45   106.35
           M_r(PCl₃) =  137.32
```

$$\therefore \%P = \frac{\text{relative mass P} \times 100}{M_r} = \frac{30.97 \times 100}{137.32} = 22.55\%$$

D *Calculate the percentage by mass of nitrogen present in ammonium nitrate.*

Formula: NH_4NO_3
```
2 × N      1 × 14.01    28.02
4 × H      4 ×  1.01     4.04
3 × O      3 × 16.00    48.00
           M_r(NH₄NO₃) =  80.06
```

$$\therefore \%N = \frac{\text{relative mass N} \times 100}{M_r} = \frac{28.02 \times 100}{80.06} = 25.00\%$$

Instead of the calculation being limited to the percentage by mass of a particular element, it can be extended to calculate the complete composition of the compound by using the figure for each element in turn. But now, all the percentages should add up to 100%. If they do not, then maybe there is something wrong with your calculation?

Worked example

A   *Find the percentage by mass composition of water.*

Formula. H_2O

$2 \times H$ 2×1.01 2.02
$1 \times O$ 1×16.00 16.00
$M_r(H_2O) = \overline{18.02}$

$\therefore \%H = \dfrac{\text{relative mass H} \times 100}{M_r} = \dfrac{2.02 \times 100}{18.02} = 11.21\%$

$\therefore \%O = \dfrac{\text{relative mass O} \times 100}{M_r} = \dfrac{16.00 \times 100}{18.02} = 88.79\%$

So, water contains 11.21% hydrogen and 88.79% oxygen.

B   *Calculate the percentage composition by mass of copper(II) carbonate.*

Formula: $CuCO_3$

$1 \times Cu$ 1×63.55 63.55
$1 \times C$ 1×12.00 12.00
$3 \times O$ 3×16.00 48.00
$M_r(CuCO_3) = \overline{123.55}$

$\therefore \%Cu = \dfrac{\text{relative mass Cu} \times 100}{M_r} = \dfrac{63.55 \times 100}{123.55} = 51.44\%$

$\therefore \%C = \dfrac{\text{relative mass C} \times 100}{M_r} = \dfrac{12.00 \times 100}{123.55} = 9.71\%$

$\therefore \%O = \dfrac{\text{relative mass O} \times 100}{M_r} = \dfrac{48.00 \times 100}{123.55} = 38.85\%$

Therefore, the percentage composition of copper carbonate, $CuCO_3$, is 51.44% Cu, 9.71% C, and 38.85% O.

Although such examples provide an introduction to calculations in chemistry, it must be realised that the formula of every known compound was found out by experiment. The data you have just used were found by experimenters and then used to work out the formulae of the various compounds. We will see how this is done later in Ch 9.4.

Sometimes it is important to know how much of a particular group of atoms is present in the compound. As an example, how much water is present in blue copper(II) sulphate, $CuSO_4.5H_2O$? All we need to do is use M for the *group* we are interested in rather than a single element.

Worked example

Calculate the percentage by mass of water present in hydrated copper(II) sulphate, $CuSO_4 .5H_2O$.

Formula: $CuSO_4.5H_2O$

$1 \times Cu$ 1×63.55 63.55
$1 \times S$ 1×32.06 32.06
$4 \times O$ 1×16.00 64.00
$5 \times H_2O$ 1×18.02 90.10
$M_r(CuSO_4.5H_2O) = \overline{249.71}$

$\therefore \%H_2O = \dfrac{\text{relative mass } H_2O \times 100}{M_r} = \dfrac{90.10 \times 100}{249.7} = 36.08\%$

SELF-ASSESSMENT QUESTIONS 9.2

1 Calculate the percentage by mass of the named substance in each of the following compounds. (We have not given the formula of those compounds which you should be able to work out or deduce from the name of the compound – see Ch 2.3 and 2.4.)

a Lithium in lithium chloride.

b Carbon in ethane, C_2H_6.

c Bromine in calcium bromide.

d Water in $MgSO_4.7H_2O$.

2 Calculate the percentage of mass of each element in:

a sodium chloride;

b sulphur dioxide;

c hydrogen chloride;

d magnesium nitrate, $Mg(NO_3)_2$.

9.3 Avogadro and the mole

Definition: One mole is the amount of substance that contains as many particles (atoms, ions, molecules, formula units) as there are carbon atoms present in 12.00000 g of the ^{12}C isotope.

Amedeo Avogadro was Professor of Physics at the University of Turin in the early nineteenth century. He stated in 1811 that equal volumes of different gases at the same temperature and pressure contain equal numbers of *molecules*. The idea gives a method of finding the molecular formula of a gas. Once that is found, it is possible to find the relative atomic masses of the elements in the gas. Avogadro's idea was the start of logical, *quantitative* chemistry, but it was not accepted until nearly 50 years later.

Avogadro's Law concerns the volumes of gases (see Ch 9.6).

The **Avogadro constant** is the number of particles in one mole of a substance:

$L = 6.022 \times 10^{23}$ mol^{-1}

This section is about *counting particles*. By this stage of the course you will have appreciated that atoms are extremely small. A grain of sand weighing approximately 0.001 g will have sufficient atoms in it to give everybody in the world today over 1000 million atoms each. To weigh masses as small as the mass of an atom is impossible. Instead, we use as our unit *a very large number* of atoms. This unit is **the mole.** You may think this is a very odd name for a unit, but it is no more odd than a *dozen* or a *hundred* – except that it is *very* much larger.

The number of carbon atoms in 12.00000 g of ^{12}C is known to be 6.022×10^{23} (to 4 significant figures). To 7 significant figures the value is 6.022045×10^{23}.

The mole (as a unit, abbreviated to **mol**) is defined in terms of **amount** of substance. The number of the specified particles in one mole of a substance is the **Avogadro constant, L**. The value of L is 6.022×10^{23} mol^{-1} (note the units!).

Why write 6.022×10^{23}?

Rather than write 602200000000000000000000, it is far easier to use *indices (as powers of 10)* to show how far we have to move the decimal point to write the number in full. So we write 6.022×10^{23}.

10^{23} means that the decimal point has to be moved 23 places to the right (since the index is positive).

1.3×10^{-4} means that the decimal point has to be moved 4 places to the left (since the index is negative) to show the number in full; i.e. this number is 0.00013.

On most calculators, the number 6.02×10^{23} is entered as:

6.02 **EXP** **23**

and the display will show 6.02^{23} or 6.02 23.

You must remember that although the display *does not show* the number 10, the figure is 6.02×10^{23} – *not* 6.02 raised to the power of 23!

On most calculators the number 1.3×10^{-4} would be entered as:

1.3 **EXP** **4** **−**

and would appear on the display as 1.3^{-4}.

Try to work out 6.02×10^{23} multiplied by 1.3×10^{-4} using a calculator. The display should show 7.826^{19}, which means an answer of 7.826×10^{19}.

In the same way as we can talk about a dozen eggs, or three dozen bottles of wine, or even half a dozen iced buns, so we can talk about a mole of sulphur atoms or two moles of methane molecules (*Figure 9.3.1*).

Each sample here contains 1 mole of the element or compound, e.g. 6×10^{23} atoms of lead, molecules of ethanol, or formula units of salt (NaCl).

Figure 9.3.1   A mole of each of several substances

One mole of sulphur atoms contains 6.022×10^{23} atoms of sulphur, and two moles of methane contain $2 \times 6.022 \times 10^{23}$ ($= 12.044 \times 10^{23}$) molecules.

However, in the laboratory, one mole of most materials is quite a large quantity, and it is much more usual to handle smaller amounts. Practical instructions often refer to 0.10 mole and even 1×10^{-4} moles. How many particles – atoms, ions, molecules, formula units – are there in a given number of moles?

How many? – worked examples

To work out how many *particles* there are, multiply the number of moles by 6.0×10^{23} (for results to two significant figures).

A   How many atoms are there in 2 moles of helium?
Atoms of He = $2 \times 6.0 \times 10^{23} = 12.0 \times 10^{23}$ atoms.

B   How many molecules are there in 0.015 moles of hydrogen sulphide, H_2S?
Molecules of $H_2S = 0.015 \times 6.0 \times 10^{23} = 0.09 \times 10^{23} = 9.0 \times 10^{21}$ molecules.

C   How many chloride ions are there in 0.10 moles of sodium chloride?
First, work out the formula of the substance, since the question does not ask about the *whole* substance.
Sodium chloride is NaCl. There is one $Cl^{\ominus}$ ion present in each formula unit of NaCl.
∴ 0.10 mole of NaCl contains 0.10 moles of $Cl^{\ominus}$ ion.
Moles of chloride ions = $0.10 \times 6.0 \times 10^{23} = 0.60 \times 10^{23} = 6.0 \times 10^{22}$.

D   How many moles of chloride ion are present in 0.25 moles of magnesium chloride?
Again, find the formula, which is $MgCl_2$.
Each mole of $MgCl_2$ contains 2 moles of chloride ions.
So 0.25 moles of $MgCl_2$ contains 0.50 moles of chloride ions.
The number of chloride ions is $0.50 \times 6 \times 10^{23} = 3.0 \times 10^{23}$.

Indeed, it is safe to assume that if in any problem you are given the *mass* of any substance, the first thing to do is to convert it into *moles*.

E   How many sodium ions are there in 2×10^{-4} moles of sodium sulphate, Na_2SO_4?
Can you work it out?
How many moles of sodium ions are there in one mole of sodium sulphate?
∴ How many moles of sodium ions in 2×10^{-4} moles of sodium sulphate?
∴ How many sodium ions are present?
Be clear in your mind that what we have done here is to calculate the number of individual particles present. Since the numbers in such calculations are so large, it is rare for us to use the actual number of particles. *Throughout the rest of the book we will nearly always use the numbers of moles of particles in any calculations involving quantities of chemicals.*

Remember, the unit of the Avogadro constant, L, is mol^{-1}. So, to calculate the number of particles in x mol of a substance, x mol $\times$ L mol^{-1}; the units cancel out and you are left with the number xL.

Molar mass and relative molecular mass

We should be clear now that 1 mole of carbon contains 6.022×10^{23} atoms, and that 1 mole of sodium chloride contains 6.022×10^{23} formula units of NaCl and 6.022×10^{23} each of sodium and chloride ions. If we take the actual mass in grams of an individual carbon atom and multiply it by the Avogadro constant (6.022×10^{23}) we will obtain what is called the **molar mass**, M. This molar mass is numerically exactly the same as the relative atomic mass we defined in Ch 1.4, **but will have the units of g mol^{-1}**.

The **molar mass**, M, of a substance is the mass of substance that contains exactly the same numbers of specified particles as there are atoms in 12.000 g of ^{12}C. So the molar mass of neon, M(Ne), is 20.18 g mol^{-1}, and for sodium chloride, M(NaCl), the molar mass is 58.34 g mol^{-1} (23.00 + 35.34).

The molar mass of any compound can be worked out by adding up the relative atomic masses (A_r values) of the atoms in the formula of the compound. All you have to be aware of is that **molar mass must have the units g mol^{-1}**.

Relative molecular mass, M_r, of a substance is obtained by adding up the A_r values of the atoms in the formula of the substance. Because it is a *relative* mass, i.e. it is relative to A_r(C) = 12.0000, M_r *has no units*. M and M_r for a substance are numerically the same.

As we have already mentioned, quantities in chemistry are measured in moles rather than in the absolute numbers of particles. **We need to be able to convert any mass of a substance into moles**. The routine in many quantitative problems in chemistry is nearly always the same – convert any *mass*, in grams, into *moles*.

The number of moles is given by the equation:

$$\text{Number of moles} = \frac{\text{mass of substance}}{\text{mass of one mole of the substance}}$$

This means that the number of moles of the materials involved in any reaction can be compared if we know their reacting masses.

We shall soon see that the same ideas can be applied when using volumes of solutions of known concentration (see Ch 10.1) or even volumes of gases (Ch 9.6).

Once you have learnt this equation – *and you need to learn and remember it* – you can use it whenever a mass appears in a question. If you know two out of the three quantities, you can calculate the third:

$$\boxed{\text{Moles} = \frac{\text{mass (in g)}}{M \text{ (in g mol}^{-1})}}$$

- mass, in grams;
- molar mass (found from values of A_r) in g mol^{-1};
- number of moles.

As long as two values are provided for you, or you can find them out, the third is readily calculated (*Figure 9.3.2*).

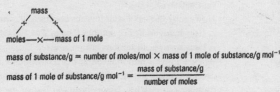

mass of substance/g = number of moles/mol × mass of 1 mole of substance/g mol^{-1}

mass of 1 mole of substance/g mol^{-1} = $\dfrac{\text{mass of substance/g}}{\text{number of moles}}$

Figure 9.3.2   Conversion triangle

Worked example

A *How many moles of phosphine, PH_3, are there in 5.30 g?*

The missing number is the mass of one mole of phosphine, $M(PH_3)$.

Data:	PH_3		
	5.30 g		
$1 \times P$	1×30.97	30.97	
$3 \times H$	3×1.01	3.03	
		$M(PH_3) = 34.00$ g mol^{-1}	

Equation: $\text{moles} = \dfrac{\text{mass}}{\text{molar mass}}$

Calculation: $\text{moles} = \dfrac{5.30}{34.00} = 0.156$ mol

B *Calculate the mass of 0.0500 moles of lithium chloride, LiCl.*

Data:	LiCl		
	0.0500 moles		
$1 \times Li$	1×6.94	6.94	
$1 \times Cl$	1×35.45	35.45	
		$M(LiCl) = 42.39$ g mol^{-1}	

Equation: $\text{moles} = \dfrac{\text{mass}}{\text{molar mass}}$ $\therefore \text{mass} = \text{moles} \times M(LiCl)$

Calculation: $\text{Mass} = 0.0500 \times 42.39 = 2.12$ g

C *0.0350 moles of an organic compound had a mass of 2.834 g. What is the molar mass of the compound?*

Data:	compound X
	2.834 g
	0.0350 moles

Equation: $\text{molar mass} = \dfrac{\text{mass}}{\text{number of moles}}$

Calculation: $\text{molar mass} = 2.834/0.0350 = 80.97$ g mol^{-1}.

1 How many particles of the named substance are there in:
   a 2 moles of argon atoms;
   b 0.004 moles carbon dioxide molecules;
   c 0.13 moles sodium iodide.

2 Calculate the number of the named particles in each of the following quantities of the named substance.
   a Atoms in 40 g argon.
   b Molecules in 27.5 g phosphorus(III) chloride, PCl_3.
   c Molecules in 3.55 g chlorine, Cl_2.
   d Atoms in 0.0355 g chlorine, Cl_2.
   e $Ca^{2\oplus}$ ions in 1.11 g calcium chloride, $CaCl_2$.
   f Chlorine atoms in 20.85 g phosphorus(V) chloride, PCl_5.
   g Bromide ions in 2.97 g barium bromide, $BaBr_2$.

3 How many moles of the named substance are there in:
   a 0.064 g sulphur atoms, S;　　　b 23 g NaCl;
   c 405 g S_2Cl_2;　　　　　　　　d 190 g KBr.

4 Silicon reacts with chlorine, when strongly heated, to form silicon tetrachloride, $SiCl_4$. In a reaction, 2.1 g of liquid silicon chloride was formed.
   a Write an equation for the formation of silicon tetrachloride.
   b Calculate the molar mass of silicon tetrachloride. (Remember: units!)
   c In a particular reaction 4.35 g of the compound was formed. How many moles did this represent?
   d How many silicon atoms are present in this prepared sample?

9.4 Empirical and molecular formulae

Empirical formulae

We are now in a position to look at how to calculate the formulae of compounds from analytical data. We can do it if we know the masses or percentages by mass of each element in the compound.

If we have prepared a new compound, or synthesised some material, or if we are checking that a material contains what it is supposed to contain – in a reduction furnace or in a food packet or whatever – the first step is to analyse the substance for the individual elements it contains. The techniques for such analysis are dealt with in Ch 30.2, but we shall concentrate here on the *interpretation* of these results.

The simplest case to start with is the analysis of a compound. Once it has been prepared and purified, analysis will give a series of figures about composition – either the actual masses or the percentages by mass of the elements in the compound. Remember, percentage tells you what mass of an element there would be if you took 100 g of the sample. So, you can use masses *or* percentages in all the following calculations.

The *problem* at this stage is to convert these figures into a formula for the compound. The process involves:

- collecting together the data;
- converting the figures into the number of moles of each element present;
- finding the ratio of the numbers of moles of elements; and
- converting this into the simplest whole number ratio possible.

Technique is important at this stage. The same sequence can be used for the great majority of chemical calculations at A level: data, moles, ratio. As you progress through the course, the routine should become easier. **But please make sure that you understand that the first step in any calculation is to convert the data – mass, volume of solution, or volume of gas – into the numbers of moles of atoms or ions or molecules or formula units**. Chemists need to compare numbers of particles, and this means comparing numbers of moles.

So, let's try an example first. A compound of nitrogen and hydrogen contains 82.4% nitrogen and 17.6% hydrogen by mass. Remember this means 100 g of the compound has 82.4 g N combined with 17.6 g H. Find the formula of the compound.

Remember:
collect data;
convert to moles;
find the mole ratio.

- To convert the mass or percentage of an element into moles we need the equation:

$$\text{Moles} = \frac{\text{mass}}{\text{mass of one mole of the element}}$$

(Remember, the mass of one mole of the element is the same as its relative atomic mass, A_r, expressed in g mol^{-1}.)

$$\therefore \text{ Moles of nitrogen} = \frac{82.4}{14.01} = 5.88 \text{ mol,}$$

$$\text{and moles of hydrogen} = \frac{17.6}{1.01} = 17.4 \text{ mol}$$

- To convert these figures into a ratio, divide by the smaller of the two numbers:

$$\text{For nitrogen } \frac{5.88}{5.88} = 1.00 \text{ and for hydrogen } \frac{17.4}{5.88} = 2.96$$

Since we cannot have anything other than *whole* atoms, formulae must contain *whole numbers of atoms*. The number 2.96 approximates to 3.00. (Do be careful *not* to round numbers up or down too early in this type of calculation. The rounding up or down should only be done in the final step **and should only involve small changes**.)

- So the ratio of N to H is 1:3, and the formula is NH_3.

Such a calculation gives the **empirical formula** of a compound.

It is much easier to use the same layout each time for these calculations. We suggest that you set them out in the form of a table.

Calculating the empirical formula of a compound

Remember, the empirical formula of a compound can be calculated from either the *masses* of the elements in the compound or from the *percentage composition* of the compound since % means g in 100 g.

For example, compound X contains 82.4% nitrogen, 17.6% hydrogen.
$(A_r(N) = 14.01; A_r(H) = 1.01)$

Element:		N	H
Data:	Percentage	82.4	17.6
	Relative atomic mass	14.01	1.01
Moles/mol	$= \dfrac{\text{percentage or mass/g}}{\text{mass of one mole/g mol}^{-1}}$	$\dfrac{82.4}{14.01}$ $= 5.88$	$\dfrac{17.6}{1.01}$ $= 17.4$
Ratio of moles (divide by the smallest value if necessary)		$\dfrac{5.88}{5.88}$ $= 1.00$ 1 N	$\dfrac{17.4}{5.88}$ $= 2.96$ 3 H_3

∴ **The empirical formula is NH_3.**

Try another worked example – this time using combining masses

A sample of compound Y contains 2.03 g sodium and 0.71 g oxygen. What is the empirical formula of Y?

- First step: look up the relative atomic masses from the Periodic Table.
  $(A_r(Na) = 23.00; A_r(O) = 16.00)$
- Draw up a table for the calculation, collecting the data you know.

Element:		Na	O
Data:	Mass (g)	2.03	0.71
	Relative atomic mass, A_r	23.00	16.00

- Convert any masses into moles using the equation
$$\text{Moles} = \frac{\text{mass}}{\text{mass of one mole}}$$

Moles/mol	$\dfrac{2.03}{23.00}$ $= 0.0883$	$\dfrac{0.71}{16.00}$ $= 0.0443$

- Work out the ratio of moles, rounding the result up or down to obtain whole numbers.

Ratio (divide by the smallest value)	$\dfrac{0.0883}{0.0443}$ $= 1.99$ 2 Na_2	$\dfrac{0.0443}{0.0443}$ $= 1.00$ 1 O

∴ **The empirical formula is Na_2O.**

This technique can be extended to three elements, and any number of elements or even groups. Work through the next two examples to give you confidence in these calculations.

Worked example

Compound Z contains 29.1% sodium, 40.5% sulphur and 30.4% oxygen. Calculate the empirical formula of Z.

$(A_r(Na) = 23.00; A_r(S) = 32.06; A_r(O) = 16.00)$

Element: Data:		Na	S	O
	Percentage	29.1	40.5	30.4
	Relative atomic mass	23.00	32.06	16.00
Moles/mol	$= \dfrac{\text{percentage or mass}}{\text{mass of one mole}}$	$\dfrac{29.1}{23.00}$ $= 1.27$	$\dfrac{40.5}{32.06}$ $= 1.26$	$\dfrac{30.4}{16.00}$ $= 1.90$
Ratio		$\dfrac{1.27}{1.26}$ $= 1.01$	$\dfrac{1.26}{1.26}$ $= 1.00$	$\dfrac{1.90}{1.26}$ $= 1.51$

In this case it would not be sensible to round the figure 1.51 up or down. The 'half atoms' can be eliminated by multiplying all the figures by 2:

		2.02	2.00	3.02

Rounding is now appropriate, with the figures being so close to whole numbers:

		2	2	3

∴ **The empirical formula is $Na_2S_2O_3$.**

Useful hint:

Whenever you find numbers containing .50, .33 or .67 (e.g. 1.50, 0.33, 2.67) appearing in the mole ratio figures, multiplying by 2 or 3 as appropriate will produce whole numbers. You should **never** round figures containing '.33', '.50' or '.67' – or figures very close to these – down or up.

Even a group of atoms within the compound, such as H_2O or the sulphate ion, $SO_4^{2\ominus}$, can be treated in the same way. In the next example you can do some of the work yourself.

Partially worked example

Compound W contains 14.3% sodium, 9.9% sulphur, 19.9% oxygen and 55.9% water. Calculate the empirical formula of W.

You will need to copy out the table and complete the columns for sodium and for water.

$(A_r(Na) = 23.00; A_r(S) = 32.06; A_r(O) = 16.00; M_r(H_2O) = 18.02)$

Element: Data:		Na	S	O	H_2O
	Percentage	14.30	9.9	19.9	55.9
	Relative atomic or molecular mass	23.00	32.06	16.00	...
Moles/mol	$= \dfrac{\text{percentage or mass/g}}{A \text{ or } M/\text{g mol}^{-1}}$	...	$\dfrac{9.9}{32.06}$ ...	$\dfrac{19.9}{16.00}$ $= 0.31$	... $= 1.24$
...					
Ratio		...	$\dfrac{0.31}{0.31}$ $= 1.00$	$\dfrac{1.24}{0.31}$ $= 4.00$	...
		...			...

∴ **The empirical formula is ....**

How many water molecules are there in each empirical formula unit of compound W?

This type of calculation always produces the **empirical formula** of a compound. This was defined earlier in this section. The empirical formula of a compound shows the simplest ratio of atoms in the formula of the compound. It is important to realise that the *empirical formula* is not necessarily the *molecular formula* of the compound; it shows us *only* the simplest ratio of atoms.

Empirical formula shows the *simplest ratio* of atoms.
Molecular formula shows the *actual numbers of atoms* in the molecule of the compound.
(However, with giant structures such as ionic compounds like sodium chloride, NaCl, we keep to the simplest ratio. There is no such thing as a molecular formula for a giant structure.)

For example, the empirical formula of water is H_2O; it is the simplest formula we can write for water, and it is also the *molecular* formula. However, hydrogen peroxide has the molecular formula H_2O_2. Hydrogen peroxide contains 5.9% H and 94.1% O. Using these figures, the empirical formula is found to be HO. The simplest ratio of hydrogen to oxygen atoms in hydrogen peroxide is $1:1$.

Molecular formulae

> **Definition:** The molecular formula of a compound shows the actual numbers of each kind of atom present in the molecule of a compound.

An empirical formula for a compound is only a first step. N.B. The term 'molecular formula' *cannot be used for giant structures*, such as those for ionic compounds.

If we are to find out the molecular formula of the compound, we need to know its relative molecular mass, M_r. We have already seen that M_r for many compounds can be measured using the mass spectrometer. However, a mass spectrometer is a complex and expensive machine with considerable running costs, and not always readily available, especially in schools and colleges.

Once M_r for the compound has been found, the molecular formula can be found by comparing M_r for the empirical formula with the value of M_r found by experiment.

An empirical formula, as we have already said, shows the *simplest ratio of atoms* in the compound. **The molecular formula must be a simple multiple of the empirical formula**.

For example, the empirical formula for a number of organic compounds is CH_2. Our ideas about bonding (see Ch 2.4) tell us that such a molecule cannot exist. However, molecules consisting of multiples of CH_2 – e.g. C_2H_4, C_3H_6, C_4H_8, etc – are all possible.

To work out the number by which the empirical formula must be multiplied to give the molecular formula, all you have to do is divide the relative molecular mass, found by experiment, by the relative mass of the empirical formula.

$$\text{Number of empirical formula units in molecular formula} = \frac{M_r \text{ found by experiment}}{M_r \text{ for empirical formula}}$$

There are one or two other techniques for measuring M_r which can be used. One important technique, which is possible even in a school laboratory, is to measure the mass of a known volume of gas. As we shall see in Ch 9.6, by measuring the mass and volume of a gas it is possible to calculate its M_r. It can even be adapted to provide M_r for volatile liquids, to a reasonable degree of accuracy. After all, much of the early determination of relative molecular masses was done using this principle. Other techniques, not really required in present A level syllabuses, use the lowering of the melting point of liquids or raising their boiling points by dissolving other materials in the liquids. The extent of elevation of the boiling point or depression of the freezing point depends on how much solute is added, and on the M_r of the dissolved substance.

Worked example

Sulphur reacts with chlorine to form a compound. A sample of this compound was found to contain 0.94 g sulphur and 1.06 g chlorine. The relative molecular mass of the compound is 135. Calculate the empirical formula and then the molecular formula of the compound.

First step, calculate the empirical formula.
($A_r(S) = 32.06; A_r(Cl) = 35.45$)

Element:		S	Cl
Data:	$\dfrac{\text{mass/g}}{\text{mass of one mole/g mol}^{-1}}$	$\dfrac{0.94}{32.06}$	$\dfrac{1.06}{35.45}$
Moles/mol		$\dfrac{0.94}{32.06}$	$\dfrac{1.06}{35.45}$
		= 0.0293	= 0.0299
Ratio (divide by the smallest value)		$\dfrac{0.0293}{0.0293}$	$\dfrac{0.0299}{0.0293}$
		= 1.00	= 1.02
		1	1
		S	Cl

∴ The empirical formula is SCl.

The relative mass of the empirical formula is 67.51; how many of these mass units are there in the M_r value of the compound (135)?

$$\frac{135}{67.51} = 2.00, \text{ so the molecular formula is } 2 \times (\text{SCl})$$

∴ The molecular formula is S_2Cl_2.

1 Calculate the empirical formulae of the compounds with the following analyses:

    a 8.0% Li, 92.0% Br;   **b** 94.1% S, 5.9% H;

    c 10.67 g Br, 1.62 g Mg   **d** 3.87 kg K, 1.39 kg N, 4.75 kg O

2 **a** 4.13 g iron combines with 1.77 g oxygen in an oxide which has M_r = 160. Calculate the formula of the oxide.

    b Potassium, when heated in oxygen, forms an oxide in which 1.46 g of potassium combines with 1.20 g oxygen. The molar mass, M, of the oxide is 71 g mol^{-1}. What are the empirical and actual formulae of the oxide?

    c A compound of aluminium and chlorine contains 20.2% aluminium and 79.8% chlorine. The relative molecular mass is 266.5. What are the empirical and molecular formulae of the compound?

3 Nitrogen in the atmosphere is unreactive. In spite of this, some nitrogen is converted into polluting gases by car engines, while a lot of money is spent by industry converting nitrogen into useful compounds for fertilisers and explosives.

    a One fertiliser compound contains 16.5% N, 56.4% O, 27.1% Na. The polluting gas has 30.4% N and 69.6% O. Calculate the empirical formula in each case.

    b The polluting gas can exist in two molecular forms, A and B; M_r(A) is 46 and M_r(B) is 92. What are the molecular formulae of the compounds A and B?

9.5 Calculations in chemical equations

We started to look at chemical equations in Ch 4.2. If you remember how to write a balanced equation for a chemical reaction, then carry on. If you are not quite sure, have a look at that section so that you can be confident of converting any word equation into a balanced chemical equation.

> **The sequence for writing a balanced equation:**
> Word equation
> Insert the formulae
> Balance the equation

In any calculations using chemical equations, you can adopt the same approach as we have been using so far in this chapter for dealing with chemical formulae.

From a balanced chemical equation you need to:

- collect the data;
- convert this information into numbers of moles;
- check the reacting ratios involved in the equation;
- calculate the quantity asked for.

If we look at an example you will see how the technique works.

How much magnesium oxide is formed when 0.240 g of magnesium is burnt in air? The balanced equation for the reaction is:

$$2Mg(s) + O_2(g) \longrightarrow 2MgO(s)$$

The mass of magnesium (M = 24.31 g mol^{-1}) has to be changed into the number of moles (0.240/24.31 = 0.00987 mol). From the equation, 2 moles of magnesium forms 2 moles of magnesium oxide, i.e. the ratio is 1:1. So 0.00987 moles of magnesium oxide will be formed. This can be converted into a mass using the now familiar equation, and the knowledge that M(MgO) = 40.3 g mol^{-1}:

$$\text{Mass} = \text{number of moles} \times \text{molar mass}.$$

0.00987 × 40.3 = 0.398 g of magnesium oxide is formed. This type of calculation can be laid out in tabular form.

Worked example

Calculate the mass of magnesium oxide that would be produced when 0.24 g magnesium is burnt in air.

Data:

Balanced equation: $2Mg$ + $O_2(g) \longrightarrow 2MgO(s)$

Mass   0.240 g

$1 \times Mg$	1×24.31	24.31
$1 \times O$	1×16.00	16.00
		40.31

Molar mass/g mol^{-1} 24.31

Moles $= \dfrac{\% \text{ or mass/mol}}{M}$ $\dfrac{0.240}{24.31}$

$= 0.00987$

Ratio from the equation   1   1

Calculation:

∴ Moles of magnesium oxide   0.00987

∴ **Mass of magnesium oxide (mass = moles × *M*)** 0.00987×40.31

$= 0.398$ g

Worked example

A   *Calculate the mass of calcium oxide that would be produced when 4.50 g chalk (calcium carbonate, $CaCO_3$) is completely decomposed by heating.*

Data:

Balanced equation: $CaCO_3 \longrightarrow CaO + CO_2$

Mass   4.50 g

Molar mass/g mol^{-1}

$1 \times Ca$	1×40.08	40.08	$1 \times Ca$	1×40.08	40.08
$1 \times C$	1×12.01	12.01	$1 \times O$	1×16.00	16.00
$3 \times O$	3×16.00	48.00			
	∴ *M* is:	100.09 g mol^{-1}		∴ *M* (CaO) is:	56.08 g mol^{-1}

Moles $= \dfrac{\% \text{ or mass}}{M}$ $\dfrac{4.50}{100.09}$

$= 0.0450$ mol

Ratio from the equation   1   1

Calculation:

∴ Moles of calcium oxide   0.0450 mol

∴ **Mass of calcium oxide (mass = moles × *M*)** 0.0450×56.08

$= 2.52$ g

B   *How many tonnes of iron oxide, Fe_2O_3, will be needed to be reduced to produce 100 tonnes of iron metal?*

Data:

Balanced equation: Fe_2O_3 +   reducing material $\longrightarrow$ 2Fe + other products

(The question can be answered even though the full balanced equation is not known – all we need to realise is that 1 mole of the oxide will form 2 moles of iron metal. 1 tonne = 10^6 grammes)

Mass   100 tonnes

Molar mass (g mol^{-1})

$2 \times Fe$	2×55.85	111.7
$3 \times O$	3×16.00	48.00
	∴ *M*(Fe$_2$O$_3$) is	159.70 g mol^{-1}

M(Fe) is 55.85/g mol^{-1}

Moles $= \dfrac{\% \text{ or mass}}{M}$ $\dfrac{100 \times 10^6}{55.85}$

$= 1.79 \times 10^6$ mol

Ratio from the equation   1   2

Calculation:

∴ Moles of iron oxide $1.79 \times 10^6 \div 2 = 0.895 \times 10^6$

∴ **Mass of iron oxide (mass = moles × *M*)** $= 0.895 \times 159.7 \times 10^6$ g

$= 143$ tonnes

Try one yourself:

Copper sulphate solution reacts with sodium hydroxide solution to precipitate copper(II) hydroxide (one possible formula for copper(II) hydroxide is Cu(OH)₂):

$$CuSO_4(aq) + 2NaOH(aq) \longrightarrow Cu(OH)_2(s) + Na_2SO_4(aq)$$

What mass of sodium hydroxide is needed to convert 20.0 g copper(II) sulphate to copper(II) hydroxide?

Data:

The balanced equation is already given.

What mass of copper(II) sulphate is being used?

Work out the value of M for copper(II) sulphate.

Moles:

How many moles of copper(II) sulphate are being used?

Ratio:

From the equation, what is the ratio of moles of copper(II) sulphate to sodium hydroxide?

So, how many moles of sodium hydroxide are required?

Calculation:

∴ Moles of sodium hydroxide = . . .

∴ **Mass of sodium hydroxide (mass = moles × M) = . . .**

The secret of any calculations involving chemical equations is to convert data into moles wherever possible. In this chapter we have always converted any masses into moles; in Ch 9.6 the volumes of gases will be converted into moles, and in Ch 10 the volumes of solutions of known concentration will be converted into moles. Remember, moles are the counting units we use in chemistry, and we **compare quantities by comparing the numbers of moles**.

Yield in a reaction

So far we have concentrated on the amounts of materials resulting from a particular chemical reaction. In all these cases we have assumed that the reaction has occurred *exactly* as the equation has shown – for example, that *all* the magnesium is converted into magnesium oxide when magnesium is burnt in air.

We have also assumed that everything is 100% what we think it is. But this is not always the case. The analytical chemist is often asked to find how pure a particular sample is. The industrial chemist making a commercial product has to be able to say how efficient the reaction is: how does the amount of material made compare with the theoretical amount?

For any reaction, the percentage yield will be the experimental yield (the actual yield) divided by the yield worked out from the balanced equation (the theoretical yield).

$$\% \text{ yield} = \frac{\text{actual yield} \times 100}{\text{theoretical yield}}$$

In the same way, the purity of a sample is calculated by comparing the measured mass of the substance under test to the mass of the impure sample.

$$\% \text{ purity} = \frac{\text{actual mass} \times 100}{\text{mass of impure sample}}$$

Calculation of the % yield in a reaction – worked example

When lithium chlorate(VII) is strongly heated, oxygen gas and lithium chloride are formed according to the equation:

$$LiClO_4(s) \longrightarrow LiCl(s) + 2O_2(g)$$

In a particular reaction, 1.06 g lithium chlorate(VII) was heated and produced 0.40 g lithium chloride. What is the yield of lithium chloride in this reaction?

Equation: $LiClO_4(s)$ $\longrightarrow$ $LiCl(s) + 2O_2(g)$

Data: mass (g)			1.06 g			0.40 g
	$1 \times Li$	1×6.94	6.94	$1 \times Li$	1×6.94	6.94
	$1 \times Cl$	1×35.45	35.45	$1 \times Cl$	1×35.45	35.45
	$4 \times O$	4×16.00	64.00			
M (g mol^{-1})			106.39			42.39
Moles/mol $= \dfrac{mass}{M}$			$\dfrac{1.06}{106.39} = 0.0100$			$\dfrac{0.40}{42.39} = 0.0094$

Ratio: (from the balanced equation)   1   1

Calculation:
The theoretical yield should be 0.0100 molecules since the ratio is 1:1.
The actual yield is 0.0094 moles.

$$\text{The \% yield} = \frac{\text{actual yield} \times 100}{\text{theoretical yield}} = \frac{0.0094 \times 100}{0.0100} = 94\%$$

Calculation of the % purity of a sample – worked example

Analytical chemistry is concerned with measuring the amount of a particular chemical present in a sample. As we have stressed in this chapter so far, the amount is determined in terms of the number of moles.

To measure the % purity of the sample, the mass of the pure chemical present has to be compared with the mass of the original sample. So, with a figure expressed as moles, it must be converted back to mass to compare with the original mass.

Analysis of an impure sample of anhydrous sodium carbonate showed that 1.20 g of it contained 0.0100 moles of sodium carbonate. Calculate the % purity of the sample.

Data:		Na_2CO_3	
	1.20 g impure solid contains 0.0100 moles		
	$2 \times Na$	2×23.00	46.00
	$1 \times C$	1×12.01	12.01
	$3 \times O$	3×16.00	48.00
M (g mol^{-1})			106.01

Moles/mol $= \dfrac{mass}{M}$ $\therefore$ Mass $=$ moles $\times M = 0.0100 \times 106.01 = 1.06$ g

Calculation:
The mass of the impure sample is 1.20 g
The actual mass of sodium carbonate is 1.06 g

$$\text{The \% purity} = \frac{\text{actual mass} \times 100}{\text{mass of impure sample}} = \frac{1.06 \times 100}{1.20} = 88\%$$

---

1 When hydrogen is reacted with oxygen, water is formed. 1.0 g hydrogen was burned with excess oxygen.

   a Write an equation for the reaction.

   b How many moles of hydrogen molecules were present?

   c How many moles of water were formed?

   d What mass of water was formed?

2 When 4.0 mol dm^{-3} aqueous potassium hydroxide is heated with solid iodine, the iodine disproportionates. Potassium iodide and potassium iodate(V) are formed. A slight excess of the potassium hydroxide solution is used so that the brown colour of the iodine solution disappears.

   a What is meant by disproportionation? (If necessary refer to Ch 6.4).

b The balanced equation for the reaction is:

$$3I_2(s) + 6KOH(aq) \longrightarrow 5KI(aq) + KIO_3(aq) + 3H_2O(l)$$

What are the oxidation states of iodine in the three iodine-containing substances?

c The mixture of iodide and iodate(V) is separated by recrystallising the crude product from aqueous solution. In a particular reaction, 1.00 g of iodine produced 0.20 g of potassium(V) iodate. Calculate the % yield in the reaction.

9.6 The gaseous state

The kinetic theory of matter describes the particles of matter as being in constant movement. The amount and type of movement determines the physical state of that matter. You should be familiar with the differences between gases, liquids and solids, and their main properties are listed in *Table 9.6.1*.

	Gas	Liquid	Solid
Arrangement of particles	Random	Random	Regular
Distance between particles	Far apart	Close together	Close together
Mobility	Fills all container	Fills container from bottom to surface of liquid	Fixed shape
Compressibility	Easily compressed	Not easily compressed	Difficult to compress
Movement of particles	Move rapidly and randomly	Roll around each other	Vibrate in fixed position

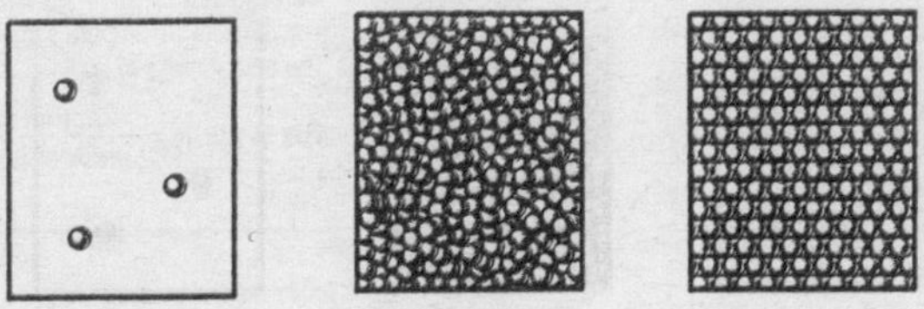

Table 9.6.1 Physical states of matter

Modern materials have extended these boundaries. Liquid crystals are now in everyday common usage (in lap-top screens, in calculator displays, etc). At temperatures above 10 000 °C some matter exists as plasma – a random array of charged ions, similar to the surface of our Sun.

1 mole of water has an approximate volume of 18 cm³. When this is changed into a vapour, at room temperature and pressure, the volume is close to 24 dm³; i.e. the water molecules occupy $(18/24000) \times 100\% = 0.075\%$ of the volume of the vapour.

Figure 9.6.1
Water as a vapour

Gases are mostly empty space. The molecules in any gas actually take up very little of the space the gas occupies, less than 1% for most gases or vapours. (The word vapour is usually used to describe the gaseous form of a substance at temperatures below its boiling point.) This is not very difficult to work out for yourself – see *Figure 9.6.1*.

The actual volume of the molecules is very small compared with the total volume occupied by the gas, and it is the rapid movement of the molecules which 'creates' the volume. Think how much more room a group of people needs if everyone is moving randomly and rapidly about, compared with standing quietly.

How fast are gas molecules moving? The speed of a hydrogen molecule at room temperature is close to 900 m s⁻¹ – the speed of a rifle bullet. Heavier molecules such as N_2 and O_2 are moving at about 250 m s⁻¹, which is about 900 km hr⁻¹.

Liquid helium, at 4K, is a
superfluid – it can flow up
the sides of the beaker.

Since there is so much empty space in a gas, gases can be compressed quite easily. The volume of a gas depends very much on its pressure and temperature. Think of a simple bicycle pump, or a syringe in the laboratory – you can easily push the piston (while holding your thumb over the exit hole!) so that the volume is only half the original. In doing so, you have doubled the pressure of the gas.

For any gas calculation work, the temperature scale used is the Kelvin scale (see Ch 1.1). This scale starts at a theoretical temperature, zero Kelvin or –273 °C, at which every gas would have zero volume. (The gases would of course have turned to liquid or solid before that temperature.) The temperature scale we have normally used in this book is Celsius, units °C. The Kelvin unit, K, has the same size as the °C. To convert °C into Kelvin:

$$K = °C + 273$$

At very low temperatures the structures of materials may change, and it is at such temperatures that superconductors (materials with no electrical resistance) and superfluids are found.

The pressure of each gas also has to be specified if we are to compare the volumes of gases. We shall normally quote gas pressures in the simplest unit – atmospheres (atm).

Standard temperatures and pressures

The atmosphere as a unit of pressure has serious limitations for accurate work. Atmospheric pressure varies from day to day: the pressure is usually higher on a clear fine day, and a lower pressure can mean a stormy day.

A more definitive way of measuring pressure is to quote the height of mercury supported in a mercury barometer at 0 °C. Normal atmospheric pressure is 760 mm Hg, and is often quoted as such (1 torr is equivalent to 1 mm Hg). The introduction of SI units led to measuring pressure in Pascals (1 Pa = 1 N m^{-2}), and 1 atm is 101.325 kPa. (More recently, standard pressure has been defined as 100 kPa – but since we are using 1 atm, it will have no effect in our calculations.)

How do we overcome the difficulties of discussing volumes of gases at differing temperatures and pressures? First, we define standard temperature and pressure. For all gas measurements, standard temperature and pressure is 0 °C (273 K) and 1 atm. These are the conditions we will use throughout the book.

s.t.p. means 'standard temperature and pressure', i.e. 0 °C (273 K) and 1 atm.

RTP means 'room temperature and pressure', i.e. –20 °C (293 K) and 1 atm.

The gas laws

The effects of temperature and pressure on gases were studied early in the development of modern science. After all, volumes of gases were relatively easy to measure with the simple apparatus available to the early scientists such as Boyle and Charles.

Boyle's work showed that doubling the pressure on a fixed mass of gas, i.e. a fixed number of moles, halved the volume, as long as *the temperature was kept the same*. This he put into an equation:

$$PV \text{ is constant, or } P \times V = k, \text{ if T stays constant}$$

and the amount of gas is constant.

Similarly, Charles showed that if *the pressure on a gas was kept constant*, doubling the temperature (in Kelvin) doubled the volume. As an equation, this looks like:

$$\frac{V}{T} = constant, \text{ if P stays constant}$$

and the amount of gas stays constant.

These two laws can be combined to provide the **gas equation**: for a given number of moles of gas:

$$\frac{PV}{T} = constant$$

Predicting the shape of a graph

Examiners are often fond of asking you to predict what the shape of the graph will be for a particular set of circumstances. How do you do it? If you can remember, then all is well. Often, the equation is known or given, but predicting can be more of a problem. Tackle it by putting some simple figures into the equation. The shape of the graph should then become obvious.

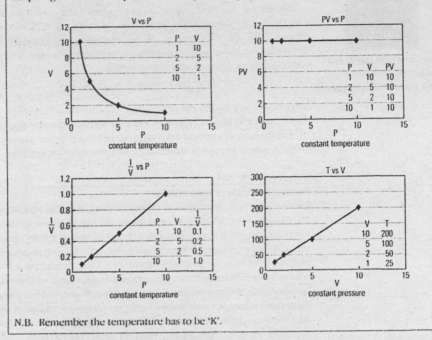

N.B. Remember the temperature has to be 'K'.

The combined gas laws of Boyle and Charles can be applied to any gas. The combined law can be stated as:

$$\frac{P_1 V_1}{T_1} = \frac{P_2 V_2}{T_2}$$

where P is the pressure of the gas;
V is the volume; and
T is the temperature (in Kelvin).

The suffixes 1 and 2 refer to each of the two sets of conditions. There are therefore six variables. If any five are known, the sixth can be calculated. As long as the units are kept exactly the same throughout, and any temperature in °C is converted into Kelvin, the calculation is straightforward.

Transforming equations

Is it worth a reminder about moving quantities in any mathematical equation?

When moving a number (or quantity) with a + or − sign in front of it to the other side of an equation, you just have to change the sign.

e.g.	$X - Y = 6$
becomes	$X = 6 + Y$
or	$l + m = n$
becomes	$l = n - m$

If the figure or symbol is part of a multiplication/division statement, the figure or quantity is moved over the equals sign by cross multiplication.

e.g.

$$\frac{A}{B} = C$$

$$\therefore A = B \times C$$

or remember it as

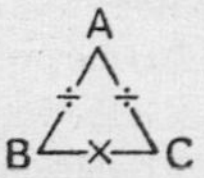

and also:

$$\frac{P_1 V_1}{T_1} = \frac{P_2 V_2}{T_2}$$

$$\therefore V_2 = \frac{P_1 V_1}{T_1} \times \frac{T_2}{P_2}$$

Worked example

What is the volume of a sample of gas at 30 °C and 1 atmosphere if its volume at 20 °C and 1 atmosphere is 25.0 cm³?

Data:

$P_1 = 1$ atm; $V_1 = 25$ cm³; $T_1 = (20 + 273) = 293$ K (it must be in K)

$P_2 = 1$ atm; $T_2 = 30$ °C $= (30 + 273) = 303$ K, and V_2 is unknown.

Check that all the units are correct (convert them if they are not).

Equation:

$$\frac{P_1 V_1}{T_1} = \frac{P_2 V_2}{T_2}$$

Calculation:

Rearranging so that V_2 is on its own:

$$\frac{P_1 V_1}{T_1} \times \frac{T_2}{P_2} = V_2$$

$$\therefore V_2 = \frac{1 \times 25}{293} \times \frac{303}{1} = 25.9 \text{ cm}^3$$

Changing the conditions in a reaction that involves gases can alter the volume of gas significantly. It is therefore very important, when doing calculations involving gases to state the temperature and pressure. (These may, of course, be s.t.p., 273 K and 1 atm, or RTP, 293 K and 1 atm.)

The ideal gas equation

One of the reasons for the similarity of all gases is the small influence the volume of the molecules has on the overall volume of the gas. This leads us to consider an **ideal gas**. Using a model of an ideal gas it is possible to derive the gas laws from a mathematical treatment of the movement of the molecules. What do we mean by an ideal gas?

An ideal gas is one in which:

- **the volume of the molecules in the gas is negligible compared with the volume of gas;**
- **all the collisions between gas molecules are perfectly elastic;**
- **the attractions between gas molecules are negligible;**
- **the pressure of the gas is due entirely to collision between gas molecules and the walls of the container.**

We do not need to consider the ideal gas any further than this. The **ideal gas equation** can be derived from the mathematical treatment of the movement of molecules, and is:

$$PV = nRT$$

where:

P is the pressure of the gas;
V is the volume of the gas;
T is the temperature, in K, of the gas;
n is the number of moles of gas; and
R is the gas constant.

Difficulties arise in ensuring that the units for the gas constant R are consistent with the units of pressure and volume. The most common situation is to measure pressure in kilopascals and volume in dm³; the value of R is then 8.3 J K⁻¹ mol⁻¹. (The same value for R applies if pressure is in N m⁻² (i.e. Pa) and the volume is in m³.)

Using this gas equation we can calculate any one of the variables if the others are known. So, if we know the volume, pressure and temperature of the gas, **then the number of moles of gas can be calculated – whichever gas we are using.** That is, of course, assuming that the gas we are handling is an ideal gas.

Most gases behave very nearly as ideal gases at normal temperatures and pressures. It is only at very high pressure or very low temperature that the assumptions we made earlier become invalid (*Figure 9.6.2*).

Note: All the gas laws only apply *exactly* to what are called **ideal gases** – i.e. gases in which there is no attraction between particles, and the particles are assumed to occupy no volume. Fortunately, *real* gases obey the gas laws closely enough for the laws to be useful in predicting behaviour.

Worked example 1

Calculate the number of moles present in 240 cm³ of a gas at 1 atm pressure and 298 K. (The gas constant, R = 0.083 atm dm³ K⁻¹ mol⁻¹.)

> **Data:** H₂
> 240 cm³ = 0.240 dm³ (since R has the units atm dm³ K⁻¹ mol⁻¹)
> R = 0.083 atm dm³ K⁻¹ mol⁻¹
> P = 1 atm
> T = 298 K
>
> **Equation:** PV = nRT
>
> **Calculation:** $n = \dfrac{PV}{RT} = \dfrac{1 \times 0.240}{0.083 \times 298} = 0.0097$ mol or 9.7×10^{-3} mol

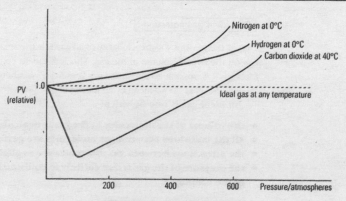

Figure 9.6.2 Non-ideal gases

At extremely high pressures, the gas molecules are pushed very close together, and then the volume of the molecules starts to become a much greater proportion of the total volume of the gas. Similarly, when the gas molecules are pushed close together the attractions between them start to become important (Ch 2.8).

For every gas, there is a particular temperature (the *critical temperature*) below which it is possible to make the gas condense to a liquid simply by compressing it. For carbon dioxide, the critical temperature is 31 °C. The carbon dioxide in any CO_2 fire extinguisher you see is probably liquid.

Gas molar volume

What is the volume of one mole of gas? The densities of gases are available as part of the physical data about any gas. Since density can be measured in $g\,dm^{-3}$, this can be related to the volume of one mole of gas?

The density of oxygen at room temperature is $1.33\,g\,dm^{-3}$, and one mole of oxygen, O_2, has a mass of $32.00\,g$ (2×16.00). Therefore:

$1.33\,g$ of oxygen has a volume of $1\,dm^{-3}$

$\therefore 32.00\,g$ of oxygen has a volume of $(1/1.33) \times 32.00 = 24.06\,dm^3$

Table 9.6.2 and *Figure 9.6.3* show the results of similar calculations for a number of other gases.

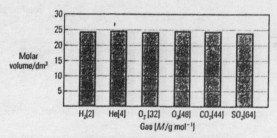

Figure 9.6.3   Gas molar volumes

Gas	Formula	Density at 25 °C or 298 K g dm^{-3}	M g mol^{-1}	Molar volume dm^3 mol^{-1}
Hydrogen	H_2	0.0824	2	24.27
Helium	He	0.163	4	24.53
Oxygen	O_2	1.33	32	24.06
Ozone	O_3	1.96	48	24.49
Carbon dioxide	CO_2	1.81	44	24.31
Sulphur dioxide	SO_2	2.68	64	23.88

Table 9.6.2   Gas molar volumes

You will see that in every case the molar volume of all these gases is very close to $24\,dm^3$ at room temperature (*Figure 9.6.3*).

Remember, it is important that all volume measurements are made under the same conditions of temperature and pressure. In the quoted examples, the densities were all measured at room temperature and pressure. If the calculation is repeated, using standard conditions, s.t.p. (0 °C and 1 atm), the result is $22.4\,dm^3$.

One mole of a gas has a volume of about $22.4\,dm^3$ ($22\,400\,cm^3$) at s.t.p., or about $24\,dm^3$ at room temperature and pressure.

i.e. The molar volume of a gas is $22.4\,dm^3$ at s.t.p. or about $24\,dm^3$ at RTP.

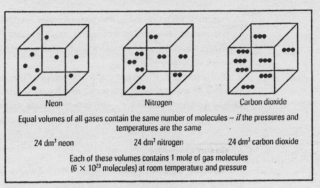

Figure 9.6.4

This quantity is called the **gas molar volume** in $dm^3 \; mol^{-1}$ or $cm^3 \; mol^{-1}$. Do not be worried about remembering the value – when you need to use the figure in a calculation, the data will be given to you.

If we turn this idea around, the number of moles of any gas can be found just by measuring the volume of the gas, converting to room temperature and pressure if needed.

$$\text{Moles of gas} = \frac{\text{volume of gas (in } cm^3)}{24\,000} \text{ at RTP}$$

Gas moles/volume calculations can be tackled using either:

$$PV = nRT$$

or:

$$\text{Moles} = \frac{\text{volume (in } cm^3)}{24\,000}$$

Which method you use depends on the data given, but *do make sure that the units are consistent throughout your calculation.*

Worked example 2 – To calculate the number of moles of a gas

How many moles of nitrogen are there in 2.4 dm^3 of the gas at room temperature and pressure? (The molar volume of a gas at room temperature and pressure is 24 $dm^3 \; mol^{-1}$.)

Data: N_2
2.4 dm^3 (keep as dm^3 since other data are also in dm^3)
Molar volume = 24 $dm^3 \; mol^{-1}$ at RTP

Equation: $\text{Moles} = \dfrac{\text{volume (in } dm^3)}{24\,dm^3 mol^{-1}}$

Calculation: $\dfrac{2.4}{24} = 0.10$ mol of nitrogen

The relative molecular mass of a gas or volatile liquid

One of the advantages of using gases is that, with the information we have so far, it is possible to calculate the molar mass of any gas (or even of a volatile liquid) without the expense of a mass spectrometer. If we measure the mass of a sample of a gas, and then measure its volume at known pressure and temperature, we can calculate the number of moles, n, by using the ideal gas equation. Since we will then know the mass of that number of moles of the gas, the molar mass can then be worked out.

The relative molecular mass (no units) or the molar mass (in g mol⁻¹) of a gas can be found by measuring the mass of a given volume of gas. One way of doing this is to use a gas syringe (*Figure 9.6.5*). The mass of liquid (and hence of vapour) is found by weighing the syringe before and after inserting the liquid.

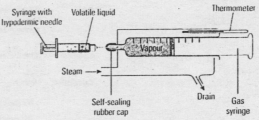

Figure 9.6.5 *M* for a volatile liquid

A number of precautions have to be taken to obtain reasonably accurate results, because the mass of the gas is so small compared with the mass of the syringe itself. As with any gas measurements, the pressure and temperature must be noted.

If the syringe is heated in a special syringe heater, the volume occupied by the vapour from a known mass of a volatile liquid can be measured (together with its pressure and temperature). The figures can then be used to calculate the molar mass of the gas or vapour; the volume of gas formed by vapourising the liquid may have to be converted to s.t.p. or RTP.

Worked example 3 – Calculating the molar mass of a gas

0.16 g of a gas at room temperature (20 °C) and pressure had a volume of 87 cm³. Calculate the molar mass of the gas. (The volume of one mole of a gas at room temperature and pressure = 24 000 cm³.)

Method A   Data:

Gas
Mass = 0.16 g
Volume = 87 cm³ = 0.087 dm³

Equation:

$$\text{Number of moles} = \frac{\text{volume (cm}^3)}{24\,000}$$

Calculation:

$$\text{Moles} = \frac{87}{24\,000} = 0.00363 \text{ mol}$$

∴ The mass of 0.00363 moles = 0.16 g

∴ The mass of 1 mole $= \dfrac{0.16}{0.00363} = 44.1$ g

∴ $M(\text{gas}) = 44$ g mol⁻¹

Method B   Using PV = nRT

The value of R to use is 0.083 atm dm³ K⁻¹ mol⁻¹, as atm is the unit used here for atmospheric pressure.
T = 25 °C = 298 K

Calculation:

$$n = \frac{PV}{RT} = \frac{1 \times 0.087}{0.083 \times 298} = 0.00352$$

∴ The mass of 0.00352 moles = 0.16 g

∴ The mass of 1 mole $= \dfrac{0.16}{0.00352} = 45$ g

∴ $M(\text{gas}) = 45$ g mol⁻¹

N.B. If you are asked to do a calculation of this kind in an examination, the appropriate value for R will always be given

(The slight difference in the result using method A and method B is because we have been using data expressed to 2 or 3 significant figures and the gas molar volume is an approximation.)

Worked example 4

0.217 g of trichloromethane was heated in a gas syringe at 100 °C so that it all vaporised. The final volume, at this temperature and 1 atm pressure, was 56 cm³. Calculate the molar mass of trichloromethane from these data.

(The volume of one mole of gas at 100 °C and 1 atm is 30.6 dm³.)

Data: Trichloromethane
Mass = 0.217 g
Volume = 56 cm³

Equation: Number of moles $= \dfrac{\text{volume (cm}^3)}{30\,600}$

Calculation: Moles $= \dfrac{56}{30\,600} = 0.00183$ mol

∴ The mass of 0.00183 moles = 0.217 g

∴ The mass of 1 mole $= \dfrac{0.217}{0.00183} = 118.6$ g

∴ Molar mass of trichloromethane is 119 g mol⁻¹

With known gas volumes we can:

convert pressure, volume, or temperature using $\dfrac{P_1 V_1}{T_1} = \dfrac{P_2 V_2}{T_2}$	calculate the number of moles using $PV = nRT$	calculate the number of moles at RTP using $n = \dfrac{\text{volume (cm}^3)}{24\,000}$	from the mass of a known number of moles, calculate the molar mass of the gas or vapour Molar mass $= \dfrac{\text{mass of gas}}{\text{number of moles}}$

N.B. If the conditions are not at RTP either the volume of gas will have to be converted to s.t.p., or the gas molar volume will have to be converted to the conditions of the experiment.

Avogadro's Law

At the start of the nineteenth century, a number of scientists were studying the way in which gases combined. The English scientist Henry Cavendish had already shown that whenever hydrogen and oxygen reacted to form water, it was always in the ratio of two volumes of hydrogen reacting with one volume of oxygen to produce two volumes of steam.

	$2H_2(g)$ +	$O_2(g)$ $\longrightarrow$	$2H_2O(g)$
e.g.	2 volumes 100 cm³	1 volume 50 cm³	2 volumes 100 cm³

Note that the total volume during the reaction decreases – since the number of molecules decreases (from 3 molecules – 2 of hydrogen and 1 of oxygen – to 2 molecules of steam).

The French scientist Gay-Lussac extended this early work and showed that, as long as the volumes of gases were measured under the same conditions of temperature and pressure, in any gas reaction the volumes of reactants and products were always in simple ratios.

Avogadro developed these ideas and in 1811 published a law (now known as **Avogadro's Law**) stating that equal volumes of all gases under the same conditions contain equal numbers of particles. Most of the work we have done so far on gases comes from the work of Avogadro.

This means that 100 cm³ of carbon dioxide (CO_2) will contain the same number of molecules as 100 cm³ of nitrogen (N_2), or 100 cm³ of argon (Ar) – in fact, the same number of molecules of any gas, held in the same volume under the same conditions. It also means that *there must be a volume, which will be the same for any gas, which should contain one mole of gas.* We have already met this as the molar volume of a gas (= 24 dm³ at RTP). In fact, as we have seen, this volume differs slightly from gas to gas, but it is close enough to 24 dm³ at RTP for us to be able to use this value.

Definition:

Avogadro's Law: Equal volumes of all gases at the same temperature and pressure contain the same number of molecules.

Avogadro's Law *allows us to use gas volumes in chemical equations.* We have already seen that most chemical equations contain simple relative numbers of reacting molecules, atoms or ions. With gas equations, the ratio of the reacting atoms or molecules can be found simply by measuring the volumes of gases involved and comparing these volumes, provided that all the volumes are measured at the same temperature and pressure.

In any reaction involving gases the volume ratio will be the same as the mole ratio of the gases. For example, the combustion of methane in oxygen produces carbon dioxide and water. The balanced equation is:

$$CH_4(g) + 2O_2(g) \longrightarrow CO_2(g) + 2H_2O(l)$$

From this equation we can already see that 2 moles of oxygen are required for each mole of methane burnt, and that 1 mole of carbon dioxide gas is produced. Avogadro's Law allows us to state that $2 \, dm^3$ of oxygen will be needed to burn $1 \, dm^3$ of methane, and will produce $1 \, dm^3$ of carbon dioxide (so long as the volumes are measured at the same temperature and pressure). Note that if *liquid* water is produced, its volume will be very small. If the reaction is done above 100 °C, each $1 \, dm^3$ of methane burned will produce $2 \, dm^3$ of steam.

One reaction we should be very familiar with by now is the reaction of hydrogen and oxygen to form water. The balanced equation is:

$$2H_2(g) + O_2(g) \longrightarrow 2H_2O(g)$$

Note that here, because the water is gaseous, the reaction has happened at a temperature higher than 100 °C.

We can now appreciate that instead of finding the masses of the gases and the water produced, the *volumes* of the gases can be measured. The ratio of hydrogen to oxygen will be 2:1. For example, $200 \, cm^3$ of hydrogen will react with $100 \, cm^3$ of oxygen, forming $200 \, cm^3$ of steam. But remember that these volumes must *all* be measured at the same temperature and pressure.

Worked example

$100 \, cm^3$ of sulphur dioxide gas reacts with exactly $50 \, cm^3$ of oxygen, to form sulphur trioxide. What is the balanced equation for the reaction?

Data:

$$SO_2(g) + O_2(g) \longrightarrow SO_3(g)$$

Volumes: $100 \, cm^3$ $50 \, cm^3$

Ratio:

    2   1

Calculation:

2 moles of SO_2 must react with 1 mole of O_2, so the equation starts as:

$$2SO_2 + O_2 \longrightarrow$$

and balancing the equation:

$$2SO_2(g) + O_2(g) \longrightarrow 2SO_3(g)$$

It is worth emphasising that volume ratios can be used to establish the equation for a gas-phase reaction. By comparing the volumes of the reacting gases and the gaseous products it is possible to work out the equation for the reaction.

The total volume will only be unchanged during the course of the reaction *if there is the same number of moles of gas on each side of the equation.* Otherwise the total volume will change during the course of the reaction. Do not confuse such a volume change with any change in mass. The total mass cannot change during a chemical reaction.

Worked example

The equation for the combustion of propane, C_3H_8, in oxygen is as follows:

$$C_3H_8(g) + 5O_2(g) \longrightarrow 3CO_2(g) + 4H_2O(g)$$

The constant temperature of the reaction is sufficient to keep the water as a vapour and the pressure is kept constant throughout.

If the initial volume of propane is 100 cm³, what volume of oxygen would be needed to ensure complete combustion?

Data: $C_3H_8(g) + 5O_2(g) \longrightarrow 3CO_2(g) + 4H_2O(g)$

Volumes: 100 m³   to be found

Ratio: 1   5

Calculation:

Volume of oxygen needed is $= 5 \times 100$ cm³ $= 500$ cm³

The total volume of reactants would then be 600 cm³.

    What volume of carbon dioxide would be produced?

    What volume of water vapour would be formed?

    The total volume of products would then be 700 cm³ (*Figure 9.6.6*).

Since air contains approximately 20% oxygen, the volume of air required to ensure the complete combustion of propane would be 500 cm³ $\times$ 5 $= 2500$ cm³ (or 2.5 dm³).

Remembering that the volume of 1 mole of any (ideal) gas is always the same, the equation for the combustion of propane, in terms of volumes involved, would look like:

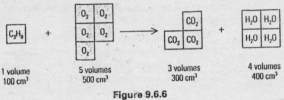

1 volume	5 volumes	3 volumes	4 volumes
100 cm³	500 cm³	300 cm³	400 cm³

Figure 9.6.6

Worked example

Carbon disulphide, CS_2, burns in air to form sulphur dioxide and carbon dioxide. When 25 cm³ gaseous CS_2 is ignited in air, 50 cm³ of SO_2 and 25 cm³ of CO_2 are formed. Calculate the volume of air needed to carry out this reaction.

Data: $CS_2(g) + O_2(g) \longrightarrow SO_2(g) + CO_2(g)$

Volumes: 25 cm³   50 cm³ 25 cm³

Ratio: 25   50   25

i.e.   1   2   1

Calculation: $CS_2(g) + O_2(g) \longrightarrow SO_2(g) + CO_2(g)$

becomes $1CS_2(g) + O_2(g) \longrightarrow 2SO_2(g) + 1CO_2(g)$

and, to balance, the equation requires $3O_2$ on the left hand side.

$$CS_2(g) + 3O_2(g) \longrightarrow 2SO_2(g) + CO_2(g)$$

So, for 25 cm³ CS_2, you would need $25 \times 3 = 75$ cm³ of oxygen.

Since air contains 20% oxygen, the volume of air needed will be 5×75 cm³ $= 375$ cm³.

SELF ASSESSMENT QUESTIONS 9.6

1   Explain in terms of kinetic theory why it is possible to compress a gas to a very small fraction of the original volume, but a liquid has very little compressibility.

2   500 cm³ H_2 was exploded with 200 cm³ oxygen and the resultant mixture returned to room temperature. (All volumes of gases were measured at room temperature and pressure.)

    a Write a balanced equation for the reaction.

    b How much hydrogen was left at the end of the reaction?

    c What volume of hydrogen would have reacted with oxygen?

d How many moles of hydrogen gas were used? (1 mole of a gas at room temperature and pressure has a volume of 24 000 cm^3.)
e Calculate the mass of water produced.
f What would be the volume of gas remaining?

3 A mixture of hydrogen and chlorine explodes in direct sunlight. The equation for the reaction is:

$$H_2(g) + Cl_2(g) \longrightarrow 2HCl(g)$$

A mixture of 40 cm^3 hydrogen and 80 cm^3 of chlorine was reacted and the volumes in the final mixture were measured.
a What is the total volume at the start of the reaction?
b What is the volume of HCl(g) produced in the reaction?
c What is the volume of chlorine that reacted?
d What is the final volume of the mixture of gases?
(All volumes are measured at the same temperature and pressure.)

Make sure you know all definitions and can do all the different types of calculation.

Summary

A_r is relative atomic mass
M_r is relative molecular mass
M is mass (in g) of 1 mole, known as molar mass
The unit chemists use is the mole; 1 mole = 6.02×10^{23} particles

$$\text{Number of moles} = \frac{\text{mass}}{M}; \quad = \frac{\text{vol (cm}^3\text{) of gas at RTP}}{24\,000}$$

$$\%\ \text{composition} = \frac{\text{relative mass of element}}{\text{mass of compound}} \times 100$$

Empirical formula shows simplest ratio of atoms in a compound
Molecular formula shows the actual number of each kind of atom in a molecule.
N.B. Giant structures are represented by their empirical formulae. Molecular formula is a whole number of empirical formula units

$$\%\ \text{yield} = \frac{\text{actual yield}}{\text{theoretical yield}} \times 100; \quad \%\ \text{purity} = \frac{\text{actual mass}}{\text{impure mass}} \times 100$$

For an ideal gas:

$$PV = nRT$$

$$\frac{P_1V_1}{T_1} = \frac{P_2V_2}{T_2}$$

S
SUMMARY 9

In this chapter the following new ideas have been covered. *Have you understood and learned them?* Could you explain them clearly to a friend? The page references will help.

- The use of mass spectrometry to measure accurate values of M_r. p219
- Molecules can fragment in a mass spectrometer. p222
- The percentage composition by mass of a compound can be worked out from its formula. p224
- The molar mass of a formula can be calculated using values of A_r. p228
- The mass of an element or compound can be converted into moles. p228
- The empirical formula of a compound can be calculated from mass or % data. p230
- The relative masses involved in a chemical equation can be calculated from the equation for the reaction. p234
- The percentage yield in a reaction and the percentage purity of a sample can be calculated. p236
- The gas laws and their uses. p239
- The use of the ideal gas equation to calculate the number of moles of gas. p242
- The use of the gas molar volume to work out the number of moles in a volume of gas. p243
- Avogadro's Law as applied to reacting volumes of gases. p246

1 The relative molecular mass of a liquid organic compound is measured by taking a weighed sample and vaporising it in a syringe at 100 °C. 0.30 g of the compound was found to have a volume of 64 cm³ at 100 °C. This converted to 50 cm³ at room temperature and pressure.

  a Show, by calculation, that a volume of 64 ml at 100 °C is equivalent to 50 cm³ at room temperature and pressure (20 °C (293 K) and 1 atm).

  b What fraction of a mole of gas is 50 cm³? (The volume of 1 mole of gas at room temperature and pressure is 24 000 cm³.)

  c What is the mass of this number of moles of gas?

  d What is the molar mass of the compound?

  e When the gas had all vaporised, the syringe was allowed to cool. The gas did not condense to liquid until 34 °C. Sketch a cooling curve to 20 °C for the gas with temperature on the vertical axis and time on the horizontal axis.

2 **a** A compound of potassium has the following analysis: 38.6% K, 13.9% N, 47.5% O. Calculate the empirical formula of the compound.

  b On heating, this compound decomposes to give oxygen. 1 mole of the compound gives 0.5 mole of oxygen gas. Calculate the volume of O_2 formed when 2.70 g of this compound is completely decomposed.

  c Calculate the volume of gas produced when 2.00 g magnesium turnings is reacted with excess aqueous hydrochloric acid.

3 In an experiment to measure a rate of reaction using a gas reaction, a student wishes to use a gas syringe.
Calculate the maximum mass of calcium carbonate, $CaCO_3$, that could be used if the volume of CO_2 produced is to be measured in a 100 cm³ syringe at room temperature and pressure. (The volume of 1 mole of gas at room temperature and pressure is 24 000 cm³.)

4 When potassium chlorate(V) is heated strongly, it decomposes into potassium chloride and oxygen.

  a What is the oxidation state of chlorine in potassium chlorate(V)?

  b Write an equation for the decomposition of potassium chlorate(V).

  c An impure sample of potassium chlorate(V) was analysed by heating the sample until no further loss in mass was recorded. The volume of gaseous oxygen collected was 24 cm³. How many moles of gas are there in 24 cm³? (The volume of 1 mole of gas at RTP is 24 000 cm³.)

  d How many moles of potassium chlorate(V) must have decomposed?

  e This calculation assumes that oxygen behaves like an ideal gas. Give two differences between a real gas and an ideal gas.

  f Calculate the mass of potassium chlorate(V) which must have decomposed in **c**.

  g This mass of potassium chlorate was contained in the 0.100 g sample which was used for this experiment. What is the % purity of the potassium chlorate(V)?

- Using **chemical equations** to represent reactions. (Ch 4.2)
- Using the **ratios** of the reactants and products represented in the equations. (Ch 9.5)
- Working out the **oxidation states** of elements in compounds. (Ch 4.3)
- Calculating the **number of moles** of compounds or elements from masses or from gas volumes. (Ch 9.3)

Many chemical reactions are carried out in solution. By working with solutions, it is possible to use very small quantities of materials. Solutions are easier to handle than solids and mixing of substances is more easily carried out. Solid phase reactions are rarely used at A level. In this chapter, a substance in solution will be assumed to be dissolved in water (i.e. an *aqueous* solution). But do remember, it is possible *to make a solution in any solvent,* and exactly the same principles will apply – the concentration of a solution is the number of moles of the substance dissolved in one dm^3 of solution.

10.1 **Concentrations in solution**
10.2 **Acid–base titrations**
10.3 **Redox titrations**

10

titration
calculations

10.1 Concentrations in solution

If we are to use a solution of a chemical to carry out a reaction, we need to know how much of the chemical is present. We need to know the **concentration** of the solution.

> **Definition**: The concentration of a solution is expressed as mol dm⁻³, i.e. the number of moles of solute in one dm³ of solution.

Note: the definition is in terms of one dm³ of *solution*, not of *solvent*. If one mole of solute is added to one dm³ of solvent, the result will *not* be one dm³ of solution! For example, the mass of one mole of sucrose, ordinary sugar, is 342 g. Adding 342 g of sugar to one dm³ of water results in more than one dm³ of solution. But doing the same thing with some ionic compounds can result in less than 1 dm³ of solution (*Figure 10.1.1*).

Dissolving ionic compounds

Ions in solution are surrounded by polar water molecules (Ch 3.4).

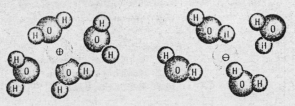

Figure 10.1.1

In some cases, because the water molecules are arranged tightly around the ions the volume of the solution can be less than the volume of water used.

> Older books refer to the 'molarity' of solutions, and you may find the symbol M used on laboratory reagent bottles. The concentration of dilute sodium hydroxide used in the laboratory is about 2 mol dm⁻³; on older bottles this will be written as '2M NaOH'.

Another way of expressing concentration is as a percentage. But using 'percent composition of the solution' tells us nothing about the number of moles present in a particular volume.
For example: 1% sodium hydroxide contains 1 g NaOH in 100 cm³, which is 0.25 mol dm⁻³, but a 1% solution of potassium chloride, KCl, will also contain 1 g in 100 cm³ – that is 0.13 mol dm⁻³, which is almost half the *concentration* of the sodium hydroxide solution.
So although the solutions are both 1% solutions, they contain different numbers of moles of solute per dm³; they have different concentrations.

Concentrations of laboratory reagents can range widely. For example, concentrated hydrochloric acid has about 10 moles of hydrogen chloride in each dm³ of solution: its concentration is about 10 mol dm⁻³. Limewater, Ca(OH)₂(aq), on the other hand, has a concentration of about 0.015 mol dm⁻³.

The highest possible concentration depends on the solubility of the **solute** (the substance dissolved) in the **solvent** (the liquid used). The lowest possible concentration depends on the sensitivity of the apparatus being used; 10⁻⁶ mol dm⁻³ is easily measured (see Ch 30.5), and modern instrumental analyses can measure concentrations much lower than this.

Units of volume

> 1 dm³ = 1000 cm³ = 1 litre = 1000 ml = 1000 cc   (We shall use cm³ and dm³.)

So: the capacity of a car engine labelled 1600 cc = 1600 cm³ = 1.6 dm³ = 1.6 litres.

Rarely do we handle 1 dm³ of solution. We are more likely to use volumes such as 25 cm³ or 100 cm³, and more likely still, volumes such as 24.3 cm³.

Therefore, we need to find out the number of moles present in the volume we are handling.

In just the same way as using **relative atomic mass** (A_r) or **relative molecular mass** (M_r) gives us a way of counting atoms or molecules (see Ch 9.3), so *the use of known volumes of solutions of known concentrations gives us a way of counting particles of solute.*

Note: the units of concentration are mol dm^{-3}, but the apparatus you use is calibrated in cm^3. Therefore, the number of moles of solute in a given volume of a solution is given by the following formula:

$$\text{Number of moles} = \frac{\text{volume in cm}^3 \text{ of solution}}{1000} \times \text{concentration in mol dm}^{-3}$$

> **Calculating the number of moles of solute**

Worked example

Calculate the number of moles of sodium hydroxide in 25.0 cm^3 of a solution of concentration 0.250 mol dm^{-3}.

As we suggested in Ch 9, it is worth collecting all the data together:

Data:

$$\text{NaOH}$$
$$\text{Volume} = 25.0 \text{ cm}^3$$
$$\text{Concentration} = 0.250 \text{ mol dm}^3$$

Relevant equation:

$$\text{Number of moles} = \frac{\text{volume}}{1000} \times \text{concentration}$$

Calculation:

$$\text{Number of moles} = \frac{25.0 \times 0.25}{1000} = 0.00625 \text{ mol}$$

Notice that the M_r of the substance is not needed in order to work out the number of moles if the concentration is given; after all, the M_r of the solute was used to define the concentration! In fact, the formula of the solute is not required for finding the number of moles in this way. But this step is only the start for many calculations, as we will show in the next two sections. *Whenever you see a concentration of a solution quoted in a question, together with the volume of that solution, it is usually worth converting the figures into the number of moles present.*

Worked examples

A *How many moles of sodium chloride are present in 100 cm^3 of 1.2 mol dm^{-3} NaCl(aq)?*

Data:

$$\text{NaCl}$$
$$\text{Volume} = 100 \text{ cm}^3$$
$$\text{Concentration} = 1.2 \text{ mol dm}^{-3}$$

Relevant equation:

$$\text{Number of moles} = \frac{\text{volume}}{1000} \times \text{concentration}$$

Calculation:

$$\text{Number of moles} = \frac{100 \times 1.2}{1000} = 0.12 \text{ mol}$$

B *How many moles of calcium nitrate are present in 23.6 cm^3 of 0.15 mol dm^{-3} Ca(NO$_3$)$_2$(aq)?*
Collect the data together and write the equation out.

Data:
Relevant equation:
Calculation:

The three figures entered into your calculator should give an answer of 0.00354 moles.

C How many moles of potassium iodide are present in 1.1 dm³ of 0.23 mol dm⁻³ KI(aq)?

Data:

KI
Volume $= 1.1$ dm⁻³ (note the volume is in dm³, so 1.1 dm³ $= 1100$ cm³)
Concentration $= 0.23$ mol dm⁻³

Relevant equation:

$$\text{Number of moles} = \frac{\text{volume (in cm}^3)}{1000} \times \text{concentration}$$

Calculation:

$$\text{Number of moles} = \frac{1100 \times 0.23}{1000} = 0.25 \text{ mol}$$

Alternatively, since the question gives a volume in dm³, the division by 1000 is not necessary (1 dm³ = 1000 cm³):

$$\text{Moles} = \text{volume in dm}^3 \times \text{concentration}$$
$$\text{Moles} = 1.1 \times 0.23 = 0.25 \text{ mol.}$$

If you cannot remember how to transform these equations, try doing it using the triangle:

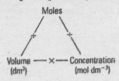

Moles = volume × concentration

$$\text{Volume} = \frac{\text{moles}}{\text{concentration}}$$

$$\text{Concentration} = \frac{\text{moles}}{\text{volume}}$$

– but remember always to get the *units* correct. The equations are illustrated for volume in dm³. Volumes measured in cm³ will need to be divided by 1000.

Calculating the concentration of a solution

We can calculate the concentration of a particular solution if we know the mass of solute in a given volume of solution. To do this we transform the 'moles' equation into:

$$\text{Concentration} = \frac{\text{number of moles}}{\text{volume in cm}^3} \times 1000$$

Notice that we need to know the **number of moles** of solute – and usually it is the **mass** of a solute which is given. Remember that mass is converted into moles using the equation (see Ch 9.3):

$$\text{Number of moles} = \frac{\text{mass of the substance}}{\text{mass of 1 mole of the substance}}$$

The equation then becomes:

$$\text{mol} = \frac{\text{mass (in g)}}{M \text{ (in g mol}^{-1})}$$

where M is the molar mass with units g mol⁻¹. Remember M is numerically the same as M_r (but M_r has no units). M_r for a substance is found from its formula, by adding up the relative atomic masses of all the atoms present in the formula of the substance.

Worked examples

In the first two examples the number of moles is known, so there is no need to work it out.

A What is the concentration of a solution of potassium hydroxide containing 0.030 mol in 100 cm³?

Data:

KOH
Volume $= 100$ cm³
Number of moles $= 0.030$ mol

Relevant equation:

$$\text{Number of moles} = \frac{\text{volume (cm}^3)}{1000} \times \text{concentration}$$

$$\therefore \quad \text{Concentration} = \frac{\text{number of moles}}{\text{volume in cm}^3} \times 1000$$

Calculation:

$$\text{Concentration} = \frac{0.030}{100} \times 1000 = 0.30 \text{ mol dm}^{-3}$$

B   What is the concentration of a solution containing 0.00235 mol caesium chloride in 45 cm^3?

Data:

Caesium chloride, CsCl
Volume $= 45$ cm^3
Number of moles $= 0.00235$ mol

Relevant equation:

Number of moles $= \dfrac{\text{volume (cm}^3)}{1000} \times \text{concentration}$

$\therefore$ Concentration $= \dfrac{\text{number of moles}}{\text{volume in cm}^3} \times 1000$

Calculation:

Concentration $= \dfrac{0.00235}{45} \times 1000 = 0.052$ mol dm^{-3}

C   Calculate the concentration of a sodium hydroxide solution which contains 4.28 g in 250 cm^3.

Remember:

The formula must be known or worked out to obtain M

Data:

NaOH
Volume $= 250$ cm^3
Mass $= 4.28$ g

The mass has to be changed into moles as the first step

Relevant equations:

Number of moles $= \dfrac{\text{mass}}{M}$

Number of moles $= \dfrac{\text{volume (cm}^3)}{1000} \times \text{concentration}$

Concentration $= \dfrac{\text{number of moles}}{\text{volume in cm}^3} \times 1000$

Calculation:

Values of A_r for the elements concerned can always be found from a Periodic Table

$M(\text{NaOH})$ $1 \times \text{Na}$ $1 \times 23.00 = 23.00$
             $1 \times \text{O}$ $1 \times 16.00 = 16.00$
             $1 \times \text{H}$ $1 \times \ 1.01 = \underline{\ 1.01}$
                              40.01 g mol^{-1}

$\therefore$ Number of moles $= \dfrac{4.28}{40.01} = 0.107$

$\therefore$ Concentration $= \dfrac{0.107}{250} \times 1000 = 0.428$ mol dm^{-3}

D   What is the concentration of a solution of copper sulphate, CuSO$_4$.5H$_2$O, which contains 23.5 g dissolved in 500 cm^3?

Data:

CuSO$_4$.5H$_2$O
Volume $= 500$ cm^3
Mass $= 23.5$ g

Relevant equations:

Number of moles $= \dfrac{\text{mass}}{M}$

Number of moles $= \dfrac{\text{volume (cm}^3)}{1000} \times \text{concentration}$

$\therefore$ Concentration $= \dfrac{\text{number of moles}}{\text{volume in cm}^3} \times 1000$

Calculation:

$M(\text{CuSO}_4.5\text{H}_2\text{O})$ $1 \times \text{Cu}$ $1 \times 63.55 = \ \ 63.55$
                       $1 \times \text{S}$ $1 \times 32.06 = \ \ 32.06$
                       $4 \times \text{O}$ $4 \times 16.00 = \ \ 64.00$
                       $5 \times \text{H}_2\text{O}$ $5 \times 18.02 = \underline{\ \ 90.10}$
                                      249.71 g mol^{-1}

$\therefore$ Number of moles $= \dfrac{23.5}{249.7} = 0.0941$

$\therefore$ Concentration $= \dfrac{0.0941}{500} \times 1000 = 0.188$ mol dm^{-3}

Calculating the mass of solute required

We should now be able to calculate the mass of a solute needed to produce a solution of known concentration in a given volume.

Using mass:

$$\text{Moles} = \frac{\text{mass}}{M}$$

$$\text{Mass} = \text{moles} \times M$$

$$M = \frac{\text{mass}}{\text{moles}}$$

For solutions:

$$\text{Moles} = \frac{\text{volume (cm}^3)}{1000} \times \text{concentration}$$

$$\text{Volume} = \frac{\text{moles}}{\text{concentration}} \times 1000$$

$$\text{Concentration} = \frac{\text{moles}}{\text{volume}} \times 1000$$

If the volumes are in dm^3, omit the 1000.

Worked example

Calculate the mass of sodium chloride needed to produce 250 cm³ of a solution of concentration 0.50 mol dm⁻³.

Data:

NaCl
Volume $= 250 \, cm^3$
Concentration $= 0.50 \, mol \, dm^{-3}$

Relevant equation:

We will need to convert volumes and concentration into moles, and then convert that into a mass, so:

$$\text{Number of moles} = \frac{\text{volume (cm}^3)}{1000} \times \text{concentration}$$

$$\text{Number of moles} = \frac{\text{mass}}{M}$$

$\therefore$ Mass $=$ number of moles $\times M$

Calculation:

$$\text{Number of moles} = \frac{250}{1000} \times 0.5 = 0.125 \, mol$$

Mass $= 0.125 \times M$

$M(\text{NaCl})$ $1 \times$ Na $1 \times 23.00 = 23.00$
               $1 \times$ Cl $1 \times 35.45 = \underline{35.45}$
                                        $58.45 \, g \, mol^{-1}$

$\therefore$ Mass $= 0.125 \times 58.45 = 7.31 \, g$

In Ch 9.3 we met the problem of finding the number of moles of one of the species (atoms or ions) in one mole of a particular formula unit. With sodium hydroxide there is no difficulty. One mole of NaOH contains 1 mole of $Na^{\oplus}$ ions and 1 mole of $OH^{\ominus}$ ions. But once there is more than one atom, ion or group in the formula you have to be careful.

One mole of $Mg(OH)_2$ contains 1 mole of magnesium ions, $Mg^{2\oplus}$, but 2 moles of $OH^\ominus$ ions, since the formula contains two $OH^\ominus$ groups. So, 0.2 moles of magnesium hydroxide contains 0.2 moles of magnesium ions and 0.4 moles of hydroxide ions. When dissolving such a compound to make a solution, exactly the same ideas apply.

A solution of calcium chloride, $CaCl_2$, of concentration 0.1 mol dm^{-3}, contains 0.1 mol dm^{-3} of calcium ions but 0.2 mol dm^{-3} of chloride ions.

Moles in a formula:

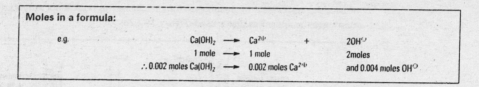

e.g.

$$Ca(OH)_2 \longrightarrow Ca^{2\oplus} \quad + \quad 2OH^\ominus$$

1 mole $\longrightarrow$ 1 mole   2moles

$\therefore$ 0.002 moles $Ca(OH)_2 \longrightarrow$ 0.002 moles $Ca^{2\oplus}$ and 0.004 moles $OH^\ominus$

Worked examples

A   *What mass of ammonium sulphate, $(NH_4)_2SO_4$, is needed to make 1.0 dm^3 of a solution containing 0.10 mol dm^{-3} of ammonium ions?*

Data:

$(NH_4)_2SO_4$
Volume = 1.0 dm^3 (which is 1000 cm^3)
Concentration of ammonium ions needed = 0.10 mol dm^{-3}

Relevant equations:

$$Moles = \frac{volume}{1000} \times concentration$$

$$Moles = \frac{mass}{M}$$

Calculation:

Since there are 2 mol of $NH_4^\oplus$ present in 1 mol of $(NH_4)_2SO_4$, for 0.10 mol dm^{-3} of $NH_4^\oplus$ ions, 0.050 mol dm^{-3} of $(NH_4)_2SO_4$ are required.

$$Moles\ of\ ammonium\ sulphate\ needed = \frac{1000}{1000} \times 0.050 = 0.050\ mol$$

$M((NH_4)_2SO_4)$		
$2 \times N$	$2 \times 14.01 =$	28.02
$8 \times H$	$8 \times 1.01 =$	8.08
$1 \times S$	$1 \times 32.06 =$	32.06
$4 \times O$	$4 \times 16.00 =$	64.00
		132.16 g mol^{-1}

$Mass = moles \times M = 0.050 \times 132.2 = 6.6\ g\ (NH_4)_2SO_4$

B   *You need to make up 250 cm^3 of a standard solution containing 0.010 mol dm^{-3} of nitrate ions, $NO_3^\ominus$. Calculate the mass of lead nitrate, $Pb(NO_3)_2$, required for this solution.*

Collect the data together from the question and write the two equations you will need (to convert volume of solution into moles, and to convert moles into mass).

Data:
Relevant equation:
Calculation:

- How many moles of nitrate ions are present in 1 mole of $Pb(NO_3)_2$?
- How many moles of nitrate ions are required in the question?
- Work out $M(PbNO_3)_2$.
- From these two figures the mass needed can be worked out as 0.414 g.

| Volumetric analysis |

Volumetric analysis is widely
used to answer the question,
'How much?'

As we have already said, the advantage of handling chemicals in solution, and certainly most common ionic substances, is that they can be more easily reacted. This has led to the development of an important area of analytical chemistry called **volumetric analysis**. Reacting solutions together in quantities measured as precisely as possible allows accurate and sensitive analyses to be carried out.

The technique of **titration** uses a particular set of apparatus with which volumes can be measured to an accuracy of greater than $0.1\ cm^3$. While this book is not a practical guide, it is worth referring to the apparatus that is used (*Figure 10.1.2*).

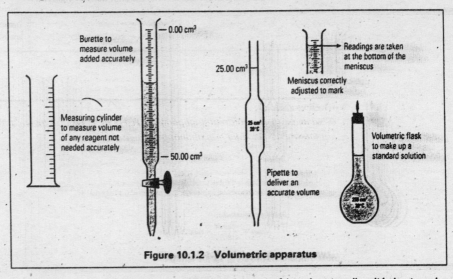

Figure 10.1.2   Volumetric apparatus

- If solutions are to be made up accurately, then the mass of the solute (usually solid – but it can be a liquid) is weighed to a suitable degree of accuracy. The substance is dissolved and thoroughly mixed in a *volumetric flask*. This flask has a long narrow neck with a mark, and it contains the stated volume when the bottom of the meniscus of the solution is just on that mark. If the exact composition of the solute is known and a solution of a known concentration is made up, we have produced a **standard solution**.

Definition: A standard
solution is one which is made up
to have a known concentration
of a known solute.

- Samples are taken from this solution using a *pipette*. A pipette, when filled properly (with the bottom of the meniscus on the marked line on the stem of the pipette), will deliver the stated volume to usually $\pm 0.05\ cm^3$. The sample is put in a conical flask.

- Finally, another solution is carefully added from a *burette* to the pipetted solution. The readings of the bottom of the meniscus at the start and at the end of the titration are noted, and the difference between the two readings gives the volume of solution that has reacted with the known volume of the first solution.

In this way, two solutions are reacted together, until the reaction is exactly completed. Often an **indicator** is needed to observe the **end point** (or **equivalence point**) of the reaction – the point at which sufficient of each reactant has been mixed together to complete the reaction. In any titration, **there are five variables, and any one can be calculated if the values of the other four variables are known or can be measured. The variables are:**

After a solution has been made
up, only part of it is actually
used for the titration (10.0 or
$25.0\ cm^3$). When doing the
final calculations, be careful to
remember that only a sample
has been taken.

- the equation for the reaction (and hence the mole ratio in which the materials react);
- the volume of reactant A;
- the concentration of reactant A;
- the volume of reactant B;
- the concentration of reactant B.

1 a How many moles of potassium hydroxide are there in 25.0 cm³ of 0.40 mol dm⁻³ potassium hydroxide solution?

   b Calculate the number of moles of magnesium nitrate in 21.7 cm³ of 0.0023 mol dm⁻³ solution.

   c How many moles of ethanol are in 250 cm³ of an aqueous solution containing 1.2 mol dm⁻³?

2 Calculate the concentration of a solution containing:

   a 2.0 g sodium hydroxide, NaOH, in 250 cm³ solution;

   b 0.0034 g silver nitrate, AgNO₃, in 100 cm³;

   c 23 g sulphuric acid, H₂SO₄, in 500 cm³.

3 a Concentrated sulphuric acid has a density of 1.84 g cm⁻³. What is the mass of 1 dm³ of the pure acid?

   b What is the relative molecular mass, M_r, of sulphuric acid?

   c Work out the number of moles present in 1 dm³ of the concentrated acid.

   d What is the apparent concentration of the concentrated acid?

   e Using the same principles, calculate the apparent concentration, in mol dm⁻³, of pure water.

4 a What mass of potassium chloride, KCl, is needed to make 500 cm³ of 0.10 mol dm⁻³ solution?

   b What mass of sodium sulphate, Na₂SO₄, is needed to make 1 dm³ of a 0.15 mol dm⁻³ solution of the salt?

   c What mass of cobalt(II) chloride, CoCl₂.6H₂O, is needed to make up a 500 cm³ solution of concentration 0.15 mol dm⁻³?

10.2 Acid–base titrations

How can you find out the concentration of an acid? How would you know, for example, that oven cleaner contains 40% sodium hydroxide solution? Acids react with alkalis and bases (and even carbonates) in a *neutralisation* reaction, and this reaction can be used, in a titration technique, to answer these questions. The acid is neutralised by reacting it in a carefully controlled way until the correct number of moles of each of the acid and alkali have been mixed to produce the neutral solution.

To find the concentration of, for example, the acid, we have to measure the volume of the acid that will react with a fixed volume of alkali of known concentration until the reaction is complete.

Acid and alkali solutions are usually colourless, and there is no obvious change when the end point has been reached. Two or three drops of an indicator that changes colour in acid or alkali solutions must be used to detect the end point. The indicator must be one that changes colour as soon as there is the slightest excess of either acid or alkali. So, using a titration technique, it is possible to work out the concentration of an acid.

Of course, the reverse is possible. By using an acid of known concentration and titrating it against an alkali solution of unknown concentration, the concentration of an alkali can be found. The choice of a suitable indicator is dealt with in Ch 23.4. For the moment, the indicators we normally use are phenolphthalein or methyl orange (*Figure 10.2.1*).

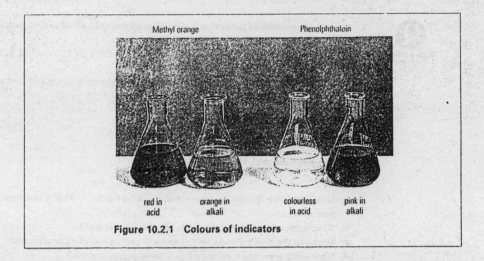

red in acid   orange in alkali   colourless in acid   pink in alkali

Figure 10.2.1 Colours of indicators

Phenolphthalein is a good indicator, changing from a pink colour in alkali solution to colourless in acid solution. But it is of no use with 'weak' alkalis such as aqueous ammonia.

Another common indicator is methyl orange, although it needs care to see the colour change. In acid solution it is red and in alkali it is yellow. If you keep a sample containing the same number of drops of indicator in roughly the same volume of distilled water alongside your titration, the colour change is easier to detect. Once the solution becomes the slightest bit acid it changes colour from yellow to red.

Methyl orange is an indicator that is of no use when using some weak acids such as ethanoic acid (found in vinegar). Litmus is *not* a good indicator for titrations, and pH or universal indicator is certainly of no use – it changes through so many colours. The apparatus used in any titration is shown in *Figure 10.2.2*.

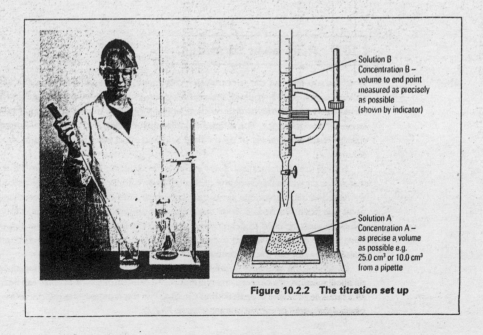

Solution B
Concentration B –
volume to end point
measured as precisely
as possible
(shown by indicator)

Solution A
Concentration A –
as precise a volume
as possible e.g.
25.0 cm^3 or 10.0 cm^3
from a pipette

Figure 10.2.2 The titration set up

Remember:

- Data
- Equation
- Moles
- Ratio

As we saw at the end of Ch 10.1, as long as four of the five variables (which include the equation for the reaction) are known or can be measured in the titration, the missing variable can be calculated. There are a variety of ways to carry out such calculations. Here we will continue to base our volumetric calculations on the same steps we have used in Ch 9:

- collecting all the given data together, including the equation;
- converting, if possible, these data into numbers of moles;
- checking the ratios in the equation;
- and then working through to the answer using this number of moles, and converting into a mass, volume or concentration as necessary.

Volumetric calculations

As we have just mentioned, in any volumetric calculations we need to have information about four out of the five variables – the concentration and volume of reactant A (to find the number of moles of A), the concentration and volume of reactant B (to find the number of moles of B), and the equation for the reaction.

When tackling any titration calculation the first step should be to write down this information: i.e. the equation for the reaction and all the available data. Since we have to convert these data into moles, *write down any equations used for converting volume and concentration into moles* (or converting mass into moles if any masses are quoted).

Reminder:

$$\text{Moles} = \frac{\text{volume (cm}^3)}{1000} \times \text{concentration}$$

$$\text{Volume (cm}^3) = \frac{\text{moles}}{\text{concentration}} \times 1000$$

$$\text{Concentration} = \frac{\text{moles}}{\text{volume (cm}^3)} \times 1000$$

In the actual calculation, the usual sequence, for a titration for which the equation is known or given, is as follows:

- Start by working out the number of moles used for one of the reactants (using the solution for which the volume **and** the concentration are known).
- Use the ratio in the equation to work out the number of moles used of the *other* reactant.
- Use the other data (amount of sample, volume of solution made up in volumetric flask, etc) to work out the concentration of the other reactant.

Worked example

A *A sample of dilute hydrochloric acid was titrated against 25.0 cm³ of 0.100 mol dm⁻³ sodium hydroxide solution. The neutralisation point was found when 23.3 cm³ of the acid had been added. What is the concentration of the hydrochloric acid?*

Calculation:

From data:

- Convert to moles
- Ratio from equation
- Moles of unknown from ratio
- Find unknown variable

(Since the volume and concentration of NaOH are known, we can find the number of moles.)

Data:

	$HCl(aq)$ +	$NaOH(aq) \longrightarrow NaCl(aq) + H_2O(l)$
Volumes	23.3 cm³	25.0 cm³
Concentrations	To be found	0.100 mol dm⁻³

All the variables, except the concentration of the acid, are known.

Relevant equations:

$$\text{Moles} = \frac{\text{volume}}{1000} \times \text{concentration}$$

With a 1:1 ratio in the equation, the number of moles of HCl used must be equal to the number of moles of NaOH when the solution is just neutralised.

Changing the relevant equation, the concentration can be calculated from the number of moles and the volume given in the question.

$$\text{Moles of NaOH in 25.0 cm}^3 = \frac{25.0}{1000} \times 0.100$$

$$= 0.00250 \text{ mol}$$

Ratio – from the balanced equation, 1 mole of HCl : 1 mole NaOH
∴ Moles of HCl in 23.3 cm³ = 0.00250 mol

$$\therefore \text{Concentration of HCl} = \frac{\text{moles}}{\text{volume}} \times 1000 = \frac{0.0025}{23.3} \times 1000$$

$$= 0.107 \text{ mol dm}^{-3}$$

B *A student had made up a solution of potassium hydroxide by weighing a given mass of the hydroxide and dissolving it in water. The hydroxide solution was titrated against 10.0 cm³ 0.0200 mol dm⁻³ sulphuric acid, and 12.5 cm³ of it were needed to neutralise the sulphuric acid. What was the concentration of the potassium hydroxide solution?*

Data:

$$H_2SO_4(aq) \quad + \quad 2KOH(aq) \longrightarrow K_2SO_4(aq) + 2H_2O(l)$$

Volumes	10.0 cm³	12.5 cm³
Concentrations	0.0200 mol dm⁻³	To be found

All the variables, except the concentration of the alkali, are known.

Relevant equations:

$$\text{Moles} = \frac{\text{volume}}{1000} \times \text{concentration}$$

Calculation:

$$\text{Moles of } H_2SO_4 \text{ in 10.0 cm}^3 = \frac{10.0}{1000} \times 0.0200$$

$$= 0.00020 = 2.00 \times 10^{-4} \text{ mol}$$

Ratio – from the balanced equation, 1 mole of H_2SO_4 : 2 moles KOH
∴ Moles of KOH in 12.5 cm³ = $2 \times (2.0 \times 10^{-4}) = 4.00 \times 10^{-4}$ mol

$$\therefore \text{Concentration of KOH} = \frac{\text{moles}}{\text{volume}} \times 1000 = \frac{4.00 \times 10^{-4}}{12.5} \times 1000$$

$$= 0.0320 \text{ mol dm}^{-3}$$

C *Calculate the volume of 0.12 mol dm⁻³ Ba(OH)₂ solution which will neutralise 10.0 cm³ of 0.16 mol dm⁻³ HNO₃ solution.*

Data:

$$2HNO_3(aq) \quad + \quad Ba(OH)_2(aq) \longrightarrow Ba(NO_3)_2(aq) + 2H_2O(l)$$

Volumes	10.0 cm³	To be found
Concentrations	0.16 mol dm⁻³	0.12 mol dm⁻³

Relevant equations:

$$\text{Moles} = \frac{\text{volume}}{1000} \times \text{concentration}$$

Calculation:

$$\text{Moles of } HNO_3 \text{ in 10.0 cm}^3 = \frac{10.0}{1000} \times 0.16$$

$$= 0.0016 \text{ mol}$$

Ratio – from the balanced equation, 2 moles of HNO₃ : 1 mole Ba(OH)₂
∴ Moles of Ba(OH)₂ $= \frac{1}{2} \times 0.0016 = 0.00080$ mol

$$\therefore \text{Volume of Ba(OH)}_2 = \frac{\text{moles} \times 1000}{\text{concentration}} = \frac{0.00080}{0.12} \times 1000$$

$$= 6.7 \text{ cm}^3$$

Remember about indices:

To convert large and small numbers into more manageable form, the number is written in standard form using powers of 10. Note that the power of 10 (the index) is the number of places you have to move the decimal point to get the original number.

e.g. $0.0025 = 2.5 \times 10^{-3}$

$1250000 = 1.25 \times 10^{6}$

$0.0000036 = 3.6 \times 10^{-6}$

On your calculator you can enter this by typing the number followed by EXP and the figure for the power of 10. For 1.26×10^{6} enter:

1.26 **EXP** **6**

and the number usually appears as 1.26^{06} or 1.26 06.

(N.B. the ($\times$ 10) part of the figure is often not shown on the display.)

For 3.6×10^{-6} enter:

3.6 **EXP** **6** **±**

and the number appears as 3.6^{-06}.

(Remember to press the minus sign, if the figure for the power of 10 is negative).

**SELF-ASSESSMENT
QUESTIONS
10.2**

1 Calculate the unknown volume in each of the following titrations.

  a What volume of 1.00 mol dm^{-3} hydrochloric acid is neutralised by 50.0 cm^3 1.50 mol dm^{-3} potassium hydroxide solution?

  b What volume of 0.160 mol dm^{-3} sodium hydroxide is neutralised by 10.0 cm^3 0.100 mol dm^{-3} sulphuric acid?

2 The concentration of a solution of limewater (saturated aqueous calcium hydroxide, Ca(OH)$_2$) was determined by titrating 25.0 cm^3 of the solution with 0.0200 mol dm^{-3} hydrochloric acid. 21.4 cm^3 of the acid was needed at the end point.

  a Write an equation for the reaction of limewater with hydrochloric acid.

  b Calculate the concentration of the limewater.

  c Limewater is used as a test for which gas?

  d Write an equation to show the reaction that occurs when limewater is used in this test.

  e Calculate the volume of carbon dioxide which can be absorbed by 1 dm^3 of limewater. (1 mole of a gas occupies 24 dm^3 at room temperature and pressure.)

  f Why is limewater not used for absorbing carbon dioxide in large quantities?

3 Phosphorus is an extremely reactive element, and the common allotropes (or polymorphs) are red phosphorus and white phosphorus. White phosphorus is kept under water since it readily starts to burn in air to form two oxides, P$_4$O$_6$ and P$_4$O$_{10}$.

  a What is the oxidation state of phosphorus in each of the oxides?

  b What is meant by the term *allotrope*?

  c P$_4$O$_{10}$ dissolves in hot water to form phosphoric acid, H$_3$PO$_4$. Write a balanced equation for the formation of H$_3$PO$_4$ from P$_4$O$_{10}$.

  d Phosphoric acid reacts with sodium hydroxide in an acid–base type reaction to form disodium hydrogen phosphate.

$$H_3PO_4(aq) + 2NaOH(aq) \longrightarrow Na_2HPO_4(aq) + 2H_2O(l)$$

  25.0 cm^3 of 0.500 mol dm^{-3} sodium hydroxide solution was used in such a reaction. How many moles of sodium hydroxide were used?

  e How many moles of phosphoric acid would react exactly with this volume of sodium hydroxide?

  f What volume of 0.100 mol dm^{-3} phosphoric acid would be needed to exactly neutralise the sodium hydroxide solution?

4 A sample of sulphuric acid, H_2SO_4, from an industrial manufacturing plant needed to be analysed to check the acid concentration. The solution had a high concentration, and the first step in the analysis was to dilute the sample by a factor of ten.

a Outline the steps that would be used, including the initial dilution, in the analysis of the acid.

b In such an analysis 25.0 cm^3 of the diluted acid reacted exactly with 25.0 cm^3 of 3.00 mol dm^{-3} sodium hydroxide solution. Calculate the concentration of the original acid, as mol dm^{-3}, and then in g dm^{-3}.

10.3 Redox titrations

Another class of reactions which is often investigated using titration techniques is redox reactions.

Remember, *redox reactions involve the transfer of electrons from one substance to another*; the oxidation state of one element decreases, while the oxidation state of another element increases (see Ch 4.3). The volumes of reacting solutions can be measured easily using titration techniques. The indicators used in these reactions are much more varied than those used in acid–base titrations, and depend on the particular reaction involved.

However, there are some such reactions that are '*self-indicating*' – colour changes during the reaction allow us to judge the 'end point'.

Reactions which involve iodide being oxidised to iodine are quite easy to follow, for example.

Iodide solutions are colourless and an iodine solution is yellow or brown. So the colourless solution would turn yellow/brown as iodine is released.

| Iodine is not very soluble in water, but dissolves freely in potassium iodide solution. Iodine titrations are often carried out in excess KI(aq). |

$$2I^{\ominus}(aq) \longrightarrow I_2(aq) + 2e^{\ominus}(aq)$$
$$\text{Colourless} \qquad \text{Brown/yellow}$$

The iodine released is titrated with standard sodium thiosulphate solution, resulting in a colourless solution. A colour change is easily observed if starch indicator is added towards the end of the titration. (If it is added too early, starch can interfere with the end point.) Close to the end point, as long as there is still some iodine left, the addition of starch will give the almost black colour of the iodine–starch complex. As soon as all the iodine has reacted, the black colour will disappear.

The routine for the calculation work is exactly as before. For example, the oxidation of sodium thiosulphate using iodine solution is shown in the box below. Notice the iodine is reduced to iodide while the sulphur atoms in thiosulphate are apparently oxidised from the +2 oxidation state to +2 (see Ch 4.3).

Worked example

Sodium thiosulphate solution reacts with iodine to produce a colourless solution according to the equation:

$$2Na_2S_2O_3(aq) + I_2(aq) \longrightarrow Na_2S_4O_6(aq) + 2NaI(aq)$$

When an iodine solution was titrated against 25.0 cm^3 0.250 mol dm^{-3} sodium thiosulphate solution, using a starch solution indicator near the end point, 22.3 cm^3 of the iodine solution was required. What is the concentration of the iodine solution?

Data:

$$2Na_2S_2O_3(aq) \ + \ I_2(aq) \longrightarrow Na_2S_4O_6(aq) + NaI(aq)$$

Volumes	25.0 cm^3	22.3 cm^3
Concentrations	0.25 mol dm^{-3}	To be found

Relevant equations:

$$\text{Moles} = \frac{\text{volume}}{1000} \times \text{concentration}$$

Calculation:

$$\text{Moles of } Na_2S_2O_3 \text{ in } 25.0 \text{ cm}^3 = \frac{25.0}{1000} \times 0.250$$

$$= 0.00625 \text{ mol}$$

Ratio – from the balanced equation, 2 moles of $Na_2S_2O_3$: 1 mole of I_2

$$\therefore \text{ Moles of } I_2 \quad = \tfrac{1}{2} \times 0.00625 = 0.00313 \text{ mol}$$

$$\therefore \text{ Concentration of } I_2 = \frac{\text{moles}}{\text{volume}} \times 1000 = \frac{0.00313}{22.3} \times 1000$$

$$= 0.140 \text{ mol dm}^{-3}$$

A variation on this method involves liberating iodine from excess aqueous potassium iodide by adding a solution of an oxidising agent, then titrating the liberated iodine with standard sodium thiosulphate solution. Provided that the equation for the reaction between the oxidising agent and iodide ions is known, the concentration of the oxidising solution can be calculated.

The available chlorine in bleach is found by using the chlorine to release iodine from excess potassium iodide solution.

$$Cl_2(aq) + 2KI(aq) \longrightarrow 2KCl(aq) + I_2(aq)$$

The quantity of iodine produced is then measured by titrating the solution with standard sodium thiosulphate solution.

$$I_2(aq) + 2Na_2S_2O_3(aq) \longrightarrow Na_2S_4O_6(aq) + 2NaI(aq)$$

Any reaction involving acidified potassium manganate(VII), $KMnO_4(aq)$, is 'self indicating'. Manganate(VII) ions are powerful oxidising reagents, especially in acid solution. When the manganate(VII) ion is reduced, the almost colourless $Mn^{2+}(aq)$ ion is formed, and so the solution will change from purple to (almost) colourless.

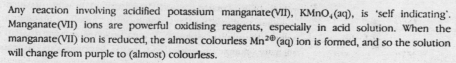

$$\underset{\text{Purple}}{MnO_4^{-}(aq)} + 8H^{+}(aq) + 5e^{-} \longrightarrow \underset{\text{Pale pink}}{Mn^{2+}(aq)} + 4H_2O(l)$$

$MnO_4^{-}(aq)$ $Mn^{2+}(aq)$

Alternatively, potassium manganate(VII) is added to a solution of a reducing agent, and the first permanent sign of the purple colour indicates that the reaction is complete.

Worked example

Potassium manganate(VII) solution can be used to analyse for iron(II) in a solution, since the manganate readily oxidises the iron(II) to iron(III) in the presence of an acid.

$$5Fe^{2+}(aq) + MnO_4^{-}(aq) + 8H^{+}(aq) \longrightarrow 5Fe^{3+}(aq) + Mn^{2+}(aq) + 4H_2O(l)$$

Calculate the concentration of a solution of iron(II) sulphate, $FeSO_4$ when 25.0 cm^3 of it decolourises 18.5 cm^3 $0.050 \text{ mol dm}^{-3}$ potassium manganate(VII) solution acidified with dilute sulphuric acid.

Data:

$$5Fe^{2\oplus}(aq) + MnO_4^{\ominus}(aq) + 8H^{\oplus}(aq) \longrightarrow 5Fe^{3\oplus}(aq) + Mn^{2\oplus}(aq) + 4H_2O(l)$$

Volumes $25.0\ cm^3$ $18.5\ cm^3$

Concentrations   To find $0.050\ mol\ dm^{-3}$

Relevant equations:

$$Moles = \frac{volume}{1000} \times concentration$$

Calculation:

$$Moles\ of\ MnO_4^{\ominus}\ in\ 18.5\ cm^3 = \frac{18.5}{1000} \times 0.050$$

$$= 0.000925\ mol$$

Ratio — from the balanced equation 5 moles of $Fe^{2\oplus}$: 1 mole of $MnO_4^{\ominus}$

$$\therefore Moles\ of\ Fe^{2\oplus} = 5 \times 0.000925 = 0.00463\ mol$$

$$\therefore Concentration\ of\ Fe^{2\oplus} = \frac{moles}{volume} \times 1000 = \frac{0.00463}{25.0} \times 1000$$

$$= 0.185\ mol\ dm^{-3}$$

$$\therefore Concentration\ of\ iron(II)\ sulphate = 0.185\ mol\ dm^{-3}$$

**SELF-ASSESSMENT
QUESTIONS
10.3**

1 Ethanedioate ions ($C_2O_4^{2\ominus}$) react at 60 °C with potassium manganate(VII) according to the equation:

$$2MnO_4^{\ominus}(aq) + 5C_2O_4^{2\ominus}(aq) + 16H^{\oplus}(aq) \longrightarrow 2Mn^{2\oplus}(aq) + 10CO_2(g) + 8H_2O(l)$$

Calculate the concentration of ethanedioate ions, both in mol dm^{-3} and in g dm^{-3}, if 10.0 cm^3 of a 0.100 mol dm^{-3} solution of potassium manganate(VII) is found to react with 12.8 cm^3 of the solution.

2 One interesting reaction of copper(II) ion is the oxidation of iodide ion to iodine – the blue solution of copper(II) sulphate changes to a 'dirty' colour as the brown colour of iodine mixes with the blue of excess copper(II) sulphate and the white of copper(I) iodide, CuI. The amount of iodine produced in the reaction can be found by titrating the solution against standard sodium thiosulphate solution. In a particular experiment 10.0 cm^3 of 0.100 mol dm^{-3} copper sulphate solution was reacted with excess potassium iodide solution.

a Write an equation for the reaction of sodium thiosulphate and iodine.

b 10.0 cm^3 of 0.100 mol dm^{-3} sodium thiosulphate was required to react with the liberated iodine. How many moles of iodine molecules are present in the solution?

c How many moles of copper(II) sulphate were used in the reaction?

d How many moles of copper(II) ions reacted to form one mole of iodine?

e Complete the following equation by considering the ratio of the answers in b and c.

$$\ldots Cu^{2\oplus}(aq) + \ldots I^{\ominus}(aq) \longrightarrow \ldots I_2(aq) + \ldots CuI(s)$$

3 A sample of iron wire, used in preparing flower displays at a florist's, was analysed to check the purity of the iron. A sample weighing 0.165 g was dissolved in excess dilute sulphuric acid to form iron(II) ions. This solution was then diluted to 50.0 cm^3.

a Complete the equation for the reaction of $Fe^{2\oplus}$ and manganate(VII) ions in acid solution.

$$\ldots Fe^{2\oplus}(aq) + MnO_4^{\ominus}(aq) + \ldots H^{\oplus}(aq) \longrightarrow 5Fe^{3\oplus}(aq) + Mn^{2\oplus}(aq) + \ldots H_2O(l)$$

b 10.0 cm^3 of the diluted solution reacted with 22.3 cm^3 of 0.00500 mol dm^{-3} potassium manganate(VII) solution. This titration is said to be self-indicating. Explain what this means.

c How many moles of manganate(VII) ions were used in the titration? How many moles of $Fe^{2\oplus}$ were present in the titrated sample?
d How many moles were present in the original 50.0 cm^3 of solution?
e What mass of iron does this correspond to?
f What mass of iron wire was used in making up the original solution?
g What is the percentage purity of the iron wire?

Summary

$$\text{Moles} = \frac{\text{mass}}{M}$$

'1 mol dm^{-3}' refers to a solution containing 1 mole of solute in 1 dm^3 of solution.
Molecular formula shows the number of each type of atom in one molecule of a compound.

$$\text{Moles} = \frac{\text{volume (cm}^3)}{1000} \times \text{concentration}$$

Acid–base equations

e.g. $HCl(aq) + NaOH(aq) \longrightarrow NaCl(aq) + H_2O(l)$

$H_2SO_4(aq) + 2KOH(aq) \longrightarrow K_2SO_4(aq) + 2H_2O(l)$

General equation: $H^{\oplus}(aq) + OH^{\ominus}(aq) \longrightarrow H_2O(l)$

Redox equations

e.g. $I_2(aq) + 2S_2O_3^{2\ominus}(aq) \longrightarrow 2I^{\ominus}(aq) + S_4O_6^{2\ominus}(aq)$

$MnO_4^{\ominus}(aq) + 5Fe^{2\oplus}(aq) + 8H^{\oplus}(aq) \longrightarrow Mn^{2\oplus}(aq) + 5Fe^{3\oplus}(aq) + 4H_2O(l)$

S
SUMMARY 10

In this chapter the following new ideas have been covered. *Have you understood and learned them?* Could you explain them clearly to a friend? The page references will help.

- The meaning of the concentration of a solution. p252
- The calculation of the number of moles in a solution of known volume and concentration. p253
- Transforming the equation which links moles, volume and concentration to calculate the volume or the concentration. p256
- The principles of volumetric analysis. p258
- How to carry out calculations involving acid–base titrations. p261
- Examples of redox titrations, and how to do the associated calculations. p264

Q
EXAMINATION QUESTIONS 10

1 Chlorine is used in swimming pools to keep the pool hygienic. Monitoring the amount of chlorine available in the water is obviously very important. One method of analysing a sample of pool water for the chlorine concentration is to treat it with excess of a solution of acidified potassium iodide. This liberates iodine, which can then be titrated against standard sodium thiosulphate solution.

a Write an equation for the reaction between chlorine and iodide ions.
b Crystalline sodium thiosulphate has the formula $Na_2S_2O_3.5H_2O$. What is the value of M_r for this compound?
c How much of the crystalline sodium thiosulphate would you need to make up 1 dm^3 of a solution of 0.0100 mol dm^{-3} concentration?

d This solution was used to titrate against the liberated iodine. Write an equation for the reaction between iodine and sodium thiosulphate solutions.

e In the analysis, a 25.0 cm³ sample of swimming pool water, after treatment with excess potassium iodide, reacted with 10.3 cm³ of the thiosulphate solution. Calculate the concentration of the iodine solution, and hence the concentration of chlorine in the pool.

f What indicator would you use to detect the end point in this titration?

2 Hydrogen peroxide is an effective bleaching agent and releases oxygen according to the equation:

$$2H_2O_2(l) \longrightarrow 2H_2O(l) + O_2(g)$$

The concentration of hydrogen peroxide has to be monitored carefully – concentrated hydrogen peroxide is a very dangerous chemical. In an analysis, 25.0 cm³ of 0.150 mol dm⁻³ solution of hydrogen peroxide were titrated against potassium manganate(VII) solution.

a How many moles of hydrogen peroxide were used?

b In a titration the volume of the aqueous manganate(VII) solution was 46.8 cm³ of 0.320 mol dm⁻³. How many moles of potassium manganate(VII) were used?

c How many moles of hydrogen peroxide reacted with one mole of potassium manganate(VII), $KMnO_4$?

d Complete the following equation

$$...MnO_4^{\ominus}(aq) + ...H_2O_2(aq) + 6H^{\ominus}(aq) \longrightarrow ...H_2O(l) + ...Mn^{2\oplus}(aq) + 5O_2(g)$$

e How would you recognise the end point of the titration?

3 Autumn treatment powders for garden lawns often contain a small amount of iron(II) sulphate, providing $Fe^{2\oplus}$ ions as a moss killer. The $Fe^{2\oplus}$ content of such a lawn treatment powder can be measured by dissolving the powder in water, and titrating the solution against standard potassium manganate(VII) solution?

a What is meant by standard solution?

b Write an ionic half equation for the oxidation of $Fe^{2\oplus}$(aq) ions to $Fe^{3\oplus}$(aq).

c A 2.00 g sample of this powder was dissolved in 50.0 cm³ of dilute sulphuric acid, and the resulting solution titrated against 0.0400 mol dm³ aqueous potassium manganate(VII). 21.3 cm³ of the manganate (VII) solution were required to complete the reaction. What colour change would be observed during the titration?

d How many moles of potassium manganate(VII) were used in the titration?

e The equation for the oxidation of $Fe^{2\oplus}$(aq) by manganate(VII) ions in acid solution is:

$$5Fe^{2\oplus}(aq) + MnO_4^{\ominus}(aq) + 8H^{\oplus}(aq) \longrightarrow 5Fe^{3\oplus}(aq) + Mn^{2\oplus}(aq) + 4H_2O(l)$$

Calculate the number of moles of $Fe^{2\oplus}$ which were present in the titrated sample.

f Therefore, what is the concentration of iron(II) ions in the original solution of the lawn treatment powder?

g What is the mass of iron in 2.00 g of the original powder?

- **Particles attract** each other. (Ch 2.8)
- **Chemical bonds** can exist between atoms. (Ch 2.2)
- Chemical reactions can be represented by **equations**. (Ch 4.2)
- The idea of the **mole**. (Ch 9.3)
- The use of the mole and equations in **chemical calculations**. (Ch 9.5)

Because particles are always attracted to one another and may be joined together by chemical bonds, energy changes **always** occur when particles are separated or brought together, and these energy changes are larger if definite chemical bonds are broken or made.

In this chapter

We shall consider the consequences of this basic idea, and show how calculations involving energy changes can be carried out.

11.1 **Introduction**
11.2 **Enthalpy changes and how they can be measured**
11.3 **Standard molar enthalpy changes; definitions**
11.4 **Hess' Law**
11.5 **Making and breaking chemical bonds: bond enthalpies**

11
energy changes

11.1 Introduction

We have already met some types of energy changes; for example, those associated with changes of state (Ch 2.9), with bonds and attractions between atoms and molecules (Ch 2), and ionisation energies (Ch 1.5). We have also met various forms of energy, e.g. thermal (heat), electrical, and light.

You will also know that one form of energy can be converted into another: e.g. chemical energy (via thermal energy and mechanical energy) to electrical energy in a power station using fossil fuels; electrical to light and thermal energy in an ordinary light bulb; gravitational (via kinetic and mechanical) to electrical energy in a hydro-electric power station. You can probably think of many others, connected for example with transport, manufacturing industry, cooking – indeed, with every aspect of humanity's physical activities.

And what about life itself? Your body, through digestion, metabolism and respiration, transforms chemical energy into an ability to run, jump, play, write, read – and think.

Energy from the Sun, through photosynthesis in green plants, provides us with our food – either as fruit, roots, tubers, leaves or stems, or as the animals, birds or fish which feed ultimately on plants. It also provides us, through combustion of the fossil fuels coal, oil and natural gas, with more than 80% of our present consumption of energy. Energy is released, always inefficiently, from these fossil fuels by combustion.

One important point to remember is that conversion of one form of energy into another is *never* 100% efficient. Some of the original energy is always lost as low grade, unusable thermal energy. The tendency of energy to dissipate was first discussed in depth by the German theoretical physicist Rudolf Clausius (1822–88), and summarised in the Second Law of Thermodynamics. One way of stating this Law is: 'Heat does not spontaneously pass from a colder body to a hotter body'. This may not seem very profound, but the Second Law is the basis for the design and development of all processes and engines involving thermal energy, e.g. all internal combustion engines, power stations, and many manufacturing methods.

The *maximum* efficiency of a coal-fired power station, in terms of converting the chemical energy in coal into electrical energy, can never be more than about 40–45%. If the heat energy released is used sensibly instead of mostly being transferred to the environment through cooling towers, the *overall* energy efficiency of the power station may be raised to 70–80%.

When a fossil fuel is burned, we are releasing solar energy which was trapped many millions of years ago. We are also releasing back into the atmosphere carbon dioxide which was taken out of it over a long period of time by photosynthesis. Because carbon dioxide is being released faster than it can be removed by photosynthesis and by passing into solution in the oceans, the proportion of carbon dioxide in the atmosphere has increased from about 280 parts per million (ppm; equivalent to 0.0280%) to about 350 ppm within the last century. This has contributed to a greater 'greenhouse effect' (see Ch 17.2).

Trapping carbon dioxide:

$$6CO_2(g) + 6H_2O(l) \underset{\text{respiration}}{\overset{\text{photosynthesis}}{\rightleftharpoons}} C_6H_{12}O_6(aq) + 6O_2(g)$$
$$\text{glucose}$$

$$\text{digestion} \updownarrow \text{ in plants}$$
$$(C_6H_{10}O_5)_n$$
$$\text{starch}$$

Forming carbon dioxide by combustion:

e.g.
$$C(s) + O_2(g) \longrightarrow CO_2(g)$$
$$\text{coal}$$

$$CH_4(g) + 2O_2(g) \longrightarrow CO_2(g) + 2H_2O(l)$$
$$\text{methane}$$
$$\text{(natural gas)}$$

Throughout the world; the living standards of ordinary people are closely linked to how much energy is available to them. In the highly industrialised countries of Europe, North America, Japan, SE Asia and Australasia, each person uses (and largely wastes) many times more energy than someone who lives in most countries of Central and South America, Asia and Africa. People in those countries see the energy consumption and living standards of those in the industrialised countries and quite naturally want the same. Yet how can this be achieved without even more rapid using up of the Earth's precious deposits of fossil fuels, and the pouring of even more carbon dioxide into the atmosphere?

As is all too often the case, it is the attitudes of individuals and communities – particularly those which are 'rich' and/or have power – which will have to change. But chemists and other scientists will also have to play their part in making our use of existing fuels more efficient, in tapping the Sun's energy more efficiently and more directly, and in developing alternative and renewable sources of energy. Nuclear fusion processes, similar to those which occur in the Sun (see Ch 1.2), *may* help us out eventually. Meanwhile there is no easy solution to the world's energy crisis, and to the injustice and inequality which result from it; and time is not on our side.

11.2   Enthalpy changes and how they can be measured

Many common chemical changes, for example burning, result in energy being released. The energy is usually, but not always, in the form of heat, but burning is often used to produce light as well. Such energy changes are described as **exothermic**; energy is lost from the reacting system to the surroundings.

Less often, chemical changes result in energy flowing *into* the system from the surroundings, usually because the reaction system has got colder. Such changes are called **endothermic**.

We shall discuss only processes in which heat energy is involved.

It is impossible to measure the total heat energy content of a system. We can, however, measure *changes* in the heat energy of a system. Because most of the chemical changes we are familiar with occur in apparatus which is open to the atmosphere – e.g. test tubes, beakers, and open calorimeters – they occur at constant pressure, *atmospheric pressure*.

The special name for the change in heat energy content of a system which occurs during a reaction at constant pressure is **enthalpy change**. Enthalpy change is given the symbol ΔH.

For exothermic reactions, the value of ΔH is negative – because heat energy is *lost* from the system. **For endothermic reactions, ΔH is positive** – because heat energy is *gained* by the system. Exothermic and endothermic changes are shown in *Figure 11.2.1*.

> Compare:
> *exothermic – exit*, the way *out*;
> *endothermic* – *entrance*, the way *in*.

> The Greek letter *delta*, written either as δ, or the capital letter Δ, is often used to indicate change in some quantity.

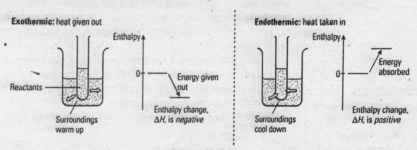

Figure 11.2.1

Enthalpy changes are often measured in **calorimeters**, which are simply devices for isolating a reacting system from the surroundings. A very simple calorimeter for measuring enthalpy changes in solution can be made from a polystyrene cup placed in a plastic mug (*Figure 11.2.2*).

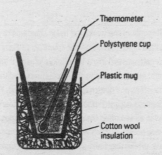

Figure 11.2.2   Simple calorimeter

Temperature changes can be measured to an appropriate level of accuracy by a thermometer.

For more accurate and complex work various types of 'bomb calorimeter' are available (*Figure 11.2.3*).

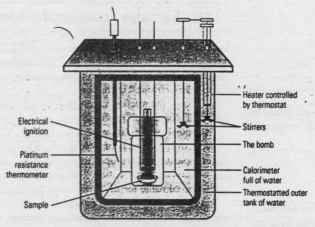

Figure 11.2.3   Bomb calorimeter

Strictly, in an exothermic reaction we should measure the heat energy which flows from the reacting system into the surroundings until the system is back at its starting temperature. But this would be very difficult to do, and anyway would take far too long. Instead, we make sure the reactants are all at room temperature; mix known amounts of the reactants; and note the maximum rise or fall in temperature.

We also need to know the **heat capacity** of the system. If we use dilute solutions, the heat capacity of the liquid will be very nearly the same as that of the same amount of water.

The heat capacity of plastic and glass is very small compared with that of water, so we can usually ignore the heat capacity of the apparatus.

The enthalpy change, ΔH, is given, with reasonable accuracy, by:

$$\Delta H = mc\Delta T$$

where:

　m is the mass of the water, in g;
　c is the heat capacity of water, in $J\,g^{-1}\,K^{-1}$;
　ΔT is the change in temperature of the water, in °C.

> Heat capacities are often expressed in the units $J\,g^{-1}\,K^{-1}$ – i.e. how much energy is needed to raise the temperature of 1 gram of the substance by one Kelvin.

The maximum temperature rise?

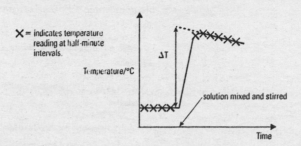

X = indicates temperature reading at half-minute intervals.

Temperature/°C

ΔT

solution mixed and stirred

Time

How do you take account of the cooling effects even as the reading on the thermometer is rising? For reactions which involve mixing substances, at least one of which is a solution, extrapolation on a graph of the temperature before and after the mixing can help.

A reminder about units

Temperature is measured in degrees Celsius, °C, or in Kelvin, K.

$K = °C + 273$ and $1 °C = 1 K.$

Energy is measured in Joules.

$1 kJ = 1000 J$ (i.e. $10^3 J$)

It takes 4.18 J to raise the temperature of 1 g of water by 1 °C (1 K).

The energy content of foods is often quoted on the packets in calories, cal. The 'calories' used for foodstuffs are actually kilocalories, kcal. If a biscuit is labelled as providing '100 calories', it is actually 100 000 calories, i.e. 418 000 J! You may, during Biology lessons, have burned one peanut and been surprised at how much hotter it can make a small beaker of water.

Exothermic changes

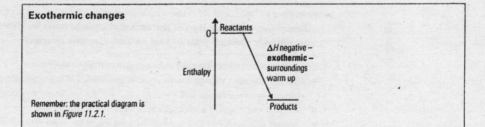

Remember: the practical diagram is shown in *Figure 11.2.1.*

Enthalpy

0 — Reactants

ΔH negative – **exothermic** – surroundings warm up

Products

Endothermic changes

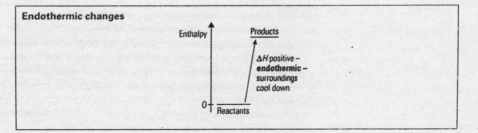

Enthalpy

Products

ΔH positive – **endothermic** – surroundings cool down

0 — Reactants

Worked example – calculating just the enthalpy change

A *Calculate the enthalpy change if 25.0 cm³ of water is warmed by 6.5 °C.*

Data:

H_2O, temperature change = 6.5 °C
Mass = 25.0 g (since 25.0 cm³ of water is used)

Equation:

$\Delta H = mc\Delta T$
$c = 4.18 \, J \, g^{-1} \, K^{-1}$ (i.e. the heat capacity of water)

Calculation:

$\therefore \Delta H = 25.0 \times 4.18 \times 6.5$
$= 679 \, J$ (remember the units of c contain Joules, J, not kJ).

Since there was a temperature rise, the enthalpy change was exothermic, its sign must therefore be negative, and the value of ΔH is -679 J.

B *Calculate the enthalpy change when 50.0 cm³ of 0.100 mol dm⁻³ HCl is heated from 18.0 °C to 29.2 °C.*

Data:

HCl
0.100 mol dm⁻³
50.0 cm³
Temperature change = 29.2 − 18.0 = 11.2 °C

Equation:

$\Delta H = mc\Delta T$
$c = 4.18 \, J \, g^{-1} \, K^{-1}$ for a dilute aqueous solution
m = 50.0 g (since it is 50.0 cm³ of a dilute aqueous solution, the mass of water will be very nearly 50 g).

Calculation:

$\therefore \Delta H = 50.0 \times 4.18 \times 11.2$
$= 2341 \, J$
$= 2.34 \, kJ$

Again, the temperature rise shows an exothermic change.

$\therefore \Delta H = -2.34 \, kJ$

Remember: In calorimetry, the equation for the energy change is $\Delta H = mc\Delta T$, where ΔH is the enthalpy change in Joules, m is the mass of material in grams, c is the heat capacity of the material in $J \, g^{-1} \, K^{-1}$, and ΔT is the temperature change in K (= the temperature change in °C).

Remember also: the sign of the enthalpy change must be clearly stated, minus (−) for an exothermic change, plus (+) for an endothermic change.

There are usually three separate stages in any enthalpy calculation:
- Find the enthalpy change for the experiment using $\Delta H = mc\Delta T$, where m is the total mass that changes in temperature and $c = 4.18 \, J \, g^{-1} \, K^{-1}$ for any water or dilute aqueous solution.
- Decide if the reaction is exothermic or endothermic, and use − or + for the enthalpy change.
- Convert this enthalpy change into a molar quantity, from the number of moles used in the experiment.

Worked example – enthalpy change of neutralisation

50.0 cm³ of sodium hydroxide solution (concentration 2.00 mol dm⁻³) are mixed with 50.0 cm³ of hydrochloric acid (2.00 mol dm⁻³). The temperature of both solutions before mixing was 20.0 °C.

The maximum temperature reached after mixing and stirring was found to be 33.1 °C. What would be the enthalpy change if one mole of sodium hydroxide (aqueous, concentration 2.00 mol dm⁻³) is completely neutralised by reacting with one mole of hydrochloric acid (aqueous, 2.00 mol dm⁻³)?

(The heat capacity of water is 4.18 J g⁻¹ K⁻¹ – i.e. it takes 4.18 Joules of energy to heat one gram of water through one Kelvin – i.e. one degree Celsius).

First, write the equation:

$$NaOH(aq) + HCl(aq) \longrightarrow NaCl(aq) + H_2O(l)$$

From the equation:

One mole of NaOH reacts completely with one mole of HCl.

But how much of the two reactants are we actually using?

50.0 cm^3 of 2.00 mol dm^{-3} NaOH(aq) = 50/1000 × 2.0 mol NaOH = 0.100 mol NaOH

Similarly, 50.0 cm^3 of 2.00 mol dm^{-3} HCl = 0.100 mol HCl.

When the two dilute solutions are mixed, there are approximately 50 + 50 = 100 cm^3 water ≡ 100 g water.

So, ignoring the heat capacity of the apparatus:

$$\Delta H = mc\Delta T$$

i.e. enthalpy change = mass of water × heat capacity of water × change in temperature.

$$\therefore \Delta H = 100 \times 4.18 \times (33.1 - 20.0)$$
$$= -5476\,J$$

As this is an *exothermic* process, a *negative* sign is used.

But this is the enthalpy change for 0.100 mol of each reactant. So, for one mole of each:

$$\Delta H = -5476 \times 10 = -54.8\ kJ\ mol^{-1}.$$

Therefore the *molar enthalpy change of neutralisation* of hydrochloric acid by sodium hydroxide, both at a concentration of 2.00 mol dm^{-3}, is -54.8 kJ mol^{-1}.

(As we shall see in the next section, the actual value of molar enthalpy changes for reactions in solution depends on the concentration of the reactants.)

Worked example – calculating the molar enthalpy change

In an experiment 50.0 cm^3 of 1.00 mol dm^{-3} HCl was exactly neutralised by 25.0 cm^3 2.00 mol dm^{-3} KOH solution. The initial temperature of both solutions was 15.3 °C and the final temperature was 24.3 °C. Calculate the molar enthalpy change of neutralisation of HCl.

Data:

$$HCl + KOH \longrightarrow KCl + H_2O$$
50.0 cm^3 25.0 cm^3
1.00 mol dm^{-3} 2.00 mol dm^{-3}
Temperature change = 24.3 − 15.3 = 9.0 °C

Equation:

$$\Delta H = mc\Delta T$$

m = (50.0 + 25.0) = 75.0 g
(since the total volume of the aqueous solution was 75.0 cm^3)

c = 4.18 J g^{-1} K^{-1} (since it is a dilute aqueous solution)

Calculation:

$$\Delta H = 75.0 \times 4.18 \times 9.0$$
$$= 2822\ J$$
$$= 2.82\ kJ$$

The temperature change shows the reaction is exothermic,

$$\therefore \Delta H \text{ must be negative.}$$
$$\therefore \Delta H = -2.82\ kJ$$

In the reaction 50.0 cm^3 of 1.00 mol dm^{-3} HCl was used.

The number of moles of HCl = $\dfrac{volume}{1000}$ × concentration

$$= \frac{50.0 \times 1.00}{1000} = 0.0500\ moles$$

∴ For one mole of HCl, the enthalpy change is $\dfrac{-2.82 \times 1}{0.0500} = -56.4$ kJ mol^{-1}

1 Calculate the enthalpy change in each of the following situations. Remember the heat capacity of any dilute solution can be assumed to be the same as that of water, i.e. $4.18\,J\,g^{-1}\,K^{-1}$.
  a 50.0 g water warmed by 25.0 °C.
  b 100 cm³ 0.100 mol dm⁻³ sodium hydroxide solution warmed from 22.3 °C to 25.7 °C.
  c 150 cm³ 0.25 mol dm⁻³ sulphuric acid heated from 20.0 °C to 40.0 °C.

2 Iron metal will react with copper sulphate solution in an exothermic reaction.

$$Fe(s) + CuSO_4(aq) \longrightarrow FeSO_4(aq) + Cu(s)$$

When 0.50 g of powdered iron is added to 100 cm³ 0.200 mol dm⁻³ copper sulphate solution the temperature rises by 3.3 °C.
  a Why is the iron metal added as a powder?
  b Calculate the enthalpy change in the reaction. (The heat capacity of the solution can be taken as $4.18\,J\,g^{-1}\,K^{-1}$. The heat capacity of the iron can be neglected.)
  c How many moles of iron were used in the reaction?
  d Calculate the enthalpy change associated with one mole of iron added to excess dilute copper sulphate solution.

3 When 1.00 g methanol was completely burned in a simple lamp burner under a container of water, the temperature of the water changed from 21.0 °C to 45.5 °C. The mass of water in the container was 220 g. The heat capacity of the water can be taken as $4.18\,J\,g^{-1}\,K^{-1}$, and the heat capacity of the container ignored.
  a How much energy was absorbed by the water in the container?
  b How much energy was given out by the methanol?
  c Is the combustion of methanol an exothermic or endothermic change?
  d How many moles of methanol, CH_3OH, were used in the reaction?
  e Calculate the enthalpy change when one mole of methanol is completely burned.

11.3 Standard molar enthalpy changes; definitions

Accurate comparisons of enthalpy changes should be measured under **standard conditions**. That is at 25 °C (298 K) and one standard atmosphere pressure (101 325 N m⁻²), with any solutions involved having a concentration of 1.00 mol dm⁻³, and with other reacting substances in their *standard states* – i.e. in their normal physical states under these conditions. (This last requirement is to take account of polymorphism – see Ch 3.2 for the meaning of *polymorphism*; e.g. graphite is slightly more stable under standard conditions than diamond.)

The amount of material also matters. Unlike, say, density or temperature, the value of ΔH depends on how much material has reacted.

It is therefore an *extensive* property – its value depends on the amount of the substance involved. Properties such as density or temperature are not dependent on the amount of substance; these are *intensive* properties.

The 'standard molar enthalpy change' in any reaction is given the symbol $\Delta H^{\ominus}_{r,\,298}$; the '298' is because of the standard temperature of 298 K, i.e. 25 °C. Note that 'r' here denotes 'reaction' of molar quantities.

Obviously, many reactions – including nearly all combustions – cannot occur at a reasonable rate at standard temperature, but the experiment can always be arranged to allow us to refer back to standard conditions. Notice the reference in the definition to 'molar quantities of reactants as stated in the equation'. The safest way of interpreting this is to take the word 'molar' to refer to moles of the *whole equation* rather than to the number of moles of any individual reactant or product.

Consider, for example, the Haber Process for making ammonia (see Ch 13.3):

$$N_2(g) + 3H_2(g) \rightleftharpoons 2NH_3(g) \qquad \Delta H^{\ominus}_{r,\,298} = -92\,kJ\,mol^{-1}$$

Two moles of methanol will give out twice as much heat energy when burned as will one mole. But both samples could have the same temperature.

Definition: The standard molar enthalpy change of any reaction, $\Delta H^{\ominus}_{r,\,298}$, is the amount of heat energy given out or absorbed when the molar quantities of reactants as stated in the equation react completely together under standard conditions to form products.

In many books an enthalpy change is often referred to as a 'heat of...' e.g. 'heat of reaction', 'heat of formation', etc.

So 92 kJ of heat energy is released, but per mole of what? Some old textbooks got into a real mess with this, quoting figures in terms of moles of nitrogen or hydrogen used up, or moles of ammonia formed. It is much safer and simpler, and makes much more sense, to **write the equation for the reaction every time**. The symbol 'mol^{-1}' then refers unambiguously to the quantities stated in the equation.

Don't forget – always show the sign of any enthalpy change. No sign means you do not know if it is positive or negative!

Other enthalpy changes

Once you have thoroughly understood the basic definition of $\Delta H^{\ominus}_{r, 298}$, the idea can be extended to all kinds of different reactions and processes, including changes of state. We shall meet several variations on $\Delta H^{\ominus}_{r, 298}$. The three most common now follow.

Definition: The standard molar enthalpy change of formation, $\Delta H^{\ominus}_{f, 298}$, of a compound is the amount of heat energy given out or absorbed when one mole of the compound is formed under standard conditions from its elements in their standard states.

e.g. $$Na(s) + \tfrac{1}{2}Cl_2(g) \longrightarrow NaCl(s) \qquad \Delta H^{\ominus}_{f, 298} = -41 \text{ kJ mol}^{-1}$$

Note that the equation is written with $\tfrac{1}{2}Cl_2$ so that 1 mole of NaCl is formed.

The formation of sodium chloride from the elements sodium and chlorine is highly exothermic. *A negative value for their standard enthalpy change of formation is typical of stable compounds.* The situation is very different, however, for the unstable and reactive gas ethyne ('acetylene').

$$2C(graphite) + H_2(g) \longrightarrow C_2H_2(g) \qquad \Delta H^{\ominus}_{f, 298} = +228 \text{ kJ mol}^{-1}$$

A positive value for their standard enthalpy change of formation is typical of unstable compounds.

For example, when methane burns:

$$CH_4(g) + 2O_2(g) \longrightarrow CO_2(g) + 2H_2O(g) \qquad \Delta H^{\ominus}_{c, 298} = -890 \text{ kJ mol}^{-1}$$

Definition: The standard molar enthalpy change of combustion, $\Delta H^{\ominus}_{c, 298}$, of a substance is the amount of heat energy given out when one mole of the substance is completely burned in oxygen under standard conditions.

Note that combustion is an *exothermic* process. It provides energy for most industrial and transport activities, as well as for much domestic heating and cooking. A simple method for measuring ΔH_c is shown in *Figure 11.3.1.*

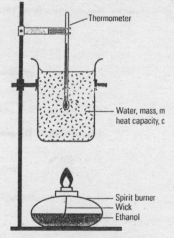

Weigh spirit burner at start and when desired temperature rise has been obtained.

Figure 11.3.1   A simple combustion calorimeter

An enthalpy change of combustion may also be an enthalpy change of formation, e.g.:

$$C(graphite) + O_2(g) \longrightarrow CO_2(g) \qquad \Delta H^{\ominus}_{f, 298} = -393.5 \text{ kJ mol}^{-1}$$

Here, -393.5 kJ mol^{-1} could be either $\Delta H^{\ominus}_{c, 298}$ for graphite or $\Delta H^{\ominus}_{f, 298}$ for carbon dioxide.

The general equation for a neutralisation reaction in solution is:

$$H^{\oplus}(aq) + OH^{\ominus}(aq) \longrightarrow H_2O(l)$$

It therefore seems sensible to define $\Delta H^{\ominus}_{n, 298}$ in terms of the amount of water produced.

For a *strong*, i.e. fully ionised, acid (see Ch 13.4) and alkali reacting together, $\Delta H^{\ominus}_{n, 298}$ has a value of about -57 kJ mol^{-1}. In our worked example in Ch 11.2, the value obtained was -55 kJ mol^{-1}. *This discrepancy is because we used solutions more concentrated than 1.00 mol dm^{-3}.* Energy is required to separate the ions in more concentrated solutions, which reduces the value of ΔH to less than $\Delta H^{\ominus}_{n, 298}$.

Also, if the reactants are *not* both fully ionised, so that energy is needed to complete the ionisation, the value of $\Delta H^{\ominus}_{n, 298}$ is again reduced to numerically less than -57 kJ mol^{-1}.

Summary of definitions (but **learn** the definitions already given)

$\Delta H^{\ominus}_{r, 298}$ – standard molar enthalpy change of a reaction *for the given equation*.

$\Delta H^{\ominus}_{f, 298}$ – standard molar enthalpy change of formation of *one mole of product* from the elements in their standard states.

$\Delta H^{\ominus}_{c, 298}$ – standard molar enthalpy change of combustion from burning *one mole of the substance*.

$\Delta H^{\ominus}_{n, 298}$ – standard molar enthalpy change of neutralisation to produce *one mole of H$_2$O*.

Remember – all of these have to happen under *standard conditions*.

Enthalpy change diagrams

One of the consequences of considering only *changes* in enthalpy during reactions and changes of state is that **the enthalpy of any element in its standard state can be assumed to be zero**. It is not possible to measure absolute enthalpies, so the assumption that elements possess zero enthalpy provides an important and useful reference point.

Simple diagrams to illustrate enthalpy changes can easily be drawn. Enthalpy diagrams for the formation of carbon dioxide and ethyne are shown in *Figures 11.3.2* and *11.3.3*.

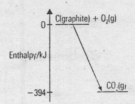

Figure 11.3.2   Diagram to show $\Delta H^{\ominus}_{f,298}$ for carbon dioxide

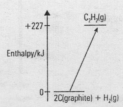

Figure 11.3.3   Diagram to show $\Delta H^{\ominus}_{f,298}$ for ethyne, C$_2$H$_2$

Notice than an *exothermic* reaction means that the total enthalpy – the energy content – of the system goes *down*.

There is a basic tendency in Nature to go to a situation of lower energy, i.e. of greater stability. *Most of the reactions we meet which are spontaneous*, i.e. which go of their own accord, *are exothermic*. Most of the compounds which don't change if you leave them in a bottle for years, the stable compounds, have an exothermic enthalpy of formation.

However, petrol and air can be put into a can for many years, and at normal temperatures there is no change at all. When one mole of the usual compounds in petrol burns in air, about 5 megajoules of energy are released – an immense amount. So, why do many exothermic reactions not occur spontaneously? The answer to that question is given in the next chapter.

**SELF-ASSESSMENT
QUESTIONS
11.3**

1 **a** Define the standard enthalpy change of formation of butane, C_4H_{10}.
   b What is meant by standard conditions?
   c Write an equation to show the formation of one mole of butane from its elements under standard conditions.

2 **a** Methanol, CH_3OH, burns to form CO_2 and water. Write a balanced equation for the reaction.
   b The standard enthalpy change of combustion of methanol is -726 kJ mol^{-1}. How much methanol must be burned to raise the temperature of 100 g of water by 10.0 °C?
   c What is meant by 'the standard enthalpy change of combustion of methanol'?

3 When 100 cm³ of 0.100 mol dm⁻³ NaOH is neutralised with 50 cm³ of 0.200 mol dm⁻³ HNO_3, the mixture warms up by 0.90 °C. Calculate the molar enthalpy change of neutralisation of sodium hydroxide, NaOH.

4 Ammonium nitrate, NH_4NO_3, is used in a chemical cooling pack. The solid ammonium nitrate is sealed in a plastic bag inside another bag containing water. When the inner bag is broken, the ammonium nitrate dissolves in the water.
   a The molar enthalpy change when one mole of the salt is dissolved in 1.00 dm³ water is $+26.0 \text{ kJ mol}^{-1}$. What would be the temperature change when 160 g of the salt is dissolved in 1.00 dm³ of water?
   b Is this an endothermic or exothermic change?

One consequence of the First Law is that no machine can be constructed which produces more energy than goes into it. All it can do, and never with 100% efficiency, is to convert energy from one form to another. If anyone tells you they have built a 'perpetual motion machine', you are dealing with a nutter! Incidentally, anything which runs on solar energy cannot be a perpetual motion machine, because the Sun is an ordinary star and will one day run out of fuel and die.

11.4 Hess' Law

Thermodynamics is the science of transforming energy from one form to another. The Laws of Thermodynamics are as well-established as anything in science, and are probably the least likely scientific laws ever to be shown to be wrong.

The *Second Law* deals with the tendency of energy to dissipate as low-grade heat energy, which we have already met in Ch 11.1.

The *First Law* can be stated very simply: **Energy can neither be created nor destroyed**.

G H Hess stated his 'law of constant heat summation' in 1840. Hess' Law follows from the First Law and, in modern wording, can be stated as follows.

For a given overall chemical reaction, the overall energy change is the same whether the reaction takes place in a single step or via a series of steps – provided that all the measurements are made under the same conditions.

One of the most important uses of Hess' Law is to 'get at' reactions which are either hard or impossible to do in the laboratory.

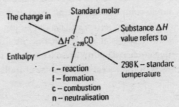

Suppose, for example, we wanted to measure $\Delta H^{\ominus}_{f, 298}$ for carbon monoxide, CO. The reaction:

$$C(graphite) + \tfrac{1}{2}O_2(g) \longrightarrow CO(g)$$

is very hard to control. Carbon, when it burns, tends to burn to carbon dioxide. But it is easy to burn carbon itself, and also carbon monoxide, to obtain the following data:

$$C(graphite) + O_2(g) \longrightarrow CO_2(g) \qquad \Delta H^{\ominus}_{c, 298} = -394 \text{ kJ mol}^{-1}$$

$$CO(g) + \tfrac{1}{2}O_2(g) \longrightarrow CO_2(g) \qquad \Delta H^{\ominus}_{r, 298} = -283 \text{ kJ mol}^{-1}$$

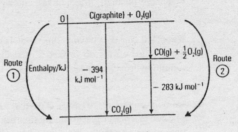

Figure 11.4.1

These data can be shown as an energy diagram (*Figure 11.4.1*), from which $\Delta H^{\ominus}_{f, 298}$ for carbon monoxide can be shown to be:

$$-394 - (-283) = -111 \text{ kJ mol}^{-1}$$

Another way of doing the calculation is to set up what is called a *Hess' Law triangle*.

$$C(graphite) + O_2(g) \xrightarrow{\;\Delta H_1\;} CO(g) + \tfrac{1}{2}O_2(g)$$

with ΔH_3 and ΔH_2 leading to $CO_2(g)$

Going around the Hess triangle, all you have to remember is:
in the direction of the arrow, put in a + sign;
in the opposite direction to the arrow, put in a − sign.
Adding all the values must then be equal to zero.
  In this triangle:

$$+ (\Delta H_1) + (\Delta H_2) - (\Delta H_3) = 0$$

By putting in the appropriate known values (including the correct signs), and rearranging the expression, the unknown quantity can be calculated.

In this, ΔH_1, ΔH_2, and ΔH_3 are the molar enthalpy changes for the three reactions. ΔH_1 must be $\Delta H^{\ominus}_{f,\,298}$ for carbon monoxide; ΔH_2 and ΔH_3 have the values given in *Figure 11.3.4* and in the equations above.

We can see at once from the triangle that, using Hess' Law,

$$+ (\Delta H_1) + (\Delta H_2) - (\Delta H_3) = 0$$
$$\therefore \Delta H_1 = \Delta H_3 - \Delta H_2$$
$$\therefore \Delta H_1 = \Delta H^{\ominus}_{f,\,298}\ CO = -394 - (-283) = -111\ kJ\ mol^{-1}$$

Worked example

To find $\Delta H^{\ominus}_{f,\,298}$ for butane, C_4H_{10}. (Call this ΔH_1).

It is not possible to make graphite and hydrogen combine together directly in the lab, but we can measure the enthalpy change of combustion for each of butane, carbon and hydrogen. Enthalpy changes such as $\Delta H^{\ominus}_{c,\,298}$, and other enthalpy changes, can be found in data books.

$$4C(graphite) + 5H_2(g) \longrightarrow C_4H_{10}(g) \qquad\qquad\qquad (\Delta H_1)$$
$$C_4H_{10}(g) + 6\tfrac{1}{2}O_2(g) \longrightarrow 4CO_2(g) + 5H_2O(g) \quad \Delta H^{\ominus}_{c,\,298} = -2877\ kJ\ mol^{-1} \quad (\Delta H_2)$$
$$C(graphite) + O_2(g) \longrightarrow CO_2(g) \qquad\quad \Delta H^{\ominus}_{c,\,298} = -394\ kJ\ mol^{-1} \quad (\Delta H_3)$$
$$H_2(g) + \tfrac{1}{2}O_2(g) \longrightarrow H_2O(l) \qquad\qquad \Delta H^{\ominus}_{c,\,298} = -286\ kJ\ mol^{-1} \quad (\Delta H_4)$$

Set up the triangle (remember to take into account the amounts of reactants in each of the equations):

From Hess' Law:

$$+ (\Delta H_1) + (\Delta H_2) - (4 \times \Delta H_3) - (5 \times \Delta H_4) = 0$$
$$\therefore \Delta H_1 = (4 \times \Delta H_3) + (5 \times \Delta H_4) - \Delta H_2$$
$$= 4 \times (-394) + 5 \times (-286) - (-2877)$$
$$\therefore \Delta H_1 = \Delta H^{\ominus}_{f,\,298} = -129\ kJ\ mol^{-1}$$

The Laws of Thermodynamics have been irreverently summarised as follows:

. You can never get something for nothing; the best you can do is break even.

. You can only break even at Absolute Zero.

. You can't get to Absolute Zero.

Carnot, Clausius, Thomson, Joule and Nernst.

Sadi Carnot (1796–1832) was a French army engineer who studied the pressure and volume changes in an idealised frictionless steam engine. His 1824 paper, *Reflections on the Motive Power of Fire*, set out the foundation of the Laws of Thermodynamics. His ideas were later developed by Rudolf Clausius and William Thomson, later Lord Kelvin, who in 1850 and 1851 respectively published what we now recognise as the Second Law.

James Joule (1818–1889) was taught at home as a child by John Dalton (see Ch 1.1). He was the first to establish, by rigorous experiment, the equivalence of heat and other forms of energy, and hence the basis for the First Law. In 1843 Joule determined the amount of mechanical work equivalent to a given amount of heat energy. Thomson and Joule collaborated in the 1850s on the Joule–Thomson effect, which is the cooling of a gas when it expands. (A carbon dioxide fire extinguisher produces some particles of solid carbon dioxide; the temperature momentarily drops about 100°C as the gas expands.) The SI unit of energy is named the Joule, to honour Joule's pioneering work.

Among Walther Nernst's (1864–1941) many contributions to science was the Third Law of Thermodynamics, one statement of which is that Absolute Zero (0 K, −273 °C) is unattainable. He was given the Nobel Prize for Chemistry in 1920 for his work in thermodynamics.

Worked example

Calculate the enthalpy of formation of carbon disulphide, CS_2. You are given the enthalpies of combustion below:

$$\Delta H^{\ominus}_{c,\,298},C = -393.5\ kJ\ mol^{-1}$$
$$\Delta H^{\ominus}_{c,\,298},S = -296.8\ kJ\ mol^{-1}$$
$$\Delta H^{\ominus}_{c,\,298},CS_2 = -1077\ kJ\ mol^{-1}$$

The first thing to do is to write out a balanced equation for each reaction.

$C(graphite) + 2S(s) \longrightarrow CS_2(l)$	$\Delta H = \Delta H^{\ominus}_{f,\,298}\ CS_2$
$C(graphite) + O_2(g) \longrightarrow CO_2(g)$	$\Delta H = -394\ kJ\ mol^{-1}$
$S(s) + O_2(g) \longrightarrow SO_2(g)$	$\Delta H = -297\ kJ\ mol^{-1}$
$CS_2(l) + 3O_2(g) \longrightarrow CO_2(g) + 2SO_2(g)$	$\Delta H = -1077\ kJ\ mol^{-1}$

Now balance all the equations with respect to each other. (In these equations the last equation contains $2SO_2$, so the combustion equation for sulphur has to be doubled. For convenience we shall leave out state symbols for the rest of the calculation.)

$C + 2S \longrightarrow CS_2$	ΔH
$C + O_2 \longrightarrow CO_2$	$\Delta H = -394\ kJ\ mol^{-1}$
$2S + 2O_2 \longrightarrow 2SO_2$	$\Delta H = 2 \times (-297)\ kJ\ mol^{-1}$
$CS_2 + 3O_2 \longrightarrow CO_2 + 2SO_2$	$\Delta H = -1077\ kJ\ mol^{-1}$

Arrange these equations in a triangle (mentally cut out each equation and align them with the same chemicals at each point of the triangle):

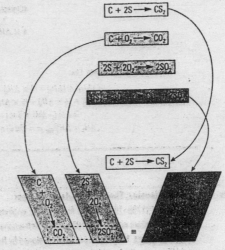

Once you get the idea this last step can be left out.

This picture can now be simplified to:

$$C + 2S \xrightarrow{\ \Delta H\ } CS_2$$
$$+ O_2 \qquad + 2O_2 \quad / + 3O_2$$
$$CO_2 + 2SO_2$$

Put in the enthalpy change values for each of the reactions, not forgetting which values have been multiplied to balance the equations with respect to each other earlier.

Now, using Hess' Law, if we go around the triangle, the total enthalpy change will be zero.

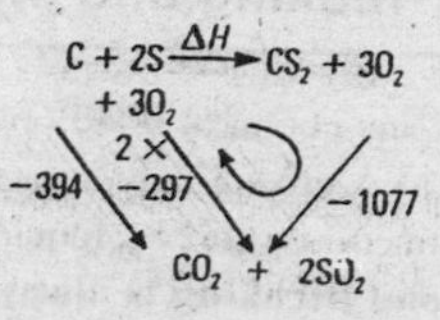

$$C + 2S \xrightarrow{\Delta H} CS_2 + 3O_2$$

$$+ 3O_2$$
$$2 \times$$
$$-394 \quad -297 \qquad -1077$$
$$CO_2 + 2SO_2$$

Remember to include the sign of the data value and to include a + when travelling in the direction of the arrow and − if going 'against' the arrow. So:

$$+ (\Delta H) + (-1077) - (-297 \times 2) - (-394) = 0$$
$$\Delta H - 1077 + 594 + 394 = 0$$
$$\Delta H = +1077 - 594 - 394$$
$$= +89 \text{ kJ mol}^{-1}$$
$$\Delta H^{\ominus}_{f, 298} CS_2 = +89 \text{ kJ mol}^{-1}$$

Would you expect carbon disulphide to be stable or unstable when heated?

1 The equation for the formation of pentane is shown below.

$$5C + 6H_2 \longrightarrow C_5H_{12}$$

a Copy the equation, and insert the state symbols in the equation.
b Use the standard enthalpy change of combustion data given below to calculate the standard enthalpy change of formation, $\Delta H^{\ominus}_f$, of pentane.

Substance	$\Delta H^{\ominus}_c$/kJ mol^{-1}
Pentane	−3509
Hydrogen	−286
Carbon	−394

2 Consider the following data:

$$H_2(g) + \tfrac{1}{2}O_2(g) \longrightarrow H_2O(l) \qquad \Delta H^{\ominus}_{f, 298} = -286 \text{ kJ mol}^{-1}$$
$$H_2(g) + \tfrac{1}{2}O_2(g) \longrightarrow H_2O(g) \qquad \Delta H^{\ominus}_{f, 298} = -242 \text{ kJ mol}^{-1}$$

a Calculate the molar enthalpy of vaporisation of water (i.e. the 'latent heat of vaporisation'). (Hint: you require the enthalpy change for the process: $H_2O(l) \longrightarrow H_2O(g)$).
b Why is it important always to write state symbols in equations which relate to energy changes in chemistry?

3 Calculate the standard molar enthalpy change of formation of methane, CH_4, given the following data:
Standard molar enthalpy change of combustion of CH_4 is −890 kJ mol^{-1};
Standard molar enthalpy changes of formation:
CO_2, −393 kJ mol^{-1}; H_2O, −286 kJ mol^{-1}.

11.5 Making and breaking chemical bonds: bond enthalpies

To break any chemical bond obviously requires energy. (Indeed, as we stated at the start of this chapter, to overcome any attraction between particles, e.g. hydrogen bonds (Ch 2.9) or van der Waals' attractions (Ch 2.8), requires energy.)

So **bond breaking is always endothermic**. What is not so obvious is that **bond making is always exothermic** (*Figure 11.5.1*).

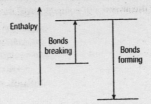

Figure 11.5.1 Bonds breaking and forming

Consider the simplest gaseous diatomic molecules, H_2.

$$H_2(g) \longrightarrow 2H(g) \qquad \Delta H^{\ominus}_{r, 298} = +436 \text{ kJ mol}^{-1}$$

If this equation is true, then two moles of gaseous hydrogen atoms combine to form one mole of H_2 molecules:

$$2H(g) \longrightarrow H_2(g) \qquad \Delta H^{\ominus}_{r, 298} = -436 \text{ kJ mol}^{-1}$$

If the value of ΔH was not *exactly* the same in each case, the First Law would be broken!

The fact that making a bond gives out *exactly* as much energy as is needed to break that bond has important consequences in free radical chain reactions (Ch 14.6 and Ch 16.7).

| Bond enthalpies |

> Bond enthalpies are often called *bond energies*.

For a singly-bonded diatomic molecule such as Cl_2 or H_2, the molar bond enthalpy is simply the energy required to break one mole of the bonds between the pairs of atoms in the gaseous molecule. Since this must be an *endothermic* process, all bond enthalpies have positive values.

If bond breaking is always endothermic, bond making must always be exothermic.

For a polyatomic molecule such as methane, CH_4, the situation is more complicated than for H_2 or Cl_2. It can be demonstrated experimentally that:

$$CH_4(g) \longrightarrow C(g) + 4H(g) \qquad \Delta H^{\ominus}_{r, 298} = +1646 \text{ kJ mol}^{-1}$$

In CH_4, there are four C—H bonds to be broken, so it seems clear that the bond enthalpy for each C—H bond is:

$$1646/4 = +411.5 \text{ kJ mol}^{-1}$$

But it is possible with methane to find the enthalpy change involved in breaking each successive bond. The results are:

	$\Delta H^{\ominus}_{r, 298}$ kJ mol^{-1}
$CH_4(g) \longrightarrow CH_3(g) + H(g)$	+425
$CH_3(g) \longrightarrow CH_2(g) + H(g)$	+470
$CH_2(g) \longrightarrow CH(g) + H(g)$	+416
$CH(g) \longrightarrow C(g) + H(g)$	+335

It is now obvious that the figure of $+411.5$ kJ mol^{-1} is merely an average: the successive bonds require different energies to break them. In other words, **the strength of an individual bond depends to some extent on its environment in the molecule.**

We shall be using only **molar bond enthalpies**. The term **bond dissociation enthalpy** is sometimes used; it is the energy in kJ needed to break one mole of a *specific* bond in a *specific* polyatomic molecule.

e.g. The bond dissociation enthalpies for

$$CH_4(g) \longrightarrow CH_3(g) + H(g)$$
$$CH_3(g) \longrightarrow CH_2(g) + H(g)$$

etc are different (see above).

The variation of bond strengths in methane is rather an extreme case. A more usual range of values is found in C—C single bonds; the energy needed to break this bond is found to vary between about 330 and 346 kJ mol^{-1}, depending on what other atoms are attached to the two carbon atoms.

This means that when you look up a table of **bond enthalpies** – as in *Table 11.5.1* – unless the bonds concerned are in simple diatomic molecules (e.g. H_2, Cl_2, HCl, HBr, N_2), then the values are averaged over those found in a large number of different compounds.

Bond	Bond length/nm	Bond enthalpy/kJ mol^{-1}
H—H	0.074	+436
F—F	0.142	+158
Cl—Cl	0.199	+243
Br—Br	0.228	+193
I—I	0.267	+151
O==O	0.121	+498
H—O	0.096	+464
H—F	0.092	+568
H—Cl	0.127	+432
H—Br	0.141	+366
H—I	0.161	+298
C—H	0.108	+413 (average)
C—F	0.138	+467 (average)
C—Cl	0.177	+346 (average)
C—Br	0.194	+290 (average)
C—I	0.214	+228 (average)
C—C	0.154	+347 (average)
C≡C	0.134	+612 (average)
C≡C	0.120	+838 (average)
C—O	0.143	+358 (average)
C==O	0.116	+805 (in CO_2)
N≡N	0.110	+945

Table 11.5.1   Molar bond enthalpies

Definition: The molar bond enthalpy of a bond between two specified atoms (which may be of the same element or two different elements) is the average energy in kJ required to break one mole of the bonds.

Because many bond enthalpies are averaged, care has to be taken using tables of bond enthalpies in calculations. The answer will only be approximate. The approximation is often very good, and such calculations are quite useful in making predictions.

Uses of bond enthalpies

The more precisely we can find out the values of bond enthalpies, the more we are able to *predict* the likely result of a particular chemical reaction – without the need of doing the experiments. After all, in any chemical reaction a definite number of bonds must be broken in the reactants before the resulting fragments join together to make a definite number of new bonds and form the products. (This is an over-simple picture in some ways, but good enough for the moment.)

If we know all the molar bond enthalpies involved, we can calculate the overall enthalpy change, ΔH, for the reaction.

For example, consider the burning of methane in air or oxygen.

$$CH_4(g) + 2O_2(g) \longrightarrow CO_2(g) + 2H_2O(g)$$

Which bonds have to break, and which have to be made? Draw out full graphical formulae for all the reactants and products:

$$\overset{\displaystyle H}{\underset{\displaystyle H}{H-C-H}} + 2O=O \longrightarrow O=C=O + 2\ \overset{H}{\underset{H}{O}}$$

Break *all* the bonds in the reactant molecules, to leave separate atoms, and add up the molar bond enthalpies involved to get the energy needed to do this (*Figure 11.5.2*).

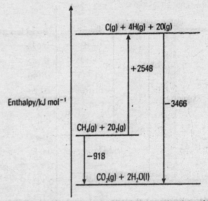

Figure 11.5.2   Enthalpy of reaction from bond enthalpies

Bonds broken:

$4 \times C-H$	$= (4 \times +413)$	$= +1652$ kJ
$2 \times O=O$	$= (2 \times +498)$	$= +\ 996$ kJ
∴ Total energy required		$= +2648$ kJ

Now add up the energy given out as the new bonds are made.

Bonds made:

$2 \times C=O$	$= (2 \times -805)$	$= -1610$ kJ
$4 \times O-H$	$= (4 \times -464)$	$= -1856$ kJ
∴ Total energy given out		$= -3466$ kJ

Therefore, for the reaction:

$$CH_4(g) + 2O_2(g) \longrightarrow CO_2(g) + 2H_2O(g)$$

$\Delta H = +2648 - 3466 = -818$ kJ mol^{-1}

The value measured by experiment is -890 kJ mol^{-1}, so our calculation is about 9% out – not a bad result!

We can now see why combustion of hydrocarbons is so widely used to provide energy for transport, space heating and the generation of electricity. The product molecules, CO_2 and H_2O, have a lower energy content, i.e. they are at a lower enthalpy than the molecules of the reactants, CH_4 and O_2. And this in turn is because the bonds in CO_2 and H_2O are stronger bonds; a great deal of energy is therefore given out as they are formed.

Enthalpy of combustion of a range of hydrocarbons

The enthalpy of combustion of a range of volatile liquid fuels, e.g. liquid alkanes, can be measured in a more sophisticated calorimeter (*Figure 11.5.3*).

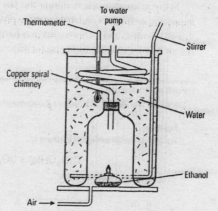

Figure 11.5.3   A flame calorimeter

To overcome the need to know the heat capacities of the various materials making up the apparatus, and the masses of each separate item, the apparatus is calibrated using a fuel with a known enthalpy of combustion or by using an electrical heater. By calibration, the value of mc in the equation $\Delta H = mc\Delta T$ is measured for the whole apparatus, and, if the temperature changes are the same in each experiment, the errors due to cooling will be minimised. The enthalpy change produced by burning a known mass of fuel is calculated, and this figure is converted to a value for the combustion of one mole of the fuel.

The molar enthalpies of the first six alkanes (CH_4, C_2H_6, C_3H_8, C_4H_{10}, C_5H_{12}, C_6H_{14} – see Ch 15.2) are shown in *Figure 11.5.4*.

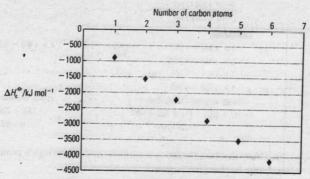

Figure 11.5.4   Enthalpy of combustion of alkanes

Can you see why the graph is a straight line? (Hint: what is the difference in formula between each successive molecule?)

In very few reactions are the reactant molecules completely broken up into atoms, which then reassemble to form the product molecules. But, from Hess' Law (Ch 11.4), the *overall* enthalpy change will be the same whether the reactants are atomised or not.

As well as helping to estimate enthalpy changes in reactions, molar bond enthalpy values also show the relative strengths of bonds, help our understanding of the structures and strengths of materials (see e.g. Ch 3), and also help us to understand the routes by which real reactions occur (see Ch 14.7).

As we mentioned at the end of Ch 11.3, just because a reaction is highly exothermic does not mean that it happens at a reasonable rate, or even that it happens at all at room temperature. The reason for this is discussed in the next chapter.

Many research workers during the last 150 years have spent much of their professional lives improving techniques and making more precise measurements of enthalpy changes in chemistry, so that all scientists and technologists may have more information to help them decide on strategies for research or for manufacturing useful materials.

Worked example

Calculate the molar enthalpy of combustion of ethanol, using bond enthalpies.

Equation:

The equation for the combustion of ethanol is:

$$C_2H_5OH(g) + 3O_2(g) \longrightarrow 2CO_2(g) + 3H_2O(g)$$

Write the equation in displayed form:

Data:

Total up the different bonds present in the displayed equation:

$1 \times$ C—C	$3 \times$ O=O	$2 \times 2 \times$ C=O	$3 \times 2 \times$ O—H
$5 \times$ C—H			
$1 \times$ O—H			
$1 \times$ C—O			

Mean bond enthalpies from *Table 11.5.1*:

C—C $+347$ kJ mol^{-1}	O=O $+498$ kJ mol^{-1}	C=O $+805$ kJ mol^{-1}	O—H $+464$ kJ mol^{-1}
C—H $+413$ kJ mol^{-1}			
O—H $+464$ kJ mol^{-1}			
C—O $+358$ kJ mol^{-1}			

Calculation:

$1 \times +347 = +347$	$3 \times +498 = +1494$	$4 \times +805 = +3220$	$6 \times +464 = +2784$
$5 \times +413 = +2065$			
$1 \times +464 = +464$			
$1 \times +358 = +358$			

∴ Energy *required* to break bonds =
$(+347 + 2065 + 464 + 358) + 1494$
$= +4728$ kJ mol^{-1}

Energy *released* on forming bonds =
$-(+3220 + 2784)$
$= -6004$ kJ mol^{-1}

Note that bond breaking is *endothermic*, so the sign is positive; bond making is *exothermic*, so the sign has become negative.

∴ $\Delta H_c = +4728 - 6004 = -1276$ kJ mol^{-1}

(How does this compare with the experimental value of -1367 kJ mol^{-1}? Why should there be a difference?)

Activity

These bond enthalpy calculations can be made simpler if you use a spreadsheet to manipulate the figures. **But do make sure you understand and can carry out the calculation without relying on a spreadsheet** – you will not have the spreadsheet to help you in the exam!

	A	B	C	D	E	F
1	bonds broken		bond	kJ/mol	bonds formed	
2		= A2*D2	H—H	436		= E2*D2
3		= A3*D3	C—H	413		= E3*D3
4		= A4*D4	C—C	347		= E4*D4
5		= A5*D5	C=C	612		= E5*D5
6		= A6*D6	C≡C	838		= E6*D6
7		= A7*D7	C—O	358		= E7*D7
8		= A8*D8	C=O	805		= E8*D8
9		= A9*D9	O=O	498		= E9*D9
10		= A10*D10	O—H	464		= E10*D10
11						
12	enthalpy change =		= (SUM(B2:B10))-(SUM(F2:F10))			
13			kJ/mol			

This spreadsheet will allow you to work out most enthalpy changes of combustion.

All you have to do is work out the number of each type of bond broken and formed in the course of the reaction – so you must write a balanced equation, with full displayed formulae before you start the calculation.

Compare the enthalpy of combustion of propane with the experimental figure of $-2219 \text{ kJ mol}^{-1}$.

'Long bonds are weak bonds'

Bond length

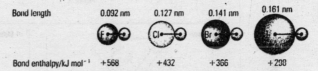

| | 0.092 nm | 0.127 nm | 0.141 nm | 0.161 nm |

Bond enthalpy/kJ mol^{-1} +568   +432   +366   +298

Figure 11.5.5   Bond lengths of the hydrogen halides

Study of the bonds involving halogens in *Table 11.5.1* shows that, in general, **long bonds are weak bonds** and are therefore likely to react. For halogen bonded either to hydrogen or carbon, the bond weakens progressively from fluorine → chlorine → bromine → iodine (*Figure 11.5.5*). Molecules containing bromine or iodine are often used in organic chemistry where reactivity is needed (see Ch 17.2 and 17.4). 'Teflon' (poly(tetrafluoroethene), $(C_2F_4)_n$), contains nothing but carbon and fluorine; it is so inert chemically that it is used in surgical replacements within the body and to coat frying pans. On the other hand, when hydrogen iodide, HI, is gently heated, it falls apart and the purple colour of iodine vapour is seen (see Ch 8.3).

The unexpected low bond enthalpy for F—F (*Table 11.5.1*) is probably caused by repulsion between the electron pairs in the small and close-together fluorine atoms. This low value, coupled with the high bond enthalpies for bonds between fluorine and other elements once the fluorine molecule has reacted, are the main reasons for the viciously reactive nature of fluorine (see Ch 8.4).

The fact that the second bond in C=C is not as strong as the first is a reason for the specific reactivity of alkenes (see Ch 16.2 and Ch 16.5). The figures in *Table 11.5.1* show that the third bond in C≡C is weaker still.

SELF-ASSESSMENT QUESTIONS 11.5

1 For this question you will need graph paper. It concerns the standard molar enthalpy change of combustion ($\Delta H^{\ominus}_{c, 298}$) of unbranched primary alcohols, where the —OH group is attached to a chain of carbon atoms (see Ch 18.1)

Alcohol	Formula	M_r	$\Delta H^{\ominus}_{c, 298}$ /kJ mol^{-1}
Methanol	CH_3OH	32	– 726
Ethanol	C_2H_5OH	46	–1367
Propan-1-ol	C_3H_7OH	60	–2021
Butan-1-ol	C_4H_9OH	74	–2676
Pentan-1-ol	$C_5H_{11}OH$	88	...

a Plot $\Delta H^{\ominus}_{c, 298}$ on the vertical axis and relative molecular mass on the horizontal axis. Use your graph to complete the table for pentan-1-ol.

b Why is the plot a straight line?

c Where does the line intersect the horizontal axis? Why does it not go through the origin?

d Write the equation for the complete combustion of ethanol.

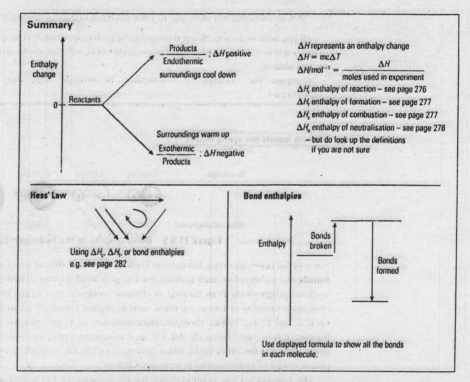

Summary

Products / Endothermic ; ΔH positive
surroundings cool down

Enthalpy change

0 — Reactants

Surroundings warm up
Exothermic ; ΔH negative
Products

ΔH represents an enthalpy change
$\Delta H = mc\Delta T$
$\Delta H/\text{mol}^{-1} = \dfrac{\Delta H}{\text{moles used in experiment}}$
ΔH_r enthalpy of reaction – see page 276
ΔH_f enthalpy of formation – see page 277
ΔH_c enthalpy of combustion – see page 277
ΔH_n enthalpy of neutralisation – see page 278
– but do look up the definitions if you are not sure

Hess' Law

Using ΔH_c, ΔH_f, or bond enthalpies
e.g. see page 282

Bond enthalpies

Enthalpy

Bonds broken

Bonds formed

Use displayed formula to show all the bonds in each molecule.

SUMMARY 11

In this chapter the following new ideas have been covered. *Have you understood and learned them?* Could you explain them clearly to a friend? The page references will help.

- The various forms of energy.   p270
- Energy conversion is never 100% efficient.   p270
- Economic and social links with energy.   p271
- Enthalpy changes and how they are measured.   p271

EXAMINATION QUESTIONS 11

1 a Butan-1-ol was burned in a simple lamp burner under a container of water. Explain in terms of bonds broken and made why the temperature of the water in the container rises.

   b When 0.60 g of butan-1-ol was burned it raised the temperature of 250 g of water by 19.4 °C. Calculate the molar enthalpy of combustion of butan-1-ol.

   c The quoted result is -2676 kJ mol^{-1}. How does this compare with your calculated value? Why should this be so?

2 A bomb calorimeter was used to measure the molar enthalpy of combustion of benzoic acid.

$$C_6H_5COOH(s) + 7\tfrac{1}{2}O_2(g) \longrightarrow 7CO_2(g) + 3H_2O(l) \quad \Delta H^{\ominus}_{c,\,298} = -3227 \text{ kJ mol}^{-1}$$

Since the enthalpy of combustion of benzoic acid is accurately known, the value is used to calibrate the calorimeter; i.e. the value of 'mc' (in the expression $\Delta H = mc\Delta T$) for the whole apparatus can be calculated.

In one series of experiments 1.9 g of the acid was completely burned and the temperature rise was 5.0 °C.

   a How many moles of benzoic acid were used?

   b How much energy is released when this amount of benzoic acid is burned?

   c What is the value of the 'mc' constant for this apparatus?

The apparatus was then used to measure $\Delta H^{\ominus}_{c,\,298}$ for a number of alkanes. The results are shown in the table. In each case 0.010 moles of the alkane was completely burned.

Alkane	Temperature rise
C_4H_{10}	2.9 °C
C_5H_{12}	3.5 °C
C_6H_{14}	4.2 °C

   d Calculate $\Delta H^{\ominus}_{c,\,298}$ for each of the alkanes.

   e From these results calculate the enthalpy of combustion associated with the $-CH_2-$ group.

3 a The chemist uses the mole to compare data for most properties of materials, but to the combustion engineer the mass of material is much more important. For instance, in designing an engine, the energy available from each kilogram of fuel is more relevant than the energy per mole. Using the following enthalpies of combustion, compare the energy provided per kilogram of fuel.

Fuel	Formula	$\Delta H^{\ominus}_{c,\,298}$/kJ mol^{-1}
Octane	C_8H_{18}	-5470
Methanol	CH_3OH	-726
Hydrogen	H_2	-286

   b Give reasons why petrol engines, which burn hydrocarbons such as octane, are still more widely used than those which burn hydrogen.

- The basic ideas about **covalent bonding**. (Ch 2.4)
- How to write balanced **chemical equations**. (Ch 4.2)
- How to convert mass into **number of moles**. (Ch 9.3)
- Why **enthalpy changes** occur during chemical reactions. (Ch 11.5)

12

rates of
reaction

12.1 Introduction

How fast can a chemical reaction happen? The slowest chemical reactions we know about must be some of those changes involving the Earth itself. The chemical changes that converted dead wood into coal have taken millions of years, and many of the chemical changes in rocks have taken similar periods of time.

At the other extreme, there are reactions that are apparently instantaneous. The neutralisation reaction between hydrochloric acid and sodium hydroxide solutions is extremely fast.

$$HCl(aq) + NaOH(aq) \longrightarrow NaCl(aq) + H_2O(l)$$

What limiting factors are there in such reactions? When the hydrochloric acid is mixed with sodium hydroxide solution, the reaction cannot go any faster than the rate at which the materials are mixed, or any faster than the rate at which particles move to meet each other. The range of time scales for chemical reactions is immense (*Figure 12.1.1*).

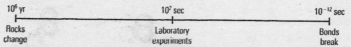

10^6 yr	10^2 sec	10^{-12} sec
Rocks change	Laboratory experiments	Bonds break

Figure 12.1.1   Time scale for reactions

Why do we need to study how fast a chemical reaction occurs? Obviously, such information is vital when changing one chemical into another on an industrial scale. Reacting fifty tonnes of iron ore with coke for five minutes or for five hours has a tremendous influence on the production costs of iron.

But information about rates of a reaction can also give us theoretical ideas about how the chemical reaction happens. Studying the rates of reactions helps us to understand how reactions occur, and how to speed up a reaction or slow it down.

Energy changes and rates of reaction

In the last chapter we saw how energy changes are important in a chemical reaction – how energy changes determine whether a chemical reaction might occur or not. A high exothermic energy change could predict that a reaction is possible, i.e. the system is *thermodynamically* unstable. But if the rate of reaction is too slow there will be no obvious chemical change, i.e. it is *kinetically* stable (*Figure 12.1.2*).

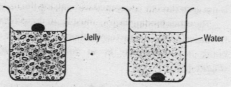

Jelly   Water

Both cases have the lowest energy with the cherry stone at the bottom of the beaker. In the case of the jelly the rate of fall of the stone is so slow that you do not see it fall – the kinetics control the process.

Figure 12.1.2   Kinetic control vs. thermodynamic control

A typical example is the 'pop' test used to test for hydrogen gas. Hydrogen mixed with air is an explosive mixture. However, at room temperature the reaction is undetectably slow.

Only when a spark or a flame such as a lighted splint is applied does the reaction mixture go 'pop'.

$$2H_2(g) + O_2(g) \xrightarrow{\text{ignite}} 2H_2O(g) \qquad \Delta H = -286 \text{ kJ mol}^{-1}$$

This is an example of a **kinetically controlled** reaction.

The same ideas apply when you try to light coal. Although graphite is *thermodynamically unstable*, it takes a long time to get a coal fire going because this reaction is *kinetically stable* at room temperature.

$$C(s) + O_2(g) \longrightarrow CO_2(g) \qquad \Delta H = -393 \text{ kJ mol}^{-1}$$

12.2 Collision theory

Before almost any chemical reaction can occur, the particles involved must meet – they have to collide. (A few reactions simply involve dissociation of molecules.) These particles could be atoms, molecules or ions, but many of the ideas are most easily discussed in terms of simple molecules which (by definition) contain covalent bonds. The following discussions will be limited to *molecules*.

An increase in the *number of collisions* should lead to an increase in the rate of reaction, and decreasing the *number of collisions* will reduce the reaction rate (*Figure 12.2.1*).

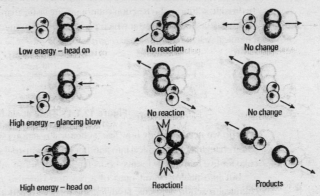

Figure 12.2.1 Collisions and reactions

But there must be more to it than this. The average speed of gas molecules can be measured and the number of collisions in a gas mixture calculated. The number of collisions is very large, and if every collision led to a chemical reaction, all the possible gas phase reactions would be over in an instant. There must be some other factors involved, since it is clear that not all possible collisions lead to reaction.

For many reactions the limiting factor in the rate of reaction is not to do with simple collision. It is not even directly concerned with mixing the reactants together.

The other limiting factors must be to do with *the way in which the molecules collide*. Although the molecules bump into each other, reaction does not occur unless there is sufficient energy for the bonds to break – there is an energy barrier to the reaction – the **activation energy, E_a** (*Figure 12.2.2*).

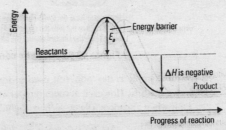

Figure 12.2.2 A reaction profile diagram for an exothermic reaction

The molecules with sufficient energy will react. Also, particularly with more complex reactants, the *right parts* of the molecules must meet in the collision – i.e. they must be *correctly oriented*. The collision has to involve the bonds that are going to break, and new bonds must be able to form.

The energy profile

Figure 12.2.2 shows the **energy profile** of a typical chemical reaction. At the start, the molecules are mixed together and are moving randomly. *When two molecules get very close, the electron clouds of each of the molecules repel each other (Figure 12.2.3).*

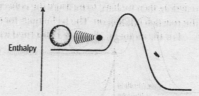

Figure 12.2.3   Electron clouds repel

To overcome this repulsion, energy is needed; and if there is enough energy the molecules get closer and the electron clouds start to interact and become distorted. The system starts to move up the energy curve (*Figure 12.2.4*).

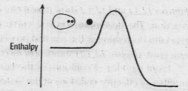

Figure 12.2.4   Electron clouds distorted

If the molecules are moving rapidly enough and therefore possess enough energy, this repulsion is overcome and the molecules get close enough to rearrange the electrons in the bonds.

The bonds in the molecules will stretch further and further and, at the top of the curve, there is some interaction between the two molecules (*Figure 12.2.5*).

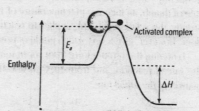

Figure 12.2.5   Bonds break and form

The original bonds are stretched to their limit, while the interaction between the molecules has become maximised. This interaction of the two species at the top of the energy profile is called an **activated complex**.

This activated complex can now move in one of two directions. The molecules can move apart again, with all the original bonds reforming. Or the bonds can break and form the new bonds of the products. Once the reaction is over the initial barrier, the reaction will continue to move forward, although there may be a number of smaller barriers as new intermediates are formed during the course of the reaction.

This picture of the progress of a chemical reaction shows how important the two factors mentioned above are. The molecules must collide with *the correct orientation* and with *sufficient energy*. Otherwise the bond breaking that could lead to reaction does not happen. The need for sufficient energy to overcome the initial repulsion of what are after all stable molecules is essential.

> **Definition:** Activation energy is the minimum energy required for colliding molecules to interact and for bonds to break.

12.3 Factors influencing the rate of reaction

What happens to the concentration of a material during the course of a reaction? How do we know what is happening in the course of a reaction? If we are going to measure the rate of a chemical reaction, then we have to measure the concentration of at least one of the reactants (or products) as the reaction is going on. The techniques for doing this will be discussed later (Ch 21.1).

For the moment, all we are concerned with is how the concentration changes with time.

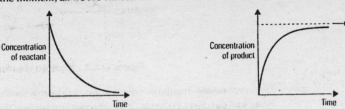

Figure 12.3.1 Typical rate curve – decomposition of a reactant

Figure 12.3.2 Typical rate curve – formation of a product

Factors which can increase the number of effective collisions
Temperature
Light
Concentration
Pressure in gas systems
Surface area for solids
Catalysts

Figures 12.3.1 and *12.3.2* show typical rate curves – of decomposition and formation – for a chemical reaction. The decomposition rate curve (*Figure 12.3.1*) shows the concentration of one of the reactants decreasing as the reaction proceeds. Looking at the formation of one of the products, the rate curve (*Figure 12.3.2*) shows how its concentration increases as the reaction proceeds.

From the idea of collisions in the last section, it is evident that any factor that increases the number of effective collisions in the system will increase the rate of reaction. But those collisions must also have sufficient energy to be effective.

The effect of temperature

The principle of the refrigerator is that at about 2 °C the rate of decomposition of the food is roughly one quarter of that at room temperature (but beware, this is not the principle behind the freezing of food).

Temperature is the factor which has most effect on the rate of a reaction. At ***higher temperatures*** molecules have a greater kinetic energy and move more rapidly, so they must collide more frequently – and, more importantly, with greater energy. And so the ***rate of reaction increases***. As a rule of thumb, an increase in temperature of 10 °C will just about double the rate of many reactions. Strictly, this relationship only applies to reactions with a particular value of the activation energy. It does, however, give a very rough guide of how rates change with temperature.

Lowering the temperature from room temperature (approximately 20 °C) to 10 °C often halves the rate of reaction, and from 10 °C to 0 °C then halves the rate again. That is a reduction to one quarter of the initial rate.

The effect of light

Another way of providing more energy for a reaction is in the form of light. A number of reactions are faster in the presence of light. You only have to think of the decomposition of silver salts on photographic film to appreciate how light can alter the rate of a reaction.

$$AgCl(s) \xrightarrow{\text{light}} Ag(s) + \tfrac{1}{2}Cl_2(g)$$

An even more fundamental reaction is photosynthesis, which just does not occur if there is no light.

$$6CO_2(g) + 6H_2O(l) \xrightarrow[\text{chlorophyll}]{\text{light}} C_6H_{12}O_6(aq) + 6O_2(g)$$

Another more dramatic way to illustrate the effect of light is the reaction between hydrogen and chlorine. In the dark the rate is extremely slow, but in sunlight, or with a photographic flash lamp, the reaction is explosive.

$$H_2(g) + Cl_2(g) \xrightarrow{\text{light}} 2HCl(g)$$

Effect of temperature on the rate of reaction

The temperature dependence of reaction rate can be shown by a number of reactions. One such reaction follows the rate at which sulphur is formed in the reaction of sodium thiosulphate with dilute acid.

$$Na_2S_2O_3(aq) + 2HCl(aq) \longrightarrow S(s) + SO_2(g) + H_2O(l) + 2NaCl(aq)$$

The solid sulphur appears as a very fine precipitate, slowly giving a milky appearance to the reaction mixture. As the amount of sulphur increases, so this milky precipitate makes the liquid more and more opaque. One way to follow the reaction is to mark a cross under the flask and time how long it takes for the cross to disappear. Repeating this with the initial mixture at different temperatures gives information about the rate at different temperatures. The composition of the mixture, and the depth of the liquid in the flask, must remain constant.

In another method, the rate curve can be obtained using a light detector and light bulb on opposite sides of the beaker.

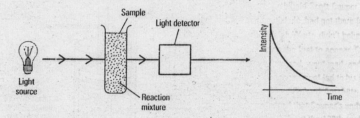

The amount of light going through the solution decreases as the sulphur precipitate builds up and can be recorded on a suitable detector or data logging equipment. The typical rate curve will be measured.

At a higher temperature the reaction should be quicker, and so the shape of the graph will alter. Although the final reading should be the same, since the initial concentrations are all the same, the final value will be achieved in much less time.

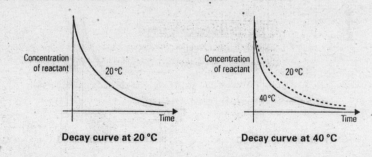

Decay curve at 20 °C   **Decay curve at 40 °C**

The effect of concentration

If you increase the number of reactant particles present in a fixed volume in a reaction, then the number of collisions must increase. This is what the effect of concentration is all about.

For solutions the *concentration (in mol dm^{-3})* is directly relevant. Any increase in concentration should also increase the rate of reaction; if there are more molecules present there will be more collisions (*Figure 12.3.3*).

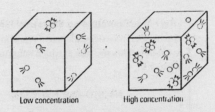

Increasing concentration leads to more collisions

Figure 12.3.3   Effect of increasing concentration

But we do need to be careful here. As we have seen already (Ch 12.2), the collisions have to be collisions that matter – **effective collisions**. Not every collision is a collision that leads to reaction.

The effect of pressure

Concentration can be increased in ways other than by increasing concentration of a solution. In a gas system, increasing *pressure* will increase the concentration of the gases (*Figure 12.3.4*).

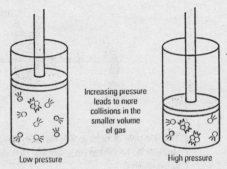

Increasing pressure leads to more collisions in the smaller volume of gas

Figure 12.3.4   Effect of increasing pressure in gas

Doubling the pressure will effectively double the concentration of the gases. So pressure has an important effect on gas reactions.

The effect of surface area

Similarly, with a reaction involving a solid surface, increasing the *surface area* of the solid increases the number of collisions with the surface (*Figure 12.3.5*).

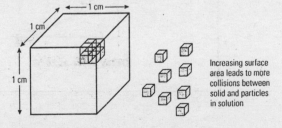

Increasing surface area leads to more collisions between solid and particles in solution

Figure 12.3.5   Effect of increasing surface area

A cube with each side 1 cm long has a total surface area of 6 cm². Halve each of the sides of the cube so that there are now eight cubes, each with sides 0.5 cm long, and the total surface area has doubled to 12 cm². Now half the size of each of these smaller cubes.

Surface area of 1 cube, sides 1 cm = 6 sides × (1 cm × 1 cm) = 6 cm²
Surface area of 8 cubes, sides 0.5 cm = 8 cubes × 6 sides × (0.5 cm × 0.5 cm) = 12 cm²
Surface are of 64 cubes, sides 0.25 = 64 cubes × 6 sides × (0.25 cm × 0.25 cm) = 24 cm²

Halving the average size of the particles, in a given quantity of solid, doubles the total surface area.

So making a solid reactant in lump form into a powder considerably increases the surface area. You only have to compare the effect, for example, of heating aluminium powder in a firework or in a Bunsen flame and heating an aluminium saucepan on a gas cooking ring. More dramatic and serious are the explosions of flour and coal powder which can happen when fine combustible powders suspended as dust in air react very rapidly. Surface area certainly makes a big difference to the reaction rate.

But, as we have stressed already, increasing the number of collisions does not always increase the rate of reaction. What matters is the number of *effective* collisions. Effective collisions have to involve the relevant molecules, in the correct orientation, with energy at least equal to the activation energy.

Clock reactions

One way of illustrating the effect of concentration on the rate of a chemical reaction is by using the so-called clock reaction. The principle of the best-known clock reaction relies on the fact that iodine and starch produce a dark blue colour, and uses a series of flasks containing different quantities of reagents. With luck and good organisation, the solutions in the flasks turn dark blue one after the other at one second intervals.

The basic reaction producing iodine is the oxidation of iodide ions by hydrogen peroxide.

$$H_2O_2(aq) + 2H^{\oplus}(aq) + 2I^{\ominus}(aq) \longrightarrow 2H_2O(l) + I_2(aq)$$

As the reaction proceeds, the brown colour of iodine increases in intensity. If starch is added to this reaction mixture, as soon as sufficient iodine is formed a dark blue colour is formed which is so intense that the increasing iodine concentration is not apparent. Once there is sufficient iodine present the dark blue colour is formed. On a rate curve, the reaction looks like this.

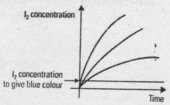

One way of delaying the formation of the iodine/starch colour is to add something to remove the iodine as it is being formed. Sodium thiosulphate reacts with iodine, re-forming iodide ions – more rapidly than the starch reacts with iodine to form the dark blue complex.

$$I_2(aq) + 2S_2O_3^{2\ominus}(aq) \longrightarrow S_4O_6^{2\ominus}(aq) + 2I^{\ominus}(aq)$$

By adding a constant amount of sodium thiosulphate solution to each of the flasks before adding the iodide solution, the delay in the formation of the iodine colour gives us a clock reaction. The dark blue colour of the starch–iodine mixture appears at different times as the effect of different concentrations of the reactants is to produce sufficient iodine (to react with all the thiosulphate present) at different rates.

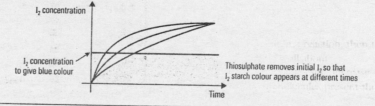

1 When 2.0 g of chalk lump is treated with excess dilute hydrochloric acid, carbon dioxide gas is given off. The equation for the reaction is:

$$CaCO_3(s) + 2HCl(aq) \longrightarrow CaCl_2(aq) + H_2O(l) + CO_2(g)$$

The volume of gas evolved can be measured, and typical results are shown in the graph by the line marked V_1. If the chalk is now ground into a powder and the experiment repeated with exactly the same quantities of chalk and acid at the same temperature, a new line, marked V_2 is produced.

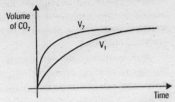

a Explain, in terms of collision theory, why the line marked V_2 is not the same as V_1.

b Copy out the graph and the two lines. Mark on your copy the shape of curve you might expect if 4.0 g of chalk powder was used with the same amount of acid (still an excess) and at the same temperature. There could be two significant differences to the shape of the graph.

2 Consider a chemical with an average particle size of 1 mm (i.e. cubes with 1 mm sides as in a coarse powder). How would you expect the surface area to change if these particles were ground into a powder of average size 1 μm – a fine powder (1 mm = 1000 μm)?

3 In an experiment to follow the rate of reaction between 0.0850 g magnesium powder and excess 1.0 mol dm^{-3} hydrochloric acid, the volume of hydrogen gas produced at set time intervals was measured using a gas syringe.

$$Mg(s) + 2HCl(aq) \longrightarrow MgCl_2(aq) + H_2(g)$$

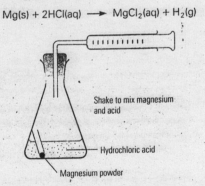

Shake to mix magnesium and acid

Hydrochloric acid

Magnesium powder

The results, at 25 °C, are listed below.

Time /min	Volume of gas /cm^3
1.0	32
2.0	57
3.0	72
4.0	80
5.0	84
6.0	83
7.0	84

a Plot a graph of volume of gas produced against time (x axis).
b Calculate the number of moles of magnesium used, and then convert that amount to the expected volume of hydrogen gas. How does this compare with the maximum volume in your graph?
   (The gas molar volume is 24 000 cm³ at 25 °C.)
c At what time was the reaction complete?
d What does *excess* hydrochloric acid mean?
e Sketch on your graph the plot you might expect if the reaction was repeated at 40 °C using the same quantities of reactants.
f The volume of acid is in excess. How would the graph change if the concentration of the acid was reduced to 0.5 mol dm⁻³ (but still an excess)?

12.4 Energy distribution of molecules

We need to consider the energy of the molecules in more detail if we are going to understand the effect of collisions on rates of reactions. You might be familiar with the normal, bell shaped, distribution curve of physical properties (*Figure 12.4.1*).

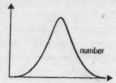

Figure 12.4.1 The normal distribution curve

The heights of students, the weight of doctors, the size of people's feet – whatever, the distribution curve has a maximum at the average value, tailing off symmetrically on either side of the maximum.

The Maxwell–Boltzmann distribution curve

On a molecular level, in the gas phase, the distribution curve is very different (*Figure 12.4.2*).

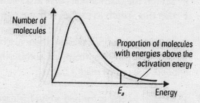

Figure 12.4.2 The Maxwell–Boltzmann distribution curve

James Clerk Maxwell (1831–1879) was a Scot who also devised the equations which successfully combined electricity and magnetism and predicted the existence of radio waves.

There might not appear to be much difference, from a casual inspection. The maximum is similar, but the energies are not distributed symmetrically. There is always a very long tail at the higher-value end of the curve. This distribution of molecular kinetic energies is called the **Maxwell–Boltzmann distribution** after the two scientists who developed, independently, the idea that, at a molecular level, energy distributions were *not normal*.

The most common energy of the molecules is represented by the maximum value, while the number of molecules with a particular energy is shown by the line. The area under the curve represents the total number of particles. The number of particles with energy greater than a value E_a in this graph is shown by the shaded area.

301

The effect of temperature

The shape of this curve changes as the temperature changes. As the temperature rises, the maximum moves to the right, the curve flattens out slightly, and the tail appears to be extended. In *Figure 12.4.3* this is shown by the two curves marked T_1 and T_2. At the higher temperature, T_2, the area under the whole curve is the same – since there are the same number of molecules present.

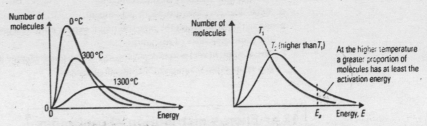

Figure 12.4.3 Energy distributions at different temperatures

However, the mean energy of the particles has increased, so that the maximum moves towards the right, and the spread of energies around this maximum is wider. We can see that at the higher temperature there are now considerably more molecules with energy greater than E_a.

In terms of reaction rates, the energy value E_a is the **activation energy** (have a look at *Figure 12.2.2* again). The molecules with energy greater than this value are the only molecules with sufficient energy for effective collisions. At the higher temperature, the number of these molecules increases and so the number of effective collisions is greater – hence a faster reaction rate.

But to be effective, collisions not only have to involve molecules with energy greater than the activation energy. In all but the simplest chemical reactions, the collisions also have to have the correct orientation. The reactive atoms in the two molecules have to be the parts that actually collide. **Steric factors,** i.e. factors which are to do with the orientation of the molecules in space, are also involved in producing effective collisions. This becomes important in the reactions of some organic molecules (see Ch 14.7).

SELF-ASSESSMENT QUESTIONS 12.4

1 a Draw a diagram to illustrate the distribution of energies in a large number of gas molecules at a particular temperature.
   b Sketch the distribution curve at a higher temperature.
   c Use this sketch to explain why the rate of reaction is increased at the higher temperature.

2 Explain, in terms of collision theory, why petrol does not burn at room temperature even though the molar enthalpy of combustion is around $-5000 \text{ kJ mol}^{-1}$, but once a spark is applied, the petrol readily burns.

3 Hydrogen iodide is formed when hydrogen and iodine are reacted together in the gas phase, although the reaction can easily be reversed. The equation for the forward reaction is:

$$H_2(g) + I_2(g) \longrightarrow 2HI(g) \qquad \Delta H = +53 \text{ kJ mol}^{-1}$$

   a What happens to the pressure in a constant volume container as the reaction proceeds?
   b Sketch a graph to show how the concentration of hydrogen iodide will change as the reaction proceeds.
   c Sketch a reaction profile for the forward reaction, showing clearly the enthalpy of formation on your diagram.
   d The activation energy for the forward reaction is $+173 \text{ kJ mol}^{-1}$. What is the activation energy for the reverse reaction? (Hint – see *Figure 12.5.4*).

12.5 Catalysis

We are all very dependent on catalysts in one way or another, even if we are not always aware of it. Many of the consumer products we use every day are only available because of the **catalysts** that are used in their manufacture. The list of examples includes the conversion of vegetable oils to margarine, crude oil into petrol for cars, and crude oil into washing up liquid. But it goes further than that – the digestion of the food we eat is only possible because of the catalysts in our bodies (known as enzymes).

Catalysts are extremely important. They provide ways of changing chemicals which are not available any other way. They can reduce the costs of changing chemicals, improve the effect of some reactions on the environment, and lower the reaction temperature or increase reaction rates.

So what are catalysts?

> **Definition:** A catalyst is a material that alters the rate of a chemical change without undergoing any permanent chemical change itself.

Obviously the catalyst must become involved in a chemical reaction in some way, even though the quantity needed is usually very small. That is what we are going to consider in this section. It helps if we consider three different types of catalysts – **heterogeneous catalysts**, **homogeneous catalysts** and **enzymes**.

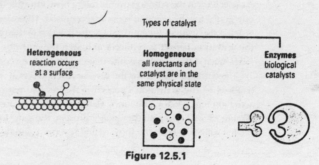

Figure 12.5.1

Catalysis and activation energy

How do catalysts alter the rates of a chemical reaction? Although there are three types of catalysts (*Figure 12.5.1*), the overall effect is the same on the reaction they catalyse – they all lower the energy barrier that controls the rate of a chemical reaction (*Figure 12.5.2*).

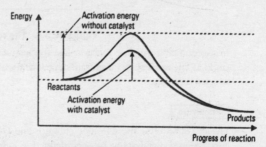

Figure 12.5.2   **Effect of catalyst on E_a**

The energy profile of a reaction is now familiar, with the formation of the **activated complex** at the top of the energy barrier, where the bonds in the reacting molecules are fully stretched. Energy is needed to overcome the repulsion of the electron clouds in the reacting molecules, which is what the energy barrier or **activation energy** is all about. If this energy barrier is lowered, the reaction will proceed more rapidly.

303

Do you remember the Maxwell–Boltzmann distribution curve for the energy of molecules (Ch 12.4)? When the activation energy is lowered, the area under the curve to the right of the E_a value increases quite dramatically, and the number of molecules with energy equal to or above the activation energy increases (*Figure 12.5.3*).

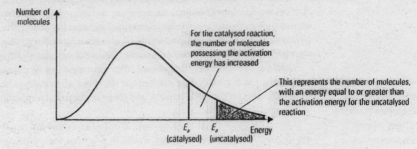

Figure 12.5.3   The effect of lowering E_a on Maxwell–Boltzmann distribution

A closer look at the reaction profile might be worthwhile at this point. In *Figure 12.4.4*, the activation energy, E_a, for the reaction is easily recognised. The value for the enthalpy change for the reaction, $-\Delta H$, is also evident (negative, since the reaction in this diagram is exothermic). If a catalyst is added the activation barrier is reduced, although the enthalpy change for the reaction is still the same, $-\Delta H$. What if the reaction is one that can be easily reversed?

The enthalpy change for the reverse reaction is now $+\Delta H$, as we found out in Ch 11.2. But what happens to the activation energy for the reverse reaction in the presence of the catalyst? The activation energy for the uncatalysed reverse reaction would be $(E_a + \Delta H)$ on our diagram (*Figure 12.5.4*). And with the catalyst present, the same lowered activation energy would apply. So a catalyst will affect both the forward and reverse reactions in any reaction that is easily reversed.

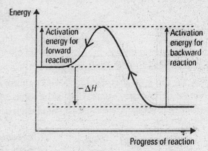

Figure 12.5.4   The reverse reaction

In a genuinely reversible reaction like the manufacture of ammonia by the Haber process (see Ch 13.3), the iron catalyst used cannot change the amount of product, but only speed up the rate at which it is obtained.

Inhibitors

The idea of lowering the activation energy is fine when increasing the rate of reaction with a catalyst is considered. Catalysts were once defined as substances which 'altered the rate of a reaction'. Materials used to decrease the rate of a reaction were called 'negative catalysts' but are now – and more correctly – called **inhibitors**. They certainly cannot work in the same sort of way as catalysts. A reaction will still continue through the lower activation energy route when an inhibitor is added.

The usual way such inhibitors work is by removing one of the reacting intermediates more rapidly than it can be used in the forward reaction. Next time you use any packaged food, have a look at the label for the antioxidants used to preserve the food. These are inhibitors which are added to stop oxidation and decay.

Heterogeneous catalysts

'Hetero' means different, and in heterogeneous catalysis the reaction proceeds with the catalysts in a different phase or state to the reactants. Typically, it is a gaseous reaction in the presence of a metal surface catalyst. Heterogeneous catalysis is often called *surface catalysis*.

Adsorption on to the surface

The metal surface provides active sites on to which the reactant gases (or liquids) can adsorb (*Figure 12.5.5*). Let us consider what happens to nitrogen oxide and carbon monoxide in the catalytic converter of a car exhaust system.

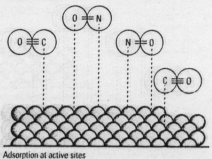

Adsorption at active sites

Figure 12.5.5   Adsorption

The **adsorption** process means that there are weak intermolecular attractions between the reactant molecules and the metal surface. (Gases are **adsorbed** on to a *surface*; water is **absorbed** into the *whole* of a sponge.) More detail of these interactions will be considered in the section on transition metals in Ch 25.3.

When the gas is adsorbed, the formation of the adsorption attraction results in a weakening of the bonds within the molecule – the bond stretches. If we think back to the energy profile (*Figure 12.1.4*), with the bond already partly stretched by the effect of the catalysts, there will be less of a 'hill' to climb to weaken the bond sufficiently for reaction to occur.

If the molecule is adsorbed on to the metal surface, it is held there for a period of time. Although this period of time might not be long, the effect is to keep the molecule close to another molecule long enough for reaction to occur, and to keep the molecules in the correct orientation for reaction to occur (*Figure 12.5.6*).

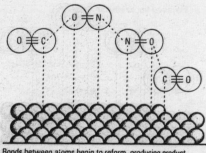

Bonds between atoms begin to reform, producing product molecules

Figure 12.5.6   Reaction

The overall effect is to lower the activation energy, E_a. As an example, the activation energy for the uncatalysed decomposition of nitrogen oxide to nitrogen and oxygen is $+240$ kJ mol^{-1}. However, in the presence of a gold catalyst this value is halved to $+120$ kJ mol^{-1}.

Desorption from the surface

Once the molecules have reacted they have to vacate the active sites on the metal surface – the products, and unreacted reactants, are **desorbed** (*Figure 12.5.7*).

$$2CO(g) + 2NO(g) \longrightarrow 2CO_2(g) + N_2(g)$$

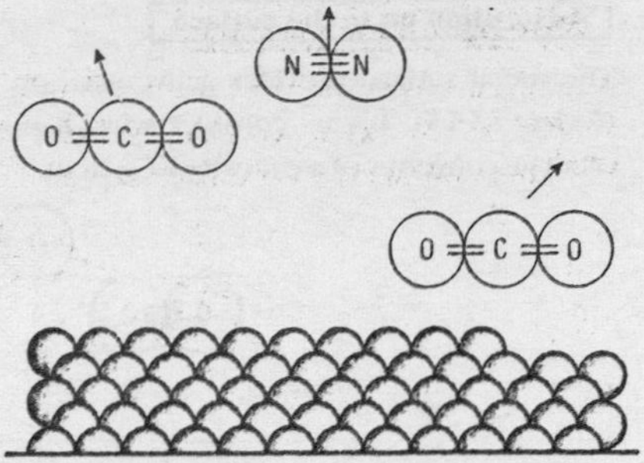

The product molecules leave the metal surface

Figure 12.5.7   Desorption

If the desorption process is not efficient, the catalyst ceases to be effective, as the reactive sites are still occupied and so are not available for any further reactions.

The surface of the metal is the critical factor. A larger surface area will provide more sites for adsorption. But there is a balance in industrial processes. Finely divided powder is very efficient, but is easily blown away, whereas a large sheet of the metal has a small surface area. The type of surface used in practice is a compromise. In the Haber Process for the production of ammonia, the iron catalyst is pea-sized.

$$N_2(g) + 3H_2(g) \xrightarrow{\text{Fe catalyst}} 2NH_3(g)$$

Iron is not an expensive metal and is relatively easy to recycle. So, in spite of this catalyst being easily poisoned by many chemicals, such as sulphur, it is more economical to use iron rather than the more efficient but more expensive alternatives.

The catalyst in the three-way catalytic converter in a car uses a mixture of very expensive platinum and rhodium. To maximise the surface area of the catalyst the metals are deposited as a thin layer on a honeycombed ceramic porous surface (see Ch 15.4).

Industrial process	Equation for chemical reaction	Catalyst	Some uses
Manufacture of ammonia (The Born–Haber Process)	$N_2(g) + 3H_2(g) \longrightarrow 2NH_3(g)$	Iron metal	Making fertilisers. nitric acid, explosives
Manufacture of sulphuric acid (The Contact Process)	$2SO_2(g) + O_2(g) \longrightarrow 2SO_3(g) \xrightarrow{H_2O} H_2SO_4(l)$	Vanadium(V) oxide	Making detergents
Catalytic converters	$2NO(g) + 2CO(g) \longrightarrow N_2(g) + 2CO_2(g)$	Rhodium and platinum	Reducing pollution from car exhausts
Hydrogenation of oils	$RCH{=}CHR'(l) + H_2(g) \longrightarrow RCH_2CH_2R'(s)$	Nickel	Making margarine
Catalytic cracking	e.g. $C_6H_{14}(g) \longrightarrow C_4H_{10}(g) + C_2H_4(s)$	Aluminium oxide	Cracking petroleum fractions

Examples of industrial heterogeneous catalysis

The use of catalysts in industry is often a compromise: balancing the efficiency of the catalyst, the conditions for catalysis (such as particle size, operating temperature and gas pressure) and environmental issues with the most cost-effective production of the desired products.

Transition metals as surface catalysts

Many transition metals make good catalysts for a large number of reactions. These metals all have d-orbitals that do not contain all the electrons possible – the d-energy sub-level is partially empty. It is these vacant d-orbitals that create the active sites for the adsorption of gas molecules on the surface of the metal.

But not all transition metals make good catalysts. In some cases the adsorption on the surface is too strong so that desorption is poor, and metals such as tungsten are rarely used as catalysts. Others, such as gold, have weakly active sites and do not often prove to be good catalysts.

One of the best catalysts is usually platinum, together with the metals close to platinum, such as rhodium and palladium. However, all these metals are rare and extremely expensive. Industrialists usually try to find cheaper, although maybe less efficient, alternatives.

Catalytic promoters

Although d-orbitals are part of the reason why transition metals are such good catalysts, the physical state of the surface, and particularly the projections from the surface, are also vitally important. **Promoters** are impurities deliberately added to the metal to produce irregularities on the surface at (on average) the correct distance apart for molecules to be absorbed on them and react.

Unfortunately, many different molecules are adsorbed on to metal surfaces. The reactants must be adsorbed and then desorbed. But other molecules and atoms may be adsorbed more effectively. Examples include lead atoms on the surface of catalytic converters.

The consequence is that the foreign atoms stay on the surface and so there are fewer sites for catalytic behaviour. The catalyst is *poisoned*. If leaded petrol is used in cars with catalytic converters, the lead is adsorbed on the catalyst surface. It does not take much leaded petrol to make the catalytic converter useless.

Remember:

- heterogeneous catalysis involves surfaces;
- transition metals provide good active sites;
- the sequence of adsorption, reaction, desorption;
- surface area is important;
- but surfaces are easily 'poisoned'.

Homogeneous catalysts

> **Definition:** A homogeneous catalyst speeds up the reaction by providing an alternative reaction mechanism of lower activation energy, with the catalyst and reactants in the same physical state.

'Homo' means the same, and the reactants and homogeneous catalysts are in the same state as the reactants – very often in solution. The catalyst again has to lower the activation energy (*Figure 12.5.8*), and the homogeneous catalyst provides an alternative reaction route for the reaction to occur. How can the activation energy be lowered?

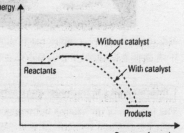

Figure 12.5.8   Reaction profile for a catalysed reaction

Even though glucose and fructose have the same molecular formulae, $C_6H_{12}O_6$, there are a number of ways of arranging these atoms (see Ch 14.4). Two examples are:

Glucose

Fructose

It is more difficult to generalise about the way in which all homogeneous catalysts work. It is easier to consider a few specific examples.

A wide range of organic reactions are catalysed in acidic solution. The hydrogen ion, $H^{\oplus}(aq)$, which is after all only a proton surrounded by water molecules, is able to attach to an electronegative element (oxygen or nitrogen) in the organic molecule and so initiate a reaction (see hydrogen bonding, Ch 2.9). The protonated species will then react readily to form the products and release $H^{\oplus}(aq)$ to continue its catalytic activity. Since the reaction will go through the easiest route possible – which is through the lower activation energy route – the reaction now proceeds more rapidly.

The conversion of cane sugar to a mixture of glucose and fructose is one such example, catalysed by dilute acid.

$$C_{12}H_{22}O_{11}(aq) + H_2O(l) \xrightarrow{H^{\oplus}(aq)} C_6H_{12}O_6(aq) + C_6H_{12}O_6(aq)$$
Sucrose Glucose Fructose

Transition metals as homogeneous catalysts

Transition metal compounds in solution are often very effective catalysts, particularly for reduction–oxidation reactions. The transition metal ion can change oxidation states readily, being oxidised or reduced, and this provides a mechanism for a reaction.

The oxidation of potassium sodium 2,3-dihydroxybutanedioate (potassium sodium tartrate, Rochelle's salt) with hydrogen peroxide provides an interesting, even amusing, example of a reaction catalysed with a transition metal compound. The tartrate is oxidised by the hydrogen peroxide to carbon dioxide, but normally the reaction is extremely slow.

However, if a little cobalt(II) chloride is added to the mixture (together with a little washing up liquid for effect), the reaction occurs at a reasonable rate. The rate is easily measured by timing the formation of the froth in a tall measuring cylinder. What is interesting is that the pink colour of the cobalt(II) chloride is obvious at the start of the reaction. As the reaction proceeds, the colour changes through purple to green. At the end of the reaction the pink colour returns.

Although it is a very complex reaction, a change of the oxidation state of cobalt from +2 to +3 and back again is involved in the catalytic activity.

Intermediates in homogeneous catalysis

The function of the catalyst in homogeneous catalysis is to form an initial reactive intermediate to help the reaction forward. This is demonstrated by the oxidation of iodide ions by potassium peroxodisulphate solution to produce iodine.

$$S_2O_8{}^{2\ominus}(aq) + 2I^{\ominus}(aq) \longrightarrow 2SO_4{}^{2\ominus}(aq) + I_2(aq)$$

The reaction is catalysed by iron(II) ions in aqueous solution, and the reaction now involves two steps. Initially, the peroxodisulphate oxidises the $Fe^{2\oplus}$ to $Fe^{3\oplus}$ and is itself reduced. In the second step the $Fe^{2\oplus}$ is regenerated when iodide ions are oxidised to iodine.

$$S_2O_8^{2\ominus}(aq) + 2Fe^{2\oplus}(aq) \longrightarrow 2Fe^{3\oplus}(aq) + 2SO_4^{2\ominus}(aq)$$

Followed by:

$$2Fe^{3\oplus}(aq) + 2I^{\ominus}(aq) \longrightarrow 2Fe^{2\oplus}(aq) + I_2(aq)$$

An easy demonstration of intermediate compound formation during a catalysed reaction

Until the 1960s the usual way of making oxygen in a school laboratory involved the thermal decomposition of sodium or potassium chlorate(V), the reaction being catalysed by manganese(IV) oxide, MnO_2 (see Ch 8.5).

e.g. $$2KClO_3(s) \longrightarrow 2KCl(s) + 3O_2(g)$$

Teachers never let their pupils smell the oxygen produced, or let them put damp blue litmus paper in the gas jar of the gas. This was because it contained a quite high concentration of chlorine. The catalysed reaction therefore cannot be as simple as the equation suggests.

It was later discovered accidentally that strongly heating a mixture of potassium chlorate(V) and manganese(IV) oxide in the ratio of 5000:1 in a test tube gives a bright pink-purple colour typical of potassium manganate(VII), $KMnO_4$ (potassium permanganate). Continued very strong heating gives a green colour, characteristic of potassium manganate(VI), K_2MnO_4. A possible sequence of reactions consistent with these observations is as follows:

$$2MnO_2(s) + 2KClO_3(s) \longrightarrow \underset{\text{purple}}{2KMnO_4(s)} + Cl_2(g) + O_2(g)$$

$$2KMnO_4(s) \longrightarrow K_2MnO_4(s) + MnO_2(s) + O_2(g)$$

$$K_2MnO_4(s) + Cl_2(g) \longrightarrow 2KCl(s) + MnO_2(s) + O_2(g)$$

Overall (cancelling out materials which occur on both sides):

$$2KClO_3(s) \longrightarrow 2KCl(s) + 3O_2(g)$$

But some of the chlorine escapes before it can react further!

Later we will see that aluminium chloride is used in many organic reactions as a catalyst because of the electron-deficient nature of the $AlCl_3$. One example is the halogenation of benzene.

$$C_6H_6 + Cl_2 \xrightarrow[\substack{\text{room} \\ \text{temperature}}]{AlCl_3} C_6H_5Cl + HCl$$

The $AlCl_3$ is able to accept electrons from the Cl_2 and so produce a very reactive species, $Cl^{\oplus}$.

$$AlCl_3 + Cl_2 \longrightarrow AlCl_4^{\ominus} + Cl^{\oplus}$$

This is possible because aluminium in $AlCl_3$ has only six outer electrons, i.e. it is electron-deficient (see Ch 2.4). But more of this later (Ch 26.8).

Autocatalysis

An interesting example of catalysis involves reactions in which the catalyst is formed in the course of the reaction. What happens then? The reaction is slow without any catalyst present at the start. As the reaction proceeds a small quantity of the catalyst is produced – which catalyses the reaction – so the reaction goes faster – to produce more catalyst, and so on. The rate curve is now somewhat different. The initial slow rate speeds up until the optimum amount of catalyst is present, and then the rate curve resumes its normal shape, as in *Figure 12.5.9*.

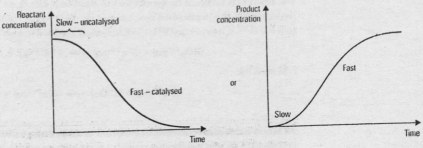

Figure 12.5.9   Autocatalysis curve

This type of catalytic activity is called **autocatalysis**, and one example is the oxidation of ethanedioate ions by potassium manganate(VII) solution. The reaction is catalysed by the $Mn^{2\oplus}(aq)$ produced in the reaction.

$$2MnO_4^{\ominus}(aq) + 16H^{\oplus}(aq) + 5C_2O_4^{2\ominus}(aq) \longrightarrow 2Mn^{2\oplus}(aq) + 8H_2O(l) + 10CO_2(g)$$

The $Mn^{2\oplus}$ ions catalyse the reaction. The rate curve shows only a slow reaction at the start, but eventually it takes on the 'usual' shape.

Enzyme catalysis

Enzymes are biological catalysts that facilitate a large number of chemical reactions in living systems. They are nearly all globular proteins and, like all catalysts, provide an alternative route for a reaction with a lower activation energy. They are very much more effective than the catalysts we have already come across; we require only minute quantities to keep us alive and well.

Hydrogen peroxide decomposes to form water and oxygen.

$$2H_2O_2(aq) \longrightarrow 2H_2O(l) + O_2(g)$$

At room temperature the reaction is very slow, but a small amount of solid manganese(IV) oxide, MnO_2, catalyses the reaction. The rate of the reaction can be followed using the apparatus shown in *Figure 12.5.10*.

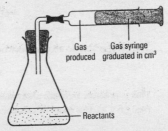

Gas produced
Gas syringe graduated in cm³
Reactants

Figure 12.5.10   Apparatus for demonstrating the decomposition of hydrogen peroxide

Once the flask is shaken in order to add the catalyst to the liquid, the volume of oxygen produced can be monitored with time.

The enzyme, catalase, found in blood, is a far more effective catalyst for this reaction than solid MnO_2. This is easily demonstrated using a small sample of chicken liver. Catalase is present in blood to remove any hydrogen peroxide produced in metabolic reactions, since hydrogen peroxide can cause mutations in DNA.

'Lock and key' enzymes

In addition to being very effective, enzymes are very specific in the reactions they catalyse. It is more than just catalysing a specific reaction; enzymes are usually specific about the shapes as well as the chemical properties of the reactants. The mechanism by which enzymes catalyse reactions is thought to involve a **'lock and key'** type of interaction. The reactant molecule (known as the **substrate**) fits, like a key, into a specific part of the complex enzyme molecule. This part of the enzyme is known as the **active site** (*Figure 12.5.11*).

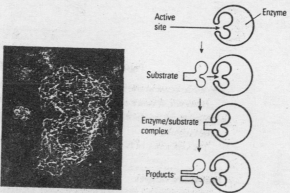

Figure 12.5.11 'Lock and key' mechanism for enzyme action

The correct atoms have to be present in the correct position before any chemical changes can occur.

The substrate molecule (or molecules) fits exactly into an active site in the enzyme, and once attached to it undergoes specific chemical changes before being released in a new form (the products). There are similarities with the pathway of heterogeneous catalysts. The substrate has to meet and interact with an active site on the enzyme and form a complex in which the bonds are rearranged, before the product molecule moves away from the enzyme.

Enzyme + Substrate → Intermediate complex → Enzyme + Product

The efficiency of an enzyme can be affected in the same way as for other catalysts. If the active sites are reduced in number, by the presence of too much substrate, or by being poisoned by a molecule that stays attached to the site, then catalytic activity is reduced.

The picture of a lock and key is sufficient at this level; but you should appreciate that although locks and keys are solid inflexible objects, the enzyme and substrate molecules are complex structures that are certainly not rigid.

Industrial enzymes

Enzymes have been used for a very long time to enable biological changes to happen on an industrial scale. The conversion of milk to cheese (using rennin), and the fermentation of sugars into alcohol using the maltase and zymase present in yeasts, have been industrial processes for thousands of years.

There is a drive to develop more enzyme-catalysed reactions for industrial chemical changes. Some of the reasons for this are that enzymes are:

- effective as catalysts;
- specific, using a 'lock and key' type mechanism;
- used at temperatures near those of the surroundings (20 °C to 40 °C);
- used at pH values close to 7;
- environmentally friendly.

Since they are 'natural' products, enzymes pose fewer pollution difficulties during and after their use and, as proteins, they are very sensitive to pH and temperature changes (see *Figure 12.5.12*).

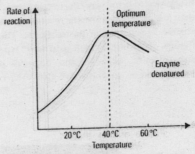

Figure 12.5.12   Effects of temperature on enzyme activity

The waste yeast in the brewing industry can be used as an animal feedstock.

However, it is not always so simple. The early attempts at adding enzymes to washing powders, to help 'break up' biological stains on clothes, led to many people suffering from allergic reactions to the enzyme.

The efficiency of biological enzymes has been further improved by 'immobilising' the enzymes.

Immobilised enzymes

In industrial processes which use soluble enzymes, the reactions are often carried out as a batch process – the substrate and enzyme are mixed in the reaction vessel and left until the reaction is 'complete'. Once the desired product is extracted, the enzyme, together with other waste products, is discarded. The cost of recovering the small quantity of enzyme is too great to be worthwhile.

With the development of the so-called *super-enzyme*, which can work over a greater temperature and pH range, the enzyme becomes much too valuable to be thrown away. By fixing the enzyme to an inert support and effectively making it insoluble, the enzyme is not lost during the reaction, and the reaction can be carried out as a continuous process. Immobilised enzymes can be used at higher concentrations (which reduces the reaction time), and allows a higher throughput of reactants, reducing the possibilities of side reactions and also reducing the cost.

The enzyme can be immobilised by adsorption on an inert resin, by trapping in a network of polymer molecules, or even by bonding the enzyme to an inert polymer support.

Examples of the use of immobilised enzymes include the conversion of the sugar glucose, obtained from the hydrolysis of starch, into the sweeter sugar fructose, and the formation of semi-synthetic penicillin – active against bacteria which have developed resistance to conventional penicillin.

But the biotechnology search does not stop. Designer enzymes are next, in which enzymes are modified so that they can operate over wider ranges of conditions for a wider range of reactions.

SELF-ASSESSMENT QUESTIONS 12.5

1 a What is meant by a catalyst?
   b Explain the term *activation energy*, and show how consideration of activation energy can be used to explain how a catalyst functions.
   c Draw a graph to show the distribution of molecular energies in the gas phase.
   d On your sketched graph show how a decrease in temperature will alter the graph.
   e Mark in a value E_a for the activation energy in an uncatalysed reaction.
   f Now mark clearly the E_a value for a catalysed reaction.

2 a Using specific examples of catalysed reactions explain what is meant by homogeneous and heterogeneous catalysis.
   b Give two examples of the use of enzymes as catalysts.
   c What are the advantages of immobilised enzymes as catalysts?

3 When starch is hydrolysed, it is changed into glucose, $C_6H_{12}O_6$. This sugar is not as sweet as fructose and so the enzyme invertase is used to change the glucose into fructose, $C_6H_{12}O_6$.

a What is an enzyme?

b Enzymes are said to work through a lock and key mechanism. Explain what is meant by *lock and key* mechanism.

c The rate of formation of the fructose was measured at different temperatures. The results are shown in the graph.

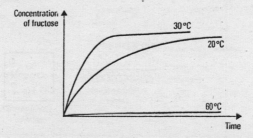

Explain the different curves at 30 °C and at 60 °C.

Summary of basic ideas

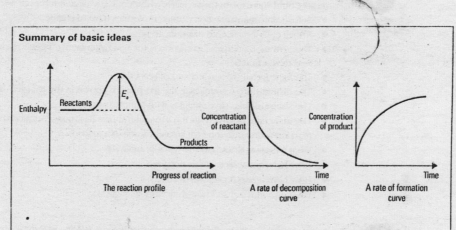

Effective collisions for molecules with energy equal to or **above** the activation energy, E_a.

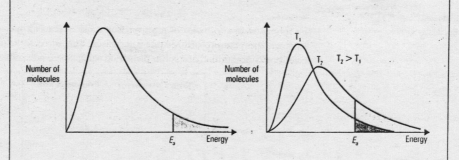

Maxwell–Boltzmann distribution curve; molecules with energy above E_a lead to effective collisions; at higher temperatures there are more molecules with energy above E_a.

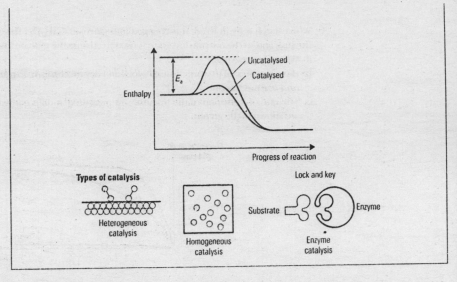

**SUMMARY
12**

In this chapter the following new ideas have been covered. *Have you understood and learned them?* Could you explain them clearly to a friend? The page references will help.

- Before almost any reaction occurs, molecules have to collide.   p294
- Not all collisions lead to reaction.   p295
- The activation energy of a reaction is the minimum energy needed before a collision can
  lead to reaction.   p295
- The factors that influence a rate of reaction.   p296
- The distribution of energies in the gas phase is shown in the Maxwell–Boltzmann curve. p302
- The shape of this curve changes with temperature.   p302
- The rate curve obtained during a reaction shows how concentration changes during the
  reaction and shows the final concentration in a reaction.   p296
- What a catalyst does, and the types of catalysts.   p303
- How heterogenous catalysis occurs.   p305
- How homogenous catalysis occurs.   p307
- What enzymes are and how they work as catalysts.   p310

**EXAMINATION
QUESTIONS
12**

1 The decomposition of sulphur dichloride dioxide, SCl_2O_2, in the gas phase occurs according to the equation:

$$SCl_2O_2(g) \longrightarrow SO_2(g) + Cl_2(g)$$

The formation of two moles of gas from one mole of reactant gas means that, in a closed container, the pressure of the system increases. This increase in pressure represents the decreasing concentration of sulphur dichloride oxide. Some rate results are given in the table.

Time / minutes	Pressure / atm
0	1.0
1	1.16
2	1.30
3	1.41
4	1.52
5	1.59

a Plot a graph of these results to show how the pressure changes with time.

b How would this graph alter if the temperature of the reaction was increased?

c Explain, using the Maxwell–Boltzmann distribution, the shape of your altered graph in **b**.

2 The reaction of nitrogen and hydrogen to form ammonia (NH_3) is an important chemical process.

a Write a balanced equation for the reaction.

b Name the catalyst that is used in this reaction.

c The reaction is exothermic with $\Delta H = -92 \text{ kJ mol}^{-1}$. Draw a reaction profile for the reaction.

d On the reaction profile mark the enthalpy change of the reaction, ΔH.

e Mark the activation energy for the forward reaction, labelling it A.

f Draw an arrow to indicate the activation energy for the reverse reaction and label it B.

g What effect will the catalyst used have on the reverse reaction?

h The activation energy for the forward reaction is $+608 \text{ kJ mol}^{-1}$. What would be the activation energy for the reverse reaction?

3 Chalk, calcium carbonate, reacts with hydrochloric acid to liberate carbon dioxide gas according to the equation:

$$CaCO_3(s) + 2HCl(aq) \longrightarrow CO_2(g) + CaCl_2(aq) + H_2O(l)$$

a Draw a suitable apparatus for following the rate of this reaction.

b 0.40 g of chalk was used in the reaction. How many moles of chalk is this?

c What volume of 2.0 mol dm^{-3} hydrochloric acid would be needed to provide a five-fold excess of acid? (Refer back to Ch 9.5 if you are unsure.)

d What volume of carbon dioxide would you expect to be produced? (The volume of one mole of gas under the conditions used in the experiment is 24 000 cm^3.)

e Draw a sketch to show the expected rate curve.

f How would adding twice the amount of hydrochloric acid alter the shape of the rate curve?

- **Changes of state**: particles **move** and particles **attract** one another. (Ch 2.8 and Ch 9.6)
- **Collision theory** and rates of reaction: particles need to meet in order to react. (Ch 12.1)
- **Chemical equations**: reactions summarised in one line. (Ch 4.2)

13

reversible reactions

13.1 Reversible reactions and dynamic equilibrium

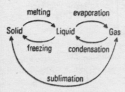

Figure 13.1.1 Changes of state

At several points in earlier chapters we have discussed **changes of state**. These, and the different states of matter, are summarised in *Figure 13.1.1*.

Imagine some water – just water, no air – in a closed flask held at a particular temperature. At first, some of the molecules of water will have sufficient energy to evaporate into the space above the water surface (*Figure 13.1.2*).

Figure 13.1.2 Before equilibrium is reached

What we are doing here is what is called a 'thought experiment' – no apparatus, but rigorous use of scientific ideas.

As soon as there are water vapour molecules in that space, some of them will collide with the surface of the liquid and re-enter it; they will condense.

After a while, the number of water molecules evaporating from the liquid in a given time will be balanced by the number of water molecules condensing back from the vapour into the liquid (*Figure 13.1.3*).

Figure 13.1.3 At equilibrium

The situation can be summed up in a simple equation:

$$\text{Water liquid} \rightleftharpoons \text{Water vapour.}$$

The double-headed arrow means that this process is reversible; it can go either way.

The rates of the two opposing processes are equal. We have reached a position of **equilibrium**: a steady state, where – if you look carefully at the system – absolutely nothing seems to be happening. In reality, as you will realise, the two processes of evaporation and condensation are going in opposite directions – **but at equal rates**, and that's why you can't see any change.

Except at absolute zero there is constant movement of particles. It is for this reason that such an equilibrium is called a **dynamic equilibrium**. (If you put a cup on the table, nothing happens. The pull of gravity on the cup is countered by the upthrust of the table on the cup. That is an example of *static equilibrium*. There is no constant movement of particles in opposite directions. The cup just sits there.)

The same sort of thing happens with a *saturated solution* (a solution which cannot dissolve any more of the solute). Consider the ionic compound, silver chloride, in contact with water at room temperature. Some of the solid will dissolve and then some of the dissolved particles will start to stick back on to the solid – they will crystallise. Eventually the rates of dissolving and crystallisation will be equal. Equilibrium will be reached (*Figure 13.1.4*).

We deliberately chose silver chloride for our example, because this was used to demonstrate dynamic equilibrium. A saturated solution of ordinary silver chloride was put in contact with solid silver chloride containing some of the radioactive isotope ^{112}Ag. Nothing appeared to happen. But within a short while a Geiger counter detected radioactivity in the solution, and it could only have come from the solid.

317

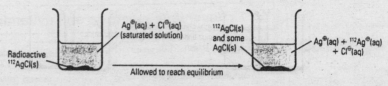

Figure 13.1.4

Common sense tells us that the double-headed arrow cannot mean that *all* the water in the flask turns to vapour, which then *all* turns back to water; or that *all* the solid silver chloride dissolves, and then *all* crystallises out again. Nature doesn't usually lurch from one side to the other!

We cannot emphasise this too strongly – *in a genuine dynamic equilibrium, on a macroscopic scale nothing seems to change.* The concentrations of the individual substances do not change. On the sub-macroscopic level you would only notice changes happening if you could see separate atoms, molecules or ions.

To return to the water/water vapour equilibrium in the vacuum flask: suppose we allowed some water vapour to escape? More water would evaporate, and eventually all the water would escape. No equilibrium would be reached.

This makes the point that *dynamic equilibrium can only be reached in a closed system*: that is, a system in which no materials can enter and leave the system, but energy can. Note that, unless gases are involved, a closed system doesn't have to be 'shut up'. The solid silver chloride/saturated solution system already mentioned would usually be in an open beaker, but – apart from some slight evaporation of water – there is no loss or gain of material.

The last point is this – *any dynamic equilibrium is affected by temperature.* Think back to the water in the flask. At higher temperatures, more water molecules would have enough energy to vaporise before equilibrium is reached. We shall discuss the effect of temperature in detail shortly.

To summarise:
- all physical equilibria of this kind are dynamic;
- no changes can be seen to occur;
- at equilibrium, the rates of the forward and reverse processes are the same; and
- all equilibria are sensitive to temperature changes.

Chemical equilibrium

So far we have discussed only equilibria involving physical changes. A **chemical equilibrium** is also dynamic. The points we have just made – about closed systems and temperature dependence – apply equally well to chemical equilibrium. But, whereas in most physical changes only comparatively weak forces between particles are affected, in any chemical reaction strong, chemical bonds are being broken and made (see Ch 11.2). And, because breaking and making bonds often involves large energy changes, temperature is obviously important in determining the balance at equilibrium.

The first main point to remember is that *all reactions are, in principle, reversible* – even dramatic and dangerous reactions like putting potassium in water – so long as the reaction happens in a closed system. When the reaction appears to have stopped there will be some amount, however small, of the reactants left. This 'drive towards equilibrium' is true of all reacting systems.

$$\text{Reactants} \rightleftharpoons \text{Products}$$

Note the use of the double headed arrow, $\rightleftharpoons$, to represent a reversible process. The top half-arrow should point from reactants to products.

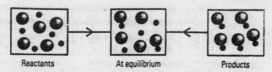

Reactants   At equilibrium   Products

Equilibrium can be reached from either side of the reaction.

But if a product of the reaction is allowed to escape, the reaction will go all the way. We say: 'The reaction goes to completion'. One familiar example involves allowing a gas to escape, e.g. the laboratory preparation of carbon dioxide from dilute hydrochloric acid and marble chips (calcium carbonate).

$$CaCO_3(s) + 2HCl(aq) \longrightarrow CaCl_2(aq) + H_2O(l) + CO_2(g)$$
carbon dioxide
escapes

Another is the formation of a precipitate when two solutions are mixed, e.g. the test for a chloride. When silver nitrate solution and dilute nitric acid are added to the solution under test, a white insoluble solid, silver chloride, forms if a chloride is present. The silver chloride precipitates as a solid which effectively removes it from the aqueous equilibrium.

$$NaCl(aq) + AgNO_3(aq) \longrightarrow AgCl(s) + NaNO_3(aq)$$
silver chloride
precipitates

An equilibrium is established between the solid and the relatively few $Ag^{\ominus}(aq)$ and $Cl^{\ominus}(aq)$ ions remaining in solution; the reaction has not *quite* gone all the way but, because of the very low solubility of silver chloride, is very nearly complete.

In this chapter we shall only discuss reactions in *homogeneous systems* – ones in which all the reactants and products are in the same state of matter. This covers many familiar reactions in solution and many very important industrial processes involving gases (as we shall soon see). In *heterogeneous systems* not all the reactants and products are in the same state, e.g. the reaction of acid on marble chips, or reactions in which reactants and products are spread over two layers of immiscible liquids (see solvent extraction, in Ch 29.2).

From now on we shall often use the symbol [X] where X is the symbol or formula of a reactant or product. Rather confusingly for us, putting square brackets round a formula is also used for complex ions (see Ch 25.4). But here, and in *any* calculation involving any kind of equilibrium, the meaning is quite straightforward and simple.

[X] means the concentration, in mol dm^{-3}, of the reactant or product X; and [X]$_{eqm}$ means the concentration of X when the system is in equilibrium. The symbol 'eqm' is often omitted if it is obvious that an equilibrium reaction is involved.

1 When chlorine gas is passed over iodine monochloride at room temperature, iodine trichloride is formed in an equilibrium reaction, according to the equation:

$$ICl(l) + Cl_2(g) \rightleftharpoons ICl_3(s)$$

Iodine monochloride is a brown liquid and iodine trichloride is a yellow crystalline solid.

a Describe what you would expect to see as the chlorine gas is passed over the iodine monochloride.

b What would you expect to see if the chlorine gas was then removed from the reaction tube?

c Sketch a graph to show how you would expect the concentration of iodine monochloride to change as equilibrium is reached in part b.

2   **a** Explain what is meant by a *dynamic equilibrium*.

   b Explain why a mixture of bromine vapour and bromine liquid in a closed flask at constant temperature is said to be in dynamic equilibrium, even though there is no colour change at all in the flask.

13.2   Factors affecting equilibrium; Le Chatelier's Principle

Remember – once equilibrium is reached, the concentrations of the reactants and products in closed system will stay constant just as long as the conditions (especially the temperature) in the system stay the same. The situation can be summed up in *Figure 13.2.1*.

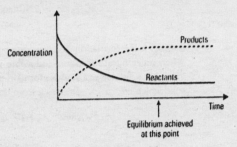

Figure 13.2.1

Consider the reaction:

$$A + B \rightleftharpoons C + D$$

The way in which the rate of a reaction depends on the concentration of reactants can *never* predicted from the equation, **but must *always* be found by experiment** (see Ch 12.2).

Let us suppose that the rate of the forward reaction is found to be proportional to t concentrations of each of the reactants A and B. We can then write:

$$\text{rate} = k_1[A][B]$$

where k_1 is the **rate constant** for the forward reaction. Similarly, *experimental evidence* shows t for the reverse reaction:

$$\text{rate} = k_2[C][D]$$

where k_2 is another rate constant, this time for the reverse reaction.

Once equilibrium is reached, the rates of the forward and reverse reactions are equal. So:

$$k_1[A]_{eqm}[B]_{eqm} = k_2[C]_{eqm}[D]_{eqm}$$

This can be rewritten like this:

$$\frac{k_1}{k_2} = \frac{[C]_{eqm}\ [D]_{eqm}}{[A]_{eqm}\ [B]_{eqm}}$$

Both k_1 and k_2 are constants, and one constant divided by another gives yet another constant. T new constant is called the **equilibrium constant**, K_c, for the reaction.

$$K_c = \frac{[C]\ [D]}{[A]\ [B]} \qquad\qquad \textbf{Equation A}$$

We can leave out the 'eqm' subscript since it is most certainly an equilibrium reaction.

N.B. The *products* always appear on the top of a K_c expression.

Now consider a slightly more complicated reaction.

$$wA + xB \rightleftharpoons yC + zD$$

Here, w, x, y, and z are the respective numbers of moles of reactants and products shown in the **stoichiometric** equation (see Ch 9.5). The numbers are usually whole numbers, but need not be; equations may be written, for example, using $\frac{1}{2}H_2$ or $\frac{1}{2}O_2$.

When equilibrium is reached, the concentrations of reactants and products are **almost always** found by experiment to fit this equation:

$$K_c = \frac{[C]^y [D]^z}{[A]^w [B]^x} \qquad\qquad \textbf{Equation B}$$

This equation (equation B) is known as the **'equilibrium law'**, or **equilibrium expression**, for this reaction.

Equilibrium and equilibrium constants

$$A \rightleftharpoons B \qquad\qquad K_c = \frac{[A]}{[B]}$$

$$A \rightleftharpoons B + C \qquad\qquad K_c = \frac{[B][C]}{[A]}$$

$$A \rightleftharpoons 2B \qquad\qquad K_c = \frac{[B]^2}{[A]}$$

The concentration term is raised to the power of the number of molecules appearing in the balanced equation.

If you are not happy with why this happens, look at the equilibrium reaction involving hydrogen, iodine and hydrogen iodide.

$$H_2(g) + I_2(g) \rightleftharpoons 2HI(g)$$

This equation can be rewritten as:

$$H_2 + I_2 \rightleftharpoons HI + HI$$

And for this equation K_c is:

$$K_c = \frac{[HI][HI]}{[H_2][I_2]}$$

This is exactly the same as:

$$K_c = \frac{[HI]^2}{[H_2][I_2]}$$

which could be derived using the first equation.

Consider this last reaction.

$$H_2(g) + I_2(g) \rightleftharpoons 2HI(g)$$

When equilibrium is reached, the concentrations of each of the reactants and products have to be found by experiment. Typical data are given in the table.

Experiment number	[H₂]	[I₂]	[HI]	[HI]²	K_c
1	1.04	0.25	3.48	12.11	46.58
2	1.29	0.14	2.90	8.41	46.57
3	0.39	0.39	2.66	7.08	46.55
4	0.87	0.35	3.77	14.21	46.66
5	0.97	0.97	6.62	43.82	46.57

By plotting a variety of graphs it is possible to show that $[HI]^2 \propto [H_2] \times [I_2]$ at equilibrium, i.e.:

$$K_c = \frac{[HI]^2}{[H_2] \times [I_2]}$$

The graph of $[HI]^2$ against $[H_2] \times [I_2]$ is the only graph that results in a straight line plot (*Figure 13.2.2*). Any other combination of the concentration values turns out to be anything but a straight line.

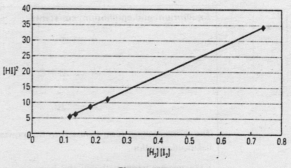

Figure 13.2.2

Alternatively, the value of K_c for each set of experimental results can be calculated. The values are shown in the table. Do you think these are constant (within experimental error)?

The equilibrium constant, K_c, for a particular reaction is found by experiment to have a constant value provided that the temperature stays constant. The c in K_c means that it is an equilibrium constant based on concentrations. In Chapters 22, 23 and 25 you will meet several more examples of equilibrium constants. All are given the symbol K, but there is a different subscript letter for each kind. In this chapter we shall use only K_c.

If you are a confident mathematician, you may already have noticed something about equations A and B. To find the value of K_c for a particular reaction you need to do a lot of experiments which involve measuring the concentrations of reactants and products in the equilibrium mixture. (You can *start* with any amounts of reactants and products: measurements on the equilibrium mixture will always give the same value for K_c at the same temperature.)

Once you have a value for K_c, provided that you know the concentrations of the reactants you start with, it becomes possible to predict accurately all the concentrations of the reactants and products in the equilibrium mixture. We shall do no calculations in this chapter. If you would like to see how to do them, see Ch 22.1.

Equilibrium constants

There are several important points to note about equilibrium constants.

- A high value of K_c usually means that in the equilibrium mixture there is much more of the products than the reactants. This is often loosely stated as: 'a large K_c means that the equilibrium is well over to the right', or: 'there is a high yield of product at equilibrium'. Statements like this should be treated with great caution.

If you look again at equation B. you will see that high values of y and z greatly boost the numerical value of K_c. So if in a reaction one or two big reactant molecules break up into lots of smaller molecules of different kinds, K_c is bound to be large even if not many of the big molecules are broken up at equilibrium. This point is developed in Ch 25.4.

- If the equilibrium constant for the forward reaction is K_c, the equilibrium constant for the reverse reaction must be $1/K_c$. Remember, the equation for K_c depends on the chemical equation it refers to. So, *always write out the equation* for the reaction before stating K_c.

Look at these examples:

$$NO(g) + \tfrac{1}{2}O_2(g) \rightleftharpoons NO_2(g) \quad K_{1c} = \frac{[NO_2]}{[NO][O_2]^{\frac{1}{2}}}$$

or:

$$2NO(g) + O_2(g) \rightleftharpoons 2NO_2(g) \quad K_{2c} = \frac{[NO_2]^2}{[NO]^2[O_2]}$$

and for the reverse reaction – perfectly valid if the reaction is an equilibrium:

$$2NO_2(g) \rightleftharpoons 2NO(g) + O_2(g) \quad K_{3c} = \frac{[NO]^2[O_2]}{[NO_2]_2}$$

Can you see that $(K_{1c})^2 = K_{2c} = 1/K_{3c}$?

Different numerical values would be obtained for K_{1c} and K_{2c} even though it is the same reaction.

- As we have already stressed, *all kinds of equilibrium constants are altered by changes of temperature*. In books of data, values for K_c and the other types of K are often given for 25 °C (298 K). You may find K_c written as $K_{c.298}$.

- A *catalyst* (see Ch 12.4) *never has any effect on K_c*. It cannot change the position of equilibrium and alter the yield in a chemical reaction. All a catalyst can do is speed up the rate at which equilibrium is reached. From this you will also realise that the value of K_c has absolutely nothing to do with how quickly equilibrium is reached. A reaction with a high value for K_c may not occur at a measurable rate at room temperature.

- The units of K_c depend on the values of w, x, y and z in equation B. If $(w + x) = (y + z)$ – i.e. if the number of moles of reactants on the left of the equation equals the number of moles of products on the right – then K_c is simply a pure number and has no units, because the units of concentration cancel out.

Units of K_c

To find out the units for any equilibrium constant, put the units of concentration into the equilibrium expression.

e.g. $\quad K_c = \dfrac{[A]}{[B]} \quad \therefore$ Units of K_c are $\dfrac{mol\ dm^{-3}}{mol\ dm^{-3}}$; i.e. K_c has no units in this case

$\quad K_c = \dfrac{[A]}{[B]^2} \quad \therefore$ Units of K_c are $\dfrac{\cancel{mol\ dm^{-3}}}{\cancel{mol\ dm^{-3}} \times mol\ dm^{-3}} = \dfrac{1}{mol\ dm^{-3}}$

$\qquad\qquad\qquad\qquad\qquad\qquad\qquad = mol^{-1}\ dm^3$

$\quad K_c = \dfrac{[B]^2[C]}{[A]} \quad \therefore$ Units of K_c are $\dfrac{(mol\ dm^{-3})^2 \times \cancel{mol\ dm^{-3}}}{\cancel{mol\ dm^{-3}}} = (mol\ dm^{-3})^2$

$\qquad\qquad\qquad\qquad\qquad\qquad\qquad = mol^2\ dm^{-6}$

No, this isn't madness. No one can visualise the inverse square of a mole or the square of a cubic decimetre, but no matter, it all works and gives correct answers which agree with experiment. Anyway, the units in physics calculations often seem even sillier. What's more, examiners seem to be very fond of finding out whether you can handle the units for K_c.

If the number of moles on each side of the equation is not equal, the units of K_c will be some multiple or reciprocal of mol dm^{-3}. So don't be surprised to see units of K_c given as, for example, mol dm^{-3}, mol^2 dm^{-6}, mol^{-1} dm^3, etc.

Remember: To define the equilibrium constant, K_c, for a reaction the concentrations of the products go on the top of the expression.

Some examples with equilibrium expressions

Define K_c for each of the following reactions and state the units in each case.

1. $H_2(g) + CO_2(g) \rightleftharpoons H_2O(g) + CO(g)$

$$K_c = \frac{[products]}{[reactants]} = \frac{[H_2O][CO]}{[H_2][CO_2]}$$

units are $\dfrac{\text{mol dm}^{-3} \times \text{mol dm}^{-3}}{\text{mol dm}^{-3} \times \text{mol dm}^{-3}}$

K_c has no units (it is *dimensionless*).

2. $2H_2S(g) \rightleftharpoons 2H_2(g) + S_2(g)$

$$K_c = \frac{[products]}{[reactants]} = \frac{[H_2]^2[S_2]}{[H_2S]^2}$$

units are $\dfrac{\text{mol dm}^{-3} \times \text{mol dm}^{-3} \times \text{mol dm}^{-3}}{\text{mol dm}^{-3} \times \text{mol dm}^{-3}}$

K_c has units of mol dm^{-3}.

3. $X + Y \rightleftharpoons XY$

$$K_c = \frac{[products]}{[reactants]} = \frac{[XY]}{[Y][X]}$$

units are $\dfrac{\text{mol dm}^{-3}}{\text{mol dm}^{-3} \times \text{mol dm}^{-3}}$

K_c has units of 1/mol dm^{-3}, which is mol^{-1} dm^3.

Le Chatelier's Principle

The usual factors which can have an effect on a chemical system in equilibrium are:

- the concentrations of reactants and products;
- pressure (especially if the reaction is in the gas phase);
- temperature.

Note carefully that, of these, **only a change in temperature actually alters the equilibrium constant**. However, changes in the others can change the relative amounts of reactants and products in the equilibrium mixture.

It would obviously be useful, especially for people in industry who use chemical reactions to make commercially valuable materials, to have some way of working out how reaction conditions can be changed in order to maximise the yield of product.

Factors affecting equilibrium reactions

Concentration changes:

K_c stays the same

Equilibrium shifts.

Pressure changes (in gas reactions):

K_c stays the same

Increase in pressure pushes reaction to side with lower number of gas molecules.

Catalyst:

K_c stays the same

Equilibrium does not shift.

Temperature changes:

K_c alters

For exothermic reactions, K_c decreases with increasing temperature

For endothermic reactions, K_c increases with increasing temperature.

Le Chatelier's Principle (1888) summarises the effects of changing the factors affecting a reacting system.

So, for example, if you increase the pressure in a gaseous reacting system, the equilibrium shifts in such a way as to tend to reduce the pressure – as we shall see very soon in the Haber Process for making ammonia.

Definition: Le Chatelier's principle states that when a system in equilibrium is subjected to a change, the composition of the equilibrium mixture will alter in such a way as to tend to counteract that change.

It's all too easy to say and write things like: 'The system tries to get back to the original state'. This is wrong! A reacting chemical system can't *try* to do anything: unlike us, it's not conscious. (Look up the word 'anthropomorphism' in a dictionary!). Le Chatelier's Principle has been thought by some people to be an aspect of Murphy's Laws. 'If you push a system in equilibrium, it pushes back'. Quite wrong!

Predicting changes in the equilibrium

A.
$$A + B \rightleftharpoons C \qquad K_c = \frac{[C]}{[A][B]}$$

Le Chatelier's Principle has been criticised as being too general. An alternative in two parts is:

- an increase in pressure favours the system which has the smaller volume;

- a rise in temperature favours the system which is formed with absorption of energy.

Adding more A will produce more C to keep K_c constant (i.e. the equilibrium will move to the right hand side).

Removing some C from the reaction vessel will produce more C.

Removing some B will produce more A and B.

B. For an exothermic reaction:

$$L + M \rightleftharpoons N + P \qquad \Delta H \text{ is negative}$$

$$K_c = \frac{[N][P]}{[L][M]}$$

Raising the temperature produces more L and M, K_c decreases.

Lowering the temperature produces more N and P, K_c increases.

C. For an endothermic reaction:

$$X \rightleftharpoons Y + Z \qquad \Delta H \text{ is positive}$$

$$K_c = \frac{[Y][Z]}{[X]}$$

Raising the temperature produces more Y and Z, K_c increases.

Lowering the temperature produces more X, K_c decreases.

1 The equilibrium constant for the formation of ethyl ethanoate from ethanol and ethanoic acid has a value of 4.0. What is the value of K_c for the reverse reaction – i.e. the hydrolysis of ethyl ethanoate?

2 Nitrogen dioxide dimerises to dinitrogen tetraoxide.

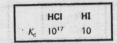

$$2NO_2(g) \rightleftharpoons N_2O_4(g)$$

Write the equilibrium expression for the reaction, and give the units for K_c.

3 The numerical values of K_c at 25 °C for the formation of hydrogen chloride and hydrogen iodide from their elements are:

	HCl	HI
K_c	10^{17}	10

a Write an equation for the formation of HCl from hydrogen and chlorine.
b Write the equilibrium expression for the formation of HI from hydrogen and iodine.
c What are the units of K_c for the formation of HI?
d Which of these two reactions goes virtually all the way to completion? How do you know?

4 Data for the equilibrium reaction between bromine and chlorine are given in the table.

$$Br_2(g) + Cl_2(g) \rightleftharpoons 2BrCl(g)$$

Experiment number	$[Br_2]$	$[Cl_2]$	$[BrCl]$
	(all concentrations are as mol dm^{-3})		
1	0.257	0.164	0.540
2	0.129	0.129	0.338
3	0.440	0.125	0.617
4	0.363	0.235	0.768

a Define K_c for this reaction.
b What are the units of K_c?
c Plot a graph of $[BrCl]^2$ against $[Br_2] \times [Cl_2]$ (on the x-axis) and explain how it shows that your expression for K_c is valid.
d Calculate a value for K_c either from the graph or by using one set of experimental results.

13.3 The Haber and Contact Processes

The Haber Process

This vitally important process is the basis for the manufacture of about 2.5 million tonnes of ammonia each year in Britain alone. Most of the ammonia (about 80%) is used for making fertilisers (see below), but much is also converted into nitric acid. Nitric acid is used for making fertilisers; it is also an essential for making the explosives used in mining and quarrying. A third major use of ammonia is in making Nylon and similar fibres for clothing and fabrics (see Ch 28.1).

It can be argued that the Haber Process, which indirectly helps to produce the food needed to keep hundreds of millions of people alive and powers the mining and quarrying needed by huge construction industries, prolonged the First World War by about two years and cost the lives of millions of soldiers and civilians. The problem of synthesising ammonia economically was solved by Fritz Haber, with the help of Carl Bosch, in 1913 – just before the war started. His process enabled Germany to make her own explosives and fertilisers when imports of nitrates from Chile were made impossible by the British naval blockade. Another example, perhaps, of the two-edged nature of science, and technology? Haber also pioneered the use of chlorine (first used by the Germans at Ypres in 1915) and other war gases, because he believed they would soon bring the war to an end and save countless lives (see Ch 8.2). Haber received the Nobel Prize for Chemistry in 1918, for his work on the synthesis of ammonia.

The starting materials for the modern Haber–Bosch process are normally air (as the source of nitrogen), with natural gas (methane) and steam, which together provide the hydrogen. In the first stages of the process, a series of catalysed reactions and removal of carbon – from the methane – in the form of carbon dioxide result in a mixture of nitrogen and hydrogen in the ratio 1:3. The flow scheme for the rest of the process is shown in *Figure 13.3.1*.

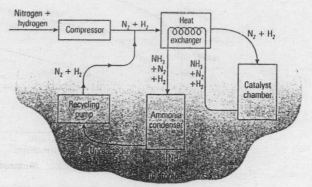

Figure 13.3.1   The Haber Process flow scheme

The equation for the main reaction of the Haber Process is:

$$N_2(g) + 3H_2(g) \underset{\text{Fe catalyst}}{\overset{450\,°C,\ 250\,atm}{\rightleftharpoons}} 2NH_3(g) \qquad \Delta H = -92\,\text{kJ mol}^{-1}$$

Notice that we have included a value for the enthalpy change, ΔH (see Ch 11.2), for the reaction. The value is negative, so energy is given out by this reaction – it is **exothermic**. Note also that the units for ΔH are kJ mol^{-1}, i.e. kilojoules per mole. Look at the equation. Per mole of what? Nitrogen used, or hydrogen used, or ammonia formed? – All different numbers of moles in the equation!

It was precisely to solve this kind of problem that, as we saw in Ch 11.3, the international scientific community decreed that kJ mol^{-1} attached to ΔH at the end of an equation means *enthalpy change per mole of the reaction equation as written.*

The conditions usually used are, as shown in the equation, a moderate temperature (usually about 450 °C), a high pressure (200–250 atmospheres), and an iron catalyst. The catalyst usually contains small amounts of potassium hydroxide and aluminium oxide as *promoters*. Promoters roughen up the surface of the catalyst in a specific way. This increases the strength of adsorption on the surface of the gases involved in the reaction, and hence the efficiency of the catalyst.

Catalysts were fully discussed in Ch 12.4; all we need to remember here is that a catalyst speeds up a particular reaction, but can always be recovered chemically unchanged at the end of the reaction. Also, that a catalyst increases the rate at which equilibrium is reached, but can never alter the composition of the equilibrium mixture.

So – why are these conditions of moderate temperature and high pressure chosen?

Choosing the temperature

Remember, this reaction is **exothermic**. This means, of course, that the reverse reaction – the splitting up of ammonia to form nitrogen and hydrogen – absorbs energy. It is **endothermic**. So what effect does temperature have on the proportion of ammonia in the gas mixture at equilibrium? We can get the answer by using Le Chatelier's Principle.

If we increase the temperature, the position of equilibrium in the system will shift and will tend to counteract the increase in the temperature – i.e. the reaction that tends to cool the system down again will be favoured. An increase in temperature will therefore mean that any ammonia formed will tend to decompose since this chemical change absorbs the extra energy. So there will be less ammonia at equilibrium; and, therefore, to get the most ammonia at equilibrium, a low temperature must be used. For example, at a pressure of 200 atm the equilibrium mixture at 200 °C would contain 88% ammonia, but at 400 °C only 39% (see *Figure 13.3.2*).

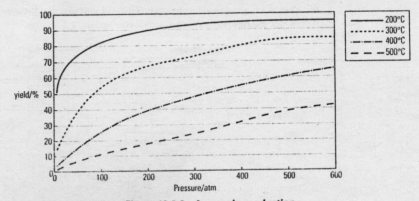

Figure 13.3.2   Ammonia production

If you know some maths, it's precisely this sort of 'optimisation' problem which simultaneous differential equations are excellent at handling. Any kind of engineering, and that includes chemical engineering, involves finding the best compromises between competing factors – e.g. strength and expense for a bridge. So, if an engineer is what you want to be – and there are few occupations more honourable – then make sure of the maths!

But there is one very big snag. Reactions go very much more slowly as the temperature drops (see Ch 12.2), so although you may get a lot of ammonia at 100 °C you have to wait a long time to get it. In the Haber Process, as in all the other industrial exothermic reactions, a compromise or 'optimum' temperature has to be found at which falling yield is best traded off against increased rate. This is also why many industrial processes need a catalyst – to get to equilibrium quicker.

A fact rarely mentioned in textbooks is that, in equilibrium processes such as the Haber Process, *equilibrium is never actually reached*. The reason why is shown in *Figure 13.3.3*.

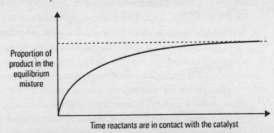

Figure 13.3.3   The usual way in which yield varies with time in a catalysed equilibrium reaction

While the reaction mixture stays in contact with the catalyst, the proportion of product increases to a constant level. When it is level, equilibrium has been reached. The rate of increase in the yield slows down as equilibrium is approached. It's simply not worth while to wait for maximum yield. Again, an optimum length of time has to be found for the gas mixture to be in contact with the catalyst.

Choosing the pressure

The reaction equation for the Haber Process shows that one mole of nitrogen molecules and three moles of hydrogen molecules combine to form two moles of ammonia. So, overall, four moles of gas molecules become two. If this happened in a closed container at constant volume, the pressure of the gas would obviously drop by a half.

We can use Le Chatelier's Principle to predict what will happen when the pressure of the gas mixture is increased. The equilibrium will shift to tend to counteract the increased pressure – so the reaction which leads to the formation of fewer gas molecules, and hence a lowering of pressure, is the one which is favoured; which is why the Haber Process uses a high pressure.

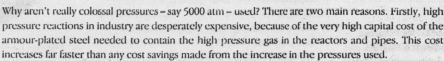

$$N_2 + 3H_2 \rightleftharpoons 2NH_3$$

1 mole   3 moles   2 moles

total 4 moles

An increase in pressure pushes the equilibrium to the right hand side (the smaller number of moles). The value of K_c remains the same.

Why aren't really colossal pressures – say 5000 atm – used? There are two main reasons. Firstly, high pressure reactions in industry are desperately expensive, because of the very high capital cost of the armour-plated steel needed to contain the high pressure gas in the reactors and pipes. This cost increases far faster than any cost savings made from the increase in the pressures used.

The second reason is the high running cost of the compressors needed to keep the gas pressure high. Both reasons come down to cost!

In many ammonia plants, the reaction mixture of hydrogen and nitrogen in the correct 3:1 ratio is passed through the catalyst chamber at, for example, 450 °C and 200 atmospheres pressure. The gas coming out of the chamber (or converter) only contains about 15% ammonia – but this is once again the result of optimisation. A small amount of ammonia produced rapidly gives more ammonia per day than a 100% yield reached very slowly.

When the gases leave the converter, they are cooled in the heat exchanger. Remember, making ammonia is an exothermic process. The catalyst chamber doesn't have to be heated – gases are passed through at the right rate to keep the chamber at the optimum temperature.

In industry, heat means money; the outgoing gases from the converter are used to heat up the incoming gases, and also to provide heat energy for the rest of the plant. The cooled gas is still under great pressure – ammonia's boiling point at 1 atm is −33 °C, but at 200 atm it will liquefy at a much higher temperature. The liquid ammonia is removed and stored. The unchanged nitrogen and hydrogen is recycled, and more hydrogen/nitrogen mixture is fed in to keep the pressure up as the ammonia is removed.

By recycling, up to 98% of the nitrogen/hydrogen mixture can be converted into ammonia. Once started, the plant runs continuously until it has to be closed down for maintenance.

Nitric acid

As we have already said, much ammonia is converted to nitric acid, which is used mainly for making fertilisers, such as ammonium nitrate, and explosives, such as nitroglycerine, TNT, etc.

- An ammonia/air mixture in the approximate ratio 12% NH_3 to 88% air is passed under mild pressure (about 5 atm) through a series of platinum/rhodium gauzes as catalyst at about 900 °C. Conversion rates of over 95% can be achieved.

$$4NH_3(g) + 5O_2(g) \xrightarrow[\text{Pt/Rh catalyst}]{\text{5 atm, 900 °C}} 4NO(g) + 6H_2O(g) \qquad \Delta H = -909 \text{ kJ mol}^{-1}$$

The exothermic reaction keeps the catalyst gauzes at the correct temperature.

- Additional air is added:

$$2NO(g) + O_2(g) \rightleftharpoons 2NO_2(g) \qquad \Delta H = -115 \text{ kJ mol}^{-1}$$

Haber converters

A small proportion of the reacting gases has to constantly be bled off and wasted. This is in order to stop a build-up of argon in the reaction loop. Argon? Yes – argon, because air is used as the source of nitrogen. You can get rid of the oxygen, the carbon dioxide, and the water vapour, but it is too costly to do anything about the 1% of the air which is argon. It goes into the reactor along with the hydrogen and nitrogen and goes round the circuit again and again completely unscathed. Once again, maths can help the operators to work out the optimum rate at which the argon bleed has to be run.

- The gases are cooled:

$$2NO_2(g) \rightleftharpoons N_2O_4(g)$$

- Before it is passed through absorption towers, the gas rises against a stream of falling water:

$$N_2O_4(g) + H_2O(l) \overset{slow}{\rightleftharpoons} HNO_2(aq) + HNO_3(aq)$$

$$4HNO_2(aq) \overset{fast}{\rightleftharpoons} 2NO(g) + N_2O_4(aq) + H_2O(l)$$

The NO and N_2O_4 are eventually reacted and absorbed so that nitric acid at about 60% concentration leaves the tower.

Structure of H_2SO_4

Herman Frasch's method, developed in the 1890s, uses three pipes, one inside the other. Sulphur melts at 119 °C; superheated water, under pressure, goes down the outer pipe; compressed air is blown down the innermost pipe, and a froth of molten sulphur and water comes up the middle pipe. On the surface the sulphur solidifies; it is insoluble in water. The sulphur is 99.5% pure.

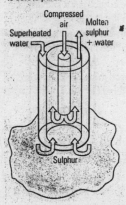

The Contact Process for sulphuric acid

Sulphuric acid is one of the most important products of all; world production is well over 1×10^8 tonnes per annum. About 30% of this amount is used in making fertilisers, including 'super-phosphate' ($Ca(H_2PO_4)_2$) and 'sulpham' (($NH_4)_2SO_4$); about 25% is used in making plastics and fibres; about 15% for making paints and pigments; and 10% for making soaps and detergents (*Figure 13.3.4*).

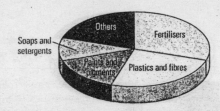

Figure 13.3.4   Uses of sulphuric acid

Sulphur is obtained from underground deposits by melting it with super-heated water under pressure (the Frasch process) and by desulphurisation of crude oil and natural gas. Large amounts of sulphur dioxide are obtained from roasting metal sulphide ores, in order to convert them to oxides before extracting the metal.

e.g. $\quad 2ZnS(s) + 3O_2(g) \overset{heat}{\longrightarrow} 2ZnO(s) + 2SO_2(g)$

If sulphur dioxide is not available, sulphur is first burned in excess dry air:

$$S(s) + O_2(g) \overset{ignite}{\longrightarrow} SO_2(g) \qquad \Delta H = -297 \, kJ \, mol^{-1}$$

The gas stream is cooled to about 420 °C and passed into the converter to produce sulphur trioxide. The energy released is used to raise steam. The catalyst is vanadium(V) oxide, V_2O_5, on silica supports. There are three or four catalyst beds; conversion to sulphur trioxide is better than 98%. Heat energy is removed by heat exchangers to raise steam.

$$2SO_2(g) + O_2(g) \overset{V_2O_5, \approx 420\,°C}{\underset{2\,atm}{\rightleftharpoons}} 2SO_3(g) \qquad \Delta H = -98 \, kJ \, mol^{-1}$$

The sulphur trioxide cannot be absorbed directly in water, as the reaction is violent and produces a mist of the acid. Instead, it is absorbed in concentrated sulphuric acid, with water being added at a rate which keeps the concentration of the acid at about 98%.

$$SO_3(g) + H_2SO_4(l) \longrightarrow \underset{\text{'oleum'}}{H_2S_2O_7(l)}$$

$$H_2S_2O_7(l) + H_2O(l) \longrightarrow 2H_2SO_4(l)$$

The temperature must be kept below 100 °C. Energy released is, as always, never wasted. Sale of the excess energy helps keep the price of sulphuric acid low.

Overall: $\quad SO_3(g) + H_2O(l) \longrightarrow H_2SO_4(l) \qquad \Delta H = -130 \, kJ \, mol^{-1}$

Application of Le Chatelier's Principle to the main step (from SO_2 to SO_3) suggests that the forward reaction is helped by:

- high pressure, because 3 moles of gas are converted to 2 moles;
- excess oxygen;
- keeping the temperature low by removing energy;
- removing sulphur trioxide from the reaction mixture.

However, as with the Haber Process, an *optimum temperature* has to be used: too low a temperature makes the reaction too slow, too high a temperature lowers the yield (*Figure 13.3.5*).

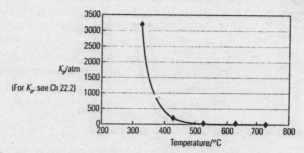

Figure 13.3.5   Equilibrium constant for sulphuric acid production

High pressure means high expenditure on armour-plated equipment. Here, a yield of nearly 100% is obtained without using more than a few atmospheres pressure to move the gas through the plant, so it is not worth using high pressures.

Excess oxygen is used, and sulphur trioxide is removed.

Note, therefore, that the conditions indicated by Le Chatelier's Principle *are not necessarily used in an industrial process*. Economic factors often decide.

The flow scheme for the Contact Process is shown in *Figure 13.3.6*.

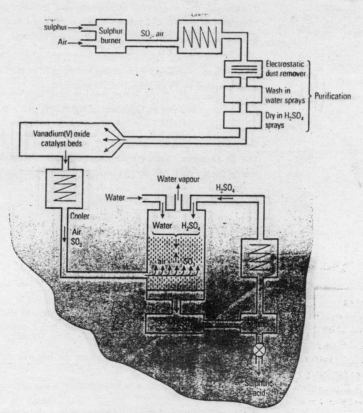

Figure 13.3.6   The Contact Process flow scheme

SELF-ASSESSMENT QUESTIONS 13.3

1   Zinc blende, ZnS, can be used to produce sulphuric acid as a by-product in the reduction of the ore. The zinc blende ore is roasted in air to produce sulphur dioxide and zinc oxide.

   a Write a balanced equation for the roasting of zinc blende in air.

   b The sulphur dioxide is then reacted with oxygen to form sulphur trioxide in a very exothermic reaction. Write an equation for the formation of sulphur trioxide in this way.

   c A catalyst is used in the reaction. Name the catalyst and briefly explain how it works.

   d Explain why the conversion of sulphur dioxide to sulphur trioxide is favoured by high pressures.

   e Why, in practice, do most industrial plants for making sulphuric acid operate close to atmospheric pressure?

2   Ammonia is produced from the reaction of hydrogen and nitrogen at a pressure of 200 atmospheres and 450 °C. The usual source of hydrogen nowadays is from methane.

   The nitrogen is obtained from air, by removing oxygen, carbon dioxide, water vapour and dust impurities.

   a Explain why the hydrogen and nitrogen composition in the reaction mixture is usually around 74% hydrogen and 24.7% nitrogen.

   b This reaction gas mixture also contains a small proportion of argon. Where has the argon come from?

   c Give one reason why the reaction is carried out at high pressure, and one disadvantage of doing so.

   d Although the reaction is exothermic and favoured by low temperature, the reaction is carried out at 450 °C. Explain why it is considered worthwhile to do this.

   e How is ammonia removed from the reacted mixture?

3   The key reactions used for obtaining hydrogen from methane and steam for the Haber Process are shown below. Both reactions use a catalyst.

$$CH_4(g) + H_2O(g) \rightleftharpoons CO(g) + 3H_2(g) \qquad \Delta H = +210 \, kJ \, mol^{-1}$$

$$CO(g) + H_2O(g) \rightleftharpoons CO_2(g) + H_2(g) \qquad \Delta H = -42 \, kJ \, mol^{-1}$$

Use Le Chatelier's Principle to suggest which conditions of pressure and temperature you would use for each reaction, in order to maximise the yield of products at equilibrium.

13.4   Acid–base theories

Definition: An acid is a substance which can give a proton (i.e an H^+ ion) to a base. A base is a substance (e.g. a metal oxide or hydroxide) which can receive a proton from an acid.

We have already used the terms 'acid', 'base' and 'alkali'; and you will know quite a lot about acids and alkalis. In this section we shall show how the behaviour of acids, bases and alkalis can be explained.

   J N Brønsted and T M Lowry in 1923 were the first to state that an **acid** is a substance which can give a proton to a base.

   An **alkali** is simply a base which is soluble in water to give $OH^\ominus(aq)$ ions, e.g. sodium hydroxide, NaOH.

   In order to be able to discuss acids and bases fully, it is necessary to learn some vocabulary.

The definitions of acid and base mean that whenever one substance is acting as an acid, another must be acting as a base. The two things can never be separated.

This should remind you of **redox** reactions (see Ch 4.3). If something is being oxidised by loss of electrons, something else must be being reduced by electron gain. The two processes can never be separated.

- Redox reactions involve the loss and gain of electrons.
- Acid–base reactions involve the loss and gain of protons.

Conjugate acids and bases

The word 'conjugate' comes from the Latin *coniungo*, to join together.

The loss or gain of a proton is all that separates an acid from a base. The species formed when an acid loses a proton *must be a* base, because under different conditions it can gain a proton once more and re-form the acid. So every acid has its **conjugate base**, and every base has its **conjugate acid**.

For example, the conjugate base of hydrochloric acid is the chloride ion; the conjugate acid of the base ammonia is the ammonium ion.

$$HCl(aq) \rightleftharpoons H^{\oplus}(aq) + Cl^{\ominus}(aq)$$

Hydrochloric acid — ACID   Chloride ion — CONJUGATE BASE

$$NH_3(aq) + H^{\oplus}(aq) \rightleftharpoons NH_4^{\oplus}(aq)$$

Ammonia — BASE   Ammonium ion — CONJUGATE ACID

> **Recognising conjugate acid–base pairs**
> Acid $\longrightarrow H^{\oplus} +$ Conjugate base
> (the acid must lose a proton)
> Base $+ H^{\oplus} \longrightarrow$ Conjugate acid
> (the base must gain a proton)

Strong and weak acids

Definition: A strong acid is one which is fully ionised when in dilute solution.

A typical **strong acid** is hydrochloric acid. In concentrated solution, the equilibrium between the *undissociated* (i.e. un-ionised) hydrogen chloride molecules and the chloride and hydrated hydrogen ions is obvious; concentrated hydrochloric acid gives off fumes of hydrogen chloride gas which can very easily be smelt! However, at a concentration of 0.1 mol dm^{-3}, hydrochloric acid is almost completely dissociated into ions with relatively few HCl molecules left in the equilibrium mixture.

The full equation can be written as:

$$HCl(g) + water \rightleftharpoons HCl(aq) \rightleftharpoons H^{\oplus}(aq) + Cl^{\ominus}(aq)$$

If more water is added, the equilibrium is shifted towards the right hand side. If there is less water, so that the concentration of the acid is greater, the equilibrium is shifted towards the left, and more molecules of the gas are formed.

> Another way of writing the right hand part of the equation for the equilibrium is:
>
> $$HCl(aq) + H_2O(l) \rightleftharpoons H_3O^{\oplus}(aq) + Cl^{\ominus}(aq)$$
>
> The $H_3O^{\oplus}$ ion is known as the **oxonium ion**. (It may also, in some books, be called the **hydronium ion** or the **hydroxonium ion**.) It has been shown that, in fact, the proton in any aqueous solution is shared over a cluster of water molecules loosely held together by hydrogen bonding, where the cluster is constantly varying in size. The most frequent size of the cluster contains four water molecules, so that the most common form of the hydrated proton is $H_9O_4^{\oplus}$. We shall write $H_3O^{\oplus}$ only when we refer specifically to the oxonium ion. Otherwise, for the hydrated proton in solution we shall write $H^{\oplus}(aq)$, to avoid suggesting that it is linked only to one water molecule.

> **Definition:** A weak acid is an acid which is only partly ionised, even in very dilute solution.

A typical **weak acid** is ethanoic acid (acetic acid), the acid found in vinegar. Other naturally occurring acids, such as citric, tartaric and lactic acids are also weak.

$$CH_3COOH(aq) \rightleftharpoons CH_3COO^{\ominus}(aq) + H^{\oplus}(aq)$$
Ethanoic acid   Ethanoate ion

A strong acid can easily lose protons, so its conjugate base is unwilling to accept a proton. **A strong acid has a weak conjugate base** (*Figure 13.4.1*).

Increasing acid strength ↑	Strong acid	Weak conjugate base	Increasing base strength
	H_2SO_4	HSO_4^-	
	HCl	Cl^-	
	H_3O^+	H_2O	
	HCOOH	$HCOO^-$	
	CH_3COOH	CH_3COO^-	
	H_2CO_3	HCO_3^-	
	H_2O	OH^-	
	Weak acid	**Strong conjugate base** ↓	

Figure 13.4.1

Conversely, the conjugate base of a weak acid such as ethanoic acid is very ready to accept a proton. **A weak acid has a strong conjugate base**.

These ideas of conjugate acids and bases, and strong and weak acids, are discussed more fully in Ch 23, and used to explain a wide range of chemical observations.

Basicity of acids

An acid such as hydrochloric acid, HCl, or nitric acid, $HNO_3(aq)$, can lose only one proton per molecule. Such acids are said to be *monobasic*. Sulphuric acid, H_2SO_4, can lose two protons per molecule. It is *dibasic*.

Acids such as sulphuric acid, which contain more than one removable proton per molecule, can ionise in more than one stage; e.g. for sulphuric acid:

$$H_2SO_4(aq) \rightleftharpoons H^{\oplus}(aq) + HSO_4^{\ominus}(aq)$$
Sulphuric acid   Hydrogensulphate ion

$$HSO_4^{\ominus}(aq) \rightleftharpoons H^{\oplus}(aq) + SO_4^{2\ominus}(aq)$$
Sulphate ion

Not all the hydrogen atoms in an acid molecule may be capable of being lost as protons. Ethanoic acid (see above) is one example. The formula is usually written as CH_3COOH. If we write out its full structure, it is:

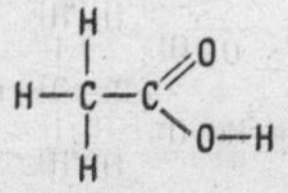

```
      H       O
      |      //
  H — C — C
      |      \
      H       O — H
```

The strong C—H bonds in the ethanoic acid molecule cannot ionise; only the weaker O—H bond (weakened even more by the polar C=O bond) can break, to release a proton.

Acid–base chemistry in non-aqueous solvents

Nearly all of the acid–base reactions which we observe, and which are used in industry, occur in aqueous solution. It was once thought that acid–base reactions could occur only in water. We now realise that the Brønsted–Lowry definitions of acid and base can apply equally well in other solvents.

The ionisation of familiar acids in non-aqueous solvents can sometimes be surprising. For example, in liquid hydrogen fluoride, nitric acid behaves as a base!

$$HNO_3 + 2HF \rightleftharpoons H_2NO_3^{\oplus} + HF_2^{\ominus}$$

When hydrogen chloride is dissolved in pure ethanoic acid, the ethanoic acid also behaves as a base:

$$HCl + CH_3COOH \rightleftharpoons CH_3C(OH)_2^{\oplus} + Cl^{\ominus}$$

Reactions in non-aqueous solvents are being increasingly studied by research teams and used in industry, but the only solvent for acid–base reactions which we shall consider here is water.

It should also be mentioned that acid–base reactions might occur when no solvent is present. For example, when ammonia gas reacts with hydrogen chloride gas, the ammonia molecule accepts a proton; white, solid ammonium chloride is formed.

$$NH_3(g) + HCl(g) \rightleftharpoons NH_4^{\oplus}Cl^{\ominus}(s)$$

Amphiprotic substances

Some substances can act, depending on the circumstances, as either acids or bases. For example, the hydrogensulphate ion, $HSO_4^{\ominus}$, can add a proton to become sulphuric acid or lose a proton to become sulphate, $SO_4^{2\ominus}$.

$$H_2SO_4 \quad \xleftarrow[\text{add proton}]{\text{add acid}} \quad HSO_4^{\ominus} \quad \xrightarrow[\text{lose proton}]{\text{add alkali}} \quad SO_4^{2\ominus}$$

Such substances are **amphiprotic**: they are capable of both donating and accepting protons.

The most important amphiprotic substance is water. A water molecule can accept a proton, by dative bond formation (see Ch 2.4), and so act as a base; or it can lose a proton, and so act as an acid. The situation can be summed up as follows (remember, $H_3O^{\oplus}$ ions are actually rare; here, they represent the hydrated protons, $H^{\oplus}(aq)$).

$$H_3O^{\oplus}(aq) \quad \xrightleftharpoons[+H^+]{-H^+} \quad H_2O(l) \quad \xrightleftharpoons[+H^+]{-H^+} \quad OH^{\ominus}(aq)$$

In pure water, there is a very small but important amount of self-ionisation. Some water molecules donate hydrogen ions to others; i.e. they act as acids, and the others act as bases.

$$2H_2O(l) \rightleftharpoons H_3O^{\oplus}(aq) + OH^{\ominus}(aq)$$

The concentrations of hydrated protons and hydroxide ions are equal; pure water is **neutral**. If a proton donor such as hydrogen chloride is present in water, more hydrated protons are formed; their concentration is now greater than the concentration of hydroxide ions, so the resulting solution is *acidic*. If a proton acceptor such as ammonia is present, it accepts protons from water molecules, so that now the concentration of hydroxide ions is the greater, and the resulting solution is *alkaline* (Figure 13.4.2).

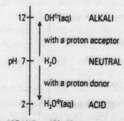

Figure 13.4.2 pH of aqueous solutions

The acid–base behaviour of water is discussed in greater depth in Ch 23.

Typical properties of acids

Unreactive metals such as copper, silver and gold do not normally react with acids. Nitric acid reacts as an oxidising agent rather than as an acid and never releases hydrogen, but when other common acids react with metals, hydrogen gas is given off.

All the common *pure* acids consist of covalent molecules. Hydrogen chloride, for example, is a gas; and in 100% sulphuric acid there are virtually no ions present. Clean dry magnesium ribbon can be dropped into pure sulphuric acid without any reaction being seen. Magnesium added to a mixture of the acid with only a very little water – and ordinary concentrated sulphuric acid is about 98% acid and 2% water – will react rapidly.

Why does this happen? As we saw above, *whenever an acid is added to water, the water acts as a base*. Acid molecules give up protons to water molecules and form hydrated protons (*Figure 13.4.3*).

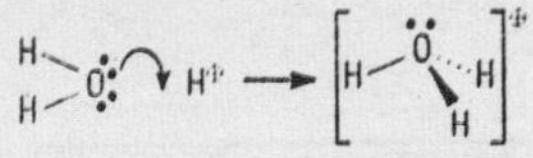

Figure 13.4.3   The hydrated proton

When a sufficiently reactive metal is put into the acid solution, the metal atoms give up electrons to the hydrated protons and hydrogen gas is formed.

e.g. $$Mg(s) + 2H^{\oplus}(aq) \longrightarrow Mg^{2\oplus}(aq) + H_2(g)$$

Essentially:

$$2H^{\oplus}(aq) + 2e^{\ominus} \longrightarrow H_2(g)$$

It is the hydrated proton, $H^{\oplus}(aq)$, which is responsible for all the typical reactions of acids.

Remember:
Acids:
- **change the colour of many indicators** (see Ch 23.2);
- **release hydrogen when in contact with reactive metals;**
- **form salts when reacted with metal oxides, hydroxides (including alkalis) and other bases;**
- **release carbon dioxide when reacted with metal carbonates and hydrogencarbonates.**

Neutralisation

As stated earlier, water can act as both an acid and a base – the water molecule can both give and accept a proton.

$$2H_2O(l) \rightleftharpoons H_3O^{\oplus}(aq) + OH^{\ominus}(aq)$$

In the same way as the hydrated proton, $H^{\oplus}(aq)$, is present in all acid solutions, so the hydroxide ion $OH^{\ominus}(aq)$ is present in all alkaline solutions. Even though $OH^{\ominus}$ ions are present in, for example, solid sodium hydroxide, $Na^{\oplus}OH^{\ominus}$, it cannot act as an alkali in the absence of water. The water releases hydroxide ions from the crystal lattice. Some substances, e.g. ammonia, actually react with water to produce hydroxide ions, so that their solutions are alkaline.

$$NH_3(g) + H_2O(l) \rightleftharpoons NH_4^{\oplus}(aq) + OH^{\ominus}(aq)$$

It is obvious that water, itself neutral, occupies a crucial position between acids and alkalis.

Neutralisation occurs when an acid and base react together to form a salt and water and nothing else. The base is often an alkali such as sodium hydroxide. For sodium hydroxide reacting with hydrochloric acid:

$$Na^{\oplus}OH^{\ominus}(aq) + H^{\oplus}Cl^{\ominus}(aq) \longrightarrow Na^{\oplus}Cl^{\ominus}(aq) + H_2O(l)$$

This equation shows that the sodium and chloride ions are not affected – they simply stay in solution. Such **spectator ions** are met in many reactions involving ionic materials in aqueous solution. In *any* reaction between an acid and an alkali, all that is happening is that hydrated protons from the acid are reacting with hydroxide ions from the alkali. So the equation which summarises *all* neutralisations in aqueous solution is:

$$H^{\oplus}(aq) + OH^{-}(aq) \longrightarrow H_2O(l)$$

The salt which is formed consists of the spectator ions which are left in solution. The water has to be evaporated if the solid salt is wanted.

Salts

Salts are all derived from a metal (or ammonium) ion and an ion from an acid. The metal ion may come from a base, an alkali, a carbonate or the metal itself.

The metal will appear first in the name and the metal ion will always carry a positive charge.

The acid part of the name will appear second and the corresponding ion will be negatively charged.

Examples:

Sodium chloride: $Na^{\oplus}Cl^{\ominus}$

Calcium sulphate: $Ca^{2\oplus}SO_4^{2\ominus}$

Potassium nitrate: $K^{\oplus}NO_3^{\ominus}$

Sodium ethanoate: $CH_3COO^{\ominus}Na^{\oplus}$

Sodium phosphate: $Na_3^{\oplus}PO_4^{3\ominus}$

Notice that when one mole of $OH^{\ominus}$ ions react with one mole of $H^{\oplus}(aq)$ ions, one mole of O—H bonds in protonated water molecules has to be broken, and one mole of new O—H bonds in new water molecules is formed (*Figure 13.4.4*).

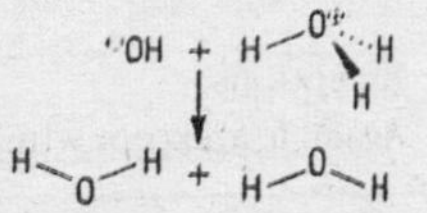

Figure 13.4.4   Neutralisation

Neutralisation is always exothermic (see Ch 11.1). Formation of a mole of water by reaction of dilute solutions of a *strong* acid and a *strong* alkali always gives the same amount of heat energy: $-57 \, kJ \, mol^{-1}$.

Neutralisation of sulphuric acid, H_2SO_4, will require two moles of $OH^{\ominus}$ for each mole of acid. This will give the **normal salt**, Na_2SO_4. If half as much alkali is used, the **acid salt**, $NaHSO_4$, is obtained. An 'acid salt' is one which can still provide a proton. Can you see how phosphoric(V) acid, H_3PO_4, can form *two* acid salts as well as a normal salt?

The base which reacts with an acid may be an insoluble oxide or hydroxide, e.g. zinc oxide. When zinc oxide reacts with sulphuric acid the equation can be written:

$$Zn^{2\oplus}O^{2\ominus}(s) + 2H^{\oplus}(aq) + SO_4^{2\ominus}(aq) \longrightarrow Zn^{2\oplus}SO_4^{2\ominus}(aq) + H_2O(l)$$

The zinc and sulphate ions are not changed; they are *spectator ions*. So the equation which summarises *all* reactions between metal oxides and acids is:

$$O^{2\ominus} + 2H^{\oplus}(aq) \longrightarrow H_2O(l)$$

For all metal hydroxides, the equation is the same as for neutralisation:

$$OH^{\ominus} + H^{\oplus}(aq) \longrightarrow H_2O(l)$$

For all metal carbonates, the equation is:

$$CO_3^{2\ominus} + 2H^{\oplus}(aq) \longrightarrow H_2O(l) + CO_2(g)$$

More about acids and bases, the self-ionisation of water, pH and buffer solutions, can be found in Ch 23.

A great deal of simple chemistry of acids can be summed up in four reactions:

- ACID + METAL ⟶ SALT + HYDROGEN

e.g. for a metal which forms a dipositive ion:

$$2H^{\oplus}(aq) + M(s) \longrightarrow M^{2\oplus}(aq) + H_2(g)$$

Exceptions:

Nitric acid never gives hydrogen.

Unreactive metals (Cu, Ag, Au) cannot displace hydrogen from an acid.

The reaction soon stops if the salt is insoluble and coats the metal.

- ACID + BASE ⟶ SALT + WATER .

   insoluble metal
   oxide or metal
   hydroxide

e.g. for an oxide:

$$2H^{\oplus}(aq) + O^{2\ominus} \longrightarrow H_2O(l)$$

Exceptions:

Very few, except when an insoluble salt is formed and hence coats the surface of the oxide or hydroxide.

- ACID + ALKALI ⟶ SALT + WATER

   soluble
   metal hydroxide

$$H^{\oplus}(aq) + OH^{\ominus}(aq) \longrightarrow H_2O(l)$$

- ACID + METAL CARBONATE ⟶ SALT + WATER + CARBON DIOXIDE

$$2H^{\oplus}(aq) + CO_3^{2\ominus} \longrightarrow H_2O(l) + CO_2(g)$$

Exceptions:

Again, few except when the salt is insoluble.

SELF-ASSESSMENT
QUESTIONS
13.4

1 Hydrogen chloride dissolves in water to form a strong acid. The formation can be pictured as the lone pair of electrons from the oxygen atom being used to form a bond with the hydrogen atom in the hydrogen chloride:

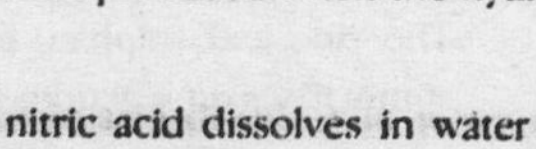

Hydrogen sulphide has a structure similar to water.

a Draw a dot and cross diagram for hydrogen sulphide.

b Which of the two elements, oxygen and sulphur, is the more electronegative?

c Show how hydrogen chloride might also dissolve in liquid hydrogen sulphide to form an acid.

d Will this acid be stronger than the acid produced when the hydrogen chloride dissolves in water?

2 a Write an equation to show how nitric acid dissolves in water to form a strongly acidic solution.

   b Which of the following acids is monobasic? HBr, H_2SO_4, CH_3COOH, H_3PO_4.

3 In the following reactions identify the acid in the reactants, and then identify the conjugate base on the product side of the reaction.

   a $CH_3COOH + H_2O \longrightarrow CH_3COO^{\ominus} + H_3O^{\oplus}$

   b $C_6H_5OH + H_2O \longrightarrow C_6H_5O^{\ominus} + H_3O^{\oplus}$

   c $OH^{\ominus} + H_3O^{\oplus} \longrightarrow 2H_2O$

Summary

For an *endothermic* reaction in dynamic equilibrium

$$\text{heat} + A + B \rightleftharpoons 2C + D$$

Adding A, equilibrium moves $\longrightarrow$
  B $\longrightarrow$ } concentration changes to keep K_c constant
  C $\longleftarrow$
  D $\longleftarrow$

$$K_c = \frac{[C]^2[D]}{[A][B]} \, \text{mol dm}^{-3}$$

Raise temperature, equilibrium moves $\longrightarrow$ } value of K_c changes
Lower temperature, equilibrium moves $\longleftarrow$

The Haber Process $N_2(g) + 3H_2(g) \underset{\text{Fe cat}}{\overset{450°C, 250\,atm,}{\rightleftharpoons}} 2NH_3(g)$ $\Delta H = -92 \, \text{kJ mol}^{-1}$

  Add N_2, equilibrium moves $\longrightarrow$
    H_2 $\longrightarrow$
    NH_3 $\longleftarrow$
    Raise temperature $\longleftarrow$ (because this reaction is *exothermic*)
    Lower temperature $\longrightarrow$ (but slower)
    Increase pressure $\longrightarrow$ (fewer gas molecules)

The Contact Process $S(g) + O_2(g) \longrightarrow SO_2(g)$

$$2SO_2(g) + O_2(g) \underset{420°C}{\overset{V_2O_5, 2\,atm,}{\rightleftharpoons}} 2SO_3(g)$$

$$SO_3(g) + H_2O(l) \longrightarrow H_2SO_4(aq) \quad \text{(but not done directly)}$$

Acids and bases Acids give protons; bases gain protons

$$\underset{\text{Acid}}{HA} \longrightarrow H^{\oplus} + \underset{\text{Conjugate base}}{A^{\ominus}}$$

A strong acid is fully ionised in sufficient water
A strong acid forms a weak conjugate base

Reactions of acid:
  Acid + Alkali $\longrightarrow$ Salt + water
  Acid + Base $\longrightarrow$ Salt + water
  Acid + Metal $\longrightarrow$ Salt + hydrogen
  Acid + Carbonate $\longrightarrow$ Salt + water + carbon dioxide

SUMMARY 13

In this chapter the following new ideas have been covered. *Have you understood and learned them?* Could you explain them clearly to a friend? The page references will help.

1 Sulphuric acid is manufactured by the Contact Process. The principal reaction involves the oxidation of sulphur dioxide.

   a Write an equation for the oxidation of sulphur dioxide. What are the oxidation states of sulphur before the reaction and after the reaction?

   b Typical conditions for the reaction are 450 °C and 2 atm pressure. Comment on the use of these conditions, assuming that the oxidation process is at equilibrium.

   c Sulphuric acid is produced by dissolving the product from the oxidation reaction in concentrated sulphuric acid, to form oleum, $H_2S_2O_7$, and diluting this mixture with water. Write an equation for the formation of oleum.

   d Write an equation for the reaction of oleum with water.

   e Why is it necessary to use concentrated sulphuric acid in the first stage of the dilution process?

2 Write the equilibrium expression, K_c, for each of the following equilibrium reactions. Make sure you clearly state the units of K_c in each case.

   a $2Fe^{3+}(aq) + 2I^-(aq) \rightleftharpoons 2Fe^{2+}(aq) + I_2(aq)$

   b $CCl_2O(g) \rightleftharpoons CO(g) + Cl_2(g)$

   c $H_2(g) + I_2(g) \rightleftharpoons 2HI(g)$

   d $2HI(g) \rightleftharpoons H_2(g) + I_2(g)$

   e $HI(g) \rightleftharpoons \frac{1}{2}H_2(g) + \frac{1}{2}I_2(g)$

   f Show how the values of K_c in each of the last three examples are related.

3 a State Le Chatelier's Principle.

   b The two polymorphs (allotropes) of carbon, diamond and graphite, are in equilibrium.

$$C(graphite) \rightleftharpoons C(diamond) \quad \Delta H = +1.88 \, kJ \, mol^{-1}$$

   What is meant by a polymorph?

   c Use Le Chatelier's Principle to decide what will happen to the graphite/diamond equilibrium as the temperature is raised.

4 Outline the Haber Process for the production of ammonia. In your outline you should include discussion of the raw materials used in the process, the conditions needed for the reaction, and an explanation of the conditions used, in the light of Le Chatelier's Principle.

5 Concentrated sulphuric and nitric acids, when they are mixed, react in an acid–base reaction according to the equation:

$$H_2SO_4 + HNO_3 \rightleftharpoons HSO_4^- + H_2NO_3^+$$

   a Which species is the acid, and what is the formula of the conjugate base of this acid?

   b Write an equation to show what happens when hydrogen iodide dissolves in water.

   c What is acting as the base in the reaction in b? What is the formula of the conjugate acid?

   d Write an expression to define K_c for the reaction in b.

Do you remember ...?

- How to draw 'dot-and-cross' diagrams for simple covalent molecules. (Ch 2.4)
- How **covalent bonds** are formed and how the numbers of pairs of electrons lead to the shapes of molecules. (Ch 2.6)
- How **electronegative atoms** affect the electron distribution in a covalent bond and give rise to **bond polarity**. (Ch 2.5)
- How to calculate the **empirical formula** of a compound given mass or percentage composition, and how the **molecular formula** can then be deduced from the relative molecular mass of the molecule. (Ch 9.4)
- The definition of **bond enthalpy** terms. (Ch 11.5)
- How the **mass spectrometer** is used to find relative molecular masses. (Ch 9.1)
- The effect of **hydrogen bonds** on the properties of molecules. (Ch 2.9)
- The structures of **diamond** and **graphite**. (Ch 3.2)

14.1 Formulae in organic chemistry
14.2 Functional groups, homologous series and frameworks
14.3 Naming organic molecules
14.4 Isomerism
14.5 Physical properties of organic molecules
14.6 Types of organic reactions
14.7 Introduction to reaction mechanisms
14.8 Writing organic equations

14
introduction to organic chemistry

Carbon is unique. Because of its electronegativity and small atomic size, carbon atoms can join with each other to form *short, strong* covalent bonds (remember, *long bonds are weak bonds, short bonds are strong bonds – Table 14.1.1*).

	nm	kJ mol⁻¹
C—H	0.108	+413
C—C	0.154	+347
Si—H	0.148	+318
Si—Si	0.235	+226
B—B	0.159	+297
B—H	0.121	+334
N—N	0.145	+158
N—H	0.101	+391

Table 14.1.1 Bond lengths and enthalpies

Why are there so many organic molecules?

1 Strong bonds
2 Four bonds
3 Catenation
4 Variety of bonds
5 Isomerism (Ch 14.4)

But, because they can each form four covalent bonds, carbon atoms can keep on joining. This is called **catenation**. We have met this already in the structures of diamond and graphite (Ch 3.2). Carbon can join with itself to form all kinds of chains, rings, and three-dimensional structures; the covalent bonds between carbon atoms can be single, double or triple.

Add to this the ability of carbon to form covalent bonds with a large number of non-metals – and even some metals – and we have the possibility of an almost infinite number of compounds. These compounds are based on frameworks or skeletons of carbon atoms which can be arranged in a large variety of ways. They range from the very small (e.g. methane) to the almost unimaginably large (e.g. DNA).

Several million compounds of carbon are already known, and hundreds more are discovered or synthesised every day. There are more known compounds of carbon than there are of all the other elements put together.

As far as we know, all forms of life in the Universe are quite likely to be based on carbon. This is simply because only carbon atoms are able to join together with covalent bonds strong enough to form the skeletons of molecules complicated enough for life to exist. Also, reactions involving these molecules have high activation energies (Ch 12.1) and therefore they do not react easily with the oxygen in air. Without proteins, without nucleic acids, even without simple sugar molecules such as glucose, no life would be possible on Earth.

Many of the materials we take for granted are based on carbon. For example, the plastics and polymers found all around the home, in shops and the workplace, and the natural or synthetic fibres used for much of our clothing, carpets and curtains, rely on the unique ability of carbon to bind to itself to form the skeletons of large, stable molecules.

It used to be thought that there was a complete and fundamental difference between **organic chemistry** and **inorganic chemistry**: that organic compounds could only be made through the 'vital force' of a living organism. For example, plants make many carbohydrates, including sugar. And ordinary alcohol is made by fermentation, in which yeasts – living organisms – act on plant carbohydrates. But in 1828 Friedrich Wöhler performed a crucial experiment, the result of which so astonished him that he repeated it many times to make absolutely sure of his results before he dared to publish them. He carefully evaporated a solution of ammonium cyanate, NH_4CNO, clearly an inorganic material, and obtained solid urea, $CO(NH_2)_2$, which had up until then only been obtained from mammalian urine. Wöhler saw that this overturned the 'vital force' theory and wrote to the great Swedish chemist, Jons Jacob Berzelius, 'I must tell you that I can prepare urea without requiring a kidney or an animal, either man or dog'.

14.1 Formulae in organic chemistry

The simplest formula for a compound is called its **empirical formula**. This shows the elements present in the compound and the simplest ratio of the atoms of those elements in that compound. For example, ethane, C_2H_6, has an empirical formula CH_3.

The empirical formula can be worked out by **qualitative analysis**, to discover which elements are present, followed by **quantitative analysis**, to find out **how much** of each element is present.

For example, if a sample of compound X is burned in oxygen, and carbon dioxide and water are both formed, then X must have contained both carbon and hydrogen. The amount of carbon dioxide and water formed will show how much carbon and hydrogen are present. Simple tests can find other elements such as the halogens, nitrogen, sulphur and even oxygen. In recent years, machines have been developed which can perform such elemental analyses quickly and accurately.

Once the mass or percentage of each of the elements which are present in the compound is found, the empirical formula can be calculated, as long as the relative atomic masses of the elements are known (Ch 9.4).

Calculating the empirical formula of an organic compound

The empirical formula of a compound can be calculated from either the masses of the elements in the compound or from the percentage composition of the compound.

For example, compound X contains 40.0% carbon, 6.7% hydrogen and 53.3% oxygen. (Relative atomic masses: C, 12.0; H, 1.0; O, 16.0) The calculation of the empirical formula can be set out like this:

Data		Element		
		C	**H**	**O**
%		40.0	6.7	53.3
Relative atomic mass		12.0	1.0	16.0
Moles $= \dfrac{\% \text{ or mass}}{A_r}$		$\dfrac{40.0}{12.0}$	$\dfrac{6.7}{1.0}$	$\dfrac{53.3}{16.0}$
		= 3.33	= 6.7	= 3.33
Ratio Ratio of moles (Divide by the smallest value)		$\dfrac{3.33}{3.33}$	$\dfrac{6.7}{3.33}$	$\dfrac{3.33}{3.33}$
		= 1	= 2	= 1
		C	H_2	O

∴ empirical formula is CH_2O

The **molecular formula** of a compound shows the actual number of each kind of atom present in a molecule of the compound. In order to work out the molecular formula, we need both the empirical formula and the relative molecular mass of the compound.

The molecular mass (M_r) can be found using a mass spectrometer. The molecular ion peak (the peak furthest to the right) in the mass spectrum gives the M_r directly (*Figure 14.1.1*) (see Ch 9.1).

Because organic molecules usually contain a large number of atoms, the molecule breaks up when bombarded by the high-speed electrons. A large number of different fragments are produced. Consequently, there are a lot of lines on most mass spectra. The relative molecular mass of the parent molecule, the molecular ion peak, is the line at the highest m/e ratio, and is marked X in *Figure 14.1.1*.

Once the M_r of the compound has been found, the molecular formula can be worked out. The molecular formula of a compound must be *either* the same as the empirical formula *or* some simple multiple of it.

To return to our compound X. We found from the analysis data that its empirical formula was CH_2O. Suppose we find that the relative molecular mass is 60. What is the molecular formula of X?

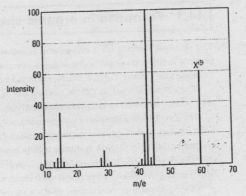

Figure 14.1.1   Mass spectrum of X

The relative formula mass of CH_2O is $(12 + (2 \times 1) + 16) = 30$. But the M_r of X is 60, which is twice the formula mass of CH_2O.

∴ The molecular formula of X must be twice the empirical formula, i.e. $2 \times (CH_2O) = C_2H_4O_2$.

The molecular formula still does not give us enough information. The molecular formula for our compound X is $C_2H_4O_2$. If you have an atomic modelling kit you will find that there are a number of ways in which two carbon, four hydrogen and two oxygen atoms can be joined together.

The **structural formula** of a compound not only shows the number of each kind of atom in the molecule, it also shows how the atoms are joined together.

In order to progress from the molecular formula to the structural formula we must do either or both of the following:

- Examine the chemical reactions of the compound. For example, our compound X reacts immediately with sodium carbonate solution to release carbon dioxide, showing that X is a carboxylic acid containing the —COOH group of atoms (see Ch 27.2).
- Record either the infra-red (IR) spectrum (*Figure 14.1.2*) or the nuclear magnetic resonance (NMR) spectrum (*Figure 14.1.3*) of the compound, or both. (For details of these spectroscopic techniques see Ch 30.5.)

The IR spectrum detects the characteristic vibrational frequencies which result when atoms are covalently bonded together. The NMR spectrum shows how many different environments there are for hydrogen atoms in the molecule, and the relative numbers of hydrogen atoms in each environment.

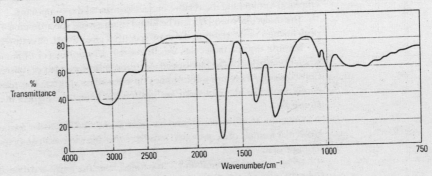

Figure 14.1.2   Infra-red spectrum of $C_2H_4O_2$

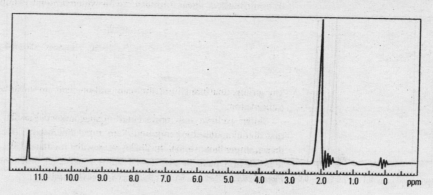

Figure 14.1.3   NMR spectrum of $C_2H_4O_2$

Using all the information, we can show that the compound X *does* contain the grouping —COOH. Because the molecular formula is $C_2H_4O_2$, this leaves only one carbon and three hydrogen atoms to be accounted for.

A formula in which all the bonds between atoms in a molecule are shown is usually called a **displayed formula** (sometimes a **graphical formula**).

The displayed formula for compound X, using the facts that carbon can form four covalent bonds, oxygen two and hydrogen only one, *must* therefore be:

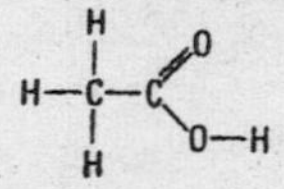

As these displayed formulae use a lot of space they are very often condensed on to one line; so the compound X becomes CH_3COOH. This kind of formula, which still shows all the important features, is often known as a **linear formula**.

Organic formulae can be represented using molecular modelling kits. These can be of three kinds: skeletal, ball and stick, and space filling. Three ways of representing compound X, CH_3COOH are shown in *Figure 14.1.4.*

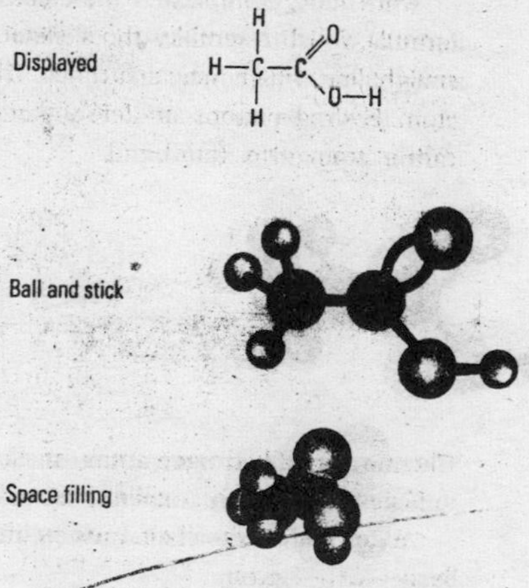

Displayed

Ball and stick

Space filling

Figure 14.1.4   Models of $C_2H_4O_2$

By using brackets, linear formulae can be written for quite complicated molecules. For example:

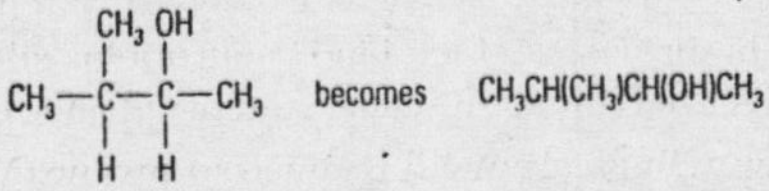

$$CH_3-\underset{\underset{H}{|}}{\overset{\overset{CH_3}{|}}{C}}-\underset{\underset{H}{|}}{\overset{\overset{OH}{|}}{C}}-CH_3 \qquad becomes \qquad CH_3CH(CH_3)CH(OH)CH_3$$

The groups that branch off the main carbon chain are included in brackets after the appropriate carbon atom.

So far, we have only represented organic molecules as flat, two-dimensional objects. But note that all the models for compound X are three-dimensional. The *majority* of organic molecules are, in fact, three-dimensional. In Ch 2.6, we saw that methane, CH_4, is tetrahedral in shape because the four electron pairs in the C—H bonds repel each other. Methane can be represented as in *Figure 14.1.5.*

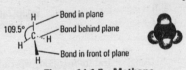

Figure 14.1.5   Methane

The ordinary lines represent bonds which are in the plane of the paper. The wedge shows that the bond is coming out of the plane of the paper towards the viewer; the dotted line shows that the bond is going into the plane of the paper, away from the viewer. All the bond angles in methane are the tetrahedral bond angle, 109.5°.

Compound X, CH_3COOH, can be represented as shown in *Figure 14.1.6.*

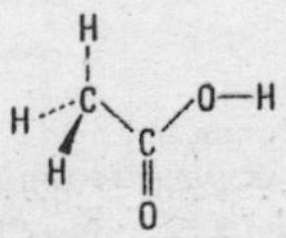

Figure 14.1.6   Compound X

For most purposes, the displayed formulae of organic molecules can be shown in two dimensions. We shall shortly see that there are times when an ability to draw three-dimensional displayed formulae for certain molecules is essential.

With more complicated molecules, a form of shorthand notation is often used – a skeletal formula, which resembles the skeletal type of model. Bonds between carbon atoms are shown as straight lines which meet at an angle. The point of the angles or the end of a line represents a carbon atom. Hydrogen atoms are left out, and carbon–hydrogen bonds are not shown. Remember, every carbon atom forms four bonds.

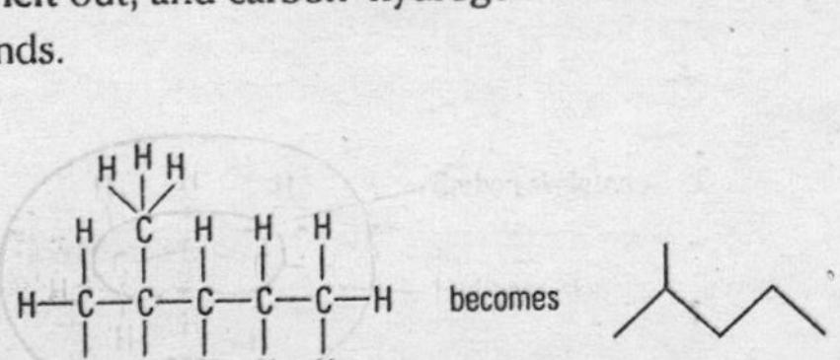

The number of hydrogen atoms on each carbon atom, at the ends of lines or the points of angles, is sufficient to make the total number of bonds on each carbon atom equal to four.

At any end of the chain must be the CH_3- group; at a simple angle in the chain, $\wedge$, there must be a $-CH_2-$ group.

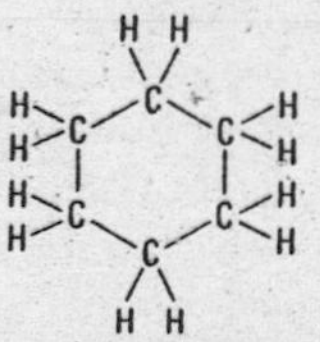

becomes

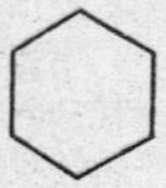

Other groups attached to the carbon skeleton are shown, so:

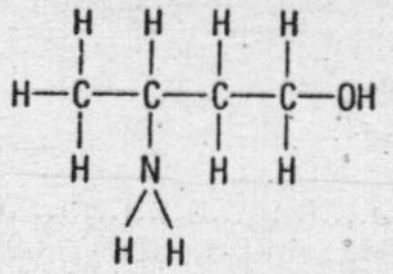

becomes

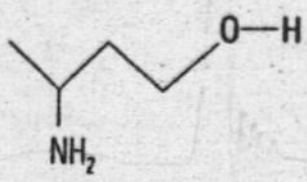

It was generally believed that the first person to have ideas about structural formulae – with a line to represent each bond between atoms, and with carbon atoms forming four bonds – was August Kekulé. Kekulé was Professor of Chemistry at Bonn in Germany, and the research paper containing his new theories was published in May 1858. A month later another paper, on chemical theory, was published by a 27-year-old unknown Scots chemist, Archibald Scott Couper. In August 1858 Kekulé attacked Couper's claim to have developed a new theory and said that he, Kekulé, had got there first.

Couper almost certainly did have the ideas before Kekulé, but his professor in Paris, Adolph Wurtz, didn't believe in them and was unhappy about letting them be published. This delay led to Kekulé's paper being the first to appear. Couper was very angry with Wurtz, who sacked him from his laboratory. Couper returned home to Scotland, went mad, and spent the rest of his life in an asylum before dying in 1892. It is at least possible that his huge disappointment led to his insanity.

The facts about Couper only came to light when Kekulé's successor as professor at Bonn, Richard Anschütz, found out about Couper's work while he was preparing a biography of Kekulé. It is now acknowledged that Couper's understanding of structure was clearer than Kekulé's, and both men were honoured and remembered in 1958 at the 100th anniversary of structural theory. One of the greatest-ever organic chemists, Nobel Prize-winner R B Woodward, said of Couper's paper of June 1858: 'In that literally astonishing paper one may see presented, for the first time, structural formulae identical with those we use today – and these are, of course, the most fundamental theoretical tools of organic chemistry'.

REMEMBER: **Working out and writing formulae – a summary**

Empirical: C_2H_4Br from % or mass composition data

Molecular: $C_4H_8Br_2$ from M_r data

Linear $CH_3CBr(CH_3)CH_2Br$

Structural CH_3CCH_2Br with Br above and CH_3 below   Functional groups from chemical or spectral data

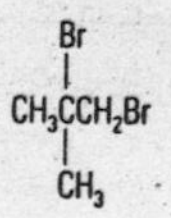

Displayed   Every bond shown

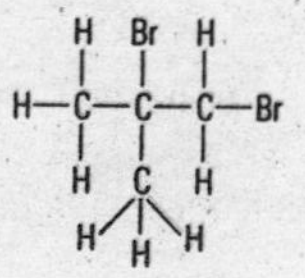

Skeletal   or in 3-D: 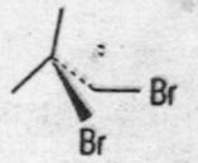   (*not* for examination answers!)

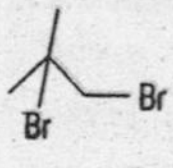

Making sense of organic chemistry

It is easy to become overawed by the vast number of organic compounds and the apparently huge range of their reactions. If all this information could not be systematised, few brains would be capable of holding it all. But have faith! *Organic chemistry at this level can be mastered if you memorise some simple rules, if you can recognise a relatively small number of arrangements of atoms and link these with particular chemical reactions, and if you are prepared to think.* Indeed, this kind of attitude and approach is still useful at higher levels of chemistry!

Four of the most helpful ideas in the learning of organic chemistry are:

1. Functional groups.
2. Homologous series.
3. Frameworks.
4. The rules for naming compounds.

If you can get on top of these unifying ideas now, you will find the study and learning of organic chemistry far easier. As we shall see, all four ideas are closely linked, but we shall discuss them in this order.

SELF-ASSESSMENT QUESTIONS 14.1

1 Calculate the empirical formula of the following compounds:
   a compound A, containing 80.0% carbon and 20.0% hydrogen;
   b compound B, containing 1.26 g carbon and 0.26 g hydrogen;
   c compound C, containing 0.303 g carbon, 0.063 g hydrogen and 0.202 g oxygen;
   d compound D, containing 23.8% carbon, 5.9% hydrogen and 70.3% chlorine.

2 Find the molecular formula of:
   a compound E, empirical formula CH_2, which has a molar mass of 56 g mol^{-1};
   b compound F, empirical formula C_2H_5, which shows a molecular ion peak at m/e = 58;
   c compound G, which has an M_r of 88 with an empirical formula of C_2H_6N;
   d compound H, which shows a molecular ion peak at 125 and has an empirical formula C_2H_3Cl.

14.2 Functional groups, homologous series and frameworks

A molecule of an organic compound can be thought of as having a 'skeleton' of carbon and a 'skin' of hydrogen (*Figure 14.2.1*).

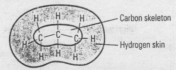

Figure 14.2.1   An organic molecule

Molecules which consist of only carbon and hydrogen atoms and in which the carbons are joined by only single bonds are not very reactive. It *is* possible to persuade them to react in interesting ways by using quite strenuous conditions (see Ch 15.5), but often we just burn them to provide energy.

Functional groups

If, however, one or more of the hydrogen atoms of the 'skin' is replaced by another atom or group of atoms which is capable of reaction, what we have is a part of the molecule which *is* able to react. Very often, at least one of the atoms involved is highly electronegative, such as a halogen, nitrogen or oxygen, causing polarity of the bond with carbon. There is, however, one common functional group which involves the skeleton. This is the alkene group, which contains a carbon–carbon double bond and is the subject of Ch 16.

A particular functional group will react in ways very different from those of any other functional group.

For example, ethane, molecular formula C_2H_6, has this structure:

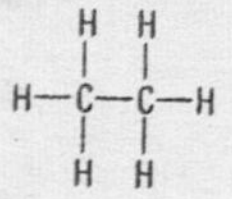

$$
\begin{array}{ccc}
 & H & H \\
 & | & | \\
H - & C - & C - H \\
 & | & | \\
 & H & H
\end{array}
$$

It is hard to do anything much with ethane except burn it.

Ethanol, however has this structure:

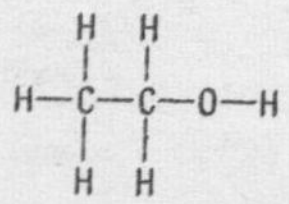

$$
\begin{array}{cccc}
 & H & H & \\
 & | & | & \\
H - & C - & C - & O - H \\
 & | & | & \\
 & H & H &
\end{array}
$$

In ethanol, the functional group is —OH. It makes the molecule react in several ways. **It is the functional group found in all the huge number of organic compounds known as alcohols: and therefore they all have very nearly the same chemical reactions.**

And that is the point about functional groups! Instead of the hopeless attempt to learn about millions of organic compounds, tackle the problem like this:

- **Learn the list of common functional groups and their structures,** and the parts they contribute to the names of the compounds (*Table 14.2.1*).
- **Always look at the structure of a molecule to see what functional group(s) it contains.** In organic chemistry, most of the molecules which we first meet contain only one functional group. The structures of the common functional groups should really be learned, and learned thoroughly, so that you can recognise them however complex the molecule appears.
- **Learn the most important reactions of the common functional groups.** At this stage, it really does mean only a relatively few reactions for each of the functional groups!

Homologous series

This is the second main idea for systematising organic chemistry. A homologous series has four features:

1. All the members of a particular homologous series have the same general formula. For example, the members of the alcohol series have the general formula, $C_nH_{2n+1}OH$, where n = 1, 2, 3, etc. For n = 1, the formula would be CH_3OH (methanol); for n = 2, C_2H_5OH (ethanol), and so on.

2. All the members of a particular homologous series have the same functional group. This means that *they have the same chemical reactions and can be made by the same general methods.* For example, all the alcohols contain the —OH group; they react with sodium to give hydrogen, and with carboxylic acids to give esters. They can all be made, for instance, by heating dilute sodium hydroxide solution with an appropriate organic halogen compound.

The only homologous series without a functional group is the **alkane hydrocarbons.** (A **hydrocarbon** is a compound, which contains *only* carbon and hydrogen). The names of all the alkanes end in *-ane*: e.g. methane, CH_4; octane, C_8H_{18}. The compounds in any other homologous series have names which all include a part which identifies the series; e.g. the names of all alcohols end in *-ol*. The name of a chemical compound can therefore tell you a great deal about its chemical reactions.

For a homologous series there is:

1. A general formula.
2. The same functional group.
3. Successive members differ by the $-CH_2-$ group.
4. There is a trend in physical properties.

3. **The successive members of any homologous series differ by $-CH_2-$.** As an example the first few alcohols are CH_3OH (methanol); CH_3CH_2OH (ethanol); $CH_3CH_2CH_2OH$ (propanol).

4. **There is a trend in the physical properties of the members of any homologous series.** As the molecules increase in size, there is a gradual change in properties, e.g. their boiling points increase.

Table 14.2.1 contains the functional groups and homologous series which you are most likely to meet.

Functional group	Identifying part of the name of the compound	Homologous series
$>C=C<$	-ene	Alkenes
$\geq C-OH$	-ol (or hydroxy-)	Alcohols
$\geq C-Cl$	chloro-	Chloroalkanes
$\geq C-Br$	bromo-	Bromoalkanes } Halogenoalkanes
$\geq C-I$	iodo-	Iodoalkanes
$-\underset{H}{C}=O$	-al	Aldehydes
$>C=O$	-one	Ketones
$-\underset{OH}{C}=O$ (often $-COOH$ or $-CO_2H$)	-oic acid	Carboxylic acids
$-\underset{Cl}{C}=O$	-oyl chloride	Acyl chlorides
anhydride structure	-oic anhydride	Acid anhydrides
$-C-O-C<$	(alkyl) -oate	Esters
$-\underset{NH_2}{C}=O$	-amide	Amides
$-C\equiv N$	-nitrile	Nitriles
$-NH_2$	-amine	Amines
$-NO_2$	nitro-	
$-N\equiv N-$	diazo-	

Table 14.2.1 Common functional groups and their structures

One other point needs to be made. Because the reactions of the homologous series depend on its functional group, a very simple way of representing this is often used. For example, any compound in the alcohol series can be written ROH, where R is the 'rest of the molecule' other than the functional group – e.g. for methanol, CH_3OH, R is CH_3—; for ethanol, C_2H_5OH, R is C_2H_5—; and so on. We shall use such simplified series formulae when writing general equations for the reactions of a homologous series. The name given to R— is **'an alkyl group'** (see Ch 14.3 on nomenclature).

Frameworks

We have mentioned that organic molecules are based on carbon skeletons. There are three basic types of skeletons, or **frameworks** as they are often known.

A framework which has *no* ring structure in it is said to be **aliphatic**. Such aliphatic frameworks are often further classified as **straight chain** (although of course chains of carbon atoms are very rarely actually straight!) or **branched chain**.

$CH_3CH_2CH_2CH_2CH_2CH_2OH$ – A 'straight-chain' six-carbon alcohol, which could be written CH_3—$(CH_2)_4$—CH_2—OH

$$CH_3CHCHCH_2OH \text{ – A 'branched-chain' six-carbon alcohol}$$

(with CH_3 branch above and CH_3 branch below)

A framework based on *flat* ring structures (most commonly the six-membered ring of benzene, see Ch 26.1) is said to be **aromatic**. The flat benzene ring contains mobile **delocalised** electrons (see for example graphite in Ch 3.2) and is usually written as a hexagon with a circle inside. Originally, the structure was written with alternate double and single bonds, but it is now usually written with the circle to represent the delocalised electrons (Ch 26.1).

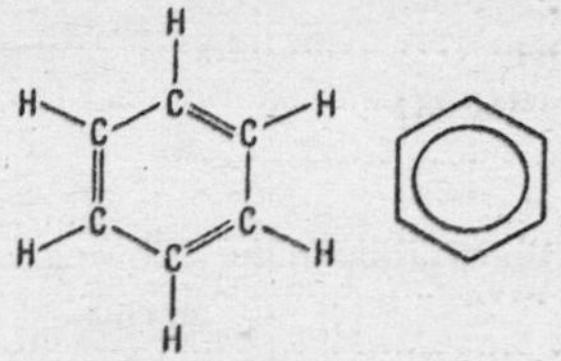

Benzene, an aromatic framework

A few functional groups have some different properties when attached to an aromatic framework; but more of this later (Ch 26.4).

Sometimes the framework involves a ring system which is not aromatic. The ring is not flat and does not contain delocalised electrons (*Figure 14.2.2*). Such a framework is said to be **alicyclic**.

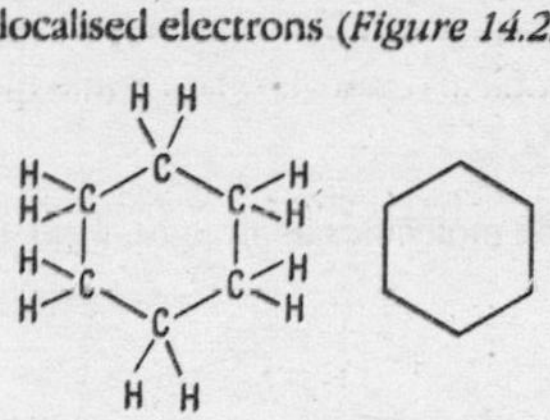

Cyclohexane; an alicyclic framework

Two possible structures of cyclohexane – try making them with a model kit. There are two extreme forms when represented in three dimensions.

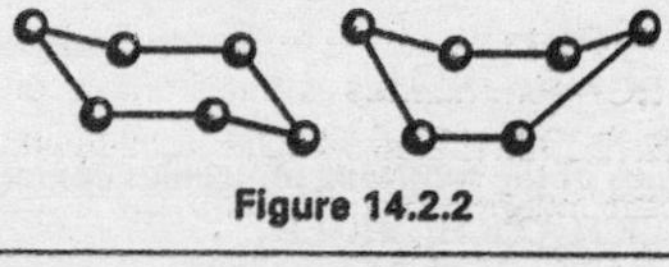

Figure 14.2.2

When functional groups are attached to an alicyclic framework, they react just as if they were attached to an aliphatic framework.

So the differences in the chemical behaviour of functional groups, where they exist, are between **aliphatic** and **alicyclic** frameworks on the one hand and **aromatic** frameworks on the other.

A

ACTIVITY

Activity

Which of the following straight-chain molecules must have at least one carbon–carbon double bond? Use molecular models to decide. Or try to draw the displayed formulae with the correct number of bonds on each atom.

C_3H_8, C_3H_6, C_5H_{10}, CH_4, C_4H_8

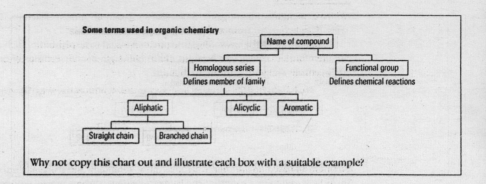

Some terms used in organic chemistry

Name of compound

Homologous series
Defines member of family

Functional group
Defines chemical reactions

Aliphatic

Alicyclic

Aromatic

Straight chain

Branched chain

Why not copy this chart out and illustrate each box with a suitable example?

Q

SELF-ASSESSMENT
QUESTIONS
14.2

1 Copy out the following formulae as displayed formulae. Name the functional group present and circle the functional group on your copy.
  a CH_3CH_2OH
  b CH_3CH_2COOH
  c $CH_3CH(CH_3)CH_2NH_2$
  d $CH_2{=}CHCH_3$

2 Identify the functional group in the following compounds as:
  A alcohol   C alkene
  B ketone   D aldehyde
  a CH_3COCH_3
  b $CH_3CH(OH)CH_3$
  c $CH_3CH_2C{=}O$
         |
         H
  d $CH_3CH{=}CHCH_3$

3 Classify the following molecules as straight chain, branched chain, alicyclic or aromatic compounds.
  a 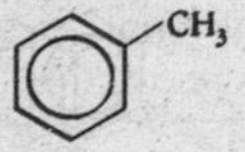
  b $CH_3C(CH_3)_2CH_3$
  c 
  d $CH_3CH_2CH_2CH_3$

4 Which of the following molecules contain an alcohol group?
  a $ClCH_2COCH(CH_3)CH_3$
  b $CH_3CH_2CH_2COOH$
  c $CH_3CH(OH)CH_2CH_3$
  d $CH_2(OH)C(CH_3)_2CH_3$

14.3 Naming organic molecules

In ancient times it was generally accepted that, if you knew someone's name, you had power over them. It is certainly true that the **systematic** name of an organic compound clearly indicates its structure, and so how it behaves chemically. For a simple compound, it should be possible to write down its systematic name from its structure, and the structure from its name.

The simplest organic molecules are the alkane hydrocarbons (see Ch 15.2). The systematic names of simple aliphatic organic compounds are based on those of the alkanes.

The simplest alkane – *meth*ane, CH_4, contains only the one carbon atom. So if you see 'meth-' as part of the main part of the name of an organic compound, you will know that the compound contains one carbon atom. The *methyl* group is CH_3—.

*Eth*ane, C_2H_6, contains two carbon atoms. So 'eth-' or 'ethan-' in a name means that the compound contains two carbon atoms. The *ethyl* group is C_2H_5—.

Similarly, *prop*ane is C_3H_8. 'Prop-' or 'propan-' indicates three carbon atoms. *But*ane is C_4H_{10} so 'but-' or 'butan-' means four carbons (*Table 14.3.1*).

C atoms	Stem name
1	Meth-
2	Eth-
3	Prop-
4	But-
5	Pent-
6	Hex-

Table 14.3.1

'Alkyl' groups come from the corresponding alkanes, so ethyl is C_2H_5—, propyl is C_3H_7—, butyl is C_4H_9—, and so on, by adding a —CH_2— each time (*Table 14.3.2*).

Alkyl groups	
CH_3—	Methyl
C_2H_5—	Ethyl
C_3H_7—	Propyl
C_4H_9—	Butyl

Table 14.3.2

Trivial names

Many important organic chemicals, especially those with more complicated molecules, are also known by so-called 'trivial' names, i.e. traditional, non-systematic names. Professional chemists have to learn to use these names because they are still used in industry. Examples are 'acetic acid' (systemically, ethanoic acid) and 'acetone' (propanone). But for teaching, and especially for learning, organic chemistry a logical systematic way of naming compounds is extremely helpful. Apart from anything else, it builds confidence. Leave the trivial names until the systematic ones are mastered!

An organic chemist therefore counts up to four like this: meth-, eth-, prop-, but-. After that, Greek prefixes are used. Some of these you will know already. 'Pent-' means five carbon atoms; 'hex-' means six; 'hept-' means seven; and 'oct-' means eight. (You might be familiar with *oct*ane ratings for petrol.)

The name of a simple aliphatic organic compound has up to three parts, and each part conveys information about the structure of the compound and therefore about its reactions.

- Meth-, eth-, oct-, etc, to tell you the **longest chain** of carbon atoms – the backbone – in the compound (*Table 14.3.1*). (It really *is* important to look for the *longest* chain, and there will be practice in this in the questions at the end of this section.)
- A short start or finish to the name, all as one word, to indicate the functional group, e.g. chloro-, nitro-, -ol, -one, -oic acid (see *Table 14.2.1*). Information about the bonding between carbon atoms in the molecule is given as part of the chain name: e.g. -ene or -en- indicates the alkene functional group, i.e. a carbon–carbon double bond. So prop*an*oic acid is CH_3CH_2COOH, prop*en*oic acid is $CH_2{=}CHCOOH$.

The short start to the name also includes the names of any side chains attached to the longest chain; $CH_2=C(CH_3)COOH$ is methylpropenoic acid, since there is now a methyl group attached to the central carbon atom.

- A number is used to show whereabouts on the carbon chain each functional group is attached. Numbers are separated from words with dashes: e.g. -2- or 2-, and the number must be as *small as possible*.

So:

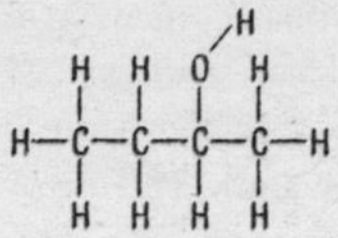

H—C—C—C—C—H   is butan-2-ol

not butan-3-ol, which it would be if you started counting from the other end! This enables us to make an important point – molecules don't line up conveniently left to right – they can rotate in space. You must be able to work out the correct name whatever the orientation of the molecule on the page. Again, practice will be given in the questions.

When there is more than one example of a particular group in a molecule, we indicate this using di-, tri-, etc (*Table 14.3.3*).

Multiple groups

1	Mono- (rarely used)
2	Di-
3	Tri-
4	Tetra-
5	Penta-
6	Hexa-

Table 14.3.3

But, do remember to put in all the numbers that show where both the groups are, and separate the numbers with a comma.

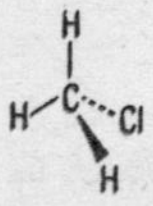

Chloromethane 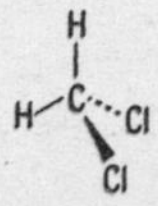   Dichloromethane 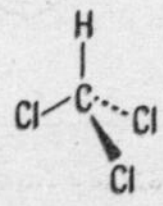   Trichloromethane 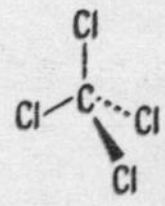   Tetrachloromethane

There is no need to use numbers with methyl compounds because there is only one carbon atom, nor with ethyl compounds unless there is more than one functional group. For example, 1,2-dibromoethane is:

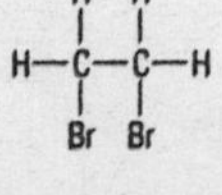

whereas 1,1-dibromoethane is:

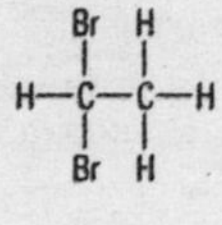

Because parts of molecules can rotate freely around single bonds as well as rotate in space, 1,2-dibromoethane could equally well be written as:

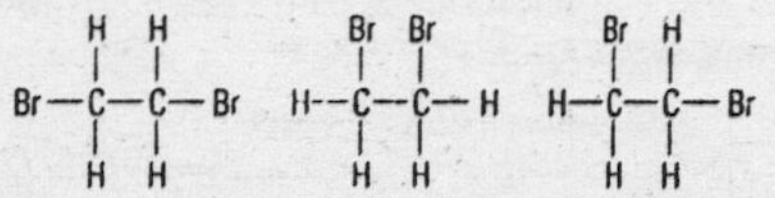

or a number of other structures. The important thing is that the numbers 1,2- mean that the two bromine atoms are attached to two *different* carbon atoms, whereas 1,1- mean that they are both attached to the *same* carbon atom.

The rules enables us to name quite complicated-looking molecules. Can you see why the molecule whose structure is shown in *Figure 14.3.1* is named 3-chloro-2-methylpropan-1-ol?

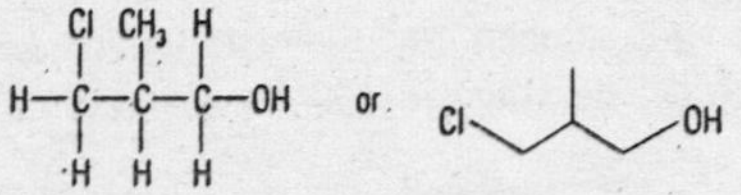

Figure 14.3.1   3-chloro-2-methylpropan-1-ol

Example

Try to justify the names of the following compounds:

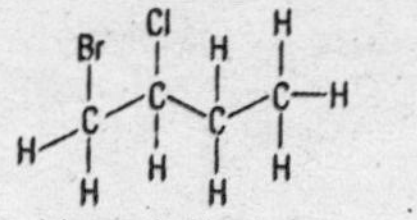

1-bromo-2-chlorobutane 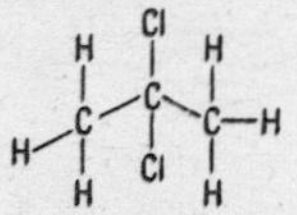   2,2-dichloropropane 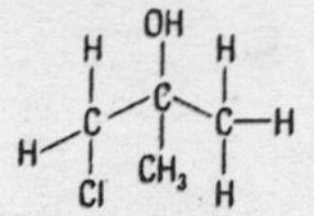   1-chloro-2-methylpropan-2-ol

Alicyclic compounds are met less frequently than aliphatic compounds. If the rings consist only of carbon atoms, naming them is simple. For example:

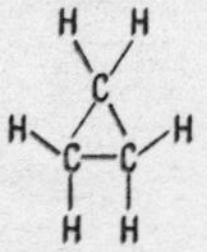

is cyclopropane, because the ring contains three carbon atoms. In shorthand it is written as:

The structure:

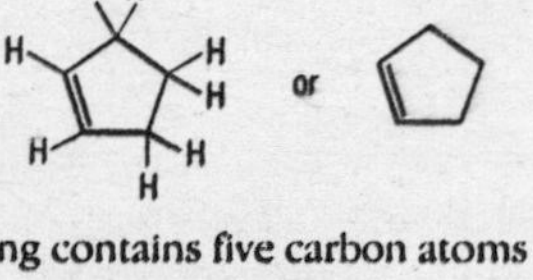

is cyclopentene, because the ring contains five carbon atoms and a carbon–carbon double bond.

The compound with the structure:

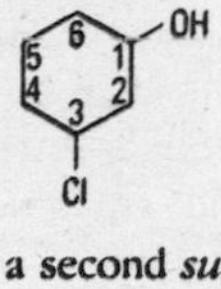

is named 3-chlorocyclohexanol. Note that a second *substituent* on a ring has a number to fix its position relative to the first substituent.

The atom or group of atoms which replaces a hydrogen atom in the hydrocarbon molecule is called a *substituent*. The hydrogen atom has been *substituted* by this newcomer.

The number is the smallest one possible; 5-chlorocyclohexanol gives the correct position but is incorrect as a name.

Aromatic compounds based on the six-membered carbon ring of benzene can be dealt with in a similar manner:

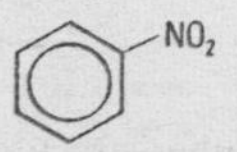

is nitrobenzene.

is 1,4-dinitrobenzene: the two numbers are needed here because both substituents are the same. (Notice here the use of 'di-' meaning two. 'Tri-' would mean three.)

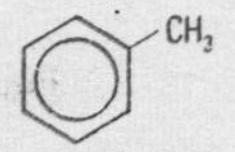

is methylbenzene.

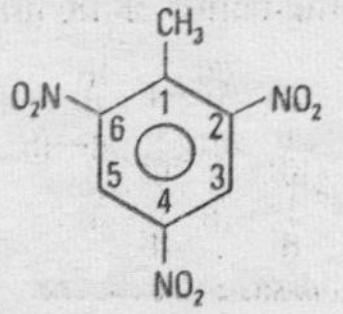

is 2,4,6-trinitromethylbenzene. Here the methyl group is the substituent which acts as the reference point. (The old name for methylbenzene is 'toluene'; the old name for this material is trinitrotoluene, TNT for short ... and probably the most notorious explosive of all.)

N.B. In some compounds, the benzene ring is represented by the prefix phen- or phenyl-.

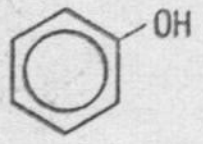

is phenol and 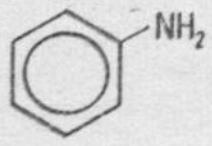   is phenylamine

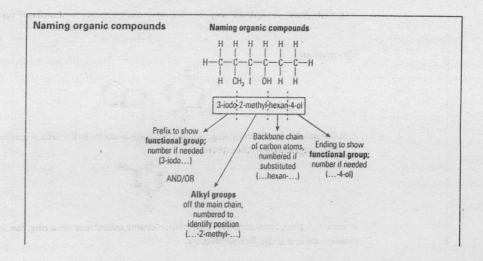

Naming organic compounds

3-iodo-2-methyl-hexan-4-ol

Prefix to show **functional group**; number if needed (3-iodo...)

AND/OR

Alkyl groups off the main chain, numbered to identify position (...-2-methyl-...)

Backbone chain of carbon atoms, numbered if substituted (...hexan-...)

Ending to show **functional group**; number if needed (...-4-ol)

Remember

1. Identify the longest chain of carbon atoms.

$$C—C—C—C—C—C \quad \therefore \text{hexan}$$

2. Number the carbon atoms from one end, minimising the numbers on substituted positions.

$$C^1—C^2—C^3—C^4—C^5—C^6$$

3. Decide on the ending to identify the functional group.

$$C^1—C^2—C^3—C^4—C^5—C^6 \quad \therefore \text{hexan ol}$$
$$\qquad\qquad\quad |$$
$$\qquad\qquad\ OH$$

4. Put in a number to show which carbon atom the functional group is attached to e.g. 2-, 3-.

$$C^1—C^2—C^3—C^4—C^5—C^6 \quad \therefore \text{hexan-4-ol}$$
$$\qquad\qquad\quad |$$
$$\qquad\qquad\ OH$$

5. Name any side chains or functional groups that use prefixes.

$$C^1—C^2—C^3—C^4—C^5—C^6 \quad \therefore \text{methyl and iodo}$$
$$\qquad\ |\qquad |\qquad |$$
$$\quad CH_3 \quad I \quad OH$$

6. Put in a number to show the carbon atom which the side chain or functional group is attached to; e.g. 2-.

$$\therefore \text{3-iodo-2-methylhexan-4-ol}$$

SELF-ASSESSMENT
QUESTIONS
14.3

1 How many carbon atoms are there in the longest chain in each of the following molecules?

a $CH_3CH_2CH_2CH_2CH_3$

b $CH_3CH(CH_3)CH(OH)CH_2CH_3$

c CH_3CHCH_3
$\quad\ |$
$\quad\ CH_2CH_2CH_3$

d $CH_3CH_2CHCH_3$
$\qquad\ |$
$\quad\ CH_3CH_2CHCH_2CH_3$

e $CH_3CHCH_2CH_2CH_3$
$\qquad |$
$\ CH_3CHCH_2CH(CH_3)CH_3$

2 Give the systematic name for the compounds with the formulae shown. (They will all end in -ane)

a $CH_3CH_2CH_2CH_2CH_3$

b $CH_3(CH_2)_4CH_3$

c $CH_3CH(CH_3)CH_2CH_3$

d $CH_3CH_2C(CH_3)_2CH_3$

3 Write the structural formulae of:

a propane

b methylpropane

c dimethylpropane

d 2,2,3-trimethylbutane

e 1,2-dimethylcyclohexane

4 To which homologous series do each of these lists of compounds belong?
  a CH$_3$CH$_2$OH,
     CH$_3$CH$_2$CH$_2$CH$_2$OH,
     CH$_3$OH,
     CH$_3$(CH$_2$)$_5$CH$_2$OH
  b CH$_3$CH=CH$_2$,
     CH$_3$(CH$_2$)$_4$CH=CH$_2$,
     CH$_3$CH=CHCH$_3$
  c CH$_3$COCH$_2$CH$_3$,
     (CH$_3$)$_2$CHCH$_2$COCH$_3$,
     CH$_3$COCH$_3$

5 Which of the following structures represent identical molecules?
  A CH$_3$CHCH$_3$
        |
        CH$_2$CH$_2$
              |
              CH$_3$
  B CH$_3$CH(CH$_3$)CH$_2$CH$_2$CH$_3$
  C CH$_3$CH$_2$
        |
        CHCH$_3$
        |
        CH$_3$CH$_2$
  D (CH$_3$)$_2$CHCH$_2$CH$_2$CH$_3$
  E CH$_2$CH$_2$CHCH$_3$
        | |
        CH$_3$ CH$_3$

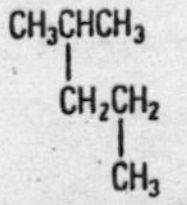

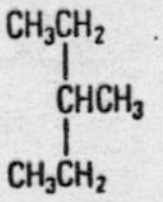

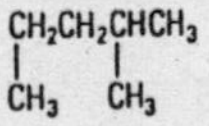

6 a What is meant by a homologous series?
   b The following molecules are all hydrocarbons. Which of them belong to the same
      homologous series?
      A CH$_3$CH$_2$CH$_3$
      B CH$_3$CHCH$_2$
      C CH$_3$CH$_2$CH$_2$CH$_3$
      D CH$_3$CH(CH$_3$)CH$_3$
      E CH$_3$CHCHCH$_3$

7 One of the waste products from bacterial activity on body secretions under the armpits is

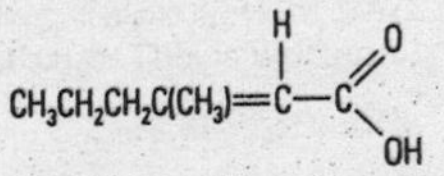

$$CH_3CH_2CH_2C(CH_3)=C{\overset{H}{\underset{OH}{\diagdown}}}C{\overset{O}{\diagdown}}$$

   a Write the molecular formula of this compound.
   b What is the empirical formula of the compound?
   c Show that the compound contains 65.6% carbon.
   d Copy out the above structure and circle the two functional groups present.
   e What are the names of these two functional groups?
   f At what value would you expect the molecular ion peak to appear in the mass spectrum of
      the compound?
   g Draw a displayed formula for this molecule.

14.4 Isomerism

We have already seen (Ch 14.1) that if compound X has the **molecular** formula $C_2H_4O_2$, there are two possible **structural** formulae:

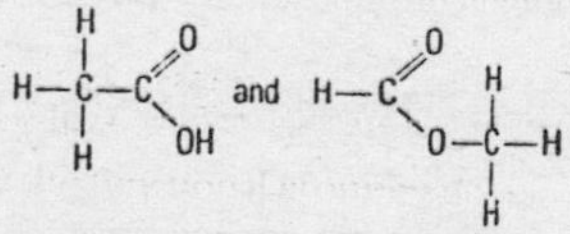

Similarly, the molecular formula C_2H_6O also has two possible structures:

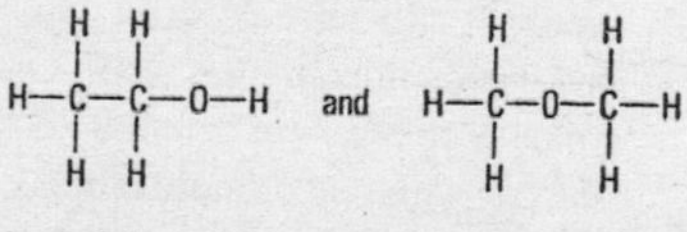

> **Definition:** Isomers are compounds which have the same molecular formula but different arrangements of the atoms.

We can see immediately that the left-hand structure has the —OH functional group and therefore represents an alcohol. (With two carbon atoms, it is our old friend ethanol.) This compound reacts immediately, as any alcohol does, with sodium to give off hydrogen gas. The right-hand structure is of a compound with a different functional group; one which does *not* react with sodium. We could test with sodium to find out which is the actual structural formula of the compound C_2H_6O.

There are a number of different types of isomers.

Structural isomers

> **Definition:** Structural isomers have the same molecular formula, but in the different isomers the atoms are joined together in a different sequence.

Structural isomerism is of three basic kinds.

Chain isomerism

When there are four or more carbon atoms in a chain, **chain branching** becomes possible. For example:

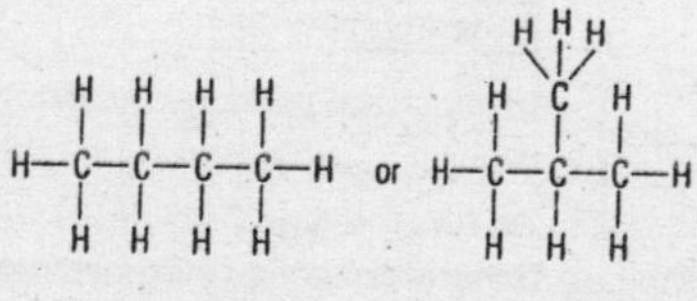

Obviously as the chain gets longer, more isomers are possible. For the alkane formula C_8H_{18} (i.e. the octanes), there are eighteen possible structures; for $C_{10}H_{22}$, there are 75 possibilities; for $C_{20}H_{42}$, there are 366 319; and for $C_{40}H_{82}$ about 6×10^{13} – i.e. 10 000 for every person in the world!

Position isomerism

This can occur when a functional group is found in a molecule which has three or more carbon atoms in its chain. For example, the linear formula C_3H_7OH should be read as propanol.

But if you draw out the full structure, or construct it from atomic models, you discover that there are two possible alcohols with the molecular formula C_3H_7OH:

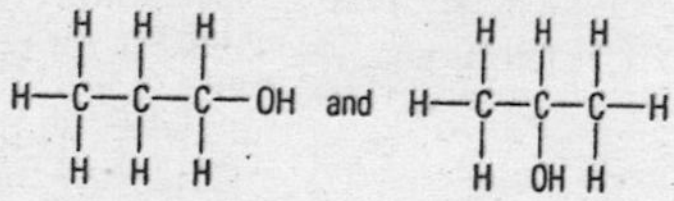

These are named using the rules in Ch 14.3.

For those functional groups which are incorporated into the skeleton, the same rules can apply; for example:

$$H-\underset{\underset{H}{|}}{\overset{\overset{H}{|}}{C}}-\underset{\underset{H}{|}}{\overset{\overset{H}{|}}{C}}-\underset{\underset{H}{|}}{\overset{\overset{H}{|}}{C}}=C\overset{H}{\underset{H}{}} \quad \text{and} \quad H-\underset{\underset{H}{|}}{\overset{\overset{H}{|}}{C}}-\underset{\underset{H}{|}}{\overset{\overset{H}{|}}{C}}=\underset{\underset{H}{|}}{\overset{\overset{H}{|}}{C}}-\underset{\underset{H}{|}}{\overset{\overset{H}{|}}{C}}-H$$

But-1-ene   But-2-ene

i.e. -1- means between carbon atoms number 1 and 2, and -2- means between 2 and 3 (not between 1 and 2).

Functional group isomerism

We have seen two examples of this kind at the start of this section. The molecular formula C_2H_6O can represent either the alcohol C_2H_5OH or the totally different compound CH_3OCH_3. In the alcohol, we find the atom sequence $\geqslant C-O-H$, typical of alcohols; in the other, we have $\geqslant C-O-C\leqslant$, typical of the class of compounds known as ethers. In general, any alcohol other than methanol is isomeric with one or more ethers.

The carboxylic acid, CH_3COOH, is, as we saw at the start of this section, isomeric with the ester, $HCOOCH_3$. Again, any carboxylic acid with two or more carbon atoms is isomeric with an ester.

Another common example of functional group isomerism is that between aldehydes (functional group $-CHO$) and ketones ($>C=O$).

For example:

$$H-\underset{\underset{H}{|}}{\overset{\overset{H}{|}}{C}}-\underset{\underset{H}{|}}{\overset{\overset{H}{|}}{C}}-C\overset{\overset{O}{\parallel}}{\underset{H}{}} \quad \text{and} \quad H-\underset{\underset{H}{|}}{\overset{\overset{H}{|}}{C}}-\underset{\underset{H}{|}}{\overset{\overset{O}{\parallel}}{C}}-\underset{\underset{H}{|}}{\overset{\overset{H}{|}}{C}}-H$$

Propanal   Propanone

ACTIVITY

Activity

Two possible molecular formulae for the empirical formula CH_2 are C_3H_6, and C_4H_8.

Use molecular models to work out the possible structure for C_3H_6 and the three (or even four) possibilities for C_4H_8.

Stereoisomers

In structural isomers, the isomers are different because the atoms are joined together in a different sequence. But it is also possible to have the same atoms joined together in the same sequence and *yet the molecules are different*, because the atoms *have a different arrangement in space*. Such isomers are called stereoisomers. (Why is a 'stereo' radio or CD player called that?)

There are two kinds of stereoisomers.

> **Definition:** Stereoisomerism occurs when the sequence of atoms joined together in the molecular structure is the same, but their arrangement in space is different.

Geometrical isomers

These are often called *cis-trans isomers*.

Consider the compound 1,2-dibromoethane. This is written:

$$H-\underset{\underset{Br}{|}}{\overset{\overset{H}{|}}{C}}-\underset{\underset{Br}{|}}{\overset{\overset{H}{|}}{C}}-H \quad \text{but, in 3-D, it is really} \quad H\cdots\underset{\underset{Br}{}}{\overset{\overset{H}{|}}{C}}-\underset{\underset{H}{}}{\overset{\overset{H}{|}}{C}}\cdots Br$$

There is nothing to stop the molecule rotating about the carbon–carbon single bond. So, in terms of position, *all three bonds around each carbon atom are the same*. We could equally well write 1,2-dibromethane as we did in Ch 14.3:

$$H-\underset{\underset{H}{|}}{\overset{\overset{Br}{|}}{C}}-\underset{\underset{Br}{|}}{\overset{\overset{H}{|}}{C}}-H \quad \text{or} \quad Br-\underset{\underset{H}{|}}{\overset{\overset{H}{|}}{C}}-\underset{\underset{H}{|}}{\overset{\overset{H}{|}}{C}}-Br \quad \text{or} \quad H-\underset{\underset{H}{|}}{\overset{\overset{Br}{|}}{C}}-\underset{\underset{H}{|}}{\overset{\overset{H}{|}}{C}}-Br$$

and all these would be structures for exactly the same compound.

But if there is a *double bond* between carbons, all is changed. The molecule is now fixed and held rigid. *It cannot rotate about the carbon–carbon double bond without the bond being broken*, which needs a lot of energy. So there are two different arrangements for 1,2-dibromoethene. (Remember that *-ene* tells you that there is a double bond in the molecule.)

cis-1,2-dibromoethene   trans-1,2-dibromoethene

The *cis* and *trans* part of the names comes from the Latin; *cis* means 'on this side of' and *trans* means 'on the other side of'.

You are able to read this because of *cis-trans* isomerism!

When light strikes a rod cell in the retina of the eye, chemical changes occur which eventually give rise to an electrical impulse in the optic nerve.

The pigment rhodopsin, or 'visual purple', is found in the retina. It was discovered in 1952 that the part of rhodopsin which absorbs the light is a compound called 11-*cis*-retinal. This joins with a protein, known as opsin, to form rhodopsin.

The 11-*cis*-retinal is bent into a curve by the *cis* formation about a particular double bond, and because of this it fits exactly into sites on the protein surface.

When rhodopsin absorbs light, the 11-*cis*-retinal is changed into the *trans* form. This form has a much 'straighter' shape and cannot fit any longer onto the opsin, so that the rhodopsin falls apart. It is this detachment which leads eventually to the nerve impulse.

in the eye

11-*cis*-retinal   rhodopsin   11-*trans*-retinal

Nerve impulse along optic nerve

light / strong light or enzyme

opsin

Enzyme in the eye converts to 11-*cis*-retinal

Unconverted 11-*trans*-retinal changed to *trans*-vitamin A

in the liver

changed to *cis*-vitamin A in the liver

cis-vitamin A   carotene from diet   trans-vitamin A

Vitamin A can be formed in the liver from carotenes absorbed in the diet. Carotenes occur in nearly all green plants and cause the orange colour of carrots. Lack of vitamin A in the diet can therefore cause problems with vision.

Optical isomers

It is possible to have optical isomers which are not caused by four different atoms or groups around a central carbon atom, but these will not be discussed here.

For these there must be *four different atoms or groups around a single carbon atom*. Labelling these a,b,c,d, we can make two different arrangements.

Notice that these two are *mirror images* of one another, and that however hard you try you cannot fit one over the other – they are *non-superimposable*. There is no symmetry around the carbon atoms. The molecule is **asymmetric**. The carbon atom with four different groups attached to it is said to be **chiral**; this is based on the Greek work for 'hand'. (Do you see how your left hand is a mirror image of your right hand, but cannot be superimposed on it?)

If we pass polarised light through one of these optical isomers, the light is rotated to the left. The other isomer rotates it the same amount to the right. The ideas about polarised light will be dealt with in Ch 28.5.

> **Definition:** A chiral carbon atom has four different atoms or groups attached to it. The presence of a chiral carbon atom gives rise to optical activity in the molecule.

ACTIVITY

Activity – Stereochemical models

For this you will need either a proper chemical modelling set, or lots of cocktail sticks, two small oranges or balls of plasticine/Blu-tack; and some soft sweets of various colours, such as Liquorice Allsorts.

Optical isomers

Assuming you are using the cheaper method, put four cocktail sticks *as far apart as possible* into one ball or orange, and *put sweets of four different colours, one on each stick.* Place the model in front of a mirror. Look carefully at the reflection, and make the model of what you can see.

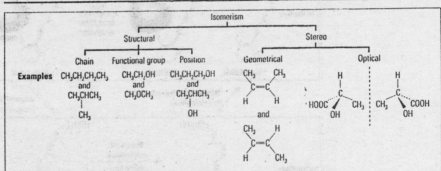

You should now have a model of both a molecule and the mirror image of that molecule. Try putting one model over the other so that they appear the same. You will find it can't be done.

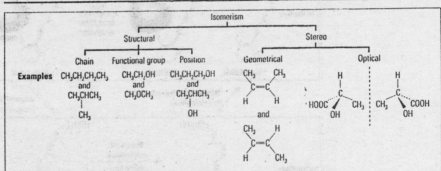

To summarise (N.B. These are *not* definitions)
Isomers have identical molecular formulae.
Structural isomers have their atoms arranged in a different sequence.
Chain isomerism involves a difference in the carbon skeleton.
Position isomerism involves a functional group located on different carbon atoms.
Functional group isomers have entirely different functional groups.
Stereoisomers have different arrangements of the atoms in space.
Geometrical isomers exist because of lack of rotation within the molecule.
Optical isomers are created in molecules with four different groups attached to a carbon atom.

ACTIVITY

Activity

Using molecular models, find out the number of different ways of arranging $C_4H_{10}O$. How many isomers are there for this molecule? It is possible to illustrate each type of isomerism except geometrical isomerism with this formula. Can you?

Draw these isomers in your notes.

*Almost all of the amino acids which build the proteins necessary for life in all the cells in your body are 'left-handed'. Optical activity seems to be associated with life. (Many such molecules, of course, contain just one chiral carbon atom, such as alanine, an **amino acid**. Others contain more than one. Glucose, which is transported by your blood to your muscles to provide energy through respiration, has four chiral carbon atoms.)*

L-(+)-alanine ('left-handed')   D-(−)-alanine ('right-handed')

Why is life on our planet based on left-handed molecules? No one really knows. But there could be, somewhere in the Universe, another planet similar to ours where all life is based on right-handed molecules. And, if you crash-landed there and tried to eat any of its fruit or vegetables, or meat from its animals, you would not be able to digest them at all and would starve! All your digestive enzymes have evolved to cope with left-handed molecules.

SELF-ASSESSMENT QUESTIONS 14.4

1   Write the possible chain isomers for the molecule C_5H_{12}, and write a systematic name for each isomer.

2   The molecular formula, C_4H_9OH, has five possible isomeric structures. Write all the isomers in displayed formulae and identify the types of isomerism present.

3   For the following compounds, identify them as three isomeric pairs of compounds.
   A $CH_3CH(CH_3)CH(CH_3)CH_3$
   B $CH_3CH_2CH{=}CH_2$
   C $CH_3CH{=}CHCH_3$
   D $CH_3C(CH_3)_2CH_2OH$
   E $CH_3CH(CH_3)CH_2CH_2CH_3$
   F $CH_3CH_2CH_2CH_2CH_2OH$

4   When fossil resin amber was distilled, in 1550, butanedioic acid (trivial name, succinic acid) was first prepared. The structure of butanedioic acid is: $HOOCCH_2CH_2COOH$.

   Nowadays, one route to this compound is by the bacterial fermentation of 2-hydroxybutanedioic acid (commonly known as malic acid).

   a   Write the displayed formula of 2-hydroxybutanedioic acid.
   b   Circle one example of each of the two functional groups present in this molecule.
   c   Mark the chiral carbon atom in this structure with an asterisk, *.
   d   Describe what particular property the presence of this chiral carbon atom gives to the molecule.
   e   A compound closely related to butanedioic acid is butenedioic acid. Write the displayed formula of butenedioic acid.
   f   Explain why there are two possible isomers of butenedioic acid, whereas there is only one structural isomer of butanedioic acid.

14.5   Physical properties of organic molecules

All kinds of particles attract one another, even if they are not polar (see Ch 2.8). This is why every material, even helium (which consists of single atoms) or hydrogen (which consists of H_2 molecules), can be liquefied. Liquids and solids could not exist if there were no attractive forces between the particles in them.

To explain the attractions between non-polar particles – such as atoms of the noble gases or the molecules of an alkane – we have to accept that the electrons in these particles are not completely fixed in their positions but can move to some small extent. This means that the distribution of electron density in a particle will not stay constant (*Figure 14.5.1*).

Reminder:

Movement of electron density   Induced movement of electron density

Figure 14.5.1   Intermolecular attractions

Even a completely covalent symmetrical molecule – or a single atom, such as argon – can develop just for a split second, a partial charge on each end. (Remember the symbol δ means very small.) The particle develops an *instantaneous dipole*.

Once more, the basic law of attraction and repulsion comes into operation. The instantaneous dipole of one particle will cause electrons to move in the opposite direction in a nearby particle, causing it to have an *induced dipole*.

The opposite partial charges on the particles then attract each other. And these two particles will have similar effects on all the particles around them, and these will be affecting each other; a tiny fraction of a second later the dipoles may all be round the other way. So our picture of a non-polar liquid or solid is that the particles are held together by very weak forces caused by the fluctuations of electron density within the particles.

Obviously, the more electrons there are in the particles, the greater the opportunity for fluctuations in electron density, and the greater the attractions between them. We saw this with the boiling points of the noble gases (single atoms) and of the halogens (molecules of the type X_2) in Ch 2.8.

The *number* of attractions between molecules can also determine the properties of materials. If we consider the simple alkane hydrocarbons (Ch 15.3), the boiling point increases with the number of carbon atoms in the molecule (*Table 14.5.1*).

> The instantaneous dipole-induced dipole attractions are often called **van der Waals forces**, after the Dutch physical chemist Johannes van der Waals. He first began to put the study of attractions between particles on a sound experimental and theoretical basis in the 1870s.

Name	Formula	b.p./°C
Methane	CH_4	−162
Ethane	CH_3CH_3	−89
Propane	$CH_3CH_2CH_3$	−42
Butane	$CH_3(CH_2)_2CH_3$	−1
Pentane	$CH_3(CH_2)_3CH_3$	36

Table 14.5.1   Boiling points of simple alkanes

As the length of the carbon chain in the molecule increases, there are more opportunities for instantaneous dipole-induced dipole attractions to occur.

For example, $C_{20}H_{42}$ is a waxy solid. The well-known plastic poly(ethene) (Polythene) consists of very long alkane chains; it can be made as a tough solid because of the great number of attractions possible along the chains.

Pentane has the molecular formula C_5H_{12}. There are three possible isomers with this formula: pentane, 2-methylbutane and 2,2-dimethylpropane.

Their structures and boiling points are shown in *Table 14.5.2*.

The three isomers of C_5H_{12} contain exactly the same number of atoms of each kind, and the three compounds might therefore be expected to have the same boiling point. But notice that the more globular and less chain-like the molecule is, the lower the boiling point. This is mainly because, in the liquid, there are fewer points of contact between the more globular molecules, such as 2,2-dimethylpropane, than there are between the longer, chain-like pentane molecules. The result is that the globular alkanes have fewer opportunities for instantaneous dipole-induced dipole interactions than the chain-like molecules, so have lower boiling points.

Name	Structure	b.p. / °C	
Pentane	CH₂ CH₂ CH₃ CH₂ CH₃	36	
2-Methylbutane	CH₂ CH₃ CH₃ CH CH₃	28	
2,2-Dimethylpropane	CH₃ C—CH₃ CH₃ CH₃	10	

Table 14.5.2   Structures and boiling points of the isomers of C_5H_{12}

Remember, attractions of this kind occur between *all* particles, whether these particles are non-polar, polar, or ionic. However, they are very weak forces when compared with hydrogen bonding (see Ch 2.9) or ionic attractions (see Ch 2.3). As we shall see, hydrogen bonding is very important in determining the properties of organic compounds with functional groups which contain oxygen and nitrogen, such as alcohols, carboxylic acids, amines, nylon and proteins.

The boiling points of the alcohols compared with those of alkanes of roughly the same relative molecular mass are shown in *Table 14.5.3*. Hydrogen bonding obviously causes a large increase in boiling point.

	Alkanes				Alcohols		
Name	Formula	M_r	b.p. /°C	Name	Formula	M_r	b.p. /°C
Ethane	CH_3CH_3	30	−89	Methanol	CH_3OH	32	65
Propane	$CH_3CH_2CH_3$	44	−42	Ethanol	CH_3CH_2OH	46	78
Butane	$CH_3(CH_2)_2CH_3$	58	−1	Propan-1-ol	$CH_3(CH_2)_2OH$	60	97
Pentane	$CH_3(CH_2)_3CH_3$	72	36	Butan-1-ol	$CH_3(CH_2)_3OH$	74	117

Table 14.5.3 The boiling points of alkanes and alcohols with similar M_r

Hydrogen bonding also leads to increased viscosity – 'stickiness' – in a liquid. As we saw in Ch 2.9, the comparative high viscosity of water means that it does not flow as easily as comparable non-polar liquids, and it is harder to stir. Molecules with more —OH groups should therefore be more viscous and have even higher boiling points. Ethane-1,2-diol (glycol) is used in antifreeze; it boils at 198 °C. Propane-1,2,3-triol (glycerol or glycerine) is well known to cooks and pharmacists as a sticky, slow-flowing liquid. Sugars contain several —OH groups per molecule. Honey, treacle and golden syrup all contain high concentrations of sugars.

SELF-ASSESSMENT
QUESTIONS
14.5

1   In each of the following groups of compounds, give a systematic name for the compound which will have the highest boiling point, and justify your choice of this compound.

   a $CH_3CH_2CH_2CH_3$,
      $CH_3CH_2CH_3$,
      $CH_3(CH_2)_6CH_3$

   b CH_3OCH_3,
      CH_3CH_2OH,
      $CH_3CH_2CH_3$

    c $CH_3C(CH_3)_2CH_2CH_3$,
        $CH_3(CH_2)_4CH_3$,
        $CH_3CH_2CH(CH_3)CH_2CH_3$

2 Look at these chemical formulae:

            $\overset{\displaystyle O}{\overset{\|}{}}$

A $CH_3CCH_2CH_3$
B $CH_3CH_2CH_2CH_2OH$
C $CH_3CH{=}CHCH_3$
D $CH_3CH(CH_3)CH_3$
  a Which is an alkane?
  b Which has the highest boiling point?
  c Which will show the largest degree of hydrogen bonding?
  d Which is an alcohol?

3 **a** From this list of data, plot a graph of boiling point (vertical axis) against the number of carbon atoms (horizontal axis).

Methanol	65
Ethanol	78
Propan-1-ol	98
2,2-Dimethylpropan-1-ol	83
Butan-1-ol	117
Pentan-1-ol	138
Hexan-1-ol	158

Boiling points for a range of alcohols/°C

  b Is there a general trend in the data?
  c Briefly explain your answer to **a**.
  d Are there any anomalous results on the graph?
  e Can you give a reason for your answer to **c**?

14.6 Types of organic reactions

In this section and the next we shall introduce the essential vocabulary which can be used t[o] describe any organic reactions. It is possible to get a long way in organic chemistry by accepting an[d] understanding the following statements:

- *There are only two ways in which a covalent bond can break.*
- *There are only three types of reagent.*
- *There are only four kinds of organic reaction.*

We shall look at each of these statements in turn.

The breaking of covalent bonds

A covalent bond consists of a shared pair of electrons (see Ch 2.4). *A covalent bond can break* [in] *only two different ways: homolysis and heterolysis.*

Homolysis ('homolytic fission')

In homolysis, the bond breaks symmetrically to form two highly reactive particles with unpaired electrons. These particles – which may be atoms or groups of atoms – are called **free radicals**.

$$A\overset{\circ}{\underset{\bullet}{}}B \longrightarrow A\bullet \quad {}^{\circ}B$$

('Homo-' means 'same'; '-lysis' means 'breaking apart').

The bond is most likely to break symmetrically if A and B have similar electronegativities (Ch 2.2), so that the electron pair is more or less dead centre between A and B. Homolysis often happens when no water or other polar solvent is present, for example when the reactant is in the gaseous state. We shall meet homolysis in reactions involving ultra-violet (UV) light (Ch 15.5) and also in the making of poly(ethene).

> **Definition:** Homolysis occurs when a chemical bond breaks by one electron from the bond pair moving to each of the atoms involved in the chemical bond.

Trouble in the ozone layer

Firstly, there is no such thing as an ozone layer. If all the ozone in the atmosphere could be compacted into a genuine gaseous layer at atmospheric pressure at sea level, it would be about 3 mm thick!

High-energy UV radiation from the sun breaks up some ordinary oxygen molecules in the stratosphere (between 15 and 50 km above sea level). This homolysis of O_2 gives oxygen radicals, i.e. oxygen atoms carrying unpaired electrons. These can combine with O_2 molecules to form trioxygen, O_3, which is usually known as ozone.

$$O_2 + O\bullet \longrightarrow O_3$$

The ozone can be broken up and converted back to O_2 by absorbing more UV radiation. It is this absorption of UV radiation which helps protect living things on Earth from being damaged and destroyed by radiation. Any failure in the ozone shield is therefore of great importance to us and to all life on the planet.

The 'ozone hole' above Antarctica was discovered in 1984 by Joe Farman and colleagues working for the British Antarctic Survey. They used simple ground-based instruments. Their results were soon confirmed by American satellites. It was then found that the satellites had observed falling ozone levels in the previous 10 years, but that the computer which processed the data had been programmed to ignore such low vales as 'impossible'. The moral of this story is obvious.

The failure to act earlier is especially strange because ozone loss and the reasons for it had been predicted during the 1970s by Paul Crutzen, Mario Molina and Sherwood Rowland, who received the Nobel Prize for Chemistry in 1995 for this work.

Under normal conditions, the rates at which ozone is created and destroyed become equal; and equilibrium is reached (see Ch 13.1). However, chlorine atoms with unpaired electrons – i.e. chlorine radicals – start a series of reactions which rapidly destroy ozone. Some chlorine radicals naturally get into the upper atmosphere because chloromethane, CH_3Cl, is given off in small amounts from oceans and through forest fires.

Homolysis by UV radiation breaks some CH_3Cl molecules into methyl and chlorine radicals. Since the 1970s it has been shown that chlorofluorocarbons (CFCs) used as aerosol propellants and in refrigeration can give chlorine radicals in the stratosphere. These compounds were invented for use as refrigerants, especially because they are chemically very stable. They last for a very long time unchanged at sea level. But, after many years, some of the CFCs released at sea level have diffused into the stratosphere, where the intense UV radiation causes homolysis of CFC molecules to form chlorine radicals, which then destroy ozone molecules faster than they can be created.

This problem is being taken very seriously indeed by the international bodies and by national governments. More than sixty countries agreed to phase out the use of CFCs, and an international fund was set up to help poorer countries to do without CFCs. But, so unreactive are the CFCs and so slow to diffuse into the stratosphere, that once their use is completely stopped it will take about another 30 years before the amount of chlorine radicals in the stratosphere goes back to what it was in the early 1980s.

> **Definition:** Heterolysis occurs when the pair of electrons in a chemical bond moves to one of the atoms involved in the chemical bond.

Heterolysis ('heterolytic fission')

The covalent bond breaks unsymmetrically – unevenly – to give ions. These ions do not normally exist for more than a tiny fraction of a second before they react.

$$A\overset{\circ}{\underset{\bullet}{}}B \longrightarrow A\overset{\circ\ominus}{\underset{\bullet}{}} B^{\oplus} \quad \text{or} \quad A\overset{\circ}{\underset{\bullet}{}}B \longrightarrow A^{\oplus} \quad \overset{\circ}{\underset{\bullet}{}}B^{\ominus}$$

Heterolysis is most likely to happen if A and B have very different electronegativities and if the A–B bond is a long one (long bonds are weak bonds – see Ch 11.5). This is particularly true if the reaction is carried out in water or some other polar solvent which interacts with any ions formed. In heterolysis, the more electronegative element in the bond will end up with both electrons and so with the negative charge.

Types of reagent

Free radicals

> **Definition:** A free radical is an atom or group of atoms with one or more unpaired electrons.

As already discussed, these are the extremely reactive species which are formed when homolysis happens. Free radicals often get involved in very rapid chain reactions (see Ch 15.5).

Nucleophiles

> **Definition:** A nucleophile is an atom or group of atoms which has an unshared (or lone) pair of electrons which it is able to donate to form a new chemical bond.

'Nucleophile' literally translated means 'nucleus loving'. A nucleophile is an electron-rich molecule or negative ion which attacks a part of the molecule where the electron density is low. (**Under any possible reaction conditions, the nucleus of an atom can never be completely bared!**) Typical nucleophiles include $OH^{\ominus}$ (hydroxide), $CN^{\ominus}$ (cyanide) and NH_3 (ammonia). Ammonia is a nucleophile because of the lone pair of electrons on the nitrogen atom (Ch 2.4).

A carbon atom can be opened up to attack by a nucleophile if an electronegative atom, such as a halogen, is attached to it. This drags the bonding pair of electrons partly away from the carbon atom – often shown by an arrow drawn on the bond – so leaving it with a partial positive charge (*Figure 14.6.1*).

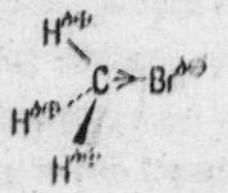

Figure 14.6.1 Polarity in CH₃Br

Electrophiles

> **Definition:** An electrophile is an atom or group of atoms that is able to accept a pair of electrons to form a new chemical bond.

'Electrophile' literally translated means 'electron loving'. An electrophile is an electron-deficient molecule or positive ion, which is therefore attracted by and attacks a part of a molecule which is rich in electrons – e.g. a carbon–carbon double bond, or a benzene ring (Ch 26.7). The electrophile is often produced as a first stage in the reaction. Typical electrophiles include $Br^{\oplus}$ (the bromonium ion) and $NO_2^{\oplus}$ (the nitronium ion).

> **Remember** (N.B. These are **not** definitions)
> **Homolytic fission** gives rise to free radicals.
> **Heterolytic fission** gives rise to charged or partially charged species.
> **Free radicals** contain at least one unpaired electron.
> **Nucleophiles** can donate a pair of electrons to form a new bond.
> **Electrophiles** are able to accept a pair of electrons to form a new bond.

Types of reaction

There are four main types of reaction associated with organic molecules, but only the first two are often met with at this level. Organic compounds can also undergo redox reactions with suitable reagents. Redox reactions will be discussed in Ch 14.8.

Substitution reactions

Definition: A substitution reaction occurs when two reactants react in such a way that an atom or group of atoms in one molecule is replaced by another atom or group of atoms. Two products must be formed.

These are the typical reactions of *saturated* compounds, i.e. those with no double or triple carbon–carbon bonds.

A carbon atom can form four covalent bonds and in saturated molecules all those bonds are single. In order to join anything else to a carbon atom in, for example, ethane, a hydrogen atom must come off. It's really just like a substitution in a football match – one player has to come off as the substitute comes on. Two examples are given below, the first involving homolysis and the second involving heterolysis. Displayed formulae are used to show which bonds are broken and made.

Ethane + Cl—Cl (diffused light) → Chloroethane + H—Cl

1-Bromopropane + OH⁻ (dilute aqueous) (heat) → Propan-1-ol + Br⁻ (aq)

Hydroxide ion from an alkali

Addition reactions

Definition: An addition reaction occurs when two reactants react to produce a single product.

These are typical of *unsaturated* compounds, such as the alkenes. The reagent adds across the double bond. A simple way of looking at it is that one of the bonds in a carbon–carbon double bond comes undone and is then used to make two new single bonds. For example:

Ethene + Br—Br (room temperature) → 1,2-Dibromoethane

Notice that an addition reaction gives only *one* product. Benzene has the molecular formula C_6H_6, which means that it is highly unsaturated. From this you might reasonably expect it to be involved in addition reactions. But the benzene ring system is so very stable that its usual reaction is substitution (Ch 26.5).

Elimination reactions

In these reactions a saturated organic molecule loses a small molecule and becomes unsaturated. Usually a halogen atom or hydroxyl group leaves together with a hydrogen atom, so that a molecule of hydrogen halide or water is lost from the original molecule. For example:

1-Bromopropane + OH⁻ (Heat concentrated in ethanol) → Propene + Br⁻ + H_2O

Any equation, almost always, summarises a reaction which occurs in more than one step. Notice that the HBr which is stripped off reacts with hydroxide to give water plus bromide ions.

$$
\begin{array}{c}
\text{H H}\\
| \ |\\
\text{H--C--C--H}\\
| \ |\\
\text{H OH}\\
\text{Ethanol}
\end{array}
\xrightarrow[\text{heat}]{\text{concentrated } H_2SO_4}
\begin{array}{c}
\text{H H}\\
| \ |\\
\text{C==C} + H_2O\\
| \ |\\
\text{H H}\\
\text{Ethene}
\end{array}
$$

In this reaction the water, which is stripped off, is absorbed by and dilutes the concentrated sulphuric acid.

The cracking process used during petroleum refining involves elimination (see Ch 15.4).

Rearrangement reactions

This is the only other type of reaction. Rearrangement reactions are rarely met with at this level. In such a reaction, the atoms in a molecule are simply rearranged, but nothing is lost or gained.

In the reforming process used during the conversion of crude petroleum into useful compounds, some of the reactions involve rearranging 'straight-chain' alkane molecules into 'branched-chain' ones by heating with a suitable catalyst (see Ch 15.4). For example:

$$
CH_3CH_2CH_2CH_2CH_2CH_3 \xrightarrow[\text{catalyst}]{\text{heat}}
\begin{array}{c}
\text{CH}_3\\
|\\
CH_3\text{---}CH\text{---}CH_2CH_2CH_3
\end{array}
$$
Hexane   2-Methylpentane

Remember

In organic chemistry, the type of reaction that is occurring can be considered in terms of the numbers of reacting particles and the number of product particles.

	Reactants	Products
Substitution:	2 molecules	2 molecules
Addition:	2 molecules	1 molecule
Elimination:	1 molecule	2 molecules
Rearrangement:	1 molecule	1 molecule

'Condensation reactions'

You may meet a few reactions in which addition is followed immediately by elimination of a small molecule such as water. A whole class of useful materials, including Nylon, is known as **condensation polymers** (Ch 28.1).

A typical condensation reaction may involve a ketone or aldehyde with a primary amine:

$$
\begin{array}{c}
\text{CH}_3\\
\diagdown\\
\quad C{=}O + H_2N\text{---}CH_2CH_3\\
\diagup\\
\text{CH}_3
\end{array}
\xrightarrow{\text{addition}}
\begin{array}{c}
\text{CH}_3 \quad \text{OH}\\
\diagdown \ | \\
\quad C \quad H\\
\diagup \ | \\
\text{CH}_3 \ \text{N---CH}_2\text{CH}_3
\end{array}
\xrightarrow{\text{elimination}}
\begin{array}{c}
\text{CH}_3\\
|\\
C{=}N\text{---}CH_2CH_3 + H_2O\\
|\\
\text{CH}_3
\end{array}
$$

Note that condensation is not a type of organic reaction. It is the result of *two* of the genuine reaction types, addition and elimination, happening together. Condensation reactions can usually be recognised because one reactant has a double bond in the molecule (usually a $>C{=}O$ double bond), and the organic product is a larger molecule containing a double bond of some kind. The only other product is water or some other small molecule.

SELF-ASSESSMENT QUESTIONS 14.6

1 Describe each of the following as electrophile, nucleophile or free radical.

   a NH_3

   b $Br\bullet$

   c H_2O

   d $NO_2^{\oplus}$

   e $Cl^{\delta\oplus}$

2 Classify each of these reactions as addition, substitution, elimination or rearrangement.

   a $CH_3CH{=}CH_2 + HBr \longrightarrow CH_3CHBrCH_3$

   b $CH_3CH_2CH_2CH_2CH_2CH_3 \longrightarrow CH_3CH(CH_3)CH_2CH_2CH_3$

   c $CH_4 + Br_2 \longrightarrow CH_3Br + HBr$

   d $CH_3CH_2CHClCH_3 \longrightarrow CH_3CH_2CH{=}CH_2 + HCl$

3 Define each of the following terms:

   a electrophile

   b free radical

   c substitution reaction

   d homolysis

   e nucleophile

14.7 Introduction to reaction mechanisms

As we have already seen (Ch 12.1), very few reactions happen in just one simple step. Careful *practical* study of a reaction, especially the influence which the concentration of each reagent has on the rate of reaction (see Ch 21.4), can give information about each individual step of the reaction.

A reaction mechanism is a detailed description of the reaction, in which the separate steps are shown. A mechanism for a reaction must fit the experimental observations.

All we shall do in this section is introduce the vocabulary which will be used for the reactions which will be met in later chapters.

First, in order to discuss mechanisms we need to introduce the 'curly arrow' idea. A curly arrow shows how a *pair of electrons* moves during a reaction. This pair could be a lone pair (as in $:OH^{\ominus}$) or a pair of electrons forming a covalent bond (as in ${\geq}C{-}Br$). *The curly arrow must start at a pair of electrons, which may already be in a bond. It must finish either between the two atoms which are to be joined by the new bond, or on one of the atoms between which the old bond has been broken (Figure 14.7.1).*

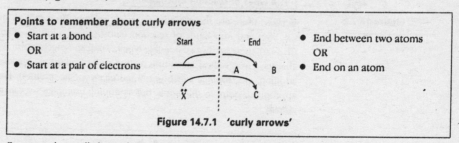

Points to remember about curly arrows

- Start at a bond
  OR
- Start at a pair of electrons

   Start   End

   A   B

   X   C

- End between two atoms
  OR
- End on an atom

Figure 14.7.1 'curly arrows'

For example, an alkali attacking a bromoalkane can be represented like this:

$$HO^{\ominus} \quad C{-}Br \longrightarrow HO{-}C \quad Br^{\ominus}$$

Here the hydroxide ion is the electron-pair donor or **nucleophile** (see Ch 14.6). The bromoalkane is the electron-pair acceptor – but, in order to accept the new pair of electrons, the carbon atom must lose a pair.

Many people get confused at this point, because they cannot link the movement of a pair of electrons with the loss of a *single* negative charge from the hydroxide ion and the gain of the negative charge by the bromine. But a covalent bond consists of a pair of electrons, usually one electron from each atom.

So when the hydroxide ion gives two electrons to form the new bond, one electron can be thought of as the extra electron which gives hydroxide its negative charge, the other as the one which would be used anyway by the oxygen to form the bond. Similarly, for the breaking of the C—Br bond, one of the pair of electrons which bonds carbon to bromine comes from the bromine anyway, so that when both the electrons end up on bromine, only one electron is extra. So the bromine only gains the one negative charge (see *Figure 14.7.2*).

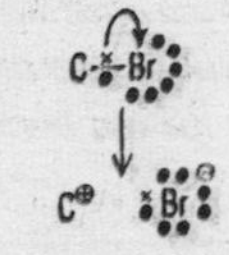

Figure 14.7.2

As one pair of electrons arrives to form the new bond from oxygen to carbon, so the other pair leaves for the bromine. The two shifts must occur together, as a carbon atom cannot cope with more than eight electrons (i.e. four pairs) in its outer electron shell.

A reaction of this kind is always helped if the *'leaving group'*, i.e. the atom or group which is expelled during the reaction, is able to accept a negative charge easily. As we have seen (Ch 8.4), bromine, chlorine and iodine readily accept the extra electrons to become negatively charged ions. Halogen atoms make good leaving groups; a good leaving group helps this type of reaction to happen.

In the example we have just used, the curly arrow represents **heterolysis** (Ch 14.6); the bond breaks unequally. In **homolysis** (Ch 14.6) the covalent bond breaks symmetrically to give very reactive **free radicals** (Ch 14.6). Homolysis can be represented by 'half arrows'. For example:

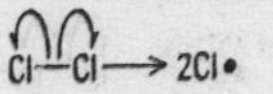

$$Cl—Cl \longrightarrow 2Cl \bullet$$

(Note how the half arrows are drawn.)

Here, the bond has broken so that one electron goes to each of the atoms involved.

Side reactions

For many *inorganic* reactions in solution such as those involved in titrations, the reaction appears to be instantaneous – all over in a split second. Furthermore, the reaction is very often complete; the yield is 100% and all the reactants are converted to the products.

Most *organic* reactions are very different. Instead of ions in solution, molecules with very strong covalent bonds are involved. Heat is often needed to force the reaction to proceed at an acceptable rate. Alternative reaction mechanisms are possible. In nearly every case, some of the reactant molecules have the 'wrong bond' broken, so that some unwanted *'by-products'* are formed.

Organic reactions *never* give a yield of 100%. There are always one or more side reactions, giving by-products. A major job after any organic reaction is separating the *wanted* product from the *unwanted* by-products in the reaction mixture. This will be dealt with in Ch 29.6.

The types of reaction mechanism

Because there are three kinds of organic reagent (free radical, nucleophile and electrophile; see Ch 14.6) and four kinds of reaction (substitution, addition, elimination and rearrangement; see Ch 14.6), there are twelve possible types of reaction mechanism. We shall meet six of them in future chapters: free radical substitution, free radical addition, nucleophilic substitution, nucleophilic addition, electrophilic substitution and electrophilic addition. Each will be fully discussed at the appropriate point in this book, but as outline examples we can use three reactions we have met already.

1. 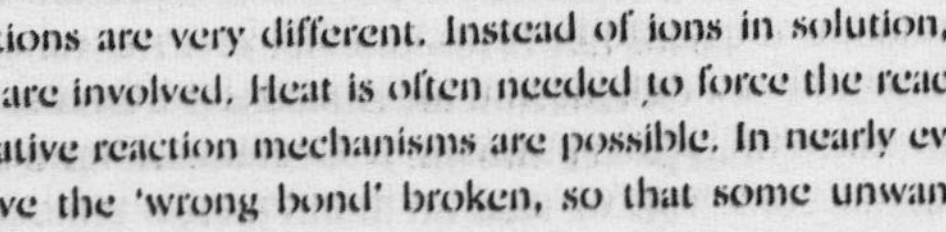
Propane + Cl—Cl → 1-Chloropropane + H—Cl

This reaction involves substitution of a chlorine atom for hydrogen. Light causes homolysis of chlorine molecules to form chlorine atoms, which are free radicals. The reaction is an example of **free radical substitution**.

2.

$$H-\underset{\underset{H}{|}}{\overset{\overset{H}{|}}{C}}-\underset{\underset{H}{|}}{\overset{\overset{H}{|}}{C}}-\underset{\underset{H}{|}}{\overset{\overset{H}{|}}{C}}-\underset{\underset{H}{|}}{\overset{\overset{H}{|}}{C}}-Br + OH^{\ominus}(\text{dilute aqueous}) \xrightarrow{\text{heat}} H-\underset{\underset{H}{|}}{\overset{\overset{H}{|}}{C}}-\underset{\underset{H}{|}}{\overset{\overset{H}{|}}{C}}-\underset{\underset{H}{|}}{\overset{\overset{H}{|}}{C}}-\underset{\underset{OH}{|}}{\overset{\overset{H}{|}}{C}}-OH + Br^{\ominus}(aq)$$

| 1-Bromobutane | Hydroxide ion | Butan-1-ol |

This is another substitution. The negatively charged hydroxide ion is an electron pair donor – a nucleophile. The reaction is therefore **nucleophilic substitution**. (N.B. In theory, it is possible to argue that 1-bromopropane is carrying out electrophilic attack on the hydroxide ion. In other words, the 1-bromopropane is *accepting* the electron pair rather than the hydroxide ion *donating* it. Both actions are, of course, happening at the same time. In practice, it is usually possible to see which *organic* molecule is being attacked and hence to classify the reaction.)

3.

$$\underset{H}{\overset{CH_3}{>}}C=C\underset{H}{\overset{H}{<}} + Br-Br \xrightarrow[\text{temperature}]{\text{room}} CH_3-\underset{\underset{H}{|}}{\overset{\overset{Br}{|}}{C}}-\underset{\underset{H}{|}}{\overset{\overset{Br}{|}}{C}}-H$$

| Propene | 1,2-Dibromopropane |

Here, addition has occurred; there is only one product. The electron-rich $>C=C<$ double bond attracts electrophiles; so the reaction is therefore **electrophilic addition**.

Examples of electron movements in reaction mechanisms

Nucleophilic substitution

$$CH_3-\underset{\underset{\ddot{O}H^{\ominus}}{|}}{\overset{\overset{H}{|}}{C}}-Br \longrightarrow CH_3-\overset{\overset{H}{|}}{\underset{\underset{H}{|}}{C}}-OH + Br^{\ominus}$$

Electrophilic addition

$$CH_2=CH_2 \quad \underset{Br}{\overset{Br}{|}} \longrightarrow {}^{\oplus}CH_2-CH_2 \longrightarrow CH_2-CH_2$$
$$\qquad\qquad\qquad :Br^{\ominus} \quad Br \qquad Br \quad Br$$

Types of chemical reactions are summarised in *Table 14.7.1*, and the *mechanisms* we will need later in the course are shown.

	Substitution	Addition	Types of bond breaking
Free radical	Alkanes with chlorine Ch 15.5	Polymerisation of alkenes Ch 16.7	Homolysis
Nucleophilic	Halogenoalkanes with hydroxide ions Ch 17.4	Carbonyls with HCN Ch 19.4	Heterolysis
Electrophilic	Nitration of arenes Ch 26.5	Alkenes with bromine Ch 16.6	Heterolysis

Table 14.7.1

SELF ASSESSMENT
QUESTIONS
14.7

1 In the following pairs of compounds, which contain pairs of structural isomers?

A $CH_3CH_2CH_2CH_2CH_3$ and $CH_3CH(CH_3)CH_2CH_3$

B $CH_3CH(CH_3)CH_2CH_3$ and $CH_3CH(CH_3)CH_2CH_2Br$

C $CH_3CH(CH_3)CH_2CH_2Br$ and $CH_3CH(CH_3)CHBrCH_3$

D $CH_3CH(CH_3)CH=CH_2$ and $CH_3C(CH_3)_2CH=CH_2$

2

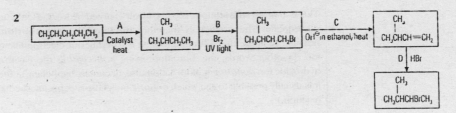

Look at this reaction sequence, and name the type of reaction corresponding to each of the steps A to D.

14.8 Writing organic equations

- The symbol R, meaning any alkyl group, can be used in equations intended to show the reactions of the **functional group.** For example:

$$R—Br + Na^{\oplus}OH^{\ominus}(aq) \xrightarrow{\text{reflux}} R—OH + Na^{\oplus}Br^{\ominus}(aq)$$

This means that *any* bromoalkane heated with dilute aqueous sodium hydroxide will give the corresponding alcohol.

Where necessary, R_1, R_2 etc can be used to indicate alkyl groups which may have different chain lengths, e.g.

$$\begin{array}{c} R_1 \\ \diagdown \\ C=0 \\ \diagup \\ R_2 \end{array}$$

would be a general formula for a ketone.

- You should have noticed that in every organic equation we have written so far, *the conditions for the reaction have been included*, written above the arrow which represents the conversion of reactants to products.

You should do this always, even if the reaction needs no heat or catalysts, etc.

$$\begin{array}{c} H \\ \diagup \\ C=C \\ \diagup \diagdown \\ H H \end{array} + Br—Br \xrightarrow{\text{room temperature}} \begin{array}{c} Br Br \\ | | \\ H—C—C—H \\ | | \\ H H \end{array}$$

An incentive for doing this can be found in most published mark schemes for A level chemistry examinations – in organic reactions, *one mark is usually given specifically for the correct conditions.*

- Oxidation and reduction reactions in organic chemistry can be represented using a form of shorthand. The reason for this is that full redox equations in organic chemistry can look very forbidding. For example, the full equation for the oxidation of ethanol to ethanal using acidified potassium dichromate(VI) is:

$$3CH_3CH_2OH(aq) + Cr_2O_7^{2\ominus}(aq) + 8H^{\oplus}(aq) \xrightarrow{\text{heat}} 3CH_3CHO(aq) + 2Cr^{3\oplus}(aq) + 7H_2O(l)$$

Oxidation can be represented by the shorthand symbol [O].
So:

$$CH_3CH_2OH + [O] \longrightarrow CH_3CHO + H_2O$$

This is far easier to comprehend and remember!
Similarly, *reduction can be represented by [H].*
So:

$$CH_3CHO + 2[H] \longrightarrow CH_3CH_2OH$$

But note that it is *very important* that such shorthand redox equations *must balance*.
- When a reagent is asked for in an equation, its formula must be given.

$$RBr + OH^\ominus(aq) \xrightarrow{heat} ROH + Br^\ominus(aq)$$

This equation is acceptable if the action of alkali on a bromoalkane is asked for; but if the reaction of sodium hydroxide solution with bromopropane is required, the equation *must* be written as:

$$CH_3CH_2CH_2Br + NaOH(aq) \xrightarrow{heat} CH_3CH_2CH_2OH + NaBr(aq)$$

A reagent must be a chemical that could be found on the shelf in a chemical store. Hydroxide is insufficient for the name of a reagent, but sodium hydroxide is fine.

SUMMARY 14

In this chapter the following new ideas have been covered. *Have you understood and learned them?* Could you explain them clearly to a friend? The page references will help.

- Calculation of an empirical formula from percentage or mass composition. p343
- Recognition of a molecular ion peak in a mass spectrum and use of the M_r to work out the molecular formula. p343
- Drawing of organic formulae in molecular, structural, linear or displayed formats. p344
- Recognition of the formulae and names of the functional groups. p350
- Identification of aliphatic, alicyclic and aromatic structures. p351
- The systematic naming of straight chain and branched chain alkanes. p353
- Identification of the isomers produced by chain, functional group and position isomerism. p359
- Understanding how geometrical and optical isomerism arise. p360
- Use of molecular structures to explain differences in physical properties. p363
- Understanding the words homolysis and heterolysis. p367
- Recognition of the three types of reagents in organic chemistry. p368
- The four types of organic reactions. p368
- The use of curly arrows to represent electron movements. p371
- The idea of organic reaction mechanisms and side reactions. p371
- The writing of organic equations, including the conditions needed. p374

At the end of each chapter in the organic chemistry section of this book, the summary of the reactions of the functional group being studied will appear. In this chapter, we have tried to show that organic chemistry is very organised and that there are a limited number of types of reactions possible. The substitution, addition and elimination reactions are three of the four types of important organic reactions. It might help you to think of a few other types of reaction classifications. Redox, condensation and polymerisation reactions are all reactions that have been mentioned so far. Cracking and reforming are considered in the next chapter.

To help remember the reactions of a particular functional group, at the end of the chapter we will try to list the reactions as:

R   Reduction
O   Oxidation
C   Condensation
A   Addition
S   Substitution
E   Elimination
P   Polymerisation

The reactions of each functional group can be remembered more easily by using a mnemonic ROCASEP diagram:

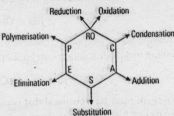

N.B. Not all functional groups are capable of the full range of these reactions. And, sometimes, some possible reactions are not important enough to be included in this book!

EXAMINATION
QUESTIONS
14

1   Campesterol is a chemical, similar in structure to cholesterol, which is extracted from soya beans and might provide a means of limiting absorption of cholesterol in the human gut. The skeletal structure is:

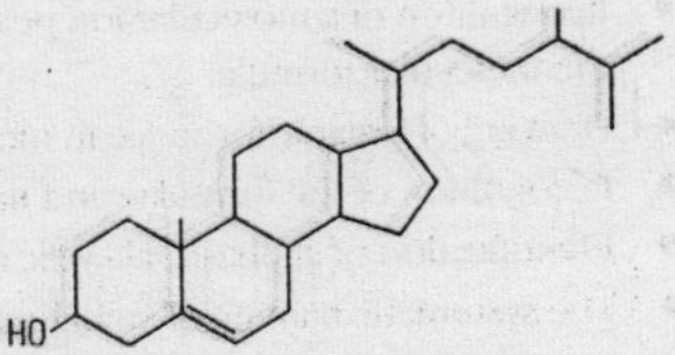

a   Identify two functional groups present in this molecule.
b   Write the molecular formula for the molecule and then calculate its relative molecular mass.
c   Show, by calculation, that the molecule contains 4.15% oxygen.
d   Part of the side chain in the molecule is $CH_3CH(CH_3)CH(CH_4)CH_3$. Draw out a full displayed formula for this fragment.
e   Draw two chain isomers of the molecule in **d**.
f   Give the systematic name for the molecule quoted in **d**.

2   A synthetic intermediate used in industrial organic chemistry is $CH_2ClCH=CH_2$.
a   Write the systematic name for this compound.
b   Draw a displayed formula for the molecule.
c   Identify the two functional groups present in this molecule.
d   The presence of an electronegative atom creates a permanent dipole in this molecule. Which is the electronegative element? On your displayed formula, mark the dipole that exists in the molecule, using the symbols $\delta^{\ominus}$ and $\delta^{\oplus}$.
e   The compound reacts at room temperature with chlorine according to the equation:

$$CH_2ClCH=CH_2 + Cl_2 \longrightarrow CH_2ClCHCl-CH_2Cl$$

Classify this reaction as substitution, elimination or addition.

3   **a** Draw a displayed formula for 2-methylbut-1-ene, C_5H_{10}.

    b Draw the structure of any chain isomers of this molecule.

    c Draw the structure of any position isomers of this molecule.

    d 2-Methylbut-1-ene reacts with bromine at room temperature to produce a new compound.

$$C_5H_{10} + Br_2 \longrightarrow C_5H_{10}Br_2$$

    Is this an addition, substitution or elimination reaction?

    e $C_5H_{10}Br_2$ exists as a number of optically active isomers. Using this compound as an example, what is the essential requirement in a molecule if it is to show optical activity?

    f $C_5H_{10}Br_2$ reacts with aqueous hydroxide ions, $OH^{\ominus}$, to form a diol. What does the term 'di-' signify?

    g Explain why the reaction in **f** can be considered as nucleophilic attack on the organic molecule.

4 One of the additives which was used in leaded petrols was 1,2-dibromoethane.

    a Write the molecular formula for 1,2-dibromoethane.

    b What is the empirical formula of this molecule?

    c When treated with aqueous sodium hydroxide, there is a nucleophilic substitution reaction. What is meant by the term nucleophilic?

    d This reaction proceeds through a heterolytic type of mechanism. What does heterolytic mean?

    e 1,1-Dibromoethane is an isomer of 1,2-dibromoethane. Explain what is meant by an isomer.

15

the
alkanes

Many of the organic chemicals used today, in a variety of products, from pain killing drugs and household detergents to polyester scarves and plastic carrier bags, are derived from fossil fuels. These same fossil remains provide three-quarters of global energy needs.

The fossil fuels – coal, natural gas and crude oil – are materials that have been formed over millions of years. Crude oil, often referred to as petroleum, and natural gas were formed from once-living materials around 200 million years ago, and coal may be up to 300 million years old. The remains of marine creatures and large fern-like plants have been subjected, over millions of years, to large temperature and pressure changes. The materials that made up the original living organisms have been changed and now appear as oil, gas, peat, lignite and coal.

This source of raw materials provides many benefits to modern society, but there is a price to pay. Fossil fuels are a finite resource – there is only a certain amount of coal, oil and gas available beneath the surface. One day these supplies will run out – these resources are *non-renewable*. While coal reserves are expected to last for a few hundred years, the gas and oil reserves are likely to run out much more quickly. New deposits are being discovered, but usually in increasingly difficult areas.

One important role for the next generation of chemists is to find alternative sources for the organic raw materials – for chemicals that contribute both to our standard of living and to providing the fuels for our ever-increasing energy demands. In less industrialised countries which do not have easy access to crude oil, the use of plants to provide energy from biomass is already established. We all have to develop the use of renewable energy sources – for example, biomass, solar and hydroelectric power – beyond their present limited applications.

15.1 Crude oil and petroleum

Crude oil, as a raw material, can benefit very many aspects of modern society. As a convenient source of fuel and as a material for further processing into chemical feedstocks (materials that are converted into commercial products), it is used for making an extensive range of everyday goods.

One reason for the dependence on crude oil is that usable materials are made in a relatively straightforward way from the crude oil. As a liquid, pumped from underground natural reservoirs created by impervious rock formations overlying porous strata containing oil and gas, crude oil itself is of very limited use. It has to be processed before it can be used – and this is where refining and refineries come in.

Crude oil is a complex mixture of compounds, consisting mainly of a variety of alkanes together with a small amount of aromatic hydrocarbons. There is also a small amount of nitrogen-, oxygen- and sulphur-containing compounds.

Most crude oils are thin, dark brown liquids, and it should be appreciated that they contain a considerable amount of petrol-like chemicals. The image of a thick, viscous material comes from pictures of the remains of crude oil spillages, where the 'light ends' – the volatile, petrol-like chemicals – have evaporated away, leaving the thick residue.

Since each source of crude oil originally came from different mixtures of marine organisms and plants, the compositions of the crude oils from different areas of the world differ slightly. For example, *Figure 15.1.1* shows the average compositions of North Sea and Middle East crude oil in terms of the fractions obtained from them.

We know that the natural gas and oil deposits were formed from the remains of animals and plankton because the structures of some of the hydrocarbons found in crude oil could only have come from these sources.

North Sea oil	Middle East oil

- petrol
- kerosene
- gas oil
- fuel oil

Figure 15.1.1

The technology of recovering the oil from the reservoirs is impressive. Engineers design rigs that can drill through thousands of feet of rock and can survive mountainous seas when the oil field is under the sea. Drills can not only go straight down but can also curve, so that the whole of an oil field can be exploited. The crude oil is transported around the world in very large crude oil carriers, carrying 100 000 tonnes or more of the crude oil at a time – which would give enough petrol to provide a car with fuel for a few thousand years. The process of refining starts once the oil is delivered from the well head.

The initial separation of the complex mixture into useful 'fractions' is achieved by primary distillation. The industrial equipment has to handle many tonnes per hour of the crude oil and provide a very efficient fractional distillation process (see Ch 29.2).

In the primary distillation process, the crude oil is heated before it enters the distillation column, which may be sixty metres or more high. The hot liquid enters the column at some point in the middle of the fractionating column. Volatile compounds rise up the column. Those materials that have not vaporised start to move down the column.

At each temperature, some compounds are changing into gas and others are condensing, depending on their boiling points. Gases rise up the column and liquids gradually fall down the column. When the liquid reaches a level corresponding to its boiling point – it boils. As the gases rise, they reach their condensation point and so condense (*Figure 15.1.2*).

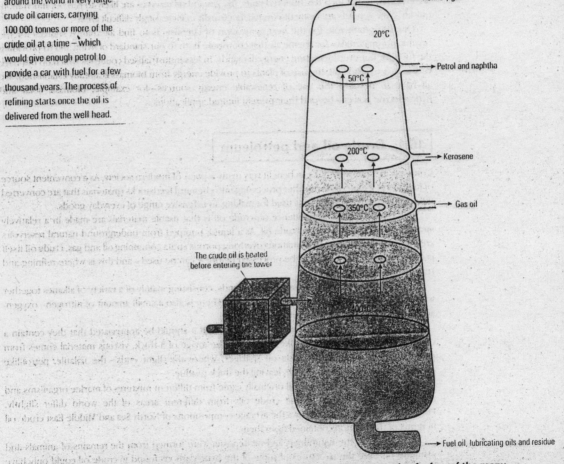

Figure 15.1.2 Primary distillation – only of a few of the many fractions produced are shown

The process makes use of the fact that there is a *temperature gradient* in the column – it g steadily cooler up the column.

The vertical stacks in many oil refineries consist of fractional distillation columns. Distillation number of different stages in the refinery process means that there are a number of columns need for the different distillation processes.

Figure 15.1.3 – An oil refinery

Fractional distillation in the laboratory uses a batch process in which each fraction is collected as it reaches the end of the apparatus. The boiling liquid rises in the fractionating column. In this column, boiling liquids and condensing vapours are in contact, and the separation of the different fractions occurs up the column. As a particular fraction reaches the top, it enters the condenser and each fraction collects in turn at the end of the condenser.

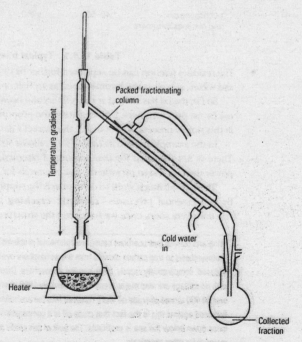

Figure 15.1.3 – A laboratory-scale fractional distillation apparatus

The industrial process needs to be continuous to cope with such large volumes of liquids. Separation has to be very carefully controlled, and a particular fraction only has a boiling range of a few degrees Celsius. By using bubble caps in the column, increased efficiency of separation is possible (*Figure 15.1.4*).

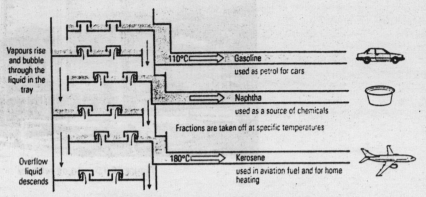

Figure 15.1.4   Inside the fractionating column

Fraction	% by mass	b.p. /°C	No. of carbon atoms	Major uses of fraction
Refinery gas (mainly CH_4)	2	<20	1–4	Domestic fuel, chemicals
Gasoline/naphtha	15–30	20–175	5–10	Solvents, petrol, cracking
Kerosene/paraffin	10–15	175–250	11–14	Aviation fuel, home heating
Gas oil	15–25	250–400	15–25	Diesel fuel, industrial heating, gasoline, cracking
Lubricating oil and residues (bitumen)	40–50	>350	20–34	Lubrication, asphalt, fuel for power stations and large ships, cracking

Table 15.1.1   Typical fractions from crude oil

The residues fraction can be separated further by distilling under vacuum, to obtain lubricating oils and waxes, as well as the bitumen used as asphalt or tar.

So far, the oil has arrived at the refinery and been separated into the different fractions ready for use by us, the customers. Unfortunately, the proportions of petrol and aviation fuel, etc, produced in this simple process do not match the market's demands.

In the example quoted in *Table 15.1.1*, almost 50% of the crude oil consists of the heavy residue. There is not the need for that amount of bitumen, and the demand for the petrol fractions – to power the cars and to provide the raw chemicals for the petrochemical industry – is much larger.

There is still much work to do to match the supply with the demand. The imbalance is addressed by two essential processes – **catalytic cracking** and **reforming**. We will consider these two reactions very soon, once we have seen the structures of the hydrocarbons involved.

Using and burning hydrocarbons cause environmental problems. The chlorophyll in the ancient plants used photosynthesis to trap carbon dioxide from the atmosphere over millions of years, and now the carbon dioxide is being released, comparatively rapidly, back into the atmosphere. Disposal of plastic waste and the pollution issues concerned with oil spillage are also major areas for concern. By 1998, there had been seven major spillages where, in each case, over 50 000 tonnes of crude oil were released into the environment.

Balanced against this is the fact that crude oil is a comparatively easy raw material to obtain and process. Even drilling miles down below the sea is profitable. The bulk of this crude oil is used for fuel, whereas only 10% is used as a raw material for other chemicals.

15.2 The alkanes

It is time now to start dealing with some real organic chemicals – real molecules to which it is possible to apply the principles of the last chapter. Where else to start but with the simplest compounds, those which contain only carbon and hydrogen?

The abundance of carbon compounds arose because of the ability of carbon atoms to form chains (catenation), the strong C—C bond, and the stability of the C—C and C—H bonds towards reaction with oxygen and water. In the previous chapter, there were a number of different frameworks which fitted these criteria – the aliphatic, alicyclic, and aromatic structures. Here, we shall discuss the **saturated** aliphatic and alicyclic compounds.

Aliphatic compounds either have straight-chain or branched-chain structures. The simplest of these hydrocarbons is methane which has a single carbon atom with four hydrogen atoms joined to it (see Ch 2.4). The structure is three-dimensional, and so the bond angle between H—C—H atoms is the tetrahedral bond angle of 109.5° (*Figure 15.2.1*).

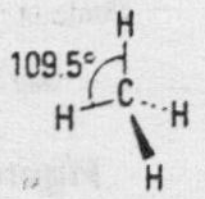

$$109.5°\quad H-\underset{\underset{H}{|}}{\overset{\overset{H}{|}}{C}}-H$$

Figure 15.2.1   Structure of methane

In the 'straight-chain' alkanes, a hydrogen on the end of the chain is replaced by a carbon atom and the correct number of hydrogen atoms to form the next member of the series (*Table 15.2.1*).

CH_4	Methane
CH_3CH_3	Ethane
$CH_3CH_2CH_3$	Propane
$CH_3CH_2CH_2CH_3$	Butane
$CH_3CH_2CH_2CH_2CH_3$ or $CH_3(CH_2)_3CH_3$	Pentane

Table 15.2.1   The first members of the alkane series

In effect, there is an extra $-CH_2-$ group added each time – to give the simplest homologous series (see Ch 14.2).

Activity

Construct models of the first four members of the alkanes series.

In particular, look at the shapes of the 'straight-chain' molecules and the possible isomers for each formula.

If we look carefully at some of these skeletons, we can see that alkane chains are *never* straight, although alkane chains with no other alkyl groups attached are usually known as 'straight-chain' alkanes. Rotation about a single carbon–carbon bond is possible, so that alkanes with more than three carbon atoms can exist in a large number of shapes, as long as each successive carbon atom is joined to just two other carbon atoms. For example some of the shapes formed by pentane might be:

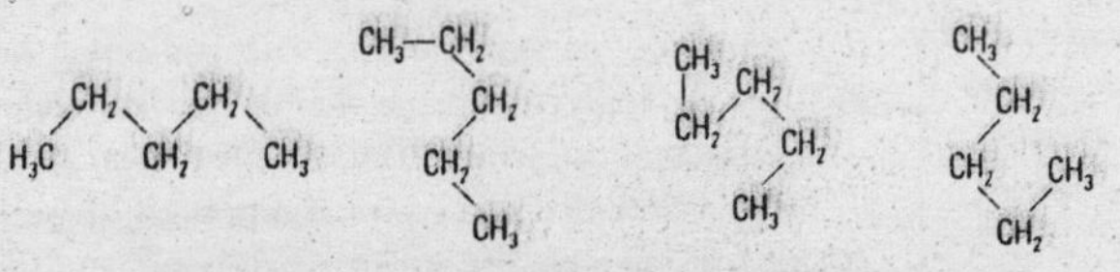

Structures of pentane – the molecule can twist

Every one of these structures represents the same compound, pentane, C_5H_{12}. The structures of pentane and its isomers are shown in *Figure 15.2.2*.

Pentane   Methylbutane   2, 2-dimethylpropane

See Ch. 14.4 to remind yourself about the different types of isomerism

Figure 15.2.2   Molecular models of pentane isomers

For each of these alkanes, a simple formula links the number of carbon atoms and hydrogen atoms:

$C_nH_{(2n+2)}$ – e.g., the alkane corresponding to C_{12} is $C_{12}H_{26}$.

So, the *general formula* for the homologous series of alkanes is $C_nH_{(2n+2)}$.

When cycloalkanes are formed, two hydrogen atoms have to be lost to form the extra C—C bond in the cyclic structure; the molecular formulae now correspond to C_nH_{2n}. So cyclobutane is C_4H_8.

The simplest of the cycloalkanes are shown in *Table 15.2.2*.

Cycloalkane	Shape	Bond angle
Cyclopropane	 △	60°
Cyclobutane	 ▢	90°
Cyclopentane	 ⬠	109.5°
Cyclohexane	 ⬡	109.5°

Table 15.2.2   Bond angles and shapes of cycloalkanes

Notice the shapes of the cycloalkanes. The bond angles in these molecules are held by the ring structure. In cyclobutane the bond angle is 90°, quite a way from the tetrahedral bond angle of 109.5°, which means the molecule is 'strained' and relatively unstable. The stable alicyclic rings have five or six carbon atoms in them. These stable rings are not planar; the carbon atoms can take up the tetrahedral configuration, so relieving the strain.

Carbon chains

Number of carbon atoms	'Stem'
1	Meth-
2	Eth-
3	Prop-
4	But-
5	Pent-
6	Hex-

Alkyl groups

CH_3—	Methyl
CH_3CH_2—	Ethyl
$CH_3CH_2CH_2$—	Propyl

Table 15.2.3

We have already looked at the naming of the branched alkanes (Ch 14.3), but it is worth refreshing our memories a little:

- decide on the longest carbon chain, and name the stem (*Table 15.2.3*);
- number the carbon atoms in the chain;
- decide on the ending – this must be -ane if it is an alkane;
- name any side chains – alkane-type side chains become 'alkyl' (*Table 15.2.3*; see also Ch 14.3);
- prefix any side chain with a number to show the carbon atom to which it is attached, minimising the numbers used;
- use di-, tri-, tetra-, penta-, if any side chain is duplicated;
- there should be a number to show where every side chain is located;
- numbers are separated from text by a hyphen, and from each other by a comma, so that we write, for example, 2,2,3-trimethylpentane;
- side chains are usually listed alphabetically, but ignoring di-, tri-.

So,

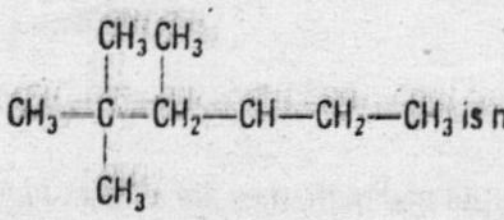

CH_3—C—CH_2—CH—CH_2—CH_3 is named 2,2,3-trimethylhexane

and

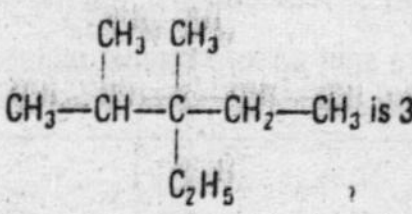

CH_3—CH—C—CH_2—CH_3 is 3-ethyl-2,3-dimethylpentane

If you do have difficulty working out the longest chain in any carbon compound, try to look at the carbon–carbon bonds as the routes through a maze – but in this maze you have to find the longest way from one end to the other. The only way to move is along a carbon–carbon bond – you cannot move across empty space to another carbon atom! All you need to draw out is the carbon skeleton. Once you have recognised the longest chain, you can number the carbon atoms, starting at the end that will give the lowest numbers in any numbering of substituents.

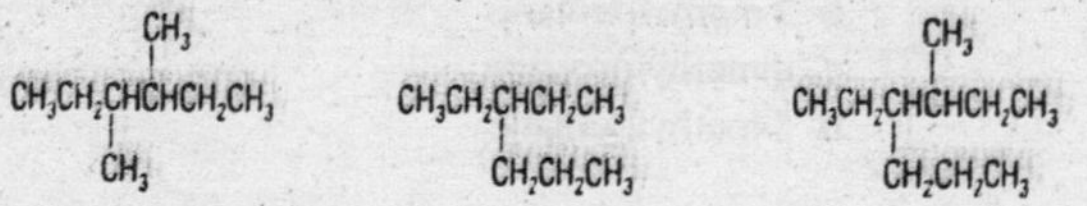

With just the carbon skeletons:

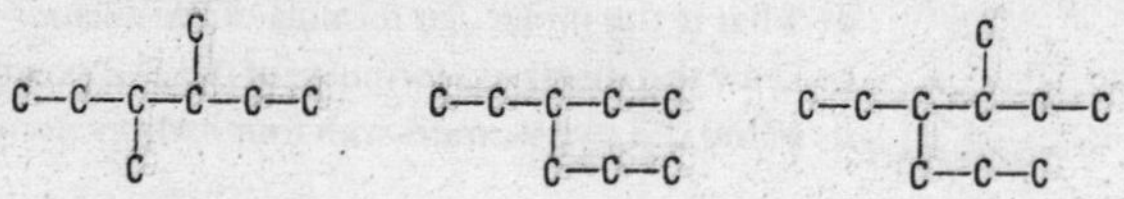

The longest way through the maze is:

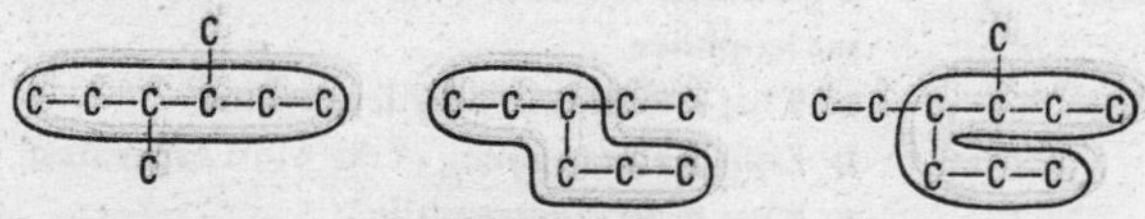

So: Six carbon atoms   Six carbon atoms   Seven carbon atoms

The main stem is:
    hex-   hex-   hept-

Name and number the side chains, and the full systematic names are:
    3,4-dimethylhexane   3-ethylhexane   4-ethyl-3-methylheptane

During and after the First World War, the famous inventor Thomas Edison experimented with the flowering plants *Euphorbia* and Goldenrod to see whether they could be used as sources of hydrocarbons – for both petrol and rubber. He was not very successful, but his ideas of looking to renewable plants for energy resources now seem increasingly sensible.

The alkanes are now familiar territory. How will they react? Don't forget, the bond enthalpies for carbon–hydrogen bonds are very high, consequently the alkanes are relatively unreactive as organic chemicals go. If the bonds are broken and new bonds formed, the resulting stability of the new compounds can provide the energy to allow the reaction to proceed.

In the case of burning, the alkanes produce CO_2 and H_2O. The resulting release of energy is large and gives us the reason for using these chemicals as fuels (Ch 15.4). However, other simple reactions do not produce such stable products, and therefore the alkanes are relatively resistant to other reactions.

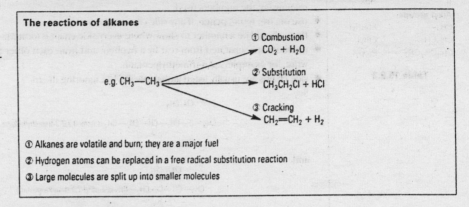

The reactions of alkanes

e.g. $CH_3—CH_3$

① Combustion
$CO_2 + H_2O$

② Substitution
$CH_3CH_2Cl + HCl$

③ Cracking
$CH_2{=}CH_2 + H_2$

① Alkanes are volatile and burn; they are a major fuel

② Hydrogen atoms can be replaced in a free radical substitution reaction

③ Large molecules are split up into smaller molecules

SELF-ASSESSMENT QUESTIONS
15.2

1 Name the following alkanes:
a $CH_3—CH_2—CH_3$
b $CH_3—CH(CH_3)CH_3$
c $CH_3(CH_2)_2CH_3$
d $CH_3—CH(CH_3)CH(CH_3)CH_3$

2 Write the linear and displayed formulae of:
a ethane
b 2-methylpentane
c dimethylpropane
d 3-methylpentane

3 An alkane hydrocarbon has the empirical formula C_3H-, and a relative molecular mass of 86.
a What is the molecular formula of the alkane?
b Draw the displayed formulae of the five possible isomers of this alkane.
c Name the two isomers with four carbon atoms in the longest chain.

4 A saturated hydrocarbon burns readily but has few chemical reactions with electrophiles or nucleophiles.
a What is meant by a **hydrocarbon**?
b Explain the meaning of the word **saturated**.
c What is an **electrophile**?
d The hydrocarbon contains 82.8% carbon and 17.2% hydrogen. Calculate the empirical formula.
e If the molecular formula is C_4H_{10}, at what m/e value would you expect to see the molecular ion peak in the mass spectrum of the hydrocarbon?

15.3 Physical properties of the alkanes

The alkanes are the first of the homologous series we encounter in organic chemistry. As with any homologous series, the chemical properties of each member are very similar, successive members differ by —CH_2—, and there is a trend in the physical properties. The extra —CH_2— group adds fourteen molecular mass units to each successive member of the series, and so, as we move along the series, there is a gradual increase in relative molecular mass.

This increase in molecular mass and the increase in the physical size of the molecules, from say, CH_3—CH_3 to CH_3—CH_2—CH_2—CH_2—CH_2—CH_3, means that the van der Waals attractions (Ch 2.8) increase as the molecules get larger. So, the values of physical properties such as boiling points, melting points and density increase as the number of carbon atoms increases. Graphs of boiling points and density are shown in *Figures 15.3.1* and *15.3.2*.

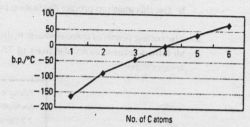

Figure 15.3.1 Boiling points of unbranched alkane hydrocarbons

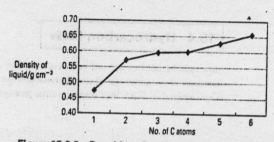

Figure 15.3.2 Densities of unbranched alkane hydrocarbons

There are two important points to remember about the alkanes. Since the molecules have very little polarity and there are no electron displacements caused by electronegative atoms in the molecule, the only intermolecular attractions are of the van der Waals type.

The increased chain length increases the van der Waals possibilities (the longer 'sausage' has more points of contact with the other 'sausages' – see Ch 14.5). The longer chains may also become more 'tangled' than the shorter chains – increasing intermolecular interaction.

The effect of branching in the chains has already been discussed (Ch 14.5) and, again, the decrease in van der Waals attractions between spheres as compared with 'sausage-like' structures is evident. The boiling points of straight chain and branched chain isomeric pentanes were given in Ch 14.5, *Table 14.5.2*.

Activity

Make up two molecular models of each of C_2H_6, C_5H_{12}, as straight chain alkanes and line them alongside each other. The extra interaction with the longer chain is very evident. Now make a model of C_5H_{12} as the dimethylpropane isomer, and look at the difference in the possible points of contact.

This gradual increase in physical properties means that the smallest alkanes are gases at room temperature; by C_5 they become volatile liquids, and by C_{20} they are waxy solids. With no possibilities for hydrogen bonds, all these alkanes are insoluble in polar solvents such as water. The liquids and solids are all less dense than water and float on the surface.

SELF ASSESSMENT QUESTIONS 15.3

1 a Plot a graph of the boiling points for the following single branched alkanes.

$CH_3CH(CH_3)CH_3$	$-12\,°C$
$CH_3CH(CH_3)CH_2CH_3$	$+28\,°C$
$CH_3CH(CH_3)CH_2CH_2CH_3$	$+60\,°C$
$CH_3CH(CH_3)CH_2CH_2CH_2CH_3$	$+90\,°C$

b Use this graph to predict the boiling point of 2-methylheptane.

2 The boiling points of three isomeric hydrocarbons are given below. Explain why, although they all have a relative molecular mass of 72, there is such a change from liquid to gas at room temperature.

Pentane	36 °C
Methylbutane	28 °C
2,2-Dimethylpropane	10 °C

15.4 Hydrocarbon fuels

Although the alkanes are not very reactive, they do have one very important reaction. The lower members of the series are all very flammable in air – a flame, or even just a spark, will cause them to burn. When they burn, they release a large amount of energy (*Table 15.4.1*).

Alkane	Formula	ΔH_c / kJ mol^{-1}
Methane	CH_4	-890
Ethane	C_2H_6	-1560
Propane	C_3H_8	-2219
Butane	C_4H_{10}	-2877
Pentane	C_5H_{12}	-3509
Hexane	C_6H_{14}	-4163
Heptane	C_7H_{16}	-4817
Octane	C_8H_{18}	-5470

Table 15.4.1 Energy released when alkanes burn

Higher members of the alkanes become more difficult to ignite, and burn with a smoky flame. When the Kuwaiti oil wells were set on fire during the 1990–91 Gulf war, the soot which was produced blackened Himalayan snows a thousand and more miles away.

It is not surprising, then, that they are the most important fuels used at the moment. Oil is relatively easy to obtain. The process of refining is straightforward, although expensive. A source of ignition, and a great deal of the energy can be released – for running motor cars, for converting into electricity, for heating systems in factories or homes, and for running machines.

The alkanes burn in a plentiful supply of oxygen. The equation for propane is:

$$C_3H_8(g) + 5O_2(g) \longrightarrow 3CO_2(g) + 4H_2O(g)$$
Propane

How much energy is released in this reaction? This can be worked out using mean bond enthalpies (Ch 11.5).

Provided that there is sufficient oxygen, any hydrocarbon will burn to form CO_2 and H_2O – each carbon atom in the hydrocarbon will form a molecule of CO_2 and every two hydrogen atoms will form a molecule of water.

$$C_3H_8(g) + \boxed{\text{oxygen}} \longrightarrow 4H_2O(g) + 3CO_2(g)$$

H_8 gives $4H_2O$

C_3 gives $3CO_2$

Complete this equation as a balanced equation by putting in the correct number of oxygen molecules.

Bond	Bond enthalpy /kJ mol^{-1}
C—C	+347
C=C	+612
C—H	+413
C—O	+358
C=O	+805
O—H	+464
O=O	+498

Table 15.4.2 Average bond enthalpies

Enthalpy of combustion using bond enthalpies

Equation:

The equation for the combustion of propane is:

$$C_3H_8(g) + \qquad 5O_2(g) \longrightarrow 3CO_2(g) + 4H_2O(g)$$

Written out in displayed form, the equation looks like:

H—C—C—C—H + 5 O=O → 3 O=C=O + 4 H—O
(with H H H above and H H H below the carbons, and H on the water)

Data:

Total up the different bonds present in the displayed equation

$2 \times$ C—C	$5 \times$ O=O	$3 \times 2 \times$ C=O	$4 \times 2 \times$ O—H
$8 \times$ C—H			

Mean bond enthalpies from data tables (*Table 15.4.2*):

C—C +347 kJ mol^{-1} O=O +498 kJ mol^{-1} C=O +805 kJ mol^{-1} O—H +464 kJ mol^{-1}
C—H +413 kJ mol^{-1}

Calculation:

$2 \times +347 = +694$	$5 \times +498 = +2490$	$6 \times +805 = +4830$	$8 \times +464 = +3712$
$8 \times +413 = +3304$			

∴ Energy *required* to break bonds =   ∴ Energy *released* on forming bonds =

$+694 +3304$ $+2490$ $-(+4830)$ $-(+712)$
                      $= +6488$ kJ mol^{-1} $= -8542$ kJ mol^{-1}

Note that bond breaking is *endothermic*, so the sign is positive; bond making is *exothermic*, so the sign has become negative (Ch 11.5).

∴ $\Delta H_c = +6488 - 8542 = -2054$ kJ mol^{-1}

(where mol^{-1} means **'per mole of the equation'** – see Ch 9.5)

There are a considerable number of alkanes, or other hydrocarbons or even alcohols, that could be used to illustrate the combustion process, and there is a separate equation for each of them. You should now be able to balance any such combustion equation.

Thus, the energy released when one mole of propane burns should be -2054 kJ mol^{-1}. It is the energy released by the formation of stable C=O and O—H bonds that provides such a favourable negative enthalpy change for the combustion reaction.

Do not forget that this calculation is based on mean bond enthalpies, which are *average* enthalpies for the individual bonds (see Ch 11.5). The measured standard enthalpy of combustion for propane is -2219 kJ mol^{-1}. The difference between the calculated value and the measured value is due to variations from the average bond enthalpy figures.

Remember:

- The formula of the alkane must be known (remember C_nH_{2n+2}), and the products are CO_2 and H_2O.
- Using butane as an example.
  Butane contains four carbon atoms. The formula must be $C_4H_{(2\times4+2)}$, and it forms CO_2 and H_2O when it reacts with oxygen:

$$C_4H_{10} + O_2 \longrightarrow CO_2 + H_2O$$

- There must be four CO_2 molecules to balance the four carbons in butane:

$$C_4H_{10} + O_2 \longrightarrow 4CO_2 + H_2O$$

- There must be 10/2 H_2O molecules to account for the ten hydrogen atoms in butane (with two hydrogen atoms in each water molecule):

$$C_4H_{10} + O_2 \longrightarrow 4CO_2 + 5H_2O$$

- Now count up the number of oxygen atoms in the products ($[CO_2]4 \times 2 = 8 + [H_2O]5 = 13$) and divide by two, since it is molecular O_2 which is reacting: $13 \div 2 = 6\frac{1}{2}$:

$$C_4H_{10} + 6\tfrac{1}{2}O_2 \longrightarrow 4CO_2 + 5H_2O$$

- The half is not often used in balanced equations, so by doubling everything, the final equation is:

$$2C_4H_{10} + 13O_2 \longrightarrow 8CO_2 + 10H_2O$$

- Putting in state symbols gives:

$$2C_4H_{10}(g) + 13O_2(g) \longrightarrow 8CO_2(g) + 10H_2O(g)$$

All chemical equations have, at some time, been determined by experimental work, and these combustion equations are no exception.

The measurement of the gas volumes before and after the combustion reaction can be used to find the balanced equation.

Using gas volumes in combustion equations (see Ch 9.6)

Example:
When 10 cm^3 of a gaseous alkane is burned in excess oxygen and the product gases are passed through sodium hydroxide solution, the volume of gas decreases by 30 cm^3. All gases are measured at room temperature and pressure. Deduce the equation for the combustion reaction.

Method:
Write the equation as far as possible:

$$C_nH_{(2n+2)} + O_2 \longrightarrow nCO_2 + H_2O$$

Alkanes all have the formula $C_nH_{(2n+2)}$ and each carbon atom in the alkane gives one molecule of CO_2.

Remember that sodium hydroxide solution removes all the CO_2 from the gases, so the volume of CO_2 produced must have been 30 cm^3. The volume of liquid water produced is negligible.

Gas volumes can be converted directly into moles (using Avogadro's Law, Ch 9.3).

$$10C_nH_{2n+2} + O_2 \longrightarrow 30CO_2 + H_2O$$

This reduces to:

$$1C_nH_{2n+2} + O_2 \longrightarrow 3CO_2 + H_2O$$

$\therefore$ n = 3, so the formula of the alkane must be $C_3H_{(2\times3)+2} = C_3H_8$, and the balanced equation is:

$$C_3H_8(g) + 5O_2(g) \longrightarrow 3CO_2(g) + 4H_2O(g)$$

Notice that in this question, only the volume ratio of gaseous alkane and carbon dioxide is important.

Problems with burning

The usefulness of alkanes can be attributed to their volatility, the ease of igniting them and the large amount of energy released when they burn. The effect on the environment when they burn also has to be considered.

- As we discussed earlier, the process of burning a fossil fuel releases, in a few minutes perhaps, the carbon dioxide that was trapped by plants, through photosynthesis, over many thousands of years. What effect does this relatively rapid increase in carbon dioxide concentration have on the environment (see *Figure 15.4.1*)? As a **greenhouse gas**, carbon dioxide is linked with **global warming**, and this is a topic we will look at later (Ch 17.2).

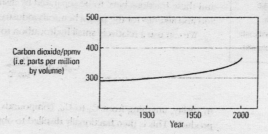

Figure 15.4.1   CO₂ in the atmosphere

- What happens when alkanes are burned in a limited amount of air? Alkanes still burn, but instead of carbon dioxide being produced, some carbon monoxide, or even carbon itself, is formed.

  As we have seen, in excess air the equation for burning propane is:

$$C_3H_8(g) + 5O_2(g) \xrightarrow{\text{ignite}} 3CO_2(g) + 4H_2O(g)$$

With less oxygen, one possibility is:

$$C_3H_8(g) + 4O_2(g) \xrightarrow{\text{ignite}} CO_2(g) + 2CO(g) + 4H_2O(g)$$

- The formation of CO instead of CO_2 is potentially very hazardous. Carbon monoxide is a toxic gas, without any smell, that interferes with respiration. It competes with oxygen for the haemoglobin in the blood. If there is a trace in the air breathed, it always wins!

In the equation above five moles of oxygen are required for the complete combustion of propane to carbon dioxide and water. If the amount of oxygen is reduced sufficiently, only carbon monoxide and water are formed.

$$C_3H_8(g) + 3\tfrac{1}{2}O_2(g) \xrightarrow{\text{ignite}} 3CO(g) + 4H_2O(g)$$

- Further reduction in the amount of oxygen leads to the formation of carbon – i.e. soot. The yellow flame in any fire, or in the Bunsen burner, is caused by particles of carbon heated to incandescence.

Oxygen is transported to where it is needed in the body by haemoglobin, the red pigment in the blood. In the lungs, where the oxygen concentration is high, O_2 molecules attach themselves to the haemoglobin; they easily detach in regions of low concentration, e.g. the muscles. Carbon monoxide molecules attach to haemoglobin much more firmly (forming carboxyhaemoglobin); they are still attached when the haemoglobin returns to the lungs (Ch 25.4). This means that a person breathing air containing a small proportion of carbon monoxide will slowly become unable to absorb oxygen, and will therefore become unconscious and eventually die. Carboxyhaemoglobin is a bright red pigment. The most obvious symptom of carbon monoxide poisoning is that the victim has cherry-red cheeks. The first-aid treatment for carbon monoxide poisoning is artificial respiration – if possible, with plenty of oxygen.

Combustion of any hydrocarbon is unsafe unless there is an ample supply of air. Every year several deaths occur in Britain because, for example, a chimney is blocked above a gas fire, or a paraffin heater is used in a badly-ventilated room.

The petrochemicals industry

Earlier, we saw that the fractional distillation of crude oil produced a wide range of very different fractions. The refinery gases are available for use immediately, but the fractions which contain the compounds with larger molecules have to be altered significantly if they are to be useful.

The petrochemicals industry requires relatively small carbon molecules, particularly the short-chain alkenes, ethene, propene and butene. These alkene molecules are used as feedstock for the production of a wide range of organic chemicals. Few alkenes are produced in the distillation of crude oil.

The market demand for the heavy fractions is also quite low; the greatest demand is for the petrol-like fractions. Consequently, after the primary distillation process at the refinery, the next step is to alter the fractions to meet the market demand.

The imbalance between supply and demand for the hydrocarbon fractions from primary distillation needs two techniques to overcome it – **cracking** and **reforming**.

Cracking

> **Definition:** Cracking is a process in which large hydrocarbon molecules are broken down, using high temperatures and catalysts, into smaller, more useful molecules.

The long-chain molecules are of limited use as a fuel, because they have higher boiling points and, consequently are more difficult to ignite. The larger molecules can be broken into simpler molecules by the process of catalytic cracking. This reaction provides fractions containing smaller molecules, and these fractions have to be separated by distillation. They provide additional supplies of petrol, and also alkenes for the petrochemicals industry.

We can use a relatively small hydrocarbon to illustrate the process:

$$C_6H_{14}(g) \xrightarrow[500\,°C]{Al_2O_3} C_4H_{10}(g) + C_2H_4(g)$$

In reality, anything from C_4 to C_{50} compounds can be cracked to produce a complex mixture of products. This is then fractionally distilled to obtain new fractions.

Activity

If we look at the bond breaking during the cracking reaction, we should be able to see that the process must produce at least one molecule of an alkene. Try using molecular models to show how hexane can be split into butane and ethene.

Now break up the hexane molecule into three molecules of ethene; there are two hydrogen atoms left, which combine to form the hydrogen molecule:

$$CH_3(CH_2)_4CH_3 \longrightarrow 3CH_2{=}CH_2 + H_2$$

Often two molecules of alkene are formed; a hydrogen molecule may be produced by an elimination reaction. Under these extreme conditions, it is difficult to control the precise composition of the final mixture, although if pushed to the limit the end products are ethene and hydrogen. These are both important chemicals; ethene in the production of plastics and hydrogen in ammonia manufacture.

e.g. $$C_6H_{14}(g) \xrightarrow[500\,°C]{Al_2O_3} 3C_2H_4(g) + H_2(g)$$

Ethene   Hydrogen

Using one more example:

e.g. $$C_{11}H_{24}(g) \xrightarrow[500\,°C]{Al_2O_3} CH_4(g) + 2C_2H_4(g) + 2CH_3CH{=}CH_2(g)$$

Methane   Ethene   Propene

Remember, in the equations for these reactions, at least one double bond is formed, together with an alkane or hydrogen.

Cracking is possible on a much smaller scale in the laboratory, although the main product in this reaction is usually ethene (see Ch 16.4). A typical apparatus is shown in *Figure 15.4.2*.

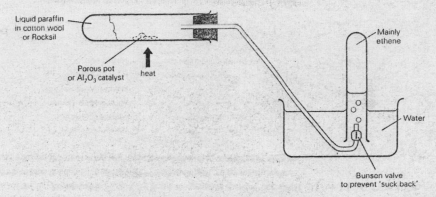

Figure 15.4.2   Cracking in the laboratory

The industrial reaction uses an aluminium oxide/silicon oxide catalyst at 500 °C, and the longer chain hydrocarbons are broken into smaller molecules. The process is often referred to as **catalytic cracking** or **cat cracking**.

Octane numbers of petrols

Different car engines require slightly different fuels to work at their maximum efficiency. One measure of the suitability of petrol for a particular car engine is the octane number. This is a number, close to 100, which shows the percentage of 2,2,4-trimethylpentane (*iso*-octane) in a mixture of 2,2,4-trimethylpentane and heptane which gives the same performance characteristics as the petrol. 95-Octane petrol will behave as would a mixture of 95% *iso*-octane and 5% heptane.

2,2,4-Trimethylpentane
(*iso*-octane)
Octane number – 100

Heptane
Octane number – 0

The principle of the internal combustion engine is that the petrol and air are mixed in a particular ratio, and then compressed by a piston in the cylinder. A spark ignites the mixture as the piston reaches its highest point in the cylinder. At these raised temperatures and pressures, it is not always easy to set the reaction off at exactly the right moment. The hydrocarbon mixture can explode prematurely. This is called pre-ignition, and causes 'knocking' in the engine. Such pre-ignition or self-ignition means that power is lost and the engine is less efficient; it may also cause severe damage in the engine.

Over the years a number of ways have been used to improve the octane number of petrols. The addition of organic lead compounds, such as tetraethyl lead (IV) ($(CH_3CH_2)_4Pb$), gives rise to 'leaded' petrol. The addition of 0.6 g of the lead compound to a litre of petrol – a concentration of about $0.002 \text{ mol dm}^{-3}$ – can raise the octane number from 87 to 93. However, lead itself is toxic and poses environmental difficulties – it is a nerve poison and a depressant and, in particular, causes brain damage to young children.

Other additives, such as metal carbonyl compounds and 2-methoxy-2-methylpropane ('methyl tertiary butyl ether', MTBE), have been used, although the modern trend is to improve the octane rating by varying the hydrocarbon mixture.

The blend of hydrocarbons in the gasoline fraction obtained directly from the fractional distillation of crude oil has a low octane number. So, high-octane hydrocarbons are added. The next section discusses the production of these alternative additives. The octane numbers of different compounds vary considerably:

		Octane number
Straight chain alkane:	Heptane	0
Straight chain alkene:	Hept-1-ene	68
Branched chain alkane:	2-Methylhexane	42
	2,2-Dimethylpentane	97
	2,2,3-Trimethylbutane	113
Alicyclic alkane:	Methylcyclohexane	84
Aromatic:	Methylbenzene	112

All these hydrocarbons have seven carbon atoms, but the increased branching alters the octane number considerably, and aromatic hydrocarbons have quite large octane numbers. By altering the petrol composition and adding more branched chain and aromatic hydrocarbons using the process of **reforming**, the octane number can be increased, even with 'unleaded' petrol. The introduction of aromatic hydrocarbons and branched chain hydrocarbons also has drawbacks, particularly with the addition of benzene, which is a carcinogen.

Summarising:

Raising octane numbers of petrols

Use lead compounds – now phased out, lead is toxic

Add aromatic hydrocarbons – benzene itself is toxic

Add branched chain hydrocarbons

Reforming

The other important process which improves the quality of the fuels we use is reforming. The octane number of a fuel can be raised by increasing the proportion of branched or cyclic hydrocarbons. Reforming does just that – it changes the arrangement of the atoms in the molecule.

An example of branching is:

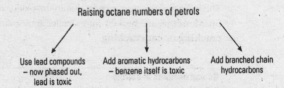

Hexane → 2-Methylpentane

Definition:
Reforming is an industrial process in which 'straight' chain hydrocarbons are converted into branched chain and cyclic hydrocarbons, using high temperature, high pressure and a catalyst.

The catalyst can be aluminium oxide at 500 °C. Typical pressures are 40 atm, when the hydrocarbons are still liquid in spite of the elevated temperatures.

In the cyclisation reaction both aromatic and alicyclic compounds are formed. The formation of an alicyclic compound releases one hydrogen molecule:

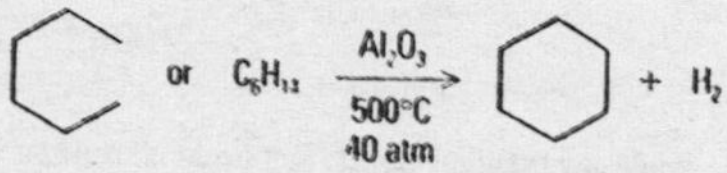

The formation of aromatic benzene rings results in four hydrogen molecules being formed:

Again, the by-product is hydrogen, so it is not surprising to find ammonia plants (which need nitrogen and hydrogen to produce ammonia) often sited close to oil refineries.

A number of catalysts are used for reforming including aluminium oxide. Platinum can also be used as a catalyst, although its cost (maybe £5 million worth of platinum in one reactor) has to be set against the saving from using a lower pressure of ten atmospheres. When platinum is used as a catalyst, reforming is sometimes referred to as 'platforming'.

Cracking

Breaks larger molecules into smaller, more useful, molecules:

- to produce more gasoline;
- to produce alkenes for petrochemicals.

Reforming

Alters the shapes of straight chain molecules to give branched and cyclic molecules:

- to improve octane ratings in petrols.

Catalytic converters

Some of the problems caused by using hydrocarbon fuels have already been mentioned. Of the many concerns about pollution caused by burning fossil fuels, some of the major points about motor car emissions are:

- Carbon dioxide is always produced by burning fossil fuels, and as it is a greenhouse gas, carbon dioxide emissions could be contributing to global warming.
- Traces of compounds containing sulphur in fossil fuels burn to form sulphur dioxide, and the high temperatures in the engine produce the nitrogen oxides NO and NO_2. All these gases give acid rain when dissolved in rain water.
- Unburned hydrocarbons contribute to photochemical smog.
- Incomplete combustion produces carbon monoxide, which is a toxic gas.
- The presence of particles in exhaust gases contributes to the deterioration of the atmosphere and, if they are from poorly adjusted diesel engines, some of these particles may be carcinogenic.
- Lead additives put into petrol to improve octane ratings cause toxic lead emissions into the environment.

One of the solutions for dealing with the exhaust gases from cars relies on the catalytic conversion of some of the problem gases into less problematic gases. The emission from car exhausts contains a complex mixture; CO_2 and H_2O are the simplest products of combustion. Other products can include CO, NO, NO_2, unburned hydrocarbons, SO_2, and solid carbon. The difficulty in dealing with some of the more hazardous pollutants is that both oxidation and reduction processes are needed simultaneously.

The modern catalytic converter in car exhaust systems is the so-called 'three-way catalytic converter'. The converter consists of platinum and rhodium metals deposited as a thin layer on a honeycombed (for a very large surface area) ceramic block. Carbon dioxide is still produced, and the catalyst only works at temperatures above 200 °C, when the engine has warmed up. The catalyst has to operate at temperatures up to 1400 °C when the engine has been working for some time.

In this converter, the carbon monoxide in the exhaust gases is used to reduce the nitrogen oxides:

$$2NO(g) + 2CO(g) \xrightarrow[200\,°C]{Pt/Rh} 2CO_2(g) + N_2(g)$$

while any residual carbon monoxide is oxidised to carbon dioxide:

$$2CO(g) + O_2(g) \xrightarrow[200\,°C]{Pt/Rh} 2CO_2(g)$$

and any unburned hydrocarbons are converted to carbon dioxide and water:

$$\text{e.g.} \quad 2C_8H_{18}(g) + 25O_2(g) \xrightarrow[200\,°C]{Pt/Rh} 16CO_2(g) + 18H_2O(g)$$

The demands for efficient working in the system are stringent, and care is needed to adjust the fuel/air mixture going into the engine to allow the catalyst to work efficiently.

Engine — Three-way catalytic converter — Silencer — Feedback from oxygen sensor

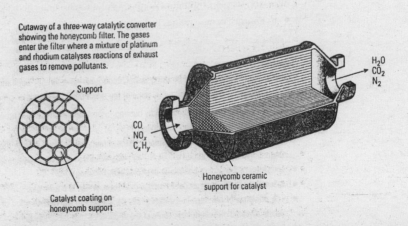

Cutaway of a three-way catalytic converter showing the honeycomb filter. The gases enter the filter where a mixture of platinum and rhodium catalyses reactions of exhaust gases to remove pollutants.

Support

Catalyst coating on honeycomb support

CO, NO_x, C_xH_y

H_2O CO_2 N_2

Honeycomb ceramic support for catalyst

SELF-ASSESSMENT QUESTIONS 15.4

1 Identify the following reactions as cracking, combustion or reforming.

a $CH_3CH_2CH(CH_3)CH_2CH_2CH_3 + 11O_2 \longrightarrow 7CO_2 + 8H_2O$

b $CH_3(CH_2)_5CH_3 \longrightarrow CH_3C(CH_3)_2CH_2CH_2CH_3$

c $CH_3CH(CH_3)CH_2CH_3 + 7O_2 \longrightarrow 3CO_2 + 2CO + 6H_2O$

d $CH_3(CH_2)_4CH_3 \longrightarrow 3CH_2{=}CH_2 + H_2$

e $CH_3CH(CH_3)CH_2CH_2CH_3 \longrightarrow CH_3CH(CH_3)CH_3 + CH_2{=}CH_2$

2 Look at the following structures:

A $CH_3(CH_2)_8CH_3$
B $CH_3C(CH_3)_2CH_2CH(CH_3)CH_3$
C $CH_3(CH_2)_{10}CH_3$

   a Which could be used as an additive to improve the octane rating of petrol?
   b Which is the most likely to be used for fuel in a paraffin lamp?
   c Which is the most likely to be used as a wax polish?

3 Copy out the following equations and balance them.

   a $CH_3CH_3 + O_2 \longrightarrow CO_2 + H_2O$
   b $CH_3C(CH_3)_2CH_2CH_3 + O_2 \longrightarrow CO_2 + H_2O$
   c $CH_3(CH_2)_3CH_3 + O_2 \longrightarrow CO + H_2O$

4 Copy out the following reactions and insert the conditions and reagents needed.

   a $CH_3CH(CH_3)CH_2CH_2CH_2CH_3 \longrightarrow CH_3C(CH_3)_2CH(CH_3)CH_3$
   b $CH_3CH_2CH_3 \longrightarrow 3CO_2 + 4H_2O$
   c $CH_3CH_2CH_2CH_3 \longrightarrow 2CH_2{=}CH_2 + H_2$

5 **a** Write a balanced equation for the reforming of C_6H_{14} to form benzene, C_6H_6.
   b Write a balanced equation to illustrate the cracking of C_6H_{14} to form ethene, C_2H_4.

6 **a** Write a balanced equation for the complete combustion of hexane in air.
   b Write a possible equation for the combustion of hexane in a limited supply of air.

15.5 Free radical substitution

Bond enthalpies

C—C	+347 kJ mol^{-1}
C—H	+413 kJ mol^{-1}
Cl—Cl	+243 kJ mol^{-1}

Electronegativities

C	2.5
H	2.1

Table 15.5.1

By now, we should have the feel for the unreactive nature of the alkane carbon skeleton. Although burning is a useful and important reaction, little else is feasible. The C—C bonds and the H—C bonds are too strong (*Table 15.5.1*).

If molecules are heated in the absence of oxygen to provide the extra energy to break the bonds, they start to break up in the cracking processes discussed in Ch 15.4.

Since these molecules have no polarity and the bond strengths are so high, the use of electrophiles or nucleophiles to cause a substitution reaction is not possible (Ch 14.6). However, substitution *is* possible with extremely reactive attacking species. Methane does react with chlorine if the reaction is carried out in ultra-violet UV light (unless the light is diffuse, an uncontrollable explosive reaction can result).

$$CH_4(g) + Cl_2(g) \xrightarrow{\text{diffuse UV light}} CH_3Cl(g) + HCl(g)$$

The reactive attacking species are **free radicals** (Ch 14.6).

We should be familiar with the dot and cross diagram for the chlorine molecule (Ch 2.4). The chlorine atom has seven outer electrons, and two atoms join to form Cl_2, sharing a pair of electrons in a covalent bond.

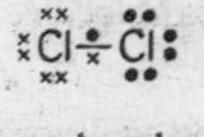

If we can reverse the process, and break this covalent bond, we produce two chlorine atoms. After **homolytic fission** of the Cl—Cl bond, each atom must have one electron unpaired, since it has seven outer electrons. Such particles are called **free radicals**. Using the half arrows, the movement of each single electron looks like:

$$Cl \overset{\frown}{\underset{\bullet\bullet}{\cdot}} Cl \longrightarrow Cl\bullet \quad Cl\bullet$$

To make the point, the unpaired electron in the free chlorine atom is written as Cl•.

The Cl_2 molecule is stable, with a bond enthalpy of $+243$ kJ mol^{-1}. It is, however, possible to break this chlorine bond with high-energy radiation in the UV region of the spectrum. The calculation in the box is not needed for examinations, but shows that the energy of UV light at 300 nm is sufficient to break the Cl—Cl bond, although not enough to break the C—H bonds in the alkane.

> **Definition:** A free radical is a species containing one or more unpaired electrons.

The energy of uv radiation

UVB radiation, the ultra-violet radiation that concerns us all when we sunbathe, has a wavelength in the region of 300 nm, which corresponds to a frequency of 10^{15} Hz. Using this value, it is possible to calculate the energy of one photon of UVB radiation by using the equation $E = h\nu$ (see Ch 1.4).

h (Plank's constant) $= 6.6 \times 10^{-34}$ J Hz^{-1}.

So $E = 6.6 \times 10^{-34} \times 10^{15}$ J for one quantum of radiation.

For a mole of photons, using the Avogadro constant 6×10^{23}, the energy would be $6.6 \times 10^{-34} \times 10^{15} \times 6 \times 10^{23}$ $= 396$ kJ mol^{-1}.

Compare this with the bond enthalpies in the reaction. There is sufficient energy in UVB radiation to break the Cl—Cl bond $(+243$ kJ mol$^{-1})$, but not enough to break the C—H bond $(+413$ kJ mol$^{-1})$.

So, the first step in a free radical reaction is the initiation step, where the UV radiation energy breaks the Cl—Cl bond. Reactions that need light to initiate them are often referred to as **photochemical reactions**.

Initiation

$$Cl \overset{\frown}{\underset{\bullet\bullet}{\cdot}} Cl \longrightarrow 2Cl\bullet$$

Once a free radical is produced, the reaction proceeds through a series of propagation steps. The free radical is sufficiently reactive to break the C—H bond, producing HCl and leaving a methyl radical – with an unpaired electron.

The movement of the single electrons is again illustrated using half arrows.

$$Cl\bullet + H{-}CH_3 \longrightarrow Cl{-}H + \bullet CH_3$$

Each free radical reacts with another molecule to form a further free radical and some stable species, setting up a **chain reaction**. One radical forms another radical.

Propagation

The methyl radical is very reactive and can react with another molecule to produce a free radical and another stable molecule.

$$CH_3\bullet + Cl_2 \longrightarrow CH_3Cl + Cl\bullet$$

Or: $$CH_3\bullet + CH_3Cl \longrightarrow CH_3CH_3 + Cl\bullet$$

The chlorine radicals can react with methane or other alkanes as before. The number of possibilities with a complex hydrocarbon is almost endless. Any propagation equation can be worked out, containing:

- one radical as a reactant;
- one radical as a product (not normally a hydrogen radical);
- HCl or chloroalkane as the product other than the radical.

In these free radical reactions involving alkanes and halogens, the radicals produced in the propagation steps are always alkyl radicals, such as $CH_3\bullet$, or halogen radicals. The hydrogen radical, $H\bullet$, is rarely formed. Simple bond enthalpy calculations will give some idea why not.

Formation of $CH_3\bullet$

$$H_3C—H + Cl\bullet \longrightarrow H_3C\bullet + HCl$$

$1 \times C—H$ broken ($+413 \text{ kJ mol}^{-1}$)

$1 \times H—Cl$ formed (-431 kJ mol^{-1})

Enthalpy change for this reaction:

$$\Delta H = -18 \text{ kJ mol}^{-1}.$$

Formation of $H\bullet$

$$H_3C—H + Cl\bullet \longrightarrow H_3C—Cl + H\bullet$$

$1 \times C—H$ broken ($+413 \text{ kJ mol}^{-1}$)

$1 \times C—Cl$ formed (-346 kJ mol^{-1})

Enthalpy change for this reaction:

$$\Delta H = +77 \text{ kJ mol}^{-1}.$$

This is endothermic, so the hydrogen free radical is rarely formed.

The energy given out when a covalent bond is formed between two radicals is exactly equal to the energy needed to break the bond again! (see Ch 11.5). So the bond cannot form unless there is another body present (the walls of the container, perhaps) to absorb some of the energy. This means that free radical reactions may go through thousands or even millions of propagation steps before they terminate. We shall meet this again in the formation of Polythene and other addition polymers (Ch 16.7).

Of course, when two radicals meet they can combine to reform a covalent bond. Any two radicals can, in principle, combine to form such a covalent bond in a reaction that finishes the chain sequence and ends the possibility of any further reaction. The reason why radicals do not immediately re-combine is discussed in the box alongside.

Termination

$$Cl\bullet + CH_3\bullet \longrightarrow CH_3Cl$$
$$Cl\bullet + Cl\bullet \longrightarrow Cl_2$$
$$CH_3\bullet + CH_3\bullet \longrightarrow CH_3CH_3$$

The reaction sequence shows that a large number of products can be formed – we have only shown three. Although the first substitution will be the most frequent, the separation of the products in free radical reactions is always a difficulty.

Remember:

The steps in any free radical reaction are:

- **Initiation:** e.g. $Cl_2 \xrightarrow{\text{UV light}} 2Cl\bullet$
- **Propagation:** e.g. $CH_4 + Cl\bullet \longrightarrow HCl + CH_3\bullet$

  $CH_3\bullet + Cl_2 \longrightarrow Cl\bullet + CH_3Cl$
- **Termination:** e.g. $CH_3\bullet + CH_3\bullet \longrightarrow C_2H_6$

  $CH_3\bullet + Cl\bullet \longrightarrow CH_3Cl$

All the halogens can substitute the hydrogen atoms in alkanes, but with different ease of reaction. Look at the overall enthalpy change involved in each of these substitutions, given here for the formation of halomethanes.

$$CH_4 + F_2 \xrightarrow{\text{room temperature}} CH_3F + HF \qquad \Delta H = -475\,kJ\,mol^{-1}$$

$$CH_4 + Cl_2 \xrightarrow{UV} CH_3Cl + HCl \qquad \Delta H = -114\,kJ\,mol^{-1}$$

$$CH_4 + Br_2 \xrightarrow{UV} CH_3Br + HBr \qquad \Delta H = -36\,kJ\,mol^{-1}$$

$$CH_4 + I_2 \xrightarrow{UV} CH_3I + HI \qquad \Delta H = +59\,kJ\,mol^{-1}$$

The reaction of methane with fluorine is explosive even in the dark, because the F—F bond is relatively weak ($+158\,kJ\,mol^{-1}$) and the C—F bond is so strong ($+467\,kJ\,mol^{-1}$). On the other hand, the reaction of iodine with methane is endothermic and easily reversed.

As pointed out above, the difficulty with free radical substitution is that the products can be very varied. Although the mono-substituted chloroalkane might be formed in the greatest yield, a variety of other products is possible and the reaction mixture has to be treated to separate the products. As an example, the chlorination of methane will produce, in differing quantities,

CH_3Cl, CH_2Cl_2, $CHCl_3$, CCl_4, C_2H_6, C_2H_5Cl, etc.

SELF-ASSESSMENT QUESTIONS 15.5

1 Which of the following reactive species are free radicals?
   Br•, $NO_2^{\oplus}$, O•, $OH^{\ominus}$, $Cl^{\ominus}$.

2 Using dot-and-cross diagrams, explain how the Br• free radical is formed from the bromine molecule in UV light.

3 Bromine reacts with ethane in UV light to form bromoethane. Write a mechanism for the formation of the mono-substituted product.

4 Write the displayed formulae of the possible mono- and di-substituted chloro-products formed when chlorine reacts with propane. Explain why UV light is necessary to carry out this reaction.

5 Which of the following reactions are free radical substitution reactions?
   a $CH_2{=}CH_2 + Cl_2 \longrightarrow CH_2Cl{-}CH_2Cl$
   b $CH_3{-}CH_2Cl + OH^{\ominus} \longrightarrow CH_3{-}CH_2OH + Cl^{\ominus}$
   c $CH_3{-}CH_3 + Cl_2 \longrightarrow CH_3{-}CH_2Cl + HCl$
   d $CH_3{-}CH_2Cl \longrightarrow CH_2{=}CH_2 + HCl$
   e $CH_4 + Cl_2 \longrightarrow CH_2Cl_2 + 2HCl$

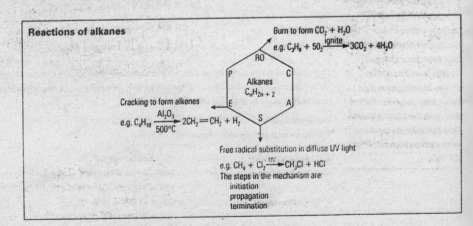

Reactions of alkanes

Burn to form $CO_2 + H_2O$
e.g. $C_3H_8 + 5O_2 \xrightarrow{\text{ignite}} 3CO_2 + 4H_2O$

Alkanes C_nH_{2n+2}

Cracking to form alkenes
e.g. $C_4H_{10} \xrightarrow[500°C]{Al_2O_3} 2CH_2{=}CH_2 + H_2$

Free radical substitution in diffuse UV light
e.g. $CH_4 + Cl_2 \xrightarrow{UV} CH_3Cl + HCl$
The steps in the mechanism are:
   initiation
   propagation
   termination

SUMMARY
15

In this chapter the following new ideas have been covered. *Have you understood and learned them?* Could you explain them clearly to a friend? The page references will help.

- Crude oil is fractionally distilled to create different fractions.   p380
- The fractions from crude oil, and their major uses, can be listed.   p382
- An alkane can be recognised and named from its molecular formula.   p383
- The boiling points of different alkanes depend on their relative molecular masses.   p387
- Balanced equations can be written for the burning of hydrocarbons.   p390
- The enthalpies of combustion of alkanes can be calculated using bond enthalpies.   p389'
- Gas volumes can be used to balance combustion equations.   p390
- Cracking is used to change large hydrocarbon molecules into smaller, more useful ones.   p392
- Fuels can be improved by reshaping their molecules.   p393
- Catalytic reforming of alkanes.   p394
- The formation of free radicals.   p397
- Free radical substitution can be represented by three steps – initiation, propagation, and termination.   p398

EXAMINATION QUESTIONS
15

1 Butane, C_4H_{10}, is used as the fuel in many small petrol lighters, in which the gas is ignited using a spark from a flint.

    a   Write a full displayed formula for butane.

    b   Write the formula, and the name, of one chain isomer of butane.

    c   The boiling point of butane is 0 °C. Would you expect the boiling point of the isomer in **b** to be higher or lower?

    d   Explain briefly why the boiling point should be different.

    e   Write a balanced equation for the complete combustion of butane.

    f   Calculate the enthalpy of combustion of butane, using the bond enthalpies given.

C—C	$+347$ kJ mol^{-1},
C—H	$+413$ kJ mol^{-1},
C=O	$+805$ kJ mol^{-1},
O—H	$+464$ kJ mol^{-1},
O=O	$+498$ kJ mol^{-1},

    g   The standard enthalpy of combustion of butane is quoted as -2880 kJ mol^{-1}. Why are the two figures not the same?

    h   The pocket lighters normally hold about 2.0 g of the liquid fuel. How much energy is potentially available from the complete combustion of all of this fuel? (C, 12.0; H, 1.0)

2 A fraction of a crude oil sample contains principally C_8H_{18}.

    a   In which fraction of crude oil would you expect to find this compound?

    b   Write an equation showing how this compound can be cracked to give ethene and hydrogen.

    c   Suggest a suitable temperature and catalyst to crack C_8H_{18}.

    d   Careful control of the cracking produces hexane as one product. Write an equation for this reaction.

3 Four of the hydrocarbons that are obtained from the fractional distillation of crude oil are:

A $CH_3CH(CH_3)CH_2CH_2CH_3$

B $CH_3CH{=}CHCH_2CH_3$

C $CH_3CH_2CH_2CH_2CH_2CH_3$

D $CH_3CH_2CH(CH_3)CH_2CH_2CH_3$

a Which two of these are a pair of isomers?

b Which molecule would be expected to have a chiral carbon atom?

c What property is normally associated with such a chiral carbon atom?

d Which molecule can exhibit geometrical isomerism?

e Write an equation for the reaction of C with bromine in UV light.

f Which compound would be expected to have the highest boiling point?

4 An alkane is burnt as a gas in excess oxygen; 20 cm^3 of the alkane produced 100 cm^3 of carbon dioxide (all volumes measured at the same temperature and pressure).

a What is the formula of the alkane?

b Draw the possible isomers of this alkane molecule.

c Give the systematic name for every isomer you have drawn.

d Write an equation to show how one of these alkane molecules might react when heated over an aluminium oxide catalyst.

e Which of the isomers will have the lowest boiling point? Explain your choice.

Do you remember ...?

- How to **name** organic molecules. (Ch 14.3)
- Understanding **geometric isomerism**. (Ch 14.4)
- Cracking **alkanes**. (Ch 15.4)
- Types of **organic reactions**. (Ch 14.6)
- Formation of **sigma bonds**. (Ch 2 4)
- Use of **curly arrows** to explain electron movements in reactions. (Ch 14.7)

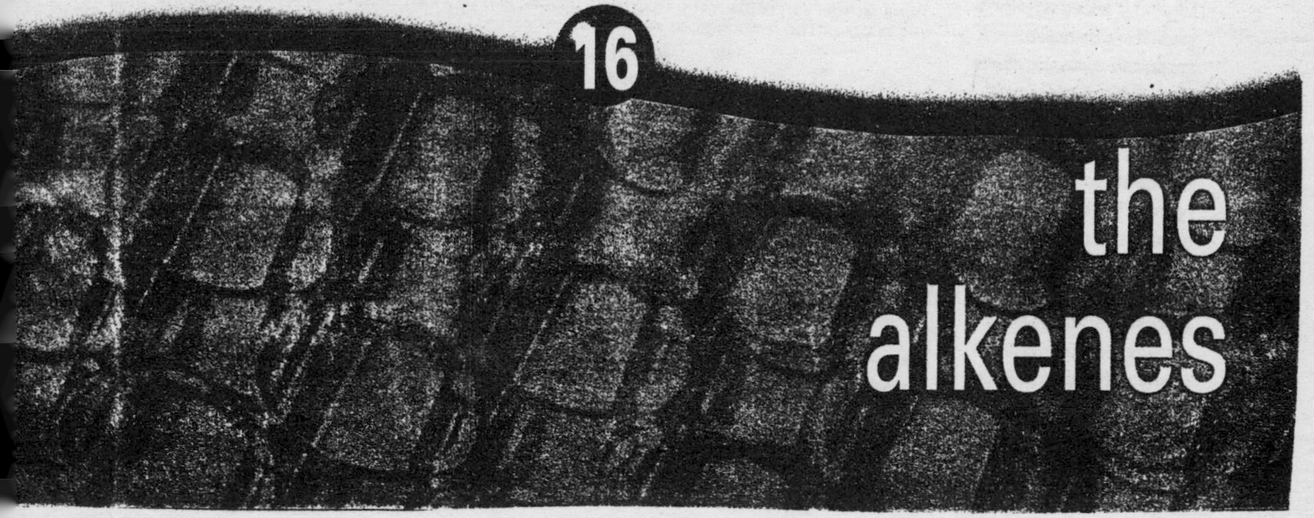

16

the alkenes

By now, the alkane skeleton should be familiar to you. Some of the alkanes are materials we come across in our day-to-day lives: natural gas (methane), butane in cylinder gases for camping or cooking, and 'octane' numbers in petrols. The alkenes are not such readily recognisable everyday chemicals. They are much more important as feedstocks for making useful products such as Polythene, polystyrene, antifreeze, refrigeration liquids, and a host of other products.

16.1 The alkenes

Any non-aromatic molecule containing a carbon–carbon double bond is an alkene. The introduction of the double bond into the carbon skeleton means that the properties of the molecule are altered significantly. The alkene functional group provides a reactive site within the carbon skeleton, which can lead to further reactions.

The alkanes, we have seen, are **saturated** molecules (see Ch 15.2). Any molecule containing a carbon–carbon double or triple bond is **unsaturated**. For each carbon–carbon double bond in the molecule there must be two less hydrogen atoms than are needed for saturation, and so the general formula for a simple alkene is C_nH_{2n}.

> **Definition:** Unsaturated compounds consist of molecules which contain one or more double or triple bonds between carbon atoms.

You are probably aware of the words 'saturated' and 'unsaturated' as applied to edible fats and oils. Animal fat, as found in pork for example, contains molecules with few carbon–carbon double bonds; vegetable oils, such as sunflower oil or olive oil, contain molecules which have one or more carbon–carbon double bonds. A diet which contains large amounts of saturated fats is considered by many doctors and dieticians to be unhealthy. The writing on the packaging of some spreadable fat mixtures (e.g. margarines) often emphasises that they contain high levels of polyunsaturates, i.e. compounds containing more than one double bond per molecule. Many people think that such spreads are less harmful than those containing saturated fats, but the question is still unresolved.

The simplest of the alkenes is ethene, $CH_2{=}CH_2$. The ethane skeleton has lost two hydrogen atoms and now contains a double bond.

Ethane   Ethene

Alkyl groups can replace the hydrogen atoms, and so a vast range of alkenes is possible. The next members of the homologous series are:

$CH_2{=}CHCH_3$ $CH_2{=}CHCH_2CH_3$ $CH_3CH{=}CHCH_3$
Propene   But-1-ene   But-2-ene

All these unsaturated molecules contain the alkene functional group. If it is the end of the molecule, the $>CH_2$ group will be present; or if in the middle, the $-CH{=}CH-$ pattern. The double bond is not always written in. The linear formulae CH_2CHCH_3 and $CH_3CHCHCH_3$ must both contain a double bond if all the carbon atoms are to have four bonds attached to them.

The ending **–ene** in the name of a compound tells us that there is a carbon–carbon double bond in the molecule.

- ethene – $CH_2{=}CH_2$
- propene – $CH_3CH{=}CH_2$ (which is the same as $CH_2{=}CHCH_3$)

Remember, the stem tells us that there are two and three carbon atoms respectively in the molecules. (Methane – with only one carbon atom – cannot possibly form a $C{=}C$ bond!)

Once the carbon skeleton contains four atoms in the straight chain, the double bond can exist in two different positions in the molecule.

> Ethene is one of the chemicals involved at the start of the ripening process of a number of fruits such as bananas and peaches. The ripening of bananas is helped by keeping the fruit in a controlled atmosphere containing ethene. The fruit is picked when still green, and transporting the fruit in a controlled ethene atmosphere controls the ripening. The fruit is ripe and ready when delivered to the shops. This is only an extension of the age old principle of ripening fruit by putting the unripe fruit in a drawer with a ripe fruit (which provides the ethene to enable ripening).

We can recognise which isomer is which by using numbers.

$$\overset{1\ \ \ 2\ 3\ 4}{CH_2{=}CHCH_2CH_3} \quad and \quad \overset{1\ 2\ \ \ 3\ 4}{CH_3CH{=}CHCH_3}$$
But-1-ene   But-2-ene

> Do not forget to keep these numbers as small as possible, counting from the other end of the longest chain if need be.

The number identifies the position of the double bond from the first of the carbon atoms forming the bond.

$$\overset{1\ \ \ 2\ 3\ \ 4}{CH_2{=}CHCH_2CH_3}$$

This is but-1-ene, not but-2-ene (or but-3-ene if you started to number from the other end of the molecule).

$$\overset{1\ \ \ 23\ \ 4\ 5}{CH_2{=}CHCH_2CH_2CH_3} \qquad \overset{5\ \ 4\ \ 3\ \ 2\ \ 1}{CH_3CH_2CH_2CH{=}CH_2}$$

Both these molecules represent pent-1-ene.

$$\overset{1\ \ 2\ \ \ 3\ 4\ 5}{CH_3CH{=}CHCH_2CH_3} \text{ is pent-2-ene and not pent-3-ene.}$$

For more than one double bond in the molecule, the name must include –*diene*, *-triene*, etc. to state the number of double bonds. The position of each double bond is identified by a number (see *Figure 16.1.1*).

$$\overset{1\ \ \ 23\ 4\ \ \ 56}{CH_2{=}CHCH_2CH{=}CHCH_3} \quad and \quad \overset{1\ \ \ 23\ \ 45\ \ \ 6}{CH_2{=}CHCH{=}CHCH{=}CH_2}$$
Hexa-1,4-diene   Hexa-1,3,5-triene

The simple aliphatic alkenes (those with only one double bond) will:

1. have a general formula C_nH_{2n};
2. have very similar chemical properties;
3. show a trend in their physical properties.

Remember, these are the typical properties of a **homologous series** (Ch 14.2).

Table 16.1.1 shows the boiling points of the straight chain alkanes and simplest alk-1-ene compounds.

Six-carbon alkenes

-ene

$$\begin{array}{ccc} CH_3 & CH & CH_2 \\ & CH & CH_2 & CH_3 \end{array}$$
Hex-2-ene

-diene

$$\begin{array}{ccc} CH_2 & CH & CH_2 \\ & CH & CH & CH_3 \end{array}$$
Hexa-1,3,-diene

-triene

$$\begin{array}{ccc} CH_2 & CH & CH \\ & CH & CH & CH_2 \end{array}$$
Hexa-1,3,5-triene

Figure 16.1.1

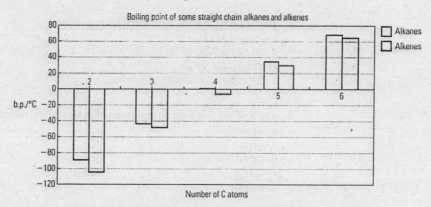

Table 16.1.1 Boiling points of straight chain alkanes and alkenes

You can see that they have slightly lower boiling points than the corresponding alkanes.

So far, all the alkenes mentioned have straight chain structures. There is no reason why the alkene functional group should not be formed in most branched chain and even cyclic molecules.

Naming such compounds follows the rules of nomenclature already established, but remember that each carbon atom must have four bonds attached to it. It is very easy to put three or five bonds around a carbon atom by mistake when you introduce the double bond into the structure. The first of the branched chain alkenes is methylpropene, $CH_2{=}C(CH_3)CH_3$. Other examples of branched chain alkenes are:

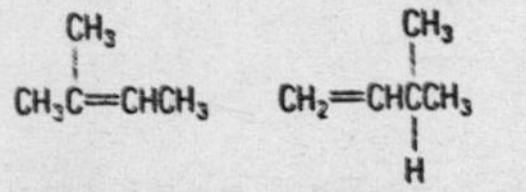

$$CH_3$$
$$|$$
$$CH_3C{=}CHCH_3 \qquad CH_2{=}CHCCH_3$$
$$|$$
$$CH_3$$
$$H$$

Methylbut-2-ene   3-Methylbut-1-ene

There is no need for a number in front of the methyl in the first example since the 2-position is the only position where a methyl group can be placed. In the second example, the 3-methyl identification is needed as another isomer would be the 2-methyl compound. In a linear format these molecules would appear as $CH_3C(CH_3){=}CHCH_3$ and $CH_2{=}CHCH(CH_3)_2$. Can you relate the two structures to the structural formulae?

Cyclic alkene molecules look like this:

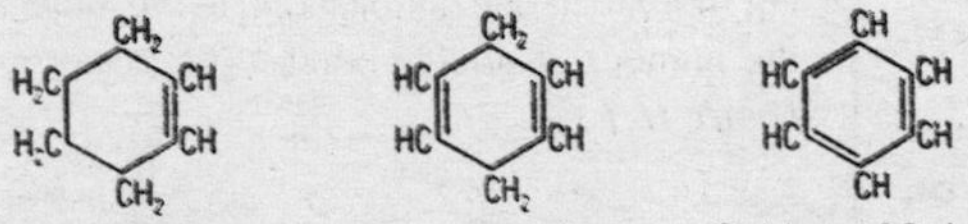

Cyclohexene   Cyclohexa-1,4-diene   Cyclohexa-1,3,5-triene

(Cyclohexa-1,3,5-triene is usually called benzene. This will be discussed in detail in Ch 26.1).

SELF-ASSESSMENT
QUESTIONS
16.1

1 Name the following alkenes.
   a $CH_3CH{=}CH_2$
   b $CH_3CH{=}CHCH_2CH_3$
   c $CH_3C(CH_3){=}CHCH_2CH_3$
   d $(CH_3)_2C{=}C(CH_3)_2$

2 Which of the following formulae represent alkenes?
   a $CH_2CHCH_2CH_3$
   b $CH_3CH(CH_3)CH_2CH_3$
   c $(CH_3)_3CCH_2CH_3$
   d $CH_3CH_2CHCHCH(CH_3)CH_3$

3 Which of the following hydrocarbons must be alkenes, if they all have straight chain skeleton structures?
   a C_5H_8
   b C_4H_8
   c C_6H_{12}
   d C_5H_{12}

4 Draw displayed formulae for the following molecules, and write their systematic names.
   a $CH_3CH_2CH(CH_3)CH_2CH_2CH_3$
   b $CH_3CH_2C(CH_3){=}CHCH_3$
   c $(CH_3)_2CHCH_2CH{=}CH_2$
   d $CH_3CH_2CH_2C(CH_3)_3$

16.2 Bonding in the alkenes and geometrical isomerism

How does the double bond form between two carbon atoms? In an earlier chapter (Ch 2.4), the formation of a single covalent bond between two atoms was shown as the sharing of a pair of electrons. Such a single bond is called a sigma (σ) bond. A σ bond is the *ordinary* covalent bond. Typical examples are chlorine, Cl_2, or methane, CH_4:

chlorine   methane

If we look carefully at ethene and try to draw the same type of *dot-and-cross* diagram, we can see that there is still one electron on each carbon atom unaccounted for in any bond. This electron on each atom can be considered as remaining in a p-orbital (see Ch 1.4), while the sigma bond clearly consists of a pair of electrons *between* the nuclei of the two atoms.

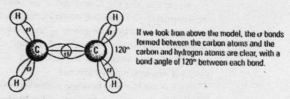

The two lobes of the p-orbitals can overlap, 'sideways on', to form the second bond between the carbon atoms (*Figure 16.2.1*). After all, a bond forms when two atomic orbitals overlap; and the pair of electrons is shared between the atoms involved.

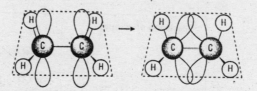

If we look from above the model, the σ bonds formed between the carbon atoms and the carbon and hydrogen atoms are clear, with a bond angle of 120° between each bond.

When we look from the side the two p-orbitals, each containing a single electron, can overlap to form the π bond.

Figure 16.2.1 σ and π bonds in ethene

The shape and structures of the two bonds in ethene are very different. The σ bond between the atoms lies along the axis between the two nuclei of the carbon atoms. This is the same as in ethane; and in all the other single bonds between carbon atoms and/or hydrogen atoms. But the 'side on' overlap in ethene is very different:

This shared electron density *both above and below* the σ bond is the π bond.

The result is a molecule that is planar (flat) about the double bond with bond angles of 120° (*Figure 16.2.2*) because the σ bonds around each carbon atom stay as far apart as possible (Ch 2.6).

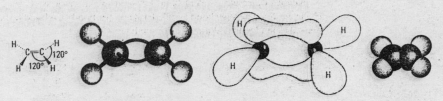

Figure 16.2.2   Models of ethene

The double bond is often represented as >C=C< or written as a dot-and-cross diagram:

$$>C \overset{\cdot}{\underset{\cdot}{\cdot}} C<$$

You must appreciate the difference between the two bonds. The carbon–carbon double bond enthalpy is +612 kJ mol⁻¹, compared with the σ C—C bond *enthalpy* of +347 kJ mol⁻¹, but the total bond enthalpy is not really significant; what *is* important is the energy needed *to break just the* π *bond*.

Notice that the C=C enthalpy is considerably less than twice the C—C value; the π bond is weaker than the σ bond. The σ bond lies along the axis joining the two carbon atoms, but the π bond lies above and below the plane of the molecule, making these bonding electrons much more accessible and more polarisable. The high electron density around the double bond enables electrophiles to attack the double bond. So **electrophilic addition reactions** become the principal reaction type for alkenes.

It is very rare for both bonds in a double bond to be broken. One way of looking at addition reactions of alkenes is this: **the weaker π bond is broken to make two new strong σ bonds** (see Ch 16.5).

Reactions of alkenes

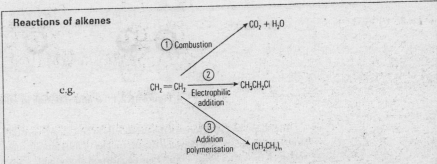

① Like all hydrocarbons, the alkenes burn in oxygen.
② The double bond is readily available for attack by an electrophile. Most of the simple reactions of alkenes involve **electrophilic addition**.
③ On the same principle of addition across the double bond, alkene molecules can join together to form long hydrocarbon chains by **addition polymerisation**.

These reactions will be fully discussed in Ch 16.5.

Non-rotation about double bonds

The other consequence of the π bond between the two carbon atoms is that there can be *no rotation about this double bond* unless the π bond is broken. Models make this much more obvious, as in *Figure 16.2.3*.

Picture a double bond

Put your middle fingers of your hands tip to tip. The other fingers on each hand represent other groups attached to the carbon atoms. It is possible to rotate one hand with respect to the other, and the atoms all move with respect to each other.

This represents a σ bond.

Now put the index and ring fingers tip to tip so that you now have three fingers tip to tip. The two outer fingers together represent the π bond. One obvious difference is that now you cannot rotate your hands, and the thumbs must both stay in the same relative position, unless you break contact between your fingers.

Figure 16.2.3   A double bond

The significance of this lack of rotation about the double bond is more obvious when the alkenes are more complex.

In molecules such as ethane and 1,2-dichloroethane, there is no restriction of rotation about the central carbon–carbon bond.

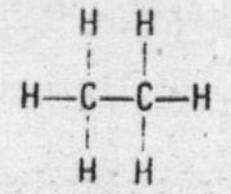

Ethane

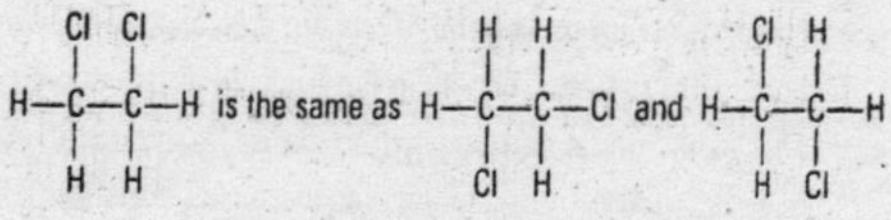

1,2-Dichloroethane

When there is a double bond between the central carbon atoms, rotation around the double bond is not possible. The presence of at least one other group, other than hydrogen, on each of the carbon atoms and the lack of rotation around the carbon–carbon double bond gives rise to stereoisomerism (see Ch 14.4) simply because the double bond holds the molecular shape.

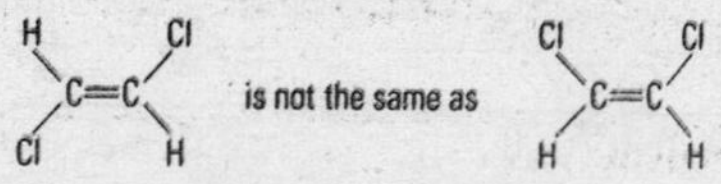

The lack of rotation keeps the chlorine atoms on the same side of the bond or on opposite sides of the double bond. This type of isomerism is called **geometrical isomerism** (Ch 14.4) and the prefixes *cis* and *trans* denote the two isomers (*trans* is the Latin word for 'across', and *cis* is the Latin word for 'on the same side').

Note that this type of isomerism can only occur when there are at least two groups, other than hydrogen, attached to the double bond.

CH_2=CH_2 and CH_2=$CHCH_3$ do not have geometrical isomers.

Ethene Propene

If the two groups on each of the carbons involved in the double bond are different, then the formation of geometrical isomers is possible.

trans-but-2-ene cis-but-2-ene

Note that the two substituents must be on different carbon atoms in the double bond. If the two substituents are on the same carbon atom, geometrical isomerism is not possible. 1,1-Dichloroethane, CCl_2=CH_2, only exists as a single isomer.

Using the butenes, C_4H_8, as another example, the possible isomers are:

But-1-ene cis-but-2-ene trans-but-2-ene Methylpropene

There is very little chemical difference between many *cis-trans* isomers and only slight differences, generally, in physical properties. The differences are, however, very important in biological systems. The importance of *cis-* and *trans*-retinal in our ability to see has already been discussed (Ch 1·i.4) and the formation of *trans*-fatty acids gives rise to much debate about healthy eating.

> Alkenes that are more complex can give rise to ambiguity over the naming of the geometrical isomers. Modern terminology refers to *E* and *Z* isomers and considers the sizes of the substituted groups, but the terms *cis* and *trans* are perfectly adequate for A level and we shall use them in this book.

---

trans-fatty acids

Vegetable oils contain triglycerides and these triglyceride esters contain long chains of alkyl groups with a varying number of double bonds.

Naturally occurring vegetable oils are unsaturated or polyunsaturated, with the *cis*-configuration around most of the double bonds. The resulting chains in the triglyceride are curved. The molecules cannot pack together and so the oil is liquid.

These oils can be partially hydrogenated (reacted with hydrogen) to reduce the number of double bonds. The production of more 'linear' saturated molecules increases the intermolecular attractions, and the melting point of the oil increases.

The conversion of olive oil to margarine (hydrogenated vegetable oil) is such a process. The process of hydrogenation also changes some of the *cis*-isomers into the 'straighter' *trans*-isomers. Typically, a vegetable oil might contain 45% *cis*-C_{18} isomer. After partial hydrogenation, the amount of this *cis*-isomer is reduced to 30%, with 15% converted to the *trans*-isomer. This is a substantial increase from the natural composition which has less than 1% of the *trans*-isomer.

Trans-fatty acids have started to be connected to the occurrence of some types of cancer and to an increased possibility of heart attack.

It is not enough to increase the unsaturated or polyunsaturated oils in your diet – they must not be *trans*-fatty acids!

ACTIVITY

Activity

Using molecular models, make up models of the following alkenes. Can you devise and illustrate a rule about how many substituents are needed around the double bond to allow geometrical isomerism to occur?

CH_2=CH_2, $ClCH$=CH_2, $ClCH$=$CHCl$, Cl_2C=CH_2, Cl_2C=$CHCl$ and Cl_2C=CCl_2

Naming alkenes – examples

Can you name the following alkenes?

a) $CH_3CH_2CHC(CH_3)CH_3$

b) CH_3CCH_2
$\quad\quad\ \|$
$\quad\quad CHCHC(CH_3)_2$

1. **Write out the carbon skeleton to find the longest chain**

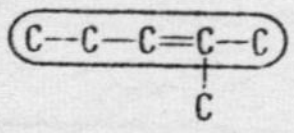

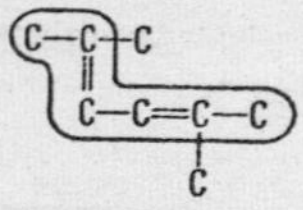

Longest chain – 5; ∴ Pent-

Longest chain – 6; ∴ Hex-

2. **Number the carbon atoms – keep the numbers as low as possible.**

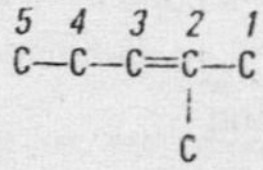

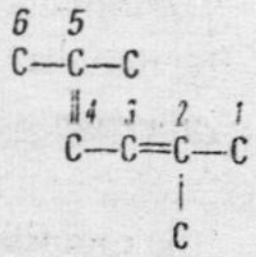

```
5  4  3  2  1
C—C—C=C—C
       |
       C
```

```
       6  5
       C—C—C
           ‖4  3  2  1
       C—C=C—C
           |
           C
```

3. **Decide on the ending – the only functional group is a double bond.**

pent- -ene

hexa- -diene

4. **Position the functional group.**

pent-2-ene

hexa-2,4-diene

5. **Name any side chains.**

methyl

methylpent-2-ene

Two methyl groups; ∴ dimethyl

dimethylhexa-2,4-diene

6. **Position the side chains.**

2-Methylpent-2-ene

2,5-Dimethylhexa-2,4-diene

SELF-ASSESSMENT
QUESTIONS
16.2

1 Name the compounds with the following formulae.

  a $CH_3CH=CHCH_3$

  b $CH_2=CHCH=CH_2$

  c $(CH_3)_2C=C(CH_3)CH_2CH_3$

  d $CH_3CH_2C(CH_3)CH_2$

2 Write the displayed formulae for both geometrical isomers of:

  a $CHCl=CHBr$

  b $CH_3CH=CHCH_2CH_3$

  c $C_6H_5CH=CHCl$

  d $(CH_3)_3CCH=CHCH_3$

3 Draw the displayed formula of each of the following isomers.

  a *cis*-$CH_3CH=CHCH_3$

  b *trans*-$CHBr=CHBr$

  c *trans*-$CH_3CH_2CH=CHCl$

  d *cis*-$CH_3CH_2CH=CHCH_2CH_3$

4 Which of the following compounds must contain π bonds?

  a CH_2CH_2

  b $CH_3CH_2CH_3$

  c $CH_2C(CH_3)_2$

  d $CH_2CHCH_2CH_2CH_3$

5 What type of isomerism does each of the following pairs of compounds show?

a $CH_3CH_2CH_2CH_2CH_2CH_3$ and $CH_3CH(CH_3)CH_2CH_2CH_3$

b $CH_2{=}CHCH_2CH_3$ and $CH_3CH{=}CHCH_3$

c

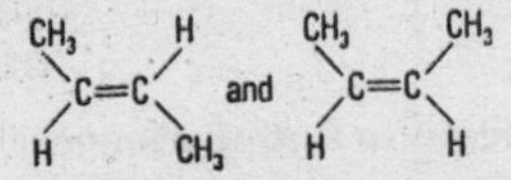

$$\underset{H}{\overset{CH_3}{}}C{=}C\underset{CH_3}{\overset{H}{}} \quad \text{and} \quad \underset{H}{\overset{CH_3}{}}C{=}C\underset{H}{\overset{CH_3}{}}$$

d $CH_2{=}CHCH_2CH_3$ and $CH_2{=}C(CH_3)_2$

16.3 Preparation of alkenes through π bond formation

The formation of an alkene requires production of a π bond from a fully saturated molecule. This can be done by cracking larger alkane molecules or by an elimination reaction – *Figure 16.3.1*. Both processes will leave two electrons to form the π bond.

The **cracking** of alkanes was introduced earlier (Ch 15.4) and, as the major industrial process for making alkenes, will be discussed in Ch 16.4.

The **elimination** of a small molecule from a saturated molecule with a suitable functional group requires a reaction which releases, or leaves behind, the electrons to form the π bond (Ch 14.6).

e.g. $CH_3CH_2OH(g) \longrightarrow CH_2{=}CH_2(g) + H_2O(g)$

Alcohols and halogenoalkanes can be made to lose the hydroxyl group (Ch 18.4) or the halogen atom (Ch 17.4), together with an adjacent hydrogen atom, to form the double bond. These **elimination** processes are often called *dehydration* or *dehydrohalogenation*, respectively.

Preparation of alkenes

Cracking   Elimination

Dehydration   Dehydrohalogenation

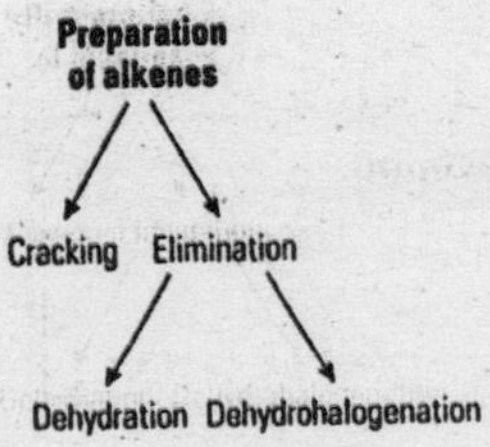

Figure 16.3.1
Preparation of alkenes

Dehydration of alcohols

Most alcohols will form the corresponding alkene if they are treated with a suitable reagent. Concentrated sulphuric or phosphoric acids can be used, or the alcohol vapour can be passed over a solid aluminium oxide (alumina) catalyst.

$$CH_3CH_2CH_2OH(g) \xrightarrow[\text{heat}]{Al_2O_3 \text{ catalyst}} CH_3CH{=}CH_2(g) + H_2O(g)$$

In the laboratory, the above reaction can be done using the apparatus shown in *Figure 16.3.2*; the ethene is collected over water.

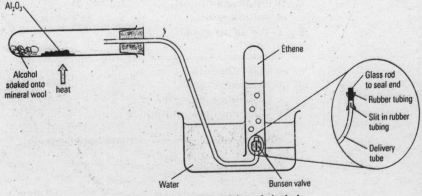

Figure 16.3.2   Dehydration of alcohols

If you carry out the experiment, do be careful not to allow the water to 'suck back' when you remove the heat source. This danger can be minimised by using a good Bunsen valve at the end of the delivery tube.

If concentrated sulphuric or phosphoric acid is used as the reagent, the reaction mixture is heated, and the alkene is distilled off or collected as a gas.

$$CH_3CH_2OH(l) \xrightarrow[\text{heat}]{\text{concentrated } H_2SO_4} CH_2{=}CH_2(g) + H_2O(l)$$

The reaction in the presence of the acid begins with the acid protonating the OH group using a lone pair of electrons from the oxygen atom. The consequent loss of water leaves the electrons to form the double bond.

Elimination of hydrogen halide

In chemical reactions, there are a number of different reagents that appear to be very similar. A **nucleophile** (see Ch 14.6) is a species capable of providing a pair of electrons to form a new bond. A **base**, similarly, has a pair of electrons able to form a new bond with an acid (especially with $H^\oplus$).

Later we will come across **ligands**, species capable of donating a lone pair of electrons to transition metals in particular. The word we use to describe the reagent in this case does depend on the reaction. $OH^\ominus$ is a base with acids, a nucleophile in organic chemistry, and a ligand in transition metal chemistry.

This is often called *dehydrohalogenation* (de- (remove) hydro- (hydrogen) halogenation (halide) – removing hydrogen halides such as HCl or HBr).

Halogenoalkanes contain an alkane chain with a halogen atom replacing one of the hydrogen atoms; e.g. CH_3CH_2Br is bromoethane (see Ch 14.3). The polarity of the C—Hal bond (Ch 2.5) within the molecule results in a relatively reactive molecule, and halogenoalkanes can react in a number of different ways.

Later, we will see that this polarity of the carbon–halogen bond, $C^{\delta\oplus}{-}X^{\delta\ominus}$, promotes nucleophilic attack by, for example, hydroxide ions, at the carbon atom resulting in substitution.

$$\text{e.g.} \quad CH_3CH_2Br + OH^\ominus \longrightarrow CH_3CH_2OH + Br^\ominus$$

Alternatively, the hydroxide ion can act as a **base**, leading to elimination. *This competition between nucleophilic substitution and elimination can be pushed in a particular direction* by altering the reaction conditions.

The **elimination** reaction is carried out in ethanol as a solvent. Potassium (or sodium) hydroxide, dissolved in ethanol, is mixed with the halogenoalkane. The mixture is heated and the alkene collected as a volatile liquid or gas.

$$CH_3CH_2Br \xrightarrow[\text{heat}]{\text{KOH in ethanol}} CH_2{=}CH_2 + HBr$$

We can picture this as the base, $OH^\ominus$, attacking hydrogen, and electron movement within the molecule releasing $Cl^\ominus$.

If the halogenoalkane contains four or more carbon atoms, and the halogen atom is not at the end of the chain, two possible alkene products are possible in the elimination reaction.

e.g. $\underset{\text{2-Chlorobutane}}{CH_3CHClCH_2CH_3} \longrightarrow \underset{\text{But-1-ene}}{CH_2{=}CHCH_2CH_3}$ or $\underset{\text{But-2-ene}}{CH_3CH{=}CHCH_3}$

413

In this example, but-2-ene is the major product, and both *cis-* and *trans-*isomers of but-2-ene will be formed (*Figure 16.3.3*).

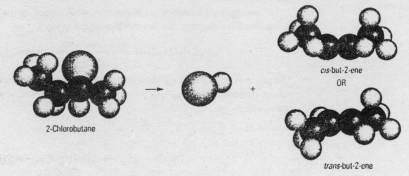

2-Chlorobutane

cis-but-2-ene

OR

trans-but-2-ene

Figure 16.3.3 Dehydrohalogenation

SELF-ASSESSMENT
QUESTIONS
16.3

1 Name the starting material and the reagents needed to prepare the following alkenes.
   a $CH_3CH_2CH\!=\!CH_2$ from an alcohol
   b $CH_2\!=\!CH_2$ from a halogenoalkane
   c $CH_3CH\!=\!CH_2$ from a halogenoalkane
   d $(CH_3)_2C\!=\!CHCH_3$ from an alcohol.

2 Complete the following equations by inserting the conditions.
   a $CH_3CH_2CH_2OH \longrightarrow CH_3CH\!=\!CH_2$
   b $CH_3CHBrCH_3 \longrightarrow CH_3CH\!=\!CH_2$

3 Identify the alkene isomers which can be prepared from the following starting materials.
   a $CH_3CH(OH)CH_3 \longrightarrow$
   b $CH_3CHBrCH_2CH_2CH_3 \longrightarrow$ 3 alkene isomers

4 Classify the following reactions as substitution or elimination reactions.
   a $CH_3CH_2CH_2CH_2CH_3 + Cl_2 \longrightarrow CH_3CHClCH_2CH_2CH_3 + HCl$
   b $CH_3CHClCH_2CH_3 \longrightarrow CH_2CH\!=\!CHCH_3 + HCl$
   c $CH_3CH(OH)CH_2CH_3 + H^\oplus \longrightarrow CH_3CH\!=\!CHCH_3 + H_2O + H^\oplus$

16.4 Industrial manufacture of alkenes

Alkenes are vital feedstocks for a whole range of organic chemicals, from pharmaceuticals to plastics. The principal source of hydrocarbons is crude oil. However, crude oil is a non-renewable resource, and one day supplies will run out. Industrial production of alkenes, ethene in particular, has diversified, and although cracking is still the most important method for the production of alkenes, other resources are being developed.

Natural gas supplies are expected to last until at least 2050, and the use of natural gas as a source of ethene is increasingly important. Renewable resources such as sugar cane can provide ethanol, and we have already seen that ethanol can be converted into ethene.

Crude oil consists mainly of saturated hydrocarbons, and the demands for the different fractions resulting from fractional distillation can vary. Alkenes are mostly produced on an industrial scale by cracking the fractions from crude oil that are in least demand (Ch 15.4).

Catalytic cracking

The naphtha fraction from crude oil distillation contains C_5 to C_{10} hydrocarbons, and can make up to 15% of the crude oil. This fraction is cracked using silicon oxide/aluminium oxide catalysts at 500 °C. These cracking conditions produce mainly ethene but can also increase the branching in the alkane chain.

e.g.
$$C_8H_{18} \xrightarrow[500\,°C]{SiO_2/Al_2O_3} CH_2{=}CH_2 + CH_3CH(CH_3)CH(CH_3)CH_3$$

Octane   Ethene   2,3-Dimethylbutane

The products containing branched alkanes are useful in increasing the octane number of petrol distillates. A single industrial plant can produce some 800 000 tons per year of ethene. The apparatus for a laboratory version of catalytic cracking is shown in *Figure 16.4.1*.

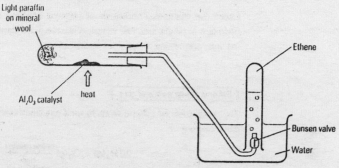

Figure 16.4.1   Cracking light paraffin (C_{10} to C_{20})

Thermal cracking

The heavy residue fraction from the fractional distillation of crude oil is also cracked to produce smaller molecules. Conditions for thermal cracking (often called 'steam cracking') are much more vigorous. The residue fraction is treated with super-heated steam at 800 °C, and this results in high yields of ethene. These high-energy conditions favour a free radical type of mechanism, and the cracking is thought to occur through the homolytic fission of a carbon–carbon bond, forming free radicals.

The **initiation** step involves the formation of two alkyl radicals.

$$RCH_2{-}CH_3 \longrightarrow RCH_2{\bullet} + CH_3{\bullet}$$

These radicals then react with other alkane chains, producing a simple alkane and another radical in a **propagation** step.

$$RCH_2{-}\overset{\displaystyle H}{\underset{\displaystyle H}{C}}{-}H + {\bullet}CH_3 \longrightarrow RCH_2CH_2{\bullet} + CH_4$$

or:

$$RC\overset{\displaystyle H}{\underset{\displaystyle H}{}}{-}\overset{\displaystyle H}{\underset{\displaystyle H}{C}}{\bullet} \longrightarrow RCH{=}CH_2 + H{\bullet}$$

The process stops in a **termination** step when two radicals combine to produce a longer chain alkane.

$$CH_3{\bullet} + CH_3{\bullet} \longrightarrow CH_3CH_3$$

Thermal cracking produces shorter chain alkanes, alkenes and hydrogen (*Figure 16.4.2*).

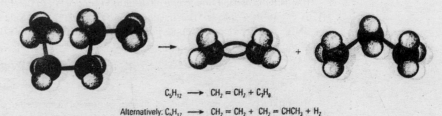

$$C_5H_{12} \longrightarrow CH_2 = CH_2 + C_3H_8$$

Alternatively: $C_5H_{12} \longrightarrow CH_2 = CH_2 + CH_2 = CHCH_3 + H_2$

Figure 16.4.2   Cracking pentane to ethene and propane

Under the vigorous conditions of thermal cracking a complex mixture of alkanes, alkenes and hydrogen is produced. The reaction mixtures are then fractionally distilled to produce the various fractions containing a variety of hydrocarbons.

Ethene from natural gas

The production of ethene from natural gas (methane) is carried out at high temperature over an oxide catalyst.

$$2CH_4(g) + O_2(g) \xrightarrow[\substack{high \\ temperature}]{oxide\ catalyst} C_2H_4(g) + 2H_2O(g)$$

Notice here that, unlike any other of the reactions so far discussed, the product has a *larger* molecule than the reactant.

Ethene is used to make a wide range of products, including Polythene, used widely for packaging, and ethane-1,2-diol, which is commercial antifreeze.

SELF-ASSESSMENT
QUESTIONS
16.4

1   Write balanced equations to show the possible products from cracking decane, $C_{10}H_{22}$, to form an alkene and an alkane.

2   Explain how crude oil is treated to produce a range of different fractions and then consequently used to prepare alkenes.

3   Butan-2-ol can be distilled with concentrated phosphoric acid to form a number of isomeric alkenes. Write the structures of the three isomers produced, and name them. Identify the isomers as pairs of position or geometrical isomers.

4   When 10 cm³ cyclohexanol, $C_6H_{13}OH$, is heated with 4 cm³ concentrated phosphoric acid, there is a reaction and a clear colourless liquid is formed which boils at around 82 °C.
   a How would you measure out the 10 cm³ and 4 cm³ of reactants?
   b Draw a diagram to show how you would separate this product from the reaction mixture.
   c Draw the structure of cyclohexanol.
   d What do you think the product of the reaction is? Draw its structure.
   e Write an equation for the reaction.

16.5 Reactions of the alkenes

Combustion

Alkenes are volatile hydrocarbons, and they burn readily to form carbon dioxide and water. One distinguishing feature about their combustion is that there is always a smoky flame when they burn in air – a helpful and simple difference between alkanes and alkenes.

e.g. $$CH_2{=}CH_2(g) + 3O_2(g) \xrightarrow{ignite} 2CO_2(g) + 2H_2O(g)$$

This equation applies when combustion is complete. But when an alkene burns under normal conditions, a small amount of carbon is always produced.

Any hydrocarbon containing a higher proportion of carbon than is present in the corresponding alkane tends to burn with a smoky flame under normal conditions. As the number of double bonds in the molecule increases, there is an increase in the smoky character of the flame.

If less oxygen is available, reactions resulting in the formation of carbon monoxide or even carbon are possible.

e.g. $$CH_2{=}CH_2(g) + 2O_2(g) \xrightarrow{ignite} 2CO(g) + 2H_2O(g)$$

$$CH_2{=}CH_2(g) + O_2(g) \xrightarrow{ignite} 2C(s) + 2H_2O(g)$$

Addition reactions

The reactions of the alkenes are dominated by the presence of the π bond. The high electron density above and below the plane of the molecule around the double bond allows an electrophile ready access to the π electrons. The π bond is broken and two new bonds are formed (*Figure 16.5.1*).

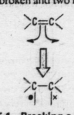

Figure 16.5.1 Breaking a double bond

These new bonds are σ bonds and are usually strong. Therefore, the enthalpy of reaction for an addition reaction is usually highly negative, which means that the reaction is likely to occur. The consequence is that there is an addition reaction across the double bond. The most important reactions of alkenes are **ADDITION REACTIONS**.

Bond	Mean bond enthalpy values/kJ mol^{-1}
C—C	+347
C=C	+612
C—H	+413
C—Br	+290
H—Br	+366
Br—Br	+193

Table 16.5.1
Some mean bond enthalpy values

The importance of addition reactions, rather than substitution reactions, for the alkenes is easily appreciated if we consider bond enthalpy changes during the reactions (*Table 16.5.1*). Note, however, that enthalpy changes will not help you decide how *rapidly* a reaction will occur (Ch 12.3).

The addition reaction of ethene and bromine can be written as:

$$CH_2{=}CH_2 + Br_2 \longrightarrow CH_2Br{-}CH_2Br$$

We can calculate the enthalpy change for this reaction by way of bond enthalpy values:

$1 \times C{=}C$	+612		$1 \times C{-}C$	+347
$4 \times C{-}H$ (4 × 413)	+1652		$4 \times C{-}H$ (4 × 413)	+1652
$1 \times Br{-}Br$	+193		$2 \times C{-}Br$ (2 × 290)	+580
	+2457			+2579
Bonds broken (so *positive*)	+2457 kJ mol^{-1}		Bonds formed (so *negative*)	−2579 kJ mol^{-1}

Overall enthalpy change for this addition reaction = 2457 − 2579 = −122 kJ mol^{-1}

For the substitution reaction of ethene by a bromine atom:

$$CH_2{=}CH_2 + Br_2 \longrightarrow CHBr{=}CH_2 + HBr$$

Enthalpy changes now would be:

$1 \times C{=}C$	$+612$		$1 \times C{=}C$	$+612$
$4 \times C{-}H$ (4×413)	$+1652$		$3 \times C{-}H$ (3×413)	$+1239$
$1 \times Br{-}Br$	$+193$		$1 \times C{-}Br$	$+290$
	$+2457$		$1 \times H{-}Br$	$+336$
				$+2477$
Bonds broken (so *positive*)	$+2457$ kJ mol^{-1}		Bonds formed (so *negative*)	-2477 kJ mol^{-1}

Overall enthalpy change for the substitution reaction = 2457 − 2477 = −20 kJ mol^{-1}

Considering these enthalpy changes, which reaction do you think is more likely?

Reaction with halogens

Alkenes react easily with most halogen molecules at room temperature, producing a di-substituted halogenoalkane. The reaction with iodine is slow, but the reaction with fluorine is explosive. With chlorine, bubbling the gas through the liquid alkene or a solution of the alkene in an organic solvent at room temperature is sufficient for the chlorine to add across the double bond.

$$CH_2{=}CH_2(g) + Cl_2(g) \xrightarrow[\substack{room \\ temperature}]{hexane} CH_2Cl{-}CH_2Cl(l)$$

With bromine, an alkene shaken with a solution of bromine in hexane will remove the brown colour of the bromine.

$$CH_3CH{=}CH_2(g) + Br_2(g) \xrightarrow[\substack{room \\ temperature}]{hexane} CH_3CHBr{-}CH_2Br(l)$$

Even a solution of bromine dissolved in water is decolorised in the presence of an alkene – although the addition product is not the same. Bromine water is less hazardous to handle than bromine liquid, or even bromine dissolved in hexane, but it still needs to be treated with care – bromine corrodes skin and flesh very rapidly!

You will remember that alkanes only react with the halogens using UV light, and so the reaction with bromine at room temperature provides a useful test for alkenes and distinguishes them from alkanes.

Reaction with hydrogen halides

The anhydrous hydrogen halides add across the double bond, if solutions in hexane or methylbenzene are reacted with the alkene. Even concentrated aqueous solutions may react.

e.g. $$CH_2{=}CH_2 + HBr \xrightarrow[\substack{room~temperature}]{\substack{solution \\ in~methylbenzene}} CH_3{-}CH_2Br$$

For the addition of hydrogen bromide to propene, see Ch 16.6.

Reaction with hydrogen

Addition of hydrogen across the double bond is achieved at high pressures and temperatures with a transition metal catalyst, often nickel or palladium.

$$CH_2{=}CH_2(g) + H_2(g) \xrightarrow[140\,°C]{Ni~catalyst} CH_3{-}CH_3(g)$$

This reaction has already been met in the conversion of oils to fats (Ch 16.2).

As we have seen (Ch 16.2), partial hydrogenation of vegetable oils forms margarines. Vegetable oils are complex molecules containing long hydrocarbon chains with differing degrees of unsaturation. The measurement of the number of double bonds present in a particular oil uses the ability of iodine (or a more reactive compound of iodine, ICl) to add across double bonds. The oil is treated with excess of the iodine reagent. ICl adds across the double bonds, and excess iodine is measured by titrating with standard sodium thiosulphate solution (Ch 10.3). The unsaturation of an oil or fat is often expressed in terms of its 'iodine number'.

Addition reactions involving oxygen

A number of different oxygen-containing products are possible.

The catalytic oxidation of ethene is achieved by reacting ethene with air in the presence of a silver catalyst. This is an important industrial reaction, since the epoxide molecule formed is very reactive.

$$CH_2{=}CH_2(g) + \tfrac{1}{2}O_2(g) \xrightarrow[170°C]{Ag\ catalyst} CH_2{-}CH_2(g)$$
$$\underset{O}{\diagup\ \diagdown}$$

Ethene   Epoxyethane

The three-membered ring containing two carbon atoms and the oxygen atom is very strained (bond angles of 60° compared to the tetrahedral bond angle of 109.5°). Epoxyethane is another feedstock chemical. One important reaction is with water.

$$CH_2{-}CH_2 + H_2O \xrightarrow[H_2SO_4\ (aq)]{60°C\ 1\ atm} CH_2OH{-}CH_2OH$$

Epoxyethane   Ethane-1,2-diol

Ethane-1,2-diol is used in antifreeze mixtures and in the preparation of polyester fibres.

When alkenes are treated with potassium manganate(VII) solution, a variety of products is possible. A cold, dilute solution in dilute sulphuric acid forms the diol, and if there is sufficient alkene the manganate(VII) purple colour will disappear. The mixture becomes almost colourless. The manganate(VII) is reduced to aqueous $Mn^{2\oplus}$, which is very weakly coloured.

$$CH_2{=}CH_2 + [O] + H_2O \xrightarrow[H^{\oplus},\ cold]{dilute\ KMnO_4} CH_2OH{-}CH_2OH$$

Ethene   Ethane-1,2-diol

However, if the solution is kept alkaline using sodium hydroxide solution, the purple colour changes to green and then a brown precipitate is formed. The manganate(VII) ion is still reduced. The green colour is due to the manganate(VI) ion, $MnO_4^{2\ominus}(aq)$, and the brown precipitate is the insoluble manganese(IV) oxide, MnO_2. Since few organic molecules produce this colour change, the reaction of alkaline potassium manganate(VII) solution with the alkene provides another test for the presence of the carbon–carbon double bond.

$$CH_2{=}CH_2 + [O] + H_2O \xrightarrow[OH^{\ominus}]{dilute\ KMnO_4} CH_2OH{-}CH_2OH$$

Ethene   Ethane-1,2-diol

Hydration

The hydration of alkenes to the corresponding alcohol is a very important industrial reaction; in effect, water is added across the double bond. If the alkene is mixed with concentrated sulphuric acid, the ethyl hydrogen sulphate formed is readily broken down by water to produce the alcohol. The sulphuric acid acts as a catalyst, providing the $H^{\delta\oplus}$ which acts as the electrophile.

$$e.g. \quad CH_2{=}CH_2 + H_2SO_4 \longrightarrow CH_3{-}CH_2 \xrightarrow[\substack{room\\ temperature}]{H_2O} CH_3CH_2OH + H_2SO_4$$
$$\underset{OSO_3H}{|}$$

Ethene   Ethyl hydrogen sulphate   Ethanol

Another route to the formation of ethanol from an alkene is to treat the alkene with high temperature steam (300 °C) at 70 atmospheres using a phosphoric acid catalyst. This is an important industrial route to many alcohols, which are used as solvents in a wide range of applications, from perfumes to paints.

e.g.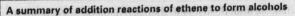

A summary of addition reactions of ethene to form alcohols

With oxygen gas and silver catalyst:	With $KMnO_4$ in alkaline solution:	With concentrated sulphuric acid:	With steam and phosphoric acid catalyst:
$CH_2{=}CH_2 + \tfrac{1}{2}O_2$ $\longrightarrow CH_2{-}CH_2$ $\qquad\quad \underset{O}{\diagdown\diagup}$ Epoxyethane which reacts with water $\longrightarrow CH_2OHCH_2OH$ Ethane-1,2-diol	$CH_2{=}CH_2 + [O] + H_2O$ $\rightarrow CH_2OH{-}CH_2OH$ Ethane-1,2-diol giving a green solution then a brown precipitate	$CH_2{=}CH_2 + H_2O$ $\longrightarrow CH_3CH_2OH$ Ethanol via the formation of $CH_2{-}CH_2{-}OSO_3H$	$CH_2{=}CH_2 + H_2O$ $\longrightarrow CH_3CH_2OH$ Ethanol at 300°C and 70 atm

---

Testing for the presence of a carbon–carbon double bond

When you have an unknown compound, you may need to find out what it is. Quantitative analysis gives you the empirical and molecular formulae, but before you can work out the structural formula, you need some information about the molecule. The organic chemist relies on some relatively simple chemical tests to start with.

If the compound contains a carbon–carbon double bond, you would expect it:

- to burn with a smoky flame;
- to decolorise bromine water;
- to turn dilute alkaline manganate(VII) green then give a brown precipitate.

None of the tests in themselves gives an unambiguous result, but when all the evidence is taken into consideration the likelihood is that the compound contains double bonds.

SELF-ASSESSMENT QUESTIONS 16.5

1 Predict the main products, and write balanced chemical equations, when:
 a propene reacts with chlorine gas;
 b propene is passed through cold concentrated sulphuric acid;
 c propene and hydrogen are passed over finely divided nickel at 250 °C;
 d propene is bubbled through an acidified potassium permanganate(VII) solution.

2 Complete the following equations:
 a $CH_2{=}CH_2 + Br_2 \longrightarrow$
 b $CH_3CH{=}CH_2 + HCl \longrightarrow$
 c $CH_3CH{=}CHCH_3 + H_2O \longrightarrow$
 d $CH_3CH_2CH{=}CH_2 + H_2O \longrightarrow$

3 The unsaturated fatty acid, *cis*-hexadec-9-enoic acid, $CH_3(CH_2)_5CH=CHCH_2(CH_2)_6COOH$, is obtained from palm oil. Hence its name palmitolenoic acid.

   a Write the molecular formula of palmitolenoic acid.

   b Write an equation for the hydrogenation of palmitolenoic acid, including the conditions needed for the reaction.

   c How would you show that palmitolenoic acid contains a carbon–carbon double bond in its molecule?

   d Naturally occurring palmitolenoic acid is always the *cis*-isomer. Use the formula of palmitolenoic acid to explain what *cis* means.

   e When palm oil is partially hydrogenated, some of this *cis*-isomer is converted into the *trans*-fatty acid. What has to happen, in terms of bonding, for this change to occur?

4 Geraniol is one of the chemicals responsible for the distinctive perfume of roses. The structure is:

$$(CH_3)_2C=CHCH_2CH_2C(CH_3)=CHCH_2OH.$$

   a Identify the two functional groups in this molecule.

   b The arrangement around the double bonds is *cis*. Explain what the term *cis* means.

   c When excess geraniol reacts with bromine in the dark, the bromine colour disappears. Explain why this happens, and write a chemical equation to explain the reaction.

   d How many moles of hydrogen gas could react with one mole of geraniol, if the gases were passed over a heated nickel catalyst?

16.6 Mechanism of electrophilic addition

The usual reactions of the alkenes are addition reactions, where an electrophile attacks the high electron density created by the pair of electrons forming the π bond. Although the symmetrical Br_2 molecule does not contain a permanent dipole, when the molecule approaches the region of high electron density, there is a partial polarisation of the bond:

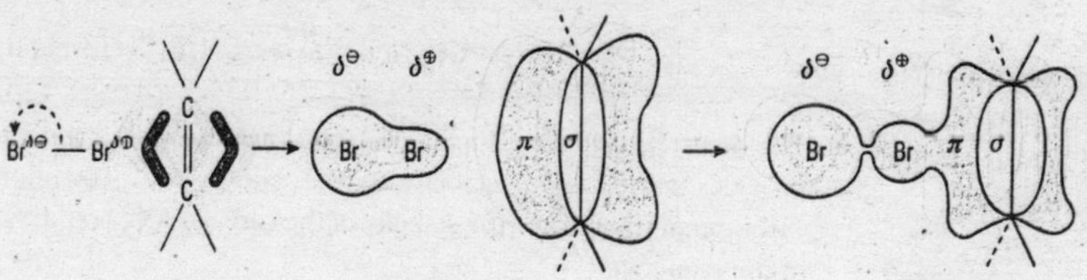

The reaction mechanism really involves two steps:

1. The interaction of the π bond and the $\delta\oplus$ end of the bromine molecule results in the formation of a carbon–bromine bond.

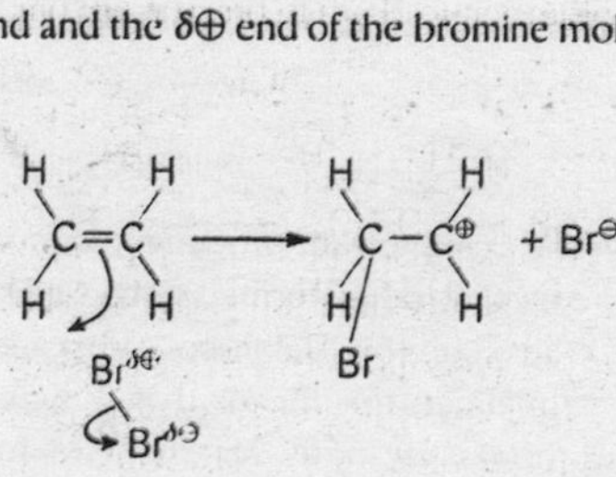

> **Definition:** A carbocation is an organic intermediate with a positive charge on one of the carbon atoms.

2. The other carbon atom of the π bond is now short of an electron and therefore has a positive charge. The $Br^{\delta\ominus}$ has become the bromide ion, $Br^{\ominus}$. The $CH_2Br—C^{\oplus}H_2$ species is called a **carbocation**, and this carbocation is now able to react with the bromide anion.

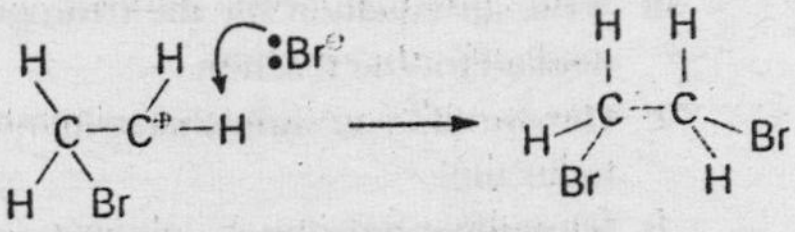

With symmetrical alkenes, there is no difficulty in working out the products of the addition reaction when a molecule such as HBr is used. Although the electrophile is permanently polarised, $H^{\delta\oplus}—Br^{\delta\ominus}$ with $H^{\delta\oplus}$ as the electrophilic end of the molecule, the possible addition products are identical.

$$CH_3CH{=}CHCH_3 + HBr \longrightarrow CH_3CH_2CHBrCH_3$$

But-2-ene   2-Bromobutane

When the alkene is not symmetrical, we must look at the stability of the carbocation formed to decide on the most likely product of addition. There are two possible products:

e.g.

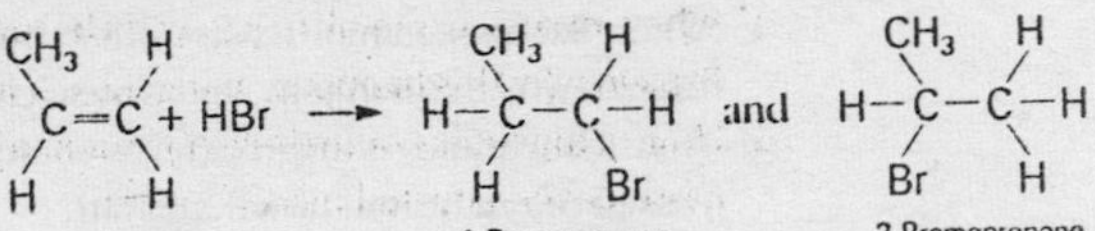

Propene   1-Bromopropane   2-Bromopropane

It is the stability of the carbocation formed from the electrophilic attack of the $H^{\delta\oplus}—Br^{\delta\ominus}$ which determines the most likely product.

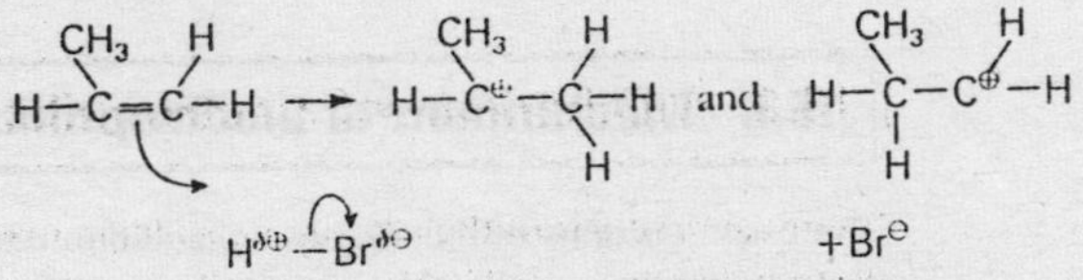

One of the compelling observations which supports such an electrophilic addition mechanism is that the carbocation formed will react with any anion present in the reaction mixture. If bromination is carried out in a solution containing hydroxide ions, $OH^{\ominus}$, two products are formed, one of which contains the hydroxyl group.

$$CH_2{=}CH_2 + Br_2 \xrightarrow{OH^{\ominus}} CH_2BrCH_2Br \text{ and } CH_2BrCH_2OH$$

> **Remember:**
> Arrows move from a pair of electrons (as a bond or lone pair). Arrows end between atoms to form a bond, or on an electronegative atom.

The methyl group has a small electron donating ability, and attached to the charged carbon atom provides a little more stability to the carbocation. The more methyl groups attached to the carbocation the greater the stability of the carbocation. So the stability of the carbocations decreases in the sequence:

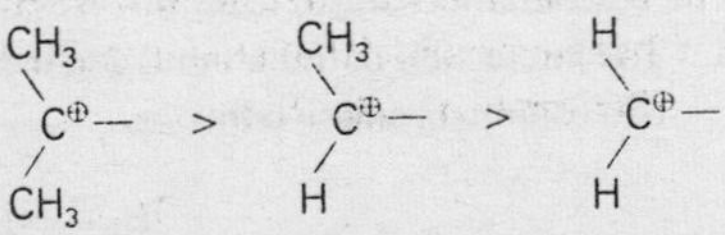

> **Definition:** According to the Markovnikov rule, when an unsymmetrical molecule containing hydrogen adds to an unsymmetrical alkene, the major product is the one in which the hydrogen atom attaches to the carbon atom which already has the more hydrogen atoms attached to it.

In the example with propene, the more stable carbocation will be $CH_3C^{\oplus}HCH_3$, and consequently the major product formed will be 2-bromopropane. The final step of the reaction involves attack by the anion on this 'stabilised' carbocation.

In effect, this means that the electrophile attaches to the carbon with the more hydrogen atoms already on it. Markovnikov formalised this as a rule for the addition to unsymmetrical alkenes.

For example:

$$CH_3 \; H$$
$$H-C=C-H + HBr \longrightarrow H-C-C-H$$

with product showing CH_3, H on top carbons, Br and H below.

$$CH_3 \; H$$
$$CH_3-C=C-H + HCl \longrightarrow CH_3-C-C-H$$

with product showing CH_3, H on carbons, Cl and H below.

Mechanism for electrophilic addition

e.g.

$$CH_3 \quad H \qquad\qquad CH_3 \quad H \qquad\qquad Br \quad H$$
$$H-C=C-H \longrightarrow H-\overset{\oplus}{C}-C-H \longrightarrow CH_3-C-C-H$$

with $:Br^{\ominus}$ and $H^{\delta\oplus}-Br^{\delta\ominus}$ shown.

Propene   2-Bromopropane

Solving a reaction sequence

Example: A hydrocarbon, L, has the empirical formula CH$_2$ and relative molecular mass of 42. When L reacts with hydrogen gas at high temperature over a nickel catalyst, it forms another hydrocarbon, M. The compound L decolorises bromine dissolved in hexane in the dark, whereas M only reacts with bromine in ultra-violet light, to form N.

Identify L, M and N using equations to explain the reactions described.

The first step in working out such a problem is to 'translate' the question into a flow diagram:

$$L \xrightarrow{H_2/Ni} M \xrightarrow{Br_2/UV} N$$
$$\downarrow Br_2$$
$$\text{Decolorises}$$

A careful look at the reagents used should suggest to you which reaction is occurring:

- H_2/Ni – adding hydrogen across some sort of double bond.
- UV light indicates free radical conditions, with bromine substitution.
- 'Decolorises bromine' could indicate a double bond.

So L has a double bond, and M has lost the double bond

Since they are both hydrocarbons, L is an alkene and M an alkane.

M as an alkane is consistent with free radical substitution by bromine in UV light.

L as an alkene is consistent with decolorising bromine in the dark.

Next step:

CH$_2$ adds up to 14; M_r of L = 42; ∴ L must contain 42/14 = 3CH$_2$ units, i.e. C$_3$H$_6$.
L is CH$_3$CH=CH$_2$.

The equations for the reactions are now:

$$CH_3CH=CH_2 + H_2 \longrightarrow CH_3CH_2CH_3$$
$$\text{L} \qquad\qquad\qquad\qquad \text{M}$$

$$CH_3CH_2CH_3 + Br_2 \longrightarrow CH_3CH_2CH_2Br + HBr$$
$$\text{M} \qquad\qquad\qquad\qquad \text{N}$$

$$CH_3CH=CH_2 + Br_2 \longrightarrow CH_3CHBrCH_2Br$$
$$\text{L}$$

1 Alkanes and alkenes react in different ways with bromine liquid, Br_2. Describe what you would expect to happen, and write a balanced equation for any chemical changes that occur, in the following cases.

 a Ethene is bubbled through bromine in the dark.
 b Ethane is bubbled through bromine in the dark.
 c A mixture of bromine vapour and ethane gas is placed in UV light.
 d What type of reaction is happening in reaction c?
 e Write a full description of the mechanism in reaction a.

2 Citronella is a part of the perfume component in many flowers and fruit. The formula is $(CH_3)_2C=CHCH_2CH_2CH(CH_3)CH_2CHO$.
 a Draw a full displayed formula for citronella.
 One of the carbon atoms in citronella is chiral. Mark that atom in your displayed formula with an asterisk, *.
 b What property does the presence of the chiral carbon atom give to citronella?
 c Citronella reacts with hydrogen bromide in the dark. Write a balanced equation for the reaction.
 d What type of reaction is occurring in this last reaction?

3 Complete the following reaction sequence, putting the missing formulae or reagents (A–F) into the scheme.

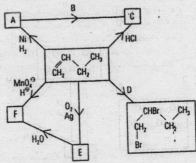

16.7 Addition polymerisation

This important reaction of the alkenes involves the breaking of the π bond.

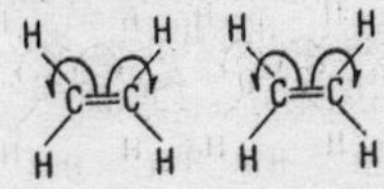

If the pair of electrons in the π bond in ethene is split up, then one of the electrons can join with an electron from the next ethene molecule to form a σ bond.

This sequence can, in theory, go on forever. The **monomer** units of ethene have added together to produce a **polymer**.

$$nCH_2=CH_2 \longrightarrow +CH_2-CH_2+_n$$

The value of n can be anything from 10 000 to over 100 000 monomer units – one molecule could have a relative molecular mass of 3×10^6. Since the molecules have simply added together, with nothing else formed, the reaction is called **addition polymerisation**.

One of the major uses of alkenes is in the production of polymers. Many of the everyday plastics that we use are based on polymers derived from alkenes. The variety of alkene-based polymers arises because other atoms or groups of atoms can replace hydrogen atoms in the ethene structure. There are a large number of possibilities. Some are shown in *Table 16.7.1*.

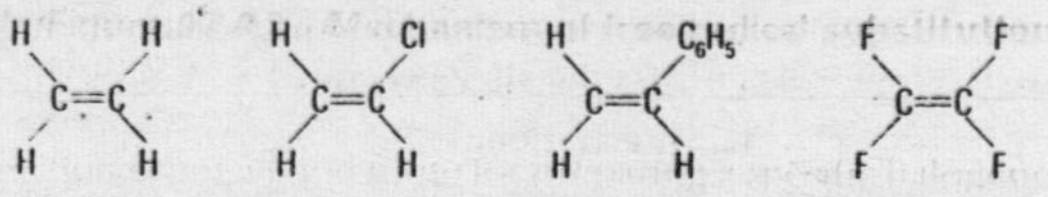

Ethene   Chloroethene   Phenylethene   Tetrafluoroethene

Polymer	Systematic name	Formula	Name of monomer	Monomer formula	Uses
Polythene	Poly(ethene)	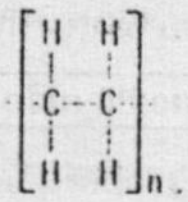	Ethene	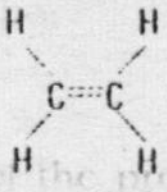	Plastic bags, 'squeezy' bottles
Polypropene	Poly(propene)	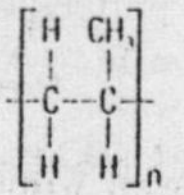	Propene	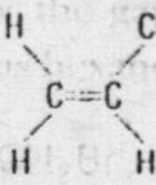	Ropes
Polybutene	Poly(but-2-ene)	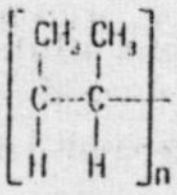	But-2-ene	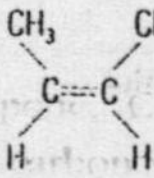	
Perspex	Poly(methyl 2-methylpropenoate)	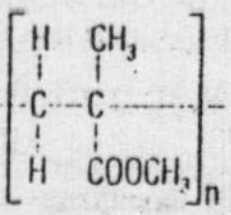	Methyl 2-methylpropenoate	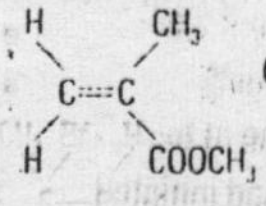	Clear sheets
PVC	Poly(chloroethene)	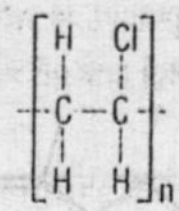	Chloroethene (vinyl chloride)	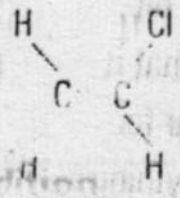	Drain pipes, upholstery
Polystyrene	Poly(phenylethene)	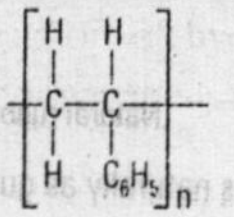	Phenylethene (styrene)	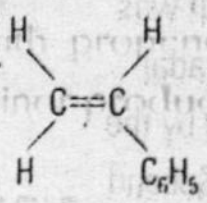	Insulation, food containers
PTFE, Teflon	Poly(tetrafluoroethene)	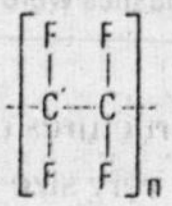	Tetrafluoroethene	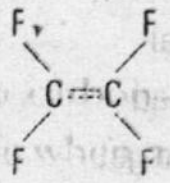	Friction-free bearings, non-stick cookware
Acrilan	Poly(propenenitrile)	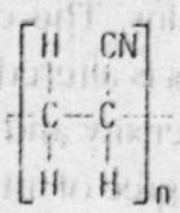	Propenenitrile, acrylonitrile	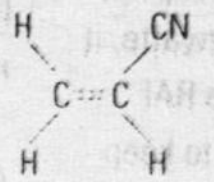	Textiles

Table 16.7.1   Some common polymers formed from alkene monomers

The systematic naming of these polymers is straightforward—'poly' followed by the name of the monomer unit in brackets. Although the double bond is lost in the formation of the polymer, the systematic name of the polymer retains the name of the alkene, including the -ene. Chloroethene polymerises to poly(chloroethene). However, many of these materials are commonly known by a less formal name, and will continue to be known by these familiar names. Poly(methyl 2-methylpropenoate) will continue to be called 'Perspex'.

The monomer unit must, of course, have a double bond, and writing out the polymer structure is simple enough if we relate the polymer to this double bond. Draw the monomer with the double bond at the centre and the four groups up and down from the double bond. It is then relatively easy to draw the polymer structure.

For poly(ethene):

Ethene
(monomer)

Poly(ethene), 'Polythene'
(polymer)
showing one repeating unit

For PVC, which is poly(chloroethene):

Chloroethene

Poly(chloroethene), 'PVC'
showing three repeating units

You must take care with writing out polymer structures and the equations to illustrate their formation. The numbers of atoms must still balance.

Remember:
- The double bond becomes a single bond (although the name, e.g. poly(ethene), still relates to the structure of the monomer).
- The brackets before and after the formula must have a bond extending beyond them.
- There is a symbol, (e.g. n), to indicate a large number.

The effect of *cis-trans* isomerism is also very significant with alkene polymers. Natural rubber is a polymer of 2-methylbuta-1,3-diene (isoprene), in which one of the C==C bonds in each monomer unit remains intact in the polymer. Because the *cis*-isomer cannot line up so closely, natural rubber is elastic.

Natural rubber (*cis*)

Gutta-percha (*trans*)

The *trans*-isomer occurs naturally as gutta-percha, which is far less elastic and was used for golf balls, and as insulation for electrical wiring, until plastics were developed.

Although the alkane structures of poly(ethene) would not be expected to have any large degree of permanent dipole, the very size of the alkane chain results in considerable van der Waals attraction between polymer chains. The extent of intermolecular attraction can vary considerably, and the manufacturing process is altered to control this.

The use of high density and low-density polythene is just one example. Low-density Polythene (LDPE) has a lesser degree of interaction between the polymer chains. The polymer chain is allowed to form with many side branches. It is more flexible, softening at 105 °C, and is used for plastic bags and collapsible 'squeezy' bottles.

The formation of the LDPE addition polymers occurs through a free radical type of reaction. The polymerisation process is started off with a free radical initiator at high temperature (200 °C) and high pressures (1500 atm).

$$nCH_2{=}CH_2 \longrightarrow {+}CH_2{-}CH_2{+}_n$$

The process goes through propagation and termination steps, just as we met with the free radical substitution of alkanes.

The first initiators used were traces of oxygen, but now organic peroxide molecules are used in the initiation step.

$$ROOR \longrightarrow 2RO{\bullet}$$

Typical organic peroxides include benzoyl peroxide:

$$C_6H_5CO{-}O{-}O{-}OC{-}C_6H_5$$

The free radical which is formed is then capable of breaking the double bond in another monomer molecule and starting off a series of propagation steps.

$$RO{\bullet} + CH_2{=}CH_2 \longrightarrow RO{-}CH_2{-}CH_2{\bullet}$$

$$RO{-}CH_2{-}CH_2{\bullet} + CH_2{=}CH_2 \longrightarrow RO{-}CH_2{-}CH_2{-}CH_2{-}CH_2{\bullet}$$

As the chain builds up, so it becomes more floppy; and when the long radical chains bend over on themselves, so the formation of side chains starts.

Eventually, two radical chains join together and terminate the reaction.

$$RO{-}(CH_2{-}CH_2)_n{-}CH_2{\bullet} + {\bullet}CH_2{-}(CH_2{-}CH_2)_n{-}OR \longrightarrow$$
$$RO{-}(CH_2{-}CH_2)_n{-}CH_2{-}CH_2{-}(CH_2{-}CH_2)_n{-}OR$$

More modern methods of polymerising these alkenes were developed by Ziegler and Natta in the early 1960s. Using triethylaluminium, $(C_2H_5)_3Al$, and titanium(IV) chloride, $TiCl_4$, as catalysts in an alkane solvent, the ethene can be polymerised at much lower pressures and just above room temperature. With these less vigorous conditions, the reaction does not involve the formation of free radicals and such conditions are much less expensive. But the structure of the polymer is different.

These different conditions result in longer chain lengths, of up to 100 000 units. There is less branching along the chains and the monomer units link up in a regular fashion.

Consequently, there are more attractions between the chains and the resulting polymer is high-density poly(ethene), HDPE (*Figure 16.7.1*), which has fewer branched chains in the structure.

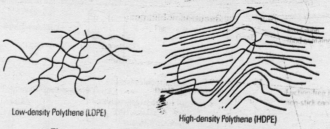

Low-density Polythene (LDPE) High-density Polythene (HDPE)

Figure 16.7.1 High and low density Polythenes

The branches are closer together, and the cross-linking possibilities along the chains are much greater. High-density Polythene softens around 135 °C, and is used for making more rigid bottles and bottle crates.

$$nCH_2{=}CH_2 \longrightarrow {+}CH_2{-}CH_2{+}$$

Problems with plastics

These polymers are very useful and have improved the quality of life in many ways. But we do need to be careful! The simple polymers are non-biodegradable, and so the plastic bag carelessly discarded stays in the environment for a considerable period of time. Even disposing of them as rubbish is not that easy.

Burning them is not always the answer – toxic products are produced when some of these plastics burn, and incomplete combustion produces carbon monoxide and carbon smoke. Even complete combustion will add to the carbon dioxide levels in the atmosphere. However, it is becoming increasingly possible to recycle plastic bottles and containers, and you should always look for the 'recyclable' symbol on a bottle. More local authorities are setting up plastics recycling bins alongside those for newspapers, metal cans and glass.

SELF-ASSESSMENT QUESTIONS 16.7

1 Write an equation to illustrate the formation of the polymer formed from tetrafluoroethene. Show three monomer units in the polymer structure.

2 Which of the following molecules can be used to form addition polymers?
   a $CH_3CH_2CH=CHCH_3$
   b $CH_3CH_2CH_2CH_3$
   c $CCl_2=CH_2$

3 Write balanced equations to illustrate the reaction of propene:
   a with bromine in hexane;
   b with hydrogen chloride in hexane;
   c in the presence of titanium(IV) chloride and triethylaluminium;
   d burning in oxygen.

4 There are three possible isomers of butene, C_4H_8, excluding the alicyclic isomer.
   a Explain the term 'isomer'.
   b Draw the three isomers and name them systematically.
   c Two of these isomers are geometrical isomers. Explain why the presence of the double bond creates these two isomers.
   d Draw a structure, containing one repeating unit, of the polymer formed from one of these isomers.
   e The formation of the polymer at high pressure and temperature occurs through a free radical type of mechanism. Explain fully this mechanism.

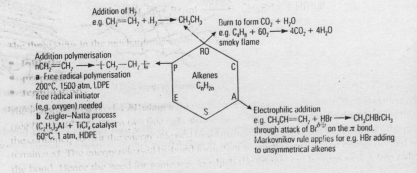

Reactions of alkenes

Addition of H_2
e.g. $CH_2=CH_2 + H_2 \longrightarrow CH_3CH_3$

Burn to form $CO_2 + H_2O$
e.g. $C_4H_8 + 6O_2 \longrightarrow 4CO_2 + 4H_2O$
smoky flame

Addition polymerisation
$nCH_2=CH_2 \longrightarrow -[CH_2-CH_2]_n-$
a Free radical polymerisation
200°C, 1500 atm, LDPE
free radical initiator
(e.g. oxygen) needed
b Zeigler–Natta process
$(C_2H_5)_3Al + TiCl_4$ catalyst
60°C, 1 atm, HDPE

Alkenes
C_nH_{2n}

Electrophilic addition
e.g. $CH_3CH=CH_2 + HBr \longrightarrow CH_3CHBrCH_3$
through attack of $Br^{\delta+}$ on the π bond.
Markovnikov rule applies for e.g. HBr adding to unsymmetrical alkenes

SUMMARY 16

In this chapter the following new ideas have been covered. *Have you understood and learned them?* Could you explain them clearly to a friend? The page references will help.

- Alkenes contain a double bond consisting of a σ bond and a π bond.   p407
- Molecules with double bonds are said to be unsaturated.   p404
- There is restricted rotation about the double bond.   p409
- Alkenes are formed by cracking alkanes or by elimination of certain groups.   p412
- The principal reaction of alkenes is electrophilic addition.   p417
- Addition across unsymmetrical double bonds favours the more stable carbocation. p422
- The Markovnikov rule predicts the addition products which are formed in unsymmetrical addition.   p422
- A variety of polymers result from addition polymerisation of alkenes.   p425

EXAMINATION QUESTIONS 16

1 3-Iodohexane can be used to form four different alkenes.
  a Draw a full displayed diagram for 3-iodohexane.
  b What reagents would you use to produce the alkene isomers from 3-iodohexane?
  c Suggest a possible mechanism for the formation of one of the alkene isomers.
  d Draw the four alkene isomers that are possible.
  e Explain why these isomers may be classed as two pairs of isomers.
  f What property in the bonding in these alkenes gives rise to these two pairs of isomers?

2 The simplest member of the alkene homologous series is ethene.
  a Explain what the term *homologous series* means.
  b Ethene can be converted into ethane. Write an equation for the reaction, including reagents and conditions necessary.
  c What is the difference in the bonding between the carbon atoms in ethane and ethene?
  d Ethane and ethene react with bromine, but under very different conditions. Write an equation for the reaction of ethane and bromine.
  e Draw a mechanism for the reaction of ethene with bromine.

3 A hydrocarbon liquid has the empirical formula CH_2 and the mass spectrum shows a molecular ion peak at m/e of 70. In air, the liquid burns readily with a smoky flame. If the liquid is dissolved in hexane and shaken vigorously with a small quantity of bromine water, the mixture becomes colourless.
  a Explain what is meant by the phrase *molecular ion peak at m/e of 70.*
  b Calculate the molecular formula of the hydrocarbon.
  c What does the smoky flame tell you?
  d Draw a possible structure for the compound.
  e Write an equation for the burning of the hydrocarbon in air.
  f Write an equation for the reaction of the molecule you have drawn in **d** with bromine.

- How **covalent bonds** are formed. (Ch 2.4)
- How the numbers of pairs of electrons leads to the **shapes of molecules**. (Ch 2.6)
- How **electronegative atoms** affect the electron distribution in a covalent bond and cause the polarity of these bonds. (Ch 2.5)
- That other groups or atoms can **replace the hydrogen atoms** attached to a carbon skeleton. (Ch 14.2)
- The types of organic **reaction mechanisms**. (Ch 14.7)
- The reactions of **alkanes** with halogens. (Ch 15.5)
- The reactions of **alkenes** with halogens. (Ch 16.5)

17
organic halogen compounds

17.1 Introduction

Organic halogen compounds are classified as those in which a halogen atom is attached *directly* to a carbon chain or ring. One such compound which is vitally important for you is thyroxine (see *Figure 17.1.1*). Halogenoalkanes are a group of compounds that contain one or more halogen atoms attached to a saturated carbon skeleton. Hydrogen atoms are replaced by halogen atoms, more commonly chlorine or bromine, but also fluorine and iodine. The general formula of a simple halogenoalkane is:

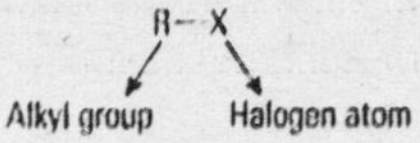

$$R — X$$

Alkyl group   Halogen atom

Thyroxine is an iodine-containing hormone, found in the thyroid gland. Anyone deficient in thyroxine over long periods may suffer from severe mental deficiency (cretinism); this is curable if treatment with thyroxine starts early enough.

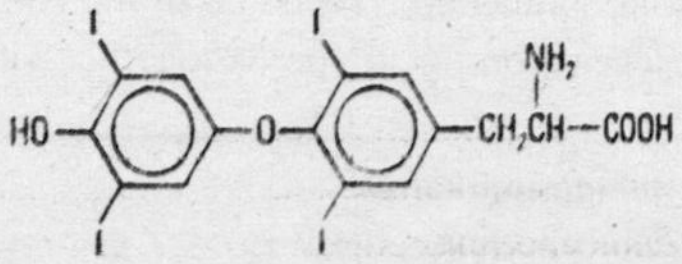

Figure 17.1.1

Some simple halogenoalkanes

CH_3Cl
Chloromethane

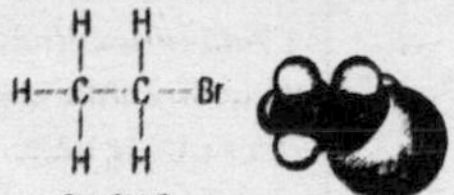

CH_3CH_2Br
Bromoethane

Most halogenoalkanes are colourless, volatile liquids that are immiscible with water. For halogenoalkanes of the same chain length, iodo-compounds have the highest boiling points, fluoro-compounds the lowest. They undergo mostly substitution reactions (see Ch 17.5). The iodo-compounds are the most reactive, fluoro-compounds the least.

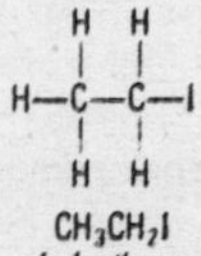

CH_3CH_2I
Iodoethane
(b.p. 72 °C)

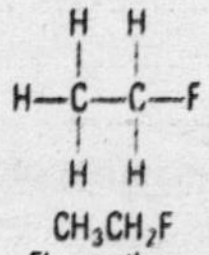

CH_3CH_2F
Fluoroethane
(b.p. −38 °C)

For example, iodoethane is more reactive and has a higher boiling point than fluoroethane.

Halogenoalkanes are important as intermediate compounds in the changing of one functional group for another. They are widely used in industrial processes and have many applications in the home. They are also common in nature.

Halogenoalkanes in nature

Simple halogenoalkanes (chloromethane, CH_3Cl; iodomethane, CH_3I) have been found in marine algae. The red algae (*Plocamium violaceum*) have been shown to contain a more complex halogenoalkane (plocamene B) with a biological activity similar to that of DDT (an insecticide very effective against mosquitoes).

Uses of halogenoalkanes

- Tetrachloromethane, CCl_4, and 1,1,1-trichloroethane, CCl_3CH_3, are widely used solvents, although CCl_4 in particular has been found to be very toxic.
- Chlorofluorocarbons (CFCs) were used as refrigerants and aerosol propellants. These are no longer used since they have been shown to damage the environment (see Ch 17.2).
- 1-bromo-1-chloro-2,2,2-trifluoroethane (halothane), $CHBrClCF_3$, is a very widely used anaesthetic.
- Poly(tetrafluoroethene) (PTFE, Teflon), is used for non-stick coatings and friction-free small bearings.

17.2 Naming halogenoalkanes

The carbon chain is named in exactly the same way as any alkane chain (Ch 14.3); the position and name of the halogen atom is prefixed to the name of the alkyl chain.

Can you see that CH_3Br is bromomethane, and CH_3CH_2Cl is chloroethane?

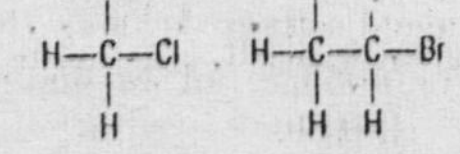

$$CH_3—Cl$$
Chloromethane

$$CH_3CH_2—Br$$
Bromoethane

These two simple halogenoalkanes do not require any numbers to show the position of the halogen atom, because there are no alternative positions for the halogen atom. However, for most halogenoalkanes, it is necessary to add to the name the position on the chain to which the halogen atom is attached. The name of the compound consists of the alkane chain with the names of the halogen atom(s) as a prefix together with position number(s).

Halogenoalkanes

*fluoro*methane, CH_3F

*chloro*ethane, CH_3CH_2Cl

2-*bromo*propane, $(CH_3)_2CHBr$

2-*iodo*-2-methylpropane, $(CH_3)_3CI$

1,1-*dichloro*ethane, $CHCl_2CH_3$

1,2-*dichloro*ethane, CH_2ClCH_2Cl

Draw the full displayed formulae of these compounds to see how the naming works. If in doubt, refer to Ch 14.3.

A few more examples of the wide range of halogenoalkanes may be helpful.

- The name of the halogen comes immediately before (and attached to) the alkyl chain.

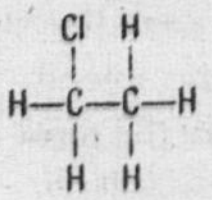

$$CH_2ClCH_3 \text{ or } CH_3CH_2Cl$$
Chloroethane

- The position of the halogen on the carbon chain is added.

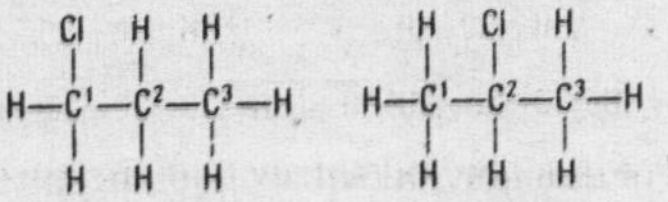

$$CH_2ClCH_2CH_3$$
1-Chloropropane

$$CH_3CHClCH_3$$
2-Chloropropane

- The numbering of the carbon chain starts at the end closest to any substituted group.

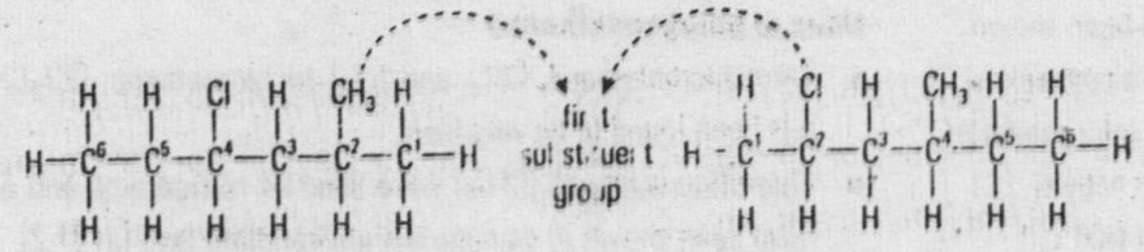

$$CH_3CH_2CHClCH_2CH(CH_3)CH_3$$
4-Chloro-2-methylhexane

$$CH_3CHClCH_2CH(CH_3)CH_2CH_3$$
2-Chloro-4-methylhexane

- The substituted groups are referred to in alphabetical order.

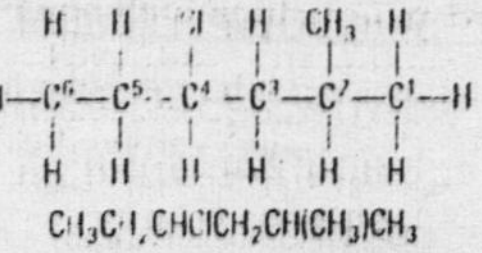

$$H—C^6—C^5—C^4—C^3—C^2—C^1—H$$

$$CH_3CH_2CHClCH_2CH(CH_3)CH_3$$

4-Chloro-2-methylhexane *not* 2-methyl-4-chlorohexane

- If there is any ambiguity the first group alphabetically takes the lowest number.

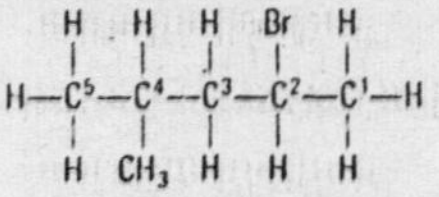

$$H—C^5—C^4—C^3—C^2—C^1—H$$

$$CH_3CH(CH_3)CH_2CHBrCH_3$$

2-Bromo-4-methylpentane *not* 4-bromo-2-methylpentane

- If more than one type of halogen atom is attached to the chain, they are written in alphabetical order.

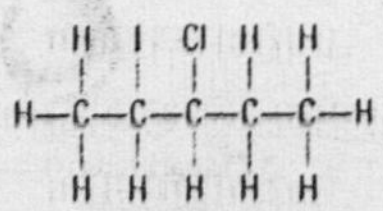

$$H—C—C—C—C—C—H$$

$$CH_3CHICHClCH_2CH_3$$

3-Chloro-2-iodopentane

- If more than one halogen atom of the same type is attached to the chain, the prefixes di-, tri- tetra-, etc are used to show how many halogen atoms there are. A number is often needed for each halogen atom in the chain.

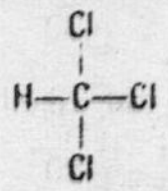

$CHCl_3$

Trichloromethane

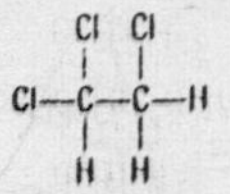

$CHCl_2CH_2Cl$

1,1,2-Trichloroethane

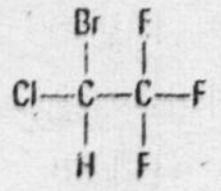

$CHBrClCF_3$

1-Bromo-1-chloro-2,2,2-trifluoroethane

Can you justify the names of the following compounds?

$$CH_3CH(CH_3)CH_2CHBrCH_2CH_3$$

4-Bromo-2-methylhexane

$$CH_3CH_2CH(CH_3)CHBrCH_3$$

2-Bromo-3-methylpentane

$$CH_3CH_2CH(CH_3)CHBrCH_2CH_2CH_3$$

4-Bromo-3-methylheptane

$$CH_3CH_2CH_2CHClCH(CH_3)CH_3$$

3-Chloro-2-methylhexane

Remember: in naming organic compounds:

- Identify the longest chain of carbon atoms and name it, as for alkanes.
- Number the carbon atoms from one end (start at the end closest to the first substituted group).
- Name the halogen atom(s) that will appear in front of the alkane chain.
- If there is more than one halogen atom of the same type use the prefixes di-, tri-, etc.
- Write the name with each substituted group in alphabetical order.
- Put in a number to show the carbon atom the functional group is attached to. e.g. 2-.

This may look clumsy, but it is completely unambiguous – which is why it is used!

Isomerism in halogenoalkanes

A halogenoalkane with molecular formula, C_4H_9Cl (*Figure 17.2.2*), can exist as a number of isomers (Ch 14.4). The isomers can be divided into three types:

- chain isomerism,
- position isomerism,
- optical isomerism.

Some halogenoalkanes

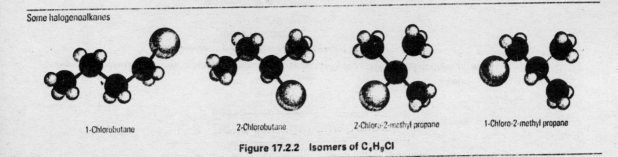

| 1-Chlorobutane | 2-Chlorobutane | 2-Chloro-2-methyl propane | 1-Chloro-2-methyl propane |

Figure 17.2.2   Isomers of C_4H_9Cl

Chain isomers have the same number of carbon atoms arranged in different chain lengths. For example:

$$CH_3CH_2CH_2CH_2Cl \quad \text{and} \quad CH_3CH(CH_3)CH_2Cl$$
1-Chlorobutane   1-Chloro-2-methylpropane

Position isomers have the same carbon skeleton but with the halogen atom attached at different positions. For example:

$$CH_3CH_2CH_2CH_2Cl \quad \text{and} \quad CH_3CH_2CHClCH_3$$
1-Chlorobutane   2-Chlorobutane

Other position isomers are possible for C_4H_9Cl:

$$CH_3CH(CH_3)CH_2Cl \quad \text{and} \quad (CH_3)_3CCl$$
1-Chloro-2-methylpropane   2-Chloro-2-methylpropane

Optical isomers: 2-chlorobutane has a chiral carbon atom (Ch 14.4). This compound will therefore exist as two optically active isomers.

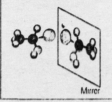

Isomerism is a very common feature of halogenoalkane molecules, so you must take care to identify the position of the halogen atom on the carbon skeleton and in naming compounds accurately.

Primary, secondary and tertiary halogenoalkanes

The environment around the C—X bond is influenced by the actual halogen (for example, chlorine is much more electronegative than bromine); but it is also affected by the structure of the carbon chain. Halogenoalkanes can be divided into three groups depending on the number of alkyl (R) groups that are attached to the carbon linked to the halogen. (The remaining bonds from that carbon atom must be to hydrogen.)

- If there is *one* R group attached, the carbon atom in question is called a **primary** carbon atom and the whole molecule is a **primary** halogenoalkane.
- If the are *two* R groups attached, the carbon atom is called a **secondary** carbon atom and the molecule is a **secondary** halogenoalkane.
- If there are *three* R groups attached to the carbon atom in question, the carbon atom is a **tertiary** carbon atom and the whole molecule is called a **tertiary** halogenoalkane.

Figure 17.2.3 shows how some of the isomers of C_4H_9Cl would be classified.

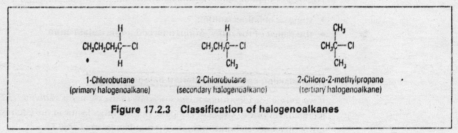

Figure 17.2.3 Classification of halogenoalkanes

1 Name the following compounds.
 a $CH_3CHClCH_3$
 b $CH_3CH_2CHBrCH_3$
 c $CH_3CH(CH_3)CH_2CH_2Cl$
 d $CH_2FCH_2CH_3$

2 Write the formula of:
 a 1-chloropropane
 b 3-bromohexane
 c 3-bromo-2-methylpentane
 d 2-bromo-2,3-dimethylbutane

3 Write the full displayed formulae and names of the isomers of:

  a C_3H_7Br

  b C_4H_9Br

4 Consider the following bromoalkanes:

  A $CH_3C(CH_3)_2Br$

  B $CH_3CH_2CH_2Br$

  C $CH_3CH_2C(CH_3)_2Br$

  Identify these as:

  a primary;

  b secondary;

  c tertiary.

17.3 Bond strength and polarity

The bond between a carbon atom and a halogen atom is polar. The electronegative halogen atom attracts electrons away from the carbon atom. This gives the halogen atom a partial negative charge and leaves the carbon atom partly positively charged, creating a permanent dipole (*Figure 17.3.1*).

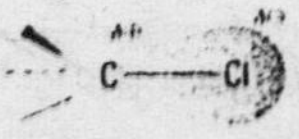

Figure 17.3.1   Polarisation of the C—Cl bond

The degree of polarisation is affected by:

- the type of halogen atom;
- the nature of the other groups attached to the carbon atom.

Polarisation caused by different halogen atoms

If we look at the electronegativities of some of the elements (*Table 17.3.1*) the most electronegative halogen is fluorine. The least is iodine. But the other factor in the chemistry of the halogenoalkanes is the difference in bond strengths (or bond lengths) (*Table 17.3.2*).

	Pauling's scale
F	4.0
Cl	3.0
Br	2.8
I	2.5
H	2.1
C	2.5

Table 17.3.1   Electronegativity list

	Bond enthalpy /kJ mol^{-1}	Bond length /nm
C—F	+467	0.138
C—Cl	+346	0.177
C—Br	+290	0.194
C—I	+228	0.214
C—H	+413	0.108

Table 17.3.2   Bond enthalpies and bond lengths

In spite of the much larger polarity in the C—F bond, the C—F bond is much stronger than even the C—H bond – and so it is no surprise to find that the simple compounds containing fluorine in an alkyl group are very unreactive. The bond strength means that the fluoroalkanes are very stable.

The opposite is true for iodoalkanes. The C—I bond is the least polar, but also the longest and weakest of all the C—halogen bonds. **Iodoalkanes are the most reactive of the halogenoalkanes. In general terms, reactivity decreases R—I > R—Br > R—Cl > R—F.**

Polarisation caused by other groups

In discussing alkenes we saw that the methyl group is weakly 'electron donating' (Ch 16.6). Three methyl groups attached to a tertiary C—Cl bond will weaken the polarity of the bond (see *Figure 17.3.2*). This means that the C—Cl bond in 2-chloro-2-methylpropane is weaker than that in chloromethane.

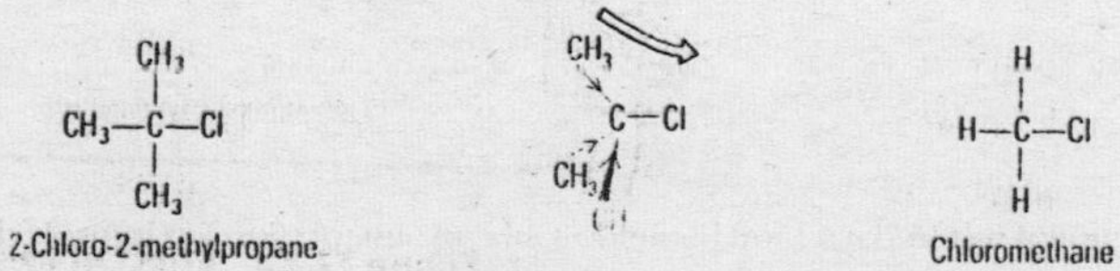

Poly(tetrafluoroethene) (PTFE) has a chemical structure very similar to Polythene in which every hydrogen atom is replaced by a fluorine atom.

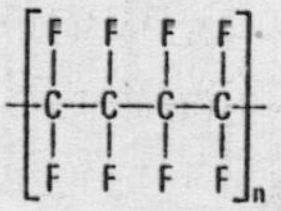

PTFE is so inert chemically, and has such a low coefficient of friction, it is used for non-stick coatings in cooking pans (as Teflon), and to cover the re-entry cones for space rockets.

Figure 17.3.2   Electron drift in 2-chloro-2-methylpropane

The former is therefore slightly more reactive. The electron donating properties of the methyl groups also help to stabilise any polar intermediates, containing a C$^{\delta+}$ atom, that are formed in the reactions of this molecule. **In general, a tertiary halogenoalkane is more reactive than a secondary one, which is more reactive than a primary one.**

Physical properties of halogenoalkanes

Chloromethane, chloroethane and bromoethane are all colourless gases at room temperature. The other low molecular mass halogenoalkanes are sweet-smelling, colourless liquids. The boiling points of a range of the halogenoalkanes with up to four carbon atoms are shown in *Table 17.3.3*. These are all immiscible with water but soluble in organic solvents.

X =	H	F	Cl	Br	I
CH₃—X ...methane	−164	−78	−24	4	44
CH₃CH₂—X ...ethane	−88	−38	12	38	72
CH₃CH₂CH₂—X 1-...propane	−42	3	47	71	103
CH₃CHXCH₃ 2-...propane			36	59	90
CH₃CH₂CH₂CH₂—X 1-...butane	0	33	79	102	131
CH₃CH₂CHXCH₃ 2-...butane			68	91	120
(CH₃)₃C—X 2-...-2-methylpropane	−12		51	73	103

Table 17.3.3   Boiling points of a range of halogenoalkanes /°C

437

For halogenoalkanes with the same chain length, iodoalkanes have the highest boiling points, fluoroalkanes the lowest (*Figure 17.3.3*). We might expect the reverse to be true, since the more polar fluoroalkanes would have larger permanent dipole–dipole interactions, which would tend to hold the molecules together.

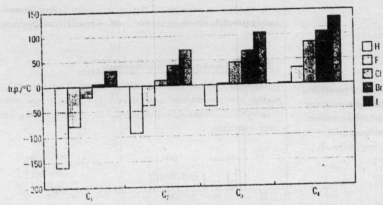

Figure 17.3.3 Effect of halogen on boiling point

Although this effect is responsible for the boiling points of halogenoalkanes being higher than those of the corresponding alkanes, the effect of the van de Waals forces (see Ch 2.8) outweighs the effect of the permanent dipole–dipole interactions. The van der Waals forces increase as the halogen atom gets bigger (and contains more electrons), and so these forces are far larger in iodoalkanes than in fluoroalkanes.

The effect on boiling point of increasing the chain length of the alkyl group has already been discussed in Ch 14.5. It is no surprise, then, to find that as the chain length increases, the boiling point increases (*Figure 17.3.4*) – as long as we consider the same halogen atoms.

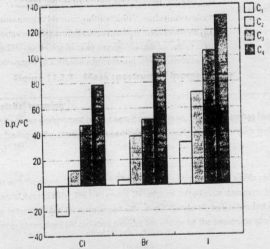

Figure 17.3.4 Effect of chain length on boiling point

The boiling points are also affected by isomerism. Moving the halogen atom towards the centre of the carbon chain alters the shape of the molecule, making it more globular. This gives the molecule a smaller surface area, and therefore reduces the van der Waals forces between the molecules. This in turn lowers the boiling point (*Figure 17.3.5*).

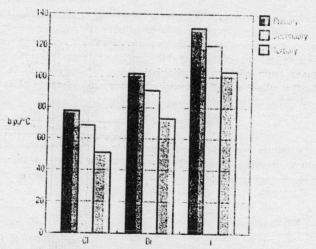

Figure 17.3.5   Effect of structure on boiling point

Remember:

the boiling point of a halogenoalkane increases with:

- a larger halogen atom (fluorine < chlorine < bromine < iodine);
- increasing number of carbon atoms in the molecule;
- a less branched (more 'linear') molecule (tertiary < secondary < primary).

Activity

Using a molecular model kit, construct models of the following formulae. Using the data in *Table 17.3.1*, compare the boiling points of these structures, and explain the trends as the alkyl chain increases and the shape of the alkyl chain changes.

$(CH_3)_3CBr$ $(CH_3)_2CH_2Br$ $CH_3CH_2CH_2Br$ $CH_3CH(CH_3)BrCH_3$

Mass spectra of halogenoalkanes

Remember:

Isotopes are *atoms* with the same number of protons but different numbers of neutrons.

Isomers are *molecules* with the same molecular formula but different arrangements of the atoms in the molecule.

Polymorphs or **allotropes** are forms of the same *element* with different structures.

Naturally-occurring chlorine and bromine both exist as a mixture of two isotopes (Ch 1.2).

Chlorine contains the ^{35}Cl and ^{37}Cl isotopes in the approximate ratio of 3:1. Since the isotopes are both present in high proportions, they have an effect on the mass spectra produced when any organic molecule containing chlorine is measured.

For example, the mass spectrum of chloromethane will show two molecular ion peaks at 50 and 52 (see *Figure 17.3.6*).

The peak at 50 will be produced by the $CH_3{}^{35}Cl^{\oplus}$ ion. The peak at 52, one third the size, is formed by the $CH_3{}^{37}Cl^{\oplus}$ ion. Other chlorine-containing fragments will show the same 3:1 ratio. The presence of ^{13}C will cause further small peaks to appear.

So, although the relative molecular mass of chloromethane would be 50.5 (using the figures in the Periodic Table at the back of the book), the molecular ion of organic molecules containing one chlorine atom will contain two peaks. These peaks will be in the ratio 3:1 and all the chlorine-containing fragments will also show the same characteristic pattern.

With bromine, the two isotopes present in naturally occurring bromine are ^{79}Br and ^{81}Br in approximately equal quantities. The mass spectrum of bromoethane, C_2H_5Br ($M_r = 109$), does not contain a molecular ion peak at 109, but two equal peaks at 108 and 110 corresponding to $C_2H_5{}^{79}Br^{\oplus}$ and $C_2H_5{}^{81}Br^{\oplus}$ (see *Figure 17.3.7*)

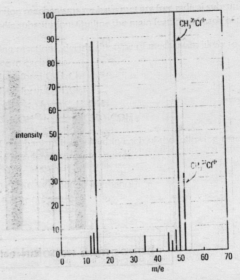

Figure 17.3.6   Mass spectrum of chloromethane

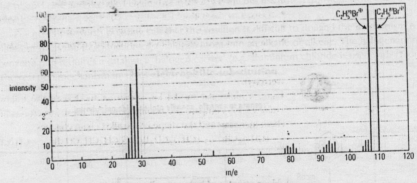

Figure 17.3.7   Mass spectrum of bromoethane

CFCs, CO$_2$, ozone and global warming

The effect of human activity on our environment is not a new issue, but in recent years there has been increasing concern about the effects we are having on the atmosphere in particular. The situation is very complex and not all the chemical ideas are fully understood.

The ozone layer

So what are the problems with the changing atmosphere? The sun radiates energy, as UV, visible and IR radiation, and the Earth receives its fair share of this radiation. The high energy UV radiation, in particular, causes problems for life on Earth; it can cause skin cancer in humans, but far more importantly, it can destroy phytoplankton and plants and so reduce photosynthesis. In the upper atmosphere UV radiation provides the energy for the conversion of some of the oxygen to ozone.

$$O_2(g) \xrightarrow{\text{UV}} 2O\bullet(g), \quad O_2(g) + O\bullet(g) \longrightarrow O_3(g)$$

The UV radiation is absorbed in the upper atmosphere by this ozone and equilibrium is established, with oxygen converted to ozone and ozone breaking up to oxygen. Simplified equations can show this.

$$O_3 \rightleftharpoons O_2 + O\bullet$$

The concentration of ozone in the upper atmosphere is only around 10 ppm (around 0.001%), but its presence is vital. Not all the UV light is absorbed and some does reach the surface of the Earth. You might have experienced sunburn from too much of this UV radiation.

Warming the atmosphere

When it reaches the surface of the Earth, sunlight is absorbed and re-emitted at different wavelengths. It warms the surface, and warm objects emit infra-red (IR) radiation. This radiation is now at a frequency that is absorbed by a number of gases in the atmosphere – H_2O vapour, CO_2, methane, other hydrocarbons, etc. As these gases absorb some of this heat energy, so the temperature rises. This is what keeps the Earth's surface temperature some 20 °C higher than it would otherwise be, and is termed the '**greenhouse effect**' (see Ch 7.4). (In a greenhouse, radiation enters through the glass, is re-emitted at a longer wavelength, and is then unable to escape through the glass, keeping the inside of the greenhouse warm.)

Everything should be in equilibrium – UV entering the upper atmosphere, ozone being formed and absorbing some UV radiation, much of the rest of the sunlight reaching the surface of the Earth, being reflected back at longer wavelength, and being absorbed by some gases in the atmosphere.

The level of CO_2 in the atmosphere is normally kept reasonably constant by photosynthesis. The increasing use of fossil fuels has led to a huge increase in CO_2 emissions, and, together with the fact that large tracts of forests and green foliage are being removed, the level of CO_2 in the atmosphere is rising. Increased amounts of CO_2, and other gases, in the atmosphere may lead to more energy being absorbed, a larger greenhouse effect, and so a slow rise in temperature at the surface of the Earth.

CFCs and ozone

So where do CFCs come into all this? What are CFCs? In CFCs (chlorofluorocarbons), all the hydrogen atoms in the alkanes have been substituted by either chlorine or fluorine. They were developed as excellent refrigerants, aerosols, extinguishers and solvents, being non-flammable, non-toxic and unreactive, with low boiling points; e.g. CCl_2F_2 (dichlorodifluoromethane) has a boiling point of −30 °C. Other common CFCs include trichlorofluoromethane ($CFCl_3$), and 1,1,2-trichloro-1,2,2-trifluoroethane ($CF_2ClCICl_2F$). It was only when their effect on the atmosphere was recognised that their use was banned.

Being so stable under normal conditions, when the gases are released they diffuse unchanged through the lower atmosphere. In the upper atmosphere UV radiation leads to the break-up of the molecules; the C—Cl bond can be broken with UV light to form free radicals. And this is where the trouble really begins. The formation of the chlorine free radicals interferes with the oxygen/ozone balance.

$$\text{e.g.} \qquad CF_3Cl \longrightarrow CF_3\bullet + Cl\bullet$$

Once the chlorine free radical is formed, it is available to react with ozone.

$$O_3 + Cl\bullet \longrightarrow O_2 + ClO\bullet$$

$$ClO\bullet + O\bullet \longrightarrow Cl\bullet + O_2$$

Artists impression of the Upper Atmosphere Research Satellite (UARS). One of its main areas of investigation will be the levels of upper atmosphere ozone and its depletion due to chlorofluorocarbons (CFCs).

The overall effect is that ozone breaks up and forms oxygen, but the chlorine radical is not consumed in the reaction. It is still available to react with another ozone molecule. It has been estimated that one chlorine radical can lead to the loss of one million ozone molecules. And so, with ozone being so rapidly removed, and the CFCs around for a considerable time (since they are so unreactive), the development of the 'hole in the ozone layer' is inevitable. The use of CFCs is now increasingly controlled, but CFCs released in the past are still diffusing and mixing in the atmosphere and will do so for many more years to come (see Ch 14.6).

CFCs act as greenhouse gases, but the main problem they cause is with the ozone in the upper atmosphere. The principal cause of recent global warming is almost certainly the extra carbon dioxide released by burning fossil fuels, which photosynthesis and increased solution in sea water cannot fully cope with.

It is best to keep the 'ozone problem' and the 'greenhouse effect' separate in your minds!

Thomas Midgley (1889–1944) was an American who invented the petrol additive tetraethyllead(IV) ($Pb(C_2H_5)_4$), TEL), which enabled more efficient high-compression petrol engines to be designed, and who also developed the idea of 'octane number' for petrols (see Ch 15.4). We now know that extra lead in the environment is dangerous, particularly for children, and 'leaded' petrol is no longer available in many parts of the world.

In 1930 Midgley also devised the first CFC, Freon-12 (CF_2Cl_2). This was the result of a search for a non-poisonous, non-flammable, non-corrosive working fluid for refrigerators and freezers. We now know that CFCs cause problems with ozone depletion.

Midgley's two major inventions were brilliant solutions to real and important technological problems. Perhaps the environmental consequences of leaded petrol could have been foreseen, but no one could have predicted the trouble which CFCs would cause.

SELF-ASSESSMENT QUESTIONS 17.3

1 The boiling points of a series of bromoalkanes are given below.

A	1-bromopentane	130 °C
B	2-bromopentane	117 °C
C	3-bromopentane	119 °C
D	1-bromo-2-methylbutane	122 °C
E	2-bromo-2-methylbutane	108 °C
F	1-bromo-2,2-dimethylpropane	106 °C
G	2-bromo-3-methylbutane	115 °C

a Draw a bar chart showing the boiling points of these compounds.
b Draw displayed formulae for compounds B, D and E.
c Explain why the pairs B and C, and E and F should have such similar boiling points.
d Explain the trend in boiling points between compounds A to B and D to F.

17.4 Preparation of halogenoalkanes

There are three main methods used to put a halogen atom on to the alkyl chain (*Figure 17.4.1*):

- Substitution of alkanes with halogens.
- Addition of a hydrogen halide to an alkene.
- Substitution of the hydroxy group in an alcohol.

R—H
R'CH═CHR" ——→ R—X
R—OH

Figure 17.4.1 Routes to halogenoalkane formation

Substitution in alkanes with halogens

Direct substitution of the hydrogen atom in an alkane is difficult to achieve. The reaction requires energetic conditions. Alkanes react with halogens only in the presence of UV light. The mechanism of substitution involves the formation of free radicals. Remember the free radical reaction mechanism from Ch 15.5? The simplest overall reaction with methane and bromine produces bromomethane. The reaction of methane with chlorine is explosive!

Bromomethane is used as an insecticide.

$$CH_4(g) + Br_2(g) \xrightarrow{\text{UV light}} CH_3Br(g) + HBr(g)$$
Methane   Bromomethane

The three steps in the mechanism involve:

- **initiation**
- **propagation**
- **termination**

The initiation step produces halogen free radicals which set up a number of propagation reactions (see *Figure 17.4.2*). When two free radicals combine – in the presence of a third body or the wall of the container, so that the energy released by formation of the bond can be removed – the process is terminated. The energy released by bond formation is exactly equal to the energy needed to break the bond. Hence the need for some way in which the energy can be removed. A free radical may survive very many propagation reactions before being terminated.

Initiation: $Br_2 \longrightarrow 2Br\bullet$

Propagation: $CH_4 + Br\bullet \longrightarrow CH_3\bullet + HBr$

$CH_3\bullet + Br_2 \longrightarrow CH_3Br + Br\bullet$

Termination: $CH_3\bullet + Br\bullet \longrightarrow CH_3Br$

$CH_3\bullet + CH_3\bullet \longrightarrow C_2H_6$

$Br\bullet + Br\bullet \longrightarrow Br_2$

Figure 17.4.2   Mechanism of free radical substitution

Unfortunately the reaction is not a good route for preparing a specific halogenoalkane. The high UV energy required can often decompose the products, and the reaction tends to produce a mixture which has to be separated in order to obtain any specific halogenoalkane.

The reason for this mixture is that successive substitution of hydrogen occurs and cannot be controlled. Methane and bromine will give a mixture of CH_3Br, CH_2Br_2, $CHBr_3$ and CBr_4. The halogen can replace any of the hydrogen atoms in an alkane, and so, for ethane and larger alkane molecules, there are many different positions available for substitution.

Addition of hydrogen halide to an alkene

This method produces a cleaner product, and the structure of the product can be predicted. Reactions can occur at room temperature between two gases, or the gaseous hydrogen halide is bubbled through the alkene dissolved in an organic solvent such as hexane.

e.g. $\qquad CH_2{=}CH_2 + HBr \xrightarrow[\text{room temperature}]{\text{hexane}} CH_3CH_2Br$

The mechanism involves **electrophilic addition**.

Addition across an unsymmetrical alkene – such as propene, $CH_3{-}CH{=}CH_2$ – follows Markovnikov's rule (see Ch 16.6) with the halogen adding to the carbon with the fewer hydrogen atoms (*Figure 17.4.3*). Do not forget to be careful exactly where you start and end your curly arrows.

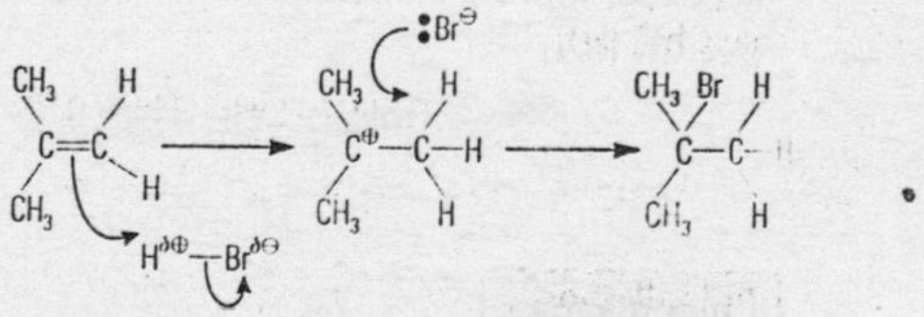

Figure 17.4.3   Electrophilic addition

The major product from the reaction of hydrogen bromide with propene is 2-bromopropane, although the formation of 1-bromopropane is still possible as a minor product.

$CH_3CH{=}CH_2 + HBr \xrightarrow[\text{room temperature}]{\text{hexane}} CH_3CHBrCH_3 \; (+ \; CH_3CH_2CH_2Br \text{ as a minor product})$

The addition of bromine is used as a test for the presence of a carbon–carbon double bond in organic compounds. The removal of the brown colour of bromine when a solution (aqueous or in an organic solvent) is added to an alkene at room temperature is a very good indication of the presence of a carbon–carbon double bond.

Substitution in an alcohol

Alcohols are readily converted into halogenoalkanes by substitution of the OH group with the halogen atom.

[margin text]

strial production of genoalkanes

rine is a readily available relatively cheap raw rial. The distillation and king of crude oil provides a y source of alkanes and es. This is an important ce of the raw materials for roduction of genoalkanes, which are ly used as a starting point dustrial syntheses and as nts. Alkanes are reacted chlorine to form a range oroalkanes, and the ion with alkenes provides fic chloroalkanes.

Treatment of an alcohol with a hydrogen halide (HCl, HBr or HI) can produce the halogenoalkane. This method works best for tertiary alcohols ($R_3C—OH$). The hydrogen halide gas is bubbled through a cold solution of the alcohol in ethoxyethane (ether).

$$R_3C—OH + H—X(g) \xrightarrow[\substack{\text{room} \\ \text{temperature}}]{\text{ethoxyethane}} R_3C—X + H_2O(l)$$

Primary alcohols ($RCH_2—OH$) and secondary alcohols (R_2CHOH) can be made to react, but require anhydrous conditions and a catalyst.

Chloroalkanes

Chloroalkanes can be produced by treatment of an alcohol with sulphur dichloride oxide ('thionyl chloride', SCl_2O). This works best for primary and secondary alcohols.

$$RCH_2—OH + SCl_2O(l) \xrightarrow{\text{reflux}} RCH_2—Cl + SO_2(g) + HCl(g)$$

One advantage of this reagent is that the other products are gases, which are easily removed from the reaction mixture.

Phosphorus trichloride (PCl_3) – a liquid – can also be used, but solid phosphorus pentachloride (PCl_5) is a more common reagent (see Ch 18.4).

$$\underset{\text{Alcohol}}{RCH_2—OH} + PCl_5(s) \xrightarrow[\text{temperature}]{\text{room}} \underset{\text{Chloroalkane}}{RCH_2—Cl} + HCl(g) + \underset{\text{Phosphorus trichloride oxide}}{PCl_3O(l)}$$

Here, the by-product, PCl_3O, has a boiling point of 103 °C and may be difficult to separate from the chloroalkane.

Bromoalkanes

A mixture of red phosphorus and bromine, used to generate phosphorus tribromide, PBr_3, can be used (PBr_3 is too unstable to be used directly). This method works best for primary and secondary alcohols.

$$P_4(s) + 6Br_2(l) \longrightarrow 4PBr_3$$

$$3RCH_2—OH + PBr_3 \xrightarrow{\text{reflux}} 3RCH_2—Br + H_3PO_3$$

Refluxing with sodium bromide and concentrated sulphuric acid will also produce a bromoalkane (see Ch 18.4).

$$RCH_2—OH + H^{\oplus} + Br^{\ominus} \xrightarrow[\substack{\text{concentrated} \\ \text{sulphuric acid}}]{\text{reflux}} RCH_2—Br + H_2O$$

Iodoalkanes

An alcohol can be reacted with a mixture of red phosphorus and iodine. PI_3 is too unstable to be used directly. Again, this method works best for primary and secondary alcohols.

$$P_4(s) + 6I_2(l) \longrightarrow 4PI_3$$

$$3RCH_2—OH + PI_3 \xrightarrow[\text{reflux}]{\text{ethoxyethane}} 3RCH_2—I + H_3PO_3$$

Table 17.4.1 summarises the different reagents that can be used to replace an —OH group with halogen.

Chlorination	Bromination	Iodination
HCl	HBr	HI (for tertiary alcohols)
PCl_5	$NaBr + H_2SO_4$	red $P + I_2$
PCl_3	red $P + Br_2$	
SCl_2O		

Table 17.4.1   Reagents for halogenation of alcohols

1 How would you prepare a sample of:

   a 2-bromopropane from propene;

   b 1-iodobutane from butan-1-ol;

   c chloroethane from ethene;

   d chloromethane from methane?

2 What type of mechanism is likely to occur in each of the following reactions? Give full equations to illustrate the reaction.

   a But-1-ene and hydrogen bromide.

   b Hexane and chlorine.

3 Consider the compound with formula $CH_3CH=CHCH_2CH_3$.

   a Give a systematic name for this compound.

   b Explain why the compound can exist in two isomeric forms.

   c When it reacts with hydrogen bromide dissolved in hexane, two products are formed. Write the formula of each compound. Which one can exist as two optical isomers and why?

   d Decide if the compounds from **c** are primary, secondary or tertiary halogenoalkanes. Explain your decision.

4 The tertiary alcohol with the formula C_5H_9OH reacts with red phosphorus and bromine.

   a Write an equation for the reaction.

   b Give the name of the product.

   c Draw a diagram of the apparatus you would use to purify the crude product.

17.5 Reactions of halogenoalkanes

Halogenoalkanes are very useful organic intermediates because the polarity of the carbon–halogen bond, $C^{\delta\oplus}-X^{\delta\ominus}$ makes the bond susceptible to nucleophilic attack. There are two main types of reaction possible: substitution involving attack from a nucleophile, and elimination involving attack from a base (proton acceptor).

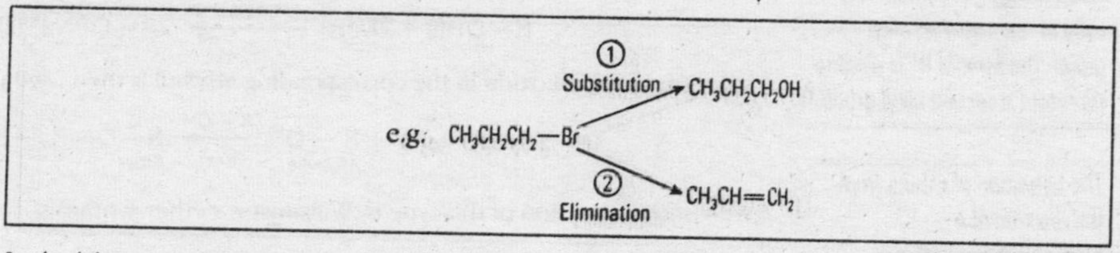

$$
\begin{array}{c}
① \\
\text{Substitution} \rightarrow CH_3CH_2CH_2OH \\
\text{e.g.} \quad CH_3CH_2CH_2-Br \\
② \qquad \rightarrow CH_3CH=CH_2 \\
\text{Elimination}
\end{array}
$$

In the laboratory, bromoalkanes are the halogenoalkanes of choice in the synthesis of other organic compounds. Chloroalkanes react fairly slowly, due to the strength of the C—Cl bond; and the weakness of the C—I bond means that iodoalkanes can be unstable (see *Table 17.5.1*). (Iodoalkanes, like all iodine compounds, are also very expensive.) However, the chloroalkanes are the cheapest to prepare (chlorine is cheap and readily available) and so these are used on an industrial scale.

ember: C—F bonds are
strong and therefore do
eact with nucleophiles
pt under extreme
.tions. This is why PTFE is
able and so very useful –
h 16.7.

	kJ mol^{-1}
C—F	+467
C—Cl	+436
C—Br	+290
C—I	+228

Table 17.5.1 C—X bond enthalpies

Substitution reactions of halogenoalkanes

The general equation for the reaction with any nucleophile (Nu) will be:

$$-\overset{|}{\underset{|}{C}} \overset{\delta+}{} -Br^{\delta-} \longrightarrow -\overset{|}{\underset{|}{C}} + Br^{\ominus}$$
$$\quad\quad :Nu^{\ominus} \quad\quad\quad\quad Nu$$

The nucleophile can be any group of atoms with a lone pair of electrons (and sometimes a negative charge) capable of forming a covalent bond with the $C^{\delta+}$ atom in the polarised bond.

Reaction with OH$^{\ominus}$

Although the water molecule can be classified as a nucleophile, the reaction of halogenoalkanes with water is very slow. Refluxing the halogenoethane in a mixture of water and ethanol (to allow the halogenoalkane to dissolve) will slowly result in the formation of the corresponding alcohol.

$$R-Br + H_2O \xrightarrow{\text{reflux}} R-OH + HBr$$

The hydroxide ion, with a negative charge as well as a lone pair of electrons, is much more able to replace the halogen atom. This reaction can also produce an alkene (see elimination reactions with OH$^{\ominus}$ below). The *substitution* reaction is best done by refluxing with *dilute aqueous* alkali.

$$R-Br + OH^{\ominus}(aq) \xrightarrow{\text{reflux}} R-OH + Br^{\ominus}(aq)$$

For example:

$$CH_3CH_2Br(l) + NaOH(aq) \xrightarrow{\text{reflux}} CH_3CH_2OH(aq) + NaBr(aq)$$

Reaction with R—O$^{\ominus}$

Halogenoalkanes react with the alkoxide ion to form ethers in which oxygen is directly bonded to two carbon atoms. The alkoxide is readily prepared by reacting sodium metal with the corresponding alcohol.

Remember: The symbol R is used to represent an alkyl group. The symbol R′ is used to represent a second alkyl group.

$$R-OH(l) + 2Na(s) \xrightarrow[\text{temperature}]{\text{room}} 2R-O^{\ominus}Na^{\oplus}(s) + H_2(g)$$

A solution of the alkoxide in the corresponding alcohol is then used to treat the halogenoalkane.

$$R-Br + R'-O^{\ominus} \xrightarrow[\text{heat}]{R'-OH} R-O-R' + Br^{\ominus}$$
$$\quad\quad\quad \text{Alkoxide} \quad\quad\quad\quad \text{Ether}$$

A well-known reaction of this type is 'Williamson's ether synthesis'.

The formation of ethers from the reaction of a halogenoalkane with an alkoxide ion is an example of the Williamson synthesis. Alexander Williamson discovered the reaction while working with Liebig in 1850, and it is one of the first 'rational' syntheses in organic chemistry. A rational synthesis is one in which the reactants are of known structure and deliberately chosen to give the 'target' molecule.

e.g.
$$CH_3O^{\ominus} \quad\quad\quad\quad\quad\quad CH_3O^{\ominus}Na^{\oplus}$$
$$\quad\quad\quad \overset{\delta+}{CH_2}-Br^{\delta-} \xrightarrow[\text{in methanol}]{} CH_3OCH_2CH_3 + Br^{\ominus}$$
$$\quad\quad\quad\quad CH_3$$
$$\quad\quad\quad \text{Bromoethane} \quad\quad\quad\quad\quad\quad \text{Methoxyethane}$$

Reaction with NH$_3$

The lone pair of electrons on the nitrogen enables ammonia to act as a nucleophile. The substitution of the halogen in the halogenoalkane forms a primary amine. The first reaction is:

$$R-Br + \ddot{N}H_3 \xrightarrow[\substack{\text{concentrated} \\ NH_3 \text{ solution}}]{\text{heat}} R-\ddot{N}H_2 + HBr$$

The primary amine itself has a nitrogen atom with a lone pair of electrons, which can also act as a nucleophile. Therefore the reaction continues and a mixture of products is formed.

$$R–Br + \overset{..}{N}H_3 \longrightarrow R–\overset{..}{N}H_2 + HBr$$

$$R\overset{..}{N}H_2 + R—Br \longrightarrow R_2\overset{..}{N}H + HBr$$

$$R_2\overset{..}{N}H + R—Br \longrightarrow R_3\overset{..}{N} + HBr$$

The tertiary amine still contains a nitrogen atom with an available lone pair of electrons. The last step in the sequence forms the quaternary ammonium salt (which is similar to the formation of $NH_4^{\oplus}$ from ammonia and $H^{\oplus}$).

$$R_3N + R–Br \longrightarrow R_4N^{\oplus}Br^{\ominus}$$

With such a mixture of products to be separated, this method is not suitable for the preparation of a particular amine.

Reaction with $CN^{\ominus}$

Halogenoalkanes react with cyanide ions to produce nitriles. The reaction is carried out using potassium cyanide in ethanol. This reaction can be an important step in an organic synthesis, as nitriles can react in a variety of ways (see Ch 27.11). The reaction can also be used to add another carbon atom to an alkyl chain (see Ch 29.3).

$$R—Br + KCN \xrightarrow[\text{ethanol}]{\text{reflux}} \underset{\text{Nitrile}}{R—CN} + KBr$$

e.g.

$$\underset{\text{bromomethane}}{CH_3Br} + KCN \xrightarrow[\text{ethanol}]{\text{reflux}} \underset{\text{ethanenitrile}}{CH_3CN} + KBr$$

Nucleophile		Product	Formula
H_2O	Water	Alcohol	R—OH
$OH^{\ominus}$	Hydroxide ion	Alcohol	R—OH
$RO^{\ominus}$	Alkoxide ion	Ether	R—O—R'
$CN^{\ominus}$	Cyanide ion	Nitrile	R—CN
NH_3	Ammonia	Primary amine	R—NH$_2$
$^{\ominus}NH_2$	Primary amine	Secondary amine	R$_2$NH
R_2NH	Secondary amine	Tertiary amine	R$_3$N

Some common nucleophiles and their products with halogenoalkanes

Elimination from halogenoalkanes

Depending on the conditions of a reaction, some nucleophiles will either:

- attack at the carbon atom and displace the halogen atom (**substitution**); or
- attack at the hydrogen and **eliminate** H—X, forming an alkene.

There is a competition between substitution and elimination. Substitution is the mechanism that normally occurs, but the conditions can be changed to favour elimination by:

- using the alkali in concentrated solution in ethanol, rather than in dilute aqueous solution;
- using more branched chain halogenoalkanes (tertiary > secondary > primary);
- increasing the temperature.

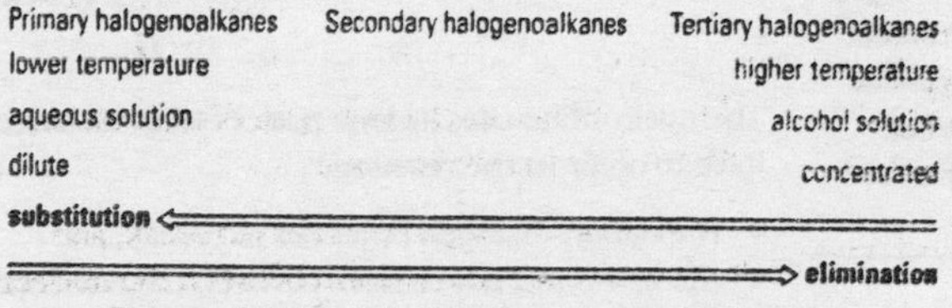

Primary halogenoalkanes	Secondary halogenoalkanes	Tertiary halogenoalkanes
lower temperature		higher temperature
aqueous solution		alcohol solution
dilute		concentrated

substitution $\Longleftarrow$

$\Longrightarrow$ elimination

Figure 17.5.1 Substitution versus elimination

In practice, whenever an elimination reaction is carried out, there will always be some substitution product; and a substitution reaction with alkali will always give some alkene by elimination. Yields can be maximised by altering the conditions, as shown in *Figure 17.4.1*.

For example, $OH^\ominus$ ions will act as strong nucleophiles in aqueous solutions. The negative charge allows the donation of the lone pair of electrons. In ethanol, the $OH^\ominus$ ion is more inclined to act as a base, accepting a proton ($H^\oplus$):

$$H-\overset{\cdots}{\underset{\cdots}{O}}\overset{\ominus}{)} \quad CH_2-CH-CH_3 \xrightarrow[\text{NaOH/ethanol}]{\text{reflux}} \quad CH_2{=}CH-CH_3 + H_2O + Br^\ominus \left(+ CH_3-\underset{OH}{\underset{|}{CH}}-CH_3\right)$$
(80%)

This mechanism is typical of elimination from primary and secondary halogenoalkanes. Tertiary halogenoalkanes react in a different way, involving, as a first step, ionisation of the C—Br bond.

SELF-ASSESSMENT
QUESTIONS
17.5

1 a Draw full displayed formulae for the two possible structural isomers of C_4H_9Br that contain a primary carbon–halogen bond.

b Both these isomers can be converted into an alkene or an alcohol under different conditions – what are the conditions needed for each reaction?

c What *type* of reaction happens in each case?

2 Write the displayed formula of 1,2,3-tribromo-2-methylbutane. Identify the primary, secondary and tertiary carbon atoms in the structure.

3 Consider the four compounds:

A C_4H_8, which can exist as two geometric isomers;
B C_4H_{10};
C C_4H_9Br;
D C_4H_9CN.

In each of the following parts, name the compounds involved, and write a balanced equation for each of the chemical changes.

a Describe how A could be obtained from C.

b How can C be converted into D?

c What reagents and conditions are needed to convert B to C?

17.6 Mechanisms of substitution

Nu is the general symbol for a nucleophile. A nucleophile is an atom or group of atoms with a lone pair of electrons (and sometimes a negative charge), capable of forming a covalent bond with a partially positive carbon atom in a polarised bond. Common examples are $OH^\ominus$, NH_3 and H_2O.

The substitution reactions of the halogenoalkanes involve nucleophiles. The reactions are therefore described as being **nucleophilic substitution reactions** (see Ch 14.7), and we use the symbol S_N to represent 'substitution nucleophilic'. The equation for the overall reaction is:

$$\overset{\cdots}{Nu}{}^\ominus \quad C-X \longrightarrow Nu-C + X^\ominus$$

The nucleophile uses its lone pair of electrons to attack the slightly positive carbon atom. Two things have to occur in the reaction:

● the carbon–halogen bond has to break; and
● the new bond has to form between the nucleophile and this same carbon atom.

Unimolecular means that only one species is involved in the slow steps (the rate determining step, see Ch 21.5) of the process; *bimolecular* means that two species are involved in the rate determining step.

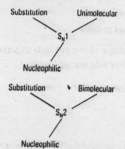

There are two ways in which this can happen.

- The C—X bond breaking and the Nu—C bond forming occur as two separate parts of the reaction. **This is the S_N1 mechanism.**

$$R-Br \longrightarrow R^\oplus \ Br^\ominus \longrightarrow R-Nu + Br^\ominus$$

i.e. Bond breaking followed by bond making (S_N1).

- The C—X bond breakage and the Nu—C bond formation occur at the same time. **This is the S_N2 mechanism.**

$$:Nu^\ominus \ R-Br \longrightarrow R-Nu + Br^\ominus$$

i.e. Simultaneous bond breaking/bond forming (S_N2).

Remember to use the curly arrows correctly when showing electron movement in reaction mechanisms (see *Figure 17.6.1*).

Reminder about curly arrows
- Start at a bond or pair of electrons.
- End on an atom or forming a bond between two atoms.

Figure 17.6.1

The S_N1 mechanism

To show this mechanism in detail, we will use as an example a tertiary bromoalkane (2-bromo-2-methylpropane) and a hydroxide ion as the nucleophile.

The first step involves the breaking of the carbon–bromine bond. This results in the formation of a bromide ion and leaves the carbon atom with a positive charge. This organic structure with a positively charged carbon atom is called a **carbocation** (or **carbonium ion**).

The stability of the carbocation is affected very much by the alkyl groups around the positive carbon atom. We have already come across the inductive effect (see Ch 16.6) of the methyl group, which involves a weak donation of electrons from the methyl group towards a carbon atom. The inductive effect of three methyl groups around the positively charged carbon atom makes this carbocation fairly stable (*Figure 17.6.2*).

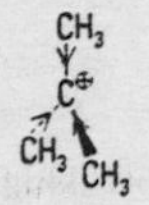

Figure 17.6.2   The inductive effect of three methyl groups

The intermediate carbocation has three groups around it, and so the shape of the group is trigonal *planar*. That means that the incoming nucleophile can attack from either side equally easily.

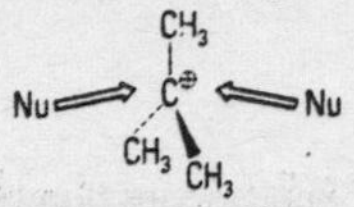

Once the carbocation has been formed, it reacts very rapidly with the nucleophile. In the example below, a hydroxide ion forms an alcohol.

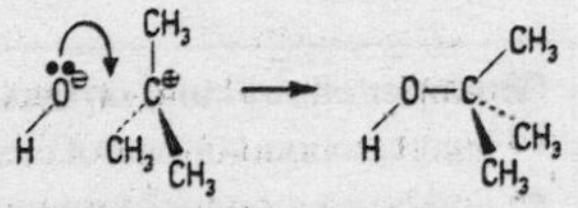

The rate of this reaction is determined by the first, slow, step; in a suitably polar solvent, such as water and ethanol, the 2-bromo-2-methylpropane dissociates spontaneously into a bromide ion and the carbocation. The nucleophilic attack happens very quickly after this. The slow (or rate-determining) step involves only one molecule – the formation of the carbocation. Therefore, the rate of the reaction will depend only on the concentration of the halogenoalkane. For this reason the mechanism is called **substitution (S) nucleophilic ($_N$) unimolecular (1) – S_N1**.

The whole mechanism can be summarised as follows.

The S_N1 mechanism

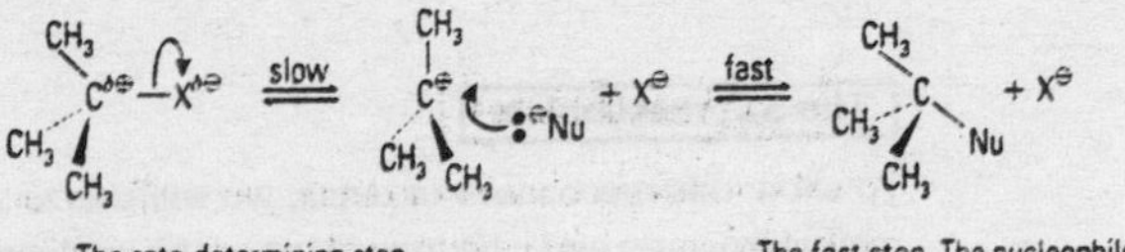

slow

fast

The rate-determining step.
The halogenoalkane dissociates to form
a halide ion and a carbocation

The fast step. The nucleophile
attacks the carbocation, equally
well from either side

The S_N2 mechanism

In this mechanism the C—X bond breaking and the Nu—C bond formation occur in one single step. As the nucleophile approaches the partially positive carbon atom, the lone pair of electrons from the nucleophile starts to form a bond with it. The start of this bond formation weakens the bond between the carbon atom and the halogen atom (C—X). Again, the reaction between a bromoalkane (bromoethane) and a hydroxide ion as nucleophile will be used to illustrate the mechanism.

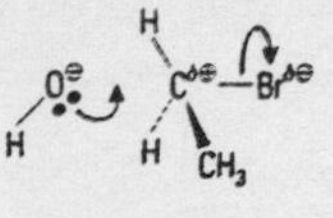

The attacking lone pair *repels* the pair of electrons in the C—Br bond. The bromine atom, which is electronegative, is ready to accept these electrons. The bromine atom is described as a *leaving group*, since it is leaving the rest of the molecule. The best leaving groups are those that are most able to accept electrons. A bromine atom is a good leaving group because it readily becomes a bromide ion.

The intermediate which is formed has five groups around the carbon atom, so the shape is trigonal bipyramidal (see Ch 2.6). This intermediate, or **activated complex**, has a negative charge overall. The structure is written enclosed in square brackets with the charge written outside these brackets.

$$\left[HO^{\delta\ominus} \cdots \overset{\displaystyle H}{\underset{\displaystyle CH_3}{\overset{|}{\underset{|}{C}}}} \cdots Br^{\delta\ominus} \right]^{\ominus}$$

From this point the reaction can go either way:

- the hydroxide ion can leave the activated complex and no reaction will have occurred; or
- the reaction can carry on further and the bromine leave as the bromide ion.

A double arrow is used in the equation to show that both outcomes are possible. Which one occurs depends on the energies of the colliding reactants (*Figure 17.6.4*).

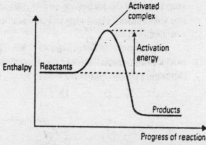

Figure 17.6.4   Energy profile for the S_N2 reaction

The forward reaction will only occur when the energy is greater than the activation energy.

In this mechanism the slow step of the reaction involves two molecules or reacting species. Both the nucleophile and the halogenoalkane are required to form the intermediate (the activated complex). The rate of the reaction will therefore depend on the concentration of both the halogenoalkane and the nucleophile. This mechanism is called **substitution (S) nucleophilic ($_N$) bimolecular (2) – S_N2**.

The S_N2 mechanism

The lone pair of electrons on the nucleophile attacks the $\delta\oplus$ carbon atom	An activated complex is formed with a partially formed Nu—C bond and a partially broken C—X bond	The halide ion leaves with the electron pair from the C—X bond

---

Substitution, S_N1 or S_N2, or elimination?

The type of mechanism that takes place in any particular reaction has to be found out by varying the reaction conditions and measuring the rates of reaction (see Ch 21.1). Often, two or more possible reaction mechanisms are competing with one another. For example, 1-bromobutane heated with sodium hydroxide solution can give both but-1-ene and butan-1-ol. Generally, elimination is favoured by higher temperatures and the use of solvents like ethanol.

Substitution is, however, the more important reaction pathway for halogenoalkanes. Of the two substitution mechanisms, tertiary halogenoalkanes use the S_N1 mechanism and primary halogenoalkanes use the S_N2 mechanism. Secondary halogenoalkanes use both mechanisms.

If we look at the shape of the tertiary carbon atom in 2-bromo-2-methylpropane, it can be seen that it is protected from attack by a nucleophile. This is because the bromine atom hinders attack on one side of the molecule, and the three methyl groups 'hide' the carbon atom on the other side (*Figure 17.6.4*).

Figure 17.6.4   2-Bromo-2-methylpropane

This means that the incoming nucleophile cannot get to the target carbon atom directly. So, the first step has to be the formation of the carbocation. The inductive effect of the methyl groups in tertiary halogenoalkanes also helps to stabilise the carbocation. For these reasons, the S_N1 mechanism is the one that occurs.

In primary halogenoalkanes the partially positive carbon atom is less protected from nucleophilic attack. There is also less of an inductive effect due to the absence of additional electron-donating groups (*Figure 17.6.5*).

Figure 17.6.5   1-Bromopropane

The experimental evidence for the differences in mechanism of two apparently similar reactions will be considered in Ch 21.3.

The rate of reaction is determined *only* by the concentration of the halogenoalkane in S_N1 reactions, but by the concentrations of *both* the halogenoalkane and the nucleophile in S_N2 reactions.

The difference in rate is easily demonstrated by reacting the bromoalkane dissolved in a little ethanol with an aqueous solution of silver nitrate. As the bromoalkane is hydrolysed (by the water in the silver nitrate solution) the formation of the halide ions can be followed by the formation of the silver bromide precipitate.

$$R—Br + H_2O(l) \longrightarrow R—OH + Br^{\ominus}(aq)$$
$$Br^{\ominus}(aq) + Ag^{\oplus}(aq) \longrightarrow AgBr(s)$$

Primary halogenoalkanes tend to react more slowly than secondary compounds, and the tertiary compounds are hydrolysed more readily through the different reaction mechanism (*Figure 17.6.6*). The electron donation effects of the alkyl groups make it easier to form the charged intermediate.

primary secondary tertiary

Figure 17.6.6   Rate of formation of AgBr(s)

Because of the C—X bond enthalpies, the overall rate of reaction will also be affected by whether the halogen is chlorine, bromine or iodine (*Table 17.6.1*). Iodoalkanes react faster than chloroalkanes.

	kJ mol^{-1}
C—F	+467
C—Cl	+346
C—Br	+290
C—I	+228

Table 17.6.1   Bond enthalpies for C-halogen

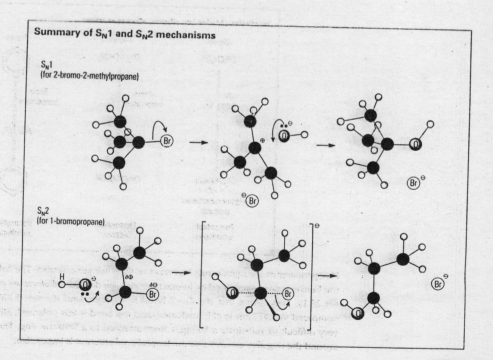

Summary of S$_N$1 and S$_N$2 mechanisms

S$_N$1
(for 2-bromo-2-methylpropane)

S$_N$2
(for 1-bromopropane)

1 So far the reaction mechanisms we have met are free radical substitution, electrophilic addition and nucleophilic substitution. Outline the main features of each of these reaction mechanisms.

2 In the following reactions, decide the type of mechanism likely to be occurring. Give reasons for your choice in each case.
  a $C_2H_6 + Cl_2$ to form C_2H_5Cl
  b $C_2H_4 + Cl_2$ to form $C_2H_4Cl_2$
  c $C_2H_5Cl + OH^{\ominus}$ to form C_2H_5OH
  d $(CH_3)_3CCl + OH^{\ominus}$ to form $(CH_3)_3COH$

3 What type of mechanism is involved in each of the following reactions? In each case give the formula of the attacking inorganic species.
  a $CH_3CH_2CH_2Br + NaOH(aq)$
  b $CH_3CH_2Br + KCN$ (in aqueous ethanol)
  c $CH_2CH_2 + HBr$ (as gases)

17.7 Arene halogen compounds

Once a benzene ring has an alkyl chain attached to it, there are effectively two parts in the molecule; the phenyl ring and the alkyl chain. In Ch 26.2, we will discuss how to introduce the chlorine atom into the benzene ring.

For now, it is worth pointing out that the reaction is difficult to do, and the mechanism is very different from the introduction of a halogen atom into an alkane or alkene. The halogenation of a benzene ring requires a catalyst, a 'halogen carrier' (e.g. aluminium chloride is used with chlorine). The reaction mechanism involves **electrophilic substitution**.

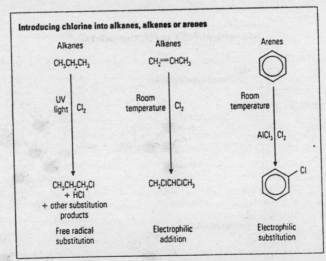

Introducing chlorine into alkanes, alkenes or arenes

Halogenoarenes are significantly less reactive than halogenoalkanes. The halogen atom attached to the benzene ring is stabilised by interaction with the delocalised electrons in the benzene ring (see Ch 26.1). This means that the C—X bond is stronger (and shorter; 0.170 nm in chlorobenzene compared to 0.177 nm in chloroalkanes) and the bond is less polarised. All of this means that it is very difficult to substitute a halogen atom attached to a benzene ring. The high electron density around the ring also ensures that nucleophilic substitution is impossible.

So the halogenation of any compound containing both an aromatic and aliphatic part will depend very much on the conditions used.

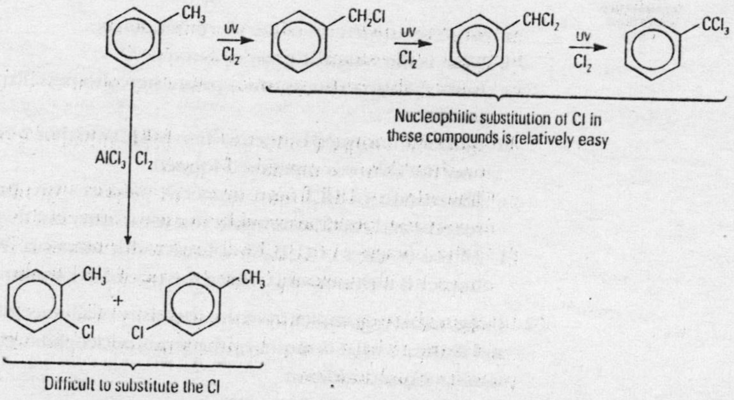

CH_3
UV
Cl_2
CH_2Cl
UV
Cl_2
$CHCl_2$
UV
Cl_2
CCl_3
Nucleophilic substitution of Cl in these compounds is relatively easy
$AlCl_3$ Cl_2
CH_3
+
CH_3
Cl
Cl
Difficult to substitute the Cl

Q
SELF-ASSESSMENT QUESTIONS 17.7

1 Write the formula of one organic product obtained from each of the following reactions.
  a Methylpropane and chlorine in UV light.
  b Methylbenzene and bromine in UV light.
  c Methylbenzene and chlorine with $AlCl_3$.

2 Ethylbenzene reacts with chlorine very differently in the presence of UV light from when an aluminium chloride catalyst is used. Write equations for the possible reactions in each of these different conditions.

Reactions of halogenalkanes

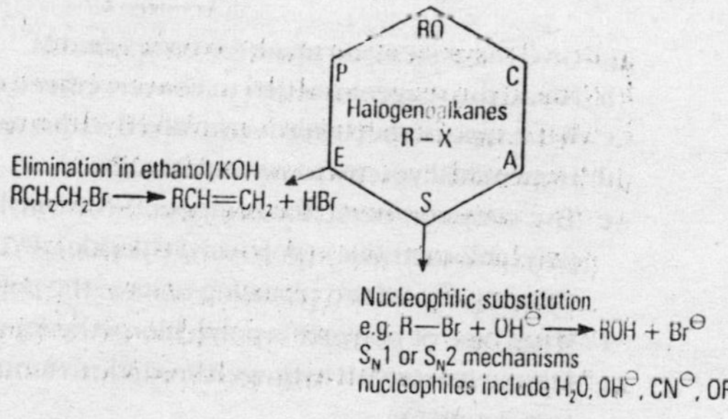

RO
P
C
Halogenoalkanes
R – X
E
A
S
Elimination in ethanol/KOH
$RCH_2CH_2Br \rightarrow RCH{=}CH_2 + HBr$
Nucleophilic substitution
e.g. $R{-}Br + OH^{\ominus} \rightarrow ROH + Br^{\ominus}$
S_N1 or S_N2 mechanisms
nucleophiles include H_2O, $OH^{\ominus}$, $CN^{\ominus}$, $OR^{\ominus}$

S
SUMMARY 17

In this chapter the following new ideas have been covered. *Have you understood and learned them?* Could you explain them clearly to a friend? The page references will help.

- Recognition of the halogen atom as a functional group. p431
- Naming the halogenoalkanes. p432
- Recognising the different types of isomerism possible with halogenoalkanes. p434
- The different bond strengths and bond polarities in the carbon–halogen bonds. p436
- The preparation of halogenoalkanes from alkanes, alkenes and alcohols. p442
- The reactions of halogenoalkanes, particularly nucleophilic substitution. p445
- The mechanisms for the substitution of halogenoalkanes with $OH^{\ominus}$. p448
- The effect of the benzene ring on the reactivity of the halogen atom. p454

1 C_4H_9Cl can exist as a number of structural isomers. The structures of two are shown.

$$CH_3CHClCH_2CH_3 \quad \text{and} \quad CH_3CCl(CH_3)CH_3$$
$$\text{A} \qquad\qquad\qquad \text{B}$$

a Write the structures of the two other isomers.

b What is the systematic name of isomer A?

c Isomer A also exists as two separate stereoisomers. Explain how two stereoisomers of A are possible.

d Isomer A is formed by the reaction of HCl with but-2-ene. Write a full mechanism to explain how the chloro compound is formed.

e Elimination of HCl from isomer A produces two products. Write the structures. What reagent and conditions would you use to carry out this elimination reaction?

f All the isomers of C_4H_9Cl will react with ammonia. Write an equation for the reaction of isomer B with ammonia. Name the type of mechanism occurring in this reaction.

2 Using suitable examples from the chemistry of alkanes, alkenes and halogenoalkanes, describe and illustrate what is meant by the terms nucleophilic, electrophilic, free radical, substitution, elimination and addition.

3 Consider the reaction scheme below.

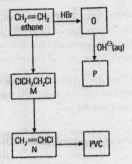

a Give the systematic name for compound M.

b Name the reagent needed to convert ethene to compound M.

c What type of mechanism is involved in this reaction?

d How would you try to convert M to N?

e The common name for compound N is vinyl chloride; it is the monomer used to prepare poly(chloroethene) (poly(vinyl chloride), PVC). Draw the structure for poly(chloroethene), showing two of the repeating units in the polymer.

f What type of polymer is poly(chloroethene) an example of?

g Ethene reacts with hydrogen bromide to form compound O. Write the displayed formula for compound O.

h O reacts with aqueous sodium hydroxide to form an alcohol, P. Give the systematic name of this alcohol.

i In view of this last reaction, suggest why poly(chloroethene) might break down in the presence of sodium hydroxide solution, whereas poly(ethene) is far more resistant to such breakdown.

4 A cyclic hydrocarbon, A, on analysis was found to contain 87.80% carbon and 12.20% hydrogen. When A was reacted with hydrogen bromide, compound B was formed, and warming B with aqueous sodium hydroxide produced compound C, $C_6H_{12}O$. If B was warmed with ethanolic potassium hydroxide solution, compound A was formed. The mass spectrum of A contained a molecular ion peak at m/e 82. Show by calculation that the empirical formula of A is C_3H_5, and hence work out the molecular formula of A. Use the data and information to deduce structures for compounds A, B, and C respectively. Explain the basis of each of your deductions, including discussions of types of reaction and reaction mechanisms where possible.

- **Functional groups, homologous series, frameworks**. (Ch 14.2)
- **Structural isomerism** and **optical isomerism**. (Ch 14.4)
- How to **name** organic compounds. (Ch 14.3)
- The factors which influence **physical properties**. (Ch 3.7)
- Factors which influence the **position of equilibrium**. (Ch 13.2)
- **Hydrogen bonding**. (Ch 2.9)
- **Types** of organic reactions. (Ch 14.6)
- **Reaction mechanisms**. (Ch 14.7)

18

alcohols

Do not confuse the hydroxy group with the hydroxide ion, OH^-, found in ionic compounds such as sodium hydroxide.

Alcohols are a group of organic compounds which contain one (or more) hydroxy groups, —OH, attached to an alkyl group in a covalent structure.

The **monohydric alcohols** contain only one —OH group per molecule; those with two or more —OH groups per molecule are called **polyhydric alcohols**. Polyhydric alcohols are more viscous and have much higher boiling points than monohydric alcohols, because of increased opportunities for hydrogen bonding (see Ch 2.9).

The general formula of a simple alcohol is:

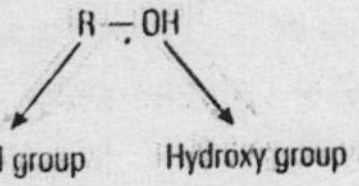

$$R - OH$$

Alkyl group   Hydroxy group

General formula: $C_nH_{2n+1}OH$

The alcohols are a very important group of compounds. The alcoholic drinks industry is well known, but there is a very wide range of alcohols used as solvents, as antifreeze and in cosmetics. After-shave lotion contains up to 70% ethanol. This volatile liquid provides a cooling effect and is mildly antiseptic.

Alcohols in nature

Vitamins A, C, D and E (retinol, ascorbic acid, calciferol and tocopherol) are all alcohols as are many hormones (e.g. testosterone, oestradiol). The peppermint plant contains menthol, . Sugars (e.g. glucose) can be considered as polyhydric alcohols

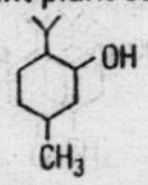

The peppermint plant

Alcohols in use

18.1 Naming and structure of alcohols

All alcohols contain the hydroxy functional group, —OH, attached to an aliphatic framework or alicyclic framework.

$CH_3CH_2CH_2CH_2CH_2CH_2OH$

$C_6H_{13}OH$
Hexan-1-ol
with an aliphatic framework

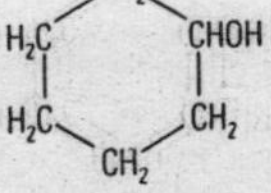

$C_6H_{11}OH$
Cyclohexanol
with an alicyclic framework

Compounds where the —OH group is attached directly to a benzene ring are called phenols, as in C_6H_5OH, and will be discussed in Ch 26.6.

Naming alcohols

The naming of alcohols follows the same rules as for alkanes (Ch 14.3). The name consists of a first part which indicates the chain length in exactly the same manner as alkanes, but without the final -e (methan-, ethan-, propan-, etc.). This stem is followed by the suffix '-ol'. So, CH_3OH is methanol; C_2H_5OH is ethanol (*Figure 18.1.1*).

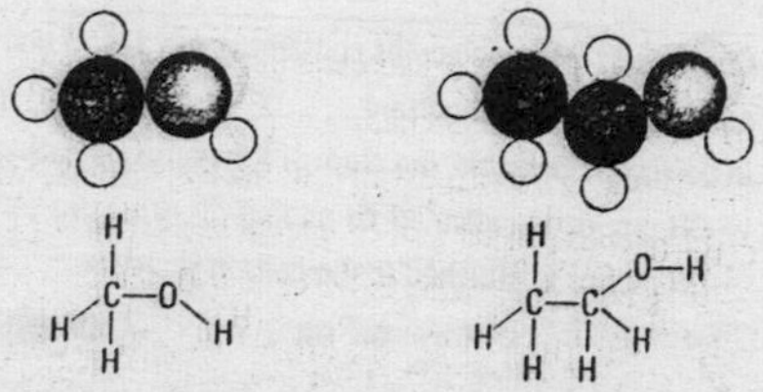

Methanol, CH_3OH Ethanol, CH_3CH_2OH

Figure 18.1.1 Simple alcohols

For chain lengths of three or more carbon atoms, the alcohols have numbers to indicate the position of the —OH group along the carbon chain. The number is placed just before the '-ol' suffix with a dash before and after the number. The numbering of the carbon chain starts at the end of the carbon chain closest to the first substituted group. So $CH_3CH_2CH_2OH$ is propan-1-ol; $CH_3CH(OH)CH_3$ is propan-2-ol.

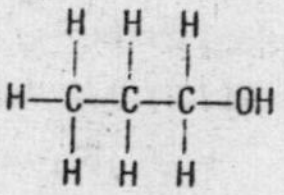

Propan-1-ol 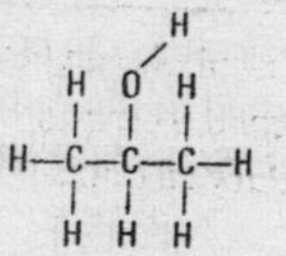   Propan-2-ol

If the numbering could begin at either end, it starts at the end nearest the substituted group that comes first alphabetically.

e.g.

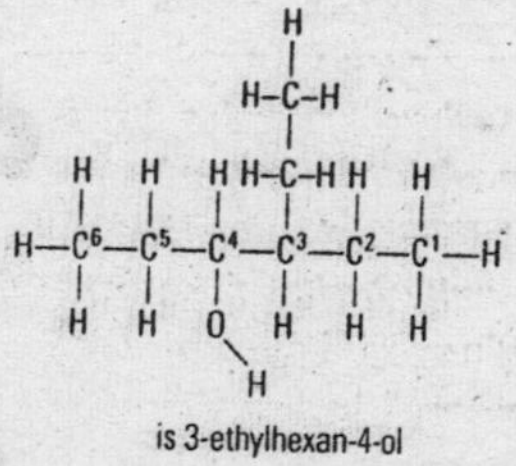

is 3-ethylhexan-4-ol

Can you justify to yourself the names of the following alcohols?

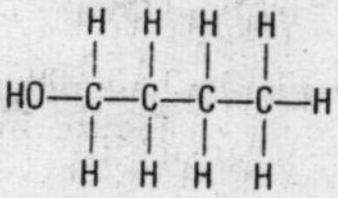

Butan-1-ol

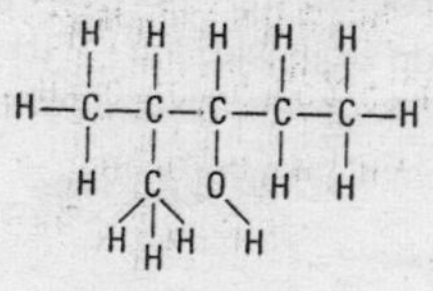

2-Methylpentan-3-ol

$CH_3CH(CH_3)CH_2OH$ 2-Methylpropan-1-ol

$CH_3CH_2C(CH_3)(OH)CH_3$ 2-Methylbutan-2-ol

Compounds with molecules containing more than one —OH group (polyhydric alcohols) are quite common and some are commercially very important.

In polyhydric alcohols, the stem of the name indicating the chain length now includes the final -e, and the total number of —OH groups is indicated by adding di-, tri- tetra-, etc: Remember, a number is needed to indicate where each of the —OH groups is attached to the carbon skeleton.

$$
\begin{array}{ccc}
\text{Ethane-1,2-diol} & \text{Propane-1,2,3-triol} & \text{Hexane-1,3,4-triol}
\end{array}
$$

The dihydric alcohol ethane-1,2-diol, CH_2OHCH_2OH, is used in antifreeze and in making the fibre Terylene. The trihydric alcohol propane-1,2,3-triol, $CH_2OHCH(OH)CH_2OH$, is also known as glycerine. It is used in cooking and in ointments and creams, and can be made into nitroglycerine – which is not only a very powerful explosive but also an effective treatment for the heart condition *angina pectoris*.

The structures of alcohols

The alcohols can be divided into three classes depending on the number of alkyl (R) groups that are attached to the carbon linked to the —OH group. The three classes are **primary, secondary** and **tertiary** alcohols (*Figure 18.1.2*).

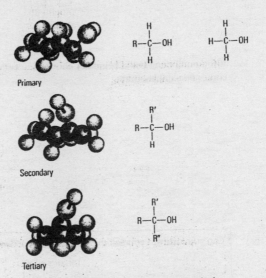

Figure 18.1.2   Structure of alcohols

Primary alcohols contain the —CH_2OH group; i.e. a maximum of one other alkyl (R) group is attached to the carbon atom with the —OH group attached. An example is ethanol, C_2H_5OH:

i.e. there are two hydrogen atoms on the C—OH carbon atom.

Methanol, CH_3OH, which contains only a single carbon atom in its molecule, is also a primary alcohol.

In **secondary alcohols,** *two* alkyl groups are attached to the carbon atom with the —OH group. An example is propan-2-ol, $CH_3CH(OH)CH_3$:

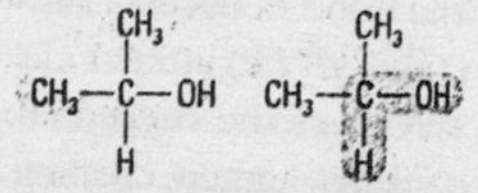

i.e. there is one hydrogen atom on the C—OH carbon atom.

In **tertiary alcohols,** *three* alkyl groups are attached to the carbon atom which is linked to the —OH group. An example is 2-methylpropan-2-ol, $(CH_3)_3COH$:

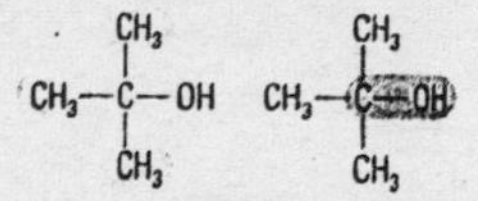

i.e. there are no hydrogen atoms on the C—OH carbon atom.

Note that a minimum of three carbon atoms is needed for a secondary alcohol and a minimum of four for a tertiary alcohol.

Table 18.1.1 classifies all the monohydric alcohols which contain up to four carbon atoms. Note that methanol and ethanol have no alcohol isomers; propanol has two and butanol five (two of which are optical isomers).

> **Remember: chain isomers** of alcohols have the same number of carbon atoms arranged in different chain lengths, with different amounts of branching; **position isomers** have the same carbon skeleton, but with the —OH group attached at different positions; **optical isomers** occur when an alcohol has a chiral carbon atom.

Alcohols with more than two carbon atoms display the usual types of isomerism.

Name	Linear formula	Structural formula	Classification	Isomerism
Methanol	CH_3OH	CH_3OH	Primary	
Ethanol	CH_3CH_2OH	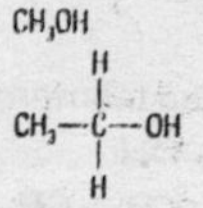 $CH_3-C(H)(H)-OH$	Primary	
Propan-1-ol	$CH_3CH_2CH_2OH$	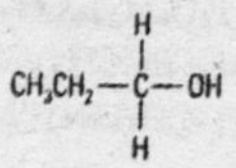 $CH_3CH_2-C(H)(H)-OH$	Primary	Position isomers
Propan-2-ol	$CH_3CH(OH)CH_3$	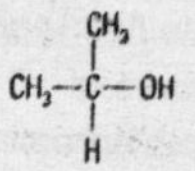 $CH_3-C(CH_3)(H)-OH$	Secondary	
Butan-1-ol	$CH_3CH_2CH_2CH_2OH$	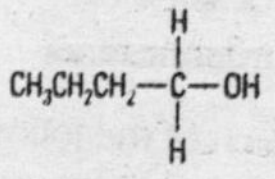 $CH_3CH_2CH_2-C(H)(H)-OH$	Primary	Position isomers
Butan-2-ol	$CH_3CH_2CH(OH)CH_3$	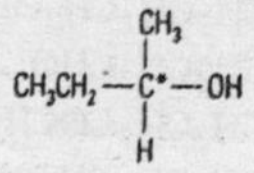 $CH_3CH_2-C^*(CH_3)(H)-OH$	Secondary	Chain isomers

*Indicates a chiral carbon atom, resulting in optical isomers

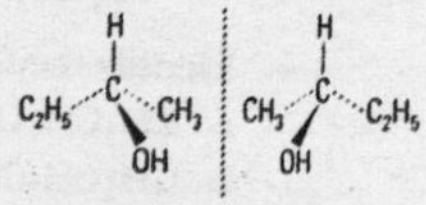

(2 optical isomers)

Name	Linear formula	Structural formula	Classification	Isomerism
2-Methylpropan-1-ol	$CH_3CH(CH_3)CH_2OH$	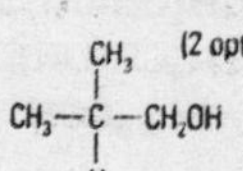 $CH_3-C(CH_3)(H)-CH_2OH$	Primary	Position isomers
2-Methylpropan-2-ol	$CH_3C(CH_3)_2OH$	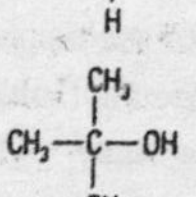 $CH_3-C(CH_3)(CH_3)-OH$	Tertiary	

Table 18.1.1   Structures of monohydric alcohols with up to four C atoms

Bond enthalpies	kJ mol^{-1}
C—H	+413
C—C	+347
C—O	+336
O—H	+464

Electronegativities	
C	2.5
H	2.1
O	3.5

**Table 18.1.2
Bond enthalpies and
electronegativities**

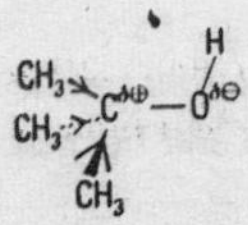

**Figure 18.1.3 Methyl
group electron donation
to partially positive
carbon**

Polarity and bond strength

The bond between a carbon atom and a hydroxy group is polar. The electronegative oxygen atom (*Table 18.1.2*) attracts electrons away from both the carbon atom and the attached hydrogen atom. This gives the oxygen atom a small negative charge and leaves the carbon and hydrogen atoms partially charged, creating a permanent dipole.

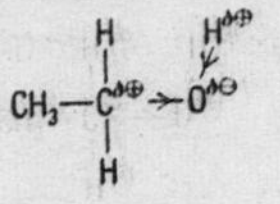

The strength of this dipole is affected by the nature of the other groups attached to the carbon atom, especially electron-donating groups, which will weaken the dipole (*Figure 18.1.3*).

Remember that the oxygen atom in an —OH group has two lone pairs of electrons and so the C—O—H bond angle in alcohols is close to 109°.

In spite of the strong covalent bonds involved in the molecule (*Table 18.1.2*), the permanent dipole present in the molecule allows either nucleophilic attack at the $C^{\delta+}$ or electrophilic attack at the oxygen atom.

Q

SELF-ASSESSMENT
QUESTIONS
18.1

1 Name the following compounds.
   a C_3H_7OH
   b $CH_3(CH_2)_4CH_2OH$
   c $CH_3CH(CH_3)CH_2OH$
   d $CH_3CH(OH)CH_2CH_3$

2 Write the full displayed formulae for:
   a pentan-1-ol
   b 2-methylpropan-2-ol
   c hexan-2-ol
   d cyclohexanol

3 Which of the following formulae represent an alcohol?
   a $CH_3CH_2CH_2OH$
   b CH_3CH_2CHO
   c $CH_3CH_2COCH_3$
   d $CH_3CH_2CH(OH)CH_2CH_3$

4 Identify the following alcohols as primary, secondary or tertiary.
   a CH_3CH_2OH
   b $CH_3CH(CH_3)CH_2OH$
   c $CH_3CH(OH)CH_3$
   d $(CH_3)_3COH$

5 a Write the displayed formulae for all the monohydric alcohols of molecular formula $C_5H_{11}OH$.
   b Name each alcohol, and mark any chiral centres with an asterisk.
   c How many of the pentanols are:
      i) primary,
      ii) secondary,
      iii) tertiary?

18.2 Physical properties of alcohols

The hydroxy, —OH, functional group in alcohols allows them to form hydrogen bonds (see Ch 2.9) with each other and with water. The smaller alcohols are colourless liquids, but alcohols with twelve or more carbon atoms are waxy solids.

The two hydrogen atoms in a water molecule can both form hydrogen bonds to oxygen atoms in two further water molecules (*Figure 18.2.1*).

Figure 18.2.1 Hydrogen bonding in ice

In alcohols there is only one hydrogen atom able to form hydrogen bonds, so there is less hydrogen bonding. As a result, the boiling points of methanol and ethanol are much lower than that of water. Propan-1-ol, with a relative molecular mass (M_r) of 60, boils at 97 °C, which is almost the same as water (M_r = 18).

However, hydrogen bonding between alcohol molecules (*Figure 18.2.2*) is still very significant in reducing volatility and thus increasing the boiling points of alcohols compared with alkanes.

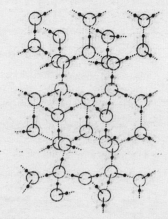

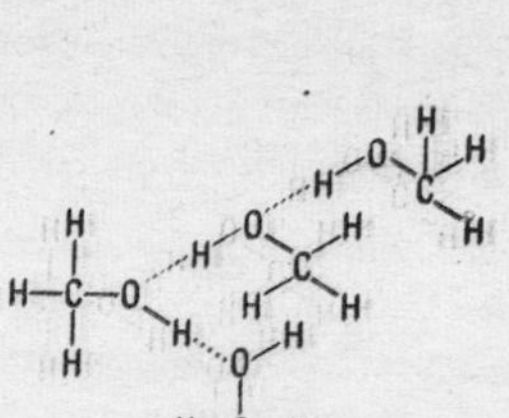

Figure 18.2.2 Hydrogen bonding in alcohols

Figure 18.2.3 compares the boiling points of the simple alcohols with those of the alkanes of similar M_r. The difference between the boiling points of propane (M_r = 44) and ethanol (M_r = 46) is 120 °C!

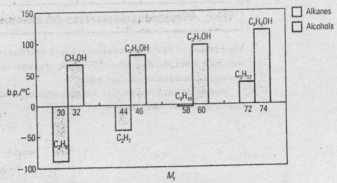

Figure 18.2.3   Boiling points of alkanes and primary alcohols

The boiling points are also affected by isomerism. Moving the —OH group towards the centre of the carbon chain alters the shape of the molecule, making it more spherical ('globular'). Adding branches (methyl groups, etc) has the same effect. Making an alcohol more spherical has two effects:

- It gives the molecule a smaller surface area. A smaller surface area reduces the van der Waals forces between the molecules.
- It makes it more difficult for —OH groups to approach one another to form hydrogen bonds.

Both these effects lower the boiling point.

Name	Structure	b.p./°C
Butan-1-ol	$CH_3CH_2CH_2CH_2OH$	117
2-Methylpropan-1-ol	CH_3CHCH_2OH $\quad\quad\mid$ $\quad\quad CH_3$	108
Butan-2-ol	$CH_3CH_2CHCH_3$ $\quad\quad\quad\mid$ $\quad\quad\quad OH$	99.5
2-Methylpropan-2-ol	CH_3 $\quad\mid$ CH_3-C-OH $\quad\mid$ $\quad CH_3$	82.5

Table 18.2.1   The structures and boiling points of the isomeric butanols, C_4H_9OH

Hydrogen bonding should allow alcohols to dissolve freely in water. Methanol, ethanol and th[e] propanols are all completely soluble (miscible) in water. However, hydrocarbon chains do not for[m] interactions with water molecules. Therefore, as the alkyl chain in the alcohol increases in lengt[h] the solubility of the alcohol decreases rapidly (*Table 18.2.2*).

Remember: hydrogen bonding in compounds such as the alcohols adds to the attractions between particles which are always present – the van der Waals' forces (see Ch 2.8 and Ch 14.5).

Name	Linear formula	Solubility in water /g per 100 g H_2O
Methanol	CH_3OH	∞
Ethanol	CH_3CH_2OH	∞
Propan-1-ol	$CH_3CH_2CH_2OH$	∞
Butan-1-ol	$CH_3(CH_2)_2CH_2OH$	7.9
Pentan-1-ol	$CH_3(CH_2)_3CH_2OH$	2.4
Hexan-1-ol	$CH_3(CH_2)_4CH_2OH$	0.6
Heptan-1-ol	$CH_3(CH_2)_5CH_2OH$	0.2
Octan-1-ol	$CH_3(CH_2)_6CH_2OH$	0.05
Nonan-1-ol	$CH_3(CH_2)_7CH_2OH$	Very small

Table 18.2.2   The solubility of simple alcohols

The viscosity of an alcohol can be demonstrated by how fast a bubble rises in it. The 'thicker' (more viscous) the liquid, the slower the bubble rises. The viscosity of alcohols increases dramatically with an increase in the number of hydroxy (—OH) groups in the molecule.

The increase in viscosity is due to an increase in hydrogen bonding. Syrup and treacle contain very high concentrations of sugars which contain many hydroxy groups. That is why they are so viscous (*Figure 18.2.4*).

Figure 18.2.4   A sugar like sucrose contains many —OH groups

Infra-red spectroscopy

So far, the bond lengths and bond enthalpies (strengths) have been given as if the bonds were stationary (e.g. the C—O bond length in CH_3—OH is 0.143 nm and the bond enthalpy is $+340 \text{ kJ mol}^{-1}$).

This is a simplification. A good analogy is to imagine atoms connected together with springs. The two atoms can move nearer to or away from each other (i.e. the bond can stretch and contract).

The atoms in the bond will usually be attached to other atoms, and therefore also able to move in other directions, giving rise to complex 'vibrations'. This will be discussed further in Ch 30.5. In polarised bonds (such as C—OH), the amount of polarisation (dipole) of the bond can change when it vibrates. If there is a change in the dipole, the bond will absorb energy in the infra-red (IR) part of the spectrum. The amount of energy that a bond absorbs depends on the mass and size of the two atoms. Therefore, different bonds will absorb different amounts of energy.

In infra-red spectroscopy, the compound is placed in a beam of infra-red light. Different wavelengths are absorbed by different bonds and the amount of each wavelength absorbed by a particular compound is plotted on a graph of wavelength (usually as wavenumber) against 'transmittance' (0–100%). A transmittance of 100% means that all the infra-red light is going through the sample; a transmittance of 0% means that all the infra-red light at that wavenumber is being absorbed by the sample.

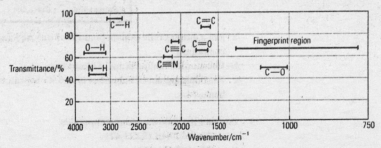

The most common unit for the scale used in infra-red spectra is wavenumber, measured in cm^{-1}. A wave number of 1500 cm^{-1} means that there are 1500 waves per centimetre.

Some wavenumbers corresponding to specific absorptions are as follows:

Bond	wavenumber/cm^{-1}
O—H	3700–3200
N—H	3500–3100
C—H	3200–2800
C≡N	2400–2200
C≡C	2300–2100
C=O	1800–1650
C=C	1700–1600
C—O	1250–1000

The infra-red spectrum of ethanol shows the O—H bond vibration around 3500 cm^{-1}. The very broad band is indicative of the extensive hydrogen bonding present in alcohols.

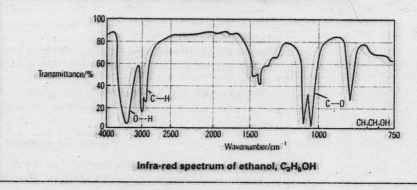

Infra-red spectrum of ethanol, C_2H_5OH

SELF-ASSESSMENT QUESTIONS 18.2

1 The boiling point of pentan-1-ol is 138 °C. Draw the structures of 3-methylbutan-2-ol and 2,2-dimethylpropan-1-ol, and estimate the boiling points of these compounds.

2 The first five members of the alcohol homologous series, with their boiling points, are:

		°C
Methanol	CH_3OH	65
Ethanol	C_2H_5OH	78
Propan-1-ol	C_3H_7OH	98
Butan-1-ol	C_4H_9OH	117
Pentan-1-ol	$C_5H_{11}OH$	138

a Plot a graph of the boiling points (vertical axis) against the number of carbon atoms in the compound.

b Estimate the boiling point of hexan-1-ol.

c The boiling point of butan-2-ol is 99.5 °C. Explain why this value is less than that quoted for butan-1-ol.

3 Consider the following four compounds:
A $CH_3CH(OH)CH_2CH_2CH_3$
B $CH_2(OH)CH_2CH_3$
C $CH_3CH_2C(CH_3)_2OH$
D $CH_3CH_2CH_2CH_2OH$

a Which are primary alcohols?
b Which are secondary alcohols?
c Which are tertiary alcohols?
d Which will have the lowest boiling point?
e Which are freely soluble in water?

4 The infra-red spectra of four compounds are shown. Which compounds contain the —OH group?

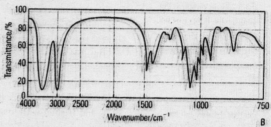

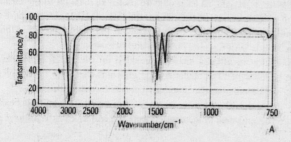

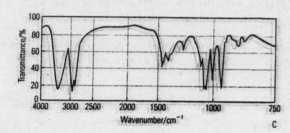

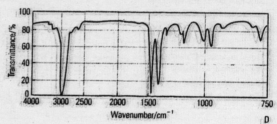

18.3 Production and preparation of alcohols

There are two important industrial methods of producing ethanol.
- Fermentation of sugars.
- Hydration of ethene.

On a smaller scale, alcohols are prepared by:
- hydration of alkenes;
- reduction of aldehydes, ketones and carboxylic acids;
- hydrolysis of halogenoalkanes;
- hydrolysis of esters.

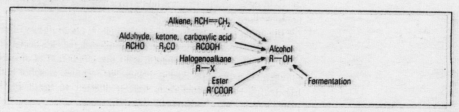

467

In 1860 Louis Pasteur first realised that fermentation was due to micro-organisms living without air. Until that time it had been thought that it was a spontaneous process. Modern theories about the evolution of life on Earth suggest that life first evolved at a time when there was no oxygen in the atmosphere. If this was the case, then most primitive organisms must have used fermentation to obtain energy.

Another common natural polymer of glucose is cellulose, $(C_6H_{10}O_5)_n$, which is the structural material in wood and grass, etc. The glucose units in starch and cellulose are joined in different ways. You can digest and ferment starch but not cellulose; a ruminant animal such as a cow or goat can digest both.

It is not only ethanol that can be produced using fermentation. The bacterium *Clostridium acetobutylicum* has been found to ferment glucose, producing butanol. Currently, there is much research into using a variety of micro-organisms to produce useful alcohols.

Industrial fermentation

About 5% of the world's production of ethanol is produced by a biological process, **fermentation**. The process is carried out by a yeast, grown on a carbohydrate food source in the absence of oxygen (*anaerobic* conditions). The weak (5–15%) ethanol solution produced can be converted into 'pure' ethanol by distillation methods, although 100% ethanol is not often required.

Anaerobic conditions are needed to prevent biochemical reactions in the yeast converting some of the carbohydrate into carboxylic acid rather than ethanol. The carbohydrate commonly used is starch.

Starch is a polymer of glucose, $C_6H_{12}O_6$. As each glucose molecule joins on to the chain in a starch molecule, a molecule of water is lost, so that starch has the formula $(C_6H_{10}O_5)_n$, where n is a large number. Starch is a food storage compound found widely in the plant kingdom. Common sources used in fermentation include barley, maize, potatoes and rice.

The first step in the production of ethanol involves breaking up the starch molecules into simpler sugars. Germinating barley ('malt') contains several enzymes that help in this process. Enzymes are natural catalysts, and the mixture contained in malt is known as *diastase*.

A carbohydrate source is 'mashed' with hot water to release the starch. This liquid is heated to 55 °C and treated with 'malt'. The enzymes in diastase convert the starch into the simpler sugar maltose.

$$2(C_6H_{10}O_5)_n + nH_2O(l) \xrightarrow[\text{no air}]{55\,°C} nC_{12}H_{22}O_{11}(aq)$$
$$\text{Starch} \qquad\qquad \text{diastase} \quad \text{Maltose}$$

The temperature of the mixture is then dropped to about 35 °C. Yeast is added. Yeast contains the enzyme, *maltase*, which breaks up maltose into glucose; at temperatures much higher than 35 °C the enzymes are 'denatured' and cannot work.

$$C_{12}H_{22}O_{11}(aq) + H_2O(l) \xrightarrow[\text{no air}]{35\,°C} 2C_6H_{12}O_6(aq)$$
$$\text{Maltose} \qquad\qquad\qquad \text{maltase} \quad \text{Glucose}$$

Another enzyme from the yeast, *zymase*, then converts glucose into ethanol and an important and useful by-product, carbon dioxide:

$$C_6H_{12}O_6(aq) \xrightarrow[\substack{\text{no air}\\ \text{zymase}}]{35\,°C} 2C_2H_5OH(aq) + 2CO_2(g)$$
$$\text{Glucose} \qquad\qquad\qquad \text{Ethanol}$$

The fermented mixture contains water, ethanol and a whole variety of other organic compounds. Purification of the ethanol requires fractional distillation. The first fractions contain the relatively volatile aldehydes and esters. The main fraction (known as rectified spirit) contains about 95% ethanol and 5% water. It is not possible to remove all the water simply by fractional distillation. Higher boiling fractions (known as 'fusel oil') have a terrible smell and consist mostly of longer-chain alcohols. Most 'industrial ethanol' is in fact 'rectified spirit'.

To produce completely pure ethanol ('absolute ethanol') from the 'rectified spirit', the water needs to be removed. This is achieved in a complex series of distillations, involving the addition of benzene. Pure ethanol distils at 78 °C.

Obtaining pure ethanol using fermentation on an industrial scale is a lengthy and expensive process. Ethene is now the preferred starting material (see next section).

The main use of fermentation is in the production of alcoholic drinks. It is also becoming important as a way of producing an alternative car fuel. 'Gasohol' contains up to 20% ethanol in lead-free petrol; car performance is scarcely affected. In Brazil, production from some sugar cane plantations is used exclusively for this purpose.

Industrially, methanol is produced from carbon monoxide and hydrogen.

$$CO(g) + 2H_2(g) \xrightarrow[\substack{catalyst \\ 15\ atm}]{400\,°C} CH_3OH(g)$$

The CO and H_2 are produced from methane and steam.

$$CH_4(g) + H_2O(g) \xrightarrow[catalyst]{heat} CO(g) + 3H_2(g)$$

Ethanol is the 'alcohol' in alcoholic drinks

The production of alcoholic drinks, using fermentation, was discovered independently by civilisations in nearly all parts of the world. The starting materials have consisted of all manner of fruits and vegetables including elderflowers, dandelions, potatoes, parsnips, grapes, apples, barley, maize and rice. Most soft fruits, like grapes, contain simpler carbohydrates than starch, and so the yeasts that occur naturally on the skins of these fruits can carry out the fermentation to form wines.

The wines and beers produced by fermentation are coloured and flavoured by substances from the starting material or by other products of the fermentation (or added, like hops for 'bitter' beer). Fermented brews can contain up to 5–15% by volume of ethanol. At this concentration, yeast is killed and fermentation stops.

'Spirits' are made by deliberate inefficient distillation of the fermented liquid, to raise the alcohol proportion to 30–45%. Whisky is obtained by distilling a mixture similar to beer, brandy is made from wine and vodka from fermented mashed potato.

The alcoholic strength of many spirits is described by the percentage of ethanol they contain. Most spirits are around 40%. Sometimes the old term 'degrees proof' is used. Originally a 'proof spirit' was one which, when poured over gunpowder, would not stop it igniting. A 100° proof spirit contains about 52% of ethanol. The term 'degrees proof' is still used in the USA, but in this case it is simply double the percentage of ethanol in the drink.

Summary

The requirements for fermentation are:

- a carbohydrate source;
- water;
- a suitable micro-organism (usually a yeast);
- absence of air;
- warmth (about 35 °C).

Industrial production from ethene

In countries which have good supplies of crude oil and/or natural gas, ethanol for industrial use is now made exclusively from ethene.

Ethene is produced by **cracking** (Ch 15.4). Ethene and steam are passed over a solid catalyst (phosphoric acid supported on silica) at about 300 °C and about 70 atm pressure.

$$CH_2{=}CH_2(g) + H_2O(g) \xrightleftharpoons[catalyst]{\substack{300\,°C/70\ atm \\ phosphoric\ acid}} C_2H_5OH(g) \quad \Delta H = -46\ kJ\ mol^{-1}$$

In this process the unreacted ethene is recycled, whereas the steam and ethanol are liquefied to produce ethanol solution. The final step is the distillation of the ethanol solution to produce 'industrial ethanol' (95% ethanol).

The enthalpy change (ΔH) for this reaction is $-46\ kJ\ mol^{-1}$. Using the ideas developed in Ch 13.2, we can see that the exothermic forward reaction is helped by low temperature, high pressure and a high concentration of steam. But low temperature means a slow reaction, so a catalyst had to be found and a compromise (optimum) temperature used. Increased pressure means expenditure on containing that pressure, so again a compromise is used. Finally, too much steam at high pressure dilutes the catalyst, so the steam:ethene ratio used is about 0.6:1. As with the Haber Process (Ch 13.3) a low yield obtained rapidly is preferred to a high yield obtained slowly and expensively.

Alcohols are poisonous

Alcohols are toxic, including ethanol. The strict meaning of the word 'intoxicated' is poisoned. With the exception of ethanol, all alcohols are very toxic: ethanol is merely toxic. In Britain the current permissible upper limit for a car driver is 80 mg ethanol per 100 cm^3 of blood, i.e. a concentration of 0.017 mol dm^{-3}. People with twice this concentration in their blood usually behave as though drunk; at four times the limit they are usually unconscious; at six times they often die.

The liver is responsible for the oxidation of ethanol. The process converts the ethanol into carbon dioxide and water with the release of energy. In the UK at present, alcohol contributes on average about 6% of the total energy in the diet. In the seventeenth century, the percentage was almost 25% – for men, women and children. Beer was then a far safer drink than water from rivers and wells, which were often polluted.

Ethanol is a 'depressant'; it reduces the functioning of nerve cells in the brain and the rest of the central nervous system. In moderate amounts it can help people to relax and feel happier for a relatively short time. But it also affects judgement and slows reaction times – which is why drinking and driving should always be kept separate.

Fermentation also produces some longer chain alcohols, particularly propanols through to hexanols, in small amounts. They are much more poisonous than ethanol, but rarely cause permanent damage because in large amounts they make people vomit. They are, however, a major cause of hangovers.

Methylated spirit, 'meths', is mainly ethanol which has deliberately had methanol and the foul-smelling and toxic pyridine added to it to make it (supposedly) undrinkable. Methanol is converted in the liver to methanal, HCHO, which is the substance used in embalming fluid. Drinking methanol or meths can cause blindness, mania and death.

Laboratory preparation of alcohols

Hydration of alkenes

This can be done directly, as in the industrial production of ethanol from ethene and steam, as discussed above.

$$RCH{=}CH_2(g) + H_2O(g) \underset{\text{catalyst}}{\overset{\text{heat/high pressure}}{\rightleftharpoons}} RCH(OH)CH_3(g)$$

The conditions and catalyst used will depend on the starting material. The use of ethene to produce ethanol is the most important reaction of this type.

Another route involves initial addition of sulphuric acid, followed by hydrolysis:

$$RCH{=}CH_2 + H_2SO_4 \xrightarrow{\substack{\text{room}\\\text{temperature}}} \underset{OSO_2OH}{RCHCH_3} \xrightarrow{\text{warm/H}_2\text{O}} \underset{OH}{RCHCH_3} + H_2SO_4$$

Notice that addition of sulphuric acid follows the Markovnikov rule (Ch 16.6), which means that the overall addition of the —OH group will also follow the rule (the —OH group adding to the carbon with the fewer number of hydrogen atoms attached to it).

The concentration of the sulphuric acid needed depends on the alkene used. Ethene reacts best with concentrated (98%) sulphuric acid, whereas 2-methylpropene will react satisfactorily with 10% acid at room temperature.

Reduction of aldehydes, ketones, carboxylic acids

Aldehydes can be reduced to **primary alcohols**, and **ketones** to **secondary alcohols** (Ch 19.3). Several methods are available. Lithium tetrahydridoaluminate(III), LiAlH$_4$, in ethoxyethane (ether) can be used. However, using LiAlH$_4$ is not for the faint hearted; it reacts violently with water and explodes at temperatures above 120 °C. It is therefore more convenient to use sodium tetrahydridoborate(III), NaBH$_4$, in water.

Although there appears to be a carbonyl group in carboxylic acids, RCOOH, and esters, RCOOR', it does not usually act as a carbonyl group (see Ch 19.1). However, carboxylic acids can be reduced to alcohols, although not quite in the same way as aldehydes and ketones.

$$R-C\underset{H}{\overset{O}{\big\|}} + 2[H] \xrightarrow[\text{room temperature}]{\text{NaBH}_4/\text{H}_2\text{O}} RCH_2OH$$

Aldehyde   Primary alcohol

$$\underset{R'}{\overset{R}{\big\backslash}}C{=}O + 2[H] \xrightarrow[\text{room temperature}]{\text{NaBH}_4/\text{H}_2\text{O}} \underset{R'}{\overset{R}{\big\backslash}}CHOH$$

Ketone   Secondary alcohol

In many cases, the reduction of aldehydes and ketones can be carried out using sodium in ethanol at room temperature.

$$R-CHO + 2Na + 2C_2H_5OH \longrightarrow R-CH_2OH + 2C_2H_5O^{\ominus}Na^{\oplus}$$

Aldehyde   Primary alcohol

Unfortunately, $NaBH_4$ cannot be used to reduce carboxylic acids. The stronger reducing agent $LiAlH_4$ must be used. A primary alcohol is always formed.

$$R-C\underset{OH}{\overset{O}{\big\|}} + 4[H] \xrightarrow[\text{room temperature}]{\text{LiAlH}_4/\text{ethoxyethane}} R-CH_2OH + H_2O$$

Carboxylic acid   Primary alcohol

Hydrolysis of halogenoalkanes

In these reactions the halogen atom is displaced by $OH^{\ominus}$ in a nucleophilic substitution reaction (see Ch 17.4). A halogenoalkane (other than fluoroalkane) is heated with a dilute aqueous solution of either sodium hydroxide or potassium hydroxide.

$$R-X + OH^{\ominus} \xrightarrow[\text{heat}]{\text{H}_2\text{O}} R-OH + X^{\ominus} \qquad \text{(where X = Cl, Br or I)}$$

Halogenoalkane   Alcohol

Hydrolysis of esters

Esters can be reduced by using sodium in ethanol and by treatment with $LiAlH_4$ in ethoxyethane. They are only slowly reduced by $NaBH_4$ in water. Reduction of esters, however, leads to a mixture of alcohol products.

The hydrolysis of an ester in water is extremely slow, but can be speeded up by refluxing with aqueous sodium hydroxide solution. This produces the salt of a carboxylic acid and a primary alcohol.

$$RCOOR' + Na^{\oplus}OH^{\ominus}(aq) \xrightarrow{\text{reflux}} RCOO^{\ominus}Na^{\oplus} + R'OH$$

Ester   Alkali   Salt of   Alcohol
                               carboxylic acid

Because this process is the basis of soap manufacture (soaps are sodium salts of long-chain carboxylic acids – see Ch 27.10) the hydrolysis of esters used to be known as *saponification*.

Figure 18.3.1 summarises all the different preparative routes for alcohols.

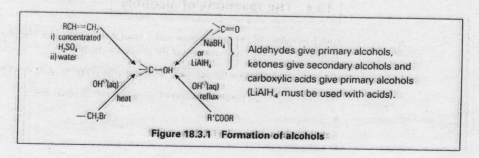

Figure 18.3.1   Formation of alcohols

Aldehydes give primary alcohols, ketones give secondary alcohols and carboxylic acids give primary alcohols ($LiAlH_4$ must be used with acids).

SELF-ASSESSMENT QUESTIONS 18.3

1 The carbohydrate found in cane sugar is sucrose. The molecular formula of sucrose is $C_{12}H_{22}O_{11}$.

   a Write an equation to show how sucrose might ferment to form ethanol and carbon dioxide.

   b Write an equation for the combustion of sucrose in a plentiful supply of air.

2 Cyclohexene (b.p. 84 °C) is converted to cyclohexanol (b.p. 161 °C) by treating with concentrated sulphuric acid at room temperature and then adding water carefully whilst refluxing.

   a Write an equation for the overall reaction.

   b How would you obtain the cyclohexanol from the reaction mixture?

   c Is cyclohexanol a primary, secondary or tertiary alcohol?

3 What starting materials, reagents and conditions could you use to make the following alcohols?

   a $CH_3CH_2CH(OH)CH_3$ from an alkene

   b $CH_3CH(OH)CH_3$ from a carbonyl compound

   c $CH_3C(CH_3)(OH)CH_3$ from a bromoalkane

   d $CH_3CH(CH_3)CH_2CH(OH)CH_3$ from an alkene

4 Which laboratory reagents and conditions would be needed to carry out the following reactions?

   a $CH_3CH_2CH{=}CH_2 \longrightarrow CH_3CH_2CH(OH)CH_3$

   b $CH_3COOCH_3 \longrightarrow CH_3OH + CH_3COOH$

   c $CH_3CH(CH_3)CHBrCH_3 \longrightarrow CH_3CH(CH_3)CH(OH)CH_3$

   d $CH_3COOH \longrightarrow CH_3CH_2OH$

5 2-Bromobutane reacts to form the corresponding alcohol by refluxing with a suitable reagent.

   a What is the name of the alcohol formed?

   b Write an equation, showing displayed formulae, for the reaction.

   c Is the alcohol primary, secondary or tertiary?

   d What reagent should be used for the reaction?

   e If there are no competing reactions and the entire product is recovered from the reaction mixture, how much of the alcohol would be obtained from 5.0 g of the bromoalkane?

18.4 The reactions of alcohols

Lower members of the homologous series burn very readily. The blue flame when a pudding is 'flambéed' is caused by the burning of the ethanol in the brandy.

$$C_2H_5OH(g) + 3O_2(g) \xrightarrow{\text{ignite}} 2CO_2(g) + 3H_2O(g) \qquad \Delta H_c = -1367 \text{ kJ mol}^{-1}$$

The three important types of reaction which occur with alcohols are:

- substitution;
- oxidation;
- elimination.

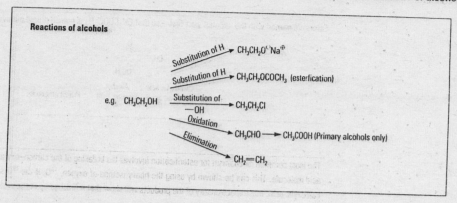

Reactions of alcohols

The general formula of an alcohol suggests that both substitution of the hydrogen in the —OH group and of the whole group is possible.

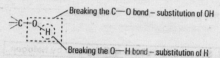

Substitution reactions can therefore be further subdivided into two types:

- reactions that involve breaking the O—H bond;
- reactions that involve breaking the C—O bond.

Substitution reactions

Substitution of the hydrogen in the —OH group

Alcohols are very weak acids (generally weaker than water). As such they will react with alkali metals like sodium, but do not show any other reactions characteristic of acids. Sodium reacts violently with water to give sodium hydroxide and hydrogen. It reacts less vigorously with alcohols to give sodium alkoxides and hydrogen.

$$2Na(s) + 2H_2O(l) \xrightarrow{\text{room temperature}} 2Na^{\oplus}OH^{\ominus}(aq) + H_2(g)$$

$$2Na(s) + 2\underset{\text{Alcohol}}{ROH(l)} \xrightarrow{\text{room temperature}} 2\underset{\substack{\text{Sodium alkoxide} \\ \text{(e.g. sodium ethoxide, } C_2H_5O^{\ominus}Na^{\oplus})}}{Na^{\oplus}OR^{\ominus}(s)} + H_2(g)$$

Substitution of the hydrogen in the hydroxy group to form an ester

Alcohols form esters when reacted with carboxylic acids in the presence of a strong acid (sulphuric or hydrochloric acid); water is the other product – a typical 'condensation' reaction.

$$\underset{\text{Carboxylic acid}}{RCOOH} + \underset{\text{Alcohol}}{R'OH} \underset{\text{reflux}}{\overset{H^{\oplus}(aq)}{\rightleftharpoons}} \underset{\text{Ester}}{RCOOR'} + H_2O$$

The mechanism for esterification – and for its reverse, the hydrolysis of an ester to give acid and alcohol – is fully considered in Ch 27.10.

Esters are named with the 'alcohol' part first – so that $CH_3COOC_3H_7$ is named propyl ethanoate.

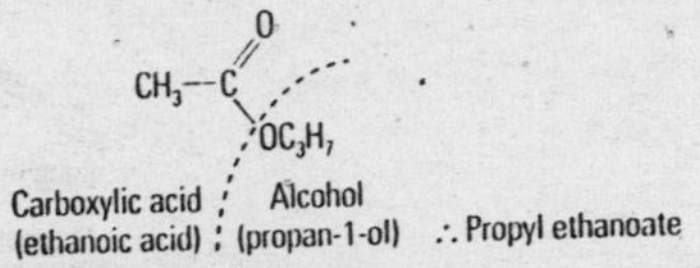

$$CH_3-C\overset{\displaystyle O}{\underset{\displaystyle OC_3H_7}{}}$$

Carboxylic acid   Alcohol
(ethanoic acid)   (propan-1-ol)   ∴ Propyl ethanoate

The most common mechanism for esterification involves the breaking of the carbon–oxygen single bond in the carboxylic acid molecule. This can be shown by using the heavy isotope of oxygen, ^{18}O. If the ^{18}O starts in the hydroxy part of the carboxylic acid, mass spectrometry of the products will show that it finishes up in the water.

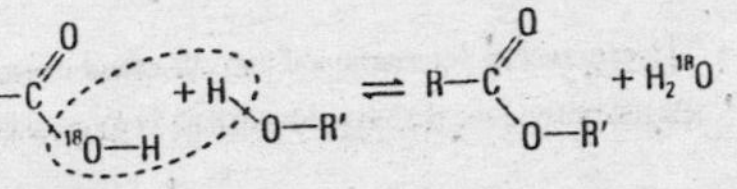

$$R-C\overset{\displaystyle O}{\underset{\displaystyle {}^{18}O-H}{}} + H\cdots O-R' \rightleftharpoons R-C\overset{\displaystyle O}{\underset{\displaystyle O-R'}{}} + H_2{}^{18}O$$

Substitution of the —OH group by halogen

The initial part of this reaction involves protonation of the alcohol. In the presence of a strong acid, the oxygen atom of the hydroxy group will accept a proton (it acts as a base). The next step is a nucleophilic attack by a negatively charged halide ion. The attack occurs at the C—O bond and results in the loss of water. For example, refluxing ethanol with sodium bromide and concentrated sulphuric acid can make bromoethane.

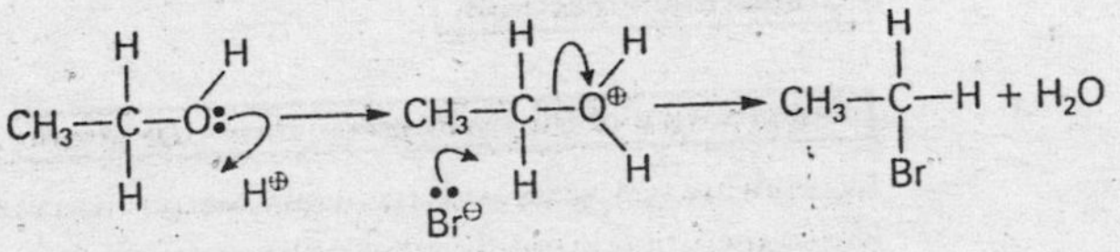

$$CH_3-\overset{\displaystyle H}{\underset{\displaystyle H}{C}}-\ddot{O}:\ \ \ \ H^{\oplus} \longrightarrow CH_3-\overset{\displaystyle H}{\underset{\displaystyle H}{C}}-\overset{\oplus}{O}\overset{H}{\underset{H}{}}\ \ \ \ Br^{\ominus} \longrightarrow CH_3-\overset{\displaystyle H}{\underset{\displaystyle Br}{C}}-H + H_2O$$

Overall:

$$CH_3CH_2OH + HBr \xrightarrow[\text{reflux}]{\text{NaBr/conc. } H_2SO_4} CH_3CH_2Br + H_2O$$

Iodoalkanes cannot be made like this, because concentrated sulphuric acid oxidises hydrogen iodide (see Ch 8.3). Red phosphorus mixed with iodine is used instead (to generate PI_3). Red phosphorus and bromine (to generate PBr_3) can be used to make bromoalkanes.

Sulphur dichloride oxide (SCl_2O, 'thionyl chloride') can be used to make chloroalkanes (see Ch 17.3).

Unlike PI_3 and PBr_3, PCl_3 (phosphorus trichloride) is a reasonably stable liquid if kept out of moist air. It can be used directly to react with alcohols to make chloroalkanes. However, the solid phosphorus pentachloride is more often used. The reaction occurs quite violently at room temperature, and a large amount of hydrogen chloride gas is rapidly given off.

$$\underset{\text{Alcohol}}{ROH} + PCl_5(s) \xrightarrow[\text{temperature}]{\text{room}} \underset{\text{Chloroalkane}}{RCl} + HCl(g) + \underset{\substack{\text{Phosphorus} \\ \text{trichloride oxide}}}{PCl_3O(l)}$$

Phosphorus trichloride oxide has a b.p. of 107 °C, so it is sometimes difficult to separate it from the chloroalkane.

The fumes of *misty* hydrogen chloride form a dense white smoke (of ammonium chloride) when concentrated ammonia solution is anywhere near. This reaction can be used to show that hydrogen chloride is a product of any of these chlorination reactions.

Because of the problem of separating the products of chlorination after the use of PCl_5, the industrial methods of chlorination often use sulphur dichloride oxide, SCl_2O. This is a liquid, b.p. 79 °C; but, other than the chloroalkane, the products of the reaction between SCl_2O and an alcohol are gases and therefore can easily be separated.

$$ROH + SCl_2O(l) \xrightarrow[\text{temperature}]{\text{room}} RCl(l) + SO_2(g) + HCl(g)$$

Chlorinating agents
(HCl), PCl_5, PCl_3, SCl_2O
for the general reaction:

Brominating agents
(HBr), Red P/Br_2, $NaBr/H_2SO_4$

Iodinating agents
(HI), Red P/I_2

Reagents for producing halogenoalkanes from alcohols for the general reaction

$$ROH + \text{'HX'} \longrightarrow RX + H_2O$$

Oxidation of alcohols

Oxidation of an alcohol produces an aldehyde, ketone or carboxylic acid depending on the alcohol used and the conditions of the reaction.

Aldehyde   Ketone   Carboxylic acid

In industry, several other oxidising agents are used, but acidified dichromate(VI) is preferred for laboratory use. Note that it should *not* be acidified with hydrochloric acid. This could react with dichromate(VI) to produce chlorine.

Heating the alcohol with aqueous potassium dichromate(VI) acidified with dilute sulphuric acid is the usual method of oxidation. As the reaction proceeds, the orange dichromate(VI) solution turns to the green of hydrated chromium(III) ions, unless there is an excess of the dichromate(VI) ion present.

A **primary alcohol** can be oxidised in two stages (see *Figure 18.4.1*). In the first stage, an **aldehyde** is formed. If the aldehyde is the required product, it can be distilled out of the mixture as it is formed. This is because the aldehyde usually has the lowest boiling point of the substances in the mixture. If excess of the oxidising agent is used and heating is continued, the aldehyde is oxidised to a **carboxylic acid**.

$$RCH_2OH + [O] \xrightarrow[\text{heat/distil}]{K_2Cr_2O_7(aq)/H^{\oplus}(aq)} RCHO + H_2O \xrightarrow[\text{[O]/reflux}]{K_2Cr_2O_7(aq)/H^{\oplus}(aq)} RCOOH$$

Primary alcohol   Aldehyde   Carboxylic acid

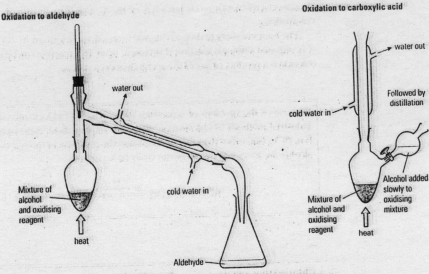

Figure 18.4.1   Oxidation of a primary alcohol

A **secondary alcohol** is oxidised in one stage to a **ketone**.

$$\underset{\text{Secondary alcohol}}{\underset{R'}{\overset{R}{\diagdown}}CHOH} + [O] \xrightarrow[\text{heat}]{K_2Cr_2O_7(aq)/H^{\oplus}(aq)} \underset{\text{Ketone}}{\underset{R'}{\overset{R}{\diagdown}}C=O}$$

The Lucas test is another way of distinguishing between primary, secondary and tertiary alcohols. It depends on substitution of chlorine for the hydroxy group. The alcohol is shaken hard in a stoppered flask with concentrated hydrochloric acid and anhydrous zinc chloride. The flask is then allowed to stand. For tertiary alcohols, substitution is rapid and the mixture rapidly becomes turbid or cloudy (or a separate layer develops) because the water-insoluble chloroalkane is formed. For secondary alcohols, turbidity takes several minutes to develop. The reaction with primary alcohols is very slow indeed.

Tertiary alcohols can only be oxidised using very vigorous conditions. Reactions like this are not very useful since the tertiary alcohol breaks up, usually into a ketone and a carboxylic acid.

Testing for primary, secondary or tertiary alcohols

Oxidation can be used to decide whether a given alcohol is primary, secondary or tertiary. Oxidation with acidified potassium dichromate(VI) solution results in a colour change from orange to green due to the formation of $Cr^{3\oplus}$ ions, as long as there is not an excess of dichromate present. A colour change will therefore only occur with primary and secondary alcohols.

To distinguish between a primary or secondary alcohol, the oxidation products can be tested to see whether they are aldehydes or ketones, using Tollens reagent or Fehling's reagent (see Ch 19.6).

Elimination reactions

If the elements of water are removed from an alcohol, an alkene is formed. (This is essentially the reverse of the manufacture of ethanol from ethene; see Ch 18.3). Ethene can be formed by passing ethanol vapour over a hot catalyst such as broken porcelain or aluminium oxide. The ethene gas is collected over water (*Figure 18.4.2*).

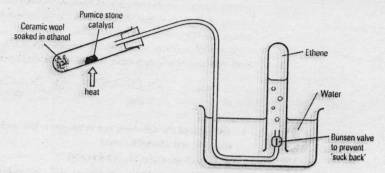

Figure 18.4.2   Elimination from an alcohol

Heating the ethanol with excess concentrated sulphuric acid at 180 °C can also carry out the 'dehydration'. In this case the ethanol is protonated by the acid. Water is then lost to form a **carbocation** (in which a positive charge is very briefly located on a carbon atom). Any **nucleophile** could now attack this carbon atom (*Figure 18.4.3*).

However, if no nucleophile is present, two electrons from a neighbouring C—H bond form another C—C bond (making a double bond) and the hydrogen is lost. An alkene is formed.

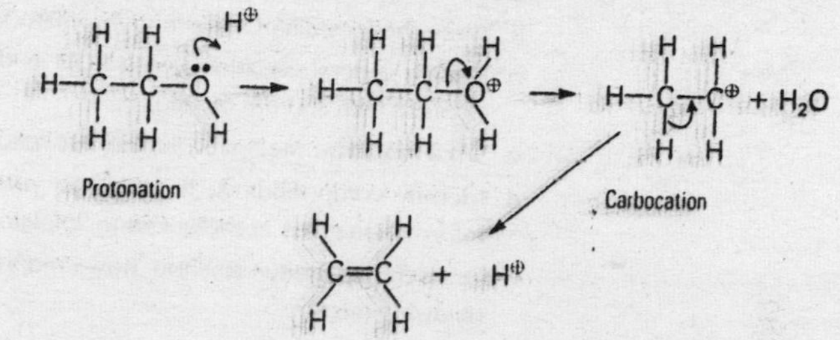

Overall:

$$RCH_2CH_2OH \xrightarrow[180°C]{H_2SO_4} RCH{=}CH_2 + H_2O$$

Figure 18.4.3   'Dehydration' of alcohols

The ethene made from ethanol by heating with excess concentrated sulphuric acid has a sickly-sweet smell. As in most organic reactions, a by-product is formed. If a molecule of water is eliminated from ethanol, ethene is formed. If water is eliminated between two molecules of ethanol, ethoxyethane ('ether' – one of the first-ever anaesthetics, used from the 1840s) is formed.

The overall reaction is:

$$2C_2H_5OH \longrightarrow (C_2H_5)_2O + H_2O$$

It is the small proportion of ethoxyethane present which gives the ethene its smell.

Tertiary alcohols are the easiest to dehydrate, primary alcohols the hardest. The reason for this is the stability of the carbocation intermediate; the electron-donating effect of three methyl groups of the tertiary alcohol stabilises the carbocation.

If alternative alkene structures are possible, the relative stabilities of carbocations means that the more substituted alkene is more likely to be formed when water is eliminated.

$$e.g. \quad CH_3CH-CH_2CH_3 \xrightarrow[180°C]{H_2SO_4} \underset{\text{Major product}}{CH_3-CH=CHCH_3} + \underset{\text{Minor product}}{CH_2=CH-CH_2CH_3}$$
$$\qquad\qquad\qquad\quad |$$
$$\qquad\qquad\qquad OH$$

SELF-ASSESSMENT QUESTIONS 18.4

1 For each of the following reactions, name the starting materials and products, and state the reagents and conditions used.

a Dehydration of $(CH_3)_2CHCH_2OH$

b Formation of $CH_3CH_2CH(Cl)CH_3$ from an alcohol

c The esterification of CH_3CH_2COOH with C_2H_5OH

d The oxidation of $CH_3CH(OH)CH_3$

e The oxidation of $CH_3CH_2CH_2CH_2OH$

2 The following tests were carried out on three compounds, A, B and C, which were all known to contain the —OH group because all three compounds react with sodium metal to form hydrogen gas. When the three alcohols were treated with acidified sodium dichromate(VI), they reacted in different ways. A reacted, and the mixture turned green after a period of time. The purified organic product showed no acid test with pH paper. After refluxing B for some time, there was still no alteration to the colour of the dichromate solution. When C was refluxed with acidified potassium dichromate(VI), the final purified product reacted with sodium carbonate solution to form carbon dioxide.

a Write the formula of sodium dichromate(VI).

b Write a general equation, using ROH as the formula of an alcohol, for the reaction of the alcohols with sodium metal.

c What would be the final colour of the reaction mixture with sample B?

d Identify compounds A, B and C as primary, secondary or tertiary alcohols, using the information given in the question. Explain your reasoning.

e Draw the apparatus to show how compound C could be added carefully to the refluxing oxidising mixture.

3 What are the major organic products formed when pentan-2-ol is reacted with the following reagents? Give the systematic names and formulae, and state the conditions necessary for the reaction to proceed.

a Sodium metal

b Potassium dichromate(VI) solution in dilute sulphuric acid

c Concentrated sulphuric acid

d Red phosphorus and iodine

Links between functional groups

How do you deal with trying to remember all these organic reactions? The functional group provides a very important 'peg' for recalling the reactions (using ROCASEP – see Ch 14.8). Try to remember each of the reactions as reactant–product pairs. Another memory aid is to look at the reagents used and identify what possible reactions involve that particular reagent.

Now that you have worked through four chapters of basic organic chemistry, it is time to try to link some of these functional groups.

Reagent	Alkanes	Alkenes	Halogenalkanes	Alcohols
Burn	Clear flame	Smoky flame		Clear flame
+ Cl_2, Br_2	UV light needed	Decolourises	–	
+ PCl_5, PBr_3	–	–		HCl fumes
HBr	–	Addition reaction	–	–
Na	–	–		H_2 gas formed
Conc. H_2SO_4	–	–		Reacts
$KMnO_4$/ H_2SO_4(aq)	–	–	–	Purple to colourless
$K_2Cr_2O_7$/ H_2SO_4(aq)	–	–	Reaction	Orange to dark green
Dilute OH⁻/reflux	–	–	Reaction	–
Concentrated OH⁻/ ethanol/heat	–	–	Reaction	–

The arrows show which other class of compound is formed in the reaction
By recognising the reagent, you can often identify the type of reactant and product in any reaction.

Worked example

Nerol is one of a number of components found in flowering plants. When nerol reacts with bromine water, the bromine solution is decolourised; and when nerol reacts with sodium metal a colourless gas, A, is formed. If the reaction is carried out quantitatively, one mole of nerol reacts exactly with two moles of bromine, Br_2. If nerol is refluxed with an acidified solution of potassium dichromate(VI), one of the products that can be isolated, B, reacts with sodium carbonate solution to produce a different colourless gas, C.

a What functional group is shown to be present because of the reaction with bromine?
b What other functional group(s) are present in nerol?
c Identify the two gases, A and C, produced in these reactions.
d How many of the groups from a) are present in one molecule of nerol?
e What functional group must be present in the product B?

The first step is to put all the information into a flow diagram.

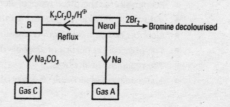

From this, we can recognise a number of familiar points.

- 'Decolourise bromine' signifies an alkene double bond.
- Since it takes two moles of Br_2 for each mole of nerol, the nerol molecule must contain two $C=C$ double bonds.
- A gas produced with sodium is usually hydrogen: gas A is hydrogen.
- Sodium reacts with alcohols to form hydrogen.
- Potassium dichromate solution is used as an oxidising agent.
- Alcohols are readily oxidised.
- Sodium carbonate solution reacts with any acid to form carbon dioxide: gas C is CO_2.
- Compound B is a carboxylic acid produced by oxidising the primary alcohol group in nerol.

Now write out the answers to parts a to e of the question.

Reactions of alcohols – summary

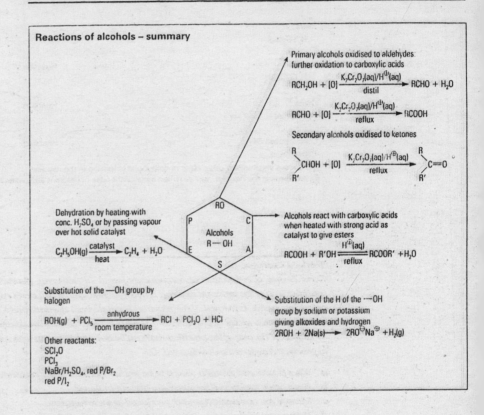

Primary alcohols oxidised to aldehydes;
further oxidation to carboxylic acids

$$RCH_2OH + [O] \xrightarrow[\text{distil}]{K_2Cr_2O_7(aq)/H^{\oplus}(aq)} RCHO + H_2O$$

$$RCHO + [O] \xrightarrow[\text{reflux}]{K_2Cr_2O_7(aq)/H^{\oplus}(aq)} RCOOH$$

Secondary alcohols oxidised to ketones

$$\underset{R'}{\overset{R}{\diagdown}}CHOH + [O] \xrightarrow[\text{reflux}]{K_2Cr_2O_7(aq)/H^{\oplus}(aq)} \underset{R'}{\overset{R}{\diagdown}}C=O$$

Alcohols react with carboxylic acids
when heated with strong acid as
catalyst to give esters

$$RCOOH + R'OH \underset{\text{reflux}}{\overset{H^{\oplus}(aq)}{\rightleftharpoons}} RCOOR' + H_2O$$

Dehydration by heating with
conc. H_2SO_4 or by passing vapour
over hot solid catalyst

$$C_2H_5OH(g) \xrightarrow[\text{heat}]{\text{catalyst}} C_2H_4 + H_2O$$

Alcohols R — OH

Substitution of the —OH group by
halogen

$$ROH(g) + PCl_5 \xrightarrow[\text{room temperature}]{\text{anhydrous}} RCl + PCl_3O + HCl$$

Other reactants:
SCl_2O
PCl_3
$NaBr/H_2SO_4$, red P/Br_2
red P/I_2

Substitution of the H of the —OH
group by sodium or potassium
giving alkoxides and hydrogen

$$2ROH + 2Na(s) \longrightarrow 2RO^{\ominus}Na^{\oplus} + H_2(g)$$

**SUMMARY
18**

In this chapter the following new ideas have been covered. *Have you understood and learned them? Could you explain them clearly to a friend?* The page references will help.

- Recognition of the alcohol functional group and naming the alcohols.   p458
- Recognition of the three types of alcohols.   p460
- Industrial methods for producing ethanol.   p468
- Alcohols can be made from compounds containing other functional groups.   p470
- The physical properties of alcohols depend largely on hydrogen bonding.   p463
- The reactions of alcohols.   p473
- The oxidation of alcohols as a way of classifying them.   p475

1 Consider the reaction:

$$CH_3CH_2CH_2CH_2OH \longrightarrow CH_3CH_2CH_2CHO$$

a How would you carry out this reaction?

b Name the starting material and the product.

c Calculate the amount of product obtained if the reaction produced a 65% yield of the pure product starting with 10.0 g of $CH_3CH_2CH_2CH_2OH$.

d Draw the apparatus you would use to carry out the reaction.

e The initial impure product is a liquid. How would you attempt to purify the mixture?

2

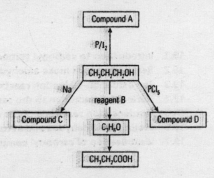

a Give the formulae of compounds A, C and D.

b What is reagent B?

c Give the systematic names for all the organic compounds in this reaction scheme.

d What type of reaction is the conversion of the starting alcohol to compound A?

3 Menthol is an alicyclic alcohol used in skin lotions because of its cooling effect on the skin. The skeletal structure is shown below.

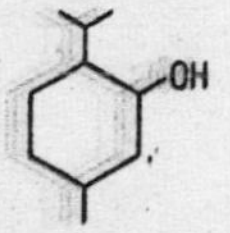

a What is the molecular formula of menthol?

b What gas would be produced if menthol was reacted with sodium metal?

c Is menthol a primary, secondary or tertiary alcohol?

d Draw the skeletal structure of the product formed when menthol is refluxed with acidified potassium dichromate(VI) solution.

Do you remember ...?

- The meaning of the terms **functional groups**, **homologous series** and **frameworks**. (Ch 14.2)
- How to recognise **structural isomers**. (Ch 14.4)
- How organic compounds are **named**. (Ch 14.3)
- The **oxidation** products of alcohols: (Ch 18.4)
- The different **types** of organic reactions. (Ch 14.6)
- How to show organic **reaction mechanisms** in equations. (Ch 14.7)
- How **electronegative** elements lead to polar bonds. (Ch 2.2)

19

the carbonyl compounds: aldehydes and ketones

19.1 Introduction to carbonyl compounds

In organic chemistry the carbonyl group is $>C=O$. Many functional groups appear to contain a carbonyl group; for example:

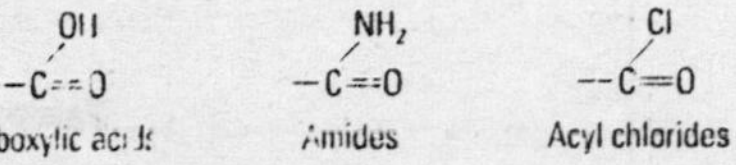

$$--\overset{\displaystyle OH}{\underset{\displaystyle }{C}}=O \qquad -\overset{\displaystyle NH_2}{\underset{\displaystyle }{C}}=O \qquad -\overset{\displaystyle Cl}{\underset{\displaystyle }{C}}=O$$

Carboxylic acids   Amides   Acyl chlorides

The presence of an electronegative atom such as O, N or Cl next to the $>C=O$ group prevents it acting as a genuine carbonyl group. As a result, the only two classes of compound in which the carbonyl group shows its true reactions are:

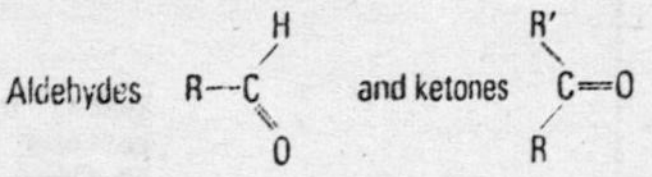

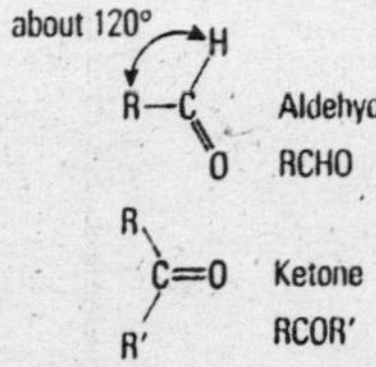

$$\text{Aldehydes} \quad R-\overset{\displaystyle H}{\underset{\displaystyle O}{C}} \qquad \text{and ketones} \quad \overset{\displaystyle R'}{\underset{\displaystyle R}{C}}=O$$

The formulae are often written as R—CHO for aldehydes and R—CO—R' for ketones (*Figure 19.1.1*).

Aldehydes and ketones are vitally important in nature. You can read this because of the aldehyde, retinal, which is found in the retina of the eye (Ch 14.4). In industry aldehydes and ketones are important as starting materials for making certain plastics, and – especially for the ketones – as solvents.

The names of aldehydes end in *-al*, and names of ketones end in *-one*. Aldehydes and ketones have a carbon atom in the functional group which is part of the carbon chain, and are named accordingly.

CH_3CHO is ethan*al*; $CH_3CH_2CH_2CH_2CHO$ is pentan*al*.

If the compound contains two functional groups and the aldehyde is not the main one, the prefix 'oxo-' is used. So:

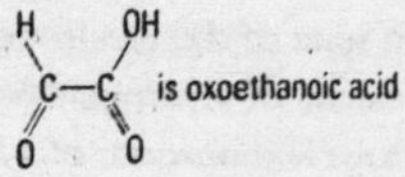

$$\overset{\displaystyle H \quad OH}{\underset{\displaystyle O \quad O}{C-C}} \quad \text{is oxoethanoic acid}$$

A ketone must contain at least three carbon atoms.

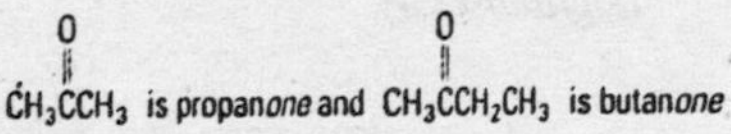

$$\overset{\displaystyle O}{\underset{\displaystyle }{CH_3\overset{||}{C}CH_3}} \text{ is propan}one \text{ and } CH_3\overset{O}{\overset{||}{C}}CH_2CH_3 \text{ is butan}one$$

Neither of these structures has an alternative, but pentanone could be either:

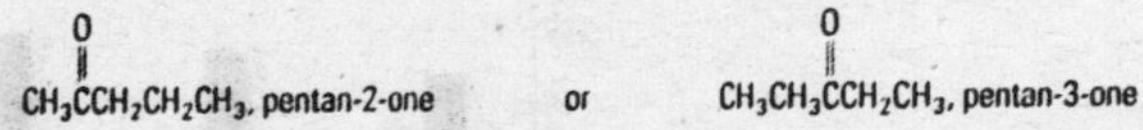

$$CH_3\overset{O}{\overset{||}{C}}CH_2CH_2CH_3, \text{ pentan-2-one} \qquad or \qquad CH_3CH_2\overset{O}{\overset{||}{C}}CH_2CH_3, \text{ pentan-3-one}$$

Again, 'oxo-' is used if the ketone is not the principal group, so that:

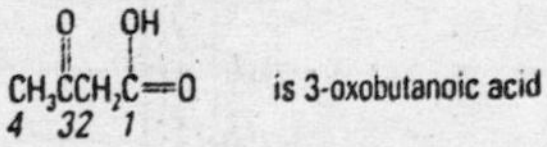

$$\underset{\displaystyle 4 \quad 32 \quad 1}{CH_3\overset{O}{\overset{||}{C}}CH_2\overset{OH}{C}=O} \quad \text{is 3-oxobutanoic acid}$$

N.B. All aldehydes with three or more carbon atoms are isomeric with ketones (*Table 19.1.1*). For example CH_3CH_2CHO, propanal, is isomeric with CH_3COCH_3, propanone.

---

about 120°

$$R-\overset{\displaystyle H}{\underset{\displaystyle O}{C}} \quad \begin{array}{l}\text{Aldehyde}\\ \text{RCHO}\end{array}$$

$$\overset{\displaystyle R}{\underset{\displaystyle R'}{C}}=O \quad \begin{array}{l}\text{Ketone}\\ \text{RCOR'}\end{array}$$

Figure 19.1.1 The functional groups in aldehydes and ketones

The carbonyl group is part of the carbon skeleton and included in it when numbering the carbon atoms.

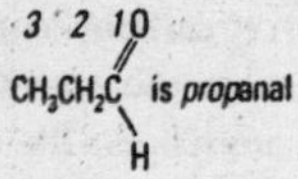

$$\underset{\displaystyle H}{CH_3CH_2\overset{3\ \ 2\ \ 10}{C}} \quad \text{is } propanal$$

Carbonyl compounds in organic chemistry should not be confused with inorganic carbonyls, which involve the bonding of molecules of carbon monoxide gas, CO, to atoms of transition metals (see Ch 25.4).

	Aldehyde	Isomeric ketone
C_1	HCHO	–
C_2	CH_3CHO	–
C_3	C_2H_5CHO	CH_3COCH_3
C_4	C_3H_7CHO	$C_2H_5COCH_3$

Table 19.1.1   Aldehydes and ketones

Reactions of carbonyl compounds

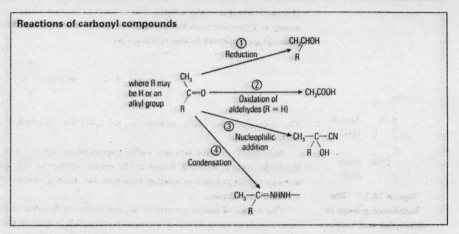

The polar double bond in the carbonyl group makes these compounds quite reactive. Aldehydes and ketones are reduced to the corresponding alcohol by strong reducing agents. Oxidation of the aldehydes is readily achieved with even quite mild oxidising agents. The polarity of the C=O double bond results in nucleophilic addition reactions. Condensation reactions of carbonyl compounds are important, but remember that these reactions proceed through an addition–elimination mechanism (see Ch 14.6).

In spite of the electronegative oxygen atom that must be present in every carbonyl compound, the extent of hydrogen bonding in these molecules is less evident than in the corresponding alcohols. The absence of a hydrogen atom attached to an electronegative atom limits the extent of possible hydrogen bonds. The boiling points of aldehydes and ketones are much higher than those of the corresponding alkanes, although less than those of the corresponding alcohols (*Figure 19.1.2*).

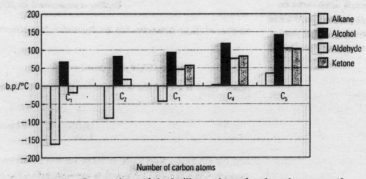

Figure 19.1.2 Comparison of the boiling points of carbonyl compounds

SELF-ASSESSMENT
QUESTIONS
19.1

1 Name the following compounds.
   a HCHO
   b $(CH_3)_2CHCOCH_3$
   c CH_3COCH_2OH
   d $CH_3CHClCH_2CHO$

2 Write full displayed formulae for:
   a butanal
   b hexan-2-one
   c 2-oxopropanoic acid
   d 2-hydroxybutanal

3 Identify the following compounds as aldehydes, ketones or alcohols.
   a $CH_3CH_3CH(OH)CH_2CH_3$
   b CH_3CHO
   c CH_3COCH_3
   d $CH_3CH_2CH(CH_3)COCH_3$

19.2 Reactions which make aldehydes and ketones

There are a number of reactions which produce carbonyl compounds. The most important ones you need to know at this point are those involving oxidation of alcohols (Ch 18.4). The oxidising agent is potassium dichromate(VI) acidified with dilute sulphuric acid (*Figure 19.2.1*). If a *primary* alcohol is heated with this, an *aldehyde* is formed. Unless the aldehyde is continuously distilled out of the reaction mixture as it is made, the aldehyde may be oxidised further to the corresponding carboxylic acid.

$$RCH_2OH + [O] \xrightarrow[\text{distil}]{K_2Cr_2O_7/H^+(aq)} RCHO + H_2O$$

Primary alcohol   Aldehyde

$$RCHO + [O] \xrightarrow[\text{reflux}]{\text{further oxidation}} RCOOH$$

Carboxylic acid

If a *secondary* alcohol is refluxed with the same oxidising agent, a ketone is formed. This cannot be oxidised any further under these conditions. More powerful oxidation breaks up the carbon chain.

$$R-\overset{R'}{\underset{}{C}}HOH + [O] \xrightarrow[\text{reflux}]{K_2Cr_2O_7/H^+(aq)} R-\overset{R'}{\underset{}{C}}=O + H_2O$$

Secondary alcohol   Ketone

In all these reactions the oxidising agent is potassium dichromate(VI) $K_2Cr_2O_7$(aq) in dilute sulphuric acid, H_2SO_4(aq) or just $H^{\oplus}$(aq).

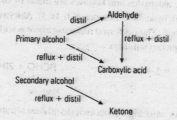

Figure 19.2.1 **Oxidation of alcohols**

SELF-ASSESSMENT QUESTIONS 19.2

1 Which carbonyl compound would be obtained by oxidation of each of the following alcohols? In each case, name the starting material and the product.
   a $CH_3CH(OH)CH_2CH_3$
   b $(CH_3)_3CCH_2OH$
   c CH_3OH
   d

   ⬡—OH

2 Classify the following alcohols as primary, secondary or tertiary. Which of them would form an aldehyde on oxidation with acidified potassium dichromate(VI) solution?
   a $CH_3CH(OH)CH_2CH_3$
   b $CH_3CH_2CH_2OH$
   c $(CH_3)_3COH$
   d CH_3CH_2OH

3 Draw and explain the apparatus you would use to prepare propanone from propan-2-ol. How would you then isolate the product from the reaction mixture? Draw a diagram of the apparatus.

19.3 The oxidation–reduction reactions of carbonyl compounds

Oxidation

Remember:
When you are writing equations for redox reactions in organic chemistry, you may write the oxidising agent as [O] or the reducing agent as [H] (Ch 14.6). Don't forget to balance the equation even when you use these symbols.

Aldehydes can easily be oxidised further to carboxylic acids. As we have just mentioned, an aldehyde has to be distilled out of the reaction vessel when a primary alcohol is being oxidised; otherwise the aldehyde will be oxidised further to a carboxylic acid.

$$RCHO + [O] \xrightarrow[\text{reflux}]{K_2Cr_2O_7(aq)/H^{\oplus}(aq)} RCOOH$$

Aldehyde   Carboxylic acid

Ketones, as we have said, cannot be oxidised any further without the molecule being broken up.

Aldehydes are so easily oxidised that they can be classified as quite powerful reducing agents. For example, they will, under suitable conditions, reduce copper in the +2 oxidation state, copper(II), to copper(I), and reduce silver(I) to metallic silver. This reducing property is used in tests to distinguish aldehydes from ketones (Ch 19.6).

Reduction

Since aldehydes and ketones are made by oxidation of alcohols, alcohols can be made by reduction of aldehydes and ketones (Ch 18.3). Aldehydes give primary alcohols, and ketones give secondary alcohols. The preferred reducing agent is sodium tetrahydridoborate(III), $NaBH_4$, which can be used in aqueous solution.

e.g. $$RCHO + 2[H] \xrightarrow[\substack{\text{room} \\ \text{temperature}}]{NaBH_4/\text{water}} RCH_2OH$$

Aldehyde   Primary alcohol

Other reducing agents include lithium tetrahydridoaluminate(III), $LiAlH_4$ (which can only be used in ethoxyethane – water must be totally excluded), sodium in ethanol, and hydrogen gas under pressure with a platinum catalyst.

1 Write balanced equations (using [H]) for the reaction of sodium tetrahydridoborate(III) in water with:

  a CH_3CHO

  b $CH_3CH_2COCH_3$

  c CH_3COCH_2CHO

2 How would you expect each of the following compounds, A, B and C, to react:

  a when refluxed with acidified potassium dichromate(VI) solution

  b with sodium tetrahydridoborate(III) in water.

  A – CH_3CH_2CHO

  B – $(CH_3)_2CHOH$

  C – CH_3COCH_3

3 2-Oxopropanal, CH_3COCHO, is thought to be one of the chemicals responsible for the smell in burnt sugar.

  a This compound contains two carbonyl groups. Explain the difference between the two groups.

  b What reagent(s) would you use to reduce this compound?

  c Draw a full displayed formula of the compound formed in the reduction reaction.

  d The product in c can be separated into two optically active isomers. What characteristic in the molecule is responsible for this isomerism?

  e If the oxopropanal is refluxed with potassium dichromate(VI) solution, a new compound is formed. Write an equation for the reaction, using [O] where appropriate.

19.4 Nucleophilic addition to the carbonyl bond

We have already seen (Ch 16.6) that the characteristic reactions of the alkene series, with the functional group $>C=C<$, involve electrophilic addition. This is because the high concentration of electron density in the double bond is attractive to electrophiles, as one would expect.

The situation is very different for the carbonyl group, $>C=O$ (*Figure 19.4.1*), even though it is unsaturated.

This is mostly because the oxygen is much more electronegative than carbon; it pulls electron density partly away from the carbon, so that the bond is polar (see Ch 2.5)

$$>C=O \longrightarrow >C^{\delta\oplus}=O^{\delta\ominus}$$

Electrophilic or nucleophilic addition?

Electrophilic addition across the carbonyl group would involve an intermediate with a positive charge on the carbon atom.

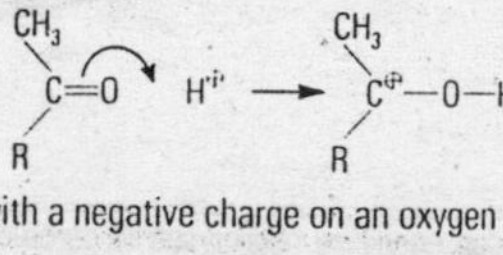

Nucleophilic addition gives rise to a species with a negative charge on an oxygen atom – a more likely occurrence.

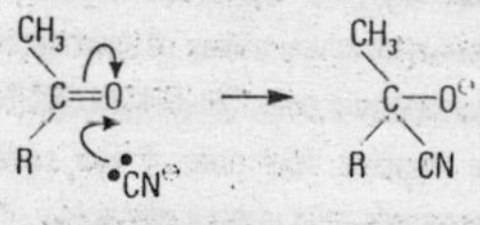

Figure 19.4.1

This partial withdrawal of electron density from the carbon exposes it to attack by negatively charged nucleophiles (*Figure 19.4.2*). Nucleophiles are often negatively charged, e.g. $CN^{\ominus}$, but may be neutral molecules with a lone pair of electrons, e.g. $:NH_3$ etc. (see Ch 14.6).

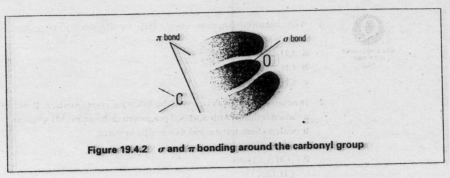

Figure 19.4.2 σ and π bonding around the carbonyl group

The process is helped by the readiness of the oxygen atom to accept a full negative charge.

Whereas bromine and hydrogen bromide add to alkenes, hydrogen cyanide does not. The exact opposite is true for carbonyl compounds. Hydrogen cyanide, HCN, adds across the carbon–oxygen double bond; bromine and HBr do not. The mechanism for the reaction of hydrogen cyanide to an aldehyde or ketone is **nucleophilic addition**. Hydrogen cyanide itself is not used, as it is an extremely poisonous gas. Instead, potassium or sodium cyanide is used, together with dilute sulphuric or hydrochloric acid.

a 2-hydroxynitrile

Other uncharged nucleophiles of the general form HNu can also add across the >C=O double bond. This always happens so that the hydrogen is attached to the oxygen atom and the $Nu^{\ominus}$ part to the carbon atom.

$$>C^{\delta\oplus}=O^{\delta\ominus} + H^{\delta\oplus}-Nu^{\delta\ominus} \longrightarrow >C-OH \atop | \atop Nu$$

Nitriles can easily be hydrolysed to carboxylic acids by refluxing with dilute hydrochloric acid. The addition of hydrogen cyanide therefore gives a way of synthesising 2-hydroxycarboxylic acids.

2-Hydroxypropanoic acid, $CH_3CH(OH)COOH$, often still known as 'lactic acid', is the product of anaerobic respiration. It is why your muscles feel sore if you over-exercise them without enough training. When isolated from muscle, 2-hydroxypropanoic acid is *optically active*. It has a chiral centre (Ch 14.4):

2-Hydroxypropanoic acid can be synthesised from ethanal in the following way:

NaCN(aq)
room temperature
dilute HCl(aq)
reflux

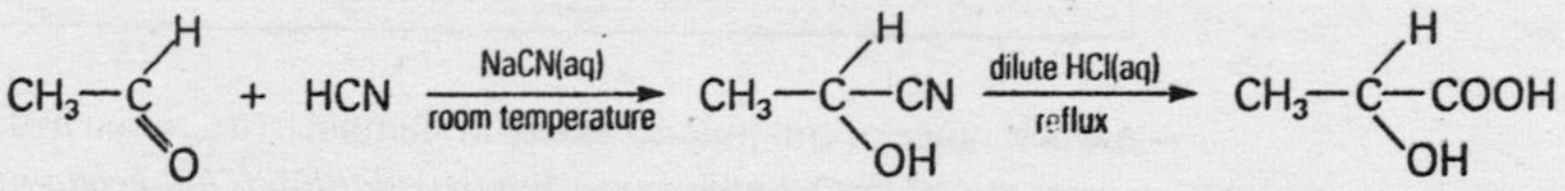

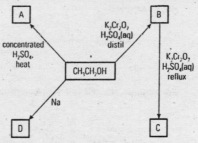

Ethanal Hydrogen cyanide 2-Hydroxypropanoic acid

The acid made in this way is *optically inactive.* It contains a 50:50 mixture of the two optical isomers. Ethanal is a *planar* molecule and the nucleophile – in this case, the cyanide ion – can attack it from either side:

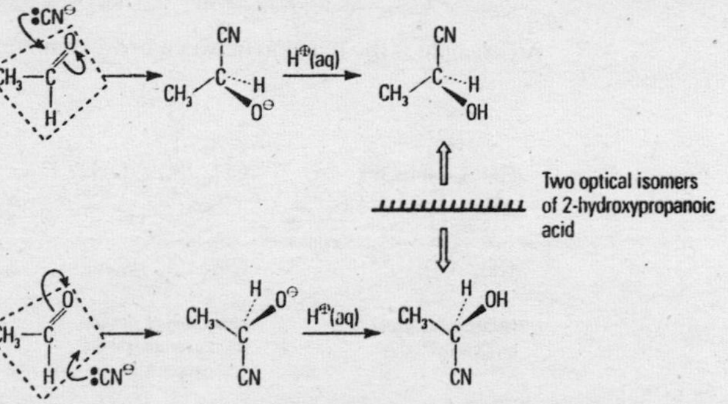

$H^{\oplus}$(aq)
Two optical isomers of 2-hydroxypropanoic acid

In general, attempts to synthesise chiral molecules from simple starting materials result in a 50:50 mixture of the optical isomers. The product is therefore optically inactive. Try 'doing' the reaction using molecular models.

Q

SELF-ASSESSMENT QUESTIONS 19.4

1 Complete the following reaction sequence by identifying compounds A, B, C and D.

A
B
C
D
CH_3CH_2OH
concentrated H_2SO_4, heat
$K_2Cr_2O_7$, H_2SO_4(aq) distil
$K_2Cr_2O_7$, H_2SO_4(aq) reflux
Na

2 Consider the following series of compounds:

A $CH_3CH_2CH_2COCH_3$
B $CH_3CH(OH)CH_2COCH_3$
C CH_3CH_2CHO
D $CH_3COCH_2COCH_3$

a Write the systematic name for each of these compounds.
b Which of these compounds are alcohols?
c Which of these compounds are aldehydes?
d Which of these compounds are ketones?
e Which could be formed by the oxidation of a primary alcohol?
f Which can be reduced to a secondary alcohol?

3 Write a mechanism to show how HCN undergoes nucleophilic addition to $CH_3CH_2COCH_2CH_3$.

19.5 Condensation reactions of the carbonyl bond

A whole range of compounds based on ammonia, NH_3, or on hydrazine, $H_2N{-}NH_2$ (N_2H_4), can add across the $>C{=}O$ double bond; *but the addition is followed immediately by loss of water*. Such an addition–elimination reaction is called a condensation reaction (Ch 14.6). (N.B. Condensation is not one of the four recognised types of reaction: it is a combination of two of those types.)

Overall, the condensation reactions considered here follow this form:

$$>C{=}O + H_2N{-} \longrightarrow\ >C{=}N{-} + H_2O$$

An example is the reaction between hydroxylamine and propanone:

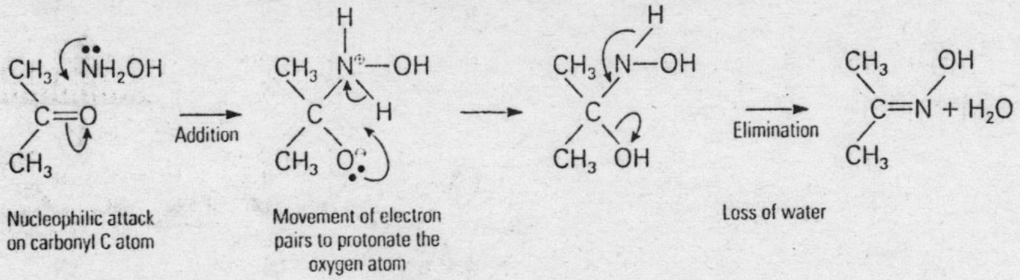

Nucleophilic attack on carbonyl C atom	Movement of electron pairs to protonate the oxygen atom		Loss of water

A very useful reagent is the impressive-sounding 2,4-dinitrophenylhydrazine (2,4-DNPH). If hydrazine is $H_2N{-}NH_2$, phenylhydrazine must be:

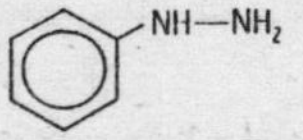

So, 2,4-DNPH is:

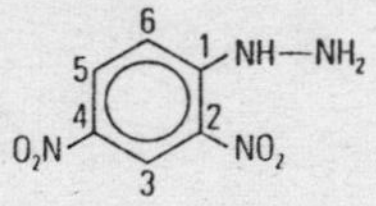

For example, the reaction between propanal and 2,4-DNPH can be written:

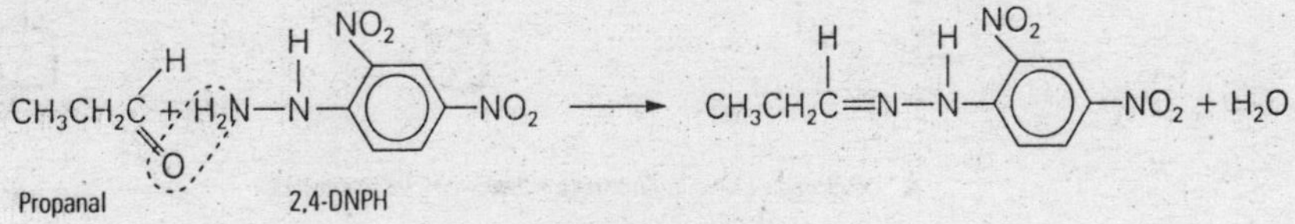

Propanal   2,4-DNPH

The usefulness of this reaction is discussed further in Ch 19.6.

SELF-ASSESSMENT
QUESTIONS
19.5

1   Write an equation for the reaction between 2,4-dinitrophenylhydrazine and $CH_3COCH_2CH_3$.

2   3-Phenylpropenal (cinnamaldehyde), $C_6H_5{-}CH{=}CHCHO$, is an important perfumery additive, with a jasmine-like odour. The *cis*-isomer is the principal isomer used.
    a   What are the two functional groups present in cinnamaldehyde?
    b   Write an equation for the reaction of cinnamaldehyde with 2,4-dinitrophenylhydrazine.
    c   When reacted with bromine, in the dark, the colour of the bromine disappears, because of the addition of bromine across the double bond. Write a mechanism to show how the bromine adds to cinnamaldehyde.

d A very different reaction occurs when the cinnamaldehyde is reacted with HCN. Write a mechanism to show how HCN might add to the cinnamaldehyde molecule.

e Explain what the term *cis-* means by drawing the full displayed formulae of the geometrical isomers of cinnamaldehyde.

3 $C_6H_{12}O$ can exist as a number of isomeric molecules. When a compound with this formula is treated with bromine dissolved in an organic solvent, the bromine colour persists, with no evidence of any reaction. If the compound is refluxed with potassium dichromate(VI) solution in dilute sulphuric acid, a ketone is formed. Indentify the structure of this molecule.

19.6 Testing for aldehydes and ketones

Testing for the presence of the carbonyl group

As shown in Ch 19.5, 2,4-dinitrophenylhydrazine (2,4-DNPH) undergoes a condensation reaction with aldehydes and ketones. This gives a convenient analytical test because the product has a characteristic colour.

Two drops, or 0.1 g, of the suspected carbonyl compound is dissolved in the minimum of methanol. To this solution is added about 3 cm³ of a solution containing 2,4-DNPH (2 g) in sulphuric acid (4 cm³ of the concentrated acid), methanol (30 cm³) and water (10 cm³). (This solution is often known as *Brady's reagent*.)

The actual quantities used are given here as an indication of how carefully analytical reagents should be made up, but you should not try to remember them! It is enough to state that a small quantity of the suspected aldehyde or ketone (in methanol) is treated with a few cm³ of a solution of 2,4-dinitrophenylhydrazine in sulphuric acid.

Using this reagent, almost all aldehydes and ketones react rapidly at room temperature to give an *orange precipitate* (although colours may range from yellow to red, depending on the carbonyl compound used). The formation of the precipitate confirms the presence of the carbonyl group (*Figure 19.6.1*). To distinguish between an aldehyde or ketone needs some more chemistry.

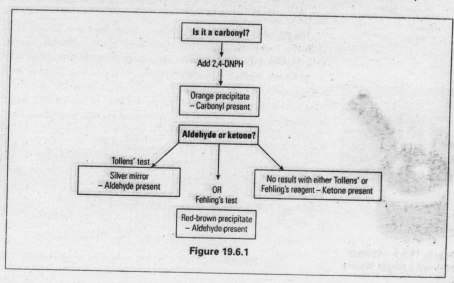

Figure 19.6.1

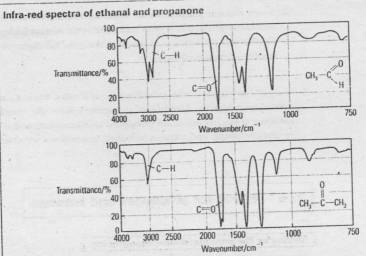

Infra-red spectra of ethanal and propanone

The >C=O group in an organic compound gives a characteristic peak in the infra-red spectrum at around 1680 to 1740 cm^{-1}. The very intense absorption in this region of the spectrum provides strong evidence for the presence of the >C=O group.

Distinguishing between aldehydes and ketones

The big difference between aldehydes and ketones – as we have discussed already in Ch 19.3 – is that aldehydes are very easily oxidised to carboxylic acids. Once you have found, by using 2,4-DNPH, that you have an aldehyde *or* ketone, the following tests only give positive results with aldehydes. Ketones have no reaction.

Tollens' test

In this reaction, silver(I) is reduced by the aldehyde to metallic silver, i.e. to silver(0). The silver is often seen as a thin reflective layer on the inside of the test tube, which is why this test is often called the 'silver mirror' test.

The Ag$^{\oplus}$ ion has to be in alkaline solution for the reaction to occur. Normally, addition of hydroxide ions to Ag$^{\oplus}$(aq) causes a precipitate of silver oxide, Ag$_2$O. The silver ions therefore have to be 'complexed' (see Ch 25.4) to keep them in solution. Ammonia is the complexing agent.

Ammonia solution is added carefully, dropwise with shaking, to about 3 cm^3 silver nitrate solution in a clean test tube, until the precipitate, which forms at first, is *just* redissolved. (**Care!** In most school laboratories, to save money the silver nitrate solution is about 0.1 mol dm^{-3}, whereas the ammonia solution may have a concentration of about 2.0 mol dm^{-3}. The ammonia should be diluted to about the same concentration as the silver nitrate before they are mixed. *Excess ammonia must be avoided* – or the reaction does not work.)

The clear solution now contains the complex ion [Ag(NH$_3$)$_2$]$^{\oplus}$. To this solution, add one or two drops (or 0.1 g) of the suspected aldehyde and shake. If there is no reaction at room temperature, place the test tube in warm (not boiling) water in a beaker. If an aldehyde is present, a shiny silver mirror will form if the test tube is clean (*Figure 19.6.2*).

In a dirty test tube, only a grey or black precipitate is formed.

$$CH_3CH_2CHO + [O] \longrightarrow CH_3CH_2COOH$$

$$2[Ag(NH_3)_2]^{\oplus}(aq) + H_2O(l) \longrightarrow 2Ag(s) + [O] + 2NH_4^{\oplus}(aq) + 2NH_3(aq)$$

Figure 19.6.2 Tollens' test (on a larger scale!)

Cleaning apparatus is always important, but is particularly so after a silver mirror test. Tollens' reagent should always be freshly prepared when required. If allowed to stand, it slowly forms explosive substances, including the detonating material silver azide. The test solution should be washed away thoroughly, and any silver mirror removed from the tube by using a little concentrated nitric acid.

Fehling's and Benedict's tests

These tests are very similar to Tollens' test, except that copper(II) is used instead of silver(I) as the oxidising agent; it is reduced to Cu(I). Again, the reaction occurs in alkaline solution, which would normally precipitate $Cu^{2\oplus}$ as the hydroxide; so, again, a complexing agent has to be used.

In *Benedict's reagent,* the complexing agent is citrate ions, from citric acid; in *Fehling's reagent* it is tartrate ions, from tartaric acid. Fehling's reagent must be freshly prepared before each use. Fehling's solution A is aqueous copper(II) sulphate; Fehling's solution B contains alkali and tartrate ions. Equal volumes of Fehling's A and B should be mixed to give a deep blue solution of copper(II) tartrate complex.

A little of the suspected aldehyde is added to a few cm³ of the Benedict's or Fehling's reagent. The mixture is warmed and, if necessary, *cautiously* boiled. An aldehyde, or a 'reducing sugar' such as glucose, gives a red-brown precipitate of copper(I) oxide, Cu_2O.

No ketone will react. The original copper(II) has oxidised the aldehyde and in turn been reduced to copper(I).

$$RCHO + [O] \longrightarrow RCOOH$$

$$2Cu^{2\oplus}(aq) + OH^{\ominus}(aq) + H_2O(l) \longrightarrow Cu_2O(s) + 3H^{\oplus}(aq) + [O]$$

In full, the redox equation is:

$$RCHO + 2Cu^{2\oplus}(aq) + OH^{\ominus}(aq) + H_2O(l) \longrightarrow RCOOH + Cu_2O(s) + 3H^{\oplus}(aq)$$

But there is no need to remember the full equation!

The triiodomethane reaction (the 'iodoform test')

Once it has been found that an unknown compound is an aldehyde or ketone, it is possible to find out more about the compound by using other tests.

One of these tests is the triiodomethane reaction, often still known as the *iodoform test*. It identifies methyl ketones, and also those with the grouping CH_3—$CH(OH)$— (which is easily oxidised to a methyl ketone). A methyl ketone contains the structure:

$$CH_3-\overset{\overset{\textstyle O}{\|}}{C}-$$

A positive result in this test is given by, among many other compounds, propanone (CH_3COCH_3), ethanal (CH_3CHO), propan-2-ol ($CH_3CH(OH)CH_3$) and even ethanol (CH_3CH_2OH).

One drop (or 0.1 g) of the compound to be tested is mixed with a few cm³ of water. To this is added 1 cm³ of a 0.5 mol dm⁻³ solution of iodine in aqueous potassium iodide. (Alternatively, a solution of potassium iodide mixed with sodium chlorate(I) may be used.) The mixture now has a dark colour.

Aqueous sodium hydroxide is then added dropwise, with shaking, until the colour becomes pale yellow or disappears. The mixture is then warmed very gently to 60 °C. A dense pale-yellow precipitate with a very characteristic 'antiseptic' smell shows that the test for CH_3CO— or $CH_3CH(OH)$— is positive.

Sugars which give a positive test with Tollens', Benedict's or Fehling's reagents are called reducing sugars. In aqueous solutions, the ring form of these sugars exists in equilibrium with the open chain form, which has an aldehyde (or 2-hydroxyketone) group.

Glucose and fructose are reducing sugars; sucrose (our ordinary sugar) is not.

D(+)-glucose – 'Ring' form

'Open chain' form

The first reaction of $CH_3CH(OH)—$ is:

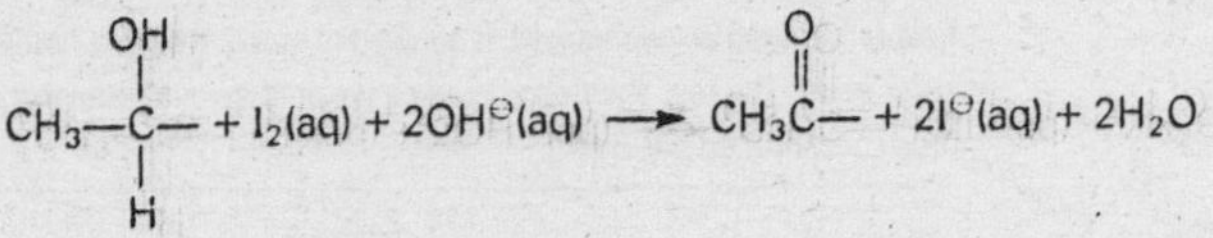

$$CH_3-\underset{\underset{H}{|}}{\overset{\overset{OH}{|}}{C}}- + I_2(aq) + 2OH^{\ominus}(aq) \longrightarrow CH_3\overset{\overset{O}{\|}}{C}- + 2I^{\ominus}(aq) + 2H_2O$$

The alcohol group is oxidised by the halogen/alkali solution to a carbonyl group.

The $CH_3-\overset{\overset{O}{\|}}{C}-$ group then reacts in two stages:

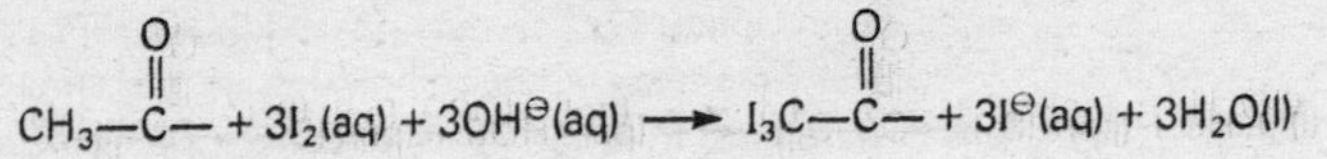

$$CH_3-\overset{\overset{O}{\|}}{C}- + 3I_2(aq) + 3OH^{\ominus}(aq) \longrightarrow I_3C-\overset{\overset{O}{\|}}{C}- + 3I^{\ominus}(aq) + 3H_2O(l)$$

Substitution on the carbon atom next to the carbonyl group occurs.

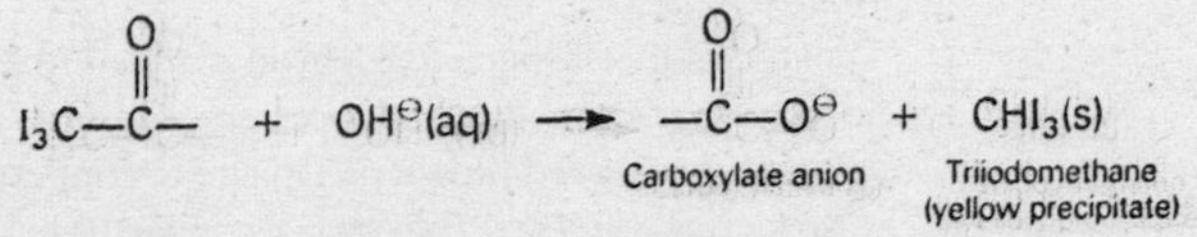

$$I_3C-\overset{\overset{O}{\|}}{C}- + OH^{\ominus}(aq) \longrightarrow -\overset{\overset{O}{\|}}{C}-O^{\ominus} + CHI_3(s)$$

Carboxylate anion   Triiodomethane
(yellow precipitate)

The C—C bond is then broken by nucleophilic attack. The bond is weakened by the polarity of the three C—iodine bonds:

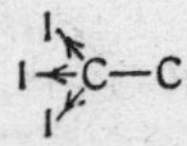

Using ethanal as an example, overall:

$$CH_3CHO + 3I_2(aq) + 4OH^{\ominus}(aq) \longrightarrow CHI_3(s) + HCOO^{\ominus}(aq) + 3I^{\ominus}(aq) + 3H_2O(l)$$

A similar reaction can be used to make trichloromethane ('chloroform') and tribromomethane ('bromoform'). The general reaction between suitable organic compounds and halogen plus alkali is known as the *'haloform reaction'*.

> Chloroform ($CHCl_3$, trichloromethane) used to be made by heating ethanol with bleaching powder, the material made from calcium hydroxide ('slaked lime', $Ca(OH)_2$) and chlorine. This was equivalent to using chlorine with alkali. Chloroform was very widely used for many decades as an anaesthetic. Queen Victoria gave birth to all her children, it is said, while anaesthetised with chloroform. However, it has been shown to cause damage to the liver and is no longer used in this way.

Tests for the carbonyl compounds

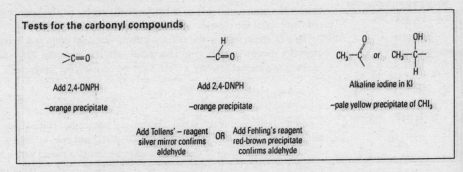

---

1 Give the displayed formulae of the organic products formed when butanal reacts with:

   a hydrogen gas and a palladium catalyst;

   b lithium tetrahydridoaluminate(III) in ethoxyethane;

   c phenylhydrazine;

   d aqueous silver nitrate mixed with the correct amount of dilute ammonia.

2 DHA is used as an artificial tan cream since the molecule reacts with protein in the dead outer skin to produce a brown pigment. The systematic name of DHA is 1,3-dihydroxypropanone.

   a Draw the structural formula of DHA.

   b The central carbon atom is part of a C=O double bond consisting of a σ and a π bond. Describe the shape of the π bond.

   c Reduction of the compound results in the formation of a triol. What reagent(s) would you use to reduce DHA?

   d Write an equation to show how DHA would react with sodium metal.

   e How would you expect DHA to react with 2,4-DNPH?

19.7 Identification of carbonyl compounds

Recrystallisation is used to purify crystalline solids, using a solvent in which the target compound is much more soluble at its boiling point than it is at room temperature (or when the flask containing the solution is put in ice/water). The orange precipitate is dissolved in the minimum of boiling solvent (often a 50:50 mixture of ethanol and water). Any insoluble impurities can be removed by rapid filtration of the hot solution. The filtrate is then cooled. The product should precipitate as crystals (the slower the precipitation, the bigger the crystals and the less impurity they will contain). The impurities will stay in solution. The crystals can be filtered off and also dried using a filter pump. Recrystallisation is also considered in Ch 29.2.

As we saw in Ch 19.6, all aldehydes and ketones form a crystalline precipitate with 2,4-dinitrophenylhydrazine (2,4-DNPH) dissolved in sulphuric acid and methanol.

Nearly all aldehydes and ketones are liquids. They are more volatile than alcohols as there is less hydrogen bonding (see Ch 19.1, *Figure 19.1.2*). It is very difficult, and often impossible, to identify a compound from its boiling point. (There are so many compounds and only a limited number of boiling points!) The solid 2,4-DNPH **derivatives** of aldehydes and ketones can easily be purified and identified by finding their *melting points*. This then identifies the original carbonyl compounds.

After the reaction between the carbonyl compound and 2,4-DNPH, the orange precipitate is filtered off and purified by recrystallisation.

Strictly, the melting point of the dry recrystallised material should then be found. No organic substance has a truly *sharp* melting point; the range of temperature between initial softening of the crystals and their final disappearance to a liquid should be noted. If, however slowly and carefully the temperature is raised, the melting range is more than 1–2 °C, *the material should be recrystallised again*.

Once the material has been purified sufficiently, its melting point can be found.

Measurement of melting points

There are several ways of doing this, but the principles are the same in each case. The dry pure solid crystals are crushed and put to a depth of 3–4 mm in a capillary tube which is closed at the bottom end. The capillary is then, for example, fixed to a thermometer by a rubber band, so that the sample in the capillary is held against the bulb of the thermometer. The thermometer is immersed in a suitable liquid (clear, colourless, high boiling point) which is heated gently, with stirring. The rate of increase of temperature near the melting point should be slow. The temperature at which the sample begins to soften is recorded, together with the temperature at which it is completely melted.

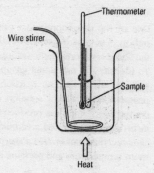

The apparatus is then allowed to cool, and the temperature at which solidification begins and finishes noted as a check on the first reading. However, there are so many solid compounds that many probably have the same melting point! How can we be sure that we have the compound we think we have? Provided that we already have an authentic sample of the solid, i.e. a sample where we are *sure* of its identity, the solution to the problem is quite simple. It depends on the fact that *any* impurity *lowers* the melting point of a solid (which is why salt sea water freezes at a lower temperature than fresh water).

Three capillaries are needed. The first contains a sample of the material which has just been made; the second contains the authentic sample; and the third contains a sample made by crushing a 50:50 mixture of the authentic compound and the just-made material. The three labelled samples are fixed as before to the thermometer and the melting points noted. If they all melt within a degree or two, the synthesised compound is the same as the authentic one. If the two materials melt at about the same temperature but the mixture melts at a lower temperature (quite possibly 20 or 30 °C lower!) then they are not the same.

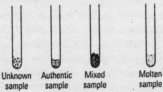

Unknown sample   Authentic sample   Mixed sample   Molten sample

An alternative method is to place three capillaries in holes in an electrically heated metal block. The samples at the bottom of the capillaries are observed through a small porthole which contains a magnifying glass.

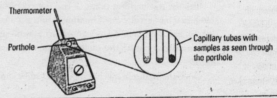

Thermometer

Porthole

Capillary tubes with samples as seen through the porthole

Identification of organic compounds using melting and boiling points

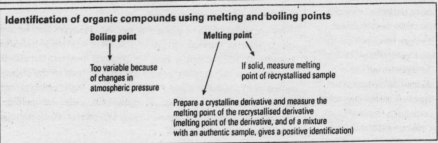

Boiling point

Too variable because of changes in atmospheric pressure

Melting point

If solid, measure melting point of recrystallised sample

Prepare a crystalline derivative and measure the melting point of the recrystallised derivative (melting point of the derivative, and of a mixture with an authentic sample, gives a positive identification)

SELF-ASSESSMENT QUESTIONS 19.7

1 One of the compounds that can be extracted from oil of bitter almonds contains 79.25% carbon, 5.66% hydrogen and 15.09% oxygen. The compound is a clear liquid. When 0.25 g of the liquid is vaporised, the volume is 56.6 cm³ (this volume is corrected to room temperature). To identify the functional groups present, a series of chemical tests was carried out. With sodium, there was no reaction; there was a similar lack of reaction with phosphorus pentachloride. With a solution of 2,4-dinitrophenylhydrazine (2,4-DNPH) in methanol an orange precipitate was produced, and oxidation of the compound with acidified potassium dichromate(VI) formed a carboxylic acid. The precipitate from the reaction with 2,4-DNPH, after recrystallisation, had a sharp melting point of 236 °C. (The volume of one mole of gas at room temperature and pressure is 24 000 cm³.)

a Calculate the empirical formula of the extracted compound.

b Calculate the relative molecular mass of the compound from the volume of vapour produced, and hence work out the molecular formula of the compound.

c Which functional group is shown not to be present because of the negative tests with sodium and phosphorus pentachloride?

d The formation of the orange precipitate identifies which possible functional group(s)?

e Describe how you would measure the melting point of the recrystallised precipitate from the reaction with 2,4-DNPH.

f The oxidation of the compound with acidified dichromate(VI), coupled with the formation of an orange precipitate with 2,4-DNPH, identifies which functional group?

g From the data you have so far, and using the table below, what is the boiling point of the compound extracted from bitter almonds?

Compound	b.p. /°C	m.p. of 2,4-DNPH precipitate /°C
Octanal	171	106
Octan-2-one	172	58
Benzaldehyde	179	237
Cycloheptanone	181	148

Solving a reaction sequence

Example: Compound A has the molecular formula $C_4H_{10}O$. When it is heated with acidified potassium dichromate(VI) solution a new compound, B, is formed. Compound A reacts readily with sodium metal to produce a colourless gas which explodes on ignition, together with compound C. When heated with excess concentrated sulphuric acid, compound A formed a mixture of products. This mixture contained compounds E and F, both of which decolorised aqueous bromine solution.

Tests to elucidate the structure of compound B were carried out and the results were as follows:

1. When added to a solution of 2,4-dinitrophenylhydrazine, an orange precipitate was formed.

2. When added to an ammoniacal solution of silver nitrate there was no apparent reaction.

3. With an alkaline solution of iodine a yellow precipitate, D, was formed, together with another organic product.

The mass spectrum of compound A showed a small peak at m/e = 74, but the two major peaks were at m/e = 29 and m/e = 45.

Use this information to work out the structures of compounds A to F, identifying the functional groups present in compounds A, B, E and F. What type of reaction is involved in the conversion of A to B and in the conversion of A to E? Suggest possible structures for the mass spectrum peaks at m/e = 29 and m/e = 45 in sample A.

The first step in working out such a problem is to translate the question into a flow diagram:

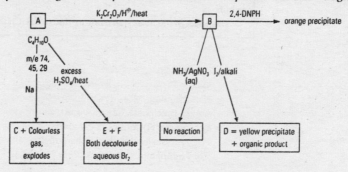

497

A careful look at the reagents used should indicate some of the reactions going on. Mark them in on your flow diagram:

- $K_2Cr_2O_7$ in acid solution is used as an oxidising reagent.
- 2,4-DNPH is used to test for the presence of a carbonyl group, $>C=O$.
- $AgNO_3$ in ammonia is a test for an aldehyde.
- 'Decolorises bromine' could indicate a $>C=C<$ double bond.
- One gas that explodes on ignition is hydrogen.
- Sodium reacts with any —OH group (possibly alcohols) to release hydrogen with sodium metal.
- E and F must both contain a $>C=C<$ double bond; and alcohols eliminate water in the presence of concentrated sulphuric acid.
- From the reaction with 2,4-DNPH, B must be a carbonyl, but since it does not react with $NH_3/AgNO_3$ it is not an aldehyde. ∴ B is a ketone.
- Ketones are formed by oxidising secondary alcohols; so, A is a secondary alcohol.

Next step:

If A is a secondary alcohol, the only possible structure is $CH_3CH_2CH(OH)CH_3$ which has an M_r of 74; this breaks up into CH_3CH_2 ($M_r = 29$) and $CH_3CH(OH)$ ($M_r = 45$) in the mass spectrometer.

Don't forget that the peaks in the mass spectrometer are only produced by positively charged ions, so 29 is due to $CH_3CH_2^{\oplus}$ and 45 is due to $CH_3CH(OH)^{\oplus}$.

Oxidation will give the ketone, $CH_3CH_2COCH_3$.

This contains the CH_3CO— grouping, which is recognised by the formation of the yellow precipitate of triiodomethane (iodoform) with alkaline iodine solution.

The equations for some of the reactions are now:

$$CH_2CH_2CH(OH)CH_3 + [O] \longrightarrow CH_3CH_2COCH_3 + H_2O$$
$$\text{A} \qquad\qquad\qquad\qquad \text{B}$$

$$CH_3CH_2CH(OH)CH_3 + Na \longrightarrow \tfrac{1}{2}H_2 + CH_3CH_2CH(O^{\ominus}Na^{\oplus})CH_3$$
$$\text{A} \qquad\qquad\qquad\qquad\qquad\qquad \text{C}$$

$$CH_3CH_2CH(OH)CH_3 \xrightarrow{-H_2O} CH_3CH=CHCH_3 + CH_3CH_2CH=CH_2$$
$$\text{A} \qquad\qquad\qquad\qquad \text{E} \qquad\qquad \text{F}$$

The answers to the questions are:

A is $CH_3CH_2CH(OH)CH_3$, containing a secondary alcohol group.

A to B is an oxidation reaction.

B is $CH_3CH_2COCH_3$, containing a ketone group.

C is $CH_3CH_2CH(O^{\ominus}Na^{\oplus})CH_3$.

D is CHI_3, iodoform.

A to E is an elimination reaction.

E and F are $CH_3CH=CHCH_3$ and $CH_3CH_2CH=CH_2$, both containing the alkene functional group.

The peaks in the mass spectrum are due to $CH_3CH_2^{\oplus}$ ($M_r = 29$) and $CH_3CH(OH)^{\oplus}$ ($M_r = 45$).

Reactions of carbonyl compounds – summary

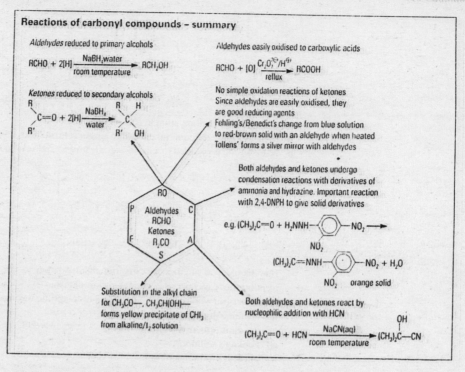

Aldehydes reduced to primary alcohols

$$RCHO + 2[H] \xrightarrow[\text{room temperature}]{\text{NaBH}_4\text{water}} RCH_2OH$$

Ketones reduced to secondary alcohols

Aldehydes easily oxidised to carboxylic acids

$$RCHO + [O] \xrightarrow[\text{reflux}]{Cr_2O_7^{2-}/H^+} RCOOH$$

No simple oxidation reactions of ketones
Since aldehydes are easily oxidised, they
are good reducing agents
Fehling's/Benedict's change from blue solution
to red-brown solid with an aldehyde when heated
Tollens' forms a silver mirror with aldehydes

Both aldehydes and ketones undergo
condensation reactions with derivatives of
ammonia and hydrazine. Important reaction
with 2,4-DNPH to give solid derivatives

Substitution in the alkyl chain
for CH₃CO—, CH₃CH(OH)—
forms yellow precipitate of CHI₃
from alkaline/I₂ solution

Both aldehydes and ketones react by
nucleophilic addition with HCN

SUMMARY 19

In this chapter the following new ideas have been covered. *Have you understood and learned them?* Could you explain them clearly to a friend? The page references will help.

- Recognition of the carbonyl functional group. p483
- The difference between aldehydes and ketones. p483
- The naming of aldehydes and ketones. p483
- Oxidation reactions which give carbonyl compounds. p485
- Aldehydes are easily oxidised; ketones are not. p486
- Nucleophiles can add to >C=O, but not >C=C<. p487
- Aldehydes and ketones can undergo condensation reactions. p490
- The condensation reaction with 2,4-dinitrophenylhydrazine can be used as a test for the presence of the carbonyl group. p491
- The easy oxidation of aldehydes in Tollens', Benedict's or Fehling's tests distinguishes aldehydes from ketones. p492
- The triiodomethane test can be used for the groupings CH₃—C=O and CH₃—CH(OH)—. p493
- Aldehydes and ketones can be identified by making the 2,4-DNPH derivative, purifying it and finding its m.p. p495

EXAMINATION QUESTIONS 19

1 The hormone testosterone is found in humans, and the optical isomer epi-testosterone is also found to be present in smaller quantities. The ratio of testosterone:epi-testosterone in normal persons is 6:1, and this fact is used as a means of determining if sports competitors have being taking extra testosterone as a performance-enhancing drug. *(continued on next page)*

Testosterone   epi-testosterone

a Mark the chiral carbon atom in testosterone.
b There are three different functional groups in testosterone. Which are they?
c Testosterone is oxidised by enzymes in the liver to androstenedione.

Androstenedione

How would you attempt to carry out this oxidation in the laboratory? Make sure you mention the reagents and apparatus you would use.

d State what you would expect to happen if testosterone was reacted with:
   i)   bromine dissolved in hexane;
   ii) 2,4-dinitrophenylhydrazine;
   iii) Fehling's solution;
   iv) phosphorus pentachloride.

2 The instructions for the preparation of propanone from propan-2-ol are as follows.
1 cm³ water and 1 cm³ of concentrated sulphuric are mixed in a flask. To this solution, add a solution containing 2.5 g potassium dichromate(VI) in 5 cm³ of water and 2 cm³ of propan-2-ol. The mixture is refluxed for 30 minutes, and then the mixture is distilled to produce propanone as a clear liquid, which boils at 56 °C.
   a Draw the apparatus you would use for the reflux reaction.
   b Write an equation for the oxidation reaction.
   c If insufficient oxidising agent had been used, what colour change would you expect to observe?
   d How many moles of potassium dichromate(VI), $K_2Cr_2O_7$, were used in the reaction?
   e What test would you use to show that the product contained a carbonyl group?
   f What test would you use to show that the product did not contain an aldehyde group?

3 One of the alarm signal pheromones given off by one species of ant is hex-2-enal, $CH_3CH_2CH_2CH\!\!=\!\!CHCHO$.
   A When this compound is reacted with ammoniacal silver nitrate solution, a new organic compound, E, is formed. E reacts with aqueous sodium carbonate to produce a colourless gas.
   B If the pheromone is added to a solution of bromine in hexane, the solution becomes almost colourless, forming F.
   C Reduction of hex-2-enal with $NaBH_4$ produces G, which does not react with sodium carbonate solution, but does give a colourless gas when sodium metal is added to it.
   D Hex-2-enal can be made to burn in air and produces two products, one of which turns lime water milky.
   Use the information given to identify E, F and G, and write balanced equations to explain each of the four reactions described.

- The meaning of **exothermic** and **endothermic**. (Ch 11.2)
- The definitions of **molar enthalpy changes**. (Ch 11.3)
- How to draw **enthalpy change diagrams**. (Ch 11.3)
- **Hess's Law** and how to use it. (Ch 11.4)
- Enthalpy changes when **bonds** are **broken** and **made**. (Ch 11.5)
- The idea of **activation energy**. (Ch 12.2)
- The ideas of **charge density** and **polarising power**. (Ch 2.5)
- Giant **ionic** structures. (Ch 3.3)

In this chapter

We shall develop further the ideas first discussed in Ch 11, and apply them to ionic lattices and the processes of solubility and hydration. In order to explain how endothermic processes can occur, we shall introduce the idea of **entropy** in a purely qualitative way – i.e. there will be no calculations here which involve entropy.

Make sure that you understand all the topics in the **Do you remember…?** section at the start of this chapter. If you can't, work through Ch 11 again before you go any further.

20.1   **Energy cycles: more uses of Hess' Law**
20.2   **Endothermic processes and a word on entropy**
20.3   **Enthalpy change of formation and the thermal stability of compounds**
20.4   **Lattice enthalpies: the Born–Haber cycle**
20.5   **More about lattice enthalpies**
20.6   **Solubility and hydration**

20

energetics

20.1 Energy cycles: more uses of Hess' Law

Hess' Law was introduced in Ch 11.4. It states that, provided all the measurements are made under standard conditions, **for a given overall chemical change, the overall energy change is the same whether the reaction takes place in a single step or via a series of steps.**

We showed how any thermochemical calculations can be structured either by using an *enthalpy change diagram* or by setting up a *Hess's Law triangle*. In both cases an *energy cycle* is shown.

Of course, as we showed in Ch 11.4, the whole point of being able to do Hess' Law calculations is that it gives a method of calculating the enthalpy changes associated with reactions which are impossible to do directly – like reacting graphite and hydrogen to form methane.

Worked example

Given the enthalpy changes of combustion of carbon, hydrogen and methane, calculate the enthalpy change of formation of methane, CH_4.

The required equation is therefore:

$$C(graphite) + 2H_2(g) \longrightarrow CH_4(g)$$

Data	$\Delta H^{\ominus}_{c, 298} (kJ\ mol^{-1})$
$C(graphite) + O_2(g) \longrightarrow CO_2(g)$	-394
$H_2(g) + \frac{1}{2}O_2(g) \longrightarrow H_2O(l)$	-286
$CH_4(g) + 2O_2(g) \longrightarrow CO_2(g) + 2H_2O(l)$	-890

Method 1: using an enthalpy change diagram (*Figure 20.1.1*)

You can draw a diagram for the enthalpy change of formation of methane in this style. If you are careful with the signs for each ΔH value, you will get the correct answer (and sign), even if you find you have drawn the corresponding arrow in the wrong direction.

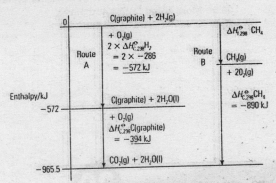

Figure 20.1.1   Enthalpy change diagram for $\Delta H^{\ominus}_{f, 298}$ CH₄

By Hess' Law, the total molar enthalpy change for the reaction:

$$C(graphite) + 2H_2(g) + 2O_2(g) \longrightarrow CO_2(g) + 2H_2O(l)$$

is the same whether the reaction goes in one stage by combustion of the carbon and hydrogen directly to carbon dioxide and water (route A), or whether the elements are converted to methane first, and this is then burned (route B).

The diagram in *Figure 20.1.1* shows that:

$$\Delta H^{\ominus}_{f, 298} CH_4 - 890 = -572 - 394 = -966$$

$$\Delta H^{\ominus}_{f, 298} CH_4 = -76\ kJ\ mol^{-1}$$

Method 2: using a Hess' Law triangle

The triangle is set out as shown in *Figure 20.1.2*, using the combustion data given above.

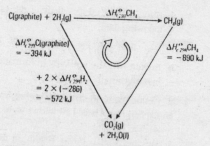

Figure 20.1.2   Hess cycle for $\Delta H_{f, 298}^{\ominus}$ CH$_4$

Figure 20.1.2 shows clearly that:

$$\Delta H_{f, 298}^{\ominus}CH_4 + \Delta H_{c, 298}^{\ominus}CH_4 - \Delta H_{c, 298}^{\ominus}(\text{graphite}) - 2 \times \Delta H_{c, 298}^{\ominus}H_2 = 0$$

$$\therefore \Delta H_{f, 298}^{\ominus}CH_4 - 890 - (-394) + 2 \times (-286) = 0$$

$$\therefore \Delta H_{f, 298}^{\ominus}CH_4 = -76 \, \text{kJ mol}^{-1}$$

The two methods given in the worked example are essentially the same, and you should use whichever of them you are more comfortable with. In our experience, more people seem to favour the triangle method of setting out these problems, so we shall continue to use it in this book.

The triangle in the worked example above is a specific case, using enthalpies of combustion, of a general principle. The general principle is shown in *Figure 20.1.3*; many calculations can be done using this method. An example is shown in *Figure 20.1.4*, using enthalpies of formation.

If all the quantities except one in a Hess' Law relationship are known or can be measured, then the missing quantity can be calculated.

> In any exam question involving a Hess' Law calculation, you will be given the values of all the molar enthalpy changes involved, except the one which you are required to calculate. If you are not given a value directly, you will be given sufficient data to work it out.

A general Hess' Law triangle

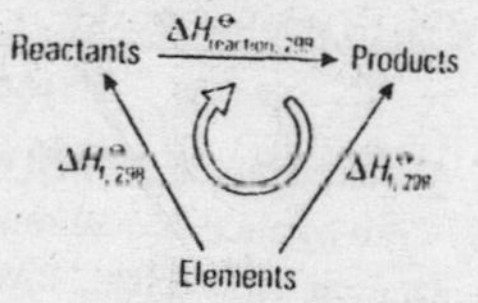

Reactants $\xrightarrow{\Delta H_{\text{reaction, 298}}^{\ominus}}$ Products

$\Delta H_{f, 298}^{\ominus}$ $\Delta H_{f, 298}^{\ominus}$

Elements

Figure 20.1.3

An example of Hess' Law in operation

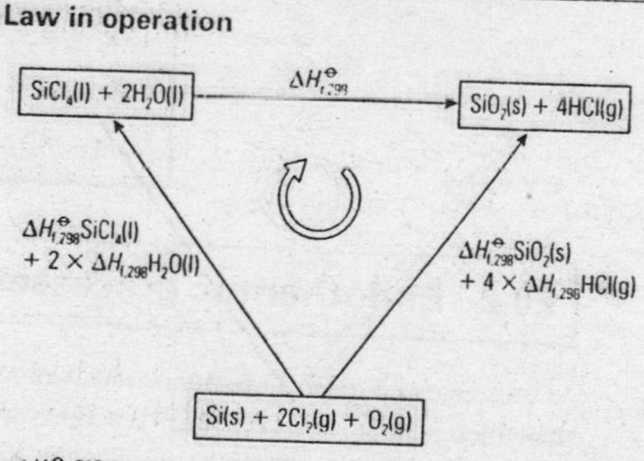

SiCl$_4$(l) + 2H$_2$O(l) $\xrightarrow{\Delta H_{r, 298}^{\ominus}}$ SiO$_2$(s) + 4HCl(g)

$\Delta H_{f, 298}^{\ominus}SiCl_4(l)$
$+ 2 \times \Delta H_{f, 298}^{\ominus}H_2O(l)$

$\Delta H_{f, 298}^{\ominus}SiO_2(s)$
$+ 4 \times \Delta H_{f, 298}^{\ominus}HCl(g)$

Si(s) + 2Cl$_2$(g) + O$_2$(g)

$$\Delta H_{r, 298}^{\ominus} - \Delta H_{f, 298}^{\ominus}SiO_2(s) - 4 \times \Delta H_{f, 298}^{\ominus}HCl(g) + \Delta H_{f, 298}^{\ominus}SiCl_4(l) + 2 \times \Delta H_{f, 298}^{\ominus}H_2O(l) = 0$$

Figure 20.1.4   Enthalpy of reaction using enthalpies of formation

Hess' Law is the basis on which all simple energy cycles are built, such as those we shall discuss in this chapter.

It is therefore essential that you understand it fully and are confident that you can use it. Practise on the questions provided.

1 a Using a Hess cycle, calculate the enthalpy of formation of carbon monoxide, given the following data:

$$\Delta H^{\ominus}_{c,\,298}CO = -283 \text{ kJ mol}^{-1}; \Delta H^{\ominus}_{f,\,298}CO_2 = -394 \text{ kJ mol}^{-1}.$$

b Explain why the enthalpy of formation of carbon monoxide cannot be determined directly by experiment.

2 a Define the term *standard enthalpy of formation*.

b Explain how Hess' Law allows us to determine the enthalpy of formation of a hydrocarbon such as hexane.

c Hexane, C_6H_{14}, burns readily in air to form carbon dioxide and water. Its enthalpy of combustion is -4163 kJ mol^{-1}. Given that the enthalpy of formation of water is -286 kJ mol^{-1} and that for carbon dioxide is -394 kJ mol^{-1}, calculate the enthalpy of formation of hexane.

3 Fermentation is the process whereby sugars such as glucose are converted to ethanol. The equation for the reaction is:

$$C_6H_{12}O_6(aq) \longrightarrow 2C_2H_5OH(aq) + 2CO_2(g)$$

a Write a balanced equation for the combustion of ethanol in oxygen.

b Glucose also burns in oxygen to form carbon dioxide and water. Write a balanced equation for the reaction.

c Draw a Hess cycle to link the fermentation reaction with the two combustion reactions, making sure that the whole cycle is balanced.

d Use the cycle to calculate the enthalpy change during the fermentation reaction.

$$(\Delta H_c C_6H_{12}O_6(s) = -2803 \text{ kJ mol}^{-1}; \quad \Delta H_c C_2H_5OH(l) = -1367 \text{ kJ mol}^{-1})$$

4 One of the gases emitted when a volcano erupts is hydrogen sulphide. The gas is oxidised in the air to sulphur dioxide.

$$...H_2S(g) + ...O_2(g) \longrightarrow ...H_2O(l) + ...SO_2(g)$$

a The equation as written is not balanced. Write the equation fully balanced.

b Write balanced equations for the enthalpy of formation of $H_2S(g)$, $H_2O(g)$ and $SO_2(g)$.

c Draw a Hess cycle, involving these equations, to enable the enthalpy change for the oxidation of hydrogen sulphide to be calculated.

d Use the cycle to calculate the enthalpy change of the oxidation of H_2S.

$\Delta H^{\ominus}_{f,\,298}$ H_2O	-286 kJ mol^{-1}
$\Delta H^{\ominus}_{f,\,298}$ H_2S	-21 kJ mol^{-1}
$\Delta H^{\ominus}_{f,\,298}$ SO_2	-297 kJ mol^{-1}

20.2 Endothermic processes and a word on entropy

As we suggested in Ch 11.5, the products of a chemical reaction are usually more energetically stable than the reactants; most chemical reactions are **exothermic**. The situation is shown in *Figure 20.2.1*,

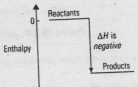

Figure 20.2.1 An exothermic reaction

If we are considering the molar enthalpy change of formation of a compound, the reactants are the elements and the product is the compound. Most compounds are energetically more stable than the elements from which they are formed; *the molar enthalpy change of formation of the compound is usually exothermic.*

But some reactions, and some molar enthalpy changes of formation of compounds, are **endothermic** (*Figure 20.2.2*). This means that the product is *less* stable than the reactants, or the compound is less stable than the elements from which it is formed. *The molar enthalpy change of formation of such a compound is endothermic.*

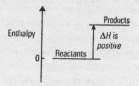

Figure 20.2.2   An endothermic reaction

How can this be? The answer is summarised in *Figure 20.2.3*.

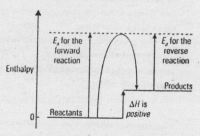

Figure 20.2.3   Energy profile for an endothermic reaction

The idea of **activation energy** was discussed in Ch 12.2. An energy barrier has to be overcome before reaction occurs; bonds have to be broken, attractions overcome, etc. The conditions which favour an endothermic *forward* reaction were discussed in Ch 13.2. Once products are formed, an energy barrier exists for the exothermic *reverse* reaction. It may not be as high as that for the forward reaction, but it may well be enough to ensure reasonable stability.

As we shall see in Ch 20.3, some quite common compounds have an endothermic molar enthalpy change of formation.

| Entropy |

In this book, we shall only consider the idea of entropy very briefly and in a non-mathematical way.

The entropy of a system depends on the number of available energy levels in the system over which the energy in the system can be spread. A highly randomised system such as a gas possesses far more energy levels than the rigorously ordered lattice of a crystalline solid.

The entropy of a system may therefore be thought of as a measure of its disorder or randomness. **Chaotic systems have high entropy.** Chaotic systems are more probable than ordered systems. Ordered systems tend towards increased disorder as time goes by; *the creation of order from disorder requires energy.*

CHAOS does rule!

We have already met the Second Law of Thermodynamics in Ch 11.1. One of the many possible statements of the Second Law is that **any spontaneous change must increase the overall entropy of the universe.** Note that this must mean that if entropy *decreases* within a system, entropy in the surroundings *must* increase to a greater extent.

Some parents have been heard to say that they never understood about entropy until they thought about a son's or daughter's bedroom!

A living system, such as you, is a highly ordered, low entropy system – and it is that way largely because it has caused and continues to cause increased chaos in the universe. For example, you eat large ordered molecules like starch and proteins, and turn them into a great number of smaller, more randomised molecules such as water, carbon dioxide and urea.

Spontaneous endothermic processes

If crystals of ammonium nitrate, NH_4NO_3, are dissolved in water, the temperature of the water falls considerably. The dissolving of ordinary salt, NaCl, is also endothermic, although not to the same extent; and the same is true for many other ionic compounds.

Here, then, is a process which is endothermic but which occurs spontaneously – the system, without any outside help, ends up at a higher level of enthalpy than it started at. We can now see the way in which this can happen. Even if the process is endothermic, it can still occur if the increase in entropy during the reaction is large enough.

When a salt dissolves, an ordered ionic lattice becomes a much more randomised solution of ions. The entropy of the system is therefore increased, and this then helps the process to occur, even though it may be endothermic.

Most spontaneous chemical reactions, i.e. those that occur at room temperature without heating or other assistance, are *exothermic* – the system gets hot, and loses energy to the surroundings. A few spontaneous reactions are *endothermic*. These usually involve large amounts of gas being given off or solids dissolving.

This reaction should *never* be done except by an experienced teacher, using small quantities in an efficient fume cupboard!

One spectacular demonstration involves pouring liquid sulphur dichloride oxide, SCl_2O, on to solid hydrated iron(III) chloride, $FeCl_3 \cdot 6H_2O$, in a beaker. The reaction is violent and immediate, vast amounts of fumes are given off, and the temperature recorded by a thermometer in the beaker may drop by up to 25 °C.

$$6SCl_2O(l) + FeCl_3.6H_2O(s) \longrightarrow 12HCl(g) + 6SO_2(g) + FeCl_3(s)$$

The reaction is driven by the production of 18 moles of highly randomised gas for every mole of crystalline solid used.

Energy conversion

As we pointed out in Ch 11.1, one consequence of the Second Law of Thermodynamics is that *any* conversion of energy from one form into another (e.g. in a power station, petrol engine, steam turbine) can **never** be 100% efficient. Some, and often *most*, of the available energy is inevitably dispersed as low grade heat energy. The First Law of Thermodynamics (see Ch 11.4) states that energy must be conserved: none can be lost.

But it is the *quality* of energy which matters. It has been said that mankind is conserving energy but increasing entropy, so destroying the availability of useful energy. It is an entropy crisis that we face, rather than an energy crisis.

Earlier, we explained how *you* create disorder and increase entropy by turning large food molecules into carbon dioxide gas and water. In his extraordinary book *The Periodic Table*, Italian chemist and Auschwitz survivor Primo Levi wrote this:
'[Life is] a parasitizing of the downward course of energy, from its noble solar form to the degraded one of low-temperature heat. In this downward course, which leads to equilibrium and thus death, life draws a bend and nests in it.'

SELF-ASSESSMENT QUESTIONS 20.2

1 In the following reactions, is there a decrease or an increase in the entropy of the system as the reaction proceeds?
a $Na_2S_2O_3(aq) \longrightarrow Na_2S_2O_3(s) + aq$
b $SCl_2O(l) + H_2O(l) \longrightarrow SO_2(g) + 2HCl(g)$
c $N_2(g) + 3H_2(g) \longrightarrow 2NH_3(g)$
d $Pb^{2\oplus}(aq) + 2I^{\ominus}(aq) \longrightarrow PbI_2(s)$

2 The molar entropy values for a range of substances are listed below. All are for 25 °C (298 K) except for water(g) and water(s).

	J K^{-1} mol^{-1}
Carbon (graphite)	+6
Helium(g)	+126
Hydrogen chloride(g)	+187
Lithium chloride(s)	+59
Silicon(s)	+19
Sodium(s)	+51
Sodium chloride(s)	+72
Water(g)	+189
Water(l)	+70
Water(s)	+48

a How do the entropy values for these substances relate to their physical states?

b What do the values for steam, water and ice tell you about their structures?

20.3 Enthalpy change of formation and the thermal stability of compounds

We would expect that the more exothermic the process in which a compound is formed from its elements, the more stable the compound. In other words, the more negative the value for its standard molar enthalpy change of formation, the more stable the compound would be expected to be. Conversely, an endothermic (positive) value should indicate an unstable compound. The situation is summed up in *Figure 20.3.1*.

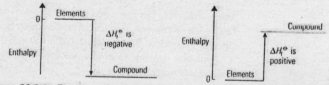

Figure 20.3.1   Exothermic and endothermic enthalpy changes of formation

Values of standard molar enthalpy changes of formation of certain compounds, compared with their thermal stability

Compound	$\Delta H_{f,298}^{\ominus}$ /kJ mol^{-1}	Thermal stability
$Al_2O_3(s)$	−1676	Very stable
$BaSO_4(s)$	−1473	Very stable
$PCl_5(s)$	−444	Unstable
$H_2O_2(l)$	−188	Unstable
$HCl(g)$	−92.3	Stable
$HBr(g)$	−36.4	Moderately stable
$HI(g)$	+26.5	Unstable
$C_6H_6(l)$	+49.0	Very stable
$C_2H_4(g)$	+52.2	Stable
$N_2O(g)$	+82.0	Unstable
$NO(g)$	+90.2	Very stable

Table 20.3.1

Table 20.3.1 shows enthalpy changes of formation of some compounds together with their thermal stability. Inspection of the table appears to show very little link between the value for $\Delta H^{\ominus}_{f, 298}$ and stability, which is not what we would expect.

Whenever we meet the unexpected, we should try first of all to find an explanation which fits with what is already known. The clue here is the definition of $\Delta H^{\ominus}_{f, 298}$ (see Ch 11.3), which we shall now repeat.

Can you now see where the problem lies? The values of $\Delta H^{\ominus}_{f, 298}$ in *Table 20.3.1* refer to *formation of the compounds from their elements*; but when they are heated, i.e. when their thermal stability is tested, **they do not necessarily break up to re-form those elements**.

Where heating *does* give the elements, the values of $\Delta H^{\ominus}_{f, 298}$ do agree with the thermal stability e.g. for the hydrogen halides:

$$\tfrac{1}{2}H_2(g) + \tfrac{1}{2}Cl_2(g) \longrightarrow HCl(g) \qquad \Delta H^{\ominus}_{f, 298} = -92.3 \text{ kJ mol}^{-1}$$

$$\tfrac{1}{2}H_2(g) + \tfrac{1}{2}I_2(g) \longrightarrow HI(g) \qquad \Delta H^{\ominus}_{f, 298} = +26.5 \text{ kJ mol}^{-1}$$

Hydrogen iodide falls apart quite easily when heated, whereas hydrogen chloride is stable at quite high temperatures (see Ch 8.3).

But, according to the figures, phosphorus(V) chloride, PCl_5, and hydrogen peroxide, H_2O_2 should both be *stable* to heat, because for both compounds the value of $\Delta H^{\ominus}_{f, 298}$ is large and negative. However, they both decompose easily on gentle heating. Why?

The answer is that *neither* of these compounds decomposes to its *elements*.

Phosphorus(V) chloride is *stable* with respect to phosphorus and chlorine; it is *unstable* with respect to phosphorus(III) chloride and chlorine. Decomposition of the solid PCl_5 to liquid PCl_3 and gaseous Cl_2 results in an increase in entropy (see Ch 20.2).

$$PCl_5(s) \underset{\text{cool}}{\overset{\text{warm}}{\rightleftharpoons}} PCl_3(l) + Cl_2(g) \qquad \Delta H^{\ominus}_{f, 298} = +124 \text{ kJ mol}^{-1}$$

The reaction is in fact reversible, and the forward reaction is endothermic. Consequently, applying Le Chatelier's Principle (see Ch 13.2), decomposition is assisted by an increase in temperature. If chlorine gas is allowed to escape, decomposition is complete.

Hydrogen peroxide does not decompose to hydrogen and oxygen, but to the highly stable molecules of water and oxygen.

$$2H_2O_2(l) \longrightarrow 2H_2O(l) + O_2(g) \qquad \Delta H = -108 \text{ kJ mol}^{-1}$$

The weak oxygen–oxygen link in hydrogen peroxide (H—O—O—H) is broken; the strong oxygen double bond in the O_2 molecule (O=O) is formed instead.

We now need to consider the other end of *Table 20.3.1*, where some compounds are 'stable' in spite of a positive value for $\Delta H^{\ominus}_{f, 298}$.

Ethene, C_2H_4, is one such compound. However, you will know that its alkene double bond makes it very reactive towards electrophiles (see Ch 16.5); it is *unstable* in terms of addition reactions.

Nitrogen oxide, NO, reacts very rapidly with the oxygen in air to give brown nitrogen dioxide, NO_2; but it is very stable with respect to its decomposition to nitrogen and oxygen. The bond between nitrogen and oxygen in NO is almost the equivalent of a triple bond. The bond enthalpy of a normal nitrogen–oxygen double bond, —N=O, is +587 kJ mol^{-1}; for the bond in NO it is considerably higher. The thermal stability of NO is therefore largely due to the *high activation energy* (see Ch 12.2) required for the decomposition reaction.

This section should leave you with the essential ideas that *there is no simple link between enthalpy change of formation of a compound and its thermal stability*, but that the stability or otherwise of a compound can usually be accounted for.

1 The enthalpy change of formation of quartz (silicon oxide, SiO_2) is $-911 \, kJ \, mol^{-1}$, whereas the comparable value for carbon dioxide is $-394 \, kJ \, mol^{-1}$. Explain why, although they are both exothermic reactions, carbon powder burns readily in air, but it is very difficult to form SiO_2 by burning silicon powder in air.

2 Elements in Group 5 of the Periodic Table can form covalent molecules with chlorine, with a general formula XCl_3.

  a How many outer electrons are there in the atom of phosphorus?

  b Draw a dot-and-cross diagram to show how the compound is formed between phosphorus and chlorine.

  c Describe the shape of the PCl_3 molecule. Explain this shape in terms of electron pair repulsion.

  d The equation for the thermal decomposition of PCl_3 is:

$$PCl_3(l) \longrightarrow P(s) + 1\tfrac{1}{2} Cl_2(g)$$

The standard enthalpy change for the reaction is $+320 \, kJ \, mol^{-1}$. The comparable value for $NCl_3(l)$ is $-230 \, kJ \, mol^{-1}$.

The equation for the decomposition of PCl_3 shows the elements in their standard states. What is meant by *standard state*?

  e Write an equation to represent the standard enthalpy of formation of $NCl_3(l)$.

  f What is the value of the standard enthalpy of formation of $NCl_3(l)$?

  g Comment on the probable stability of the molecule NCl_3 at 25 °C.

20.4 Lattice enthalpies: the Born–Haber cycle

The Born–Haber cycle is a special case of Hess' Law, as applied to ionic lattices.

As we saw in Ch 3.3, there is no such thing as a molecule of an ionic solid like sodium chloride, NaCl. The material exists as a giant lattice of ions, held together by the fact that the electrostatic *attraction* between the oppositely charged ions is at a shorter distance, and therefore stronger, than the *repulsion* between ions of the same charge.

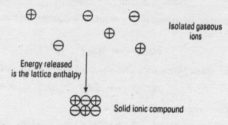

Isolated gaseous ions

Energy released is the lattice enthalpy

Solid ionic compound

Lattice enthalpies cannot be found directly by experiment. They can be estimated quite closely by finding the sizes and relative positions of ions from X-ray crystallography (see Ch 3.9), knowing the charges on the ions, and summing up the attractions and repulsions throughout the lattice.

Such a calculated value is known as a **theoretical lattice enthalpy value**.

Hess' Law can be used to find the lattice enthalpy of an ionic compound by using five other quantities *which can be found by experiment*, and arranging these together with the lattice enthalpy in a logical sequence – **the Born–Haber cycle**.

We have met some of these quantities before, but one further definition is needed.

e.g. $\quad$ $C(graphite) \longrightarrow C(g) \quad \Delta H^{\ominus}_{at, 298} = +717 \, kJ \, mol^{-1}$

$\quad\quad\quad$ $\tfrac{1}{2} H_2(g) \longrightarrow H(g) \quad \Delta H^{\ominus}_{at, 298} = +218 \, kJ \, mol^{-1}$

The energy needed to break one mole of a molecule up into gaseous atoms is often referred to as its *molar enthalpy change of atomisation*.

e.g. $$CH_4(g) \longrightarrow C(g) + 4H(g) \qquad \Delta H = +1662 \text{ kJ mol}^{-1}$$

The process is, in effect, breaking the four C—H bonds in methane to produce isolated gaseous atoms (see Ch 11.2).

The Born–Haber cycle for sodium chloride, NaCl, is shown in *Figure 20.4.1*.

The Born–Haber cycle for sodium chloride, NaCl

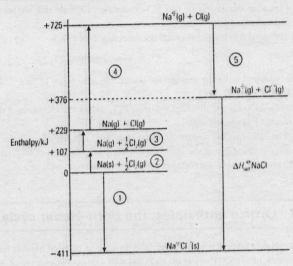

Figure 20.4.1

- ① represents the molar enthalpy of formation (see Ch 11.3) of sodium chloride.

$$Na(s) + \tfrac{1}{2}Cl_2(g) \longrightarrow NaCl(s) \qquad \Delta H^{\ominus}_{f,\,298} = -411 \text{ kJ mol}^{-1}$$

Notice that the *direction of the arrow* is important, as it determines the *sign* of the quantity.

- ② is $\Delta H^{\ominus}_{at}$ for sodium.

$$Na(s) \longrightarrow Na(g) \qquad \Delta H^{\ominus}_{at,\,298} = +107 \text{ kJ mol}^{-1}$$

- ③ is $\Delta H^{\ominus}_{at}$ for chlorine. (Notice that, for a diatomic molecule like Cl_2 or H_2, $\Delta H^{\ominus}_{at} = \tfrac{1}{2} \times$ bond enthalpy.)

$$\tfrac{1}{2}Cl_2(g) \longrightarrow Cl(g) \qquad \Delta H^{\ominus}_{at,\,298} = +122 \text{ kJ mol}^{-1}$$

- ④ is the first ionisation energy (see Ch 1.5) for sodium.

$$Na(g) \longrightarrow Na^{\oplus}(g) + e^{\ominus} \qquad \Delta H^{\ominus}_{1st\,I.E.} = +496 \text{ kJ mol}^{-1}$$

- ⑤ is the first electron affinity (see Ch 1.5) for chlorine.

$$Cl(g) + e^{\ominus} \longrightarrow Cl^{\ominus}(g) \qquad \Delta H^{\ominus}_{1st\,E.A.} = -349 \text{ kJ mol}^{-1}$$

Figure 20.4.1 clearly shows that the enthalpy change on *one* side of the diagram must equal that on the *other* – this can be used for checking your calculation. Also, it shows that we must *change the sign* on the values for ②, ③ and ④.

$$\Delta H^{\ominus}_{latt}\, NaCl + ⑤ = ① - ② - ③ - ④$$
$$\therefore \Delta H^{\ominus}_{latt}\, NaCl = ① - ② - ③ - ④ - ⑤$$
$$= -411 - 107 - 122 - 496 - (-349)$$
$$= -787 \text{ kJ mol}^{-1}$$

Worked example

Calculate the molar lattice enthalpy change for magnesium chloride, MgCl$_2$.

Quantities required:

		kJ mol^{-1}
$\Delta H^{\ominus}_{at} Mg$	=	+148
1st ionisation energy of Mg	=	+738
2nd ionisation energy of Mg	=	+1451
$\Delta H^{\ominus}_{at} Cl$	=	+122
$\Delta H^{\ominus}_{f} MgCl_2$	=	−641
1st electron affinity of Cl	=	−349

Draw out the cycle as in *Figure 20.4.2*

Note that, as the formula is MgCl$_2$, the cycle requires *twice* the molar enthalpy change of atomisation of chlorine and *twice* the first electron affinity of chlorine, as well as *both* ionisation energies for magnesium.

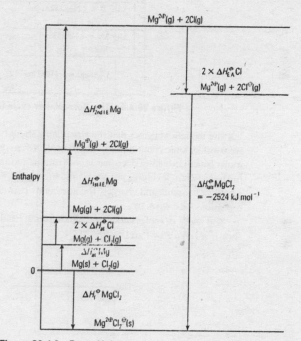

Figure 20.4.2   Born–Haber cycle for magnesium chloride, MgCl$_2$

Notice that the molar lattice enthalpy for magnesium chloride is considerably more than that for sodium chloride. This is because the magnesium ion is *doubly-charged*, so that it attracts negative ions more strongly, and therefore the forces between the ions are greater.

A moment's thought should convince you that if the anion is *also* doubly charged, the lattice enthalpy of the resulting compound should be even higher. And this is the case: the value for magnesium oxide, Mg$^{2\oplus}$O$^{2\ominus}$, is about −3800 kJ mol^{-1}.

The Born–Haber cycle diagram for MgO will look a little different (*Figure 20.4.3*), *because the second electron affinity for oxygen is endothermic* (see Ch 2.2).

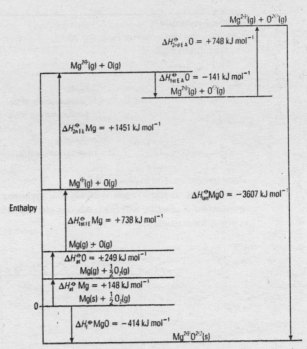

$Mg^{2+}(g) + O^{2-}(g)$

$\Delta H^{\ominus}_{2nd\,E.A.}O = +748 \text{ kJ mol}^{-1}$

$Mg^{2+}(g) + O(g)$

$\Delta H^{\ominus}_{1st\,E.A.}O = -141 \text{ kJ mol}^{-1}$

$Mg^{2+}(g) + O^{-}(g)$

$\Delta H^{\ominus}_{2nd\,I.E.}Mg = +1451 \text{ kJ mol}^{-1}$

$Mg^{+}(g) + O(g)$

$\Delta H^{\ominus}_{latt}MgO = -3607 \text{ kJ mol}^{-1}$

Enthalpy

$\Delta H^{\ominus}_{1st\,I.E.}Mg = +738 \text{ kJ mol}^{-1}$

$Mg(g) + O(g)$

$\Delta H^{\ominus}_{at}O = +249 \text{ kJ mol}^{-1}$

$Mg(g) + \frac{1}{2}O_2(g)$

$\Delta H^{\ominus}_{at}Mg = +148 \text{ kJ mol}^{-1}$

$Mg(s) + \frac{1}{2}O_2(g)$

0 ─

$\Delta H_f^{\ominus}MgO = -414 \text{ kJ mol}^{-1}$

$Mg^{2+}O^{2-}(s)$

Figure 20.4.3   The Born–Haber cycle for magnesium oxide, MgO

Further thought suggests that the *size of ions* should also affect the value of the lattice enthalpy. If we consider ionic compounds of the form $M^{\oplus}X^{\ominus}$, so that all the ions are singly charged, we find that molar lattice enthalpies decrease as both cations and anions increase in size. The bigger the ions, the further apart their centres are, and consequently the weaker the forces between them. (Remember, whatever the sizes and charges, the attractions between ions in ionic lattices always outweigh the repulsions – see Ch 3.3.)

The values of molar lattice enthalpies for the halide salts of lithium, sodium and potassium are shown in *Table 20.4.1* and *Figure 20.4.4*.

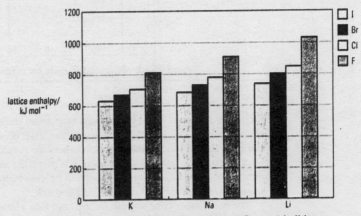

Figure 20.4.4   Lattice enthalpies of some Group 1 halides

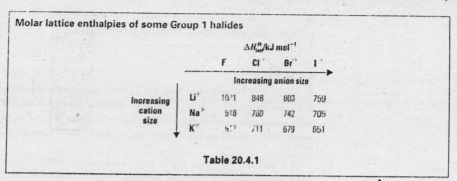

Molar lattice enthalpies of some Group 1 halides

		F	Cl	Br	I
			Increasing anion size $\rightarrow$		
Increasing cation size $\downarrow$	Li$^+$	10?1	848	803	759
	Na$^+$	918	760	742	705
	K$^+$	6??	711	679	651

$\Delta H^{\ominus}_{lat}/kJ\ mol^{-1}$

Table 20.4.1

1 The lattice enthalpy of an ionic solid cannot be measured directly by experiment.
  a Write an equation to define what is meant by lattice enthalpy.
  b Name all the enthalpy changes that need to be measured to calculate the lattice enthalpy of a simple ionic compound MX. In each case write an equation to represent the change that occurs.

2 Calculate the lattice enthalpy of magnesium bromide from the following data (all values are in kJ mol^{-1}):

$$\Delta H^{\ominus}_{at,\ 298}Mg \qquad\qquad = +148$$
$$\Delta H^{\ominus}_{at,\ 298}Br \qquad\qquad = +112$$
$$\text{1st ionisation enthalpy Mg} = +738$$
$$\text{2nd ionisation enthalpy Mg} = +1451$$
$$\text{First electron affinity Br} \quad = -325$$
$$\Delta H^{\ominus}_{f,\ 298}MgBr_2 \qquad\qquad = -524$$

3 a Draw the Born–Haber cycle for CaO given the following data (all values are in kJ mol^{-1}):

For calcium:	Atomisation enthalpy	+177
	1st ionisation enthalpy	+590
	2nd ionisation enthalpy	+1145
For oxygen:	Atomisation enthalpy	+249
	1st electron affinity	−141
	2nd electron affinity	+798
	Enthalpy of formation for CaO	−635

  b Explain why the second ionisation enthalpy of calcium is greater than the first ionisation enthalpy.
  c How does the first ionisation enthalpy of calcium compare with the first ionisation enthalpy of magnesium? Why should this be so?
  d The second electron affinity of oxygen is endothermic. Why should this be so?

4 a The enthalpy of atomisation of methane, CH$_4$, is +1662 kJ mol^{-1}. Write an equation for the atomisation of methane.
  b What is the average bond enthalpy for the C—H bond in methane?
  c The related compound, CCl$_4$, has an enthalpy of atomisation of +1308 kJ mol^{-1}. What is the average bond enthalpy for the C—Cl bond?
  d Calculate the enthalpy of atomisation of chloromethane, CH$_3$Cl, using your answers to **b** and **c**.

20.5 More about lattice enthalpies

Why MgCl₂?

In *Figure 20.4.2* we showed how the molar lattice enthalpy of magnesium chloride, $MgCl_2$, and hence any ionic salt, could be found from experimental measurements of enthalpies of atomisation and formation, ionisation energies and electron affinities. We can see that $MgCl_2$ is energetically very stable. But we would expect it to be – the formula follows from simple ideas of ionic bonding (see Ch 2.3).

But why not $MgCl_3$ or MgCl? If we look at the energy changes involved, can we find an explanation in terms of the molar enthalpy changes of formation, $\Delta H^{\ominus}_{f,\,298}$ MgCl₃ and $\Delta H^{\ominus}_{f,\,298}$ MgCl?

In order to complete a Born–Haber cycle we need to calculate a theoretical value for the lattice enthalpy. As indicated above, this can be done from knowledge of ion sizes, charges, and relative positions.

To set up a cycle for $MgCl_3$, we need the figures given for *Figure 20.4.2*, plus the third ionisation energy of Mg and the 'theoretical' molar lattice enthalpy for $MgCl_3$.

3rd ionisation enthalpy of magnesium = $+7733 \text{ kJ mol}^{-1}$

'Theoretical' lattice enthalpy for $MgCl_3 = -5440 \text{ kJ mol}^{-1}$

Figure 20.5.1 shows the resulting Born–Haber cycle for $MgCl_3$. Note that we now need $3 \times \Delta H^{\ominus}_{at,\,298}$ Cl and $3 \times$ first electron affinity for chlorine.

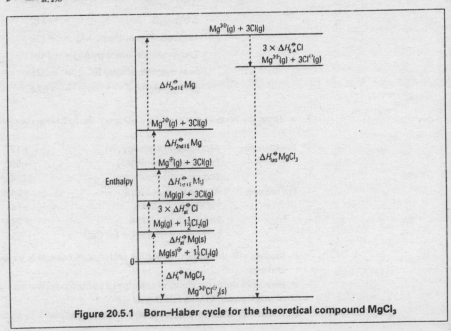

Figure 20.5.1 Born–Haber cycle for the theoretical compound MgCl₃

So, from the diagram:

$$\Delta H^{\ominus}_{latt}\, MgCl_3 + 3 \times (1st\ EA\ Cl) - \Delta H^{\ominus}_f\, MgCl_3 = -\Delta H^{\ominus}_{at}\, Mg - 3 \times (\Delta H^{\ominus}_{at}\, \tfrac{1}{2}Cl_2) - 1st\ IE\ Mg$$
$$- 2nd\ IE\ Mg - 3rd\ IE\ Mg$$

$$\therefore -5440 + 3 \times (-349) - \Delta H^{\ominus}_f\, MgCl_3 = -148 - 3 \times (122) - 738 - 1451 - 7733$$

$$\therefore -\Delta H^{\ominus}_f\, MgCl_3 = -10436 + 6487$$

$$\therefore \Delta H^{\ominus}_f\, MgCl_3 = +3949 \text{ kJ mol}^{-1}$$

So *Figure 20.5.1* as we have drawn it is not correct in its bottom part. *The molar enthalpy change of formation of MgCl₃ is not exothermic as we have shown; it is massively endothermic.* MgCl₃ would be an extremely unstable compound! And the most important reason for this is the huge value of the third ionisation energy of magnesium.

We can explain that, because magnesium has only two electrons in its outer electron energy level, so the third electron would have to be torn out from an inner level closer to the attraction of the nucleus (see Ch 1.5). MgCl₂, with Mg³⁺ ions, would certainly have a large lattice enthalpy. The energy required to get those Mg³⁺ ions would be even greater, and ensures that MgCl₃ cannot exist.

ACTIVITY

Activity

Set up a Born–Haber cycle for the theoretical compound MgCl, using the appropriate values from those provided for *Figure 20.4.2* and the calculated value of the molar lattice enthalpy of MgCl, −753 kJ mol⁻¹.

Show, by putting the necessary values in the cycle, that $\Delta H^{\ominus}_{f, 298}$ MgCl is slightly *exothermic*.

Why, then, does MgCl not exist?

(Hint: consider $\Delta H^{\ominus}_{f, 298}$ MgCl₂ and remind yourself of the meaning of disproportionation – see Ch 8.4.)

Theoretical and experimental values for lattice enthalpies

Table 20.5.1 shows some theoretical and experimental values for molar lattice enthalpies for sodium and silver halides.

Some theoretical and experimental (Born–Haber) values of molar lattice enthalpies

$$M^{\oplus}(g) + X^{\ominus}(g) \longrightarrow MX(s)$$

Compound	Theoretical $\Delta H^{\ominus}_{latt}$ / kJ mol⁻¹	Experimental $\Delta H^{\ominus}_{latt}$ / kJ mol⁻¹
NaCl	−770	−780
NaBr	−735	−742
NaI	−687	−705
AgCl	−833	−905
AgBr	−816	−891
AgI	−778	−889

Table 20.5.1

It is easily seen that the two values are very similar for the sodium halides, but are very different for the silver halides (*Figure 20.5.2*).

The theoretical value is calculated on the assumption that each ion is spherical and separate, with its charge distribution evenly around it. The close agreement between the two values for the sodium halides is very good evidence that the simple model of an ionic crystalline lattice is valid; and equally, that the sodium halides are very largely ionic in character (see Ch 3.3).

For the silver halides, however, the disagreement must be because they do not conform to a near-perfect ionic model. The bonding in them cannot be fully ionic, but partially covalent instead. This can be linked to the smaller differences between the *electronegativities* (see Ch 2.2) of silver and the halogens compared with sodium and the halogens (*Figure 20.5.3*).

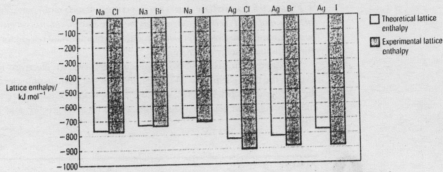

Figure 20.5.2   Theoretical and experimental (Born–Haber) lattice enthalpies

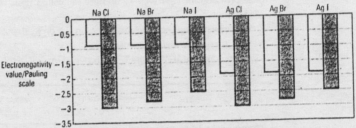

Figure 20.5.3   Pauling electronegativities

In general, *the smaller the cation the larger its charge*, i.e. *the higher its charge density and hence its polarising power* (see Ch 2.5), the more lattice enthalpies diverge from the theoretical value.

SELF-ASSESSMENT QUESTIONS 20.5

1   The lattice enthalpies for the Group 2 oxides and sulphides are shown in the table. All values are quoted in kJ mol^{-1}.

MgO	−3791	MgS	−3299
CaO	−3401	CaS	−3013
SrO	−3223	SrS	−2848
BaO	−3054	BaS	−2725

   a   Plot a bar chart of the lattice enthalpies for both series of compounds.
   b   Explain the general trend in both series.
   c   Explain why the lattice enthalpy for the sulphide of each element is lower than for the corresponding oxide.

2   By comparing the measured lattice enthalpy (via a Born–Haber cycle) with the theoretical lattice enthalpy in each of the following compounds, what conclusions can you make about the structures of the compounds? All values are in kJ mol^{-1}.
   a   Calcium bromide, $CaBr_2$:
       Born–Haber value, −2176; theoretical value, −2132.
   b   Copper bromide, $CuBr_2$:
       Born–Haber value, −976; theoretical value, −870.
   c   Sodium bromide, NaBr:
       Born–Haber value, −742; theoretical value, −735.

20.6 Solubility and hydration

The state symbol (aq) simply means a sufficient quantity of water to give a **suitably** dilute solution.

As we mentioned while briefly discussing entropy in Ch 20.2, dissolving ionic salts in water often involves an endothermic enthalpy change. If we apply Le Chatelier's Principle (see Ch 13.2) to the dissolving of an ionic salt in water, an increase in temperature should favour more salt dissolving if the process is endothermic. And, indeed, many salts are more soluble in hot water than they are in cold. The situation is complicated by entropy effects (crystalline solid to widely dispersed ions in solution), and by changes of hydration as temperature increases.

For example, for sodium chloride:

$$NaCl(s) + aq \longrightarrow Na^{+}(aq) + Cl^{\ominus}(aq) \qquad \Delta H^{\ominus}_{soln} = +4 \text{ kJ mol}^{-1}$$

Dissolving an ionic compound can be considered as two stages, ① and ②:

① Disruption of the lattice to form gaseous ions (the *opposite* of the lattice enthalpy).

② **Hydration** of these ions, as polar water molecules (see Ch 2.5) are attracted to the ions. *Figure 20.6.1* represents the process for sodium chloride and sodium hydroxide.

③ is the overall enthalpy change of solution for NaCl. The diagram for NaOH is set out more fully.

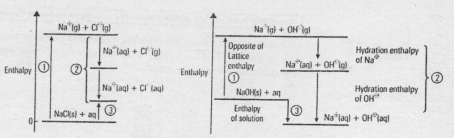

Figure 20.6.1   Dissolving an ionic compound

Some standard molar enthalpy changes of hydration are given in *Table 20.6.1*.

Molar enthalpy changes of hydration for some ions	
Ion	$\Delta H^{\ominus}_{hyd}$ /kJ mol^{-1}
Li^{+}	-499
Na^{+}	-390
K^{+}	-305
Mg^{2+}	-1891
Al^{3+}	-4613
Cl^{-}	-381
Br^{-}	-351
I^{-}	-307

Table 20.6.1

The table shows that, for the same charge, the smaller the ion the greater the value of $\Delta H^{\ominus}_{hyd}$.

An increased charge on the ion leads to a far higher value of $\Delta H^{\ominus}_{hyd}$. An important idea is the charge density on the ion (see Ch 2.5). The smaller the ion and the greater its charge, the higher the charge density, and the more attractive to the polar water molecules it will be – with a corresponding higher value of $\Delta H^{\ominus}_{hyd}$.

If we discount entropy effects, the solution of an ionic solid can be considered to be in large part a competition between *breaking up the lattice* (endothermic) and *hydration of the ions* which are thereby released (exothermic). High charge density will mean high hydration enthalpy, but it will also mean high lattice enthalpy.

Some helpful rules about the solubility of common ionic compounds are given in the box below.

Solubility rules (at room temperature)
- All common sodium, potassium and ammonium salts are soluble.
- All nitrates are soluble.
- Lead compounds are insoluble (except the nitrate and ethanoate).
- Hydroxides are insoluble (except for Group 1 and barium hydroxides; calcium hydroxide is slightly soluble).
- Carbonates are insoluble (except for sodium, potassium and ammonium carbonate).
- Common chlorides are soluble (except silver and lead chloride).
- Common sulphates are soluble (except barium sulphate; calcium sulphate is slightly soluble).

Variation in solubility of the sulphates and hydroxides of Group 2

In Ch 6.5 we gave figures to show that the solubility of the *hydroxides* of Group 2 elements *increases* from magnesium hydroxide, $Mg(OH)_2$, through calcium and strontium hydroxides to barium hydroxide, $Ba(OH)_2$ (*Figure 20.6.2*).

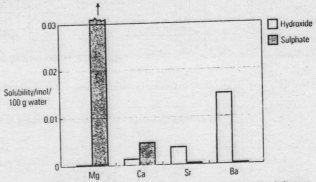

Figure 20.6.2   Solubility of Group 2 hydroxides and sulphates

The solubility of the Group 2 *sulphates*, however, *decreases* from magnesium sulphate, $MgSO_4$, to barium sulphate, $BaSO_4$.

Why is this?

As the cations increase in size from Mg^{2+} through Ca^{2+} and Sr^{2+} to Ba^{2+}, *both the hydration and lattice enthalpies will decrease*. However, with the *doubly-charged* sulphate ion, SO_4^{2-}, the drop in the lattice enthalpy is less significant than the drop in hydration enthalpies, so that hydration of the ions becomes less able to compensate for the break-up of the ionic lattice. So the Group 2 sulphates become *less* soluble from $MgSO_4$ to $BaSO_4$.

For the hydroxides, however, where the lattices contain the *singly-charged* hydroxide ion, OH^-, the drop in lattice enthalpy from $Mg(OH)_2$ to $Ba(OH)_2$ as the cations increase in size is more significant than the drop in hydration enthalpy, so that hydration becomes better able to compensate for the break-up of the lattice. So the Group 2 hydroxides become *more* soluble from $Mg(OH)_2$ to $Ba(OH)_2$.

Q

SELF-ASSESSMENT
QUESTIONS
20.6

1 Draw a Born–Haber type of cycle to show the connection between lattice enthalpy, enthalpies of hydration and enthalpy of solution for:

   a potassium bromide

   b calcium bromide.

2 On graph paper draw, to scale, a Born–Haber type of cycle to calculate the molar enthalpy of solution for magnesium hydroxide. The necessary data are:

Lattice enthalpy, $Mg(OH)_2$,	$-3006 \, kJ \, mol^{-1}$
Enthalpy of hydration, Mg^{2+},	$-1891 \, kJ \, mol^{-1}$
Enthalpy of hydration, OH^-,	$-460 \, kJ \, mol^{-1}$

Summary of lattice enthalpies

Lattice enthalpy of MX is given by the equation:

$$M^+(g) + X^-(g) \longrightarrow MX(s) \qquad \Delta H^\ominus_{latt} \text{ is negative}$$

It has to be calculated using the Born–Haber cycle:

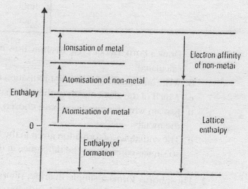

If Born–Haber cycle values are close to the theoretical value, the ionic model for the structure of the compound is a good model, i.e. the structure has a high proportion of ionic character.

Lattice enthalpy and solution enthalpy – dissolving ionic compounds

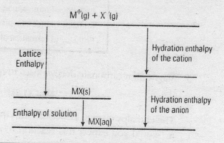

S

SUMMARY
20

In this chapter the following new ideas have been covered. *Have you understood and learned them?* Could you explain them clearly to a friend? The page references will help.

- Hess' Law is the basis of all simple energy cycles.   p502
- Why endothermic processes are possible.   p505
- The entropy of a system depends on the number of available energy levels.   p505
- Spontaneous endothermic processes can occur because of an increase in entropy in the system.   p506

- The link between the molar enthalpy change of formation of a compound and its thermal stability depends on a number of factors. p507
- The Born–Haber cycle is a special case of Hess' Law. p509
- The definitions of molar lattice enthalpy and molar enthalpy change of atomisation. p509
- How to find molar lattice enthalpies using cycles. p510
- Born-Haber cycles and 'theoretical' compounds. p514
- Factors influencing the solubility of ionic compounds. p517
- The reasons for the variations of solubility in Group 2 sulphates and hydroxides. p518

EXAMINATION
QUESTIONS
20

1 The enthalpies of formation (in kJ mol⁻¹) of the alkali metal chlorides are shown in the table.

LiCl	−408
NaCl	−411
KCl	−437
RbCl	−435
CsCl	−443

a Draw a Born–Haber cycle to show how the lattice enthalpy for rubidium chloride could be calculated.
b Write an equation for the first ionisation energy of potassium.
c Which of the alkali metals in the table has the lowest first ionisation energy?
d Explain why the metal you have chosen in c would be expected to be the most reactive of the metals.
e The enthalpies of formation given in the table are similar in value. Why do you think they are so similar considering the difference in the reactivity of metals?

2 Phosphorus forms a simple hydride, phosphine, which has the formula PH_3. Using the data below calculate the atomisation enthalpy of phosphine, and from the value obtained give an estimate of the P—H bond enthalpy. One quoted value is +321 kJ mol⁻¹. Explain why your value does not agree with the quoted value.

Enthalpy of atomisation of red phosphorus,	+332 kJ mol⁻¹
Enthalpy of atomisation of hydrogen,	+218 kJ mol⁻¹
Enthalpy of formation of phosphine,	+5.4 kJ mol⁻¹

3 Magnesium carbonate decomposes to release carbon dioxide gas.

$$MgCO_3(s) \longrightarrow MgO(s) + CO_2(g)$$

The enthalpy change associated with this reaction is +100 kJ mol⁻¹.
a Would you expect this reaction to occur spontaneously at room temperature? If so, give your reasons.
b If the temperature is raised would you expect the reaction to be more or less likely to occur?
c How does the thermal stability of magnesium carbonate compare with the thermal stability of calcium carbonate?
d Why would you expect there to be a difference in the thermal stability of the two compounds?

Do you remember ...?

- How to calculate the **numbers of moles** for solids, gases, and solutions. (Ch 9)
- That **titrations** can be used to measure quantities of substances in solutions. (Ch 10)
- The **collision theory** to explain rates of reaction. (Ch 12.1)
- Which **factors** affect how fast a chemical reaction occurs. (Ch 12.2)
- What is meant by the **activation energy** of a reaction, and its position on the **Maxwell–Boltzmann distribution curve** of particle energies. (Ch 12.3)
- What is meant by a **reaction mechanism**, and the use of **curly arrows** in explaining mechanisms in organic chemistry. (Ch 14.7)

21.1 **Introduction to reaction kinetics**
21.2 **Experimental methods**
21.3 **Order of reaction**
21.4 **Rate equations and rate constants**
21.5 **Reaction rates and reaction mechanisms**
21.6 **Activation energy and reaction rates**

21

reaction kinetics

21.1 Introduction to reaction kinetics

Two of the most important questions about a chemical reaction are:

● Is it likely to happen?
● If so, how fast will it happen?

The first involves **chemical energetics** (see Ch 11 and Ch 20.1); the second involves **chemical kinetics**. To find ΔH° for a reaction will tell us whether the reaction is feasible, and may tell us what sort of equilibrium will result. *But it cannot tell us the rate at which the reaction will occur.* Some reactions occur almost instantaneously, e.g. the neutralisation of an acid, or – more dramatically – an explosion. Others happen very slowly at room temperature.

For example, a mixture of petrol vapour and air in a container at room temperature will show no sign of reaction if left for many years, even though ΔH°_{298} for the combustion of octane is an immense -5498 kJ mol^{-1}. But pass a spark through that mixture to overcome the activation energy of the reaction (see Ch 12.2), and the reaction is extremely rapid – as happens in petrol-driven car engines.

Under a specific set of conditions, all reactions occur at a definite rate. Nature has several ways of speeding up reactions, and so do the manufacturers of useful chemical substances.

In Chapter 12 we looked at the principles underlying ideas about the rates of chemical reactions. Very briefly, before any two molecules can react together, they have to collide with sufficient energy and in the correct orientation.

The factors that can influence the number of collisions, and therefore the rate of a chemical reaction, include temperature, and concentration effects (*Figure 21.1.1*). Concentration effects might involve a change in the surface area of a solid reactant, or of pressure in a reaction involving gases. Catalysts reduce activation energy and so can influence rate greatly, and so may the energy provided by light.

Temperature
Light
Concentration
Gas pressure
Solid surface area
Catalysts

**Figure 21.1.1 Factors affecting rates of
chemical reactions**

So far, we have given a *qualitative* treatment of reaction rates. Here, we are going to look at a *quantitative* treatment. To do this, we have to obtain reliable data to plot graphs to show the disappearance of reactants or the formation of products over time, and use these graphs to gain information about rates. We then have to interpret the information over time to understand how the reaction we are studying may be proceeding.

What is meant by the term 'rate of reaction'?

A reaction may have the form:

$$X + 2Y \longrightarrow Z$$

Obviously, as the reaction proceeds X and Y are being used up, so their concentrations decrease. Z is being produced, so its concentration rises from zero to a maximum when one of the reactants has been completely used up. The curves for X and Y are **decay curves**; the curve for Z is a **growth curve**. The changes in concentration of reactants and products can be shown as in *Figure 21.1.2*.

Reminder: We use square brackets to represent concentration. So [X] means 'the concentration of X in mol dm^{-3}'.

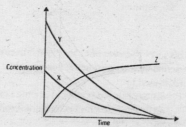

Figure 21.1.2 Decay and growth curves

Notice that the slope is greatest and therefore that Z is produced most rapidly at the *start* of the reaction, when there is plenty of X and Y to react. As they get used up, so the rate at which Z is produced drops. When there is no more X or Y to react, [Z] stays constant.

> This idea of a limiting reactant is an important one and should be fully understood.

We have drawn the graph so that both X and Y are used up completely. It will often be the case that the reaction will stop because *one* of the reactants has been used up, and some of the other reactant will remain. The reactant which is used up first is known as the *limiting reactant*. If, for example, all Y had been used up before X, the line for X would become horizontal at the same time as the line for Z, because [X] would remain constant.

If we consider a small interval of time Δt, we find that the concentration of, say, Z changes during that time by Δ[Z] (*Figure 21.1.3*). So, during that time, the rate of change of [Z] is Δ[Z]/Δt.

$$\text{Rate of reaction} = \frac{\text{change in concentration of chemical}}{\text{time}}$$

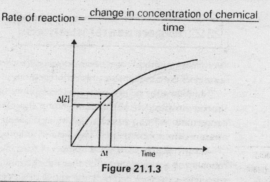

Figure 21.1.3

If you have used the branch of mathematics called calculus, you will know what comes next. We represent an infinitely small period of time by dt, and the corresponding change in [Z] by d[Z]. Therefore, at time t the rate of change of [Z] is d[Z]/dt.

This value can be found by measuring the *gradient of the tangent* to the curve at time t. (When doing this, always take care to get the slope as accurate as possible, and draw good long lines in order to get as accurate a value for the tangent as possible.) From *Figure 21.1.4*, we can see that the tangent to the curve at time t is a/b.

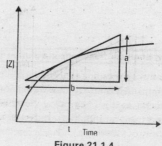

Figure 21.1.4

So, at time t,

$$\frac{d[Z]}{dt} = \frac{a}{b}$$

This gives the rate of change of concentration of the product, Z, at time t. The equation for the reaction we are considering is, remember, X + 2Y → Z. So, for every mole of Z produced, one mole of X is used up.

So, if d[Z]/dt = a/b, d[X]/dt at time t must be −a/b.

And, because each new mole of Z means that two moles of Y are consumed, d[Y]/dt must be −2a/b.

What is the *rate of the reaction* X + 2Y → Z? You can now see that this is an impossible question to answer, *unless we specify which product or reactant we are referring to.*

Factors affecting the rates of reactions

These have been discussed already in Ch 12. The factors involved are **temperature**, **concentration**, **pressure** (in reactions involving gases), **surface area** of solids, **light** and **catalysis** (see *Figure 21.1.1*). If you are still uncertain about how and why these factors operate, you should work though the appropriate sections of Ch 12 now.

21.2 Experimental methods

Very many methods have been developed for following the rates of reactions. We shall limit ourselves here to discussing only those methods which might be met at A level.

To follow the rate of a particular reaction we have to measure the concentration of one of the species involved in the reaction during the course of the reaction. A rate curve shows the concentrations of a particular species at different times. So, to obtain such a curve we need to measure the concentration of one material at known time intervals.

In any reaction, there are a number of reactants and products. Any one of these can be used to monitor the reaction, provided that the chemical equation for the reaction is known. Unless the reaction is extremely complex, the rate of decay of one reactant must be proportional to the rate of decay of any other reactant, and proportional to the rate of formation of any of the products.

For example, the rate at which hydrochloric acid is used up in the reaction of magnesium and acid:

$$Mg(s) + 2HCl(aq) \longrightarrow MgCl_2(aq) + H_2(g)$$

must be related to the rate of formation of hydrogen gas.

$$\therefore \frac{-d[HCl]}{dt} = 2 \times \frac{d[H_2]}{dt}$$

It is unlikely that we would worry about studying the rate of a chemical reaction before we knew the overall equation for that reaction.

'Wet analysis': reactions in solution

One way of measuring concentration is to set up the reaction under investigation, and then take a relatively small sample of known volume from the reaction mixture at set time intervals. This sample, or *aliquot* as it is often called, can then be used to measure the concentration of a reactant or product using an appropriate chemical reaction – very often involving a titration of some kind. This basic way of following a chemical reaction is often termed '*wet analysis*'.

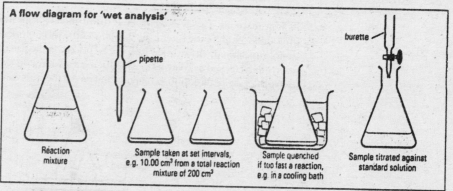

A flow diagram for 'wet analysis'

Reaction mixture

pipette

Sample taken at set intervals, e.g. 10.00 cm³ from a total reaction mixture of 200 cm³

Sample quenched if too fast a reaction, e.g. in a cooling bath

burette

Sample titrated against standard solution

If the reaction is slow (i.e. it takes a day or more to complete), the time needed to take the sample and do the titration poses no difficulty. If, however, the reaction is complete in a much shorter time (anything less than 2 hours), the time taken to sample and titrate becomes significant in the time scale for the complete reaction. During the process of sampling and titrating, the reaction is still going on. What time do you record for any particular sample – the time at which the sample was taken or when the titration was complete?

One way of overcoming this problem is to *quench* the sample – further reaction is stopped at the time of sampling. Quenching can be done by cooling a sample from room temperature to almost 0 °C (but not to 0 °C, when the sample might freeze). An alternative way of quenching is to remove one of the reactants immediately. In any reaction involving an acid, the sample can be run immediately into a solution that will neutralise the acid (e.g. sodium carbonate solution) but will not interfere with the reaction being studied.

Titration is useful for reactions between solutions in which the concentration of one of the reactants or products can be measured using a standard solution of a suitable reagent. For example, the alkaline hydrolysis of an ester (see Ch 27.10) such as methyl methanoate, $HCOOCH_3$, can be followed. The equation is:

$$HCOOCH_3(aq) + OH^{\ominus}(aq) \longrightarrow HCOO^{\ominus}(aq) + CH_3OH(aq)$$

The reacting solutions are mixed in a flask, usually immersed in a thermostated water-bath to keep the temperature constant at 25 °C (298 K), and a clock started. A magnetic stirrer can be used to keep the mixture completely homogeneous. At fixed intervals a small amount of the mixture can be removed using, say, a 10.0 cm³ pipette. Note that removing a sample does not change the composition of the mixture. The only problem is that the reaction keeps on going *in the sample*. The sample is therefore immediately put into a known excess amount of some reagent which stops the reaction (this technique is known as *quenching* – see above). Here, an excess of dilute acid would do. The resulting mixture is then titrated at leisure with standard alkali.

As this hydrolysis reaction goes on, $OH^{\ominus}(aq)$ ions are used up; so the concentration of the alkali in the reaction mixture goes down. Each successive sample taken out of the mixture will therefore react with less of the quenching acid, which means that more of the standard alkali will be needed in the titration.

A look at the equation shows that every mole of $OH^{\ominus}(aq)$ ions takes a mole of ester with it when it goes – and produces a mole each of methanol and methanoate ions. So, if you know the concentration of each of the ester and alkali at the start, *each measurement of the concentration of alkali left in the mixture in a sample tells you the concentration of the three other species at the time the sample was taken*. It is therefore possible to follow precisely the course of the reaction.

Volume or pressure of a gas

In many reactions, a gas is given off – e.g. the reaction of many metals or marble chips with dilute hydrochloric acid. In the latter case, the *mass* of the relatively dense carbon dioxide given off can easily be measured. The reaction flask can be placed on the pan of an electronic balance, and the mass of the flask and its contents recorded at regular intervals (*Figure 21.2.1*). It may be possible to couple the balance directly to a computer and record the mass continuously.

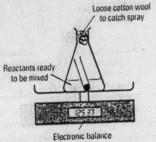

Figure 21.2.1 **Measuring the rate of change of mass**

The rate of formation of the gas, and so the rate of reaction, can therefore be found. The information can be plotted as shown in *Figure 21.2.2*. Remember that the loss in mass will be small in comparison with the mass of the flask and reactants: the mass plotted on the vertical scale should not therefore start at zero!

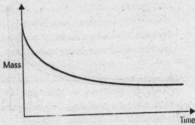

Figure 21.2.2

Alternatively, the volume of the gas given off can be measured using either an inverted burette or a gas syringe. Again, it may be possible to couple the syringe to a computer and make a continuous recording. The basic arrangement using a syringe is shown in *Figure 21.2.3*. Care should be taken that the final volume of gas given off does not exceed the volume of the gas syringe.

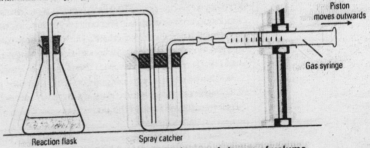

Figure 21.2.3 **Measuring rate of change of volume**

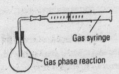

Figure 21.2.4 Volume change in gas phase reaction

In some reactions which involve only gases, the number of molecules of the products is different from the number of molecules of reactants. As the reaction proceeds in a vessel at a fixed pressure, the total gas volume will change. This change in volume can be followed using a gas syringe (*Figure 21.2.4*).

For example, in the reaction of hydrogen with nitrogen oxide at more than 100 °C:

$$2NO(g) + 2H_2(g) \longrightarrow N_2(g) + 2H_2O(g)$$

four moles of reactants produce three moles of products. So, as the reaction proceeds in a container of fixed volume, there is a reduction in pressure which is directly proportional to the number of moles of nitrogen formed.

Instrumental methods

'Wet analysis' measures the concentration of a reactant or product directly using chemical methods. There are other properties of some reaction mixtures which can be used to measure concentration. As long as a particular property is proportional to the concentration of one or more of the species being investigated, the rate of reaction can be followed using that property.

Colorimetry

As its name implies, this method involves colour intensities, and is suitable for coloured solutions containing e.g. iodine or potassium manganate(VII). The experiment is set up as in *Figure 21.2.5*. The colour of the filter is chosen to match the colour in the reaction mixture.

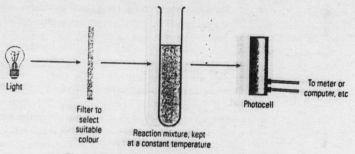

Figure 21.2.5   The principle of the colorimeter

As the reaction proceeds and the reactants are used up, so the intensity of colour in the reaction mixture changes, which means that the intensity of light getting through to the photocell changes (*Figure 21.2.6*). The current generated by the photocell is proportional to the intensity of light falling on it. This can be read off from the meter or recorded continuously by chart recorder, data logger or computer.

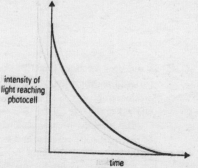

Figure 21.2.6   A typical intensity curve for a reaction in which the colour in the reaction mixture becomes more intense

Conductimetry

This method uses measurements of the conductivity of a solution. In many ionic reactions in solution, the total number of ions of products is different from that of the reactants. An extreme example is given by the reaction of manganate(VII) and ethanedioate ions in acid solution.

$$2MnO_4^{\ominus}(aq) + 16H^{\oplus}(aq) + 5C_2O_4^{2\ominus}(aq) \longrightarrow 2Mn^{2\oplus}(aq) + 10CO_2(g) + H_2O(l)$$

Here there are 23 moles of ions of reactants which will end up as just 2 moles of ions among the products. As it is the ions in a solution which enable it to conduct electricity, the conductivity of the solution is bound to change as the reaction proceeds, and this can be followed using suitable apparatus. This last reaction could be studied using all five methods we have mentioned so far – can you see why?

Other methods

As we have said, any property of the reacting mixture which varies with the extent to which the reaction has proceeded can be used to follow the rate of that reaction. Any spectroscopic technique (see Ch 30.5) can in principle be used; the reaction can be set up to occur in the sample chamber of the machine at constant temperature and followed continuously. Examples include the appearance or disappearance of characteristic absorption peaks caused by products or reactants in the infra-red (IR), ultra-violet (UV) or nuclear magnetic resonance (NMR) spectra.

For some reactions involving chiral molecules (see Ch 14.4), the reaction can be followed by using a polarimeter to follow changes in the rotation of the plane polarised light as it passes through the reaction solution (see Ch 28.5).

'Clock' methods

These methods measure the time taken for a definite, *small* amount of product to be formed *at the start of the reaction*, i.e. when the reaction is at its most rapid. Provided that no more than 10–20% of the total amount of product is formed when the measurement is taken, the portion of the curve up to that point is virtually a straight line (*Figure 21.2.7*). The rate at the start of the reactions can therefore be taken as proportional to the initial concentration of reactant.

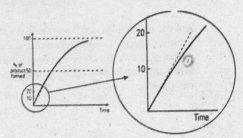

Figure 21.2.7

A series of reactions could be done in which the *concentration of* only *one of the reactants varied*, the others being kept constant, and the time noted in each case for a definite amount product to be formed. (The *rate* of formation of the product is inversely proportional to the time rate $\propto$ 1/time.)

This gives information about the way in which the rate of reaction depends on the concentration of that particular reactant and hence about the **order of the reaction** (see next section); and possibly the **mechanism** of the reaction (see Ch 21.5).

The difficulty with this method is how to define clearly the amount of product to be formed the time to be measured. An outline of two methods is given here.

The thiosulphate reaction

The equation for the reaction is:

$$S_2O_3^{2\ominus}(aq) + 2H^{\oplus}(aq) \longrightarrow S(s) + SO_2(g) + H_2O(l)$$

Thiosulphate
ion

As the reaction proceeds, solid sulphur is produced, as very fine particles. This will cause the initially transparent colourless solution to gradually become opaque.

A cross is drawn on a piece of card and the reaction flask containing one reactant is placed over it (*Figure 21.2.8*). The other reactant is added and a clock started as the flask is swirled briefly to ensure good mixing. The time taken for the cross just to become invisible, as the mixture becomes more cloudy, is noted.

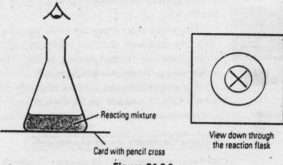

Reacting mixture

Card with pencil cross

View down through
the reaction flask

Figure 21.2.8

A variation of this reaction uses exactly the same amounts and concentrations of reactants each time but changes the temperature in the reaction flask. This gives information about the link between reaction rate and temperature, and can be used to calculate the activation energy for the reaction (see Ch 21.5).

The experiment can be repeated using different initial concentrations of reactants. The temperature must be held constant, and so must the total volume and hence depth of the liquid mixture in the flask. Information about the link between the rate of reaction and the initial concentration of the reactants is obtained (*Figure 21.2.9*).

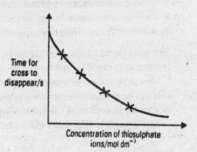

Time for
cross to
disappear/s

Concentration of thiosulphate
ions/mol dm^{-3}

Figure 21.2.9   Results graph

The iodine clock

With an aqueous oxidising agent, iodide ions can be converted into iodine. A yellow-brown solution of iodine in aqueous iodide is formed. Iodine reacts with aqueous thiosulphate to form a colourless solution.

Suitable oxidising agents are peroxodisulphate, $S_2O_8^{2\ominus}(aq)$, or hydrogen peroxide, H_2O_2. We shall consider the reaction with hydrogen peroxide, acidified with a small amount of dilute sulphuric acid.

$$H_2O_2(aq) + 2H^{\oplus}(aq) + 2I^{\ominus}(aq) \longrightarrow 2H_2O(l) + I_2(aq)$$

A mixture containing:

- a known volume of one of the reactant solutions, plus
- a definite small volume of standard sodium thiosulphate solution, plus
- another fixed small amount of dilute sulphuric acid, and
- 1 cm^3 of freshly prepared starch solution

is made up. A known volume of the other reactant solution is added and the flask swirled gently as a clock is started.

At first any iodine formed reacts very rapidly with thiosulphate ions:

$$2S_2O_3^{2\ominus}(aq) + I_2(aq) \longrightarrow 2I^{\ominus}(aq) + S_4O_6^{2\ominus}(aq)$$

Thiosulphate ion   Tetrathionate ion

The solution therefore stays colourless. As soon as the iodine made by the reaction has used up all the thiosulphate, any further iodine produced reacts at once with the starch to give a deep blue-black colour.

The overall effect is that, once the reactants are mixed, the contents of the flask stay colourless until – very suddenly and completely – they turn blue-black! This gives the time needed to form a definite amount of the product iodine. The rate of reaction corresponding to the initial concentration of reactants can be found.

The temperature, the volumes of standard thiosulphate solution and starch indicator, and the total volume of liquid in the flask are all kept constant. The effect on reaction rate of varying the concentrations of each of the reactants in turn can then be investigated.

mix solutions   18s   19s

A 'clock' reaction

Using the results of rate experiments

In a rate experiment such as those described above, you are measuring *the change with time' in the amount of a reactant or product*. This is often done by measurement of a property which is *related* to the amount of the substance present – e.g. colour intensity, volume of gas, etc.

So, to measure the rate of a reaction:

- decide on a suitable property (of a reactant or product) which can be easily measured;
- ensure that all other variables are kept the same initially;
- measure the change in the property in a certain time;
- find the rate, as change of property over time.

The rates found in this way will be in a variety of units (although they all will contain the unit s^{-1}), but the units can all, if necessary, be converted to mol s^{-1}.

Suppose we are investigating the decomposition of hydrogen peroxide solution using a suitable catalyst. Keeping the temperature, the amount of catalyst, and the volume of solution all constant, graphs are plotted of the volume of oxygen given off against time for different initial concentrations of H_2O_2, $[H_2O_2]$. The result will look like *Figure 21.2.10*.

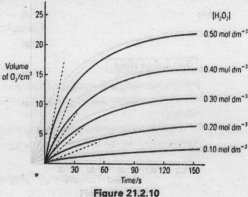

Figure 21.2.10

The **initial rate of reaction** for each concentration is then found by drawing a tangent to each curve at time zero and finding the gradient of each tangent. The gradients are then plotted against initial $[H_2O_2]$. Typical results are given in *Table 21.2.1*; the resulting graph is given in *Figure 21.2.11*.

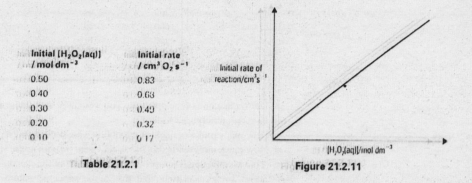

Initial $[H_2O_2(aq)]$ / mol dm^{-3}	Initial rate / cm^3 O_2 s^{-1}
0.50	0.83
0.40	0.68
0.30	0.49
0.20	0.32
0.10	0.17

Table 21.2.1

Figure 21.2.11

Because the graph is a straight line, we can write:

$$\text{Rate} = \text{constant} \times [H_2O_2(aq)]$$

In the next section, we see how to use such results. The value of the constant is given by the gradient of the line in *Figure 21.2.11*.

SELF-ASSESSMENT QUESTIONS 21.2

1 Decide on and describe the appropriate apparatus you would use to measure the rates of the following chemical reactions.

  a The reaction of granulated zinc with dilute hydrochloric acid.

  b The gas phase decomposition of N_2O_5 forming NO_2 gas and oxygen.

  c The reaction of copper(II) sulphate solution with granulated zinc metal.

  d The reaction of bromoethane with sodium hydroxide solution to produce ethanol and sodium bromide solution.

2 In the reaction of granulated zinc metal with 1.0 mol dm^{-3} nitric acid at 20 °C, how will the rate of reaction alter by using the following conditions?

  a Using 0.5 mol dm^{-3} nitric acid.

  b The reaction is carried out at 15 °C.

  c Using copper(II) sulphate solution as a catalyst.

  d Using powdered zinc metal.

3 Magnesium metal reacts with dilute hydrochloric acid according to the equation:

$$Mg(s) + 2HCl(aq) \longrightarrow MgCl_2(aq) + H_2(g)$$

In an experiment to follow the rate of this reaction, 1.0 g of magnesium ribbon was added to 50 cm^3 of 0.12 mol dm^{-3} acid and the hydrogen gas collected in a gas syringe. The volume of gas collected at set time intervals was recorded as follows. (One mole of gas at room temperature and pressure occupies 24 000 cm^3.)

Time / s	Volume of hydrogen / cm³
20	25
40	43
60	55
80	64
100	69
120	71
140	72
160	72

a Draw a diagram of the apparatus you would use to monitor this reaction.

b Plot a graph of these results.

c How many moles of magnesium ribbon were used in this reaction?

d How many moles of hydrochloric acid were present at the start of the reaction?

e What volume of gas was produced by the end of the reaction?

f Show, by a calculation, that this is the expected volume of gas when the reaction is complete.

g How long did it take for half of the hydrochloric acid to be used up?

21.3 Order of reaction

Measurements as in Ch 21.2 enable us to find out how the rate of a reaction is affected by the concentration of each of the reactants.

Consider the reaction:

$$A + B \longrightarrow products$$

> The rate equation *cannot be predicted from the equation* – it must be found by experiment.

It is found by experiment that the rate equation for this reaction is:

$$Rate = k[A]^x \times [B]^y$$

where *k* is the **rate constant** for the reaction and x or y are the order with respect to the reactants A or B, respectively.

The overall order of a reaction is the sum of the powers of the concentrations of the individual reactants shown in the rate equation.

So, the overall order of the reaction $A + B \rightarrow$ products above is $(x + y)$. The order with respect to reagent A is x, and with respect to reagent B is y.

If $x = 1$, the reaction is said to be **first order** with respect to reagent A. If $y = 2$, the reaction is **second order** with respect to reagent B.

> **Definition:** The order of a reaction with respect to a particular reactant equals the power to which the concentration of the reactant must be raised when the experimentally found rate equation is written.

For example, for the alkaline hydrolysis of 1-bromobutane (a primary halogenoalkane) with alkali (see Ch 17.5), the rate equation is found to be:

$$Rate = k[\text{1-bromobutane}][OH^{\ominus}]$$

So, this reaction is *first order* with respect to each of 1-bromobutane and hydroxide ion, and *second order* overall.

However, for the alkaline hydrolysis of 2-bromo-2-methylpropane (a tertiary halogenoalkane), the rate equation is:

$$Rate = k[\text{2-bromo-2-methylpropane}]$$

Obviously, *some* hydroxide ion is needed to achieve the hydrolysis; but changing its concentration does not alter the rate of reaction. We say that this reaction is **zero order** with respect to hydroxide ion, **first order** with respect to 2-bromo-2-methylpropane, and first order overall. As will be seen, this sort of investigation of a reaction can be used to find out the **mechanism** of a reaction. What we have just discussed are the two alternative mechanisms for the hydrolysis of halogenoalkanes (see S_N1 and S_N2 – Ch 17.5).

Reactions may commonly be **zero order**, **first order** or **second order** with respect to a particular reactant. If zero order, the rate of reaction does not change if the concentration of reactant is changed; if first order, the rate is directly proportional to the concentration of reactant; if second order, the rate is proportional to the square of the reactant concentration (*Figure 21.3.1*). These different relationships are summarised in *Figure 21.3.2*. Obviously, plotting a graph of rate against reactant concentration quickly reveals the order.

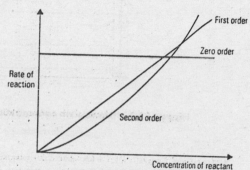

Figure 21.3.2   How reaction rate varies with reactant concentration for different orders

Some gas phase decomposition reactions are zero order (the rate of reaction does not depend on the amount of gas present). The rate of decomposition depends on the diffusion of the gaseous products from a surface.

Orders higher than 2 for specific reactants are uncommon. Fractional orders are occasionally met with but are unlikely at AS and A level.

First order reactions: half life, $t_{\frac{1}{2}}$

Advanced textbooks on physical chemistry usually include the theory of second order and other reactions, but here we shall discuss only reactions which are first order overall.

Remember, for a first order reaction: rate = k[reactant] – the rate depends solely on the concentration of this single reactant. (There may be another reactant taking part in the reaction; if so, as we said above, its order will be zero.) If we plot rate vs. the concentration of reactant we get a straight line going through the origin (well, it has to – at zero concentration of reactant the rate of reaction must be zero!). Plotting the graph and getting such a line is one test for a first order reaction.

Another test involves the idea of half-life.

Now, after one half-life, half the reactant will be left. Suppose we find that after another half-life, half of that will remain – i.e. one quarter of the original reactant, and so on – see *Figure 21.3.3*. Wherever you start on the curve, you find it takes $t_{\frac{1}{2}}$ for the concentration to fall to half the concentration at the start. This is another test for a first order reaction – **all first order reactions have constant half-lives.**

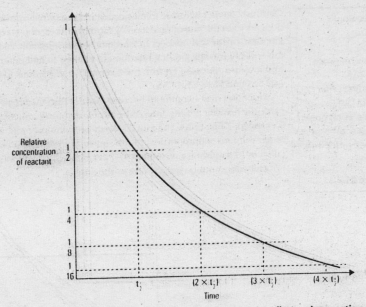

Figure 21.3.3   How relative concentration decreases in a first order reaction

At first sight *Figure 21.3.3* looks like an ordinary decay curve for a reactant. In *any* reaction, concentration of reactant falls with time. But in a first order reaction, the curve is *precisely* as in *Figure 21.3.3*, with a constant half-life.

An example of a first order process is the decay of radioactive nuclei. The idea of radioactive half-life, and the use of this in archaeological dating using $^{14}_{6}C$, has already been discussed (see Ch 1.2). Carbon-14 has a half-life of about 5600 years. For geological dating, a longer lived isotope is needed; $^{40}_{19}K$ has $t_{\frac{1}{2}} = 1.3 \times 10^{9}$ years. It decays to argon which is trapped in the rock and can be measured; it is therefore often used for this purpose.

Summary: effects of concentration on rates

Order of reaction	Double the concentration	Treble the concentration
Zero	No change	No change
1st	×2	×3
2nd	×4	×9

Table 21.3.1

Worked example

From lists of concentrations and reaction rates the order of reaction with respect to a particular reactant can be found by looking at pairs of results (see Table 21.3.1). What is the effect of doubling only one concentration figure on the overall rate of reaction?

Example A
In the reaction of $A + B \rightarrow C$, the following rate results were obtained in three experiments with different starting concentrations of A and B.

Experiment number	[A] / mol dm^{-3}	[B] / mol dm^{-3}	Rate of reaction / mol dm^{-3} s^{-1}
1	0.010	0.10	0.0032
2	0.010	0.20	0.0032
3	0.020	0.20	0.0064

What is the order of reaction for A and B and overall? Write the rate equation for the reaction.

Calculation:

Using the results of experiments 1 and 2, [B] has doubled and [A] is the same. The rate of reaction has not altered.

∴ The order of reaction is zero with respect to B.

With experiments 2 and 3, [A] has doubled and [B] is the same. The rate of reaction has doubled.

∴ The rate is first order with respect to A.

The overall rate equation is: rate = k[A], and the overall order of reaction is 1.

Example B

The aldehyde, ethanal, CH_3CHO, dimerises in dilute alkaline solution according to the equation:

$$2CH_3CHO \xrightarrow{OH^-} CH_3CH(OH)CH_2CHO$$

In a study of the rate of this reaction, a series of three experiments with different starting concentrations of ethanal and hydroxide ions were carried out. The results of the relative initial rates of these reactions are shown in the table.

Experiment number	[OH$^-$] / mol dm^{-3}	[CH$_3$CHO] / mol dm^{-3}	Relative rate
1	0.01	0.15	1
2	0.02	0.15	2
3	0.01	0.30	2

What is the order of reaction for ethanal, hydroxide ion and overall?

Calculation:

In experiments 1 and 2, [OH$^-$] has doubled and the rate of reaction has doubled.

∴ The reaction is first order for [OH$^-$].

In experiments 1 and 3 (in experiments 2 and 3 both concentrations change), the concentration of ethanal has doubled and the rate of reaction has doubled.

∴ The reaction is first order for ethanal.

Overall the reaction is second order and the rate equation is:

$$Rate = k[CH_3CHO][OH^-]$$

Example C

The rate of the gas phase oxidation of NO was followed and a series of results obtained. The equation for the reaction is:

$$2NO(g) + O_2(g) \longrightarrow 2NO_2(g)$$

Experiment number	Initial NO concentration / mol dm^{-3}	Initial O$_2$ concentration / mol dm^{-3}	Initial rate / mol dm^{-3} s^{-1}
1	1	1	4
2	2	1	16
3	3	1	36
4	3	3	108

What is the order of reaction for each reactant and overall?

Calculation:

In experiments 1 and 2 the concentration of NO has doubled and $[O_2]$ is the same; the rate has increased by a factor of 4.

The order of reaction must be second order for NO.

In experiments 3 and 4 (only one concentration can be changed) [NO] is the same but the concentration of oxygen has trebled.

The rate of reaction has also trebled (×3).

So the order of reaction with respect to oxygen must be first order.

The overall reaction is third order and rate $= k[NO]^2[O_2]$

N.B. You will see why we said at the beginning of this section that the rate equation cannot be predicted from the chemical equation.

Remember:

- In any reaction which is zero order with respect to a particular reactant, changing the concentration of that reactant has no effect on the rate of reaction.

    Rate $= k[reactant]^0$ or Rate $= k$

- In a first order reaction, doubling the concentration of the reactant doubles the rate of reaction.

    Rate $= k[reactant]^1$ or Rate $= k[reactant]$

- In a second order reaction, doubling the concentration of the reactant quadruples (×4) the rate of reaction.

    Rate $= k[reactant]^2$.

**SELF-ASSESSMENT
QUESTIONS
21.3**

1 For the reaction:

$$A + 2B \longrightarrow 2C$$

it is found that the rate is multiplied by 4 if [A] is doubled and [B] is kept constant, and is doubled if [B] is doubled and [A] is kept constant. (Remember – square brackets represent *concentration in mol dm^{-3}*.)

a Write the equation which relates the rate of reaction to [A] and [B].

b What would happen to the rate of reaction if [B] is increased by a factor of 4?

c What is the order of reaction:

   i) with respect to A;

   ii) overall?

2 In a study of the reaction:

$$CH_3CSNH_2 + 2OH^\ominus \longrightarrow CH_3COO^\ominus + HS^\ominus + NH_3$$

the rate equation for the reaction was found to be:

$$Rate = k[CH_3CSNH_2][OH^\ominus]$$

a How many molecules of reactant are involved in the balanced equation?

b How many molecules of reactant are involved in the rate equation?

c What is the order of reaction with respect to each reactant and overall?

d What would be the effect on the rate of reaction of halving the hydroxide concentration?

e What would be the effect on the rate of reaction of doubling the CH_3CSNH_2 concentration?

3 Aqueous hydrogen peroxide, H_2O_2, decomposes in the presence of many catalysts.

a Write an equation for the decomposition of hydrogen peroxide.

b What is meant by a catalyst?

c The rate of reaction can be followed by measuring the volume of oxygen evolved as the reaction proceeds. Draw a suitable apparatus for following the reaction rate.

d The reaction is an exothermic reaction. How could you minimise the effect of the reaction mixture getting warmer during the course of the reaction?

e The time taken for half the final volume of oxygen to be produced was measured at different starting concentrations of hydrogen peroxide. In each case the time taken ($t_\frac{1}{2}$) was 20 minutes. What does this tell you about the order of reaction?

f Write a rate equation for the decomposition of hydrogen peroxide.

4 The results of some rate experiments with the following reaction are shown in the table.

$$2H_2(g) + 2NO(g) \longrightarrow 2H_2O(g) + N_2(g)$$

Initial [H_2] / 10^{-3} mol dm^{-3}	Initial [NO] / 10^{-3} mol dm^{-3}	Initial rate / 10^{-3} mol dm^{-3} s^{-1}
0.5	6.0	1.5
1.0	6.0	3.0
1.5	6.0	4.5
3.0	1.0	0.25
3.0	2.0	1.0
3.0	3.0	2.25

a What is the order of reaction with respect to H_2?

b What is the order of reaction with respect to NO?

c What is the order of reaction overall?

d Write a rate equation for the reaction.

e What would the initial rate be if the concentration of H_2 was 3.0×10^{-3} mol dm^{-3} and the initial concentration of NO was 4.0×10^{-3} mol dm^{-3}?

f Plot a graph of initial rate against the concentration of H_2.

21.4 Rate equations and rate constants

As we have seen in Ch 21.3, evidence from experiments can be used to write rate equations for reactions. For a typical reaction

$$A + B \longrightarrow products$$
$$Rate = k[A]^x[B]^y$$

where k is the **rate constant** for the reaction.

The value of k can be found by experiment. For example, if you refer back to *Figure 21.2.11*, in which the initial rate of decomposition of hydrogen peroxide is plotted against the various concentrations of hydrogen peroxide, k is given by the gradient of the resulting straight line. This is because, experimentally, rate = $k[H_2O_2]$.

The value of the rate constant, k, can vary greatly with temperature; hence the need to keep the temperature constant when investigating the variation of reaction rates with concentration of reactants. The dependence of k on temperature is discussed in Ch 21.6.

Remember:
For a zero order reaction:
Rate = k
Units of k are mol dm^{-3} s^{-1}.
For a 1st order reaction:
Rate = k[reactant]
Units of k are s^{-1}.
For a 2nd order reaction:
Rate = k[reactant]^2
Units of k are mol^{-1} dm^3 s^{-1}.

The units of k depend of the order of reaction. For example, consider a first order process: rate = k[X].

Here, the units of *rate* are based on the change of concentration of X with time: the units are therefore, mol dm^{-3} s^{-1}. The units of concentration are mol dm^{-3}. *In any equation involving units, the units on each side must be the same overall.*

So for   Rate = k[X];
$$\text{mol dm}^{-3}\text{s}^{-1} = k \times \text{mol dm}^{-3}$$

For the two sides to balance the units for k must be s^{-1}. **The units of k for any first order process are s^{-1}.**

The first order rate equation is rate = k[reactant], which means that a graph of concentration of reactant against rate of reaction should be a straight line, going through the origin. Also, the gradient of the line will give the value of k (in s^{-1}).

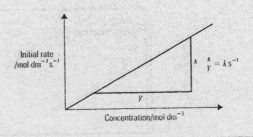

For a second order process,

e.g. rate = k[X][Y];
$$\text{mol dm}^{-3}\text{s}^{-1} = k \times (\text{mol dm}^{-3})^2$$

The units for k for a second order process are therefore:
$$\text{mol dm}^{-3}\text{s}^{-1} / (\text{mol dm}^{-3})^2 = \text{mol}^{-1}\text{dm}^3\text{s}^{-1}$$

The units of k for any second order process are mol^{-1} dm^3 s^{-1}.

There is no simple graphical solution to a second order reaction which produces a straight line.

The units of k for a zero order reaction are mol dm^{-3} s^{-1}. Since rate = k, the units of k must be the same as for rate of reaction (mol dm^{-3} s^{-1}).

For a zero order reaction, rate = k, and so a graph of rate against concentration will produce a horizontal straight line. The value of the rate constant will be the same as the rate of reaction (in mol dm^{-3} s^{-1}).

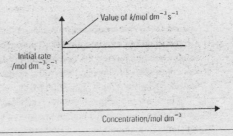

Worked example

Although inspection of a series of rates results can lead to information concerning the order of reaction, it is much more usual to use a graphical method to determine the order of reaction – after all, inspection of one pair of figures can often be misleading, especially if there are significant errors in the readings themselves.

The investigation of the decomposition of a compound C at a fixed temperature gave the following series of results.

Experiment number	Initial concentration of C / mol dm^{-3}	Initial rate / mol dm^{-3} s^{-1}
1	0.0194	1.01×10^{-3}
2	0.0404	1.99×10^{-3}
3	0.0590	3.04×10^{-3}
4	0.0806	3.95×10^{-3}
5	0.1178	6.04×10^{-3}

Use a graphical method to determine the order of reaction with respect to C and hence the overall order of reaction. From your graph calculate the value of the rate constant, k. Use your results to estimate the value of the rate of reaction when the initial concentration is 0.10 mol dm^{-3}.

Simple inspection of the results might not be so obvious with such a variation in figures, but if a graph of rate against concentration of the reactant is plotted then we can identify if the reaction is first order or not. We can also use the gradient of such a graph to calculate the value of the rate constant.

Calculation:

First, plot the graph (*Figure 21.4.1*):

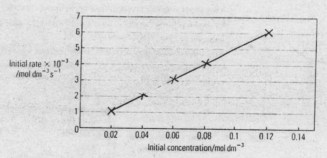

Figure 21.4.1 Initial rate of reaction versus initial concentration

The straight line shows clearly that rate ∝ concentration.

∴ Rate = $k[C]$ and the reaction is first order with respect to C.

From this equation k = rate/[C]

So from the graph we can find the value of $k = 5.1 \times 10^{-2}$ s^{-1} (Do not include too many significant figures – after all, you are estimating the position of the straight line on the graph.)

If we check by looking at one set of data (from experiment 3):

Rate = $k[C]$

$3.04 \times 10^{-3} = k \times 0.590$

$k = 5.1 \times 10^{-2}$ s^{-1}.

When the concentration is 0.10 mol dm^{-3}, the rate of reaction, from the graph, is 5.0×10^{-3} mol dm^{-3} s^{-1}.

1 **a** The half-life of any radioactive isotope has a constant value. What does this tell you about the order of reaction in any radioactive decay?

   b The half-life for ^{60}Co is 5.26 years. How long would it take for an 8 g sample of ^{60}Co to break down to leave only 1 g of ^{60}Co?

   c Tritium, a radioactive isotope of hydrogen, $^{3}_{1}$H, has a half-life of 12.3 years. How long would it take the concentration of tritium in a sample to decrease to 25% of its original value?

   d What fraction of a sample of the ^{220}Rn isotope, half-life 56 s, will be left after 168 s?

2 Sulphur dichloride dioxide, SCl_2O_2, is an important reagent in industrial organic chemistry. It breaks down according to the equation:

$$SCl_2O_2(g) \longrightarrow SO_2(g) + Cl_2(g)$$

The rate of the decomposition was investigated at a fixed temperature. The first experiment measured the concentration of SCl_2O_2 as the reaction proceeded and the results are given in the table.

Time / s	$[SCl_2O_2]$ / mol dm^{-3}
0	0.050
1000	0.041
2000	0.035
3000	0.029
4000	0.025
5000	0.019

a Plot a graph of $[SCl_2O_2]$ against time.

This graph shows a typical decomposition curve for the compound, and was used to determine the initial rate of reaction by measuring the gradient at the very start of the decomposition. The value was 1.04×10^{-5} mol dm^{-3} s^{-1}.

The experiment was repeated with a range of starting concentrations of SCl_2O_2. The gradient at the start of each of these graphs was measured to find the rate of reaction in each case. The results were:

Initial $[SCl_2O_2]$ / mol dm^{-3}	Initial rate / mol dm^{-3} s^{-1}
0.050	1.04×10^{-5}
0.039	0.81×10^{-5}
0.032	0.67×10^{-5}
0.027	0.56×10^{-5}
0.024	0.50×10^{-5}
0.018	0.37×10^{-5}

b Plot a graph of $[SCl_2O_2]$ against reaction rate.

c What does the shape of the graph tell you about the reaction kinetics of the reaction?

d What is the order of the reaction overall?

e Write a rate equation for the decomposition reaction of sulphur dichloride dioxide.

f Using your graph, calculate the value of the rate constant at the temperature of the experiment.

21.5 Reaction rates and reaction mechanisms

Nearly all the chemical reactions we meet are thought to occur in a series of simple steps, each step usually involving no more than two reacting species. A reaction mechanism is this series of steps for a particular reaction.

Reaction mechanisms are man-made, mental models. They are always worked out from detailed experimental observations; they are rarely obvious; and published ones may well turn out to be wrong. (Max Bodenstein spent many years working out a mechanism for the reaction:

$$H_2(g) + I_2(g) \longrightarrow 2HI(g)$$

He was proved wrong about 70 years later.)

Reaction mechanisms often give us a very detailed understanding of what goes on at the molecular level. Finding them out can be complicated, particularly if two or more of the simple steps occur at about the same rate. The situation is simplified if one of the steps is much slower than any of the others. This step will fix the overall rate of the reaction, and is called the **rate-determining step**.

Identification of the rate-determining step can normally be done by finding out, by experiment, the *order* of the reaction with respect to each of the reactants.

For example, the reaction of iodine with the ketone propanone in acid solution is (see Ch 19.1):

$$CH_3COCH_3(aq) + I_2(aq) \longrightarrow CH_2ICOCH_3(aq) + H^{\oplus}(aq) + I^{\ominus}(aq)$$

The reaction is found to be first order with respect to each of propanone and $H^{\oplus}(aq)$ ions, but *zero order* with respect to iodine.

i.e. Rate $= k[CH_3COCH_3(aq)][H^{\oplus}(aq)]$

This shows that the rate determining step in the reaction involves propanone and acid, but not iodine. This suggests that protonation of the propanone by the acid starts the reaction. It then becomes an intellectual puzzle to construct a series of subsequent steps which corresponds to chemical common sense and also fits the data from any further experiments.

The suggested mechanism is:

The term **molecularity of a reaction** is sometimes used. It should not be confused with **order of reaction**. Molecularity refers to the number of species involved in the rate-determining step of the reaction. In the iodination of propanone discussed above, two species are involved in the slow step.

The chances of three or more particles colliding all at once are extremely small.

The overall rate at which a motorway journey is completed may well depend on a single-line section resulting from repairs or an accident. A 20 mile journey on a motorway has more than once been in three *steps* – 10 miles in 10 minutes, then the next 5 miles taking one hour, and the last 5 miles in five minutes. The slowest step determines the overall rate.

The iodination of propanone is not a mechanism which you will normally be required to remember. You should check your syllabus to find out which mechanisms you do need to know. The list would normally include examples of the types introduced in Ch 14.6, and especially:

- free radical substitution (Ch 15.5);
- electrophilic addition to alkenes (Ch 16.6), addition polymerisation (Ch 16.7);
- nucleophilic substitution of halogenoalkanes (Ch 17.5);
- nucleophilic addition to carbonyl compounds (Ch 19.4)
- electrophilic substitution on the benzene ring (Ch 26.6).

S_N1 is shorthand for
Substitution Nucleophilic
unimolecular.
S_N2 means Substitution
Nucleophilic bimolecular.

it is therefore said to be **bimolecular**. If only one species is involved (e.g. in the S_N1 mechanism, Ch 17.5), the reaction is **unimolecular**. It is probably better to use the terms unimolecular and bimolecular for the separate steps in the reaction rather than as a descriptor of the overall reaction. In the iodination of propanone discussed above, it can be seen that steps ① and ③ are bimolecular and ② and ④ are unimolecular.

An observed rate equation may be the result of a complicated series of unimolecular and bimolecular steps which may take a great deal of hard work and ingenuity to sort out.

For a reaction $A + 2B \rightarrow AB_2$, the rate equation has to be found by experiment. If the rate equation is: rate $= k[B]^2$, the reaction is a second order reaction.

This information tells us that the first step in the reaction (the slowest step or the **rate determining step**) involves two molecules of B, which collide to produce some reactive intermediate.

$$B + B \xrightarrow{\text{rate determining step}} [B - B]$$

The sequence that follows this first step will eventually produce the final product.

In this example the next step in the reaction sequence could simply be the collision of this reactive intermediate with one molecule of A to form the product in a fast reaction:

$$[B - B] + A \xrightarrow{\text{fast}} AB_2$$

These two reaction steps are the **reaction mechanism** for the reaction.

SELF-ASSESSMENT
QUESTIONS
21.5

1 For each of the following reactions write an equation for the rate determining step and suggest possible reaction sequence(s) to produce the final product.

a $P + Q \longrightarrow R + S$
Rate $= k[P][Q]$

b $A + B \longrightarrow D$
Rate $= k[A]$

c $2X + Y \longrightarrow W$
Rate $= k[Y]^2$

2 The kinetics of the hydrolysis of a halogenoalkane, RX, with aqueous hydroxide ions were determined as follows.

Rate / 10^{-3} mol dm^{-3} s^{-1}	[RX] / mol dm^{-3}	[OH$^-$] / mol dm^{-3}
0.78	0.103	0.100
1.56	0.204	0.100
1.24	0.163	0.100
6.23	0.823	0.100
3.12	0.200	0.199
6.21	0.200	0.400
4.33	0.200	0.277
0.56	0.200	0.036

a Draw a graph of rate against [RX] and a graph of rate against [OH$^-$]. Make sure that you only use the values where the concentration of the other reactant is the same throughout.

b What is the order of reaction with respect to the halogenoalkane, and with respect to the hydroxide ion?

c Calculate the value of the rate constant at this temperature.

d Suggest a possible mechanism for the hydrolysis of the halogenoalkane. Make sure that the rate determining step is clearly marked.

3 Write a mechanism for each of the following reactions:

a $CH_3CH_3 + Cl_2 \longrightarrow CH_3CH_2Cl + HCl$

b $CH_2=CH_2 + Cl_2 \longrightarrow CH_2ClCH_2Cl$

c $CH_3CHO + HCN \longrightarrow CH_3CH(OH)CN$

4 The gas phase reaction of iodine monochloride with hydrogen is shown in the equation.

$$2ICl(g) + H_2(g) \longrightarrow 2HCl(g) + I_2(g)$$

The results of a kinetics investigation into the reaction were:

[ICl] / mol dm^{-3}	[H$_2$] / mol dm^{-3}	Initial rate / mol dm^{-3} s^{-1}
0.200	0.100	0.041
0.200	0.200	0.167
0.400	0.100	0.083

a What is the order of reaction with respect to ICl, H$_2$ and overall?

b Write the rate equation for the reaction.

c Draw a dot and cross diagram to show the bonding in ICl.

d What is the polarity in the ICl molecule?

e Calculate an average value of the rate constant using these results. What are the units of the rate constant, k?

21.6 Activation energy and reaction rates

Rate constants for reactions are usually dependent on temperature. In this section, we provide a way of linking rate to temperature.

A very simple picture of any chemical reaction is that chemical bonds in the reactants have to be broken before new bonds form to make the products. This idea was discussed in Ch 12, and can be shown in the form of an energy profile diagram as in *Figures 21.6.1* and *21.6.2*.

As we saw in Ch 12.4, catalysts speed up reactions by lowering the activation energy.

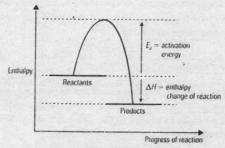

Figure 21.6.1 Energy profile for an exothermic reaction

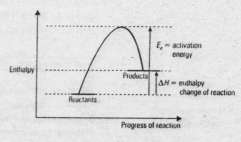

Figure 21.6.2 Energy profile for an endothermic reaction

The activation energy of a reaction is the minimum energy needed for a reaction to occur. (N.B. It does *not* necessarily correspond to the bond enthalpies of the bonds broken; the collision and rearrangement of two rapidly-vibrating molecules is a far from simple process.) The higher the activation energy, the greater the 'energy barrier' to reaction, and therefore the slower the reaction at a given temperature.

Reactions with an activation energy of about 80 kJ mol^{-1} will occur reasonably rapidly at room temperature; with double that activation energy, a temperature of about 400 °C will be needed to achieve the same rate.

As we saw in Ch 12, the existence of this energy barrier explains why exothermic reactions don't happen spontaneously; e.g. why a petrol/air mixture needs a spark. It also explains why compounds with an *endothermic* enthalpy change of formation, i.e. for which $\Delta H_f^{\ominus}$ is positive, can exist at room temperature. Once it is formed, energy has to be applied to the endothermic compound to break it up again.

The peak of the energy profile is often called the **activated complex** – the most unstable arrangement in which the old bonds are not quite broken or the new bonds formed. We met examples when discussing nucleophilic substitution in halogenoalkanes (Ch 17.5).

Arrhenius, in 1889, linked the reaction rate (as measured by the rate constant, k) with the activation energy and temperature in a single equation – the **Arrhenius equation**:

$$k = Ae^{-E_A/RT}$$

where

k is the rate constant;
A is a constant for the particular reaction;
e has the usual mathematical meaning;
E_a is the activation energy;
R is the gas constant; and
T is the temperature in Kelvin.

The right-hand side of the Arrhenius equation can be considered in two parts.

- The constant A (often called the *pre-exponential factor*) can be considered as summarising two influences on the rate of reaction – the *collision frequency* (obviously molecules must collide before they can react) and the *orientation (or steric) factor* (the right bits of molecules have to collide, moving in the right direction, for the correct bonds to break – for an example, see the S_N2 mechanism in Ch 17.5).
- The $e^{-E_a/RT}$ part can be called the *activation state factor*; it is determined by the fraction of molecules with the necessary energy to be able to react (see Ch 12.1).

Some people say that reactions go faster at higher temperature because the molecules move faster and so collide more frequently. However, it can be shown that for a 10 K rise in temperature the average speed of molecules increases by less than 2%, but if the E_a for the reaction is about 50 kJ mol^{-1}, a 10 K rise in temperature *doubles* the rate of reaction. So it's not the speed with which molecules move which matters so much; particles can't necessarily react when they collide. It's the *energy* with which they smash into each other that matters; if it exceeds E_a and the molecules are correctly orientated, reaction can then occur.

Measuring activation energies

For this, you must know a little about logarithms, and especially that $\log_{10}xy = \log_{10}x + \log_{10}y$. Also, that e.g. $\log_{10}100 = \log_{10}10^2 = 2$; and $\log_e e^x = x$.

To return to $\log_{10}xy = \log_{10}x + \log_{10}y$. For example, $10000 = 100 \times 100$. $\log_{10}10000 = 4$; $\log_{10}100 = 2$. Can you see it now? If not, consider the following story.

Noah – once the Ark had safely grounded – said to all the animals 'Go forth and multiply'. A pair of snakes came back and said they'd tried very hard to multiply but couldn't. Fortunately, Noah found some logs among the driftwood from the Flood, and using the logs, the snakes found they could happily multiply. Because they were adders, of course ...

N.B. $\log_e x$ is often written $\ln(x)$, as logarithms to base e are often called 'natural logarithms'; or, since they were invented by the Scots mathematician John Napier in the early 1600s, 'Napierian' logarithms.

Write the Arrhenius equation, $k = Ae^{-E_a/RT}$, and take logs to the base e.
Then:

$$\log_e k = \log_e A - E_a/RT$$

This means that a graph of $\log_e k$ (vertical axis) against $1/T$ should be a straight line with gradient $-E_a/R$ (Figure 21.6.3). The gas constant, R, has the value 8.3 J K^{-1} mol^{-1}; the units of E_a will then be J mol^{-1}.

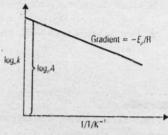

Figure 21.6.3

If logs to the base 10 are used, the above equation becomes:

$$\log_{10} k = \log_{10} A - E_a/2.303RT$$

and the gradient of the graph becomes $-E_a/2.303R$.

Practical methods for measuring the variation of rates with temperature can be found in textbooks of physical chemistry, but we shall not consider them here.

The distribution of molecular energies at a particular temperature is discussed in Ch 12.3. The curves are called Maxwell–Boltzmann curves (Figure 21.6.4). Maxwell and Boltzmann showed that the fraction of molecules with energy greater than E_a – and therefore capable of reacting – is given by $e^{-E_a/RT}$. At any temperature, T, the rate of reaction – and so the rate constant, k – is proportional to $e^{-E_a/RT}$. The Arrhenius equation follows from this.

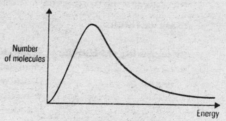

Figure 21.6.4   Maxwell–Boltzmann distribution curve

SELF ASSESSMENT QUESTIONS 21.6

1   a   Sketch the Maxwell–Boltzmann distribution curve for the molecular energy distribution in a gas at room temperature.

    b   On the sketch, draw the line that would represent the distribution of molecular energies at a lower temperature.

c Mark on your sketch the activation energy for a reaction that is extremely slow at room temperature.

d Explain why at the higher temperature the reaction rate is increased, although the activation energy of the reaction does not change at higher temperature.

2 2-Bromo-2-methylbutane is hydrolysed by aqueous potassium hydroxide solution. The rate law for the reaction is given by:

$$\text{rate} = k[\text{2-bromo-2-methylbutane}]$$

a Write the full displayed formula for 2-bromo-2-methylbutane.

b What is the order of reaction in this hydrolysis reaction?

c What type of reaction occurs in this reaction (e.g. addition, substitution, elimination)?

d Write the mechanism for the hydrolysis reaction that is consistent with the rate law quoted above.

Summary

All rates of reaction have to be determined by experiment – the rate equation has no relationship to the balanced equation.

Experimental methods include: wet analysis (titration, gas volumes)

instrumental (colorimetry)

A rate experiment follows the change of concentration with time.

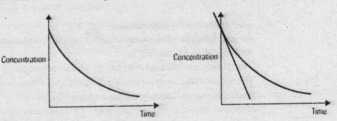

The gradient at the start gives the initial rate of reaction.

Using the results

By measuring the half-life
- If constant then first order.

Comparing effects of changing concentration
Double the concentration:
- no change in rate, zero order;
- rate doubles, 1st order;
- rate ×4, 2nd order.

Plot rate vs. concentration
- Horizontal line – zero order, rate = k.
- Straight line through origin – 1st order, gradient gives k.
- Otherwise results inconclusive.

Order of reaction

Rate equation is rate = $k\,[X]^x\,[Y]^y$

Overall order of reaction is $x + y$.

For zero order – units of k are $\text{mol dm}^{-3}\,\text{s}^{-1}$

1st order – units of k are s^{-1}

2nd order – units of k are $\text{mol}^{-1}\,\text{dm}^3\,\text{s}^{-1}$

SUMMARY
21

In this chapter the following new ideas have been covered. *Have you understood and learned them? Could you explain them clearly to a friend?* The page references will help.

- The meaning of the term rate of reaction.   p522
- Decay curves and growth curves.   p523
- The idea of a limiting reactant.   p523
- The gradient of the tangent to a decay or growth curve gives the rate of change.   p523
- Experimental methods for finding reaction rates.   p524
- The meaning of order of a reaction.   p532
- Half-life.   p533
- The meaning of rate equation and rate constant.   p537
- How to find the units of k.   p538
- What is meant by rate-determining step and reaction mechanism.   p541
- How rate is linked to activation energy and temperature through the Arrhenius equation.   p543
- The principle of measuring E_a for a reaction.   p544

EXAMINATION
QUESTIONS
21

1 The reaction of hydrogen and iodine produces hydrogen iodide.

$$H_2(g) + I_2(g) \longrightarrow 2HI(g)$$

The rate of the reaction was followed by measuring the concentration of iodine as the reaction proceeded. A number of experiments were carried out with different starting concentrations of hydrogen and of iodine. The results are shown in the table.

Initial [H_2] / mol dm^{-3}	Initial [I_2] / mol dm^{-3}	Relative initial rate
0.010	0.010	1.0
0.040	0.010	4.0
0.010	0.030	3.0

a Describe a method by which the concentration of iodine could be measured during the course of the reaction.

b Write a rate equation for the reaction of hydrogen and iodine, including the units for the rate constant.

c At a particular temperature, the initial rate was 1.4×10^{-9} mol dm^{-3} s^{-1}, with initial concentrations of 0.001 mol dm^{-3} for both hydrogen and iodine. Calculate the value of the rate constant at this temperature.

2 In the reaction of $2X + Y$ to form products, a series of kinetic experiments gave the following results.

[X] / mol dm^{-3}	[Y] / mol dm^{-3}	Relative initial rate
1	1	1
3	1	9
3	2	9

a What is the order of reaction overall?
b Write the rate equation for the reaction.
c What are the units of the rate constant?
d Write the rate-determining step of the reaction.
e Sketch the distribution curve for the molecular energies in the reaction mixture.
f Using this sketch explain why the reaction is not completed instantaneously when the two reactants are mixed.

3 One of the catalysed reactions in the modern car exhaust is to remove the carbon monoxide in the exhaust gas by oxidising with nitrogen monoxide and oxygen.

$$CO(g) + NO(g) + O_2(g) \longrightarrow NO_2(g) + CO_2(g)$$

a Why is the removal of CO so important from car exhaust emissions?
b In the equation there are two products, both of which contribute to atmospheric problems. What are the problems associated with each of these products?

A gas phase kinetics experiment obtained the following results for the reaction:

[CO] / 10^{-3} mol dm^{-3}	[NO] / 10^{-3} mol dm^{-3}	[O$_2$] / 10^{-3} mol dm^{-3}	Initial rate / 10^{-3} mol dm^{-3} s^{-1}
0.50	0.50	0.50	2.0
1.0	0.50	0.50	2.0
1.0	1.0	0.50	8.0
1.0	1.0	1.0	8.0

c Why can you not use the balanced chemical equation to work out information about the rate of reaction?
d What is the order of reaction with respect to each of the reactants?
e Write a rate equation for the reaction.
f Calculate the value of the rate constant at the temperature of the experiment.

- The meaning of **reversible reaction** and **dynamic equilibrium**. (Ch 13.1)
- The **factors** which affect equilibrium, especially **temperature**. (Ch 13.2)
- The **equilibrium constant**, K_c. (Ch 13.2)
- **Le Chatelier's Principle.** (Ch 13.2)
- How Le Chatelier's Principle can be used in **industrial** processes. (Ch 13.3)

In this chapter

We shall first make sure that you fully understand the ideas met in Ch 13. (If the checklist in the box above makes you feel at all uncertain, it would be sensible to work briefly through Ch 13 again before proceeding. The first section of this chapter will include a lot of questions!)

We shall then move from general equilibria in solution to equilibria in gaseous systems and in solid/gas systems. The final section of this chapter will remind you of the use of equilibrium ideas in working out the best conditions for industrial reactions, and also look at equilibria involving sparingly soluble solids. The next two chapters will be devoted to two very important special cases of equilibria in solution – acid–base equilibria and redox equilibria.

22.1 Equilibria in solution and K_c: a reminder
22.2 Gaseous equilibria: K_p
22.3 Some applications of equilibria
22.4 Sparingly soluble salts: K_{sp}

22

equilibria

22.1 Equilibria in solution and K_c: a reminder

In Ch 13.1 we pointed out that any chemical equilibrium is a **dynamic equilibrium**: nothing appears to be happening, but in reality two reactions are occurring in opposite directions at exactly equal rates. We also pointed out that the relative amounts of products and reactants at equilibrium, i.e. what is often called the **position of equilibrium**, depend on the conditions. Because chemical equilibrium involves making and breaking bonds, and that usually involves a great deal of heat energy, temperature often greatly affects the position of equilibrium.

In Ch 13.2 we showed how the effect of changing conditions for an equilibrium system can be predicted using **Le Chatelier's Principle**, and in Ch 13.3 we discussed how the Principle is applied in industrial methods such as the Haber Process for making ammonia and the Contact Process for sulphuric acid. We also explained that chemical equilibria can only be reached in a *closed system*, i.e. one into which nothing enters and from which nothing leaves.

> For many reactions in solution, indeed for all those in which reactants and products are soluble and involatile, an ordinary beaker or flask is a perfectly adequate closed system.

For any chemical equilibrium:
- the equilibrium is dynamic;
- the equilibrium constant is temperature dependent;
- the changes in the equilibrium position can be predicted using Le Chatelier's Principle.

Le Chatelier's Principle tells us that when a system in equilibrium is subjected to a change the equilibrium changes in such a way as to tend to counteract this change.

If we look at the reaction of dinitrogen tetraoxide, N_2O_4, breaking up into nitrogen dioxide, we can use Le Chatelier's Principle to predict the outcome of any changes on the equilibrium mixture.

$$N_2O_4(g) \rightleftharpoons 2NO_2(g) \qquad \Delta H = +58 \text{ kJ mol}^{-1}$$

Changing concentrations

Adding a reactant (in this case N_2O_4) pushes the reaction forward (although the value of the equilibrium constant, K_c, is unchanged).

$$N_2O_4(g) \rightleftharpoons 2NO_2(g)$$

More NO_2 is formed.
Or removing a reactant pulls the reaction back:

$$N_2O_4(g) \rightleftharpoons 2NO_2(g)$$

More N_2O_4 is formed.
The same ideas will apply as a product is added or removed from the equilibrium mixture.

Changing temperature

Exothermic reactions release energy in the course of the reaction and endothermic reactions take in energy. Altering the temperature of an equilibrium reaction alters the value of the equilibrium constant. Raising the temperature of an endothermic reaction at equilibrium raises the value of the equilibrium constant, K_c, so that more of the product is formed (*Figure 22.1.1*).

$$\xrightarrow{\text{raise temperature}}$$
$$N_2O_4(g) \rightleftharpoons 2NO_2(g) \qquad \Delta H = +58 \text{ kJ mol}^{-1}$$
$$\text{yellow} \qquad \text{brown}$$

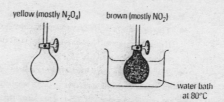

Figure 22.1.1   Temperature affects an equilibrium

Lowering the temperature will lower the value of the equilibrium constant so that more reactant, N_2O_4, is produced in this example.

$$\xleftarrow{\text{lower temperature}}$$

$$N_2O_4(g) \rightleftharpoons 2NO_2(g) \qquad \Delta H = +58\,\text{kJ mol}^{-1}$$

Changing pressure

In most equilibrium reactions involving gases, changing the pressure will alter the position of equilibrium – although the equilibrium constant does not change. To predict the effect of pressure changes, you need to note how many reactant gaseous molecules and product gaseous molecules there are. Increasing pressure will push the equilibrium to the side with the fewer gas molecules.

$$\xleftarrow{\text{increase pressure}}$$

$$N_2O_4(g) \rightleftharpoons 2NO_2(g)$$
$$\text{one gas} \qquad \text{two gas}$$
$$\text{molecule} \qquad \text{molecules}$$

Decreasing pressure in this example will produce more NO_2 (*Figure 22.1.2*).

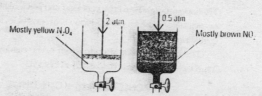

Figure 22.1.2   Pressure affects a gaseous equilibrium

Pressure changes will have no effect on gas reactions with the same number of gas molecules on each side of the equilibrium.

The equilibrium constant

We considered in Ch 13.2 a general reversible reaction which reaches equilibrium:

$$wA + xB \rightleftharpoons yC + zD$$

We showed that such a reaction is found experimentally to give rise to an equilibrium constant based on concentrations, K_c, such that:

$$K_c = \frac{[C]^y_{\text{eqm}}\,[D]^z_{\text{eqm}}}{}$$

where $[\ ]_{\text{eqm}}$ represents the concentration in mol dm^{-3} of a component in the equilibrium mixture.

For convenience, we shall omit 'eqm' from the concentration $[\ ]$ symbols used in equilibrium constant expressions.

Points to note

- The *products* of the reaction are always on the upper line of the K_c expression. This usually means that a large value for K_c indicates that there is a large concentration of products in the equilibrium mixture (but see Ch 13.2).
- The *units of K_c depend on the values of w, x, y, and z in the equation above*. The units will always be a multiple of mol dm^{-3} (e.g. mol dm^{-3}, mol^2 dm^{-6}, mol^{-1} dm^3, etc).
- No catalyst (see Ch 12.4) can alter the position of equilibrium, i.e. it cannot change the value of K_c for a reaction.
- Changing the temperature has a considerable effect on nearly all equilibria of any kind.

A large value of K_c tells us *how far* a reaction has gone. For example, in the reaction of hydrogen and oxygen to form water:

$$2H_2(g) + O_2(g) \longrightarrow 2H_2O(l)$$

K_c is approximately 9×10^{80} dm^5 mol^{-1} at 25 °C – which means that virtually all the $H_2(g)$ and $O_2(g)$ has reacted. We would not normally write an equilibrium double arrow for the reaction.

Equilibrium calculations

For any reaction which reaches equilibrium at a particular temperature, the following information may be available or can be found by experiment.

- The *equation* for the reaction.
- The initial amount(s) or concentration(s) of the starting material(s).
- The concentrations of all the components in the equilibrium mixture.

The amounts *put in* to the reaction can obviously be controlled, and the *equation* is usually known or available from a book. How is it possible to find the concentration of *all* components at equilibrium?

The key to solving this problem is the reaction equation. Consider the 'esterification' reaction which we have already discussed in Ch 13.2:

$$CH_3COOH(aq) + C_2H_5OH(aq) \rightleftharpoons CH_3COOC_2H_5(aq) + H_2O(l)$$

Ethanoic acid   Ethanol   Ethyl ethanoate

$$\therefore K_c = \frac{[CH_3COOC_2H_5(aq)]\ [H_2O(l)]}{[CH_3COOH(aq)]\ [C_2H_5OH(aq)]}$$

> Ethyl ethanoate is one of a class of compounds called *esters* – see Ch 18.4 and Ch 27.9.

If we are doing the experiment, we will know how much ethanoic acid and ethanol we have put in.

The equation shows that for every mole of ethanoic acid which reacts, one mole of ethanol is used up, and one mole each of ethyl ethanoate and water are formed.

$$CH_3COOH(aq) + C_2H_5OH(aq) \rightleftharpoons CH_3COOC_2H_5(aq) + H_2O(l)$$

1 mole   reacts with 1 mole   produces 1 mole   produces 1 mole

So, if we know how much acid we have put in, and we can find by titration (see Ch 10.2) how much acid is left in the equilibrium mixture, we can easily calculate how much ethanol has been used up and how much ethyl ethanoate and water has been formed. We can then put all the concentrations into the K_c expression and work out the value of K_c.

Note that for many reactions the *volume* of the system at equilibrium will need to be known so that the *concentrations* of components can be worked out. For the esterification reaction, the concentration terms cancel out, so the volume of the system at equilibrium is not needed. This also means that K_c for the esterification reaction *has no units*.

This reaction is relatively easy to deal with because of the 1:1 ratios and the simple method of finding the equilibrium concentration of one product. But other reactions can be treated in a similar way using other methods, and the general idea can be stated like this:

For any equilibrium reaction for which:

- **the equation is known, and**
- **the initial concentrations of the reactants are also known,**
- **if the concentration of *one* of the components in the equilibrium mixture can be found;**
- **then the equilibrium constant K_c for the reaction can be determined.**

Worked example

Consider the reverse of the esterification reaction, in which ethyl ethanoate reacts with water to form ethanoic acid and ethanol.

A   *In an experiment, 44 g of ethyl ethanoate are mixed and stirred with 9.0 g water until equilibrium is reached. 0.167 mol ethanoic acid was found, by titration, to be present in the equilibrium mixture. All measurements were made at 25 °C (298 K). Calculate the equilibrium constant, K_c, for the reaction.*

$$CH_3COOC_2H_5(aq) \ + \ H_2O(l) \ \rightleftharpoons \ CH_3COOH(aq) \ + \ C_2H_5OH(aq)$$

Data:

	$CH_3COOC_2H_5$	H_2O	CH_3COOH	C_2H_5OH
Mass at start	44 g	9 g	0	0
Mass of 1 mole / g	88	18	–	–
Number of moles at start	$\frac{44}{88} = 0.50$	$\frac{9}{18} = 0.50$	–	–
Number of moles at equilibrium	$(0.50 - x)$	$(0.50 - x)$	x	x

– where x is the number of moles of each component used up, or produced, according to the equation.

K_c expression:

$$K_c = \frac{[CH_3COOH(aq)]\,[C_2H_5OH(aq)]}{[CH_3COOC_2H_5(aq)]\,[H_2O(l)]}$$

Calculation: 0.167 mol ethanoic acid was produced at equilibrium, so $x = 0.167$

∴ Moles at equilibrium $(0.50 - 0.167)$ $(0.50 - 0.167)$ 0.167 0.167

∴ Assuming the total volume of the system at equilibrium is V dm³, the concentration of the components at equilibrium are:

$$[CH_3COOC_2H_5] = [H_2O] = \frac{0.333}{V}$$

$$[CH_3COOH] = [C_2H_5OH] = \frac{0.167}{V}$$

$$K_c = \frac{[CH_3COOH]\,[C_2H_5OH]}{[CH_3COOC_2H_5]\,[H_2O]}$$

$$= \frac{\frac{0.167}{V} \times \frac{0.167}{V}}{\frac{0.333}{V} \times \frac{0.333}{V}} = 0.25 \quad \text{(to 2 significant figures)}$$

N.B. Because the mol dm⁻³ units at the top and bottom lines of the expression for K_c cancel out, K_c here has *no units*. K_c for the *hydrolysis* of ethyl ethanoate to the acid and alcohol is precisely $1/K_c$ for the formation of ethyl ethanoate from the same acid and alcohol (see Ch 13.2); so K_c for the *formation* of ethyl ethanoate is $1/0.25 = 4.0$ at 25 °C.

Worked example

B   *We shall repeat the worked example above, but this time using different quantities of starting materials: 0.50 mol of ethyl ethanoate and 1.00 mol of water. At equilibrium at 25 °C, 0.228 mol of ethanoic acid is found in the equilibrium mixture.*

$$CH_3COOC_2H_5(aq) + H_2O(l) \rightleftharpoons CH_3COOH(aq) + C_2H_5OH(aq)]$$

Data:

Moles at start	0.50	1.00	0	0
Moles at equilibrium			0.228	

K_c expression:
$$K_c = \frac{[CH_3COOH(aq)]\,[C_2H_5OH(aq)]}{[CH_3COOC_2H_5(aq)]\,[H_2O(l)]}$$

Calculation: If one mole of CH_3COOH was produced then the same number of moles of C_2H_5OH would be formed. Similarly, since all ratios in the equation are 1:1, the same number of moles of ester and water must have reacted

∴ Moles at equilibrium

$(0.50 - 0.228)$	$(1.00 - 0.228)$	0.228	0.228
$= 0.272$	$= 0.772$		

If the total volume is V dm³:

then
$$[CH_3COOC_2H_5] = \frac{0.272}{V}$$

$$[H_2O] = \frac{0.772}{V}$$

$$[CH_3COOH] = [C_2H_5OH] = \frac{0.228}{V}$$

$$\therefore K_c = \frac{\frac{0.228}{V} \times \frac{0.228}{V}}{\frac{0.272}{V} \times \frac{0.772}{V}} = \frac{0.0520}{0.210} = 0.25 \quad \text{(to 2 significant figures)}$$

Note that the value of K_c is independent of the amount of starting materials used, so long as the temperature remains constant.

Note also that *increasing the amount of one of the starting materials increases the yield of product.* This is to be expected from Le Chatelier's Principle (Ch 13.2).

Worked example

Working the problem like this the other way round, i.e. using the value of K_c at 25 °C together with the amounts of starting materials to find the composition of the equilibrium mixture, involves solving a *quadratic equation.*

Check your A level exam syllabus. Some state explicitly that you will not be required to solve a quadratic equation. Just in case yours does not, and for the sake of completeness, we show how such a problem can be solved.

Problem: using the same reaction and quantities as in worked example B and given that K_c at 25 °C = 0.25, find the amounts of the substances present in the equilibrium mixture.

Let x mol be the amount of ethanoic acid produced.

$$CH_3COOC_2H_5(aq) + H_2O(l) \rightleftharpoons CH_3COOH(aq) + C_2H_5OH(aq)$$

Data:

Number of moles at start	0.500	1.00	0	0
Number of moles at equilibrium	$(0.500 - x)$	$(1.00 - x)$	x	x

Equation:
$$K_c = \frac{[CH_3COOH]\,[C_2H_5OH]}{[CH_3COOC_2H_5]\,[H_2O]} = 0.25$$

Calculation: If the total volume = V dm³

then
$$[CH_3COOH] = [C_2H_5OH] = \frac{x}{V}$$

and
$$[CH_3COOC_2H_5] = \left(\frac{0.500 - x}{V}\right)$$

$$[H_2O] = \left(\frac{1.00 - x}{V}\right)$$

$$\therefore K_c = 0.25 = \frac{\frac{x}{V} \times \frac{x}{V}}{\frac{(0.500 - x)}{V} \times \frac{(1.00 - x)}{V}}$$

From this equation, and solving the resulting quadratic equation (see *Figure 22.1.3*)

$$x = 0.228 \text{ mol}$$

∴ The amounts of each substance present at equilibrium are:

Ethanoic acid and ethanol: each 0.228 mol

Ethyl ethanoate: $(0.500 - 0.228) = 0.272$ mol

Water: $(1.00 - 0.228) = 0.772$ mol

$$K_c = \frac{0.228 \times 0.228}{0.272 \times 0.772} = \frac{0.0520}{0.210} = 0.25$$

(Check to 2 significant figures which agrees with the values of K_c given earlier.)

For a quadratic equation:

$ax^2 + bx + c = 0$,

$$x = \frac{-b \pm \sqrt{b^2 - 4ac}}{2a}$$

So: $\quad 0.25 = K_c = \dfrac{x^2}{0.5 - 1.5x + x^2}$

$$\therefore x^2 = 0.25(0.5 - 1.5x + x^2)$$

$$\therefore 4x^2 = 0.5 - 1.5x + x^2$$

$$\therefore 3x^2 + 1.5x - 0.5 = 0$$

$$\therefore x = \frac{-1.5 \pm \sqrt{(1.5)^2 + (4 \times 3 \times 0.5)}}{2 \times 3}$$

$$= \frac{1.5 \pm 2.87}{6}$$

$$= -0.728 \text{ (which is impossible)}$$

or $+0.228$

Figure 22.1.3   Solving the quadratic equation

1 **a** State Le Chatelier's Principle.

b Show how Le Chatelier's Principle can be used to predict the effects of changes in concentration in the following reactions.

i) $Cr_2O_7^{2-}(aq) + H_2O(l) \rightleftharpoons 2CrO_4^{2-}(aq) + 2H^+(aq)$

ii) $Fe^{3+}(aq) + SCN^-(aq) \rightleftharpoons [Fe(SCN)]^{2+}(aq)$

iii) $X(s) \rightleftharpoons Y(s) + Z(g)$

(Remember to think about the concentration of a solid.)

c Use Le Chatelier's Principle to predict the effect of lowering the temperature on the following equilibrium reactions:

i) $2NO_2(g) \rightleftharpoons N_2O_4(g)$, ΔH is negative

ii) $CO(g) + 2H_2(g) \rightleftharpoons CH_3OH(g)$, ΔH is negative

2 The equilibrium constant, K_c, for the reduction of iron(III) ions using tin(II) ions, at 25 °C, is 1.0×10^{10}. The equation for the reaction is:

$$Sn^{2+}(aq) + 2Fe^{3+}(aq) \rightleftharpoons Sn^{4+}(aq) + 2Fe^{2+}(aq)$$

a At 25 °C, to which side of the equation does the position of equilibrium lie?

b Write an expression for K_c.

c What is the value of K_c for the reverse reaction?

d What would happen to the value of K_c if more $Sn^{2+}(aq)$ was added to the equilibrium mixture?

e Write an expression for K_c for the reaction:

$$\tfrac{1}{2}Sn^{2+}(aq) + Fe^{3+}(aq) \rightleftharpoons \tfrac{1}{2}Sn^{4+}(aq) + Fe^{2+}(aq)$$

f How does this value compare with the value given at the start of the question?

3 Ethanoic acid reacts with ethanol in an esterification reaction, to form ethyl ethanoate and water.

a Write an equation for this reaction.

b If the reactants are mixed together they eventually reach what is called a position of *dynamic equilibrium*. What does the term *dynamic equilibrium* mean?

c Write the expression for the equilibrium constant, K_c.

d The equilibrium mixture was analysed and found to contain 0.090 mol ethanoic acid, 0.037 mol ethanol, 0.18 mol of the ester and 0.074 mol water. Calculate the value of the equilibrium constant at the temperature of the experiment.

e The experiment was repeated starting with 0.020 mol of the acid and 0.050 mol ethanol. At equilibrium, there were 0.010 mol of acid present. Calculate the amounts of the other species in the equilibrium mixture.

f Explain why the total volume of the system need not be known in the calculations involving K_c for this reaction.

22.2 Gaseous equilibria: K_p

Equilibrium reactions involving only gases

Many reactions involve only gases: for example, the manufacture of ammonia by the Haber Process (see Ch 13.3).

$$N_2(g) + 3H_2(g) \rightleftharpoons 2NH_3(g)$$

The equilibrium constant, K_c, in terms of the concentrations at equilibrium can be written:

$$K_c = \frac{[NH_3]^2}{[N_2][H_2]^3} \text{ mol}^{-2}\text{ dm}^6$$

> Below is *Dalton's Law of partial pressures*, first stated by John Dalton in 1802. This is the same Dalton who introduced the Atomic Theory – see the introduction to Ch 1.

The units for K_c here are $\text{mol}^{-2}\text{ dm}^6$ because:

$$\frac{(\text{mol dm}^{-3})^2}{(\text{mol dm}^{-3})(\text{mol dm}^{-3})^3} = (\text{mol dm}^{-3})^{-2} = \text{mol}^{-2}\text{ dm}^6$$

However, it is probably easier to use gas pressures rather than concentrations. In order to be able to do this, we need to understand the meaning of **partial pressure** of a gas.

From this it follows that **the total pressure of a mixture of gases is the sum of the partial pressures of the different gases in the mixture**.

> **Definition:** The partial pressure of each gas in a mixture of gases is the pressure which that gas would exert if it alone occupied the whole volume occupied by the gas mixture.
>
> Partial pressure of a gas = mole fraction × total pressure
> = % by volume × total pressure
>
> **The mole fraction of a gas**
> mole fraction =
> $\dfrac{\text{number of moles of the gas}}{\text{total number of moles}}$

The symbol p is used for the partial pressure of a gas. So, for air at a total pressure of 100 kPa:

$$pO_2 = 20/100 \times 100 \text{ Pa} = 20 \text{ kPa}$$

This is because air contains about 20% oxygen. Similarly, $pN_2 = 80$ kPa, because air is about 80% nitrogen (*Figure 22.2.1*).

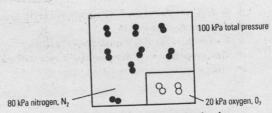

Figure 22.2.1   Partial pressures in air

1. Like all gas laws (see Ch 9.6), the idea of partial pressure depends on the assumption that gases behave as *ideal gases* – there is no attraction between the gas particles, i.e. the collisions are perfectly elastic.

2. **The units of gas pressure**. Gas pressures have often been expressed in *atmospheres*, atm. One standard atmosphere has been defined as that pressure able to support a column of mercury 760 mm high in a barometer. Consequently, pressures have also often been expressed in terms of millimetres of mercury, mm Hg.

   However, the internationally recognised (SI) unit of pressure is *newtons per square metre*, Nm^{-2}. A pressure of 1 Nm^{-2} is *one Pascal*, Pa.

   Using the Pascal as a unit, *standard atmospheric pressure* is 101 325 Pa, more often written as 101.325 kPa. Because atmospheric pressure at sea level can vary by several percent either way from this value, and also for convenience in calculation, *a pressure of one atmosphere is often taken to be approximately 100 kPa*. This is the unit we shall use here, although gas pressures may be quoted in atm occasionally. (N.B. In industrial processes in earlier chapters we have used 'atm' as a unit of pressure.)

$$1 \ Nm^{-2} = 1 \ Pa$$
$$1 \ atm \approx 10^5 \ Pa$$
$$1 \ kPa = 10^3 \ Pa$$

In meteorology, atmospheric pressure is often expressed in *millibars*. In Britain, sea level air pressures can range from about 950 (deep depression) to about 1040 millibars (anticyclone).

Considering once more the equation for the Haber Process:

$$N_2(g) + 3H_2(g) \rightleftharpoons 2NH_3(g)$$

The partial pressures of the gases in the equilibrium mixture (*Figure 22.2.2*) can be written as pN_2, pH_2 and pNH_3.

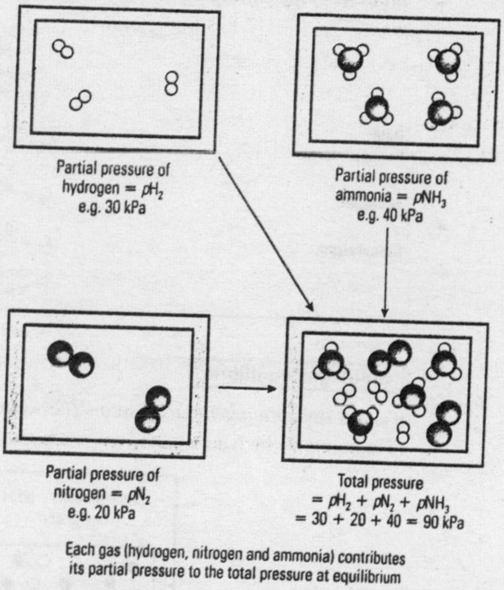

Partial pressure of hydrogen = pH_2 e.g. 30 kPa

Partial pressure of ammonia = pNH_3 e.g. 40 kPa

Partial pressure of nitrogen = pN_2 e.g. 20 kPa

Total pressure = $pH_2 + pN_2 + pNH_3$ = 30 + 20 + 40 = 90 kPa

Each gas (hydrogen, nitrogen and ammonia) contributes its partial pressure to the total pressure at equilibrium

Figure 22.2.2 Partial pressures in the Haber process

In Figure 22.2.2

mole fraction of $H_2 = \frac{30}{90} = 0.33$

mole fraction of $N_2 = \frac{20}{90} = 0.22$

mole fraction of $NH_3 = \frac{40}{90} = 0.44$

The equilibrium constant, K_p, in terms of the partial pressures at equilibrium is therefore:

$$K_p = \frac{(pNH_3)^2}{(pN_2) \times (pH_2)^3} \ Pa^{-2}$$

The units of K_p here are:

$$\frac{(Pa)^2}{Pa \times (Pa)^3} = Pa^{-2}$$

For gas reactions in which the number of moles of reactants is the same as the number of moles of products, K_p and K_c have the same value. Otherwise there is a difference between K_c and K_p, but we shall not discuss it here.

The importance of linking the equation and the equilibrium constant expression

Consider the equilibrium between hydrogen, iodine vapour and hydrogen iodide:

$$H_2(g) + I_2(g) \rightleftharpoons 2HI(g)$$

$$K_p = \frac{(pHI)^2}{(pH_2) \times (pI_2)}$$

(Note that here K_p has no units as the Pa terms cancel out.)

But the equation for the reaction could also have been written as:

$$\tfrac{1}{2}H_2(g) + \tfrac{1}{2}I_2(g) \rightleftharpoons HI(g)$$

In that case,

$$K_p = \frac{(pHI)}{(pH_2)^{\frac{1}{2}} \times (pI_2)^{\frac{1}{2}}} = \frac{(pHI)}{\sqrt{(pH_2)} \times \sqrt{(pI_2)}}$$

which would have a different value, although still no units.

Remember: **Always write the equilibrium equation before writing the expression for K_c or K_p, and always check the units for K_c or K_p.** Remember, whether the values for K_c or K_p have units or not depends on whether the units in the equilibrium constant expression cancel out.

Worked example

Calculate the gas phase equilibrium constant, K_p, for the following reaction:

$$COCl_2(g) \rightleftharpoons CO(g) + Cl_2(g)$$

Analysis of the equilibrium mixture of gases gave the following partial pressures for each of the gases:

$$pCOCl_2 = 0.64 \ Pa$$
$$pCO = 0.28 \ Pa$$
$$pCl_2 = 0.46 \ Pa$$

Data:
Partial pressures

$$COCl_2(s) \rightleftharpoons CO(g) + Cl_2(g)$$
$$0.64 \ Pa \qquad 0.28 \ Pa \quad 0.46 \ Pa$$

Equation:

$$K_p = \frac{(pCO) \times (pCl_2)}{(pCOCl_2)} \ Pa$$

Calculation:

$$K_p = \frac{0.28 \times 0.46}{0.64}$$

$$= 0.20 \ Pa$$

Solid–gas equilibria

It is not only *homogeneous* systems (i.e. systems in which all the components are in the same state of matter) in which an equilibrium can be set up (*Figure 22.2.3*).

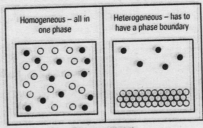

| Homogeneous – all in one phase | Heterogeneous – has to have a phase boundary |

Figure 22.2.3

One example of a *heterogeneous* equilibrium system involves the thermal decomposition of chalk or limestone (calcium carbonate, $CaCO_3$).

The action of heat on calcium carbonate is of great industrial importance (see Ch 6.5). If you live or have been on holiday in a limestone area, you will have seen the remains of old lime-kilns built into hillsides, often with the furnace entrance facing the prevailing wind.

Why were the kilns built to ensure a good draught of air through them? The equation is:

$$CaCO_3(s) \rightleftharpoons CaO(s) + CO_2(g)$$

$$\therefore K_p = \frac{(pCaO) \times (pCO_2)}{(pCaCO_3)}$$

Even apparently involatile solids are in equilibrium with vapour, even though the partial pressure may be *very* small. (Any solid which smells must have a vapour pressure.) So long as *any* solid is present, even if only a very tiny amount, the partial pressure of its vapour will stay constant at constant temperature. So, in the K_p expression above, $pCaO$ and $pCaCO_3$ are constant and can be incorporated in the overall constant.

$$\therefore K_p = (pCO_2)_{eqm} \text{ Pa}$$

If calcium carbonate is heated in a closed container, there must at equilibrium be a definite pressure of carbon dioxide. This pressure will depend on the temperature, and is known as the *dissociation pressure* for calcium carbonate. It has a value of 100 kPa (i.e. one atmosphere) at about 900 °C – red heat.

Obviously, because the reaction is reversible, large amounts of limestone (calcium carbonate) cannot completely decompose unless the carbon dioxide is removed from the contact with the calcium oxide which is also formed. And that is why a good draught of air is needed through a lime-kiln.

'Lime' (calcium hydroxide, $Ca(OH)_2$) was – and still is – produced from limestone by heating it to form 'quicklime', CaO, and then adding water. Lime is used to make acid moorland soils more productive.

SELF-ASSESSMENT
QUESTIONS
22.2

1 In each of the following reactions explain what will happen to the position of equilibrium if the pressure on the system in equilibrium is reduced.
   a $N_2(g) + 3H_2(g) \rightleftharpoons 2NH_3(g)$
   b $C(s) + O_2(g) \rightleftharpoons CO_2(g)$
   c $PCl_3(g) + Cl_2(g) \rightleftharpoons PCl_5(g)$
   d $CH_4(g) + CO_2(g) \rightleftharpoons 2CO(g) + 2H_2(g)$

2 One of the steps in the commercial manufacture of nitric acid is the oxidation of ammonia, according to the equation:

$$4NH_3(g) + 5O_2(g) \rightleftharpoons 4NO(g) + 6H_2O(g) \qquad \Delta H = -908 \text{ kJ mol}^{-1}$$

How would the position of equilibrium change in the following circumstances?
   a The addition of a catalyst.
   b The addition of ammonia to the reaction mixture.
   c An increase in pressure.
   d A decrease in temperature.
   e The removal of water vapour from the equilibrium mixture.

3 Methanol is a simple alcohol with potential uses as a fuel. It can be manufactured by the reaction of carbon monoxide and hydrogen.

$$CO(g) + 2H_2(g) \rightleftharpoons CH_3OH(g) \qquad \Delta H = -92 \text{ kJ mol}^{-1}$$

The value of K_p for the reaction at 500 K is 6.25×10^{-3} atm^{-2}.

a Write an expression for the equilibrium constant, K_p, for the gas phase reaction.
b Explain why high pressure is used to carry out the reaction.
c The reaction is normally carried out at 500 K, even though the value for ΔH would indicate that a lower temperature should increase the yield of methanol. Explain why the higher temperature is used.

4 a A mixture of hydrogen and nitrogen, at a total pressure of 100 kPa, contained 15% by volume of hydrogen. Calculate the partial pressures of hydrogen and nitrogen in the mixture.
b The products from a reaction chamber in the production of ammonia contained equal amounts of nitrogen and hydrogen, together with 24% ammonia. The total pressure of the gas mixture was 120 kPa. Calculate the partial pressures of each of the gases in the mixture.
c Dinitrogen tetraoxide dissociates on heating into nitrogen dioxide.

$$N_2O_4(g) \rightleftharpoons 2NO_2(g)$$

Calculate the partial pressures of dinitrogen tetraoxide and nitrogen dioxide in a sample in which the tetraoxide is 30% dissociated and the total pressure is 12 atm.

22.3 Some applications of equilibria

Industrial processes

We have already seen how the relative amounts of the reactants and products in a reversible reaction can be altered by changing the conditions (*Figure 22.3.1*), and how the result of changing the conditions can be predicted by using Le Chatelier's Principle (*Figure 22.3.2* and Ch 13.2).

Figure 22.3.1 shows that the value of K_c for a reaction can change considerably by changing the temperature. A temperature difference of just 100 °C can alter the value of K_c more than a factor of 10 – up or down depending on whether the reaction is endothermic or exothermic.

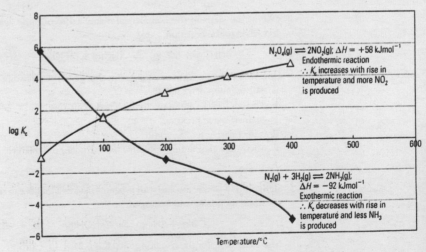

Figure 22.3.1 Effect of temperature on the value of K_c

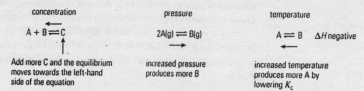

Figure 22.3.2   Using Le Chatelier's Principle

These ideas were then applied to the manufacture of ammonia by the Haber Process (*Figure 22.3.3*) and sulphuric acid by the Contact Process.

(If you cannot remember *how* these equilibria are manipulated in terms of temperature and pressure to maximise the yield of product, this would be a good time to review Ch 13.3).

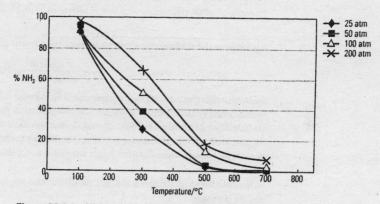

Figure 22.3.3   Yields of ammonia at different temperatures and pressures

---

Remember: changing conditions and the value of the equilibrium constant.

Change	Value of K_p or K_c
Temperature	Changes
Concentration	No change
Pressure	No change
Catalyst	No change but equilibrium achieved more rapidly

One general principle widely used to improve the yield in industrial methods which involve equilibrium processes is *removal of a product as it is formed*. For a general equilibrium system:

$$A + B \rightleftharpoons C + D$$

removal of either C or D as it is formed *stops the reverse reaction occurring*. The forward reaction will continue, and eventually conversion of reactants to products will be complete.

We have just met an example of this in the use of lime-kilns to make lime (Ch 22.2).

$$CaCO_3(s) \rightleftharpoons CaO(s) + CO_2(g)$$

Limestone   'Quicklime'
                 (calcium oxide)

561

Removal of carbon dioxide by a draught of air through the kiln prevents the reverse reaction, and ensures almost complete conversion of calcium carbonate to calcium oxide.

This is an example of a **batch process**. Once all the limestone has been converted the solid product has to be removed from the kiln and a fresh charge of limestone loaded into it.

Removal of product can also be used with a **continuous process**, as for example in the Haber Process.

$$N_2(g) + 3H_2(g) \quad \overset{\overset{\sim 200\ atm}{Fe\ catalyst,\ \sim 450\ °C}}{\rightleftharpoons} \quad 2NH_3(g) \qquad \Delta H = -92\ kJ\ mol^{-1}$$

As explained in Ch 13.3, the hot outgoing gas mixture from the converter contains only about 15% ammonia. The hot gas is cooled by being used to heat up incoming gas, and is then cooled further. Because the pressure inside the plant is so great, the ammonia (but not the nitrogen and hydrogen) condenses to a liquid and can be run off. Unchanged nitrogen and hydrogen are recycled, and more H_2/N_2 mixture fed in to maintain pressure. The forward reaction is therefore continuously maintained while the reverse reaction is inhibited as much as possible.

It is instructive also to consider the K_p expression for the Haber Process (see Ch 22.2).

$$K_p = \frac{(pNH_3)^2}{(pN_2) \times (pH_2)^3}\ Pa^{-2}$$

Inspection reveals that increasing the partial pressure of hydrogen has more effect in increasing the yield of ammonia than does increasing the partial pressure of nitrogen. This is because pH_2 in the K_p expression is *cubed*.

Doubling the partial pressure of hydrogen, pH_2, will increase the value of the bottom line of K_p by 2^3, i.e. by 8. K_p has to stay constant, so $(pNH_3)^2$ must also increase by a factor of 8; pNH_3 will therefore increase by a factor of $\sqrt{8}$, i.e. almost 2.8.

Doubling pN_2 however, means that $(pNH_3)^2$ must also double, so that pNH_3 must increase only by a factor of $\sqrt{2}$, i.e. about 1.4.

So increasing the partial pressure of hydrogen fed into the Haber reaction has double the effect of increasing the partial pressure of nitrogen.

Worked examples

A   *At high temperatures, iodine molecules dissociate:*

$$I_2(g) \rightleftharpoons 2I(g)$$

In a particular experiment at atmospheric pressure (10^5 Pa), the equilibrium mixture contained 25% by volume of iodine atoms. Calculate K_p for the equilibrium at the particular temperature.

Data: $\qquad\qquad\qquad\qquad\qquad\qquad I_2(g) \rightleftharpoons 2I(g)$

$$25\%\ of\ iodine\ atoms$$

Total pressure $= 10^5$ Pa (i.e. 1.0×10^5 Pa)

Equation: $\qquad\qquad\qquad\qquad\qquad K_p = \frac{(pI)^2}{pI_2}\ Pa$

Calculation: If the equilibrium mixture contained 25% iodine atoms, I, then there must be 75% of I_2 molecules present.

The partial pressure of I atoms $= 25\%$ of 10^5 Pa

$$= \frac{25}{100} \times 10^5 = 0.25 \times 10^5\ Pa$$

The partial pressure of I_2 molecules $= 75\%$ of 10^5 Pa

$$= \frac{75}{100} \times 10^5 = 0.75 \times 10^5\ Pa$$

$$\therefore K_p = \frac{(pI)^2}{pI_2} = \frac{(0.25 \times 10^5)^2}{0.75 \times 10^6} = 8.3 \times 10^3\ Pa$$

B   In the Haber Process, the equilibrium mixture at 2×10^4 kPa and 400 °C is found to contain 39% by volume of ammonia. The ratio of nitrogen to hydrogen at the start was 1:3. The gas pressure is kept at 2×10^4 kPa (200 atm) during the reaction. What is K_p for this reaction at 400 °C?
Data:

$$N_2(g) + 3H_2(g) \rightleftharpoons 2NH_3(g)$$

At equilibrium 39% of mixture is NH_3

$\therefore$ remaining 61% contains N and H in the ratio 1:3

$\therefore$ at equilibrium mixture contains

15.25% N_2 45.75% H_2 and   39% NH_3

since the system is all gaseous, the molar ratios will be the same as % ratios.

So the mole fraction of each gas is:

$$\frac{15.25}{100} (N_2) \quad \frac{45.75}{100} (H_2) \quad \frac{39}{100} (NH_3)$$

Since total pressure is 2×10^4 kPa the partial pressure of each gas is

$$\frac{15.25}{100} \times 2 \times 10^4 \quad \frac{45.75}{100} \times 2 \times 10^4 \quad \frac{39}{100} \times 2 \times 10^4$$

$$= 3050 \text{ kPa } (N_2) \qquad = 9150 \text{ kPa } (H_2) \qquad = 7800 \text{ kPa } (NH_3)$$

$$K_p = \frac{(pNH_3)^2}{(pH_2) \times (pH_2)^3} = \frac{(7800)^2}{(3050) \times (9150)^3} \text{ kPa}^{-2}$$

$$= 2.6 \times 10^{-8} \text{ kPa}^{-2}$$

C   Carbon dichloride oxide, CCl_2O, produces an equilibrium mixture when heated at 200 °C containing CCl_2O, CO and Cl_2. The equation for the reaction is:

$$CCl_2O(g) \rightleftharpoons CO(g) + Cl_2(g)$$

In a particular experiment starting with 9.89 g CCl_2O, carried out in a 2.00 dm³ flask, the equilibrium mixture contained 0.355 g of Cl_2. The total pressure in the flask at equilibrium was 0.100 kPa.

Calculate:
a   the number of moles of CCl_2O present at the start of the reaction

$M_r CCl_2O = (12.0 + 70.9 + 16.0) = 98.9$

$\therefore$ number of moles of $CCl_2O = \frac{9.89}{98.9} = 0.100$ mol

b   the number of moles of Cl_2 present in the equilibrium mixture

$M_r Cl_2 = (35.45 \times 2) = 70.9$;

$\therefore$ number of moles of $Cl_2 = \frac{0.355}{70.9} = 0.0050$ mol

c   the concentration of Cl_2 and CO in the flask at equilibrium

The flask has a volume of 2.00 dm³;

$\therefore$ concentration of $Cl_2 = \frac{0.0050}{2.00} = 0.0025$ mol dm⁻³;

for every mole of Cl_2 produced in the reaction, the same number of moles of CO are produced;

the number of moles of CO is 0.00500 and the concentration of CO is also $\frac{0.0050}{2.00} = 0.0025$ mol dm⁻³

d   the concentration of CCl_2O in the equilibrium mixture

0.0050 moles of chlorine has been produced from 0.100 moles of CCl_2O;

the moles of CCl_2O in the equilibrium mixture is $(0.100 - 0.0050) = 0.0950$ mol, in the 2.00 dm³ flask;

$\therefore$ concentration is $\frac{0.0950}{2.00} = 0.0475$ mol dm⁻³

e   a value for K_c for the reaction

For this reaction, $K_c = \frac{[CO] \times [Cl_2]}{[CCl_2O]} = \frac{0.0025 \times 0.0025}{0.0475} = 0.00013$ mol dm⁻³

f   the total number of moles of gases present in the equilibrium mixture and the mole fraction of each of the gases in the mixture

$$\text{Mole fraction} = \frac{\text{number of moles}}{\text{total number of moles present}}$$

Total number of moles at equilibrium = 0.0950 + 0.0050 + 0.0050 = 0.105 mol

Mole fraction of $CCl_2O = \frac{0.0950}{0.105} = 0.905$

Mole fraction of $Cl_2 = \frac{0.0050}{0.105} = 0.0477$

Mole fraction of $O = \frac{0.0050}{0.105} = 0.0477$

g *the partial pressure of each of the gases in the equilibrium mixture*

Partial pressure of each gas = mole fraction of each gas × total pressure

$pCCl_2O = 0.905 \times 0.100 = 0.0905$ kPa

$pCl_2 = 0.0477 \times 0.100 = 0.00477$ kPa

$pCO = 0.0477 \times 0.100 = 0.00477$ kPa

h *The value of K_p at 200 °C*

$$K_p = \frac{pCO \times pCl_2}{pCCl_2O} = \frac{0.00477 \times 0.00477}{0.0905} = 0.00025 \text{ kPa}$$

Equilibria in nature

As we noted in Ch 7.4, conditions on the Earth's surface and life on it are maintained by a range of reversible processes which exist in approximate equilibrium. Among these processes are photosynthesis and respiration, which can be summarised in one equation:

$$6CO_2(g) + 6H_2O(l) \underset{\substack{\text{enzymes in living cells,} \\ \text{temperatures around 35 °C}}}{\overset{\substack{\text{sunlight, range of} \\ \text{temperatures around 20 °C}}}{\rightleftharpoons}} C_6H_{12}O_6(aq) + 6O_2(g)$$

Note that this equation is a *summary* of quite a large number of individual reactions. Many natural equilibrium systems are far from simple, and – although they are vital to the well being of the planet Earth – they do not come within the scope of this book.

The same is true for equilibrium systems within living organisms, including ourselves. *Homoeostasis*, the maintenance of a constant environment within the body even when external conditions (such as temperature) change, involves complex networks of interlinked equilibria. (For example, the pH of human blood is maintained very close to 7.4, or else death results; this 'buffering' is achieved through acid–base equilibria – see Ch 23.6.)

Beginning in 1969 the British environmental scientist James Lovelock proposed and developed the Gaia Hypothesis. (This is named after the Greek Earth goddess, Gaia, daughter of Chaos.) Earth can be thought of as a single system in which a multitude of complex equilibrium systems interact and provide feedback mechanisms which control and regulate the environment. Life flourishes on Earth because these feedback mechanisms have developed a complicated and interlinked pattern of life, which not only makes the best use of its environment but also is capable of altering the environment to support life better.

This idea was, and still is, controversial. But Lovelock and others have used it to make specific predictions (for example, about aspects of the carbon cycle) which have been tested and verified by observation. Lovelock's work, both in his invention of devices able to detect environmental contaminants in low concentrations and through his Gaia Hypothesis, has increased our understanding of global pollution and helped to establish environmental science as a major scientific field.

**SELF-ASSESSMENT
QUESTIONS
22.3**

1 The partial oxidation of natural gas, CH_4, is used to manufacture hydrogen for use in the Haber Process. The overall reaction is endothermic and the equation for the reaction is:

$$CH_4(g) + H_2O(g) \rightleftharpoons CO(g) + 3H_2(g)$$

a Write the equation to define K_p for the reaction.

b The reaction can be catalysed by a variety of catalysts. What effect does a catalyst have on:
  i) the value of K_p;
  ii) the position of equilibrium in the reaction;
  iii) the time taken to reach equilibrium?

   c How does increasing the pressure in the reactor change the position of equilibrium?

   d How does raising the temperature of the reaction alter the position of equilibrium?

2 When heated, PCl_5 dissociates into PCl_3 and chlorine gas. In an experiment, with all reactants and products in the gas phase, the dissociation of 20.8 g of PCl_5 in a 10 dm^3 flask produced 2.84 g chlorine. The total pressure in the flask at equilibrium was 300 kPa.

   a Write a balanced equation for the dissociation of PCl_5 in the gas phase.

   b Write an expression for the equilibrium constants K_c and K_p for this reaction.

   c How many moles of PCl_5 were present at the start of the reaction?

   d How many moles of chlorine, Cl_2, were present at equilibrium?

   e What was the concentration of Cl_2 and PCl_3 in the flask when equilibrium had been reached?

   f Work out the concentration of PCl_5 present in the equilibrium mixture.

   g Calculate the value of K_c for the reaction at the temperature of the reaction.

   h Using the number of moles of PCl_5 present at equilibrium, and the total number of moles present in the equilibrium mixture, work out the partial pressure of PCl_5 in the reaction flask (total pressure 300 kPa).

   i Calculate the partial pressure of PCl_3 and Cl_2 in the equilibrium mixture.

   j Calculate a value for K_p at the reaction temperature.

22.4 Sparingly soluble salts: K_{sp}

Provided that an ionic solid is only sparingly soluble in water, so that the ions in the solution are sufficiently far apart on average not to interfere with each other, it is possible to describe the behaviour of the system accurately.

A **saturated solution** of an ionic solid is another example of a *heterogeneous* equilibrium system. Undissolved solid is in equilibrium with the solution above it; the solid is dissolving and precipitating at the same rate, and nothing appears to be happening (*Figure 22.4.1*).

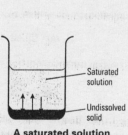

Figure 22.4.1   **A saturated solution**   **PbI_2 crystals – soluble in hot water, insoluble in cold**

With all equilibria, a change in temperature can have a dramatic effect on the position of equilibrium, and the solubility of many salts increases sharply with temperature. For example, the solubility of potassium nitrate, KNO_3, increases tenfold – from about 13 g/100 g water to about 137 g/100 g water – between 0 °C and 70 °C; see *Figure 22.4.2*. Sparingly soluble salts may vary even more surprisingly. The solubility of copper(II) sulphide, CuS, can increase a thousand-fold from 18 °C (solubility 3.0×10^{-19} g dm^{-3}) to 25 °C (solubility 2.4×10^{-16} g dm^{-3}).

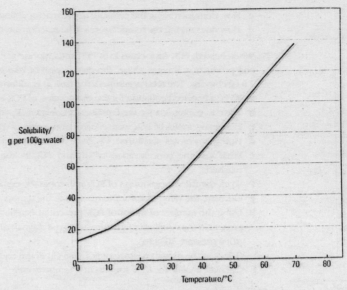

Figure 22.4.2   Solubility of KNO₃

Solubility product

Silver chloride, AgCl, is sparingly soluble and will be used as an example. If solid silver chloride is stirred with pure water at 25 °C (298 K) until equilibrium is reached, the concentration of silver chloride in the resulting saturated solution is found to be 1.41×10^{-5} mol dm⁻³; i.e. $[\text{AgCl}]_{\text{eqm}} = 1.41 \times 10^{-5}$ mol dm⁻³.

Solid AgCl consists of $Ag^{\oplus}$ and $Cl^{\ominus}$ ions; when the solid dissolves, the ions are hydrated and separated. Each mole of solid AgCl which dissolves gives one mole of each of $Ag^{\oplus}$ and $Cl^{\ominus}$ ions, so in the saturated solution:

$$[\text{Ag}^{\oplus}(\text{aq})]_{\text{eqm}} = [\text{Cl}^{\ominus}(\text{aq})]_{\text{eqm}} = 1.41 \times 10^{-5} \text{ mol dm}^{-3}$$

> Remember, 'eqm' is omitted from the equilibrium constant expression in order to make it less clumsy and easier to write.

The equilibrium equation can be written as:

$$\text{Ag}^{\oplus}\text{Cl}^{\ominus}(\text{s}) \rightleftharpoons \text{Ag}^{\oplus}(\text{aq}) + \text{Cl}^{\ominus}(\text{aq})$$

The equilibrium constant can therefore be written:

$$K_c = \frac{[\text{Ag}^{\oplus}(\text{aq})]_{\text{eqm}} [\text{Cl}^{\ominus}(\text{aq})]_{\text{eqm}}}{[\text{AgCl}(\text{s})]_{\text{eqm}}}$$

However, provided that there is some solid present at equilibrium – and for a solution to be *saturated*, there *must* be solid present – [AgCl] will be constant.

> You can vary the *amount* of a solid, but – except for a slight change caused by expansion due to increasing temperature – you *can't* vary its *concentration*. For a solid, concentration would be expressed in mol dm⁻³; since the volume can only change slightly, the concentration may be taken as constant, whatever the volume or amount of solid.

We can therefore incorporate the two constants K_c and $[\text{AgCl}(\text{s})]_{\text{eqm}}$ into a fresh constant. This is called the solubility product, K_{sp}, of silver chloride.

$$K_{sp} \text{ for AgCl at 25 °C} = [\text{Ag}^{\oplus}(\text{aq})] \times [\text{Cl}^{\ominus}(\text{aq})]$$
$$= (1.41 \times 10^{-5} \text{ mol dm}^{-3}) \times (1.41 \times 10^{-5} \text{ mol dm}^{-3})$$
$$= 2.00 \times 10^{-10} \text{ mol}^2 \text{ dm}^{-6}$$

A few values of K_{sp} for sparingly soluble salts are shown in *Table 22.4.1*.

The solubility products of some sparingly soluble ionic solids

Substance	Formula	K_{sp} at 25 °C (298 K)
Lead sulphide	PbS	1.25×10^{-28} mol² dm⁻⁶
Silver bromide	AgBr	5.0×10^{-13} mol² dm⁻⁶
Silver iodide	AgI	8.0×10^{-17} mol² dm⁻⁶
Silver chromate(VI)	Ag_2CrO_4	3.0×10^{-12} mol³ dm⁻⁹
Iron(III) hydroxide	$Fe(OH)_3$	8.0×10^{-40} mol⁴ dm⁻¹²
Bismuth sulphide	Bi_2S_3	1.0×10^{-97} mol⁵ dm⁻¹⁵

Table 22.4.1

The units of K_{sp} always relate to the total number of ions in the compound.

2 ions: units are $(\text{mol dm}^{-3})^2 = \text{mol}^2 \, \text{dm}^{-6}$

5 ions: units are $(\text{mol dm}^{-3})^5 = \text{mol}^5 \, \text{dm}^{-15}$

Some of the values for K_{sp} in the table are breathtakingly small. (Try to visualise 10^{-97}!) The reason is, as we discussed in Ch 13.2, the powers to which the concentrations are raised in the solubility product expression.

e.g. $\qquad K_{sp}$ for $Bi_2S_3 = [Bi^{3+}(aq)]^2[S^{2-}(aq)]^3$

If e.g. $\qquad x = 10^{-10}, \; x^3 = (10^{-10}) \times (10^{-10}) \times (10^{-10}) = 10^{-30}$

If a number is very small already, squaring or cubing it is going to make it *very* small indeed.

Worked example

A *If for zinc sulphide, ZnS, the value for K_{sp} at 25 °C is 1.0×10^{-24} mol² dm⁻⁶, find the solubility (in mol dm⁻³) of zinc sulphide in water at 25 °C.*

Data: $\quad ZnS(s) \rightleftharpoons Zn^{2+}(aq) + S^{2-}(aq)$

$\qquad\qquad K_{sp} = 1.0 \times 10^{-24}$ mol² dm⁻⁶

Equation: $K_{sp} = [Zn^{2+}(aq)][S^{2-}(aq)]$

Calculation: $\qquad$ 1 mole of dissolved ZnS will give 1 mole each of Zn^{2+} and S^{2-} ions in solution

$\qquad\qquad \therefore [ZnS] = [Zn^{2+}] = [S^{2-}]$

$\qquad\qquad$ since the equation shows a 1:1:1 ratio

$\qquad\qquad \therefore K_{sp} = 1.0 \times 10^{-24} = [Zn^{2+}][S^{2-}]$

$\qquad\qquad\qquad\qquad = [Zn^{2+}]^2$

$\qquad\qquad \therefore [ZnS] = [Zn^{2+}] = \sqrt{1.0 \times 10^{-24}}$ mol dm⁻³

$\qquad\qquad\qquad\qquad = 1.0 \times 10^{-12}$ mol dm⁻³

$\therefore$ solubility of ZnS in water is 1.0×10^{-12} mol dm⁻³ at 25 °C.

Note: for any simple salt MX, the solubility in mol dm⁻³ $= \sqrt{K_{sp}}$

B *With salts consisting of more than two ions, the calculation is a little less straightforward, and may require working out cube roots, or more.*

e.g. K_{sp} for silver chromate(VI), Ag_2CrO_4, at 25 °C is 3.0×10^{-12} mol³ dm⁻⁹. Calculate its solubility at 25 °C.

$$Ag_2CrO_4(s) \rightleftharpoons 2Ag^{+}(aq) + CrO_4^{2-}(aq)$$

Data: $K_{sp} = 3.0 \times 10^{-12} \, mol^3 \, dm^{-9}$

Equation: $K_{sp} = [Ag^{\oplus}(aq)]^2 [CrO_4^{2\ominus}(aq)]$

Calculation: If the solubility of Ag_2CrO_4 is x mol dm^{-3}, then:

$[Ag^{\oplus}] = 2x$ and $[CrO_4^{2\ominus}] = x$ from the ratios in the equation

$$\therefore K_{sp} = 3.0 \times 10^{-12} = [Ag^{\oplus}]^2 [CrO_4^{2\ominus}]$$
$$= (2x)^2 \times x$$
$$= 4x^3$$
$$\therefore x^3 = 7.5 \times 10^{-13}$$
$$x = 9.1 \times 10^{-5} \, mol \, dm^{-3}$$

If $x^n = y$, to find the nth root of y:
Find the $\log_{10}$ of y, divide this value by n, and then antilog that new value. Anything more than a simple cube is unlikely! On your calculator:

NUMBER then:

LOG **÷** **n** **=** **SHIFT** **10^x**

gives the nth root of the initial number

C   If you feel confident with the maths, use the values in Table 22.4.1 to calculate the solubility in mol dm^{-3} of iron(III) hydroxide and bismuth sulphide.

Write the equation for iron(III) hydroxide dissolving to form $Fe^{3\oplus}(aq)$ and $OH^{\ominus}(aq)$.

Define K_{sp} for this reaction.

When x mol dm^{-3} of $Fe(OH)_3$ dissolves, how much $Fe^{3\oplus}(aq)$ and how much $OH^{\ominus}(aq)$ is formed?

Put these concentrations into the K_{sp} expression.

Calculate the value of x (see the box to work out the 4th root).

You should obtain a value of $9.7 \times 10^{-11} \, mol \, dm^{-3}$.

Now repeat the procedure with bismuth sulphide.

The 'common ion' effect

The solubility product for an ionic solid represents a *maximum* at the temperature involved. In the saturated solution, K_{sp} **cannot be exceeded**. If for a moment it is, the salt precipitates until equilibrium is restored. **The relative amounts of ions present do not matter.**

For example, suppose some solid silver chloride is stirred into 0.0100 mol dm^{-3} hydrochloric acid, HCl, at 25 °C. The dilute acid can be assumed to be fully ionised, so that $[Cl^{\ominus}(aq)]$ from the acid is $1.0 \times 10^{-2} \, mol \, dm^{-3}$. The very tiny amount of silver chloride which dissolves to give chloride ions can be ignored; so, in the solution, $[Cl^{\ominus}(aq)] = 1.0 \times 10^{-2} \, mol \, dm^{-3}$.

But:

$K_{sp}AgCl = [Ag^{\oplus}(aq)] \times [Cl^{\ominus}(aq)]$
$= 2.0 \times 10^{-10} \, mol^2 \, dm^{-6}$ at 25 °C
In the solution, $[Cl^{\ominus}(aq)]$ now $= 1.0 \times 10^{-2} \, mol \, dm^{-3}$.
$\therefore [Ag^{\oplus}(aq)] \times 10^{-2} \, mol \, dm^{-3} = 2.0 \times 10^{-10} \, mol \, dm^{-3}$
$\therefore [Ag^{\oplus}(aq)] = 2.0 \times 10^{-8} \, mol \, dm^{-3}$.

And, since 1 mole AgCl gives rise to 1 mole $Ag^{\oplus}$ ions, the solubility of AgCl must also be 2.0×10^{-8} mol dm^{-3}.

We see from this calculation that the solubility of silver chloride in 0.0100 mol dm^{-3} hydrochloric acid is only $2.0 \times 10^{-8} \, mol \, dm^{-3}$, compared with $1.41 \times 10^{-5} \, mol \, dm^{-3}$ in water at the same temperature. *Its solubility has therefore been reduced almost a thousandfold*.

Remember: Any salt is less soluble in a solution which already contains one of the ions present in the salt; and if a solution containing one of its ions is added to a saturated solution of the salt, some of the salt must precipitate.

The salt and the other solution involved must have an ion in common; the effect is known as the **common ion effect**.

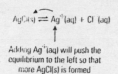

$AgCl(s) \rightleftharpoons Ag^+(aq) + Cl^-(aq)$

Adding $Ag^+(aq)$ will push the equilibrium to the left so that more $AgCl(s)$ is formed

Figure 22.4.3

The common ion effect is yet another example of Le Chatelier's Principle: if the concentration of one of the ions of the ionic solid in the equilibrium mixture (the saturated solution) is increased, the position of equilibrium will shift to reduce that concentration (*Figure 22.4.3*).

And the only way available for it to do so is by removal of some of the extra ions by precipitation of the ionic solid.

Complex ion formation

Sometimes an ionic solid will dissolve if a high concentration of its anion is added to the saturated solution.

For example, lead(II) chloride is sparingly soluble in water at 25 °C ($K_{sp} = 2.0 \times 10^{-5}$ mol³ dm⁻⁹). If concentrated hydrochloric acid is added to the saturated solution, at first – as expected – some solid lead(II) chloride is precipitated because of the extra chloride ions from the acid. If more of the hydrochloric acid is slowly added with stirring, all the lead(II) chloride will eventually dissolve. This is because an ion containing the metal in the anion, one of the types of complex ion (see Ch 25.4), is formed. At high values of $[Cl^-(aq)]$, hydrated Pb^{2+} ions form the $PbCl_4^{2-}$ complex ion which is more soluble in water.

$$PbCl_2(s) \rightleftharpoons Pb^{2+}(aq) + 2Cl^-(aq)$$
$$+ 4Cl^-(aq) \text{ (from the added hydrochloric acid)}$$
$$\parallel$$
$$PbCl_4^{2-}(aq)$$

As lead ions are removed from the original equilibrium, the value for $[Pb^{2+}(aq)][Cl^-(aq)]^2$ is momentarily less than K_{sp}. So more lead(II) chloride dissolves to restore equilibrium. But if enough chloride ions are added, the Pb^{2+} ions will continue to be removed by formation of the $PbCl_4^{2-}$ complex ion. Addition of sufficient hydrochloric acid will therefore 'dissolve' all the solid lead(II) chloride.

SELF-ASSESSMENT QUESTIONS 22.4

1 Write an expression for the solubility product, K_{sp}, for the following ionic compounds. In each case clearly identify the units of K_{sp}.

  a Mercury(II) sulphide.

  b Lead(II) chloride.

  c Chromium(III) hydroxide.

2 Magnesium hydroxide has a solubility of 9.9×10^{-3} g dm⁻³ at 25 °C.

  a What is the formula of magnesium hydroxide?

  b Write an expression for the solubility product, K_{sp}, of magnesium hydroxide.

  c What are the units of K_{sp} in this expression?

  d What is the concentration, in mol dm⁻³, of magnesium ions in the saturated solution?

  e What is the concentration of hydroxide ions in the saturated solution?

  f Calculate the value for K_{sp} for magnesium hydroxide at 25 °C.

3 The solubility product of calcium sulphate is 2.4×10^{-5} mol² dm⁻⁶ at 25 °C.

  a Explain why the units quoted for K_{sp} are mol² dm⁻⁶.

  b Calculate the solubility of calcium sulphate, in both mol dm⁻³, and g dm⁻³, at 25 °C.

Summary of equilibria calculations

For the reaction of nitrogen and hydrogen:

$$N_2(g) + 3H_2(g) \rightleftharpoons 2NH_3(g)$$

$K_c = \dfrac{[NH_3]^2}{[N_2] \times [H_2]^3}$; units are $\dfrac{(\text{mol dm}^{-3})^2}{(\text{mol dm}^{-3}) \times (\text{mol dm}^{-3})^3} = \text{mol}^{-2}\,\text{dm}^6$

$K_p = \dfrac{(pNH_3)^2}{(pN_2) \times (pH_2)^3}$; units are $\dfrac{\text{Pa}^2}{\text{Pa} \times \text{Pa}^3} = \text{Pa}^{-2}$ (or in kPa or atm)

When x moles of N_2 are used up in the reaction, 3x moles of H_2 are used and 2x moles of NH_3 are formed. At equilibrium there are (initial moles of N_2 − x) moles,

and the concentration is $\left(\dfrac{\text{initial moles} - x}{\text{volume of container}} \right)$.

Partial pressure, $pN_2 = \dfrac{\text{pressure of } N_2}{\text{total pressure}}$

Temperature changes alter the values of K_c and K_p.

With changes in concentration, K_c or K_p do not change – the equilibrium moves to reduce the concentration of an added substance or increase the concentration of the substance that has been removed.

Pressure changes only affect equilibria involving gases:

- if there are the same number of gas molecules on each side, there is no effect;
- otherwise, an increase in pressure moves the equilibrium towards the side with the lower number of gas molecules

$K_{sp} = [\text{cation}]^x\,[\text{anion}]^y$

**SUMMARY
22**

In this chapter the following new ideas have been covered. *Have you understood and learned them?* Could you explain them clearly to a friend? The page references will help.

**EXAMINATION
QUESTIONS
22**

1   Consider the equilibrium reaction:

$$N_2(g) + O_2(g) \rightleftharpoons 2NO(g) \qquad \Delta H = +180\,\text{kJ mol}^{-1}$$

a   Write the equilibrium expression for K_p for this reaction. What are the units of K_p in this case?

b   The equilibrium is said to be *dynamic*. What does the term dynamic equilibrium mean for this reaction?

c   What effect would a catalyst have on the position of equilibrium in this reaction?

d Changing the pressure and temperature alters the percentage yield of NO(g).

   i) Which of the curves below shows how *increasing pressure* affects the yield of product?

   ii) Which of the curves below shows how *increasing temperature* affects the yield? Explain your reasoning in each case.

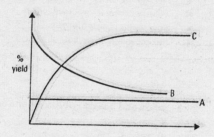

2 In the Contact Process for the manufacture of sulphuric acid, the oxidation step involves converting sulphur dioxide to sulphur trioxide.

   a Write an equation for the gas phase oxidation.

   b What is the name of the catalyst normally used in industry for this reaction? What effect has the catalyst on the percentage yield in the reaction?

   c Explain why, even though more sulphur trioxide is produced at a lower temperature, the reaction is carried out at a temperature around 450 °C.

   d The reaction is an equilibrium reaction. Write an expression for the gas phase equilibrium constant.

   e The value of this constant at 25 °C is 4.0×10^{19} Pa^{-1} and at 200 °C is 2.5×10^5 Pa^{-1}. Using this information, decide if the reaction is endothermic or exothermic.

   f The value of the equilibrium constant at 800 °C is 1.3×10^{-6} Pa^{-1}. Estimate the value of the constant at 400 °C.

   g A 90% yield of sulphur trioxide was obtained when 20 mol of SO_2 and 10 mol of oxygen reacted at 450 °C and 200 kPa pressure. How many moles of SO_2, of oxygen and of SO_3 were present in the equilibrium mixture?

   h Calculate the partial pressure of each gas in the equilibrium mixture.

   i Calculate the value of the equilibrium constant at the temperature of this particular experiment.

3 When a mixture containing 0.015 moles of each of hydrogen and iodine is heated in a sealed tube an equilibrium mixture is formed. The equation for the reaction is:

$$H_2(g) + I_2(g) \rightleftharpoons 2HI(g)$$

   a Write expressions for the equilibrium constants, K_c and K_p, for this reaction.

   b Explain why the values of K_p and K_c in this particular reaction are the same.

   c The concentration of the iodine in the mixture can be estimated by titration of the reaction mixture with 0.300 mol dm^{-3} sodium thiosulphate solution. Write an equation for the reaction of iodine and aqueous sodium thiosulphate.

   d In a particular reaction, the total iodine in the equilibrium mixture was found to be equivalent to 24.5 cm^3 of the thiosulphate solution. How many moles of iodine were present in the equilibrium mixture?

   e From the equation and the data given, how many moles of hydrogen and hydrogen iodide were present at equilibrium?

   f Calculate the value of K_c for the reaction.

23

acid–base equilibria

There is a helpful summary of the maths ideas used in this chapter in the **Mathematical Appendix**. Do look at it if you find the maths difficult.

Whenever chemical equilibrium is discussed (as in Chs 13, 22, 24, 25), it is very helpful to be able to write very large and very small numbers accurately in standard form, using powers of ten. (Examples of large and small numbers: the Avogadro Constant – see Ch 9.3 – is 6.02×10^{23}; the mass of a proton – see Ch 1.2 – is 1.7×10^{-27} kg.) Compare this way of writing such numbers with having to put in the noughts if they were written fully … Powers of ten are very useful!

We have already met **logarithms** briefly in discussing successive ionisation energies (Ch 1.5) and in discussing activation energy. In this chapter a basic understanding of logarithms is essential if we are to use ideas about pH intelligently. It is our belief that if you can understand powers of ten, you can understand ordinary logarithms, which are based on powers of ten. If you need help, you will find it in the Appendix.

23.1 Acids and bases: reinforcement of essential ideas

The vocabulary of acid–base chemistry was introduced in Ch 13.4. No doubt you have looked through the 'Do you remember…?' box at the start of this chapter, and made an honest assessment of your understanding of ideas listed there before you started this chapter.

Even so, knowing, understanding, and *being able to use* the correct vocabulary is so important that we suggest you try a brief test before you go on. The answers are at the end of the chapter. Unless you can get nearly all the correct answers, you should have *another* careful read of Ch 13.4 *now*, before you work through the rest of this chapter.

Acid–base diagnostic test

1. Define a) **acid**, b) **base**.
2. a) What is an **alkali**?
   b) Which ion is always found in high concentration in an alkaline solution?
3. a) What is the **conjugate base** of ethanoic acid, CH_3COOH?
   b) What is the **conjugate acid** of ammonia, NH_3?
4. What is meant by a **weak acid**?
5. Why is sulphuric acid, H_2SO_4, called a **dibasic acid**?
6. a) What is meant by **neutralisation**?
   b) What is the ionic equation for the neutralisation of an acid by an alkali?
7. Write equations to show how water is **amphiprotic**.

If you did well and feel confident, we can now press on!

23.2 K_w and pH

The self-ionisation of water

Pure water ionises very slightly.

$$H_2O(l) \rightleftharpoons H^{\oplus}(aq) + OH^{\ominus}(aq)$$

Hydrated proton   Hydroxide ion
(hydrogen ion)

Water therefore always contains some hydrated protons and hydroxide ions, however pure it is.

This can also be written as:

$$2H_2O(l) \rightleftharpoons H_3O^{\oplus}(aq) + OH^{\ominus}(aq)$$

Oxonium ion   Hydroxide ion

Notice that this *self-ionisation of water* is the exact opposite of neutralisation (Ch 13.4). Neutralisation is always *exothermic* (see Ch 11.3) as bonds are being made. For a strong acid reacting with a strong base in dilute solution, ΔH_n is always the same:

$$H^{\oplus}(aq) + OH^{\ominus}(aq) \rightleftharpoons H_2O(l) \qquad \Delta H_n = -57 \text{ kJ mol}^{-1}$$

If neutralisation is always exothermic, the ionisation of water must always be *endothermic* to exactly the same extent, as the same bonds are being broken rather than made (see Ch 11.5).

$$H_2O(l) \rightleftharpoons H^{\oplus}(aq) + OH^{\ominus}(aq) \qquad \Delta H_n = +57 \text{ kJ mol}^{-1}$$

The ionisation of water can be discussed in terms of Le Chatelier's Principle (see Ch 13.2). If ionisation of water is endothermic, raising the temperature will encourage the process which tends to lower the temperature again – i.e. the endothermic breaking of bonds in ionisation.

At 25 °C (298 K) the concentration of hydrated protons (hydrogen ions) in pure water has been found to be $1.0 \times 10^{-7} \text{ mol dm}^{-3}$. The concentration of hydroxide ions must be the same as the concentration of $H^{\oplus}(aq)$.

Using the symbol $[X]_{eqm}$ to represent the concentration in mol dm^{-3} of a species in solution at equilibrium, the equilibrium constant, K_c (see Ch 13.2), for the ionisation of water must be:

$$K_c = \frac{[H^{\oplus}(aq)]_{eqm}[OH^{\ominus}(aq)]_{eqm}}{[H_2O(l)]_{eqm}}$$

As the amount of ionisation is very small, $[H_2O]_{eqm}$ is almost constant – it is *very* nearly the same as it would be if no ionisation occurred. We *can* therefore include $[H_2O]_{eqm}$ in the overall equilibrium constant. This new constant is called the **ionic product of water**, K_w.

At 25 °C (298 K),

$$K_w = [H^{\oplus}(aq)]_{eqm} \times [OH^{\ominus}(aq)]_{eqm}$$
$$= (1.0 \times 10^{-7} \text{ mol dm}^{-3}) \times (1.0 \times 10^{-7} \text{ mol dm}^{-3})$$
$$= 1.0 \times 10^{-14} \text{ mol}^2 \text{ dm}^{-6}$$

Definition: The ionic product of water,
$K_w = [H^{\oplus}(aq)][OH^{\ominus}(aq)]$
$= 1.0 \times 10^{-14} \text{ mol}^2 \text{ dm}^{-6}$
at 25 °C (298 K).

The important thing to remember about K_w is that, like any other equilibrium constant, it *is* constant so long as the temperature stays constant.

Whatever the relative concentrations of H^+(aq) and OH^-(aq), in any aqueous solution at a particular temperature the value of K_w stays the same.

- If the solution is acidic, $[H^{\oplus}(aq)]_{eqm}$ will be high, and $[OH^{\ominus}(aq)]_{eqm}$ low; but $[H^{\oplus}(aq)]_{eqm} \times [OH^{\ominus}(aq)]_{eqm}$ at 25 °C will *always* be $1 \times 10^{-14} \text{ mol}^2 \text{ dm}^{-6}$.
- For alkaline solutions, $[OH^{\ominus}(aq)]_{eqm}$ will be high and $[H^{\oplus}(aq)]_{eqm}$ low, but *still* the product of the two concentrations will be $1 \times 10^{-14} \text{ mol}^2 \text{ dm}^{-6}$ (*Table 23.2.1*).

solution	pH	$[H^+]$/mol dm^{-3}	$[OH^-(aq)]$/mol dm^{-3}	K_w/mol^2 dm^{-6}
1.0 mol dm^{-3} HCl(aq)	0	1.0	1.0×10^{-14}	1.0×10^{-14}
1.0×10^{-4} mol dm^{-3} HCl(aq)	4.0	1.0×10^{-4}	1.0×10^{-10}	1.0×10^{-14}
2.0 mol dm^{-3} NaOH(aq)	14.3	5.0×10^{-15}	2.0	1.0×10^{-14}
0.10 mol dm^{-3} NaCl(aq)	7	1.0×10^{-7}	1.0×10^{-7}	1.0×10^{-14}
0.10 mol dm^{-3} C$_6$H$_5$COOH(aq)	3.1	7.9×10^{-4}	1.3×10^{-11}	1.0×10^{-14}
0.10 mol dm^{-3} NH$_3$(aq)	11.1	7.9×10^{-12}	1.3×10^{-3}	1.0×10^{-14}

Table 23.2.1 $K_w = 1.0 \times 10^{-14} \text{ mol}^2 \text{ dm}^{-6}$ at 25 °C

As discussed above, water ionises more when it is hot; this means that K_w must also increase with temperature – see *Table 23.2.2* and *Figure 23.2.1*.

Temperature / °C	0	10	25	50	100
$K_w \times 10^{14}$ / mol^2 dm^{-6}	0.114	0.19	1.08	5.476	51.3

Table 23.2.2 K_w at different temperatures

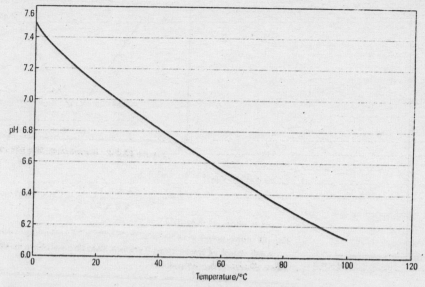

Figure 23.2.1   Variation of pH of water with temperature (but the water is still neutral)

pH

The 'p' in pH (note: *small* p, *capital* H) simply means 'the negative logarithm to base 10 of …'. We shall meet this use again during this chapter.

You will almost certainly have met the term pH already, particularly if you study biology. It is a convenient way to refer to the acidity or alkalinity of a particular solution.

The pH of an aqueous solution is determined by the following mathematical relationship:

Definition:
$pH = -\log_{10}[H^{\oplus}(aq)]$

- In pure water at 25 °C,

  $[H^{\oplus}(aq)] = 1 \times 10^{-7} \text{ mol dm}^{-3}$
  $\log_{10}[H^{\oplus}(aq)] = -7$
  $-\log_{10}[H^{\oplus}(aq)] = 7$
  $pH = 7$

As, once again, you will almost certainly know, **pH 7 defines a neutral aqueous solution**.

But remember (see *Table 23.2.1*) that a pH of 7 only indicates a neutral solution if the temperature is about 25 °C (298 K). The pH of boiling pure water at 100 °C is about 6.15 – but the water is still neutral.

- Now consider dilute hydrochloric acid with a concentration of 0.010 mol dm^{-3} of hydrogen chloride, HCl. Assuming that *all* the HCl is ionised:

$$HCl(aq) \longrightarrow H^{\oplus}(aq) + Cl^{\ominus}(aq)$$

$[H^{\oplus}(aq)] = 0.010 = 1 \times 10^{-2} \text{ mol dm}^{-3}$
$pH = -\log_{10}[H^{\oplus}(aq)] = -\log_{10}(10^{-2}) = 2$

So the pH of this solution of hydrochloric acid is 2, whereas the pH of pure water is 7.

A pH value of less than 7 means that the solution is acidic. The lower the pH, the more acidic the solution (*Figure 23.2.2*).

- Finally, consider a dilute solution of sodium hydroxide, NaOH, with a concentration of 0.0010 mol dm^{-3}. Assuming that ionisation is complete, $[OH^{\ominus}(aq)] = 0.0010 = 1.0 \times 10^{-3}$ mol dm^{-3}. We know the value of $[OH^{\ominus}(aq)]$; how can we calculate $[H^{\oplus}(aq)]$?

Remember: $K_w = [H^{\oplus}(aq)] \times [OH^{\ominus}(aq)] = 1.0 \times 10^{-14} \text{ mol}^2 \text{ dm}^{-6}$ for any aqueous solution at 25 °C.

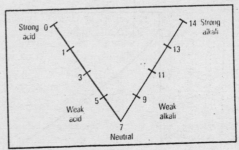

Figure 23.2.2   Remember the pH chart?

So, if $[OH^\ominus(aq)] = 10^{-3}$ mol dm^{-3}

$$[H^\oplus(aq)] = \frac{1.0 \times 10^{-14}}{1.0 \times 10^{-3}} = 1.0 \times 10^{-11}\text{ mol dm}^{-3}$$

$$pH = -\log_{10}(1.0 \times 10^{-11}) = 11$$

The pH of this solution of sodium hydroxide is 11, compared with pH 7 for pure water.

A pH value of more than 7 means that the solution is alkaline. The higher the pH, the more alkaline the solution.

Worked example

A   *Calculate the pH of 0.015 mol dm^{-3} nitric acid, assuming the acid is completely dissociated.*

Data: $HNO_3(aq) \longrightarrow H^\oplus(aq) + NO_3^-(aq)$
0.015 mol dm^{-3}

Equation: $pH = -\log_{10}[H^\oplus(aq)]$

Calculation: Since the concentration of HNO_3 is 0.015 mol dm^{-3},
the $[H^\oplus(aq)]$ is also 0.015 mol dm^{-3} as the acid is completely ionised.

$$\therefore pH = -\log_{10}(0.015)$$
$$= 1.82$$

On your calculator:

0.015 `log` `+/-` gives 1.8239 ...

Remember, don't write down all the figures – *make sure you use the correct number of significant figures.*

B   *Calculate the pH of 0.0250 mol dm^{-3} potassium hydroxide solution, assuming that the alkali is completely ionised at this concentration.*

Data: $KOH(aq) \longrightarrow K^\oplus(aq) + OH^\ominus(aq)$
0.0250 mol dm^{-3}

Equation: $pH = -\log_{10}[H^\oplus(aq)]$
We will also need $K_w = [H^\oplus(aq)][OH^\ominus(aq)] = 1.00 \times 10^{-14}$ mol^2 dm^{-6}.

Calculation: Since the concentration of KOH is 0.0250 mol dm^{-3} and
it is completely ionised, $[OH^-(aq)]$ is also 0.0250 mol dm^{-3}.
To find $[H^\oplus(aq)]$ we need to use K_w.

$$[H^\oplus(aq)] = \frac{K_w}{[OH^\ominus(aq)]} = \frac{1.00 \times 10^{-14}}{0.0250} = 4.00 \times 10^{-13}$$

$$\therefore pH = -\log_{10}(4.00 \times 10^{-13}) = 12.4$$

On your calculator:

1.00 `EXP` `14` `+/-` `÷` 0.025 `=`

then

`log` `+/-` gives 12.3979

You may need to use the calculator keys in a slightly different sequence. But do make sure you find out the exact sequence needed on *your* calculator.

pOH and pK_w

We can, of course, do the same with hydroxide ions as we did with hydrogen ions.

$$pOH = -\log_{10}[OH^{\ominus}(aq)]$$
So if $[OH^{\ominus}(aq)] = 1 \times 10^{-2} \, mol \, dm^{-3}$
$$pOH = 2$$

And, again with K_w:

$$pK_w = -\log_{10}K_w = -\log_{10}(1 \times 10^{-14}) = 14, \text{ at } 25\,°C$$
Therefore, if $K_w = [H^{\oplus}(aq)] \times [OH^{\ominus}(aq)]$
$$pK_w = pH + pOH = 14$$

At 25 °C (298 K), therefore, **the sum of the pH and pOH of any aqueous solution must be 14.** And, as we saw earlier, **the product $[H^{\oplus}(aq)] \times [OH^{\ominus}(aq)]$ for any aqueous solution at 25 °C must be $1 \times 10^{-14} \, mol^2 \, dm^{-6}$.**

The pH scale

It has often been stated in textbooks in the past that the pH range for aqueous solutions is 0–14. This is wrong. In theory, the pH of hydrochloric acid of concentration $10 \, mol \, dm^{-3}$, i.e. ordinary concentrated hydrochloric acid, HCl(aq), should be −1, and that of sodium hydroxide, NaOH(aq), of concentration $10 \, mol \, dm^{-3}$, should be 15 – see *Figure 23.2.3*.

In fact, because ions at such concentrations interfere with each other and are not completely free to move independently, pH can fall a little below 0 for acids and rise slightly above 14 for alkalis.

The pH of a solution can be estimated using suitable indicators (see Ch 23.4) or measured using an electrical instrument called a pH meter.

N.B. Many students fail to remember that the pH scale is a **logarithmic** one, i.e. based on powers of ten. This means that $[H^{\oplus}(aq)]$ in a solution of pH 2 is a *thousand times* more than in a solution of pH 5 (because $5 - 2 = 3$, and $10^3 = 1000$). So rainwater can easily be of pH 5 and you would never notice unless you measured it carefully; whereas at pH 2 an acid is very obviously acidic.

There is, in the laboratory, little difference in practice between solutions in the range pH 6–8, but in some circumstances a small variation in pH can become very serious. For example, the pH of human blood is normally 7.4; a variation of about 0.4 either way could cause death. We shall discuss this further in Ch 23.5. The pH values of some solutions are at 25 °C shown in *Figure 23.2.4*.

We stated above that water ionises more at higher temperatures. This means that both K_w and $[H^{\oplus}(aq)]$ will increase as the temperature increases; so, since the symbol 'p' in pX means $-\log_{10}X$, **both pK_w and pH for pure water decrease with increasing temperature.** As we have already pointed out, the pH of pure water at 100 °C (373 K) is about 6.15. You might think that this means boiling water is slightly acidic. In fact, it is still neutral – *because it must contain equal concentrations of $H^{\oplus}$ and $OH^{\ominus}$ ions*, and pK_w at 100 °C is pH + pOH = 6.15 + 6.15 = 12.3.

$pH = -\log_{10}[H^{\oplus}(aq)]$
$\therefore pH = -\log_{10}(10)$
$\therefore pH = -1$ for the 10 mol dm^{-3} acid.
For the 10 mol dm^{-3} alkali:
$pOH = -\log_{10}[OH^{\ominus}(aq)]$
$\quad = -\log_{10}(10)$
$\quad = -1$
But $pOH + pH = 14$
$pH = 14 - pOH = 14 - (-1)$
So pH should = 15 for the alkali.
Figure 23.2.3

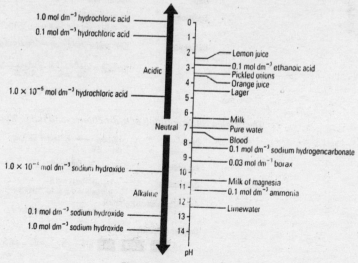

Figure 23.2.4   pH values

Worked example

A *What is the pH of 0.050 mol dm^{-3} aqueous hydrochloric acid, HCl(aq)?*

Assuming complete ionisation,

Data: $HCl(aq) \longrightarrow H^{\oplus}(aq) + Cl^{-}(aq)$

$[H^{\oplus}(aq)] = 0.050$ mol dm^{-3}

Equation: $pH = -\log_{10}[H^{\oplus}(aq)]$

Calculation: $pH = -\log_{10}(0.050)$

$= 1.3$

B *What is the pH of 0.010 mol dm^{-3} aqueous sulphuric acid, H$_2$SO$_4$?*

Assuming complete ionisation (remembering that sulphuric acid is a dibasic acid giving two hydrated protons per molecule – see Ch 13.4):

Data: $H_2SO_4(aq) \longrightarrow 2H^{\oplus}(aq) + SO_4^{2-}(aq)$

0.010 mol dm^{-3}

Equation: $pH = -\log_{10}[H^{\oplus}(aq)]$

Calculation: $[H^{\oplus}(aq)] = 2 \times [H_2SO_4(aq)]$

$= 2 \times 0.010$ nol dm^{-3}

$= 0.020$ mol dm^{-3}

$\therefore pH = -\log_{10}(0.020)$

$= 1.7$

C *What is the pH of 0.0050 mol dm^{-3} aqueous sodium hydroxide, NaOH?*

Assuming complete ionisation, $[OH^{\ominus}(aq)] = 0.0050$ mol dm^{-3}

Data: $NaOH(aq) \longrightarrow Na^{\oplus}(aq) + OH^{-}(aq)$

0.0050 mol dm^{-3}

Equations: $pOH = -\log_{10}[OH^{-}(aq)]$

$pK_w = pOH + pH = 14$

Calculation: $[OH^{\ominus}(aq)] = [NaOH(aq)] = 0.0050$ mol dm^{-3}

$pOH = -\log_{10}(0.0050) = 2.3$

$pH = pK_w - pOH$

So the pH of this solution $= 14 - 2.3 = 11.7$

Alternatively, example C can be solved as follows – without using pOH

Data: 0.0050 mol dm^{-3} NaOH

Equations: $pH = -\log_{10}[H^{\oplus}(aq)]$

$K_w = [H^{\oplus}(aq)][OH^{-}(aq)] = 1.0 \times 10^{-14}$ mol dm^{-6}

Calculation: $[OH^{\ominus}(aq)] = [NaOH(aq)] = 0.0050$ mol dm^{-3}

$[H^{\oplus}(aq)] = \dfrac{K_w}{[OH^{-}(aq)]} = \dfrac{1.0 \times 10^{-14}}{0.0050} = 2.0 \times 10^{-12}$

$pH = -\log_{10}(2.0 \times 10^{-12}) = 11.7$

D *The pH of a dilute solution of hydrochloric acid is 3.6. What is the concentration of the acid?*

Data: $HCl(aq) \longrightarrow H^{\oplus}(aq) + Cl^{-}(aq)$

pH = 3.6

Equation: $pH = -\log_{10}[H^{\oplus}(aq)]$

Calculation: $-\log_{10}[H^{\oplus}(aq)] = 3.6$

$\therefore [H^{\oplus}(aq)] = 2.5 \times 10^{-4}$ mol dm^{-3}

Assuming complete ionisation:

$[HCl(aq)] = 2.5 \times 10^{-4}$ mol dm^{-3}

On your calculator:

3.6 +/-   shift   log   gives 2.51 ... $-04 = 2.5 \times 10^{-4}$ mol dm^{-3}

Summary of equations and definitions

$$pH = -\log_{10}[H^{\oplus}(aq)]$$

$$pOH = -\log_{10}[OH^{\ominus}(aq)]$$

$$K_w = [H^{\oplus}(aq)][OH^{\ominus}(aq)] = 1.00 \times 10^{-14} \text{ mol}^2 \text{ dm}^{-6} \text{ at } 25\,°C$$

$$pK_w = pH + pOH = 14$$

SELF-ASSESSMENT
QUESTIONS
23.2

1 Calculate the pH of the following:
  a 0.020 mol dm^{-3} hydrochloric acid solution
  b 1.5 mol dm^{-3} nitric acid solution
  c 1.2×10^{-4} mol dm^{-3} hydrochloric acid solution
  d 0.10 mol dm^{-3} sulphuric acid (assuming complete dissociation of the dibasic acid)

2 Calculate the pH of the following solutions:
  a 1.0 mol dm^{-3} sodium hydroxide solution
  b 0.050 mol dm^{-3} potassium hydroxide solution
  c 1.5×10^{-5} mol dm^{-3} sodium hydroxide solution.

3 Define the following terms:
  a the ionic product of water
  b pH
  c pOH
  d a dibasic acid
  e a weak acid

4 **a** Write an equation to show the dissociation of water.
  b Given that this is an endothermic change, will the dissociation increase or decrease as the temperature is raised?
  c Write equilibrium expressions, K_c and K_w, for this reaction.
  d What is the value of the equilibrium constant, K_w, for this expression at 25 °C?

5 **a** The pH of a solution of hydrochloric acid was 1.5. What is the concentration in mol dm^{-3} of the HCl?
  A solution of potassium hydroxide was found to have a pH of 12.0.
  b What is the concentration of potassium hydroxide in mol dm^{-3}?
  c What is the concentration of potassium hydroxide in g dm^{-3}?

23.3 K_a, pK_a, K_b, pK_b

The strengths of acids

As we have already discussed in Ch 13.4:
- a **strong** acid is one which **ionises completely** in water when in dilute solution;
- a **weak** acid is one which **does not ionise completely** in water, however dilute the solution.

The terms **strong base** and **weak base** can be used in exactly the same way as the comparable terms for acids.

N.B. You must never confuse a *strong* acid or base with a *concentrated* acid or base, or a *weak* acid or base with a *dilute* one. Hydrochloric acid of concentration 0.001 mol dm^{-3} is certainly dilute, but it is still a strong acid. Ethanoic acid of concentration 10 mol dm^{-3} may be very concentrated, but it is still a weak acid.

Hydrochloric acid is a **strong** acid.
2.0 mol dm^{-3} is quite concentrated, and 0.01 mol dm^{-3} is dilute.
Ethanoic acid is a **weak** acid.
2.0 mol dm^{-3} is quite concentrated, and 0.01 mol dm^{-3} is dilute.

In aqueous solution, a weak acid exists as molecules in equilibrium with the ions from the acid. For ethanoic acid, CH_3COOH:

$$CH_3COOH(aq) \rightleftharpoons H^{\oplus}(aq) + CH_3COO^{\ominus}(aq)$$

The equilibrium constant for this reaction is given a special name and symbol: the **acid dissociation constant, K_a.**

> Notice the units of K_a (mol dm^{-3}) – see Ch 22.1.

$$K_a = \frac{[H^{\oplus}(aq)]_{eqm}[CH_3COO^{\ominus}(aq)]_{eqm}}{[CH_3COOH(aq)]_{eqm}} \text{ mol dm}^{-3}$$

And $pK_a = -\log_{10}K_a$.

> From this it follows that **a smaller value of pK_a means a stronger acid, and a larger value of pK_a means a weaker acid**. Remember the meaning of 'p' – see Ch 23.2.

- **A larger value of K_a means a stronger acid**: the equilibrium mixture contains more of the ions and fewer of the molecules.
- **A smaller value of K_a means a weaker acid**: the equilibrium mixture contains less of the ions and more of the molecules.

With diabasic acids there will be two acid dissociation constants, e.g.

$$H_2CO_3 \rightleftharpoons H^{\oplus} + HCO_3^{\ominus}$$

$$K_1 = \frac{[H^{\oplus}][HCO_3^{\ominus}]}{[H_2CO_3]}$$

and

$$HCO_3^{\ominus} \rightleftharpoons H^{\oplus} + CO_3^{2\ominus}$$

$$K_2 = \frac{[H^{\oplus}][CO_3^{2\ominus}]}{[HCO_3]}$$

The constants might be very different:

for H_2CO_3,

$K_1 = 4.5 \times 10^{-7}$ mol dm^{-3}
$K_2 = 4.8 \times 10^{-11}$ mol dm^{-3}

H_3PO_4 will have three acid dissociation constants of very different values:

$K_1 = 7.9 \times 10^{-3}$ mol dm^{-3}
$K_2 = 6.2 \times 10^{-8}$ mol dm^{-3}
$K_3 = 4.4 \times 10^{-13}$ mol dm^{-3}

Strictly, in considering the equilibrium which results when the weak acid HA dissociates, we should include the water. So, accepting for the moment that the ion $H_3O^{\oplus}(aq)$ is formed (see Ch 13.4), the equation for dissociation is:

$$HA(aq) + H_2O(l) \rightleftharpoons H_3O^{\oplus}(aq) + A^{\ominus}(aq)$$

The equilibrium constant, K_c (see Ch 22.1) for this is therefore:

$$K_c = \frac{[H_3O^{\oplus}(aq)]_{eqm}[A^{\ominus}(aq)]_{eqm}}{[HA(aq)]_{eqm}[H_2O(l)]_{eqm}}$$

But, because a weak acid is only slightly dissociated and therefore comparatively few water molecules are involved, $[H_2O(l)]$ is very nearly constant. If we assume that it *is* constant, we can include it in a *new* constant, which is called K_a. For this, we can go back to writing the correct formula for the hydrated proton, $H^{\oplus}(aq)$.

$$K_a = \frac{[H^{\oplus}(aq)]_{eqm}\,[A^{\ominus}(aq)]_{eqm}}{[HA(aq)]_{eqm}} \text{ mol dm}^{-3}$$

Look back at the equation:

$$HA(aq) + H_2O(l) \rightleftharpoons H_3O^{\oplus}(aq) + A^{\ominus}(aq)$$

By Le Chatelier's Principle (see Ch 13.2), addition of more water will push the position of equilibrium towards the right: i.e. **diluting a weak acid increases its dissociation into ions**.

But, however dilute it becomes, a weak acid is *never* completely ionised.

The 'concentration' of pure water can be calculated to be 55.6 mol dm^{-3} (see Ch 10.1). When a dilute solution of a weak acid, HA, is made, some of the acid molecules dissociate by 'reacting' with the water.

$$HA(aq) + H_2O(l) \rightleftharpoons H_3O^{\oplus}(aq) + A^{\ominus}(aq)$$

If 0.01 mol dm^{-3} of the acid dissociates, then 0.01 mol dm^{-3} of water must have reacted – leaving $(55.6 - 0.01) = 55.59$ mol dm^{-3} of water. So the 'concentration' of the water is effectively constant.

Table 23.3.1 shows values of K_a and pK_a for some weak acids.

K_a and pK_a values for some weak acids

Acid	Formula	K_a / mol dm^{-3}, at 25 °C	pK_a
Methanoic	HCOOH	1.6×10^{-4}	3.8
Ethanoic	CH$_3$COOH	1.7×10^{-5}	4.8
Chloroethanoic	CH$_2$ClCOOH	1.3×10^{-3}	2.9
Dichloroethanoic	CHCl$_2$COOH	5.0×10^{-2}	1.3
Trichloroethanoic	CCl$_3$COOH	2.3×10^{-1}	0.7
Benzoic	C$_6$H$_5$COOH	6.3×10^{-5}	4.2

Table 23.3.1

The reasons for the variations in acid strength which are shown in the table are discussed in Ch 27.6. For the moment, note that a halogen atom attached to the carbon atom next to the carboxylic acid group, —COOH, helps ionisation to occur. This is shown by the rapid increase in K_a, and consequent decrease in pK_a, as the methyl hydrogen atoms in CH$_3$COOH are replaced by chlorine.

Calculating pH from K_a for solutions of known concentration

A useful way of looking at the dissociation of weak acids – and one which enables us to do worthwhile calculations – is as follows.

Consider once more a solution of a weak acid, HA, of concentration 1.0 mol dm^{-3}.

$$HA(aq) \rightleftharpoons H^{\oplus}(aq) + A^{\ominus}(aq)$$

Concentration before
ionisation / mol dm^{-3} 1.0   0   0

Because every mole of HA molecules which ionise gives 1 mole of $H^{\oplus}(aq)$ and one mole of $A^{\ominus}(aq)$, the equilibrium concentration of HA(aq) must be $1 - [H^{\oplus}(aq)]_{eqm}$, and the equilibrium concentration of $A^{\ominus}(aq)$ must be the same as $[H^{\oplus}(aq)]_{eqm}$.

$$HA(aq) \rightleftharpoons H^{\oplus}(aq) + A^{\ominus}(aq)$$

∴ Concentration at
equilibrium / mol dm^{-3}

$$1 - [H^{\oplus}(aq)] \qquad [H^{\oplus}(aq)] \qquad [H^{\oplus}(aq)]$$

$$\therefore K_a = \frac{[H^{\oplus}(aq)]^2}{1 - [H^{\oplus}(aq)]}$$

But HA is a *weak* acid, so the amount of ionisation is small. So $[H^{\oplus}(aq)]_{eqm}$ is small, and therefore $1 - [H^{\oplus}(aq)]_{eqm}$ is very nearly equal to 1.

$$\therefore K_a \simeq \frac{[H^{\oplus}(aq)]^2}{1}$$

$$\therefore [H^{\oplus}(aq)] \simeq \sqrt{K_a}$$

∴ **For a 1.0 mol dm^{-3} solution of a weak acid, the value of $[H^{\oplus}(aq)]_{eqm}$ is very nearly equal to the square root of K_a.**

For ethanoic acid, CH$_3$COOH, of concentration 1.0 mol dm^{-3}

$K_a = 1.7 \times 10^{-5}$ mol dm^{-3} (see *Table 23.3.1*)

$\therefore \sqrt{K_a} = 4.12 \times 10^{-3}$ mol dm$^{-5} \simeq [H^{\oplus}(aq)]$

$\therefore$ pH $= -\log_{10}(4.12 \times 10^{-3}) \simeq 2.38$

∴ The pH of a 1.0 mol dm^{-3} solution of ethanoic acid is, to a near approximation, 2.4.

Consider now ethanoic acid of concentration 0.10 mol dm^{-3}.

$$K_a = \frac{[H^{\oplus}(aq)]^2}{0.1 - [H^{\oplus}(aq)]} = 1.7 \times 10^{-5} \text{ mol dm}^{-3}$$

But, because $[H^{\oplus}(aq)]_{eqm}$ is very small,

$$0.1 - [H^{\oplus}(aq)] \simeq 0.1$$

$$\therefore \frac{[H^{\oplus}(aq)]^2}{0.1} \simeq 1.7 \times 10^{-5}$$

$$\therefore [H^{\oplus}(aq)] \simeq \sqrt{1.7 \times 10^{-6}} \simeq 1.3 \times 10^{-3} \text{ mol dm}^{-3}$$

$$\therefore \text{pH} \simeq 2.88$$

∴ The pH of a 0.10 mol dm^{-3} solution of ethanoic acid, to a near approximation, is 2.9.

The concentration of $H^{\oplus}(aq)$ ions in 0.10 mol dm^{-3} aqueous solution of an acid would be expected to be only 1/10 of what it is in a 1.0 mol dm^{-3} solution. But, as you can see above for ethanoic acid, $[H^{\oplus}(aq)]_{eqm}$ for the 0.10 mol dm^{-3} solution has only dropped to about 1/3 of what it was in the 1.0 mol dm^{-3} solution. Even though the acid concentration has been reduced to 1/10 (1.0 to 0.1 mol dm^{-3}) the $H^{\oplus}(aq)$ concentration has only reduced to 1/3 (4.12×10^{-3} to 1.3×10^{-3} mol dm^{-3}). This is because, as shown previously, the ionisation of a weak acid increases with dilution; and this partly compensates for the decrease in concentration of the acid (*Table 23.3.2*).

Concentration / mol dm^{-3}	Hydrochloric acid pH	Ethanoic acid pH
1.0	0.0	2.4
0.1	1.0	2.9
0.01	2.0	3.4
1×10^{-3}	3.0	3.9
1×10^{-5}	5.0	4.9

Table 23.3.2   The pH of various concentrations of HCl and CH$_3$COOH

The two examples for ethanoic acid above give us a general method for finding the pH of a solution of a weak acid of known K_a. Let the concentration of the acid be X mol dm^{-3}.

$$K_a = \frac{[H^+(aq)]^2}{x - [H^+(aq)]}$$

$$K_a \simeq \frac{[H^+(aq)]^2}{x}$$

$$\therefore [H^+(aq)] \simeq \sqrt{K_a \times x}$$

$$\therefore pH \simeq -\log_{10}\sqrt{K_a \times x}$$

$\therefore$ **The pH of a solution of a weak acid is, to a good approximation, equal to the negative logarithm of the square root of the product of the K_a value for the acid and the concentration of the acid.**

Worked example

What is the pH of a 0.010 mol dm^{-3} solution of benzoic acid, C_6H_5COOH? (K_a(benzoic acid) = 6.3 × 10^{-5} mol dm^{-3})

Data: Benzoic acid, C_6H_5COOH
Concentration 0.010 mol dm^{-3}
K_a = 6.3 × 10^{-5} mol dm^{-3}

Equation: $[H^+(aq)] = \sqrt{K_a \times \text{(concentration of acid)}}$

Calculation: $[H^+(aq)] = \sqrt{6.3 \times 10^{-5} \times 0.010}$
$\therefore pH = -\log_{10}\sqrt{6.3 \times 10^{-5} \times 0.010} = 3.1$

Alternatively, the calculation can be solved working from first principles. *You need be able to use only one of these methods.*

Data: Benzoic acid, C_6H_5COOH
Concentration 0.010 mol dm^{-3}
K_a = 6.3 × 10^{-5} mol dm^{-3}

Equation: $C_6H_5COOH(aq) \rightleftharpoons H^+(aq) + C_6H_5COO^-(aq)$
$pH = -\log_{10}[H^+(aq)]$

$$K_a = \frac{[H^+(aq)][C_6H_5COO^-(aq)]}{[C_6H_5COOH(aq)]}$$

Calculation: Since benzoic acid is a weak acid the equilibrium concentration of the acid, $[C_6H_5COOH(aq)]$ is very close to the initial concentration of the acid = 0.010 mol dm^{-3}.

$$[C_6H_5COO^-(aq)] = [H^+(aq)]$$

$$K_a = \frac{[H^+(aq)]^2}{[C_6H_5COOH]} \quad \therefore [H^+] \simeq \sqrt{K_a \times [C_6H_5COOH]}$$

$$= \sqrt{6.3 \times 10^{-5} \times 0.010} = 7.94 \times 10^{-4}$$

$$pH = -\log_{10}(7.94 \times 10^{-4}) = 3.1$$

N.B. **These calculations are only reasonably accurate for dilute solutions and genuinely weak acids.** The lower the value of pK_a for the acid (and so the greater its strength), the greater the inaccuracy of the approximate pH obtained by this method.

Calculating K_a from the pH of a solution of known concentration

This is the exact opposite of what we have just been doing, and once again only works for dilute solutions of weak acids.

As we have seen:

$$K_a \simeq \frac{[H^+(aq)]^2}{\text{concentration of acid}}$$

So, if $[H^+(aq)]_{eqm}$ can be measured and the concentration is known, K_a can be calculated.

Worked example

The pH of 0.15 mol dm^{-3} propanoic acid, C_2H_5COOH, is 2.85. Calculate K_a, and hence pK_a, for propanoic acid.

Data: Propanoic acid, C_2H_5COOH
Concentration = 0.15 mol dm^{-3}
pH = 2.85

Equation: $pH = -\log_{10}[H^+(aq)]$

$$K_a = \frac{[H^+(aq)]^2}{[HA(aq)]}$$

Since propanoic acid is a weak acid.

Calculation: Using $pH = -\log_{10}[H^+(aq)] = 2.85$ and 'undoing' logs;

i.e. 10^x or `+/-` `shift` `log` on your calculator

$[H^+(aq)] = 1.41 \times 10^{-3}$ mol dm^{-3}

$$K_a = \frac{(1.41 \times 10^{-3})^2}{0.15} = 1.3 \times 10^{-5} \text{ mol dm}^{-3}$$

$\therefore pK_a = 4.88$

Weak bases: K_b and pK_b

The arguments used above for weak acids can also be applied to weak bases.

Consider a weak base, B, in aqueous solution.

$$\begin{array}{cc} B(aq) + H_2O(l) \rightleftharpoons & BH^+(aq) + OH^-(aq) \\ \text{Base} & \text{Conjugate} \\ & \text{acid} \end{array}$$

$$K_c = \frac{[BH^+(aq)]_{eqm}\,[OH^-(aq)]_{eqm}}{[B(aq)]_{eqm}\,[H_2O(l)]_{eqm}}$$

But, as for weak acids, $[H_2O(l)]$ is hardly affected by the ionisation of the base, so $[H_2O(l)]$ can be considered to be constant. If we incorporate it into K_c, we obtain a new constant, the **base dissociation constant, K_b**.

$$K_b = \frac{[BH^+(aq)]_{eqm}\,[OH^-(aq)]_{eqm}}{[B(aq)]_{eqm}}$$

Notice that the units of K_b are mol dm^{-3}, as for K_a.

However, like K_a, **the larger the value of K_b, and the smaller the value of pK_b, the stronger the base**. *Table 23.3.3* lists some values of K_b.

K_b and pK_b values for some weak bases

Base	Formula	K_b / mol dm^{-3} at 25 °C	pK_b
Ammonia	NH_3	1.8×10^{-5}	4.8
Methylamine	CH_3NH_2	4.4×10^{-4}	3.4
Dimethylamine	$(CH_3)_2NH$	5.2×10^{-4}	3.3
Phenylmethylamine	$C_6H_5CH_2NH_2$	2.0×10^{-5}	4.7

Table 23.3.3

The relationship between pK_b and pK_a

Consider a solution of ammonia in water:

$$\begin{array}{cc} NH_3(aq) + H_2O(l) \rightleftharpoons & NH_4^+(aq) + OH^-(aq) \\ \text{Base} & \text{Conjugate} \\ & \text{acid} \end{array}$$

From *Table 23.3.3*, for this equilibrium at 25 °C:

$K_b = 1.8 \times 10^{-5}$ mol dm^{-3} and p$K_b = 4.8$

We can also find K_a for NH$_4^{\oplus}$ acting as an acid:

$$NH_4^{\oplus}(aq) \rightleftharpoons H^{\oplus}(aq) + NH_3(aq)$$

It is found that $K_a = 5.6 \times 10^{-10}$ mol dm^{-3}, and, therefore, p$K_a = 9.2$

$\therefore K_a \times K_b = (5.6 \times 10^{-10}) \times (1.8 \times 10^{-5}) = 1.0 \times 10^{-14}$ mol^2 dm^{-6}

and pK_a + p$K_b = 9.2 + 4.8 = 14$ (*Table 23.3.4*).

Acid		Equilibrium		K_a at 25 °C	pK_a	pK_b
	Acid		Conjugate base	/ mol dm^{-3}		
Sulphuric acid	H_2SO_4	$\rightleftharpoons H^{\oplus} +$	$HSO_4^{\ominus}$	Very large	–	–
Nitric acid	HNO_3	$\rightleftharpoons H^{\oplus} +$	$NO_3^{\ominus}$	40	–1.6	15.6
Trichloroethanoic acid	CCl_3CO_2H	$\rightleftharpoons H^{\oplus} +$	$CCl_3CO_2^{\ominus}$	0.23	0.7	13.3
Sulphurous acid	H_2SO_3	$\rightleftharpoons H^{\oplus} +$	$HSO_3^{\ominus}$	0.015	1.8	12.2
Hydrofluoric acid	HF	$\rightleftharpoons H^{\oplus} +$	$F^{\ominus}$	0.00056	3.3	10.7
Methanoic acid	HCO_2H	$\rightleftharpoons H^{\oplus} +$	$HCO_2^{\ominus}$	0.00016	3.8	10.2
Ethanoic acid	CH_3CO_2H	$\rightleftharpoons H^{\oplus} +$	$CH_3CO_2^{\ominus}$	1.7×10^{-5}	4.8	9.2
Phenylammonium ion	$C_6H_5^{\oplus}NH_3$	$\rightleftharpoons H^{\oplus} +$	$C_6H_5NH_2$	2.4×10^{-5}	4.6	9.4
Carbonic acid	H_2CO_3	$\rightleftharpoons H^{\oplus} +$	$HCO_3^{\ominus}$	4.5×10^{-7}	6.3	7.7
Hydrogen sulphide	H_2S	$\rightleftharpoons H^{\oplus} +$	$HS^{\ominus}$	8.9×10^{-8}	7.0	7.0
Ammonium ion	$NH_4^{\oplus}$	$\rightleftharpoons H^{\oplus} +$	NH_3	5.6×10^{-10}	9.2	4.8
Phenol	C_6H_5OH	$\rightleftharpoons H^{\oplus} +$	$C_6H_5O^{\ominus}$	1.3×10^{-10}	10.3	3.7
Hydrogen peroxide	H_2O_2	$\rightleftharpoons H^{\oplus} +$	$HO_2^{\ominus}$	2.4×10^{-12}	11.6	2.4
Water	H_2O	$\rightleftharpoons H^{\oplus} +$	$OH^{\ominus}$	1.0×10^{-14}	14.0	0

Increasing strength of acid (left side, upward arrow)

Increasing strength of conjugate base (right side, downward arrow)

Table 23.3.4 Trends and strengths of acids and their conjugate bases

For any pair of weak acid/conjugate base, or weak base/conjugate acid, in aqueous solution, $K_a \times K_b = 1.0 \times 10^{-14}$ mol^2 dm^{-6} at 25 °C, and pK_a + p$K_b = 14$. (You should, of course, recognise 1×10^{-14} mol^2 dm^{-6} as the ionic product of water, K_w – see Ch 23.2.)

Calculating pH from K_b for solutions of known concentration

Consider a base, B, in aqueous solution of concentration X mol dm^{-3}. (N.B. Don't forget, [H$_2$O] stays very nearly constant, so we can ignore it.)

$$B(aq) + H_2O(l) \rightleftharpoons BH^{\oplus}(aq) + OH^{\ominus}(aq)$$

Concentrations before ionisation (mol dm^{-3})	X	–	0	0
Concentrations at equilibrium (mol dm^{-3})	X – [OH$^{\ominus}$(aq)]	–	[OH$^{\ominus}$(aq)]	[OH$^{\ominus}$(aq)]

The reasoning is exactly the same as for the weak acid. One mole of B(aq) must form one mole of BH$^{\oplus}$(aq) and one mole of OH$^{\ominus}$(aq); so the concentrations of BH$^{\oplus}$(aq) and OH$^{\ominus}$(aq) must be equal, and the equilibrium concentration of B(aq) must be reduced from its original concentration by [OH$^{\ominus}$(aq)]$_{eqm}$.

$$\therefore K_b = \frac{[OH^{\ominus}(aq)]^2}{X - [OH^{\ominus}(aq)]}$$

But [OH$^{\ominus}$(aq)]$_{eqm}$ for a weak base will be very small compared with X, the concentration of B(aq).

$$\therefore K_b \simeq \frac{[OH^{\ominus}(aq)]^2}{(\text{concentration of base})}$$

$$\therefore [OH^{\ominus}(aq)] \simeq \sqrt{K_b \times (\text{concentration of base})}$$

$$\therefore pOH \simeq -\log_{10}\sqrt{K_b \times (\text{concentration of base})}$$

But, because $[H^{\oplus}(aq)]_{eqm} \times [OH^{\ominus}(aq)]_{eqm} = 1.0 \times 10^{-14}$ mol dm^{-3} for any aqueous solution at 25 °C, as we have seen (Ch 23.2), then:

pH + pOH = 14 for any aqueous solution.

So, once pOH has been found: pH = 14 − pOH.

Worked example

What is the pH of a solution of ammonia, NH$_3$, of concentration 0.10 mol dm^{-3}, given that for ammonia $K_b = 1.7 \times 10^{-5}$ mol dm^{-3}?

Data: Ammonia, NH$_3$

$NH_3(aq) + H_2O(l) \rightleftharpoons NH_4^{\oplus}(aq) + OH^{\ominus}(aq)$

Concentration = 0.10 mol dm^{-3}

$K_b = 1.7 \times 10^{-5}$ mol dm^{-3}

Equation: $K_b \cong \dfrac{[OH^{\ominus}(aq)]^2}{\text{concentration of base}}$

Since $[NH_4^{\oplus}(aq)] = [OH^{\ominus}(aq)]$ and the equilibrium concentration of the base is approximately the same as the initial concentration of the base if it is a weak base.

$pOH = -\log_{10}[OH^{\ominus}(aq)]$

$pOH + pH = 14$

Calculation: $K_b = 1.7 \times 10^{-5} = \dfrac{[OH^{\ominus}(aq)]^2}{0.10}$

$\therefore [OH^{\ominus}(aq)] = \sqrt{1.7 \times 10^{-5} \times 0.10}$

$= 1.3 \times 10^{-3}$

$\therefore pOH = -\log_{10}(1.3 \times 10^{-3})$

$= 2.9$

$\therefore pH = 14 - pOH$

$= 14 - 2.9$

$= 11.1$

A method for finding K_a for a weak acid

For a weak acid, HA, as we have seen:

$$K_a = \frac{[H^{\oplus}(aq)]_{eqm}\,[A^{\ominus}(aq)]_{eqm}}{[HA(aq)]_{eqm}}$$

From this expression, it follows that if we can make $[A^{\ominus}(aq)]$ equal to $[HA(aq)]$, then $K_a = [H^{\oplus}(aq)]$, and $pK_a = pH$.

A given volume of the aqueous weak acid is titrated with standard alkali and the volume of alkali needed for complete neutralisation is noted. The same volume of the acid is then again used, but this time only *half* the volume of alkali; the acid is *half-neutralised*. If half the HA molecules have reacted with alkali to form A$^{\ominus}$ anions, then $[A^{\ominus}(aq)] = [HA(aq)]$.

The pH of the half-neutralised solution is measured with a pH meter – and, since $pK_a = pH$ of the half-neutralised solution, pK_a and hence K_a for the acid are easily obtained.

Svante Arrhenius (1859–1927) was Swedish and Freidrich Ostwald (1853–1932) was German. Together with the Dutchman Jacobus van't Hoff (1852–1911) they founded that major branch of chemistry which is known as physical chemistry. All three won Nobel Prizes, and all three worked within a wide range of topics. But Arrhenius was the first person to investigate closely solutions of acids, bases and salts so that ideas of ionisation could be developed and increasingly understood. Ostwald contributed a great deal to the understanding of the ionisation of water and the behaviour of weak acids. The mathematical expression of the link between K_a, $[H^{\oplus}(aq)]$ and concentration for weak acids was for very many years known as *Ostwald's Dilution Law*.

pH and pOH equations

For a weak acid:

$HA(aq) \rightleftharpoons H^+(aq) + A^-(aq)$

$pH = -\log_{10}[H^+(aq)]$

$K_a = \dfrac{[H^+(aq)][A^-(aq)]}{[HA(aq)]}$ mol dm^{-3}

$pK_a = -\log_{10}(K_a)$

$pH = -\log_{10}\sqrt{K_a \times [HA(aq)]}$

For a weak base:

$B(aq) + H_2O(aq) \rightleftharpoons BH^+(aq) + OH^-(aq)$

$pOH = -\log_{10}[OH^-(aq)]$

$K_b = \dfrac{[BH^+(aq)][OH^-(aq)]}{[B(aq)]}$ mol dm^{-3}

$pK_b = -\log_{10}(K_b)$

$pOH = -\log_{10}\sqrt{K_b \times [B(aq)]}$

In any solution: $K_w = [H^+(aq)][OH^-(aq)] = 1.0 \times 10^{-14}$ mol^2 dm^{-6}

$pK_w = -\log_{10}(K_w) = 14$

$pH + pOH = 14$

For any conjugate acid–base pair:

$pK_a + pK_b = 14$

$K_a \times K_b = 1.0 \times 10^{-14}$ mol^2 dm^{-6}

SELF-ASSESSMENT QUESTIONS 23.3

1 a Define the pH of a dilute solution.

b What is the pH of 0.01 mol dm^{-3} sulphuric acid (assuming complete dissociation)?

c What is the pH of 0.01 mol dm^{-3} sodium hydroxide solution?

d Explain what is meant by a *weak* acid, a *strong* acid, a *weak* base and a *strong* base. Give an example in each case.

e Benzoic acid, C_6H_5COOH is a weak acid ($K_a = 6.3 \times 10^{-5}$ mol dm^{-3} at 25 °C). What is the pH of a solution of 0.01 mol dm^{-3} of benzoic acid at 25 °C?

2 Water dissociates according to the equation:

$$H_2O(l) \rightleftharpoons H^+(aq) + OH^-(aq)$$

a Define the term K_w, and state its value at 25 °C. What is this constant called?

b Explain in which way the value of this constant varies with change in temperature.

c What do the terms pH and pOH mean? What is the significance of the sum of these two values in pure water at 25 °C?

d 1.0 dm^3 of a strong monobasic acid was exactly neutralised by 6.00 g of solid NaOH. How many moles of NaOH are present in 6.00 g of NaOH?

e Write an equation for the neutralisation reaction between sodium hydroxide and a monobasic acid.

f What is the concentration of the acid?

g Calculation the initial pH of the acid, before the reaction.

3 Citric acid is one of the chemicals present in all citrus fruits. It is a weak acid, $K_a = 7.4 \times 10^{-4}$ mol dm^{-3}. It acts as a monoprotic (monobasic) acid:

$$C_5H_7O_5COOH(aq) + H_2O(l) \rightleftharpoons H_3O^+(aq) + C_5H_7O_5COO^-(aq)$$

a Write an expression for K_a for this acid.

b Calculate the value of pK_a for the citric acid.

c Calculate the pH of lemon juice containing 0.18 mol dm^{-3} citric acid.

d The pH you have calculated is an approximation. What assumptions have you made to calculate this approximate pH?

e In the dissociation equation above, there are two conjugate acid–base pairs. Identify the two acid–base pairs.

f Acids other than citric which are found in or used in foods are ethanoic acid (pK_a = 4.8) and benzoic acid (pK_a = 4.2). Put the three acids in order of acid strength, with the strongest acid first.

23.4 Titration curves and indicator theory

Titration curves

For the technique of titration, see Ch 10.2.

If a dilute strong acid, such as 0.100 mol dm⁻³ hydrochloric acid, HCl, is titrated with a dilute strong alkali, e.g. 0.100 mol dm⁻³ sodium hydroxide, NaOH, we can easily follow how the pH changes (*Figure 23.4.1*).

① If we start with 25.0 cm³ of the hydrochloric acid in the flask, then:

$$pH = -\log_{10}[H^{\oplus}(aq)] = -\log_{10}(0.100) = 1$$

② After 25.0 cm³ of the sodium hydroxide solution has been added the acid is exactly neutralised and the pH of the mixture is 7.

③ After 50.0 cm³ of the alkali has been added, the pH will be about 12.5.

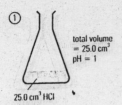
② total volume = 50.0 cm³
pH = 7

25.0 cm³ HCl
+ 25.0 cm³ NaOH

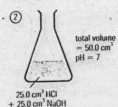
① total volume = 25.0 cm³
pH = 1

25.0 cm³ HCl

The pH of 0.100 mol dm⁻³ NaOH(aq) is 13. This is because $[OH^{\ominus}(aq)] = 0.100 = 10^{-1}$, so pOH = 1. But, since pOH + pH = 14, the pH of 0.100 mol dm⁻³ NaOH is therefore 13.

However, when 50.0 cm³ of 0.100 mol dm⁻³ NaOH(aq) has been added to 25.0 cm³ of 0.100 mol dm⁻³ HCl(aq), the excess alkali – 25.0 cm³ of it – is now spread over 75 cm³ of solution. Its concentration is therefore 0.033 mol dm⁻³; so $[OH^{\ominus}(aq)]$ is 0.033; pOH = $-\log_{10}(0.033)$, therefore pOH = 1.5, and so pH (14.0 − 1.5) = 12.5.

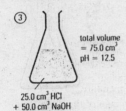
③ total volume = 75.0 cm³
pH = 12.5

25.0 cm³ HCl
+ 50.0 cm³ NaOH

Figure 23.4.1

The variation of pH with the volume of alkali added is shown in *Figure 23.4.2*.

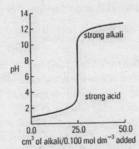

Figure 23.4.2 Titration curve for strong acid/strong alkali

Notice that the pH changes very rapidly at the 'end-point' of the titration.

Assuming that one drop from the burette is 0.05 cm³, then when 24.95 cm³ of alkali have been added, we have 0.05 cm³ of the 0.100 mol dm⁻³ hydrochloric acid left un-neutralised in 49.95 cm³ of solution. For simplicity, let's say the volume is 50.0 cm³.

∴ The concentration of the un-neutralised acid is:

$$\frac{0.05}{1000} \times 0.100 \text{ mol HCl in 50 cm}^3$$

$$= \frac{0.05}{1000} \times 0.100 \times \frac{1000}{50} \text{ mol dm}^{-3}$$

$$= 1 \times 10^{-4} \text{ mol dm}^{-3}$$

∴ The pH is $-\log_{10}(1 \times 10^{-4}) = 4$

A similar calculation shows that *after the addition of just two more drops of alkali*; i.e. a total of 25.05 cm³, *the pH rises to 10*.

So the pH of the solution changes by 6 units, i.e. *the hydrogen ion concentration changes by a factor of one million*, for just two extra drops of alkali!

Since most acid-base reactions involve colourless solutions, how do we know the reaction is neutralised?

- Measure pH, but that is difficult without a pH meter.
- Use indicators that change from one colour to another over a particular pH change.

Phenolphthalein is colourless in acid, pink in alkali solutions. Methyl orange is red in acid, yellow in alkali.

Any acid–base indicator which changes colour between pH 4 and pH 10 will give a reasonably accurate end-point for the titration of a *strong* acid with a *strong* base. The indicator will go through the full colour change for just a drop of added reagent at the end-point. Both methyl orange and phenolphthalein could be used.

As you can see from *Figures 23.4.3* and *23.4.4*, the situation is very different for titrations involving either a weak acid or a weak alkali. For a weak acid/strong alkali titration, the rapid change of pH occurs above pH 7, so to get a good-end point an indicator which changes colour at a high pH has to be used, e.g. phenolphthalein.

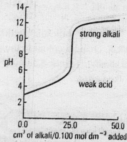

Figure 23.4.3   Titration curve for weak acid/strong alkali

For a strong acid/weak alkali titration, the situation is reversed, and an indicator which changes colour at a pH below 7 must be used, e.g. methyl orange.

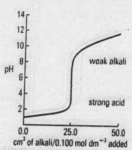

Figure 23.4.4   Titration curve for strong acid/weak alkali

Titration of a weak acid with a weak base using an indicator is not usually possible. As you can see from *Figure 23.4.5, no part of the titration curve is vertical.* In other words, at no point does the pH change rapidly when one or two drops of alkali are added.

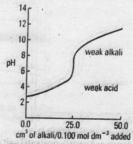

Figure 23.4.5   Titration curve for weak acid/weak alkali

ACTIVITY

The theoretical change in pH of the reaction mixture as a strong acid is titrated against a strong alkali can be calculated using a spreadsheet such as EXCEL.

Acid–base indicators

An **acid–base indicator** is a substance which changes colour depending on whether it is in acid or alkaline solution. You have probably used litmus paper or solution to test whether a solution is acid or alkaline. Litmus is extracted from lichen, and its composition varies. It turns red in acids and blue in alkalis, but it is useless for finding a sharp end-point in a titration as it changes colour over a wide pH range.

Familiar indicators in titrations (*Figure 23.4.6*) include methyl orange (red in acid, yellow in alkali; changes colour in the pH range 3.1–4.4) and phenolphthalein (colourless in acid, red/purple in alkali; pH range 8.3–10.0).

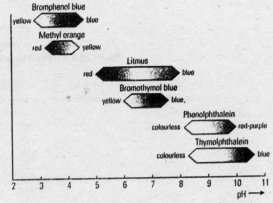

Figure 23.4.6   Colours and pH range of indicators

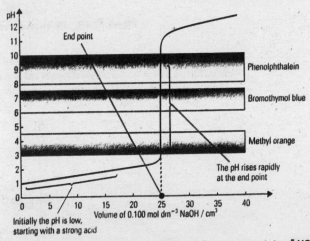

Figure 23.4.7   Titrating a strong acid (25.0 cm³ of 0.100 mol dm⁻³ HCl) and a strong alkali (0.100 mol dm⁻³ NaOH)

Figure 23.4.7 shows that both of these indicators could be used for a strong acid/strong alkali titration. But, for a weak acid/strong alkali titration (*Figure 23.4.8*), phenolphthalein can be used but not methyl orange; and for a strong acid/weak alkali (*Figure 23.4.9*), methyl orange can be used but not phenolphthalein.

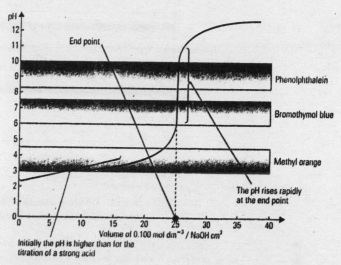

Figure 23.4.8   Titrating a weak acid (25.0 cm³ of 0.100 mol dm⁻³ ethanoic acid) and a strong alkali (0.100 mol dm⁻³ NaOH)

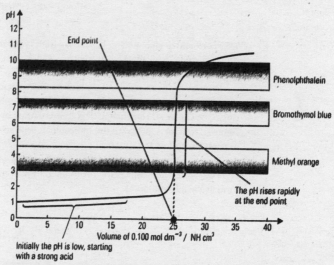

Figure 23.4.9   Titrating a strong acid (25.0 cm³ of 0.100 mol dm⁻³ HCl) and a weak alkali (0.100 mol dm⁻³ NH₃)

Figure 23.4.10 shows that, in this case, no indicator will change colour with the addition of 1 or 2 drops of alkali. Weak acids cannot be titrated against weak alkalis using indicators.

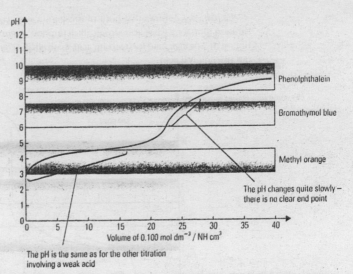

Figure 23.4.10   Titrating a weak acid (25.0 cm³ of 0.100 mol dm⁻³ ethanoic acid) and a weak alkali (0.100 mol dm⁻³ NH₃)

A mixture of different indicators can be made which gives a series of colours across the whole range of pH. Such a mixture is called 'Universal Indicator' or 'Full Range Indicator'. A few drops are added to the solution being tested, and the resulting colour is compared with a chart which relates pH to colour. Perhaps you can see why Universal Indicator cannot be used in a titration.

An acid–base indicator (In) *is itself a weak acid*:

$$\text{HIn(aq)} \rightleftharpoons \text{H}^{+}\text{(aq)} + \text{In}^{\ominus}\text{(aq)}$$

Acid   Conjugate
                      base

The acid and its conjugate base have different colours. In acid solution the acid form HIn predominates; in alkali, $H^{\oplus}$(aq) ions are removed from the equilibrium, more HIn(aq) molecules ionise, and the conjugate base $In^{\ominus}$(aq) predominates.

In order for us to see them, indicators must absorb light in the visible part of the spectrum. To do this, they need *delocalised systems* – fairly complicated organic structures containing benzene rings (see Ch 26.1). The change of colour is caused by changes in molecular structure as a proton is removed from or attached to the molecule. It is these changes of structure which cause the changes in light absorption.

The structures of the *protonated* and *deprotonated* forms of methyl orange are shown in *Figure 23.4.11.* (Remember, the protonated form dominates in acid solution, and the deprotonated form in alkaline solution.)

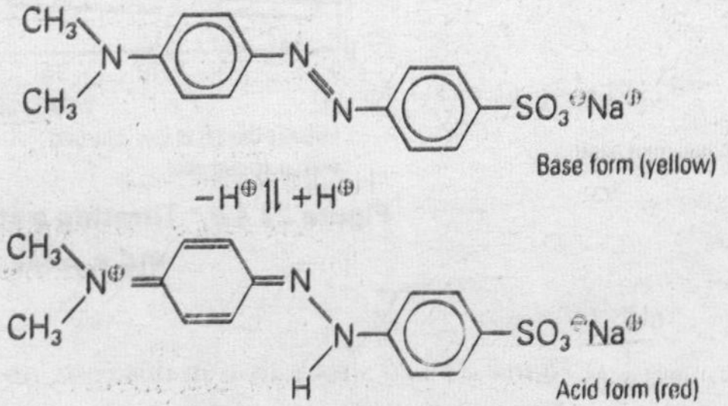

Figure 23.4.11   The protonated and deprotonated forms of methyl orange

From the indicator ionisation equation above, K_a for an indicator can be written as:

$$K_a = \frac{[H^+(aq)]_{eqm} \, [In^-(aq)]_{eqm}}{[HIn(aq)]_{eqm}}$$

If we assume that $HIn(aq)$ and $In^-(aq)$ absorb light to about the same extent (although at different wavelengths), at the end-point there must be roughly equal concentrations of the differently coloured $HIn(aq)$ and $In^-(aq)$. So $[In^-(aq)]$ is roughly equal to $[HIn(aq)]$, therefore:

$$\frac{[In^-(aq)]}{[HIn(aq)]} \simeq 1$$

$$\therefore K_a = [H^+(aq)]_{eqm}$$

In words – K_a for an indicator is approximately equal to ‘the hydrogen ion concentration in the solution at the end-point.

$$\therefore pK_a(\text{indicator}) \simeq pH(\text{at end-point}).$$

SELF-ASSESSMENT QUESTIONS 23.4

1 a Sketch a graph of the change in pH which happens when 0.100 mol dm^{-3} sodium hydroxide solution is added to 25.0 cm^3 of 0.100 mol dm^{-3} benzoic acid ($K_a = 6.3 \times 10^{-5}$ mol dm^{-3}) solution until the alkali is in excess.

   b Calculate the value of pK_a for benzoic acid.

   c The pK_a values, and colours, of some indicators are:

Thymol blue	1.7	red in acid – yellow in alkali
Bromocresol green	4.7	yellow in acid – blue in alkali
Phenolphthalein	9.3	colourless in acid – red in alkali

Which indicator would be used to determine the end-point of the titration of benzoic acid and sodium hydroxide? Explain why you have chosen this particular indicator.

   d What colour change would you expect to see at the end-point of the titration with the indicator you have selected?

2 a Calculate the pH of 0.050 mol dm^{-3} hydrochloric acid.

   b Calculate the pOH of 0.010 mol dm^{-3} potassium hydroxide solution, and hence calculate the pH of the solution.

   c What is the pH of the solution produced by mixing 10.0 cm^3 of this hydrochloric acid solution with 50.0 cm^3 of the potassium hydroxide solution?

   d What would be the pH of a solution produced by mixing 10.0 cm^3 of this hydrochloric acid with 100 cm^3 of the potassium hydroxide solution?

   e Sketch a graph of the change in pH as 0.010 mol dm^{-3} potassium hydroxide solution is titrated with 10.0 cm^3 of 0.050 mol dm^{-3} hydrochloric acid until 100 cm^3 of potassium hydroxide has been added.

   f Suggest a suitable indicator for the end-point of this reaction.

23.5 Salt hydrolysis and displacement of weak acids

Salt hydrolysis

As we have seen, every acid has a **conjugate base** and every base a **conjugate acid**. The link is the proton. Remember that, according to the Brønsted–Lowry idea, every acid is a proton donor and every base a proton acceptor.

Which ions are present in a solution of sodium chloride?

Hydrated sodium ions , $Na^{\oplus}(aq)$, and chloride ions, $Cl^{\ominus}(aq)$, obviously. But also the ions from water – $H^{\oplus}(aq)$ and $OH^{\ominus}(aq)$. Never forget that these ions are present in *all* aqueous solutions.

In dilute sodium chloride solution, none of the ions interferes with any of the others, and the solution is neutral. The possible combinations of ions are sodium chloride, sodium hydroxide and hydrochloric acid. All these substances are completely ionised in dilute solution. **All salts made from a strong acid and a strong alkali give a neutral solution in water** (i.e. pH 7); sodium chloride is such a salt (*Figure 23.5.1*).

Remember: An **alkali** is simply a **base** which is soluble in water to give plenty of OH ions – see Ch 13.4.

Mixture of	In solution	N.B.
NaCl(s)	$Na^{\oplus}(aq)$	NaOH and HCl
and	$Cl^{\ominus}(aq)$	are not formed
$H_2O(l)$	$H^{\oplus}(aq)$	since these both
	$OH^{\ominus}(aq)$	ionise completely
	$H_2O(l)$	in water

Figure 23.5.1   Solution of a salt of a strong acid and a strong alkali

Now consider a solution of sodium ethanoate. The aqueous ions present in the solution are $Na^{\oplus}(aq)$, $CH_3COO^{\ominus}(aq)$, $OH^{\ominus}(aq)$, and $H^{\oplus}(aq)$. Of the possible combinations of ions, sodium hydroxide and sodium ethanoate are completely ionised; but the ethanoate ion is a *strong conjugate base* because ethanoic acid is a weak acid, and so it can seize protons from the water to form molecules of ethanoic acid (*Figure 23.5.2*).

Mixture of	In solution
$CH_3COO^{\ominus}Na^{\oplus}(s)$	$Na^{\oplus}(aq)$
and	$CH_3COO^{\ominus}(aq)$
$H_2O(l)$	$H^{\oplus}(aq)$
	$OH^{\ominus}(aq)$
	$H_2O(l)$
	and $CH_3COOH(aq)$

Figure 23.5.2   Solution of a salt of a weak acid and a strong alkali

The situation can be summarised like this:

Ions from sodium ethanoate $Na^{\oplus}(aq)$ $CH_3COO^{\ominus}(aq)$
+

Ions from water $H_2O(l) \rightleftharpoons OH^{\ominus}(aq) + H^{\oplus}(aq)$
$\|$
$CH_3COOH(aq)$

As ethanoate ions take $H^{\oplus}(aq)$ ions away from the water equilibrium, so more water must ionise to restore K_w (see Ch 23.2). This must mean that there will be more $OH^{\ominus}(aq)$ ions in the solution than $H^{\oplus}(aq)$ ions; the solution will therefore become alkaline.

Ammonium hydroxide, NH$_4$OH, does not exist. The ammonia molecule forms a hydrogen bond with water

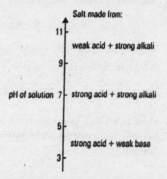

Sodium ethanoate can be made from sodium hydroxide, which is a strong alkali, and ethanoic acid, a weak acid. **All salts made from a strong alkali and a weak acid give an alkaline solution in water** (i.e. pH greater than 7).

If an aqueous solution of a salt does not have a pH of 7, i.e. if it is not neutral, salt hydrolysis has occurred. Salt hydrolysis is one kind of *hydrolysis* – which means the breaking up of a compound by the action of water.

Ammonium chloride is a salt of a weak base and a strong acid. In solution there are NH$_4^\oplus$(aq), Cl$^\ominus$(aq), H$^\oplus$(aq) and OH$^\ominus$(aq) ions.

Ions from NH$_4$Cl	NH$_4^\oplus$(aq) + Cl$^\ominus$(aq)
	+
Water partially dissociates	H$_2$O $\rightleftharpoons$ OH$^\ominus$(aq) + H$^\oplus$(aq)
	$\Updownarrow$
	NH$_3$.H$_2$O(aq)

Because ammonia is a weak base, the ammonium ion is a strong conjugate acid. It therefore takes hydroxide ions from the solution. More water ionises, so eventually there must be more H$^\oplus$(aq) ions than OH$^\ominus$(aq) ions. **All salts made from a weak base and a strong acid give an acid solution in water** (pH less than 7).

If the salt is made from a weak base and a weak acid, the pH depends on the relative strengths of the acid and base – i.e. which of them ionises more in solution. If the base is stronger than the acid – more completely ionised – the solution will be alkaline; if the acid is stronger, the solution will be acidic (*Figure 23.5.3*).

```
                                    Salt made from:

                              11

                                   weak acid + strong alkali

                               9

          pH of solution       7   strong acid + strong alkali

                               5

                                   strong acid + weak base

                               3
```

Figure 23.5.3   Summary of salt hydrolysis and pH of solutions

For the salt ammonium ethanoate, both the acid and the base are almost equally weak, so the pH of a solution of ammonium ethanoate is nearly 7.

Displacement of weak acids by strong acids

Similar arguments explain what is going on when, say, sulphuric acid releases the vinegar smell of ethanoic acid from sodium ethanoate crystals, or solid or aqueous carbonates fizz with practically any acid.

The anion from the weak acid is a strong conjugate base, and will therefore take protons from the strong acid. In other words, the weak acid is released (or 'displaced') from its salt by any strong acid.

$$CH_3COO^\ominus Na^\oplus(s) + H^\oplus(aq) \rightleftharpoons CH_3COOH(aq) + Na^\oplus(aq)$$

Sodium ethanoate   Ethanoic acid

In the same way, a strong base can displace a weaker one, e.g.

$$NH_4Cl(aq) + NaOH(aq) \longrightarrow NaCl(aq) + NH_3(g) + H_2O(l)$$

The formation of $NH_3(g)$ in this reaction is used to test for ammonium salts.

A general statement is this: **a strong acid will displace a weak acid from a salt of the weak acid**.

All carbonates react with acids to give carbon dioxide. The carbonate ion comes from carbonic acid, H_2CO_3, which is a weak acid. The carbonate ion is therefore a strong conjugate base. When a strong acid is added, carbonate ions seize hydrogen ions to form carbonic acid. As well as being weak, carbonic acid is very unstable, and falls apart to give carbon dioxide and water. If the CO_2 is allowed to escape, the reaction goes to completion.

$$CO_3{}^{2-} + 2H^+(aq) \rightleftharpoons H_2CO_3(aq) \longrightarrow CO_2(g) + H_2O(l)$$

Allowing a product to escape like that, and therefore preventing any equilibrium from being established, can mean that apparently unlikely reactions can occur. For example, the standard laboratory preparation of hydrogen chloride gas is by dropping concentrated sulphuric acid on to sodium chloride.

$$NaCl(s) + H_2SO_4(l) \xrightarrow{RT} NaHSO_4(s) + HCl(g)$$

This doesn't mean that sulphuric acid is a stronger acid than hydrochloric acid – simply that sulphuric acid has a higher boiling point, and stays in the reaction mixture while the hydrogen chloride gas is allowed to escape. If the reaction is done in a closed container an equilibrium is set up.

Hydrogen iodide is easily oxidised (see Ch 8.3), and reacts with concentrated sulphuric acid to give iodine, sulphur dioxide and even hydrogen sulphide. This means that, unlike hydrogen chloride, hydrogen iodide gas cannot be prepared in the laboratory by dropping concentrated sulphuric acid on to a solid salt.

$$\text{e.g.} \quad 2NaI(s) + 2H_2SO_4(l) \xrightarrow{RT} I_2(s) + SO_2(g) + 2H_2O(l) + Na_2SO_4(aq)$$

Instead, the iodide (e.g. NaI) is warmed with 100% phosphoric(V) acid, which although a weak acid has a high boiling point and does not oxidise the HI as it is produced.

$$NaI(s) + H_3PO_4(l) \longrightarrow HI(g) + NaH_2PO_4(s)$$

Hydrogen iodide gas is given off. In this instance, a weak acid is actually displacing a strong acid, because the weak acid is comparatively involatile. In a closed container the reaction would not happen.

SELF-ASSESSMENT QUESTIONS 23.5

1 Estimate a value of the pH of aqueous solutions of the following salts. If your estimate is not at pH 7, explain why.
   a Potassium chloride
   b Ammonium sulphate
   c Sodium benzoate
   d Caesium nitrate

2 Ethanoic acid, CH_3COOH, reacts with sodium hydroxide in a neutralisation reaction.
   a Write an equation for the reaction.
   b If 25.0 cm³ of a solution of ethanoic acid, 0.100 mol dm⁻³, is titrated with 0.100 mol dm⁻³ sodium hydroxide, the end-point is reached when 25.0 cm³ of the alkali has been added. The pH of the solution at this point is not, however, pH 7. Explain why this should be so.

23.6 Buffer solutions

Consider a solution containing both ethanoic acid, CH_3COOH, and sodium ethanoate, $CH_3COO^- Na^+$ (*Figure 23.6.1*). The ions present are:

- $Na^+(aq)$ and $CH_3COO^-(aq)$ from the salt;
- $OH^-(aq)$ from the water; and
- $H^+(aq)$ from the water and from the acid.

Also present are molecules of CH_3COOH.

$$CH_3COOH(aq) \rightleftharpoons CH_3COO^-(aq) + H^+(aq)$$

$$CH_3COO^- Na^+(s) \longrightarrow CH_3COO^-(aq) + Na^+(aq)$$

$$H_2O(l) \rightleftharpoons H^+(aq) + OH^-(aq)$$

Note that two reactions are equilibrium reactions

Figure 23.6.1   A solution of ethanoic acid and sodium ethanoate

Because ethanoic acid is weak, the CH_3COO^- ion is a **strong conjugate base** (see Ch 13.4). *If an acid contaminates the buffer solutions, the $H^+(aq)$ ions from that acid are mopped up by CH_3COO^- ions to form molecules of CH_3COOH, and the pH of the solution stays roughly the same. Similarly, if alkali contaminates the buffer solution, the $OH^-(aq)$ ions from the alkali cause more CH_3COOH molecules to ionise and react to form water and CH_3COO^- ions. Again, the pH stays roughly the same (see Figure 23.6.2).*

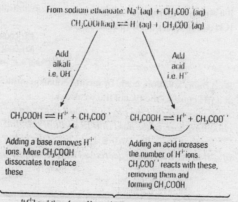

From sodium ethanoate: $Na^+(aq) + CH_3COO^-(aq)$

$CH_3COOH(aq) \rightleftharpoons H^+(aq) + CH_3COO^-(aq)$

Add alkali i.e. OH^-

Add acid i.e. H^+

$CH_3COOH \rightleftharpoons H^+ + CH_3COO^-$

$CH_3COOH \rightleftharpoons H^+ + CH_3COO^-$

Adding a base removes H^+ ions. More CH_3COOH dissociates to replace these

Adding an acid increases the number of H^+ ions. CH_3COO^- reacts with these, removing them and forming CH_3COOH

$[H^+]$ and therefore pH remain constant (or very nearly so)

Figure 23.6.2   A buffer in action

Obviously, if too much acid or alkali is added the CH_3COO^- ions or CH_3COOH molecules will be used up and the pH will rapidly change.

To sum up:

- the buffer action depends on the equilibrium between the weak acid, CH_3COOH, and its conjugate base, CH_3COO^-, in a mixture of the acid and one of its salts;
- extra $H^+(aq)$ ions from added acid will push the equilibrium to the left, and most of the extra $H^+(aq)$ ions will be removed;
- extra $OH^-(aq)$ ions from added alkali will remove $H^+(aq)$ ions from the equilibrium, pushing the equilibrium to the right;
- either way, the pH stays reasonably constant until the buffering capacity of the mixture is used up.

An aqueous acidic buffer (i.e. one with a pH rather less than 7) contains a weak acid together with a salt of that acid.

Once that is grasped, it is fairly easy to see that an aqueous alkaline buffer (i.e. one with a pH rather greater than 7) contains a weak base with a salt of that base.

A typical alkaline buffer consists of aqueous ammonia, NH_3, together with ammonium chloride, $NH_4^{\oplus}Cl^{\ominus}$.

The ions present are:

$NH_4^{\oplus}$ (aq), $Cl^{\ominus}$(aq), $H^{\oplus}$(aq), and $OH^{\ominus}$(aq).

If any alkali is added the $OH^{\ominus}$(aq) ions from this alkali react with the $NH_4^{\oplus}$(aq) ions and are removed.

$$NH_4^{\oplus}(aq) + OH^{\ominus}(aq) \longrightarrow NH_3(aq) + H_2O(l)$$

If any acid is added the $H^{\oplus}$(aq) ions from this acid react with NH_3(aq) molecules to give $NH_4^{\oplus}$(aq) ions. In both cases, so long as the buffering capacity of the solution is not overwhelmed by adding too much acid or alkali, the pH stays roughly the same.

Calculation of the pH of buffer solutions

It would be very useful to have an equation which linked the pH of a buffer solution to the acid dissociation constant and to the relative concentrations of the acid and of the anion from the acid.

Consider the K_a expression for ethanoic acid (see Ch 23.3).

$$CH_3COOH(aq) \rightleftharpoons CH_3COO^{\ominus}(aq) + H^{\oplus}(aq)$$

$$K_a = \frac{[CH_3COO^{\ominus}(aq)]_{eqm} [H^{\oplus}(aq)]_{eqm}}{[CH_3COOH(aq)]_{eqm}}$$

Rearranging: $[H^{\oplus}(aq)] = K_a \times \dfrac{[CH_3COOH(aq)]}{[CH_3COO^{\ominus}(aq)]}$

If a solution is made up containing *equal* concentrations of CH_3COOH and $CH_3COO^{\ominus}Na^{\oplus}$, so that $[CH_3COOH(aq)] = [CH_3COO^{\ominus}(aq)]$:

then $\dfrac{[CH_3COOH(aq)]}{[CH_3COO^{\ominus}(aq)]} = 1$

$\therefore [H^{\oplus}(aq)] = K_a$ and $pH = pK_a$

So, a buffer solution, made by mixing equal quantities of the weak acid and its sodium salt, will buffer at a pH equal to the pK_a of the weak acid.

Remember that the pK_a of a weak acid can be found by making a solution containing equal concentrations of the acid and its sodium salt and measuring the resulting pH (see also Ch 23.4).

Further, since $\log_{10}(10) = 1$, the pH of the buffer can be raised by one unit by using a 10:1 ratio of salt to acid. It can be decreased by one unit by using a 1:10 ratio of salt to acid.

Worked example

If K_a for ethanoic acid at 25 °C is 1.8×10^{-5} mol dm^{-3}, what is the pH of a solution which contains 1.0 mol dm^{-3} sodium ethanoate and 0.10 mol dm^{-3} ethanoic acid?

Data:	$CH_3COOH(aq) \rightleftharpoons H^{\oplus}(aq) + CH_3COO^{\ominus}(aq)$
	Concentration $= 0.10$ mol dm^{-3}
	$K_a = 1.8 \times 10^{-5}$ mol dm^{-3}
	$CH_3COONa \longrightarrow CH_3COO^{\ominus}(aq) + Na^{\oplus}(aq)$
	Concentration $= 1.0$ mol dm^{-3}
Equations:	$pH = -\log_{10}[H^{\oplus}(aq)]$
	$K_a = \dfrac{[H^{\oplus}(aq)] [CH_3COO^{\ominus}(aq)]}{[CH_3COOH(aq)]}$
Calculation:	Since ethanoic acid is a weak acid, the concentration of CH_3COOH at equilibrium is almost the same as the initial concentration.

The concentration [CH_3COO^-(aq)] is almost entirely due to the ionisation of sodium ethanoate.

$$K_a = 1.8 \times 10^{-5} = \frac{[H^+(aq)][CH_3COO^-(aq)]}{[CH_3COOH(aq)]}$$

$$= \frac{[H^+(aq)] \times 1.0}{0.10}$$

$$\therefore [H^+(aq)] = \frac{1.8 \times 10^{-5} \times 0.10}{1.0}$$

$$= 1.8 \times 10^{-6}$$

$$\therefore pH = -\log_{10}(1.8 \times 10^{-6}) = 5.7$$

The pH of buffer solutions – the Henderson–Hasselbach equation

The equation which we have derived for ethanoic acid and ethanoate can be made into a general equation which can be used for *any* weak acid and its conjugate base.

$$HA(aq) \rightleftharpoons H^+(aq) + A^-(aq)$$

$$K_a = \frac{[A^-(aq)]_{eqm}[H^+(aq)]_{eqm}}{[HA(aq)]_{eqm}}$$

Rearranging:

$$[H^+(aq)] = K_a \times \frac{[HA(aq)]}{[A^-(aq)]}$$

Take $-\log_{10}$ of everything (see Ch 23.2):

$$pH = pK_a - \log_{10}\frac{[HA(aq)]}{[A^-(aq)]}$$

This is called the *Henderson–Hasselbach equation*.

This equation is written in some books:

$$pH = pK_a + \log_{10}\frac{[A^-(aq)]}{[HA(aq)]}$$

But this is exactly the same!

Remember (see Appendix), $\log_{10}\frac{x}{y} = -\log_{10}\frac{y}{x}$

$$\left(Try\ x = 10, y = 1; \log_{10}\frac{10}{1} = 1; \log_{10}\frac{1}{10} = -1\right)$$

In either case, if [HA(aq)] = [$A^\ominus$(aq)], pH = pK_a.

And if [HA(aq)] > [$A^\ominus$(aq)], i.e. if there is a greater concentration of acid than of conjugate base, the pH, not surprisingly, drops below the pK_a value.

If [$A^\ominus$(aq)] > [HA(aq)], the pH rises above the pK_a value.

Now, the example we used above can be worked out as follows.

If K_a for ethanoic acid at 25 °C is 1.8×10^{-5} mol dm^{-3}, what is the pH of a solution which contains 1.0 mol dm^{-3} sodium ethanoate and 0.10 mol dm^{-3} ethanoic acid?

$$pH = pK_a - \log_{10}\frac{[CH_3COOH(aq)]}{[CH_3COO^\ominus(aq)]}$$

$$pH = -\log_{10}(1.8 \times 10^{-5}) - \log_{10}\left(\frac{0.1}{1.0}\right)$$

$$= -(-4.74) - \log_{10}0.1 = +4.74 + 1$$

$$pH = 5.74$$

pH and living organisms

Living things, including you, depend greatly on constancy of pH in their body fluids. Human blood has a pH of 7.4; this is essential for efficient respiration. If the pH drops below 7 or rises to near 8 the result could be death. Details of the buffering of blood and other body fluids can be found in AS and A Level biology textbooks.

Many aquatic organisms, including fish, are very sensitive to the pH of their environment. Over recent years many lakes in Northern Europe and Scandinavia, and also North America, have experienced a rapid drop in pH and in some cases total loss of their fish populations. This has been linked to 'acid rain' caused by industrial activity and internal combustion engines, but the topic is complex and will not be further discussed here.

Acid–base chemistry in non-aqueous solvents

A lot of modern chemistry involves acid–base reactions in solvents other than water. As we have seen (Ch 13.4), water molecules can act both as acids, i.e. proton donors, or bases, i.e. proton acceptors. Some other solvents can also do this. An example is liquid ammonia.

$$NH_4^{\oplus} \underset{+H^+}{\overset{-H^+}{\rightleftharpoons}} NH_3(l) \underset{+H^+}{\overset{-H^+}{\rightleftharpoons}} NH_2^{\ominus}$$

It is perfectly possible (though difficult without special apparatus!) to titrate ammonium chloride, $NH_4^{\oplus}Cl^{\ominus}$, as the acid with sodamide, $Na^{\oplus}NH_2^{\ominus}$, as the base in solution in liquid ammonia.

$$NH_4^{\oplus} + NH_2^{\ominus} \longrightarrow 2NH_3$$

When hydrogen chloride is dissolved in pure ethanoic acid, an equilibrium results:

$$CH_3COOH + HCl \rightleftharpoons CH_3COOH_2^{\oplus} + Cl^{\ominus}$$

Here, ethanoic acid receives the proton, and therefore acts as a base.

Remember: whether a substance acts as an acid or a base may depend on what else is involved in the reaction.

With ammonia, for example, water acts as an acid.

$$NH_3(g) + H_2O(l) \longrightarrow NH_4^{\oplus}(aq) + OH^{\ominus}(aq)$$

With hydrogen chloride, however, water acts as a base; it accepts the proton from the HCl:

$$HCl(g) + H_2O(l) \longrightarrow H_3O^{\oplus}(aq) + Cl^{\ominus}(aq)$$

(But remember that the hydrated proton is best represented as $H^{\oplus}(aq)$ rather than $H_3O^{\oplus}(aq)$.)

Lewis acids and bases

When a proton adds to, say, ammonia the new bond is formed by the lone pair on the nitrogen being donated to the proton.

G N Lewis in 1923 defined an acid as an electron pair acceptor and a base as an electron pair donor. So, in the example above, the proton is a Lewis acid and the ammonia a Lewis base. In this case, the Lewis idea corresponds to the Brønsted–Lowry definition (see Ch 13.4). But Lewis's definition is not limited solely to protons. For example, ammonia will donate a lone pair to boron trifluoride to form an addition compound (see Ch 2.4):

$$H_3N{\colon} BF_3 \longrightarrow H_3N—BF_3$$

Here, the ammonia is acting as a Lewis base and the boron trifluoride as a Lewis acid.

It is important not to confuse Lewis's idea with the simple idea of oxidation and reduction (see Ch 4.3). The difference is that in acid–base chemistry the donation or acceptance of a pair of electrons does not cause a change in oxidation state (see Ch 4.3).

For nearly all everyday chemical purposes the Brønsted–Lowry theory of acids and bases will serve very well.

1 0.10 mol of an acid HX was mixed with 0.10 mol of the sodium salt of X and the mixture made up to 100 cm^3 with water. The pH of the resulting solution at 25 °C was found to be 4.8. What is the value of K_a at 25 °C for the acid HX?

2 **a** If K_a for methanoic acid at 25 °C is 1.8×10^{-4} mol dm^{-3}, find the pH of a solution which contains 1.00 mol dm^{-3} of each of methanoic acid and sodium methanoate.

  b How could you use methanoic acid and sodium methanoate to make up a solution with a pH value one unit less than your answer for **a**?

3 **a** What is meant by a buffer solution?

  b Explain how a mixture of sodium butanoate and butanoic acid acts as a buffer when dilute hydrochloric acid is added.

  c Why should buffer solutions be so important in living systems?

  d Calculate the pH of a solution containing 0.10 mol dm^{-3} sodium butanoate and 0.40 mol dm^{-3} butanoic acid (K_a for butanoic acid, C_3H_7COOH, is 1.5×10^{-5} mol dm^{-3}).

Answers to diagnostic test on acids (p 573)

1. **a** An acid is a proton donor.
   b A base is a proton acceptor.

2. **a** An alkali is a soluble base.
   b OH^-(aq) is present in alkaline solutions.

3. **a** The conjugate base of CH_3COOH is CH_3COO^-.
   b The conjugate acid of NH_3 is NH_4^+.

4. A weak acid only partially ionises in aqueous solution.

5. Sulphuric acid is dibasic because it can donate two protons per molecule of H_2SO_4.

6. **a** Neutralisation occurs when an acid and a base react to form only a salt and water.
   b H^+(aq) + OH^-(aq) $\longrightarrow$ H_2O(l)

7. H_3O^+(aq) $\underset{+H^+}{\overset{-H^+}{\rightleftharpoons}}$ H_2O(l) $\underset{+H^+}{\overset{-H^+}{\rightleftharpoons}}$ OH^-(aq)

Summary

$K_w = [H^+(aq)][OH^-(aq)] = 1.0 \times 10^{-14}$ mol^2 dm^{-6} at 25°C

$pH = -\log_{10}[H^+(aq)]$ $pOH = -\log_{10}[OH^-(aq)]$

$pH + pOH = 14$

For a weak acid, HA (only slightly ionised in water): HA $\rightleftharpoons$ H^+ + A^-

$K_a = \dfrac{[H^+(aq)][A^-(aq)]}{[HA(aq)]}$ $H^+(aq) \simeq \sqrt{K_a \times [HA]}$ for a weak acid

For a weak base: B + H_2O $\rightleftharpoons$ BH^+ + OH^-

$K_b = \dfrac{[BH^+][OH^-]}{[B]}$

$pK_a + pK_b = 14$

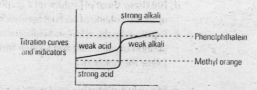

Titration curves and indicators

strong alkali — Phenolphthalein
weak acid   weak alkali — Methyl orange
strong acid

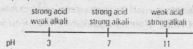

Hydrolysis of salts:

strong acid weak alkali	strong acid strong alkali	weak acid strong alkali

pH   3   7   11

Buffer solutions: resist change in pH and are made from weak acid and salt of the acid

SUMMARY 23

In this chapter the following new ideas have been covered. *Have you understood and learned them?* Could you explain them clearly to a friend? The page references will help.

EXAMINATION QUESTIONS 23

1 When carbon dioxide dissolves in water, some of the gas is thought to dissolve as 'H_2CO_3'. When the pressure is released in bottles containing carbonated drinks, the 'fizz' is due to the release of some of the CO_2 gas.

  a Write an equation to show how carbon dioxide dissolves in water, producing 'H_2CO_3'.

  b In such a solution an equilibrium exists as follows:

$$H_2CO_3(aq) + H_2O(l) \rightleftharpoons HCO_3^{\ominus}(aq) + H_3O^{\oplus}(aq)$$

Explain how H_2CO_3 is acting as an acid in this reaction.

  c Define the term *acid dissociation constant*, and write an expression for the first acid dissociation constant of carbonic acid, 'H_2CO_3'.

  d What is meant by pK_a?

  e The K_a value for carbonic acid is 4.5×10^{-7} mol dm^{-3} What is the value of pK_a?

  f Is carbonic acid a stronger acid or weaker acid than ethanoic acid, $pK_a = 4.8$?

2 **a** The pH of 50.0 cm^3 solution of 0.0200 mol dm^{-3} nitric acid is 1.7. When 50.0 cm^3 of a 0.0200 mol dm^{-3} solution of sodium hydroxide is added, the pH changes to pH 7. Explain why there is a pH change when the solutions are mixed.

  b Define K_w for an aqueous solution. What is the value of pK_w at 25 °C?

  c What will be the pH of the solution from **a** if another 50 cm^3 of 0.020 mol dm^{-3} sodium hydroxide is added to the mixture?

  d Plot these three pH values on a graph, and show how you would expect the points to join up as the volume of sodium hydroxide is gradually added to the nitric acid solution.

  e Explain why the acid–base indicator methyl violet ($pK_a = 0.8$) cannot be used to follow the titration of nitric acid and sodium hydroxide, whereas phenol red ($pK_a = 8$) can be used to follow the reaction.

3 **a** The pH of swimming pool water is kept around 7.3. Calculate the $H^{\oplus}(aq)$ concentration at this pH.

  b The reason why chlorine is used in the upkeep of swimming pools is because of the disinfectant action of HClO. Write an equation for the formation of HClO when chlorine dissolves in water.

   c HClO is a weak acid. Write an equation for the dissociation reaction in water.

   d Define K_a for HClO.

   e The free acid, HClO(aq), rather than the anion $ClO^{\ominus}$(aq), is the bactericide. If K_a for the acid is 3.7×10^{-8} mol dm^{-3}, and the pH of the water is 7.3, Calculate the ratio of HOCl(aq):$ClO^{\ominus}$(aq) in the swimming pool water.

4 a Using hydrochloric acid and propanoic acid, C_2H_5COOH ($K_a = 1.3 \times 10^{-5}$ mol dm^{-3}) explain what is meant by a weak acid and a strong acid.

   b Calculate the pH of 0.010 mol dm^{-3} aqueous hydrochloric acid.

   c Calculate the pH of 0.010 mol dm^{-3} aqueous sulphuric acid (assuming full dissociation).

   d Calculate the pH of 0.010 mol dm^{-3} aqueous propanoic acid.

   e Sodium hydroxide reacts with propanoic acid to form a salt and water. Calculate the pH of a solution containing 0.010 mol dm^{-3} sodium propanoate and 0.010 mol dm^{-3} propanoic acid.

   f How will the pH of this last solution change if 5.0 cm^3 of 1.0×10^{-6} mol dm^{-3} aqueous hydrochloric acid is added to the solution?

   g Sketch a graph of the changes in pH as dilute sodium hydroxide solution, 0.010 mol dm^{-3}, is added to 25.0 cm^3 of 0.010 mol dm^{-3} propanoic acid.

- The ideas of **reduction** and **oxidation**. (Ch 4.3)
- Reactions that can **change in both directions**. (Ch 13.1)
- The meaning of a **dynamic equilibrium**. (Ch 13.1)
- The use of the **mole**. (Ch 9.5)
- The **chlor-alkali industry**. (Ch 8.8)

24

redox
equilibria

We have already met oxidation and reduction in terms of adding or removing oxygen (or an electronegative element), and in terms of electrons (Ch 4.3). Oxidation is the loss of electrons, and reduction occurs when an element gains electrons.

Oxidation Is Loss, Reduction Is Gain (of electrons): hence, OILRIG

We can see this in the reaction of magnesium and oxygen to form magnesium oxide.

$$Mg(s) + \tfrac{1}{2}O_2(g) \longrightarrow MgO(s)$$

There are two processes occurring. The first involves oxidation: the magnesium atom loses two electrons to form the magnesium ion.

$$Mg \longrightarrow Mg^{2\oplus} + 2e^{\ominus}$$

In the second process, the oxygen atom gains these two electrons. It is reduced to form the oxide ion.

$$\tfrac{1}{2}O_2 + 2e^{\ominus} \longrightarrow O^{2\ominus}$$

OILRIG
Oxidation Is Loss, Reduction Is Gain
(of electrons)
Oxidising agents are easily reduced.
Reducing agents are easily oxidised.

24.1 Half-equations and oxidation states

In Ch 4.3, we divided redox reactions into two halves to show oxidation and reduction processes. For example, the reaction of sodium and chlorine:

$$Na(s) + \tfrac{1}{2}Cl_2(g) \longrightarrow NaCl(s)$$

Here the sodium atom is oxidised to the sodium ion by losing its outer electron.

$$Na(s) \longrightarrow Na^{\oplus} + e^{\ominus}$$

At the same time a chlorine atom gains this electron, and so it becomes reduced to a chloride ion.

$$\tfrac{1}{2}Cl_2(g) + e^{\ominus} \longrightarrow Cl^{\ominus}$$

Any redox reaction can be divided into an oxidation half-equation and a reduction half-equation.

A reduction process must always accompany any oxidation process and vice versa. The two processes combined are called a *redox* reaction.

The reaction of copper with chlorine can be broken down into the oxidation of copper:

$$Cu(s) \longrightarrow Cu^{2\oplus} + 2e^{\ominus} \quad \text{(the \textbf{oxidation} half-equation)}$$

and the reduction of the chlorine:

$$Cl_2(g) + 2e^{\ominus} \longrightarrow 2Cl^{\ominus} \quad \text{(the \textbf{reduction} half-equation)}$$

If we add these two half equations together, the overall equation is:

$$Cu(s) + Cl_2(g) \longrightarrow CuCl_2(s)$$

It is interesting to compare the redox process with the acid–base reactions we have already covered (Ch 23.1), where the essential process is the loss or gain of protons. In redox processes, electrons are lost or gained.

Displacement reactions

You are probably familiar with the displacement reaction between copper sulphate and zinc, where the more reactive metal, zinc, replaces the copper.

$$CuSO_4(aq) + Zn(s) \longrightarrow ZnSO_4(aq) + Cu(s)$$

When zinc powder is added to a 1.0 mol dm^{-3} solution of copper sulphate, the blue colour of the copper sulphate solution starts to fade, and if sufficient zinc is added the blue colour disappears completely. At the same time a brown precipitate of copper metal is formed, and the temperature of the solution rises because the reaction is exothermic. If we split this reaction up into half-equations, we can see that copper ions are reduced in the reaction and zinc is oxidised.

$$Cu^{2\oplus}(aq) + 2e^{\oplus} \longrightarrow Cu(s) \quad \text{(the reduction process)}$$

$$Zn(s) \longrightarrow Zn^{2\oplus}(aq) + 2e^{\ominus} \quad \text{(the oxidation process)}$$

By adding these two half equations together (and cancelling out the two electrons on each side of the equation) the full ionic equation becomes:

$$Cu^{2\oplus}(aq) + Zn(s) \longrightarrow Cu(s) + Zn^{2\oplus}(aq)$$

A similar reaction, this time involving non-metals, occurs when chlorine gas is bubbled through a solution containing bromide ions (see *Figure 24.1.1*):

$$Cl_2(g) + 2e^{\ominus} \longrightarrow 2Cl^{\ominus}(aq) \quad \textbf{(reduction)}$$

$$2Br^{\ominus}(aq) \longrightarrow Br_2(aq) + 2e^{\ominus} \quad \textbf{(oxidation)}$$

These half equations can be combined to give the ionic equation:

$$Cl_2(g) + 2Br^{\ominus}(aq) \longrightarrow 2Cl^{\ominus}(aq) + Br_2(aq)$$

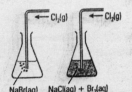

Figure 24.1.1 Chlorine displaces bromine

The redox chart

A way to keep a check on what is going on in any redox reaction is to recognise the changes in oxidation states of the elements involved. In Ch 4.3 we showed that the **oxidation states** of elements can be worked out by using some assumptions about the common oxidation states of certain elements, and by assuming that the unreacted element always has a zero oxidation state. The other 'rules' for assigning oxidation states are summarised in *Figure 24.1.2*.

Assigning oxidation states

	Oxidation state
Uncombined elements	0
Sodium, potassium	+1
Magnesium, calcium	+2
Aluminium	+3
Fluorine	−1
Hydrogen, usually	+1
Oxygen, usually	−2
Chlorine, often	−1

Figure 24.1.2

Examples of working out oxidation states are contained in Ch 4.3. If you do not remember how the following examples are worked out, then look back to Chapter 4 to make sure. Remember, the sum of the oxidation states of the elements in a compound must be zero. In a multi-atom ion, the sum of the oxidation states must equal the charge on the ion.

Examples of working out oxidation state

S in K_2SO_3	: $2 \times K(+1)$, $3 \times O(-2)$; total $= -4$;	$\therefore S = +4$ oxidation state
S in S_2Cl_2	: $2 \times Cl(-1)$; total $= -2$;	$\therefore S = +2/2 = +1$ oxidation state
Br in $NaBrO_3$	: $1 \times Na(+1)$, $3 \times O(-2)$; total $= -5$;	$\therefore Br = +5$ oxidation state
S in SO_4^{2-}	: $4 \times O(-2)$; total $= -8$; and overall charge is -2;	$\therefore S = +6$ oxidation state

The redox changes in any reaction can be displayed on a redox chart. If one element is oxidised then another element has to be reduced.

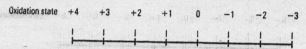

Oxidation is the process of increasing the oxidation number (moving left on the chart) and so losing electrons, and reduction is the reverse process.

For the oxidation of magnesium metal by oxygen:

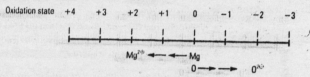

When sodium burns in chlorine:

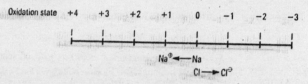

And the displacement reaction with copper ions and zinc is:

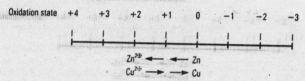

In these three examples, the oxidation process has involved the same number of electrons as the reduction process. There are the same number of atoms in each half-equation. In all these processes, exactly the same number of electrons must be involved in each direction, so that overall the electrons gained and lost cancel out.

When copper reacts with chlorine:

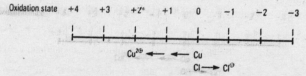

In this case, two chlorine atoms have to be reduced for each copper atom oxidised, if the numbers of electrons are to balance out. The balanced equation for this reaction becomes:

$$Cu(s) + Cl_2(g) \longrightarrow CuCl_2(s)$$

Worked examples with balancing redox equations

The reduction and oxidation half-equations can be used to work out the balanced equation for a redox reaction. Just remember the number of electrons provided by the oxidation half-equation must be the same as the number of electrons accepted by the reduction half-equation.

A *Work out the equation for the reaction of potassium burning in bromine vapour.*

The reduction process is: $$Br_2(g) + 2e^\ominus \longrightarrow 2Br^\ominus$$

The oxidation process is: $$K(s) \longrightarrow K^\oplus + e^\ominus$$

The oxidation half-equation has to be doubled to provide the two electrons needed in the bromine half-equation:

$$2K(s) \longrightarrow 2K^\oplus + 2e^\ominus$$

Adding the reduction and oxidation half-equations together will give the overall balanced equation.

$$Br_2(g) + 2e^{\ominus} \longrightarrow 2Br^\ominus$$
$$2K(s) \longrightarrow 2K^\oplus + 2e^\ominus$$
$$\overline{2K(s) + Br_2(g) \longrightarrow 2K^\oplus + 2Br^\ominus} \text{ (which is equivalent to 2KBr)}.$$

What we have done is:

- divide the reaction into the reduction and oxidation half-equations;
- balance up the number of electrons in the two half-equations;
- add the half-equations together.

B *Iron metal reacts with dry chlorine to form iron(III) chloride. What is the balanced equation for the reaction?*

The oxidation half-equation is: $$Fe(s) \longrightarrow Fe^{3\oplus} + 3e^\ominus$$

The reduction half-equation is: $$Cl_2(g) + 2e^\ominus \longrightarrow 2Cl^\ominus$$

To balance the number of electrons in each half-equation, the iron half-equation needs to be multiplied by 2 and the chlorine half-equation by 3.

$$2Fe(s) \longrightarrow 2Fe^{3\oplus} + 6e^\ominus$$
$$3Cl_2(g) + 6e^\ominus \longrightarrow 6Cl^\ominus$$

Adding the half-equations gives the balanced redox equation.

$$2Fe(s) + 3Cl_2(g) \longrightarrow 2FeCl_3(s)$$

C *Write the balanced redox equation for the oxidation of $Fe^{2\oplus}(aq)$ to $Fe^{3\oplus}(aq)$ using acidified potassium manganate(VII). The half-equation for the reduction of potassium manganate(VII) is:*

$$MnO_4^\ominus(aq) + 8H^\oplus(aq) + 5e^\ominus \longrightarrow Mn^{2\oplus}(aq) + 4H_2O(l)$$

The oxidation half equation is: $$Fe^{2\oplus}(aq) \longrightarrow Fe^{3\oplus}(aq) + e^\ominus$$

And the reduction equation is: $$MnO_4^\ominus(aq) + 8H^\oplus(aq) + 5e^\ominus \longrightarrow Mn^{2\oplus}(aq) + 4H_2O(l)$$

The oxidation half-equation needs to produce 5 electrons, so becomes:

$$5Fe^{2\oplus}(aq) \longrightarrow 5Fe^{3\oplus}(aq) + 5e^\ominus$$

Adding the two half-equations, the final equation is:

$$5Fe^{2\oplus}(aq) + MnO_4^\ominus(aq) + 8H^\oplus(aq) \longrightarrow 5Fe^{3\oplus}(aq) + Mn^{2\oplus}(aq) + 4H_2O(l)$$

Notice that this particular reaction will depend on the presence of hydrogen ions. If there is insufficient acid present a very different reaction occurs, resulting in precipitation of black manganese(IV) oxide, MnO_2.

1 What is the oxidation state of the halogen atom in each of the following compounds?
   a KIO_4
   b $NaClO_2$
   c $HBrO$
   d Na_5IO_6

2 Which has happened, oxidation or reduction, to the named elements in each of the following reactions?
   a Magnesium metal forming $MgCl_2$.
   b Sulphur reacting to form Na_2S.
   c Iron, when FeO reacts to form Fe_2O_3.
   d Carbon, when CO is burned to form CO_2.

3 Write the two half-equations for each of the following reactions, and identify the oxidation and reduction steps:
   a $4K(s) + O_2(g) \longrightarrow 2K_2O(s)$
   b $Ca(s) + Cl_2(g) \longrightarrow CaCl_2(s)$
   c $2Al(s) + 3Cl_2(g) \longrightarrow 2AlCl_3(s)$
   d $2Na(s) + Br_2(l) \longrightarrow 2NaBr(s)$

24.2 Half-cells (electrodes)

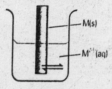

Figure 24.2.1
A simple half-cell

A redox process can be split into two half-reactions (see Ch 24.1). The redox process which occurs in an electrochemical cell (such as a battery) can be understood by considering the reactions occurring in the separate oxidation and reduction half-cells. We can develop our understanding of how half-cells work by building up ideas from a simple starting point. Consider a strip of a suitable metal (*not* sodium or potassium – see Ch 6.4) placed in an aqueous solution of its ions (*Figure 24.2.1*).

The metal atoms tend to lose electrons to become ions in solution; the electrons lost from those atoms build up on the surface of the metal. The positive metal ions in solution tend to gain electrons from the metal surface to re-form metal atoms which deposit on that surface. An equilibrium is eventually established:

$$M^{n+}(aq) + ne^{\ominus} \rightleftharpoons M(s)$$

The metal becomes charged by overall loss or gain of electrons, so that a potential difference (p.d.) exists between the metal and the solution. The *sign* of the p.d. will depend on whether, at equilibrium, the metal has gained electrons (negative) or lost electrons (positive). The *value* of this p.d. depends on, among other things, the readiness of the metal to ionise (see Ch 2.3) and the hydration energy of the resulting ions (see Ch 20.6). The p.d. is given the symbol E.

> The value of a potential difference is measured in *volts*.

Suppose we decrease the concentration of $M^{n+}(aq)$ ions in the equilibrium solution in *Figure 24.2.1*. More atoms of the metal M will ionise until equilibrium is re-established. If we increase the concentration of $M^{n+}(aq)$ ions, atoms of M will deposit on the metal strip until equilibrium is re-established. In other words, **the value of the potential difference between a metal and a solution of its ions at equilibrium depends on the concentration of the solution.**

Because the value of E depends on an equilibrium and is a measure of the equilibrium constant for the reaction:

$$M^{n+}(aq) + ne^{\ominus} \rightleftharpoons M(s)$$

it will – like every other equilibrium constant (see Ch 13.2) – vary with **temperature.**

The position of equilibrium, as shown by the sign and value of E, gives information about how easily the metal *loses* electrons – and also about how hard it is for the metal ion in solution to *gain* electrons. A metal which loses electrons easily, such as magnesium, will have an ion, $Mg^{2\oplus}(aq)$, which will not easily accept electrons. The opposite will be true for a metal which does not lose electrons easily, such as silver and its ion, $Ag^{\oplus}(aq)$. Silver ions readily accept electrons.

It is easy to visualise a metal/metal ion half-cell, as in *Figure 24.2.1*. As we shall see in Ch 24.3, it is possible to construct half-cells which involve gaseous elements in contact with solutions of their ions. For example, half-cells involving $H_2(g)/H^{\oplus}(aq)$ and $Cl_2(g)/Cl^{\ominus}(aq)$ are possible. It is also possible to construct half-cells containing a mixture of ions of the same element in different oxidation states, e.g. $Fe^{2\oplus}(aq)/Fe^{4\oplus}(aq)$ and $MnO_4^{\ominus}(aq)/Mn^{2\oplus}(aq)$.

24.3 Electrochemical cells

Joining two half-cells and finding the potential difference

It is impossible to measure a potential difference for a half-cell on its own. (It would be similar to the Zen problem: what is the sound of one hand clapping?) The half-cell you wish to study must be compared with a *standard half-cell of known potential*. From the p.d. between the half-cells, the potential of the half-cell being studied can be worked out.

When two half-cells are connected together, the result is an **electrochemical cell** (*Figure 24.3.1*).

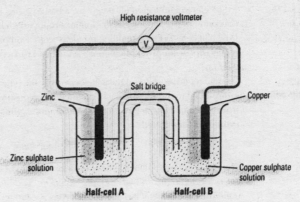

Figure 24.3.1 An electrochemical cell

If you don't believe this, park a car facing a wall with headlights full on. See what happens to the lights when you start the engine.

Two important points must be noted:

- The p.d. is measured with a voltmeter which must have a *very high resistance* so that the current in the circuit is as near as possible to zero. Only then will it be possible to measure the maximum potential difference of the cell. When current is taken from any cell or battery, its voltage drops (see box).
- The **salt bridge** completes the circuit. (A metal strip would not do because it would interact with solutions at both ends.) In the electrodes and around the external circuit conduction occurs through a flow of *electrons*. But between the electrodes, within the half-cells, conduction is by movement of *ions*.

Because very little current is flowing, there is only a very small movement of ions; but, even so, a path for conduction of ions must be provided. The salt bridge is often a U-tube with porous sintered glass ends and filled with a salt-impregnated jelly. But for ordinary laboratory use, a strip of filter paper soaked in saturated potassium nitrate solution can be used. Obviously, the salt used in the salt bridge must contain ions which do not cause chemical reactions in the half-cells. Potassium nitrate is usually used because all nitrates are soluble, so are nearly all potassium salts.

The standard hydrogen electrode (SHE)

The half-cell which is used as the **primary standard**, i.e. the standard on which all other measurements of electrode potentials are based, is the **standard hydrogen electrode**. The potential for this half-cell is assumed to be *zero*.

The standard hydrogen electrode is constructed as shown in *Figure 24.3.2*, which also shows the completed cell for measuring the potential for copper. Note carefully the labelling, which is often asked for in examinations. The reasons why the pressure of gas, temperature of the cell, and concentration of solutions are specified are discussed below.

Apparatus for measuring the standard electrode potential of copper

The right-hand half-cell (inside the dotted lines) is the **standard hydrogen electrode (SHE)**.

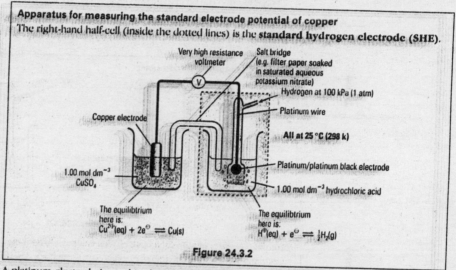

Figure 24.3.2

A platinum electrode is used in the standard hydrogen electrode because of its very low reactivity with acid and because it is a very good catalyst in many reactions involving hydrogen. It is often coated with 'platinum black' – i.e. platinum freshly deposited on the electrode by electrolysis of a solution of a platinum salt. This adsorbs hydrogen gas, and helps the equilibrium:

$$2H^{\oplus}(aq) + 2e^{\ominus} \rightleftharpoons H_2(g)$$

to become established more rapidly. To provide a basis for measurements, *the standard hydrogen electrode is assumed to have a potential of zero:*

i.e. $$H^{\oplus}(aq) + e^{\ominus} \rightleftharpoons \tfrac{1}{2}H_2(g), \quad E^{\ominus} = 0.00 \text{ V}$$

Standard electrode potentials (SEP), $E^{\ominus}$

As we have already seen in Ch 24.2, the potential (E) developed in a half-cell depends on the position of equilibrium in the reaction occurring in the half-cell. E is a measure of the equilibrium constant for that reaction, and will therefore vary with temperature (see Ch 13.2). In Ch 24.2 we also showed why the value of E depended on the concentration of ions in the solution. In order to make measurements which can be reproduced anywhere in the world, *all the conditions must be standardised.*

The effect of concentration on $E^\ominus$ was studied by Walther Nernst (1864–1941) who developed the Nernst Equation. This is beyond the scope of this book.

The point about the metal being 'in its standard state' is that some metals are polymorphic, and this could make a slight difference.

The standard conditions are:

- All solutions must have a concentration of 1.00 mol cm^{-3}.
- All gases must be at a pressure of 1 atm (about 100 kPa).
- Metals must be in their *standard state*.
- All materials and apparatus must be at 25 °C (298 K).

Measurements of potential differences against a standard half-cell, such as the SHE, are known as **standard electrode potentials, $E^\ominus$**. As we shall see, $E^\ominus$ values are very important in understanding and predicting many chemical reactions and industrial processes.

The apparatus in *Figure 24.3.2* is suitable for most metals, except the very reactive ones such as the Group 1 metals.

Definition: The standard electrode potential $(E^\ominus)$ of an element is the potential developed between a half-cell which consists of the element immersed in a 1.00 mol dm^{-3} solution of its ions and another half-cell which consists of a standard hydrogen electrode, all being at 25 °C (298 K).

This definition can be extended to include the SEP of a half-cell containing any redox pair.

For a redox ionic pair such as $Fe^{3+}(aq)/Fe^{2+}(aq)$ or $MnO_4^{-}(aq)/Mn^{2+}(aq)$, or a gas/ion pair such as $Cl_2(g)/Cl^{-}(aq)$, the half-cell being tested against the hydrogen electrode can be modified as shown in *Figure 24.3.3*.

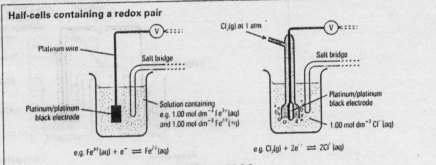

Half-cells containing a redox pair

Platinum wire

Salt bridge

Platinum/platinum black electrode

Solution containing e.g. $1.00 \text{ mol dm}^{-3} Fe^{3+}(aq)$ and $1.00 \text{ mol dm}^{-3} Fe^{2+}(aq)$

e.g. $Fe^{3+}(aq) + e^- \rightleftharpoons Fe^{2+}(aq)$

$Cl_2(g)$ at 1 atm

Salt bridge

Platinum/platinum black electrode

$1.00 \text{ mol dm}^{-3} Cl^-(aq)$

e.g. $Cl_2(g) + 2e^- \rightleftharpoons 2Cl^-(aq)$

Figure 24.3.3

Note the use of a platinum electrode. This is always used when the half-cell does not contain a metal and is used for the same reasons as it is used in the hydrogen electrode. Platinum is unreactive, and allows electrons to enter or leave the mixture of the oxidised and reduced ions, e.g. $Fe^{3+}(aq)$ and $Fe^{2+}(aq)$. **These should normally both be present at a concentration of 1.00 mol dm^{-3}**. If the salts are not soluble enough to reach this concentration, then a mixture containing **equal concentrations** of each of the two ions is necessary.

N.B. If gases are involved, they should be at a pressure of 100 kPa (1 atm).

As we shall soon see, values of standard electrode potentials range from about +3 V to about −3 V. This apparently small range of voltage encompasses everything from the most reactive non-metal (e.g. fluorine) to the most reactive metals (Group 1).

Intensive properties such as $E^\ominus$, density and colour, do not depend on the amount of substance involved. Properties which do depend on the amount of substance present (e.g. volume, energy content, mass) are called *extensive* properties.

N.B. $E^\ominus$ is an **intensive property**. Unlike ΔH in calculations involving energy changes (Ch 11.3), values of $E^\ominus$ do not depend on the number of moles appearing in the equation or the amounts of substances involved in the reaction.

Remember: The potential of the standard hydrogen electrode is taken to be zero.

$$2H^+(aq) + 2e^\ominus \rightleftharpoons H_2(g) \qquad E^\ominus = 0.00 \text{ V}$$

Here, there is an important international convention to be remembered:

All the half-cell reaction equations written in connection with $E^\ominus$ values are reduction processes.

So, in the equilibrium equations representing SEPs, **electrons are always written on the left. The product is always the reduced form of the reactant.**

For example, in the half-cell equation:

$$Mg^{2\oplus}(aq) + 2e^{\ominus} \rightleftharpoons Mg(s) \qquad E^{\ominus} = -2.37 \text{ V}$$

oxidised form   reduced form

The standard electrode potential, SEP, always refers to the reduction equation; i.e. the electrons in the half-equation are always on the left hand side of the equilibrium arrow.

$E^{\ominus}$, the standard electrode potential, is measured at:
- 25 °C (298 K);
- 1 atm pressure;
- using 1.00 mol dm^{-3} solutions.

Half-cell equation		$E^{\ominus}$/V
$Li^+(aq) + e^-$	$\rightleftharpoons$ Li(s)	−3.03
$K^+(aq) + e^-$	$\rightleftharpoons$ K(s)	−2.92
$Na^+(aq) + e^-$	$\rightleftharpoons$ Na(s)	−2.71
$Mg^{2+}(aq) + 2e^-$	$\rightleftharpoons$ Mg(s)	−2.37
$Al^{3+}(aq) + 3e^-$	$\rightleftharpoons$ Al(s)	−1.66
$Zn^{2+}(aq) + 2e^-$	$\rightleftharpoons$ Zn(s)	−0.76
$2H^+(aq) + 2e^-$	$\rightleftharpoons$ H_2(g)	0.00
$Cu^{2+}(aq) + 2e^-$	$\rightleftharpoons$ Cu(s)	+0.34
$O_2(g) + 2H_2O(l) + 4e^-$	$\rightleftharpoons$ $4OH^-$(aq)	+0.40
$I_2(aq) + 2e^-$	$\rightleftharpoons$ $2I^-$(aq)	+0.54
$Fe^{3+}(aq) + e^-$	$\rightleftharpoons$ Fe^{2+}(aq)	+0.77
$Ag^+(aq) + e^-$	$\rightleftharpoons$ Ag(s)	+0.80
$Br_2(aq) + 2e^-$	$\rightleftharpoons$ $2Br^-$(aq)	+1.09
$Cl_2(aq) + 2e^-$	$\rightleftharpoons$ $2Cl^-$(aq)	+1.36
$MnO_4^-(aq) + 8H^+(aq) + 5e^-$	$\rightleftharpoons$ Mn^{2+}(aq) + $4H_2O$(l)	+1.51
$F_2(g) + 2e^-$	$\rightleftharpoons$ $2F^-$(aq)	+2.87

Increasing reducing strength ↑   Increasing oxidising strength ↓

Table 24.3.1   Some standard electrode potentials

Look now at the $E^{\ominus}$ values in *Table 24.3.1*, and remember OILRIG (see introduction to this chapter).
- The strongest **reducing** agents have the highest **negative** values of $E^{\ominus}$ because they have the greatest tendency to lose electrons. For example, sodium is reactive because of its tendency to lose electrons. It takes energy to force electrons back on to sodium ions in aqueous solution.
- The strongest **oxidising** agents have the highest **positive** values of $E^{\ominus}$, because they have the greatest tendency to gain electrons. For example, fluorine or acidified potassium manganate(VII).
- A positive $E^{\ominus}$ value therefore, means that at equilibrium the *reduced* form predominates. This means that **a positive value of $E^{\ominus}$ shows that the reaction tends to go from left to right as written in the half-equation.** Conversely, **a negative value of $E^{\ominus}$ means that the reaction tends to go from right to left as written.** (We shall soon examine the predictive power of these ideas as applied to full chemical equations – see Ch 24.7.)
- One result of this is that **the sign of $E^{\ominus}$ for a half-cell – whether it is $\oplus$ or $\ominus$ – is also the polarity of the electrode of that half-cell when coupled with the standard hydrogen electrode.** So, for example, a zinc half-cell would be negative (*Figure 24.3.4*). Electrons tend to flow *from* the zinc around the external circuit *to* the hydrogen half-cell. For a copper half-cell, the copper is positive; electrons therefore tend to flow *from* the hydrogen half-cell *to* the copper (*Figure 24.3.5*).

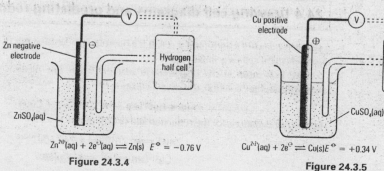

$Zn^{2+}(aq) + 2e^-(aq) \rightleftharpoons Zn(s)$ $E^{\ominus} = -0.76$ V

Figure 24.3.4

$Cu^{2+}(aq) + 2e^- \rightleftharpoons Cu(s)$ $E^{\ominus} = +0.34$ V

Figure 24.3.5

A word of warning about $E^\ominus$ values is required. Many people think of a table of such values as if it was the old 'reactivity series', probably learned in early secondary school. Table 23.3.1 shows that the halogens – fluorine, chlorine, bromine, and iodine – are indeed in the expected order, with fluorine having by far the most positive $E^\ominus$ value of the four. But look at the alkali metals! Lithium is certainly the least reactive of the alkali metals (see Ch 6.4), but it has the highest negative $E^\ominus$ value of the Group 1 elements.

The point here is that $E^\ominus$ values involve reactions in solution, and not just simple electron loss or gain. Other factors help to determine the overall value of $E^\ominus$, notably the hydration enthalpy of the ion (see Ch 20.6). Lithium has a very small ion, which as a result has high polarising power (see Ch 2.5) and a high enthalpy of hydration. It is largely for this reason that lithium beats sodium and potassium in the $E^\ominus$ table.

SELF-ASSESSMENT QUESTIONS 24.3

1 **a** Draw a fully labelled diagram to illustrate the construction of the standard hydrogen electrode.

   b Draw a fully labelled diagram of the apparatus you would use to measure the standard electrode potential of the $Cu^{2+}(aq)/Cu(s)$ half-cell.

   c Draw a fully labelled diagram of the apparatus needed to measure the e.m.f. produced by a cell containing the $Cu^{2+}(aq)/Cu(s)$ half-cell and the $Br_2(aq)/Br^\ominus(aq)$ half-cell.

   d Write half-equations for the reactions in each of the half-cells in **c**.

   e There are two possible reactions involving the half-equations in **c** – one where $Cu^{2+}(aq)$ is reduced and the other where $Cu(s)$ is oxidised by the $Br_2(aq)/Br^-(aq)$ system. Write balanced equations for the two possible reactions.

2 Consider the following half-equations.

$Cu^{2+}(aq) + e^\ominus \rightleftharpoons Cu^+(aq)$	$E^\ominus = +0.15\ V$
$Ce^{4+}(aq) + e^\ominus \rightleftharpoons Ce^{3+}(aq)$	$E^\ominus = +1.70\ V$
$Cr^{3+}(aq) + 3e^\ominus \rightleftharpoons Cr(s)$	$E^\ominus = -0.74\ V$
$Br_2(aq) + 2e^\ominus \rightleftharpoons 2Br^\ominus(aq)$	$E^\ominus = +1.09\ V$
$Mg^{2+}(aq) + 2e^\ominus \rightleftharpoons Mg(s)$	$E^\ominus = -2.37\ V$

   a Which is likely to be the strongest reducing system?

   b Which is the best oxidising system?

   c Draw a diagram to show how you would set up the cerium, Ce, half cell.

3 **a** What is meant by the standard electrode potential of magnesium?

   b What are the standard conditions used to measure the SEP of magnesium?

   c Write a half-equation for the reaction that tends to occur when the magnesium half-cell is connected to a standard hydrogen electrode.

24.4 Drawing cell diagrams and predicting redox reactions

A cell diagram is *not* a drawing of a cell. It is a rigorous standardised way of representing a complete electrochemical cell on a single line.

If we look again at the displacement reaction when zinc metal is placed in a solution of copper(II) sulphate solution, the overall equation is:

$$Zn(s) + Cu^{2+}(aq) \longrightarrow Zn^{2+}(aq) + Cu(s)$$

The two half-equations for the reduction and oxidation processes are:

$$Zn(s) \longrightarrow Zn^{2+}(aq) + 2e^\ominus \text{ and}$$
$$Cu^{2+}(aq) + 2e^\ominus \longrightarrow Cu(s)$$

The **cell diagram** for the electrochemical cell involving two half-cells made up with these metals, in equilibrium, with their ions is written as:

$$Zn(s) \mid Zn^{2\oplus}(aq) \; \vdots \vdots \; Cu^{2\oplus}(aq) \mid Cu(s)$$

This is a typical cell diagram, which is analysed in detail in *Figure 24.4.1*.

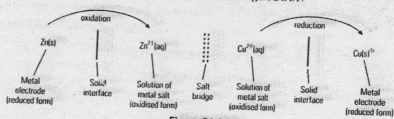

Figure 24.4.1

Some books refer to 'anodes' and 'cathodes' in electrochemical cells. These terms should only be used for electrolytic (or electrolysis) cells (see Ch 24.8).
For cells in which chemical reaction leads to a flow of electrons, always refer to negative and positive terminals.

- A firm vertical line represents a phase boundary, i.e. between solid and liquid, solid and gas, etc.
- The dashed double vertical line represents the salt bridge.
- The reduced species in each half-cell is written furthest from the salt bridge (in this case, the two metals).

A closer look at this cell diagram shows that it is written in the direction of the overall chemical reaction that occurs, i.e. $Zn(s)$ oxidised to $Zn^{2+}(aq)$, then $Cu^{2+}(aq)$ being reduced to $Cu(s)$.

Even when the direction of flow of electrons in the external circuit is not obvious, the E° values for the half-cells can always be used to work out which terminal is which under standard conditions, and also to work out E_{cell}.

Cell diagrams for non-metal half-cells

Remember here that a platinum or platinum/platinum black electrode is usually used for such half-cells (see Ch 24.3). The half-cell equation for the standard hydrogen electrode is:

$$2H^+(aq) + 2e^- \rightleftharpoons H_2(g) \qquad E^\circ = 0.00\,V$$

To take account of the phase boundary, in which the gas is adsorbed on the platinum, the half-cell can be written:

$$2H^\oplus(aq) \mid [H_2(g)]Pt$$

Suppose a standard hydrogen electrode, i.e. a hydrogen half-cell, is coupled with the manganate(VII)/manganate(II) half-cell (i.e. acidified $MnO_4^\ominus(aq)/Mn^{2\oplus}(aq)$; see *Table 24.3.1*).

$$MnO_4^\ominus(aq) + 8H^\oplus(aq) + 5e^\ominus \rightleftharpoons Mn^{2\oplus}(aq) + 4H_2O(l)$$

The complete cell can be written:

$$Pt[H_2(g)] \mid 2H^\oplus(aq) \; \vdots\vdots \; [MnO_4^\ominus(aq) + 8H^\oplus(aq)], [Mn^{2\oplus}(aq) + 4H_2O(l)] \mid Pt$$

Two extra rules apply to this type of cell diagram.

- In each half-cell, the more reduced species (i.e. $H_2(g)$ or $Mn^{2\oplus}(aq)$) is written nearer to the electrode (i.e. again furthest from the salt bridge).
- The two pairs of reactants ($MnO_4^\ominus(aq) + 8H^\oplus(aq)$ and $Mn^{2\oplus}(aq) + 4H_2O(l)$) are written in square brackets and separated by a comma.

Where the half-cell reaction is simpler, there is no need for the square brackets around the pairs of reactants.

e.g.
$$Pt[H_2(g)] \mid 2H^\oplus(aq) \; \vdots\vdots \; Br_2(aq), 2Br^\ominus(aq) \mid Pt$$

Predicting redox reactions

In this section we shall consider in detail what happens in the electrochemical cell, and provide simple rules for calculating E_{cell}. We shall then show how E_{cell} can be used in predicting reactions carried out in a normal chemical environment.

Consider again the simple zinc-copper cell diagram (*Figure 24.4.1*):

$$Zn(s) \mid Zn^{2+}(aq) \; \| \; Cu^{2+}(aq) \mid Cu(s)$$

We already know the chemistry that takes place.

On the left of the cell diagram:

$$\text{Zinc metal} \longrightarrow \text{zinc ions in solution}$$

Looking at table 24.3.1, we find:

$$Zn^{2+}(aq) + 2e^- \rightleftharpoons Zn(s) \qquad E^\ominus = -0.76\,V$$

But the process is the exact opposite of what we know the chemistry to be. So turn the half-equation around:

$$Zn(s) \rightleftharpoons Zn^{2+}(aq) + 2e^- \qquad E = +0.76\,V$$

It is very important to realise that **if you reverse the equation for a half-reaction, you simply reverse the sign for the e.m.f.** (N.B. We can no longer use the symbol $E^\ominus$ – that can only be used for *reduction* processes.)

On the right of the cell diagram:

$$\text{Copper ions in solution} \longrightarrow \text{copper metal}$$

This is the same as for the half-cell reaction for the standard electrode potential.

$$Cu^{2+}(aq) + 2e^- \rightleftharpoons Cu(s) \qquad E^\ominus = +0.34\,V$$

The cell reaction:

Now we have the two half-reactions, we simply add the e.m.f. values:

$$Zn(s) \rightleftharpoons Zn^{2+}(aq) + 2e^- \qquad E = +0.76\,V$$
$$Cu^{2+}(aq) + 2e^- \rightleftharpoons Cu(s) \qquad E^\ominus = +0.34\,V$$
$$\overline{Zn(s) + Cu^{2+}(aq) \rightleftharpoons Zn^{2+}(aq) + Cu(s) \qquad E_{cell} = +1.10\,V}$$

This brings us to a very important point: **a positive overall e.m.f. value means that the reaction tends to go from left to right as written**.

Does this agree with experiment? As we saw in Ch 24.1, when metallic zinc is put into copper(II) sulphate solution, the blue colour disappears, copper is precipitated, and heat energy is produced. Zinc has 'displaced' the copper and the complete cell diagram becomes:

$$Zn(s) \mid Zn^{2+}(aq) \; \| \; Cu^{2+}(aq) \mid Cu(s) \qquad E_{cell} = +1.10\,V$$

> Note that if the number of electrons lost in one half-cell equation is not equal to the number gained in the other, the two half-equations will need to be balanced in order to obtain the overall equation for the redox reaction. But there is no need to change the $E_{half-cell}$ values to take account of the number of electrons lost or gained. Electrode potentials are an **intensive** property – and do not depend on the amount of substance involved.

A rule for finding E_cell

E_{cell} can always be worked out by the method shown above, but a straightforward rule can shorten the process.

The rule consists of a simple statement:

$$E_{cell} = E_{right-hand\ half-cell} - E_{left-hand\ half-cell}$$

Using this for the reaction of zinc and copper sulphate solution in the last paragraph:

$$E_{cell} = E_{RHS} - E_{LHS}$$
$$= +0.34 - (-0.76)$$
$$= +1.10 \text{ V}$$

Does anything happen if metallic copper is put into a solution of zinc sulphate? This would be equivalent to writing the cell diagram the other way around:

$$Cu(s) \mid Cu^{2\oplus}(aq) \; \vdots \; Zn^{2\oplus}(aq) \mid Zn(s)$$

Reactants   Possible products

i.e. copper metal placed in a solution of zinc ions.

The answer from our own knowledge of the reactivity of metals is 'no'. But now we can give a better explanation of why this is so in terms of redox potentials.

We can use the $E_{cell} = E_{RHS} - E_{LHS}$ rule for the copper/zinc sulphate reaction. The cell diagram is:

$$Cu(s) \mid Cu^{2\oplus}(aq) \; \vdots \; Zn^{2\oplus}(aq) \mid Zn(s)$$

The equation shows the zinc half-cell on the right, and the copper half-cell on the left.

$$Cu^{2\oplus}(aq) + 2e^{\ominus} \rightleftharpoons Cu(s) \qquad E^{\ominus} = +0.34 \text{ V}$$
$$Zn^{2\oplus}(aq) + 2e^{\ominus} \rightleftharpoons Zn(s) \qquad E^{\ominus} = -0.76 \text{ V}$$
$$E_{cell} = E_{RHS} - E_{LHS}$$
$$\therefore E_{cell} = -0.76 - (+0.34) = -1.10 \text{ V}$$

The negative value of E_{cell} indicates that the reaction as written in the cell diagram will not occur.

> It does not matter if you have drawn the cell diagram the wrong way round. In that case E_{cell} will be negative, and redrawing it the other way round will represent the feasible reaction.

Worked examples of calculating E_{cell}

Calculate the value of E_{cell} for the following cell diagrams.

A *$Mg(s) \mid Mg^{2\oplus}(aq) \; \vdots \; Ag^{\oplus}(aq) \mid Ag(s)$*
The two half-cells in this cell diagram are:

$$Mg^{2\oplus}(aq) + 2e^{\ominus} \rightleftharpoons Mg(s) \qquad E^{\ominus} = -2.37 \text{ V}$$
$$Ag^{\oplus}(aq) + e^{\ominus} \rightleftharpoons Ag(s) \qquad E^{\ominus} = +0.80 \text{ V}$$
$$E_{cell} = E_{RHS} - E_{LHS}$$
$$= +0.80 - (-2.37)$$
$$= +3.17 \text{ V}$$

B *$Mg(s) \mid Mg^{2\oplus}(aq) \; \vdots \; Zn^{2\oplus}(aq) \mid Zn(s)$*
The two half-cells in this cell diagram are:

$$Mg^{2\oplus}(aq) + 2e^{\ominus} \rightleftharpoons Mg(s) \qquad E^{\ominus} = -2.37 \text{ V}$$
$$Zn^{2\oplus}(aq) + 2e^{\ominus} \rightleftharpoons Zn(s) \qquad E^{\ominus} = -0.76 \text{ V}$$
$$E_{cell} = E_{RHS} - E_{LHS}$$
$$= -0.76 - (-2.37)$$
$$= +1.61 \text{ V}$$

C *$Pt \mid 2I^{\ominus}(aq), I_2(aq) \; \vdots \; Al^{3\oplus}(aq) \mid Al(s)$*
The two half-cells in this cell diagram are:

$$I_2(aq) + 2e^{\ominus} \rightleftharpoons 2I^{\ominus}(aq) \qquad E^{\ominus} = +0.54 \text{ V}$$
$$Al^{3\oplus}(aq) + 3e^{\ominus} \rightleftharpoons Al(s) \qquad E^{\ominus} = -1.66 \text{ V}$$
$$E_{cell} = E_{RHS} - E_{LHS}$$
$$= -1.66 - (+0.54)$$
$$= -2.20 \text{ V}$$

Making predictions

We now know that, **for a reaction to be feasible, the overall e.m.f., E_{cell}, must be positive**.

So, in any redox process, if we know $E^\ominus$ for each of the two half-reactions, we can find E_{cell} and hence whether the reaction is feasible in a laboratory situation.

Worked example

A *Will a solution of iron(III) chloride, $FeCl_3$, cause iodine to form from a solution of potassium iodide, KI?*

Data:

$$Fe^{3+}(aq) + e^\ominus \rightleftharpoons Fe^{2+}(aq) \qquad\qquad E^\ominus = +0.77\,V$$
$$I_2(aq) + 2e^\ominus \rightleftharpoons 2I^\ominus(aq) \qquad\qquad E^\ominus = +0.54\,V$$

Method:

The reaction we wish to look at is $Fe^{3+}(aq)$ (as $FeCl_3$) oxidising iodide ions to iodine. If we write the cell diagram with the oxidation half-cell ($I^\ominus$ to I_2) as the left-hand half-cell we will write:

$$Pt \mid 2I^\ominus(aq), I_2(aq) \;\|\; Fe^{3+}(aq), Fe^{2+}(aq) \mid Pt$$
$$E_{cell} = E_{RHS} - E_{LHS}$$
$$E_{cell} = +0.77 - (+0.54) = +0.23\,V$$

Overall, E_{cell} is positive – so this reaction will tend to go from left to right as written. Therefore, iron(III) compounds in solution should oxidise iodide ions to iodine; i.e. Fe^{3+} will change to Fe^{2+} as $I^\ominus$ changes to I_2.

So, the two half-equations become:

$$2Fe^{3+}(aq) + 2e^\ominus \rightleftharpoons 2Fe^{2+}(aq)$$

and doubled up to provide $2e^\ominus$ (but $E^\ominus$ stays the same):

$$\underline{2I^\ominus(aq) \rightleftharpoons I_2(aq) + 2e^\ominus}$$
$$2Fe^{3+}(aq) + 2I^\ominus(aq) \rightleftharpoons 2Fe^{2+}(aq) + I_2(aq)$$

N.B. You must, however, read on to the end of this section to understand the limitations of these redox calculations!

B *Will a solution of iron(III) chloride, $FeCl_3$, cause bromine to form from a solution of potassium bromide, KBr?*

Data:

$$Fe^{3+}(aq) + e^\ominus \rightleftharpoons Fe^{2+}(aq) \qquad E^\ominus = +0.77\,V$$
$$Br_2(aq) + 2e^\ominus \rightleftharpoons 2Br^\ominus(aq) \qquad E^\ominus = +1.09\,V$$

Method:

The reaction we wish to look at is $Fe^{3+}(aq)$ (as $FeCl_3$) oxidising bromide ions to bromine. If we write the cell diagram with the **oxidation** half-cell ($Br^\ominus$ to Br_2) as the **left-hand** half-cell we will write:

$$Pt \mid 2Br^\ominus(aq), Br_2(aq) \;\|\; Fe^{3+}(aq), Fe^{2+}(aq) \mid Pt$$
$$E_{cell} = E_{RHS} - E_{LHS}$$
$$E_{cell} = +0.77 - (+1.09) = -0.32\,V$$

For $Fe^{3+}(aq)$ attempting to oxidise $Br^\ominus(aq)$ to $Br_2(aq)$, $E_{cell} = -0.32\,V$; *the reaction is not feasible.*

> When two half-cells are combined the feasible forward reaction is the reaction represented by the cell diagram with the more negative (or less positive) $E^{\ominus}$ half cell forming the LHS of the cell diagram.

Limitations of making predictions from $E^{\ominus}$ values

Remember that $E^{\ominus}$ values refer to solutions which are all $1.00 \ mol \ dm^{-3}$, gases at 100 kPa (1 atm), all temperatures at 25 °C. Most reactions are carried out under different conditions. **To be sure of a reaction being feasible at *any* reasonable concentration, etc, the overall value of E_{cell} should be greater than about +0.5 V.**

Also remember – **the overall value of E_{cell} tells us whether a reaction is feasible, i.e. whether it tends to proceed; it does *not* tell us whether it occurs at a detectable rate.** In this way, the situation is the same as with $\Delta H_{reaction}$ (see Ch 11.3).

Finally, for many redox processes $E_{half-cell}$ depends on the pH of the solution. For example, for the reduction of nitrate ion, $NO_3^{\ominus}(aq)$, to ammonium ion, $NH_4^{\oplus}(aq)$, at pH = 0, $E_{half-cell}$ is about +0.75 V, but at pH = 14 (when the reduction product is the ammonia molecule, NH_3) it is nearly −2.9 V. This means that reducing agents such as aluminium metal, in alkaline solution, can reduce nitrogen in the +5 oxidation state (as in nitrate ion) all the way to the −3 state in ammonia.

The reaction of metals with acids

Consider the general process:

$$M(s) + nH^{n\oplus}(aq) \rightleftharpoons M^{n\oplus}(aq) + \tfrac{1}{2}nH_2(g)$$

The cell diagram is:

$$M(s) \mid M^{n\oplus}(aq) \; \vdots \; 2H^{\oplus}(aq) \mid [H_2(g)]Pt$$

and

$$E_{cell} = E_{RHS} - E_{LHS}$$

Since E_{RHS} is *zero* for the hydrogen half-cell, then *for E_{cell} to be positive, E_{LHS} must be negative*.

This means that **for a metal to be able to displace hydrogen from an acid, $E^{\ominus}$ for the metal must be negative.** Otherwise E_{cell} for the metal/acid reaction will not be positive, and therefore the reaction will not be feasible. This is why copper and silver, which have positive values for $E^{\ominus}$ (see *Table 24.3.1*) cannot release hydrogen from acids.

$$Cu(s) \mid Cu^{2\oplus}(aq) \; \vdots \; 2H^{\oplus}(aq) \mid [H_2(g)]Pt \qquad E_{cell} = -0.34 \ V$$

Metals which should, according to their $E^{\ominus}$ value, release hydrogen from an acid, sometimes do not because of surface effects. An example is aluminium ($E^{\ominus} = -1.66$ V) which is protected by a thin but very tough layer of aluminium oxide.

SELF-ASSESSMENT QUESTIONS 24.4

(Data for these questions will be found in *Table 24.3.1*.)

1 **a** Draw the apparatus that would be used to measure the e.m.f. of the cell made up from the $Zn/Zn^{2\oplus}$ half-cell and the $Fe^{2\oplus}/Fe^{3\oplus}$ half-cell, showing which is the negative terminal and the direction in which electrons would flow.

   b Construct a cell diagram for the overall cell, and calculate the e.m.f. of this cell.

   c Write a balanced equation for the reaction that would occur in such a cell.

   d Explain why you have included a salt bridge in the apparatus in **a**.

   e How would you make up a salt bridge?

2 Write cell diagrams for the cells produced with the following pairs of half-cells. Calculate the e.m.f. for each cell, including the correct sign.

   a Copper and aluminium half-cells

   b Magnesium and chlorine half-cells.

   c Oxygen and zinc half-cells.

24.5 Redox processes in electrochemical cells

As we have already seen in Ch 4.3 and Ch 24.1, oxidation is loss of electrons and reduction is gain of electrons (OILRIG), and reduction and oxidation *always* occur together. This is why we use the term redox to describe such processes and reactions.

Consider an electrochemical cell consisting of a zinc half-cell and a copper half-cell (*Figure 24.5.1*). In which of the half-cells will reduction and oxidation each occur?

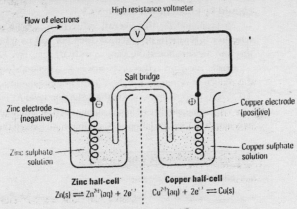

Figure 24.5.1

The two half-equations linked with the relevant values for $E^\ominus$ are:

$$Zn^{2\oplus}(aq) + 2e^\ominus \rightleftharpoons Zn(s) \qquad E^\ominus = -0.76\ V$$
$$Cu^{2\oplus}(aq) + 2e^\ominus \rightleftharpoons Cu(s) \qquad E^\ominus = +0.34\ V$$

In Ch 24.2, we saw that a system consisting of a metal immersed in a solution of its own ions will eventually reach equilibrium. The atoms of a relatively reactive metal like zinc will tend to lose electrons and go into solution as zinc ions, until equilibrium is reached. Electrons will build up on the zinc; it will develop a *negative polarity*.

On the other hand, copper ions will take electrons from the copper strip and plate on to it as copper metal until equilibrium is reached. Because electrons are being removed from it, the copper strip will develop a *positive polarity*.

If the zinc and copper are now linked by an external circuit, and the circuit is completed with a salt bridge or other suitable contact between the two half-cells, *electrons will flow in the external circuit from the zinc to the copper.* Zinc atoms lose electrons; i.e. *oxidation occurs in the zinc half-cell.* Copper ions gain electrons; i.e. *reduction occurs in the copper half-cell.*

The cell diagram is:

$$Zn(s) \mid Zn^{2\oplus}(aq) \; \vdots \vdots \; Cu^{2\oplus}(aq) \mid Cu(s)$$
$$E_{cell} = E_{RHS} - E_{LHS}$$
$$\therefore E_{cell} = +0.34 - (-0.76) = +1.10\ V$$

and the reaction is:

$$Zn(s) + Cu^{2\oplus}(aq) \longrightarrow Zn^{2\oplus}(aq) + Cu(s)$$

In the zinc/copper cell, one $E^\ominus$ value is negative and the other positive, so it is easy to follow what is happening. But the principle still applies even if *both* $E^\ominus$ values are negative or both are positive. For example, consider the following cell diagram:

$$Zn(s) \mid Zn^{2\oplus}(aq) \; \vdots \vdots \; Pb^{2\oplus}(aq) \mid Pb(s)$$

In the lead half-cell, the equilibrium is:

$$Pb^{2\oplus}(aq) + 2e^{\ominus} \rightleftharpoons Pb(s) \qquad E^{\ominus} = -0.13 \text{ V}$$

So:

$$E_{cell} = E_{RHS} - E_{LHS}$$
$$= -0.13 - (-0.76)$$
$$= +0.63 \text{ V}$$

As the value for E_{cell} is positive, the reaction should occur; i.e.

$$Zn(s) + Pb^{2\oplus}(aq) \longrightarrow Zn^{2\oplus}(aq) + Pb(s)$$

Although lead atoms tend to lose electrons when in contact with their aqueous ions, zinc has a greater tendency to do so. Therefore, electrons flow from zinc to lead around the external circuit, and lead ions are forced to discharge on the lead strip and become lead atoms.

The reaction of copper sulphate solution with zinc formed the basis for one of the earliest batteries – **the Daniell cell,** first used in 1836. When zinc is added to copper sulphate solution, the mixture becomes hot. By **not** allowing the reagents to mix, the reaction can be controlled. Instead of the energy from the reaction being liberated as thermal energy, the energy is available as an electric current.

A zinc rod is immersed in a solution of a zinc salt (zinc sulphate) inside a porous pot. The porous pot allows electrical contact between the two solutions inside the battery without allowing them to mix. This porous pot is then placed in a solution of copper sulphate inside a copper beaker (*Figure 24.5.2*). The *voltage* produced by such a battery stays constant at about 1.1 V for a long time. Daniell cells were used for many years for telegraphy, for door bells, and for electrolysis.

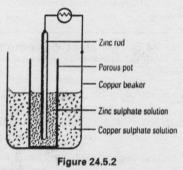

- Zinc rod
- Porous pot
- Copper beaker
- Zinc sulphate solution
- Copper sulphate solution

Figure 24.5.2

Many books refer to the 'anode' and 'cathode' of a cell or battery. *This is confusing and wrong.* **The terms anode and cathode should be used only when discussing electrolysis** – see Ch 24.8.
- In a cell or battery, electrons flow *from* the **negative terminal** (or **negative pole**) *to* the **positive terminal** (or **positive pole**), as we have shown. Oxidation occurs in the half-cell which provides the negative terminal, where electrons are released into the external circuit. In a cell or battery, **chemical energy is converted into electrical energy** – a chemical reaction is used to produce a flow of electrons, i.e. an electric current.
- In electrolysis, electrons in the external circuit flow *into* the **cathode** and *out* of the **anode**. Reduction occurs at the cathode and oxidation at the anode. Electrolysis is the exact opposite of what happens in a cell or battery (see Ch 24.8). In electrolysis, **electrical energy is converted into chemical energy** – an electric current is used to cause a chemical reaction.

Remember:

- An electrical cell always consists of two **half-cells**. In one of these **oxidation** occurs and in the other, **reduction**. Overall, a **redox** reaction happens.
- Oxidation happens in the half-cell with the *more negative* value of $E^\ominus$. This becomes the *negative terminal* of the cell.
- Electrons flow in the external circuit from the negative terminal to the positive terminal.

SELF-ASSESSMENT QUESTIONS 24.5

1 a Draw and label the apparatus you would use to construct a cell containing the magnesium and iron half-cells.

b Which of these two metals is the more reactive?

c Mark on your drawing of the apparatus the direction of electron flow in the external circuit.

d At which metal electrode will the oxidation process occur?

2 What will be the cell potential made up from the following electrode combinations? Write a balanced equation for the chemical reaction that would occur if the cell were allowed to provide a current. The values for $E^\ominus_{\text{half-cell}}$ are listed below.

	$E^\ominus_{\text{half-cell}}$
$Ag^+(aq) \mid Ag(s)$	$+0.80$ V
$Cu^{2+}(aq) \mid Cu(s)$	$+0.34$ V
$Mg^{2+}(aq) \mid Mg(s)$	-2.37 V
$Pb^{2+}(aq) \mid Pb(s)$	-0.13 V
$Zn^{2+}(aq) \mid Zn(s)$	-0.76 V

a Magnesium and lead half-cells.

b Silver and zinc half-cells.

c Zinc and magnesium half-cells.

d Lead and copper half-cells.

24.6 The electrochemical series

A list of $E^\ominus$ values is given in *Table 24.3.1*, with the most negative potentials at the top and the most positive at the bottom. The list arranged in this way is called the **electrochemical series**.

Because the equilibria are all listed with the electrons on the left, with the *reduced form* on the right, $E^\ominus$ values are often known as **reduction potentials**.

$$Mg^{2+}(aq) + 2e^- \rightleftharpoons Mg(s) \qquad E^\ominus = -2.37\text{ V}$$

A large negative value of $E^\ominus$ means that the species on the left hand side of the equilibria will not easily *accept* electrons but that the species on the right hand side can easily *lose* electrons. In this example, magnesium readily loses electrons to form $Mg^{2+}(aq)$.

Remember:
The half-equation with the more positive potential gains electrons (as negative electrons are attracted to the positive ion more strongly), so that reaction tends to go in the forward direction. For example:

$$Ag^+(aq) + e^- \longrightarrow Ag(s) \qquad E^\ominus = +0.80\text{ V}$$

Since electron donors are reducing agents (see Ch 4.3), it follows that *a negative $E^\ominus$ value means that the species on the left of the $E^\ominus$ equilibrium equation is a reducing agent*. Conversely, *a positive value of $E^\ominus$ indicates an oxidising agent.*

Notice that the most active metals – those which most readily lose their outermost electrons – are at the top of the electrochemical series with the highest negative values of $E^\ominus$. The most reactive non-metals and most powerful oxidising agents in solution (e.g. $MnO_4^-(aq)$) are at the bottom with the most positive $E^\ominus$ values.

Predicting reactions using the electrochemical series

The electrochemical series gives a rough measure of relative reducing or oxidising power and so can be used to predict reactivity and the result of redox reactions, such as the displacement of one element by another.

For example, zinc will displace lead from $Pb(NO_3)_2(aq)$. Zinc has a more *negative* $E^\ominus$ value than lead, so zinc atoms will *lose* electrons which are *gained* by $Pb^{2\oplus}$ ions:

$$Zn(s) + Pb^{2\oplus}(aq) \longrightarrow Zn^{2\oplus}(aq) + Pb(s)$$

Chlorine will displace bromine from KBr(aq). Bromine has a *less positive* (i.e. more negative) $E^\ominus$ value than chlorine, so bromide ions will *lose* electrons which are *gained* by chlorine:

$$Cl_2(aq) + 2Br^\ominus(aq) \longrightarrow 2Cl^\ominus(aq) + Br_2(aq)$$

Some $E^\ominus$ values

$Zn^{2\oplus}(aq) + 2e^-$ ⇌ $Zn(s)$		-0.76 V
$Pb^{2\oplus} + 2e^-$ ⇌ $Pb(s)$		-0.13 V
$Cu^{2\oplus}(aq) + 2e^-$ ⇌ $Cu(s)$		$+0.34$ V
$Br_2(aq) + 2e^-$ ⇌ $2Br^\ominus(aq)$		$+1.09$ V
$Cl_2(aq) + 2e^-$ ⇌ $2Cl^\ominus(aq)$		$+1.36$ V

It is perhaps unfortunate that the alkali metals, e.g. sodium and potassium, are described as highly *electropositive* and the halogens, e.g. fluorine and chlorine, as highly *electronegative*. But this can be explained – a sodium atom easily loses an electron to become a positive ion, and a chlorine atom gains an electron to leave a negative ion.

But care is needed when using the series in this way. As we saw earlier (Ch 24.3), the series is *not* the same as the 'reactivity series', because all the reactions occur in solution. Also, it cannot predict the *rate* at which reactions will happen. For example, dilute sulphuric acid does not react with aluminium, although the $E^\ominus$ value for aluminium (-1.66 V) indicates that aluminium should displace hydrogen ($E^\ominus = 0.00$ V). This is because of a tough, thin coherent film of oxide on the aluminium.

The fact that $E^\ominus$ values cannot predict rates should remind you that $\Delta H_{reaction}$ values also cannot be used to predict reaction rates (see Ch 12.4).

Disproportionation reactions

Disproportionation occurs when the same element is both oxidised and reduced during the same reaction. We have met disproportionation already in the reactions between halogens and alkalis (Ch 8.4). For example:

$$Cl_2(g) + 2OH^\ominus(aq) \longrightarrow Cl^\ominus(aq) + OCl^\ominus(aq)$$

In the reaction above, two chlorine atoms (oxidation state zero) become a chloride ion (oxidation state of chlorine, -1) and a chlorate(I) ion (oxidation state of chlorine, $+1$).

Copper(I) cannot exist in aqueous solution – it disproportionates to metallic copper and aqueous copper(II) ions:

$$2Cu^\oplus(aq) \rightleftharpoons Cu(s) + Cu^{2\oplus}(aq)$$

The two half-equations involved are:

$$Cu^{2\oplus}(aq) + e^\ominus \rightleftharpoons Cu^\oplus(aq) \qquad E^\ominus = +0.15 \text{ V}$$

$$Cu^\oplus(aq) + e^\ominus \rightleftharpoons Cu(s) \qquad E^\ominus = +0.52 \text{ V}$$

The most negative (least positive) half-cell will be the left-hand electrode in the cell diagram if E_{cell} is going to be positive. So,

$$Pt \mid Cu^\oplus(aq), Cu^{2\oplus}(aq) \; \vdots \; Cu^\oplus(aq) \mid Cu(s)$$

$$\begin{aligned} E_{cell} &= E_{RHS} - E_{LHS} \\ &= +0.52 - (+0.15) \\ &= +0.37 \text{ V} \end{aligned}$$

and the disproportionation of $Cu^\oplus(aq)$ occurs according to the equation:

$$2Cu^\oplus(aq) \rightleftharpoons Cu^{2\oplus}(aq) + Cu(s) \qquad E_{cell} = +0.37 \text{ V}$$

Electrochemical cells and diagrams – a summary

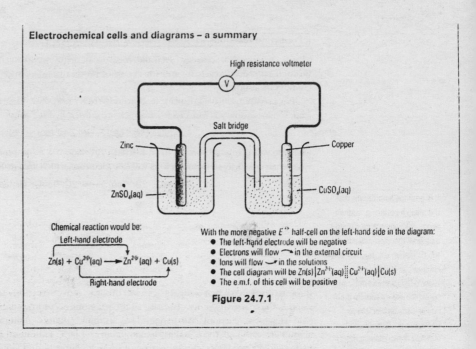

Figure 24.7.1

1 The SEPs for two half-equations are:

$$PbO_2(s) + 4H^{\oplus}(aq) + 2e^{\ominus} \rightleftharpoons Pb^{2\oplus}(aq) + 2H_2O(l) \quad E^{\ominus} = +1.46 \text{ V}$$

$$Cl_2(g) + 2e^{\ominus} \rightleftharpoons 2Cl^{\ominus}(aq) \quad E^{\ominus} = +1.36 \text{ V}$$

a Use these data to determine if PbO_2 will react with $Cl^{\ominus}$ or if Cl_2 will react with $Pb^{2\oplus}$.
b Write a balanced equation for the reaction.
c Calculate the cell potential, E_{cell}.

2 The standard electrode potential of cadmium is found using the following electrochemical cell.

$$Cu(s) \mid Cu^{2\oplus}(aq) \parallel Cd^{2\oplus}(aq) \mid Cd(s) \quad E_{cell} = -0.74 \text{ V}$$

a Draw a diagram of apparatus to set up this cell.
b Calculate the SEP for the $Cd^{2\oplus}(aq)/Cd$ electrode, given that $E^{\ominus}$ for $Cu^{2\oplus}(aq) \mid Cu(s)$ is +0.34 V.
c Mark in your diagram the direction of the movement of the electrons in the circuit.
d In which direction would the ions in the cell move?
e Write an equation for the reaction which could occur.
f Why is there no evidence for the precipitation of a metal when the cell is set up?

3 a Using a fully labelled diagram, show how you would measure the standard electrode potential of the $Zn^{2\oplus}(aq)/Zn$ electrode. Make sure you state the conditions and all the chemicals you would use for the measurement. Explain the function of each of the parts of the apparatus.
b The standard electrode potential for the $Zn^{2\oplus}(aq)/Zn$ electrode is −0.76 V. What would you expect to happen if zinc was placed in a dilute solution of aqueous hydrogen ions, $H^{\oplus}(aq)$? Explain your prediction by referring to the relevant half-equations.

24.7 Practical examples of electrochemical cells

So far we have discussed this topic in terms of reactions which might be seen occurring in a test tube or beaker. Like so much else in chemistry, the ideas in this topic can be used for important industrial purposes or to explain natural processes.

The pH electrode

The variation of e.m.f. with concentration is the basis of the pH electrode, as used in the pH meter. Because a standard hydrogen electrode (including a source of hydrogen gas!) is so cumbersome, a glass electrode is used in connection with a suitable reference electrode. The principle is shown in *Figure 24.7.1*.

In practice, all the components of the apparatus except the solution under test are fitted into a single probe (*Figure 24.7.2*).

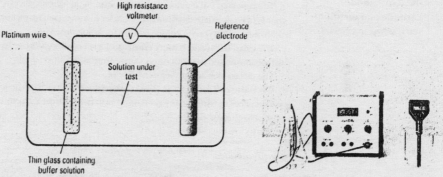

Figure 24.7.1   The pH electrode   **Figure 24.7.2   pH probes**

The platinum wire is sealed into a thin-walled glass tube (the glass contains lithium). This glass tube contains a buffer solution, i.e. a solution which has a fixed concentration of $H^{\oplus}(aq)$ ions. The glass of the tube is *permeable* to protons. The e.m.f. of this electrode will depend on the $[H^{\oplus}(aq)]$ value of the solution around it, and hence on the pH of the solution. The p.d. between the glass electrode and the reference electrode is measured on a high resistance voltmeter. Since the p.d. developed depends on the pH of the solution, the voltmeter can be calibrated to read in pH units. The reference electrode is a suitable compact *secondary standard* electrode, such as a silver/silver chloride electrode, which has a known potential compared with the standard hydrogen electrode.

Storage cells, batteries and fuel cells

In a cell, redox chemical reactions are used to produce electrical energy. Electricity generating companies use fossil fuels, nuclear reactors, or renewable resources to produce electricity. It is highly uneconomical to switch power stations on and off frequently, especially nuclear power stations.

One of the major problems in using electrical energy is storing it for use when it is needed.

As 'alternative', renewable, sources of electrical energy – e.g. solar panels, wind and wave power – become more important, the problem of energy storage becomes more acute. The Sun does not always shine (particularly at night!) nor does the wind always blow. A great deal of research is now being devoted to trying to solve the storage problem.

There are two types of storage cell:

Primary cells, once used, cannot normally be recharged. A common primary cell is the 'dry' cell, as used in torches, personal stereos, etc.

Secondary cells (or 'accumulators') are rechargeable and can be used many times. These include batteries in vehicles, and NiCad (nickel/cadmium) cells for use in place of dry cells.

We shall consider here the lead–acid battery and the dry cell.

The lead-acid battery

An example of intermittent use of electrical energy is provided by a car battery. Such a battery still usually consists of six lead–sulphuric acid cells.

The battery is very heavy for the energy it can store. Its great virtue is that it can deliver a current of about 200 amps to get a vehicle engine started on a cold winter morning; and carry on doing it a few times a day for years on end, recharging from an alternator while the engine is running, and also being used for lights, wipers, demisters, etc as required.

The so-called NiFe cells (nickel and iron in an alkaline electrolyte) have similar virtues. Several other lighter alternative batteries are under investigation by motor companies, but the conventional lead battery is likely to continue in cars for some time yet.

The usual lead–acid battery consists of a set of six cells in series. Each cell contains an anode and cathode immersed in moderately concentrated sulphuric acid.

> For the correct use of the terms **anode**, **cathode** and **terminal** see the box in Ch 24.5

The electrodes are grids made of lead.

When the cell is discharged – i.e. when it is flat – both of the lead grids are covered with lead(II) sulphate, $PbSO_4$. The battery is charged by passing direct current (DC) through it from an alternator coupled to the car engine.

The following changes occur.

At each cathode:

$$Pb^{2+}SO_4^{2-}(s) + 2e^- \longrightarrow Pb(s) + SO_4^{2-}(aq).$$

At each anode:

$$Pb^{2+}SO_4^{2-}(s) + 2H_2O(l) \longrightarrow PbO_2(s) + 4H^+(aq) + SO_4^{2-}(aq) + 2e^-$$

When the battery has been charged by electrolysis, the lead grid cathodes are covered with spongy lead and the anodes with dark brown lead(IV) oxide, PbO_2. Note that as the battery becomes charged, hydrogen and sulphate ions go into solution – i.e. the concentration of sulphuric acid increases. The state of charge of the battery can be tested by measuring the density of the acid electrolyte.

If the battery is now connected to an external circuit, the battery discharges. The spongy lead cathode of the end cell becomes the *negative terminal* of the battery, and the PbO_2-coated anode at the other end is the *positive terminal*.

At each negative terminal in the cells:

$$Pb(s) \longrightarrow Pb^{2+}(aq) + 2e^-$$

The $Pb^{2+}(aq)$ ions join with $SO_4^{2-}(aq)$ ions from the electrolyte and precipitate on to the lead grid as solid $PbSO_4$.

At each positive terminal:

$$PbO_2(s) + 4H^+(aq) + 2e^- \longrightarrow Pb^{2+}(aq) + 2H_2O(l)$$

Again, $Pb^{2+}(aq)$ ions join with $SO_4^{2-}(aq)$ ions from the electrolyte and deposit as $PbSO_4$ (*Figure 24.7.3*).

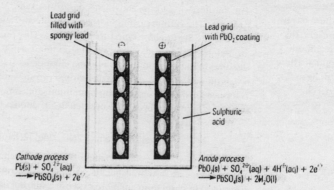

Figure 24.7.3   A summary of the reactions in the lead–acid battery

Each cell produces about 2 V; the battery gives about 12 V.

When all the PbO_2 has gone and all the lead grids are coated with lead(II) sulphate, $PbSO_4$, the battery is flat and no more electric current can be produced. Notice that as PbO_2 is used up and $PbSO_4$ deposited, $H^+(aq)$ and $SO_4^{2-}(aq)$ ions are used up. This means that sulphuric acid is removed from the electrolyte, so that its *density decreases*.

Lead–acid batteries are also used to power vehicles which have to stop frequently, do not need to travel fast, and cover a relatively short distance daily – e.g. milk floats. Lighter batteries with greater energy densities, which can be used on a wider range of vehicles, are under active development.

'Dry cells'

Georges Leclanché (1839–1882) was a French railway engineer who devised his cell in 1866 for use in telegraphy.

The commonest and cheapest form of primary cell is the Leclanché 'dry cell', as used in torches, radios, toys, etc.

Note that the cell contains ammonium chloride paste; there is no such thing as a really *dry* cell! The construction of a typical cell is shown in *Figure 24.7.4*.

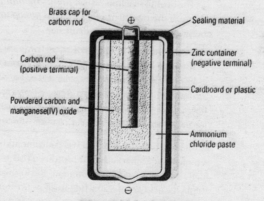

Figure 24.7.4   A dry cell

The negative terminal is zinc, which of course loses electrons:

$$Zn(s) \longrightarrow Zn^{2+}(aq) + 2e^-$$

At the positive terminal (the carbon rod), ammonium ions are reduced, using up electrons:

$$2NH_4^+(aq) + 2e^- \longrightarrow 2NH_3(g) + H_2(g)$$

627

Note that there has to be water present; dry ammonium chloride cannot conduct. There is no such thing as a genuine *dry* cell.

The mixture of carbon and manganese(IV) oxide, MnO_2, around the carbon rod serves two purposes. The carbon granules increase the effective surface area of the carbon rod, and the MnO_2 oxidises the hydrogen formed to give water. This prevents the formation of hydrogen bubbles around the carbon rod, which would stop the reaction. The ammonia formed is very soluble and dissolves in the water in the ammonium chloride paste.

As the cell is discharged, the zinc casing corrodes away as it forms Zn^{2+}(aq) ions, and more water is formed inside it. Eventually the case disintegrates and a horrible, corrosive goo emerges. (As you will have discovered if you have ever left dead batteries in a torch or radio for too long.)

The cell diagram is:

$$Zn(s)\,|\,Zn^{2\oplus}(aq)\,\|\,2NH_4^{\oplus}(aq),[2NH_3(aq) + H_2(g)]\,|\,C(graphite)$$

More expensive, longer-life variations on the Leclanché cell – some using alkaline electrolyte in a steel casing – are now available. Rechargeable cells containing nickel and cadmium and an alkaline electrolyte in paste form are more expensive still, but can be recharged hundreds of times and save money in the long run.

Fuel cells

In a conventional power station, fuel is burned to heat water to make steam to pass through turbines to turn the dynamo rotors to produce electricity. This is a grossly inefficient process, for thermodynamic reasons as well as mechanical ones.

The most efficient fossil fuel power stations can convert only about 45% of the energy available from the fuel into electrical energy. (Nuclear fission power stations are not very much better overall, if the energy needed to produce the fuel and construct the core and plant is taken into account; they have been described as a very silly way of boiling water.) Fossil fuel power stations also produce a great deal of carbon dioxide, and often sulphur dioxide.

In the fuel cell, chemical energy is converted directly into electrical energy in a device which has no moving parts. This is obviously more efficient than the conventional method. The first fuel cell, using hydrogen and oxygen, was made in 1842 by Sir William Grove. It became a practical device in 1959, thanks largely to another Englishman, Francis Bacon. Hydrogen–oxygen alkaline fuel cells have often been used to power electrical systems in spacecraft, because the only product of the reaction is pure water: so that two problems are neatly solved with one process!

The overall reaction, which occurs *spontaneously*, in the hydrogen/oxygen fuel cell is the exact reverse of the overall reaction which is *forced* to happen during the electrolysis of water by electrical energy supplied from outside the cell.

Sir William Grove (1811–1896) was a famous lawyer and judge who also contributed greatly to science and technology in his spare time. Among other contributions, in 1845 he made the first electric lamp which used a filament. His 'gas battery' of 1842 consisted of two platinum strips dipping into dilute sulphuric acid. The top of one strip was surrounded with hydrogen gas, the other with oxygen. A current flowed between the strips when they were connected by a wire.

Francis Bacon (1904–1982) was a descendant of another Francis Bacon, the great early 17th-century statesman and philosopher who developed 'scientific induction' – the basis of modern scientific method.

Recent fuel cell development has concentrated on cells which use a solid polymer electrolyte, which is a membrane permeable to protons. (This is usually called a *proton exchange membrane*, PEM; fuel cells using such a membrane are termed PEMFCs.) The polymer is based on Teflon, with sulphonic acid groups attached to allow proton conduction.

The electrodes are made of porous carbon cloth and covered with extremely finely divided platinum, perhaps as little as 0.1 mg Pt per cm^2. The membrane-electrode assembly is less than 1 mm thick.

Hydrogen is oxidised at the 'anode', or negative terminal; electrons are removed from the hydrogen atoms and go round the external circuit, and protons pass through the membrane. Oxygen is reduced at the 'cathode' (positive terminal), and with the protons water is made as the only product. The general principles are shown in *Figure 24.7.5*.

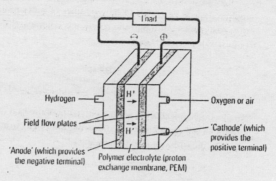

Figure 24.7.5 The hydrogen fuel cell

The reactions occurring in the cell are:

At the 'anode' electrons are released, so oxidation occurs:

$$2H_2(g) \longrightarrow 4H^+(aq) + 4e^-$$

At the 'cathode', reduction occurs:

$$O_2(g) + 4H^+(aq) + 4e^- \longrightarrow 2H_2O(l)$$

Overall:

$$2H_2(g) + O_2(g) \longrightarrow 2H_2O(l)$$

Figure 24.7.6 A fuel cell powered car

Fuel cells can be put together in stacks; sufficient stacks provide enough power for many possible uses.

Buses powered by hydrogen fuel cells have been running in Vancouver and Chicago since 1997 and more recently in Germany, and rapid development is expected.

Demonstration cars have been running for several years and seven of the ten largest car manufacturers in the world are collaborating with fuel cell firms and intend to mass produce fuel cell-powered cars before the year 2010 (*Figure 24.7.6*). Fuel cells are already in use for providing domestic electrical supplies, and also for factories and hospitals. There are still problems to be overcome, especially in securing a 'clean' way of producing hydrogen in large quantities, but it looks as though a decentralised method of producing power for transport and domestic use is becoming increasingly possible. And the only waste product from a hydrogen fuel cell is water!

If ever nuclear fusion (see Ch 1.2) is harnessed as a means of generating electrical power, and if it is as 'clean' in terms of radioactivity as is hoped, it might become possible to generate hydrogen by electrolysis on a large scale near coasts, pipe it to where it is needed, and then use it in fuel cells to power industry, run homes, etc. in many parts of the world. The water produced could be vital for plants and people.

A highly efficient, non-polluting, easily controllable and widely available source of electricity would make a very welcome change.

Systems based on solar energy are also being developed with pilot plants running at 200 kW or more. The solar cells (which are based on silicon, see Ch 3.8) generate electricity, which is used to electrolyse water. The hydrogen produced is stored until required, then used to generate energy in a fuel cell. The overall energy efficiency of the system is potentially higher than that of any combustion method.

Corrosion and rusting

Iron is one of the most important metals in modern society – in Britain alone more than 7 500 000 tonnes are produced each year. It is used, usually as steel, for making a wide range of construction materials, from girders to reinforcement of concrete and sheeting for cladding buildings. It is converted into steels and alloys, with small amounts of carbon and/or a wide range of other transition metals added, to improve the strength and corrosion resistance of the steel.

The rusting of iron causes many problems. Normally, pure iron corrodes fairly rapidly when exposed to water and oxygen. Many millions of pounds are spent in trying to minimise the effects of rusting.

The rusting process is a redox reaction involving atmospheric oxygen and water to form the hydrated oxide $Fe_2O_3.xH_2O$. The number of water molecules present in the brown residue when iron rusts varies. Although the reaction is complex, the first step in the process involves the metal being oxidised to Fe^{2+}(aq).

Half-cell	E°/V
Fe^{2+}(aq)∣Fe(s)	– 0.44
[O_2(g) + 2H_2O(l)],4OH^-(aq)∣Pt	+ 0.40

An electrochemical cell is set up with parts of the iron metal becoming negative. Any stress point in the metal will cause an uneven distribution of electrons in the metal; and so will a point or sharp edge, or the stress caused by bending.

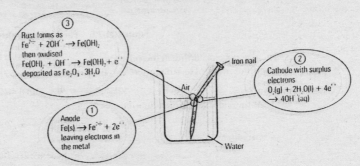

Figure 24.7.7   A rusting iron nail

① The iron metal is oxidised to Fe^{2+}(aq).

$$Fe(s) \longrightarrow Fe^{2\oplus}(aq) + 2e^\ominus$$

② The electrons supplied by this process will then allow the reduction of oxygen where there is moisture and oxygen present. This is usually as oxygen dissolved in the water. More oxygen from air is available to replace oxygen used up by the reaction.

$$O_2(g) + 2H_2O(l) + 4e^\ominus \longrightarrow 4OH^\ominus(aq) \quad E^\ominus = +0.40 \text{ V}$$

The cell diagram for the reaction becomes:

$$Fe(s) ∣ Fe^{2\oplus}(aq) \vdots [O_2(g) + 2H_2O(l)],4OH^\ominus(aq) ∣ Pt$$

For the overall process, E_{cell} involving the oxidation of Fe to $Fe^{2\oplus}$(aq) can be calculated.

$$E_{cell} = E_{RHS} - E_{LHS}$$
$$= +0.40 - (-0.44)$$
$$= +0.84 \text{ V}$$

The rusting is more rapid in salt water areas. Sodium chloride in solution is an electrolyte, and the solution is a better conductor than pure water.

③ Once the $Fe^{2\oplus}$ and $OH^\ominus$ ions have been formed, the formation of $Fe(OH)_2$(s) is an example of simple combination of ions present to make an insoluble compound.

$$Fe^{2\oplus}(aq) + 2OH^\ominus(aq) \longrightarrow Fe(OH)_2(s)$$

Oxidation to $Fe(OH)_3$ in the presence of hydroxide ions is also possible, and it is the formation of hydroxide ions from water and oxygen that drives both the redox reactions involved in the rusting of iron.

$$Fe(OH)_2(s) + OH^\ominus(aq) \longrightarrow Fe(OH)_3(s) + e^\ominus$$

The E value for this second reaction is positive. The half cells and cell diagram can be used to show this.

$$[O_2(g) + 2H_2O(l)], 4OH^-(aq) \mid Pt \qquad E^\ominus = +0.40 \text{ V}$$

$$Fe(OH)_3(s), [Fe(OH)_2(s) + OH^-(aq)] \mid Pt \qquad E^\ominus = -0.56 \text{ V}$$

$$Pt \mid [Fe(OH)_2(s) + OH^-(aq)], Fe(OH)_3(s) \vdots [O_2(g) + 2H_2O(l)], 4OH^-(aq) \mid Pt$$

$$E_{cell} = E_{RHS} - E_{LHS}$$
$$= +0.40 - (-0.56)$$
$$= +0.96 \text{ V}$$

Iron(III) hydroxide, $Fe(OH)_3$, is more accurately represented as hydrated iron(III) oxide, $Fe_2O_3.3H_2O$; but this can lose some water, hence $Fe_2O_3.xH_2O$. And so we have rust.

Can we prevent rusting?

Yes, at a cost. Obvious methods include the physical exclusion of air and/or moisture, and the feeding of electrons into the iron or steel to counter the oxidation process. The prevention of rusting is discussed in Ch 25.7.

SELF-ASSESSMENT QUESTIONS 24.7

1 Explain what is meant by, and give an example of:
   a a primary cell;
   b a secondary cell.

2 The simplest nickel–cadmium cell could be based on their standard electrode potentials.

$$Ni^{2+}(aq) \mid Ni(s) \qquad E^\ominus = -0.25 \text{ V}$$

$$Cd^{2+}(aq) \mid Cd(s) \qquad E^\ominus = -0.40 \text{ V}$$

   a Write the cell diagram incorporating these two half cells, and deduce the e.m.f. of the cell diagram you have written.
   b The rechargeable nickel–cadmium battery is based on a more complex series of reactions:

$$Cd(s) + 2OH^-(aq) \longrightarrow Cd(OH)_2(s) + 2e^-$$

$$NiO_2(s) + 2H_2O(l) + 2e^- \longrightarrow Ni(OH)_2(s) + 2OH^-$$

   Write an equation for the reaction that occurs when the battery is being used.
   c What are the oxidation states of both metals before and after the reaction?
   d Write the equation for the reaction that takes place when the battery is recharged.

3 Outline the principles of the hydrogen fuel cell. Why do you think this type of cell is likely to prove useful in the future?

4 When the price of tin metal increased significantly, chromium was looked at as an alternative coating of iron for 'tin' cans. The standard redox potentials for iron, tin and chromium are given.

	$E^\ominus / V$
$Cr^{3+}(aq) \mid Cr(s)$	−0.74
$Fe^{2+}(aq) \mid Fe(s)$	−0.44
$Sn^{2+}(aq) \mid Sn(s)$	−0.14

   a In which of these two cases is iron more likely to rust if the metal covering the iron is scratched? Explain your answers.

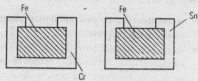

b Although both tin and chromium electrode potentials are negative, comment on the fact that chromium can react with 1.0 mol dm⁻³ hydrochloric acid whereas tin has very little reaction with the same acid.

5 The oxidation process in the presence of water can involve the reaction:

$$O_2(g) + 2H_2O(l) + 4e^{\ominus} \longrightarrow 4OH^{\ominus}(aq)$$

The standard electrode potential for the half-cell is +0.40 V.

a Would you expect the three metals listed below to oxidise in damp air? Justify your answer by referring to the SEPs.

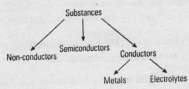

$$Ag^+(aq) \mid Ag(s) \quad +0.80\ V$$
$$Fe^{2+}(aq) \mid Fe(s) \quad -0.44\ V$$
$$Al^{3+}(aq) \mid Al(s) \quad -1.66\ V$$

b Write a cell diagram using the oxygen and iron half-cells.

c Calculate the e.m.f. produced by this cell.

d Write a balanced equation for the reaction between iron, oxygen and water.

e Explain why rusting of a car is much more likely in the seaside town of Teignmouth in Devon than in Phoenix in the Arizona desert.

f Aluminium cooking utensils are commonly used, and although they are used at temperatures up to 100 °C with aqueous solutions, there is rarely any significant sign of corrosion on the cooking pan. Why should there be so little evidence of corrosion in spite of the large negative standard electrode potential for aluminium?

24.8 Electrolysis

In terms of conducting electricity, substances can be classified as follows:

Substances
→ Non-conductors
→ Semiconductors
→ Conductors
→ Metals
→ Electrolytes

In metallic conduction, electrons move through the body of the metal (see Ch 2.7). *There is no chemical change.* The metal stays exactly as it is, except that it might get hotter.

Conduction through metals involves a flow of electrons and there is no chemical change.

However, the word **electrolysis** implies that a substance is being broken up by electricity.

In the electrical cells we have discussed so far, **a chemical reaction produces electrical energy** when the circuit is complete.

The word-ending '-lysis' generally means 'breaking up'; e.g. 'hydrolysis' means breaking-up of a material by water.

- **In an electrolytic cell, electrical energy from outside the cell causes chemical reaction to occur in the cell.**
- A crystal of sodium chloride does NOT conduct electricity. Yet we know that it is a giant structure of charged ions, $Na^{\oplus}$ and $Cl^{\ominus}$, which should respond to a potential difference. If the salt is melted or dissolved in water, the melt or solution DOES conduct electricity (*Figure 24.8.1*).
- **For electrolysis to be able to happen, ions must be able to move.**

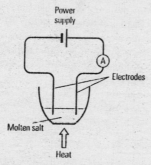

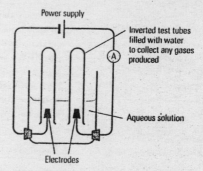

Figure 24.8.1 Electrolysis of a molten salt   **Electrolysis of an aqueous solution**

- In electrolysis , the chemical changes are caused by ions moving to the oppositely charged electrodes, where they pick up or lose electrons.
- For electrolysis, direct current (DC) is necessary. Electrons enter the electrolyte at the **negative** electrode, which is called the **cathode**; electrons leave the electrolyte at the **positive** electrode, called the **anode**.

The usual household supply of electricity is alternating current (AC) at 50 Hz (50 cycles per second). If AC current was used in electrolysis the ions would be changing direction 50 times a second!

- The **positive** ions in the electrolyte go to the **negative** electrode, the **cathode**. Because of this, positive ions are called **cations** (N.B. not cathions!).
- The **negative** ions go to the **positive** electrode, the **anode**; negative ions are therefore called **anions**.
- When cations arrive at the cathode, they pick up electrons from the surface of the cathode. These electrons neutralise the positive charge on the cations; the ions are *discharged*.
- When anions arrive at the anode, they give up their extra electrons to the surface of the anode, and so the anions are *discharged*.
- *The electric current therefore passes through the electrolyte by ions moving in the electrolyte to electrodes where they are discharged by losing or gaining electrons.*

The situation is shown in *Figure 24.8.2*.

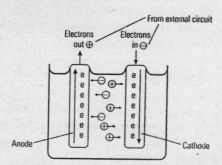

Figure 24.8.2 Electrons and ions in electrolysis

- As in a storage cell, the electrodes must be made of material which is both a good conductor and which does not react with the electrolyte. (Graphite, which is comparatively cheap, is used where possible.)

Strictly speaking, a *solid* salt should not be called an electrolyte, as the ions *cannot move*. Similarly, an anhydrous acid – such as $H_2SO_4(l)$ or $HCl(g)$ – is not an electrolyte, as it does not contain enough ions until water is present. But aqueous *solutions* of all acids, alkalis and salts are electrolytes. *Molten* alkalis and salts are also electrolytes, but not molten pure acids. Pure acids consist of molecules, and do not contain any ions.

For a substance to be an electrolyte:

● it must contain a significant proportion of ions;
● the ions must be able to move.

Reactions at electrodes

Examination of *Table 24.3.1* shows that values of standard electrode potentials, $E^{\ominus}$, rarely exceed ± 3 V. Conditions during electrolysis are usually far from standard, but in even the most vigorous industrial methods *the potential difference between anode and cathode is only a few volts*. This is an important point to remember. In the manufacture of aluminium, for example, the current used may be 100 000 A – but the voltage is in single figures.

Molten salts

Description of electrode reactions for molten simple salts, e.g. molten sodium chloride, is usually quite straightforward. Positive metal ions go to the cathode.

During electrolysis of a molten salt, at the cathode electrons are forced on to the metal cations to form the metal atoms.

Similarly, the negative non-metal anions lose their extra electrons.

During electrolysis of a molten salt, at the anode electrons are removed from non-metal anions to form atoms (and eventually molecules) of the non-metal.

So, for molten sodium chloride:

At the cathode:

$$Na^{\oplus} + e' \longrightarrow Na(l)$$

> A *cathode* supplies electrons, so it *reduces cations*. Conversely, an *anode oxidises anions*. Electrolysis is *a redox process* (see Ch 4.3).

At the anode:

$$Cl^{\ominus} \longrightarrow (Cl) + e^{\ominus}$$
$$2(Cl) \longrightarrow Cl_2(g)$$

Obviously, the *same number of electrons must enter the anode from the anions as are collected from the cathode by the cations*. The same current, i.e. the same flow of electrons, must occur in all parts of the same circuit.

By looking at the electrode half-equations above, we can see that to form one mole of sodium *atoms* needs one mole of electrons; but one mole of electrons given up at the anode will only give half a mole of chlorine *molecules*. We shall shortly see how amounts of substances liberated during electrolysis can be worked out.

Solutions

Electrolysis of solutions is a little more complicated, because water always contains some hydrated protons, $H^{\oplus}(aq)$, and hydroxide ions, $OH^{\ominus}(aq)$, through self-ionisation (see Ch 13.4).

In a solution of sodium chloride, therefore, the ions present are:

> Notice that the 'hydrogen ion', $H^{\oplus}(aq)$ is positively charged. Hydrogen is the only non-metal element which can be formed at the cathode during electrolysis of an aqueous solution.

From the salt: $\qquad Na^{\oplus}(aq)$ and $Cl^{\ominus}(aq)$

From water: $\qquad H_2O(l) \rightleftharpoons H^{\oplus}(aq) + OH^{\ominus}(aq)$

In this solution there are two cations and two anions. Which ions discharge first? (N.B. Remember that the word discharge here means exactly what it says – an ion is *discharged* when it loses its charge.)

The order of discharge of ions during electrolysis of a solution depends on three main factors.

- The **standard electrode potentials**, $E^\ominus$, of the elements.
- The **relative concentrations** of the ions.
- **What the electrode is made of.**

We shall discuss these factors in turn.

Standard electrode potential, $E^\ominus$

The more reactive a metal, the easier it should be for its atoms to lose electrons, and the harder it will be to force electrons back on to its ions. The more positive its $E^\ominus$ value, the less reactive a metal is likely to be, and the easier it should be to discharge its cation.

For hydrogen, by definition, $E^\ominus = 0.00$ V; so metals with positive values of $E^\ominus$ should discharge before hydrogen does. So, in electrolysis of a solution of copper(II) sulphate, $CuSO_4$, copper ($E^\ominus = +0.34$ V) will plate on the cathode.

However, for solutions of metals with a negative $E^\ominus$ value, $H^\oplus(aq)$ should discharge before the metal cations, and hydrogen gas should bubble off at the cathode. For example, electrolysis of zinc sulphate solution, $ZnSO_4$, results in hydrogen gas at the cathode.

As we have pointed out earlier, the situation is complicated to some extent because $E^\ominus$ values are measured in 1.00 mol dm^{-3} aqueous solutions and at 25 °C, and other factors such as hydration of the ions are also important (see box in Ch 24.3). The order of $E^\ominus$ values is not exactly the same as the order of reactivity (see *Table 24.3.1*). But it is near enough to make some helpful rules.

For cations, the order of discharge of common ions in aqueous solution is:

$$Ag^\oplus \longrightarrow Cu^{2\oplus} \longrightarrow H^\oplus \longrightarrow Fe^{2\oplus} \longrightarrow Zn^{2\oplus} \longrightarrow Mg^{2\oplus} \longrightarrow Na^\oplus$$

most positive $E^\ominus$ value – discharge first ... most negative $E^\ominus$ value – discharge last

For anions:

$$OH^\ominus \longrightarrow I^\ominus \longrightarrow Br^\ominus \longrightarrow Cl^\ominus \longrightarrow NO_3^\ominus \longrightarrow SO_4^{2\ominus}$$

most negative $E^\ominus$ value – discharge first ... most positive $E^\ominus$ value – discharge last

In the electrolysis of aqueous sodium chloride, therefore, hydrated hydrogen ions should discharge at the cathode before sodium ions: hydrogen should bubble off, and sodium ions stay in solution. At the anode, if $E^\ominus$ is the only factor, hydroxide ions should discharge before chloride ions to produce oxygen gas.

We now write the ion-electron half-equations so that the number of electrons balance:

At the anode:

$$OH^\ominus(aq) \longrightarrow (OH) + e^\ominus$$

Hence:

$$4OH^\ominus(aq) \longrightarrow 4(OH) + 4e^\ominus$$

$$4(OH) \longrightarrow 2H_2O(l) + O_2(g)$$

At the cathode:

$$H^\oplus(aq) + e^\ominus \longrightarrow (H)$$

Hence:

$$4H^\oplus(aq) + 4e^\ominus \longrightarrow 2H_2(g)$$

Notice that every four moles of electrons gives two moles of hydrogen molecules but only one mole of oxygen molecules, and so the hydrogen:oxygen ratio of volumes is also 2:1 (see Ch 9.6).

This 2:1 ratio of hydrogen and oxygen is always found to be the case when only hydrogen and oxygen are produced in an electrolysis.

Notice also that if the water in the solution is being split apart and lost as hydrogen and oxygen, the solution must be getting more concentrated.

Note: as $H^\oplus(aq)$ and $OH^\ominus(aq)$ ions discharge, more water ionises very rapidly to restore K_w (see Ch 23.2). Eventually the water will all be used up.

Concentration

As shown above, electrolysis of *dilute* aqueous sodium chloride gives hydrogen at the cathode and oxygen at the anode. If the experiment is repeated with *concentrated* solution, hydrogen is once more produced at the cathode, but *chlorine* (not oxygen) is released from the anode.

The value of $E^\ominus$ for sodium is -2.71 V; the difference between hydrogen ($E^\ominus = 0.00$ V) and sodium is so large that purely by altering concentrations it is impossible to reverse the order of discharge. (In any case, sodium cannot exist as the metal when water is present – see Ch 6.4.) For oxygen and chlorine, however, the $E^\ominus$ values are much closer:

$$Cl_2(aq) + 2e^\ominus \rightleftharpoons 2Cl^\ominus(aq) \qquad \bar{E}^\ominus = +1.36 \text{ V}$$

$$O_2(g) + 2H_2O(l) + 4e^\ominus \rightleftharpoons 4OH^\ominus(aq) \qquad E^\ominus = +0.40 \text{ V}$$

In addition, the concentration of $OH^\ominus(aq)$ in a solution of sodium chloride is very low. (The pH is about 7, so $[OH^\ominus(aq)] = [H^\oplus(aq)] = 1.0 \times 10^{-7}$ mol dm^{-3} – see Ch 23.2.) A high concentration of chloride ion will therefore cause chlorine to be produced at the anode.

There is no sharp cut-off point for this switch from oxygen to chlorine during electrolysis. However dilute the solution of sodium chloride, a little chlorine will be produced with the oxygen, and in saturated sodium chloride solution, a little oxygen will be mixed with the chlorine.

To manufacture chlorine economically, the amount of chlorine which is obtained from a given amount of (expensive) electricity must be maximised by using the highest possible concentration of sodium chloride (see next section).

The nature of the electrode

Occasionally, the order of discharge of the ions is totally disrupted by using an electrode which itself is involved in the process. Normally, an inert electrode such as platinum or carbon (graphite) is used. Carbon electrodes can flake badly when gases – particularly hydrogen – are produced, and carbon electrodes at which oxygen is released tend to oxidise to carbon dioxide. But carbon is cheap and platinum is not.

The two best-known examples of electrodes getting involved are the use of a mercury cathode during the electrolysis of brine, and of a copper anode during the purification of copper. Each of these is the basis for a massive commercial process and is discussed in the next section.

N.B. We have used the electrolysis of aqueous sodium chloride to illustrate the factors which affect the order of discharge of ions. The same ideas apply to many other electrolytic systems.

SELF-ASSESSMENT QUESTIONS 24.8

1 Which of the following substances are electrolytes?
  a Dilute aqueous potassium hydroxide, KOH.
  b Solid lead nitrate, $Pb(NO_3)_2$.
  c Dilute aqueous nickel sulphate, $NiSO_4$.
  d Dilute aqueous sucrose, $C_{12}H_{22}O_{11}$.
  e Anhydrous sulphuric acid, H_2SO_4.
  f Dilute aqueous barium nitrate, $Ba(NO_3)_2$.

2 Write a half-equation for the reaction occurring at each of the electrodes when a current is passed through the following substances, using carbon electrodes.
  a Molten sodium iodide.
  b Dilute aqueous copper(II) chloride.
  c 1.0 mol dm^{-3} aqueous potassium bromide solution.
  d Molten calcium oxide.
  e Dilute aqueous silver nitrate solution.

24.9 The Faraday constant: quantitative electrolysis

Many important industrial processes involve electrolysis to form a more useful product. Aluminium is produced effectively by the electrolysis of molten aluminium oxide. It is obviously useful to know *how much* electricity you need for a given quantity of your desired product.

- Electric current is measured in ampères (amps, symbol A).
- The quantity of electricity when a current of 1 amp flows for one second is called the coulomb (symbol C).

(Coulombs)
= amps × seconds

Although no A level syllabus at the present time includes quantitative electrolysis, we feel that a brief mention is needed as the ideas are so obviously useful in industry – and it is industrial electrolysis which we discuss in the next section.

The Faraday constant is named after Michael Faraday, who may well be the finest experimental scientist who ever lived, in terms of what he did with the resources he had available to him. He was born in 1791, son of a poor blacksmith, and aged 13 became a bookbinder's apprentice.

He read some of the scientific books he worked on. He was given tickets for Sir Humphry Davy's lectures at the Royal Institution in London; he wrote up his notes and sent them to Davy, who took on Faraday as a temporary helper. He stayed on as Davy's assistant, succeeded him as Director of the Royal Institution, and was showered with honours by scientific organisations, the British government, and other nations.

He discovered electromagnetic induction, and built primitive motors, transformers and dynamos. He discovered benzene, synthesised the first known chlorocarbons, liquefied chlorine, and developed alloy steels. He was a brilliant lecturer, and in 1826 started the Royal Institution Christmas lectures for young people, which still continue. For many years, up until 1999, his portrait, and a picture of him lecturing at the RI, were on the back of British £20 notes. He died in 1867.

Among his discoveries were the laws of electrolysis, known as Faraday's Laws. In *modern* terms, these state that:

1. **The mass of a substance produced at an electrode during electrolysis is directly proportional to the quantity of electricity passed through the electrolytic cell.**
2. **The number of moles of different substances produced during electrolysis by the same quantity of electricity is inversely proportional to the number of charges (⊕ or ⊖) on each of the ions concerned.**

Definition: The Faraday constant (symbol F) is the quantity of electricity which corresponds to one mole of electrons – i.e. Le, where L is the Avogadro constant (see Ch 9.3) and e is the charge on the electron.

The value of F is 9.6465×10^4 C mol^{-1} (this is often rounded up to 96 500 C mol^{-1}).

The Faraday constant (symbol F) provides the key to **quantitative electrolysis**. If we can write ion-electron half-equations for the two electrode reactions in any electrolysis, *we can use the Faraday constant to relate the quantity of electricity required to the amount of the substances produced.*

For example, consider these reactions at a cathode:

$$Cu^{2+}(aq) + 2e^{\ominus} \longrightarrow Cu(s)$$

$$Ag^{\oplus}(aq) + e^{\ominus} \longrightarrow Ag(s)$$

You can see at once that to get one mole of copper atoms needs two moles of electrons, i.e. $2F$, whereas one mole of silver atoms requires only one mole of electrons.

We met this point earlier, in Ch 24.8, while discussing the order of discharge of ions.

Consider the equations for the electrolysis of dilute aqueous sulphuric acid (remembering that hydroxide ions, OH^-(aq) discharge before sulphate ions, SO_4^{2-}(aq)).

At the cathode: $$2H^+(aq) + 2e^- \longrightarrow H_2(g)$$

At the anode: $$4OH^-(aq) \longrightarrow 2H_2O(l) + O_2(g) + 4e^-$$

Again, you can see that to get *one mole of oxygen gas needs four moles of electrons, i.e. 4F, whereas the same amount of electricity will give two moles of hydrogen gas.* Remembering that one mole of any gas occupies about $24\,000$ cm³ (24 dm³) at 25 °C and 100 kPa (1 atm), we can work out the volumes of gases produced during electrolysis.

SELF-ASSESSMENT QUESTIONS 24.9

1 **a** Strontium bromide is an ionic solid, melting point 543 °C. It contains 35.41% by mass of strontium. Calculate the empirical formula of strontium bromide. (Remember: values of A_r are in the Periodic Table at the back of the book.)

b Strontium is a reactive metal, and to isolate the metal strontium bromide can be electrolysed. Explain what temperature is needed to carry out the electrolysis, and why it is needed.

c Draw a simple electrolysis circuit to show how the electrolysis might be carried out, and a diagram to show the affinities required.

Remember – strontium is so reactive that the electrode where the strontium collects needs to be surrounded by a test tube and bromine is so toxic that this experiment must be done in a fume cupboard.

d Write an equation for the discharge at the cathode.

24.10 Industrial uses of electrolysis

We shall limit discussions to three extremely important industrial processes.

The manufacture of sodium hydroxide and chlorine by electrolysis of brine (saturated aqueous sodium chloride)

One major method uses the **diaphragm cell** (*Figure 24.10.1*), which is fully described in Ch 8.8. This should be reviewed before you progress further.

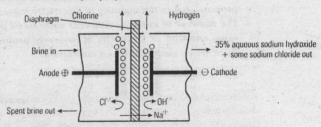

Figure 24.10.1 The diaphragm cell (summary)

An alternative method, which produces rather purer aqueous sodium hydroxide but involves high capital cost, is the **mercury cell**.

When brine is electrolysed, sodium hydroxide solution is produced at the cathode and chlorine at the anode. As we saw in Ch 8.8, because these two products react they have to be kept apart in some way which still allows electrolysis to happen. In the cell discussed in Ch 8.8, the cathode and anode compartments are separated by a porous asbestos diaphragm. Here, the cell uses mercury to achieve separation – and it does it by using some peculiar properties of mercury.

Brine is obtained from underground rock salt deposits. It is electrolysed in the double cell shown in *Figure 24.10.2*.

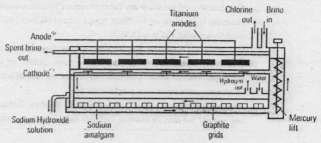

Figure 24.12.2   The mercury cell

In the top compartment, the anodes are now made from titanium (for many years graphite was used). The cathode, however, is a thin layer of mercury running across the gently sloping floor. The anodes are arranged to be as close as possible to the mercury.

Hydrogen does not easily discharge on some clean metal surfaces, probably because of the high activation energy of the process in which the hydrogen atoms formed on the surface combine to form hydrogen molecules.

$$H^{\oplus}(aq) + e^{\ominus} \longrightarrow (H); \quad 2(H) \longrightarrow H_2(g)$$

This difficulty does not occur with platinum, but platinum is extremely expensive.

The extra voltage needed to get hydrogen gas to form on a metal cathode is called the *overpotential* of the metal. *Mercury has a high overpotential.* It is this fact which is the key to this process. Normally, because $E^{\ominus}$ for sodium is –2.71 V there would be no possibility of sodium ions discharging at the cathode before hydrogen. But, in this case, *because of hydrogen's high overpotential on mercury and also because sodium is soluble in mercury to form an amalgam*, the sodium ions do discharge first. No metallic sodium is formed, which would instantly react with water.

As discussed in the box above, sodium ions are discharged at the liquid mercury cathode, and the sodium atoms formed dissolve in the mercury to form an amalgam.

Chlorine is, as expected, liberated at the anode.

At the cathode:	$Na^{\oplus}(aq) + e^{\ominus} \longrightarrow (Na)$
	$(Na) + Hg(l) \longrightarrow Na/Hg(l)$
At the anode:	$Cl^{\ominus}(aq) \longrightarrow (Cl) + e^{\ominus}$
	$2(Cl) \longrightarrow Cl_2(g)$

An **amalgam** is a mixture (or solution) of a metal or metals in mercury. You quite possibly have some very unreactive amalgam in your mouth as tooth filling.

The sodium/mercury amalgam then runs into the lower compartment, where it reacts with water to produce sodium hydroxide solution and hydrogen gas. The water is in contact with graphite grids which are held at a small negative potential with respect to the amalgam. (So, the mercury is the *cathode* in the upper cell, but the *anode* in the lower one.) This encourages the sodium in the amalgam to react with the water, so forming aqueous sodium hydroxide and hydrogen.

The mercury is continuously recycled to the upper cell. As discussed in Ch 8.8, all three products of this process – sodium hydroxide solution, chlorine and hydrogen – are widely used in industry.

Comparison of mercury and diaphragm cells

	Mercury	Diaphragm
Raw materials	Brine	Brine
Overall energy requirements	Similar	Similar
Capital costs	Expensive	Relatively cheap
Operational details	Run at 4.5 V.	Run at 3.8 V
	Highly toxic mercury must be removed from effluent	Diaphragms have to be replaced frequently
Product	High purity; 50% concentration as required by industry	Less pure, needs concentration by evaporation

The absolute need to remove any traces of mercury from the effluent from mercury cells was tragically demonstrated at Minamata in Japan during the 1950s. Here, mercury cells were used mostly to make chlorine for the manufacture of poly(chloroethene), PVC (see Ch 16.7).

Cats were the first to suffer: they became paralysed, then comatose and eventually died. The health authorities were slow to react until human deformities, paralysis and death resulted. The investigations showed that mercury compounds were in the spent effluent which ran directly into the sea; the mercury entered the food chain and became concentrated in fish, which were eaten by the local people (and cats).

There are now extremely strict controls worldwide about the discharge of any heavy metals and their compounds (not only mercury, but also, for example, cadmium and lead) into any seas, rivers or lakes.

The purification of copper

The common copper ore, copper pyrites ($CuFeS_2$), can be heated at a controlled temperature in the correct amount of air to form impure 'blister' copper and sulphur dioxide.

Slabs of this crude copper are used as anodes in cells containing copper(II) sulphate as an electrolyte with a little sulphuric acid. The cathodes are thin sheets of pure copper. As expected from its $E^{\ominus}$ value of $+0.34$ V, copper is deposited on the cathode. **At the anode, however, instead of oxygen being given off, the least energetic process is the dissolving of the copper.**

At the cathode: $\qquad Cu^{2+}(aq) + 2e^{\ominus} \longrightarrow Cu(s)$

At the anode: $\qquad Cu(s) \longrightarrow Cu^{2+}(aq) + 2e^{\ominus}$

Overall, copper passes from the anodes to the cathodes. The cathodes grow through deposition of nearly 100% pure copper, which has many uses (see Ch 25.7). Any impurities, for example, severely reduce the conductivity of copper wiring.

The impure copper contains some valuable materials, including silver. As the copper ionises and dissolves, these impurities fall to the bottom of the cell or can be caught in a porous bag around the anodes. Separating and refining this 'anode sludge' contributes to the overall profitability of the process.

A source of 'cheap' electricity (e.g. a major hydroelectric power scheme) is advisable for copper purification, because two moles of electrons are needed for every mole of copper atoms deposited – i.e. 1.93×10^5 C (A × sec) for every 63.5 g copper. Much of the energy for the 'copper belt' in Zambia, for example, is supplied from the Kariba Dam on the Zambesi river.

The extraction of aluminium

The most common ore of aluminium is bauxite, $Al_2O_3.2H_2O$. It contains oxides of silicon, titanium and iron as impurities. It is first roasted in air to convert any iron(II) compounds to iron(III), then heated at 160 °C under pressure with aqueous sodium hydroxide. The aluminium oxide is **amphoteric** – i.e. it dissolves in alkali as well as in acid – and so it reacts and dissolves.

$$Al_2O_3(s) + 6OH^{\ominus}(aq) + 3H_2O(l) \longrightarrow 2[Al(OH)_6]^{3-}(aq)$$

Iron and titanium oxides do not dissolve, although a little silica may react to form sodium silicate.

The undissolved impurities are filtered off, the disposal of this highly alkaline 'red mud' poses severe environmental problems.

The remaining liquid is diluted and stirred, and 'seeded' with freshly made hydrated aluminium oxide ('aluminium hydroxide'). The $[Al(OH)_6]^{3-}(aq)$ complex breaks down and hydrated alumina precipitates. Any silicate stays in solution.

$$2[Al(OH)_6]^{3-}(aq) \longrightarrow Al_2O_3.3H_2O(s) + 6OH^{\ominus}(aq)$$

The precipitate is filtered, washed, dried and heated to give highly pure aluminium oxide, Al_2O_3, a white powder.

Note that in this process the *concentration of the electrolyte does not change*. In contrast, if you electrolyse aqueous copper(II) sulphate using platinum or carbon electrodes, the blue colour of the electrolyte gets paler and eventually disappears altogether as copper ions are discharged and copper atoms are deposited on the cathode.

Aluminium oxide has a *very high* melting point (2072 °C). Electrolysis was impossible until Hall and Héroult independently discovered in 1886 that the melting point could be greatly lowered by adding the minerals cryolite, Na_3AlF_6, and fluorspar, CaF_2 (*Figure 24.10.3*).

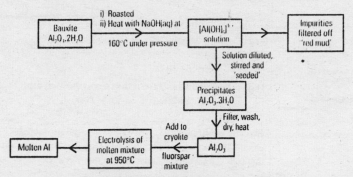

Figure 24.10.3   The extraction of aluminium

Electrolysis is carried out in cells of the type shown in *Figure 24.10.4*. The cell operates at about 950 °C.

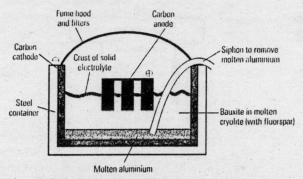

Figure 24.10.4   Manufacture of aluminium

The electrolyte is mainly cryolite, with fluorspar to lower the melting point, and contains approximately 5% of alumina, Al_2O_3, which is frequently replenished. The fluorides stay unchanged; only the oxide decomposes. The cells are connected in series. Typically, a current of 120 000 A at a voltage of 5 V is used.

The molten aluminium (about 99% pure) is siphoned off at intervals. The carbon anodes are specially made on site from baked bitumen. At the high temperatures of the cell they are oxidised quite rapidly by the oxygen given off and are frequently replaced; but this still works out to be more economical than any other possible anode.

The reactions which occur in the cell are still not fully understood, but the *overall* electrode reactions are:

At the cathode: $2Al^{3+} + 6e^- \longrightarrow 2Al(l)$

At the anode: $3O^{2-} \longrightarrow 1\tfrac{1}{2}O_2(g) + 6e^-$

Each mole of aluminium atoms produced has a mass of only 27 g, but requires 3 moles of electrons (i.e. 3 F, nearly 300 000 C). To produce the large amounts of aluminium required worldwide (about 1.5×10^7 tonnes annually), huge quantities of cheap electricity are necessary. Aluminium production plants are usually sited in mountainous areas, to make use of hydroelectric energy.

Because of the energy requirements for extraction of aluminium, recycling makes particularly good economic sense. To recycle aluminium needs only 5% of the energy necessary for extracting the same amount of metal from the ore.

Until 1886, aluminium could only be made in reasonable quantities by the Deville method (1855). This used sodium to reduce aluminium chloride, $AlCl_3$. Not surprisingly, aluminium was extremely expensive, and because it does not corrode (because of a coherent, tough surface layer of oxide) it was used extensively in France for jewellery. The Emperor Napoleon III of France and Empress Eugénie impressed their royal dinner guests by serving food on aluminium plates.

In 1886, two 23-year-olds, Charles M Hall of America and Paul-Louis-Toussaint Héroult of France, independently discovered the electrolytic method for aluminium. Economical mass production was soon possible, and the jewellery and tableware became worthless.

Summary

Half-equations e.g.

$$K(s) \longrightarrow K^+ + e^- \quad \text{oxidation}$$
$$I_2(s) + 2e^- \longrightarrow 2I^- \quad \text{reduction}$$

To balance, both equations need to involve 2 electrons

$$\therefore \quad 2K(s) \longrightarrow 2K^+ + 2e^-$$

adding the reduction and oxidation equations gives the redox equation:

$$2K(s) + I_2(s) \longrightarrow 2KI(s)$$

Half-cells contain both the reduced and oxidised form

$$e.g. \quad Pb^{2+}(aq) + 2e^- \rightleftharpoons Pb(s)$$
$$\underset{\text{oxidised form}}{} \qquad \underset{\text{reduced form}}{}$$

Electrochemical cells – two half-cells combined produce an electrochemical cell. The potential difference between the two half-cells is measured against a reference electrode.

The standard hydrogen electrode

- Platinum electrode
- $H_2(g)$ at 1 atm pressure
- 1.0 mol dm^{-3} HCl(aq)
- 25 °C

Labels: H_2, Platinum electrode, 1.0 mol dm^{-3} HCl(aq)

Diagram: High resistance voltmeter (V), flow of electrons, Salt bridge (KNO$_3$), Mn, Cl$_2$, 1.0 mol dm^{-3} MnSO$_4$, 1.0 mol dm^{-3} NaCl, flow of ions

$$Mn^{2+}(aq) + 2e^- \rightleftharpoons Mn(s) \qquad Cl_2(g) + 2e^- \rightleftharpoons 2Cl^-(aq)$$
$$E^\ominus = -1.19\,V \qquad\qquad\qquad E^\ominus = +1.36\,V$$

Cell diagram: $Mn(s)\,|\,Mn^{2+}(aq)\;\|\;Cl_2(g), 2Cl^-(aq)\,|\,Pt$

$$E_{cell} = E_{RHS} - E_{LHS}$$
$$= +1.36 - (-1.19)$$
$$= +2.55\,V$$
$$\therefore \text{ reaction is feasible}$$

Remember the half-cell with the most negative $E^\ominus$ is:

- the negative terminal
- on the LHS of the cell diagram for a feasible reaction

E_{cell} should be positive for a feasible reaction.

SUMMARY 24

In this chapter the following new ideas have been covered. *Have you understood and learned them?* Could you explain them clearly to a friend? The page references will help.

- The meaning of electrolyte and electrolysis.
- Why the electrolysis of solutions differs from that of molten salts.   p632
- The factors affecting the order of discharge of ions at an electrode.   p634
- The Faraday constant: how to work out quantities in electrolysis.   p635
- The mercury cell in the chlor-alkali industry.   p637
- The purification of copper.   p638
- The extraction of aluminium.   p640
                                                                          p640

EXAMINATION QUESTIONS 24

1 a Define the term *standard electrode potential*.

b Use the following list of standard electrode potentials to answer the following questions.

		$E^\ominus$/V
A	$Fe^{3+}(aq) + 3e^- \rightleftharpoons Fe(s)$	$+0.20$
B	$Sn^{2+}(aq) + 2e^- \rightleftharpoons Sn(s)$	-0.14
C	$In^{3+}(aq) + 3e^- \rightleftharpoons In(s)$	-0.34
D	$Zn^{2+}(aq) + 2e^- \rightleftharpoons Zn(s)$	-0.76

Write the cell diagram that would be formed between half-cells A and B.

c What would you use as a salt bridge in this cell and what is the purpose of the salt bridge?

d Calculate the e.m.f. of the cell formed using half-cells C and D. How would you measure this e.m.f.?

e Zinc reacts with 2.0 mol dm^{-3} hydrochloric acid. By reference to the e.m.f. of a standard zinc-hydrogen cell, explain why the reaction would be expected to occur.

f From the standard electrode potentials the reaction between tin and 2.0 mol dm^{-3} hydrochloric acid would be expected to occur. Can you explain why the reaction does not occur in practice?

2 One of the cheapest and most convenient types of battery in common use is the Leclanché dry cell. The cell consists of a central electrode containing a carbon rod in a carbon-manganese(IV) oxide mixture. This electrode is placed in an ammonium chloride paste. The other half of the cell consists of a zinc metal casing. The two half-cells involved in making the battery are the zinc-zinc ion system:

$$Zn^{2+}(aq) + 2e^- \rightleftharpoons Zn(s)$$

and the ammonia-ammonium ion half-cell

$$2NH_4^+(aq) + 2e^- \rightleftharpoons 2NH_3(g) + H_2(g)$$

a Why is ammonium chloride paste used to complete the half-cell rather than the dry powdered ammonium chloride?

The standard electrode potentials of the two half-cells are:

$$Zn^{2+}(aq) + 2e^- \rightleftharpoons Zn(s) \qquad E^\ominus = -0.76 \text{ V}$$
$$2NH_4^+(aq) + 2e^- \rightleftharpoons 2NH_3(g) + H_2(g) \qquad E^\ominus = +0.74 \text{ V}$$

b Define the term standard electrode potential.

c Write the cell diagram using these two half-cells and calculate the standard e.m.f. of the overall cell.

d What are the changes in oxidation states of the elements in the two half cells?

Zinc changes from ____ to ____ ;

Hydrogen changes from ____ to ____ .

e Deduce the overall equation for the reaction in this cell.

3 Outline the commercial production of aluminium

Your account should include chemical equations and take into consideration the electricity needed to produce the aluminium.

- Electron energy levels and **electron spin.** (Ch 1.4)
- **Dative covalent bonds.** (Ch 2.4)
- The reasons for the **shapes of molecules.** (Ch 2.6)
- **Coordination number.** (Ch 3.3)
- **Oxidation states**, and how to work them out. (Ch 4.3)
- The meaning of **amphoteric.** (Ch 5.5)
- **Equilibrium constants.** (Ch 22.1)
- **Catalysis**, and how it happens. (Ch 12.4)
- **Acids and bases**, and acid–base equilibrium. (Ch 13.4 and Ch 23.1)
- **Standard electrode potentials** and how to use them. (Ch 24.3)

25
the periodic table: the d-block

25.1 Transition elements and the d-block

The term *transition elements* probably came into use, after the Periodic Table had been established, because chemists looked on these elements as having properties in between those of the very reactive metals in Groups 1 and 2 and the much less reactive metals, and then non-metals, further to the right in each Period. They are part of the *transition*, or change, in properties from metallic to non-metallic from left to right in each Period.

A glance at a Periodic Table reveals that the d-block, together with the f-block, forms a very large part of the Table. Most elements are metals; the great majority of the metals are found in the d- and f-blocks.

This chapter deals with metals which have immense industrial importance. Indeed, industry as we now know it – and transport, too – would be inconceivable without them. Some of them are essential for life, and collectively they are of great interest from the point of chemical theory.

Although there are many important metals (e.g. silver, gold, mercury, platinum, tungsten) in the second and third rows of the d-block, we shall concentrate on the first row, from scandium to zinc. This is often called the first transition series.

The first transition series

Sc Ti V Cr Mn Fe Co Ni Cu Zn

Figure 25.1.1

Definition: A transition metal is an element which forms at least one ion in which there is an incomplete d-sub level of electrons.

One definition of a transition metal is given in the box.

This definition would exclude scandium and zinc, which differ from the other elements in the first transition series by existing in only one common oxidation state in their respective compounds. They are, however, included here so that the first series of 'd-block elements' (Sc → Zn), which corresponds to the filling of the 3d sub-level, may be complete (*Figure 25.1.1*).

The d-block elements generally have the usual physical properties of metals – they are all, for example, good conductors of heat and electricity, and all have shiny surfaces when freshly cut. The metals of the first series also have some properties which are characteristic of the transition metals.

- They can exist in more than one oxidation state in their compounds.
- Many of their compounds are coloured and/or *paramagnetic* – i.e. they are weakly attracted by a magnet. (*Ferromagnetic* materials, such as iron, respond much more strongly.)
- Many d-block elements and their compounds are used as catalysts in major industrial processes.
- All readily form **complexes**.

These properties (*Figure 25.1.2*) do not usually apply for scandium and zinc, and are not entirely confined to d-block elements – see, for example, tin and lead (Ch 7.3).

All these properties are discussed later in this chapter. All ultimately depend on the electron configurations of the atoms of these elements. These are shown in *Table 25.1.1*. Note carefully that only the 3d and 4s sub-levels are shown in the table; all these elements possess the *argon core*, i.e. the electron configuration $1s^22s^22p^63s^23p^6$.

Transition metal properties

- Variable oxidation state
- Coloured compounds, often paramagnetic
- Good catalysts
- Form complex ions or molecules because of incomplete d-sub levels

Figure 25.1.2

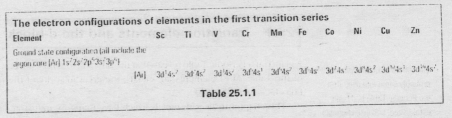

The electron configurations of elements in the first transition series

Element	Sc	Ti	V	Cr	Mn	Fe	Co	Ni	Cu	Zn
Ground state configuration (all include the argon core [Ar] $1s^2 2s^2 2p^6 3s^2 3p^6$) [Ar]	$3d^1 4s^2$	$3d^2 4s^2$	$3d^3 4s^2$	$3d^5 4s^1$	$3d^5 4s^2$	$3d^6 4s^2$	$3d^7 4s^2$	$3d^8 4s^2$	$3d^{10} 4s^1$	$3d^{10} 4s^2$

Table 25.1.1

So, using 'electrons in boxes' (Ch 1.4), the electron configuration of manganese is:

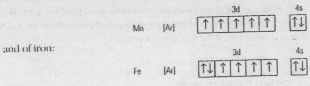

		3d	4s
Mn	[Ar]	↑ ↑ ↑ ↑ ↑	↑↓

and of iron:

		3d	4s
Fe	[Ar]	↑↓ ↑ ↑ ↑ ↑	↑↓

Notice in particular the odd structures for chromium and copper. This is caused by the extra stability of the 3d sub-level – with its *five* orbitals – when it is exactly half full or when it is completely full. Remember *all* the orbitals in a sub-level must contain one electron before *any* of them is allowed to contain two (see Ch 1.4).

For chromium, the configuration in the **ground state** (i.e. when the atom is unexcited) is:

		3d	4s
Cr	[Ar]	↑ ↑ ↑ ↑ ↑	↑
and not	[Ar]	↑ ↑ ↑ ↑	↑↓

For copper, the configuration is:

		3d	4s
Cu	[Ar]	↑↓ ↑↓ ↑↓ ↑↓ ↑↓	↑
and not	[Ar]	↑↓ ↑↓ ↑↓ ↑↓ ↑	↑↓

Q
SELF-ASSESSMENT QUESTIONS 25.1

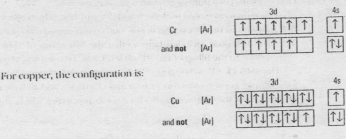

1 Write the electron configurations of the following d-block metals. You can simplify the configurations by using [Ar] and [Kr] for the electron structures of the noble gases where appropriate.

a Vanadium
b Manganese
c Copper
d Zirconium

2 'Transition metals' and 'd-block elements' are terms that can be easily confused.

a By describing the distinctive properties of the transition metals, show that scandium and zinc are not transition metals.

b Write the electron configurations of scandium and zinc. What would you expect to be the oxidation states of these metals in their most common compounds?

3 For the last 8000 years silver has been used as an ornamental metal. It has the highest electrical conductivity of the d-block elements.

a Explain why silver is not used for the wires in the National Grid to carry electrical power across the country.

b With substances which contain high sulphur concentrations, such as boiled eggs and mustard dressing, silver quickly tarnishes, according to the equation:

$$4Ag(s) + 2H_2S(g) + O_2(g) \longrightarrow 2Ag_2S(s) + 2H_2O(l)$$

What is the oxidation state of silver in the black compound formed?

c Which element has been reduced in this reaction?

d Silver is also important in the photographic industry because silver halides decompose readily in sunlight to form silver metal and the free halogen. Write an equation for the formation of silver when silver iodide decomposes.

25.2 Physical properties of the elements

The relevant properties are shown in *Table 25.2.1*.

Some physical properties of the first transition series

Element	Sc	Ti	V	Cr	Mn	Fe	Co	Ni	Cu	Zn
Atomic (metallic) radius / nm	0.164	0.147	0.135	0.129	0.137	0.126	0.125	0.125	0.128	0.137
Melting point / °C	1541	1660	1890	1857	1244	1535	1495	1455	1083	420
Boiling point / °C	2731	3287	3380	2670	1962	2750	2870	2730	2567	907
Ionisation energies / kJ mol^{-1}										
First	+631	+658	+650	+653	+717	+759	+758	+737	+746	+906
Second	+1235	+1310	+1414	+1592	+1509	+1561	+1646	+1753	+1958	+1733
Third	+2389	+2653	+2828	+2987	+3249	+2958	+3232	+3394	+3554	+3833

Table 25.2.1

Because no new electron energy levels are being occupied as the atomic number increases from Sc → Zn, the atomic radii only vary by a small percentage, especially after Ti (*Figure 25.2.1*).

> As nuclear charge increases, with no new energy level being started, one would expect atomic size to decrease. But don't forget that extra electrons cause extra repulsion inside the atom. Here, the two opposing effects just about cancel out.

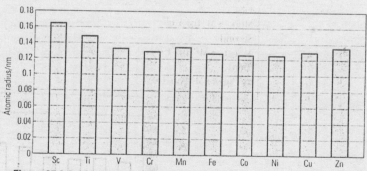

Figure 25.2.1   Atomic (metallic) radius of the first row transition metals

Because their atoms have very similar sizes, these metals form many **alloys**: e.g. 'cupronickel' for coinage (copper and nickel); brass (copper and zinc); and a variety of steels (iron, with manganese, chromium, nickel, etc), all with different properties and uses.

These elements also form so-called **interstitial compounds**. These have no fixed whole number ratio between the numbers of the different kinds of atoms in the substance; i.e. they are *non-stoichiometric*, so they are not true compounds. These materials involve small atoms – e.g. H, B, C, N – occupying the gaps, or *interstices*, in between the metal atoms in the crystal lattice.

Many such materials are important in industry; for example, all steels contain interstitial carbon. The cheapest and most common type of steel, *mild steel*, contains just iron with about 1% carbon. The carbon atoms prevent slippage of the planes of metal atoms past one another (see Ch 2.7), so making the iron very much tougher.

Table 25.2.1 also shows that the melting and boiling points of the first transition series are generally much higher than those of the s-block metals. However, when the d sub-level is completely filled, the melting points and boiling points become much lower (*Figure 25.2.2*).

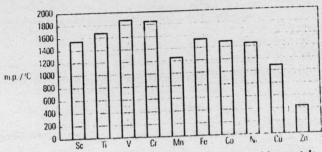

Figure 25.2.2   Melting point of the first row transition metals

The elements zinc, cadmium and mercury (which form a vertical mini-group within the d-block) are much more volatile than the other members of the d-block, and mercury is, of course, liquid at room temperature.

The high melting points of d-block metals cause problems during their extraction from ores and subsequent purification; but they are also one of the main reasons why these metals are in such wide commercial use. (Consider, for example, the properties required of a material for making stoves, engines, exhausts, etc.)

Because nuclear charge increases across a Period, the outer electrons of the d-block elements are more firmly held than those of the s-block; their first ionisation energies are therefore higher than those of the corresponding s-block elements (*Figure 25.2.3*).

> Iron and steel can be protected from rusting by 'galvanising' – by dipping objects into a bath of molten zinc (see Ch 25.7).

Equations for ionisation energies

First
$$M(g) \rightarrow M^{\oplus}(g) + e^{\ominus}$$
Second
$$M^{\oplus}(g) \rightarrow M^{2\oplus}(g) + e^{\ominus}$$
Third
$$M^{2\oplus}(g) \rightarrow M^{3\oplus}(g) + e^{\ominus}$$

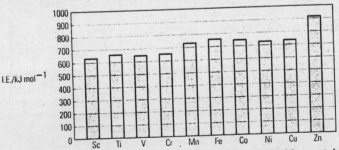

Figure 25.2.3   First ionisation energies of the first row transition metals

The ionisation energies are still relatively low, however, so that the metals of the d-block form cations relatively easily.

1 The metallic radii of some metals are given below, together with data about the atoms.

	Atomic number Z	Mass number A	Radius / nm
Sodium	11	23	0.191
Potassium	19	39	0.235
Scandium	21	45	0.144
Copper	29	64	0.117

Although the difference in numbers of protons and the difference in mass numbers between sodium and potassium and between vanadium and copper are similar, the difference in metallic radii is in complete contrast. Explain why this should be so.

2 The formation of the Fe^{3+} ion is relatively easy compared to the elements on either side of it in the Periodic Table.

a What are the names and symbols of the elements either side of iron in the Periodic Table?

b Write an equation to represent the third ionisation energy of iron.

c The first three ionisation energies of these elements are (in $kJ\ mol^{-1}$):

Element	X	Fe	Y
1st I.E.	+717	+759	+758
2nd I.E	+1509	+1651	+1646
3rd I.E.	+3249	+2958	+3232

Justify the initial statement in this question by explaining why the total energy required to form the Fe^{3+} ion from the neutral atom is less than the comparable value for the formation of the $3+$ ions of the elements either side of it in the Periodic Table.

25.3 Characteristic chemical properties of d-block elements

Oxidation states of the first transition series

Table 25.3.1 shows the oxidation states; the most common and important are shown in bold type. It is important to *know* the most usual oxidation states of the elements most frequently met with: iron, copper, chromium and manganese. Some of their detailed chemistry, and that of vanadium, cobalt, nickel and zinc, is discussed in Ch 25.7.

Oxidation states of the first transition series

Oxidation states (other than zero); the important ones are in bold type

Sc	Ti	V	Cr	Mn	Fe	Co	Ni	Cu	Zn
				+7					
		+6	+6	+6	+6				
		+5	+5	+5	+5	+5			
	+4	+4	+4	+4	+4	+4	+4		
+3	+3	+3	+3	+3	+3	+3	+3	+3	
	+2	+2	+2	+2	+2	+2	+2	+2	+2
	+1	+1	+1	+1	+1	+1	+1	+1	

Table 25.3.1

In general, the compounds of the d-block metals in their higher oxidation states are *covalent* and act as *oxidising agents*. The higher oxides, because they are covalent like the oxides of non-metals, behave like them and are *acidic*. For example, chromium(III) oxide, Cr_2O_3, is an ordinary, basic metal oxide; chromium(VI) oxide, CrO_3, melts at a low temperature, ignites ethanol as soon as it comes into contact with it, is soluble in some other organic solvents, and reacts with aqueous alkalis to form e.g. sodium chromate(VI), Na_2CrO_4.

Don't forget the difference between **ionic charge** and **oxidation state**. For example, $Mn^{2\oplus}$ and $Cr^{3\oplus}$ are perfectly good ions, and correspond to the manganese(II) and chromium(III) oxidation states. But, in higher oxidation states, *the corresponding ions simply can't exist* – the ionisation energies would be *far* too great (see Ch 1.5). So, in sodium chromate(VI), Na_2CrO_4, the chromium(VI) is *covalently* bonded to oxygen in the ion $CrO_4^{2\ominus}$.

> The zero oxidation state in compounds is encountered in the **carbonyls**, e.g. tetracarbonylnickel(0), $Ni(CO)_4$, which is used in purifying nickel (see Ch 7.4).

> If you have forgotten the use of Roman numerals for oxidation states, refer to Ch 4.3.

Oxoanions

The higher oxidation states of these elements are only found when they are combined with very electronegative elements – oxygen or fluorine. The anions containing the metal and oxygen are often termed **oxoanions** (oxyanions). It is the electronegative oxygen that stabilises the formation of the high oxidation state.

e.g.

$$MnO_4^{\ominus} - \text{contains Mn(VII)}$$
$$CrO_4^{2\ominus} - \text{contains Cr(VI)}$$

Figure 25.3.1

The two most common ions of chromium to show the +6 oxidation state are the yellow chromate(VI) ion, $CrO_4^{2\ominus}$, and the orange dichromate(VI) ion, $Cr_2O_7^{2\ominus}$. It is this second ion you will have met (in acid solution) as an oxidising reagent in organic chemistry. The two ions are in equilibrium in solution, and in acid solution it is the dichromate(VI) ion that predominates.

$$2CrO_4^{2\ominus}(aq) + 2H^{\oplus}(aq) \rightleftharpoons Cr_2O_7^{2\ominus}(aq) + H_2O(l)$$

The $Cr^{6\oplus}$ ion does not exist under normal chemical conditions. The same applies to manganese(VII) in the well-known manganate(VII) ion, $MnO_4^{\ominus}$ ('permanganate'); the metal is covalently bonded to oxygen; and there is *no* $Mn^{7\oplus}$ ion present, or indeed possible! *Figure 25.3.2* shows common ions, oxoanions and some typical compounds of the metals of the first transition series.

	+1	+2	+3	+4	+5	+6	+7
						mostly as oxoanions	
Sc			$Sc^{3\oplus}$				
Ti			$Ti^{3\oplus}$	TiO_2			
V		$V^{2\oplus}$	$V^{3\oplus}$	$VO^{2\oplus}$	$VO_2^{\oplus}$		
Cr		$Cr^{2\oplus}$	$Cr^{3\oplus}$		CrF_5	$CrO_4^{2\ominus}$	
Mn		$Mn^{2\oplus}$	$Mn^{3\oplus}$	MnO_2		$MnO_4^{2\ominus}$	$MnO_4^{\ominus}$
Fe		$Fe^{2\oplus}$	$Fe^{3\oplus}$			$FeO_4^{2\ominus}$	
Co		$Co^{2\oplus}$	$Co^{3\oplus}$				
Ni		$Ni^{2\oplus}$					
Cu	$Cu^{\oplus}$	$Cu^{2\oplus}$					
Zn		$Zn^{2\oplus}$					

Figure 25.3.2   Ions, oxoanions and some other compounds of the first transition series

The electron configuration of the ions of the first transition series

When ionisation of an atom of a d-block element does occur, **the s-electrons are removed first**. The extra stability of the half-filled 3d sub-level, with one electron in each of the five orbitals, also becomes important. For example, the configuration of the manganese atom in the ground state (i.e. when unexcited) is $[Ar]3d^54s^2$:

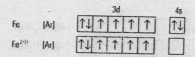

The ion Mn^{2+} therefore has the configuration $[Ar]3d^5$, i.e. the 3d sub-level is exactly half full:

This means that, as we shall see, manganese(II), $[Ar]3d^5$, is more stable than, for example, vanadium(II), $[Ar]3d^3$, or chromium(II), $[Ar]3d^4$, or iron(II), $[Ar]3d^6$.

Iron(II) gives us another good example. If Fe^{2+} has the configuration $[Ar]3d^6$, then one of its five 3d orbitals must have two electrons in it:

If it now loses one more electron, the more stable $[Ar]d^5$ configuration results. This means that it should be possible to remove one electron from Fe^{2+}, i.e. to *oxidise* it to Fe^{3+} (*Figure 25.3.3*). And this is what we find experimentally.

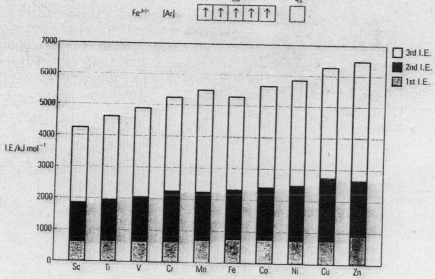

Figure 25.3.3   1st, 2nd and 3rd ionisation energies of the first row transition metals

Catalysis and the d-block

A large number of d-block elements and their compounds are used as catalysts in major industrial processes, for example:

- iron in the Haber Process for making ammonia (see Ch 13.3);
- vanadium(V) oxide in the Contact Process for sulphuric acid (see Ch 13.3);
- a platinum/rhodium alloy in the manufacture of nitric acid from ammonia; and
- nickel in the hydrogenation of vegetable oils to make margarine (Ch 16.2).

Platinum and rhodium are also used in the catalytic converters which are fitted to car exhausts.

How catalysts work is
discussed in Ch 12.4

Whether the catalyst's action occurs by surface adsorption (*heterogeneous catalysis*, Figure 25.3.4) or by formation of unstable intermediates, the ability of d-block elements to flip between oxidation states is involved (see Ch 12.4).

Heterogeneous catalysis

Vacant d-orbitals provide sites for reactants to combine

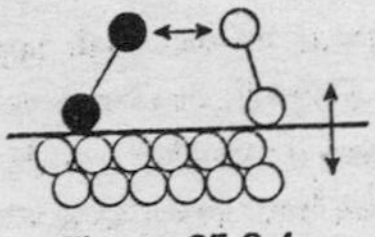

Figure 25.3.4

Homogeneous catalysis
Variable oxidation states
facilitate redox reactions.

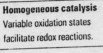

**SELF-ASSESSMENT
QUESTIONS
25.3**

1 What are the oxidation states of the transition metal in each of the following compounds?
  a $KMnO_4$
  b Na_2CrO_4
  c CrO_3
  d MnO_2
  e $Na_2Fe_2O_4$
  f $Mn_2(CO)_{10}$

2 The following transition metals or their compounds are used to catalyse important industrial reactions. Which reactions are they? Write balanced equations to illustrate the reaction involved in each case.
  a Nickel metal
  b Vanadium(V) oxide
  c Iron metal

3 Write the electron configuration of the following atoms or ions.
  a Fe
  b Mn^{2+}
  c V^{3+}
  d Cu
  e Cu^{+}
  f Cu^{2+}

25.4 Complexes

Ligands can donate lone pairs of electrons into vacant orbitals in the metal atom or ion.

Definition: A complex
contains a central metal atom or
ion surrounded by oppositely
charged ions or neutral
molecules, which are called
ligands.

Ligands

In transition metal chemistry a ligand donates a pair of electrons to a central metal atom. The bond is, in effect, a dative covalent bond (see Ch 2.4).

Examples of ligands include:
F^{-}, Cl^{-}, Br^{-}, I^{-}
CN^{-}, OH^{-}, SCN^{-}
H_2O, NH_3, CO

Some of these species also act as *nucleophiles* in organic chemistry, where the lone pair is the key to starting a reaction.
In acid–base chemistry the lone pair donation is typical of a *Lewis base* (see Ch 23.6).

Most of the complexes we shall meet contain d-block elements, since these often have partly filled electron energy levels into which lone pairs may be donated.

The total number of donations of lone pairs from the ligands in a complex is the **coordination number** of the metal atom or ion in the complex (*Figure 25.4.1*).

(We have already met the term 'coordination number' to describe the number of nearest neighbours to a given ion or atom in an ionic or metallic structure – see Ch 3.3 and 3.5.)

Formation of complexes
- Vacant d-orbitals
- Ligand (with lone pair of electrons)
- Dative covalent bond
- Coordination number of mainly 2, 4 or 6

Figure 25.4.1

Some complexes are **anionic**, e.g. $[Fe(CN)_6]^{4\ominus}$; in these, a metal is found in an *anion*, which is not where one might expect to find a metal. Other complexes are **cationic**; e.g. $[Cu(NH_3)_4(H_2O)_2]^{2\oplus}$, which is responsible for the deep royal blue colour when excess aqueous ammonia is added to copper(II) sulphate solution.

Complexes should not be confused with what are known as *double salts*, such as the 'alums', which contain two different cations. Double salts, such as potash alum – potassium aluminium sulphate, $KAl(SO_4)_2.12H_2O$ – show the typical reactions of *all* the components in them; here, the $K^\oplus$, $Al^{3\oplus}$ and $SO_4^{2\ominus}$ ions all act separately in solution.

Complexes, however, usually break up in water into a **complex ion** together with, usually, a simple ion. For example, potassium hexacyanoferrate(II):

$$K_4[Fe(CN)_6](s) \longrightarrow 4K^\oplus(aq) + [Fe(CN)_6]^{4\ominus}(aq)$$

N.B. *A complex is usually signified by means of a square bracket [ ] around the formula of the complex species.*

Stability constants

The stabilities of complexes range very widely; the stability of a particular complex with respect to its constituents can be expressed by a **stability constant**.

A stability constant is yet another case of an equilibrium constant (see Ch 13.1 and Ch 22.1), applied this time to the formation of complex ions.

Consider the equilibrium between a metal ion, $M^{n\oplus}$, with n positive charges, and neutral ligands, L:

$$M^{n\oplus}(aq) + 2L(aq) \rightleftharpoons [ML_2]^{n\oplus}(aq)$$

Unfortunately, it is just one of those awkward and annoying things that the symbol [ ] for a complex is also the symbol which means 'concentration in mol dm^{-3} at equilibrium' in equilibrium constant expressions. Bear this in mind as we write the equilibrium constant for the above reaction; which, since it relates to the stability of the complex, is known as the stability constant, K_{stab}:

$$K_{stab} = \frac{[[ML_2]^{n\oplus}]}{[M^{n\oplus}][L]^2} \text{ mol}^{-2} \text{ dm}^6$$

Notice that, as before with other equilibrium constants (*Figure 25.4.2*), the overall units for the constant depend on the relative number of ions on either side of the equilibrium equation.

So, for example, for the tetraamminecopper(II) complex, $[Cu(NH_3)_4]^{2\oplus}$:

$$Cu^{2\oplus}(aq) + 4NH_3(aq) \rightleftharpoons [Cu(NH_3)_4]^{2\oplus}(aq)$$

$$K_{stab} = \frac{[[Cu(NH_3)_4]^{2\oplus}]}{[Cu^{2\oplus}][NH_3]^4}$$

$$= 1.4 \times 10^{13} \text{ mol}^{-4} \text{ dm}^{12} \text{ at } 298 \text{ K}$$

Do you see how the 'mol^{-4} dm^{12}' happens? Put just the concentration units into the K_{stab} expression:

$$\frac{(\text{mol dm}^{-3})}{(\text{mol dm}^{-3})(\text{mol}^4 \text{ dm}^{-12})} = \text{mol}^{-4} \text{ dm}^{12}$$

(Strictly the complex is $[Cu(NH_3)_x(H_2O)_2]^{2\oplus}$, as we shall see shortly; but we shall ignore the water molecules here, since, in dilute solution, the concentration of water is effectively constant.)

For *complexes of similar structure only*, the greater the value of K_{stab} the more stable the complex. Some stability constants are shown in *Table 25.4.1*. You will see that tetracyanocopper(I), $[Cu(CN)_4]^{3\ominus}$, is more stable than tetraamminecopper(II), $[Cu(NH_3)_4]^{2\oplus}$, and hexaamminecobalt(III), $[Co(NH_3)_6]^{3\oplus}$, is more stable than hexaamminecobalt(II), $[Co(NH_3)_6]^{2\oplus}$. But it is difficult to compare the stabilities of $[Cu(CN)_4]^{3\ominus}$ and $[Co(NH_3)_6]^{3\oplus}$, as the concentration units are raised to different powers in the expression for K_{stab}.

Do you remember how to read the names in organic chemistry? The name appears as one long word, and the skill is in recognising the individual parts of the name. The same applies in reading the names of complexes. You need to recognise the number (hexa-, tetra-, etc), then the name of the ligand(s), then the metal and finally the oxidation state of the metal in brackets. We shall look at the details of naming complexes in the next section.

The stability constants of some complexes

Complex	K_{stab}	Units	$\log_{10} K_{stab}$
$[Cu(CN)_4]^{3\ominus}$	2.0×10^{27}	mol^{-4} dm^{12}	27.3
$[Cu(NH_3)_4]^{2\oplus}$	1.4×10^{13}	mol^{-4} dm^{12}	13.1
$[Co(NH_3)_6]^{3\oplus}$	4.5×10^{33}	mol^{-6} dm^{18}	33.7
$[Co(NH_3)_6]^{2\oplus}$	7.7×10^{4}	mol^{-6} dm^{18}	4.9

Table 25.4.1

Ligand exchange and substitution

Since we are dealing with an equilibrium, it follows that if a more powerful complexing agent (i.e. ligand which makes a more stable complex) is added to a solution of an existing complex, the existing ligands will be replaced.

e.g.   for $[Ag(NH_3)_2]^{\oplus}$ $K_{stab} = 1.7 \times 10^7 \text{ mol}^{-1} \text{ dm}^3$

but for $[Ag(CN)_2]^{\ominus}$ $K_{stab} = 1.0 \times 10^{21} \text{ mol}^{-1} \text{ dm}^3$

Because the two complexes are similar in structure, we can see at once from the figures that the cyanide complex is far more stable than the ammine complex. So, if aqueous potassium cyanide added to a solution of the ammine complex, the ammonia molecules are displaced as ligands by the more powerful cyanide ions.

$$[Ag(NH_3)_2]^{\oplus}(aq) + 2CN^{\ominus}(aq) \rightleftharpoons [Ag(CN)_2]^{\ominus}(aq) + 2NH_3(aq)$$

The formation of the diamminesilver(I) ion explains why silver chloride dissolves in excess aqueous ammonia. This was one way of distinguishing a chloride from a bromide (Ch 8.10).

$$Cl^{\ominus}(aq) + Ag^{\oplus}(aq) \rightleftharpoons AgCl(s)$$

Although silver chloride is almost insoluble, there are sufficient silver ions to react with ammonia and form the ammine complex:

$$Ag^{\oplus}(aq) + 2NH_3(aq) \rightleftharpoons [Ag(NH_3)_2]^{\oplus}(aq)$$

Silver ions are therefore removed from the silver chloride solubility equilibrium, more silver chloride dissolves to restore equilibrium, and so all the silver chloride eventually dissolves in excess ammonia solution.

Stepwise reactions

Where the overall reaction can occur in two or more steps, the product of the equilibrium constants of the individual steps is equal to the overall value for the stability constants.

Consider the formation of the deep-blue copper–ammonia complex, $[Cu(NH_3)_4(H_2O)_2]^{2\oplus}$, from the hydrated copper ion, $[Cu(H_2O)_6]^{2\oplus}$, which is responsible for the paler blue colour of copper salts in dilute solution.

i) $[Cu(H_2O)_6]^{2+}(aq) + NH_3(aq) \rightleftharpoons [Cu(NH_3)(H_2O)_5]^{2\oplus}(aq)$
$$K_1 = 2.0 \times 10^4 \text{ mol}^{-1} \text{ dm}^3 \ (\log_{10}K_1 = 4.3)$$

ii) $[Cu(NH_3)(H_2O)_5]^{2+}(aq) + NH_3(aq) \rightleftharpoons [Cu(NH_3)_2(H_2O)_4]^{2\oplus}(aq)$
$$K_2 = 4.2 \times 10^3 \text{ mol}^{-1} \text{ dm}^3 \ (\log_{10}K_2 = 3.6)$$

iii) $[Cu(NH_3)_2(H_2O)_4]^{2\oplus}(aq) + NH_3(aq) \rightleftharpoons [Cu(NH_3)_3(H_2O)_3]^{2\oplus}(aq)$
$$K_3 = 1.0 \times 10^3 \text{ mol}^{-1} \text{ dm}^3 \ (\log_{10}K_3 = 3.0)$$

iv) $[Cu(NH_3)_3(H_2O)_3]^{2\oplus}(aq) + NH_3(aq) \rightleftharpoons [Cu(NH_3)_4(H_2O)_2]^{2\oplus}(aq)$
$$K_4 = 1.7 \times 10^2 \text{ mol}^{-1} \text{ dm}^3 \ (\log_{10}K_4 = 2.2)$$

Overall:

$$[Cu(H_2O)_6]^{2\oplus}(aq) + 4NH_3(aq) \rightleftharpoons [Cu(NH_3)_4(H_2O)_2]^{2\oplus}(aq)$$

$$K_{stab} = K_1 \times K_2 \times K_3 \times K_4 = 1.4 \times 10^{13} \text{ mol}^{-4} \text{ dm}^{12}$$

(So, $\log_{10}K_{stab} = \log_{10}K_1 + \log_{10}K_2 + \log_{10}K_3 + \log_{10}K_4$
$$= 4.3 + 3.6 + 3.0 + 2.2 = 13.1)$$

The reversibility of **ligand exchange** processes can be demonstrated with a solution of copper(II) sulphate and concentrated hydrochloric acid or brine (NaCl(aq)). The blue colour of $CuSO_4$(aq) is due to the $[Cu(H_2O)_6]^{2\oplus}$ complex. If a high concentration of chloride ions is added, the chloride ion ligands displace some of the water molecules to form the yellow–green complex $[CuCl_4(H_2O)_2]^{2\ominus}$:

$$[Cu(H_2O)_6]^{2\oplus}(aq) + 4Cl^{\ominus}(aq) \longrightarrow [CuCl_4(H_2O)_2]^{2\ominus}(aq) + 4H_2O(l)$$

This can be returned to the original blue $[Cu(H_2O)_6]^{2\oplus}$ simply by diluting with water and so reducing the chloride concentration:

$$[CuCl_4(H_2O)_2]^{2\ominus}(aq) + 4H_2O(l) \longrightarrow [Cu(H_2O)_6]^{2\oplus}(aq) + 4Cl^{\ominus}(aq)$$

Adding more chloride ions turns the solution yellow–green once more; and so on.

The solubility of d-block 'hydroxides' in ammonia or alkali

With the exception of those of Group 1 and some of Group 2, metal hydroxides (or hydrated metal oxides, as they often are) are insoluble in water. This means that they can be precipitated from aqueous solution by adding sodium hydroxide solution, $Na^{\oplus}OH^{\ominus}(aq)$, or ammonia solution (as we saw in Ch 23.1, aqueous ammonia contains hydroxide ions):

$$NH_3(aq) + H_2O(l) \rightleftharpoons NH_4^{\oplus}(aq) + OH^{\ominus}(aq)$$

Some metal oxides and 'hydroxides', notably those of aluminium, zinc, and lead, are **amphoteric** – they can dissolve in excess alkali. Some can dissolve in ammonia solution, especially those of cobalt(II), nickel(II), copper(II) and again zinc.

In the previous section, the formation of the deep blue $[Cu(NH_3)_4(H_2O)_2]^{2\oplus}$ ion was described in terms of stepwise replacement of water molecules in the $[Cu(H_2O)_6]^{2\oplus}$ complex. In fact, it is not quite so simple, at least to begin with. If ammonia solution (or, indeed, sodium hydroxide solution) is added, the hydrated cation, $[Cu(H_2O)_6]^{2\oplus}$, is **deprotonated** (i.e. protons are removed by the alkali from water molecules in the cation) *until the charge on the cation complex is neutralised*. The neutral particles can then clump together and precipitate, as they no longer repel each other because of their positive charge (*Figure 25.4.3*).

$$[Cu(H_2O)_6]^{2\oplus}(aq) + 2OH^{\ominus}(aq) \rightleftharpoons [Cu(OH)_2(H_2O)_4](s) + 2H_2O(l)$$

Pale blue precipitate

or

$$[Cu(H_2O)_6]^{2\oplus}(aq) \rightleftharpoons [Cu(OH)_2(H_2O)_4](s) + 2H^{\oplus}(aq)$$

This precipitate is not easily soluble in excess sodium hydroxide, but easily dissolves in excess ammonia solution by replacement of the hydroxide ligands and two more of the water molecules, to give the deep blue $[Cu(NH_3)_4(H_2O)_2]^{2\oplus}$ (see *Figure 25.4.3*).

Copper sulphate solution and ammonia

The simplest form of the equation for this reaction is:

$$Cu^{2\oplus}(aq) + 2OH^{\ominus}(aq) \longrightarrow Cu(OH)_2(s)$$

Pale blue precipitate

or with excess aqueous ammonia:

$$Cu^{2\oplus}(aq) + 4NH_3(aq) \longrightarrow [Cu(NH_3)_4]^{2\oplus}(aq)$$

Deep blue solution

But the reality is more complex. The several steps can be summarised as:

$$[Cu(H_2O)_6]^{2\oplus}(aq) + 2NH_3(aq) \longrightarrow [Cu(OH)_2(H_2O)_4](s) + 2NH_4^{\oplus}(aq)$$

$$[Cu(OH)_2(H_2O)_4](s) + 2NH_3(aq) + 2NH_4^{\oplus}(aq) \longrightarrow [Cu(NH_3)_4(H_2O)_2]^{2\oplus}(aq) + 4H_2O(l)$$

Figure 25.4.3

Similar processes occur for cobalt(II), nickel(II) and zinc with aqueous ammonia. The sequence can be reversed by adding acid and thus removing the ammonia ligands.

The zinc precipitate, $[Zn(OH)_2(H_2O)_2]$, and aluminium precipitate, $[Al(OH)_3(H_2O)_3]$, are soluble in excess sodium hydroxide *because in high concentrations the $OH^{\ominus}$ deprotonates more water molecules in the complex*, effectively replacing water molecules with hydroxide ions.

e.g. $$[Zn(OH)_2(H_2O)_2](s) \rightleftharpoons [Zn(OH)_4]^{2\ominus}(aq) + 2H^{\oplus}(aq)$$

The resulting anionic particles can no longer clump together because they are now negatively charged and repel. The precipitate therefore dissolves. Again, the process is easily reversible via precipitate to the hydrated cations by adding acid (*Figure 25.4.4*).

Hydration and precipitation

$$[M(H_2O)_6]^{n\oplus}(aq) + nOH^-(aq) \xrightarrow{\text{deprotonation}} [M(H_2O)_{6-n}(OH)_n](s) + nH_2O(l)$$
ionic, soluble   Neutral, insoluble

Further deprotonation $\downarrow$ + (6−n)OH$^-$(aq) (excess OH$^-$ ions)

$$[M(OH)_6]^{(6-n)\ominus} + (6-n)H_2O$$
ionic, soluble

Figure 25.4.4

SELF-ASSESSMENT QUESTIONS 25.4

1 Iron(III) chloride contains the complex ion $[Fe(H_2O)_6]^{3\oplus}$ in aqueous solution. However, the complex is deprotonated and eventually produces a species $[Fe(H_2O)_3(OH)_3]$, which is brown.

  a Write an equation for the reaction where one of the water molecules is deprotonated, producing $H^\oplus$.

  b Why does a solution of iron(III) chloride have a pH in the region 1–3?

  c Explain why there is always a brown precipitate present in any solution of iron(III) chloride if it is left for some time.

2 When copper(II) sulphate is reacted with sodium hydroxide solution a gelatinous pale blue precipitate is formed. When the reaction is repeated with aqueous ammonia (ammonium hydroxide solution), the pale blue precipitate dissolves as soon as excess ammonia is added. Write equations to explain the reactions with sodium hydroxide and then with ammonia and excess ammonia.

3 Write the expressions for the stability constants for the following complexes. Make sure the units are clearly stated in each case.

  a $[Cu(CN)_4]^{3\ominus}$

  b $[Ag(NH_3)_2]^\oplus$

  c $[CoCl_4]^{2\ominus}$

4 Three copper-containing complexes have the formulae:

  A $[CuCl_4(H_2O)_2]^{2\ominus}$ – yellow–green

  B $[Cu(H_2O)_4(OH)_2]$ – pale blue

  C $[Cu(NH_3)_4(H_2O)_2]^{2\oplus}$ – deep blue

  a What is the oxidation state of copper in each of these complexes?

  b If samples of all three complexes were dotted on a piece of moist filter paper and an electric current passed across, as in the diagram, what would you expect to see when the current had been on for ten minutes?

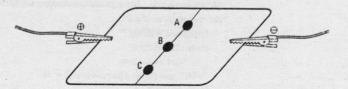

25.5 The structure and naming of complexes

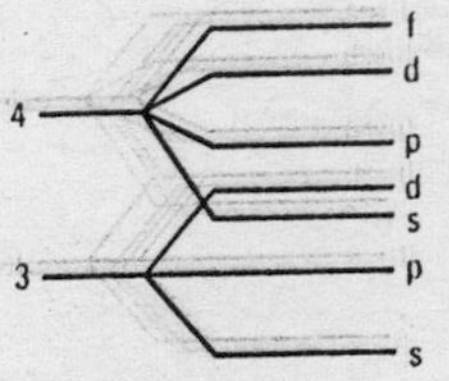

**Figure 25.5.1
Electron energy levels**

At the start of Ch 25.4 we described a complex as a central metal atom or ion surrounded by ligands, with the ligands able to donate lone pairs of electrons into vacant orbitals in the central atom or ion. For the first d-block series (Sc → Zn), vacant orbitals are so close in energy to occupied or partly-occupied ones that they can easily accept donated electron pairs (*Figure 25.5.1*).

For example, in the complex $[Fe(CN)_6]^{4\ominus}$ each cyanide ligand donates one lone pair, and the coordination number of the $Fe^{2\oplus}$ ion is 6.

But in the complex $[Fe(H_2NCH_2CH_2NH_2)_3]^{2\oplus}$, the coordination number is also 6. Why is this?

In the second example the ligand is ethane-1,2-diamine, $H_2NCH_2CH_2NH_2$.

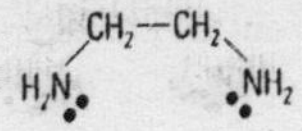

Definition: The coordination number of the metal atom or ion in a complex is the total number of lone pairs donated by the ligands in that complex. (See also Ch 3.3.)

Each nitrogen atom can donate one lone pair, so each diamine ligand can donate *two*. So the total number of donations in the complex is *six* (*Figure 25.5.2*).

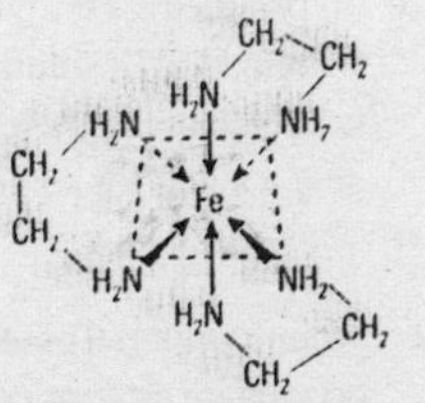

Figure 25.5.2   A bidentate ligand around iron

(Notice that the **dative covalent bond** is shown here by an arrow – but it is no different from *any* covalent bond.)

Ligands such as ethane-1,2-diamine can form ring-like structures which give extra stability to a complex. These ring structures are known as **chelates**, from the Greek word for 'a claw'. The ligands which are involved in **chelation** are said to be **polydentate**: ethane-1,2-diamine is a *bidentate* ligand (because its 'claw' has two 'teeth'!)

An extremely powerful and important chelating agent is 1,2-*bis*[*bis*(carboxymethyl)amino]ethane, which is still known by its old name of 'ethylenediaminetetraacetic acid', EDTA. This is a *hexadentate* ligand and can wrap itself completely around a metal ion to form an exceptionally stable complex.

The lone pairs which can be donated are shown in the structure:

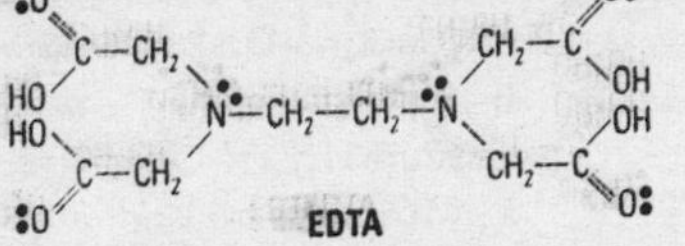

EDTA

For the first transition series (Sc → Zn), a 36-electron rule was proposed in the 1920s and 1930s. Krypton, the next noble gas, has 36 electrons. If we take the atomic number of the central atom, take away the electrons needed to make the metal ion, and then add two from each lone pair donation, we find that many common complexes have 36 electrons associated with the central metal atom.

For example, $[Fe(CN)_6]^{4\ominus}$:

Iron has 26 electrons,

∴ $Fe^{2\oplus}$ has 24 electrons;

Six $CN^{\ominus}$ each donate a lone pair $(6 \times 2) = 12$

So, the total is 36.

But there are so many exceptions that it is best not to think of it as a 'rule'.

Alfred Werner was a German–Swiss inorganic chemist who established the theory of inorganic complexes. He started work on them in 1892, and over the next 20 years developed the idea of a central metal ion with other ions or molecules arranged in space around it. His work systemised a huge section of inorganic chemistry, to the extent that the theory of coordination complexes was capable of making predictions and providing explanations in activities ranging from industrial catalysis to plant biochemistry. Werner was awarded the Nobel Prize for Chemistry in 1913.

The naming of complexes

So far we have used some names for complexes without explaining how these names work. The International Union of Pure and Applied chemistry (IUPAC) rules for naming complexes can be summarised as follows.

1. The cation is always named before the anion.
2. The names of ligands come before the name of the central ion or atom.
3. The oxidation state of the central metal is shown (in Roman numerals) in brackets immediately after its name (see Box below to work out this oxidation state).
4. Metals which are in complex *anions* are given the word ending *-ate*; complex cations and neutral molecules are given their usual names.
5. The order for naming ligands is anions, neutral molecules, cations. Within each category of ligand, if there is more than one type they are named in alphabetical order.

    The names of negatively charged *anionic ligands* end in *-o*, e.g. $Cl^\ominus$, chloro; $CN^\ominus$, cyano. *Neutral ligands* usually keep their normal names; exceptions include H_2O, *aqua*; NH_3, *ammine* (note the double 'm' – *not* to be confused with amine, NH_2); and CO, *carbonyl*. The much rarer *cationic ligands* end in *-ium*.
6. When a number prefix (e.g. di-, tri-) is already part of the name of a ligand, the name of the ligand is put into a bracket with the prefix *bis-, tris-, tetrakis-*, instead of di-, tri-, or tetra-.

Ammine – NH_3 as a ligand.
Amine – NH_2 as a functional group.

Examples of names:

$[Cu(NH_3)_4(H_2O)_2]^{2+} \cdot SO_4^{2-}$ – diaquatetraamminecopper(II) sulphate

$[Cr(H_2O)_6]^{3+} \cdot Cl_3^\ominus$ – hexaaquachromium(III) chloride

$[Co(H_2NCH_2CH_2NH_2)_3]^{3+} \cdot Cl_3^{--}$ – tris(ethane-1,2-diamine)cobalt(III) chloride

$[CuCl_4]^{3\ominus}$ – tetrachlorocuprate(I) ion

$[Fe(CN)_6]^{4\ominus}$ – hexacyanoferrate(II) ion

Calculating oxidation states in complexes

Charge on ligand × number of that type of ligand	=	
Charge on the complex ion	=	
Subtract (bottom from top – careful with the sign)	=	
Change the sign		

Which is the oxidation state of the metal

e.g. $[Fe(CN)_6]^{3\ominus}$

Charge on ligand × number of that type of ligand	$= 1\ominus \times 6$	$= -6$
Charge on the complex ion	$= 3\ominus$	$= -3$
Subtract (bottom from top – careful with the sign)	$= -6 - (-3)$	-3
Change the sign		$+3$

$\therefore$ Fe(III)

e.g. $[Pt(NH_3)_3Cl_3]^\ominus$

Charge on ligand × number of that type of ligand (NH_3)	$= 0 \times 3$	$= 0$
Charge on ligand × number of that type of ligand (Cl)	$= 1\ominus \times 3$	$= -3$
Charge on the complex ion	$= \ominus$	$= -1$
Subtract (bottom from top – careful with the sign)	$= -3 - (-1)$	-2
Change the sign		$+2$

$\therefore$ Pt(II)

e.g. $[Cr(H_2O)_4Cl_2]^\oplus$

Charge on ligand × number of that type of ligand (H_2O)	$= 0 \times 4$	$= 0$
Charge on ligand × number of that type of ligand (Cl)	$= 1\ominus \times 2$	$= -2$
Charge on the complex ion	$= \oplus$	$= -1$
Subtract (bottom from top – careful with the sign)	$= -2 - (+1)$	-3
Change the sign		$+3$

$\therefore$ Cr(III)

Metal carbonyls are complexes in which neutral carbon monoxide molecules are the only ligands and the metal atom is therefore in the oxidation state of *zero* – which is very unusual in compounds. Nickel carbonyl, $Ni(CO)_4$, which is an intermediate in the purification of nickel should therefore be named tetracarbonylnickel(0).

Magnetic properties

In the introduction to this chapter we stated that many compounds of the d-block elements have magnetic properties and/or are coloured.

If all the electrons in an ion or molecule are *paired* – i.e. if all the orbitals with electrons are fully occupied, and each contain a pair of electrons of opposite spin (see Ch 1.4) – the material is *diamagnetic*.

But if any electrons in the ion or molecule are *unpaired*, the material is *paramagnetic* – it is weakly attracted by a magnet. (*Ferromagnetic* materials, which of course include iron – Latin name 'ferrum' – are much more intensely magnetic.) A paramagnetic material's response to a magnetic field depends directly on the number of unpaired electrons in its molecule or ion.

Colour

Many complexes which contain transition metals are coloured. This is a result of the comparatively small differences between the orbital energies in the complex. Visible light is not very energetic, and the energy of some of its wavelengths may be sufficient to excite electrons into *higher energy orbitals* (see *Figure 25.5.3*). If the difference in energy between any two levels is ΔE:

$$\Delta E = h\nu$$

where h is Plank's constant and ν is the frequency of the absorbed radiation (see Ch 1.4). When white light shines on many complexes, therefore, visible frequencies are absorbed. The other frequencies are reflected or transmitted, so that the materials appear coloured.

The colour of the complex will be the *complementary colour* to the light absorbed – e.g. copper(II) sulphate solution absorbs the red–orange part of the visible spectrum, and so we see the solution as blue. Potassium dichromate(VI) solution absorbs in the blue–violet region and so the solution appears yellow–orange in colour.

The origin of the colour in these complexes is through the **absorption** of visible light. Do not confuse this with the *emission (line) spectra* produced when electrons are excited and emit energy as they fall back to lower energy levels.

> Virtually all compounds of metals in Groups 1 and 2 contain no unpaired electrons and are diamagnetic. This means that they are virtually unaffected by a magnetic field.

> It can be argued that it was the work of inorganic chemists such as Vauquelin and Thénard at the turn of the 18th/19th centuries – work which was often subsidised by the French Government – which gave painters greater freedom and encouraged the Impressionist revolution. Their experiments produced a range of new, brilliantly coloured, stable pigments based on chromium, cobalt, copper, arsenic and cadmium.

For a gaseous metallic ion, the five 3d orbitals all have the same energy. (They are said to be '*degenerate*' – an apparently odd use of the word!) But in a complex, ligands surround the central metal ion or atom. Because the ligands are negative ions or lone pair donors, they affect the energy levels in the metal. The 3d orbitals which are close to the ligands have their energies slightly raised; the others become rather lower in energy. The result is that the d-orbital energies are no longer the same – they are *split*. It is this splitting which makes a major contribution to the absorption of light of visible frequency and hence to the colour.

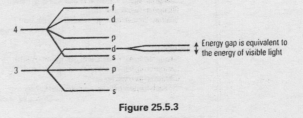

Figure 25.5.3

The colour of the complex depends on the number of electrons in the d sub-level, the arrangement of the ligands, and on what the ligands are.

This d-orbital splitting, resulting in what is called 'd-d transition', is the cause of the colour in several pigments much used by painters: for example, Prussian blue ($Fe_4[Fe(CN)_6]_3.14–16H_2O$); malachite ($CuCO_3.Cu(OH)_2$, green); azurite ($2CuCO_3.Cu(OH)_2$, blue); cobalt blue ($CoO.Al_2O_3$); and viridian ($Cr_2O_3.2H_2O$, green).

Colour of some common transition metal aqueous ions

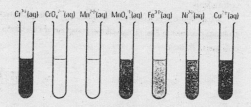

Cr^{3+}(aq) CrO_4^{2-}(aq) Mn^{2+}(aq) MnO_4^-(aq) Fe^{3+}(aq) Ni^{2+}(aq) Cu^{2+}(aq)

Coordination number and the shapes of complexes

Some common complexes and their shapes are shown in *Table 25.5.1*.

Some complexes and their shapes

Coordination number	Shape	Examples
2	Linear	$[Ag(NH_3)_2]^+$, $[Ag(CN)_2]^-$
4	Square planar	$[Ni(CN)_4]^{2-}$, $[Pt(NH_3)_4]^{2+}$
4	Tetrahedral	$[Cd(CN)_4]^{2-}$, $Ni(CO)_4$
6	Octahedral	$[Cr(H_2O)_6]^{2+}$, $[Co(NH_3)_6]^{3+}$ $[Fe(CN)_6]^{3-}$, $[Fe(CN)_6]^{4-}$
6	Distorted octahedral	$[Cu(NH_3)_4(H_2O)_2]^{2+}$

Table 25.5.1

The shapes of these complexes can be accounted for by the mutual repulsion of electron pairs (see Ch 2.6). We shall consider here only tetrahedral, square planar and octahedral complexes. N.B. In the diagrams below, ⇅ represents a donated electron pair. 'Strong' ligands can *push electrons into the lowest available orbitals*, pairing them up where necessary. 'Weak' ligands cannot do this so easily.

Tetrahedral

An example is tetracyanocuprate(I), $[Cu(CN)_4]^{3\ominus}$, *Figure 25.5.4*.

$$\begin{bmatrix} CN \\ | \\ CN\!-\!Cu\!-\!CN \\ | \\ CN \end{bmatrix}^{3-}$$ Tetracyanocuprate(I)

Figure 25.5.4

		3d	4s	4p
Cu ground state	[Ar]	↑↓ ↑↓ ↑↓ ↑↓ ↑↓	↑	
$Cu^{\oplus}$ ion (remember, s electrons are lost first)	[Ar]	↑↓ ↑↓ ↑↓ ↑↓ ↑↓		
$Cu^{\oplus}$ ion in complex (4 donated lone pairs)	[Ar]	↑↓ ↑↓ ↑↓ ↑↓ ↑↓	⇅	⇅ ⇅ ⇅

There are no unpaired electrons in this complex, so it is *diamagnetic*. Note that it has the full 36 electrons. It is also colourless, since there are no d–d electron transitions possible.

Another example is tetracarbonylnickel(0).

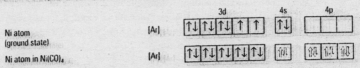

Ni atom (ground state)   [Ar]

Ni atom in Ni(CO)₄   [Ar]

Notice the 'packing' of electrons into the 3d orbitals. This molecule is also diamagnetic and colourless.

Weak ligands such as $Cl^{\ominus}$ in $[NiCl_4]^{2\ominus}$ can give *paramagnetic* tetrahedral complexes. Here, the paramagnetism is the result of two unpaired electrons. Also, because of possible d–d transitions, it is coloured (blue).

e.g. $Ni^{2\oplus}$ ion in $[NiCl_4]^{2\ominus}$ [Ar]

Square planar

Check with your syllabus whether any knowledge of square planar complexes is required.

Although $[NiCl_4]^{2\ominus}$ is tetrahedral and paramagnetic, the corresponding cyanide complex, $[Ni(CN)_4]^{2\ominus}$, is square planar and diamagnetic. The 'stronger' cyanide ligand causes electrons to pair up.

e.g. $Ni^{2\oplus}$ ion in $[Ni(CN)_4]^{2\ominus}$ [Ar]

It is often difficult to decide whether a complex with four donated pairs will be tetrahedral or square planar. One square planar complex, used as an anti-cancer drug, is $[Pt(NH_3)_2Cl_2]$, commonly known as *cisplatin* (*Figure 25.5.5*).

At this level, tetrahedral complexes are the more likely of the two kinds.

Figure 25.5.5   Square planar

Octahedral

An example is hexacyanoferrate(II), $[Fe(CN)_6]^{4\ominus}$ (*Figure 25.5.6*):

Fe ground state   [Ar]

$Fe^{2\oplus}$ ion   [Ar]

$Fe^{2\oplus}$ ion as in $[Fe(CN)_6]^{4\ominus}$ [Ar]

There are no unpaired electrons here, so the complex is diamagnetic.

The corresponding hexacyanoferrate(III), $[Fe(CN)_6]^{3\ominus}$, has one electron less because it contains the $Fe^{3\oplus}$ ion.

$Fe^{3\oplus}$ ion as in $[Fe(CN)_6]^{3\ominus}$ [Ar]

This has one unpaired electron and is therefore paramagnetic. Also, it can easily accommodate another electron to give the 'full' 36 electron structure. It is no surprise therefore that $[Fe(CN)_6]^{4\ominus}$ is used as a mild oxidising agent.

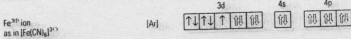

Figure 25.5.6 An octahedral iron complex ion

Hexacyanoferrate(II)

The familiar $[Cu(NH_3)_4(H_2O)_2]^{2\oplus}$ complex contains ammonia molecules in square planar positions around the $Cu^{2\oplus}$ ion, with water molecules above and below (*Figure 25.5.7*).

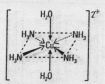

Figure 25.5.7 $[Cu(NH_3)_4(H_2O)_2]^{2+}$

Because the lengths of the copper–ammonia and copper–water bonds are not quite the same, the structure is described as a 'distorted octahedron'.

SELF-ASSESSMENT
QUESTIONS
25.5

1 Consider the following complexes:

A $[Cr(CO)_6]$

B $[Cu(CN_4)]^{3\ominus}$

C $[CuCl_4(H_2O)_2]^{2\ominus}$

D $[Fe(CN)_6]^{4\ominus}$

a Which of these are octahedral in shape?

b Which have a tetrahedral shape?

c What is the oxidation state of the metal atom in each of the complexes?

2 When sodium chloride solution is tested with dilute nitric acid and silver nitrate solution, a dense white precipitate is formed. On addition of aqueous ammonia solution, this precipitate dissolves.

a Complete the following electron-in-boxes diagrams for the ground states of the silver atom and the $Ag^{\oplus}$ ion.

		4d	5s
Ag	[Kr]	☐☐☐☐☐	☐
$Ag^{\oplus}$	[Kr]	☐☐☐☐☐	☐

b Write an ionic equation for the formation of the dense white precipitate when silver nitrate is added to sodium chloride solution.

c Explain why the precipitate dissolves when excess dilute ammonia is added.

d What is the formula and shape of the complex ion formed?

3 Nickel has an atomic number of 28, and appears in the transition metals in the Periodic Table.

a Write the electron configuration, using electrons in boxes, of the ground state of the nickel atom and of the nickel(II) ion.

b Give the formula and draw the shape of a tetrahedral complex of nickel.

c Explain briefly what happens to nickel when it is used to catalyse the reaction of hydrogen with unsaturated oils.

4 a Explain why many of the compounds of the d-block elements are coloured.

b Iron has properties typical of a transition metal, including a variety of oxidation states in its compounds. Using electrons in boxes, explain why $Fe^{2\oplus}$ might be expected to be less stable than $Fe^{3\oplus}$.

c What colour change would you expect to observe if a pale green solution of $Fe^{2\oplus}$(aq) was allowed to stand for several days? Why should this reaction occur?

25.6 The biological importance of some transition metals

Only a brief treatment of the role which transition elements play in biology is given here.

Titanium

Titanium is important in medicine precisely because it does **not** react with body fluids! It is non-toxic, is not rejected by the body, and can be bonded to bone. It is used in hip and knee replacements; bone plates, screws and pins; and in heart pacemakers.

Vanadium

Vanadium is thought to be essential in the human diet, but has not yet been proved to be so. It may be a regulator of an enzyme which controls the role of sodium in the body. The average intake is about 2 mg per day; the human body contains about 20 mg.

Chromium

Again, chromium has not so far been proved to be essential in the human diet; there is a fall in the (very small) amount of chromium in the body with increasing age, but the significance of this is not yet known. Chromium compounds in moderate amounts are toxic; chromate(VI) is corrosive to skin and probably carcinogenic (i.e. it may cause cancer). Chromium is found in, for example, brewer's yeast, barley and wheat germ.

Manganese

Because bacteria are susceptible to oxidising agents, people with sore throats used to be made to gargle with potassium manganate(VII) solution!

All plants and animals need manganese. What exactly it does is not fully known, although it has been shown to be involved with glucose metabolism and the effectiveness of vitamin B_1, and it seems to be associated with RNA. Small amounts of low oxidation state manganese are added to fertilisers and animal feedstuffs. The daily human intake in the diet averages 4 mg; above 20 mg per day can be dangerous.

Manganese in a high oxidation state, e.g. manganate(VII), $MnO_4^{\ominus}$, ('permanganate'), is very toxic.

Some manganese compounds have been shown to have cancer-inducing properties. Manganese is found in, for example, peanuts, almonds, olives, corn, wheat, rice, and tea.

Iron

Iron is in the active part of haemoglobin, which is found in the red corpuscles in blood (*Figure 25.6.1*).

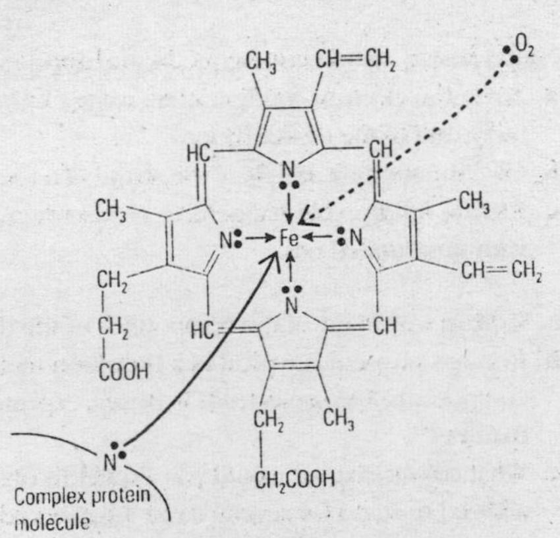

Figure 25.6.1   Haemoglobin

Without iron, the red corpuscles would be unable to accept oxygen from the lungs and carry it to where it is needed in the body. Without iron, anaemia results – a failure to produce enough red corpuscles. (But there are also other, more intractable, causes of anaemia than lack of iron.)

Iron is involved in enzymes responsible for DNA synthesis, and those connected with glucose metabolism. It is also essential for normal brain function. The average man needs 10 mg of iron per day; the average woman, mostly because of menstruation, needs about 18 mg. A good diet should provide all that is needed – rich sources of iron include liver, baked beans, peanut butter, bread, raisins, eggs, bran and curries.

Much of the iron taken in from the diet is not absorbed; only about 2 mg passes into the bloodstream to compensate for the 2 mg lost daily. If this balance is tipped one way, anaemia may result (about half a billion people are anaemic to some extent); if it tips the other way, iron overload in the body can cause mental and physical degeneration and an increased likelihood of cancer.

Many parts of the ocean have little or no iron in their surface layers. In a recent experiment by a joint American–British team, iron(II) sulphate was spread over 60 square miles in the Pacific. In less than a week the sea turned green with plankton, the basic foodstuff for fish. This raises the possibility that the sea could be fertilised and farmed in the same way as land.

Cobalt

Cobalt is an essential element for us, as it is a vital part of vitamin B_{12}.

If we lack vitamin B_{12}, pernicious anaemia is the often fatal result. We probably have only about 1 mg of the vitamin in our body. Too much cobalt may damage the heart and the thyroid gland, and cobalt is suspected as a carcinogen. Foods which contains vitamin B_{12} include sardines and some other fish, peanuts, butter and bran.

Nickel

The role of nickel in the human body is still not fully understood, but it has been proved to be essential for some animal species, and is involved in growth. The average human body contains about 1 mg. One source of nickel is tea, which contains 8 ppm of nickel in the dried leaves.

Copper

Some respiratory enzymes depend on the presence of copper. We need 1–2 mg per day, from foods such as liver, kidney, almonds, soya bean, wheat germ, margarine, mushrooms, and bran. The problem is that we may get too much, particularly through copper water pipes in a soft water area. Copper can displace iron and zinc from active sites in enzymes. There have been fatal accidents with copper(II) sulphate; 30 g has been known to kill, and following one fatal accident it has been banned from children's chemistry sets.

Copper is found in haemocyanin, which is the respiratory pigment in spiders, snails and octopuses; they are literally blue-blooded!

Zinc

Zinc is present in many enzymes in the body, and is involved in the proteins which mediate the synthesis of RNA from DNA. It is especially involved in enzymes which affect growth, maturity, and fertility.

Some areas of the world have low zinc levels in soil; many people there may be stunted in growth. The nature and causes of zinc deficiency in humans were first established by Dr Ananda Prasad working in Iran and Egypt between 1968 and 1972, studying men who had been rejected for army service because of inadequate development.

The daily intake for men should be about 7.5 mg and for women 5.5 mg. Liver is a good source, as are most cheeses and herring, sunflower seeds, brewer's yeast, maple syrup and – once again – bran.

Alcohol is dealt with by a zinc-containing enzyme in the liver. Cirrhosis of the liver is a result of excessive drinking; cirrhosed livers contain low zinc levels. Minor damage can be helped by zinc supplementation of the diet.

Zinc is non-toxic in reasonable amounts and 'zinc pills' can be bought from pharmacies and health shops. Studies by Professor Derek Bryce-Smith at Reading University and others have shown that some infections, and some conditions such as anorexia and post-natal depression, lead to lowering of zinc levels in the body and may respond to zinc supplements.

SELF-ASSESSMENT QUESTIONS 25.6

1 Most plants require a small amount of iron if they are to grow healthily. Lack of iron leads to yellowing of the leaves and reduced yields of crops. In alkaline soils the iron(II) present in the soil is precipitated.
   a What is the formula of the hydrated complex of $Fe^{2\oplus}$?
   b Show how this complex ion is deprotonated to form an insoluble compound in alkaline conditions.
   c Explain why the addition of EDTA in the form of the salt Na[Fe(EDTA)] overcomes the effect of the alkaline soil.

2 Carbon monoxide is a very dangerous gas since the molecule has lone pairs available for donation as a ligand.
   a What is meant by the term ligand?
   b Draw a dot-and-cross diagram to show the bonding in carbon monoxide.
   c Explain why CO might be dangerous as a poison in the presence of haemoglobin.

3 Summarise the involvement of d-block elements in biological systems. You should try to use a wide variety of resources in your research, including the Internet, CD-ROMs and encyclopaedias. Limit your summary to 500 words.

25.7 Aspects of the chemistry of some d-block elements

In this section we consider briefly some of the interesting chemistry of the first transition series of elements.

Vanadium

Vanadium is used in steel for hardening; it is especially good for high speed machine tools. Vanadium can have the oxidation states +2, +3, +4, +5. The two higher states are normally only reached in oxo-cations, i.e. when vanadium is combined with oxygen.

All four oxidation states can be observed in succession if small pieces of zinc metal are dropped into a solution of a vanadate(V) compounds acidified with sulphuric acid. (Hydrochloric acid should not be used as it can be oxidised.) Zinc acts as a reducing agent by providing electrons.

Species:	$VO_2^{\oplus}(aq)$			$[V(H_2O)_6]^{2\oplus}(aq)$
Oxidation state:	+5			2
Colour:	Yellow–orange			lilac

This is a delightfully colourful experiment, but it is easy to be misled into believing that there are five oxidation states, two of them appearing green. Can you see why?

Air will convert vanadium(II) to vanadium(III); any reasonable oxidising agent will convert vanadium(III) to vanadium +4 as in vanadate(IV); but because vanadium in the +5 oxidation state is itself a fairly powerful oxidising agent, only the most powerful oxidising agents (e.g. acidified potassium manganate(VII), $KMnO_4$) will oxidise vanadium +4 to vanadium +5, as in vanadate(V), $VO^{2\oplus}$.

Standard electrode potentials (see Ch 24.3) can be used to predict and explain these reactions, but it should be remembered that standard electrode potentials apply accurately only to *standard conditions* and that by changing the relative concentrations of reagents actual electrode potentials can be altered by up to about 0.4 V. The standard electrode potentials for the different vanadium species are shown in *Table 25.7.1*.

Electrode potentials for vanadium species

Oxidation state change	Electrode process	$E^{\ominus}/V$
$+3 \longrightarrow +2$	$V^{3+}(aq) + e^- \rightleftharpoons V^{2+}(aq)$	-0.26
$+4 \longrightarrow +3$	$VO^{2+}(aq) + 2H^+(aq) + e^- \rightleftharpoons V^{3+}(aq) + H_2O(l)$	$+0.34$
$+5 \longrightarrow +4$	$VO_2^+(aq) + 2H^+(aq) + e^- \rightleftharpoons VO^{2+}(aq) + H_2O(l)$	$+1.00$

Table 25.7.1

How can we reduce vanadium(V)?

With all these possible oxidation states for vanadium, how is it possible to stop at any particular oxidation state? We have already mentioned that vanadium(V) can be reduced through all the oxidation states down to vanadium(II) using zinc metal in hydrochloric acid. If we look at the standard electrode potential for zinc we can see why. For the reduction of zinc ions:

$$Zn^{2+}(aq) + 2e^{\ominus} \rightleftharpoons Zn(s) \qquad E^{\ominus} = -0.76\,V$$

The zinc system is able to reduce all the oxidation states down to the +2 state. How is this?

The lowest oxidation state of vanadium is +2.

The half equation for the reduction to $V^{2+}(aq)$ is:

$$V^{3+}(aq) + e^{\ominus} \rightleftharpoons V^{2+}(aq) \qquad E^{\ominus} = -0.26\,V$$

The half equation for the *oxidation* of zinc becoming:

$$Zn(s) \rightleftharpoons Zn^{2+}(aq) + 2e^{\ominus} \qquad E = +0.76\,V$$

Therefore, the overall value of E for *oxidation* of zinc and the *reduction* of $V^{3+}(aq)$ to $V^{2+}(aq)$ is +0.5 V, which means that the reaction has a high tendency to happen (see Ch 24.4):

$$
\begin{aligned}
2V^{3+}(aq) + 2e^{\ominus} &\rightleftharpoons 2V^{2+}(aq) & E^{\ominus} &= -0.26\,V \\
Zn(s) &\rightleftharpoons Zn^{2+}(aq) + 2e^{\ominus} & E &= +0.76\,V \\
\hline
2V^{3+}(aq) + Zn(s) &\rightleftharpoons 2V^{2+}(aq) + Zn^{2+}(aq) & E &= +0.50\,V
\end{aligned}
$$

However, the standard electrode potential for another reducing metal, tin, is:

$$Sn^{2+}(aq) + 2e^{\ominus} \rightleftharpoons Sn(s) \qquad E^{\ominus} = +0.14\,V$$

When we add this value to the redox table (*Table 25.7.2*) we can see, using the method we have just used for zinc, that tin will reduce vanadium(V) to vanadium(IV) and even down to vanadium(III). Under standard conditions, tin will not reduce the vanadium any further.

Some potentials for reducing vanadium compounds

Electrode process	$E^{\ominus}/V$
$Zn^{2+}(aq) + 2e^{\ominus} \rightleftharpoons Zn(s)$	-0.76
$V^{3+}(aq) + e^{\ominus} \rightleftharpoons V^{2+}(aq)$	-0.26
$Sn^{2+}(aq) + 2e^{\ominus} \rightleftharpoons Sn(s)$	$+0.14$
$VO^{2+}(aq) + 2H^+(aq) + e^{\ominus} \rightleftharpoons V^{3+}(aq) + H_2O(l)$	$+0.34$
$VO_2^+(aq) + 2H^+(aq) + e^{\ominus} \rightleftharpoons VO^{2+}(aq) + H_2O(l)$	$+1.00$

Table 25.7.2

To obtain vanadium(II), a more powerful reducing agent such as zinc is needed. This reinforces the point that with redox reactions it is the reducing power or oxidising power of the reactants that has to be considered in deciding if any reaction is possible.

Chromium

Chromium is used in making 'stainless' steel, in making hardened steels, and also as chromium plating on steel external body items and trim for cars.

The important oxidation states of chromium are +3 and +6; +2 is also quite easy to obtain. If zinc is added to an acid solution of dichromate(VI), all three oxidation states can be seen in turn (compare vanadium, above).

Species:	$Cr_2O_7^{2\ominus}(aq) \longrightarrow$	$Cr^{3\oplus}(aq) \longrightarrow$	$Cr^{2\oplus}(aq)$
Oxidation state:	+6	+3	+2
Colour:	Orange	Green	Blue

Chromium(II) is unstable in aqueous solution and a powerful reducing agent. The standard electrode potential for the conversion of $Cr^{3\oplus}$ to $Cr^{2\oplus}$ is -0.41 V. The aqueous $Cr^{3\oplus}$ ion is stable, but chromium(VI) as in chromate(VI), $CrO_4^{2\ominus}$, or dichromate(VI), $Cr_2O_7^{2\ominus}$, is a powerful oxidising agent ($E^{\ominus} = +1.33$ V). Aqueous sodium or potassium dichromate(VI) acidified with dilute sulphuric acid is much used as an oxidising agent in organic chemistry.

A test for sulphur dioxide depends on it being the only common gas which will turn the colour of a strip of filter paper soaked in acidified aqueous potassium dichromate(VI) from orange to green – and indeed, often to the blue colour associated with chromium(II).

In many laboratories bottles of yellow potassium chromate(VI) and orange potassium dichromate(VI) solutions stand next to one another, and this can cause some confusion. They are very easily interconverted: note that the oxidation state of the chromium does not change during the reaction. The $Cr_2O_7^{2\ominus}$ ion is, in effect, two $CrO_4^{2\ominus}$ tetrahedra joined together at one of the oxygen atoms (*Figure 25.7.1*).

Figure 25.7.1 The $Cr_2O_7^{2\ominus}$ ion

$$2CrO_4^{2\ominus}(aq) + 2H^{\oplus}(aq) \underset{alkali}{\overset{acid}{\rightleftharpoons}} Cr_2O_7^{2\ominus}(aq) + H_2O(l)$$

Yellow   Orange

These two ions can be considered to be in equilibrium in aqueous solution (see Ch 13.1). If alkali is added, $H^{\oplus}$ ions are removed from the left hand side of the equation; the position of equilibrium is therefore shifted to the left, so that the yellow $CrO_4^{2\ominus}$ ion is present in high concentration. If however, acid is added, the concentration of $H^{\oplus}(aq)$ ions on the left hand side is momentarily increased. The equilibrium is therefore shifted to the right, so that now the orange $Cr_2O_7^{2\ominus}$ ion predominates.

The standard electrode potentials for chromium species are shown in *Table 25.7.3*

Standard electrode potentials for chromium species	
Electrode process	$E^{\ominus}/$ V
$Cr^{3\oplus}(aq) + e^{\ominus} \rightleftharpoons Cr^{2\oplus}(aq)$	-0.41
$Cr_2O_7^{2\ominus}(aq) + 14H^{\oplus}(aq) + 6e^{\ominus} \rightleftharpoons 2Cr^{3\oplus}(aq) + 7H_2O(l)$	$+1.33$
Table 25.7.3	

Potassium dichromate(VI) crystals were used in the original 'breathalyser' for testing the breath of suspected drunk-drivers. Any alcohol in the driver's breath was oxidised by the orange crystals in the tube which led to an inflatable bag; the orange dichromate(VI) was reduced to green chromium(III). Once the bag was inflated – i.e. a certain volume of breath had been passed – the distance along the tube for which the crystals had turned green showed how much alcohol was in the breath, and hence how much was in the body.

Manganese

Manganese is also used in steel making, especially for tough, hard-wearing and elastic steels such as that used for railway tracks. Manganese can exist in all oxidation states from +1 to +7, but +1, +3 and +5 are not important. Manganese(II) compounds are ionic. The electron configuration of $Mn^{2\oplus}$ is $[Ar]3d^5$, which is stable (see Ch 25.3). The most common and most stable oxidation state of manganese is therefore +2. Manganese in the +4 state can act as an oxidising agent. For example, for the black compound manganese(IV) oxide, MnO_2, in acid, $E^\ominus = +1.23$ V.

$$MnO_2(s) + 4H^\oplus(aq) + 2e^\ominus \longrightarrow Mn^{2\oplus}(aq) + 2H_2O(l) \qquad E^\ominus = +1.23 \text{ V}$$

An old method for preparing chlorine gas was by heating concentrated hydrochloric acid with manganese(IV) oxide; the hydrogen chloride was oxidised to chlorine. Alternatively, concentrated sulphuric acid, sodium chloride and manganese(IV) oxide were heated together. The hydrogen chloride gas produced from the reaction between the chloride and the concentrated sulphuric acid was mostly oxidised by the manganese(IV) oxide.

A more convenient method, which works rapidly and well at room temperature, is to drip concentrated hydrochloric acid (*definitely not* concentrated sulphuric acid) from a tap funnel onto potassium manganate(VII) crystals.

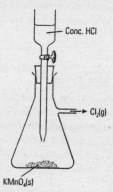

An immediate, vigorous reaction gives chlorine gas. The rate of production can be controlled through the rate of adding the concentrated hydrochloric acid.

The reaction works even though the $E^\ominus$ values are so close (*Figure 25.7.6*).

$$MnO_4^\ominus(aq) + 8H^\oplus(aq) + 5e^\ominus \rightleftharpoons Mn^{2\oplus}(aq) + 4H_2O \qquad E^\ominus = +1.52 \text{ V}$$
$$2Cl^\ominus(aq) \longrightarrow Cl_2(g) + 2e^\ominus \qquad E^\ominus = +1.36 \text{ V}$$

The reason, of course, is that *concentrated* hydrochloric acid can never be considered as 'standard conditions'.

Manganese in the +7 state is a very powerful oxidising agent. For an acid solution of manganate(VII), $MnO_4^\ominus$, $E^\ominus = +1.52$ V.

Even more powerful oxidising agents (e.g. lead(IV) oxide, PbO_2, or even sodium bismuthate(V), $NaBiO_3$, in nitric acid) are able to oxidise the almost colourless $Mn^{2\oplus}(aq)$ all the way to the intensely purple manganate(VII), $MnO_4^\ominus(aq)$.

Black manganese(IV) oxide can be converted to purple manganate(VII) in two stages. First, the oxide is mixed with potassium hydroxide solution and a source of oxygen, such as potassium chlorate(V), and heated strongly:

$$2MnO_2(s) + 4OH^\ominus(aq) + O_2(g) \xrightarrow{\text{heat}} 2MnO_4^{2\ominus}(aq) + 2H_2O(g)$$
$$\text{Black} \qquad\qquad\qquad\qquad\qquad\qquad \text{Green}$$

The green manganate(VII), $MnO_4^{2\ominus}$, which is formed can be made to *disproportionate* (see Ch 8.4) in acid to give a purple solution of manganate(VII), $MnO_4^\ominus$, and a black precipitate of manganese(IV) oxide, MnO_2.

$$3MnO_4^{2\ominus}(aq) + 4H^\oplus(aq) \longrightarrow 2MnO_4^\ominus(aq) + MnO_2(s) + 2H_2O(l)$$

Oxidation state:	+6	+7	+4
	Green	Purple	Black

669

Notice again that, as always in a disproportionation reaction (see Ch 8.4), the oxidation states of manganese add up to the same total on both sides of the equation. We start with three manganese atoms in the +6 state; one has been reduced to +4; to compensate for this, the other two are oxidised to +7.

Iron

Iron, used mostly as steel – i.e. iron alloyed with small amounts of carbon and/or d-block elements such as vanadium, chromium, manganese, cobalt and nickel – still provides the structural basis for an industrial civilisation. Without it, large buildings, bridges, and all transport of large volumes of passengers and freight by land and sea would not be possible.

The chemistry of the extraction of iron from iron(III) oxide in the blast furnace has already been discussed in Ch 7.4.

An old, unrusted, iron pillar in Delhi

Rusting

Most metals are energetically unstable with respect to the formation of an oxide, and so tend to *corrode* by oxidation. (An obvious exception is gold; its use in jewellery and ornament is directly linked with its lack of reactivity.) The corrosion of iron is known as *rusting*.

The rusting of iron and steel, and its prevention, is a major field of study. Rust prevention, and the financial consequences of failure to do so, costs Britain billions of pounds each year. We can only touch the topic briefly here. The electrochemistry of rusting is discussed fully in Ch 24.7.

- Rusting requires the presence of both oxygen and liquid water. Water vapour is not enough. Rusting is speeded up by the presence of most electrolytes – which is why car owners don't know whether to applaud or deplore the use of salt to keep roads free of ice in winter – but is slowed down by alkalis. Iron without any defects or impurities does not rust.
- The surface of an iron or steel object is never completely uniform, either because of impurities in the material or stress on the object. If water is present, a potential develops between *cathodic areas* (where electrons are gained) and *anodic areas* (where electrons are lost).

The characteristic reactions are shown in *Figure 25.7.2*.

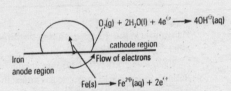

Figure 25.7.2 The rusting process

At cathodic areas:

$$O_2(g) + 2H_2O(l) + 4e^\ominus \longrightarrow 4OH^\ominus(aq)$$
$$2H^\oplus(aq) + 2e^\ominus \longrightarrow H_2(g)$$

At anodic areas:

$$Fe(s) \longrightarrow Fe^{2\oplus}(aq) + 2e^\ominus$$

Electrons flow *through the metal* from anodic areas to cathodic areas. If the water in contact with the metal contains an electrolyte such as sodium chloride – or even carbonic acid, from dissolved carbon dioxide – corrosion will be speeded up as a current flows more easily in the circuit. The $Fe^{2\oplus}$ ions from the anodic corrosion and the $OH^\ominus$ ions from cathodic areas form iron(II) hydroxide, which oxidises rapidly in air to give red–brown hydrated iron(III) oxide – i.e. rust.

The point here is that rust is a result of a *reaction in solution*. Furthermore, study of the equations above suggests that iron undergoes anodic corrosion faster when it is not in contact with oxygen but a nearby area is. If a drop of water is placed on a clean steel surface, rusting occurs most rapidly under the middle of the drop.

Obviously, iron which has been painted over is protected from both water and oxygen. But if the paint is scratched through, rusting occurs rapidly *not* at the scratch, but *under the paint*, because there is a large difference in the availability of oxygen for the freshly-exposed iron and the paint-covered iron.

- Corrosion can be prevented by covering the iron or steel to exclude air and water with a protective layer of oil, paint, plastic or a metal such as tin ('tinplate').
- An alternative to this *physical* exclusion of air and water is to supply electrons to the iron or steel object. This is why it is the *negative* terminal of a car battery which must be earthed to the chassis. (If the *positive* terminal is connected by mistake, corrosion occurs *more* rapidly. Can you see why?) If corrosion involves oxidation by loss of electrons, a supply of electrons will slow it down or stop it. Underground or undersea metal objects can be connected to the negative terminal of a low voltage electrical supply.
- Another, and often more effective, way of supplying electrons is to join the iron or steel to a metal more electropositive than iron, either in the form of a surface coating or by bolting a lump of the other metal firmly to it.

For example, galvanised iron is coated with zinc by dipping it in a bath of molten zinc. The zinc loses electrons more readily than iron when in contact with air; it *corrodes preferentially*, releasing electrons which go to protect the iron. Because the zinc acts as the anode in the cell, it is described as a *sacrificial anode*. The protection still occurs when the zinc layer is scratched through. The zinc continues to corrode and provide extra electrons to keep the iron at a small negative potential. If, however, the tin coating on tinplate is scratched through, the iron is now the more electropositive of the two metals, and corrodes rapidly, providing electrons to protect the tin!

Tin and zinc both provide physical protection against rusting; only zinc provides *electrochemical protection*. Damage to tinplate means that rusting of the underlying iron or steel is much faster than it would have been without the tin covering.

- Yet another, but more expensive, way of preventing the corrosion of steel is to alloy the iron with chromium (about 15%) to form so-called *stainless steel*. This forms a tough, coherent film of oxide on the surface, which prevents further oxidation. Stainless steel is used for cutlery and long-life car exhaust systems, for example.

The protective surface oxide film on stainless steel works in a way similar to that on aluminium. Unlike iron, aluminium is a comparatively reactive metal. If a piece of aluminium foil is dipped for a few seconds in an aqueous mercury(II) salt, shaken dry, and placed on a protective sheet, it will become very hot, steam and crumble into a grey powder of aluminium oxide. But normally, if the protective oxide layer is scratched off aluminium or stainless steel, it immediately re-forms and continues to protect.

Iron in catalysis

Iron is used as the catalyst in the Haber Process for the manufacture of ammonia (see Ch 13.3). This obviously involves interaction between the surface of the iron and gas molecules adsorbed on it; it is an example of *heterogeneous catalysis* (see Ch 12.4).

Because of its ability to change oxidation states, iron can also become involved in *homogeneous catalysis* by formation of intermediate compounds (se Ch 12.4). An example is the oxidation of aqueous iodide ions to iodine by peroxodisulphate ions, $S_2O_8^{2\ominus}$.

$$S_2O_8^{2\ominus}(aq) + 2I^{\ominus}(aq) \longrightarrow 2SO_4^{2\ominus}(aq) + I_2(aq)$$

The uncatalysed reaction is slow since two negative ions must collide to react. The rate of the reaction is increased greatly by addition of $Fe^{3\oplus}(aq)$. This can oxidise $I^{\ominus}$ ions to iodine (see *Table 25.7.4*); the $Fe^{2\oplus}$ formed is then oxidised back to $Fe^{3\oplus}$ by the $S_2O_8^{2\ominus}$ ions. The reaction can therefore proceed in two steps which, because their activation energy is less than that of the original reaction (see Ch 12.3), means that the overall reaction is speeded up.

Standard electrode potentials

Electrode process	$E^{\ominus}$ / V
$Fe^{3\oplus}(aq) + e^{\ominus} \rightleftharpoons Fe^{2\oplus}(aq)$	+0.77
$I_2(aq) + 2e^{\ominus} \rightleftharpoons 2I^{\ominus}(aq)$	+0.54
$S_2O_8^{2\ominus}(aq) + 2e^{\ominus} \rightleftharpoons 2SO_4^{2\ominus}(aq)$	+2.01

Table 25.7.4

$$2Fe^{3\oplus}(aq) + 2I^{\ominus}(aq) \longrightarrow 2Fe^{2\oplus}(aq) + I_2(aq)$$

$$2Fe^{2\oplus}(aq) + S_2O_8^{2\ominus}(aq) \longrightarrow 2Fe^{3\oplus}(aq) + 2SO_4^{2\ominus}(aq)$$

Overall: $\quad S_2O_8^{2\ominus}(aq) + 2I^{\ominus}(aq) \longrightarrow 2SO_4^{2\ominus}(aq) + I_2(aq)$

Redox chemistry of iron

Aqueous iron(III) is an oxidising agent. The value of $E^{\ominus}$ for the reduction of $Fe^{3\oplus}(aq)$ is considerably positive.

$$Fe^{3\oplus}(aq) + e^{\ominus} \rightleftharpoons Fe^{2\oplus}(aq) \qquad E^{\ominus} = +0.77\,V$$

Yellow $\qquad\qquad$ Pale green

Any reasonable reducing agent, e.g. sulphur dioxide gas, reduces the characteristic yellow solution of $Fe^{3\oplus}(aq)$, to the very pale green of $Fe^{2\oplus}(aq)$. A *strong* oxidising agent, such as hydrogen peroxide or chlorine gas, is needed to oxidise $Fe^{2\oplus}(aq)$ to $Fe^{3\oplus}(aq)$.

If steel wool is used as the anode in the electrolysis of cold concentrated aqueous sodium hydroxide, a purple colour develops. This is caused by the unstable ferrate(VI) ion, $FeO_4^{2\ominus}(aq)$, containing iron in the +6 state. It is a *very powerful* oxidising agent!

$$FeO_4^{2\ominus}(aq) + 8H^{\oplus}(aq) + 3e^{\ominus} \rightleftharpoons Fe^{3\oplus}(aq) + 4H_2O(l) \qquad E^{\ominus} = +2.20\,V$$

The value of $E^{\ominus}$ for the reduction of iron(III) is changed greatly by the presence of different ligands (see *Table 25.7.5*).

The ion represented as $Fe^{3\oplus}(aq)$ is in fact the complex $[Fe(H_2O)_6]^{3\oplus}(aq)$. Replacement of the water molecules as ligands by the stronger ligand cyanide ion stabilises the iron(III) state; the hexacyanoferrate(III) ion, $[Fe(CN)_6]^{3\ominus}$, is a much weaker oxidising agent than $Fe^{3\oplus}(aq)$ ($E^{\ominus} = +0.36\,V$ compared with $+0.77\,V$).

Standard electrode potentials for iron(III) with different ligands

Half-equation	$E^{\ominus}$ / V
$[Fe(OH)_3(H_2O)_3](s) + H_2O(l) + e^{\ominus} \rightleftharpoons [Fe(OH)_2(H_2O)_4](s) + OH^{\ominus}$	−0.56
$[Fe(CN)_6]^{3\ominus}(aq) + e^{\ominus} \rightleftharpoons [Fe(CN)_6]^{4\ominus}(aq)$	+0.36
$Fe^{3\oplus}(aq) + e^{\ominus} \rightleftharpoons Fe^{2\oplus}(aq)$	+0.77

Table 25.7.5

A high pH causes partial replacement of water molecules by $OH^{\ominus}$ ions in the $[Fe(H_2O)_6]^{3\oplus}$ ion until charges are neutralised and precipitation occurs. We can now see that, in alkaline solutions, iron(II) acts as a quite *strong reducing agent*. (If $E^{\ominus}$ for the iron(III) hydroxide is −0.56 V, then E for the conversion of $Fe(OH)_2(H_2O)_4$ to $Fe(OH)_3(H_2O)_3$ must be +0.56 V).

Tests for $Fe^{2\oplus}(aq)$ and $Fe^{3\oplus}(aq)$

Addition of aqueous alkali causes precipitation of the 'hydroxides' or hydrated oxides, as just discussed. The iron(II) compound is a dirty green colour, which rapidly oxidises to brown in air; the iron(III) compound is a muddy brown.

Aqueous iron(II), $Fe^{2\oplus}$(aq), reacts immediately to give a deep blue precipitate with aqueous potassium hexacyanoferrate(III), $K_3[Fe(CN)_6]$.

$$4Fe^{2\oplus}(aq) + 3[Fe(CN)_6]^{3\ominus}(aq) + aq \longrightarrow Fe_4[Fe(CN)_6]_3 \cdot xH_2O$$

Deep blue   (where x = 14–16)

A similar reaction occurs between aqueous iron(III), $Fe^{3\oplus}$(aq), and aqueous potassium hexacyanoferrate(II), $K_4[Fe(CN)_6]$, to give a blue precipitate which has been shown by X-ray diffraction analysis to be exactly the same as that from $Fe^{2\oplus}$(aq) and the hexacyanoferrate(III).

The blue precipitate from this reaction is a widely used artist's pigment, *Prussian blue*. It was discovered by accident in 1707, and by the 1720s was widely used by painters. It was not originally made by precipitation, but by methods involving the burning of dried blood, or hooves and horns! Those materials provided the nitrogen to make the cyanide complex. Prussian blue was much cheaper than the existing major blue pigment, ultramarine. It was also, in comparison with many of the mercury, lead and arsenic compounds then being used as pigments, relatively non-toxic. In early advertisements for Prussian blue, artists were told they could lick their brushes to a point and not die as a result.

Although it contains cyanide, the stability constant (see Ch 25.4) of the $[Fe(CN)_6]^{4\ominus}$ ion is so large that few free cyanide ions are available. In any laboratory in which cyanide ions are being used, bottles of aqueous iron(II) sulphate and dilute ammonia solution are available and clearly labelled. In the event of accidental cyanide poisoning, the contents are rapidly mixed and the resulting slurry poured into the victim. It reacts with cyanide to form the stable $[Fe(CN)_6]^{4\ominus}$ complex, and gives time to rush the victim to hospital. (First aiders should be aware that, in such circumstances, they have about 90 seconds to make and administer this treatment before it is too late. But there are well-documented accounts of its success.)

A very sensitive test for aqueous iron(III) is to add aqueous potassium thiocyanate, KSCN. This gives a blood-red solution.

$$[Fe(H_2O)_6]^{3\oplus}(aq) + SCN^{\ominus}(aq) \longrightarrow [Fe(SCN)(H_2O)_5]^{2\oplus}(aq) + H_2O(l)$$

Blood-red

Tests for iron ions	$Fe^{2\oplus}$(aq)	Fe^{2+}(aq)
NaOH(aq)	Dirty green precipitate	Brown precipitate
Hexacyanoferrate(III)	Deep blue precipitate	–
Hexacyanoferrate(II)	–	Deep blue precipitate
Thiocyanate	–	Blood-red colour

Cobalt

Cobalt is used for making temperature-resistant alloys for the blades of gas turbines and for making Alnico, a steel which contains iron alloyed with aluminium, nickel and cobalt. This is used in permanent magnets which are more than 20 times as powerful as ordinary magnets.

The common oxidation states of cobalt are +2 and +3. The $Co^{2\oplus}$(aq) ion is pink; $Co^{3\oplus}$(aq) is green. Aqueous cobalt(III) is a powerful oxidising agent; cobalt(II) is stable.

$$Co^{3\oplus}(aq) + e^{\ominus} \rightleftharpoons Co^{2\oplus}(aq) \qquad E^{\ominus} = +1.81 \text{ V}$$

Cobalt(II) chloride is used as a test for the *presence* of water (*not* as a test for pure water). Anhydrous $CoCl_2$ is blue; the hydrated salt, $CoCl_2 \cdot 6H_2O$ is red. Strips of filter paper are soaked in a solution of cobalt(II) chloride. When placed in an oven, they become blue. They stay blue if kept dry. When in contact with a liquid containing water, they turn pink.

The solution of cobalt(II) chloride turns from pink to blue in colour at about 50 °C. The blue colour can also be made by adding a high concentration of chloride ions, e.g. as concentrated hydrochloric acid. Either way the octahedral pink $[Co(H_2O)_6]^{2\oplus}$ complex turns into the tetrahedral blue $[CoCl_4]^{2\ominus}$ complex (*Figure 25.7.3*).

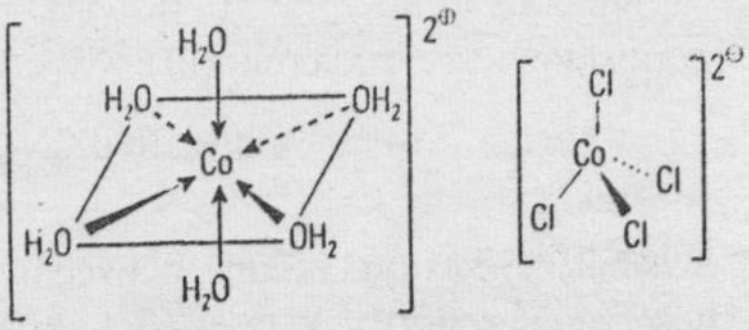

Figure 25.7.3   Cobalt complexes

The ability of cobalt(III) to act as an oxidising agent is, as with iron(III), much reduced when the ligands are more powerful than water (*Table 25.7.6*).

Standard electrode potentials for some cobalt complexes	
Half-equation	$E^\ominus$ / V
$[Co(CN)_6]^{3\oplus}(aq) + H_2O(l) + e^\ominus \rightleftharpoons [Co(CN)_5(H_2O)]^{3\ominus}(aq) + CN^\ominus$	-0.80
$[Co(NH_3)_6]^{3\oplus}(aq) + e^\ominus \rightleftharpoons [Co(NH_3)_6]^{2\oplus}(aq)$	$+0.11$
$[Co(H_2O)_6]^{3\oplus}(aq) + e^\ominus \rightleftharpoons [Co(H_2O)_6]^{2\oplus}(aq)$	$+1.81$

Table 25.7.6

Nickel

The production of nickel via the tetracarbonylnickel(0) complex has already been discussed in Ch 7.4.

Nickel is widely used in alloys, for example, stainless steel, Alnico (see cobalt, above) and cupronickel (25% Ni, 75% Cu) for coinage. Invar steel is 35% Ni – it has a very small coefficient of thermal expansion, and is therefore used in surveying, measuring, and timekeeping instruments. Monel metal (70% Ni, 30% Cu) is very resistant to corrosion and is therefore used in chemical plants. Nichrome (80% Ni, 20% Cu) resists oxidation and is used in electrical heating elements. (You may have also used Nichrome wire for flame tests for s-block elements – see Ch 1.4, Ch 6.3 and Ch 30.3.)

Nickel is also used as a catalyst for the hydrogenation of alkenes, for example, in the making of margarine from unsaturated vegetable oils (Ch 16.2). This is an example of heterogeneous catalysis.

Copper

Copper is also much used in alloys (see, for example, cupronickel and Monel metal, above). Bronze, in the ancient world, was copper with about 10% tin. Until quite recently British bronze coinage contained Cu 95%, Sn 4%, Zn 1%. Brass is copper with 20–30% zinc.

Because copper is relatively inert chemically, as well as being very malleable and ductile (i.e. it can be shaped easily and drawn into wire) and a good conductor of electricity, it is used in large amounts in housing for water pipes, hot water cylinders, gas pipes and electrical wiring.

- The common oxidation states of copper are +1 and +2. The solutions of $Cu^{2\oplus}(aq)$ are blue; this is caused by the $[Cu(H_2O)_6]^{2\oplus}$ complex. Copper(I) compounds are unstable in aqueous solution; they disproportionate to copper(II) and metallic copper. This can be explained in terms of the standard electrode potentials in *Table 25.7.7*.

Bronze was the first metallic material to be produced in quantity, because of the relative ease of reduction of copper and tin ores. A whole Age of humanity was named after bronze. Iron ore is much harder to reduce to metallic form. Eventually, the Iron Age succeeded the Bronze Age, once methods of extracting iron were developed.

Standard electrode potentials	
Half-equation	$E^\ominus$ / V
$Cu^{2\oplus}(aq) + e^\ominus \rightleftharpoons Cu^{\oplus}(aq)$	$+0.15$
$Cu^{2\oplus}(aq) + 2e^\ominus \rightleftharpoons Cu(s)$	$+0.34$
$Cu^{\oplus}(aq) + e^\ominus \rightleftharpoons Cu(s)$	$+0.52$

Table 25.7.7

From these values, it follows that $Cu^{\oplus}(aq)$ being reduced to Cu can oxidise $Cu^{\oplus}$ to $Cu^{2\oplus}$.

$$2Cu^{\oplus}(aq) \longrightarrow Cu(s) + Cu^{2\oplus}(aq) \qquad E^{\ominus} = +0.37\,V$$

If an attempt is made to dissolve, say, white Cu_2SO_4, in water, a blue solution and a brown solid result.

$Cu^{\oplus}(aq) + e^{\ominus} \rightleftharpoons Cu(s)$	$E^{\ominus} = +0.52\,V$
$Cu^{\oplus}(aq) \rightleftharpoons Cu^{2\oplus}(aq) + e^{-}$	$E = -0.15\,V$
$\therefore 2Cu^{\oplus}(aq) \rightleftharpoons Cu(s) + Cu^{2\oplus}(aq)$	$E = +0.37\,V$

- Copper(I) is stable in insoluble compounds, e.g. CuCl, CuI, CuCN. All these are white. Copper(I) oxide, Cu_2O, is red. Copper(I) can also be stable in solution if it forms a very stable complex; e.g. $[CuCl_2]^{\ominus}$ can be formed if CuCl is treated with concentrated hydrochloric acid, or if $CuCl_2$ and copper metal are heated together with the concentrated acid.
- Insoluble, white, copper(I) iodide (CuI) is formed if aqueous copper(II) compounds are mixed with aqueous iodide.

$$2Cu^{2\oplus}(aq) + 4I^{\ominus}(aq) \longrightarrow 2CuI(s) + I_2(aq)$$

A check on oxidation states in this reaction shows that in order for two copper($+2$) to be reduced to the two copper($+1$), two iodine(-1) must be oxidised to two iodine(0).
- In alkaline solution, $Cu^{2\oplus}(aq)$ precipitates as the 'hydroxide', $[Cu(OH)_2(H_2O)_4]$ (see Ch 25.4). However, if, for example, tartrate ions (or the anions of some other organic acids) are present, these complex preferentially with the $Cu^{2\oplus}(aq)$ so that it is not precipitated by alkali.

   Such an alkaline solution of complexed copper(II) is able to oxidise aldehydes and – as Benedict's or Fehling's reagent – is used as a test for aldehydes (see Ch 19.3 and Ch 30.4). When the reactants are warmed or gently heated together, a red precipitate of copper(I) oxide is formed.

$$2Cu^{2\oplus}(aq) + 2OH^{\ominus} + 2e^{\ominus} \longrightarrow Cu_2O(s) + H_2O(l)$$

Zinc

Zinc is used in 'galvanising' iron or steel (see the section on iron, above). It is also used in brass (see copper, above), and is the cathode in 'dry' batteries.

Because the 3d orbitals are full in the atom and in the $Zn^{2\oplus}$ ion, zinc can only form compounds in which it is in the $+2$ oxidation state.

Zinc reacts with acids as a moderately reactive metal, but also reacts with alkali. Hydrogen is given off in both instances.

$$Zn(s) + 2H^{\oplus}(aq) \longrightarrow Zn^{2\oplus}(aq) + H_2(g)$$
$$Zn(s) + 2OH^{\ominus}(aq) + 2H_2O(l) \longrightarrow [Zn(OH)_4]^{2\ominus}(aq) + H_2(g)$$

The oxide and hydroxide of zinc react with aqueous alkali to give a solution of $[Zn(OH)_4]^{2\ominus}$; they are **amphoteric** (see Ch 23.1 and Ch 25.4).

Zinc compounds are usually colourless (because of the full d-sub-level) but zinc oxide is *thermochromic*: its colour depends on its temperature. When heated in air it turns from white to yellow; the change is reversed on cooling. The change is caused by the reversible loss of some oxygen on heating.

One ore of zinc is *calamine*, white zinc carbonate, $ZnCO_3$. *Calamine lotion* is used for treating irritated or sunburnt skin. Zinc oxide is mildly antiseptic and is used in plasters for covering scratches and other minor wounds. The biological importance of zinc has already been discussed in Ch 25.6.

SELF-ASSESSMENT QUESTIONS 25.7

1 Although iron is used extensively in the construction industry, it is very susceptible to rusting.
   a Explain the principles involved in rusting, showing clearly how the redox potentials of the reactants allow the rusting process to occur.
   b Outline the principal ways of preventing rusting.

2 Neptunium (Np) can be prepared by the reaction of neptunium fluoride with lithium vapour. Neptunium has an oxidation state of $+3$ in the compound.

a What is the formula of neptunium fluoride?

b Write an equation for the reduction with lithium.

c Would you expect neptunium fluoride to be colourless, or not? Explain your reasoning.

d What other properties might you expect for neptunium, given that it is in the transition metals region of the Periodic Table?

3 Two complex ions have the following percentage compositions:

$[A]^{2\ominus}$: 29.36% Co; 70.64% Cl

$[B]^{2\oplus}$: 35.28% Co; 64.72% H_2O

a Calculate the empirical formulae of the two complex ions.

b Work out the oxidation state of cobalt in both the complexes.

c Describe the shape of both of the complexes.

d What is the coordination number of cobalt in each complex?

e When concentrated hydrochloric acid is added to an aqueous solution of the complex $[B]^{2\oplus}$, the colour changes from pink to blue. Write an equation for the change.

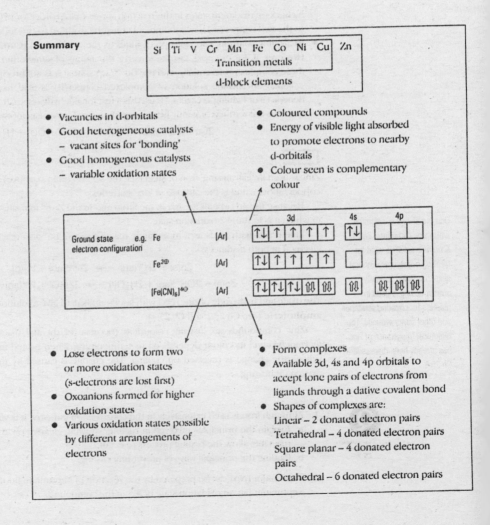

Summary

| Si | Ti | V | Cr | Mn | Fe | Co | Ni | Cu | Zn |

Transition metals

d-block elements

- Vacancies in d-orbitals
- Good heterogeneous catalysts
  – vacant sites for 'bonding'
- Good homogeneous catalysts
  – variable oxidation states

- Coloured compounds
- Energy of visible light absorbed to promote electrons to nearby d-orbitals
- Colour seen is complementary colour

- Lose electrons to form two or more oxidation states (s-electrons are lost first)
- Oxoanions formed for higher oxidation states
- Various oxidation states possible by different arrangements of electrons

- Form complexes
- Available 3d, 4s and 4p orbitals to accept lone pairs of electrons from ligands through a dative covalent bond
- Shapes of complexes are:
  Linear – 2 donated electron pairs
  Tetrahedral – 4 donated electron pairs
  Square planar – 4 donated electron pairs
  Octahedral – 6 donated electron pairs

SUMMARY 25

In this chapter the following new ideas have been covered. *Have you understood and learned them?* Could you explain them clearly to a friend? The page references will help.

EXAMINATION QUESTIONS 25

1 **a** What is a ligand?
   b What type of bonding occurs between a ligand and a transition metal ion?
   c Which of the following can act as a ligand?
   H_2O, $Br^{\ominus}$, $Al^{3\oplus}$, $AlCl_3$.
   Justify your answers.

2 Ammonium dichromate(VI) is an orange solid which on heating decomposes to form a green 'volcano'. The equation for the reaction is:

$$(NH_4)_2Cr_2O_7(s) \longrightarrow Cr_2O_3(s) + N_2(g) + 4H_2O(g)$$

   a In this redox reaction, what has been oxidised and what has been reduced?
   b What properties of a transition metal are evident in this reaction?

3 Transition metal complexes can show a variety of types of isomerism, comparable to the isomerism found in organic chemistry. (*Cis-trans* isomerism is characteristic of a double bond, where the groups can be on the same side or on opposite sides of the double bond.) One square planar complex of platinum is $[PtCl_2(NH_3)_2]$.
   a Draw the square planar structure of this complex.
   b Show that two isomers are possible, *cis* and *trans*, in such a planar structure.
   c The *cis* isomer is *cisplatin*, an anticancer drug. What is the oxidation state of platinum in this compound?
   d What feature of the NH_3 molecule makes it able to form a bond with the platinum ion?

4 Aqueous copper(II) sulphate is a blue colour. The stability constants for some copper complexes are:

$\log_{10}K_{stab}$	Formula	Colour
5.6	$[CuCl_4(H_2O)_2]^{2-}$	Yellow–green
13.1	$[Cu(NH_3)_4(H_2O)_2]^{2+}$	Deep blue
18.8	$[Cu(EDTA)]^{2-}$	Pale blue

a What ion is responsible for the colour in aqueous copper(II) sulphate?

b Explain how such colour arises in transition metal complexes.

c Describe what you would see if fairly concentrated hydrochloric acid is added to the aqueous copper(II) sulphate solution.

d What is the formula of the ion formed?

e If EDTA solution is now added to the solution from d, what would you expect to see?

f Finally, if aqueous ammonia is now added to the solution from e, what do you expect to happen? Justify your answer using the K_{stab} values given.

Do you remember ...?

- How **empirical** and **molecular formulae** are related. (Ch 14.1)
- That double bonds contain a **sigma** and **pi** bond. (Ch 16.2)
- Many of the reactions of alkenes are a consequence of the double bond breaking through **electrophilic addition**. (Ch 16.6)
- The reactions of **alcohols**. (Ch 18.4)
- How to use **curly arrows** to describe electron movements. (Ch 14.7)
- How to calculate the **energy of a reaction** and **bond enthalpy** terms. (Ch 11.4)

26.1   The benzene problem
26.2   The reactions of arenes
26.3   Sources of benzene and its uses
26.4   Isomerism around the benzene ring
26.5   Substitution reactions of benzene
26.6   Nitration and halogenation of the benzene ring
26.7   Phenols
26.8   The Friedel–Crafts reaction
26.9   Addition reactions of benzene
26.10 Alkyl derivatives of benzene
26.11 Summary of hydrocarbon reactions

26
hydrocarbons
– the arenes

The only hydrocarbon framework left to consider is that of the arenes. The simplest and typical arene is benzene. Its empirical formula is CH, and the relative molecular mass of 78 means that the molecular formula must be C_6H_6 (*Figure 26.1.1*).

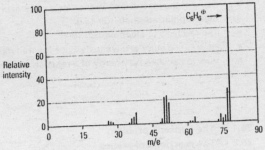

Figure 26.1.1   Mass spectrum of benzene

In the early days of organic chemistry a large number of highly unsaturated compounds were obtained from materials such as balsams, resins and 'aromatic oils'. Most had a pleasant smell or aroma. These compounds were therefore classified as aromatic. It was only later that their structures were found to be based on benzene.

Any model we try to make with such a formula must show considerable unsaturation (Ch 16.1). Benzene was a puzzle for a long time – unsaturated, yet showing few of the addition reactions which are typical of unsaturation. The term **aromatic** is often used to describe compounds containing a benzene-type structure.

26.1 The benzene problem

The nature of the problem can be appreciated if we approach it in the way that the chemists of the early 19th century did – by combustion analysis and physical measurement.

A clear, colourless hydrocarbon isolated from the distillation products of crude oil contained 92.3% carbon and 7.7% hydrogen. When 0.250 g of this liquid was vaporised at 100 °C, the vapour had a volume of 98 cm^3. The boiling point of 80 °C and melting point of 5 °C indicated that the liquid had a simple covalent structure with weak intermolecular attractions, although the early chemists would not have thought in those terms. The empirical formula and molecular formula can be calculated (as we did in Ch 14.1) in *Table 26.1.1*.

Calculating the empirical formula of an organic compound

		C	H
Element:		C	H
Data:	%	92.3	7.7
	Relative atomic mass (A_r)	12.0	1.0
Moles:	$= \dfrac{\% \text{ or mass}}{A_r}$	$\dfrac{92.3}{12.0}$	$\dfrac{7.7}{1.0}$
		$= 7.69$	$= 7.7$
Ratio:	Ratio of moles	$\dfrac{7.69}{7.69}$	$\dfrac{7.7}{7.69}$
	(Divide by the smallest value)		
	∴ Ratio of atoms is	1 :	1
		C	H

∴ *Empirical formula* is CH ($M_r = 13$)

Calculating the molecular formula of an organic compound

The volume of one mole of gas is 24 dm^3 at room temperature, 20 °C (293 K).

At 100 °C (373 K) this volume would be $24 \times \dfrac{373}{293} = 30.6$ dm^3

(gases expand on heating, see Ch 9.6)

98 cm^3 of the sample has a mass of 0.25 g at 100 °C

1 mole of the sample (30.6 dm$^3 = 30600$ cm^3) would have a mass of $0.25 \times \dfrac{30600}{98} = 78$ g

∴ The relative molecular mass of the sample is 78
∴ The molecular formula must contain $6 \times$ CH units (78/13 = 6)
∴ The *molecular formula* is C_6H_6

Table 26.1.1

A number of isomers are possible for the molecular formula C_6H_6. Two of these are:

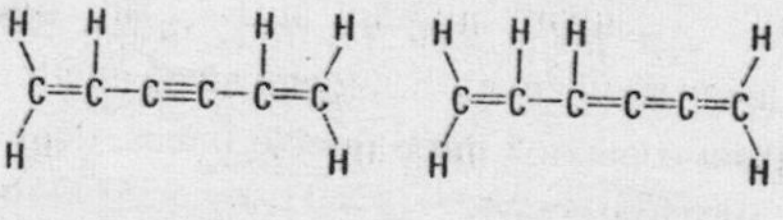

$$\underset{H}{\overset{H}{C}}=\underset{H}{\overset{H}{C}}-C\equiv C-\underset{H}{\overset{H}{C}}=\underset{H}{\overset{H}{C}} \qquad \underset{H}{\overset{H}{C}}=\underset{H}{\overset{H}{C}}-\underset{H}{\overset{H}{C}}=C=C=\underset{H}{\overset{H}{C}}$$

With so many double bonds present, these compounds should react readily with HBr. From your knowledge of alkene chemistry (Ch 16.5) a large number of monobromo products are possible from each of these isomers;

e.g: $CH_3CHBr-C\equiv C-CH=CH_2$ and $CH_2=CH-CHBr-CH=C=CH_2$

When benzene reacts with bromine there is only one monobromo- derivative, and that is not easy to make, unlike the reaction of Br_2 with the alkenes. And benzene has no reaction at all with HBr.

So far, the chemical reactions of benzene are inconsistent with those of a highly unsaturated molecule. Closer study shows that the typical reaction of benzene is substitution, not addition as for alkenes.

If we consider the problem from a different angle, the energy released when an alkene reacts with hydrogen has already been measured (Ch 16.5).

$$CH_2=CH_2(g) + H_2(g) \longrightarrow CH_3CH_3(g) \qquad \Delta H = -208\ kJ\ mol^{-1}$$

With any of the structures already proposed for the C_6H_6 molecular formula, with as many as four carbon–carbon double bonds, the enthalpy change for the hydrogenation reaction should be up to about $-800\ kJ\ mol^{-1}$.

But the measured value for the hydrogenation of benzene is only $-207\ kJ\ mol^{-1}$. The arrangement of six carbon atoms and six hydrogen atoms in benzene is much more stable than any of the formulae we have come across so far.

Kekulé proposed, in 1865, a structure for the benzene molecule with six carbon atoms arranged in a hexagonal ring.

His structure, with alternating double and single bonds in a ring of six carbon atoms, only started to solve the difficulties with benzene. This structure of six carbon atoms in a σ bonded ring, with three alternating π bonds, is still apparently very unsaturated.

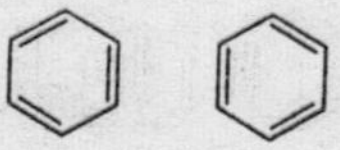

Although benzene had been isolated by Michael Faraday as early as 1825, no one had considered a structural explanation for the lack of reactions of benzene. Kekulé, forty years later, is supposed to have seen snakes in his open fire biting their own tails. This inspiration became the carbon atoms of benzene arranged in a ring.

Why should such a structure be so unreactive? After all, it still has three double bonds, which might be expected to react with bromine in the dark at room temperature – something that does not happen with benzene.

When techniques were developed to measure the bond lengths in molecules, the bond length of all the bonds in the benzene ring turned out to be identical (*Figure 26.1.2*).

Bond lengths	/ nm
C—C cyclohexane	0.154
C=C cyclohexene	0.134
C⁓C benzene	0.140

Figure 26.1.2

Furthermore, the distances between the carbon atoms in the ring were intermediate between the bond lengths of carbon–carbon single bonds and carbon–carbon double bonds. This did not fit in with Kekulé's idea of alternating double and single bonds around the ring.

Another look at the hydrogenation enthalpies of related compounds also shows that there is more to understanding the structure of benzene.

The hydrogenation of cyclohexene is well recorded.

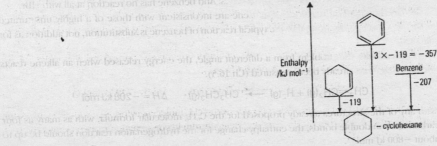

$$\bigcirc + H_2 \longrightarrow \bigcirc \qquad \Delta H = -119 \text{ kJ mol}^{-1}$$

If it contained three double bonds, the comparable reaction of benzene with hydrogen should therefore liberate $-(3 \times 119) = -357$ kJ mol^{-1}.

But:

$$\bigcirc \longrightarrow \bigcirc \qquad \Delta H = -207 \text{ kJ mol}^{-1}$$

benzene   cyclohexane

So benzene is about $(357 - 207) = 150$ kJ mol^{-1} more stable than it would be if it contained three ordinary C=C double bonds (*Figure 26.1.3*).

Enthalpy /kJ mol^{-1}

$3 \times -119 = -357$
Benzene
-207

-119

– cyclohexane

Figure 26.1.3 Hydrogenation enthalpy diagram

Delocalisation

Instead of alternating double and single bonds (*Figure 26.1.4*), the evidence therefore suggests that the carbon–carbon bonds in benzene are something between double and single bonds. Each carbon atom is joined by a strong σ bond, but whereas in an ordinary alkene there is overlap of p orbitals to form a π bond (see Ch 16.2), in benzene the π-bonded system is **delocalised** (*Figure 26.1.5*).

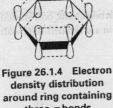

Figure 26.1.4 Electron density distribution around ring containing three π bonds

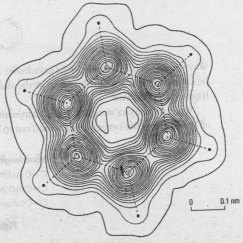

0   0.1 nm

Figure 26.1.5 Electron density map of benzene at −3 °C to show equal bond lengths in benzene

The orbital overlap of the π orbitals is around the whole of the ring (*Figure 26.1.6*), so that the electrons in these orbitals are not fixed on any one atom, but can move around the whole ring. This

Definition: Delocalised electron systems in organic molecules involve π bonds with all the carbon (or other) atoms in the system in a single plane and in which overlap of p orbitals extends over a significant number of carbon (or other) atoms.

Figure 26.1.6   Delocalisation model

delocalisation gives the molecules a much more stable structure. As seen above, the benzene molecule has an extra stability, the **delocalisation enthalpy**, of about 150 kJ mol^{-1}.

(Note that all the other atoms which are directly attached to the carbon atoms of the delocalised system must also be in the same plane.)

Delocalised ring systems in organic chemistry are very important. The extra stability provided by delocalisation is equivalent to the energy normally associated with a single covalent bond (*Figure 26.1.3*). You can recognise a delocalised system by looking for the alternate single and double bonds in a structure drawn using conventional bonding ideas.

There are a number of different ring structures which contain delocalised electrons. These ring structures are all *planar*, and the bond lengths between all the carbon atoms are the same as in benzene (*Figure 26.1.5*). The flat structure means that a higher degree of intermolecular attraction is possible, with the planar molecules able to pack closely together.

Also, the delocalised rings lead to greater dipole-induced dipole interactions and therefore stronger attractions between molecules (see Ch 2.8). This is reflected in the relatively high melting point of 6 °C for benzene. Although benzene is the most important example of a delocalised system, straight chain compounds can also contain delocalised bonds.

For example: hexa-1,3,5-triene, $CH_2{=}CH{-}CH{=}CH{-}CH{=}CH_2$, contains a delocalised π bond system.

Such alternate double and single bonds are often called **conjugated**.

The evidence for delocalisation in benzene

- Carbon–carbon bond lengths are equal in the delocalised system.
- Hydrogenation energies are lower than expected (see *Figure 26.1.3*).
- Although delocalised systems are highly unsaturated, their reactions (particularly those of the arenes) are often substitution rather than addition.

The cyclic structure of benzene, molecular formula C_6H_6, might be written as:

To indicate the delocalised structure of the ring and the ability of the π electrons to move round it, the double bonds are not shown individually. Instead, the three pairs of electrons involved in the π bond, which are the electrons capable of moving round the ring, are shown as a circle. The structure of benzene is therefore written as a hexagon with a ring in it:

Never draw benzene as a simple hexagon, without anything inside it. Such a diagram represents a molecule of cyclohexane, C_6H_{12} (*Figure 26.1.7*), which has no delocalised electrons and is far from flat.

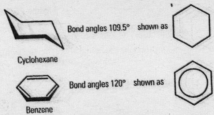

Bond angles 109.5°   shown as

Cyclohexane

Bond angles 120°   shown as

Benzene

Figure 26.1.7

Activity
Use a molecular model kit to make the structures of benzene, cyclohexene and cyclohexane.
Note the planar structure of benzene, contrasted with the puckered rings of cyclohexene and cyclohexane.

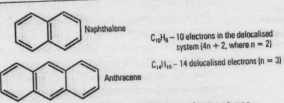

Naphthalene

$C_{10}H_8$ – 10 electrons in the delocalised system ($4n + 2$, where $n = 2$)

$C_{14}H_{10}$ – 14 delocalised electrons ($n = 3$)

Anthracene

Figure 26.1.8   Other simple aromatic structures

Where two or more benzene units are joined together, rings can be used in the diagram. However, the bonds are often drawn as shown in this box, because the electrons in the bonds where the benzene units join are largely localised.

- A molecule will be aromatic if there is a π electron system containing $(4n+2)\pi$ electrons (*Figure 26.1.8*).
- When the benzene ring is attached, as a group, to an aliphatic skeleton, it is called the *phenyl* group. The formula of the phenyl group is C_6H_5.
- Any compound in which the C:H ratio is close to one, and which contains six or more carbon atoms, is likely to contain a benzene ring.

SELF-ASSESSMENT
QUESTIONS
26.1

1 Draw displayed formulae for:
    a methylbutane
    b benzene
    c *trans*-pent-2-ene
    d cyclopentane

2 Explain why the carbon–carbon bond lengths in cyclohexane are 0.154 nm, whereas the carbon–carbon bond lengths in benzene are 0.140 nm.

3 A sample of an aromatic hydrocarbon, naphthacene, contains 0.432 g carbon and 0.024 g hydrogen.

   a Calculate the empirical formula of naphthacene.

   b If the relative molecular mass is 228, calculate the molecular formula of the compound.

   c The structure of the compound is composed entirely of six-membered carbon rings. Draw a possible structure of naphthacene, clearly showing any double bonds in the structure.

   d How many double bonds are present in this structure?

   e What do you notice about the double and single bonds in naphthacene? What term is used to describe such a structure?

26.2 The reactions of arenes

The stability of the delocalised ring makes the arenes relatively unreactive. Any reactions usually involve quite vigorous conditions including ultra-violet light or catalysts.

Like all volatile hydrocarbons, arenes burn very easily. However, the presence of so many double bonds – and therefore a high C:H ratio – means that the flame is extremely smoky. Many of the dangers from burning materials such as polystyrene, poly(phenylethene), come from the smoke.

When benzene burns in excess oxygen, the equation for the combustion of benzene is:

$$2C_6H_6(l) + 15O_2(g) \xrightarrow{ignite} 12CO_2(g) + 6H_2O(l)$$

With far less oxygen, one possible equation is:

$$2C_6H_6(l) + 3O_2(g) \xrightarrow{ignite} 12C(s) + 6H_2O(l)$$

With the benzene ring stabilised by about -150 kJ mol^{-1} addition reactions, although not impossible, are difficult to achieve (because of the *stabilisation* or *delocalisation energy* – *Figure 26.1.3*). *The principal reactions of benzene are those that end up with the delocalised ring intact* – and that means that the main reactions of benzene involve **substitution**. With such a large number of π electrons above and below the plane of the benzene molecule, the substitution reaction has to proceed through an electrophilic mechanism (see Ch 14.7).

Reactions of benzene

① Electrophilic attack by powerful electrophiles, such as $NO_2^{\oplus}$ (see Ch 26.6), results in substitution of the hydrogen atoms on the benzene ring.

② In the presence of UV light, or with transition metal catalysts at high temperatures, it is possible to break the delocalisation in the benzene ring (equivalent to three double bonds), giving an addition product.

ACTIVITY

Activity

Any spreadsheet, such as EXCEL, can be used to set up a simple table to calculate the atomisation energy of any molecule. If the atomisation energies of both reactants and products are considered, then the theoretical enthalpy change for the reaction, based on bond enthalpy terms, can be calculated. Try it like this, inserting the number of each type of bond broken and formed in columns C and E respectively. Once set up, the spreadsheet can be used for any such calculation.

	A	B	C	D	E	F
1						
2	Bond	Enthalpy	Number broken		Number formed	
3	C—C	347		=B3*C3		=B3*E3
4	C=C	612		=B4*C4		=B4*E4
5	C—H	413		=B5*C5		=B5*E5
6	C—O	358		=B6*C6		=B6*E6
7	C≡C	838		=B7*C7		=B7*E7
8	C—Cl	346		=B8*C8		=B8*E8
9	O—H	464		=B9*C9		=B9*E9
10	H—H	436		=B10*C10		=B10*E10
11	H—Cl	432		=B11*C11		=B11*E11
12	Cl—Cl	243		=B12*C12		=B12*E12
13	O=O	498		=B13*C13		=B13*E13
14	Atomisation of:		Reactants	=SUM(D3:D13)	Products	=SUM(F3:F13)
15	Enthalpy of reaction (from bond enthalpies) is					=D14−F14

Use your spreadsheet to calculate the theoretical enthalpy of combustion of benzene.

$$C_6H_6(l) + 7\tfrac{1}{2}O_2(g) \longrightarrow 6CO_2(g) + 3H_2O(l)$$

Draw out a full displayed formula, so that all the bonds broken and formed are counted, and insert the figures into the spreadsheet.

How does your result compare to the quoted value of -3267 kJ mol^{-1}?

What is the reason for this discrepancy?

SELF-ASSESSMENT QUESTIONS 26.2

1 Identify the following molecules as alkane, alkene, arene or cycloalkane.
   a $CH_3CH_2CH_2CH_2CH_3$
   b $C_6H_5CH_3$
   c $CH_3CH=CHCH_2CH_3$
   d $CH_3CH(CH_3)CH_2CH_3$
   e

2 Benzene is classified as an aromatic compound because of the *delocalised system of electrons* around the benzene ring.
   a Explain what is meant by the phrase *'delocalised system of electrons'*.
   b Why is the knowledge of bond lengths used in the discussion of the structure of benzene?
   c What thermochemical data supports the idea of a stable ring structure for benzene, with delocalised electrons?

26.3 Sources of benzene and its uses

Benzene is one of the many hydrocarbons obtained when crude oil is fractionally distilled. This provides a ready source of a chemical that is used extensively as a raw material for a wide range of products – but it is toxic! (*Figure 26.3.1*).

Aromatic hydrocarbons have been connected to the development of some types of cancer. Benzene is particularly dangerous.

However, the demand for benzene as a feedstock chemical (around 10^6 tonnes/year in the UK) is so large that it also has to be made by reforming straight chain hydrocarbons. Hexane, from the naphtha fraction in crude oil distillation, can be converted into benzene using oxide catalysts at high temperatures. This is part of the **catalytic reforming** of the C_6 to C_8 fractions we discussed in the chemistry of the alkanes (Ch 15.4).

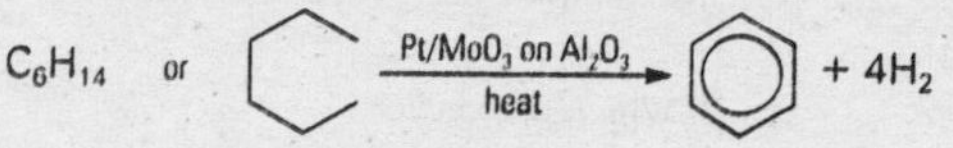

$$C_6H_{14} \quad \text{or} \quad \text{[cyclohexane]} \xrightarrow[\text{heat}]{\text{Pt/MoO}_3 \text{ on Al}_2\text{O}_3} \text{[benzene]} + 4H_2$$

The benzene ring is an important part of a wide range of products (*Figure 26.3.1*). The aromatic ring, reacted with concentrated sulphuric acid, and attached to a long chain alkane, forms the non-soapy detergents in washing-up liquid. The aromatic ring is present in polystyrene. Benzene compounds are added to petrol to raise its octane number.

Other common materials containing the benzene ring or made from it

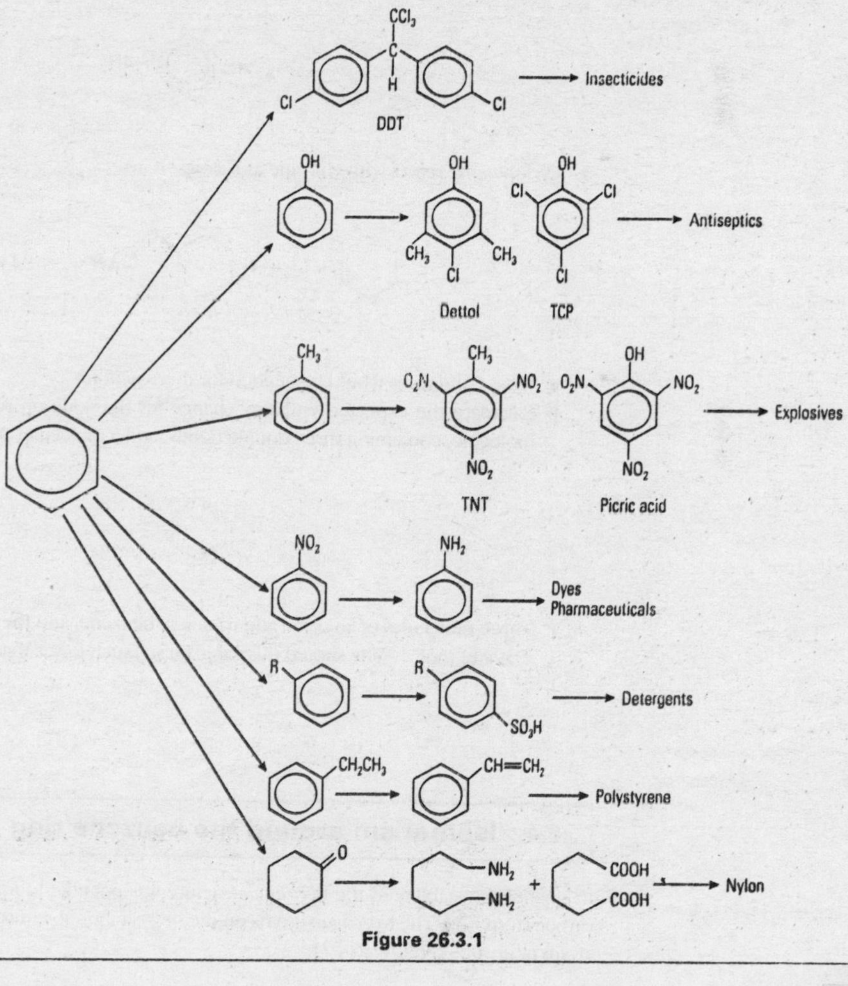

- DDT → Insecticides
- Dettol, TCP → Antiseptics
- TNT, Picric acid → Explosives
- → Dyes, Pharmaceuticals
- → Detergents
- → Polystyrene
- → Nylon

Figure 26.3.1

SELF-ASSESSMENT QUESTIONS 26.3

1 Consider the following molecules:

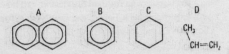

a Which are cyclic?
b Which are aromatic?
c Which are planar?
d Which might undergo electrophilic substitution?
e Which will have all the C—C bond lengths equal?
f Which will react with bromine in the dark by an electrophilic mechanism?

2 Using bond enthalpy values (values are in the activity in Ch 26.2), calculate the enthalpy change for the reaction:

$$C_6H_{14} \longrightarrow \bigcirc + 4H_2$$

3 Cyclohexene reacts with chlorine as follows:

$$\bigcirc + Cl_2 \longrightarrow \begin{array}{c} Cl \\ Cl \end{array} \qquad \Delta H = -184 \text{ kJ mol}^{-1}$$

a What conditions would be needed for this reaction?
b Calculate the expected enthalpy change for the addition reaction between the following molecule, containing three double bonds, and 3 molecules of chlorine.

c Three molecules of chlorine add to one of benzene, and the enthalpy of reaction is actually −399 kJ mol⁻¹. Why should this value be so much lower than that calculated in b?

26.4 Isomerism around the benzene ring

In spite of the stability of the benzene structure it is possible to introduce other groups on to the carbon atom ring. The hydrogen atoms on the ring can be substituted by a large number of different atoms or groups (*Figure 26.4.1*).

Naming substituted benzene rings

The name of the substituent is usually added in front of 'benzene':

_____benzene. One or two trivial names (such as phenol) are still commonly used, and a few functional groups are placed after 'benzene'

Before	After
bromo- benzene	
chloro-	-sulphonic acid
nitro-	-carboxylic acid
methyl-	-carboxylate

When a long alkyl chain with other substituents is present, think of the benzene as substituted on to the chain, using 'phenyl' and a number to position it on the chain.

2-Chloro-3-phenylbutane is

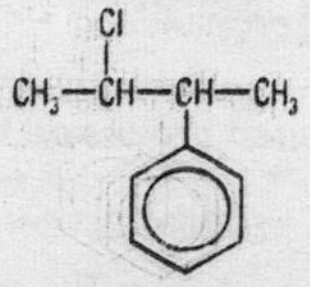

Figure 26.4.1

Since the benzene ring is planar and symmetrical, there is only one possible structure for a monosubstituted benzene ring. Introduction of one chlorine atom forms C_6H_5Cl – chlorobenzene.

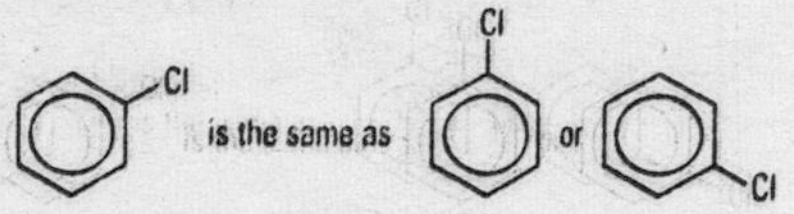

These structures are all equivalent and identical.

You should appreciate that, in the case of the benzene ring, what you draw on paper in two dimensions is a close approximation for the real structure, since the benzene ring is planar. (Beware – this is *not* true when you draw out the structures of alkyl chains – see Ch 15.2.)

It is worth recalling some of the conventions in writing formulae. For chlorobenzene, C_6H_5Cl would be written in text, but the displayed formula (often required in examination answers) should be written as:

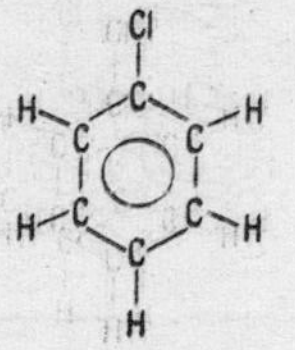

with all the carbon and hydrogen atoms and all the bonds shown. However, in other circumstances the formula is often written as:

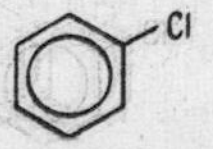

But do remember that the six carbon atoms and the five hydrogen atoms not shown are still there in the molecule.

As for naming these compounds, the term *aryl group* is sometimes used for any arene, whereas *phenyl* is specific for the C_6H_5 group, derived from benzene.

Methylbenzene is $C_6H_5CH_3$, and phenylethene has the formula $C_6H_5—CH=CH_2$.

689

A number of substituted benzene compounds are shown in *Table 26.4.1*. The two situations for naming most benzene-containing compounds can be summarised as:

- -benzene, when substituted or the alkyl side chain is CH_3.
- Phenyl-, when the aliphatic chain has *two* carbon atoms or more and the chemical properties of the side chain are more dominant in the molecule.

Chlorobenzene Benzenecarboxylic acid (benzoic acid) 1-Phenylpropane

Remember to divide the names of these compounds up mentally into the functional groups when reading the names. Nitrobenzene contains nitro | benzene and 1-phenylpropane is in fact 1-phenyl | propane.

Substituted groups on the benzene ring

Formula	Systematic name	Common name	Examples of use
OH	Hydroxybenzene (rarely used)	Phenol	Phenolic resins Antiseptics
CH₃	Methylbenzene	Toluene	TNT
NO₂	Nitrobenzene	Nitrobenzene	
COOH	Benzenecarboxylic acid	Benzoic acid	Food additive
CHO	Benzenecarbaldehyde (rarely used)	Benzaldehyde	Almond flavouring
Cl	Chlorobenzene	Chlorobenzene	
NH₂	Phenylamine	Aniline	Aniline dyes Pharmaceuticals

Table 26.4.1

Disubstituted benzene compounds

Once a second atom or group is introduced into the benzene ring, position isomerism becomes possible (Ch 14.4). To identify the particular isomer, and to provide an unambiguous name, the carbon atoms in the ring are numbered from 1 to 6. You must always start the numbering at one of the substituents on the ring.

The sequence around the ring is arranged to keep the numbers as low as possible. Examples are:

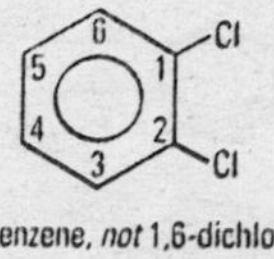

1,2-Dichlorobenzene, *not* 1,6-dichlorobenzene

The chlorine atoms can be placed on different carbon atoms around the ring and produce position isomers.

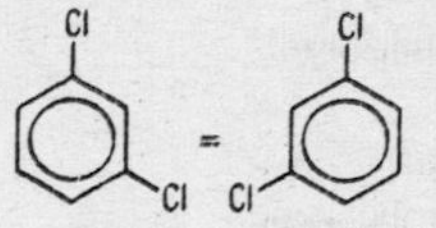

1,3-Dichlorobenzene 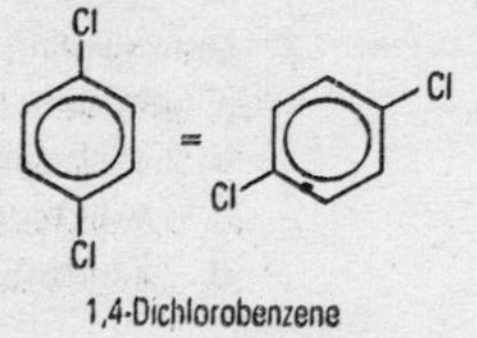   1,4-Dichlorobenzene

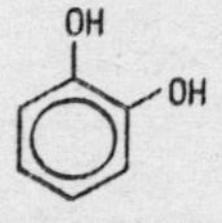

1,2-Dihydroxybenzene 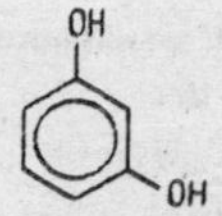   1,3-Dihydroxybenzene 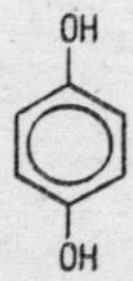   1,4-Dihydroxybenzene

Remember that, if there are two substituents on the benzene ring, there are only four hydrogen atoms attached to the ring.

It is possible to replace more of the hydrogen atoms, although the variety of isomers so formed is not needed at this level (*Figure 26.4.2*).

Some of the **common** compounds in *Table 26.4.1* with further substitution include:

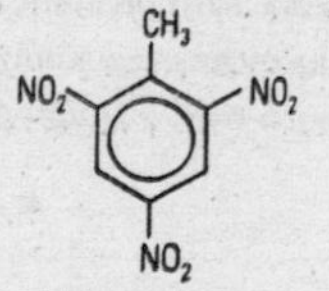

TNT   2,4,6-Trinitromethylbenzene   Explosive
('trinitrotoluene')

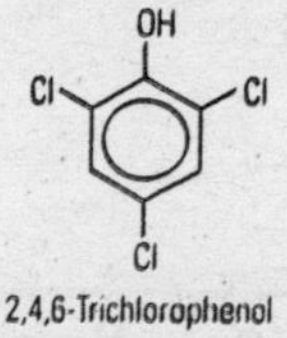

TCP   2,4,6-Trichlorophenol   Antiseptic

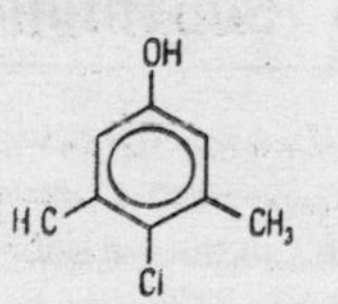

Dettol   4-Chloro-3,5-dimethylphenol   Antiseptic

Figure 26.4.2

SELF-ASSESSMENT QUESTIONS 26.4

1 Name the following structures:

a Br—⟨ring⟩—CH₃

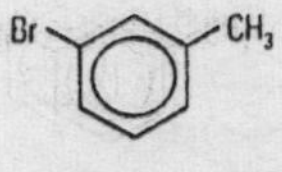

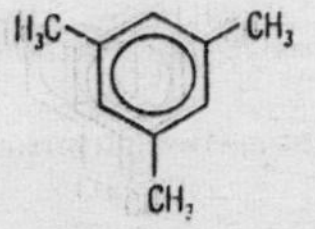

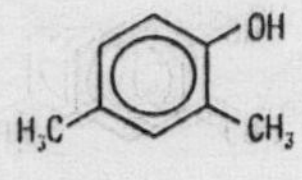

b H₃C—⟨ring⟩—CH₃ , CH₃

c ⟨ring⟩—OH , H₃C , CH₃

d ⟨ring⟩—CH=CHCH₃

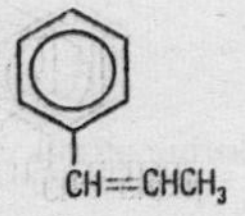

2 Draw displayed formulae for:
  a ethylbenzene
  b 2,4-dibromophenol
  c 3,5-dichloromethylbenzene
  d 2,4,6-trimethylbenzoic acid

3 How many position isomers are possible for the following molecules? Draw their structures and name them systematically.
  a C_6H_5—Br
  b CH_3—C_6H_4—Cl
  c $(CH_3)_2C_6H_4$

4 The following compounds all react with a mixture of concentrated HNO_3/H_2SO_4, producing eventually a tri-substituted product.

  $C_6H_5OH, C_6H_5CH_3, C_6H_6$.

  a Draw a displayed formula for C_6H_5OH.
  b Draw displayed formulae for the tri-substituted product for each of the compounds, and give the systematic names of each of the products.
  c Write an equation for the reaction with $C_6H_5CH_3$.

5 State which of the following compounds, A, B, C and/or D, would:
  a react with bromine in the dark;
  b decolourise acidified potassium manganate(VII);
  c burn with a smoky flame.

A ⟨ring⟩—NO₂

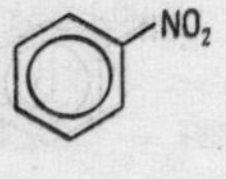

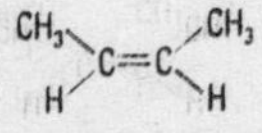

B CH₃—C=C—CH₃ , H , H

C ⟨ring⟩

D ⟨ring⟩

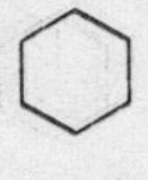

26.5 Substitution reactions of benzene

As we have seen, the delocalisation of the π electrons around the benzene ring gives extra stability to the compound. Consequently, it is very difficult to achieve addition reactions with benzene, where this additional stability would be lost.

As shown in *Figure 26.1.4*, there are rings of electron density above and below the plane of the benzene ring. Electrophiles can attack this region of high electron density, leading to the substitution of one of the hydrogen atoms attached to the carbon atoms (*Table 26.5.1*).

The benzene ring is so unreactive that the electrophile has to be quite powerful; otherwise a catalyst that helps the formation of an electrophilic intermediate is necessary. The mechanisms for some of these reactions will be discussed in the next section. Here, we shall merely list the more important reactions. Remember, in each of these substitution reactions *a hydrogen atom is lost* from the ring, although it is not shown on the usual hexagon symbol for benzene.

With halogens

Benzene does not react with halogens unless there is a catalyst present. A variety of catalysts can be used; all are essentially halogen carriers, i.e. they activate the halogen molecule, producing a high degree of polarity in the halogen–halogen bond. The most common catalyst used is aluminium chloride.

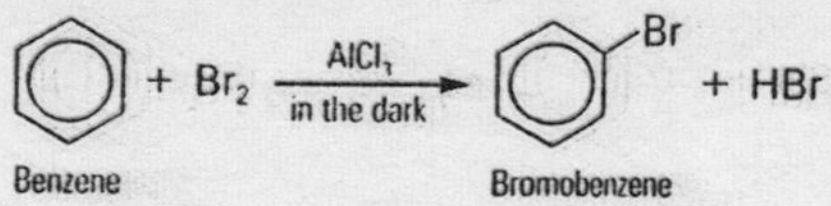

$$\text{Benzene} + Br_2 \xrightarrow[\text{in the dark}]{AlCl_3} \text{Bromobenzene (Br)} + HBr$$

With sulphuric acid

This important reaction in the production of non-soapy detergents uses fuming sulphuric acid or even sulphur trioxide to substitute the —SO_3H group on to the aromatic ring.

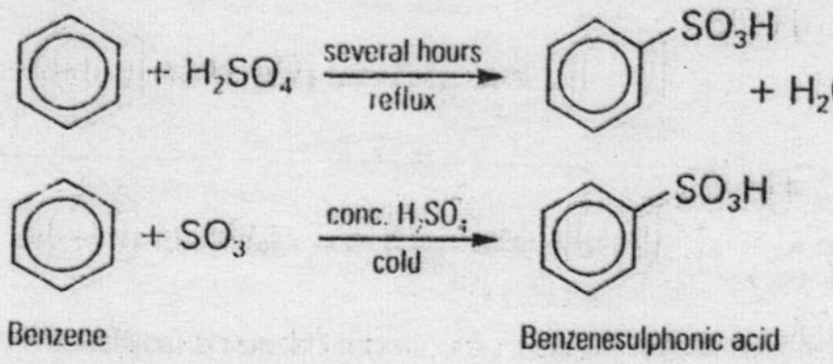

$$\text{Benzene} + H_2SO_4 \xrightarrow[\text{reflux}]{\text{several hours}} SO_3H + H_2O$$

$$\text{Benzene} + SO_3 \xrightarrow[\text{cold}]{\text{conc. } H_2SO_4} \text{Benzenesulphonic acid } (SO_3H)$$

With nitric acid and sulphuric acid

Nitric acid itself has no effect on the benzene ring. The mixture of concentrated nitric acid with concentrated sulphuric acid reacts to substitute the —NO_2 group on to the benzene ring.

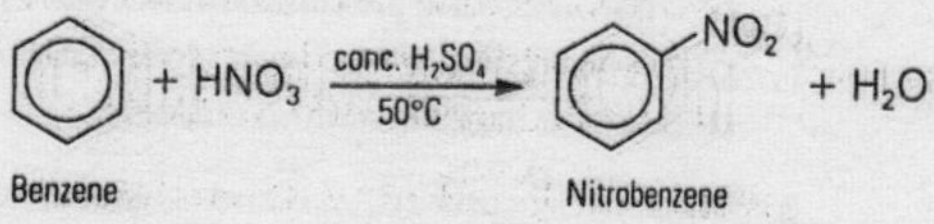

$$\text{Benzene} + HNO_3 \xrightarrow[50°C]{\text{conc. } H_2SO_4} \text{Nitrobenzene } (NO_2) + H_2O$$

With organic groups

An organic group can be introduced onto the benzene ring by using a halogen derivative of the organic group and an aluminium chloride catalyst. The mechanism of this *Friedel–Crafts* reaction will be considered later (Ch 26.8).

e.g.

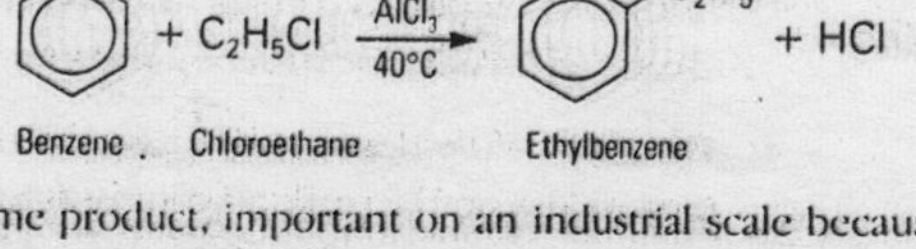

$$\text{Benzene} + C_2H_5Cl \xrightarrow[40°C]{AlCl_3} \text{Ethylbenzene } (C_2H_5) + HCl$$

Another route to the same product, important on an industrial scale because of the availability of ethene, uses ethene with an aluminium chloride catalyst in the presence of an acid.

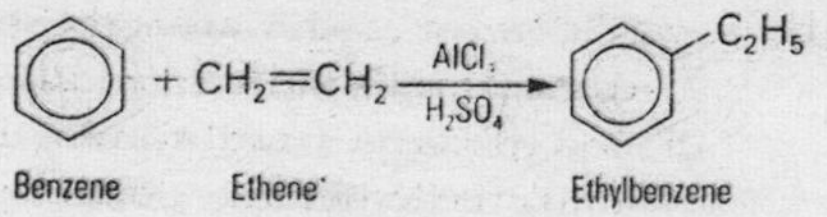

$$\text{Benzene} + CH_2{=}CH_2 \xrightarrow[H_2SO_4]{AlCl_3} \text{Ethylbenzene } (C_2H_5)$$

A reaction similar to that with chloroethane can be achieved with an acyl chloride to form a ketone (Ch 19.1).

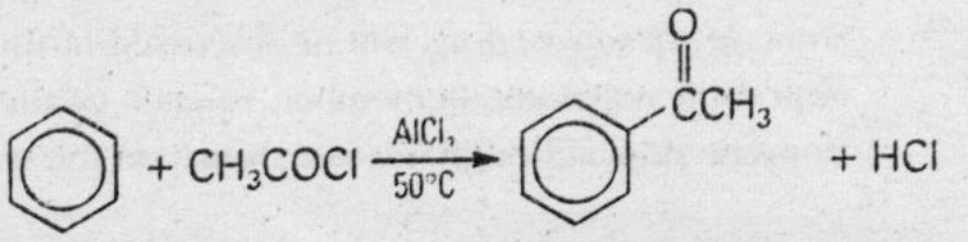

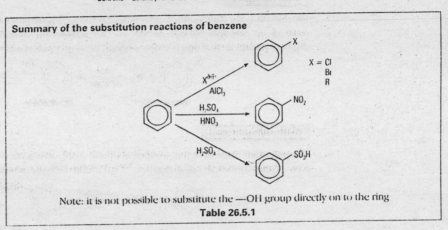

Benzene   Ethanoyl chloride   Phenylethanone

Summary of the substitution reactions of benzene

Note: it is not possible to substitute the —OH group directly on to the ring

Table 26.5.1

SELF-ASSESSMENT
QUESTIONS
26.5

1 a Describe, with suitable diagrams, the σ and/or π bonding in $CH_3CH_2CH_3$, $CH_3CH{=}CH_2$ and C_6H_6.

   b Predict in each of these examples the C—C bond angles.

   c Outline the different reaction conditions needed for the reaction of $CH_3CH{=}CH_2$ with bromine and of C_6H_6 with bromine.

2 Write a balanced chemical equation for the reaction of methylbenzene, $C_6H_5CH_3$, with:

   a concentrated sulphuric acid;

   b chloromethane and aluminium chloride;

   c sulphuric acid mixed with nitric acid (both concentrated);

   d bromine and aluminium chloride.

3 Three hydrocarbons, A, B and C, gave the following series of reactions.

   A burnt with a very smoky flame and reacted with bromine in the presence of a catalyst to form a substitution product.

   B reacted with bromine in the dark, giving an addition product; it burnt with a fairly smoky flame.

   C burnt with an almost clear flame and only reacted with bromine in UV light in a substitution-type reaction.

   All three compounds contain six carbon atoms and have cyclic frameworks. Decide which is an alkane, an alkene, or an arene, and draw a structure for each of the compounds A, B and C.

4 The bulk of the benzene produced in the catalytic reforming process during the refining of crude oil is used to make alkyl-substituted benzene compounds.

   a Write an equation for the reaction of 2-chloropropane with benzene.

   b What catalyst is used in this reaction?

   c Ethylbenzene, used in making polystyrene, is produced from ethene and benzene. Write an equation to show this reaction and include the catalysts needed.

   d Dodecylbenzene is used in making detergents. What reactants and catalysts would you use to make dodecylbenzene? Dodecylbenzene has the formula $CH_3(CH_2)_{11}C_6H_5$.

26.6   Nitration and halogenation of the benzene ring

The π complex corresponds to the highest point on the energy profile diagram (see Ch 21.6).

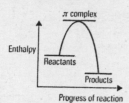

Figure 26.6.1

Under the right conditions, one of the hydrogen atoms around the carbon ring can be replaced by the NO_2 group to form nitrobenzene.

$$C_6H_6 + NO_2^{\oplus} \xrightarrow[\text{reflux}]{60\,^{\circ}C} C_6H_5NO_2 + H^{\oplus}$$

The electrophile in this reaction is the nitryl (or nitronium) cation $NO_2^{\oplus}$, produced by reacting concentrated nitric acid with concentrated sulphuric acid.

$$HNO_3 + 2H_2SO_4 \longrightarrow H_3O^{\oplus} + 2HSO_4^{\ominus} + NO_2^{\oplus}$$

The benzene is warmed with a mixture of the two concentrated acids – rather vigorous conditions, but not so vigorous that more of the dinitro- and trinitro-compounds are formed. The $NO_2^{\oplus}$ acts as the electrophile and forms what is called a π *complex* with the benzene ring (*Figure 26.6.1*).

From this complex, the $NO_2^{\oplus}$ could be lost or the hydrogen ion lost.

Mechanism for electrophilic substitution

π complex

Take care in writing the formula of the π complex. The positive charge in the structure is now only delocalised over five of the carbon atoms in the ring, and the partial circle inside the ring must show this.

Chlorination, i.e. substitution of chlorine on to the benzene ring, is also possible.

$$C_6H_6 + Cl_2 \xrightarrow[\text{temperature}]{AlCl_3 \atop \text{room}} C_6H_5Cl + HCl$$

Although the temporary polarity of the chlorine molecule is sufficient for reaction with alkenes (Ch 16.6), for the aromatic reaction a catalyst is needed to produce a strong enough electrophile to attack the delocalised π system. A number of catalysts are possible. The need is for an atom that is capable of accepting electrons from the chlorine molecule – a *halogen carrier*. Iron(III) bromide, or even an iron nail in the presence of bromine, will do, although one of the most commonly used is aluminium chloride.

Mechanism for chlorination of benzene by electrophilic substitution

The first step in both these *substitution* mechanisms could conceivably be used to form *addition* products. So why does substitution nearly always result? The answer is, there is nothing to be gained by losing the delocalisation of the benzene ring, as shown below.

695

Addition or substitution?

Substitution:

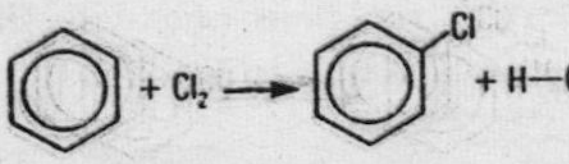

Bonds broken		Bonds formed	
1 × C—H	+ 413	1 × C—Cl	– 346
1 × Cl—Cl	+ 243	1 × H—Cl	– 432
	+ 656		– 778

Enthalpy change for substitution reaction = 656 – 778
$$= – 122 \text{ kJ mol}^{-1}$$

Addition:

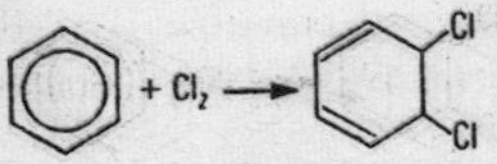

Bonds broken		Bonds formed	
6 × C—C (6 × 520)		2 × C=C (2 × 612)	– 1224
(for benzene)	+ 3120	4 × C—C (4 × 347)	– 1388
1 × Cl—Cl	+ 243	2 × C—Cl (2 × 346)	– 692
	+ 3363		– 3304

Enthalpy change for addition reaction = 3363 – 3304
$$= +59 \text{ kJ mol}^{-1}$$

Which reaction is more likely to occur?

The directing influence of substituents

Any atom or group of atoms substituted on to the benzene ring must influence the π electron clouds above and below the plane of the carbon atoms. Carbon and hydrogen have similar electronegativities; the electrons are equally shared and there is very little polarity in the C—H bond.

In the C—Cl bond, the electronegative chlorine atom attracts electrons and so the bond is polarised. The electronegativity of the elements and the presence of any lone pairs of electrons will therefore distort the symmetry of the π clouds (*Figure 26.6.2*).

The effect caused by differences in electronegativity is called the **inductive effect**, and significantly influences the reactivity of the molecule.

An electronegative atom pulls electron density away from the π clouds, and such an effect is a negative inductive effect, –I. The effect is to reduce the susceptibility of the π cloud to electrophilic attack, and so the compound is less reactive towards electrophiles (*Figure 26.6.3*).

Figure 26.6.2 Inductive effect with chlorobenzene (–I)

Definition: The inductive effect is the electron drift in a molecule caused by the atoms or groups of atoms in the molecule.

Groups with +I effect	Groups with –I effect
Activate – substitution mostly in the 2-, 4- and 6- positions	Deactivate – substitution normally in the 3- and 5-positions
—OH	—Cl
—CH$_3$	—COOH
—NH$_2$	—NO$_2$

Figure 26.6.3

Chlorobenzene is such a compound, where the electronegative chlorine atom withdraws electrons from the central ring, and so deactivates the molecule. This effect overcomes the effect of the lone pairs on the chlorine atom.

Figure 26.6.4   Phenol

When the substituent group is able to push electron density into the ring, the molecule is activated, and becomes more reactive towards electrophiles. This is a positive inductive effect, +I. This is particularly true with groups which possess one or more lone pairs of electrons. The lone pairs repel the bond pair towards the ring, and the delocalised system of the ring extends to include the lone pairs. Phenol (hydroxybenzene) is activated by the presence of the —ÖH group (*Figure 26.6.4*), and phenylamine by the —NH₂ group.

The activation of the benzene ring does not mean a change in the chemical reaction, but rather a change in the conditions needed to achieve the same chemical result. The nitration of benzene and methylbenzene with nitric acid and sulphuric acid is typical. The methyl group has a +I effect. The activated methylbenzene will react at room temperature, whereas benzene has to be refluxed for an hour before the reaction is significantly completed.

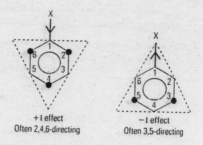

Although the detail of *how* and *why* there are the push and pull effects on the benzene ring is not needed, the diagrams below should help you to remember the effect of changes of electron density by electron-withdrawing and electron-donating groups around the benzene ring. The higher electron density around different carbon atoms results in further substitution occurring preferentially at specific carbon atoms.

+I effect
Often 2,4,6-directing

−I effect
Often 3,5-directing

Chromatography – TLC stands for thin layer chromatography

One of the techniques used to separate and identify the wide range of organic chemicals is chromatography. Other forms of chromatography, such as paper, column, gas–liquid and high performance liquid chromatography are discussed in Ch 29.2.

The problem posed by the nitration of methylbenzene ('toluene') illustrates the usefulness of **thin layer chromatography**. The methyl group on the benzene ring activates the ring towards electrophilic attack. In the nitration of methylbenzene, the first step in nitration can produce two mono-substituted isomers. The introduction of a second nitro group can lead to another two isomers and the third nitrating group produces just the one isomer – 2,4,6-trinitromethylbenzene ('trinitrotoluene' – TNT). Refluxing phenol with concentrated sulphuric and nitric acids will form a complex mixture of products.

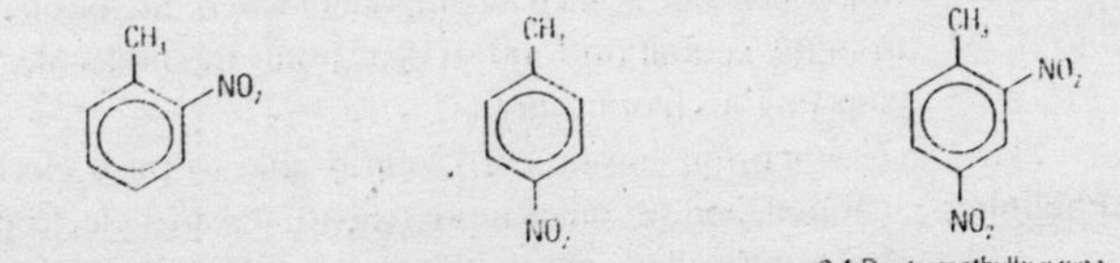

2-Nitromethylbenzene   4-Nitromethylbenzene   2,4-Dinitromethylbenzene

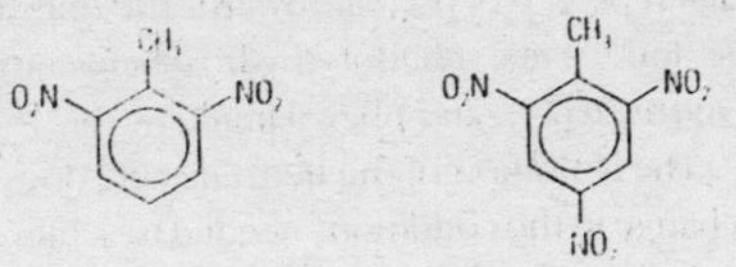

2,6-Dinitromethylbenzene   2,4,6-Trinitromethylbenzene

The reaction mixture is spotted close to the bottom of a plate (glass, plastic or aluminium) coated with a very thin layer of an inert powder such as silica gel. Samples of each of the possible isomers are also placed, as spots on a line near the bottom of the plate. The plate is placed in a beaker, with a layer of solvent sufficient to reach the layer of powder on the plate, but not enough to be in direct contact with the sample spots. The solvent moves up the silica gel layer, by capillary action, and when it reaches the spots there is competition. The samples may dissolve in the solvent or stay weakly attracted to the powder. Each molecule will have different attractions to the solvent and the layer of powder. Consequently, the samples will start to move up the plate at different rates. In this way the compounds are separated and the products of the reaction mixture identified by comparing them to the known samples.

Watch glass (to minimise evaporation of solvent)

Solvent front

Separated sample

Reference spots

Starting line

SELF-ASSESSMENT
QUESTIONS
26.6

1   Chlorine reacts with benzene and cyclohexene in different ways. Explain why there is a substitution reaction with benzene and an addition reaction with cyclohexene. Make sure you refer to the different conditions needed for these two reactions and the types of mechanisms involved.

2   Benzocaine is used as a local anaesthetic. The first step in its preparation involves nitrating methylbenzene.

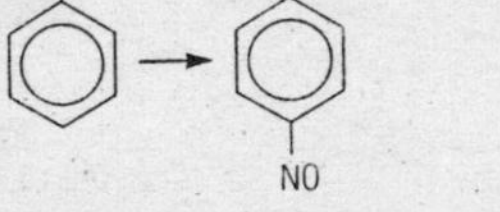

a   What reagents would you use to nitrate the methylbenzene?
b   The last step in the production of the benzocaine is a reduction of the nitro group to an amine

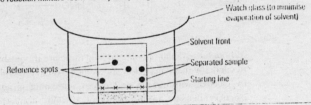

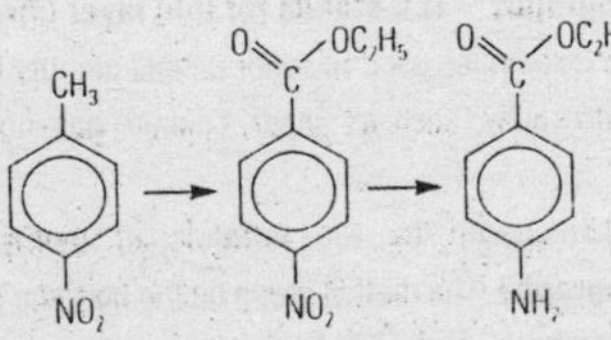

The reducing agent is tin and hydrochloric acid, where tin(II) chloride is one of the product formed. Write an equation for the reaction of tin with hydrochloric acid.
c   Write an equation for the reduction (using the symbol [H]) of the —NO$_2$ group to for benzocaine.

d In this three-step preparation 5.0 g methylbenzene produces 5.0 g of benzocaine. What is the percentage yield in the overall reaction?

3 C_7H_7Br is an aromatic compound and can exist as four isomeric structures. One isomer is produced by reacting methylbenzene with bromine in UV light, whereas two of the other isomers are formed when methylbenzene is reacted with bromine and aluminium chloride in the dark. Outline the different mechanisms of these two reactions.

26.7 Phenols

At first sight, the compound C_6H_5OH looks very much like an alcohol, with an —OH group attached directly to a carbon atom. However, the benzene ring influences the hydroxy group, and the hydroxy group influences the benzene ring. We can therefore consider any molecule consisting of a delocalised system attached to an —OH group as belonging to a separate group – the phenols.

Phenol itself is an important feedstock chemical for the preparation of Nylon, and for plastics, polymers and epoxy resins, as well as the starting material for a range of antiseptic and disinfectant products.

Although phenol cannot be made directly from benzene (the benzene ring is inert to attack by the nucleophilic $OH^{\ominus}$ ion), the industrial manufacture of phenol still starts with benzene. In the so-called 'cumene' process, benzene and propene react, using aluminium chloride as a catalyst, to form (propan-2-yl)benzene ('cumene'). This is then oxidised and hydrolysed in acid to give phenol and propanone.

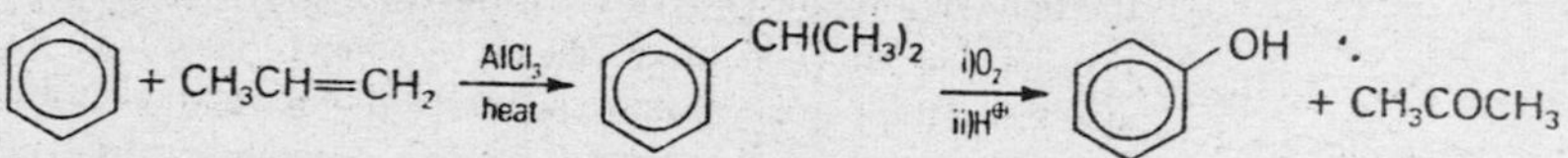

Benzene | Propene | (Propan-2-yl)benzene | Phenol | Propanone

Phenol is the common name for hydroxybenzene. This latter name is rarely used, and compounds based on phenol are named as derivatives of phenol, with the positions of the substituents numbered relative to the -OH group. Remember to keep the numbers as low as possible by counting around the ring in the appropriate direction.

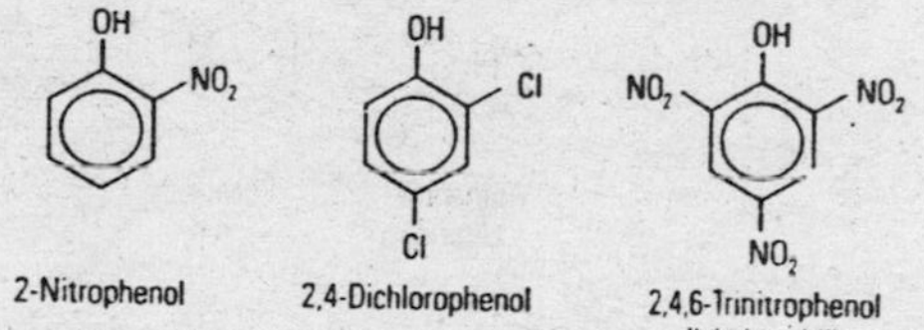

2-Nitrophenol | 2,4-Dichlorophenol | 2,4,6-Trinitrophenol ('picric acid')

The —OH group attached to the benzene ring activates the ring towards electrophilic substitution. The positive inductive effect of the —OH group pushes electrons towards the benzene ring and so effectively increases the electron density around the ring, particularly at the 2, 4, and 6 positions (*Figure 26.7.1*).

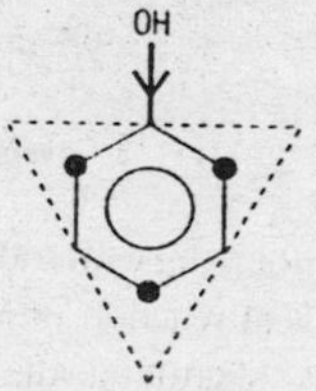

Figure 26.7.1 Inductive effect in phenol

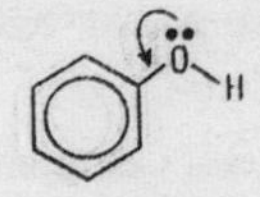

**Figure 26.7.2
Delocalisation in phenol**

This inductive effect arises because one of the lone pairs of electrons on the oxygen atom becomes involved with the delocalised electron system around the benzene ring; the p orbital containing the lone pair becomes part of the delocalised system (*Figure 26.7.2*).

This is borne out by bond length measurements (*Figure 26.7.3*).

C—O bond lengths	/ nm
Ethanol	0.143
Phenol	0.136
Ethanal ($>$ C $=$ O)	0.122

Figure 26.7.3

As we have already mentioned, the presence of the benzene ring has some effect on the chemical reactivity of the —OH group. The main reactions of alcohols are (Ch 18.4):

- oxidation of primary and secondary alcohols, not tertiary;
- condensation reaction to form esters;
- substitution of H$^{\oplus}$ in the —OH group;
- substitution of the whole —OH group;
- elimination of H_2O to form an alkene.

Looking at the phenol structure, there is no hydrogen atom attached to the carbon in the $>$C—OH part of the molecule – effectively, it is a tertiary carbon atom. Consequently, phenols cannot be oxidised directly, and no elimination reaction is possible.

The stability provided by the delocalisation of the lone pair of electrons on the oxygen atom into the benzene ring means that the condensation and substitution reactions of the —OH group are possible, but generally more difficult to achieve. Since both these reactions require nucleophilic attack, and the partially positive polarisation of the carbon atom is reduced by the delocalisation of electrons, the reactions need more vigorous conditions to succeed (*Figure 26.7.4*).

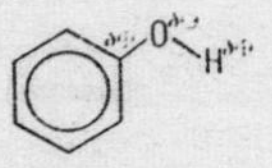

The esters are only formed if the acyl chloride is used (*not* the free organic acid), and if the phenoxide ion is used, rather than phenol itself. This is because $C_6H_5O^{-}$ is a better nucleophile for the reaction with the acyl chloride. A typical equation for the formation of a phenol ester is:

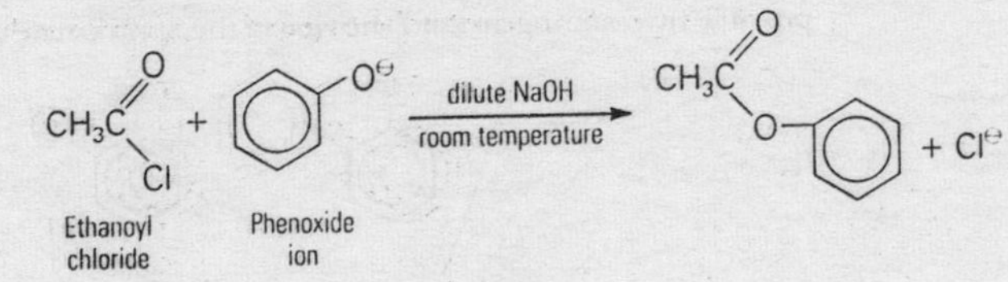

**Figure 26.7.4
Polarity in phenol**

An important difference between phenol and ethanol involves substitution of the hydrogen in the —OH group. Delocalisation allows H$^{\oplus}$ to be formed much more readily by phenol. The phenoxide ion, $C_6H_5O^{\ominus}$, is much more stable than the ethoxide ion, $C_2H_5O^{\ominus}$, since the negative charge is delocalised over the whole of the benzene ring and the oxygen atom. With sodium, the reaction of phenol is quite vigorous:

$$2C_6H_5OH + 2Na \xrightarrow[\substack{room \\ temperature}]{ethoxyethane} 2C_6H_5O^{\ominus}Na^{\oplus} + H_2$$

Phenol   Sodium phenoxide

Alcohols do not react with sodium hydroxide solution, but phenols do react with it to form a salt and water – a typical reaction of an acid. Phenol is more acidic than water. The pK_a values (Ch 23.3) confirm this slightly acidic character of phenols compared with water and the alcohols (*Figure 26.7.5*).

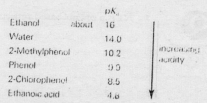

		pK$_a$
Ethanol	about	16
Water		14.0
2-Methylphenol		10.2
Phenol		9.9
2-Chlorophenol		8.5
Ethanoic acid		4.8

increasing acidity

Figure 26.7.5 pK$_a$ values

$$C_6H_5OH + NaOH(aq) \xrightarrow[\text{room temperature}]{\text{aqueous}} C_6H_5O^\ominus Na^\oplus(aq) + H_2O(l)$$

However, it must be appreciated that the phenols are **weak acids**. Although they react with reactive metals and alkalis, they do not react with sodium carbonate solution. This allows phenols to be distinguished from carboxylic acids.

The other important reaction which distinguishes phenols, from alcohols is the reaction with neutral aqueous iron(III) chloride. Phenols produce a coloured complex with this reagent. The colour is often purple, but any colour other than yellow may indicate a phenol.

So far, the effect of the benzene ring on the —OH group has been considered. The —OH group activates the benzene ring towards electrophilic substitution. Typical electrophilic substitution reactions of phenol are illustrated in the following equations. Note that, unlike the —OH group in alcohols (Ch 18.4), the —OH group in phenol does not react with PCl$_5$ or SOCl$_2$.

With bromine

With dilute nitric acid

With concentrated nitric and sulphuric acids

With concentrated sulphuric acid

The chemical reactions of phenols can be summarised as:

- no direct oxidation reaction;
- condensation reactions to form esters, but using an acyl chloride rather than a carboxylic acid;
- weakly acidic, reacting with sodium hydroxide solution;
- electrophilic substitution on the benzene ring is preferred to nucleophilic substitution of the —OH group;
- no direct elimination reactions.

1 TCP is a phenol-based antiseptic. The formula is

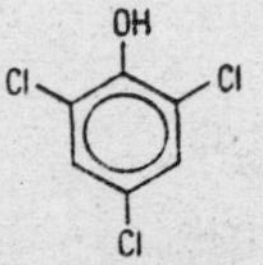

and the systematic name is 2,4,6-trichlorophenol (hence TCP). The commercial antiseptic liquid is sold as a 0.68% solution, equivalent to a 0.034 mol dm^{-3} solution. Another phenol-based antiseptic is Dettol, which is approximately ten times more effective than TCP. Dettol has the structure

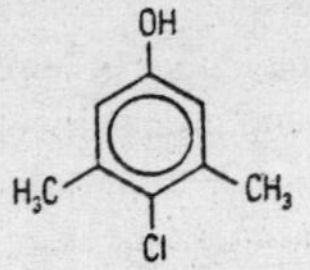

a Write a systematic name for Dettol.
b What concentration of Dettol would be required to provide a solution of similar effectiveness to that of commercial TCP? Calculate as a molar concentration, and then give a figure as g dm^{-3}.
c Explain why the reaction of Dettol with sodium hydroxide solution results in the formation of a sodium salt, but does not substitute any of the chlorine atoms.

2 The following isomeric compounds A and B both contain an —OH group.

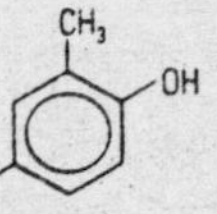

A and B

a Name the compound A.
b To which classes of compounds do each of A and B belong?
c How would each of these compounds react with concentrated sulphuric acid? Give balanced equations to show the reactions.
d What reaction would there be if each of A and B was reacted with sodium metal?
e How, and under what conditions, would you expect each of these compounds to react with a 2 mol dm^{-3} solution of sodium hydroxide?

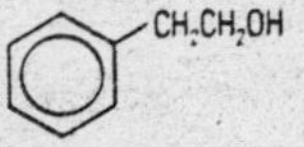

3 The hormone adrenaline has the structural formula:

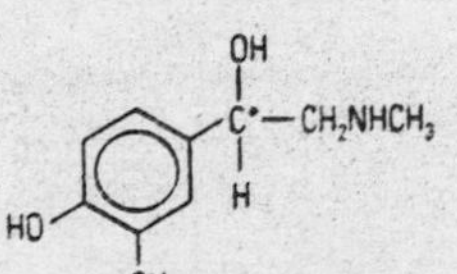

a What type of isomerism is observed in this compound?
b The —OH group on the marked (*) carbon atom is different from the —OH groups on the benzene ring. Explain the difference in the reactions of these apparently similar groups.
c How many moles of hydrogen gas, H_2, will be formed if one mole of adrenaline is reacted with excess sodium metal?
d How many moles of adrenaline will react with 1 dm^{-3} of 1 mol dm^{-3} sodium hydroxide solution?

26.8 The Friedel–Crafts reaction

One of the important reactions in the chemistry of benzene is the substitution of an alkyl chain on to the benzene ring. The stability of the delocalised benzene ring makes the reaction difficult, so it requires the use of a catalyst. The reaction is important, since many of our everyday chemicals contain such an aromatic ring attached to an alkyl chain. Soapless detergents (*Figure 26.8.1*) and polystyrene are two such materials.

Non-soapy detergents are prepared from benzene and long chain alkenes in the presence of a catalyst.

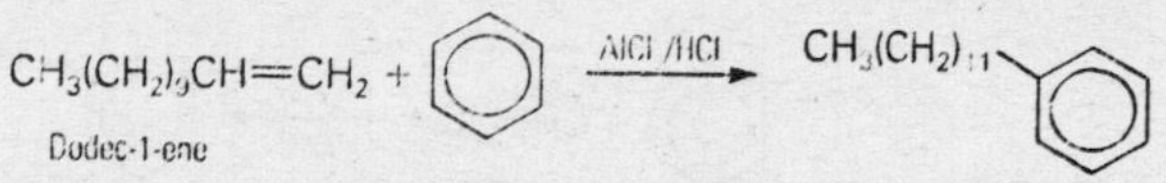

$$CH_3(CH_2)_9CH{=}CH_2 + \bigcirc \xrightarrow{AlCl_3/HCl} CH_3(CH_2)_{11}\bigcirc$$

Dodec-1-ene

The benzene ring is then sulphonated, using SO_3 rather than H_2SO_4, to form the alkyl benzene sulphonate.

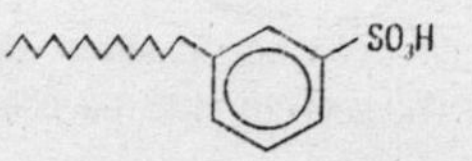

SO_3H

This acid now reacts with NaOH forming the sodium salt, which, dissolved in water, is used in washing-up liquid and makes up 10–20% of the bulk in washing powder.

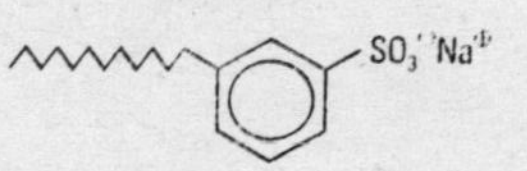

$SO_3^- \ ^-Na^+$

Figure 26.8.1

There are a number of different Friedel–Crafts catalysts. The most important types are electron pair acceptors (Lewis acids). The most common of these is aluminium chloride. Large quantities of this catalyst are used in industry (10^6 tonnes per year). It is the most effective catalyst, although it can cause problems of disposal.

Other electron pair acceptor catalysts include $FeCl_3$.

The Friedel–Crafts reaction involves benzene reacting with a halogenoalkane. The catalyst is normally anhydrous aluminium chloride, which reacts in the same way as the 'halogen-carrier' does when chlorine reacts with benzene (see Ch 26.6).

The mechanism must still involve the electrophilic substitution of one of the aromatic hydrogens.

The overall reaction could be considered as a 'condensation' reaction (see Ch 14.6). Two reactants combine by addition. A simple molecule such as HCl or H_2O is then lost from the addition compound to form the new product. But it is best remembered as an *electrophilic substitution*.

$$\bigcirc + CH_3CH_2Cl \xrightarrow[\text{room temperature}]{AlCl_3} \bigcirc^{CH_2CH_3} + HCl$$

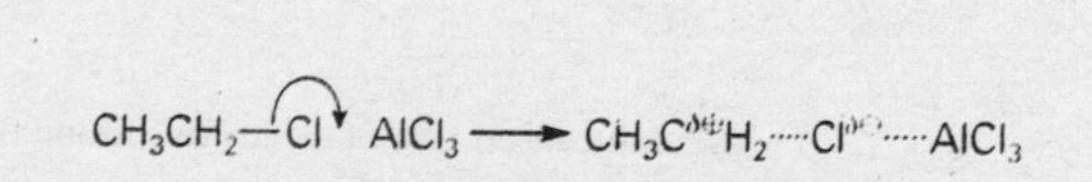

Benzene   Halogenoalkane   Alkyl-substituted benzene

The aluminium chloride promotes the formation of the electrophilic reactant. The initial step involves the halogenoalkane becoming polarised in the presence of the aluminium chloride, where the electron deficient aluminium chloride accepts electrons from the chlorine and so polarises the halogenoalkane.

$$CH_3CH_2{-}Cl \ AlCl_3 \longrightarrow CH_3C^{\delta+}H_2{\cdots}Cl^{\delta-}{\cdots}AlCl_3$$

This creates a partial positive charge on the alkyl chain, and so promotes attack by it on the high electron density around the benzene ring. It is possible to write this intermediate polarised complex as $CH_3C^{\oplus}H_2[AlCl_4]^{\ominus}$, although the existence of the free $CH_3C^{\oplus}H_2$ carbocation is not thought to be very likely. The process is called **alkylation**, since an alkyl group is substituted on to the benzene ring.

Definition: Alkylation is the electrophilic substitution of an alkyl chain onto a benzene ring using, for example, aluminium chloride as a catalyst

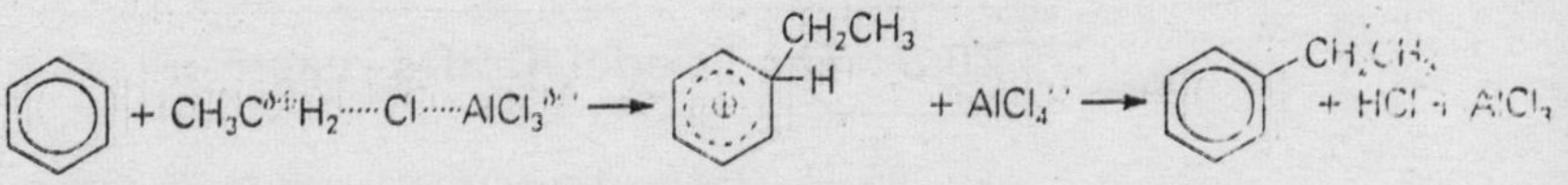

Here, the attacking group is the alkyl carbocation, and the leaving group is the proton.

In industry, the formation of phenylethene (styrene) involves the reaction of benzene with ethene. Ethene is a major product in the cracking of crude oil fractions and is a readily available raw material. In the presence of hydrogen chloride and with aluminium chloride as a catalyst, phenylethane is formed, and this is heated over zinc metal to form phenylethene ('styrene'), the monomer for poly(phenylethene) (polystyrene).

Phenylethene

The Friedel–Crafts reaction can also be used to introduce the R—C=O group on to the benzene ring. The acyl chloride reacts in a way very similar to the halogenoalkane.

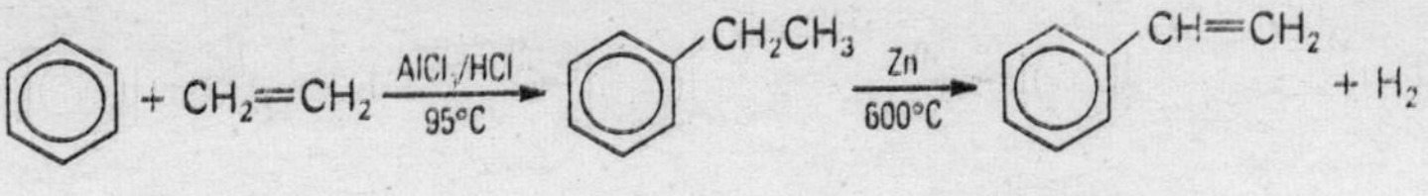

Definition: Acylation involves the electrophilic substitution of an acyl group onto a benzene ring using, for example, aluminium chloride as a catalyst

Phenylethanone ('acetophenone') an aromatic ketone

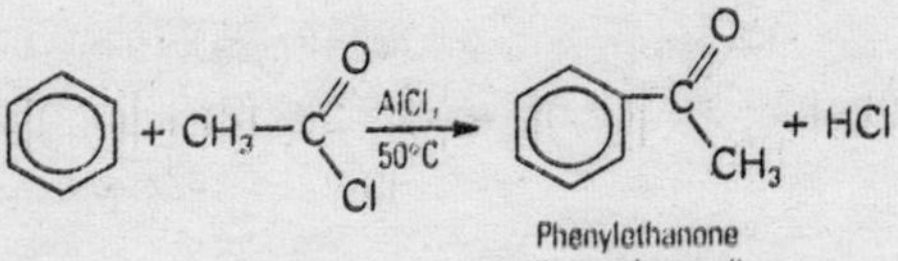

In this case the electrophile is the $CH_3C^{\delta+}{=}O$ group.

Q SELF-ASSESSMENT QUESTIONS 26.8

1 Which of the following reagents might be considered as halogen carriers in the electrophilic substitution of benzene?
PCl_3, $AlCl_3$, $FeBr_3$, BCl_3

2 Give the name and formula of the mono-substituted product from the following reactants.

a + HNO_3 + H_2SO_4 at 50 °C

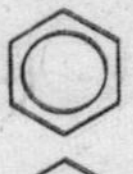

b + CH_3Br with $AlCl_3$

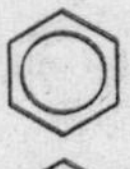

c + Cl_2 with $FeCl_3$

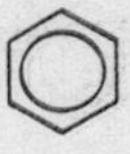

3 The delocalised ring of electrons around the benzene molecule means that benzene reacts by electrophilic substitution.
a Explain what is meant by *delocalised*
b What is meant by an *electrophile*?
c Describe the mechanism for the reaction of benzene with CH_3Cl in the presence of $AlCl_3$.

4 The pain-relieving drug Ibuprofen has the formula:

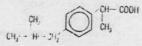

a Calculate the relative molecular mass of Ibuprofen.
b Write the formula of 2-phenylpropanoic acid.
c How would you convert 2-phenylpropanoic acid into Ibuprofen using the Friedel–Crafts reaction?

26.9   Addition reactions of benzene

In spite of the resistance of benzene to addition reactions, it is possible to achieve addition products – but it is always addition across all three of the theoretical 'double bonds' in the molecule. It is impossible to add across just one of the double bonds and leave a cyclic diene product, because then the delocalisation stability would be lost.

With hydrogen, cyclohexane is formed when benzene is reacted with hydrogen over very finely divided nickel as a catalyst.

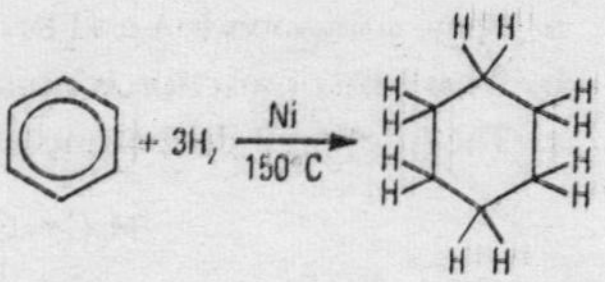

With chlorine, the reaction has to be carried out with ultra-violet light. A chlorine atom is added to each of the six carbon atoms in the ring.

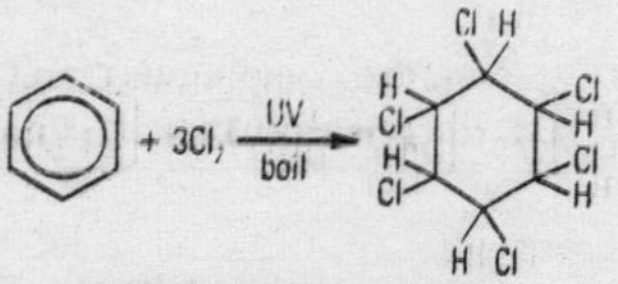

1,2,3,4,5,6-Hexachlorocyclohexane

This molecule can exist as a number of geometrical isomers.

The ring of six carbon atoms inhibits rotation about the carbon–carbon bonds – like the restricted rotation about the π bond in alkenes – and so it is possible to have the chlorine atoms above the ring, below the ring, or some up and some down (*Figure 26.9.1*).

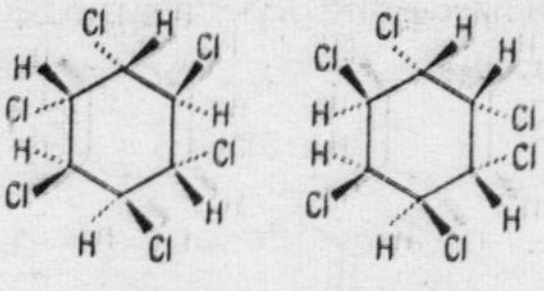

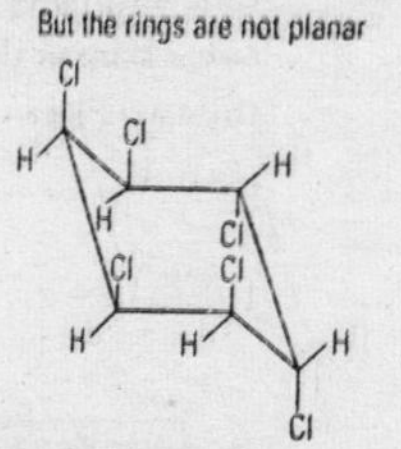

But the rings are not planar

Figure 26.9.1   Two isomers of hexachlorocyclohexane

When chlorine is reacted with benzene in UV light, up to six of the eight possible isomers are formed. One of these isomers was widely used as an insecticide under the names of Lindane, Gammexane or BHC. Since the toxic effects of the compound, on fish particularly, became known, its use has been widely banned.

1 Write an equation for the reaction of $C_6H_5NH_2$ with:
   a the nitryl ion, $NO_2^{\oplus}$;
   b bromine;
   c 2-chloropropane and aluminium chloride.

2 Classify these reactions as electrophilic substitution, electrophilic addition, or free radical substitution.
   a $CH_3CH{=}CHCH_3 + HBr \longrightarrow CH_3CH_2CHBrCH_3$
   b $C_6H_5C_2H_5 + I{-}Cl \longrightarrow I{-}C_6H_4C_2H_5 + HCl$
   c $C_6H_5CH_3 + Cl_2 \longrightarrow C_6H_5CH_2Cl + HCl$
   d $CH_3CH_2CH_2CH_3 + Br_2 \longrightarrow CH_3CH_2CHBrCH_3 + HBr$

3 (Propan-2-yl)benzene (cumene) is an important intermediate product in the synthesis of aromatic compounds. It is formed in the reaction:

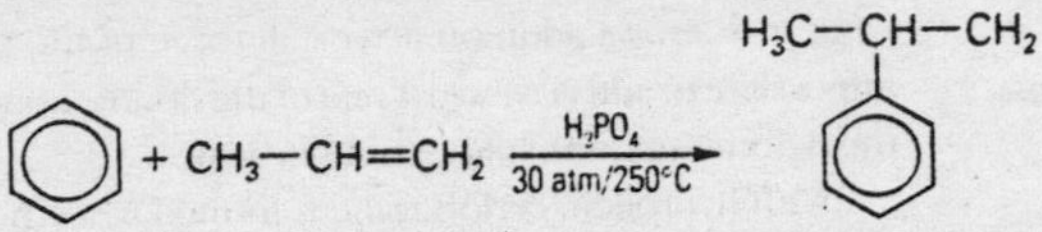

$$\text{A} + CH_3{-}CH{=}CH_2 \xrightarrow[\text{30 atm/250°C}]{H_3PO_4} \text{(Propan-2-yl)benzene}$$

   a Name compounds A and B.
   b What type of reaction is occurring?
   c The (propan-2-yl)benzene reacts, in a series of steps, to produce two products, C and D:

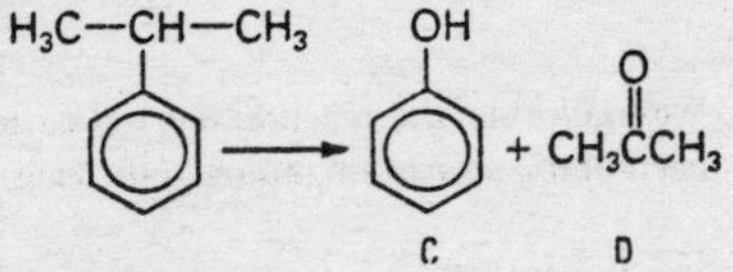

$$\text{C} + CH_3\overset{O}{\overset{\|}{C}}CH_3$$

   Name the compounds C and D.
   d C reacts with chlorine to form a trichloro derivative. Draw the structure of this product.

26.10 Alkyl derivatives of benzene

In molecules that contain both the aromatic benzene ring and an alkyl chain, there are two very different parts of the molecule, and the reactions reflect this. Chlorination can occur in either the side chain or the benzene ring depending on the conditions. The reactions of methylbenzene show the two types of reaction.

e.g.

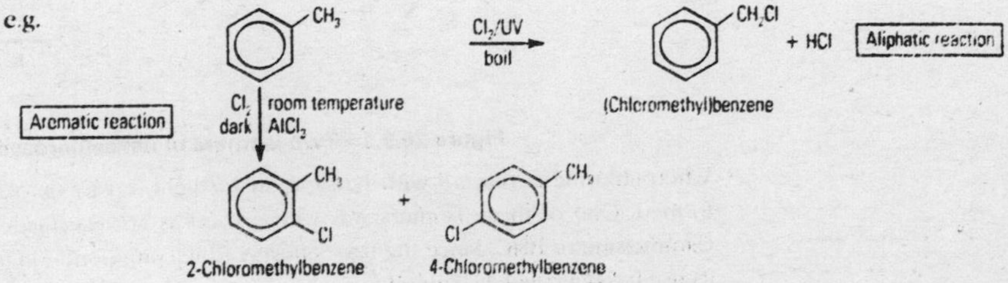

Aromatic reaction

Cl_2/UV boil

(Chloromethyl)benzene   + HCl   Aliphatic reaction

Cl_2 room temperature
dark $AlCl_3$

2-Chloromethylbenzene   +   4-Chloromethylbenzene

The alkyl side chain attached to a benzene ring is oxidised by manganate(VII) ions and dichromate(VI) ions in acid solutions, although the benzene ring itself is very difficult to oxidise. Any alkyl side chain is oxidised to benzenecarboxylic acid (benzoic acid), but you only need to know the reaction for the methyl group.

$$\text{C}_6\text{H}_5\text{CH}_3 + 3\,[\text{O}] \quad \text{e.g.} \quad \xrightarrow[\text{reflux}]{\text{K}_2\text{Cr}_2\text{O}_7/\text{H}^+} \quad \text{C}_6\text{H}_5\text{COOH} + \text{H}_2\text{O}$$

SELF-ASSESSMENT QUESTIONS 26.10

In the following series of reactions identify the products.

A ← C$_2$H$_5$Cl / AlCl$_3$

B ← HNO$_3$ / H$_2$SO$_4$

C ← KMnO$_4$ / H$_2$SO$_4$

D ← Cl$_2$ / AlCl$_3$

(central structure: CH$_3$ on benzene ring)

26.11 Summary of hydrocarbon reactions

Points to remember about hydrocarbon reactions

	Alkanes	Alkenes	Arenes
• Burning forming $CO_2 + H_2O$	Clear flame	Smoky flame	Very smoky flame
• With bromine	Substitution in UV light	Addition in the dark	Substitution with $AlCl_3$ Addition in UV light
• With alkaline manganate(VII)	No reaction	Oxidation – solution turns green	No reaction
• With acid manganate(VII)	No reaction	Oxidation – solution turns colourless	No reaction
• Hydrogen/Ni	No reaction	Addition of hydrogen	Addition of hydrogen
• Conc. sulphuric acid	No reaction	Addition; intermediate gives —OH when water is added	Substitution if refluxed

Comparison of chloro-derivatives of alkanes and arenes

1-Chlorohexane, CH$_3$(CH$_2$)$_5$—Cl
- Nucleophilic substitution of —Cl possible with OH$^-$, CN$^-$, NH$_2$
- Elimination to form an alkene is possible

Chlorobenzene
- Nucleophilic substitution of —Cl extremely difficult
- Electrophilic substitution on the benzene ring is difficult because of electron withdrawal by —Cl, but occurs in preference to addition reactions
- Elimination is not possible

Comparison of hydroxy-derivatives of alkanes and arenes

Hexan-1-ol, CH$_3$(CH$_2$)$_5$—OH
- Readily oxidised to carbonyl group
- Elimination possible to an alkene
- Nucleophilic substitution of —OH is possible
- Formation of esters using carboxylic acid is possible
- Reacts with Na to give H_2
- Does not react with NaOH

Phenol
- Impossible to oxidise the —OH group directly
- Elimination impossible with the ring structure
- Nucleophilic substitution of the —OH group is impossible
- Electrophilic substitution on the benzene ring is easily achieved because of electron donation by —OH group
- Formation of esters using acyl chloride is possible, but not using carboxylic acids
- Reacts with Na to give H_2
- Does react with NaOH to form a salt

Using the C_6 hydrocarbons as examples, compare the structures and reactions of alkanes, alkenes and arenes. Make sure you discuss the bonding in these molecules, and how the difference in bonding leads to differences in their chemical reactions. You might like to consider the reactions with oxygen, hydrogen bromide, chlorine, sulphuric acid and hydrogen. Balanced equations should be given where appropriate.

Summary – Reactions of arenes

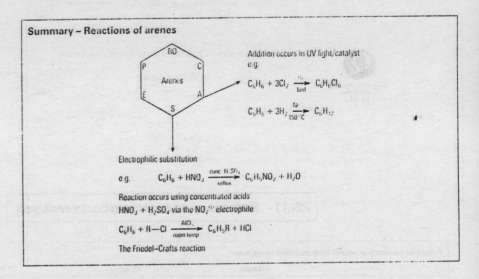

Addition occurs in UV light/catalyst
e.g.

$$C_6H_6 + 3Cl_2 \xrightarrow[\text{boil}]{uv} C_6H_6Cl_6$$

$$C_6H_6 + 3H_2 \xrightarrow[150\,°C]{Ni} C_6H_{12}$$

Electrophilic substitution

e.g. $C_6H_6 + HNO_3 \xrightarrow[\text{reflux}]{\text{conc. } H_2SO_4} C_6H_5NO_2 + H_2O$

Reaction occurs using concentrated acids
$HNO_3 + H_2SO_4$ via the $NO_2^{\oplus}$ electrophile

$C_6H_6 + R—Cl \xrightarrow[\text{room temp}]{AlCl_3} C_6H_5R + HCl$

The Friedel–Crafts reaction

In this chapter the following new ideas have been covered. *Have you understood and learned them?* Could you explain them clearly to a friend? The page references will help.

- The delocalised structure of benzene.   p682
- The dangers of benzene.   p687
- Electrophilic substitution on the benzene ring.   p685
- Position isomerism, with two substituents on the benzene ring.   p690
- Activation and deactivation of the ring towards electrophilic attack.   p696
- The reactions of phenol compared with alcohols.   p699
- The mechanism for electrophilic substitution with $NO_2^{\oplus}$ and Cl_2.   p695
- Alkylation and acylation of the benzene ring.   p703
- Addition reactions of benzene and why addition is less common.   p705
- The reactions of the alkyl chain attached to the benzene ring.   p706

1   The —CH_3 group attached to the benzene ring activates the ring towards electrophilic attack by the $NO_2^{\oplus}$ group. The incoming groups mainly occupy the 2, 4 and 6 positions. The reaction of methylbenzene with $NO_2^{\oplus}$ produces two mono-, two di- and one tri-substituted compounds. Draw the structure of these five isomers. Explain why, under normal conditions, the presence of the trisubstituted compound cannot be detected.

2   One of the compounds used as a sun screen in sun tan lotions has the formula:

a How would you expect this compound to react with sodium metal?

b What would you expect to see when the compound is burned in air?

c What would happen if the compound is reacted with 2,4-dinitrophenylhydrazine?

d The strong absorption at around 290 nm in the UV region of the spectrum – the wavelengths which can cause damage to the skin – is due to the extensive delocalisation within the molecule. Explain what delocalisation means.

3 a Write an equation for the complete combustion of 1,3-dimethylbenzene in excess oxygen.

b The catalytic reforming of hexane to produce aromatic ring structures is carried out at high temperature and uses an oxide catalyst. What is the advantage of converting hexane to an aromatic ring structure?

c What is the evidence for the extra stability of aromatic ring structures such as benzene?

4 'BHT' is used, at about 0.02% levels, as an antioxidant in many packaged foods. It can be made by reacting 4-methylphenol with methylpropene in a Friedel–Crafts alkylation.

a Write the displayed formula for 4-methylphenol.

b What is the formula of methylpropene?

c What catalyst is used in the alkylation reaction?

d The formula of BHT is:

What is the molecular formula of this compound?

- The basic ideas in organic chemistry including **homologous series** and **functional group**. (Ch 14.2)
- The **naming** of organic compounds. (Ch 14.3)
- The principles of organic **reaction mechanisms**. (Ch 14.7)
- **Primary alcohols**, R—CH$_2$OH, and **aldehydes**, R—CHO, can be oxidised to carboxylic acids, R—COOH. (Ch 18.4)
- **Equilibrium calculations**. (Ch 22.1)
- **Weak acids** are only partly broken up ('dissociated') into ions when in dilute aqueous solution. (Ch 23.1)
- The extent to which a weak acid dissociates at a particular concentration in solution is related to its **acid dissociation constant**, K_a. (Ch 23.3)

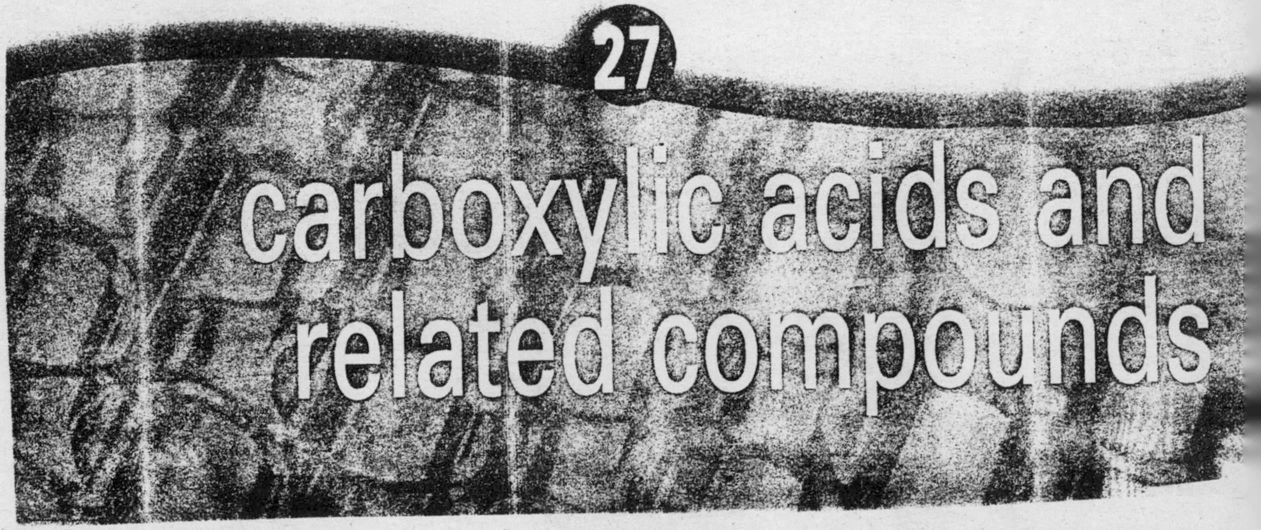

27

carboxylic acids and related compounds

The sharp taste of a lemon or of vinegar, and the stings of ants, are all caused by the presence of a range of acidic organic compounds. These compounds all contain the **carboxylic acid** functional group, —COOH. Sometimes this is called simply the *carboxyl* group.

27.1   Naming carboxylic acids

The usual *rules* of organic nomenclature are followed (Ch 14.3). The name of the compound is based on the longest chain of carbon atoms. However, as with the aldehydes (Ch 19.1), in the carboxylic acid series the carbon atom in the functional group counts as part of the chain.

The carbon atoms in the chain are numbered so that the carbon atom in the functional group is number one. The names of all the carboxylic acids end in '-*oic acid*'. The first member of the series contains only the one carbon atom, which is in the *carboxylic acid group*, —COOH. As there is only one carbon atom, the name of the acid is based on methane.

The first member is, therefore, *methanoic acid*, H—COOH; the second is ethanoic acid (because ethane is the alkane with two carbon atoms), CH_3—COOH; the third is propanoic acid, C_2H_5—COOH; and so on (*Figure 27.1.1*).

Examples of organic acids

Formula	Name
HCOOH	Methanoic acid
CH_3COOH	Ethanoic acid
C_2H_5COOH	Propanoic acid
C_3H_7COOH	Butanoic acid

Figure 27.1.1

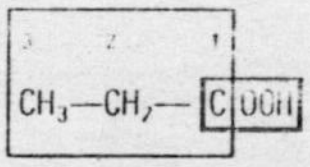

It will help if you try to recognise the two different parts of the molecule:

- the —COOH is the functional group responsible for most of the chemical reactions of the molecule;
- the part in the bigger box is the carbon chain and gives the compound a specific name.

Substituted carboxylic acids are named in the usual way (see Ch 14.3), so that:

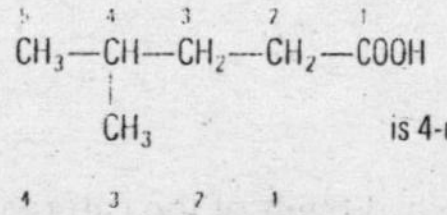

$$CH_3—CH—CH_2—CH_2—COOH$$
$$|$$
$$CH_3 \qquad \text{is 4-methylpentanoic acid}$$

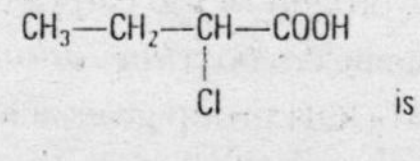

$$CH_3—CH_2—CH—COOH$$
$$|$$
$$Cl \qquad \text{is 2-chlorobutanoic acid}$$

However, you should make sure you do recognise the longest carbon chain, even if it is not an obvious straight chain. So:

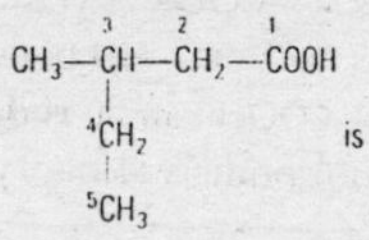

$$CH_3—CH—CH_2—COOH$$
$$|$$
$$^4CH_2 \qquad \text{is 3-methylpentanoic acid}$$
$$|$$
$$^5CH_3$$

1 Which of the following formulae represent carboxylic acids?

   a CH_3—CH_2—$CH(CH_3)$—$COOH$

   b $(CH_3)_2CH$—CHO

   c CH_3—CH_2OH

   d C_6H_5—CH_2—$COOH$

2 Draw displayed formulae for:

   a propanoic acid

   b 2-methylbutanoic acid

   c 3,3-dimethylbutanoic acid

   d 2-methylbenzoic acid

3 Name the following acids:

   a CH_3—CH_2—CH_2—CH_2—$COOH$

   b CH_3—CH_2—CH_2—$COOH$

   c CH_3—$CH(CH_3)$—CH_2—$COOH$

   d $ClCH_2$—CH_2—CH_2—CH_2—$COOH$

27.2 The carboxylic acid functional group

The carboxylic acid group is most frequently written as the formula —$COOH$ (and less frequently in the form —CO_2H).

The displayed formula of the group shows why it is important to appreciate that the two oxygen atoms in the formula are very different.

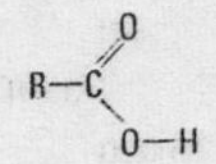

$$R-C\overset{O}{\underset{O-H}{}}$$

Drawing out the full structure of the carboxyl group enables us to see what chemical reactions it can do.

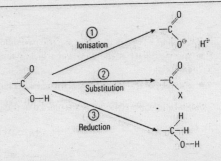

① The hydrogen of the carboxylic group can be lost by **ionisation**; this means that metal salts containing *carboxylate anions*, R—$COO^⊖$, can be formed. And, of course, it is the reason why the —$COOH$ group gives acidic properties to the series of carboxylic acids.

② The entire —OH part of the —$COOH$ group can be replaced in a **substitution** reaction. The replacement can be an atom, as for the acyl chlorides, R—$COCl$; or a group of atoms, as for the esters, R—$COOR'$. (As we shall see in Ch 27.9, esterification can be considered a substitution as it is the $>C$—OH bond in the carboxylic group which is broken.)

③ The —$COOH$ can be **reduced**, by replacing one oxygen atom with two hydrogen atoms, to form the primary alcohol group, R—CH_2OH.

You may meet the older names for a few organic compounds which were called 'acids', although they do not contain the —COOH group. Two of the most likely to be met are 'carbolic acid', for phenol (C_6H_5OH), and 'picric acid', for 2,4,6-trinitrophenol.

'Carbolic acid' – a solution of phenol in water – was widely used as a disinfectant. It was first used during surgery in 1865 by Joseph Lister, and reduced the death rate from infections following surgery by two-thirds.

'Picric acid' was a high explosive used in bombs and shells. It is still sometimes used as a stain in biology.

The carboxyl group —COOH contains both a *hydroxyl part*, —OH, and a *carbonyl part*, $>C{=}O$. The carbonyl group in *aldehydes* and *ketones* makes them behave the way they do (Ch 19.1). But, except for reduction, carboxylic acids show **none** of the typical reactions of aldehydes or ketones.

The carbonyl ($>C{=}O$) part of the carboxyl group is largely 'hidden' and does not contribute to the list of carboxyl group reactions. The reactivity of the $>C{=}O$ depends on its polarisation as $>C^{\delta+}{=}O^{\delta-}$. Any atom or group which donates electrons, or has an available lone pair which can be donated towards the carbon of the carbonyl group, will reduce the carbonyl reactivity if that atom or group is attached to the carbon of the carbonyl group.

So carbonyl reactivity reduces in the sequence:

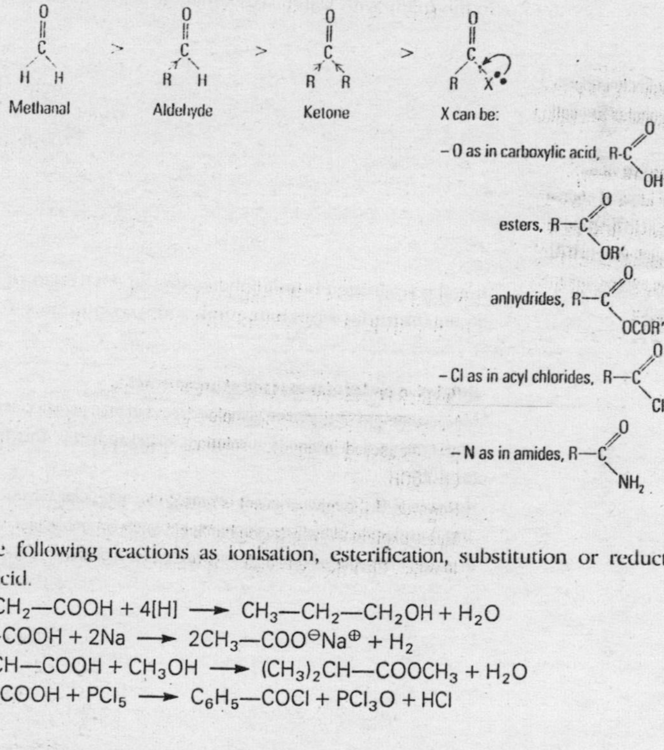

Identify the following reactions as ionisation, esterification, substitution or reduction of the carboxylic acid.

SELF-ASSESSMENT
QUESTIONS
27.2

a $CH_3{-}CH_2{-}COOH + 4[H] \longrightarrow CH_3{-}CH_2{-}CH_2OH + H_2O$

b $2CH_3{-}COOH + 2Na \longrightarrow 2CH_3{-}COO^{\ominus}Na^{\oplus} + H_2$

c $(CH_3)_2CH{-}COOH + CH_3OH \longrightarrow (CH_3)_2CH{-}COOCH_3 + H_2O$

d $C_6H_5{-}COOH + PCl_5 \longrightarrow C_6H_5{-}COCl + PCl_3O + HCl$

27.3   The physical properties of carboxylic acids

Because every carboxyl group has two electronegative oxygen atoms, one of which is attached to a hydrogen atom, carboxylic acids readily form *hydrogen bonds* (Ch 2.9). Carboxylic acid molecules can hydrogen bond with each other, so that carboxylic acids have unusually high melting and boiling points (*Figure 27.3.1*) – for the same reason that water has, for a compound with a small molecule, a high melting point.

Ethanoic acid (old name 'acetic acid') has a melting point of 17°C.
This means that in a cold laboratory or in an outside store it solidifies in the bottle; it then looks rather like ice, although it has needle-shaped crystals. It is for this reason that you will often see bottles labelled 'glacial acetic acid' or 'glacial ethanoic acid'. If any water is present in the acid, the melting point is lowered.

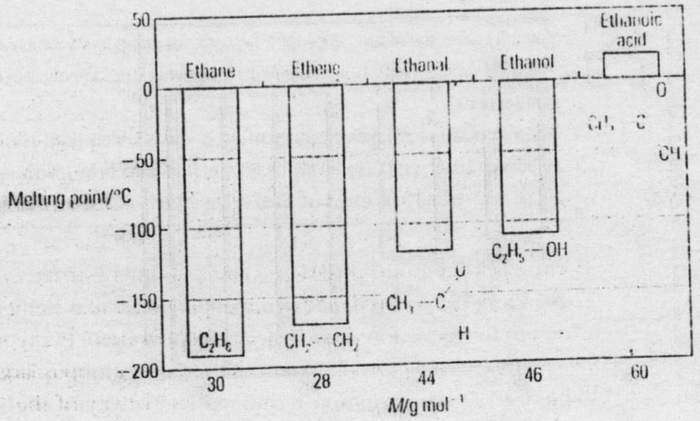

Figure 27.3.1   Melting points of common compounds with two carbon atoms

Carboxylic acids can hydrogen bond with water, which allows the smaller carboxylic acid molecules to mix freely with water – e.g. ethanoic acid and water form vinegar.

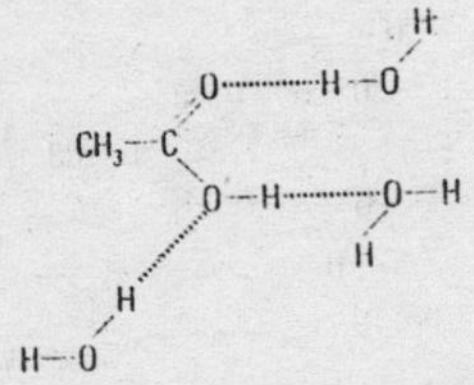

Some biologically important organic molecules are called 'acids'.
Examples are vitamin C, 'ascorbic acid', and of course ribonucleic acid (RNA) and deoxyribonucleic acid (DNA). None of these contains the —COOH group.

It is this hydrogen bonding between the —OH part of the —COOH and a water molecule which is an important factor in ionisation, and so contributes to the acid properties of carboxylic acids.

Relative molecular masses of organic acids

Measurement of the relative molecular mass of many organic acids gives some surprising results. The M_r of ethanoic acid, measured in aqueous solution, is found to be 60. This fits in with the formula we have been using so far, CH_3COOH.

However, if the measurement is repeated in a non-polar solvent, such as benzene, M_r is found to be 120.

This is explained by hydrogen bonding between two molecules of ethanoic acid, leading to the formation of a **dimer**. In water, the hydrogen bonding with the water molecules is enough to prevent dimerisation.

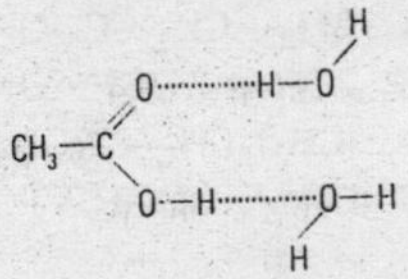

In benzene, there is no possible hydrogen bonding with the solvent, C_6H_6, and consequently two acid molecules combine together by hydrogen bonds.

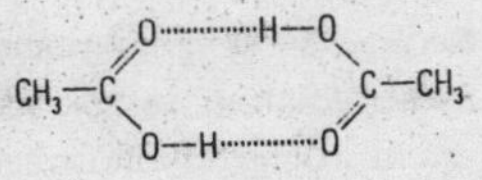

1 Which of the following molecules might be expected to show appreciable hydrogen bonding in the liquid state?
   a C_6H_6
   b $(CH_3)_3C—C(CH_3)_3$
   c C_4H_9OH
   d $C_5H_{11}COOH$

2 The relative molecular mass of the carboxylic acid, $C_6H_5—COOH$, is measured as 244 (in a non-aqueous solvent) although the formula suggests that it should be 122. Draw a diagram to show why the measured relative molecular mass is double that expected.

3 The boiling points of a series of compounds with three carbon atoms are shown in the table.

		b.p. / °C
A	Propan-1-ol	98
B	Propanoic acid	141
C	Propanone	56
D	Propan-2-ol	83
E	Propanal	49

   a Name the functional group present in each molecule.
   b Plot a bar chart of boiling points of these compounds.
   c Explain why some values of the boiling points are significantly higher than others.

27.4   Preparation of carboxylic acids

Summary of routes:

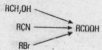

By oxidation

Carboxylic acids can often be made by **oxidation** of primary alcohols and aldehydes. The starting material is refluxed (*Figure 27.4.1*) with excess aqueous potassium dichromate(VI) acidified with dilute sulphuric acid. If the starting material is a primary alcohol and insufficient oxidising solution is used (or if the product is distilled off as it is formed), the product will contain large amounts of aldehyde (see Ch 19.2). Hence the need for oxidation under reflux with excess oxidising agent if the carboxylic acid is required.

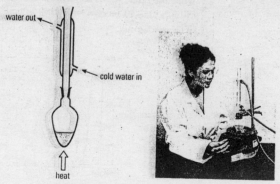

Figure 27.4.1   Reflux apparatus

Oxidation in organic chemistry normally involves the gain of oxygen, or the loss of hydrogen to form water. The symbol [O] is often included in the equation. It is, however, perfectly possible to write balanced redox equations for organic reactions, although you do not need to learn them!

The oxidation of the primary alcohol can be shown as happening in two steps:

e.g.

$$CH_3-CH_2OH + [O] \xrightarrow[\text{distil}]{\text{K}_2\text{Cr}_2\text{O}_7/\text{dil. H}_2\text{SO}_4} CH_3-CHO + H_2O$$

$$CH_3-CHO + [O] \xrightarrow[\text{reflux}]{\text{K}_2\text{Cr}_2\text{O}_7/\text{dil. H}_2\text{SO}_4} CH_3-COOH$$

Do remember that it is worth trying to include the reagents and the conditions necessary for any organic reaction – include them above and below the arrow in the equation (see Ch 14.8).

The fully balanced equation for producing ethanal from ethanol is:

$$3CH_3CH_2OH(aq) + Cr_2O_7^{2-}(aq) + 8H^+(aq) \xrightarrow[\text{and distil}]{\text{heat}} 3CH_3CHO(aq) + 2Cr^{3+}(aq) + 7H_2O(l)$$

It is not necessary to learn such equations, although you should eventually be able to work them out when necessary. After completion of the oxidation reaction to carboxylic acid, an aqueous solution of the acid can be obtained by distilling the mixture left in the reaction flask (*Figure 27.4.2*).

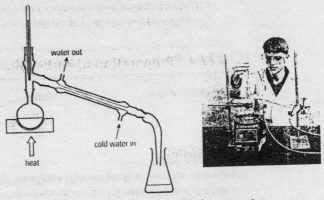

Figure 27.4.2   Distillation apparatus

The oxidation of the alcohol is clearly shown in the infra-red spectra of the reactant and product. While the broad —OH band, around 3500 cm^{-1}, in both the alcohol and the acid is still evident, the appearance of the >C=O band around 1700 cm^{-1} in the product clearly shows the formation of the acid group.

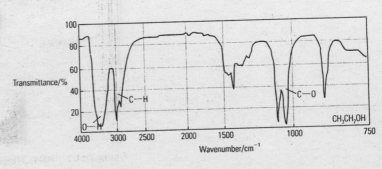

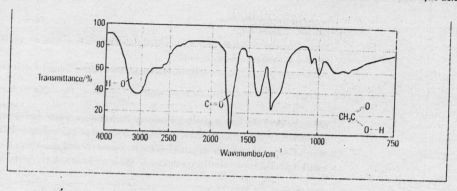

By hydrolysis

Carboxylic acids can be formed by *hydrolysis* of the nitrile or amide. This is, in fact, part of a reversible series of reactions linking the nitrile, the amide and the ammonium salt of the acid:

$$R\text{---}C\equiv N \underset{-H_2O}{\overset{+H_2O,\ P_4O_{10}}{\rightleftharpoons}} R\text{---}\underset{NH_2}{\overset{O}{C}} \underset{-H_2O}{\overset{+H_2O}{\rightleftharpoons}} R\text{---}\underset{O^-\ NH_4^+}{\overset{O}{C}} \underset{+NH_3(aq)}{\overset{+H^+(aq)}{\rightleftharpoons}} R\text{---}\underset{OH}{\overset{O}{C}}$$

Nitrile Amide Ammonium salt Carboxylic acid

Overall, this can be written as:

$$R\text{---}CN(l) + 2H_2O(l) + HCl(aq) \xrightarrow{heat} R\text{---}COOH(aq) + NH_4^+Cl^-(aq)$$

The hydrolysis can be done by heating the nitrile with either dilute aqueous hydrochloric acid or dilute aqueous sodium hydroxide. If the acid is used, the intermediate ammonium salt is converted directly into the carboxylic acid. If sodium hydroxide solution is used, it reacts with the ammonium salt to release ammonia gas and form the sodium salt of the acid.

$$CH_3COO^-\ NH_4^+(aq) + OH^-(aq) \longrightarrow CH_3COO^-Na^+(aq) + NH_3(g) + H_2O(l)$$

This sodium salt is then reacted with a 'strong acid', such as hydrochloric acid or sulphuric acid, to release the carboxylic acid.

$$CH_3COO^-Na^+(aq) + H^+(aq) \longrightarrow CH_3COOH(aq) + Na^+(aq)$$

Using a Grignard reagent

Another route starts by adding powdered magnesium to a solution of a bromoalkane in dry ethoxyethane. This forms the so-called *Grignard reagent* (see Ch 17.4).

$$RBr + Mg \longrightarrow \text{'}RMgBr\text{'}$$

Grignard reagent

By stirring solid carbon dioxide, 'dry ice', into this solution and treating the resulting mixture with hydrochloric acid, the carboxylic acid RCOOH is formed.

$$RMgBr \xrightarrow[\text{cold solution in ethoxyethane}]{\text{i) solid } CO_2,\ \text{ii) } H_2O/H^+} R\text{---}COOH + Mg(OH)Br$$

Other methods

There are other individual ways of preparing specific carboxylic acids, particularly in the preparation of the industrially important acids.

Benzoic acid is prepared by oxidising methylbenzene using catalysts and air.

$$C_6H_5-CH_3 + 3[O] \xrightarrow[150\,°C]{O_2 + catalyst} C_6H_5-COOH + H_2O$$

Ethanoic acid is a very important industrial chemical used as a starting material for many different products. Aerobic fermentation is still used as a way of producing vinegar, but more important routes include the preparation from crude oil fractions and natural gas. There is no need to remember the industrial conditions, which are included to illustrate what is done on the industrial scale.

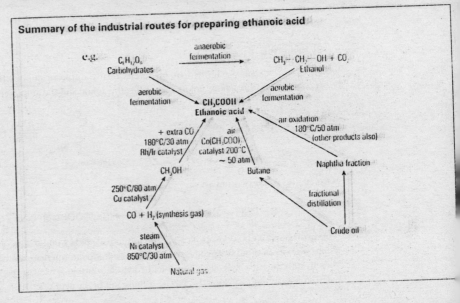

Summary of the industrial routes for preparing ethanoic acid

SELF-ASSESSMENT
QUESTIONS
27.4

1. Name suitable starting materials and the reagents needed to prepare the following compounds. Use a different method for each preparation.
   a. Propanoic acid
   b. 4-phenylbutanoic acid
   c. 2-ethylpentanoic acid

2. State the reagents and conditions needed to carry out the following reactions:
   a. $CH_3-CH_2-CH_2OH \longrightarrow CH_3-CH_2-COOH$
   b. $CH_3-CHO \longrightarrow CH_3-COOH$
   c. $CH_3-CH_2CN \longrightarrow CH_3-CH_2-COOH$
   d. $CH_3CH_2CONH_2 \longrightarrow CH_3CH_2COOH$

3. Outline how you would attempt to carry out the following chemical conversions. Remember that there might be a number of steps involved in any particular conversion. Make sure that you mention the reagents and conditions needed for each step.
   a. $CH_3-CH_2-CH_2Cl \longrightarrow CH_3-CH_2-COOH$
   b. $C_6H_6 \longrightarrow C_6H_5-COOH$
   c. $CH_3-CH_2Br \longrightarrow CH_3-CH_2-COOH$

27.5 Acidic properties of the carboxyl group

Carboxylic acids all ionise to some extent in water (see Ch 23.1). This is helped considerably by the polarity of the $>C=O$ bond. The electron movements are shown below:

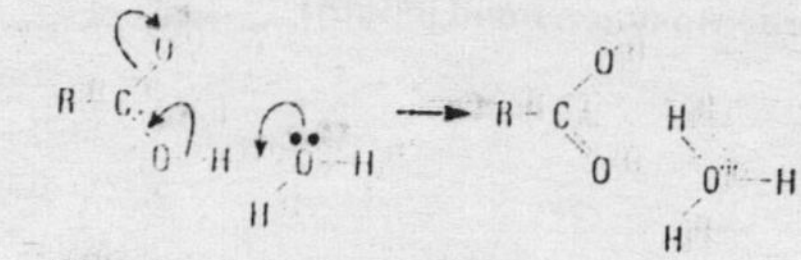

(The curly arrows represent the movement of a pair of electrons – see Ch 14.7 for the correct use of these curly arrows.)

It is the release of the hydrogen ion to the water molecule, to form $H^{\oplus}$(aq) or $H_3O^{\oplus}$(aq), which allows these compounds to act as acids.

The number of protons (which is what $H^{\oplus}$ ions are) released from one molecule of the carboxylic acid gives the *basicity* of the acid.

Ethanoic acid, $CH_3—COOH$, is *monobasic*;

ethanedioic acid, COOH
      |
     COOH is *dibasic*;

and citric acid, (2-hydroxypropane-1,2,3-tricarboxylic acid)

$H_2C—COOH$
   |
$HO—C—COOH$
   |
$H_2C—COOH$ is *tribasic*.

<table>
<tr><td>

The sodium salt of ethanoic acid has an unusual use. When a saturated solution of the salt crystallises the process is exothermic. The saturated solution is prepared by gently warming the solid crystals, $CH_3COO^{\ominus}\,Na^{\oplus}.3H_2O$, when the salt will dissolve in the water already present in the crystals. As this solution cools, heat is slowly given out as the crystals form. Packets of the saturated solution are sold as hand warmers which can be reused by gently warming the packet to re-dissolve the solid.

</td><td>

Because a carboxylic acid releases protons when in water, the solution will have a pH of less than 7 and will give the usual reactions of an acid:

● Neutralisation with an alkali (such as sodium hydroxide solution) or a base (such as copper(II) oxide) to form a salt:

e.g.

$$CH_3—COOH(aq) + OH^{\ominus}(aq) \xrightarrow[\text{temperature}]{\text{room}} CH_3—COO^{\ominus}(aq) + H_2O(l)$$

$$2CH_3—COOH(aq) + O^{2\ominus}(aq) \xrightarrow[\text{temperature}]{\text{room}} 2CH_3—COO^{\ominus}(aq) + H_2O(l)$$

● Release of hydrogen with a moderately reactive metal such as magnesium (very reactive metals such as sodium would be too dangerous):

e.g.

$$2CH_3—COOH(aq) + Mg(s) \xrightarrow[\text{temperature}]{\text{room}} 2CH_3—COO^{\ominus}(aq) + Mg^{2\oplus}(aq) + H_2(g)$$

● Release of carbon dioxide with any carbonate:

e.g.

$$2CH_3—COOH(aq) + CO_3^{2\ominus}(aq) \xrightarrow[\text{temperature}]{\text{room}} 2CH_3—COO^{\ominus}(g) + CO_2(g) + H_2O(l)$$

Examination of the structure of the resulting anion helps us to understand this ability of the carboxyl group to release a proton:

</td></tr>
</table>

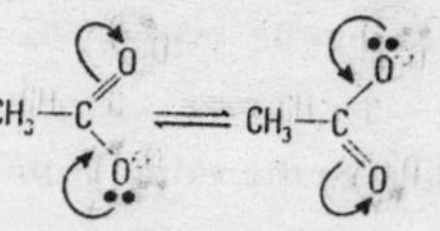

Note that the extra negative charge can be on either oxygen atom: it is partly *delocalised* and this adds to the stability of the anion, which, in turn, means that there is more likelihood of it being formed.

Bond lengths can be measured by the diffraction of X-rays by crystals. Measurements on sodium methanoate crystals show that the carbon–oxygen bond length is intermediate between the single and double bond lengths, and both carbon–oxygen bonds are the same length (*Figure 27.5.1*).

Carbon–oxygen bond lengths

C—O 0.122 nm

C—O 0.143 nm

C=O 0.127 nm

(in sodium methanoate)

Figure 27.5.1

This is physical evidence for delocalisation, and any carboxylate ion can be represented as:

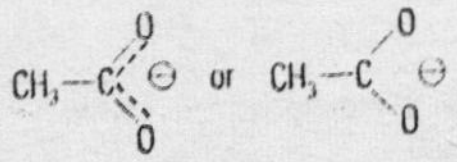

Write balanced chemical equations for the following reactions (you may use the symbols [H] and [O] in the reduction and oxidation reactions):

a ethanol and potassium dichromate in acid solution;

b ethanoic acid and sodium carbonate solution;

c propanal and potassium dichromate in acid solution;

d 2-methylpropanoic acid and aqueous potassium hydroxide solution;

e aqueous propanoic acid and magnesium metal.

SELF-ASSESSMENT
QUESTIONS
27.5

27.6   Carboxylic acids as weak acids

Carboxylic acids are *weak acids*. They only dissociate (break up into ions) to a small extent in water; the dissociation increases with increasing dilution.

e.g. For ethanoic acid:

$$CH_3COOH(aq) \overset{H_2O}{\rightleftharpoons} CH_3COO^-(aq) + H^{\oplus}(aq)$$

The acid dissociation constant, K_a (Ch 23.3) gives a measure of the tendency of an acid to ionise:

$$K_a = \frac{[CH_3COO^-(aq)]_{eqm}[H^{\oplus}(aq)]_{eqm}}{[CH_3COOH(aq)]_{eqm}} = 1.8 \times 10^{-5} \text{ mol dm}^{-3}$$

(The 'eqm' subscript is used in this equation to emphasise the need to use concentrations at equilibrium in all K_a expressions.)

A general expression for any monobasic acid, HA, is:

$$K_a = \frac{[H^{\oplus}(aq)]_{eqm}[A^-(aq)]_{eqm}}{[HA(aq)]_{eqm}}$$

Looking at this expression, it becomes clear that a small value of K_a means a low concentration of both the anion and $H^{\oplus}$ at equilibrium: i.e. that dissociation has only happened to a small extent.

The smaller the value of K_a, the weaker the acid.

The relationship between K_a and pK_a is the same as that between [$H^{\oplus}$] and pH (Ch 23.2).

$$pK_a = -\log_{10}K_a$$

So: **the higher the value of pK_a, the weaker the acid.**

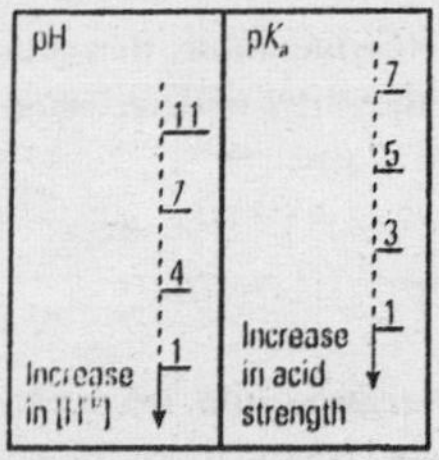

Remember, that as with pH, pK_a values are logarithmic. A difference of one in the pK_a value means a difference of 10 in the K_a value; a difference of two in pK_a means a difference of 100 in the K_a value.

Acid	Formula	K_a / mol dm^{-3}	pK_a
Methanoic	H—COOH	1.7×10^{-4}	3.8
Ethanoic	CH$_3$—COOH	1.7×10^{-5}	4.8
Propanoic	CH$_3$—CH$_2$—COOH	1.3×10^{-5}	4.9
Butanoic	C$_3$H$_7$—COOH	1.5×10^{-5}	4.8
Chloroethanoic	CH$_2$Cl—COOH	1.3×10^{-3}	2.9
Dichloroethanoic	CHCl$_2$—COOH	5.0×10^{-2}	1.3
Trichloroethanoic	CCl$_3$—COOH	2.3×10^{-1}	0.6
2-Chlorobutanoic	CH$_3$—CH$_2$—CHCl—COOH	1.6×10^{-3}	2.8
3-Chlorobutanoic	CH$_3$—CHCl—CH$_2$—COOH	7.9×10^{-5}	4.1
Benzoic	C$_6$H$_5$—COOH	6.3×10^{-5}	4.2

Table 27.6.1   Acid dissociation constants for some organic acids

The strengths of individual carboxylic acids vary quite widely (see *Table 27.6.1*). Inspection of the table indicates that the strength depends on the atom or group of atoms attached to the carbon atom which is **next to the carboxyl group**.

The acidity of the carboxyl group is due to the tendency to lose its proton by ionisation. Any *electron releasing group* (such as the methyl group) attached to the key carbon atom (marked * in *Figure 27.6.1*) will cause a drift of electrons in the C—C bond towards the carboxyl group, which will hinder ionisation and make the acid weaker. So ethanoic acid, $pK_a = 4.8$, is a weaker acid than methanoic acid, $pK_a = 3.8$, since ethanoic acid has the methyl group attached to the key carbon atom.

On the other hand, an *electron withdrawing group* (such as one or more chlorine atoms) will have the opposite effect. Chloroethanoic acid has a chlorine atom attached to the key carbon atom, which withdraws electrons and makes the acid much stronger.

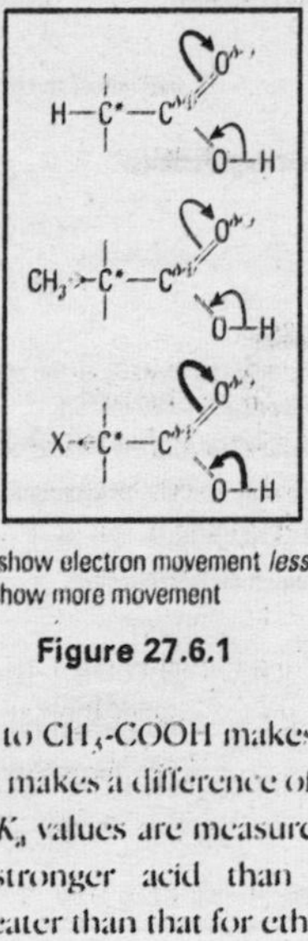

Thinner arrows show electron movement *less* than usual,
thicker arrows show more movement

Figure 27.6.1

Notice that changing from H-COOH to CH$_3$-COOH makes a difference of one in the pK_a value; but one chlorine on the key carbon atom makes a difference of nearly two, and three chlorine atoms makes a difference of more than four. pK_a values are measured on a logarithmic scale. This means that CCl$_3$-COOH is a **very** much stronger acid than CH$_3$-COOH – the value of K_a for trichloroethanoic acid is 10 000 times greater than that for ethanoic acid.

A closer look at the pK_a values in *Table 27.6.1* shows that the more chlorine atoms that are substituted on the 'key' carbon atom the more acidic is the compound. So the order of decreasing acidity is:

$$\xrightarrow{\hspace{2cm}} \text{decreasing acid strength}$$

$$CCl_3-COOH \; > \; CHCl_2-COOH \; > \; CH_2Cl-COOH \; > \; CH_3-COOH$$

pK_a	0.6	1.3	2.9	4.8

You will also see from the table that acidity decreases as the position of the chlorine atom becomes moved further away from the carboxylic acid group. Hence, 2-chlorobutanoic acid is more acidic than 3-chlorobutanoic acid.

$$\xrightarrow{\hspace{2cm}} \text{decreasing acid strength}$$

$$CH_3-CH_2-CHCl-COOH \; > \; CH_3-CHCl-CH_2-COOH \; > \; CH_2-CH_2-CH_2-COOH$$

pK_a	2.8	4.1	4.8

Calculating the pH of organic acid solutions (Ch 23.2)

Since these acids are weak acids, you will need to know the acid dissociation constant, K_a, of the acid as well as the concentration of the acid.

For ethanoic acid, which dissociates:

$$CH_3COOH(aq) \; \rightleftharpoons \; CH_3COO^-(aq) + H^+(aq)$$

K_a is defined as:

$$K_a = \frac{[CH_3COO^-(aq)][H^+(aq)]}{[CH_3COOH(aq)]} \quad \text{(for concentrations at equilibrium)}$$

Worked example

As an example, calculate the pH of a 0.020 mol dm^{-3} solution of 3-chlorobutanoic acid, for which $K_a = 7.9 \times 10^{-5}$ mol dm^{-3}

Equation: $\quad CH_3CHClCH_2COOH(aq) \; \rightleftharpoons \; CH_3CHClCH_2COO^-(aq) + H^+(aq)$

Data:

Initial concentrations	0.020 mol dm^{-3}	0	0
Ratio:	1	1	1

(remember these numbers are in front of each formula in the equation)

Relevant expressions: $\quad K_a = \dfrac{[CH_3CHClCH_2COO^-(aq)][H^+(aq)]}{[CH_3CHClCH_2COOH(aq)]}$

$$= 7.9 \times 10^{-5} \text{ mol } dm^{-3}$$

$$pH = -\log_{10}[H^+(aq)]$$

Calculation:
Since the acid is a weak acid, the amount of the free acid dissociated is very small and so the concentration of the free acid is still very nearly 0.020 mol dm^{-3}.

$$\therefore [CH_3CHClCH_2COOH] = 0.020 \text{ mol } dm^{-3}$$

Since the ratio of anion to hydrogen ion is 1:1, the concentration of both is the same.

$$\therefore [CH_3CHClCH_2COO^-] = [H^+]$$

The K_a equation then becomes:

$$K_a = \frac{[H^+][H^+]}{[CH_3CHClCH_2COOH]} = \frac{[H^+]^2}{[CH_3CHClCH_2COOH]}$$

$$\therefore \; [H^+]^2 = K_a \times [CH_3CHClCH_2COOH]$$

so $\quad [H^+] = \sqrt{(K_a \times [CH_3CHClCH_2COOH])}$

$$= \sqrt{7.9 \times 10^{-5} \times 0.020} = 1.265 \times 10^{-3} \text{ mol } dm^{-3}$$

$$pH = -\log_{10}(1.26 \times 10^{-3}) = 2.9$$

Some points to remember:

- Write out the relevant equation.
- Collect the data together: the initial concentrations and K_a value.
- Define the terms K_a and pH.
- Since it is weak the acid does not dissociate appreciably:
  $$\therefore [HA]_{eqm} \approx [HA]_{initial}$$
- For a monobasic acid, $[R-COO^-] = [H^+]$.
- For a weak acid, $[H^+] = \sqrt{K_a \times (\text{concentration of the acid})}$.

Activity

ACTIVITY

If you have access to a spreadsheet, such as EXCEL, enter K_a as 'A1' and concentration as 'B1'. In 'C1' enter the following formula, which will give the pH of a weak acid.

= -LOG(SQRT(A1*B1))

Some compounds found in nature have more than one carboxyl group – for example, ethanedioic acid ('oxalic acid'), a poisonous compound found in rhubarb leaves, and 'citric acid', which is found in citrus fruit. Other very important compounds have a carboxyl group plus at least one other functional group. One example is the amino acid series (Ch 28.5), the building blocks of proteins; the simplest amino acid is 'glycine' (aminoethanoic acid). Another example is 'lactic acid' (2-hydroxypropanoic acid).

Organic acids in nature

A large number of organic acids are found in nature, where they are often the cause of a sharp or sour taste.
Ant stings are mainly due to the presence of methanoic acid, $H-COOH$.
The flavour of vinegar (ethanoic acid, CH_3-COOH) is familiar in any pickled food and is very evident in wine or beer that has 'gone off'. 'Malic acid' is found in unripe apples.
Sour milk contains 'lactic acid', and the smell of rancid butter is partially due to butanoic acid. Lactic acid is also formed by anaerobic respiration in muscles which are not supplied with sufficient oxygen; this is why you feel stiff and sore after unaccustomed exercise.

| 2-Hydroxypropane-1,2,3-tricarboxylic acid (citric acid) | Ethanedioic acid (oxalic acid) | 2-Hydroxybutanedioic acid (malic acid) | 2-Hydroxypropanoic acid (lactic acid) | Butanoic acid |

SELF-ASSESSMENT QUESTIONS 27.6

1 Calculate the pH of the following solutions of organic acids. The K_a data are available in *Table 27.6.1*.
  a 0.025 mol dm^{-3} ethanoic acid
  b 0.040 mol dm^{-3} 3-chlorobutanoic acid
  c 0.5 mol dm^{-3} methanoic acid
  d 0.20 mol dm^{-3} benzoic acid

2 a Write full displayed formulae for butan-1-ol, $C_4H_{10}O$, and butanoic acid, $C_4H_8O_2$.

b Write the structures of the other isomers of $C_4H_{10}O$ containing the alcohol group. Give the systematic name for each of these isomers.

c Explain why these other isomers cannot be used to prepare butanoic acid.

d Show, by means of a chemical equation, including reagents and conditions, how you might convert butan-1-ol to butanoic acid. (Remember, use [O] for oxidation equations in organic chemistry.)

e Butanoic acid dissociates when dissolved in water. Write an equation to illustrate this dissociation, and use this equation to define K_a, the acid dissociation constant of butanoic acid.

f Calculate the pH of a 0.15 mol dm^{-3} solution of butanoic acid ($K_a = 1.5 \times 10^{-5}$ mol dm^{-3}).

27.7 Substitution reactions of the OH group

The loss of a proton from the carboxyl group has already been discussed (Ch 27.5). In the same way that –OH in alcohols is replaced by halogen using PCl_3, PBr_3, PCl_5, or SCl_2O, so the –OH in carboxyl acids can be replaced by halogen. Sulphur dichloride oxide, 'thionyl chloride', SCl_2O, is often preferred for this reaction as the only by-products are gases.

e.g.

$$CH_3COOH(l) + SCl_2O(l) \xrightarrow{\text{room temperature}} CH_3COCl(l) + SO_2(g) + HCl(g)$$

With phosphorus pentachloride, PCl_5, the equation is:

$$CH_3COOH(l) + PCl_5(s) \xrightarrow{\text{room temperature}} CH_3COCl(l) + PCl_3O(l) + HCl(g)$$

Whichever way this reaction is carried out, the fumes of hydrogen chloride are always very evident, but when PCl_5 is used it is quite difficult to separate the liquid products. The special case of esterification is considered in Ch 27.9.

27.8 Reduction reactions

Carboxylic acids are hard to reduce. The simple carbonyl group in aldehydes and ketones can be reduced with comparatively gentle reagents, including sodium tetrahydrido-borate(III), $NaBH_4$, in water (Ch 19.3). Carboxylic acids require the much more powerful reagent, lithium tetrahydridoaluminate(III), $LiAlH_4$, in anhydrous ethoxyethane. No water must be present. It is not possible to stop the reduction at the halfway stage to get the aldehyde.

$$CH_3COOH + 4[H] \xrightarrow[\substack{\text{anhydrous} \\ \text{ethoxyethane}}]{LiAlH_4} CH_3CH_2OH + H_2O$$

1 Compare the reactions of ethanol and of ethanoic acid with the following reagents; include balanced equations and the conditions for any reactions that do occur.

   a Sodium metal

   b Sodium hydroxide solution

   c Lithium tetrahydridoaluminate(III)

2 Identify the lettered conditions, reagents and compounds in the following reaction sequence:

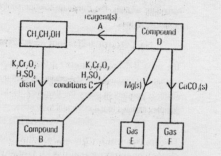

3 2-Hydroxypropanoic acid (lactic acid) is the chemical responsible for the sour taste and texture in yoghurt. The lactose in milk is converted into lactic acid, and the low pH coagulates the protein and so thickens the yoghurt.

   a Analytical data for 2-hydroxypropanoic acid show 40.0% carbon, 6.7% hydrogen and 53.3% oxygen. Show, by calculation, that this corresponds to the empirical formula CH_2O.

   b Work out the relative molecular mass of 2-hydroxypropanoic acid if the molecular formula is $C_3H_6O_3$.

   c The structural formula is CH_3—CH(OH)—COOH, which can exist as two isomers. What type of isomerism is possible with this structure? Draw three-dimensional diagrams to represent these two isomers.

   d What functional group, other than the carboxylic acid group, is present in this molecule?

   e Is 2-hydroxypropanoic acid, $pK_a = 3.87$, a stronger acid than propanoic acid, $pK_a = 4.9$? Explain this difference in acidic property.

   f Calculate the pH of a solution containing 9.0 g dm^{-3} of 2-hydroxypropanoic acid, using the data in **e**.

   g Write an equation to illustrate the reaction of 2-hydroxypropanoic acid with sulphur dichloride oxide, SCl_2O.

   h In the presence of lithium tetrahydridoaluminate(III) in ethoxyethane, the acid can be reduced. Write a displayed formula of the product. What is the systematic name of this compound?

27.9   Esterification

An ester is formed by heating a carboxylic acid with an alcohol in the presence of an acid catalyst.

e.g. $CH_3-COOH + C_2H_5OH \xrightarrow[\text{catalyst}]{\text{acid}} \rightleftharpoons CH_3-COOC_2H_5 + H_2O$

Semi-micro preparation of esters

A small-scale preparation of an ester can be done in a test tube using the top end of the test tube as an air condenser to reflux the reaction mixture.

$$R\text{---}COOH + R'OH \longrightarrow R\text{---}COOR' + H_2O$$

A tube containing 1 cm³ of the organic acid, mixed with 1 cm³ of an alcohol and 2 drops of concentrated sulphuric acid, is carefully heated over a small Bunsen flame so that the vapours can be seen condensing on the side of the test tube. After refluxing for a few minutes, the mixture is poured into a beaker containing 25 cm³ of 1 mol dm⁻³ sodium carbonate solution. The carbonate solution neutralises the acids (sulphuric and excess organic acid) thus:

$$Na_2CO_3(aq) + 2H^+(aq) \longrightarrow 2Na^+(aq) + CO_2(g) + H_2O(l)$$

Test tube hand held via an extended clothes peg

Gentle heat

The smell of the particular ester is then released. Many esters are used in flavouring and essences, both natural and synthetic. For example, pentyl ethanoate was used to flavour 'pear drop' sweets. Synthetic flavourings are made from mixtures of esters to mimic the flavour of the fruit (cherry, raspberry, etc). They can, however, never exactly match the complex mixtures found in the real fruit.

The mechanism for the acid-catalysed esterification reaction starts with nucleophilic attack by the alcohol on the protonated acid. The most common mechanism for esterification is as follows:

The displayed formula of an ester is:

e.g.

Acid stem | Alcohol stem

The reverse sequence is the mechanism for hydrolysis of an ester (Ch 27.10).

Esters are named as follows:
- **the alcohol stem comes at the start of the ester name;**
- **the acid stem provides the second part of the name;**
- **the name of the ester usually ends with *-anoate*.**

The example above is therefore *ethyl ethanoate*.

Another example is:

Acid from pentanoic acid | Alcohol was methanol

Esters all must contain the pattern:

In linear format:
—COOC⊂

The name of this ester is methyl pentanoate.

It is important to be able to distinguish between esters and ketones, particularly when they are written in linear formulae: $CH_3\text{---}CO\text{---}CH_3$ is the linear structure of a ketone, whereas $CH_3\text{---}COOC_2H_5$ is an ester. You must look for the —COOC— group to recognise the ester (it may sometimes still be written as —CO₂C—).

Even the catalysed reaction between carboxylic acid and alcohol is low, and it always reaches an equilibrium so that high yields are not possible (see Ch 22.1). For this reason, the commercial synthesis of esters usually involves reacting an alcohol with a more reactive compound such as the acyl chloride or acid anhydride (see Ch 27.11).

Equilibrium constants in esterification (see Ch 22.1)

Although initial concentrations may be given, you must remember that the equilibrium constant is calculated using the concentrations *at equilibrium*. As long as the equation is known, the final concentrations can be deduced, provided that the initial concentrations and one of the concentrations at equilibrium are known.

For the reaction:

$$CH_3COOH(l) + C_2H_5OH(l) \rightleftharpoons CH_3COOC_2H_5(l) + H_2O(l)$$

The equilibrium constant, K_c, is defined as:

$$K_c = \frac{[CH_3COOC_2H_5(l)][H_2O(l)]}{[CH_3COOH(l)][C_2H_5OH(l)]}$$

(for concentrations at equilibrium)

Can you see why K_c has no units? The concentration terms in this equation should all be calculated from (moles/volume). In this reaction, because the number of moles of reactants equals the number of moles of products, the moles/volume terms all cancel out, so K_c *has no units.* The number of moles can be used directly, rather than converted into concentration.

As an example, calculate the equilibrium constant for the esterification of ethanoic acid with ethanol from the following data. A mixture containing 1.0 moles of ethanoic acid and 1.4 moles of ethanol is mixed with a small volume of acid catalyst. The temperature is kept at 25 °C throughout the reaction. At equilibrium, 0.77 moles of ethyl ethanoate are found to be present. The total volume of the reaction mixture is V dm³.

Equation:

$$CH_3COOH(l) + C_2H_5OH(l) \rightleftharpoons CH_3COOC_2H_5(l) + H_2O(l)$$

Data:

Initial concentration
$\frac{1.0}{V}$ mol dm⁻³ $\frac{1.4}{V}$ mol dm⁻³   0   0

Equilibrium 'concentration'
$\frac{0.77}{V}$ mol dm⁻³

Ratio:
1   1   1   1

(remember these are the numbers in front of each formula)

Relevant expressions:

$$K_c = \frac{[CH_3COOC_2H_5][H_2O]}{[CH_3COOH][C_2H_5OH]}$$

Calculation:

If 0.77 mol of ethyl ethanoate is produced, then 0.77 moles of water must also be produced. Also, to make 0.77 moles of each product, 0.77 moles of each of the ethanoic acid and ethanol must be used. (This is all because of the 1:1 ratio throughout the reactants and products.)

Therefore, the concentrations at equilibrium are:

Ethanoic acid, 1.0 − 0.77 $= \frac{0.23}{V}$ mol dm⁻³

Ethanol, 1.4 − 0.77 $= \frac{0.63}{V}$ mol dm⁻³

Ethyl ethanoate and water $= \frac{0.77}{V}$ mol dm⁻³

$$K_c = \frac{[CH_3COOC_2H_5(l)][H_2O(l)]}{[CH_3COOH(l)][C_2H_5OH(l)]} = \frac{0.77/V \times 0.77/V}{0.23/V \times 0.63/V}$$

The volumes cancel out, so:

$$K_c = \frac{0.77 \times 0.77}{0.23 \times 0.63} = 4.1$$

Examples of esters and their uses

Ethyl 2-methylbutanoate	$CH_3CH_2CHICH_3)COOCH_2CH_3$	Apple flavour
3-Methylbutyl ethanoate	$CH_3COOCH_2CH_2CH(CH_3)_2$	Pear flavour
1-Methylpropyl ethanoate	$CH_3COOCH(CH_3)CH_2CH_3$	Banana flavour
Ethyl methanoate	$HCOOCH_2CH_3$	Raspberry flavour
Butyl butanoate	$CH_3CH_2CH_2COOCH_2CH_2CH_2CH_3$	Pineapple flavour
Phenylmethyl ethanoate	$CH_3COOCH_2C_6H_5$	Oil of jasmine
2-Ethanoyloxybenzoic acid ('acetylsalicylic acid')		Aspirin
Methyl 2-hydroxybenzoate ('methyl salicylate')		Muscle rub
Ethyl ethanoate	$CH_3COOCH_2CH_3$	Glue solvent
Methyl 2-cyanopropenoate	$CH_2=C(CN)COOCH_3$	Superglue
Ethenyl ethanoate	$CH_3COOCH=CH_2$	PVA glue

In this table you can see that certain esters are used in adhesives, as perfumes and flavourings, and even in painkillers.

Esters which are more complex occur as insect attractants (pheromones). These can sometimes be synthesised and used to control insect pests. The *cis,trans-* and the *cis,cis-*isomers of 7,11-hexadecadienyl ethanoate:

$$CH_3COO(CH_2)_6CH=CHCH_2CH_2CH=CH(CH_2)_3CH_3$$

are both released by the pink bollworm moth as attractants.

Q

SELF-ASSESSMENT QUESTIONS 27.9

1 Identify the following as acids, esters, alcohols or ethers:
   a $CH_3CH(CH_3)CH(OH)CH_2CH_3$
   b $CH_3CH_2COOCH(CH_3)CH_3$
   c $CH_3CH_2OCH_3$
   d $CH_3CH_2CH_2COOCH_3$

2 Draw the full structures of these named esters:
   a methyl butanoate
   b propyl 2-methylpropanoate
   c butyl ethanoate
   d 1-methylethyl propanoate

3 Name these acids and esters:
   a $CH_3CH_2CH_2COOH$
   b $CH_3(CH_2)_5COOH$
   c $CH_3C(CH_3)_2COOCH_2CH_3$
   d $CH_3CH_2COOCH_3$

4 Write balanced equations for the following reactions, including the conditions necessary to carry out the reactions. (Use [H] and [O] in any redox reactions.)

a Propanoic acid and phosphorus pentachloride.

b Butanoic acid and lithium tetrahydridoaluminate(III).

c 2-Methylbutanal and acidified potassium dichromate(VI) solution.

d Ethanol and 2-methylpropanoic acid.

e Butanoic acid and sodium hydroxide solution.

27.10 Hydrolysis of esters

Esters can be hydrolysed by both acids and alkalis. In hydrolysis, the overall result is that a water molecule is added for each ester linkage which is broken.

$$CH_3-C\!\!\begin{array}{c}O\\\\O-C_2H_5\end{array} + H_2O \xrightleftharpoons{reflux} CH_3-C\!\!\begin{array}{c}O\\\\OH\end{array} + C_2H_5OH$$

Acid hydrolysis leads to equilibrium (Ch 27.9) and the yield of products is never 100%. Alkaline hydrolysis breaks up the ester completely.

If the hydrolysis is done using dilute hydrochloric or sulphuric acid, the carboxylic acid itself is produced. If an alkaline solution is used, the sodium or potassium salt of the carboxylic acid is made. This has then to be treated with hydrochloric or sulphuric acid to free the carboxylic acid.

During hydrolysis, which of the two C—O bonds breaks?

$$CH_3-C\!\!\begin{array}{c}O\\\\O\!\!\xi\!\!-CH_2CH_3\end{array}$$

This problem was solved using water containing the heavy oxygen isotope, ^{18}O. The products were purified, and then examined using a mass spectrometer (see Ch 1.3).

$$CH_3-C\!\!\begin{array}{c}O\\\\O-CH_2CH_3\end{array} + H\!\!-\!\!\overset{18}{O}\!\!-\!\!H \rightleftharpoons CH_3-C\!\!\begin{array}{c}O\\\\^{18}OH\end{array} + H-OCH_2CH_3$$

The heavy oxygen appears only in the acid, showing that it must be the C—O bond next to the >C=O group in the ester which is broken.

Application of hydrolysis of esters

Natural fats or vegetable oils are **triglycerides** – esters of 'glycerol', propane-1,2,3-triol. The carboxylic acids, which are joined to the triol by the ester linkages, contain twelve to eighteen carbon atoms – always an even number (*Figure 27.10.1*).

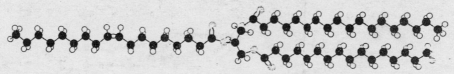

Figure 27.10.1   Model of a triglyceride

They may be saturated (i.e. with no carbon–carbon double bonds in the chain), monounsaturated (with one >C=C< bond), or polyunsaturated (two or more >C=C< bonds). Examples are given in the table.

Sources of some natural acids from glycerides

	Common name	Systematic name of acid	Source
Saturated acids			
$CH_3-(CH_2)_{16}-COOH$	Stearic	Octadecanoic	Animal fats
Mono-unsaturated acids			
$CH_3-(CH_2)_7-CH=CH-(CH_2)_7-COOH$	Oleic	cis-octadec-9-enoic	Olive oil
$CH_3-(CH_2)_5-CH=CH-(CH_2)_7-COOH$	Palmitoleic	cis-hexadec-9-enoic	Sunflower oil
Poly-unsaturated acids			
$CH_3-(CH_2-CH=CH)_3-(CH_2)_7-COOH$	Linolenic	cis,trans,trans-octadeca-9,11,13-trienoic	Linseed oil

N.B. Do not try to remember the detailed names or structures.

Soaps are the sodium or potassium salts of these carboxylic acids. They are made by prolonged boiling of fats or oils with sodium or potassium hydroxide solution. This hydrolyses the triglyceride esters. The products are propane-1,2,3-triol ('glycerol') and soap. 'Glycerol' is also known as 'glycerine', which is commercially quite valuable; it is used, for example, in making cake icing easier to shape – and also used for making explosives such as gelignite!

It is because of this process of boiling fats with alkalis that the old name for the alkaline hydrolysis of *any* ester was 'saponification' – the making of soap.

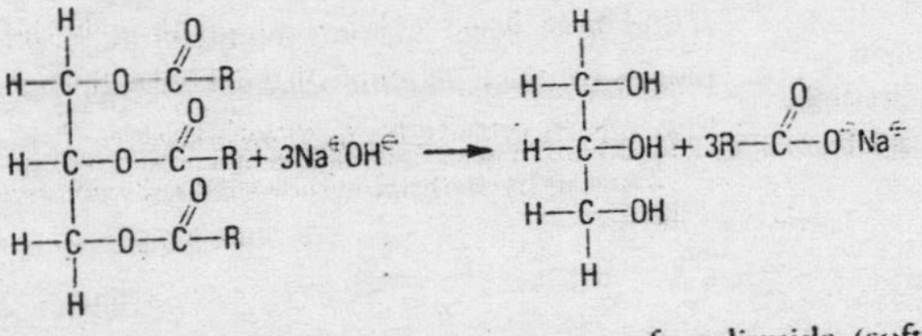

Sodium soaps are usually solids, but potassium soaps are often liquids (soft soap). As they are sodium and potassium salts, they are soluble in water.

Mechanisms for the hydrolysis of an ester

(The mechanism for ester formation is given in Ch 27.9. Try to see how the formation mechanism is the reverse of the hydrolysis mechanism.)

Remember, it is the $-C \underset{O-R}{\overset{O}{<}}$ bond which is broken.

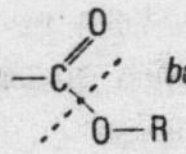

In acid conditions ($H^{\oplus}$ catalyses the reaction):

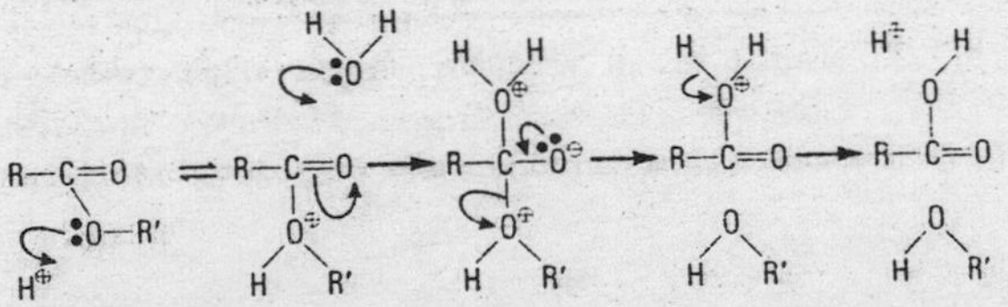

In alkaline conditions:

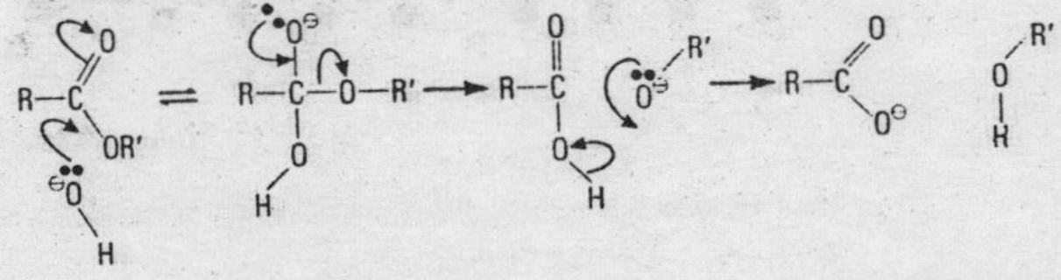

As soaps are salts of weak acids, they are slightly alkaline in solution (see Ch 23.5). They do not work well at low pH, because the carboxylate anion will then gain a proton to become the acid molecule, which is much less soluble.

$$RCOO^{\ominus}(aq) + H^{\oplus}(aq) \longrightarrow RCOOH(s)$$

Soaps are also much less effective in 'hard water', which contains calcium and magnesium ions and is found in chalk and limestone areas. This is because the carboxylate anions combine with the calcium or magnesium ions and precipitate out as an insoluble scum.

$$2RCOO^{\ominus}(aq) + Ca^{2\oplus}(aq) \longrightarrow (RCOO)_2Ca(s)$$

Detergents destroy the surface tension of water and improve its ability to wet surfaces and fabrics. All soaps are detergents, but not all detergents are soaps; most washing powders are non-soapy. **Non-soapy detergents** do not form scum in hard water, and use sulphonate or phosphate anions rather than carboxylate. A typical non-soapy detergent is:

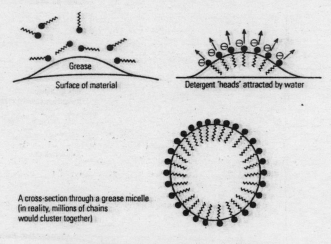

A detergent can be thought of as a long hydrocarbon tail with an ionic head. The tail does not mix with water but mixes easily with fat or grease – it is 'lipophilic'. The head mixes easily with water but not with fat or grease – it is 'hydrophilic'.

The ionic head enables the detergent to dissolve in water, but when it does so it forms spherical groups of carboxylate chains called *micelles*. The ionic ends are on the outside, exposed to the water; the lipophilic chains are hidden inside.

However, when the micelle comes into contact with a grease spot, the tails dissolve in the greasy material, and the ionic heads are left on the surface. These are attracted by the polar water molecules; the grease is helped off the surface and forms small spheres – micelles again – surrounded by negative charge (*Figure 27.10.2*).

Grease
Surface of material

Detergent 'heads' attracted by water

A cross-section through a grease micelle (in reality, millions of chains would cluster together)

Figure 27.10.2   Grease micelle in water

Next time you are washing up and confronted with a greasy or oily pan, pour hot water into the pan. You will see the oily material floating on top of the water. Now add a little soap or detergent, and swirl the contents around with your hand. You will see the liquid go cloudy as the water and oil 'mix'.

Because they carry the same charge, the spheres repel each other – so that they cannot rejoin and precipitate. Overall, the detergent helps the grease and water to mix, so that the dirt can easily be washed away.

Activity

Use molecular models to show the structure of the di-ester, dimethyl propandioate, $CH_3OOC.CH_2.COOCH_3$.

Use this model to show how the molecule is hydrolysed by two molecules of water to form a di-acid and two molecules of methanol.

SELF-ASSESSMENT QUESTIONS 27.10

Complete the following reaction sequence by putting in the missing names, formulae and conditions:

1 $BrCH_2CH_2CH_2CH_3 \xrightarrow[\text{conditions B}]{\text{reagent A}} HOCH_2CH_2CH_2CH_3$

2 $HOCH_2CH_2CH_2CH_3 \xrightarrow[\text{reflux}]{K_2Cr_2O_7/H_2SO_4(aq)}$ compound C + H_2O

3 $C_6H_5CH_2CH_2OH + CH_3COOH \underset{\text{reflux}}{\overset{\text{HCl(aq)}}{\rightleftharpoons}}$ compound D

4 $CH_3CH_2CN \xrightarrow[\text{conditions F}]{\text{reagent E}} CH_3CH_2COOH$

5 $(CH_3)_2CHCH_2COOH \xrightarrow[\text{conditions H}]{SCl_2O}$ compound G + SO_2 + HCl

6 compound I $\xrightarrow[\text{conditions J}]{LiAlH_4} CH_3CH(CH_3)CH_2OH$

7 $CH_3COOC_2H_5 + H_2O \overset{H^{\oplus}/\text{reflux}}{\rightleftharpoons}$ compound K + compound L

27.11 Derivatives of carboxylic acids

The carboxylic acids give rise to four closely-related homologous series:

Carboxylic acids
$$R-C\overset{O}{\underset{OH}{}}$$

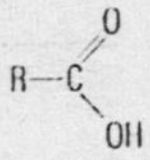

Esters
$$R-C\overset{O}{\underset{OR'}{}}$$

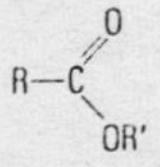

Acyl chlorides
(sometimes called acid chlorides)
$$R-C\overset{O}{\underset{Cl}{}}$$

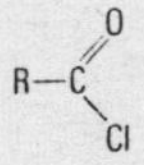

Acid anhydrides
$$R-C\overset{O}{\underset{O}{}}$$
$$R-C\overset{O}{\underset{O}{}}$$

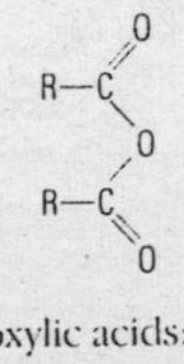

Two other homologous series are also related to the carboxylic acids:

Amides
(sometimes called acid amides)
$$R-C\overset{O}{\underset{NH_2}{}}$$

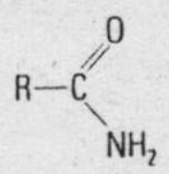

Nitriles
(sometimes called cyanides)
$$R-C\equiv N$$

These both contain nitrogen and will be discussed in Ch 28, as will the biologically important amino acids.

The effect of replacing the hydroxy group in the acid with other electronegative atoms or groups changes the reactivity of the molecule. The more electronegative the substituted atoms or groups, the greater the general reactivity of the new compound.

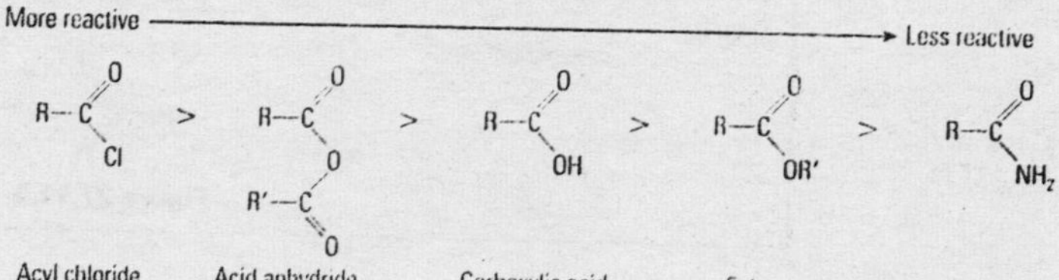

More reactive ———————————————————————→ Less reactive

| Acyl chloride | Acid anhydride | Carboxylic acid | Ester | Amide |

As we have already discussed (Ch 27.2), the $\overset{X}{\underset{}{>}}C^{\delta+}=O^{\delta-}$ group is polarised. The degree to which the carbon atom is positive, and therefore able to be attacked by a nucleophile, depends on whether X attracts or releases electrons.

In an acyl chloride, polarity in $>C=O$ is increased by the presence of chlorine:

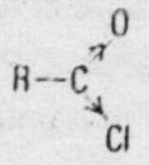

In the acid anhydride, polarity in $C=O$ is also increased:

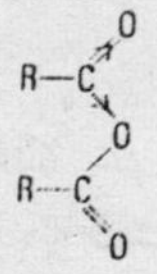

The polarisation is less when hydrogen is attached to the second oxygen, to form a carboxylic acid:

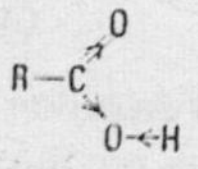

It is even less in an ester:

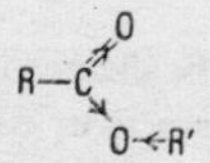

and is less still in the amide when oxygen is replaced by the less electronegative nitrogen:

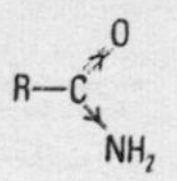

Acyl chlorides

Acyl chlorides are formed by reacting the carboxylic acid with PCl_5 or SCl_2O. The name of the acyl chloride is based on the name of the starting acid, with the -oic of the acid changing to -oyl. Ethanoyl chloride, CH_3—$COCl$, is formed from ethanoic acid, CH_3—$COOH$:

$$CH_3COOH(l) + PCl_5(s) \xrightarrow{\text{room temperature}} CH_3COCl(l) + PCl_3O(l) + HCl(g)$$

Or, in a general equation using sulphur dichloride oxide:

$$RCOOH(l) + SCl_2O(l) \xrightarrow{\text{room temperature}} RCOCl(l) + SO_2(g) + HCl(g)$$

The increased polarity in acyl chlorides (*Figure 26.11.1*) allows them to be very useful in making other compounds.

Chlorine withdraws electrons and makes the carbon atom more susceptible to nucleophilic attack.

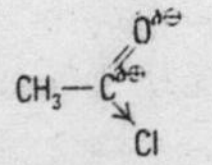

$$CH_3 - C\begin{smallmatrix} O^{\delta-} \\ \\ Cl \end{smallmatrix}$$

Figure 27.11.1

e.g.

$$CH_3COCl + C_2H_5OH \xrightarrow{\text{room temperature}} CH_3COOC_2H_5 + HCl$$
Alcohol $\qquad$ Ester

$$CH_3COCl + C_6H_5OH \longrightarrow CH_3COOC_6H_5 + HCl$$
Phenol $\qquad$ Ester

It is worth noting that esters of phenol **cannot** be made directly using the carboxylic acid; they have to be made by reacting phenol with an acyl chloride or an acid anhydride.

You will meet two more reactions of acyl chlorides in Ch 28.1:

$$CH_3COCl + NH_3(aq) \xrightarrow[\substack{\text{violent} \\ \text{reaction}}]{\substack{\text{room} \\ \text{temperature}}} CH_3CONH_2 + HCl(g)$$

Acyl chloride + Ammonia $\longrightarrow$ Amide

$$CH_3COCl + C_2H_5NH_2 \longrightarrow CH_3CONHC_2H_5 + HCl$$

Acyl chloride + Amine $\longrightarrow$ Substituted amide $\longrightarrow$ further substitution

The acyl chlorides are rapidly **hydrolysed** when reacted with water:

$$CH_3COCl(l) + H_2O(l) \longrightarrow CH_3COOH(aq) + HCl(g)$$

The very rapid evolution of hydrogen chloride gas can be seen as steamy white fumes in moist air, and these fumes produce a dense white smoke if placed near a source of ammonia gas.

$$HCl(g) + NH_3(g) \longrightarrow NH_4Cl(s)$$

The effect of the electronegative oxygen atom on the C—Cl bond in the acyl chlorides results in a much greater reactivity compared to halogenoalkanes.

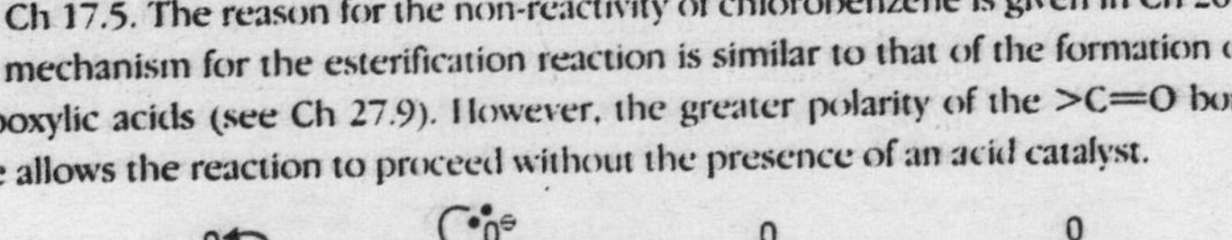

The reasons for the gradation of reactivity for primary, secondary and tertiary halogenoalkanes are given in Ch 17.5. The reason for the non-reactivity of chlorobenzene is given in Ch 26.11.

The mechanism for the esterification reaction is similar to that of the formation of esters using the carboxylic acids (see Ch 27.9). However, the greater polarity of the >C=O bond in the acyl chloride allows the reaction to proceed without the presence of an acid catalyst.

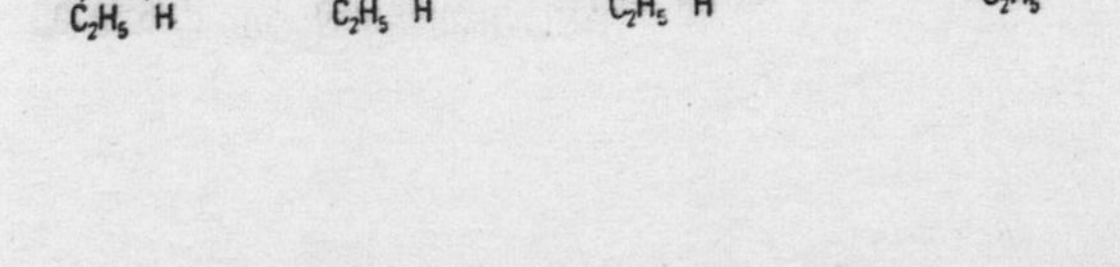

Acid anhydrides

Elimination of a water molecule from two carboxylic acid molecules produces a new acid derivative – an acid anhydride.

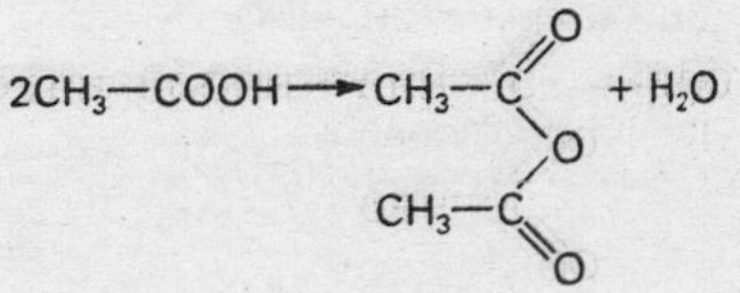

$$2CH_3-COOH \longrightarrow CH_3-C \overset{O}{\underset{O}{\big\langle}} \quad + H_2O$$
$$CH_3-C \overset{O}{}$$

However, this reaction is difficult under normal conditions. The linear formula for an anhydride is often written as $(RCO)_2O$.

Acid anhydrides are easily prepared by heating an acyl chloride with the anhydrous sodium salt of a carboxylic acid:

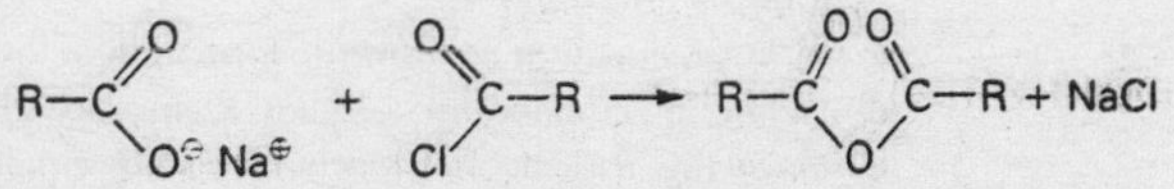

$$R-C \overset{O}{\underset{O^\ominus Na^\oplus}{\big\langle}} \quad + \quad \overset{O}{\underset{Cl}{\big\rangle}}C-R \longrightarrow R-C \overset{O \quad O}{\underset{O}{\big\langle \quad \big\rangle}} C-R + NaCl$$

The acid anhydrides are less reactive than the acyl chlorides, but they are more reactive than the parent acid. One advantage of the anhydrides is that hydrogen chloride is not produced in the reactions of the anhydride.

e.g.
$$(CH_3CO)_2O + CH_3CH_2OH \longrightarrow CH_3COOCH_2CH_3 + CH_3COOH$$

Ethanoic anhydride   Ethanol   Ethyl ethanoate   Ethanoic acid

---

Calculation of the yield in a reaction (see Ch 9.5)

Aspirin can be prepared by the esterification of 2-hydroxybenzoic acid using ethanoic anhydride according to the equation below. Typically, the reaction of 2.5 g of the acid produces 2.8 g of aspirin after recrystallisation from ethanol. To work out the percent yield you have first to calculate the theoretical yield.

Equation:

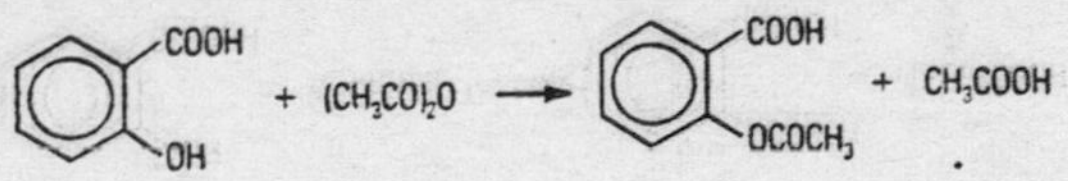

	2-Hydroxybenzoic acid	Ethanoic anhydride	Aspirin	Ethanoic acid
Data:	2.5 g		2.8 g	
M_r	138		180	
Moles: ($= mass/M_r$)	0.0181		0.0156	
Ratio:	1		1	

Calculation:
The theoretical yield should be 0.0181 moles since the ratio is 1:1
The actual yield is 0.0156 moles

The % yield is therefore $\dfrac{actual\ yield \times 100}{theoretical\ yield} = \dfrac{0.0156 \times 100}{0.0181} = 86\%$

One of the insect attractants (pheromones) used by the Greater Grain Borer (*Prostephanus truncatus*) is an ester with the systematic name 1-methylethyl-2-methylpent-2-enoate. Its alternative name is 'trunc-call', and it is used in grain stores to control the pest. The synthetic route used to prepare the compound starts with an unsaturated aldehyde, 2-methylpent-2-enal, $CH_3CH_2CH=C(CH_3)CHO$.

The final product has the linear formula: $CH_3CH_2CH=C(CH_3)COOCH(CH_3)CH_3$, and hence the displayed formula:

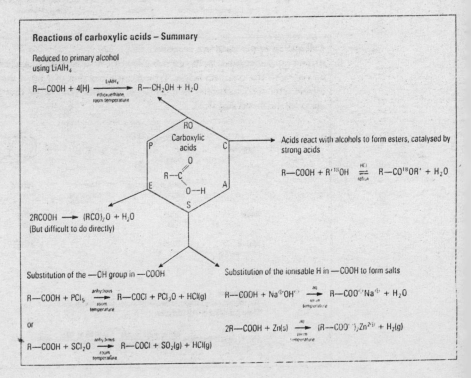

a   Since the aldehyde is quite reactive, the oxidation of the unsaturated aldehyde to the corresponding carboxylic acid has to be carried out using silver oxide. Write the displayed formula of the acid obtained.

b   Write the displayed formula and name of the alcohol that needs to be reacted with this acid to form the pheromone.

c   Write an equation to show the formation of the ester.

d   Define the equilibrium constant, K_c, for this reaction.

e   Since the acid–alcohol reaction reaches equilibrium, in the chosen synthetic route the acyl chloride is reacted with the alcohol. How would you prepare the acyl chloride from the free acid prepared in **a**?

f   This acyl chloride is then reacted with the appropriate alcohol to form the pheromone. Write an equation to show this reaction.

g   When 2.5 g of the aldehyde was used in this preparation, 3.0 g of the ester was recovered. Calculate the % yield in the reaction.

SUMMARY
27

In this chapter the following new ideas have been covered. *Have you understood and learned them?* Could you explain them clearly to a friend? The page references will help.

- Recognition of the carboxylic acid functional group and the naming of organic acids. p711
- The physical properties of carboxylic acids are dominated by hydrogen bonds. p713
- How acids are prepared in the laboratory. p715
- There is a variety of industrial routes for making ethanoic acid. p718
- The acid properties of carboxylic acids. p719
- The use of K_a and pK_a to recognise the strengths of acids. p720
- Explanation of the stability of the carboxylate anion. p719
- The general reactions of carboxylic acids. p712
- How esters are prepared. p725
- Recognition and naming of esters. p726
- The wide range of uses of esters. p728
- How to write the structural formula of a triglyceride. p729
- The usefulness of the hydrolysis of natural triglycerides. p730
- Writing equations for the reactions of acid chlorides and acid anhydrides. p733

A few earlier ideas have been revised in this topic. Are you sure about them now? The calculation of:

- The equilibrium constant for an esterification reaction. Ch 22.1
- The pH of a weak organic acid. Ch 23.2
- % yield for a reaction. Ch 9.5

EXAMINATION
QUESTIONS
27

1 a An organic compound has the composition 54.5% carbon, 36.4% oxygen and 9.1% hydrogen. Calculate the empirical formula of the compound.

b In the mass spectrum of the molecule, the molecular ion peak appears at m/e corresponding to 88. What is the relative molecular mass of the compound?

c What is the molecular formula of the compound?

d In the vapour phase at high pressure the relative molecular mass is twice the figure found in the mass spectrum above. Can you account for this?

e When the compound is reacted with lithium tetrahydridoaluminate(III) the infra-red spectrum changes and the peak present at $1700\ cm^{-1}$ in the original sample disappears. What reaction accounts for the disappearance of this peak?

2 a Nettle rash is an irritation caused by stinging nettles, and is partly due to the presence of methanoic acid in the hairs on nettle leaves. Write the full displayed formula of methanoic acid.

b Crushing nettle leaves in warm water gives a weak solution of methanoic acid. Write an equation to show how methanoic acid dissociates in water.

c Since the acid is a weak acid ($K_a = 1.7 \times 10^{-4}\ mol\,dm^{-3}$), this dissociation reaches equilibrium. Write the expression for the acid dissociation constant of methanoic acid.

d Calculate the pH of the solution if the concentration of methanoic acid in the nettle extract is $0.15\ mol\,dm^{-3}$.

e How would you expect methanoic acid to react with ethanol? Write a balanced equation which includes the conditions needed for the reaction to occur.

3 **a** Show how propanoyl chloride can be considered an electrophile when the compound is reacted with ethanol, although the reaction is normally explained as nucleophilic attack on the acyl chloride group.

b Explain why the reactivity of propanoyl chloride is much greater than the reactivity of the parent acid.

c Write an equation for the reaction of propanoic acid with sulphur dichloride oxide, SCl_2O.

4

For each of the organic compounds A to F, give the displayed formula and the name of the compound.

Do you remember ...?

- **Functional groups, homologous series, frameworks.** (Ch 14.2)
- Ammonia as a **weak base.** (Ch 23.3)
- The reactions of **arenes**. (Ch 26.2)
- The definition of a **nucleophile**. (Ch 14.7)
- The reactions of the **carboxylic acid derivatives**. (Ch 27.11)

In this chapter

We introduce several functional groups which contain nitrogen atoms, and also discuss the amino acids – a group of compounds which contain two functional groups and without which no life would be possible.

Amides	–	$R—CONH_2$
Nitriles	–	$R—C\equiv N$
Amines	–	$R—NH_2$
Diazonium salts	–	$ArN_2^{\oplus}Cl^{\ominus}$

28.1 **Amides**
28.2 **Nitriles**
28.3 **Amines**
28.4 **Diazonium salts**
28.5 **Amino acids**

28
organic nitrogen
compounds

28.1 Amides

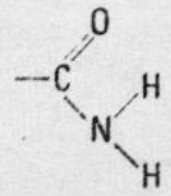

Figure 28.1.1 The amide group

Amides contain the functional group —CO—NH$_2$. As with esters, acyl chlorides and acid anhydrides, the >C=O (carbonyl) part of the functional group shows none of its usual properties because it is next to a very electronegative atom, in this case nitrogen, as shown in *Figure 28.1.1* (see Ch 19.1).

The names of amides take into account the fact that, as for the carboxylic acids which they resemble, the functional group contains a carbon atom which is part of the chain (see Ch 14.3). So:

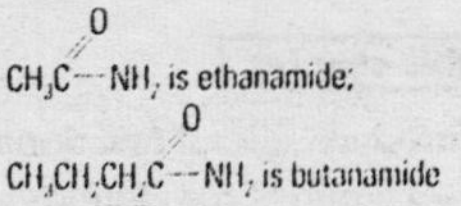

$$CH_3C{-}NH_2 \text{ is ethanamide;}$$

$$CH_3CH_2CH_2C{-}NH_2 \text{ is butanamide}$$

– see *Figure 28.1.2*.

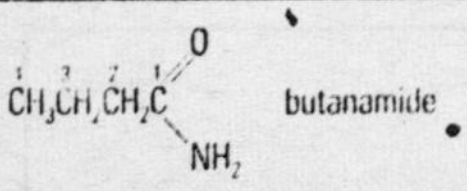

CH$_3$CH$_2$CH$_2$C   butanamide

NH$_2$

The —CONH$_2$ carbon atom is part of the longest chain

Figure 28.1.2 Naming amides

Amides are considered to be derivatives of carboxylic acids because they can be made fairly easily from carboxylic acids, and because the —OH part of the carboxylic acid group has been replaced by —NH$_2$. (But please note that this replacement cannot be done in one step!) Because of the link with carboxylic acids, amides are still sometimes known as 'acid amides'.

It might be expected that the amide group would have quite strong basic properties, similar to those of ammonia (see Ch 13.4) or the amines (see Ch 28.3). The polarity of the >C=O part of the amide group greatly reduces the likelihood that the lone pair on the nitrogen can be donated, so that amides show little basicity.

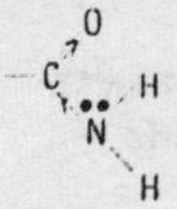

Preparation of amides

The most important routes for the preparation of amides start with either the acid or acyl chloride (*Figure 28.1.3*).

Preparation of amides by substitution

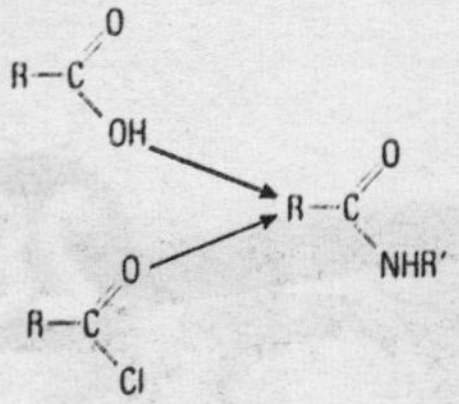

R' = hydrogen – simple amide
R' = alkyl group – N-substituted amide

Figure 28.1.3

Amides can be prepared by dehydration of the ammonium salts of the corresponding carboxylic acids; the salt is usually heated for some hours under reflux.

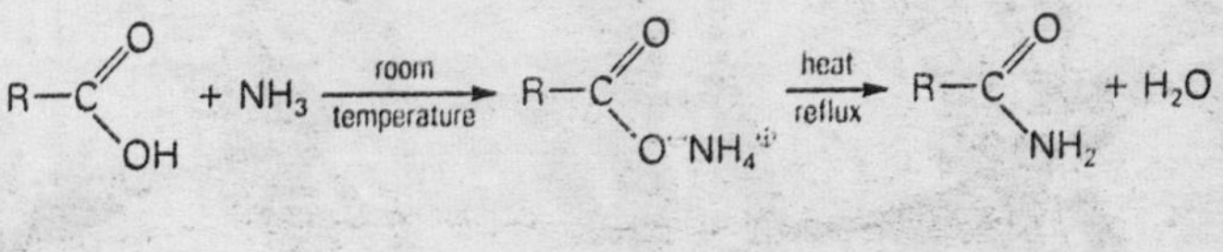

Substituted amides, in which one or both of the hydrogen atoms of the —$CONH_2$ groups are replaced by alkyl groups, can be made by direct reaction of an acyl chloride with a primary or secondary amine (see Ch 28.3 for the mechanism).

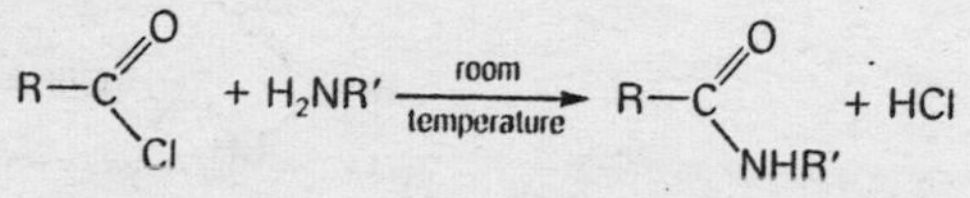

$$R-\underset{Cl}{\overset{O}{\underset{\|}{C}}} + H_2NR' \xrightarrow[\text{temperature}]{\text{room}} R-\underset{NHR'}{\overset{O}{\underset{\|}{C}}} + HCl$$

Such substituted amides are named according to the usual rules except for the group attached to the nitrogen atom, e.g.:

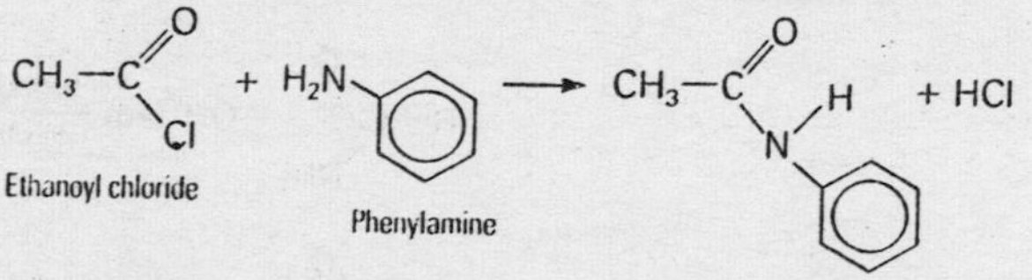

$$CH_3-\underset{Cl}{\overset{O}{\underset{\|}{C}}} + H_2N-\bigcirc \longrightarrow CH_3-\underset{\underset{\bigcirc}{N}}{\overset{O}{\underset{\|}{C}}}H + HCl$$

Ethanoyl chloride   Phenylamine   N-phenylethanamide

Groups attached to the nitrogen atom are recognised in the name of the compound by writing N- (e.g. $CH_3CONHC_6H_5$ is called *N-phenyl*ethanamide).

The substituted amide group, —CONH—, is found in polyamide polymers (see below) and also in all peptides and proteins (see Ch 28.5).

Reaction of an acyl chloride with ammonia itself gives the simple amide:

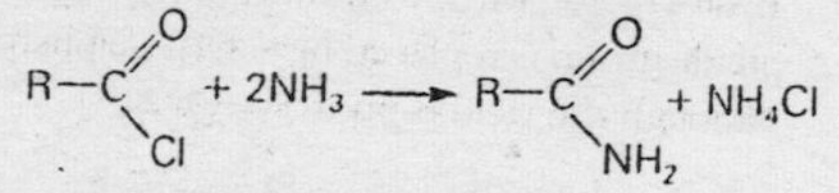

$$R-\underset{Cl}{\overset{O}{\underset{\|}{C}}} + 2NH_3 \longrightarrow R-\underset{NH_2}{\overset{O}{\underset{\|}{C}}} + NH_4Cl$$

Reactions of amides

The amides are reactive compounds, and their reactions are summarised in the box.

Reactions of amides

e.g.

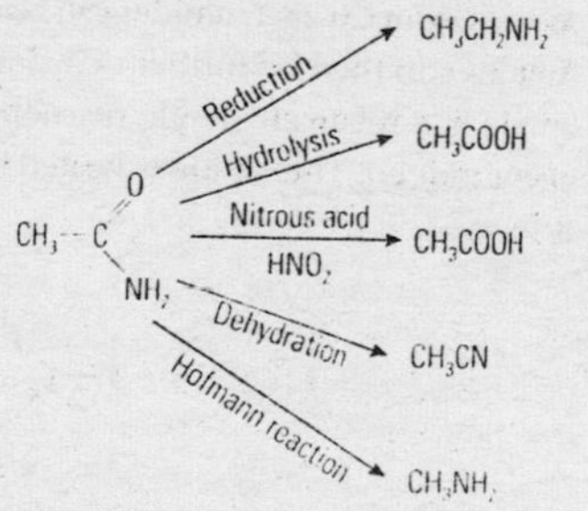

Remember:

When reading the names of organic compounds, you must recognise the different groups in the name, and divide the name up mentally.

N-methylpropanamide has three distinct parts, even though it appears as one word:

N-methyl   —N—CH_3
propan CH_3CH_2C—
amide   —$CONH_2$

But one hydrogen of the amide group has been replaced by the methyl group, and the carbon of the amide group is also part of the propane chain.
So the formula is:

   $CH_3CH_2CONHCH_3$

Reduction

Like carboxylic acids, amides can be reduced with lithium tetrahydridoaluminate(III) in ethoxyethane (ether). Amides give primary amines.

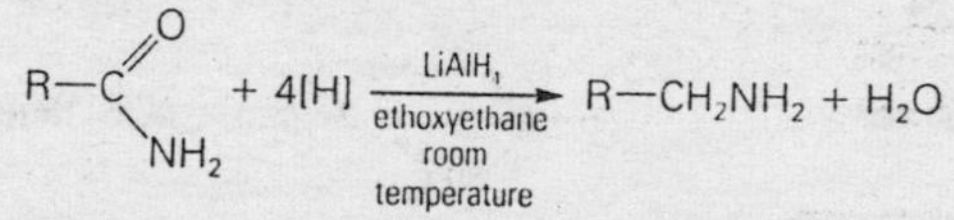

$$R-\underset{NH_2}{\overset{O}{\underset{\|}{C}}} + 4[H] \xrightarrow[\substack{\text{room}\\\text{temperature}}]{\substack{\text{LiAlH}_4\\\text{ethoxyethane}}} R-CH_2NH_2 + H_2O$$

Hydrolysis

Amides can be hydrolysed by heating under reflux with dilute hydrochloric acid, rather like esters.

$$R-\overset{\displaystyle O}{\underset{\displaystyle NH_2}{C}} + H_2O(l) + H^+(aq) \xrightarrow[HCl(aq)]{heat/reflux} R-\overset{\displaystyle O}{\underset{\displaystyle OH}{C}} + NH_4^+$$

If an alkali is used instead of an acid, ammonia is given off and the salt of the carboxylic acid is formed. The carboxylic acid is formed if the salt is treated with dilute hydrochloric acid.

$$R-\overset{\displaystyle O}{\underset{\displaystyle NH_2}{C}} + OH^-(aq) \xrightarrow[reflux/heat]{NaOH(aq)} R-\overset{\displaystyle O}{\underset{\displaystyle O^-}{C}} + NH_3(g)$$

$$R-\overset{\displaystyle O}{\underset{\displaystyle O^-}{C}} + H^+(aq) \xrightarrow[\substack{room \\ temperature}]{HCl(aq)} R-\overset{\displaystyle O}{\underset{\displaystyle OH}{C}}$$

Reaction with nitrous acid

Amides react with nitrous acid, HNO_2 or HONO (produced by reacting sodium nitrite and dilute hydrochloric acid), to form carboxylic acids with nitrogen being given off. The $-NH_2$ part of the amide group is replaced by $-OH$. (Aliphatic primary amines react in the same way to give alcohols, although the yield is poor; see Ch 28.3).

$$R-\overset{\displaystyle O}{\underset{\displaystyle NH_2}{C}} + HONO \xrightarrow[\substack{room \\ temperature}]{NaNO_2/HCl} R-\overset{\displaystyle O}{\underset{\displaystyle OH}{C}} + N_2(g) + H_2O(l)$$

Dehydration

As we saw in Ch 28.1, amides can be made by dehydration of the ammonium salt of a carboxylic acid. Amides can then be further dehydrated to make nitriles, RCN. Whereas the dehydration to form an amide is a relatively gentle reaction, dehydration to a nitrile requires a very powerful dehydrating agent indeed. The amide is heated with phosphorus(V) oxide, P_4O_{10}. The nitrile is distilled off as it is formed.

$$R-\overset{\displaystyle O}{\underset{\displaystyle NH_2}{C}} \xrightarrow[distil]{P_4O_{10}} R-C\equiv N + H_2O(l)$$

August Wilhelm von Hofmann (1818–92) studied under Liebig at Giessen in Germany, but came to London in 1845 as the first Professor of the Royal College of Chemistry in Oxford Street, London. He stayed for 20 years; he and his students started systematic organic chemistry in England. He returned to Germany in 1865. He discovered the 'Hofmann reaction' in 1881.

Loss of a carbon atom to form a primary amine (the Hofmann reaction)

Amides warmed with a solution of bromine in aqueous sodium hydroxide (i.e. aqueous sodium bromate(I)) give a primary amine with one carbon atom less. *This reaction can be used as the key step in descending a homologous series.*

$$R-\overset{\displaystyle O}{\underset{\displaystyle NH_2}{C}} + Br_2(aq) + 4OH^-(aq) \xrightarrow{warm} RNH_2 + 2Br^-(aq) + CO_3^{2-}(aq) + 2H_2O(l)$$

Polyamide polymers

Alkenes form polymers by *addition* reactions (see Ch 16.7); the double-bonded monomer molecules join together by new single carbon–carbon bonds, and the unsaturation is destroyed. **Polyamide** polymers, on the other hand, are formed by **condensation** reactions: that is, addition followed immediately by elimination (*Figure 28.1.4*; see also Ch 14.6).

Addition polymerisation

$$nCH_2{=}CH_2 \longrightarrow {-}[CH_2{-}CH_2]{-}_n$$

Monomer $\longrightarrow$ Polymer

One type of monomer, giving only a single product

Condensation polymerisation

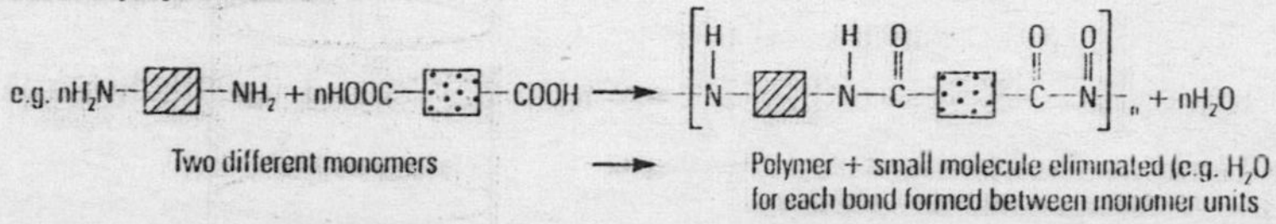

e.g. $nH_2N{-}\!\!\!\diagup\!\!\!\diagup\!\!\!{-}NH_2 + nHOOC{-}\boxed{\vdots}{-}COOH \longrightarrow \left[\begin{smallmatrix}H\\|\\N\end{smallmatrix}\!\!-\!\!\diagup\!\!\!\diagup\!\!\!-\!\!\begin{smallmatrix}H&&O\\|&&\|\\N&-&C\end{smallmatrix}\!-\boxed{\vdots}-\!\begin{smallmatrix}O&&O\\\|&&\|\\C&-&N\end{smallmatrix}\right]_n + nH_2O$

Two different monomers $\longrightarrow$ Polymer + small molecule eliminated (e.g. H_2O or HCl) for each bond formed between monomer units

Two types of monomer (or functional groups) are present, forming the polymer and another product. A single monomer can be used if it has two suitable functional groups, one at each end of the molecule: e.g. amino acids (see Ch 28.5) can form chains, as in proteins or in Nylon – see below.

Figure 28.1.4

In polyamides, the linkage between monomer units is —CONH—, as in substituted amides. The same linkage is found in peptides and proteins, where it is known as the **'peptide' link**. The differences between these compounds and the polyamides will be discussed in Ch 28.5.

Several types of the polyamide fibre Nylon have been made commercially. Their names depend on the lengths of the carbon chains in the monomers. A very common type is Nylon-6,6. This was made in industry by condensing a dicarboxylic acid with a diamine:

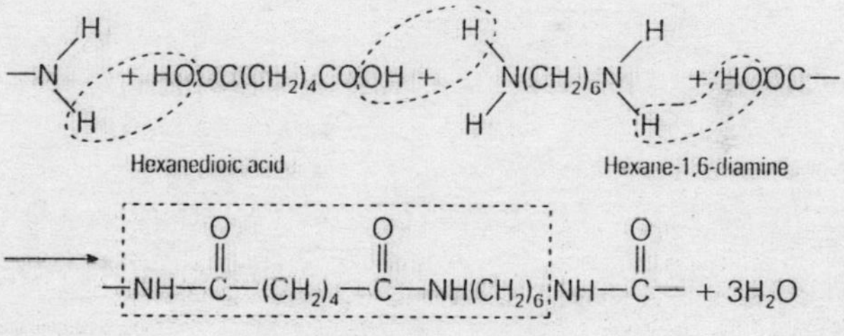

Hexanedioic acid   Hexane-1,6-diamine

the repeating unit

For every amide link, —CONH—, which is formed, a molecule of water is eliminated.

Poly(ethene) was discovered by accident (see Ch 16.7), but the polyamide polymers were the result of a deliberate search for a synthetic material to replace very expensive silk in parachute canopies. Wallace Carothers worked for the American company Du Pont. He was the first to classify polymers as 'addition' or 'condensation' polymers. He began his systematic studies of polymerisation in 1930, and his team successfully developed a method for making artificial rubber. Rubber, however, is cross-linked; the structure is three-dimensional. Any substitute for silk has to have a structure similar to that of silk – a linear fibre. Carothers discovered Nylon in 1935. Nylon was the first synthetic fibre to be made by a condensation reaction involving small units. Carothers died in 1937 at the tragically early age of 41, three years before Nylon went into full-scale commercial production.

In the laboratory, the reaction can be done at room temperature using the acyl chloride rather than the acid. An aqueous solution of the diamine is placed on top of a solution of the acyl chloride dissolved in trichloromethane (chloroform), which is denser than water and only dissolves slightly in it. Nylon-6,6 forms at the interface between the two layers and can be pulled out steadily as a thick thread using a pair of tongs (*Figure 28.1.5*).

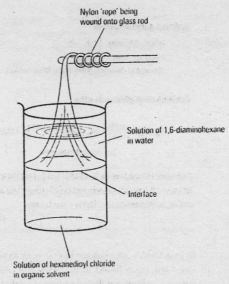

Nylon 'rope' being wound onto glass rod

Solution of 1,6-diaminohexane in water

Interface

Solution of hexanedioyl chloride in organic solvent

Figure 28.1.5

In this reaction a molecule of hydrogen chloride is eliminated for every amide link which is formed.

the repeating unit

Nylon-6 was originally prepared from 6-aminohexanoic acid; i.e. the one monomer molecule had both the reactive groups on it.

This monomer is, strictly speaking, an amino acid. However, it is very different from the amino acids found in nature, as we shall see in Ch 28.5.

SELF-ASSESSMENT QUESTIONS 28.1

1 Name the following amides.
  a $CH_3CH_2CONH_2$
  b $CH_3(CH_2)_3CONH_2$
  c $CH_3CH(CH_3)CH_2CONH_2$
  d $CH_3CH_2CONHC_2H_5$

2 Write out the displayed formulae of:
  a hexanamide
  b 2-methylpropanamide
  c N-methylpropanamide
  d N-ethyl-2-methylpropanamide

3 Write balanced equations for the following reactions:
  a ethanoyl chloride + ammonia
  b ethanoyl chloride + methylamine
  c ethanamide + hot dilute hydrochloric acid
  d butanamide + lithium tetrahydroaluminate(III)

4 Propanal can be converted into ethylamine in the following series of reactions. For each reaction, insert the conditions and reactants needed.
  a $CH_3CH_2CHO \rightarrow CH_3CH_2COOH$
  b $CH_3CH_2COOH \rightarrow CH_3CH_2COCl$
  c $CH_3CH_2COCl \rightarrow CH_3CH_2CONH_2$
  d $CH_3CH_2CONH_2 \rightarrow CH_3CH_2NH_2$

28.2 Nitriles

Nitriles contain the functional group $—C\equiv N$. As with carboxylic acids and their other derivatives, nitriles possess a functional group which contains a carbon atom and is therefore part of the chain. So, CH_3CN is named ethanenitrile, CH_3CH_2CN is propanenitrile, and so on.

Because they contain the $—CN$ group, nitriles used to be called 'cyanides'. However, the first members of the nitrile series are covalent liquids and not particularly poisonous, unlike the solid, water-soluble inorganic cyanides which contain the $CN^\ominus$ ion.

Preparation of nitriles

Formation of nitriles

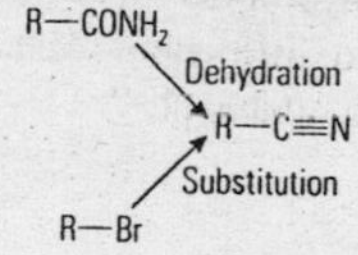

$$R—CONH_2$$
Dehydration
$$R—C\equiv N$$
Substitution
$$R—Br$$

Figure 28.2.1

By powerful dehydration of amides

A summary linking the ammonium salts of carboxylic acids through amides to nitriles is given below. The dehydration reactions have already been met in Ch 28.1.

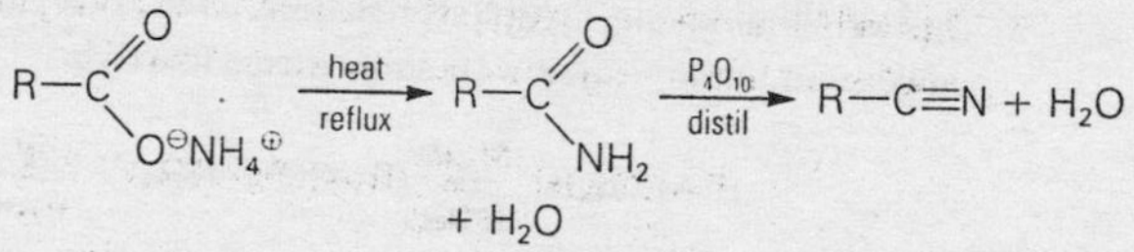

$$R—C\underset{O^\ominus NH_4^\oplus}{\overset{O}{\big\|}} \xrightarrow[\text{reflux}]{\text{heat}} R—C\underset{NH_2}{\overset{O}{\big\|}} + H_2O \xrightarrow[\text{distil}]{P_4O_{10}} R—C\equiv N + H_2O$$

745

By substitution of the halogen atom in a halogenoalkane with —CN

Usually the bromoalkane is used (see Ch 17.4). The halogenoalkane is heated under reflux with potassium cyanide in solution in aqueous ethanol. The nucleophile, $CN^{\ominus}$, attacks the carbon in the polarised carbon–halogen bond (*Figure 28.2.1*).

$$R{-}CH_2Br + CN^{\ominus} \xrightarrow[\text{neat}]{\substack{\text{KCN}\\\text{ethanol/water}}} R{-}CH_2CN + Br^{\ominus}$$

N.B. *This reaction can be used to increase the length of aliphatic carbon chains* and is therefore important in planning syntheses. It should be known together with the Hofmann reaction (Ch 28.1), which is a way of decreasing the length of a carbon chain.

The reactions of nitriles

Reactions of nitriles

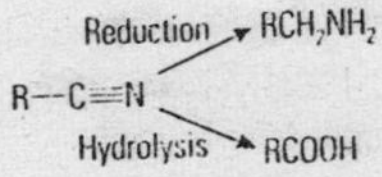

$$R{-}C{\equiv}N \quad \begin{array}{l} \xrightarrow{\text{Reduction}} RCH_2NH_2 \\ \xrightarrow{\text{Hydrolysis}} RCOOH \end{array}$$

Figure 28.2.2

Reduction

Nitriles are reduced to primary amines by reaction with sodium in ethanol, or by lithium tetrahydridoaluminate(III) in anhydrous ethoxyethane (ether). (This is usually the next step in any chain-lengthening sequence involving a nitrile, as the amine can then be converted into other functional groups – see Ch 28.3.)

$$R{-}C{\equiv}N + 4[H] \xrightarrow[\text{ethoxyethane}]{\text{LiAlH}_4} R{-}CH_2NH_2$$

Hydrolysis

Nitriles can be hydrolysed to amides, and hence to carboxylic acids (see Ch 28.1), by heating under reflux with dilute hydrochloric acid. If dilute alkali is used, ammonia is given off and the salt of the carboxylic acid is obtained.

$$RCN + H_2O \xrightarrow[\text{reflux}]{\text{HCl(aq)}} RCONH_2$$

$$RCONH_2 + H_2O(l) + H^{\oplus}(aq) \xrightarrow[\text{reflux}]{\text{heat}} RCOOH + NH_4{}^{\oplus}(aq)$$

If alkali is used: $\qquad RCONH_2 + OH^{\ominus}(aq) \xrightarrow[\text{reflux}]{\text{heat}} RCOO^{\ominus} + NH_3(aq)$

The dehydration and hydrolysis reactions which link the ammonium salts of carboxylic acids with amides and nitriles can now be summarised like this:

$$R{-}COOH \underset{\text{HCl(aq)}}{\overset{\text{NH}_3\text{(aq)}}{\rightleftharpoons}} R{-}COO^{\ominus}NH_4{}^{\oplus} \underset{+H_2O}{\overset{-H_2O}{\rightleftharpoons}} R{-}CONH_2 \underset{+H_2O}{\overset{-H_2O}{\rightleftharpoons}} R{-}CN$$

1 Identify the following as amides, nitriles or ammonium salts. Give the systematic names of each compound.

   a $CH_3CH_2CH_2CONH_2$

   b CH_3CH_2CN

   c $CH_3CH_2CH(CH_3)CONH_2$

   d $CH_3CH(CH_3)CN$

   e $CH_3CH_2CH_2COO^{\ominus}NH_4^{\oplus}$

2 Write the displayed formulae for:

   a ethanamide

   b butanenitrile

   c 3-methylbutanenitrile

   d N-methylbutanamide

3 What conditions and reagents are needed to carry out the following reactions?

   a $(CH_3)_2CH_2CONH_2 \longrightarrow (CH_3)_2CH_2CN$

   b $(CH_3)_2CH_2CONH_2 \longrightarrow (CH_3)_2CH_2COOH$

   c $(CH_3)_2CH_2CONH_2 \longrightarrow (CH_3)_2CH_2NH_2$

   d $(CH_3)_2CH_2CONH_2 \longrightarrow (CH_3)_2CH_2CH_2NH_2$

   e $CH_3CH(CH_3)CH_2COO^{\ominus}NH_4^{\oplus} \longrightarrow CH_3CH(CH_3)CH_2CONH_2$

28.3 Amines

Although molecules of primary and secondary amines can hydrogen bond to one another, because of the lower electronegativity of nitrogen compared with oxygen the hydrogen bonds in amines are weaker than those in alcohols. This means that the volatility of amines is higher than that of the corresponding alcohols, so that their boiling points are lower: e.g. ethanol, 78 °C; ethylamine, 17 °C.

Like ammonia, the lower amines have very powerful smells. The smell of rotting fish is caused largely by a mixture of secondary and tertiary amines released by decomposing proteins. The notorious smell of maggot factories producing bait for anglers is said to be due to triethylamine.

Amines are essentially molecules of ammonia in which one or more of the hydrogen atoms has been replaced by an alkyl group. If one hydrogen has been replaced, the result is a primary amine; replacement of two hydrogens gives a secondary amine, and of three gives a tertiary amine. Ammonia can form ammonium salts containing the ammonium ion, $NH_4^{\oplus}$; so it is possible to obtain quaternary ammonium salts, with all four hydrogens in the ammonium ion replaced by alkyl groups.

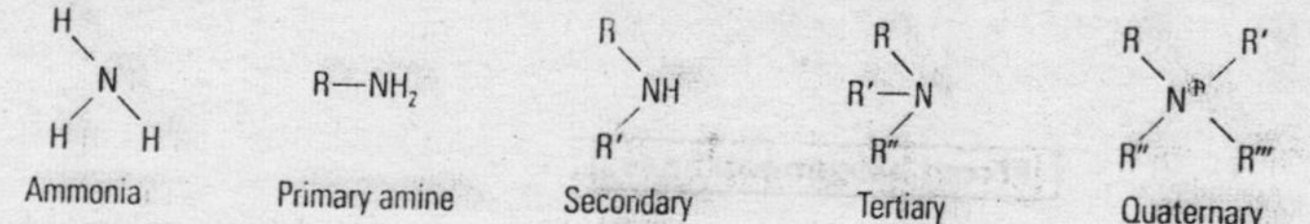

Ammonia   Primary amine   Secondary amine   Tertiary amine   Quaternary ammonium salt

Examples are:

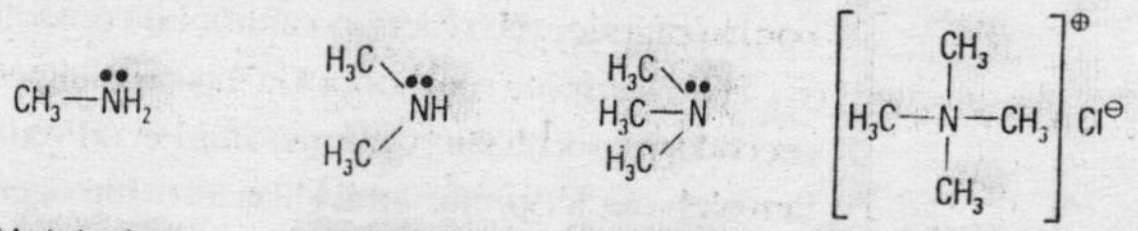

Methylamine   Dimethylamine   Trimethylamine   Tetramethylammonium

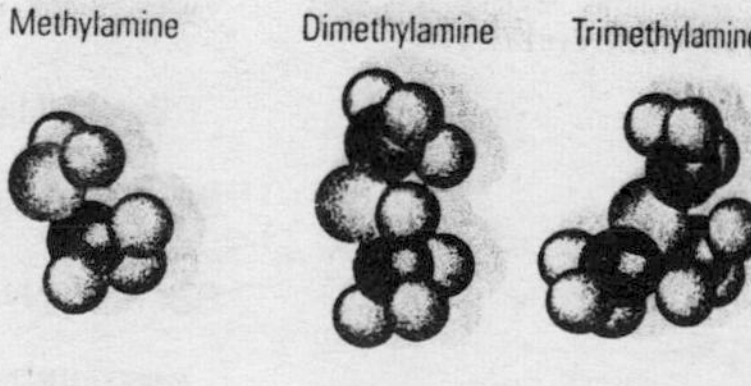

As with —OH and halogen, attachment of the $-NH_2$ group to a benzene ring leads to chemical properties which are often different from those found when the group is attached to an aliphatic or alicyclic framework. Some reactions of phenylamine will be discussed in the next section (Ch 28.4).

Primary, secondary or tertiary?

The differences between primary, secondary or tertiary structures lie in the availability of hydrogen atoms close to the functional group, and also the inductive effect of alkyl (particularly methyl) groups in making the lone pair of electrons in the amine functional group more available.

	Amines	Alcohols	Halogenoalkanes
Primary	$-NH_2$	$-CH_2-OH$	$-CH_2-X$

Two hydrogen atoms on the nitrogen atom (for amines) or the adjacent carbon atom (three if $-CH_3$)

Secondary	$>NH$	$>CH-OH$	$>CH-X$

One hydrogen atom on the nitrogen atom (for amines) or the adjacent carbon atom

Tertiary	$>N$	$>C-OH$	$>C-X$

No hydrogen atoms on the nitrogen atom (for amines) or the adjacent carbon atom

Notice that, in the amines, *the functional group itself* changes between primary, secondary and tertiary. With the alcohols, it is the *environment* of the functional group which changes. It is perfectly possible to obtain a tertiary alcohol from a primary amine (see 'Reaction of primary amines with nitrous acid', below).

Preparation of amines

Preparation of amines

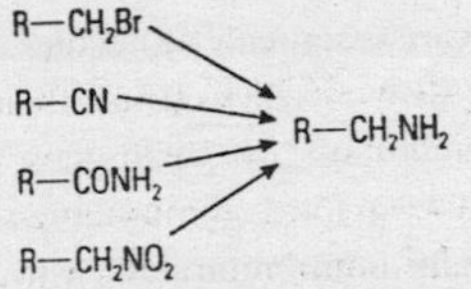

$$R-CH_2Br$$
$$R-CN$$
$$\searrow R-CH_2NH_2$$
$$R-CONH_2$$
$$R-CH_2NO_2$$

From halogenoalkanes

This method can, in principle, be used to prepare any kind of amine. A halogenoalkane should react with ammonia to give a primary amine, and with a primary amine to give a secondary amine, and so on. In practice, the reaction cannot be controlled, and a mixture results.

For example, reaction of a bromoalkane with ammonia will give a mixture of the primary, secondary and tertiary amines, and even the quaternary ammonium salt. In every case the reaction involves nucleophilic attack by the nitrogen of the amine on the carbon atom of the polar carbon–halogen bond.

$$RBr + NH_3 \longrightarrow RNH_2 + HBr$$
$$RBr + RNH_2 \longrightarrow R_2NH + HBr$$
$$RBr + R_2NH \longrightarrow R_3N + HBr$$
$$RBr + R_3N \longrightarrow R_4N^{\oplus}Br^{\ominus}$$

If this reaction is attempted, the mixture which results is hard to separate.

By reduction of amides and nitriles

Using lithium tetrahydridoaluminate(III) in ethoxyethane, amides and nitriles are reduced to primary amines (see Ch 28.1 and Ch 28.2).

$$R-C{\overset{O}{\underset{NH_2}{}}} + 4[H] \xrightarrow[\text{ethoxyethane}]{\underset{\text{in}}{LiAlH_4}} RCH_2NH_2 + H_2O$$

$$R-C{\equiv}N + 4[H] \xrightarrow[\text{ethoxyethane}]{\underset{\text{in}}{LiAlH_4}} RCH_2NH_2$$

By reduction of nitro compounds

Compounds which contain the $-NO_2$ group can be reduced using hydrogen and a nickel catalyst, or by using lithium tetrahydridoaluminate(III) in ethoxyethane. However, the specific reduction of nitrobenzene to phenylamine is often done using metallic tin and concentrated hydrochloric acid. A complex tin(IV) salt is formed from which the phenylamine can be liberated with excess alkali. Overall:

$$R-NO_2 + 6[H] \longrightarrow RNH_2 + 2H_2O$$

e.g.
$$\underset{\text{Nitrobenzene}}{C_6H_5NO_2} + 6[H] \xrightarrow[\text{ii) NaOH(aq)}]{\text{i) HCl/Sn}} \underset{\text{Phenylamine}}{C_6H_5NH_2} + 2H_2O$$

By the Hofmann reaction

With bromine and alkali the amide forms an amine. For details, see Ch 28.1. Remember, this reaction shortens the chain by one carbon atom.

Reactions of amines

Reactions of amines

The reactivity of amines of any kind depends on the availability of the 'lone pair' of electrons on the nitrogen atom (see Ch 2.4). Since nitrogen is less electronegative than oxygen, the lone pair of electrons on the nitrogen is more available than it is with the alcohols. Amines are better bases and nucleophiles than the alcohols. The donation of the lone pair on the nitrogen atom in amines results in amines acting:

- as quite good **bases** (donating the lone pair to an $H^{\oplus}$ ion);
- as excellent **ligands** (with transition metals, see Ch 25.4); and
- as good **nucleophiles** (being able to attack at the positive end of a polarised bond).

Amines as bases

Bases are proton 'acceptors' (see Ch 13.4). They do not in fact 'accept' protons: a lone pair of electrons is donated to the proton to form a dative covalent bond. Ammonia and the bases can do this with any suitable acid to form a salt, e.g.:

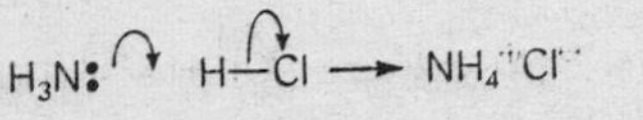

$$H_3N: \quad H—Cl \longrightarrow NH_4^+Cl^-$$

An alkyl group is slightly 'electron donating', because the electron pairs around the carbon atom slightly repel the electron pair in the bond from the carbon to the functional group.

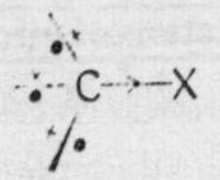

$$:C \longrightarrow X$$

So, replacement of a hydrogen atom in ammonia to form a primary amine causes some electron donation towards the nitrogen in the $\equiv C—N$ bond, which polarises it so that the nitrogen becomes partially negative. This in turn means that the lone pair on the nitrogen is slightly repelled and can therefore be donated to a proton more easily. The result is that a primary amine is more basic than ammonia.

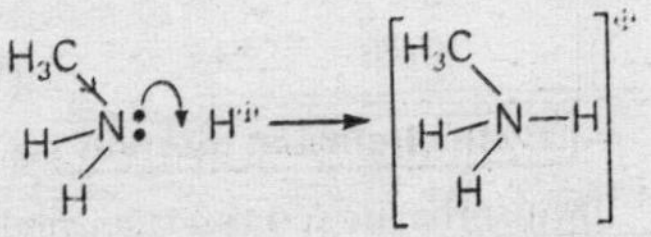

$$\begin{bmatrix} H_3C \\ H—N: \quad H^+ \longrightarrow \quad H—N—H \\ H \end{bmatrix}^+$$

Using the same argument, a secondary amine should be more basic than a primary one, and a tertiary one more basic still. In phenylamine ($C_6H_5NH_2$), however, the lone pair becomes part of the benzene ring's delocalised system (see Ch 26.1), and is less readily donated to a proton.

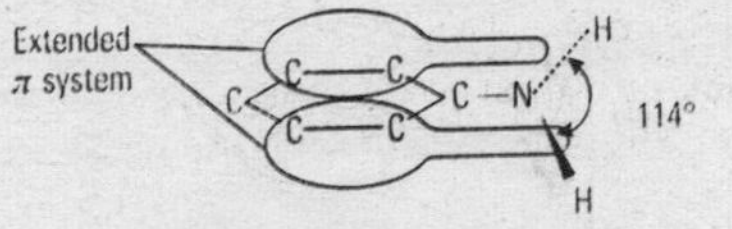

Extended π system ... 114°

Figure 28.3.2   The shape of the phenylamine molecule

Phenylamine is therefore less basic than ammonia.

The free amine can be regenerated from the protonated form by reacting the salt formed with dilute alkali.

$$R—NH_3^{\oplus}Cl^{\ominus} + NaOH(aq) \xrightarrow{\text{room temperature}} R—NH_2 + NaCl(aq) + H_2O(l)$$

The pK_a and pK_b values (see Ch 23.3) of some amines are shown in *Table 28.3.1*. The higher the pK_a – or the lower the pK_b – the more basic the compound. Our predictions are not quite borne out by the figures; the pK_a value for triethylamine is slightly lower than for diethylamine or ethylamine.

(It is worth remembering that, for a particular substance: $pK_a + pK_b = 14.0$.)

The relative basicities of some amines

Name	Type	Structure	pK_a	pK_b	Relative basicity
Ammonia		NH_3	9.25	4.75	1
Ethylamine	Primary	$C_2H_5NH_2$	10.73	3.27	30
Diethylamine	Secondary	$(C_2H_5)_2NH$	10.93	3.07	48
Triethylamine	Tertiary	$(C_2H_5)_3N$	10.64	3.36	25
Phenylamine	Aromatic	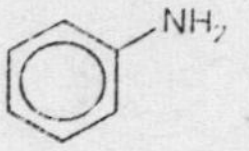 $C_6H_5NH_2$	4.62	9.38	2×10^5

Table 28.3.1

Amines as nucleophiles

Because of the lone pair on the nitrogen atom, ammonia and the amines attack regions of low electron density and partial positive charge. They are therefore usually quite powerful nucleophiles. As discussed in Ch 28.1 and Ch 27.11, ammonia reacts with acyl chlorides to give amides, and primary and secondary amines react with acyl chlorides to give substituted amides. Such reactions involve a nucleophilic substitution mechanism:

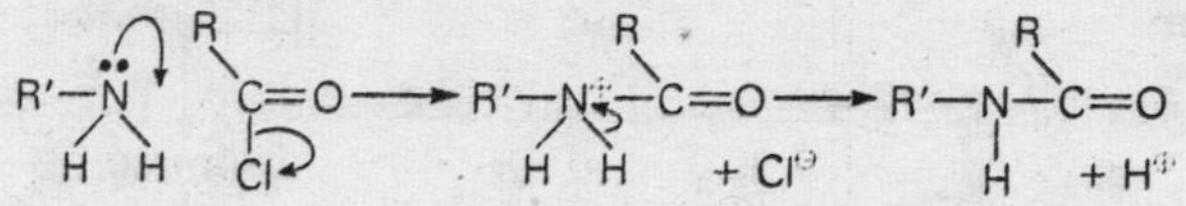

A very similar reaction occurs when ammonia or a primary or secondary amine reacts with an acid anhydride. In this case a molecule of the carboxylic acid is produced as well as the amide or substituted amide.

e.g. $RNH_2 + R'—C(=O)—O—C(=O)—R' \longrightarrow RNHCOR' + R'COOH$

In inorganic chemistry, ammonia and amines can donate electron pairs into vacant orbitals in the ions of the d-block elements to form coloured complexes (see Ch 25.4)

Reaction of primary amines with nitrous acid

Nitrous acid is very unstable and has to be made by adding dilute hydrochloric or sulphuric acid to sodium nitrite. Its formula is often written as HNO_2, but the atoms are joined as H—O—N=O. It reacts at room temperature with primary aliphatic amines to replace the —NH_2 group with —OH, although the yield is often very poor. A driving force for this reaction is the great stability of the nitrogen molecule, which is one of the products (see Ch 28.4). Note that it is perfectly possible by using this reaction to get a tertiary alcohol from a primary amine (see Ch 18.4).

$$CH_3CH_2CH_2NH_2 \xrightarrow[\text{room temperature}]{NaNO_2/H^+(aq)} CH_3CH_2CH_2OH$$

Propylamine → Propan-1-ol

$$CH_3—\underset{\underset{CH_3}{|}}{\overset{\overset{CH_3}{|}}{C}}—NH_2 \xrightarrow[\text{room temperature}]{NaNO_2/H^+(aq)} CH_3—\underset{\underset{CH_3}{|}}{\overset{\overset{CH_3}{|}}{C}}—OH$$

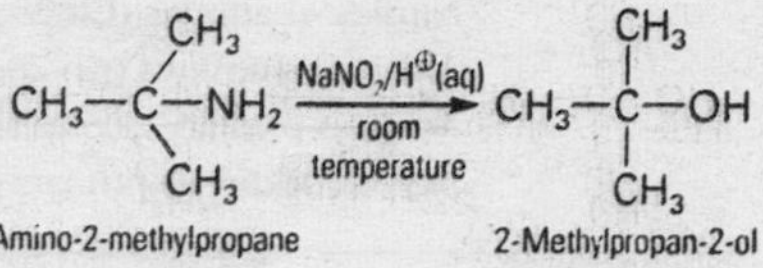

2-Amino-2-methylpropane → 2-Methylpropan-2-ol

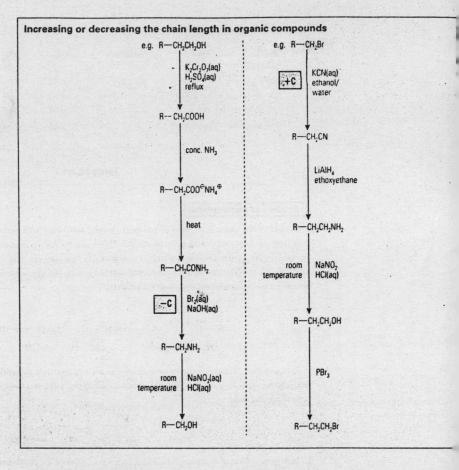

Increasing or decreasing the chain length in organic compounds

As this is the last organic chapter to deal with functional groups, this may be a good place for revision guide to reducing and oxidising agents in organic chemistry.

Reducing agents in organic chemistry

Hydrogen gas + transition metal catalyst (Ni)
   C≡C double bonds → C—C single bonds (Ch 16.5)

Tin + concentrated hydrochloric acid
   Aromatic nitro compounds → aromatic amines (Ch 28.3)

Sodium + ethanol
   Carbonyls → alcohols (Ch 19.3)

Sodium tetrahydridoborate(III) – aqueous
   Aldehydes → primary alcohols (Ch 19.3)
   Ketones → secondary alcohols (Ch 19.3)

Lithium tetrahydroaluminate(III) – anhydrous (in ethoxyethane)
   Nitriles → amines (Ch 28.3)
   Amides → amines (Ch 28.3)
   Acids → primary alcohols (Ch 27.8)
   Acyl chlorides → primary alcohols (Ch 27.11)

> **Oxidising agents in organic chemistry**
> *Oxygen gas + silver catalyst*
>   Alkene → 'alkene oxide' (epoxide) (Ch 16.5)
> *Potassium manganate(VII) + dilute sulphuric acid*
>   Alkene → 'diol' (Ch 16.5).
> *Potassium dichromate(VI) + dilute sulphuric acid*
>   Primary alcohol → aldehyde (distil as formed) (Ch 18.4)
>   Primary alcohol → carboxylic acid (reflux) (Ch 18.4)
>   Secondary alcohol → ketone (Ch 18.4)

SELF-ASSESSMENT QUESTIONS 28.3

1 Identify the following compounds as amines, nitriles or amides.
  a $CH_3CH_2CONH_2$
  b $CH_3CH(CN)CH_2CH_3$
  c $CH_3CH_2CH_2NH_2$
  d $CH_3CH(NH_2)CH_3$

2 Write balanced equations to illustrate the following reactions for preparing amines.
  a Propanamide to propylamine
  b 2-Nitromethylbenzene to 2-aminomethylbenzene
  c 2-Bromopropane to 2-aminopropane

3 Give the reagents and conditions necessary to change 1-aminobutane into:
  a butan-1-ol;
  b N-butylethanamide;
  c 1-aminobutane hydrochloride.

28.4   Diazonium salts

'Diazonium' means that there are two nitrogen atoms joined together in the positive ion. In French, nitrogen is still called by its old name 'azote', which came from the Greek for 'not live', i.e. unable to support life.

Because of the delocalisation in the benzene ring the primary amine phenylamine, $C_6H_5NH_2$, reacts with nitrous acid at low temperature (usually 5 °C) in a way very different from that of aliphatic primary amines. Both these and phenylamine react with nitrous acid to give a diazonium ion, $R-N^{\oplus}{\equiv}N$.

This ion is normally very unstable and rapidly breaks up to give the exceptionally stable molecule of nitrogen gas plus a carbocation, $R^{\oplus}$, which can react with any nucleophilic species in the solution. When $R^{\oplus}$ reacts with a water molecule, the alcohol ROH is obtained (see Ch 28.3).

Diazonium ions from aromatic amines are much more stable. Even so, they have to be kept below about 10 °C. They must also only be used in solution; they explode when solid! Their extra stability, compared with the aliphatic diazo compounds, is explained by delocalisation of the positive charge over the aromatic ring, due to overlap of p-orbitals in the diazo group with the π system in the ring (*Figure 28.4.1*).

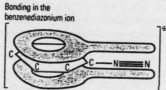

Bonding in the benzenediazonium ion

Figure 28.4.1 Delocalisation in the diazonium ion

For example, phenylamine gives benzenediazonium chloride:

$$C_6H_5NH_2 + HONO(aq) + HCl(aq) \xrightarrow{5\,°C} C_6H_5N_2{}^{\oplus} + 2H_2O(l) + Cl^{\ominus}(aq)$$

Reactions of aromatic diazonium salts

Reactions of aromatic diazonium salts

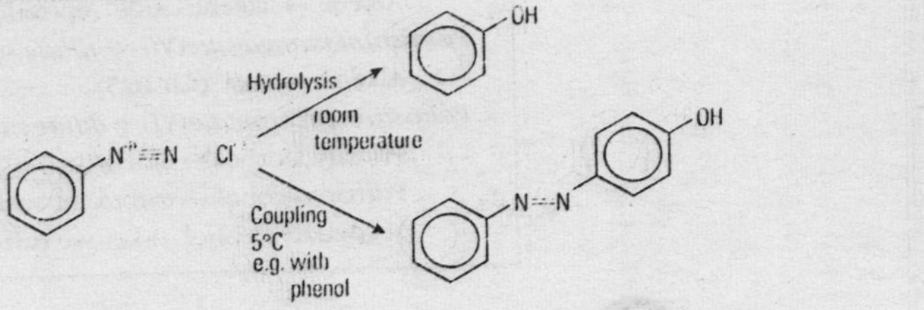

Hydrolysis

If a solution of a diazonium salt is heated, nitrogen is lost, and as with the aliphatic compounds the resulting carbocation can react with negative ions and other nucleophilic species in solution. The decomposition of the carbocation results in the loss of the very stable N_2 molecule. Using potassium halide or cyanide, together with a copper(I) compound as catalyst, gives halogeno- or cyanoarenes.

However, the only simple reaction of this kind which concerns us here is the formation of phenol by allowing an aqueous solution of benzenediazonium chloride to warm up to room temperature.

$$C_6H_5N_2^+Cl^-(aq) + H_2O(l) \longrightarrow C_6H_5OH + N_2(g) + HCl(aq)$$

Coupling reactions

The cation of an aryldiazonium salt is a powerful electrophile, and will attack phenols or aromatic amines, e.g.:

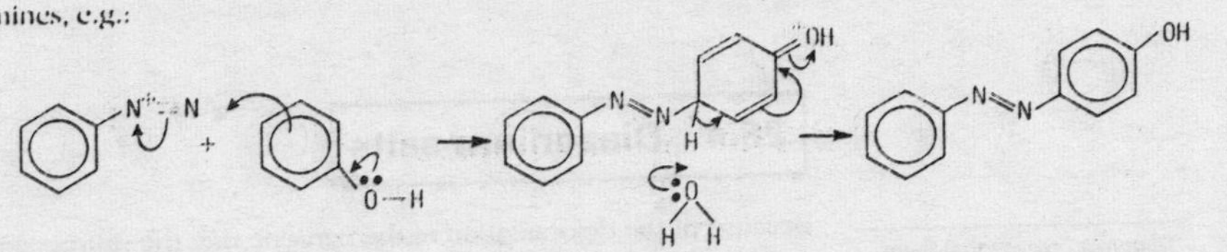

The resulting compounds contain benzene rings linked by the —N=N— group; this is itself unsaturated, and enables the delocalised system to extend over the entire new molecule (*Figure 28.4.2*).

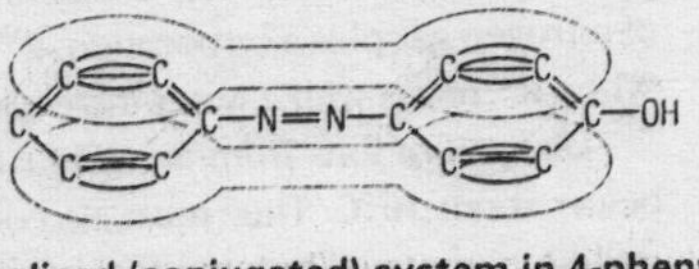

Figure 28.4.2   The delocalised (conjugated) system in 4-phenylazophenylamine

Such extended delocalised (or 'conjugated') systems absorb light well into the visible region. They are often intensely coloured and many are used as dyes for fabrics and fibres – and some, controversially, are used as food dyes. A group of atoms which is responsible for colour in a series of compounds is known as a chromophore. The **chromophore** in azo dyes is:

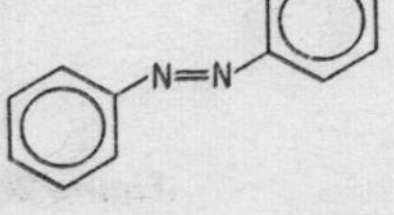

The colour produced is modified by attaching functional groups to the chromophore system, or by extending the delocalised system through use of a naphthalene-based compound, or by doubling up the diazo chromophore.

A general method for making azo dyes in small amounts would be as follows. First, make the diazonium salt by adding a cold aqueous solution of sodium nitrite slowly (with cooling and stirring) to a cold solution of the amine compound in excess hydrochloric acid. The temperature must not rise above 5 °C. Then add this cold solution of the diazonium salt slowly and with stirring to a cold solution of the other compound – the 'coupling agent'. (If the coupling agent is a phenol, the reaction occurs most rapidly in slightly alkaline solution.) Again, the temperature must be kept at about 5 °C.

Figure 28.4.3 shows some examples of coupling reactions leading to azo dyes – and in one case to a common acid–base indicator.

Some coupling reactions leading to azo dyes

Chrysoidine G (orange)

Methyl orange (yellow) in the deprotonated form

Naphthalen-2-ol

Orange II

Congo red

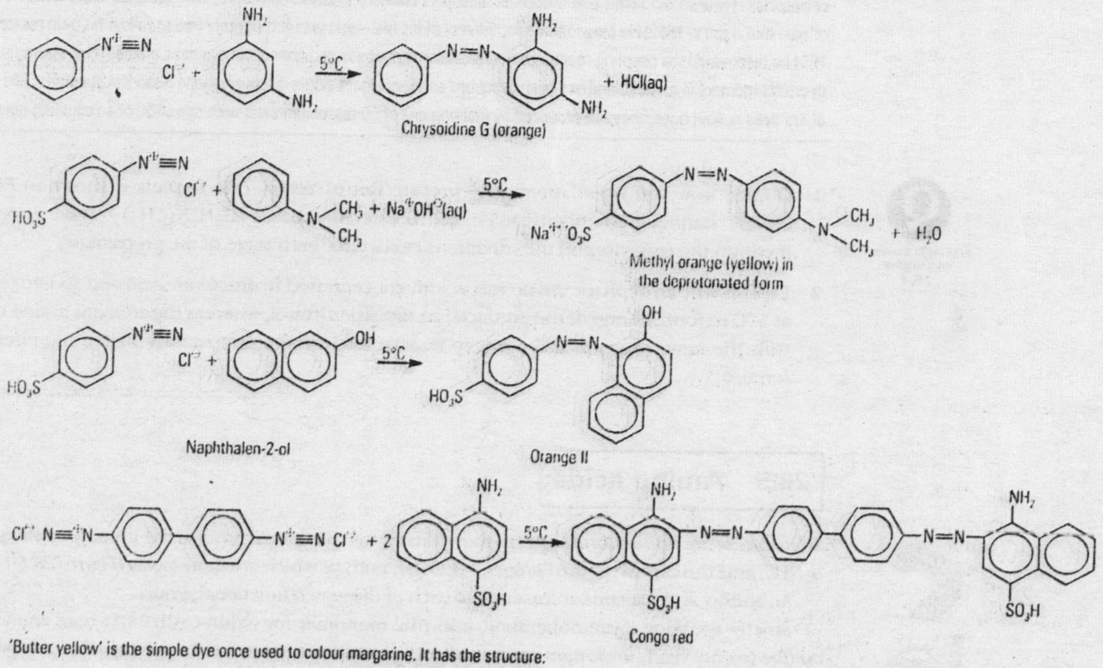

'Butter yellow' is the simple dye once used to colour margarine. It has the structure:

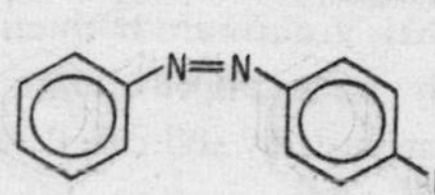

Butter Yellow was shown to be carcinogenic (i.e. it can cause cancer) and its use in food is now forbidden.

Figure 28.4.3

Phenylamine ('aniline'), W H Perkin and Peter Griess

The early history of the dyestuffs industry in Britain is closely connected with two young men, Peter Griess and William Henry Perkin, Senior. (The 'senior' is to distinguish him from his son, whom he inconsiderately christened William Henry, and who followed his father in becoming a distinguished organic chemist and a Fellow of the Royal Society.)

Perkin was only 15 when he began to study chemistry at the Royal College of Chemistry in London, under A W Hofmann, in 1853. He also had a fully equipped laboratory in his own home, where he discovered what was probably the first azo dye, although he did not patent it until 1863. In 1856, while trying to synthesise the anti-malarial drug quinine, he heated phenylamine with potassium dichromate(VI), and thus discovered 'Mauve'. This was the beginning of a vast dyestuffs industry, mostly developed in Germany, which almost completely replaced vegetable dyes – such as indigo and alizarin – with those manufactured from compounds, such as phenylamine and phenol, which could cheaply be made from coal tar.

The fabrics and clothing industries were transformed, as was the economy of rural areas in southern Europe and elsewhere where vegetable dyes had been produced. For example, in 1868, 70 000 tonnes of madder root was grown and processed to give about 750 tonnes of alizarin. Five years later no madder fields were left; all the alizarin used was synthetic.

Perkin was 20 when he opened his dyestuffs factory at Greenford. He discovered other dyes, made a great deal of money, and sold up in 1873 – aged just 35! – to spend the rest of his life in chemical research.

Peter Griess was born in Germany in 1829, and during his student days was noted for his unruliness and large debts. However, he eventually began to work hard at chemistry and discovered the diazotisation reaction with nitrous acid in 1858. In the same year he came to London to work with Hofmann, who commissioned him to continue with diazo compounds. He also did some other work for Allsopp's brewery in Burton-on-Trent, who were so impressed that they offered him a job in 1862. He stayed there for the rest of his life – and was left entirely free to follow his own research. In 1864 he discovered the coupling reaction between diazo compounds and phenols or aromatic amines. This started the azo dyestuffs industry in earnest, and has been described as the greatest single discovery in the history of dyestuffs. The range of azo dyes is now huge; they are produced by reacting one of 50 diazonium salts with one of 50 or so coupling agents.

SELF-ASSESSMENT QUESTIONS 28.4

• 1 Outline how you would attempt to prepare Butter Yellow (the formula is shown in *Figure 28.4.3*) starting from phenylamine and the tertiary amine $C_6H_5N(CH_3)_2$. Make sure you mention the reagents and the conditions needed for each stage of the preparation.

2 Explain why an aliphatic amine reacts with concentrated hydrochloric acid and sodium nitrite at 5 °C to form, as one of the products, an aliphatic alcohol, whereas the aromatic amine reacts with the same reagents and needs to be warmed to room temperature before the phenol is formed.

28.5 Amino acids

Amino acids are *bifunctional*: that is, they contain two functional groups, the primary amino group, —NH_2, and the carboxylic acid group, —COOH, both of which are quite polar (*Figure 28.5.1*).

An amino acid contains at least one of each of these two functional groups.

Strictly speaking, 6-aminohexanoic acid (the monomer for Nylon-6, Ch 28.1) is an amino acid, but the twenty vitally important amino acids which are found in nature in polypeptides and proteins are all 2-amino acids, i.e. the —COOH and —NH_2 groups are both attached to the same carbon atom. All of them can be named systematically, for example alanine is 2-aminopropanoic acid, $CH_3CH(NH_2)COOH$ (*Figure 28.5.2*); but they are usually still called by their old names. Indeed, 2-amino acids are often still referred to as α-amino acids.

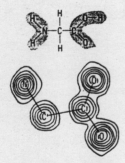

Figure 28.5.1 Electron density map of glycine

Figure 28.5.2 2-aminopropanoic acid (alanine)

The general formula of a 2-amino acid is:

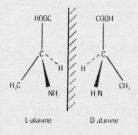

Figure 28.5.3 The mirror images of L- and D-amino acids

L-alanine   D-alanine

If the general formula for an amino acid is H_2N—$CH(R)COOH$, then, unless R is —H, —NH_2, or —COOH, there will be four different atoms or groups around the central carbon atom.

e.g.

$$H_2N\!-\!\overset{\displaystyle H}{\underset{\displaystyle CH_3}{C}}\!-\!\overset{\displaystyle O}{C}\!\diagdown OH$$

The molecule will therefore be chiral and possess optical activity (*Figure 28.5.3*). As discussed in Ch 14.4, the twenty or so amino acids found in proteins are, with the exception of glycine, all chiral – *and they are all left-handed*.

Some common natural amino acids are shown in *Table 28.5.1*; they differ only in the R group.

Some amino acids

Name of amino acid	R group	Abbreviation
Glycine	—H	gly
Alanine	—CH_3	ala
Phenylalanine	—CH_2—⬡	phe
Serine	—CH_2OH	ser
Cysteine	—CH_2SH	cys
Aspartic acid	—CH_2COOH	asp
Lysine	—$CH_2CH_2CH_2CH_2NH_2$	lys
Tryptophan	—CH_2—(indole)	try

Proline is different: it cannot have a normal 'R' group as it is a *secondary amine*.

pro

Table 28.5.1

Optical activity

Ordinary light consists of waveforms, with waves vibrating in all directions. Certain materials, such as Polaroid, act as filters to normal light, allowing through only the light waves which vibrate in one direction – plane polarised light. This is exactly what happens with most sunglasses.

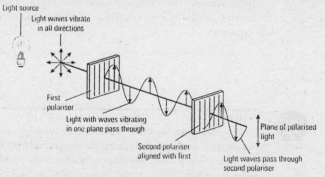

When plane polarised light meets a second piece of Polaroid, for example, then only light vibrating in one plane will pass through. If the polarised light is vibrating at right angles to this 'grid', no light will get through.

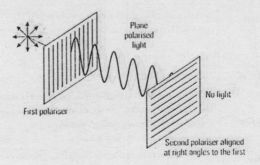

Plane polarised light

First polariser

No light

Second polariser aligned at right angles to the first

Optically active molecules (see Ch 14.4) can rotate the plane of polarisation. If plane-polarised light enters a solution of an optically active substance, interaction with the molecules in the solution rotates the plane of polarisation.

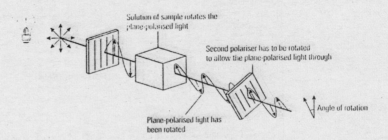

Solution of sample rotates the plane-polarised light

Second polariser has to be rotated to allow the plane-polarised light through

Plane-polarised light has been rotated

Angle of rotation

The degree of rotation will depend on the number of molecules present – i.e. the concentration and the length of the tube which contains the solution. With molecules that do not contain a chiral carbon atom, i.e. that are symmetrical, the polarised light will not be affected. Unsymmetrical molecules rotate polarised light to the right or to the left. The degree of rotation varies with the molecule. The mirror image of the unsymmetrical molecule – its *optical isomer* (see Ch 14.4) – will rotate polarised light in the opposite direction, to exactly the same amount. Louis Pasteur was the first to notice the effect, when he found that crystals of tartaric acid he had recrystallised from the residue in wine casks had two mirror image structures.

ACTIVITY

Activity

Using a molecular model kit make up one molecule of dibromomethane, CH_2Br_2. Look at the structure and make another model of this molecule that is a mirror image of the first.

Put the two models alongside each other. Show that molecules without a chiral carbon atom, such as CH_2Br_2, have mirror images that are identical since the two molecules can be superimposed on each other. Now make a model of bromochlorofluoromethane, CHBrClF, and then make a model of the mirror image. Show that in this case the mirror image cannot be superimposed over the original. Such molecules are optically active.

Some examples of organic nitrogen compounds used as drugs

Pain killers	Paracetamol	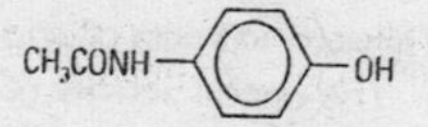 CH_3CONH—⬡—OH
Stimulant drugs	Amphetamine	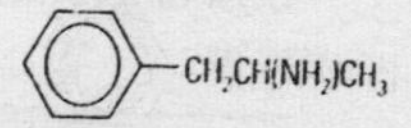 ⬡—$CH_2CH(NH_2)CH_3$
	Ephedrine	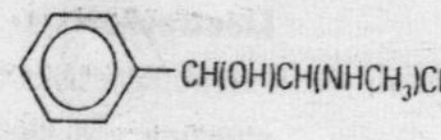 ⬡—$CH(OH)CH(NHCH_3)CH_3$
Antibacterial	Sulphanilamide	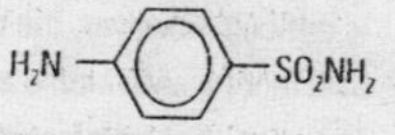 H_2N—⬡—SO_2NH_2

The physical and acid–base properties of 2-amino acids

Amino acids are white solids with relatively high melting points. (The simplest one, glycine, has a melting point of 235 °C.) Usually, they are readily soluble in water but almost insoluble in non-polar organic solvents. In these ways they are rather like ionic compounds.

The reason for this behaviour is that they *are*, very largely, ionic compounds. The carboxyl group can lose a proton and the amine group can gain one; the result is a **zwitterion** (from the German for hermaphrodite, hybrid or mongrel!)

For example, glycine exists mainly as H_3N^+—CH_2—CO_2^-. It is the strong attractions between these charges in the crystal which cause the high melting point. In aqueous solution, an amino acid with an equal number of amino and carboxyl groups will be very roughly neutral. In strongly acid solution, the carboxylate ions will be protonated to carboxylic acid groups. The positive charge will stay on the nitrogen, so the amino acid will have become a cation. In strongly alkaline solution, the proton will be taken off the nitrogen of the zwitterion, leaving the negative charge on the carboxylate; so the amino acid will have become an anion.

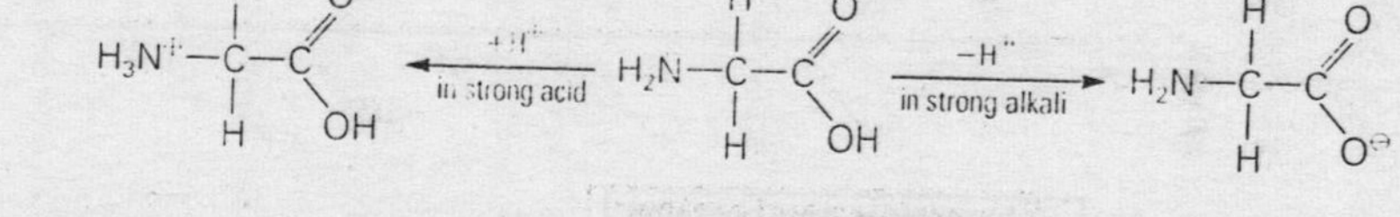

For each amino acid there exists a definite pH, the **isoelectric point**, at which the acidic and basic ionisations are equal – i.e. the molecule is effectively neutral because it carries equal and opposite charges. The isoelectric point is rarely very near pH 7, even for glycine, because the tendencies for the amino group to gain a proton and the carboxyl to lose one are affected by the rest of the molecule and are not exactly equal anyway.

If the amino acid contains more carboxyl groups than amino, e.g. aspartic acid, it will be acidic in aqueous solution. And, of course, a compound such as lysine – with more amino than carboxyl – will be alkaline (*Figure 28.5.4*).

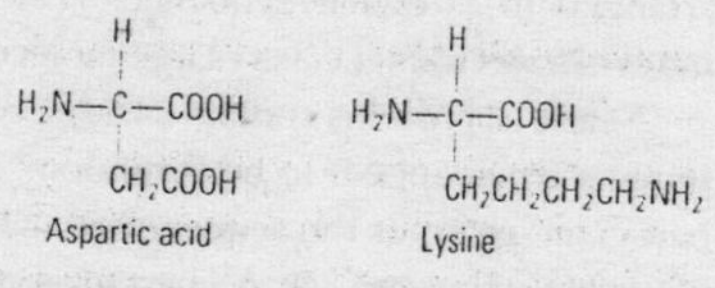

$$H_2N-\overset{\overset{\displaystyle H}{|}}{C}-COOH$$
$$CH_2COOH$$
Aspartic acid

$$H_2N-\overset{\overset{\displaystyle H}{|}}{C}-COOH$$
$$CH_2CH_2CH_2NH_2$$
Lysine

Figure 28.5.4   Aspartic acid and lysine

One of the consequences of amino acids being able to act as both proton donors and acceptors is that their solutions can act as buffer solutions, and maintain a reasonably constant pH in the face of limited addition of acid or alkali (see Ch 23.6).

The optical activity of almost all the naturally occurring amino acids has been discussed at the beginning of this section.

Electrophoresis

Electrophoresis is the technique used to separate and identify different amino acids. Since the acids are so similar in structure, with the $H_2N—C(R)H—COOH$ group present in each amino acid, the separation using normal chromatography is difficult. However, the presence of a different R group in each amino acid means that the pH at the isoelectric point is different for each amino acid.

The mixture of amino acids is placed in a gel or on filter paper soaked in a solution at a specific pH. A potential difference is established across the filter paper – i.e. one side is negative, the other positive. Depending on the charge on the particular amino acids at the pH of the solution, the acids move towards the cathode or towards the anode. If the pH is at the value for the zwitterion to be formed, the amino acid does not move at all. In this way different amino acids can be separated and, using reference standards, identified.

The technique is adapted in genetic fingerprinting, when genes are separated into bands, after the DNA is broken up into small fragments.

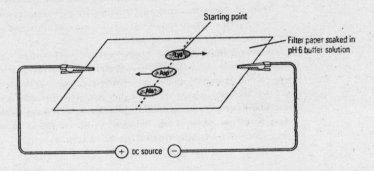

Polypeptides and proteins

Introduction to proteins

The characteristic breakdown product of protein in mammals is urea, $CO(NH_2)_2$, which you excrete in urine. Kidney failure prevents this excretion and, without dialysis treatment or a kidney transplant, can rapidly cause death.

About 15% of your bodyweight consists of protein. Protein occurs in muscle, and also as collagen in all connective tissue and skin, in the keratin of hair and nails, in blood plasma, and as the enzymes which enable a vast range of chemical reactions to occur in every living cell. Every animal needs protein in food, not only for growth but also for the replacement of old, used-up, discarded protein.

Proteins consist of many amino acid units joined together. They are not polymers. Unlike a polyalkene such a poly(ethene) or a Nylon, they do not consist of one or two monomer units arranged in a regular sequence. Proteins, together with DNA, are perhaps best described as **macromolecules**, i.e. very large molecules which do not contain a regular sequence of units.

Nearly all proteins contain a large proportion of the possible twenty or so amino acids, and their sequence may appear to be at random. In fact, for the functioning of many natural proteins and, in particular, enzymes the sequence must be *absolutely* right. If a single amino acid unit in a sequence of several thousand in an enzyme is missed out or wrongly positioned, the enzyme's ability to function may be completely destroyed.

Proteins contain carbon, hydrogen, oxygen and nitrogen, and almost always sulphur, from the amino acids cysteine and methionine. The stench of decomposing protein is caused not only by amines (see Ch 28.3) but also alkyl derivatives of hydrogen sulphide and other organosulphur compounds.

Test for proteins

Any compound containing the *peptide link* (see p 743) will give a characteristic colour with an alkaline solution of copper(II) sulphate. Proteins contain a large number of peptide bonds and form a deep blue—violet colour with the reagent. This is the **biuret test.**

The shapes of proteins

The shapes of proteins are very important for their function. The primary structure of a protein is fixed by the sequence of amino acids along the protein chain. X-ray diffraction investigation of protein molecules shows that they usually have a helical (spiral) structure, rather as DNA has, but this time without a double strand. This helical arrangement is by far the most usual in natural proteins, and is called the secondary structure. Usually, eighteen amino acids make up every five turns of the helix; the spiral turns are held together by hydrogen bonds between C—O and H—N groups in amino acids on successive turns. The helix then can fold or twist into its final shape – its tertiary structure.

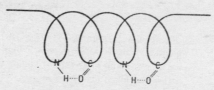

The structure is also determined by the position of cysteine units in the amino acid sequence. Whenever the —SH groups of two cysteine units come together, oxidation can result in the formation of a 'disulphide bridge', —S—S—.

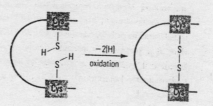

It doesn't matter if the cysteine units are widely separated by other amino acids; a 'loop' is formed. It is loops of this kind which help determine the properties of the hormone insulin. (A deficiency of insulin causes diabetes in humans.)

The weak hydrogen bonds which largely determine the detailed structure of a protein are sensitive to temperature. This is why the catalytic action of enzymes is destroyed by heating. They are 'denatured' by alteration of their structures. A good example is boiling an egg: the egg white is largely the protein albumin, and when heated it solidifies.

Hair contains the protein keratin, which contains a quite high proportion of cysteine and hence many disulphide bridges. To 'perm' hair, the —S—S— bonds are broken using a mild reducing agent. The hair is then curled and shaped, before the —S—S— bonds are re-made using a mild oxidising agent (e.g. dilute hydrogen peroxide). This fixes the hair in its new style.

How amino acid units join together

The carboxyl group of one amino acid can react with the amino group of another to form a substituted amide (see Ch 28.1):

$$H_2N-\overset{\overset{\displaystyle H}{|}}{\underset{\underset{\displaystyle R}{|}}{C}}-COOH + H_2N-\overset{\overset{\displaystyle H}{|}}{\underset{\underset{\displaystyle R'}{|}}{C}}-COOH \longrightarrow H_2N-\overset{\overset{\displaystyle H}{|}}{\underset{\underset{\displaystyle R}{|}}{C}}-CONH-\overset{\overset{\displaystyle H}{|}}{\underset{\underset{\displaystyle R'}{|}}{C}}-COOH + H_2O$$

A dipeptide

Where this kind of bond is formed between two 2-amino acids, it is called a peptide bond or peptide link (*Figure 28.5.4*).

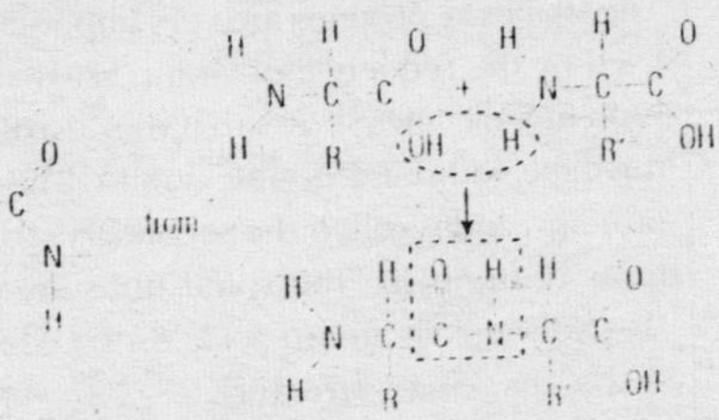

Figure 28.5.4   The peptide bond

If the resulting compound contains two amino acid units, it is called a dipeptide; if three, a tripeptide; if many, a polypeptide. There is no specific point at which a polypeptide becomes a protein, some authorities put the change at about 40 amino acids.

However long the chain of amino acid units, there will be a free —NH₂ group at one end and a free —COOH group at the other. This makes it very difficult to form peptide bonds to create a specific amino acid sequence in the laboratory, although enzymes do it at astonishing rates in any living cell.

The problem was solved by the American chemist Bruce Merrifield, who anchored one end of the growing chain on tiny resin beads. The synthesis of medium-sized proteins can now be automated. Merrifield received the Nobel Prize for Chemistry in 1984.

Proteins are a vital component of your diet. They contain the same type of link between units as is found in polyamides (see Ch 28.1). Why, then, can you not digest Nylon? The main reason is that in a peptide or protein there is only one carbon atom between each —CONH— link, because it is 2-amino acids which are joined together. In polyamides there is a chain of a few carbon atoms between the —CONH— links. Your digestive system can deal with the first but not the second.

Acid hydrolysis of proteins

If a polypeptide or protein is refluxed for 24 hours with moderately concentrated (6 mol dm⁻³) hydrochloric acid, all the peptide bonds are broken. After neutralisation, the mixture contains all the amino acids which were in the original material. These can be separated and identified using **paper chromatography** (Ch 29.2).

Column chromatography using cation-exchange resins can be used to find the relative amounts of each amino acid. This information, coupled with knowledge of the target substance relative molecular mass, gives the total number of amino acid units in the molecule. Finding the sequence of amino acids is much more difficult, and is not considered here. Several methods are now available.

The breaking-up of the peptide bonds is the exact opposite of their formation. A molecule of water is, in effect, added across each linkage to regenerate the original amino and carboxyl groups. This is why the reaction is called hydrolysis.

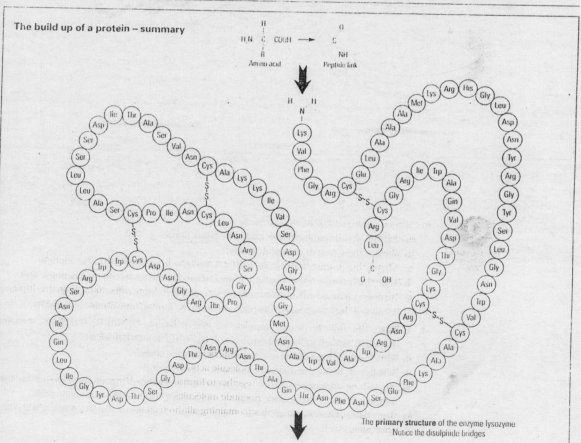

Diabetes is caused by a deficiency in production of the protein insulin by the human pancreas

Knowledge of the detailed structure of insulin would help understanding of the disease and how it might be tackled, and could lead to a synthesis of pure material for use by sufferers. Fred Sanger and his research group at Cambridge developed methods for finding the amino acid sequence, and established the 51-unit primary structure in 1955. For this work, Sanger received the Nobel Prize for Chemistry in 1958. The full three-dimensional structure of insulin was found later, using X-ray diffraction methods, by Dorothy Hodgkin's team at Oxford. (She won the Nobel Prize for Chemistry in 1964, largely for her work on vitamin B_{12}, the anti-anaemia vitamin.) Insulin can now be synthesised, but it is cheaper to produce it by biological methods.

Sanger won a second Nobel Prize for Chemistry in 1980, this time for pioneering the sequencing of DNA. He became the second person to be awarded two Science Nobel Prizes of the same kind. (The other double Nobelists are Marie Curie, Physics and Chemistry; Linus Pauling, Chemistry and Peace; and John Bardeen, Physics in 1956 and 1972.)

The build up of a protein – summary

The **primary structure** of the enzyme lysozyme
Notice the disulphide bridges

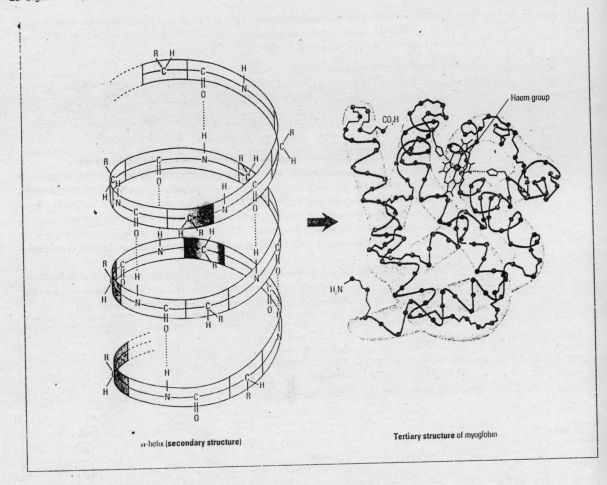

α-helix **(secondary structure)**

Tertiary structure of myoglobin

**SELF-ASSESSMENT
QUESTIONS
28.5**

1 Alanine is one of the simplest amino acids.
   a Explain why the amino acid is said to be bifunctional.
   b What is the systematic name of alanine?
   c What is the product formed when alanine reacts with sodium hydroxide solution?
   d Write the formula of the structure formed when alanine reacts with hydrochloric acid.
   e Alanine combines with another amino acid molecule to form a dipeptide. Write the displayed formula of the two possible dipeptides that can be formed from alanine and glycine.

2 a Write the full displayed formulae of phenylalanine, $H_2NCH(C_6H_5)COOH$; cysteine, $H_2NCH(CH_2SH)COOH$; and aspartic acid, $H_2NCH(CH_2COOH)COOH$.
   b Which part of the aspartic acid molecule behaves as a base?
   c Which part of the phenylalanine molecule acts as an acid?
   d These three amino acids can join together to form a number of tripeptides. Write the displayed formula of one of the possible tripeptide molecules containing all three amino acids.
   e How many different tripeptides (containing all three amino acids) are possible with these three amino acids?

Summary of reactions

Formation	Reaction

Amide

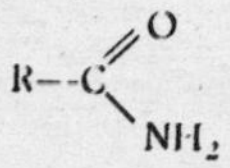

$R-\overset{\displaystyle O}{\underset{\displaystyle NH_2}{C}}$

$RCOOH + NH_3 \xrightarrow[\text{ii) reflux}]{\text{i) } NH_4 / RT} RCONH_2 + H_2O$

$RCONH_2 + 4[H] \xrightarrow[\text{ethoxyethane}]{LiAlH_4 / RT} RCH_2NH_2 + H_2O$

$RCOCl + NH_3 \xrightarrow{RT} RCONH_2 + HCl$

$RCONH_2 + H_2O + H^{\oplus} \xrightarrow[\text{HCl(aq)}]{\text{reflux}} RCOOH + NH_4^{\oplus}$

$RCONH_2 + HONO \xrightarrow[RT]{NaNO_2/HCl} RCOOH + N_2 + H_2O$

$RCONH_2 \xrightarrow[\text{distil}]{P_4O_{10}} RCN + H_2O$

$RCONH_2 + 4OH^{\ominus}(aq) + Br_2 \xrightarrow{\text{warm}}$
$RNH_2 + CO_3^{2\ominus} + 2H_2O + 2Br^{\ominus}$

Nitrile
$R-C\equiv N$

$RCONH_2 \xrightarrow[\text{distil}]{P_4O_{10}} RCN + H_2O$

$RCN + 4[H] \xrightarrow[\text{ethoxyethane}]{LiAlH_4} RCH_2NH_2$

$RBr + CN^{\ominus} \xrightarrow[\substack{\text{ethanol}\\\text{water / heat}}]{KCN} RCN + Br^{\ominus}$

$RCN + H_2O \xrightarrow[\text{reflux}]{HCl(aq)} RCONH_2$

Amine
$R-NH_2$

$RBr + NH_3 \xrightarrow{\text{warm}} RNH_2 + HBr$

$RNH_2 + HCl \xrightarrow{RT} RNH_3^{\oplus}Cl^{\ominus}$

$RCONH_2 + 4[H] \xrightarrow[Et_2O]{LiAlH_4} RCH_2NH_2 + H_2O$

$RNH_2 + R'COCl \xrightarrow{RT} R'CONHR + HCl$

$RCN + 4[H] \xrightarrow[Et_2O\ RT]{LiAlH_4} RCH_2NH_2$

$RNH_2 + HONO \xrightarrow[RT]{NaNO_2/HCl} ROH + N_2 + H_2O$

$ArNO_2 + 6[H] \xrightarrow[\text{conc. HCl}]{Sn} ArNH_2 + 2H_2O$

$RCONH_2 + 4OH^{\ominus}(aq) + Br_2 \xrightarrow{\text{warm}}$
$RNH_2 + CO_3^{2\ominus} + 2H_2O + 2Br^{\ominus}$

Diazonium salts

$RN^{\oplus}H_3Cl^{\ominus} + NaOH(aq) \xrightarrow{\text{warm}} RNH_2 + NaCl + H_2O$

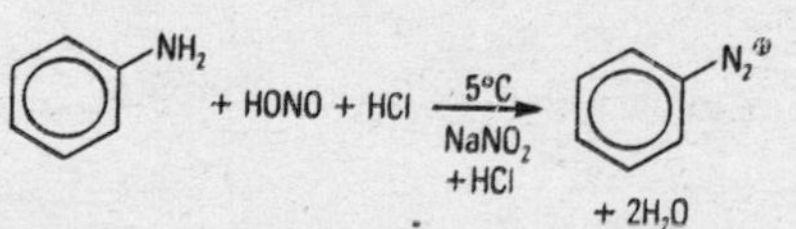

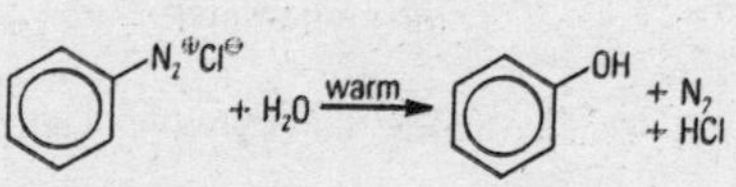

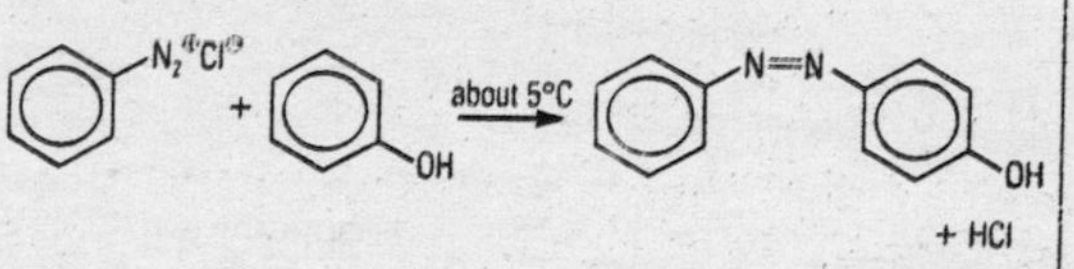

**SUMMARY
28**

In this chapter the following new ideas have been covered. *Have you understood and learned them?* Could you explain them clearly to a friend? The page references will help.

- Recognition of the amide, nitrile and amine functional groups.   p739
- The preparation and reactions of compounds containing the amide group.   p740
- The preparation and reactions of compounds containing the nitrile group.   p745
- The preparation and reactions of compounds containing the amine group.   p747
- The formation of polyamide polymers, such as the nylons.   p743
- Primary, secondary and tertiary amines, and quaternary ammonium salts.   p747
- Reactions for shortening and lengthening chains.   p752
- The formation and reaction of diazonium salts.   p753
- Coupling reactions for making dyestuffs.   p754
- The structure and properties of amino acids.   p756
- Amino acids in nature, and their chirality.   p757
- Zwitterions; the isoelectric point.   p759
- Peptides and proteins; the peptide link.   p760
- Proteins as macromolecules; their structures and shapes.   p761

**EXAMINATION
QUESTIONS
28**

1 In each of the following cases give the name and displayed formula of a simple molecule that will react in the way described.

   a An organic nitrogen compound that is hydrolysed in dilute acid, to form an organic compound which does not contain nitrogen and which has a pH of 3 in aqueous solution.

   b An aliphatic organic nitrogen compound that is reduced to an amine using LiAlH₄ in ethoxyethane.

   c An aromatic compound that can be reduced to an amine by tin and concentrated hydrochloric acid.

   d A compound that will dissolve both in acid and in alkaline solutions, two molecules of which join together with formation of a —CO—NH— linkage.

2 Paracetamol and aspirin are two common pain-relieving drugs. Their structures are:

Paracetamol   Aspirin

   a Name the functional groups present in each structure.

   b What compound would you need to react with ethanoyl chloride to form paracetamol?

   c What compound would you need to react with ethanoyl chloride to produce aspirin?

   d Explain why you might expect paracetamol to decolourise an aqueous solution of bromine, 'bromine water'.

3 Serine is an example of an organic molecule with three functional groups.

$$H_2N-\overset{\displaystyle H}{\underset{\displaystyle CH_2OH}{\vphantom{|}}C}-COOH$$

It is optically active.

   a Which class of organic compound is serine put into because of two of the functional groups present in the molecule?

b One mole of serine reacts with excess sodium metal to produce one mole of hydrogen gas. Write a balanced equation for this reaction.

c Explain what is meant by *optically active*.

d What feature of the serine molecule makes it optically active?

e Draw full structural diagrams to show clearly the two optical isomers of serine.

f At the isoelectric point, the serine molecule exists as a zwitterion. Draw a diagram to show clearly the structure of the zwitterion.

4

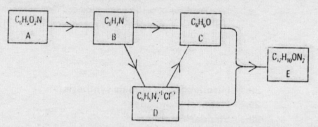

In this reaction sequence you are asked to identify the five compounds, A to E. The following questions should help you.

a The ratio of carbon:hydrogen atoms in all these molecules should indicate something about the structure of these molecules. What is it?

b Compound C reacts with sodium metal to produce a colourless gas. What functional group is present in C?

c The conversion of A to B is a reduction reaction. What reagents do you need for this reaction?

d What reagents and conditions do you need to convert B to D?

e How would you convert compound B to C?

f E is a solid with a bright orange/yellow colour. What type of group exists in the molecule to produce this colour?

g Finally, draw full structures for the compounds A to E.

- The idea of a **functional group**. (Ch 14.2)
- Simple **reaction mechanisms**. (Ch 14.7)
- The reactions of **alkenes**. (Ch 16.5)
- The reactions of **halogenoalkanes**. (Ch 17.4)
- The reactions of **alcohols**. (Ch 18.4)
- The reactions of **aldehydes and ketones**. (Ch 19)
- The reactions of **arenes**. (Ch 26.2)
- The reactions of **carboxylic acids** and derivatives. (Ch 27)
- The reactions of **organic nitrogen** compounds. (Ch 28)

29.1   **Introduction to organic synthesis**
29.2   **Practical techniques**
29.3   **Standard methods**
29.4   **Aliphatic sequences**
29.5   **Aromatic sequences**
29.6   **Yields in organic reactions**

29

organic synthesis

29.1 Introduction to organic synthesis

In *synthesis*, smaller units are combined together to make something more complex. The word is often used to describe a sequence of reactions which results in a substance which is very different from the starting material, although not necessarily more complex.

A synthesis of the first kind *may* occur in a simple single step. A synthesis of the second kind *always* needs more than one step.

For example, the reaction of ethanoyl chloride with phenylamine can legitimately be called a *synthesis*:

Ethanoyl chloride   Phenylamine   *N*-Phenylethanamide

The hydrolysis of ethanoyl chloride to ethanoic acid is simply a *reaction*:

Ethanoyl chloride   Ethanoic acid

A 'rational' synthesis is one in which the structure of the target compound is known, and the chemist deliberately sets out to build it from starting materials of known structure.

Probably the first ever rational synthesis of an organic molecule was carried out in 1850 by the British chemist, Alexander Williamson. His target was ethoxyethane, known as 'ether'. This was then, and for many years afterwards, used as a general anaesthetic, so that an easy synthesis of the pure material from cheap starting materials could be profitable. His starting materials were bromoethane (made from ethanol, Ch 18.4) and sodium ethoxide (also from ethanol).

$$C_2H_5Br + C_2H_5O^{\ominus}Na^{\oplus} \xrightarrow{\text{warm}} C_2H_5OC_2H_5 + Na^{\oplus}Br^{\ominus}$$
Bromoethane   Sodium ethoxide   Ethoxyethane

As more was understood about the ways in which organic molecules could react, syntheses that are much more elaborate became possible. The great American chemist Robert Woodward was awarded the Nobel Prize for Chemistry in 1965 for his contributions to the *art* of organic synthesis. In 1971, he achieved the total synthesis of vitamin B_{12}, which has the molecular formula $C_{63}H_{90}O_{14}N_{14}PCo$. This molecule was targeted because lack of vitamin B_{12} in humans causes pernicious anaemia, which is often fatal. Synthesis of the vitamin confirmed the structure already found using X-ray crystallography by Dorothy Hodgkin, working in Oxford, in 1956. (Hodgkin was awarded the 1964 Nobel Prize for Chemistry for this work.) Knowledge of the precise structure and stereochemistry of a biologically active molecule develops understanding of how it works in living organisms. In the case of, for example, antibiotics or hormones, it can lead to deliberate synthesis of compounds which do the job better and have fewer side effects.

29.2 Practical techniques

The organic chemist works with covalent compounds, in which strong covalent bonds are often broken and made. Such reactions can involve large energy changes and may be very slow; and they will almost always involve *side-reactions* which produce unwanted by-products from which the intended products must be separated and purified (see Ch 14.7).

The only techniques which you are expected to know in any detail are simple ones:

- for temperature control;
- for the separation of mixtures of liquids;
- for purification of solids;
- for measuring the melting point and hence determining the purity of a crystalline organic solid; and
- for checking the purity of a sample using chromatography.

You should also be able to choose apparatus suitable for mixing and containing the reactants for the required contact time.

You have met some of these earlier in this book.

Control of temperature: heating and cooling

Heating

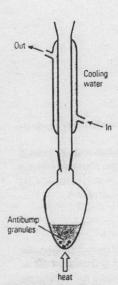

Figure 29.2.1 Boiling under reflux

A rise in temperature increases the rate of reaction (see Ch 12.2). A reaction which is very slow or not observable at room temperature, such as the alkaline hydrolysis of primary halogenoalkanes (see Ch 17.4), can often be made to react at a reasonable rate at about 100 °C.

A useful technique for increasing the efficiency and safety of such a reaction is *boiling under reflux* (*Figure 29.2.1*).

The reagents are heated together in a flask which has a Liebig condenser fitted vertically above it. Vapour from the boiling mixture enters the condenser, is cooled and condenses, and flows back ('refluxes') into the flask. For smooth boiling, the flask should contain a few *anti-bump granules*. For safety, the level at which condensation occurs should not be more than about one-third up the condenser.

Boiling under reflux allows the reaction to take place at or near the boiling point of the reactants without losing any of the reactants or products by evaporation. Another advantage of boiling under reflux is that immiscible reactants are constantly agitated and brought into more intimate contact – the contents of the flask not only splash about gently during boiling but constantly have liquid dropping from above. In the hydrolysis of halogenoalkanes, already mentioned, the alkali is in aqueous solution and does not mix with the halogenoalkane. The general equation can be written:

$$R—X + OH^{\ominus}(aq) \xrightarrow{\text{heat under reflux}} R—OH + X^{\ominus}(aq)$$

N.B. Care is needed in heating the reaction flask. It is inadvisable to heat the flask, even a small pear-shaped flask, directly with a Bunsen flame. The flask can be seated in a sand bath in a metal tray which is on a gauze heated from underneath; or it can be heated in a hot air-bath, by being supported above a heated gauze. Or, of course, it may be placed in an electrical heating mantle, so that close control of the heat input into the flask is possible.

For reactions involving a flammable organic solvent such as ethanol, hexane or ethoxyethane, a flame should *never* be used. (Imagine the scene if the flask should crack!) A heating mantle should be used, or – since these solvents all boil at less than 100 °C – the flask can be immersed in an electrically heated water bath.

Cooling

Cooling decreases the rate of a reaction. Reactions which are strongly **exothermic** (see Ch 11.2), i.e. in which formation of the product releases energy which heats up the reaction mixture, can often be dangerous.

As the reaction begins, the mixture heats up, so the reaction goes faster. This means the product is formed at a greater rate, which releases energy at a greater rate, which increases the temperature in the flask at a greater rate ... The rate may accelerate very rapidly to explosion.

Such a reaction can be controlled by adding one reactant to the other slowly and with stirring, with a thermometer in the flask and with the flask surrounded by cold water or ice.

A reaction which is notorious for its violence is Skraup's synthesis of quinoline,

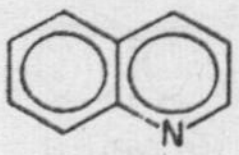

Phenylamine is heated under reflux with nitrobenzene, propane-1,2,3-triol (glycerol), concentrated sulphuric acid and iron(II) sulphate. For some unknown reason, this reaction was a frequent part of university courses in practical organic chemistry, and the ceilings of teaching laboratories bore many circular brown stains as a result.

Separation of liquid mixtures

Mixtures of liquids which do not in fact mix, i.e. which are *immiscible* or only dissolve in each other to a small extent, can be separated using a tap funnel (see *Figure 29.2.2*).

Provided that the liquids have different densities, they will separate into layers in the funnel. With the stopper taken out, the bottom layer is slowly released through the tap until the interface between the layers is just going through the tap. After discarding the interface, the top layer is then run out into another receiver flask.

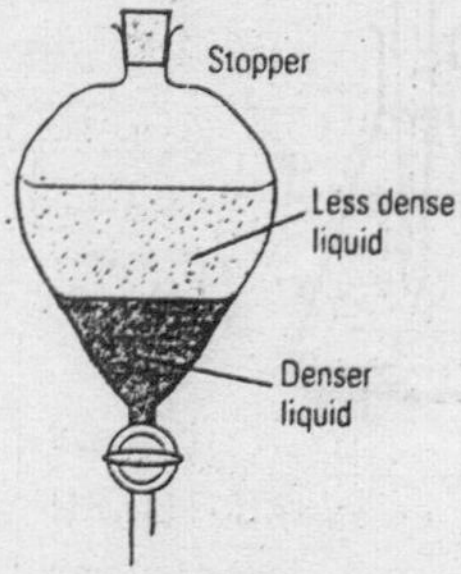

This technique is particularly useful in separating organic products from an aqueous reaction mixture. The cold reaction mixture is shaken with *successive amounts* of a water-immiscible organic solvent. (Each time, the organic layer is run off and collected before fresh solvent is added.) The organic material will distribute itself between the water and the organic solvent until equilibrium is reached, but more will dissolve in the organic layer. Two or three *extractions* with a suitable solvent will remove most of the product. Typical solvents are ethoxyethane (ether) and trichloromethane (chloroform). But be careful about which layer is kept – ethoxyethane is less dense than water, but trichloromethane is denser than water. (Densities (g cm^{-3}): water, $\rho = 1.00$; ethoxyethane, $\rho = 0.71$; trichloromethane, $\rho = 1.59$)

Stopper

Less dense liquid

Denser liquid

**Figure 29.2.2
Separating (tap) funnel**

Distillation of liquids

Mixtures of liquids which do mix (i.e. are *miscible*) can usually be separated, provided that they have different boiling points. The technique of fractional distillation is of crucial importance in the first stage of refining crude oil and is discussed in Ch 15.1. Here we shall look at the simple theory of fractional distillation.

If two liquids have boiling points that differ by at least 10 °C, it is usually possible to separate them using **distillation**. The mixture of the two liquids is heated so that the mixture boils. The lower boiling liquid will start to boil first and the vapour will contain more of this lower boiling liquid. This enriched vapour will rise from the liquid mixture. The vapour is led into a cooler part of the apparatus, the condenser, and in this the vapour condenses to a liquid which contains a much higher proportion of the liquid with the lower boiling point (*Figure 29.2.3*).

The normal (Liebig) condenser has cold water flowing through it, to provide the cooling effect – although liquids with high boiling points can easily be condensed using an air condenser, without the need for the flow of water through the condenser.

By repeating this process of distillation two liquids can be separated. However, repeating the distillation a number of times is time consuming. Introducing a *fractionating column* into the apparatus (*Figure 29.2.4*) allows the process of distillation to be repeated a large number of times all at once, so to speak.

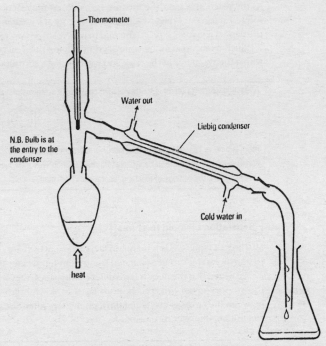

Figure 29.2.3   Simple distillation

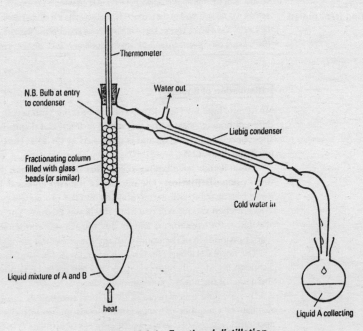

Figure 29.2.4   Fractional distillation

The vapour rises in the column and, as it does so, the higher-boiling liquid begins to condense and run down. With suitable adjustment of the heating rate, only the vapour of the lower-boiling liquid reaches the side arm and enters the condenser. All the way up the column, liquid is condensing and evaporating. As the vapour goes up the column, it becomes richer in the lower-boiling component as the higher-boiling one condenses.

As the liquid runs down the column, it becomes richer in the higher-boiling component as the lower-boiling one vaporises. After a time, a steady *temperature gradient* is established in the column, with the boiling point of the lower-boiling component registering on the thermometer.

This process of **fractional distillation** can repeat many times the process of boiling and condensing within the single piece of apparatus. The beads or other packing in the column provide a large surface area for the boiling and condensing process. As we have said, fractional distillation on a huge scale is used in refining crude oil (see Ch 15.1); in the oil industry the process is continuous and the different fractions run off at different levels of the column.

Purification of solids by recrystallisation

An important technique used to purify solids is recrystallisation, already mentioned in Ch 19.7.

The crude product is dissolved in the *minimum* quantity of *hot solvent*, filtered hot to remove *insoluble impurities*, allowed to cool to *crystallise the product*, and filtered to remove the solvent together with any *soluble impurities*. The technique relies on a number of assumptions – that the desired product is the main material in the mixture (impurities are only present in relatively small quantities) and that the solubility of the desired product increases with increasing temperature in a particular solvent.

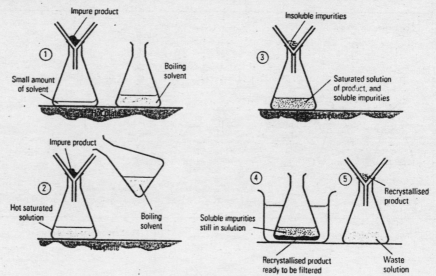

Figure 29.2.5   Recrystallisation

One method of recrystallisation involves the following steps (*Figure 29.2.5*).

- The crude product is placed in a filter paper in a filter funnel.
- This funnel is placed in the neck of a conical flask containing a small quantity of the solvent (to prevent the glass flask getting too hot).
- A second flask containing boiling solvent is also available – with both flasks on a hotplate. ①
- Hot solvent is slowly poured into the filter funnel to dissolve as much of the solid as possible – impurities that are insoluble in the solvent are held in the filter paper. A saturated solution of the product, with soluble impurities, passes through the filter paper. ② and ③

- The saturated solution of the product is cooled slowly to room temperature (or even to 0 °C by using ice if needed). ④
- The product should crystallise out, leaving behind the soluble impurities, since there should be insufficient of the impurities to crystallise out.
- The recrystallised product is filtered off and dried (between filter papers, in an oven below the m.p. of the product, or in a desiccator). ⑤
- If necessary, the recrystallisation is repeated to produce as pure a product as possible.

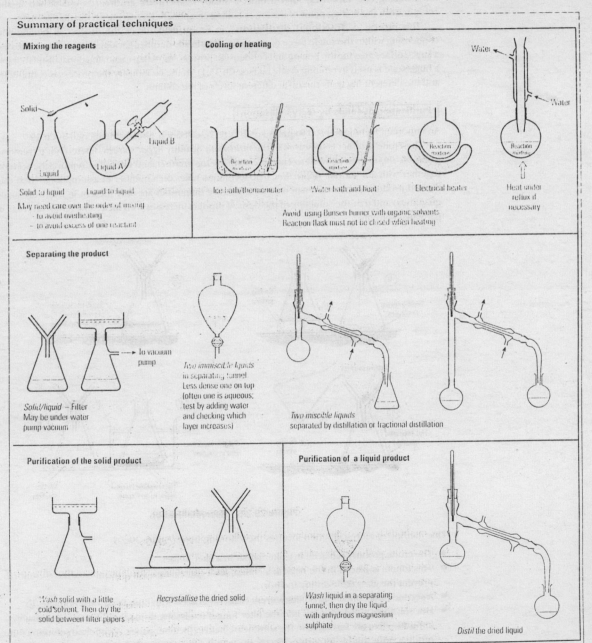

Summary of practical techniques

Mixing the reagents

Solid

Liquid B

Liquid A

Liquid

Solid to liquid   Liquid to liquid

May need care over the order of mixing
 – to avoid overheating
 – to avoid excess of one reactant

Cooling or heating

Water

Water

Reaction mixture

Reaction mixture

Reaction mixture

Reaction mixture

Ice bath/thermometer   Water bath and heat   Electrical heater   Heat under reflux if necessary

Avoid using Bunsen burner with organic solvents
Reaction flask must not be closed when heating

Separating the product

To vacuum pump

Solid/liquid – Filter May be under water pump vacuum

Two immiscible liquids in separating funnel Less dense one on top (often one is aqueous; test by adding water and checking which layer increases)

Two miscible liquids separated by distillation or fractional distillation

Purification of the solid product

Wash solid with a little cold solvent. Then dry the solid between filter papers

Recrystallise the dried solid

Purification of a liquid product

Wash liquid in a separating funnel, then dry the liquid with anhydrous magnesium sulphate

Distil the dried liquid

Measuring melting points

The melting point of a solid is a useful indication of the purity of a product, and can even provide a method of confirming the identity of the product. We have already met the technique briefly in Ch 19.7 when we discussed the identification of carbonyl compounds. A pure solid will melt over a very small temperature range – normally less than one degree. If there are impurities present then the solid will gradually melt over a range of quite a few degrees.

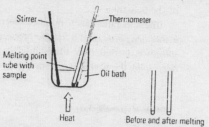

Figure 29.2.6   Melting point apparatus

Electrical melting point apparatus

A number of methods can be used to measure the melting point (e.g. *Figure 29.2.6*), but all require that a very small sample of the product is slowly heated and the temperature raised very gradually. The temperature at which the solid melts, i.e. the temperature when the crystalline solid structure collapses and a meniscus is formed in the sample tube, is recorded as the melting point of the solid.

The solid must be dry. The solid is placed in a melting point tube – a very fine glass tube, 0.5 mm internal diameter, sealed at one end by heating the very end in a hot flame. This tube is then placed in or on the apparatus to be used.

The simplest apparatus contains a boiling tube of liquid paraffin (or even water, if the melting point of the sample is below 100 °C), and the sample tube attached to a thermometer is placed in this oil. The temperature of the tube is raised slowly with constant stirring by heating the boiling tube. Improvements on this use a side tube to allow better circulation of the oil, or a metal block that is heated electrically and the thermometer and melting point tube observed through small holes drilled in the block.

The accurate measurement of a melting point is a slow and painstaking exercise, but it does provide a clear indication of the purity, and possibly the identity, of a compound. When the identity of a solid product is thought to be known, determining the melting point of the product mixed with a sample of the 'real' compound will confirm the identity.

Since the sample is now a mixture – of the unknown and the real compound – the melting point will still be sharp if the two are the same compound. If they are not identical, the mixture will contain two different compounds, and even if the two pure samples melt at the same temperature, the mixed melting point will occur over a range and also be lowered. For more details, see Ch 19.7.

Liquid crystals

Most crystalline solids melt at a sharp 'melting point'. Amorphous solids melt over a wide range of temperatures. There are a small group of compounds that appear to have two distinct melting points. The solid melts at one sharp melting point to form a *liquid crystal*. This liquid crystal has another sharp 'melting point' when it changes into a liquid. In the liquid crystal phase there is sufficient order in the material to resemble a crystal.

Picture a large box of pencils containing packets of pencils. The whole tidily packed box resembles a solid liquid crystal. If the box is opened, the packets of pencils will be spread randomly, but each packet of pencils is still very much in order – the liquid crystal state. When all the pencils are removed from the packets and are all randomly spread out, we have the true liquid state.

Liquid crystal display units, in your watch etc, contain compounds that are in the liquid crystal state at room temperature. When they become overheated, the liquid crystal melts – becomes a true liquid – and no longer exhibits the properties that enable it to be used as a display.

Introduction to chromatography

Chromatography is a very widely used technique in modern chemistry. It is used to separate the components of mixtures as well as testing for the purity of a wide range of materials. Most of you will have tried to separate coloured inks using filter paper, and modern chromatographic techniques allow the separation of optically active isomers in drug manufacture. We have mentioned chromatography briefly in Ch 26.6 when separating and identifying the isomers from the nitration of the benzene ring.

The apparatus has a number of common features. A *support* medium is used to hold the *stationary phase*, and a *mobile phase* (or solvent) containing the materials to be separated (the *solute*) is passed over the stationary phase. Most systems need some way of identifying or *developing* the resulting *chromatogram* (*Figure 29.2.7*).

Thin-layer chromatography

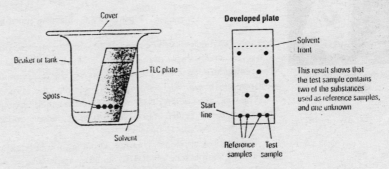

Column chromatography

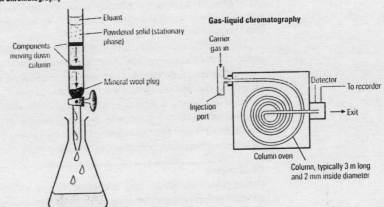

Figure 29.2.7 Chromatography

This chromatogram can appear as a sheet containing a number of 'spots', as 'bands in a column', or even as peaks recorded from an electrical signal.

The principles of the different methods are all very similar. **Partition chromatography** relies on the difference in solubility of the solute between the stationary phase and the mobile phase. The slight differences in solubility for each component of a mixture means that the components separate out as the chromatograph runs (*Figure 29.2.8*).

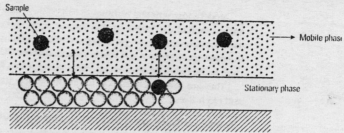

Figure 29.2.8   Partition and adsorption chromatography

Adsorption chromatography relies on the slight differences in attraction of the solute for the stationary phase.

These principles have been developed into a wide range of techniques from gas–liquid chromatography (GLC), thin layer chromatography (TLC), 2-D thin layer chromatography, column chromatography, paper chromatography, high performance liquid chromatography (HPLC), electrophoresis (see Ch 28.5) and ion-exchange.

In each of these techniques, the components of a mixture can be separated out using the appropriate stationary phases. However, this does not tell you what each component is.

Chromatography with coloured compounds is relatively simple – you can see the separation of the compounds as it happens. Often the final chromatograms are apparently clear sheets – they have to be developed using some reagent that reacts with the hidden spots to produce a coloured spot. Or on a chart the peaks are just peaks – with no indication as to what material they represent (*Figure 29.2.9*).

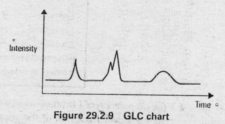

Figure 29.2.9   GLC chart

The area under the peak will give us information about the amount present, but not what it is.

The most basic way to identify a spot or peak is to repeat the experiment with a pure sample of what you think the material is. It should appear in exactly the same place in the chromatogram (*Figure 29.2.10*).

Here, X is the unknown substance: A, B, C and D are the trial substances. It can be seen that X is most likely to be substance B.

$$R_f = \frac{x}{y}$$

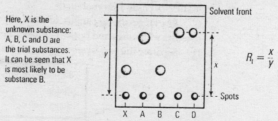

Figure 29.2.10   Paper chromatogram

Alternatively, since the experiment should produce exactly the same results if repeated exactly, we can measure the retention factor, the R_f value for the sample. The R_f value is given by the equation:

$$R_f = \frac{\text{distance of spot of compound from the base line}}{\text{distance of solvent front from the baseline}}$$

Modern instrumental techniques can connect up a chromatograph with a second instrument to analyse each of the 'peaks' as they appear at the end of the chromatograph. A typical set up uses a mass spectrometer at the end of a gas chromatography machine (GC–MS). Each peak is analysed on the mass spectrometer as it reaches the end of the GLC. Such machines provide powerful analytical tools in the hands of research chemists in the search for new drugs.

They are also widely used in testing for minute traces of drugs in the urine or blood of athletes and racehorses.

The first use of chromatography was by a Russian botanist, Mikhail Tsvett, in 1906. It was greatly developed and extended as a method of separating complex mixtures of amino acids by Archer Martin and Richard Synge, working at the Wool Industries Research Institution in Leeds in the early 1940s. By 1944 they were using paper chromatography routinely. Martin later joined the Medical Research Council and worked extensively on the early development of GLC. Martin and Synge were awarded the Nobel Prize for Chemistry in 1952 for their work on chromatography.

SELF-ASSESSMENT
QUESTIONS
29.2

1 a At the end of a particular organic preparation the product was a white crystalline powder. Explain how this powder would be recrystallised to purify it and remove both soluble and insoluble impurities.

  b To establish the identity of the powder, mixed melting points were carried out using the pure product and two possible identities of the product. Explain the principle of mixed melting points using the following results with an unknown compound, which might be either cholesterol or glucose.

Compound	Melting point / °C
Unknown	150
Cholesterol	149
Glucose	150
Mixture of unknown + cholesterol	149
Mixture of unknown + glucose	123

2 The preparation of 1-bromobutane from butan-1-ol can be carried out by the following method. Draw a labelled diagram of the apparatus you would use at each of the stages A–G of this preparation.

  A Mix 5 g sodium bromide, 3.0 g butan-1-ol and 5 cm³ water in a 50 cm³ reaction flask, fitted with a reflux condenser. Slowly add 5 cm³ concentrated sulphuric acid, keeping the reaction flask cool during the addition.

  B When the addition is complete, rearrange the apparatus and reflux for 0.5 hr to produce a mixture of crude 1-bromobutane.

  C Rearrange the apparatus to distil the crude product, which contains 1-bromobutane together with unreacted butan-1-ol and water in two separate layers.

  D Separate the two layers keeping the bromobutane layer. Relevant densities are:

1-Bromobutane	1.3 g cm⁻³
Water	1.0 g cm⁻³
Butan-1-ol	0.8 g cm⁻³

  E Wash the 1-bromobutane layer successively with concentrated hydrochloric acid (to remove any unreacted alcohol), and with sodium hydrogen carbonate solution (to remove residual hydrochloric acid).

  F Dry the almost pure 1-bromobutane with anhydrous sodium sulphate.

  G Distil this product, collecting the fraction boiling between 101 °C and 103 °C.

29.3   Standard methods

Organic chemistry very often involves the change of one functional group into another. The most common types of reactions in the organic chapters of this book involve oxidation, reduction, dehydration and hydrolysis, and in this section we will pull all these types of reactions together (*Figure 29.3.1*).

Oxidising agents
$K_2Cr_2O_7$ dilute H_2SO_4
$KMnO_4$ dilute H_2SO_4
$KMnO_4$ OH^-

Reducing agents
$LiAlH_4$ in ether
$NaBH_4$ in water (milder)
Na/C_2H_5OH
Sn/HCl (concentrated)
H_2/Ni

Dehydrating agents
P_4O_{10} or H_3PO_4
Concentrated H_2SO_4

Hydrolysis agents
HCl (aq)
NaOH (aq)

Halogenating agents
PCl_5
PCl_3
SCl_2O
Red P/Br_2
Red P/I_2

Figure 29.3.1

The development of organic chemistry still continues, faster than ever, and new techniques for achieving chemical changes are continually being discovered.

Oxidising reactions

One oxidising agent that is usually kept away from most organic reactions is oxygen itself. Too many organic chemicals burn in oxygen to form carbon dioxide and water – not very useful chemicals in the formation of new organic molecules. The controlled industrial oxidation of alkenes uses a silver catalyst. In schools, and quite often in industry, oxidation is carried out with potassium dichromate(VI) in dilute sulphuric acid, or potassium manganate(VII) usually in acid solution.

Mild oxidising agents which are used in detecting aldehyde groups are Tollens, Benedict's and Fehling's reagents. Details of these can be found in Ch 19.3.

Some oxidation reactions

$$C_4H_8 + 6O_2 \xrightarrow{\text{ignite}} 4CO_2 + 4H_2O$$

$$CH_2{=}CH_2 + \tfrac{1}{2}O_2 \xrightarrow[\text{170 °C}]{\text{Ag catalyst}} CH_2{-}CH_2$$ (with O bridging)

$$CH_2{=}CH_2 + [O] + H_2O \xrightarrow[\text{H}^+, \text{cold}]{\text{dil. } KMnO_4} CH_2OHCH_2OH$$

$$CH_3CH_2OH \xrightarrow[\text{H}_2SO_4, \text{distil}]{K_2Cr_2O_7} CH_3CHO$$

$$CH_3CHO \xrightarrow[\text{H}_2SO_4 \text{ reflux}]{K_2Cr_2O_7} CH_3COOH$$

Reduction reactions

Reduction of alkenes using hydrogen at high temperatures, high pressures, and a transition metal catalyst such as nickel is an important reaction in the hydrogenation of vegetable oils to form partially saturated fats such as margarine.

A more important reagent in the laboratory is lithium tetrahydridoaluminate(III), $LiAlH_4$, in ethoxyethane as a solvent (N.B. no water at all must be present.) A less vigorous reducing agent is sodium tetrahydridoborate(III), $NaBH_4$, which can be used in aqueous solution. Other reducing agents include tin in fairly concentrated hydrochloric acid (specifically for reducing aromatic nitro groups) or sodium in ethanol.

Some reduction reactions

$$CH_3CHO + 2[H] \xrightarrow[RT]{NaBH_4/H_2O} RCH_2OH$$

$$R-\overset{\displaystyle O}{\underset{\displaystyle OH}{C}} + 4[H] \xrightarrow[RT]{LiAlH_4/ethoxyethane} RCH_2OH + H_2O$$

$$CH_2 = CH_2 \xrightarrow[150°C/5atm]{H_2/Ni\ cat} CH_3CH_3$$

(benzene ring with NO_2) $\xrightarrow[\substack{\text{ii) NaOH(aq), conc.}\\ \text{iii) Steam distil}}]{\text{i) Sn, conc HCl}}$ (benzene ring with NH_2)

Dehydrating reactions

Concentrated sulphuric acid is the most frequently used reagent to remove the elements of water from an organic compound, although phosphorus(V) oxide (often as phosphoric acid, H_3PO_4) o catalysts such as aluminium oxide (with the hot vapour of the compound) can be used.

Some dehydration reactions

$$CH_3CH_2OH \xrightarrow[heat]{conc.\ H_2SO_4} CH_2{=}CH_2 + H_2O$$

$$CH_3CH_2OH \xrightarrow[heat]{Al_2O_3} CH_2{=}CH_2 + H_2O$$

Hydrolysis reactions

Water as a reagent in itself is not very useful in organic synthesis. After all, most organic compoun have very little affinity for the very polar water molecule. Hydrolysis reactions need something to s the reaction off – an acid or alkali group to initiate the reaction. Industrially, reaction with water possible, but only with superheated steam together with a catalyst and high pressure.

Some hydrolysis reactions

$$CH_3Br + OH^\ominus(aq) \xrightarrow{warm} CH_3OH + Br^\ominus$$

$$RCOOR' + H_2O \xrightarrow[\text{ii) } H^\oplus(aq).]{\text{i) reflux}} RCOOH + R'OH$$

$$RC\overset{\displaystyle O}{\underset{\displaystyle NH_2}{}} + H^\oplus(aq) + H_2O \longrightarrow RCOOH + NH_4^\oplus(aq)$$

Grignard reactions

Victor Grignard of France introduced these reactions in 1901. They are widely used in industry. He was awarded the Nobel Prize for Chemistry in 1912.

When a bromoalkane is reacted with magnesium metal in dry ethoxyethane, a new compound, RMgBr, is formed. Experiments with radioactive Mg suggest that there is an equilibrium between RMgBr and the dialkylmagnesium and magnesium bromide, $R_2Mg + MgBr_2$. We shall use the simple formula RMgBr here.

$$R\!-\!Br + Mg \longrightarrow RMgBr$$
A Grignard compound

This is an example of a very large group of compounds that cross the artificial boundary between organic and inorganic chemistry – *organometallic* compounds. Essentially the Grignard compound can be considered a source of R^- ions (compare the Pauling electronegativities of carbon = 2.5 and magnesium = 1.2 – see Ch 2.2).

The advantage of these reagents is that they provide a short route from the halogenoalkane to many different functional groups – to carboxylic acid, different types of alcohols, or even the alkane.

Some Grignard reactions

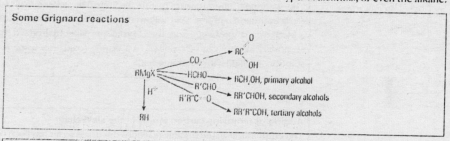

Starting compounds into target compounds

The problem facing any preparative chemist is to convert one compound into another specific compound. The starting compounds used are governed by availability, cost, and safety and environmental considerations. The desired compound is the **target** compound – the compound which is required for further research, for testing as a potential pharmaceutical drug, for development into a new consumer product, or whatever.

As a chemist, you have to look at what you can do with the possible starting compound and at the same time look at possible routes for preparing your target compound. At this level, hopefully, this will provide an answer which is either a single- or two-stage synthesis. Try these two examples.

1. How would you prepare propanoic acid starting from propan-1-ol?

The first thing to do is to list what is possible with propan-1-ol, and possible ways to prepare propanoic acid.

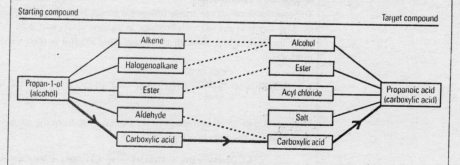

This example is quite simple and a single stage oxidation of the alcohol to the carboxylic acid is possible (refluxing with acidified potassium dichromate(VI)).

2. *How would you convert propene into propanone?*
Again, what is possible – from both sides of the problem.

Starting compound   Target compound

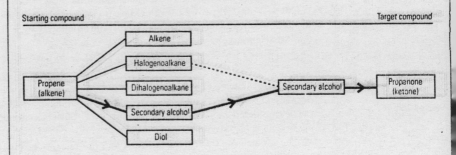

The common intermediate is the alcohol. The alkene will form the secondary alcohol (concentrated sulphuric acid followed by reaction with water); or hydrogen bromide can be added to the propene; and the 2-bromopropane hydrolysed to the secondary alcohol. A secondary alcohol can be oxidised to a ketone (oxidation using hot potassium dichromate(VI) solution acidified with sulphuric acid).

Adding or removing carbon atoms in the alkyl chain

The emphasis in all the organic reactions discussed so far has been on the reaction of a functional group, while the carbon skeleton has remained unchanged. If we need to make the carbon chain longer or shorter, various methods can be used.

Adding a carbon atom

A useful reaction for adding a carbon atom to the chain is to substitute a $CN^{\ominus}$ group into the molecule at the appropriate position. The nitrile group is then reduced to an amine or hydrolysed to a carboxylic acid.

e.g. $CH_3CH_2OH \xrightarrow[\text{heat}]{\text{NaBr/H}_2\text{SO}_4} CH_3CH_2Br \xrightarrow[\substack{\text{reflux} \\ \text{in H}_2\text{O/} \\ \text{C}_2\text{H}_5\text{OH}}]{\text{KCN}} CH_3CH_2CN \xrightarrow[\substack{\text{ethoxyethane} \\ \text{RT}}]{\text{LiAlH}_4} CH_3CH_2CH_2NH_2$

$\downarrow \substack{\text{H}^{\oplus}\text{(aq)} \\ \text{boil}}$ $\text{RT} \downarrow \substack{\text{NaNO}_2 \\ \text{HCl}}$

                     CH_3CH_2COOH $CH_3CH_2CH_2OH$

Removing a carbon atom

The reaction involves the amide functional group, but in a rather complex reaction.

The amide is reacted with bromine in alkaline solution (Ch 28.1). The product is an alcohol with one less carbon atom in the chain. This alcohol is then used to produce the required product.

e.g. $CH_3CH_2OH \xrightarrow[\text{reflux}]{\substack{\text{K}_2\text{Cr}_2\text{O}_7 \\ \text{H}_2\text{SO}_4\text{(aq)}}} CH_3COOH \xrightarrow[\substack{\text{ii) heat}}]{\text{i) NH}_3} CH_3CONH_2 \xrightarrow[\substack{\text{ii) NaOH(aq)} \\ \text{warm}}]{\text{i) Br}_2} CH_3NH_2 \xrightarrow[\text{RT}]{\text{NaNO}_2, \text{HCl}} CH_3OH$

The reaction involved in the loss of the carbon atom from the alkyl chain is called the *Hofmann degradation*. The full equation is:

$$CH_3CONH_2 + Br_2 + 4NaOH \xrightarrow{\text{warm}} CH_3NH_2 + Na_2CO_3 + 2NaBr + 2H_2O$$

– but you do not need to learn this equation!

SELF-ASSESSMENT QUESTIONS 29.3

1 What reagent(s) would you need to carry out the following chemical changes?

a $CH_3CHO \longrightarrow CH_3COOH$

b $C_6H_5NO_2 \longrightarrow C_6H_5NH_2$

c $CH_3CH_2OH \longrightarrow CH_2{=}CH_2$

d $CH_3CHO \longrightarrow CH_3CH_2OH$

2 All the following chemicals can be prepared from pentan-1-ol. Are they formed in an oxidation, reduction or dehydration reaction?

a $CH_3CH_2CH_2CH_2CHO$

b CO_2 and H_2O

c $CH_3CH_2CH_2CH{=}CH_2$

d $CH_3CH_2CH_2CH_2COOH$

3 Outline a reaction scheme to prepare:

a ethanoic acid starting from bromomethane;

b 2-methylpropanoic acid starting from 2-bromopropane;

c 1-aminobutane starting from pentanoic acid.

4 How would you carry out the following chemical conversions?

a Bromoethane into ethanoic acid, in two stages.

b Chloroethane into propanoic acid, in two stages.

c Bromoethane into propanoic acid, in two stages.

In each case state clearly the reagents and conditions needed, and write equations to show the reactions involved.

29.4 Aliphatic sequences

Summary of aliphatic sequences

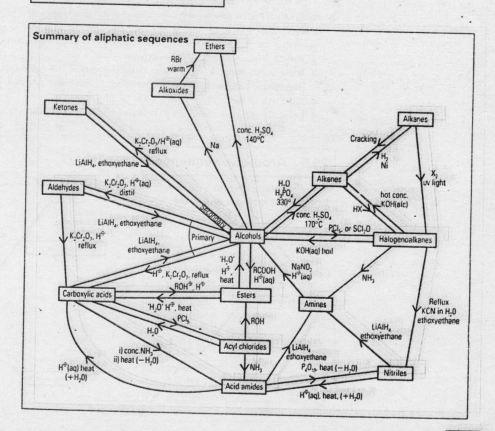

1 Construct a flow chart to show the relationship between alcohols, halogenoalkanes and nitriles. Make sure you include all the reagents and conditions for each reaction used in the diagram.

2 How would you carry out the following conversions? Make sure you include the reagents and conditions needed.

   a Propan-2-ol to 2-bromopropane.
   b Ethanoic acid to ethanol.
   c Ethanol to ethanal.
   d Cyclohexanol to cyclohexene.

3 Tetraethyl-lead(IV), $(C_2H_5)_4Pb$ can be made by reacting a lead/sodium alloy with chloroethane. It used to be used in petrol, together with 1,2-dibromoethane. The principal starting chemical for both these organic compounds is ethene, produced by cracking crude oil fractions.

   a Write an equation to show what cracking means.
   b What is meant by an alloy?
   c How would you covert ethene to chloroethane?
   d How would you convert ethene to 1,2-dibromoethane?
   e Tetraethyl-lead(IV) is a very volatile liquid at room temperature. What type of bond do you think exists between the ethyl groups and the lead atom?
   f Draw a diagram to show the shape of the molecule around the lead atom.

4 a Show how you would prepare ethanoic acid from ethanol. Describe the apparatus you would use.
   b How would you separate the ethanoic acid from the reaction mixture? Illustrate your answer with a diagram of suitable apparatus.
   c How would you convert this ethanoic acid into methylamine? (Hint: the chain length has to be shortened.)

5 The plastic, PVC, is widely used in guttering on modern houses. It is prepared by polymerising chloroethene, which, in turn, is obtained from ethene in a two stage synthesis. Ethene is converted into 1,2-dichloroethane, which is then converted into chloroethene.

   a Write a display formula for chloroethene.
   b Write a two-stage flow diagram connecting ethene to chloroethene. Include on your diagram the reagents and conditions needed to carry out each stage of the sequence.
   c What type of mechanism is involved in the first stage?
   d Draw a displayed structure for PVC, showing clearly the repeating unit in the polymer.

29.5 Aromatic sequences

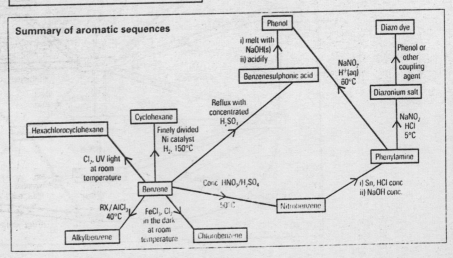

Summary of aromatic sequences

1 Part of the flow chart showing the inter-relationship between aromatic compounds looks as follows:

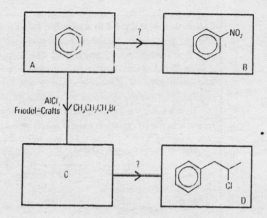

a Write the formula that should be placed in the blank box for C.

b What reagent(s) and conditions are needed to convert compound A to compound B?

c What reagent(s) and conditions are needed to convert compound C to compound D?

d How would you purify the product B (b.p. 211 °C, m.p. 6 °C)?

2 Outline how you would attempt to prepare phenylamine starting from benzene in a two-stage synthetic route. Make sure you include the reagents and conditions needed for each stage of the synthesis.

3 What reagents and conditions would be needed to carry out the following chemical changes?

a Nitrobenzene to phenylamine

b Methylbenzene to 4-nitromethylbenzene

c Phenylamine to phenol

d Benzene to ethylbenzene

4 In the following reaction sequence, what are the conditions and reagents necessary to carry out each step, **a** to **e**, of the sequence? What type of reaction mechanism is involved in stage **a**?

785

29.6 Yields in organic reactions

Calculation of the yield in a reaction (see Ch 9.5 and Ch 27.11)

The azo dye 4-hydroxyazobenzene is prepared by diazotising phenylamine using hydrochloric acid and sodium nitrite below 5 °C. The mixture is then reacted with a solution of phenol in 2.5 mol dm⁻³ sodium hydroxide solution. The equation for the overall reaction is given below. The yellow crystals of the dye separate out. Typically, the reaction of 2.5 g of the phenylamine produces 3.3 g of the dye after recrystallisation.

To work out the percent yield you have first to calculate the theoretical yield, which is given by the scheme or equation for the reaction.

Equation:

	Phenylamine		4-Hydroxyazobenzene
Data:	2.5 g		3.3 g
M_r	93		198
Moles:	0.0269		0.0167
(= mass/M_r)			

Ratio: 1 ◄─────────────────────► 1

(N.B. The equation need only show the ratio of the original material to product)

Calculation:

The theoretical yield should be 0.0269 moles since the ratio is 1:1

The actual yield is 0.0167 moles

The % yield is therefore $\dfrac{\text{actual yield} \times 100}{\text{theoretical yield}} = \dfrac{0.0167 \times 100}{0.0269} = 62\%$

The preparation of many organic chemicals often involves a multistage synthesis – with a yield in each step that might not be 100%. The simplest way to find the overall yield is to multiply the % yield at each step together and divide by $100^{(n-1)}$, for an n stage synthesis. For a two stage synthesis with 80% yield in stage 1 and 60% yield in stage 2, the overall yield will be $(80 \times 60)/100 = 48\%$.

Q

SELF-ASSESSMENT QUESTIONS 29.6

The preparation of 3-amino-4-methylbenzenesulphonic acid involves a three stage synthesis starting with methylbenzene.

a What inorganic reagents would you require to carry out stage 1?

b What type of reaction is occurring in stage 1?

c In stage 1, another product, isomeric with the given compound, is possible. Name this second isomer.

d What reagents do you need for the stage 3 reduction reaction?

e 5.0 g methylbenzene was used at the start of this reaction sequence. How many moles are there in 5.0 g methylbenzene?

f If 4.6 g of product was collected at the end of stage 1, what would the percentage yield be for the stage 1 reaction?

g The overall yield in this reaction sequence is usually around 35%. Given that the yield is 35%, what mass of 3-amino-4-methylbenzenesulphonic acid would be prepared in this sequence, starting with 5.0 g methylbenzene?

**SUMMARY
29**

In this chapter the following new ideas have been covered. *Have you understood and learned them?* Could you explain them clearly to a friend? The page references will help.

- The techniques for heating and cooling reaction mixtures.
- How to separate mixtures of compounds.   p770
- How to purify a liquid or solid product.   p771
- Measuring melting points and the use of such measurements.   p773
- Chromatographic methods.   p775
- Common reactions in organic synthesis.   p776
- Flow charts for common reaction sequences of aliphatic compounds.   p779
- Flow charts for common reaction sequences of aromatic compounds.   p783
- How to work out a single stage or two stage synthesis.   p784
- Lengthening or shortening a carbon chain.   p781
- Calculating the percent yield in a two-stage synthesis.   p782
    p786

**EXAMINATION
QUESTIONS
29**

1 Paracetamol can be made from phenol by the following synthetic route.

$$\text{A} \xrightarrow[\text{dilute HNO}_2]{\text{nitration}} \text{B} \xrightarrow{\text{Zn/H}^+\text{(aq)}} \text{C} \xrightarrow[\text{CH}_3\text{COCl}]{\text{ethanoylation}} \text{4-(N-ethanoylamino)phenol}$$

a What is the significance of the circle in each of the benzene hexagons?

b Write a displayed formula for paracetamol.

c In the reaction to form compound B, there is often contamination of the product with another isomer. Suggest a possible formula for a mono-substituted isomer.

d How would you test the product to see if there was any of this contaminating isomer present?

e 5.0 g of phenol was used in the first step of this reaction sequence. What mass of Paracetamol would be formed if the yield in *each* of the three steps was 60%?

f What type of reaction is the conversion of compound B to compound C?

g Compound C can react with an acid and with an alkali. Write balanced equations to show these reactions with sodium hydroxide and with hydrochloric acid.

2 Sorbitol is a polyhydroxy compound used in moisturising cream. It is occasionally used as an alternative sweetener in confectionery and wines, since the compound is not metabolised or fermented, although it can lead to diarrhoea if too much is consumed. The structure of sorbitol is:

$$\underset{\text{CH}_2-\text{CH}-\text{CH}-\text{CH}-\text{CH}-\text{CH}_2}{\overset{\text{OH} \quad \text{OH} \quad \text{OH} \quad \text{OH} \quad \text{OH} \quad \text{OH}}{\rule{0pt}{1em}}}$$

a Explain why sorbitol is used in cosmetics as a cream to retain moisture.

b Esters of long-chain carboxylic acids with the triol, propan-1,2,3-triol (glycerol), are the major constituent of vegetable and animal fats (the triglycerides). Since sorbitol is not metabolised, there is a possibility that esters of sorbitol with long chain carboxylic acids could provide alternatives to the use of animal fats. How many carboxylic acid molecules will react with one mole of sorbitol?

c Draw the formula of the ester formed between sorbitol and hexanoic acid, showing clearly the bonding in one of the ester linkages.

d How would you attempt to esterify sorbitol, using hexanoyl chloride as the source of the long chain acid?

e Calculate the percent yield if 20 g of sorbitol ($M_r = 182$) was esterified with excess hexanoyl chloride and 25 g of the ester ($M_r = 542$) was produced.

3

a Outline a two-stage synthesis to carry out the reaction outlined above. Explain clearly the conditions and reagents needed for each of the stages.

b The benzoic acid can be esterified using ethanoyl chloride. Write the formula of the ester formed.

c The ester methyl benzoate can be nitrated, putting a nitro group in the 3-position of the benzene ring. Draw a displayed formula for this methyl 3-nitrobenzoate.

d Write a mechanism, including a clear formula for the attacking species, for the nitration reaction.

- The origin of **flame colours** for compounds of s-block elements. (Ch 6.3)
- Chemical tests for **halogens** and for **halide ions**. (Ch 8.10)
- The theory and use of the **mass spectrometer**. (Ch 1.3 and 9.1)
- How to find **empirical formulae**. (Ch 9.4)
- How to use **equations**. (Ch 9.5)
- How to use **titrations**. (Ch 10.2)
- Tests for **acids** and **alkalis**. (Ch 13.4)
- What is meant by the term **'functional group'**. (Ch 14.2)
- Tests for **alkenes**. (Ch 16.5)
- Tests for **halogenoalkanes**. (Ch 17.4)
- Tests for **alcohols**. (Ch 18.4)
- Tests for **aldehydes** and **ketones**. (Ch 19.6)
- Tests for **arenes**. (Ch 26.2)

30.1 Introduction to analysis
30.2 Quantitative analysis
30.3 Inorganic qualitative analysis
30.4 Organic qualitative analysis
30.5 The use of instruments in analysis

30

analytical techniques

30.1   Introduction to analysis

If **synthesis** involves the combination of smaller units in order to make something which is more complex, **analysis** can be thought of as the breaking up of something into its constituent parts. More usefully, however, chemical analysis is described as a process which leads to the identification of the component parts of a substance, and the determination of the relative proportions of those parts.

Analytical chemistry is the science of finding out, with the highest possible accuracy, what is in a material, and is a subject in its own right. Many major textbooks of chemical analysis have been written, but we shall deal here only with those methods of analysis which you may need to know at this level.

In this topic, we have tried to gather together all the tests and techniques which are listed by the A-level examination boards. All are useful and interesting, but you may wish to check what is and what is not required by the particular syllabus which you may be working to.

Chemical analysis can be divided into two main sections.

- **Qualitative analysis.** In this, the analyst is concerned only to identify what is present (e.g. elements, ions, water, gases, etc). So, for example, the analysis of a well-known blue crystalline compound will show that it contains copper(II) ions, sulphate ions, and water.
- **Quantitative analysis.** Here, analysis is not only carried out to find what is present, but also how much. Results are often expressed as percentage by mass. For example, the blue crystalline compound may be found to contain: copper, 25.5%; sulphate ion, 38.5%; water, 36.1%. These figures, with the necessary relative atomic masses, can be used to find the formula: $CuSO_4.5H_2O$.

Modern analytical techniques can detect some materials in tiny amounts – parts per million (ppm) is commonplace; parts per billion (ppb, where one billion = 10^9) is often reliably and repeatedly achieved.

Methods of this degree of sensitivity are needed when, for example, testing for traces of drugs in the blood or urine of athletes or racehorses.

In the heroic age of chemistry – the first seventy years of the nineteenth century – many advances in our understanding of chemistry were achieved on the basis of analytical results. Indeed, it could be said that accumulated information provided by the analysts was the springboard for our understanding of both inorganic and organic chemistry.

During the last fifty years or so the test tubes, platinum wires, burettes and pipettes of the pioneers – still in general use in schools, and still with applications in research and industry – have largely been replaced by instrumental methods. The machines concerned can be very expensive – a gas chromatograph coupled to a mass spectrometer (GC–MS) is extremely powerful in analysing traces of drugs, for example, but may cost £1 million.

We shall discuss three instrumental methods later (Ch 30.5) – mass spectrometry (MS), infra-red spectroscopy (IR) and nuclear magnetic resonance spectroscopy (NMR).

Qualitative analysis
What is there?
Quantitative analysis
How much is there?

Some commercially available mass spectrometers can routinely measure to parts per trillion (ppt, where one trillion = 10^{12}).

30.2   Quantitative analysis

Combustion analysis of organic compounds

This can only be used for carbon, hydrogen and nitrogen – and by difference, oxygen. The first action is to use qualitative analysis to make sure that there are no other elements (e.g. sulphur and the halogens) present in the compound.

A weighed sample of the compound is then burned in excess oxygen mixed with helium. The gas stream carries the carbon dioxide, water vapour and nitrogen oxides which result from combustion of the sample. The stream is then passed over a hot mesh of metallic copper; the copper removes the excess oxygen from the stream and also reduces any oxides of nitrogen to nitrogen. The remaining gas stream therefore consists of the inert carrier gas helium, plus carbon dioxide, water vapour and (if there is any nitrogen in the compound) nitrogen gas.

The old method involved absorbing water (in anhydrous magnesium chlorate(VII)) and carbon dioxide (in soda-lime – a mixture of calcium and sodium hydroxides), and weighing the two absorption tubes before and after in order to find the masses of water and carbon dioxide produced. In modern equipment, the thermal conductivity of the gas stream is measured before and after each absorption, which gives the amount of water, carbon dioxide, and nitrogen in the gas stream (*Figure 30.2.1*). However, calculations in examinations are set using data obtained from the masses of water and carbon dioxide absorbed, and also possibly the volumes of nitrogen gas obtained.

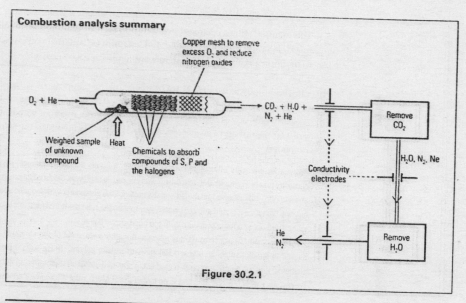

Figure 30.2.1

Worked example 1

1.00 g of a compound X, which contains carbon, hydrogen, nitrogen and oxygen only, is burned in excess oxygen to give 1.49 g of carbon dioxide, 0.765 g water, and 203 cm³ nitrogen, measured at room temperature and pressure. (Under these conditions one mole of gas occupies 24 000 cm³.) Find the empirical formula of X.

Method:

Collect the **data** together with an 'equation':

$$X + O_2 \longrightarrow CO_2 + H_2O + N_2$$
(C, H, N, O)

(From Periodic Table, $A_r(C) = 12.0$; $A_r(O) = 16.0$; $A_r(H) = 1.0$; $A_r(N) = 14.0$)

Data:	1.00 g	1.49 g	0.765 g	203 cm³
	Relative molecular mass	$12.0 + (2 \times 16.0) = 44.0$	$(2 \times 1.0) + 16.0 = 18.0$	1 mol = 24000 cm³
Moles:	$= \dfrac{\text{mass}}{\text{mass of 1 mole}}$	1.49/44.0	0.765/18.0	203/24000
		= 0.0339	= 0.0425	= 0.00846

So far we have the moles of C, H and N in the compound X. To find the moles of oxygen in X we need to find out the mass of oxygen. To find the mass of oxygen we have to convert the moles so far calculated into mass, to find the mass of each of carbon, hydrogen and nitrogen. The mass of oxygen will then be found by difference.

		1 C atom in CO_2	2 H atoms in H_2O	2 N atoms in N_2
		A_r 12.0	A_r 1.0	A_r 14.0
Mass	= moles of element	0.0339×12.0	$2 \times 0.0425 \times 1.0$	$2 \times 0.00846 \times 14.0$
	× mass of 1 mole		= 0.407 g	= 0.0850 g = 0.237 g

∴ Mass of oxygen = 1.00 − (0.407 + 0.0850 + 0.237)
= 0.271 g

Calculating the empirical formula of compound X

Element:		**C**	**H**	**N**	**O**
Data:	mass (g)	0.407	0.085	0.237	0.271
	Relative atomic mass	12.0	1.0	14.0	16.0
Moles: $= \dfrac{\% \text{ or mass}}{\text{mass of 1 mole}}$		$\dfrac{0.407}{12.0}$ = 0.0339	$\dfrac{0.085}{1.0}$ = 0.085	$\dfrac{0.237}{14.0}$ = 0.0169	$\dfrac{0.271}{16.0}$ = 0.0169
Ratio: Ratio of moles (divide by the smallest value)		$\dfrac{0.0339}{0.0169}$ = 2.0 C_2	$\dfrac{0.085}{0.0169}$ = 5.0 H_5	$\dfrac{0.169}{0.0169}$ = 1.0 N	$\dfrac{0.0169}{0.0169}$ = 1.0 O

∴ Empirical formula is C_2H_5NO

This example shows each and every step in the calculation. With some practice such calculations quickly become routine and rapid. But you must be sure that you use your calculator correctly, and to the appropriate number of significant figures at each stage.

N.B. The empirical formula is usually only the first step. Knowledge of M_r (the relative molecular mass) for a compound means that the empirical formula can be converted into a molecular formula (see Ch 9.4). Study of the reactions of the compound, together with instrumental methods to detect particular groupings of atoms (see Ch 30.5), leads eventually to the structure of the molecule of the compound.

Analysis for elements other than carbon, hydrogen, nitrogen and (by difference) oxygen used to be part of an A-level chemistry practical course. One test which was commonly used was the Lassaigne test, which detected halogens, sulphur and nitrogen. This involved heating a small sample of the compound with metallic sodium in a soda-glass tube. The red-hot tube was then plunged into cold water, whereupon the tube shattered (and any excess sodium exploded). The process was carried out in a fume cupboard with the hood drawn down.

This fusion with sodium converts any nitrogen (together with some of the carbon in the compound) to sodium cyanide, sulphur to sodium sulphide, and any halogens to sodium halides. These salts dissolve in the water when the tube is shattered, and the anions can be detected using standard inorganic qualitative analysis methods.

For obvious safety reasons, this test is no longer used! Much safer alternatives are available.

Quantitative inorganic analysis

Very large textbooks have been written on this subject, and indeed many methods of analysis have been used, some of which have now been fully automated.

The more common of these methods can be classified into five main types.

- **Volumetric**, usually involving titrations of known materials in solution with reagents of known concentration.
- **Gravimetric**, involving the measurement of mass, often of a suitable precipitate formed by reaction of ionic solutions, or the change in mass caused by loss of a volatile material (water or carbon dioxide, for example) when the test substance is heated.
- **Methods involving the absorption or emission of light of particular wavelengths**, e.g. colorimetric or spectroscopic methods. These methods can be used for many ions other than the obvious coloured ones of the d-block metals.
- **Methods involving electrical measurements**, e.g. potentiometric, conductometric or polarographic methods.
- **Gas analysis**, for mixtures of gases.

Quantitative volumetric analysis involves calculations. It is essential that you understand how to use the mole (Ch 9.3) and stoichiometric equations (Ch 9.5), and are able to do titration calculations (Ch 10.2 and 10.3). You may want to check through these sections before proceeding.

Now carefully study the worked examples, and then try the self-assessment questions.

Worked example 2

To determine the purity of a sample of iron wire

1.250 g of clean iron wire was reacted with warm sulphuric acid in a flask fitted with a valve so that the hydrogen produced in the reaction could escape, but no air could get in. (Under these conditions only $Fe^{2+}(aq)$ ions are produced, and no $Fe^{3+}(aq)$.)

The reaction mixture was cooled, then filtered and washed into a 250 cm^3 graduated flask, made up to the mark with distilled water, and thoroughly mixed by shaking.

25.00 cm^3 samples of this solution were pipetted into a flask, acidified with about 25 cm^3 of dilute sulphuric acid, and titrated with 0.0200 mol dm^{-3} potassium manganate(VII), $KMnO_4(aq)$, until the first appearance of a permanent pink colour in the titration flask. The average of three good titres was 21.70 cm^3. Calculate the percentage purity of the iron wire.

The ionic equation for the titration is:

$$MnO_4^-(aq) + 5Fe^{2+}(aq) + 8H^+(aq) \longrightarrow Mn^{2+}(aq) + 5Fe^{3+}(aq) + 4H_2O(l)$$

Method:

Collect all the data together with an 'equation'.

($A_r(Fe)$, from Periodic Table = 55.8)

$$Fe \xrightarrow{H_2SO_4} Fe^{2+}$$

1.250 g → 250 cm^3 flask
wire → 25.00 cm^3 sample

$$MnO_4^-(aq) + 5Fe^{2+}(aq) + 8H^+(aq) \longrightarrow Mn^{2+}(aq) + 5Fe^{3+}(aq) + 4H_2O(l)$$

Data: 0.0200 mol dm^{-3}
21.70 cm^3 25.00 cm^3 25 cm^3

Moles: of MnO_4^- $= \dfrac{0.0200 \times 21.70}{1000}$

$= 4.34 \times 10^{-4}$ ①

Ratio: 1   5 ②

Moles: of Fe^{2+}
$5 \times 4.34 \times 10^{-4}$
$= 0.00217$ moles in 25.00 cm^3
$= 0.0217$ moles in 250 cm^3 ③

Mass: of Fe
0.0217×55.8 g in 250 cm^3
$= 1.211$ g of iron in a 1.250 g sample ④

% purity of the iron wire $= \dfrac{1.211 \times 100}{1.250} = 96.9\%$ ⑤

Notes on the working in example 2

(The sequence here is to find the number of moles of MnO_4^- ions in 21.70 cm^3 ①; then use the equation to find the number of moles of Fe^{2+} ions in 25.00 cm^3 of the solution, and hence the moles of Fe^{2+} in the 250 cm^3 flask and so the number of moles of iron in the sample of wire ③. This is multiplied by the relative atomic mass of iron to find the mass of iron in the sample ④, which is compared with the mass of the sample to find the percentage of iron in it. ⑤)

Volume of 0.0200 mol dm^{-3} $MnO_4^-(aq)$ used = 21.70 cm^3

$$\therefore \text{number of moles of } MnO_4^-(aq) = \frac{21.70}{1000} \times 0.0200$$

$$= 0.000434 \text{ mol} ①$$

The reaction equation shows that one mole of MnO_4^- reacts with 5 moles Fe^{2+}. ②
$\therefore 0.000434$ mol MnO_4^- reacts with 5×0.000434 mol of Fe^{2+}.

But 21.70 cm³ of the $MnO_4^{\ominus}$ reacted with 25.00 cm³ of the iron solution.

∴ 25.00 cm³ of the iron solution contains 0.00217 mol $Fe^{2\oplus}$

∴ the 250 cm³ flask contains 0.0217 mol $Fe^{2\oplus}$ ③.

∴ the mass of iron from the wire which is in the flask
$$= 55.8 \times 0.0217 = 1.211 \text{ g } ④$$

The mass of iron wire used was 1.250 g

This has been shown to contain 1.211 g of iron

∴ The percentage purity of the wire is $\dfrac{1.211}{1.250} \times 100$ ⑤

$$= 96.9\%$$

Points to note

This looks rather drawn out because each step is fully explained, and the strategy of the calculation was stated beforehand. Also, the practical method is given in detail to illustrate how the reaction is carried out.

When you have practised some calculations of this kind, you will find that you can work out *in your mind*, before you start, the sequence you will follow. **However, you should *always* write down clearly what you are doing at each stage of the calculation, as shown here.**

This not only keeps your mind clear, but helps anyone who is trying to follow your argument (who could, of course, be an A-level or similar examiner; and don't forget that a clear correct method gains marks even if it contains a slip in arithmetic).

Finally, notice the precision with which the intermediate quantities are reported and the final result stated. The mass of the sample was given to four figures; the pipette and burette were accurate to ±0.05 cm³; but the titration reagent concentration was limited to three figures. In these circumstances, a result to three figures is suitable.

Worked example 3

To find the relative atomic mass of a Group 2 element, given a sample of its chloride.

0.954 g of the anhydrous chloride, BCl_2, of a Group 2 element was dissolved in water. A slight excess of aqueous silver nitrate was added and a suitable method followed to obtain the dry precipitate of silver chloride, AgCl. The mass of precipitate obtained was 1.725 g.

Find the relative atomic mass of the metal, B.

Method:

(From Periodic Table, $A_r(Cl) = 35.5$; $A_r(Ag) = 107.9$)

Data:	BCl_2(aq) +	silver nitrate →	2AgCl +	'B nitrate'
	since it is in Group 2			
Mass:	0.954 g	excess	1.725 g	
M_r:	X + (2 × 35.5)		107.9 + 35.5 = 143.4	
	(where X = A_r of B)			
Moles:	$\dfrac{0.954}{X + 71}$		$\dfrac{1.725}{143.4} = 0.0120$	
Ratio:	1		2	

Moles of BCl_2: Since 1 mole BCl_2 gives 2 moles of AgCl:

$0.0120/2 = 0.0060$

$0.0060 = \dfrac{0.954}{X + 71}$

$0.0060(X + 71) = 0.954$

$0.0060 \times X + 0.0060 \times 71 = 0.954$

$0.0060 \times X = 0.954 - 0.0060 \times 71$

$= 0.954 - 0.426$

$= 0.528$

mass of 1 mole $= \dfrac{\text{mass}}{\text{moles}}$ ∴ $X = \dfrac{0.528}{0.0060}$

$X = 88$

The relative atomic mass of B is 88.0   (Incidentally, B is strontium. $A_r(Sr) = 87.6$)

Alternatively:

Moles AgCl = 1.725/(molar mass of AgCl)

= 1.725/143.4 = 0.0120

0.0120 mol Cl = 0.0120 × 35.5 = 0.426 g of Cl present in BCl_2

∴ Mass of B in 0.954 g = (0.954 − 0.426) = 0.528 g.

But, since 0.0120 mol AgCl comes from 0.0060 mol BCl_2,

A_r(B) = 0.528/0.0060 = 88.0

Points to note

The mass of the precipitate gives the mass of chlorine in the sample of BCl_2, and hence the mass of B. The mass of chlorine is converted into moles, which gives – using the formula of the compound – the number of moles of B.

This, with the mass of B, gives the relative atomic mass of B. See Ch 9.2: mass of element = number of moles × mass of 1 mole, so that:

$$\text{Mass of 1 mole of B} = \frac{\text{mass of B}}{\text{number of moles of B}}$$

The relative atomic mass, A_r, is numerically equal to the mass of one mole of atoms of the element but has no units.

SELF-ASSESSMENT QUESTIONS 30.2

1 1.80 g of a compound Y which contains carbon, hydrogen, nitrogen and oxygen only, is burned in excess oxygen and gives 4.682 g carbon dioxide, 1.077 g water and 160 cm^3 nitrogen, measured at room temperature and pressure. Find the empirical formula of Y.

2 Hydrated ammonium iron(II) sulphate has the formula $(NH_4)_2SO_4.FeSO_4.xH_2O$. 8.73 g of the salt was made up to 250 cm^3 of solution using water and dilute sulphuric acid. 25.0 cm^3 portions of this solution reacted with an average 22.3 cm^3 of 0.0200 mol dm^{-3} potassium manganate(VII). Find the value of x.

Data: the equation for the reaction is:

$$MnO_4^{\ominus}(aq) + 5Fe^{2\oplus}(aq) + 8H^{\oplus}(aq) \longrightarrow Mn^{2\oplus}(aq) + 5Fe^{3\oplus}(aq) + 4H_2O(l)$$

(*Helping hand*: 1 mol of ammonium iron(II) sulphate contains 1 mol Fe; so find the number of moles of $Fe^{2\oplus}$ in the 250 cm^3 flask, and hence the relative formula mass of $(NH_4)_2SO_4.FeSO_4.xH_2O$.)

3 A solution contains a mixture of hydrochloric acid and sodium chloride. 25.0 cm^3 portions of this solution were titrated with 0.100 mol dm^{-3} aqueous sodium hydroxide, using phenolphthalein as indicator, and required an average 17.3 cm^3.

Separate 25.0 cm^3 portions of the solution were titrated with 0.100 mol dm^{-3} aqueous silver nitrate, after addition of about 1 g calcium carbonate to neutralise the acid, using potassium chromate(VI) as indicator. The average titre was 26.8 cm^3.

Calculate the concentration (in g dm^{-3}) of the sodium chloride in the solution.

Data: the equation for the second titration is:

$$Ag^{\oplus}(aq) + Cl^{\ominus}(aq) \longrightarrow AgCl(s)$$

4 A solution contained sodium nitrate contaminated with calcium nitrate. An analyst wished to find the concentration (in mol dm^{-3}) of calcium ions in the mixture.

50.0 cm^3 of the solution was taken, acidified with hydrochloric acid, heated with added ammonium ethanedioate and slowly neutralised with aqueous ammonia. Hydrated calcium ethanedioate ($CaC_2O_4.H_2O$) was precipitated. This was converted quantitatively to calcium carbonate, $CaCO_3$, by careful heating in an oven. The mass of calcium carbonate obtained was 0.356 g.

Calculate the concentration of calcium ions (in mol dm^{-3}) in the original solution.

30.3 Inorganic qualitative analysis

Very large textbooks have been written about this topic. For exam purposes, it is necessary for you to know a number of tests. If your exam pattern includes a practical exam, you will probably be allowed to take in books – indeed, the exam board may provide a handbook for candidates which includes, among other things, the standard qualitative analytical tests.

However, many of these analytical tests are found as parts of questions on theory papers. Hence the need to know them. We have tried to include here all the tests found in current exam syllabuses for A-level chemistry.

Before proceeding further, it would be a good idea to examine the syllabus for your exam carefully to find which tests you need to know. (This does not, of course, stop you from knowing the others as well!)

Tests for the presence of water

- Add a drop or two of the liquid to a little white powdered, anhydrous copper sulphate. If a blue colour results, the liquid contains water.
- Add a drop of the liquid to blue cobalt chloride paper. (This is prepared by soaking strips of filter paper in pink aqueous cobalt(II) chloride, and drying them carefully in an oven at 80 °C until they turn blue. The blue strips must be kept in a desiccator – the paper will change colour with the small quantity of water vapour in the air.) If a pink colour results, the liquid contains water.

Both these tests depend on the differences in colour between hydrated and anhydrous ions of d-block metals (*Figure 30.3.1*; see also Ch 25.3)

Equations for water tests

$$CuSO_4(s) + 5H_2O(l) \longrightarrow CuSO_4.5H_2O(l)$$

White   Blue

$$CoCl_2(s) + 6H_2O(l) \longrightarrow CoCl_2.6H_2O(s)$$

Blue   Pink

Figure 30.3.1

N.B. Note that these tests merely show that water *is present* in a liquid, not that the liquid is *pure* water. The way of testing for reasonably pure water would be to measure the freezing point of the liquid. If it freezes completely at 0 °C, it is likely to be water of high purity.

Tests for the presence of an acid or alkali

Acids and alkalis are usually tested for in aqueous solution by using strips of indicator paper or pH paper (*Figure 30.3.2*), or a pH meter.

However, all but the very weakest acids can be tested for by adding the suspected acid in aqueous solution to either solid or aqueous sodium carbonate or sodium hydrogencarbonate. Effervescence ('fizzing') shows that an acid is present, as carbon dioxide is given off (Ch 13.4).

$$2H^{\oplus}(aq) + CO_3^{2\ominus}(aq) \longrightarrow CO_2(g) + H_2O(l)$$

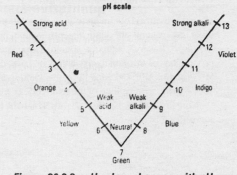

pH scale

Figure 30.3.2   pH colour changes with pH paper

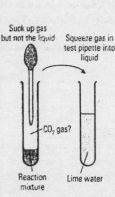

Figure 30.3.3   Testing for CO$_2$

Tests for common gases

Several of these tests have been met with already, but they are tabulated here (*Table 30.3.1*) for convenient reference. Many of the tests require the gas to be tested with an aqueous solution. In these cases the technique used is to sample the gas using a teat pipette and then bubble the gas through the solution. The process of sampling and bubbling can be repeated a number of times (*Figure 30.3.3*).

Gas	Test
Hydrogen, H$_2$	With *lighted splint* or Bunsen flame, gives an audible 'pop'. (N.B. a relatively pure sample of the gas, i.e. one containing only a little air, burns smoothly. Also, collect in an inverted tube.) $$2H_2(g) + O_2(g) \xrightarrow{\text{ignite}} 2H_2O(g)$$
Oxygen, O$_2$	Relights a *glowing splint*. (N.B. When the glowing splint is put into a tube of oxygen, a faint 'pop' is often heard because the hot wood gives off flammable gases. This can cause confusion with hydrogen, if the difference between a *lighted* splint and a *glowing* splint is not understood.)
Carbon dioxide, CO$_2$	Gives a *white precipitate* when bubbled through, or shaken with, *lime water*. (N.B. Do not state that the lime water turns milky.) $$CO_2(g) + Ca(OH)_2(aq) \xrightarrow{\text{RT}} CaCO_3(s) + H_2O(l)$$
Ammonia, NH$_3$	Turns *damp red litmus paper* or *pH paper* blue. (N.B. *Dry* paper doesn't work.)
Sulphur dioxide, SO$_2$	With *aqueous potassium dichromate(VI)* (either when bubbled through a little of the solution or tested with a strip of filter paper moistened with the solution), sulphur dioxide causes the colour of the solution to turn from orange to green (and perhaps through green to blue). $$3SO_2(g) + Cr_2O_7^{2\ominus}(aq) + 2H^{\oplus}(aq) \longrightarrow 3SO_4^{2\ominus}(aq) + 2Cr^{3\oplus}(aq) + H_2O(l)$$ Orange Green
Chlorine, Cl$_2$	Turns *damp blue litmus paper red and then bleaches it* colourless. (N.B. The bleaching action may be so fast that the red stage is missed. Again, dry litmus paper doesn't work.) $$Cl_2(aq) + H_2O(l) \xrightarrow{\text{RT}} HOCl(aq) + HCl(aq)$$ 'Bleach' 'Acid'
Hydrogen chloride, HCl	Fumes with moist air, dense white fumes with *ammonia gas*. $$HCl(g) + NH_3(aq) \xrightarrow{\text{RT}} NH_4Cl(s)$$ Forms a white precipitate with *aqueous silver nitrate* acidified with dilute nitric acid. The precipitate dissolves readily in aqueous ammonia. This distinguishes from HBr and HI. $$HCl(g) + Ag^{\oplus}(aq) \xrightarrow{\text{RT}} AgCl(s) + H^{\oplus}(aq)$$
Ethene, C$_2$H$_4$	When bubbled through dilute *aqueous potassium manganate(VII)* acidified with dilute sulphuric acid, ethene causes the purple solution to become colourless. $$CH_2{=}CH_2(g) + [O] + H_2O(l) \xrightarrow[\text{RT}]{\text{KMnO}_4\text{(aq)/H}^{\oplus}\text{(aq)}} CH_2OH{-}CH_2OH(aq)$$

Table 30.3.1   Tests for gases

Tests for common cations in solution

All tests are carried out in dilute aqueous solution at room temperature, unless otherwise stated. The usual test reagents are dilute aqueous sodium hydroxide, NaOH(aq), and dilute aqueous ammonia, NH_3(aq). However, aqueous sodium carbonate, Na_2CO_3(aq) is sometimes specified. This gives precipitates of insoluble carbonates, or basic carbonates, with most $M^{2\oplus}$(aq) ions; but with the $M^{3\oplus}$(aq) ions $Al^{3\oplus}$, $Cr^{3\oplus}$ and $Fe^{3\oplus}$, the sodium carbonate solution acts as aqueous sodium hydroxide does – giving a precipitate of the hydroxide or hydrated oxide.

Note in *Table 30.3.2* the ease with which the hydroxides of manganese(II), iron(II) and cobalt(II) are oxidised by air.

Cation	Result of adding: Aqueous sodium hydroxide, NaOH(aq)	Result of adding: Aqueous ammonia, NH_3(aq)
Magnesium, $Mg^{2\oplus}$(aq) (Group 2)	White precipitate, insoluble in excess reagent	White precipitate, insoluble in excess reagent (Ch 6.5)
Calcium, $Ca^{2\oplus}$(aq) (Group 2)	White precipitate, if concentration of $Ca^{2\oplus}$(aq) relatively high	No precipitate (Ch 6.5)
Barium, $Ba^{2\oplus}$(aq) (Group 2)	No precipitate	No precipitate (Ch 6.5)
Aluminium, $Al^{3\oplus}$(aq) (Group 3)	White precipitate, soluble in excess reagent	White precipitate, insoluble in excess reagent (Ch 25.4)
Lead(II), $Pb^{2\oplus}$(aq) (Group 4)	White precipitate, soluble in excess reagent (To distinguish from $Al^{3\oplus}$(aq), add dilute HCl to the suspected $Pb^{2\oplus}$(aq). A white precipitate is formed.)	White precipitate, insoluble in excess reagent (Ch 25.4)
Chromium(III), $Cr^{3\oplus}$(aq) (d-block)	Grey–green precipitate soluble in excess reagent to give dark green solution	Grey–green precipitate, only slightly soluble in excess reagent to give a violet or pink solution (Ch 25.4)
Manganese(II), $Mn^{2\oplus}$(aq) (d-block)	White precipitate, insoluble in excess reagent, which rapidly oxidises in air to brown colour	White precipitate, insoluble in excess reagent, which rapidly oxidises in air to brown colour (Ch 25.4)
Iron(II), $Fe^{2\oplus}$(aq) (d-block)	Dirty green precipitate insoluble in excess reagent, which rapidly oxidises in air to red–brown colour	Dirty green precipitate, insoluble in excess reagent, which rapidly oxidises in air to red–brown colour (Ch 25.7)
Iron(III), $Fe^{3\oplus}$(aq) (d-block)	Red–brown precipitate, insoluble in excess reagent	Red–brown precipitate, insoluble in excess reagent (Ch 25.7)
Cobalt(II), $Co^{2\oplus}$(aq) (d-block)	Blue precipitate which may go pink with excess reagent, but is insoluble. Slowly turns brown–black in air	Blue precipitate soluble in excess reagent to brown–yellow solution, which slowly turns red in air (Ch 25.4)
Copper(II), $Cu^{2\oplus}$(aq) (d-block)	Pale blue precipitate, insoluble in excess reagent	Pale blue precipitate soluble in excess reagent to intense blue solution (Ch 25.4)
Zinc, $Zn^{2\oplus}$(aq) (d-block)	White precipitate, soluble in excess reagent	White precipitate, soluble in excess reagent (Ch 25.7)
Ammonium, $NH_4^{\oplus}$(aq)	When the mixture is warmed with aqueous or solid alkali, ammonia gas is produced and can be detected (see *Table 30.3.1*) (N.B. This test can be performed on the solid)	

Table 30.3.2   Tests for aqueous cations

Flame test for s-block metals

The flame test uses a suitable metal wire (platinum, fused into a glass rod; or much more cheaply, but not so satisfactorily, a length of Nichrome wire, one end of which is pushed into a cork), hydrochloric acid in a small beaker or watch glass, a roaring Bunsen flame, and a small powdered sample of the solid compounds to be tested (*Figure 30.3.4*).

Figure 30.3.4

The wire is first cleaned by dipping it in the hydrochloric acid and holding the end of it in the flame to the side of the cone and near the top of the cone. Repeat the cleaning until the wire does not colour the flame. Moisten the end of the wire with the acid, then dip it in the powdered sample. A small amount of the powder will stick to the wire. Place this in the flame as before and note the colour produced (do not put the wire into the cone). Finally, clean the wire as before.

Compounds of s-block elements are used in fireworks to produce colour (*Table 30.3.3*).

In particular, strontium compounds are added to fireworks to produce the red colour, and barium gives the green colours often seen in such flares.

The reason why several elements in Groups 1 and 2 cause colour in the Bunsen flame is discussed in Ch 6.3. Colour photographs of the flames are also in Ch 6.3.

Element	Colour
Lithium, Li	Rich scarlet red
Sodium, Na	Bright yellow (often wrongly described as orange)
Potassium, K	Lilac (this is much less intense than the sodium colour; so because potassium compounds are usually contaminated with small amounts of sodium, a potassium flame should be observed through blue cobalt glass which absorbs the sodium colour)
Rubidium, Rb	Red
Caesium, Cs	Blue
Calcium, Ca	Brick-red
Strontium, Sr	Crimson
Barium, Ba	Pale green

Some other elements also give flame colours, but you are not required to know these.

Table 30.3.3 · Flame colours produced by s-block elements and their compounds

---

Murderers convicted by simple test

Not very many years ago, A-level students regularly did analytical practical work with poisonous compounds of arsenic, antimony and cadmium, and used the extremely poisonous gas hydrogen sulphide as a common laboratory reagent.

You probably know that arsenic was for some centuries the preferred method of poisoning people. The element itself was not used, but the white powder of arsenic(III) oxide, As_2O_3. The fatal dose was often less than 0.1 g, although resistance can be built up by regular small doses. It is undoubtedly true that many murders committed in this way went undetected, so that it was a great advance when tests were developed which enabled arsenic to be detected in stomach contents and in other organs. One such test was Marsh's test, and the guilty verdict in several famous Victorian murder trials depended greatly on the evidence of the analyst who performed the test.

Hydrogen was produced in a flask by reacting arsenic-free zinc with dilute pure sulphuric acid. (The hydrogen produced had to be rigorously tested to ensure that it contained no arsenic compounds.) The solution containing the suspected arsenic was then added a little at a time to the reacting zinc and acid. If arsenic is present under these conditions, the highly poisonous gas arsine, AsH_3, is produced. As the gases left the flask, they were dried by passing through anhydrous calcium chloride and then entered a horizontal glass tube heated at one point by a small flame. The arsine was decomposed by heat to form a dark shining mirror of arsenic on the inside of the tube. The deposit of arsenic was soluble in aqueous sodium chlorate(I).

Antimony compounds are generally almost as poisonous as those of arsenic, and can also be detected by Marsh's test. A similar 'mirror' is formed, but it does not react with aqueous sodium chlorate(I).

Tests for common anions in solution

All tests are carried out in dilute aqueous solution at room temperature unless otherwise stated. In some instances, a single reaction does not give sufficiently strong evidence, so a second reaction is also used (*Table 30.3.4*).

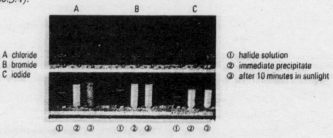

A chloride
B bromide
C iodide

① halide solution
② immediate precipitate
③ after 10 minutes in sunlight

Figure 30.3.5   Halide precipitates with silver nitrate solution

Ion	Test and observation
Carbonate, $CO_3^{2\ominus}$(aq)	Add dilute hydrochloric to the solution or solid. Effervescence is seen (and heard!); carbon dioxide is given off and can be detected (Table 30.3.1)

$$CO_3^{2\ominus}(aq) + 2H^{\oplus}(aq) \xrightarrow{RT} CO_2(g) + H_2O(l)$$

Hydrogencarbonate, $HCO_3^{\ominus}$(aq)	Boil the solution, or heat the solid. Carbon dioxide is given off and can be detected (see Table 30.3.1)

$$\text{e.g.} \qquad 2HCO_3^{\ominus}(aq) \xrightarrow{heat} CO_2(g) + CO_3^{2\ominus}(aq) + H_2O(l)$$

Chloride, $Cl^{\ominus}$(aq) (Figure 30.3.5)	Add dilute aqueous silver nitrate(I) acidified with dilute nitric acid. A white precipitate is formed which dissolves readily when excess dilute aqueous ammonia is added.

$$Cl^{\ominus}(aq) + Ag^{\oplus}(aq) \xrightarrow{RT} AgCl(s)$$

And in ammonia:

$$AgCl(s) + 2NH_3(aq) \xrightarrow{RT} [Ag(NH_3)_2]^{\oplus}(aq) + Cl^{\ominus}(aq)$$

(N.B. this precipitate darkens on standing in daylight, rapidly if in direct sunlight)

$$2AgCl(s) \xrightarrow[uv]{RT} 2Ag(s) + Cl_2(g)$$

Bromide, $Br^{\ominus}$(aq) (Figure 30.3.5)	As for chloride. The precipitate is off-white, and does not dissolve in dilute aqueous ammonia, although it will in concentrated aqueous ammonia. (Again, the precipitate darkens in daylight)

$$Br^{\ominus}(aq) + Ag^{\oplus}(aq) \xrightarrow{RT} AgBr(s)$$

Iodide, $I^{\ominus}$(aq) (Figure 30.3.5)	As for chloride. The precipitate is pale yellow and does not dissolve, even in concentrated ammonia.

$$I^{\ominus}(aq) + Ag^{\oplus}(aq) \xrightarrow{RT} AgI(s)$$

Sulphate, $SO_4^{2\ominus}$(aq)	Add dilute acidified $Ba^{2\oplus}$(aq) (e.g. $BaCl_2$(aq) with dilute hydrochloric acid). A dense white precipitate is formed, insoluble in excess acid. (The same result can be obtained with $Pb^{2\oplus}$(aq), e.g. $Pb(NO_3)_2$(aq) with dilute nitric acid.)

$$SO_4^{2\ominus}(aq) + Ba^{2+}(aq) \xrightarrow{RT} BaSO_4(s) + H^{\oplus}(aq)$$

Hydrogensulphate, $HSO_4^{\ominus}$(aq)	As for sulphate. However, the solution of a hydrogensulphate is quite strongly acidic, and also gives a positive test for an acid.

$$HSO_4^{\ominus}(aq) + Ba^{2\oplus}(aq) \xrightarrow{RT} BaSO_4(g) + H^{\oplus}(aq)$$

Sulphite, $SO_3^{2\ominus}$(aq)	Warm the solution (or better, the solid) gently with dilute hydrochloric acid or sulphuric acid. Sulphur dioxide is given off and can be detected (Table 30.3.1)

$$SO_3^{2\ominus}(aq) + 2H^{\oplus}(aq) \xrightarrow{RT} SO_2(g) + H_2O(l)$$

Nitrate, $NO_3^{\ominus}$(aq)	1. Heat gently with dilute aqueous sodium hydroxide and aluminium foil. Ammonia is given off and can be detected (Table 30.3.1). 2. Add dilute hydrochloric acid or sulphuric acid to the test solution (or, better, the solid). There is no reaction.
Nitrite, $NO_2^{\ominus}$(aq)	1. As for nitrate, test 1 – the result is the same. 2. As for nitrate, test 2 – the solution turns pale blue and a gas is given off which turns brown on contact with air.
Chromate(VI), $CrO_4^{2\ominus}$(aq)	1. The solution is yellow; add dilute sulphuric acid, and it turns orange.

$$2CrO_4^{2\ominus}(aq) + 2H^{\oplus}(aq) \xrightarrow{RT} Cr_2O_7^{2\ominus}(aq) + H_2O(l)$$

2. Add $Pb^{2\oplus}$(aq) to the test solution. A bright yellow precipitate is formed.

$$CrO_4^{2\ominus}(aq) + Pb^{2\oplus}(aq) \xrightarrow{RT} PbCrO_4(s)$$

Manganate(VII), $MnO_4^{\ominus}$(aq)	The solution is an intense purple colour. When acidified with dilute sulphuric acid, it is rapidly decolorised by, e.g. aqueous hydrogen peroxide, $SO_3^{2\ominus}$(aq), SO_2(g) or $Fe^{2\oplus}$(aq)

$$\text{e.g.} \qquad 2MnO_4^{\ominus}(aq) + 5H_2O_2(aq) + 6H^{\oplus}(aq) \xrightarrow{RT} 2Mn^{2+}(aq) + 5O_2(g) + 8H_2O(l)$$

Table 30.3.4 Tests for aqueous anions

1 Identify the gas in each of the following series of experimental results.

 a Pale green gas; puts out a lighted splint; no reaction with lime water; turns pH paper colourless.

 b Colourless gas; small explosive noise with a lighted splint; no reaction with lime water; pH paper remains neutral colour.

 c Colourless gas; burns with a smoky flame when a lighted splint is applied; no change with lime water; no colour change with pH paper.

 d Colourless gas; lighted splint goes out; forms a white precipitate with lime water; turns litmus paper an orange colour.

2 In the early days of flour production, unscrupulous manufacturers used to increase the bulk of the flour by adding much cheaper, unreactive materials. In one such sample, tests on the flour sample showed the following results.

A   A sample of the adulterated flour, when treated with hydrochloric acid, produced a colourless gas.

B   This gas when bubbled through lime water gave a white precipitate.

C   The solution, when tested in a flame, gave a brick red coloration to the flame.

What information do these tests give you about the possible material used to adulterate the flour?

3 After an accident there was some concern about a clear liquid found in a puddle at the site. Initial tests were as follows:

A   pH 1.

B   There was no coloration with a flame test.

C   Addition of silver nitrate solution – no precipitate.

D   Did not ignite.

E   Addition of $Ba^{2\oplus}$ ions with dilute nitric acid – white precipitate.

 a What does the pH value tell you?

 b Which group of anions does the silver nitrate test show to be *not* present?

 c Which anion does the test with barium ions show to be present?

 d Suggest a possible identity of the main chemical present in the spillage.

30.4 Organic qualitative analysis

All that is required here is knowledge of tests for a relatively few functional groups, plus a test for a high degree of unsaturation.

Test for a high degree of unsaturation

Ignite a small amount of solid or a drop of liquid on a piece of broken porcelain or on a spatula held in a Bunsen flame. A very smoky flame (quite often producing flakes of carbon in the air!) shows that the material contains a high carbon:hydrogen ratio and is therefore highly unsaturated, and probably aromatic (see Ch 26.2). Unsaturated gases, such as ethene, $CH_2{=}CH_2$, and especially ethyne, $CH{\equiv}CH$ (acetylene) burn with smoky flames, but may be dangerous to test (*Figure 30.4.1*).

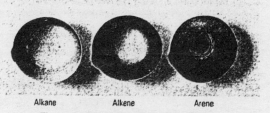

Alkane   Alkene   Arene

Figure 30.4.1 Burning hydrocarbons

| Tests for common functional groups |

All these reactions have been met with earlier. The table includes references to more detailed discussion of the reactions involved in the tests (*Table 30.4.1*). Reactions are carried out at room temperature unless otherwise stated.

Functional group | **Test**

Alkene, $>C=C<$

1. Shake with a little aqueous bromine (bromine water), which is brown. The reagent is decolorised. (This is not a very good test, as it is sometimes impossible to distinguish between the addition and substitution reactions of bromine. It is better to dissolve a little of the test material in trichloromethane, and then add a 1% solution of bromine in trichloromethane dropwise. Genuine alkenes will decolourise the bromine but not liberate hydrogen bromide.) (Ch 16.5)

e.g. $CH_2=CH_2 + Br_2 \xrightarrow[CHCl_3]{RT} CH_2BrCH_2Br$

2. To a little of the test material dissolved in water or propanone add a solution of 0.01 mol dm^{-3} potassium manganate(VII) acidified with dilute sulphuric acid until the pink colour persists. The immediate decolorisation of more than 1 cm^3 of the manganate(VII) solution indicates an alkene. (Ch 16.5)

e.g. $CH_2=CH_2 + [O] + H_2O \xrightarrow[dilute,acidified,RT]{KMnO_4} CH_2OHCH_2OH$

Chloroalkane, $>C-Cl$
Bromoalkane, $>C-Br$
Iodoalkane, $>C-I$

Heat gently with aqueous sodium hydroxide plus ethanol (to ensure that the alkali and test material can mix) for a few minutes. Cool, dilute with water, and make the solution acid by adding dilute nitric acid. Then add aqueous silver nitrate, and identify the halide ions as in *Table 30.3.4*. (Ch 17.4)

e.g. $CH_3Br + NaOH \xrightarrow{heat} CH_3OH + NaBr$

$Ag^+ + NaBr \xrightarrow[RT]{HNO_3} AgBr(s) + Na^+$

Aliphatic alcohol, $>C-OH$

The test material must be completely free from water.
Check for the presence of the —COOH (carboxylic acid) group by adding the test material to aqueous sodium carbonate (see below).
If no —COOH is present, add phosphorus pentachloride to the test material. If $>C-OH$ is present, hydrogen chloride is given off and can be identified (*Table 30.3.1*). (Ch 18.4)

e.g. $CH_3OH + PCl_5 \xrightarrow{RT} CH_3Cl + PCl_3O + HCl(g)$

To distinguish between primary, secondary and tertiary alcohols see Ch 18.4

Phenol, —OH

The —OH group attached to a benzene ring, or other delocalised system, gives a colour, other than yellow – and often purple – with aqueous iron(III) chloride. (Ch 26.7)

Ketone, $>C=O$

To a small amount of test material add a few cm^3 of a solution of 2,4-dinitrophenylhydrazine (2,4-DNPH) in sulphuric acid and methanol. An orange precipitate is formed. (Ch 19.6)

e.g. $(CH_3)_2C=O + H_2NNH$— ⟶ $(CH_3)_2C=NNH$— $+ H_2O$

The test material does not give a positive Tollens, Fehling's or Benedict's test (see below).

Aldehyde, —C$\stackrel{O}{\scriptstyle H}$

The test material gives a positive result with 2,4-DNPH (see above)
The test material does give a positive Tollens, Fehling's or Benedict's test.

e.g. Tollens: $CH_3CHO + 2Ag^+ + H_2O \xrightarrow{RT} CH_3COOH + 2Ag(s) + 2H^+$

The details of at least one of these tests should be known. (Ch 19.6)

Carboxylic acid, —C$\stackrel{O}{\scriptstyle OH}$

Add the test material to aqueous or solid sodium carbonate.
Carbon dioxide is given off and can be identified (*Table 30.3.1*). (Ch 27.5)

$2CH_3COOH + Na_2CO_3 \xrightarrow{RT} 2CH_3COONa + CO_2(g) + H_2O$

Table 30.4.1 Tests for functional groups

1 An organic liquid compound, Q, was tested with the following reagents, with the results shown. What do these tests tell you about the compound? (Do not forget that negative results still tell you something about the compound.)

   a Addition of sodium carbonate solution to Q produced no gas.

   b When bromine water was added to a solution of Q in ethanol the brown colour of the bromine disappeared.

   c When PCl_5 was added to pure anhydrous Q, a gas was evolved, which produced dense white fumes in the presence of ammonia fumes.

   d When Q was ignited in air a dense black smoke was formed.

   e A few cm^3 of a solution of 2,4-DNPH was added to a few drops of Q. The reaction mixture remained a yellow colour and no precipitate was formed.

2 What tests would you carry out to identify the functional group present in the following molecules?

   a $CH_3CH_2COCH_3$

   b C_6H_5OH

   c CH_3CH_2COOH

   d $CH_3CH_2CH(CH_3)CHCHCH_3$

3 Describe the chemical test(s) you would carry out to distinguish between the following pairs of compounds.

   a $CH_3COCH_2CH_3$ and $CH_3CH_2CH_2CHO$

   b $C_6H_{11}OH$ and C_6H_5OH

   c CH_3CH_2Br and CH_3CH_2OH

   d C_6H_6 and C_6H_{12}

30.5 The use of instruments in analysis

The use of automated methods has grown dramatically in the last few decades, particularly mass spectrometry, chromatography of all kinds, and methods involving the emission or absorption of radiation of all wavelengths from γ-rays to microwaves.

The only methods discussed here are mass spectrometry (MS), infra-red spectroscopy (IR), and nuclear magnetic resonance spectroscopy (NMR). The mass spectrometer has already been discussed in some detail (Ch 1.3 and Ch 9.1), but you are not expected to know details of the apparatus needed for IR and NMR.

A major advantage of these methods is that they only require very small samples of material (Figure 30.5.1).

Instrumental analysis

Advantages:

- small amounts of sample
- non-destructive, or only tiny quantities are needed
- extremely sensitive

Figure 30.5.1

IR and NMR are non-destructive – the sample can be recovered.

Mass spectrometry

Early mass spectrometers were used to measure isotopic masses (see Ch 1.3). When used with organic compounds, MS can give information not only about the relative molecular mass of the compound, but also about its structure.

The interior of the mass spectrometer is kept at high vacuum. Molecules in the vaporised sample are ionised, usually by impact with electrons generated by a heated filament. The ionised particles are accelerated through an electric field, and then deflected in a magnetic field before being detected (*Figure 30.5.2*).

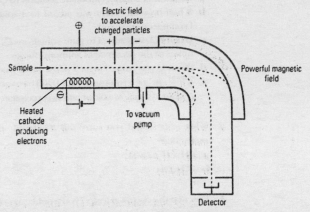

Figure 30.5.2   The mass spectrometer

Many of the ionised molecules **fragment**, i.e. break into pieces. The molecular structure is destabilised by loss of an electron, and as it vibrates, various bonds may break. The fragments give information about the structure of the original molecule. Normally, some of the ionised molecules survive their flight through the machine. These will, of course, cause the peak with the highest mass. For example, a simplified mass spectrum for butane, C_4H_{10}, $M_r = 58$, is shown in *Figure 30.5.3*.

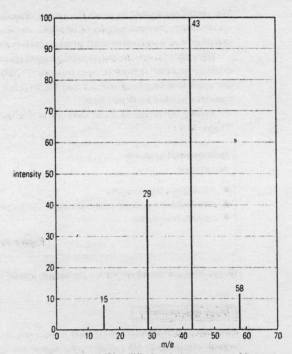

Figure 30.5.3   Simplified mass spectrum of butane

The most abundant fragment (here, $M_r = 43$) causes the strongest signal on the detector. This is called the **base peak**. This peak is set to 100% by the software in the machine and the others measured against it. (Strictly, the horizontal axis should be mass/charge ratio, but it is assumed that all the particles have one positive charge.)

The main butane peaks are 58, 43, 29 and 15. The peak at 58 is obviously due to the molecular ion. (If we do not already know M_r for the compound concerned, the peak with the highest mass – the **molecular ion peak** – will very often give it to us. However, sometimes – particularly with large, floppy molecules – no measurable amount of the original molecules survives.) What causes the other peaks?

Consider the structure of butane. The molecule can break in two different places, which explains the peaks (*Figure 30.5.4*).

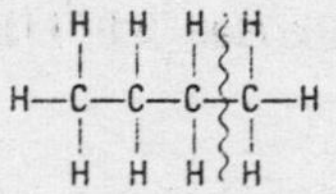

15	$CH_3^{\oplus}$
29	$C_2H_5^{\oplus}$
43	$C_3H_7^{\oplus}$
Molecular ion peak at: 58	$C_4H_{10}^{\oplus}$

Figure 30.5.4
Fragments from C_4H_{10}

Fragments of mass 43 and 15

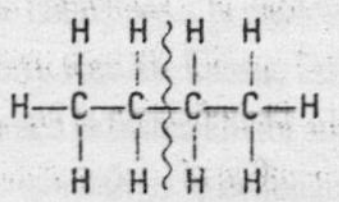

Fragments of mass 29

The fragmentation pattern depends on the structure of the compound. The cleavage of a particular bond does not necessarily give a 50:50 mix of the fragments, as the fragments may have very different stabilities. Specific groupings of atoms in a compound often give specific fragmentation patterns, so establishing the pattern for an unknown compound can often lead to its structure.

Conversely, given the molecular structure of a compound, it is possible to 'chop it up' and predict what peaks will appear in its mass spectrum.

High resolution mass spectrometry

A good modern MS machine can easily measure masses to several decimal places (*Figure 30.5.5*), and this can help decide between compounds. For example, propanal, CH_3CH_2CHO apparently has the same M_r as butane – i.e. 58. But to an appropriate degree of accuracy, the relevant atomic masses are ^{12}C, 12.00000 (this is the standard for relative atomic masses); ^{1}H, 1.00782; ^{16}O, 15.99492.

So for butane, C_4H_{10}, $M_r = 58.0782$

But for propanal, C_3H_6O, $M_r = 58.0418$

High resolution MS can easily decide between these two.

^{1}H	1.00782
^{12}C	12.00000
^{16}O	15.99492
^{14}N	14.00307

Figure 30.5.5 Accurate isotope masses

Modern MS machines can be coupled to a computer database containing all probable combinations of elements to be found in molecular formulae together with their M_r values to several decimal places, enabling the rapid determination of molecular formulae (*Figure 30.5.6*).

Accurate isotope masses to give formulae

The mass spectra of C_2H_4 and N_2 both have a molecular ion peak at 28. With a high resolution mass spectrometer it is possible to distinguish between them:

e.g. a molecular ion peak at 28.03128 must be from $C_2H_4^{\oplus}$.

Using the data from *Figure 30.5.5*:

C_2 $2 \times 12.0000 = 24.00000$

H_4 $4 \times 1.00782 = \dfrac{4.03128}{28.03128}$

for N_2:

N_2 $2 \times 14.00347 = 28.00614$

Figure 30.5.6

Real mass spectra

The mass spectra presented in textbooks and exam questions are often much more simple than 'real' ones, for two main reasons.

- Fragments of molecules under the extreme conditions inside a mass spectrometer behave in odd ways. For example, the peak at $M_r = 29$ in the butane spectrometer is caused by the ion with structure $CH_3—CH_2^{\oplus}$. This could lose a hydrogen to become an ion of ethene, $CH_2=CH_2^{\oplus}$. A quite intense peak is found at $M_r = 28$ in the butane spectrum.
- Many elements exist as a mixture of isotopes (Ch 1.2). For example, 1.1% of carbon atoms are actually ^{13}C rather than ^{12}C. This will cause extra peaks in any organic mass spectrum. Also the mass spectrum of any compound containing chlorine will reflect the fact that chlorine has two isotopes, with the ratio $^{35}Cl{:}^{37}Cl$ being 3:1.

The detailed mass spectrum of a compound acts as a fingerprint, which can be used in the identification of the compound. Databases of such mass spectra are now available, and help in, for example, the identification of drugs taken by athletes or given to racehorses; the identification of compounds responsible for giving foods and drinks their characteristic tastes and smells; and the determination of contaminants in food and drinking water.

Real mass spectra – summary

Mass spectrum of dodecane, $C_{12}H_{26}$

Relative molecular mass 170

A – All particles have $\oplus$ charge
  – Molecular ion gives the relative molecular mass of the molecule
  – Fragmentation pattern can help to identify the structure of the molecule

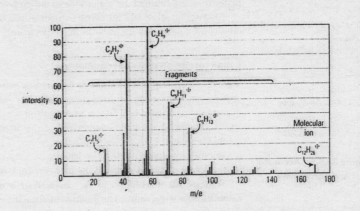

Mass spectrum of chlorobenzene, C_6H_5Cl

B – The presence of the 3:1 ratio of ^{35}Cl and ^{37}Cl helps to identify fragments containing Cl

  – The fragments at m/e = 112 and 114 are in a 3:1 ratio

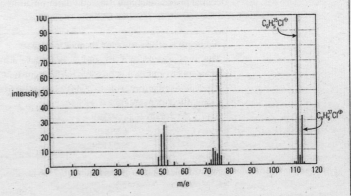

Mass spectrum of cyanobenzene, C₆H₅CN

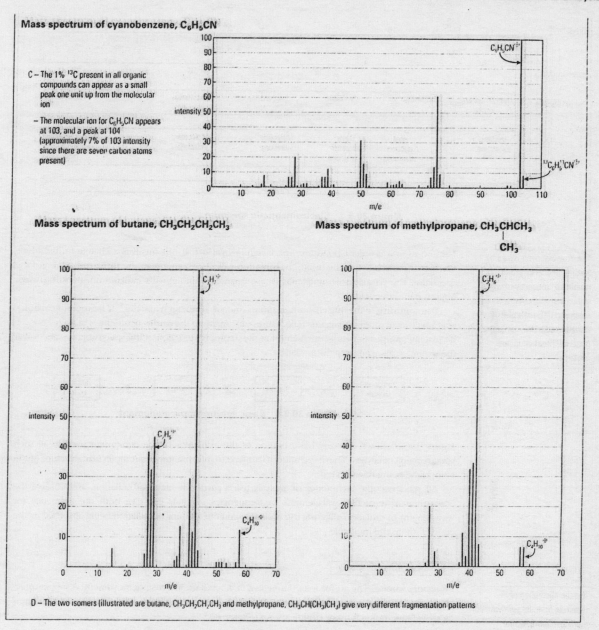

C – The 1% ^{13}C present in all organic compounds can appear as a small peak one unit up from the molecular ion

– The molecular ion for C_6H_5CN appears at 103, and a peak at 104 (approximately 7% of 103 intensity since there are seven carbon atoms present)

Mass spectrum of butane, CH₃CH₂CH₂CH₃

Mass spectrum of methylpropane, CH₃CHCH₃
CH₃

D – The two isomers (illustrated are butane, $CH_3CH_2CH_2CH_3$ and methylpropane, $CH_3CH(CH_3)CH_3$) give very different fragmentation patterns

---

Spectroscopic techniques

In the first chapter of this book we saw that the radiation in the electromagnetic spectrum contains a wide range of energies (*Figure 30.5.7*).

Some of these energies can be related to specific energy changes in the particles around us. The nucleus is associated with γ radiation, the inner electron energy levels with X-rays, and the outer electron energy levels with the emission and absorption spectra of atoms and molecules in the UV and visible regions of the spectrum.

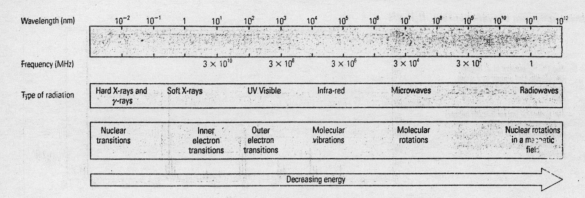

Figure 30.5.7   Electromagnetic spectrum

A microwave oven works by feeding in energy at the rotational frequency of water molecules. The energy of the spinning H_2O molecules is transferred to the rest of the material, heats it up and cooks it.

The atoms in a covalent bond are in constant **vibration**. If this involves a change in the dipole moment associated with the bond, the vibration will absorb energy in the **infra-red** region of the spectrum. The **rotation** of molecules is also quantised and absorbs radiation in the **microwave** region of the spectrum.

The **spinning** of the nuclei of certain atoms (notably **hydrogen** and the ^{13}C isotope of carbon), when held in an intense magnetic field, causes absorption in the **radio** part of the spectrum.

By placing samples in beams of radiation in the appropriate region of the spectrum, we can look at the extent of absorption of this radiation.

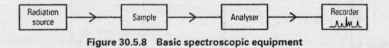

Figure 30.5.8   Basic spectroscopic equipment

Figure 30.5.8 shows the very basic outline of the principles of the equipment used in all such spectroscopic analyses. These techniques can lead to information about molecular structure of the compound being investigated.

All spectroscopic techniques of analysis use a particular range of radiation, and observe the absorption of particular wavelengths or frequencies. Analysis can give both the frequency (or wavelength) of radiation affecting the molecule and the amount of radiation being absorbed by the activity within the molecule.

Infra-red spectroscopy

It is the absorption of IR radiation from the sun-warmed surface of earth by the asymmetrical stretching and bending modes of vibration of CO_2 which is largely responsible for global warming.

Molecules vibrate. The atoms joined together in a bond are in constant movement. A simple three-atom molecule, such as carbon dioxide, can vibrate in the three ways shown. Molecules with many atoms vibrate in much more complex ways (*Figure 30.5.9*).

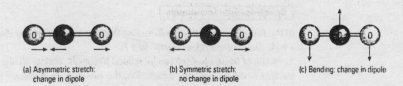

(a) Asymmetric stretch: change in dipole

(b) Symmetric stretch: no change in dipole

(c) Bending: change in dipole

Figure 30.5.9   Molecules vibrate

In vibration, energy is quantised (Ch 1.4). The frequency of vibration in a bond, and hence the energy associated with it, depends on the masses of the atoms and the strength of the bond between them. So particular bonds are associated with particular frequencies and energies. These frequencies occur in the infra-red (IR) part of the spectrum, so that if IR radiation is passed through a sample of a compound, frequencies are absorbed which correspond to the vibrational frequencies of the bond in the compound (*Figure 30.5.10*).

But not *all* the bonds which vibrate at IR frequencies actually absorb IR radiation. First the bond must have a dipole, that is it must be *polar* (Ch 2.5). Although a bond might be polar, the molecule does not necessarily have a permanent dipole. The C=O bond in carbon dioxide is polar but the *symmetrical* stretch in the CO_2 molecule does not produce a signal in the infra-red spectrum. A *change* in dipole on vibration is required for any infra-red activity.

The infra red spectrum of carbon dioxide is shown in *Figure 30.5.11*.

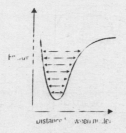

Figure 30.5.10
Vibrational energies of a
bond

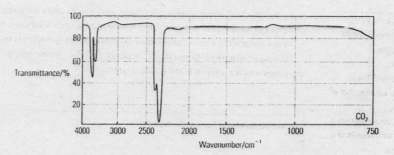

Figure 30.5.11   The infra red spectrum of carbon dioxide

The above discussion means that hydrogen, H—H, does not absorb IR, whereas hydrogen chloride, H—Cl, does. Secondly, in the vibration the overall polarity of the molecule must change in some way. So, the symmetrical stretching in O=C=O does not absorb IR, whereas the asymmetrical stretching and bending does (*Figure 30.5.9*).

In an IR machine (*Figure 30.5.12*), the sample is subjected to IR radiation which varies steadily over a period of minutes from about 2.5×10^{13} Hz to about 10^{14} Hz. A chart recorder registers the frequencies at which the IR radiation is absorbed. (These absorptions, resulting in dips in the trace, are mysteriously called 'peaks'.) An example of the kind of plot obtained is shown (*Figure 30.5.13*).

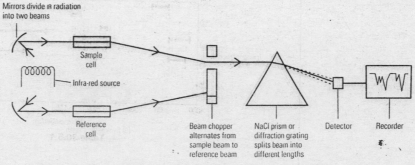

Figure 30.5.12   Simplified diagram of a conventional infra-red spectrometer

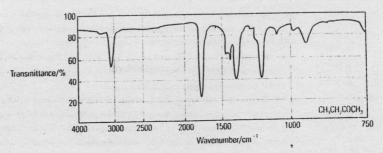

Figure 30.5.13   The infra-red spectrum of butanone

For radiation, $c = \lambda \nu$
where c = velocity of light,
λ = wavelength,
ν = frequency
∴ $1/\lambda = \nu/c$, and $1/\lambda$ is
known as the wavenumber of
the radiation.
For example, an absorption at
5×10^{13} Hz corresponds to a
wavelength of
$3 \times 10^{8} / 5 \times 10^{13} =$
6×10^{-6} m $= 6 \times 10^{-4}$ cm
(where $c = 3 \times 10^{8}$ ms^{-1})
The unit for wavenumber is
cm^{-1}, so here the wavenumber
$= 1/(6 \times 10^{-4}) = 1067$ cm^{-1}.

Note that the horizontal scale is *not linear*, and that its units are *wavenumbers*. This unit is retained in IR spectroscopy, and nowhere else! The unit is derived as shown below.

The characteristic wavenumbers of common bonds are shown in *Table 30.5.1*.

Because vibration frequencies depend on the surroundings of the bond, the range of wavenumbers for a given bond is often quite wide.

IR wavenumbers for some bonds

N.B. The strength of the absorption can vary greatly. The C==O absorption is usually very strong.

Bond	Wavenumber /cm^{-1}
C—H (aromatic)	3200–3000
C—H (aliphatic)	3000–2800
O—H	3700–3200
N—H	3500–3100
C==O	1800–1650
C—O	1250–1000
C—Cl	800–600
C==C	1700–1600

Infra-red correlation chart

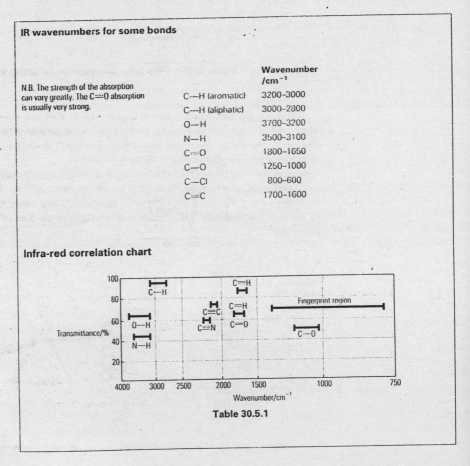

Table 30.5.1

Worked example with spectra

Which bonds give rise to the major peaks in these spectra? (Table 30.5.1 might help). Do you agree with the assignments marked?

Infra-red spectrum of benzoyl chloride

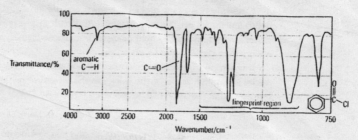

Infra-red spectrum of hexane

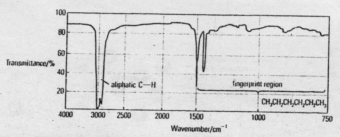

Infra-red spectrum of cyclohexene

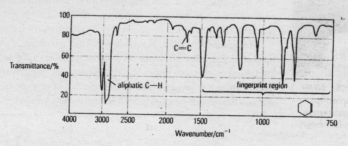

Infra-red spectrum of benzene

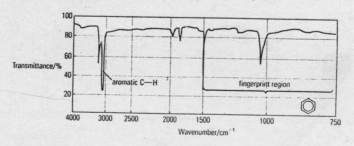

Nuclear magnetic resonance spectroscopy (NMR)

NMR is a technique which can be used with several nuclei with odd mass numbers, especially ^{13}C, ^{19}F and ^{31}P; but we shall be concerned here only with ^1H. (NMR as applied to ^1H is often called proton magnetic resonance, PMR.)

Electrons possess a property known as *spin* (see Ch 1.4); so do protons. Because of this, hydrogen nuclei act like tiny magnets. If placed in a very strong magnetic field, they are aligned either *with* the magnetic field or *against* it. Those aligned against the field are at a *slightly higher energy* than those aligned with it.

If radiation of a frequency corresponding to the energy difference between the two groups of nuclei is applied, some of the nuclei in the lower energy level will move up to the higher energy level, and that radiation will be absorbed (*Figure 30.5.14*).

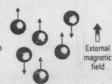
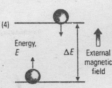

Figure 30.5.14 How protons behave when an external magnetic field is applied

Figure 30.5.15 NMR machine used for magnetic resonance imaging (MRI) in hospital

For very intense magnetic fields, the radiation needed to cause hydrogen nuclei to 'flip' to the higher level is at radio frequencies (*Figure 30.5.15*). The nuclei then 'relax' back to the lower energy and emit radiation of the same frequency. NMR machines used for diagnostic scanning have very powerful magnets, often involving superconductors cooled by liquid helium (*Figure 30.5.15*).

The reason why NMR is such a powerful technique is that the energy, and hence frequency, needed to flip different 'kinds' of hydrogen nuclei to the higher level is slightly different for each kind. Further, the area under the absorption peak in the recorder trace which corresponds to each kind of proton is proportional to the number of protons of that kind. What is meant by 'kind' of proton? Consider a simplified NMR spectrum for methanol, CH_3OH (*Figure 30.5.16*).

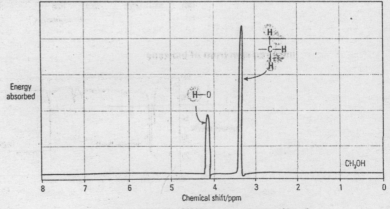

Figure 30.5.16 NMR spectrum – methanol

Chemical shifts

The protons in a particular molecule do not experience exactly the same magnetic field – some protons in the hydrogen atom are **shielded** by the electrons surrounding it and this affects the magnetic field the proton 'feels'. More electronegative elements withdraw electrons from the proton – the proton is *deshielded* and the result is that the proton absorbs radiation at a higher frequency than the normal shielded —CH_3 proton.

The *environments* of the four protons in methanol are different. One proton is attached to oxygen; three are attached to carbon in a methyl group, —CH_3. The relative areas under the peaks are 1:3. The lower electron density around the —OH proton (because of the electronegative oxygen atom) results in this proton absorbing at a higher frequency.

The proton signal from the —CH_3 group in tetramethylsilane, $Si(CH_3)_4$, is taken as a reference point. The resonance signal is compared with this signal and recorded as a chemical shift (the δ scale).

An NMR spectrum shows absorption of energy on the vertical axis with, on the horizontal axis, the **chemical shift**, which increases from right to left. Chemical shifts are measured in parts per million (ppm).

A standard reference material has to be used. This is usually tetramethylsilane, TMS, $Si(CH_3)_4$; it is used because it is chemically inert, non-toxic, gives an NMR signal well away from any other proton environments in organic compounds, and gives an intense single peak because all twelve protons are in exactly the same environment.

The simplified NMR spectrum of ethanol is shown in *Figure 30.5.18*. Notice that the protons in —O—CH_2— give a signal at a chemical shift very different from those in —CH_3.

The high resolution NMR spectrum of ethanol shows much more detail. The multiple peaks are caused because of 'coupling' – interaction between protons attached to nearby atoms. Some of the more usual chemical shifts are shown in *Table 30.5.2* and *Figure 30.5.17*.

Some chemical shifts

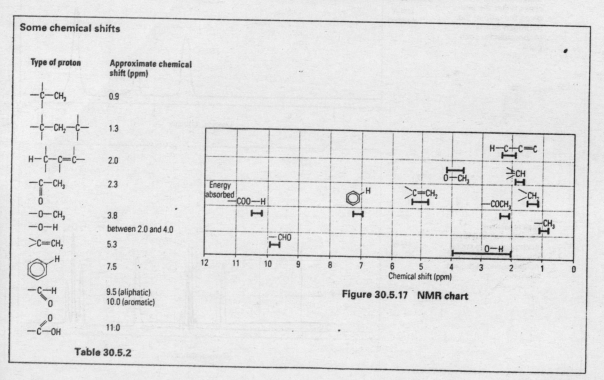

Type of proton	Approximate chemical shift (ppm)
—$\overset{\mid}{\underset{\mid}{C}}$—$CH_3$	0.9
—$\overset{\mid}{\underset{\mid}{C}}$—$CH_2$—$\overset{\mid}{C}$—	1.3
H—$\overset{\mid}{\underset{\mid}{C}}$—$\overset{\mid}{C}$=$\overset{\mid}{C}$—	2.0
—$\overset{\mid}{\underset{\overset{\parallel}{O}}{C}}$—$CH_3$	2.3
—O—CH_3	3.8
—O—H	between 2.0 and 4.0
>C=CH_2	5.3
⌬—H	7.5
—$\overset{\mid}{C}$—H, $\overset{\parallel}{O}$	9.5 (aliphatic) 10.0 (aromatic)
—$\overset{\overset{O}{\parallel}}{C}$—OH	11.0

Table 30.5.2

Figure 30.5.17 NMR chart

Figure 30.5.18   The NMR spectrum of pure ethanol, CH₃CH₂OH

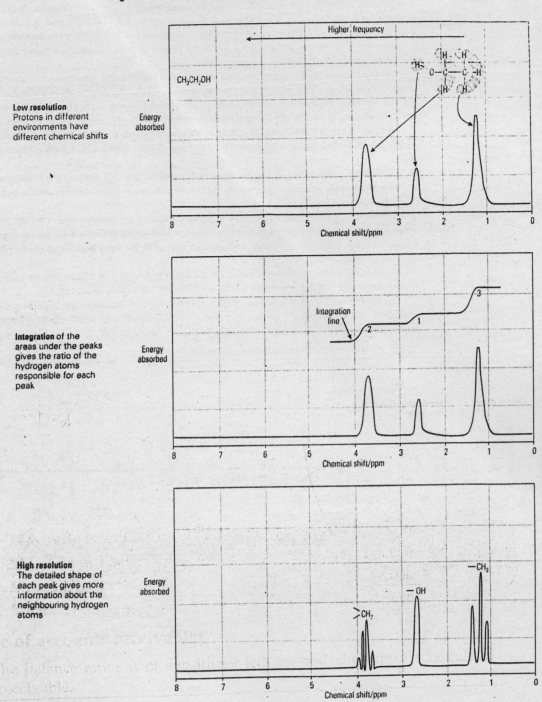

Low resolution
Protons in different
environments have
different chemical shifts

Integration of the
areas under the peaks
gives the ratio of the
hydrogen atoms
responsible for each
peak

High resolution
The detailed shape of
each peak gives more
information about the
neighbouring hydrogen
atoms

MRI scans in medicine

MRI means 'magnetic resonance imaging'. Protons in water, carbohydrates, proteins, lipids, collagen and most tissues in the body give different signals. For example, a cancer tumour can be distinguished from healthy tissue. CAT (computerised axial tomography) scanners using MRI can examine a 'slice' of the brain or liver which is only 1 mm or so thick, and produce an image very much clearer than can be obtained using X-rays.

Further, there seem to be no ill effects from subjecting patients to intense magnetic fields and radio frequency radiation; this is very different from using high-energy X-rays. This means that patients can be monitored regularly using MRI.

An image of a slice through a human head taken with an MRI machine

The technique of CAT scanning using X-rays was developed independently by Allan Cormack, a South African working in the USA, and Godfrey Hounsfield at EMI in Britain. Hounsfield also played a major part in developing MRI–CAT scanning. Hounsfield (now Sir Godfrey) and Cormack shared the 1979 Nobel Prize for Medicine or Physiology.

SELF-ASSESSMENT QUESTIONS 30.5

1 When the hydrocarbon C_5H_{12} is tested in a mass spectrometer, there are five main peaks observed – at m/e values corresponding to 15, 29, 43, 57 and 72. Identify the species responsible for each of these peaks.

2 Name the bond most likely to be responsible for each of the peaks at the following wavenumbers in an infra-red spectrum. (You might like to look at *Table 30.5.1*.)
   a 1750 cm^{-1}
   b 3100 cm^{-1}
   c 3600 cm^{-1}
   d 685 cm^{-1}

3 The high resolution mass spectrum of a particular gas produced an m/e peak at 27.99492.
   a Using the data provided in *Figure 30.5.5*, calculate the accurate relative molecular masses of the following species: N_2, CO, C_2H_4.
   b Use this information to decide the identity of the gas being analysed.

Summary
Analysis of any material requires a number of sequential steps.

- Purify the material
  - distillation
  - recrystallisation
  - chromatography, etc.
- Check the purity of the material
  - chromatography
  - sharpness of melting point if organic solid, etc.
- % composition analysis
  - automated chemical analysers
  - possibly using a mass spectrometer
- Identify the groups present
  - cations
  - anions
  - organic functional groups

For organic molecules

- Mass spectrometry
  - fragmentation patterns
- Infra-red spectroscopy
  - recognising functional groups
- Nuclear magnetic resonance spectroscopy
  - identifying the environment of protons

Both infra-red and nuclear magnetic resonance spectroscopy can also be useful for studying transition metal complexes.

In this chapter the following new ideas have been covered. *Have you understood and learned them?* Could you explain them clearly to a friend? The page references will help.

**SUMMARY
30**

- What is meant by analytical chemistry.   p790
- The difference between qualitative and quantitative analysis.   p790
- Quantitative analytical calculations.   p790
- Tests for water, and acids and alkalis.   p796
- Tests for common gases.   p797
- Tests for common cations in solution.   p798
- Tests for common anions in solution.   p799
- Tests for common functional groups.   p802
- Fragmentation of organic molecules in a mass spectrometer.   p804
- Recognition of a molecular ion peak and use of the M_r to work out the molecular formula.   p805
- Molecular vibrations and infra-red (IR) spectroscopy.   p808
- Proton spin and nuclear magnetic resonance (NMR) spectroscopy.   p812

**EXAMINATION
QUESTIONS
30**

1 In each of the following tests, identify the molecule or ion which would give a positive test.
   a A dense white precipitate is formed when a solution of barium nitrate is added.
   b A dense white smoke is formed when the gas is placed near an open bottle of ammonia.
   c An apple green colour is produced if a small amount of solid is heated in a strong flame.
   d A red-brown precipitate is formed when sodium hydroxide solution is added.

2 One of the important chemicals found in oil of peppermint is menthol. The white solid can be recrystallised from ethanol, and has a melting point of 43 °C. The infra-red and mass spectra of the purified product are shown in the diagrams.

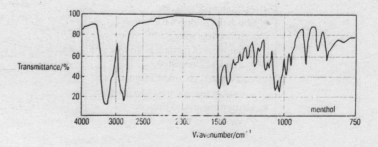

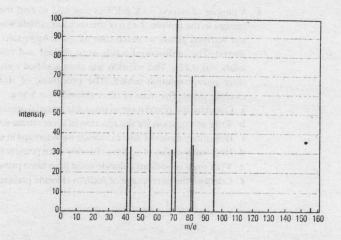

a Describe how you would recrystallise the impure extract of menthol, and how you would measure the melting point of the pure product.

b In the mass spectrum, identify the molecular ion peak.

c Decide which type of bond in the molecule can be assigned to the peaks in the infra-red spectrum in the 4000–1600 cm^{-1} region.

d Which functional group do you think is present in menthol? What test(s) would you carry out to confirm the presence of this functional group?

3 A mass spectrometer can be used to analyse organic compounds. The fragmentation pattern produced in the spectrometer can lead to information about the structure of the molecule.

a Describe the principles of the working of a mass spectrometer.

b The simplified mass spectrum of propenoic acid, CH_2=CHCOOH, is shown below. What is the formula of the molecular ion peak at m/e = 72?

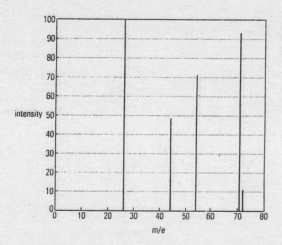

c Write the formulae of the fragments that produce peaks at m/e = 27 and m/e = 45.

d Describe one chemical test you would carry out to show that propenoic acid contained a carbon–carbon double bond.

4 A sample of mined rock salt was analysed to find the percentage of salt, sodium chloride, present in the sample. 2.43 g of the rock salt sample was dissolved in 100 cm^3 of distilled water, and warmed gently to make sure all the soluble salts had dissolved. The mixture was then filtered. The undissolved solids were washed and carefully dried. The mass of undissolved solids was 0.12 g. The solution was then acidified with dilute aqueous nitric acid, and excess silver nitrate solution added. The precipitate of silver chloride was filtered, washed and carefully dried. The mass of silver chloride was 5.58 g.

a Calculate the percentage of insoluble material present in the rock salt sample.

b Write an equation for the reaction of silver nitrate solution with sodium chloride solution.

c How many moles of silver chloride were formed in the precipitation reaction?

d How many moles of sodium chloride were present in the 'rock salt' solution?

e What mass of sodium chloride must have been present in the 'rock salt' solution?

f Calculate the percentage of sodium chloride present in the original sample.

Mathematical Appendix

Notation

A brief reminder of some of the notations used (the quantity mol is used as an example):

$mol \times mol = (mol)^2$

$\dfrac{1}{mol}$ is the same as mol^{-1}

$(mol)^{\frac{1}{2}}$ is $\sqrt{(mol)}$

Arithmetic

We assume that readers have access to a calculator with all the usual functions, including log x and memory. We also assume that the basic common-sense rules of using a calculator are followed, especially:

- always perform the calculation twice;
- do not write down what the calculator tells you without thinking first about how many significant figures are required in your answer;
- always check your answer against a mental estimate; and
- write down every stage of your calculation.

Large and small numbers should be represented in standard form wherever possible, i.e. as a number between 0 and 10 together with the appropriate power of ten.

e.g. $1234 = 1.234 \times 10^3$; $0.00043 = 4.3 \times 10^{-4}$

[For 0.00043 move the decimal point four places to the right; 0.00043 – gives 10^{-4}; for 1234, move the decimal point three places to the left; 1234 – gives 10^3.]

So: 236.5 in standard form is 2.365×10^2.

Many calculations involve **simple proportion**. In these there are usually three known quantities and one unknown.

For example: if 51 g aluminium oxide gives 27 g aluminium, how many tonnes of the oxide are needed to make 100 tonnes of aluminium?

- Write a statement of the known proportion, with aluminium oxide on the right hand side (always keep the unknown quantity on the RHS):

  27 g aluminium ① from 51 g aluminium oxide ②

- Write a second statement starting with the third known quantity:

$$100 \text{ tonnes aluminium } ③ \text{ from } \frac{51}{27} \times 100 \text{ tonnes}$$

of aluminium oxide i.e. $\dfrac{②}{①} \times ③$ tonnes

Algebra

Moving over an equals sign, the sign in front of the quantity changes:

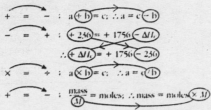

One way of remembering this for calculations involving reacting quantities is:

mass = moles × M; ∴ mass

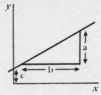

Logarithms (to base 10, i.e. log x or $\log_{10} x$)

Logarithms are very useful for handling very large and very small numbers in terms of powers of ten.

$\log_{10} 10^x = x$

$\log_{10} 1 \times 10^{-3} = -3$

$\log_{10} 1 \times 10^7 = 7$

If the numbers are more complicated, we use

$\log_{10} xy = \log_{10} x + \log_{10} y$.

∴ $\log_{10} 5 \times 10^4 = \log_{10} 5 + \log_{10} 10^4$

= 0.699 + 4.00 = 4.699

It is much simpler on your calculator.

$\log_{10} 5 \times 10^4 =$ 5 EXP 4 log = 4.699

$\log_{10} 2.3 \times 10^{-4} =$ 2.3 EXP 4 +/- log = −3.638

To reverse the process, i.e. to convert a log number back to standard form:

Insert the number (remember to include the sign), then press 10^x (on some calculators for 10^x, you will need SHIFT log).

e.g. to reverse from log number 4.699, enter:

4.699 SHIFT log (or 10^x) = 5.00...,

which is 5.0×10^4

and from −3.638:

3.638 +/- SHIFT log (or 10^x) = 2.3...

which is 2.3×10^{-4}.

Graphical solutions

If two quantities can be related in an equation of the type $y = mx + c$ (i.e. where y is proportional to x), the values of m and c can be found by plotting a graph of x (horizontal axis) against y.

The gradient of the straight line is the value of m and the intercept will be the value of c.

The constant m is a÷b on the graph.

Answers

Self-Assessment Questions – 1.2

1 a 3p, 3e, 4n   b 18p, 18e, 22n
   c 53p, 53e, 74e   d 26p, 26e, 30n
2 a $^{24}_{12}Mg$ b $^{1}_{1}H$ c $^{11}_{5}H$ d $^{14}_{7}Si$
3 a See p4   b See p4   c See p4
4 a 16   b 8   c 8
   d $^{16}_{8}O–8$; $^{17}_{8}O–9$; $^{18}_{8}O–10$; $^{20}_{8}O–12$
   e See p5   f $^{20}_{9}O \longrightarrow {}^{0}_{-1}\beta + {}^{20}_{9}F$

Self-Assessment Questions – 1.4

1 a $1s^2 2s^2 2p^6$ b $1s^2 2s^2 2p^6 3s^2 3p^6$
   c $1s^2 2s^2 2p^6 3s^2 3p^5$ d $1s^2 2s^2 2p^1$
2 a 1s 2s
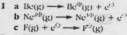
   b 1s 2s 2p 3s
   c 1s 2s 2p 3s 3p
   d 1s 2s 2p

3 a s-block; Group 1   b p-block; Group 7
   c s-block; Group 2   d p-block; Group 6

Self-Assessment Questions – 1.5

1 a $Be(g) \longrightarrow Be^{\oplus}(g) + e^{\ominus}$
   b $Ne^{2\oplus}(g) \longrightarrow Ne^{3\oplus}(g) + e^{\ominus}$
   c $F(g) + e^{\ominus} \longrightarrow F^{\ominus}(g)$
2 a There is a large increase between the first and second ionisation energies.
   b The outer electrons are in the p energy sub-level.
   c The largest increase occurs between the fourth and fifth ionisation energies.
3 a See p26.   b $He(g) \longrightarrow He^{\oplus}(g) + e^{\ominus}$
   c The 1s orbital is closest to the nucleus; there are two protons in the helium nucleus, so the first ionisation energy is higher than for hydrogen.
   d Decreases.
   e Outer electrons are shielded from the nucleus by completed electron energy levels; more shielding means lower first ionisation energy
4 a and b
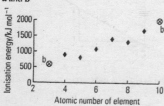
   c Lower
   d Exactly half-filled p-orbitals are relatively stable; nitrogen has a half-filled p-sub-level; with one more electron, the stability is slightly reduced for oxygen.
5 a 1s 2s 2p 3s 3p

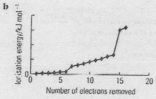

   c Increases as more electrons are removed; it is harder to remove electrons as they become closer to the nucleus and the charge on the ion increases; there are large 'jumps' as new main energy levels, closer to the nucleus, are entered; the much smaller 'jumps' indicate sub-levels.
   d 6   e $S(g) + e^{\ominus} \longrightarrow S^{\ominus}(g)$
   f The first electron affinity of nearly all elements is negative (exothermic), as attraction from the nucleus is still experienced even in the outermost energy level; the second electron affinity is always positive (endothermic) as the incoming electron is repelled by the first extra electron.

Self-Assessment Questions – 1.6

1 a Na   b Si   c Br⁻   d I
2 a $1s^2 2s^2 2p^6$ b $1s^2 2s^2 2p^6 3s^2 3p^6$
   c $1s^2 2s^2 2p^6 3s^2 3p^6 4s^2$ d $1s^2 2s^2 2p^6 3s^2 3p^6$
3 a The three 2p orbitals all have the same energy; none of the orbitals can contain two electrons until all are half full.
   b Spin   c 2p, 2s, 1s
   d Larger; more electron–electron repulsion.
   e Oxygen is slightly smaller.
4 a
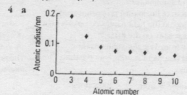
   b
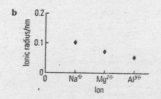
   c
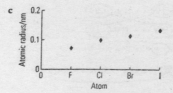

Self-Assessment Questions – 2.2

1 a Ionic   b Covalent
   c Ionic   d Metallic
2 a Na ⋮ I   b Rb ⋮ Cl
   c H ⋮ Cl   d Cl ⋮ Cl
3 a 3.3 – ionic   b 0.4 – covalent
   c 0 – covalent   d 0.2 – covalent

Self-Assessment Questions – 2.3

1 a Na, F → Na⁺ [F]⁻
   b Ca, O → Ca²⁺ [O]²⁻
   c 2K, O → 2K⁺ [O]²⁻
   d Mg, 2Br → Mg²⁺ 2[Br]⁻
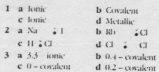
2 a [Ne] $3s^2 3p^5$ b [Ar] $4s^2 3d^{10} 4p^6$
   c [Ne] $3s^2 3p^1$ d $1s^2 2s^2 2p^6$
3 a Sodium loses one electron, sulphur gains two electrons; hence Na_2S.
   b Calcium loses two electrons, bromine gains one electron; hence $CaBr_2$.
   c Aluminium loses three electrons, oxygen gains two; hence, to balance, formula must be Al_2O_3.
   d Magnesium loses two electrons, nitrogen gains three electrons; hence Mg_3N_2.

Self-Assessment Questions – 2.4

1 a H:C:H (with H above and below)   b H:H
   c H:S:H   d H:I:
2 a Br:C:Br (with Br above and below)
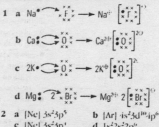
   b Half the distance between the centres of the two chlorine atoms, which are joined by a single covalent bond.
   c It increases with atomic number as the number of complete electron energy levels increases.
   d It forms no bonds
3 a H:O:H   [H:O:H]⁺ (with H below)
   b An electron pair is formed using one electron from oxygen and the electron from hydrogen; the shared pair of electrons is the bond.
   c A non-bonded pair of electrons (a 'lone pair') from the oxygen atom in a water molecule is donated to a proton to make $H_3O^{\oplus}$.

   H:O: + H⁺ → [H:O:H]⁺ (with H below)

Self-Assessment Questions – 2.5

1 a $^{\delta+}H-Cl^{\delta-}$ b $^{\delta+}H-S^{\delta}-H^{\delta+}$

c $^{\delta-}Cl-S^{\delta+}-Cl^{\delta-}$ d $^{\delta-}F-N$

2 a H_2O b HF c NF_3 d H_2O
3 a Ionic b Covalent
 c Polar covalent d Ionic

Self-Assessment Questions – 2.6

1 a 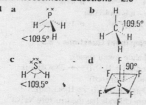 b

c d

2 a Three bonding pairs, one non-bonding pair, hence trigonal pyramid.

b Four bonding pairs, as two double bonds, hence linear.

c Three bonding pairs, one non-bonding pair, hence trigonal pyramid.

d Two bonding pairs, two non-bonding pairs, hence crooked, as for H_2O.

3 a Covalent – no metal present; small difference in electronegativity.
 b Covalent – no metal present; small difference in electronegativity.
 c Ionic – metal and non-metal present; large difference in electronegativity.
 d Ionic – metal and non-metal present; large difference in electronegativity.

Self-Assessment Questions – 2.7

1 The outer electrons of metal atoms can move within the metal lattice, and hence metals and alloys conduct electricity; in the diamond form of carbon, all the outer electrons of the atoms are fixed in covalent bonds.

2 a Ionic

b Polar covalent

c Polar covalent $H^{\delta+}-Br^{\delta-}$

d Metallic

Self-Assessment Questions – 2.8

1 a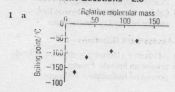

b Larger molecules have more electrons and hence more instantaneous dipole-induced dipole forces.

2 a HF b HF
3 a and c; SO_2 and HBr have permanent dipoles.
4 c and d; CCl_4 and BF_3 have polar bonds, but are symmetrical molecules with no overall dipole.
5 There are only very weak van der Waals (instantaneous dipole-induced dipole) forces between argon atoms; molecules of C_2H_6 contain more electrons and therefore have stronger intermolecular forces between the molecules; LiCl is an ionic substance, with strong forces of attraction between the ionic particles.

Self-Assessment Questions – 2.9

1 d, with a slight effect noticed (in terms of boiling point) in b.
2 Hydrogen atoms attached to a very electronegative element.
3 a 120.
 b Hydrogen bonding.
 c The ethanoic acid molecules are no longer hydrogen bonded to each other.
4 a

b The boiling point increases with increasing atomic number.
 c Larger molecules have more electrons and stronger van der Waals attractions.
 d Hydrogen bonding between the water molecules.
5 Neon has weak van der Waals attractions between atoms; methane has slightly stronger intermolecular attractions – a larger molecule (than neon), therefore more chance of instantaneous dipole-induced dipole attractions; ammonia has hydrogen bonding between molecules.

Self-Assessment Questions – 3.2

1 a Atoms joined by strong covalent bonds in a very large regular structure.
 b Forms of the same element in the same state but with different structures.
 c Electrons which are not fixed in position (in a covalent or ionic bond) but can move around a structure containing at least three atoms.
2 Graphite is a lubricant.
3 a $C(s) + O_2(g) \longrightarrow CO_2(g)$;
  $Si(s) + O_2(g) \longrightarrow SiO_2(s)$
 b CO_2 – gas; SiO_2 – solid
 c To change from solid to gas (gas to solid) without passing through the liquid state.
 d CO_2 has a molecular structure, SiO_2 has giant atomic structure.

Self-Assessment Questions – 3.3

1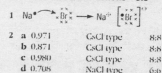

2 a 0.971 CsCl type 8:8
 b 0.871 CsCl type 8:8
 c 0.980 CsCl type 8:8
 d 0.708 NaCl type 6:6

3 a

b Opposite charges are closer together than are like charges – attractions > repulsions.

4

Self-Assessment Questions – 3.4

3 a Yes b No c No d Yes

2 a

b High attractive forces between oppositely charged ions which are close together.
 c Hydration of ions aids solution.
 d Ions cannot move in the solid structure.
 e Ions can move in the liquid.

3 a

b A regular 3-D arrangement of ions in the solid.
 c See *Figure 3.3.2*, p84.
 d 6
 e The smaller cations have to fit between the larger anions; below a radius ratio of about 0.732 the NaCl structure is possible
 f See *Figure 3.3.7*, p86.
 g Attractions between oppositely charged ions outweigh repulsions between similarly charged ions because they are closer together.

Self-Assessment Questions – 3.6

1 a The outer electrons in each atom are delocalised and can move throughout the metal.
 b See p92.
2 Gold and graphite have delocalised electrons; in molten caesium bromide the ions can move.
3 a See *Figure 3.6.4*, p94; movement of planes of atoms prevented by the impurity alloying atoms.
 b See *Figure 3.6.2*, p94.
 c Ions can only move in molten material.

Self-Assessment Questions – 3.7

1 a Trigonal pyramidal; bond angle $< 109.5°$.
 b $PCl_4^{\oplus}$ – tetrahedral; $PCl_6^{\ominus}$ – octahedral
2 i) a $1s^2$
  b $1s^2 2s^2 2p^3$
  c $1s^2 2s^2 2p^6 3s^2 3p^2$
  d $1s^2 2s^2 2p^6 3s^2 3p^4$
  e $1s^2 2s^2 2p^6 3s^2 3p^6 3d^{10} 4s^2 4p^3$

ii) a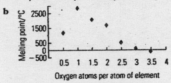

3 Diamond has a giant atomic lattice and all the outer electrons are used in covalent bonds; sodium chloride has a giant ionic lattice and can conduct when the ions are free to move.

4 A Giant atomic   B Metal
  C Molecular   D Ionic

5 a Forms of the same element with different structures.
  b Structural units (molecules) fit together differently; different amounts of intermolecular forces.
  c Molecular formula of sulphur is S_8; for oxygen, O_2; greater intermolecular forces possible for S_8.

Self-Assessment Questions – 4.1

1 a 1 Ca, 2 Br   b 2 B, 3 O
  c 1 Ca, 2 N, 6 O   d 2 N, 8 H, 1 S, 4 O
  e 2 C, 6 H, 1 O
2 a MgS   b Na_2O c KI
  d $CaCl_2$ e $(NH_4)_3PO_4$
3 a NaBr: $23.00 + 79.91 = 102.91$;
  b K_2S: $(2 \times 39.10) + 32.06 = 110.26$;
  c SiO_2: $28.09 + (2 \times 16.00) = 60.09$;
  d $(NH_4)_2SO_4$: $(2 \times 14.01) + (8 \times 1.01) + 32.06 + (4 \times 16.00) = 132.16$;
  e $MgSO_4$: $24.31 + 32.06 + (4 \times 16.00) = 120.37$;
  f SO_2: $32.06 + (2 \times 16.00) = 64.06$

Self-Assessment Questions – 4.2

1 a caesium + oxygen $\longrightarrow$ caesium oxide;
  b aluminium + chlorine $\longrightarrow$ aluminium chloride;
  c magnesium sulphate + barium chloride $\longrightarrow$ barium sulphate + magnesium chloride.
2 a $2Sr(s) + O_2(g) \longrightarrow 2SrO(s)$
  b $2Al(s) + Cr_2O_3(s) \longrightarrow 2Cr(s) + Al_2O_3(s)$
  c $CH_4(g) + 2O_2(g) \longrightarrow CO_2(g) + 2H_2O(l)$
  d $Fe_2O_3(s) + 3C(s) \longrightarrow 3CO(g) + 2Fe(s)$
3 a $Ba^{2+}(aq) + 2NO_3^{-}(aq) + Mg^{2+}(aq) + SO_4^{2-}(aq) \longrightarrow BaSO_4(s) + Mg^{2+}(aq) + 2NO_3^{-}(aq)$
  b $2Ag^{+}(aq) + 2NO_3^{-}(aq) + Ca^{2+}(aq) + 2Cl^{-}(aq) \longrightarrow 2AgCl(s) + Ca^{2+}(aq) + 2NO_3^{-}(aq)$

Self-Assessment Questions – 4.3

1 a Al is oxidised; Cl is reduced
  b Ba is oxidised; oxygen reduced
  c O is oxidised; Hg is reduced
2 a +3   b −4   c +6
  d +6   e +5

3 a oxidation   b reduction   c oxidation
4 a K oxidised, Cl reduced
  b Fe oxidised, Cu reduced

Self-Assessment Questions – 5.1

1. Your account should include the similarity of the reactions of some elements, attempts to organise based on these properties, using relative atomic masses to arrange the elements. Confirmation of the validity of the Table came with the discovery of neutrons and protons (and atomic number) and the arrangement of the electrons in different energy levels around the nucleus.
2 a LiCl–Li +1 Cl −1, $BeCl_2$–Be +2 Cl −1, BCl_3–B +3 Cl −1, CCl_4–C +4 Cl −1, NCl_3–N +3 Cl −1, Cl_2O–Cl +1 O −2, ClF–Cl +1 F −1.
  b Lithium loses an electron: the oxidation state increases from 0 to +1; chlorine gains one electron: the oxidation state goes from 0 to −1.
3 a p-block; oxygen, O
  b p-block; argon, Ar
  c s-block; lithium, Li
  d d-block; chromium, Cr.

Self-Assessment Questions – 5.2

1 a NaO; MgO; $AlO_{1\frac{1}{2}}$; SiO_2; $PO_{2\frac{1}{2}}$; SO_3; $ClO_{3\frac{1}{2}}$
  b

Melting point/°C (graph, y-axis values: −500, 0, 500, 1500, 2500; x-axis: Oxygen atoms per atom of element 0.5 1 1.5 2 2.5 3 3.5 4)

  c Melting points are high for giant ionic compounds (Na, Mg, Al) and giant atomic structures (Si, possibly Al); values are low for molecular structures.
2 Ionic structures have high boiling points (Groups 1, 2); then the boiling points of the molecular hydrides increase across the rest of the Period as M_r increases. Down a Group the boiling point increases with M_r, except first members in Groups 5, 6, 7 – e.g. HF in Group 7 – because of hydrogen bonding.
3 a The ionisation energies tells us that B is in Group 1, so A must be a noble gas.
  b $B^{2+}(g) \longrightarrow B^{3+}(g) + e^{-}$
  c Approximately +3500 kJ mol^{-1}
  d Peaks are at noble gases, troughs at Group 1, with a similar pattern across each Period (allowing for d- and f-blocks in Periods 4 onwards).

Self-Assessment Questions – 5.3

1 a non-metal   b non-metal
  c metal   d metal
2 a ionic   b ionic
  c covalent   d ionic
3 a The relative molecular mass of the P_4 units is less than that of S_8.
  b Aluminium has three outer (delocalised electrons) compared to magnesium (two outer electrons).

c Metallic (Na, Mg, Al) and giant atomic structures (Si) usually have very high boiling points; all the other elements have molecular structures.
d Chlorine exists as Cl_2, argon is monatomic. Cl_2 has many more electrons, hence more opportunity for intermolecular forces.

Self-Assessment Questions – 5.4

1 a S   b K   c K, Mg   d K   e K, Mg
2 a +4   b ionic
  c no – it is molecular
  d $1s^2 2s^2 2p^6 3s^2$ e $1s^2 2s^2 2p^6 3s^2$

Self-Assessment Questions – 5.5

1 a Alkaline colour (blue).
     $Na_2O(s) + H_2O(l) \longrightarrow 2NaOH(aq)$
  b Acid colour (red).
     $SO_2(g) + H_2O(l) \rightleftharpoons H_2SO_3(aq)$
2 Sodium chloride has the structure $Na^{+}Cl^{-}$. In water, the ions separate and are hydrated: $NaCl(s) + (aq) \longrightarrow Na^{+}(aq) + Cl^{-}(aq)$. PCl_5 has a molecular structure (or, more accurately, $PCl_4^{+}.PCl_6^{-}$). The P—Cl bonds are relatively weak. One product is volatile. $PCl_5(s) + H_2O(l) \longrightarrow PCl_3O(l) + 2HCl(g)$.
3 a NaCl   b $SiCl_4$ c MgO
  d Na_2O e $SiCl_4$ f P_4O_6

Self-Assessment Questions – 6.2

1 a 2.8.1   b 2.8.8.2   c 2.8.8   d 2.2
2 Potassium atoms (2.8.8.1) easily lose an electron.
3 a Decrease.   b Decrease.
  c Increase from Group 1 to Group 2.
4 a First ionisation energy (or first ionisation enthalpy).
  b Rubidium atoms are larger than sodium; the outer electron is more shielded from the nucleus.
  c Smaller.
  d The loss of the electron decreases the electron-electron repulsions in the atom; the number of protons remains the same; the remaining electrons are therefore closer to the nucleus.

Self-Assessment Questions – 6.3

1 a sodium + water $\longrightarrow$ sodium hydroxide + hydrogen
  b $Na + H_2O \longrightarrow NaOH + H_2$
  c $2Na(s) + 2H_2O(l) \longrightarrow 2NaOH(aq) + H_2(g)$
  d 2.8.1 (or $1s^2 2s^2 2p^6 3s^1$)
  e See p156.
2 a The colour is not a pure colour but a mixture of lines at a number of wavelengths, because more than one electron transition is involved.
  b calcium + oxygen $\longrightarrow$ calcium oxide
  c $2Ca(s) + O_2(g) \xrightarrow{heat} 2CaO(s)$
  d Calcium is oxidised by electron loss, oxygen reduced by electron gain

Self-Assessment Questions – 6.4

1 a rubidium + chlorine $\longrightarrow$ rubidium chloride
  b potassium hydride + water $\longrightarrow$ potassium hydroxide + hydrogen

c lithium carbonate ⟶ lithium oxide
        + carbon dioxide

d No reaction.

2 a $2K(s) + 2H_2O(l) \xrightarrow{RT} 2KOH(aq) + H_2(g)$

b $4LiNO_3(s) \xrightarrow{heat} 2Li_2O(s) + 4NO_2(g) + O_2(g)$

c $4Cs(s) + O_2(g) \xrightarrow{heat} 2Cs_2O(s)$

d $2KNO_3(s) \xrightarrow{heat} 2KNO_2(s) + O_2(g)$

3 a Relative atomic mass, A_r, of an element is the weighted mean of the isotopic masses of its natural isotopes on a scale on which $^{12}_6C$ has $A_r = 12.0000$.

b Lilac.

c $2K(s) + Cl_2(g) \xrightarrow{heat} 2KCl(s)$

d $K^\bullet \rightsquigarrow \overset{\times\times}{\underset{\times\times}{Cl}} \longrightarrow K^\oplus \left[\overset{\times\times}{\underset{\times\times}{Cl}} \right]^-$

e $39.10 + 35.45 = 74.55$

f Potassium is oxidised by electron loss to chlorine; chlorine is reduced by gaining the electron.

4 a $RbNO_3$.

b The ionic lattice breaks up as ions are hydrated.

c The solid melts; bubbles of gas are eventually seen in the liquid.

d The gas would relight a glowing splint.

e $2RbNO_3(s) \xrightarrow{heat} 2RbNO_2(s) + O_2(g)$

Self-Assessment Questions – 6.5

1 a $Sr(s) + 2H_2O(l) \xrightarrow{RT} Sr(OH)_2(s) + H_2(g)$

b $2Ba(s) + O_2(g) \xrightarrow{heat} 2BaO(s)$

c $Mg(s) + Cl_2(g) \xrightarrow{heat} MgCl_2(s)$

d $BaO(s) + H_2O(l) \xrightarrow{RT} Ba(OH)_2(s)$

e $2Ca(NO_3)_2(s) \xrightarrow{heat} 2CaO(s) + 4NO_2(g) + O_2(g)$

f $BaCO_3(s) \xrightarrow{heat} BaO(s) + CO_2(g)$

g $Ca(OH)_2(s) \xrightarrow{heat} CaO(s) + H_2O(l)$

2 a Solubility decreases from $MgSO_4$ to $BaSO_4$. $BaSO_4$ is so insoluble that it passes through the gut unchanged.

b More vigorously. Barium atoms are larger than calcium (and lose the outer electrons more readily).
$Ba(s) + 2H_2O(l) \longrightarrow Ba(OH)_2(s) + H_2(g)$

c $SrCO_3$:
$87.62 + 12.00 + (3 \times 16.00) = 147.62$;
it is more stable.

Self-Assessment Questions – 6.6

1 a See p167.   **b** See p167.

2 a $CaCO_3(s) \xrightarrow{heat} CaO(s) + CO_2(g)$

b $CaO(s) + H_2O(l) \xrightarrow{RT} Ca(OH)_2(s)$;
pH is high (above 12), showing an alkaline solution.

c Water which does not lather easily with soap.

d Chalk slowly dissolves in rain water because of the carbon dioxide in the rain water; $CaCO_3(s) + H_2O(l) + CO_2(aq) \rightleftharpoons Ca(HCO_3)_2(aq)$

Self-Assessment Questions – 7.2

1 a

The atomic radius increases as there are more complete electron energy levels.

b Smaller.

c A greater number of completed energy levels means more shielding of the outer electrons from the attraction of the nucleus;
$Pb(g) \longrightarrow Pb^+(g) + e^-$;
$Pb^{2+}(g) \longrightarrow Pb^{2+}(g) + e^-$

d Lead has a high density; therefore a close-packed structure (i.e. more atoms per unit volume is expected).

Self-Assessment Questions – 7.3

1 a $SiCl_4(l) + 4H_2O(l) \longrightarrow$
$Si(OH)_4(s) + 4HCl(g)$;
H_4SiO_4 or $SiO_2.2H_2O$ are alternate formulae for $Si(OH)_4$:
approximately pH 1 because of the HCl produced.

b Covalent.

c

d $SiH_4(g) + 2O_2(g) \longrightarrow SiO_2(s) + 2H_2O(l)$

2 a Carbon dioxide, CO_2.

b Carbon monoxide, CO.

c Lead(IV) oxide, PbO_2; or tin(IV) oxide, SnO_2.

d Lead(II) oxide, PbO.

e Tetrachlormethane, CCl_4.

f Lead(II) chloride, $PbCl_2$.

g Carbon dioxide, CO_2; or carbon monoxide, CO.

3 a Immediate evolution of gas – forms steamy fumes in moist air and leaving a white solid; $GeCl_4(l) + 4H_2O(l) \longrightarrow$
$Ge(OH)_4(s) + 4HCl(g)$

b The yellow solid forms a colourless solution;
$PbO(s) + 2HNO_3(aq) \longrightarrow$
$Pb(NO_3)_2(aq) + H_2O(l)$

c Silvery metal forms a red/orange solid;
$6Pb(s) + 4O_2(g) \longrightarrow 2Pb_3O_4(s)$

Self-Assessment Questions – 7.4

1 a A white smoke forms as the silane burns;
$SiH_4(g) + 2O_2(g) \longrightarrow SiO_2(s) + 2H_2O(l)$

b A white suspension is formed, which dissolves as more CO_2 is bubbled through the mixture to form a clear, colourless solution;
$Ca(OH)_2(aq) + CO_2(g) \longrightarrow$
$CaCO_3(s) + H_2O(l)$;
$CaCO_3(s) + H_2O(l) + CO_2(g) \rightleftharpoons$
$Ca(HCO_3)_2(aq)$

c No reaction.

d Immediate reaction producing steamy fumes;
$SnCl_4(l) + 4H_2O(l) \longrightarrow$
$SnO_2.2H_2O(s) + 4HCl(g)$

2 SiO_2 has a giant structure; CO_2 molecular; CCl_4 has no orbitals into which the non-bonding pairs of electrons in water can donate – so there is no reaction with water; $SiCl_4$ has vacant 3d orbitals; it is rapidly attacked by water; CH_4 has no reaction with water; nor does pure SiH_4.

3 a $2CO(g) + O_2(g) \longrightarrow 2CO_2(g)$

b $PbO_2(s) + 4HCl(conc\ aq) \longrightarrow$
$PbCl_2(aq) + Cl_2(g) + H_2O(l)$

c $CO_2(g) + H_2O(l) \rightleftharpoons H_2CO_3(aq)$

Self-Assessment questions – 8.1

1 a $Br^{\delta-}—P^{\delta+}$ **b** $Br^{\delta+}—Cl^{\delta-}$

c $Cl^{\delta+}—F^{\delta-}$ **d** $F^{\delta-}—H^{\delta+}$

2 a Ionic   **b** Covalent

c Covalent   **d** Ionic

e Covalent

3 a Chlorine   **b** NaX

c Ionic   **d** $Na^\oplus \overset{\times\times}{\underset{\times\times}{X}}^\ominus$

Self-Assessment questions – 8.2

1 a Half the distance between centres of atoms of the same element joined by a single covalent bond.

b Difficult to see where one ion starts and the other finishes; the distance between ions can be measured – so self-consistent data can be built up.

c Larger.

d The bromide ion has an extra electron, so more internal repulsion of electrons leads to expansion of the ion.

2 See p195.

3 a Atoms of the same element with the same number of protons but with different mass numbers.

b Solid; it probably sublimes.

c Black

d $At_2(s) + 2e^\ominus \longrightarrow 2At^\ominus$

e Larger

f $At(g) \longrightarrow At^\oplus(g) + e^\ominus$

g More completed electron energy levels shield the outer electrons from the attraction of the nucleus.

Self-Assessment Questions – 8.3

1 a H—H; H—Cl; H—F

b Only weak intermolecular forces between H_2 or HCl molecules; much stronger hydrogen bonds between H—F molecules.

c H—I molecules dissociate into H_2 and I_2 vapour, which is purple;
$2HI(g) \longrightarrow H_2(g) + I_2(g)$; no observable reaction with hydrogen chloride.

2 a SO_2, S, H_2S

b NaCl would produce misty fumes of HCl; NaBr would produce misty fumes of HBr and brown fumes of Br_2;
NaI would produce a mixture of products including yellow sulphur and purple fumes of $I_2(g)$.

c $H_2SO_4(l) + 2Br^- + 2H^+(aq) \longrightarrow$
$$SO_2(g) + Br_2(g) + 2H_2O(l)$$
$H_2SO_4(l) + 2I^- + 2H^+(aq) \longrightarrow$
$$SO_2(g) + I_2(g) + 2H_2O(l)$$
$H_2SO_4(l) + 6I^- + 6H^+(aq) \longrightarrow$
$$S(s) + 3I_2(g) + 4H_2O(l)$$
$H_2SO_4(l) + 8I^- + 8H^+(aq) \longrightarrow$
$$H_2S(g) + 4I_2(g) + 4H_2O(l)$$

3 a Ionic; giant ionic lattice.
b The white crystalline solid dissolves.
c Hydrogen fluoride evolved;
$$NaF(s) + H_2SO_4(l) \longrightarrow NaHSO_4(s) + HF(g)$$

Self-Assessment Questions – 8.4

1 a Simultaneous oxidation and reduction of the same element during a reaction.
b $Cl_2(g) + 2OH^-(aq) \longrightarrow$
$$Cl^-(aq) + ClO^-(aq) + H_2O(l)$$
c NaCl and NaClO.
d Dissolved chlorine – i.e. $Cl_2(aq)$.

2 a $Br_2(l) + H_2O(l) \rightleftharpoons Br^-(aq) + H^+(aq)$
$$+ HBrO(aq)$$
b $I_2(s) + 2OH^-(aq) \longrightarrow$
$$I^-(aq) + IO^-(aq) + H_2O(l)$$
c $Cl_2(g) + 2OH^-(aq) \longrightarrow$
$$Cl^-(aq) + ClO^-(aq) + H_2O(l)$$
d $3Cl_2(g) + 6OH^-(aq) \longrightarrow$
$$5Cl^-(aq) + ClO_3^-(aq) + 3H_2O(l)$$

3 See p201/2.
a e.g. $2Br^-(aq) + Cl_2(Aq) \longrightarrow$
$$Br_2(aq) + 2Cl^-(aq)$$
b e.g. Br^- is oxidised from -1 to 0;
$$Cl_2 \text{ is reduced from } 0 \text{ to } -1.$$

4 a $I_2(s) + Cl_2(g) \longrightarrow 2ICl(s)$
b
c ICl $-+1$; ICl_3, $-+3$
d $2ICl(s) \longrightarrow I_2(s) + Cl_2(g)$

Self-Assessment Questions – 8.5

1 a $+7$ **b** $+1$ **c** $+5$ **d** $+3$
2 a Sodium chlorate(V).
b Iodic(VII) acid.
c Iodine(V) oxide.
d Potassium bromate(V).

Self-Assessment Questions – 8.6

1 a White precipitate, darkens with time and sunlight.
b Brief appearance of a red colour, quickly bleached to colourless solution.
c Off-white precipitate, darkens rapidly in light.

2 a $KClO_3$
b $2KClO_3(s) \longrightarrow 2KCl(s) + 3O_2(g)$
c A reaction involving simultaneous oxidation and reduction; here, chlorate(V) is reduced to chloride (i.e. chlorine(-1)) and oxygen (-2 in the chlorate(V) ion) is oxidised to oxygen (0) – i.e. the element oxygen.
d The solid dissolves to form a clear colourless solution;
$$KCl(s) + (aq) \longrightarrow K^+(aq) + Cl^-(aq)$$

Self-Assessment questions – 8.7

1 a

---

b The structure is a trigonal pyramid to accommodate the non-bonding pair of electrons.
c $PBr_5 - +5$; $PCl_3 - +3$
d Iodine atoms are too large to fit five I atoms around the P atom.
e $PCl_5(s) + H_2O(g) \longrightarrow PCl_4O(l) + 2HCl(g)$

2 a All exist as molecules; dipole-dipole and other intermolecular interactions are too weak to hold molecules together.

b

c $+6$
d Oxygen atoms cannot form six covalent bonds – oxygen does not have orbitals available to accommodate six electron pairs around it.

Self-Assessment Questions – 8.8

1 a See p210.
b Hydrogen and chlorine.
c Hydrogen; $2H^+(aq) + 2e^- \longrightarrow H_2(g)$
d Chlorine; $2Cl^-(aq) \longrightarrow Cl_2(g) + 2e^-$
e Sodium chloride, NaCl, and sodium chlorate(I), NaClO.
f The products are sodium chloride, NaCl, and sodium chlorate(V), $NaClO_3$.

Self-Assessment Questions – 8.9

1 a $Cl_2(g) + H_2O(l) \longrightarrow HCl(aq) + HClO(aq)$
b HClO, chloric(I) acid.
c Chlorine is put in to remove the bacteria; the HClO formed is what actually destroys the bacteria.

2 a $2AgBr(s) \longrightarrow 2Ag(s) + Br_2(l)$
b Iodine is a covalent molecular material; KI is an ionic solid, and its ions can be hydrated by water and so KI dissolves.
c The only oxidation state is -1; HF is not a strong acid; C—F bonds cannot easily be hydrolysed.

Self-Assessment Questions – 8.10

1 A is HCl; B is NaCl; C is HClO; D is AgCl; E is Ag.
2 Chloride: forms a white precipitate with silver nitrate/nitric acid solutions, easily soluble in dilute ammonia solution; bromide: off-white precipitate with silver nitrate/nitric acid solutions (not easily distinguished from chloride), dissolves with difficulty in ammonia solution; iodide: yellowish precipitate with silver nitrate/nitric acid solutions, insoluble even in concentrated ammonia solution.

Self-Assessment Questions – 9.1

1 a See p219.   **b** See p219.
c The peak with the highest mass number, corresponding to M^+, the molecule carrying a single positive charge.

2 a 46(100)   **b** 72(72)
c 44(51)   **d** 64(100)

3 $15-CH_3^+$, $127-I^+$, $142-CH_3I^+$, $141-CH_2I^+$
4 a 121.8   **b** 63.6   **c** 6.93
5 a $C_2H_6^+$ **b** $C_3H_6O^+$

Self-Assessment Questions – 9.2

1 a LiCl; $6.94 + 35.45 = 42.39$;
$$Li = 6.94 \times 100/42.39 = 16.4\%$$
b $24.02 + 6.06 = 30.08$;
$$C = 24.02 \times 100/30.08 = 79.9\%$$
c $CaBr_2$: $40.08 + (2 \times 79.90) = 199.88$;
$$Br = 159.8 \times 100/199.88 = 79.9$$
d $24.32 + 32.00 + (4 \times 16.00)$
$$+ (7 \times 18.02) = 246.51;$$
$$H_2O = 126.1 \times 100/246.51 = 51.2\%$$

2 a NaCl = $22.99 + 35.45 = 58.44$;
$$39.3\% \text{ Na}, 60.7\% \text{ Cl}$$
b $SO_2 = 32.06 + (2 \times 16.00) = 64.06$;
$$50.0\% \text{ S}, 50.0\% \text{ O}$$
c HCl = $1.01 + 35.45 = 36.46$;
$$2.76\% \text{ H}, 97.2\% \text{ Cl}$$
d $24.3 + (2 \times 14.01) + (6 \times 16.00) = 148.3$;
$$16.4\% \text{ Mg}, 18.9\% \text{ N}, 64.7\% \text{ O}$$

Self-Assessment Questions – 9.3

1 a $2 \times L = 1.20 \times 10^{24}$ atoms
b $0.004 \times (6.02 \times 10^{23})$
$$= 2.41 \times 10^{21} \text{ molecules}$$
c $0.13 \times L = 7.8 \times 10^{22}$ ions of each of Na^+ and I^-

2 a 1 mole $= 6.02 \times 10^{23}$
b $30.97 + (3 \times 35.45) = 137.32$;
$$27.5/137.32 = 0.200 \text{ mol};$$
$$0.200 \times L = 1.20 \times 10^{23} \text{ molecules}$$
c $Cl_2 = 70.9$; $3.55/70.9 = 0.0500$ mol
$$= 3.01 \times 10^{22} \text{ molecules}$$
d Cl = 35.45; $0.0355/35.45 = 0.00100$ mol
$$= 6.02 \times 10^{20}$$
e $CaCl_2 = 40.08 + (2 \times 35.45) = 111.0$;
$1.11 \text{ g} = 0.010 \text{ mol } CaCl_2 = 6.02 \times 10^{21}$
Ca^{2+} ions
f $PCl_5 = 30.97 + (5 \times 35.45) = 208.22$;
$20.85 \text{ g} = 0.10 \text{ mol } PCl_5 = 0.50 \text{ mol Cl}$
$$= 3.0 \times 10^{23} \text{ Cl}$$
g $BaBr_2 = 137.3 + (2 \times 79.9) = 297.1$;
$2.97 \text{ g} = 0.01 \text{ mol } BaBr_2 = 0.02 \text{ mol } Br^-$
$$= 1.2 \times 10^{22} \text{ Br}^- \text{ ions}$$

3 a 0.0020 mol   **b** 0.39 mol
c 3.00 mol   **d** 1.60 mol
4 a $Si(s) + 2Cl_2(g) \longrightarrow SiCl_4(l)$
b $28.09 + (4 \times 35.45) = 170 \text{ g mol}^{-1}$
c $4.35/170 = 0.0256$
d 1.54×10^{22}

Self-Assessment Questions – 9.4

1 a $8.0/6.94 = 1.15$;
$$92/79.9 = 1.15; \text{ so } 1:1, \text{ LiBr}$$
b $94.1/32.1 = 2.93$;
$$5.9/1.01 = 5.84; \text{ so } 1:2, H_2S$$
c $10.61/79.9 = 0.134$; $1.62/24.3 = 0.067$;
$$\text{so } 2:1, MgBr_2$$
d $3.87/39.1 = 99.0$; $1.39/14.0 = 99.2$;
$$4.75/16.0 = 297; \text{ so } 1:1:3, KNO_3$$

2 a $4.13/55.85 = 0.0739$ mol;
$1.77/16.0 = 0.111$ mol; so $2:3$, Fe_2O_3;
$$M_r = (2 \times 55.85) + (3 \times 16.00) = 159.7$$
b $1.46/39.1 = 0.0373$ mol;
$1.20/16.0 = 0.0750$ mol; so $1:2$, KO_2;
$M = 39.1 + (2 \times 16.0) = 71.1 \text{ g mol}^{-1}$;
empirical formula = molecular formula
$$= KO_2$$
c $20.2/27.0 = 0.750$; $79.8/35.45 = 2.25$;
so $1:3$, $AlCl_3$; empirical formula is $AlCl_3$;

---

c Propane;
$$CH_3CH=CH_2(g) + H_2(g) \longrightarrow$$
$$CH_3CH_2CH_3(g)$$

d Propan-1,2-diol;
$$CH_3CH=CH_2(g) \longrightarrow CH_3CH(OH)CH_2OH$$

2 a CH_2BrCH_2Br
b $CH_3CHClCH_3$
c $CH_3CH(OH)CH_2CH_3$
d $CH_3CH_2CH(OH)CH_3$

3 a $C_{16}H_{30}O_2$
b $C_{16}H_{30}O_2(l) + H_2(g) \xrightarrow[Ni]{H_2} C_{16}H_{32}O_2(l)$
c Decolorisation of Br_2 (in $CHCl_3$) or $MnO_4^-(aq)/H^+(aq)$.
d Both H on the same side — or both bulky groups on the same side.
e π bond is broken, rotation of molecule around a single bond and the π bond reforms.

4 a Alkene and a (primary) alcohol.
b Both bulky groups on the same side of the double bond — in this case
$—C(CH_3)=CHCH_2OH$.
c The bromine adds across each of the double bonds
$(CH_3)_2C=CHCH_2CH_2C(CH_3)$
$=CHCH_2OH + 2Br_2(in\ CHCl_3) \longrightarrow$
$(CH_3)_2CBrCHBrCH_2CH_2CBr(CH_3)$
$CHBrCH_2BrCH_2OH$
d Two, since there are two C=C groups in the molecule.

Self-Assessment Questions – 16.6

1 a Bromine is decolorised;
$C_2H_4(g) + Br_2(l) \longrightarrow CH_2BrCH_2Br(l)$
b No reaction.
c Bromine decolorised; first step;
$C_2H_6(g) + Br_2(g) \longrightarrow C_2H_5Br(l) + HBr(g)$
d Free radical substitution.
e Electrophilic addition; see p421.

2 a
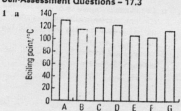
b Optical activity/rotation of the plane of plane polarised light.
c $(CH_3)_2C=CHCH_2CH_2CH(CH_3)CH_2CHO(l)$
$+ HBr(g) \longrightarrow (CH_3)_2CBr=CH_2CH_2CH_2$
$CH(CH_3)CH_2CHO(l)$
d Electrophilic addition.

3 A is $CH_3CH_2CH_2CH_3$.
B is $Cl_2(g)$ in UV light.
C is $CH_3CHClCH_2CH_3$.
D is Br_2 in an organic solvent.
E is $CH_3CHCH_2CH_3$.
  $\underset{O}{\diagdown}$
F is $CH_2(OH)CH(OH)CH_2CH_3$.

Self-Assessment Questions – 16.7

1
$$3F_2C=CF_2 \longrightarrow \begin{bmatrix} F & F & F & F & F & F \\ | & | & | & | & | & | \\ -C-C-C-C-C-C- \\ | & | & | & | & | & | \\ F & F & F & F & F & F \end{bmatrix}$$

---

2 a Yes. b No. c Yes.

3 a $CH_3CH=CH_2(g) + Br_2(l) \longrightarrow$
$CH_3CHBrCH_2Br(l)$
b $CH_3CH=CH_2 + HCl(hexane) \longrightarrow$
$CH_3CHClCH_3(l)$
c
$$nCH_2CH=CH_2 \longrightarrow \begin{bmatrix} CH_3 & H \\ | & | \\ -C-C- \\ | & | \\ H & H \end{bmatrix}_n$$

d $2CH_3CH=CH_2(g) + 9O_2(g) \longrightarrow$
$6CO_2(g) + 6H_2O(l)$

4 a Substances with the same molecular formula but different arrangements of atoms in space.
b $CH_3CH_2CH=CH_2$; but-1-ene.

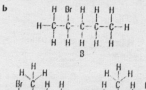
cis-but-2-ene

trans-but-2-ene

c There can be no rotation about a double bond.

d
$$\begin{bmatrix} CH_3 & CH_3 \\ | & | \\ -C-C- \\ | & | \\ H & H \end{bmatrix}$$
  e See p427.

Self-Assessment Questions – 17.2

1 a 2-Chloropropane. b 2-Bromobutane.
c 1-Chloro-3-methylbutane.
d 1-Fluoropropane.

2 a $CH_2ClCH_2CH_3$
b $CH_3CH_2CHBrCH_2CH_2CH_3$
c $CH_3CH(CH_3)CHBrCH_2CH_3$
d $CH_2Br(CH_3)CH(CH_3)CH_3$

3 a 1-Bromopropane   2-Bromopropane
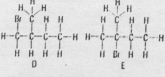

b 1-Bromobutane
H H H H
H-C-C-C-C-Br
H H H H

2-Bromobutane (two optical isomers)
H H Br H
H-C-C-C-C-H
H H H H

1-Bromo-2-methylpropane
H H H
H-C-C-C-Br
H H H (with CH_3)

2-Bromo-2-methylpropane
H H H
H-C-C-C-H
H Br H (with CH_3 groups)

4 A tertiary; B primary; C tertiary.

Self-Assessment Questions – 17.3

1 a
(bar chart: Boiling point/°C vs A B C D E F G)

b
H Br H H H H
H-C-C-C-C-C-C-H
H H H H H H
B

D   E
(structures)

c They are isomeric molecules of similar shape and very similar intermolecular forces.
d Increasingly spherical molecules hence decreasing intermolecular forces.

Self-Assessment Questions – 17.4

1 a Add HBr gas to propene dissolved in hexane.
b Heat butan-1-ol with red phosphorus and iodine.
c Add HCl gas to ethene dissolved in hexane.
d Mix the gases in diffused UV light and separate the mixture.

2 a Electrophilic addition;
$$CH_3CH_2CH=CH_2 + HBr \xrightarrow{RT}$$
$$CH_3CH_2CHBrCH_3$$
b Free radical substitution;
$$C_6H_{14} + Cl_2 \xrightarrow{UV} C_6H_{13}Cl + HCl \longrightarrow$$
further substitution products

3 a Pent-2-ene.
b It can exist as cis and trans isomers because there is no rotation about the C=C double bond.
c $CH_3CH_2CHBrCH_2CH_3$ and
$CH_3CHBrCH_2CH_2CH_3$
$CH_3CHBrCH_2CH_2CH_3$ can exist as two optical isomers since it contains a chiral carbon atom.
d Both compounds are secondary since they both contain a $C—CHBr—C$ structure.

4 a $3C_5H_{11}OH + \ 'PBr_3' \xrightarrow{RT} 3C_5H_{11}Br + H_3PO_3$
b 2-Bromo-2-methylbutane.
c See p381.

Self-Assessment Questions – 17.5

1 a
H H H H
H-C-C-C-C-Br
H H H H

H H H
H-C-C-C-Br
H H H (with CH_3)

Answers

b The alkene is formed by treating with hot concentrated hydroxide ion in ethanol; the alcohol is formed by treating with hot dilute aqueous hydroxide.

c The alkene is formed by elimination; the alcohol is formed by substitution.

2

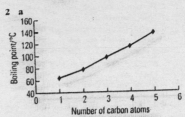

3 a $C_4H_9Br \xrightarrow[\text{OH}^-, \text{EtOH}]{\text{hot conc}} C_4H_8 + HBr$
2-Bromobutane · But-2-ene

b $C_4H_9Br + KCN \xrightarrow[\text{ethanol}]{\text{reflux KCN}} C_4H_9CN + KBr$
1-(or 2-)Bromobutane · 1-(or 2-)Pentanenitrile or 2-methylbutanenitrile

c $C_4H_{10}(g) + Br_2(g) \xrightarrow{UV} C_4H_9Br(l) + HBr(g)$
Butane · 1-(or 2-)Bromobutane

Self-Assessment Questions – 17.6

1 Free radical substitution, see p397; electrophilic addition, see p421, nucleophilic substitution, see p448.

2 a Free radical substitution for halogenation of alkanes.
b Electrophilic addition in the addition to an alkene.
c Nucleophilic substitution (S_N2) for replacement of Cl in a primary halogenoalkane.
d Nucleophilic substitution (S_N1) for replacement of Cl in a tertiary halogenoalkane.

3 a If dilute aqueous solution, nucleophilic substitution; OH$^{\ominus}$(aq).
b Nucleophilic substitution, CN$^-$(aq).
c Electrophilic addition; H$^{\delta+}$—Br$^{\delta-}$.

Self-Assessment Questions – 17.7

1 a $CH_3CH(CH_3)CH_2Cl$

c CH₂Br [structure] **b** [structure with CH₃ and Cl]

2 [benzene structures with reactions]
+ Cl₂ UV → + HCl
+ Cl₂ AlCl₃ → + HCl

Self-Assessment Questions – 18.1

1 a Propan-1-(or -2-)ol. **b** Hexan-1-ol.
c 2-Methylpropan-1-ol. **d** Butan-2-ol.

2 a [structure: H–C–C–C–C–C–OH chain]

b [structure]

---

c [structure: H–C–C–C–C–C–H with OH]

d [ring structure with OH]

3 a Yes. **b** No. **c** No. **d** Yes.
4 a Primary. **b** Primary.
c Secondary. **d** Tertiary.

5 [structure] Pentan-1-ol, primary

[structure] Pentan-2-ol, secondary

[structure] Pentan-3-ol, secondary

[structure] 2,2-Dimethylpropanol, primary

[structure] 3-Methylbutan-1-ol, primary

[structure] 2-Methylbutan-1-ol, primary

[structure] 2-Methylbutan-2-ol, tertiary

Self-Assessment Questions – 18.2

1 [structure] ~125 °C

[structure with CH₃, OH, CH₃] ~110 °C

2 a

[graph: Boiling point/°C vs Number of carbon atoms, plotted points rising from ~65 at 1 to ~145 at 5]

b 158 °C
c The more spherical the isomer, the less intermolecular interaction, and hence the lower the boiling point.

3 a B, D **b** A **c** C
d B **e** B
4 B and C

Self-Assessment Questions – 18.3

1 a $C_{12}H_{22}O_{11}(aq) + H_2O(l) \longrightarrow$
$4C_2H_5OH(aq) + 4CO_2(g)$
b $C_{12}H_{22}O_{11}(s) + 12O_2(g) \longrightarrow$
$12CO_2(g) + 11H_2O(l)$

2 a $C_6H_{10}(l) + H_2O(l) \longrightarrow C_6H_{11}OH(l)$
b Fractional distillation.
c Secondary.

3 a Butene-1-(or-2-)ene with HBr gas at room temperature.
b Propanone reacted with NaBH₄(aq) at room temperature.
c Heat 2-bromo-2-methylpropane with dilute aqueous sodium hydroxide.
d Bubble HBr through solution of 2-methylpent-1-ene in hexane at room temperature; then reflux with dilute NaOH(aq).

4 a Add HBr at room temperature, reflux with dilute NaOH(aq)
b Reflux with dilute HCl(aq).
c Reflux with dilute NaOH(aq).
d Add LiAlH₄ in ethoxyethane at room temperature, followed by water.

5 a Butan-2-ol.

b [structure] H–C–C–C–C–H + OH$^{\ominus}$ →

[structure] H–C–C–C–C–H + Br$^{\ominus}$

c Secondary.
d Dilute NaOH(aq).
e 5.0 × 74/137 = 2.7 g

Self-Assessment Questions – 18.4

1 a Dehydration of $(CH_3)_2CHCH_2OH$

b Butan-2-ol + PCl_5 (or PCl_3 or SCl_2O) at room temperature to form 2-chlorobutane.

c Propanoic acid + ethanol refluxed with hydrochloric acid to form ethyl propanoate.

d Heat propan-2-ol with acidified aqueous dichromate solution to form propanone.

e Heat butan-1-ol with acidified aqueous dichromate solution to form butanal; or refluxing the mixture will form butanoic acid.

2 a $Na_2Cr_2O_7$

b $2ROH(l) + 2Na(s) \longrightarrow$
$$2RO^- Na^\oplus(s) + H_2(g)$$

c Orange.

d A is secondary; no acid formed although oxidation occurs; B is tertiary; no reaction with the oxidising agent; C is primary; the final product is an acid.

e See p476.

3 a Sodium pentan-2-oxide; $CH_3CH_2CH_2CH(O^- Na^\oplus)CCH_3$ at room temperature.

b Pentan-2-one; $CH_3CH_2CH_2COCH_3$; reflux.

c Pent-1-ene (or pent-2-ene); $CH_3CH_2CH_2CH=CH_2$; heat.

d 2-Iodopentane; $CH_3CH_2CH_2CHICH_3$; heat.

Self-Assessment Questions – 19.1

1 a Methanal. **b** 3-Methylbutanone.

c 2-Oxopropan-1-ol.

d 3-Chlorobutanal.

2 a

b

c

d

3 a Alcohol.   **b** Aldehyde.

c Ketone.   **d** Ketone.

Self-Assessment Questions – 19.2

1 a butan-2-ol ⟶ butanone

b 2-methylpropan-1-ol ⟶ 2-methylpropanal

c methanol ⟶ methanal

d cyclohexanol ⟶ cyclohexanone

2 a Secondary; no. **b** Primary; yes.

c Tertiary; no. **d** Primary; yes.

3 See p476.

Self-Assessment Questions – 19.3

1 a $CH_3CHO + 2[H] \longrightarrow CH_3CH_2OH$

b $CH_3CH_2COCH_3 + 2[H] \longrightarrow$
$$CH_3CH_2CH(OH)CH_3$$

---

c $CH_3COCH_2CHO + 4[H] \longrightarrow$
$$CH_3CH(OH)CH_2CH_2OH$$

2 a A ⟶ CH_3CH_2COOH;
B ⟶ CH_3COCH_3;
C ⟶ no reaction

b A ⟶ $CH_3CH_2CH_2OH$;
B ⟶ no reaction;
C ⟶ $(CH_3)_2CHOH$

3 a Ketone, $>C=O$ and aldehyde $-CHO$ groups; aldehyde can be oxidised further.

b $NaBH_4(aq)$, or $LiAlH_4$ in ethoxyethane

c

d Four different atoms or groups of atoms attached to the same carbon atom.

e $CH_3COCHO + [O] \longrightarrow CH_3COCOOH$

Self-Assessment Questions – 19.4

1 A is $CH_2=CH_2$; B is CH_3CHO; C is CH_3COOH; D is $CH_3COO^- Na^\oplus$.

2 a A is pentan-2-one;
B is 4-hydroxypentan-2-one;
C is propanal;
D is penta-2,4-dione

b B   **c** C   **d** A, B, D

e C   **f** A, B, D

3

Self-Assessment Questions – 19.5

1

2 a Alkene and aldehyde.

b

c

d

---

e

cis   *trans*

3 Cyclohexanol

Self-Assessment Questions – 19.6

1 a and b

c

d

2 a $HOCH_2COCH_2OH$

b Two sausage shaped regions of high electron density above and below the sigma bond.

c Sodium tetrahydridoborate(III) in water.

d $HOCH_2COCH_2OH(l) + 2Na(s) \longrightarrow$
$$Na^\oplus {}^- OCH_2COCH_2O^- Na^\oplus(s) + H_2(g)$$

e In a condensation reaction to form a $>C=N-$ bond.

Self-Assessment Questions – 19.7

1 a $C = 79.25/12.0 = 6.60$; $H = 5.66$;
$O = 15.09/16.00 = 0.94$; C_7H_6O

b 0.25 g forms 56.6 cm³ of gas;
$\therefore M_r = 24000/56.6 \times 0.25 = 106$;
$C_7H_6O = 106$; $\therefore M_r = 106$

c $-OH$ **d** $-CHO$ or $>C=O$

e See p495.   **f** $-CHO$

g 179 °C

Self-Assessment questions – 20.1

1 a
$$C(s) + \tfrac{1}{2}O_2(g) \xrightarrow{\Delta H} CO(g)$$
-394 $+\tfrac{1}{2}O_2$ -283
$$CO_2(g)$$
$\Delta H = +(-394) - (-283) = -111$ kJ mol⁻¹

b When carbon burns it produces carbon monoxide and carbon dioxide.

2 a See p277.

b Hexane cannot be formed directly from $C(s)$ and $H_2(g)$, but hexane, carbon and hydrogen all burn to form CO_2 and H_2O.

c $6C(s) + 7H_2(g) \xrightarrow{\Delta H} C_6H_{14}(l)$

$+9\tfrac{1}{2}O_2$ -286×7 $+9\tfrac{1}{2}O_2$
-394×6 -4163
$$6CO_2(g) + 7H_2O(l)$$
$\Delta H = +(-394 \times 6) + (-286 \times 7)$
$- (-4163) = -203$ kJ mol⁻¹

3 a $C_2H_5OH(l) + 3O_2(g) \longrightarrow$
$$2CO_2(g) + 3H_2O(l)$$

b $C_6H_{12}O_6(s) + 6O_2(g) \longrightarrow$
$$6CO_2(g) + 6H_2O(l)$$

c $C_6H_{12}O_6(s) \xrightarrow{\Delta H} 2C_2H_5OH(aq) + 2CO_2(g)$

$+9O_2$ $\qquad$ $+9O_2$
-2803 $\qquad$ -1367×2

$$6CO_2(g) + 6H_2O(l)$$

d $\Delta H = +(-2803) - (-1367 \times 2)$
$= -69 \text{ kJ mol}^{-1}$

4 a $2H_2S(g) + 3O_2(g) \longrightarrow 2H_2O(l) + 2SO_2(g)$

b $H_2(g) + S(s) \longrightarrow H_2S(g)$;
$H_2(g) + \frac{1}{2}O_2(g) \longrightarrow H_2O(l)$;
$S(s) + O_2(g) \longrightarrow SO_2(g)$

c $H_2S(g) + 1.5O_2(g) \xrightarrow{\Delta H} H_2O(l) + SO_2(g)$

-21 $\qquad$ -286 $\qquad$ -297

$$H_2(g) + S(s) + 1.5O_2(g)$$

d $\Delta H = -(-21) + (-286) + (-297)$
$= -562 \text{ kJ mol}^{-1}$

Self-Assessment Questions – 20.2

1 a Decrease. **b** Increase.
c Decrease. **d** Decrease.

2 a Gases have a high molar entropy value; solids have a low molar entropy value.
b The low molar entropy of ice tells us the structure is very organised; in water liquid there is less organisation to the structure; the much higher value for steam shows that there is no overall structure.

Self-Assessment Questions – 20.3

1 The activation energy for the formation of $SiO_2(s)$ is much higher than for $CO_2(g)$ and so the reaction is much slower in the case of Si.

2 a 5

b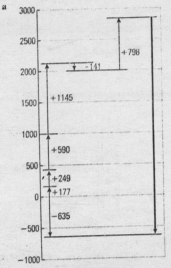

c Trigonal pyramid; 4 pairs of electrons around the P atom; one pair of electrons is a non-bonding pair, with 3 bonding pairs arranged in a trigonal pyramid structure.
d The structure of the element at 25 °C and 1 atm pressure.
e $N_2(g) + 1\frac{1}{2}Cl_2(g) \longrightarrow NCl_3(g)$.
f $+230 \text{ kJ mol}^{-1}$
g It is likely to decompose into the elements.

Self-Assessment Questions – 20.4

1 a For NaCl, enthalpy change for
$Na^{\oplus}(g) + Cl^{\ominus}(g) \longrightarrow Na^{\oplus}Cl^{\ominus}(s)$
b Enthalpy of atomisation for M;
$M(s) \longrightarrow M(g)$; enthalpy of atomisation of X (assume a solid); $X(s) \longrightarrow X(g)$;
1st ionisation energy of M;
$M(g) \longrightarrow M^{\oplus}(g) + e^{\ominus}$; 1st electron affinity of X; $X(g) + e^{\ominus} \longrightarrow X^{\ominus}$;
enthalpy of formation of MX;
$M(s) + X(s) \longrightarrow MX(s)$

2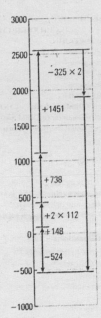

$+(-524) - (+148) - (+112 \times 2) - (+738)$
$- (+1451) - (-325 \times 2) = -2435 \text{ kJ mol}^{-1}$

3 a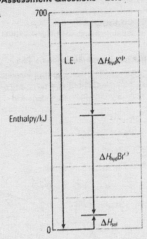

b The removal of an electron from a positive ion requires much more energy.
c Less since Ca has more electron energy levels and the outer electrons are more shielded from the nucleus.
d Energy has to be put in to add an electron to a negative ion.

4 a $CH_4(g) \longrightarrow 4C(g) + 4H(g)$
b $1662/4 = +413 \text{ kJ mol}^{-1}$
c $1308/4 = +327 \text{ kJ mol}^{-1}$
d $(+413 \times 3) + 327 = +1566 \text{ kJ mol}^{-1}$

Self-Assessment Questions – 20.5

1 a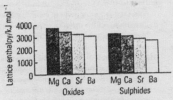

b The lattice enthalpies decrease down the group; in each case it is the decreasing 1st and 2nd ionisation enthalpies that give rise to this trend, since the other values in the cycle are all very similar.
c The sum of the electron affinities for S is less positive than for O, so the resulting lattice enthalpy is lower.

2 a The ionic model is a valid model for $CaBr_2$.
b The ionic model is a poor model for $CuBr_2$.
c The ionic model is a valid model for NaBr.

Self-Assessment Questions – 20.6

1 a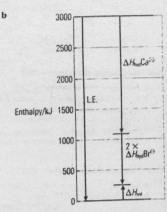

b

2 $\Delta H = -(-3006) + (-1891) + (-460 \times 2)$
$= +195\ \text{kJ mol}^{-1}$

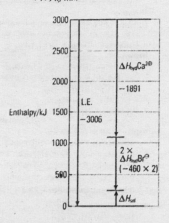

Self-Assessment Questions – 21.2

1 a See Figure 21.2.3, p526, using a gas syringe.
 b See Figure 21.2.4, p526, using a closed flask and monitoring the volume change.
 c See Figure 21.2.5, p527, following the change in colour intensity.
 d Take samples at intervals from the reaction mixture, kept in a thermostatic bath; titrate against standard acid using a suitable indicator; if the reaction is too rapid, quench by adding to known volume of standard acid solution; titrating using standard alkali and an acid/base indicator.
2 a Slower by a factor of $\frac{1}{2}$.
 b Slightly slower.
 c More rapid.
 d More rapid.
3 a See Figure 21.2.4, p526.
 b

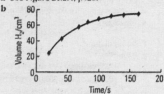

 c $1.0/24.0 = 0.417$ mol.
 d $50 \times 0.12/1000 = 0.0060$ mol.
 e $72\ \text{cm}^3$.
 f Mg is in excess, so HCl is limiting reagent; 0.0060 mol HCl produces 0.0030 mol $H_2 = 0.0030 \times 24\,000 = 72\ \text{cm}^3$.
 g 140 s.

Self-Assessment Questions – 21.3

1 a Rate $= k[A]^2[B]$.
 b Increase by a factor of 4.
 c i 2   ii 3
2 a 3   b 2
 c First with respect to each reactant, second overall.

d Halving the rate.
e Doubling the rate.
3 a $2H_2O_2(aq) \longrightarrow 2H_2O(l) + O_2(g)$
 b See p303.
 c See Figure 21.2.4, p526.
 d Dip the flask into iced water as necessary.
 e Rate depends on $[H_2O_2]$, and since the half life is constant, is first order with respect to $[H_2O_2]$.
 f Rate $= k[H_2O_2]$.
4 a 1   b 2
 c 3   d Rate $= k[H_2][NO]^2$.
 e $4.0\ \text{mol dm}^{-3}\ \text{s}^{-1}$; from the table of results, for $[H_2] = 3.0 \times 10^{-3}\ \text{mol dm}^{-3}$ and $[NO] = 2.0 \times 10^{-3}\ \text{mol dm}^{-3}$ the rate was $1.0\ \text{mol dm}^{-3}\ \text{s}^{-1}$; if $[NO]$ is doubled the rate is increased by $\times 4$ since the rate equation contains $[NO]^2$.
 f

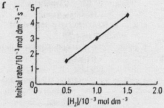

Self-Assessment Questions – 21.4

1 a First.   b $3 \times t_\frac{1}{2} = 15.78\ \text{y}$
 c $2 \times t_\frac{1}{2} = 24.6\ \text{y}$ d $1/2^3 = 1/8 = 0.125$
2 a

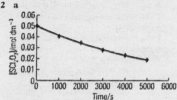

 b

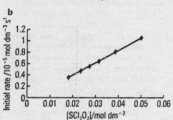

 c First order.   d First.
 e Rate $= k[SCl_2O_2]$.
 f $2.1 \times 10^{-4}\ \text{s}^{-1}$.

Self-Assessment Question – 21.5

1 a Slow step: P + Q ⟶ 'PQ'; fast step: 'PQ' ⟶ R + S (where 'PQ' is an unstable intermediate).
 b Slow step: A ⟶ A*; fast step: A* + B ⟶ D (where A* is an unstable intermediate).
 c Slow step: 2Y ⟶ Y_2^*; fast step: Y_2^* + 2X ⟶ W (where X_2^* is a reactive intermediate).

2 a

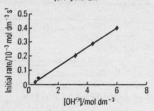

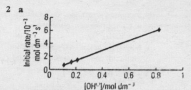

 b First order in halogenoalkane; first order in $[OH^-]$.
 c Rate $= k[RX][OH^-]$; using the first set of results: $0.78 = k \times 0.103 \times 0.100$; $\therefore k = 75.7 \times 10^{-3}\ \text{mol}^{-1}\ \text{dm}^3\ \text{s}^{-1}$.
 d S_N2; see p453.
3 a Free radical substitution; see p399.
 b Electrophilic addition; see p423.
 c Nucleophilic addition; see p488.
4 a First order with respect to ICl; second order with respect to H_2 and third order overall.
 b Rate $= k[ICl][H_2]^2$.
 c

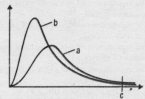

 d $^{69}ICl^{69}$
 e $0.041 = k \times 0.200 \times (0.100)^2$; $\therefore k = 20.5$
 $0.167 = k \times 0.200 \times (0.200)^2$; $\therefore k = 20.9$
 $0.083 = k \times 0.400 \times (0.100)^2$; $\therefore k = 20.75$
 average is $20.7\ \text{mol}^{-2}\ \text{dm}^6\ \text{s}^{-1}$

Self-Assessment Questions – 21.6

1 a, b and c

 d At the higher temperature a greater proportion of the molecules will possess energy equal to or greater than the activation energy.
2 a

 b First order.   c Substitution.
 d S_N1 mechanism; see p450.

Self-Assessment Questions – 22.1

1 a See p325.
 b i Adding acid (increasing H^+ ions) produces more $Cr_2O_7^{2-}(aq)$.

ii Increased concentration of either reagent produces more of $[Fe(SCN)]^{2\oplus}(aq)$.

iii Allowing the gas to disperse means total conversion of solid X to solid Y.

c i More NO_2 is formed.

ii Greater yield of CH_3OH.

2 a On the right hand side (products).

b $K_c = \dfrac{[Sn^{4\oplus}(aq)]_{eqm}[Fe^{2\oplus}(aq)]^2_{eqm}}{[Sn^{2\oplus}(aq)]_{eqm}[Fe^{3\oplus}(aq)]^2_{eqm}}$

c 1.0×10^{-10}

d Nothing.

e $K_c = \dfrac{[Sn^{4\oplus}(aq)]^{\frac{1}{2}}_{eqm}[Fe^{2\oplus}(aq)]_{eqm}}{[Sn^{2\oplus}(aq)]^{\frac{1}{2}}_{eqm}[Fe^{3\oplus}(aq)]_{eqm}}$

f Value for K_c in part e is $\sqrt{K_c}$ in part b.

3 a $CH_3COOH(l) + C_2H_5OH(l)$
$\rightleftharpoons CH_3COOC_2H_5(l) + H_2O(l)$

b See p317.

c $K_c = \dfrac{[CH_3COOC_2H_5(l)]_{eqm}[H_2O(l)]_{eqm}}{[CH_3COOH(l)]_{eqm}[C_2H_5OH(l)]_{eqm}}$

d $K_c = \dfrac{(0.18) \times (0.074)}{(0.090) \times (0.037)} = 4.0$.

e 0.010 mol of acid present at equilibrium; $\therefore (0.20 - 0.10) = 0.10$ mol acid was used up; at equilibrium $(0.050 - 0.010) = 0.040$ mol alcohol present in equilibrium mixture; 0.010 mol of ester and 0.010 mol water also present at equilibrium.

f K_c has no units, and the volumes used to convert number of moles into a concentration term will all cancel out in the equilibrium expression.

Self-Assessment Questions – 22.2

1 a Less product formed.
b No change.
c Less product formed.
d More product formed.

2 a No change.
b The equilibrium moves to the right (more product formed).
c The equilibrium moves to the left (more reactants formed).
d The equilibrium moves to the right (more product formed).
e The equilibrium moves to the right (more product formed).

3 a $K_p = pCH_3OH/(pCO \times (pH_2)^2)$
b The number of moles of gaseous products is less than the number of moles of gaseous reactants; hence, applying Le Chatelier's Principle, high pressure will move the equilibrium to the side with the smaller number of gaseous molecules.
c At a lower temperature, although the yield is higher, the rate is slower, so a compromise temperature is used.

4 a 15 kPa H_2, 85 kPa N_2
b $pNH_3 = 120 \times 24/100 = 28.8$ kPa; $pH_2 = pN_2 = 120 \times 38/100 = 45.6$ kPa
c 70% N_2O_4 is unreacted, and each mole of N_2O_4 produces 2 moles NO_2; 30 moles N_2O_4 produces 60 moles NO_2; ratio N_2O_4:NO_2 will be 70:60; $\therefore pN_2O_4 = 12 \times 70/130 = 6.46$ atm; $pNO_2 = 12 \times 60/130 = 5.54$ atm.

Self-Assessment Questions – 22.3

1 a $K_p = pCO \times (pH_2)^3/pCH_4 \times pH_2O$
b i No effect.
ii No effect.
iii Less time.
c Less product is formed because there are more moles of product in the reaction.
d More product is formed because the reaction is endothermic.

2 a $PCl_5(g) \rightleftharpoons PCl_3(g) + Cl_2(g)$
b $K_c = [PCl_3]_{eqm}[Cl_2]_{eqm}/[PCl_5]_{eqm}$
$K_p = pPCl_3 \times pCl_2/pPCl_5$
c $20.8/208.5 = 0.100$
d $2.84/71 = 0.0400$
e $[Cl_2] = [PCl_3] = 0.0400/10$
$= 0.00400$ mol dm^{-3}
f 0.040 mol has dissociated; $[PCl_5] = (0.100 - 0.0400)/10$
$= 0.00600$ mol dm^{-3}
g $K_c = (0.00400 \times 0.00400)/0.00600$
$= 0.0027$ mol dm^{-3}
h $pPCl_5 = 300 \times 0.0600/(0.0600 + 0.0400 + 0.0400)$
$= 129$ kPa
i $pCl_2 = pPCl_3 = 300 \times 0.0400/(0.0600 + 0.0400 + 0.0400)$
$= 85.7$ kPa
j $K_p = (85.7 \times 85.7)/129 = 56.9$ kPa

Self-Assessment Questions – 22.4

1 a $K_{sp} = [Hg^{2\oplus}(aq)][S^{2\ominus}(aq)]$ mol^2 dm^{-6}
b $K_{sp} = [Pb^{2\oplus}(aq)][Cl^{\ominus}(aq)]^2$ mol^3 dm^{-9}
c $K_{sp} = [Cr^{4\oplus}(aq)][OH^{\ominus}(aq)]^4$ mol^5 dm^{-12}

2 a $Mg(OH)_2$
b $K_{sp} = [Mg^{2\oplus}(aq)][OH^{\ominus}(aq)]^2$
c mol^3 dm^{-9}
d $M_r(Mg(OH)_2) = 24 + (2 \times 17) = 58$; solubility is 0.0099 g dm^{-3}; $\therefore [Mg(OH)_2] = 0.0099/58$
$= 0.000171$ mol dm^{-3}.
e $2 \times 0.000171 = 0.000342$ mol dm^{-3}
f $K_{sp} = (1.71 \times 10^{-4}) \times (3.42 \times 10^{-4})^2$
$= 2.00 \times 10^{-11}$ mol^3 dm^{-9}

3 a $K_{sp} = [Ca^{2\oplus}(aq)][SO_4^{2-}(aq)]$; both concentrations are in mol dm^{-3}, neither are raised in power, so units of K_{sp} are (mol dm^{-3} × mol dm^{-3}) = mol^2 dm^{-6}.
b $M_r(CaSO_4) = 40 + 32 + (4 \times 16) = 136$; $K_{sp} = 2.4 \times 10^{-5}$ mol^2 dm$^{-6} = [Ca^{2\oplus}(aq)][SO_4^{2\ominus}(aq)]$; $\therefore [CaSO_4(aq)] = [Ca^{2\oplus}(aq)] = [SO_4^{2\ominus}]$
$= \sqrt{(2.4 \times 10^{-5})} = 4.9 \times 10^{-3}$ mol dm^{-3}; $\therefore$ solubility $= 4.9 \times 10^{-3} \times M_r$
$= 0.67$ g dm^{-3}.

Self-Assessment Questions – 23.2

1 a 1.7   b 0.18   c 3.9   d 0.70
2 a pOH = 0; pH = 14
b pOH = 1.3; pH = 12.7
c pOH = 2.8; pH = 11.2
3 a $K_w = [H^{\oplus}(aq)][OH^{\ominus}(aq)]$
$= 1.0 \times 10^{-14}$ mol^2 dm^{-6} at 25 °C (298 K)
b For any aqueous solution, pH $= -\log_{10}[H^{\oplus}(aq)]$.
c For any aqueous solution, pOH $= -\log_{10}[OH^{\ominus}(aq)]$.

d An acid which has a molecule containing two ionisable (or 'replaceable') hydrogen atoms.
e An acid which is not completely ionised even in very dilute aqueous solution.

4 a $H_2O(l) \rightleftharpoons H^{\oplus}(aq) + OH^{\ominus}(aq)$
b Increase.
c $K_c = [H^{\oplus}(aq)][OH^{\ominus}(aq)]/[H_2O(l)]$; but $[H_2O(l)]$ stays very nearly constant, so $K_w = [H^{\oplus}(aq)][OH^{\ominus}(aq)]$.
d 1.0×10^{-14} mol^2 dm^{-6}

5 a $[HCl(aq)] = [H^{\oplus}(aq)] = 10^{-1.5}$
$= 0.032$ mol dm^{-3}
b pH = 12.0 so pOH = 2.0; $\therefore [KOH(aq)] = 10^{-2} = 0.010$ mol dm^{-3}
c $M_r(KOH) = (39 + 16 + 1) = 56$; $\therefore [KOH(aq)] = 0.010 \times 56 = 0.56$ g dm^{-3}

Self-Assessment Questions – 23.3

1 a For any aqueous solution, pH $= -\log_{10}[H^{\oplus}(aq)]$.
b $[H^{\oplus}(aq)] = 0.02$ mol dm^{-3}, so pH = 1.7
c $[OH^{\ominus}(aq)] = 0.01 = 10^{-2}$; pOH = 2, so pH = 12
d See p579; examples: ethanoic acid is a weak acid; hydrochloric acid is a strong acid; aqueous ammonia is a weak base; aqueous sodium hydroxide is a strong base.
e $K_a = [H^{\oplus}(aq)]^2/0.01$; $\therefore [H^{\oplus}(aq)] = \sqrt{(0.01 \times 6.3 \times 10^{-5})}$
$= 7.9 \times 10^{-4}$; $\therefore$ pH = 3.1.

2 a $K_w = [H^{\oplus}(aq)][OH^{\ominus}(aq)] = 1.0 \times 10^{-14}$ at 25 °C; the ionic product of water.
b The dissociation of water is endothermic, therefore K_w increases with increasing temperature.
c pH $= -\log_{10}[H^{\oplus}(aq)]$; pOH $= -\log_{10}[OH^{\ominus}(aq)]$; pH + pOH = 14.0 = p$K_w$.
d $M_r(NaOH) = 40$; $6.00/40 = 0.15$ mol
e $NaOH(aq) + HX(aq)$
$\longrightarrow H_2O(l) + NaX(aq)$
f 0.15 mol dm^{-3}
g 0.82

3 a $K_a = \dfrac{[H_3O^{\oplus}(aq)]_{eqm}[C_5H_7O_5COO^{\ominus}(aq)]_{eqm}}{[C_5H_7O_5COOH(aq)]_{eqm}[H_2O(l)]_{eqm}}$
b pKa $= -\log_{10}(7.4 \times 10^{-4}) = 3.1$
c $K_a = [H^{\oplus}(aq)]^2/0.18$; $\therefore [H^{\oplus}(aq)] = \sqrt{(0.18 \times 7.4 \times 10^{-4})}$
$= 1.2 \times 10^{-2}$; $\therefore$ pH = 1.9
d The concentration of citric acid stays nearly constant.
e $C_5H_7O_5COOH(aq) + H_2O(l) \rightleftharpoons$
acid A   base B
$H_3O^{\oplus}(aq)$ + $C_5H_7O_5COO^{\ominus}(aq)$
conjugate acid B   conjugate base A
f Citric acid (pK_a = 3.1) > benzoic acid (pK_a = 4.2) > ethanoic (pK_a = 4.6).

Self-Assessment Questions – 23.4

1 a See Figure 23.4.3, p589.
b p$K_a = -\log_{10}(6.3 \times 10^{-5}) = 4.2$

c Phenolphthalein; the graph shows that a rapid change of pH in titration occurs above pH7, phenolphthalein will change colour when one drop NaOH(aq) is added.

d Colourless to red.

2 a $pH = -\log_{10}[H^+(aq)] = -\log_{10}0.050$
$= 1.3$

b $pOH = -\log_{10}0.010 = 2.0$;
$\therefore pH = 12.0$, since $pH + pOH = 14.0$

c $10.0\ cm^3$ of $0.050\ mol\ dm^{-3}$ HCl is exactly neutralised by $50.0\ cm^3$ $0.010\ mol\ dm^{-3}$ KOH $\therefore$ the solution will be neutral at pH7 since both the acid and the alkali are strong.

d There will be $50\ cm^3$ $0.010\ mol\ dm^{-3}$ excess KOH in $(10.0 + 100) = 110\ cm^3$ solution; $\therefore[OH^-(aq)] = 0.010 \times 50/110$
$= 0.00455\ mol\ dm^{-3}$;
$\therefore pOH = -\log_{10}0.00455 = 2.3$;
$\therefore pH = 11.7$.

e See *Figure 23.4.2*, p588.

f Methyl orange.

Self-Assessment Questions – 23.5

1 a 7

b The ammonium ions remove $OH^-(aq)$ from water equilibrium; a new equilibrium is established with excess $H^+(aq)$;
$\therefore pH < 7$.

c Benzoate ions remove $H^+(aq)$ from the water equilibrium; a new equilibrium is established with excess $OH^-(aq)$;
$\therefore pH > 7$.

d 7

2 a $CH_3COOH(aq) + NaOH(aq) \longrightarrow$
$CH_3COO^-Na^+(aq) + H_2O(l)$

b The salt formed is $CH_3COO^-Na^+$; this will be fully dissociated; however CH_3COO^- is a strong conjugate base and removes $H^+(aq)$ from the water equilibrium and more water ionises; so that OH^- is in excess; hence pH > 7.

Self-Assessment Questions – 23.6

1 $K_a = [H^+(aq)][X^-(aq)]/[HX(aq)]$;
concentration of HX and NaX is
$0.10 \times 10 = 1.0\ mol\ dm^{-3}$;
$[H^+(aq)] = 10^{-4.8} = 1.6 \times 10^{-5}\ mol\ dm^{-3}$;
$[X^-(aq)] = 1.0\ mol\ dm^{-3}$ since the sodium salt is completely dissociated and provides nearly all the anion concentration;
undissociated acid $[HX(aq)] = 1.0\ mol\ dm^{-3}$ assuming it to be a weak acid;
$\therefore K_a = (1.6 \times 10^{-5}) \times 1.0/1.0$
$= 1.6 \times 10^{-5}\ mol\ dm^{-3}$.

2 a $pK_a = -\log_{10}(1.8 \times 10^{-4}) = 3.7$;
$pH = pK_a - \log_{10}([HA]/[A^-])$
$= 3.74 - \log_{10}(1.0/1.0) = 3.74$.

b $pH = pK_a - \log_{10}([HA]/[A^-])$; so use $[methanoic\ acid] = 10 \times [methanoate\ ion]$; e.g. $1.0\ mol\ dm^{-3}$ methanoic acid and $0.10\ mol\ dm^{-3}$ sodium methanoate.

3 a A solution made up to have a particular pH, which maintains that pH closely when small amounts of alkali or acid are added.

b The added HCl provides protons; butanoate ions react with the added protons;
$C_3H_7COO^-(aq) + H^+(aq) \rightleftharpoons$
$C_3H_7COOH(aq)$,
removing them from solution as undissociated acid molecules.

c Blood and cellular fluids need constant pH for enzymes to function properly.

d $K_a = [H^+(aq)][A^-(aq)]/[HA(aq)]$;
$[H^+(aq)] = 1.5 \times 10^{-5} \times 0.40/0.10$
$= 6.0 \times 10^{-5}$; pH = 4.2.

Self-Assessment Questions – 24.1

1 a +5   **b** +3   **c** +1   **d** +7

2 a Oxidation.   **b** Reduction.

c Oxidation.   **d** Oxidation.

3 a $K(s) \longrightarrow K^+ + e^-$ is oxidation;
$O_2(g) + 4e^- \longrightarrow 2O^{2-}$ is reduction.

b $Ca(s) \longrightarrow Ca^{2+} + 2e^-$ is oxidation;
$Cl_2(g) + 2e^- \longrightarrow 2Cl^-$ is reduction.

c $Al(s) \longrightarrow Al^{3+} + 3e^-$ is oxidation;
$Cl_2(g) + 2e^- \longrightarrow 2Cl^-$ is reduction.

d $Na(s) \longrightarrow Na^+ + e^-$ is oxidation;
$Br_2(l) + 2e^- \longrightarrow 2Br^-$ is reduction.

Self-Assessment questions – 24.3

1 a See p611.

b See p611.

c See p611 and p612.

d $Cu(s) \longrightarrow Cu^{2+}(aq) + 2e^-$;
$Br_2(aq) + 2e^- \longrightarrow 2Br^-(aq)$

e $Cu(s) + Br_2(aq) \longrightarrow Cu^{2+}(aq) + 2Br^-(aq)$;
$Cu^{2+}(aq) + 2Br^-(aq) \longrightarrow Cu(s) + Br_2(aq)$

2 a Mg/Mg^{2+}

b $Ce^{4+}(aq)/Ce^{3+}(aq)$

c See p612.

3 a See p611.

b 298 K, 1 atm, $1.0\ mol\ dm^{-3}$ $Mg^{2+}(aq)$

c $Mg(s) \longrightarrow Mg^{2+}(aq) + 2e^-$

Self-Assessment Questions – 24.4

1 a See p611 and 612 for a diagram; zinc is the negative terminal; electrons flow in the external circuit from zinc to iron.

b $Zn(s)|Zn^{2+}(aq) \| Fe^{3+}(aq),Fe^{2+}(aq)|Pt$;
$E_{cell} = E_{RHS} - E_{LHS} = +0.77 - (-0.76)$
$= +1.53\ V$

c $Zn(s) + 2Fe^{3+}(aq) \longrightarrow$
$Zn^{2+}(aq) + 2Fe^{2+}(aq)$

d To ensure ionic conduction between the half-cells and so complete the circuit.

e Use a strip of filter paper soaked in saturated potassium nitrate solution.

2 a $Cu(s)|Cu^{2+}(aq) \| Al^{3+}(aq)|Al(s)$.
$E_{cell} = E_{RHS} - E_{LHS} = -1.66 - (+0.34)$
$= -2.0\ V$

b $Mg(s)|Mg^{2+}(aq) \| Cl_2(g),2Cl^-(aq)|Pt$;
$E_{cell} = E_{RHS} - E_{LHS} = +1.36 - (-2.37)$
$= +3.73\ V$

c $Pt|4OH^-(aq),O_2(g)$
$+ 2H_2O(l) \| Zn^{2+}(aq)|Zn(s)$;
$E_{cell} = E_{RHS} - E_{LHS} = -0.76 - (-0.40)$
$= -1.16\ V$

Self-Assessment Questions – 24.5

1 a See p620.

b Mg

c From Mg to Fe.

d Mg

2 a $Mg(s)|Mg^{2+}(aq) \| Pb^{2+}(aq)|Pb(s)$;
$E_{cell} = E_{RHS} - E_{LHS} = -0.13 - (-2.37)$
$= +2.24\ V$;
$Mg(s) + Pb^{2+}(aq) \longrightarrow Mg^{2+}(aq) + Pb(s)$

b $Ag(s)|Ag^+(aq) \| Zn^{2+}(aq)|Zn(s)$;
$E_{cell} = -0.76 - (+0.80) = -1.56\ V$;
$Zn(s) + 2Ag^+(aq) \longrightarrow$
$Zn^{2+}(aq) + 2Ag(s)$

c $Zn(s)|Zn^{2+}(aq) \| Mg^{2+}(aq)|Mg(s)$;
$E_{cell} = -2.37 - (-0.76) = -1.61\ V$
$Mg(s) + Zn^{2+}(aq) \longrightarrow Mg^{2+}(aq) + Zn(s)$

d $Pb(s)|Pb^{2+}(aq) \| Cu^{2+}(aq)|Cu(s)$;
$E_{cell} = +0.34 - (-0.13) = +0.47\ V$
$Pb(s) + Cu^{2+}(aq) \longrightarrow Pb^{2+}(aq) + Cu(s)$

Self-Assessment Questions – 24.6

1 a $PbO_2(s)$ will oxidise $Cl^-(aq)$ because $E^{\ominus}$ for $PbO_2(s) \longrightarrow Pb^{2+}(aq)$ is more positive than $E^{\ominus}$ for $Cl_2(g) \longrightarrow Cl^-(aq)$.

b $PbO_2(s) + 4H^+(aq) + 2Cl^-(aq) \longrightarrow$
$Pb^{2+}(aq) + Cl_2(g) + 2H_2O(l)$

c $E_{cell} = +1.36 - (+1.46) = -0.10\ V$

2 a See p620.

b $E_{cell} = E_{RHS} - E_{LHS}$;
$\therefore -0.74 = E_{RHS} - (+0.34)$;
$\therefore E_{RHS} = -0.40\ V$ which is $E^{\ominus}$ for $Cd^{2+}(aq)/Cd(s)$.

c Electrons flow from Cd $\longrightarrow$ Cu in the external circuit.

d Towards the Cu.

e $Cd(s) + Cu^{2+}(aq) \longrightarrow Cd^{2+}(aq) + Cu(s)$

f Because use of a very high resistance voltmeter means virtually no electrons flow.

3 a See p611.

b $Zn^{2+}(aq) + 2e^- \rightleftharpoons Zn(s)$; $E^{\ominus} = -0.76\ V$;
$2H^+(aq) + 2e^- \rightleftharpoons H_2(g)$; $E^{\ominus} = 0.00\ V$;
Zn will displace hydrogen from acids; for Zn and $H^+(aq)$,
$E_{cell} = E_{RHS} - E_{LHS} = 0.00 - (-0.76)$
$= +0.76\ V$;
i.e. the reaction is feasible.

Self-Assessment Questions – 24.7

1 a See p627.   **b** See p626.

2 a $Cd(s)|Cd^{2+}(aq) \| Ni^{2+}(aq)|Ni(s)$;
$E_{cell} = E_{RHS} - E_{LHS} = -0.25 - (-0.40)$
$= +0.15\ V$

b $Cd(s) + NiO_2(s) + 2H_2O(l) \longrightarrow$
$Cd(OH)_2(s) + Ni(OH)_2(s)$

c Before: Cd, 0; Ni, +4; after: Cd, +2, Ni, +2.

d $Cd(OH)_2(s) + Ni(OH)_2(s) \longrightarrow$
$Cd(s) + NiO_2(s) + 2H_2O(l)$

3 See p628.

4 a Iron rusts in contact with tin (provided air can get at the iron) because it corrodes by electron loss in preference to tin as it has a more negative value of $E^{\ominus}$.

b As chromium has a more negative value for $E^{\ominus}$ than tin, it would be expected to react better with $1.0\ mol\ dm^{-3}$ HCl(aq).

With Al the situation is complicated by surface effects, so that although $E^{\ominus} = -1.66$ V for Al, it does not normally react with dilute acids.

5 a Iron and aluminium, yes; silver, no; $E^{\ominus}$ for Ag is more positive than the $E^{\ominus}$ value for O_2/H_2O.

b $Fe(s) | Fe^{2\oplus}(aq)$ ⫶
$$[O_2(g)+2H_2O(l)],4OH^{\ominus}(aq) | Pt$$

c $E_{cell} = E_{RHS} - E_{LHS} = +0.40 - (-0.44)$
$= +0.84$ V

d $O_2(g) + 2H_2O(l) + 2Fe(s) \longrightarrow$
$$4OH^{\ominus}(aq) + 2Fe^{2\oplus}(aq)$$

e Sea-side atmosphere often contains salt spray and is wet; electrolytes help corrosion.

f Aluminium is protected by a very tough, coherent, thin layer of Al_2O_3; although $E^{\ominus}$ is large and negative, the film excludes oxygen and so Al does not corrode. ●

Self-Assessment Questions – 24.8

1 a Yes. **b** No.
c Yes. **d** No.
e No. **f** yes.

2 a Cathode: $Na^{\oplus} + e^{\ominus} \longrightarrow Na(s)$;
anode: $2I^{\ominus} \longrightarrow I_2(s) + 2e^{\ominus}$.

b Cathode: $Cu^{2\oplus} + 2e^{\ominus} \longrightarrow Cu(s)$;
anode: $2Cl^{\ominus} \longrightarrow Cl_2(g) + 2e^{\ominus}$.

c Cathode: $H^{\oplus} + 2e^{\ominus} \longrightarrow H_2(g)$;
anode: $2Br^{\ominus} \longrightarrow Br_2(l) + 2e^{\ominus}$.

d Cathode: $Ca^{2\oplus} + 2e^{\ominus} \longrightarrow Ca(s)$;
anode: $2O^{2\ominus} \longrightarrow O_2(g) + 4e^{\ominus}$.

e Cathode: $Ag^{\oplus} + e^{\ominus} \longrightarrow Ag(s)$;
anode: $4OH^{\ominus} \longrightarrow 2H_2O(l) + O_2(g) + 4e^{\ominus}$

Self-Assessment Questions – 24.9

1 a $Sr = 35.41/87.7 = 0.404$;
$Br = 64.59/80.0 = 0.807$;
$\therefore$ 1:2; formula is $SrBr_2$

b Temperature above 543 °C, $SrBr_2$ must be molten, so that ions can move.

c
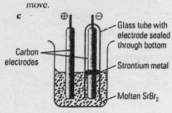
Glass tube with electrode sealed through bottom
Carbon electrodes
Strontium metal
Molten $SrBr_2$

d $Sr^{2\oplus} + 2e^{\ominus} \longrightarrow Sr$

Self-Assessment Questions – 25.1

1 a $[Ar]3d^84s^2$ **b** $[Ar]3d^54s^2$
c $[Ar]3d^{10}4s^1$ **d** $[Kr]4d^55s^2$

2 a See p643.
b $Sc - [Ar]3d^14s^2$, oxidation state +3;
$Zn - [Ar]3d^{10}4s^2$, oxidation state +2.

3 a Too expensive.
b +1
c Sulphur.
d $2AgI(s) \longrightarrow 2Ag(s) + I_2(s)$

Self-Assessment Questions – 25.2

1 The electron structures of Na and K are Na – 2.8.1, K – 2.8.8.1; K has one extra electron energy level; the electron structures for Sc and Cu are Sc – $[Ar]3d^14s^2$, Cu – $[Ar]3d^{10}4s^1$; there is no new electron energy level to accommodate the extra electrons, and the increased nuclear attraction for electrons counteracts increased electron-electron repulsion.

2 a Mn, manganese; Co, cobalt.
b $Fe^{2\oplus}(g) \longrightarrow Fe^{3\oplus}(g) + e^{\ominus}$
c The electron configurations of the three elements are:

		3d					4s
Mn [Ar]	↑	↑	↑	↑	↑		↑↓
Fe [Ar]	↑↓	↑	↑	↑	↑		↑↓
Co [Ar]	↑↓	↑↓	↑	↑	↑		↑↓

The 4s electrons are lost first; loss of 3e⁻ in each case leaves $3d^44s^0$ for Mn, $3d^54s^0$ for Fe and $3d^64s^0$ for Co; the $3d^5$ configuration in Fe (exactly half-filled) has extra stability.

Self-Assessment Questions – 25.3

1 a +7 **b** +6 **c** +6
d +4 **e** +3 **f** 0

2 a Hydrogenation, e.g. of vegetable oils $\longrightarrow$ fats;
$$>C=C< + H_2(g) \xrightarrow{H_2/Ni\ catalyst} >CH-CH<$$

b Contact process for sulphuric acid;
$$2SO_2(g) + O_2(g) \xrightleftharpoons[\ ]{450\ °C/V_2O_5\ catalyst} 2SO_3(g)$$

c Haber process for ammonia;
$$N_2(g) + 3H_2(g) \xrightleftharpoons[\ ]{450\ °C/200\ atm/Fe\ catalyst} 2NH_3(g)$$

3 a $[Ar]3d^64s^2$ **b** $[Ar]3d^5$ **c** $[Ar]3d^2$
d $[Ar]3d^{10}4s^1$ **e** $[Ar]3d^{10}$ **f** $[Ar]3d^9$

Self-Assessment Questions – 25.4

1 a $[Fe(H_2O)_6]^{3\oplus}(aq) \rightleftharpoons$
$$[Fe(H_2O)_5(OH)]^{2\oplus}(aq) + H^{\oplus}(aq)$$
b Loss of protons into solution increases $[H^{\oplus}(aq)]$ – the solution becomes acidic.
c Eventually neutral $[Fe(H_2O)_3(OH)_3]$ is formed; particles no longer repel, and can aggregate and precipitate.

2 With $OH^{\ominus}(aq)$, from NaOH(aq) and $NH_3(aq)$:
$Cu^{2\oplus}(aq) + 2OH^{\ominus}(aq) \longrightarrow Cu(OH)_2(s)$ or
$[Cu(H_2O)_6]^{2\oplus}(aq) + 2OH^{\ominus}(aq) \longrightarrow$
$$[Cu(H_2O)_4(OH)_2](s);$$
with excess $NH_3(aq)$:
$[Cu(H_2O)_4(OH)_2](s) + 4NH_3(aq) \rightleftharpoons$
$[Cu(H_2O)_2(NH_3)_4]^{2\oplus}(aq) + 2H_2O(l) +$
$$2OH^{\ominus}(aq)$$

3 a $K_{stab} = [Cu(CN)_4]^{3\ominus}/[Cu^{\oplus}][CN^{\ominus}(aq)]^4$
$$mol^{-4}\ dm^{12}$$
b $K_{stab} = [Ag(NH_3)_2]^{\oplus}/[Ag^{\oplus}][NH_3]^2\ mol^{-2}\ dm^6$
c $K_{stab} = [CoCl_4]^{2\ominus}/[Co^{2\oplus}][Cl^{\ominus}]^4\ mol^{-4}\ dm^{12}$

4 a A: +2; B: +2; C: +2.
b A – yellow–green spot moved towards the $\oplus$ electrode.
B – the pale blue spot would not have moved.
C – the deep blue spot would have moved towards the $\ominus$ electrode.

Self-Assessment Questions – 25.5

1 a $[Cr(CO)_6]$, $[CuCl_4(H_2O)_2]^{2\ominus}$, $[Fe(CN)_6]^{4\ominus}$
b $[Cu(CN)_4]^{3\ominus}$
c A: 0; B: +1; C: +2; D: +2.

2 a
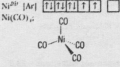

b $Ag^{\oplus}(aq) + Cl^{\ominus}(aq) \longrightarrow AgCl$s)
c $Ag^{\oplus}$ ions are removed as $[Ag(NH_3)_2]^{\oplus}(aq)$; more AgCl dissolves to restore the equilibrium; enough NH_3 leads to all the AgCl dissolving.
d $[Ag(NH_3)_2]^{\oplus}$, linear

3 a

		3d					4s
Ni [Ar]	↑↓	↑↓	↑↓	↑	↑		↑↓

		4d					5s
$Ni^{2\oplus}$ [Ar]	↑↓	↑↓	↑↓	↑	↑		

b $Ni(CO)_4$;
```
        CO
        |
 CO— Ni —CO
        |
        CO
```

c Ni forms weak 'surface' bonds with the reactants, allowing reaction to occur between the reactants; after reaction, the weak bonds are broken and the product leaves the surface of the Ni catalyst.

4 a See p660.
b Electron configurations: $Fe^{2\oplus} - [Ar]3d^64s^0$; $Fe^{3\oplus} - [Ar]3d^54s^0$; there is an extra stability of a half-filled sub level in $Fe^{3\oplus}$.
c Goes brown, probably with a brown precipitate, caused by oxidation of $Fe^{2\oplus}(aq)$ by O_2 in the air.

Self-Assessment Questions – 25.6

1 a $[Fe(H_2O)_6]^{2\oplus}$
b $[Fe(H_2O)_6]^{3\oplus}(aq) \longrightarrow$
$[Fe(H_2O)_5(OH)]^{2\oplus}(aq) + H^{\oplus}(aq)$;
eventually $[Fe(H_2O)_3(OH)_3]$ is formed; since this last complex has no charge to repel the particles, the particles aggregate and form a precipitate.
c EDTA complexes any iron, so that it cannot precipitate as the hydrated hydroxide.

2 a A species (usually molecule or ion) able to donate a non-bonding electron pair to a central metal atom.
b
$$:C:::O:$$
c Complexes to iron in haemoglobin and prevents transport of O_2 by haemoglobin.

Self-Assessment Questions – 25.7

1 a See p629. **b** See p671.
2 a NpF_3
b $NpF_3(s) + 3Li(g) \longrightarrow Np(s) + 3LiF(s)$
c Np is in the f-block and would be expected to contain partially-filled electron sub-levels; this would result in a coloured compound.
d Variable oxidation state; catalytic properties; complex ion formation.

3 a Co, 29.36/58.93 = 0.498;
Cl, 70.64/35.45 = 1.99; ∴ $[CoCl_4]^{2-}$;
Co, 35.28/58.93 = 0.599;
H_2O, 64.72/18.02 = 3.59; ∴ $[Co(H_2O)_6]^{2+}$
b $[CoCl_4]^{2-}$: Co is +2;
$[Co(H_2O)_6]^{2+}$: Co is +2.
c $[CoCl_4]^{2-}$ is tetrahedral;
$[Co(H_2O)_6]^{2+}$ is octahedral.
d Coordination number in $[CoCl_4]^{2-}$ is 4;
in $[Co(H_2O)_6]^{2+}$ is 6.
e $[Co(H_2O)_6]^{2+}$(aq) + $4Cl^-$(aq) ⟶
$[CoCl_4]^{2-}$(aq) + $6H_2O$(l)

Self-Assessment Questions – 26.1

1 a

b

c

d

2 See p682.
3 a C, 0.432/12.0 = 0.036; H, 0.024; C_3H_2;
b C_3H_2 = 38.0; 228/38.0 = 6; molecular
formula is $C_{18}H_{12}$
c

d 9
e The alternate double and single bonds are termed conjugated.

Self-Assessment Questions – 26.2

1 a Alkane. **b** Arene. **c** Alkene.
d Alkane. **e** Cycloalkane.
2 a See p682.
b Because the carbon–carbon bond lengths in benzene are intermediate between those in alkanes and in alkenes.
c Enthalpy change of hydrogenation of benzene is ≈150 kJ mol^{-1} less negative than if it contained three alkene C=C bonds; hence benzene is ≈150 kJ mol^{-1} more stable than if it had alternate C—C and C=C bonds.

Self-Assessment Questions – 26.3

1 a A, B, C. **b** A, B.
c A, B, D. **d** A, B.
e A, B. **f** D
2 (5 × C—C)+(14 × C—H) – (3 × C—C)
– (3 × C=C) – (6 × C—H) – (4 × H—H)
= (5 × 347) + (14 × 413) – (3 × 347)
– (3 × 612) – (6 × 413) – (4 × 436) = 1735
+ 5782 – 1041 – 1836 – 2478 – 1744
= +418 kJ mol^{-1}
3 a Room temperature in the dark.
b (6 × 413) + (3 × 347) + (3 × 612)
+ (3 × 243) – (6 × 413) – (6 × 347)
– (6 × 346) = – 552 kJ mol^{-1}
c Because delocalised benzene is 150 kJ mol^{-1} more stable than the six membered ring with 3 C=C bonds would be (here actually 552 – 399 = 153 kJ mol^{-1}).

Self-Assessment Questions – 26.4

1 a 1-Bromo-3-methylbenzene.
b 1,3,5-Trimethylbenzene.
c 2,4-Dimethylphenol.
d 1-Phenylprop-1-ene.
2 a

b

c

d

3 a One;

bromobenzene;
b Three;

2-Chloromethylbenzene

3-Chloromethylbenzene

4-Chloromethylbenzene
c Three;

1,2-Dimethylbenzene   1,3-Dimethylbenzene

1,4-Dimethylbenzene

4 a

b

2,4,6-Trinitrophenol

2,4,6-Trinitromethylbenzene

1,3,5-Trinitrobenzene
c $C_6H_5CH_3 + 3NO_2^+$ ⟶
$(NO_2)_3C_6H_2CH_3 + 3H^+$

5 a B
b B
c A, B, C

Answers

Self-Assessment Questions – 26.5

1 a

π bond

σ bonds
delocalised
π bonds

b $CH_3CH_2CH_3$ – bond angle 109.5°; $CH_3CH=CH_2$ – 120°; C_6H_6 – 120°.

c Propene reacts with bromine vapour or liquid at room temperature; benzene reacts with bromine liquid at room temperature using an iron catalyst.

2 a $C_6H_5CH_3(l) + H_2SO_4(l) \longrightarrow$
$\qquad CH_3C_6H_4SO_3H + H_2O(l)$

b $C_6H_5CH_3(l) + CH_3Cl(g) \xrightarrow{AlCl_3}$
$\qquad (CH_3)_2C_6H_4(l) + HCl(g)$

c $C_6H_5CH_3(l) + NO_2^{\oplus} \longrightarrow$
$\qquad CH_3C_6H_4NO_2(l) + H^{\oplus}$

d $C_6H_5CH_3(l) + Br_2(l) \xrightarrow{AlCl_3}$
$\qquad CH_3C_6H_4Br(l) + HBr(g)$

3 A is an arene, e.g.

B is a cycloalkene, e.g.

C is a cycloalkane, e.g.

4 a $CH_3CHClCH_3(l) + C_6H_6(l) \longrightarrow$
$\qquad (CH_3)_2CHC_6H_5(l) + HCl(g)$

b $AlCl_3$

c $C_2H_4(g) + C_6H_6(l) \xrightarrow{AlCl_3/H^{\oplus}} C_2H_5C_6H_5(l)$

d $CH_3(CH_2)_{11}CH_2Cl$ and benzene, C_6H_6, with an aluminium chloride catalyst.

Self-Assessment Questions – 26.6

1 Substitution reactions are common with benzene because of the great stability of the benzene ring; addition reactions are typical of alkenes as a π bond is broken and two stronger σ bonds are formed; with benzene the reaction is electrophilic substitution and a catalyst (Fe or $AlCl_3$) is needed; with cyclohexene, electrophilic addition results when the gases are mixed at room temperature in the dark.

2 a A mixture of concentrated sulphuric and nitric acids at room temperature.

b $Sn(s) + 2HCl(aq) \longrightarrow SnCl_2(aq) + H_2(g)$

c $-NO_2 + 6[H] \longrightarrow -NH_2 + 2H_2O(l)$

d $M_r(C_6H_5CH_3) = 92$;
$M_r((NH_2)C_6H_4COOC_2H_5) = 165$;
theoretical yield = $5.0 \times 165/92 = 8.97$ g;
% yield = $5.0/8.97 = 56\%$.

3 With bromine in uv light – a free radical substitution in the methyl group; see p399; with bromine and $AlCl_3$ in the dark – electrophilic substitution on the benzene ring occurs, see p695.

Self-Assessment Questions – 26.7

1 a 4-chloro-3,5-dimethylphenol

b 0.0034 mol dm^{-3},
$0.0034 \times 156.5 = 0.53$ g dm^{-3}

c The —OH in a phenol can act as a weak acid but a Cl group attached to the benzene group is very unreactive.

2 a 2,5-Dimethylphenol.

b A – phenol; B – alcohol.

c A would be sulphonated:
$(CH_3)_2C_6H_4OH + H_2SO_4 \longrightarrow$
$\qquad (CH_3)_2C_6H_2(SO_3H)OH + H_2O$
B would give an alkene resulting from dehydration:
$C_6H_5CH_2CH_2OH \longrightarrow$
$\qquad C_6H_5CH=CH_2 + H_2O$

d A and B would react to produce $H_2(g)$;
for A, $(CH_3)_2C_6H_3OH(s) + Na(s) \longrightarrow$
$\qquad (CH_3)_2C_6H_3O^{\ominus}Na^{\oplus}(s) + \frac{1}{2}H_2(g)$
for B, $C_6H_5CH_2CH_2OH(l) + Na(s) \longrightarrow$
$\qquad C_6H_5CH_2CH_2O^{\ominus}Na^{\oplus}(s) + \frac{1}{2}H_2(g)$

e A reacts as a weak acid in aqueous solution at room temperature to form $(CH_3)_2C_6H_3ONa$. B would not react with aqueous sodium hydroxide solution.

3 a Optical.

b —OH attached to a benzene ring is a phenol, and has some acidic properties; the —OH in the side chain is an alcohol

c $1\frac{1}{2}$

d $\frac{1}{2}$ – from just the phenolic —OH groups.

Self-Assessment Questions – 26.8

1 $AlCl_3$, $FeBr_3$, BCl_3.

2 a Nitrobenzene, $C_6H_5NO_2$

b Methylbenzene, $C_6H_5CH_3$

c Chlorobenzene, C_6H_5Cl

3 a See p682.

b See p368.

c See p704 for electrophilic substitution.

4 a 206

b

CH_3CHOOH

c Using the Friedel–Crafts reaction, reacting 2-phenylpropanoic acid with 1-chloro-2-methylpropane using $AlCl_3$ catalyst.

Self-Assessment Questions – 26.9

1 a

b

c

2 a Electrophilic addition.

b Electrophilic substitution.

c Electrophilic substitution.

d Free radical substitution.

3 a A – benzene; B – propene.

b Electrophilic substitution.

c C – phenol; D - propanone

d

Self-Assessment Questions – 26.10

A is $CH_3C_6H_4C_2H_5$; B is $CH_3C_6H_4NO_2$;
C is C_6H_5COOH;
D is $CH_3C_6H_4Cl$.

Self-Assessment Questions – 27.1

1 a Yes.　**b** No.
c No　**d** Yes.

2 a

b

c

d

3 a Pentanoic acid.

b Butanoic acid.

c 3-methylbutanoic acid.

d 5-chloropentanoic acid

Self-Assessment Questions – 27.2

a Reduction. **b** Ionisation.
c Esterification. **d** Substitution.

Self-Assessment Questions – 27.3

1 a No. **b** No.
c Yes. **d** Yes.

2

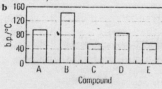

3 a A – primary alcohol; B – carboxylic acid;
C – ketone; D – secondary alcohol;
E – aldehyde.

b

```
160
120
 80
 40
  0
    A   B   C   D   E
       Compound
```
b.p./°C

c Hydrogen bonding in propan-1-ol,
propan-2-ol and propanoic acid; hydrogen
bonding is not possible in propanone and
propanal.

Self-Assessment Questions – 27.4

1 a b and c. By oxidation of the suitable alcohol
using $Cr_2O_7^{2-}$ in acid solution and refluxing; by
refluxing the suitable nitrile or amide with
HCl(aq) or with NaOH(aq) followed by
acidification / starting with a suitable
halogenoalkane and forming the Grignard
reagent, stirring with solid CO_2 and acidifying.

2 a Reflux with $K_2Cr_2O_7$ in acid solution.
b Reflux with $K_2Cr_2O_7$ in acid solution.
c Reflux with HCl(aq) or reflux with
NaOH(aq) and acidify.
d Reflux with HCl(aq) or reflux with
NaOH(aq) and acidify.

3 a Reflux with alkali to give primary alcohol,
then acidify with $Cr_2O_7^{2-}$ in acid solution.
b React benzene with CH_3Cl and $AlCl_3$ catalyst
to form methylbenzene; this is then
oxidised using, for example, concentrated
nitric acid to form benzoic acid.
c Reflux with KCN in aqueous ethanol to
form the nitrile, then reflux with HCl(aq).

Self-Assessment Questions – 27.5

a $C_2H_5OH(aq) + [O] \longrightarrow CH_3CHO + [O]$
$+ H_2O \longrightarrow CH_3COOH$
b $2CH_3COOH(aq) + Na_2CO_3(aq) \longrightarrow$
$2CH_3COO^-Na^+(aq) + CO_2(g) + H_2O(l)$
c $CH_3CH_2CHO(aq) + [O] \longrightarrow CH_3CH_2COOH$
d $CH_3CH(CH_3)COOH(aq) + KOH(aq) \longrightarrow$
$CH_3CH(CH_3)COO^-K^+(aq) + H_2O(l)$
e $2CH_3CH_2COOH(aq) + Mg(s) \longrightarrow$
$(CH_3CH_2COO^-)_2Mg^{2+} + H_2(g)$

Self-Assessment Questions – 27.6

1 a $[H^+] = \sqrt{(0.025 \times 1.7 \times 10^{-5})}$
$= 6.5 \times 10^{-4}$; pH = 3.2
b $[H^+] = \sqrt{(0.040 \times 7.9 \times 10^{-5})}$
$= 1.8 \times 10^{-3}$; pH = 2.8

c $[H^+] = \sqrt{(0.50 \times 1.7 \times 10^{-4})}$
$= 9.2 \times 10^{-3}$; pH = 2.0
d $[H^+] = \sqrt{(0.20 \times 6.3 \times 10^{-5})}$
$= 3.5 \times 10^{-3}$; pH = 2.4

2 a

```
      H  H  H
      |  |  |       O
 H----C--C--C--C
      |  |  |       \
      H  H  H        O----H
```
Butan-1-ol

```
      H  H  H
      |  |  |       O
 H----C--C--C--C
      |  |  |       \
      H  H  H        O----H
```
Butanoic acid

b

```
      H  H     H
      |  |     |
 H----C--C--C--C--O
      |  |  |  |
      H  H  O  H
            |
            H
```
Butan-2-ol (2 optical isomers)

```
      H  H  H
      |  |  |
 H----C--C--C--O----H
      |  |  |
      H  C  H
         |
         H
```
2-Methylpropan-1-ol

```
      H     H
      |  O  |  H
      |  |  |  |
 H----C--C--C--H
      |  |  |
      H  H  H
```
2-Methylpropan-2-ol

c Oxidation of butan-2-ol produces a
ketone; 2-methylpropan-1-ol forms 2-
methylpropanoic acid, the oxidation of 2-
methylpropan-2-ol is very difficult and
even would result in products other
than butanoic acid.

d $CH_3CH_2CH_2CH_2OH(l) + 2[O] \longrightarrow$
$CH_3CH_2CH_2COOH(aq) + H_2O(l)$
e $CH_3CH_2CH_2COOH(aq) \rightleftharpoons$
$CH_3CH_2CH_2COO^-(aq) + H^+(aq)$
$K_a = \dfrac{[CH_3CH_2CH_2COO^-(aq)]_{eqm}[H^+(aq)]_{eqm}}{}$

f $[H^+] = \sqrt{(0.15 \times 1.5 \times 10^{-5})}$
$= 1.5 \times 10^{-3}$; pH = 2.8

Self-Assessment Questions – 27.8

1 a $2C_2H_5OH(l) + 2Na(s) \xrightarrow{RT}$
$2C_2H_5O^-Na^+(s) + H_2(g)$
$2CH_3COOH(s) + 2Na(s) \longrightarrow$
$2CH_3COO^-Na^+(s) + H_2(g)$
b $C_2H_5OH(aq)$ – no reaction under normal
conditions.
$CH_3COOH(aq) + NaOH(aq) \longrightarrow$
$CH_3COO^-Na^+(aq) + H_2O(l)$
c C_2H_5OH – no reaction
$CH_3COOH(l) + 4[H] \xrightarrow[\text{in ethoxyethane}]{RT}$
$CH_3CH_2OH(l) + H_2O(l)$

2 A – $LiAlH_4$ in ethoxyethane; B – CH_3CHO;
C – reflux; D – CH_3COOH; E – H_2; F – CO_2

3 a C:40/12 = 3.33; H:6.7/1 = 6.7;
O:53/16 = 3.31; ratio = 1:2:1;
empirical formula is CH_2O.
b 90
c Optical isomerism

```
      CH_3                CH_3
      |                   |
 H----C----COOH    HOOC---C----H
      |                   |
      OH                  HO
```

d Alcohol.
e Yes, the lower pK_a means a stronger acid;
the electron-withdrawing —OH group on
the C atom next to the —COOH group
assists ionisation.
f 9.0/90 = 0.10 mol; $[H^+]$
$= \sqrt{(0.10 \times 1.35 \times 10^{-4})}$
$= 3.67 \times 10^{-3}$;
∴ pH = 2.4.
g $CH_3CH(OH)COOH(s) + SCl_2O(l) \longrightarrow$
$CH_3CH(Cl)COCl(l) + SO_2(g) + 2HCl(g)$
h Propane-1,2-diol

```
         H
         |
      O  H  H
      |  |  |
 H----C--C--C--O----H
      |  |  |
      H  H  H
```

Self-Assessment Questions – 27.9

1 a (Secondary) alcohol. **b** Ester.
c Ether. **d** Ester.

2 a

```
      H  H  H
      |  |  |       O       H
 H----C--C--C--C          |
      |  |  |       \      |
      H  H  H        O----C--H
                          |
                          H
```

b

```
         H
         |
         C--H
         |
      H  H        O     H  H  H
      |  |        |     |  |  |
 H----C--C--------C-----O--C--C--C--H
      |  |                 |  |  |
      H  H                 H  H  H
```

c

```
      H
      |
 H----C----C
      |      \  O
      H       \
               O----C--C----H
               |    |
```

d

```
      H  H
      |  |       O
 H----C--C-------C
      |  |        \
      H  C         O----C--C----H
         |         |    |
         H         H    H
                   |
                   C--H
                   |
                   H
```

3 a Butanoic acid.
b Heptanoic acid.
c Ethyl-2,2-dimethylpropanoate.
d Methylpropanoate.

4 a $CH_3CH_2COOH(l) + PCl_5(s) \xrightarrow{RT}$
$CH_3CH_2COCl(l) + PCl_3O(l) + HCl(g)$
b $CH_3CH_2CH_2COOH(l) + 4[H] \xrightarrow[\text{in ethoxyethane}]{RT}$
$CH_3CH_2CH_2CH_2OH(l) + H_2O(l)$

c $CH_3CH_2CH(CH_3)CHO(aq) + [O] \xrightarrow{reflux}$
$CH_3CH_2CH(CH_3)COOH(aq)$

d $CH_3CH(CH_3)COOH(l) + C_2H_5OH(l) \xrightarrow[with\ H_2SO_4]{reflux}$
$CH_3CH(CH_3)COOC_2H_5(l) + H_2O(l)$

e $CH_3CH_2CH_2COOH(l) + NaOH(aq) \xrightarrow{RT}$
$CH_3CH_2CH_2COO^-Na^{\oplus}(aq) + H_2O(l)$

Self-Assessment Questions – 27.10

1 A – aqueous potassium hydroxide; B – reflux.
2 C – HOOCCH$_2$CH$_2$CH$_3$, butanoic acid.
3 D – CH$_3$COOCH$_2$C$_6$H$_5$, phenylethyl ethanoate.
4 E – dilute aqueous HCl; F – reflux.
5 G – (CH$_3$)$_2$CHCH$_2$COCl; H – room temperature.
6 I – CH$_3$CH(CH$_3$)COOH, 2-methylpropanoic acid; J – ethoxyethane at room temperature.
7 K and L – CH$_3$COOH, ethanoic acid and C$_2$H$_5$OH, ethanol.

Self-Assessment Questions – 27.11

a

b Propan-2-ol;
or

c $CH_3CH_2CH=C(CH_3)COOH$
$+ HOCH(CH_3)CH_3 \longrightarrow$
$CH_3CH_2CH=C(CH_3)COOCH(CH_3)CH_3$
$+ H_2O$

d $K_c = \dfrac{[ester]}{[acid][alcohol]}$
$= \dfrac{[CH_3CH_2CH=C(CH_3)COOCH(CH_3)CH_3]_{eqm}}{[CH_3CH_2CH=C(CH_3)COOH]_{eqm}[HOCH(CH_3)CH_3]_{eqm}}$

e Add SCl$_2$O or PCl$_5$ at room temperature.
f $CH_3CH_2CH=C(CH_3)COCl$
$+ HOCH(CH_3)CH_3 \longrightarrow$
$CH_3CH_2CH=C(CH_3)COOCH(CH_3)CH_3 + HCl$

g $M_r(aldehyde) = 98; \therefore 2.5\ g = 0.0255\ mol;$
$M_r(ester) = 156; \therefore 3.0\ g = 0.0192\ mol;$
% yield $= 100 \times 0.0192/0.0255\ g = 75\%$.

Self-Assessment Questions – 28.1

1 **a** Propanamide.
  b Pentanamide.
  c 3-methylbutanamide.
  d N-ethylpropanamide.
2 **a**

b

c

d

3 **a** $CH_3COCl(l) + NH_3(g) \longrightarrow$
$CH_3CONH_2(s) + HCl(g)$

b $CH_3COCl(l) + H_2NCH_3 \longrightarrow$
$CH_3CONHCH_3(s) + HCl(g)$

c $CH_3CONH_2(s) + H_2O(l) + H^{\oplus}(aq) \longrightarrow$
$CH_3COOH(aq) + NH_4^{\oplus}(aq)$

d $CH_3CH_2CH_2CONH_2(s) + 4[H] \longrightarrow$
$CH_3CH_2CH_2CH_2NH_2(l) + H_2O(l)$

4 **a** $+ K_2Cr_2O_7(aq)/H^{\oplus}(aq)$; reflux.
  b $+ SCl_2O$ or PCl_5 or PCl_3; room temperature.
  c $+ NH_3(g)$; room temperature.
  d $+ Br_2/OH^-(aq)$; warm.

Self-Assessment Questions – 28.2

1 **a** Amide; butanamide.
  b Nitrile; propanenitrile.
  c Amide; 2-methylbutanamide.
  d Nitrile; 2-methylpropanenitrile.
  e Ammonium salt; ammonium butanoate.
2 **a**
b

c

d

3 **a** Reflux with $H^{\oplus}$(aq) or with OH^-(aq) and then acidify.
  b Reflux with $H^{\oplus}$(aq) or with OH^-(aq) and then acidify.
  c Warm with Br$_2$ in aqueous alkali.
  d LiAlH$_4$ in ethoxyethane at room temperature.
  e Heat with P$_4$O$_{10}$.

Self-Assessment Questions – 28.3

1 **a** Amide. **b** Nitrile.
  c Amine. **d** Amine.
2 **a** $CH_3CH_2CONH_2 + 4[H] \xrightarrow[room\ temperature]{LiAlH_4/ethoxyethane}$
$CH_3CH_2NH_2 + H_2O$

b

c $CH_3CH(Br)CH_3 + NH_3 \xrightarrow{heat}$
$CH_3CH(NH_2)CH_3 + HBr$

3 **a** NaNO$_2$ in aqueous HCl at room temperature.
  b Ethanoyl chloride at room temperature.
  c Concentrated HCl at room temperature.

Self-Assessment Questions – 28.4

1
; then

2 See p753.

Self-Assessment Questions – 28.5

1 **a** Two functional groups on the same molecule: —COOH and —NH$_2$.
  b 2-Aminopropanoic acid.
  c Sodium 2-aminopropanoate.
  d

  e

2 **a**

  b —NH$_2$
  c —COOH
  d

  e 6

Self-Assessment Questions – 29.2

1 **a** See p773.
  b Any impurity lowers the melting point – so if the unknown is mixed with cholesterol with no lowering of m.p., the unknown must be cholesterol; this is confirmed by the lower m.p. when the unknown is mixed with glucose.
2 Stage A – Cooling in a large beaker of ice-water.
  Stage B – Reflux; see p770.
  Stage C – Simple distillation; see Figure 29.2.3, p772.
  Stage D – Using a separating funnel; see p771.
  Stage E – Using a separating funnel; see p771.
  Stage F – In a small flask.
  Stage G – Fractional distillation; see Figure 29.2.4, p772.

Self-Assessment Questions – 29.3

1 a $K_2Cr_2O_7$ and sulphuric acid and reflux.
 b Tin and concentrated hydrochloric acid; then react with aqueous NaOH.
 c Heat with concentrated sulphuric acid.
 d NaBH$_4$ in water at room temperature.

2 a Oxidation.   **b** Oxidation.
 c Dehydration.   **d** Oxidation.

3 a $CH_3Br + KCN \longrightarrow CH_3CN + KBr$;
   $CH_3CN \longrightarrow CH_3COOH$
 b $CH_3CH(CH_3)Br + KCN \longrightarrow$
   $CH_3CH(CH_3)CN + KBr$;
   $CH_3CH(CH_3)CN \longrightarrow CH_3CH(CH_3)COOH$
 c $CH_3(CH_2)_3COOH \longrightarrow$
   $CH_3(CH_2)_3COO^-NH_4^{\oplus} \longrightarrow$
   $CH_3(CH_2)_3CONH_2 \longrightarrow CH_3(CH_2)_3NH_2$

4 a $C_2H_5Br \longrightarrow C_2H_5OH \longrightarrow CH_3COOH$
 b $C_2H_5Cl \longrightarrow C_2H_5CN \longrightarrow C_2H_5COOH$
 c $C_3H_7Br \longrightarrow C_3H_7CN \longrightarrow C_3H_7COOH$

Self-Assessment Questions – 29.4

1
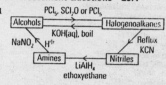
PCl$_5$, SCl$_2$O or PCl$_5$
Alcohols → Halogenoalkanes
KOH(aq), boil
NaNO$_2$, H$^{\oplus}$ Reflux / KCN
Amines → Nitriles
LiAlH$_4$ ethoxyethane

2 a Reflux with $K_2Cr_2O_7$ in aqueous sulphuric acid.
 b LiAlH$_4$ in ethoxyethane.
 c Heat with $K_2Cr_2O_7/H^{\oplus}$(aq) and distil off as formed.
 d Heat with concentrated H_2SO_4.

3 a e.g. $C_6H_{14} \xrightarrow{\text{heat, catalyst}} C_4H_{10} + CH_2{=}CH_2$
 b An apparently homogeneous mixture of metals.
 c React with hydrogen chloride.
 d Add Br$_2$ vapour.
 e Covalent.
 f
   C$_2$H$_5$
   Pb — C$_2$H$_5$
   C$_2$H$_5$ C$_2$H$_5$

4 a Reflux with $K_2Cr_2O_7/H^{\oplus}$(aq); see p770.
 b Using fractional distillation, see Figure 29.2.4, p772.
 c $CH_3COOH \xrightarrow{NH_3} CH_3COO^-NH_4^{\oplus} \xrightarrow{heat}$
   $\xrightarrow{Br_2/OH^-}$ warm
   $CH_3CONH_2 \longrightarrow CH_3NH_2$

5 a
   H   Cl
    C={=}C
   H   H

---

b $CH_2{=}CH_2 + Cl_2 \xrightarrow{RT} CH_2Cl{-}CH_2Cl$
   $\xrightarrow[\text{in ethanol}]{\text{hot, conc. NaOH}} CH_2{=}CHCl$

c Electrophilic addition.

d
   H H
   | |
   C—C
   | |
   H Cl $_n$

Self-Assessment Questions – 29.5

1 a
   CH$_2$—CH$_3$
   CH$_2$ (on benzene ring)

b Concentrated H_2SO_4 and concentrated HNO_3 at 55 °C.
 c Cl$_2$ in UV light.
 d Distillation.

2 $C_6H_6 \xrightarrow[\text{warm}]{\text{conc. }H_2SO_4/\text{conc. }HNO_3} C_6H_5NO_2 \xrightarrow[\text{in NaOH(aq)}]{\text{Sn-HCl conc. reflux}} C_6H_5NH_2$

3 a Tin with concentrated HCl followed by reacting with aqueous NaOH.
 b concentrated H_2SO_4/HNO_3
 c NaNO$_2$/H$^{\oplus}$(aq) above 10 °C.
 d C_2H_5Br with AlCl$_3$.

4 a Concentrated H_2SO_4/HNO_3; electrophilic substitution.
 b Further nitration with conc. H_2SO_4/HNO_3; possibly by heating.
 c Reflux with Sn/HCl(conc) followed by aqueous NaOH.
 d NaNO$_2$/H$^{\oplus}$(aq), below 5 °C.
 e Allow reaction mixture to warm above 10 °C.

Self-Assessment Questions – 29.6

a Warm with a mixture of concentrated H_2SO_4/HNO_3.
 b Electrophilic substitution.
 c 4-Nitromethylbenzene.
 d Reduce with tin in concentrated HCl followed by aqueous NaOH.
 e $M_r(C_6H_5CH_3) = 92$; $\therefore 5.0/92 = 0.054$ mol.
 f $M_r(C_6H_4(CH_3)NO_2) = 137$;
   $4.6/137 = 0.034$ mol;
   % yield $= 100 \times 0.034/0.054 = 63\%$.
 g 5.0 g $C_6H_5CH_3 = 0.054$ mol;
   35% of 0.054 mol $= 0.019$ mol;
   $M_r(C_6H_3(CH_3)(NH_2)(SO_3H)) = 185$;
   mass $= 0.019 \times 185 = 3.5$ g.

Self-Assessment Questions – 30.2

1 C: $12/44 \times 4.682$ g $= 1.28/12 = 0.107$ mol;
   H: $2/18 \times 1.077 = 0.120$ g $= 0.120$ mol;

N: $160/24000$ mol $= 0.0133$ mol;
N atoms $= 0.187$ g;
O: mass $= 1.80 - 1.587 = 0.213$ g
   $= 0.213/16 = 0.0133$ mol;
empirical formula is C_8H_9ON.

2 Mol $MnO_4^{-} = 22.3 \times 0.0200/1000$
   $= 0.000446$ mol;
 $\therefore$ mol $Fe^{2\oplus} = 0.000446 \times 5 \times 10$
   $= 0.0223$ mol;
 $\therefore M_r$ of salt $= 8.73/0.0223 = 391.5$;
 $M_r((NH_4)_2SO_4 \cdot FeSO_4) = 284$; $\therefore xH_2O =$
 $(391.5 - 284)/18 = 6.0 = x$.

3 Concentration of HCl acid
   $= 17.35 \times 0.100/25 = 0.069$ mol dm^{-3};
 total Cl$^-$ concentration
   $= 26.85 \times 0.100/25 = 0.107$ mol dm^{-3};
 $\therefore$ [NaCl] $= 0.038$ mol dm^{-3} $= 0.038 \times 58.5$
   $= 2.22$ g dm^{-3}.

4 $M_r(CaCO_3) = 100$;
 $\therefore 0.356$ g $CaCO_3 = 0.00356$ mol;
 $\therefore 0.00356$ mol $Ca^{2\oplus}$ in 50 cm^3 of original solution;
 $\therefore$ [Ca$^{2\oplus}$] $= 0.00356 \times 1000/50$
   $= 0.0712$ mol dm^{-3}.

Self-Assessment Questions – 30.3

1 a Chlorine.   **b** Hydrogen.
 c Ethene.   **d** Carbon dioxide.
2 Calcium carbonate, i.e. chalk.
3 a Acid.   **b** Halide.
 c Sulphate.   **d** Sulphuric acid.

Self-Assessment Questions – 30.4

1 a Not acid.
 b Unsaturated.
 c Alcohol group (not acidic from the test in part a).
 d Probably aromatic.
 e Not carbonyl.

2 a Add 2,4-dinitrophenylhydrazine.
 b Add FeCl$_3$(aq).
 c Add Na$_2$CO$_3$.
 d Add Br$_2$.

3 a Tollens' or Fehling's tests.
 b Add FeCl$_3$(aq).
 c Add SCl$_2$O, PCl$_5$ or PCl$_3$.
 d Burn the samples.

Self-Assessment Questions – 30.5

1 $15 - CH_3^{\oplus}$; $29 - C_2H_5^{\oplus}$; $43 - C_3H_7^{\oplus}$; $57 - C_4H_9^{\oplus}$; $72 - C_5H_{12}^{\oplus}$

2 a C={=}O   **b** C—H
 c O—H   **d** C—Cl

3 a N$_2$: 28.00614; CO: 27.99492; C_2H_4: 28.03128.
 b CO.

Index